2022 24th European Conference on Power Electronics and Applications (EPE'22 ECCE Europe)

Hanover, Germany
5-9 September 2022

Pages 1-672

IEEE Catalog Number: CFP22850-POD
ISBN: 978-1-6654-8700-9

Copyright © 2022, The European Power Electronics and Drives Association
All Rights Reserved

*** *This is a print representation of what appears in the IEEE Digital Library. Some format issues inherent in the e-media version may also appear in this print version.*

IEEE Catalog Number: CFP22850-POD
ISBN (Print-On-Demand):   978-1-6654-8700-9
ISBN (Online):   978-9-0758-1539-9

Additional Copies of This Publication Are Available From:

Curran Associates, Inc
57 Morehouse Lane
Red Hook, NY 12571 USA
Phone: (845) 758-0400
Fax: (845) 758-2633
E-mail: curran@proceedings.com
Web: www.proceedings.com

2022 24th European Conference on Power Electronics and Applications (EPE'22 ECCE Europe)

Hanover, Germany
5-9 September 2022

Pages 1-672

IEEE Catalog Number: CFP22850-POD
ISBN: 978-1-6654-8700-9

TABLE OF CONTENTS

Dynamic Power Analysis of Inverter-Fed Drives Based on the Switching Period of the Power Electronics 1
Alexander Stock

Stability Analysis in an Inverter-Dominant Microgrid Facing In-Rush Current of an Induction Machine 11
Nastaran Fazli, David Hammes, Sidney Gierschner, Hans-Gunter Eckel

Self-Oscillating Capacitive Power Transfer with Multiple Receiver Capability and Coupling Path Adaption 22
Norbert Seliger

An Electrically Driven Gas Compressor for Hydrogen Refueling Stations with Active Power Smoothing 30
Alfred Rufer

Unsymmetrical Fault Behavior of PLL Based Grid-Connected Converters 39
Philipp Hackl, Ziqian Zhang, Robert Schuerhuber

Stability Assessment and Optimization of MMC Energy Balancing for Drive Applications at Standstill using an Averaging Approach 49
Qiuye Gui, Hendrik Fehr, Albrecht Gensior

Turn-On Losses Optimization for Medium Power SiC MOSFET Half-Bridge Module 59
Pham Ha Trieu To, Felix Kayser, Hans-Günter Eckel

Oscillation Damping in a 500kW Hybrid Si/SiC Three-Level ANPC Inverter with Decoupling Capacitor 70
Pham Ha Trieu To, Hans-Günter Eckel

Multi Busbar Sub-Module Modular Multilevel STATCOM with Partially Rated Energy Storage Configured in Sub-Stacks 80
Chuantong Hao, Wenhao Ma, Michael Merlin, Paul Judge, Stephen Finney

Three-Phase ZVS Inverter with Variable and Fixed Frequency Operation Based on GaN Semiconductors 88
Benedikt Kohlhepp, Michael Lutsch, Thomas Dürbaum

Influences of Conductor Positions and Fast Rising Impulse Voltages on the Line-End Coil Based on a Three-Phase High-Frequency Model 97
Ting Helmholdt-Zhu, Volker Grabs

Simulation Tool for Optimization of Digital Active Gate Drive Sequence using Genetic Algorithm 108
Hajime Takayama, Shuhei Fukunaga, Takashi Hikihara

Analysis of Balancing Algorithms for Quasi- Two/Three-Level Single Phase Operation of a Flying Capacitor Converter 115
Stefan Mersche, Markus Bayer, Kai Rickert, Marc Hiller

Instability in Active Balancing Control of Dc Bus Voltages in VSC Converters Interconnected via Multi-Winding Transformers 125
Duro Basic, Sami Siala

Online Learning-Based Islanding Detection Scheme for Grid-Connected Systems 135
Mohammed Ali Khan, V S Bharath Kurukuru, Rupam Singh

Difference in the Design Process of LCL Filters for Grid Connected VSI When using SiC/GaN Instead of Si Semiconductors 145
Dennis Kampen, Lukas Fräger, Niklas Badenhop, Arthur Mambetow

Analysis and Design of a Resonant DC/DC Transformer in Modular Operation 152
Abraham López, Manuel Arias, Pablo F. Miaja, Arturo Fernández

Predictive Braking Algorithm for Soft Starter Driven Induction Motors 160
Hauke Nannen, Heiko Zatocil, Gerd Griepentrog

Ambient Electromagnetic Energy Harvesting Circuit using Rectennas Manufactured with Stereolithography Resin 169
Xuan Viet Linh Nguyen, Tony Gerges, Jacques Verdier, Philippe Lombard, Michel Cabrera, Bruno Allard, Jean-Marc Duchamp, Philippe Benech

Boost/Buck-Boost Based Grid Connected Solar PV Micro-Inverter with Reduced Number of Switches and Having Power Decoupling Capability 178
Arup Ratan Paul, Arghyadip Bhattacharya, Kishore Chatterjee

Operation and Selection of Multilevel Power Converters for Doubly Fed Induction Generator-Based Wind Turbines 187
Kapil Jha, Joseph Banda, Hridya I, Arvind Tiwari

A Detailed View on the Trapezoidal Operation for MMC Type Braking Chopper in Medium Voltage Application 195
Patrick Hofstetter, Viktor Hofmann, Dennis Karwatzki

Influence of Operating Frequency on High-Power Medium-Voltage Medium-Frequency Transformers 203
Thomas B. Gradinger, Ralph M. Burkart, Marko Mogorovic

Output Power Characteristics of Isolated Secondary-Resonant SAB DC-DC Converter for Output Voltage Variation 213
Shota Yamashita, Kohei Budo, Takaharu Takeshita

Hardware and Control Design of a High Precision Modular Power Converter Based on GaN Technology for Particle Accelerator Magnets 223
Thomas Margreiter, Ivan De Cesaris, Maurizio Incurvati, Sebastien Pelletier, Martin Schiestl, Ronald Stärz

Battery Cycler to Generate Open Li-Ion Cell Aging Data and Models 232
Matthias Luh, Thomas Blank

Function Blocks of a Highly-Integrated All-In-GaN Power IC for DC-DC Conversion 242
Michael Basler, Richard Reiner, Stefan Moench, Patrick Waltereit, Rüdiger Quay

Comparison of Redundancy Requirements for Modular Multilevel Converter Considering Manufacturer Reliability Inputs and Mission Profile 251
Diego Velazco, Guy Clerc, Emmanuel Boutleux, Francois Wallart

Impact of Insulation and Cooling on Performance Due to Reliability-Oriented Design of Electrical Machines 261
Lucas Vincent Hanisch, Jonas Franzki, Markus Henke

Long Switching Horizon Model Predictive Controller for High-Speed Integrated Modular Motor Drives 268

    Martin Schiestl, Maurizio Incurvati, Ronald Starz, Markus Schmid

Standalone Power Management System for Flexible Piezo Electric Nano Generators (PENG) Based on the Co-Polymer P(VDF:TrFE) 279

    Alexander Wölk, Mahmoud Shousha, Shashank Shekhawat Singh, Martin Haug, Lorandt Fölkel, Michael Brooks, Asier Alvarez, Andreas Petritz, Philipp Schäffner, Jonas Groten, Andreas Tschepp, Barbara Stadlober

Analysis and Estimation of Neutral-Point Voltage Balancing Ability of an Optimized Balancing Algorithm for Grid Connected Active-NPC Converter 289

    Joseph Banda, Kapil Jha, Hridya Ittamveettil, Arvind Kumar Tiwari, Fernando Ramirez

A Direct Model Predictive Control Strategy of Back-To-Back Modular Multilevel Converters using Arm Energy Estimation 297

    Akseli Hakkila, Antonios Antonopoulos, Petros Karamanakos

Study on Commutation Loop Inductance and Current Distribution to DC-Link Capacitors in a GaN Half-Bridge 307

    Benedikt Kohlhepp, Samuel Faber, Jeremias Kaiser, Thomas Dürbaum

Cooperative Control of Online Impedance Spectroscopy Monitoring Method and Maximum Power Point Tracking Method for Photovoltaic Panels 315

    Xin Wang, Zhixue Zheng, Michel Aillerie, Alexandre De Bernardinis, Jean–paul Sawicki, Marie-Cécile Péra, Daniel Hissel

Benefits of Switching from Si to SiC Modules with Further Converter Optimization 325

    Antxon Arrizabalaga, Mikel Mazuela, Iosu Aizpuru, June Urkizu, Jon Aztiria

On the Reduction of Output Capacitance in Two-Level Three Phase PFC Boost Rectifier for Pulsating Loads 335

    Tania C. Cano, Douglas Pedroso, Alberto Rodríguez, Ignacio Castro, Diego G. Lamar

Cognitive Insights into Metaheuristic Digital Twin Based Health Monitoring of DC-DC Converters 344

    Abdul Basit Mirza, Kushan Choksi, Sama Salehi Vala, Krishna Moorthy Radha, Madhu Sudhan Chinthavali, Fang Luo

A Three-Phase Isolated Secondary-Resonant Single-Active-Bridge DC-DC Converter with a Delta-Star Connected Transformer 351

    Atsushi Nishio, Kohei Budo, Mai Van Tuan, Takaharu Takeshita

A Novel Concept to Optimize Core Loss in Planar Magnetic Based on an Unbalanced-Flux-Approach 361

    Sobhi Barg, Kent Bertilsson, Grover Torrico

Model Reduction using Singular Perturbation Methods for a Microgrid Application 370

    Lasse Gnärig, Albrecht Gensior, Saioa Burutxaga Laza, Miguel Carrasco, Carsten Reincke-Collon

Drive Level Parameter Identification of an Induction Motor 380

    Andreas Bünte, Alex Hald, Andreas Kirsch

Impedance Stability of Single-Phase LCL Grid-Connected Voltage Source Inverters with Wideband Gap Devices Under Different Control Approaches 390

    Ramy Ali, Terence O'Donnell

Design and Modulation Optimization of an MMC Based Braking Chopper 400
  Viktor Hofmann, Patrick Hofstetter

Modeling the Arrangement of Drill Holes for Orthogonal Biasing in Controllable Inductors for
Power Electronic Converters 411
  Jonas Pfeiffer, Christoph Drexler, Pierre Küster, Peter Zacharias, Michael Schmidhuber

A Sectorized FCS-MPC Transformerless SST for Power Transmission Application 421
  *Gabriel Gaburro Bacheti, Renner Sartório Camargo, Emilio José Bueno, Marco Liserre,
  Lucas Frizera Encarnação*

Inductance Estimation for Square-Shaped Multilayer Planar Windings 432
  Theofilos Papadopoulos, Antonios Antonopoulos

Cost and Efficiency Considerations in On-Board Chargers 442
  *Marija Jankovic, Christian Felgemacher, Kevin Lenz, Aly Mashaly, Abdelmouneim
  Charkaoui*

A Novel Combined Control of Ground Current and DC-Pole-To-Ground Voltage in Symmetrical
Monopole Modular Multilevel Converters for HVDC Applications 451
  Pablo Briff, Amit Kumar

A PFC Boost Converter with Reduced Switching Losses Operating at a Fixed Switching Frequency 459
  Burkhard Ulrich

Predictive Control of Power Electronics Autotransformer for Mitigating Three-Phase Grid Current
Unbalance in Railway Supply Systems 468
  Tabish Nazir Mir, Faysal Hardan, Masood Hajian, Tamer Kamel, Pietro Tricoli

Parameter Sensitivity of a MRAS-Based Sensorless Control for AFPMSM Considering Speed
Accuracy and Dynamic Response at Multiple Parameter Variations 474
  Michael Brüns, Christian Rudolph, Tankred Müller

Synchronization Stability of a Grid Forming Converter Under the Effect of Current Limit in
Voltage Dips with VI Based Current Limiting Method: Analysis and Solution 484
  *Siam Hasan Khan, Markel Zubiaga Lazkano, Pedro Izurza, Alain Sanchez-Ruiz, Javier Cañas
  Aceña, Joseba Arza*

Analytic Calculation of Touch and Leakage Currents of Non-Isolated EV Chargers using a Fast
Common Mode Calculation Method and Non-Ideal Passive Component Models 493
  Christian Stutz, Sebastian Nielebock, Martin März

Triple-Phase-Shift Controlled Dual Active Bridge Converter with Variable Input Voltage in
Auxiliary Railway Supply 504
  Martin Scohier, Olivier Deblecker, Carlos Valderrama

Loss Characterization Methodology for Soft Magnetic Nano-Crystalline Tape Materials in Coupled
Inductors 514
  David Bohne, Valentin Wagner, Patrick Deck, Christian P. Dick

Substitution of Nanocrystalline Toroid by Laminated Ferrite Toroid in the Application of a
Common-Mode Choke 525
  Lukas Reißenweber, Fritz Wohlrath, Alexander Stadler

Direct Active Stabilization of the DC-Link in Voltage-Source Converters 534
  Matthieu Bertin, Mohamad Koteich

Hardware-In-The-Loop Control of a Modular Induction Motor Drive in Power Electronics Education...... 544
Jens Peter Kaerst

Design and Efficiency Analysis of an LCL Capacitive Power Transfer System with Load-Independent ZPA...... 554
Francesco Musolino, Ahmed Abdullah, Mario Pavone, Fabio Ferreyra, Paolo Crovetti

A Pulse Generator Based on Transmission Line Transformer for Insulation Aging Test...... 562
Xiao Yu, Khanh-Hung Nguyen, Peter Zacharias

Design of a Single-Phase Common Mode and Differential Mode Inductor for Interleaved Converters...... 572
Jonathan Robinson, Gopal Mondal, Stefan Hänsel, Matthias Neumeister

Steady-State Analysis and Comparison of SSFB, SDFB and DSFB MMC-Based STATCOM...... 582
Mohamed Moez Belhaouane, Pierre Vermeerch, François Gruson, Pierre Rault, Sébastien Dennetiere, Xavier Guillaud

Current Distribution Control in Parallel Connected Power Converters with Continuous Output Voltage...... 593
Sabrina Ulmer, Andreas Brunner, Philipp Czerwenka, Gernot Schullerus, Ertugrul Sönmez

Optimized Pulse Pattern with Half-Wave Symmetry for 5-Level Converter...... 604
Jonas Weires, Pedro Leal Dos Santos, Steven Liu

Characterization of Si-IGBT Crosstalk with a Concentration on Power Circuit Parasitic Elements and the Device Operation Point...... 614
Amir Azam Rajabian, Sadegh Mohsenzade, Javad Naghibi, Kamyar Mehran

Impact of Higher Current Harmonics on Component Current Stress and Conduction Losses of Half-Bridge-Series-Resonant-Converters in Discontinuous Conduction Mode for High-Power Applications...... 624
Daniel Haake, Anton Grodnichev, Fabian Schnabel, Marco Jung

Control of a Zero-Voltage Switching Isolated Series-Resonant Power Circuit for Direct 3-Phase AC to DC Conversion...... 634
Yusuf Kosesoy, Remco Bonten, Henk Huisman, Jan Schellekens

Design of a Robust Voltage Control for Inverters with LC Filter Based on the Internal Model Control...... 641
Frederik Stallmann, Axel Mertens, Lukas Fräger

Influence of Power Semiconductor Device Variations on Pulse Shape of Nanosecond Pulses in a Solid-State Linear Transformer Driver...... 651
Raffael Risch, Anliang Hu, Jürgen Biela

Optimal Design of Integrated Motor Drives - Comparison of Topologies (2L/3L/Modular), PWM Variants, and Switch Technologies (Si/SiC/GaN)...... 662
Thilo Bringezu, Jürgen Biela

Distribution Transformer Voltage Control using a Single-Phase Matrix Converter...... 673
Rui Wang, Henk Huisman, Korneel Wijnands

Influence of Carrier-Based PWM Techniques on the Common-Mode Voltage and Common-Mode Current of Six-Phase Full-Bridge Inverters...... 681
Juris Arrozy, Esin Ilhan Caarls, Henk Huisman, Jorge L. Duarte, Lorenzo Ceccarelli

Mitigation of Dead-Time Effects on Transient DC Bias Elimination in Dual Active Bridge Link Current 689

MK Kharabela Mohanta, Dipankar De, Silpashree Sahu, Alberto Castellazzi

Generalized Automated Tool for Analysis and Design of Multiphase Coupled Inductor Buck Converters 698

Rana Asad Ali, Mahmoud Shousha, Martin Haug

Experimental Study of a Directly Oil-Cooled Electrical Machine for a Full-Electric Vehicle by using Low Viscosity Oil 709

Huihui Xu, Georg Tobias Götz, Shimin Zhang, Rik W. De Doncker

Development of a Family of High Voltage Gain Step-Up Multi-Port DC-DC Converters for Fuel Cell-Based Hybrid Vehicular Power Systems 719

Pouya Zolfi, Sina Vahid, Ayman El-Refaie

Bidirectional DC Circuit Breaker with Improved Performance During Commissioning and Reclosing 730

Aditya Pogulaguntla, Venkata Raghavendra I, Satish Naik Banavath, Andrii Chub, T Sreekanth, Harish Sarma Krishnamoorthy

Modeling Method for Conducted Noise Flowing in Power Lines of DC/DC Converter 739

Takato Hattori, Wataru Kitagawa, Takaharu Takeshita

High-Bandwidth Power Hardware-In-The-Loop for Motor and Battery Emulation at High Voltage Levels 749

Manuel Fischer, Philipp Kemper, Johannes Herbold, Daniel Epping, Frank Puschmann

Analysis and Discussion of Different Three-Phase dv/dt Filter Topologies and the Influences of Their Filter Parameters on Losses and EMC 758

Eric Fritze, Michael Meissner, Klaus F. Hoffmann, Kai-Uwe Rathjen, Stefan Dickmann, Oliver Woywode

State of Charge Prediction of Lithium-Ion Batteries Based on Artificial Neural Networks and Reduced Data 767

Sebastian Pohlmann, Ali Mashayekh, Dominic Karnehm, Manuel Kuder, Antje Gieraths, Thomas Weyh

Investigation for Condensation Test Condition of HVIGBT Modules 777

Kenji Hatori, Keiichi Nakamura, Wakana Noboru, Nils Soltau, Eugen Wiesner

Three Phase PV Inverter LCOE Optimization Considering Technological Choice 787

Morteza Tadbiri Nooshabadi, Jean-Luc Schanen, Shahrokh Farhangi, Hossein Iman-Eini

Square Wave Operation to Reduce Pulsating Power in Isolated MMC-Based Ultrafast Chargers 798

Ygor Pereira Marca, Maurice G. L. Roes, Jorge L. Duarte, Korneel Wijnands

Surge Current Protection for Railway Traction Applications 805

Michael Gleissner, Mark-M. Bakran

Impedance-Based Analysis of HVDC Converter Control for Robust Stability in AC Power Systems 814

André Schön, Andreas Lorenz, Rodrigo Alonso Alvarez Valenzuela

Class-E Push-Pull Resonance Converter with Load Variation Robustness for Industrial Induction Heating 825

Janus Dybdahl Meinert, Benjamin Futtrup Kjærsgaard, Thore Stig Aunsborg, Asger Bjorn Jorgensen, Stig Munk-Nielsen, Sune Bro Duun

Review of Power Converter Topologies for Electrochemical Impedance Spectroscopy of Lithium-Ion Batteries 833

Hamzeh Beiranvand, Julius M. Placzek, Marco Liserre, Giorgia Zampardi, Doriano Constantino Brogioli, Fabio La Mantia

Design and Experimental Validation of a Voltage Sensing-Current Cancellation Common Mode Linear Active Filter 843

B. Mohamed Nassurdine, PE Lévy, D. Labrousse, JL Schanen, X. Maynard, S. Carcouet

Partial Discharges of Insulated Wires Under Impulses from Wide Bandgap Power Electronics 854

Ting Helmholdt-Zhu, Vivien Grau, Urs Obernolte

Analysis of a Droop-Based Power Controller for Three-Phase Microgrids 865

Andrea Lauri, Hossein Abedini, Davide Biadene, Tommaso Caldognetto, Paolo Mattavelli

Efficiently Paralleling GaN-Transistors for High Current and High Frequency Applications using a Butterfly Layout 873

Martin Wattenberg, Oscar Lorenz, Juan Sanchez

Data-Driven Decentralized Volt/Var Control for Smart PV Inverters in Distribution Systems 883

Yizhou Lu, Qianwen Xu, Lars Nordström

Study of Current Ripple Generators for Accelerated Ageing of Capacitors 891

Robert Keilmann, Hendrik Schefer, Regine Mallwitz

Intra-Arm Balancing Control of Cascaded Multi-Port Converter for Whole Power Unbalance Conditions 902

Takumi Yasuda, Jun-Ichi Itoh

Investigation of Creepage Distances on Printed Circuit Boards for Avionic Applications 912

Hendrik Schefer, Zhongqing Xu, Tobias Kopp, Regine Mallwitz, Michael Kurrat

A 20 kW, 3-Level Flying Capacitor 1500 V Inverter with Characterized GaN Devices for Grid-Tie Applications 922

Van Sang Nguyen, Anthony Bier, Hajar Es-Seghier, Ulrich Soupremanien, Gérard Delette, Stephane Catellani

New Analytical Model for Calculating HF-Losses in Litz Wire Regions Located Outside the E/U-Core Window of Transformers 933

Qingchao Meng, Jürgen Biela

Fast and Accurate Soft-Switching and Hard-Switching Losses Estimation for Power Converter, Application to the Dual Active Bridge (DAB) Converter 944

Francois Boige, Nicolas Videau, Adel Ziani, Bruno Guerrero, Julien Laclaverie

Influence of an Electrical Machine on the Dimension and Packaging of Multi-Machine Systems 952

Thomas Stöckl, Hans-Georg Herzog

Design of a Serial Impingement Cooling Heatsink for a 30 kW PV String Inverter 960

Paul Bruyere, Guillaume Piquet Boisson, Gaëtan Perez

Online Junction Temperature Measurement of SiC-MOSFETs via Gate Impedance using the Gate-Signal Injection Method 971

David Hirning, Luca Bauer, Johannes Ruthardt, Jörg Haarer, Philipp Ziegler, Jörg Roth-Stielow

Powercycling Test Bench with Realistic Loss Distribution and Temperature Ripples 980
Till-Mathis Plötz, Jan Fuhrmann, Hans-Günter Eckel

Design, Implementation and Characterization of an Integrated Current Sensing in GaN HEMT
Device by using the Current-Mirroring Technique 990
*Van-Sang Nguyen, René Escoffier, Stéphane Catellani, Murielle Fayolle-Lecocq, Jérémy
Martin*

GaN-Based Modular Multilevel Converter for Low-Voltage Grid Enables High Efficiency 999
Philip Kiehnle, Patrick Himmelmann, Marc Hiller

Energy Management of Smart Homes with Electric Vehicles using Deep Reinforcement Learning............. 1006
Xavier Weiss, Qianwen Xu, Lars Nordström

Simple and Low-Computational Losses Modeling for Efficiency Enhancement of Differential
Inverters with High Accuracy at Different Modulation Schemes.. 1015
Ahmed Shawky, Mokhtar Aly, Emad M. Ahmed, Samir Kouro, José Rodriguez

Estimation of Battery Parameters in Cascaded Half-Bridge Converters with Reduced Voltage
Sensors 1025
Nima Tashakor, Bita Arabsalmanabadi, Elham Hosseini, Kamal Al-Haddad, Stefan Goetz

Method to Analyze the Influence of Switching Behavior in Hard Switching Half Bridge Topologies
for Traction Application 1036
Dominik Nehmer, Michael Gleissner, Lukas Bergmann, Mark-M. Bakran

Impact of Aluminum Casing on High-Frequency Transformer Leakage Inductance and AC
Resistance.. 1046
*Reda Bakri, Xavier Margueron, Wendell Da Cunha Alves, Xavier Cimetiere, Frédéric Gillon,
Antoine Bruyere, Lucian Vatamanu*

Neural Networks-Generalized Predictive Control for MIMO Grid-Connected Z-Source Inverter
Model 1056
Navid Salehi, Herminio Martinez-Garcia, Guillermo Velasco-Quesada

Voltage Estimation for Diode-Clamped MMCs Based on a Simplified Neural Network 1064
Nima Tashakor, Davood Keshavarzi, Shady Banana, Stefan Goetz

A Non-Cooperative Game-Theoretic Distributed Control Approach for Power Quality
Compensators 1074
*Claudio Burgos-Mellado, Victor Bucarey, Helmo K. Morales-Paredes, Diego Muñoz-
Carpintero*

A Comparative Analysis of Power Converter Topologies for Integration of Modular Batteries in
Electric Vehicles.. 1083
*Alberto Cárcamo, Aitor Vázquez, Alberto Rodriguez, Diego G. Lamar, Marta M. Hernando,
Daniel Remón*

Design of a High-Dynamic Test Bench for Accelerated Dielectric Lifetime Testing with Adjustable
Voltage Slopes and Temperatures 1094
Hendrik Schefer, Lucas Hanisch, Tim-Hendrik Dietrich, Regine Mallwitz, Markus Henke

Novel Modulation Method for Common-Mode Noise Reduction in Solid-State Transformer Based
on ISOP Configuration 1104
Naoto Kikuchi, Hiroki Watanabe, Keisuke Kusaka, Jun-Ichi Itoh

Modular STATCOM for Compensation of Reactive Power and Voltage Asymmetry in Medium-Voltage Distribution Power Grids 1114
Josef Štengl, Tomáš Kormska, Jakub Talla, Zdenek Peroutka

Novel Method for Active Short Circuit (ASC) Tests of Power Module in Automotive Traction Application 1121
Tobias Appel, Arne Bieler

Short Circuit Performance and Current Limiting Mode of a Monolithically Integrated SiC Circuit Breaker for DC Applications Up to 800 V 1128
Norman Boettcher, Taro Takamori, Keiji Wada, Wataru Saito, Shin-Ichi Nishizawa, Tobias Erlbacher

Application of a HV Bipolar Square-Wave Voltage Generator for Qualification and Assessment of Energy Equipment 1137
Rico Fischer-Baeumer, Kai Gohrmann, Konrad Domes, Benjamin Sahan, Christian Staubach

A Decentralized and Communication-Free Control Algorithm of DC Microgrids for the Electrification of Rural Africa 1147
Lucas Richard, David Frey, Marie-Cecile Alvarez-Herault, Bertrand Raison

Universal Real-Time Model for Active Rectifiers in Versatile Totem-Pole PFC Configurations 1157
Axel Kiffe, Thorben Hoffstadt

Investigation of Core-Loss Mechanisms in Large-Scale Ferrite Cores for High-Frequency Applications 1167
Michael Baumann, Christoph Drexler, Jonas Pfeiffer, Jens Schueltzke, Erwin Lorenz, Michael Schmidhuber

Generation of Methodology for Making Benchmark Microgrids and Application in ESUSCON Microgrid 1177
Oscar Dorner, Patricio Mendoza-Araya

An Overview of Grid-Connection Requirements for Converters and Their Impact on Grid-Forming Control 1187
Paul Imgart, Mebtu Beza, Massimo Bongiorno, Jan R. Svensson

Modular Battery-Integrated Power Electronics-Modelling, Advantages, and Challenges 1197
Nima Tashakor, Jan Kacetl, Tomas Kacetl, Stefan Goetz

Design of Triple-Active Bridge Converter with Inherently Decoupled Power Flows 1207
Dong-Uk Kim, Byengjoo Byen, Byunghwang Jeong, Sungmin Kim

Application of a Multi-Winding Magnetic Component Characterization Method to Optimize Cross-Regulation Performances in DCM Flyback Converters 1216
Denis Motte-Michellon, Brahim Ramdane, Yves Lembeye, Bruno Cogitore

Application of an Electrostatic Machine in a Low-Voltage Microgrid 1226
Gabriel Ramos Huerta, Patricio Mendoza-Araya

Influences of Parasitic Capacitances in Wide Bandwidth Rogowski Coils for Commutation Current Measurement 1237
Philipp Ziegler, Tobias Festerling, Jorg Haarer, Philipp Marx, David Hirning, Jorg Roth-Stielow

Systematic Analysis of Oscillations in DC-Links of Fast Switching Power Electronics 1247
Tobias Fricke, Regine Mallwitz

EMI Mitigation Induced by an IGBT Driver Based on a Controlled Gate Current Profile 1256
  Daniel S. Martinez-Padron, Nicolas Patin, Eric Monmasson

An Accurate and Fast Model of Three-Level Three-Phase Dual-Active Bridge Converters in Real-
Time Simulation 1266
  Ming Jia, Philipp Joebges, Rik W. De Doncker

A Calorimetric and Electrical Method for Measuring Loss Energies of Half-Bridges 1277
  *Jörg Haarer, Mattea Eckstein, Philipp Ziegler, Philipp Marx, David Hirning, Jörg Roth-
  Stielow*

Condition Monitoring Approach of a SiC Power Semiconductor using Turn-Off Delay with an
Integration in a SiC Driver 1286
  Victor Golev, Ulf Schümann, Rando Raßmann, Jan Bockholt

Measurement Results of Multilevel Hysteresis Control for Paralleled Two-Level Converters 1294
  Magdalena Gierschner, Yves Hein, Hans-Günter Eckel, Christian Heien

Design and Development of a Short-Circuit Test Bench for Low-Voltage Direct Current Protection
Devices 1300
  Simon Ravyts, Thomas Vandenbussche, Koen Stul, Jan Cappelle

A Novel Modified-TOGI Based PLL for the Three-Phase Unbalanced and Distorted Grid
Conditions 1309
  Khanh-Hung Nguyen, Ahmad Ali Nazeri, Xiao Yu, Peter Zacharias

Comparison of Two and Three-Level AC-DC Rectifier Semiconductor Losses with SiC MOSFETs
Considering Reverse Conduction 1319
  Guangyao Yu, Thiago Batista Soeiro, Jianning Dong, Pavol Bauer

Measurement Method for Simple Determination of Sinusoidal Large Signal Losses in Inductive
Components 1328
  Peter Zacharias, Alejandro Aganza-Torres

A Novel Technique for the Suppression of the Displacement Current Through Power Module Base-
Plate Capacitance 1336
  Mahmoud Saeidi, Ahmad Ali Nazeri, Rufad Zilic, Peter Zacharias

Analysis and Implementation of Effective Placement of EMC Capacitors for WBG Modules 1343
  Mahmoud Saeidi, Ahmad Ali Nazeri, Firas Jenhani, Peter Zacharias

Power Hardware-In-The-Loop Verification of a Cold Load Pickup Scenario for a Bottom-Up Black
Start of an Inverter-Dominated Microgrid 1350
  Mina Mirzadeh, Robin Strunk, Tobias Erckrath, Axel Mertens

Detection of Incipient Inter-Turn Short-Circuit Faults by Artificial Intelligence Classifiers 1361
  Osman Örgüt, Ilker Sahin, Ece Olcay Günes

Modeling the Impact of Grid-Forming E-STATCOMs on Inter-Area System Oscillations 1371
  A. Bolzoni, N. Johansson, J. P. Hasler

Combining Schwarz-Christoffel Mappings and Biot-Savart Law to Calculate the High-Frequency
Current Distribution Inside a Single Slot 1381
  Torben Fricke, Phil Leon Pickert, Babette Schwarz, Bernd Ponick

Standardised Switching Cell Building Block for Converter Design Optimisation with Detailed Electro-Thermal Model 1391
Georgios Papadopoulos, Jürgen Biela

Design Procedure for Transformer-Based Solid-State Pulse Modulators with Damping Network 1402
Spyridon Stathis, Juergen Biela

DC Bias Impact on Magnetic Core Losses at High Frequency 1413
Bima Nugraha Sanusi, Ziwei Ouyang

Investigation of the Short-Circuit Type II Safe Operating Area of IGBTs.................................... 1424
Madhu Lakshman Mysore, Mohamed Alaluss, Abhishek Maitra, Thomas Basler, Roman Baburske, Franz-Josef Niedernostheide, Hans-Joachim Schulze

Single Transformer, MMC Based MV Power Electronic Traction Transformer 1434
Simon Fuchs, Simon Beck, Jürgen Biela

A New Power MOSFET Technology Achieves a Further Milestone in Efficiency 1445
Ralf Siemieniec, Michael Hutzler, Cesar Braz, Tomasz Naeve, Elias Pree, Heimo Hofer, Ingmar Neumann, David Laforet

Experimental Evaluation of Battery Impedance and Submodule Loss Distribution for Battery Integrated Modular Multilevel Converters 1456
Arvind Balachandran, Tomas Jonsson, Lars Eriksson, Anders Larsson

Constant DC Power Infeed Grid Forming with Improved Ability to Ride-Through Unbalanced Low-Voltage Faults 1466
Tayssir Hassan, Malte Eggers, Huoming Yang, Peter Teske, Sibylle Dieckerhoff

Constrained Long-Horizon Direct Model Predictive Control for Grid-Connected Converters with LCL Filters 1476
Mattia Rossi, Petros Karamanakos, Francesco Castelli-Dezza

Performance Evaluation of SiC-Based Isolated Bidirectional DC/DC Converters for Electric Vehicle Charging... 1486
Kaushik Naresh Kumar, Rafal Miskiewicz, Przemyslaw Trochimiuk, Jacek Rabkowski, Dimosthenis Peftitsis

Impact of Threshold Voltage Shifting on Junction Temperature Sensing in GaN HEMTs........................... 1497
Burhan Etoz, Jose Ortiz Gonzalez, Arkadeep Deb, Saeed Jahdi, Olayiwola Alatise

Comparison of Power Cycling Results of Discrete GaN Cascodes for Automotive Power Electronics with High Temperature Swings 1506
Florian Lippold, Philipp Hauenschild, Regine Mallwitz

Current Distortion Study for Hybrid Multi-Level Grid Inverter with Active Neutral-Point-Clamped 4-Leg Topology 1515
Jonas Steffen, Matthias Klee, Fabian Schnabel, Axel Seibel, Marco Jung

Dynamic Maximum Power Point Tracking Method Including Detection of Varying Partial Shading Conditions for Photovoltaic Systems 1525
Rosalie Rouphael, Nezha Maamri, Jean-Paul Gaubert

Novel Operation Mode of the Modular Multilevel Matrix Converter Based on a Dimensioning Algorithm 1533
Rebecca Dierks, Axel Mertens

On the Cosmic Ray Influence on the Electronics Design of a High Altitude Electric Aircraft 1543
  Philippe Morey, Mauro Carpita

DC-Bus Control Considerations of Asymmetrical Multilevel Inverters with Embedded Buck-Boost
Converter 1551
  Theodoros P. Mouselinos, Emmanuel C. Tatakis

A Seamless Modulation Strategy for Step-Up/Down Partial Power Processing Converter (SUD-
P3C) 1561
  *Chao Liu, Zhe Zhang, Ziwei Ouyang, Jiasheng Huang, Michael A. E. Andersen, Tiberiu
  Gabriel Zsurzsan*

Performances Analysis of Non-Model-Based Speed Estimation Algorithms for Motor Drives 1569
  *Gaetano Turrisi, Luigi Danilo Tornello, Giacomo Scelba, Giulio De Donato, Giuseppe
  Scarcella*

A Method to Design Power Control System of Wayside Energy Storage System for Energy Saving
in DC-Electrified Railway 1580
  Kota Sato, Keiichiro Kondo, Hiroyasu Kobayashi, Makoto Chida

A Reconfigurable Single-Stage Three-Phase Electric Vehicle DC Fast Charger Compatible with
Both 400V and 800V Automotive Battery Packs 1590
  Mojtaba Forouzesh, Yan-Fei Liu, Paresh C. Sen

Efficiency Improvement of Single-Stage AC-DC LLC Converter using a Line Cycle Synchronous
Rectifier (SR) Driving Strategy 1601
  Mojtaba Forouzesh, Yan-Fei Liu, Paresh C. Sen

Influence of DC Supply Voltage Unbalances on the Performance of ARCP Inverters 1611
  Gholamreza Tabrizi, Sebastian Sprunck, Marco Jung

Grid-Forming Control for Enhanced Microgrid Interconnection 1620
  Tobias Erckrath, Christian Bendfeld, Peter Unruh, Axel Seibel, Marco Jung

Low Phase Shift Filter for Current Sensing Based on the Difference Between AC Machine Models
with and Without Iron Losses 1631
  Niklas Himker, Marcel Krümpelmann, Axel Mertens

Design and Analysis of a Voltage Clamping Active Delay Control Method for Series Connected
SiC MOSFETs 1641
  Rui Wang, Asger Bjørn Jørgensen, Hongbo Zhao, Stig Munk-Nielsen

Practical Implementation of a Concept for In-Situ Detection of Humidity-Related Degradation of
IGBT Modules 1649
  Benedikt Kostka, Axel Mertens

Design for Enhanced Noise Immunity of PCB Coils Used for Sensing Current Through Power
Devices 1658
  Aamir Rafiq, Sumit Pramanick

Measurement Principle for Measuring High Frequency Bearing Currents in Electric Machines and
Drive Systems 1665
  Benjamin Knebusch, Lennart Junemann, Pauline Holtje, Axel Mertens, Bernd Ponick

Climatically Induced Insulation Degradation in Power Semiconductor Modules of Wind Turbines 1674
  Timo Lichtenstein, Sören Fröhling, Bernd Tegtmeier, Katharina Fischer

Comparison of Magnetic Noise Compensation Techniques for Dual Three-Phase Electrically Excited Synchronous Machines...1684
    Jonas Henkenjohann, Jan Andresen, Axel Mertens

PCB Technology Comparison Enabling a 900V SiC MOSFET Half Bridge Design for Automotive Traction Inverters 1692
    Matthias Spieler, Che-Wei Chang, Ayman El-Refaie, Muhammad H Alvi, Dong Dong, Rolando Burgos

Desaturated Turn-Off of Low-Saturation IGBTs with Clamping Method to Reduce Turn-Off Energy Losses...1703
    Vishwas Acharya Nayampalli, Hans-Günter Eckel

Impact of Bond Wire Configuration on the Power Cycling Capability of Discrete SiC-MOSFET Devices 1713
    Patrick Heimler, Nick Thönelt, Josef Lutz, Thomas Basler

A Low-Leakage, Low-Loss Magnetic Transformer Structure for High-Frequency Applications.................1722
    Allen Nguyen, Ajinkya Phanse, Michael Solomentsev, Alex J. Hanson

Temperature Distribution of an IGBT Chip During Repetitive Switching Events Under Consideration of Front-Side Ageing...1733
    Christian Bäumler, Bo Zhang, Maximilian Goller, Xing Liu, Thomas Basler

Boosting Pilot-Diode Reverse-Conducting IGBTs Turn-ON and Reverse-Recovery Losses with a Simple Gate-Control Technique...1744
    Daniel Lexow, Hans-Günter Eckel

Modeling of an Interleaved DC-DC Boost Converter for a Direct Model Predictive Control Strategy...1754
    Thomas Effenberger, Hannes Böorngen, Eyke Liegmann, Michael Hoerner, Petros Karamanakos, Ralph Kennel

Static Analysis and Control Strategies of the Single Active Bridge Converter 1765
    Alexis A. Gómez, Alberto Rodríguez, Marta M. Hernando, Diego G. Lamar, Javier Sebastián, Ibán Ayarzaguena, Jose Manuel Bermejo, Igor Larrazabal, David Ortega, Francisco Vázquez

Multi-Port Inductive Power Transfer System Considering Charging Auxiliary Battery in EVs...................1776
    Zhuoqi Zhang, Ryosuke Ota, Ryohei Okada, Nobukazu Hoshi

Influence of IGBT and Diode Parameters on the Current Sharing and Switching-Waveform Characteristics of Parallel-Connected Power Modules...1785
    Y. Ando, J. Sakai, K. Hatori, N. Soltau, E. Wiesner

Innovative Driving Scheme for Electrical Generators in More Electric Aircrafts Employing Series Active Filtering...1796
    Nena Apostolidou, Nick Papanikolaou

Field-Measurement Based Hygrothermal Modelling of the Converter-Cabinet Climate in Wind Turbines...1804
    Katharina Fischer, Katherina Gohler

A Multi-Mode Control Based Asymmetrical Dual-Active-Bridge Series-Resonant DC-DC Converter (DABSRC) 1815
    M. Yaqoob, Grover Torrico, Wang Shuqin

Extended Balancing and Dimensioning of Capacitors in MMC Double Submodules 1824
Ali Sharaf Addin, Christopher Dahmen, Thomas Brückner

Saliency Extraction and Torque Sharing Estimation of Dual Motor Drive using Special Current Sensor Configuration... 1834
E. Rodriguez Montero, M. Vogelsberger, T. Wolbank

Soft-Switching Converter for Inductive Power Transfer System with Double-Sided LCC Resonant Network 1844
Ryohei Okada, Ryosuke Ota, Nobukazu Hoshi

Ultra Low Loss - MMC Submodules Favorable for SiC-FET Enabling High Functional Safety 1855
Christopher Dahmen, Rainer Marquardt

Control of an Active Gate Driver for an Electric Vehicle Traction Inverter using Artificial Neural Networks 1865
Julius Wiesemann, Jacob Dumtzlaff, Axel Mertens

Cascaded H-Bridge Converter Designs for Future Short-Range All-Electric Aircraft Propulsion 1875
Maximilian Hagedorn, Malte Lorenz, Axel Mertens

Overview and Evaluation of Energy Balancing Techniques for MMCs with Various Input and Output Frequencies... 1885
Gyanendra Kumar Sah, Michael Schütt, Hans-Günter Eckel

Comparative Lifetime Estimations for IGBT Modules in Wind Turbine Converters 1895
Christian Neumann, Hans-Gunter Eckel

Single-Phase, Five-Level Inverter with SPWM-Based Neutral Point Voltage Balancing Scheme 1906
Dmytro Kondratenko, Arkadiusz Lewicki, Charles Odeh

Magnetic Core Evaluation Kit for the Comparison of Core Losses 1914
Wilmar Martinez, Xiaobing Shen, Siqi Lin, Jens Friebe

Multi-Objective Optimization of Modular Multilevel Converter Systems... 1923
Nikolaus Patzelt, Christian Schlegel, Michail Vasiladiotis

Sizing of Hybrid Energy Storage System for Residential PV Applications 1933
Xiangqiang Wu, Zhongting Tang, Tamas Kerekes

DC Bias Currents in Full-Bridge DC-DC Converters in Context of WBG Semiconductors and High Switching Frequencies... 1939
Niklas Badenhop, Lukas Fräger, Dennis Kampen, Sascha Langfermann, Michael Owzareck

Parameter Tuning Method for Class Φ_2 Converters for High-Frequency Wireless Power Transfer Applications.. 1947
Yining Liu, Prasad Jayathurathnage, Jorma Kyyrä

Inductor Design Optimization using FEA Supervised Machine Learning 1955
D. Cajander, I. Viarouge, P. Viarouge, D. Aguglia

Enabling Large-Scaled MMC EMT-RMS Co-Simulation by Data Exchange in the Loop (DXiL)............... 1966
Xiong Xiao, Soham Choudhury, Martin Coumont, Jutta Hanson

Advanced Low-Voltage System-In-Package Half-Bridge MOSFET with Added Protection Features.......... 1975
S. Musumeci, V. Barba, F. Scrimizzi, C. Mistretta

Evaluation of Common-Mode Leakage Current of Aalborg-Type Transformerless PV Inverters 1985
Georgios I. Orfanoudakis, Eftychios Koutroulis, Georgios Foteinopoulos, Weimin Wu

Multi-Frequency Traction-To-Auxiliary Integrated EV Drivetrain: Eliminating the Need for an
Auxiliary Power Module 1995
Caniggia Viana, Mehanathan Pathmanathan, Peter W. Lehn

Potentials to Improve the Post-Fault Performance of a Fault-Tolerant Inverter System in Electrified
Aircraft Propulsion System 2003
Yongtao Cao, Leon Fauth, Jens Friebe, Axel Mertens

Model Predictive Control-Enabled Fault Ride Through Operation Strategy for High Power Wind
Turbine 2011
Pedro Catalán, Yanbo Wang, Zhe Chen, Joseba Arza

A Theoretical Comparison of Different Virtual Synchronous Generator Implementations on
Inverters 2021
Patrick Körner, Andrea Reindl, Hans Meier, Michael Niemetz

Linear Flux-Switching Machine Design - A Multiobjective Optimization 2030
Hendrik Marks, Henning Schillingmann, Sridhar Balasubramanian, Markus Henke

Single-Arm MMC-Based Converter for Transformerless Rail Interties 2038
Simon Beck, Simon Fuchs, Jürgen Biela

Medium Voltage Diode Rectifier Design for High Step-Up DC-DC Converter 2049
Pierre Le Métayer, Cyril Buttay, Drazen Dujic, Piotr Dworakowski

Fast Switching Planar Inductance Current Source ZETA Converter with Integrated Common Mode
Filter 2058
Benjamin H. Zacher, Christian Schumann

System Level Simulation of Moisture Propagation and Effects in Wind Power Converters 2066
Johannes C. Wenzel, Axel Mertens

PWM-Based Optimization-Free Active Voltage-Balancing Control of 7-Level Active Neutral-
Point-Clamped Flying-Capacitor Multicell Inverters 2073
Vahid Dargahi

Model Predictive Power Sharing Algorithm for Fuel Cell Integration in a Dual Inverter Electric
Vehicle Drivetrain 2084
Mehanathan Pathmanathan, Caniggia Viana, Sukhjit Singh, Peter W. Lehn

Comparative Evaluation of the 5-Phase Vienna and the 5-Phase PWM Rectifiers Under DC
Voltage Control 2092
A. Dieng

Modelling and Control of a 50kW SiC-Based Isolated DAB Converter for Off-Board Chargers of
Electric Vehicles 2101
*Haaris Rasool, Manh Tuan Tran, Sajib Chakraborty, Joeri Van Mierlo, Thomas Geury,
Mohamed El Baghdadi, Omar Hegazy*

Impact of Cyber Attacks on Cost Oriented Power Routing Schemes in Microgrids 2110
Kirti Gupta, Subham Sahoo, Bijaya Ketan Panigrahi, Frede Blaabjerg

Response of IGBT Chip Characteristics Due to Critical Stress 2119
Kohei Yamauchi, Rik W. De Doncker

Mega-Hertz High-Power WPT System with Parallel-Connected Inverters using Current Balance Circuit 2127
    Masamichi Yamaguchi, Keisuke Kusaka, Jun-Ichi Itoh

Investigation and Mitigation of Common-Mode Voltage in Four-Level NPC Converters Modulated by Redundant Level Modulation 2136
    Jun Wang, Wei Xu, Xibo Yuan, Lihong Xie

Ferrite Optimization for a Three-Phase Wireless Power Transfer System for Electric Vehicles 2145
    Shuang Nie, Mehanathan Pathmanathan, Peter W. Lehn

Frequency and Modulation Index Related Effects in Continuous and Discontinuous Modulated Y-Inverter for Motor-Drive Applications 2156
    Hamzeh J. Jaber, Alberto Castellazzi

Performance Evaluation of Sinusoidal-Flux Reluctance Machine for Improving Power Density with Reduced Torque and Input-Current Ripples 2164
    Kiwa Nagayasu, Masaki Iida, Kazuhiro Umetani, Mastaka Ishihara, Eiji Hiraki

Power Hardware-In-The-Loop Test of Low-Voltage Battery for a Plug-In Hybrid Electric Vehicle 2175
    Ronan German, Florian Tournez, Alain Bouscayrol, Aurelien Lievre, Betty Lemaire-Semail

Stability Analysis of DFIG System Connected with High-Frequency Capacitive Grid Based on Closed-Loop Current Control and Direct Power Control 2182
    Bin Hu, Heng Nian, Subham Sahoo, Frede Blaabjerg, Yaqian Zhang, Zixiao Xu

Full-Bridge Modular Multilevel Converter for the Four-Quadrant Supply of High Power Magnets in Particle Accelerators 2189
    Manuel Colmenero, Ricardo Vidal-Albalate, Francisco R. Blanquez, Ramon Blasco-Gimenez

Deep Neural Network for Magnetic Core Loss Estimation using the MagNet Experimental Database 2197
    Xiaobing Shen, Hans Wouters, Wilmar Martinez

Hybrid Circuit Board Structure for Power Electronics 2205
    Gerrit Braun, Deniz-Heinz Moldenhauer

Active Control of Gear Mesh Vibration using a Permanent-Magnet Synchronous Motor and Simultaneous Equation Method 2211
    Dominik Reitmeier

Research Laboratory for Testing Grid Connected Devices Under Grid Voltage / Grid Impedance Variations and Microgrid Conditions 2219
    Swen Bosch, Jochen Staiger, Heinrich Steinhart

Reducing the Impact of Skin Effect Induced Measurement Errors in M-Shunts by Deliberate Field Coupling 2230
    Hauke Lutzen, Jonas Müller, Vladimir Polezhaev, Till Huesgen, Nando Kaminski

Grid Forming Control for HVDC Systems: Opportunities and Challenges 2241
    Adil Abdalrahman, Ying-Jiang Häfner, Malaya Kumar Sahu, Khirod Kumar Nayak, Ashkan Nami

A Highly Integrated and Modular High Speed Electric Drive for Lightweight Electric Mountain Bikes 2251
    Matthias Hofer, Mario Nikowitz, Manfred Schrödl

Performance Enhancement of Power Conditioning Systems in More Electric Aircrafts 2257
Nick Rigogiannis, Nick Papanikolaou, Yongheng Yang

Steady State Simulations of a Hybrid HVAC/HVDC Network using OS Based ARM Devices 2266
Ioan Catalin Damian, Mircea Eremia

Experimental Comparison of FPGA-Implemented Model Predictive Voltage Control to Cascaded
Proportional Resonant Control for a Three-Phase Four-Wire Three-Level Grid-Forming Inverter of
250 kVA 2276
Jarren Lange, Dominik Schmies, Karl Stephan Stille, Joachim Böcker, Oliver Wallscheid

Experimental Study of Interleaved Y-Inverter Performance 2285
Yusuke Endo, Masataka Minami, Hamzeh J. Jaber, Alberto Castellazzi

Design of a GaN-Based Reconfigurable Resonant Converter for High Frequency On-Board
Charger of Battery Electric Vehicles 2293
*Manh Tuan Tran, Haaris Rasool, Dai Duong Tran, Mohamed El Baghdadi, Philippe Lataire,
Omar Hegazy*

Transient Liquid Phase Bond Reliability Evaluation of Die-Attach for Power Module Packaging 2301
Laxma R. Billa, Yangang Wang, Thomas Grant, Xiang Li, Harley Neal, Muhammad Morshed

Experimental Evaluation on Observer-Based Delay-Compensating Active Damping for LC-Filters............ 2308
Michael Schütt, Hans-Günter Eckel

Influence of Static Rotor Imbalance on the Roller Bearing Damage Due to Inverter-Induced
Bearing Currents... 2316
Martin Weicker, Omid Safdarzadeh, Andreas Binder

Novel Current Balancing Method for HF Interleaved Converters with Reduced Control Effort 2327
Christian Beckemeier, Jens Friebe

dV/dt-Based Filter Design for Motor Inverters with Continuous Output Voltage 2334
Sabrina Ulmer, Stevan Bugarski, Gernot Schullerus, Ertugrul Sönmez

Evaluation of Core Losses in Transformers for Three-Phase Multi-Level DAB Converters........................ 2344
Babak Khanzadeh, Yuriy Serdyuk, Torbjörn Thiringer

A Quasi-Offline Condition Monitoring Method of DC-Link Capacitor Banks in Accelerator Power
Converters 2355
*Timm Felix Baumann, Konstantinos Papastergiou, Raul Murillo Garcia, Dimosthenis
Peftitsis*

Minimizing Voltage Stress in Auxiliary Resonant Commutated Pole Inverters using Saturable
Inductors... 2366
Markus Zocher, Norbert Grass, Ralph Kennel

Adaptive Dead-Time Control in a Resonant Wireless Power Transfer System 2375
Tim Krigar, Martin Pfost

Multilevel Battery Converter with Cascaded H-Bridges on Cell Level-Battery Management System
Or a Renewed Attempt for Power Electronic Building Blocks? 2383
*Max Rothenburger, Markus Horn, Xiao Yu, Gerold Schulze, Koenraad Muyllaert, Peter
Zacharias, Ludwig Brabetz, Hartmut Hillmer*

Design and Potential of EMI cm Chokes with Integrated DM Inductance..................................... 2392
Mohammad Ali, Rehnuma Bushra, Jens Friebe, Axel Mertens

Implementation Options of a Fully SiC Buck-CSI for Advanced Motor Drive Application........................ 2402
    Yonghwa Lee, Alberto Castellazzi

Optimized Control Scheme to Achieve ZVS for the Complete Pre-Charging Phase of
Supercapacitors with a 500 kHz SiC- And GaN-Based Dual Active Bridge 2413
    Patrick Lenzen, Martin Pfost

Fault Blocking Capability in the DC-MMC with Reduced Number of Sub-Modules................................... 2422
    J. D. Páez, F. Morel, S. Bacha, P. Dworakowski

An Open-Source FEM Magnetic Toolbox for Calculating Electric and Thermal Behavior of Power
Electronic Magnetic Components 2432
    Nikolas Förster, Jonas Hölscher, Till Piepenbrock, Philipp Rehlaender, Oliver Wallscheid,
    Frank Schafmeister, Joachim Böcker

Comparison of Dual-Active-Bridge-Based Topologies for Single-Phase Single-Stage EV On-Board
Chargers 2441
    Daniel Gaona, Denis Pauls, Eduardo Facanha De Oliveira

Design Concepts for Medium Voltage DC Networks Supplying the Future Circular Collider (FCC)........... 2451
    Manuel Colmenero, Francisco R. Blanquez, Ramon Blasco-Gimenez

A Novel Dual CC-CV Output Wireless EV Charger with Minimal Dependency on Both Coil
Coupling and Load Variation 2462
    Subhranil Barman, Kishore Chatterjee

A High-Performance EMI Filter Based on Laminated Ferrite Ring Cores 2470
    Marcin Kacki, Marek S. Rylko, John G. Hayes, Charles R. Sullivan

Investigation of the Static Performance and Avalanche Reliability of High Voltage 4H-SiC
Merged-PiN-Schottky Diodes 2477
    Chengjun Shen, Saeed Jahdi, Phil Mellor, Juefei Yang, Erfan Bashar, Jose Ortiz-Gonzalez,
    Olayiwola Alatise

On Chain-Link Based Multi-Port Converters Able to Connect HVDC and MVDC to AC
Transmission Network... 2486
    Daniele Falchi, Oriol Gomis-Bellmunt, Eduardo Prieto-Araujo, Olivier Despouys

Voltage Control Scheme for Multilevel Interfacing PV Application: Real-Time MRAC-Based
Approach 2496
    Mohammad Sadegh Orfi Yeganeh, Mehdi Rahmani, Nenad Mijatovic, Tomislav Dragicevic,
    Frede Blaabjerg, Pooya Davari

Control Principles for Island Operation and Black Start by Offshore Wind Farms Integrating Grid-
Forming Converters... 2504
    Daniela Pagnani, Lukasz Kocewiak, Jesper Hjerrild, Frede Blaabjerg, Claus Leth Bak

Experimental Study of the Reduction and Removal of Turn-On Snubber for IGCT Based MMC
Submodule using Fast Silicon Diodes 2515
    Arthur Boutry, Cyril Buttay, Besar Asllani, Bruno Lefebvre, Eric Vagnon, Dong Dong

Characterisation of a Ferrite-Polymer Based Magnetic Material 2526
    Johan Le Leslé, Guillaume Lefevre, Julien Morand, Rémi Perrin, Pierre-Yves Pichon,
    Guillaume Regnat

Model Predictive-Based Control Technique for Fault Ride-Through Capability of VSG-Based Grid-Forming Converter.. 2537
    Mobina Pouresmaeil, Amir Sepehr, Basit Ali Khan, Jafar Adabi, Edris Pouresmaeil

Grounding Points in HV/MV Hybrid Transformer Auxiliary Converters.. 2544
    Adrian Wiemer, Jürgen Biela

Non-Parasitic Induced Transient Overvoltage in ANPC Topology Due to Critical Switching Sequences 2554
    Michael Geiss, Robert Kragl, Jürgen Thoma, Benjamin Volzer

Open-Delta SBC: A New Converter Topology with Low Number of Sub-Modules for MV Applications... 2564
    D. Lanzarotto, P. B Steckler, K. Vershinin, F. Morel

Characterising the Effect of an Inverter on the Regulation of the AC Voltage using a Frequency Response Identification Technique 2574
    Mohamed Aldarmon, Joan Marc Rodriguez, Adria Junyent-Ferre

Artificial-Intelligence Based DC-DC Converter Efficiency Modelling and Parameters Optimization 2581
    Fanghao Tian, Diego Bernal Cobaleda, Wilmar Martinez

Analysis of the Loss Distribution of a 6 kW Two Stage Power Supply for 600 V DC Applications............. 2588
    Lukas Fräger, Sascha Langfermann, Michael Owzareck, Dennis Kampen, Jens Friebe

Study on the Gate Loop Design and Its Impact on Switching Characteristics of GaN Transistors................ 2596
    Xiaomeng Geng, Carsten Kuring, Oliver Hilt, Mihaela Wolf, Joachim Würfl, Sibylle Dieckerhoff

Analysis of Current Sharing in the Parallel Connection of GaN Transistors 2607
    Frederik Stalleicken, Sibylle Dieckerhoff, Karsten Handt, Sebastian Nielebock

Verification of GaN-HEMT Spice Models using an S-Parameters Approach 2618
    Alonso Gutierrez, Nasri Said, Emmanuel Marcault, Mathieu Gavelle

Power Loss Modelling of GaN HEMT-Based 3L-ANPC Three-Phase Inverter for Different PWM Techniques.. 2628
    Salvatore Mita, Arjun Sujeeth, Giuseppe Aiello, Dario Patti, Francesco Gennaro, Giacomo Scelba, Mario Cacciato

Generalized Core and Winding Area Ratio - Trends for Inductors and Transformers in Power Electronics with High Switching Frequencies... 2638
    Siqi Lin, Leon Fauth, Wilmar Martnez, Jens Friebe

Active Substrate Termination of Discrete and Monolithic Bidirectional GaN HEMTs in a T-Type Inverter 2644
    Carsten Kuring, Yannic Lange, Xiaomeng Geng, Oliver Hilt, Mihaela Wolf, Joachim Würfl, Sibylle Dieckerhoff

Transformer Design Optimization and Comparison for a DC-DC Converter Used in PV Micro-Inverters... 2655
    Tobias Manthey, Meriem Khader, Jens Friebe

Automated Gate Impedance Network Design for SiC MOSFETs using SPICE Solver Interfaced with MATLAB Environment 2661
    Pawel Piotr Kubulus, Szymon Michal Beczkowski, Stig Munk-Nielsen, Asger Bjørn Jørgensen

An Improved Multi-Loop Resonant and Plug-In Repetitive Control Schemes for Three-Phase
Stand-Alone PWM Inverter Supplying Non-Linear Loads 2670
  Ahmad Ali Nazeri, Peter Zacharias

High Switching Frequency Operation of a Single-Phase Five-Level Hybrid Active Neutral Point
Clamped Inverter with a Model Predictive Control Approach 2682
  Mohammad Najjar, Mahdi Shahparasti, Rasool Heydari, Morten Nymand

Design of Planar Coupled Inductor Applied to Zero-Current Switching Clamped Current Converter 2689
  Vinicius Freire Bezerra, Tobias Manthey, Montiê Alves Vitorino, Jens Friebe

Characterization of Online Junction Temperature of the SiC Power MOSFET by Combination of
Four TSEPs using Neural Network 2698
  Kanuj Sharma, Simon Kamm, Kevin Muñoz Barón, Ingmar Kallfass

Novel Extended Robust Disturbance Observer for Improved Cogging Force Compensation in
Permanent Magnet Linear Motors 2706
  Franz Luckert, Axel Mertens

Improvement of a Self-Powered Gate Driver Power Supply 2715
  *Mariana Raya, Oriol Aviñó, Sergio Busquets-Monge, Xavier Perpiñá, Miquel Vellvehi, Xavier
  Jordà*

Optimization and Scaling of a Compact High-Power IGCT Capacitor Charger Based on Simulation
and Measurements with a 300 kW/3.3 kV Demonstrator 2726
  Felix Haag, Fabian Albrecht, Volker Brommer, Oliver Liebfried, Klaus F. Hoffmann

Multilayer Busbars for Medium Voltage ANPC Converter Dedicated to Battery Energy Storage
Systems 2736
  Mamadou Lamine Beye, Luc Bimmel, Anthony Bier, Jérémy Martin

A Simulation Model for SiC MOSFET Switching Transients Controlled by an Adaptive Gate
Driver with the Capability of Reducing Switching Losses and EMI Across the Full Operating
Range 2744
  Zheming Li, Robert W. Maier, Mark-M. Bakran, Franz-J. Niedernostheide, Daniel Domes

Phase-Shift Modulation for Flying-Capacitor DC-DC Converters 2754
  Philipp Rehlaender, Frank Schafmeister, Joachim Böcker

An EV Integrated Isolated DC Charger using a Six-Phase Synchronous Machine 2763
  Sukhjit S Ghumman, Mehanathan Pathmanathan, Peter W Lehn

Configurable ISOP-IPOP DC-DC Converter for Universal Solid-State Transformer 2773
  Pramod Apte, Jens Friebe, Lukas Fräger

Using System-On-Chip Boards for the Deployment of Controller for Verification and Prototyping 2780
  Adeel Jamal, Gerd Griepentrog

Utilizing the Reactive Current Control Capability of an MMC-Fed AC/DC Converter for Volt-
Second Balancing in Medium Frequency Transformers 2788
  *Kaveh Pouresmaeil, Maurice Roes, Jorge Duarte, Korneel Wijnands, Nico Baars, George
  Papafotiou*

Cost Comparison for Different PV-Battery System Architectures Including Power Converter
Reliability 2795
  *Martijn Deckers, Leander Van Cappellen, Glenn Emmers, Fereshteh Poormohammadi, Johan
  Driesen*

Insulation Design and Analysis of a Medium Voltage Planar PCB-Based Power Bus Considering Interconnects and Ancillary Circuit Integration 2806
Joshua Stewart, Rolando Burgos, Dushan Boroyevich

Modular Multilevel Converter Control with using a General Space Vector PWM Method in Medium Voltage Hydro Power Application 2813
Chengjun Tang, Torbjörn Thiringer

A Technical Overview of Single-Stage Three-Port DC-DC-AC Converters 2824
Sebastian Neira, Zoe Blatsi, Michael M. C. Merlin, Javier Pereda

Common-Mode EMI Noise Modeling of Three-Level T-Type Inverter for Adjustable Speed Drive Systems 2835
Vefa Karakasli, Abdelmoumin Allioua, Gerd Griepentrog

A Condition Monitoring Scheme for Semiconductor Devices in Modular Multilevel Converters with Cascaded H-Bridge Submodules 2843
Mohsen Asoodar, Mehrdad Nahalparvari, Christer Danielsson, Hans-Peter Nee

Particular Requirements on Drive Inverters for Safe and Robust Operation on an Open Industrial DC Grid 2852
Simon Puls, Jan-Niklas Koch, Martin Ehlich, Holger Borcherding

Investigation About Operation and Performance of Gate Drivers for Power Electronics Converters for Cryogenic Temperatures 2860
Mustafeez-Ul-Hassan, Yuxuan Wu, Vyacheslav Solovyov, Fang Luo

Synchronization Angle Determination in DVCSFO of DFIM Naval Propulsion 2869
Youssef Drimizi, Maria Pietrzak-David, Pascal Maussion

Power Control of LCR-DAB Converter with Phase Shift in Fixed Switching Frequency 2877
Seung-Hyuk Baek, Jaehong Lee, Seung-Hwan Lee, Sungmin Kim

A Simplified Braking Method for Direct Matrix Converter-Fed PMSM Drives with Consideration of Avoiding Regenerative Energy 2885
Jun Xie, Dustin Henneberg, Martin Suberski, Thomas Ellinger, Uwe Radel, Jürgen Petzoldt

Inverter-Machine Parametric Co-Design for Energy Efficient Electric Drives 2893
Jaedon Kwak, Alberto Castellazzi

Bidirectional Cuk Converter in Partial-Power Architecture with Current Mode Control for Battery Energy Storage System in Electric Vehicles 2903
J. S. Artal-Sevil, J. Anzola, V. Ballestín-Bernad, I. Aizpuru

Design Space Exploration for a Capacitive 36V, 4A, 4:1 DCDC Converter with GaN Switches using a Performance-Cost-Matrix Including Uncommon Topologies 2912
Adrian Gehl, Malte Kempchen, Simon Disselkamp, Markus Olbrich, Bernhard Wicht

A Fast Control for a Three-Switch Multi-Input DC-DC Converter 2919
Simone Cosso, Andrea Formentini, Mario Marchesoni, Massimiliano Passalacqua, Luis Vaccaro

Impact on the Torque and on the Copper Losses Under Fault-Tolerant Control of 5-Phase PMSG 2930
A. Dieng

Weighting Factor Design for FS-MPC in VSCs: A Brain Emotional Learning-Based Approach 2939
Mohammad Sadegh Orfi Yeganeh, Arman Oshnoei, Saeed Peyghami, Nenad Mijatovic,
Tomislav Dragicevic, Frede Blaabjerg

A Strategy for Smooth Microgrid Transitions Without Phase Misalignment and Voltage Mismatch 2948
Gabriel Silva Rocha, Amiron Wolff Dos Santos Serra, Cesar Augusto Santana Castelo
Branco, Hercules Araujo Oliveira, Jose Gomes De Matos, Luiz Antonio De Souza Ribeiro

Subtle Design and Performance Comparison of WF-FSM and DC-VRM for Large-Scale Direct-
Drive Wind Power Generation 2958
Udochukwu B. Akuru, Maarten J. Kamper, Zi-Qiang Zhu

Analysis and Implementation of Different Non-Isolated Partial-Power Processing Architectures
Based on the Cuk Converter.. 2967
J. S. Artal-Sevil, J. Anzola, V. Ballestín-Bernad, J. L. Bernal-Agustín

GaN HEMT and SiC Diode Commutation Cell Based Dual-Buck Single-Phase Inverter with
Premagnetized Inductors and Negative Gate Driver Turn-Off Voltage 2977
Tobias Brinker, Hendrik Gräber, Jens Friebe

Determination of Optimal Associated Discrete Circuit Switch Model Parameters for Real-Time
Simulation of Dual-Active Bridge Converters 2985
Marija Stevic, Ravinder Venugopal

Integrated Motor Drive: A Multidisciplinary Approach... 2996
Betty Lemaire-Semail, Nadir Idir, Eric Semail, Souad Harmand

Hardware in the Loop Test of an Electric Aircraft Powertrain.. 3005
Sebastian Mönninghoff, Moritz Scholjegerdes, Kay Hameyer

A Multi-Port Smart Transformer for Green Airport Electrification 3014
Giampaolo Buticchi, Giovanni De Carne, Thiago Pereira, Kangan Wang, Xiang Gao, Jiajun
Yang, Youngjong Ko, Zhixiang Zou, Marco Liserre

Improvement of EMI Filter Attenuation using Shielding.. 3022
Mohammad Ali, Rehnuma Bushra, Jens Friebe, Axel Mertens

Implementation of Onsite Junction Temperature Estimation for a SiC MOSFET Module for
Condition Monitoring.. 3031
Farzad Hosseinabadi, Shahid Jaman, Sachin Kumar Bhoi, Md. Mahamudul Hasan, Sajib
Chakraborty, Mohamed El Baghdadi, Omar Hegazy

Energy Storage Systems for Airborne Wind Generators.. 3037
Bakr Bagaber, Axel Mertens

Design Interactions of AC- And DC-Side Filters for Traction Drives with SiC Inverters 3048
Hedieh Movagharnejad, Benjamin Knebusch, Axel Mertens, Bernd Ponick

Investigation of an Interleaved Current-Fed Single Active Bridge DC-DC Converter for PV
Applications.. 3059
Lucas Vinícius De Araújo Gomes, Tobias Manthey, Montiê Alves Vitorino, Jens Friebe

Real-Time Thermal Characterization of Power Semiconductors using a PSO-Based Digital Twin
Approach 3067
Johannes Kuprat, Younn Pascal, Marco Liserre

Self-Sensing Design and Control for an Induction Machine with an Additional Short-Circuited Rotor Coil 3075
Stefan Luecke, Axel Mertens

Calculating the Tractive Power and Power Conversion Efficiency of Battery Electric Vehicles using a Global Navigation Satellite System and a Road Elevation Database 3084
Shinichi Domae, Alberto Castellazzi, Hamzeh J. Jaber, Tenghui Dong, Taketsune Nakamura

PCB Layer Optimization of Planar Medium Frequency Transformer for On-Board EV Chargers 3092
Fabian Groon, Hamzeh Beiranvand, Thiago Pereira, Görkem Can, Marco Liserre

Fault Current Capability Assessment of Low-Voltage Side Inverters in Smart-Transformers 3101
Thiago Pereira, Luis Camurca, Francisco Santos, Marco Liserre

Adaptive Resonant-Valley Switching for a GaN HEMT Direct AC-AC Auxiliary Resonant Commutated Pole Converter 3112
Kyle Steyn, Johan Beukes

The Variation of Core Loss in High-Frequency Transformers Under Different Load Conditions 3120
Navid Rasekh, Jun Wang, Xibo Yuan

A Complete PFC Inductor Design for Lighting Equipment Applications 3130
Wai Keung Mo, Kasper M. Paasch, Thomas Ebel

Automatic Generation Control-Based Charging/Discharging Strategy for EV Fleets to Enhance the Stability of a Vehicle-To-Weak Grid System 3140
Majid Mehrasa, Mehrdad Gholami, Reza Razi, Khaled Hajar, Antoine Labonne, Ahmad Hably, Seddik Bacha

Model-Based Converter Control for the Emulation of a Wind Turbine Drive Train 3149
Alexander Ernst, Wilfried Holzke, Dawid Koczy, Nando Kaminski, Bernd Orlik

A Novel Grid-Demanded Power Point Tracking (GPPT) Control Method for Wind Turbines to Preserve Grid Stability with High Wind Energy Penetration 3159
David Matthies, Alexander Ernst, Henning Sauerland, René Reimann, Wilfried Holzke, Bernd Orlik

Extension and Implementation of a Model-Based Lifetime Monitoring System with Parallel Calculation of Multiple Power Semiconductors 3169
Steffen Menzel, Wilfried Holzke, Michael Hanf, Holger Groke, Bernd Orlik, Nando Kaminski

Smart Charging Strategy for Electric Vehicles using an Optimized Fuzzy Logic System 3179
M. Gholami, M. Mehrasa, R. Razi, K. Hajar, A. Hably, S. Bacha, A. Labonne

Analysis and Discussion of a Concept for an Adjustable Inductance Based on an Impact of an Orthogonal Magnetic Field 3188
Guido Schierle, Michael Meissner, Klaus F. Hoffmann

A Field Programmable and Dynamic Configurable Power Electronic Converter Concept 3198
Bjarte Hoff

DAB Converter Discrete ADRC Control into Real-Time CHIL Simulation of a MVDC/LVDC Power Grid 3206
Alessio Clerici, Riccardo Chiumeo, Diego Raggini, Alessandro Veroni

SNNFT: Sequential Neural Network-Fuzzy Thermal Early Warning System for Lithium-Ion Batteries.. 3215
Marui Li, Chaoyu Dong, Yunfei Mu, Qian Xiao, Jingming Cao, Hongjie Jia

Fine-Grained Dynamics Representation and Stability Analysis for MMC-Based Hybrid AC/DC Power Systems 3225
Jingming Cao, Chaoyu Dong, Qian Xiao, Marui Li, Xiaodan Yu, Hongjie Jia

Adaptive Pontryagin's Minimum Principle-Inspired Supervised-Learning-Based Energy Management for Hybrid Trains Powered by Fuel Cells and Batteries 3235
Hujun Peng, Feifei Li, Zhu Chen, Kai Deng, Sebina Jeschke, Kay Hameyer

A Case Study of Pole-Phase Changing Induction Machine Performance 3246
Konstantina Bitsi, Sjoerd G. Bosga

New Topology of Superconducting Fault Current Limiter with Bypass Resistor 3254
D. Baimel, Eli Barbi, S. Bronstein, N. Baimel, A. Kuperman

A Pre- And Discharge Unit for Capacitive DC-Links Based on a Dual-Switch Bidirectional Flyback Converter 3262
Madlen Hoffmann, Martin März

Control and Integration of a Multiphase Brushless Wounded Synchronous Motor Drive 3272
Remi Perrin, Guilherme Bueno-Mariani

A Way Forward to Achieve Interoperability in Multi-Vendor HVDC Systems 3282
Adil Abdalrahman, Ying-Jiang Häfner, Philippe Maibach, Christoph Haederli

Model Predicitve Position Control of Electrical Drives on an Industrial PC 3292
Fabian Karau, Michael Leuer

Bidirectional Active EMC Filter for Industrial Power Converters 3301
Bernhard Wunsch, Stanislav Skibin, Ville Forsstrom

A General Method to Measure Parasitic Capacitance of Transformer using Guarding Technique 3309
Shaokang Luan, Stig Munk-Nielsen, Bruce Wakelin, Magnus Hortans, Jan Schupp, Hongbo Zhao

Inductance Analysis of Electric Machines by Classical and Numerical Methods 3318
J. J. Germishuizen, T. J. E. Miller

Dynamic Wireless Power Transfer DWPT Time Domain Model: Xyz Position and Speed Coupling Effect 3327
Iosu Aizpuru, Eneko Agirrezabala, Mikel Mazuela, Unai Iraola, Estanis Oyarbide, Carlos Bernal

Dynamic Average Small Signal Model of the SAB Converter 3336
Alexis A. Gómez, Alberto Rodríguez, Marta M. Hernando, Diego G. Lamar, Javier Sebastián, Ibán Ayarzaguena, Jose Manuel Bermejo, Igor Larrazabal, David Ortega, Francisco Vázquez

Algorithm for Optimal Selection of Drive Motor Transmission Combination............................ 3344
Santiago Ramos Garces, Dries Jacques, Stijn Derammelaere, Simon Houwen, Nick Van Oosterwyck, Bart Vanwalleghem

Evaluation of Drain-Source Voltage in Switch Transient Time Intervals as Gate Oxide Degradation Precursor of SiC Power MOSFETs.. 3353
Javad Naghibi, Sadegh Mohsenzade, Kamyar Mehran, Martin P. Foster

Active Output LLC Converter Topology 3362
  Hannes Börngen, Eyke Liegmann, Sriram Jagannath, Ralph Kennel

Short Circuit Type II and III Behavior of 1.2 kV Power SiC-MOSFETs.. 3373
  Xing Liu, Xupeng Li, Thomas Basler

Analog MPPT Comparison for Interplanetary Small Satellites Missions 3382
  C. Torres, A. Garrigós, J. M. Blanes, P. Casado, D. Marroquí, C. Orts

Feasibility Assessment of Variable-Speed Generator Set Concepts with Focus on Rating of Power
Electronic Equipment 3391
  Hendrik Fehr, Albrecht Gensior, Andreas Möckel, Frank Atzler, Tilo Roß, Carsten Reincke-Collon

Bus Voltage Regulation using Sequentially Switched ZVZCS Converters for Spacecraft Power
Systems.. 3401
  A. Garrigós, C. Orts, D. Marroquí, J. M. Blanes, C. Torres, P. Casado

A Standardized and Modular Power Electronics Platform for Academic Research on Advanced
Grid-Connected Converter Control and Microgrids 3411
  Frank S. R., Schulz D., Stefanski L., Schwendemann R., Hiller M.

Gate Input Capacitance Characterization for Power MOSFETs using Turn-On and Turn-Off
Switching Waveforms 3420
  Yota Nishitani, Michiko Inoue, Takashi Sato, Michihiro Shintani

AC Battery: Modular Layout with Cell-Level Degradation Control 3429
  Claudio Burgos-Mellado, Marcos Orchard, Diego Muñoz-Carpintero, Tomislav Dragicevic, Lorenzo Reyes-Chamorro, Jacqueline Llanos

Analysis of Test Methods for Measurement of Leakage and Magnetising Inductances in Integrated
Transformers 3440
  Sajad A. Ansari, Jonathan N. Davidson, Martin P. Foster, David A. Stone

A Topology-Morphing Series Resonant Converter for Photovoltaic Module Applications.................... 3450
  Grigorios Sergentanis, Liliana De Lillo, Lee Empringham, C. Mark Johnson

A Novel Parameter for the Evaluation of Protective Circuits for IGBT Explosion Protection in
Submodules of MMC 3460
  Christoph Junghans, Hans-Guenter Eckel

Sub-Modules Switching Algorithms for Dual Active Bridge Modular Multilevel Converters to
Optimize Capacitor Voltage Deviation Versus Power Efficiency 3470
  Peizhou Xia, Chuantong Hao, Stephen Finney, Michael Merlin

Systematic Adaptive Robust State Feedback Control for Active Front-End Rectifiers 3480
  Aidar Zhetessov, Giri Venkataramanan

An Optimized Compensation Strategy of Direct Matrix Converter-Fed PMSM Drives with Field
Weakening Under Unbalanced Supply Conditions 3491
  Jun Xie, Dustin Henneberg, Martin Suberski, Manuel Kusebauch, Uwe Rädel, Jürgen Petzoldt

Double Inverter Concept for High-Speed Drives Without Motor Filters 3501
  Henning Kasten, Stephan Beineke, Matthias Bachmann

A Universal Single Stage Current-Fed Bidirectional Converter with Both AC and DC Input Power Source Compatibility 3511
Manish Kumar, Sumit Pramanick, Bijaya Ketan Panigrahi

Optimization of Electric Vehicle Charge Scheduling with Consideration of Battery Degradation 3518
Raka Jovanovic, Sertac Bayhan, Islam Safak Bayram

Onboard ESU Sizing and Dynamic IPT Charging Scenarios for a Tramway Application 3529
Endika Bilbao Muruaga, Irma Villar, Florian Legay, Pierre Prenleloup, Jean-François Reynaud

Investigations on the Active Reduction of Common Mode Noise with Opposing Noise Sources 3536
Philipp Marx, Felix Seybold, Philipp Ziegler, David Hirning, Jörg Roth-Stielow

Knowledge Based Grey Box Modeling of Inaccessible Circuits for System EMC-Simulation in Time Domain 3545
Jan-Philipp Roche, Jens Friebe, Oliver Niggemann

Novel Quasi-Direct Rotor Position Estimator for Permanent Magnet Synchronous Machines Based on the Back-Electromotive Force using Current Oversampling 3555
Georg Lindemann, Viktor Willich, Axel Mertens

Design Considerations for Fast On-State Voltage Measurement Circuits 3565
Mathias C. J. Weiser, Manuel Rueß, Ingmar Kallfass

Analytical, FEM and Experimental Study of the Influence of the Airgap Size in Different Types of Ferrite Cores 3574
Asier Arruti, Francisco Jose Perez-Cebolla, Jon Anzola, Iosu Aizpuru, Mikel Mazuela

Design Method of a High Frequency GaN-Based Half-Bridge with Bottom-Side Cooled Transistors using Multi-PCB Assembly 3582
Loris Pace, Florian Chevalier, Thierry Duquesne, Nadir Idir

A 30 kW Dynamic Wireless Inductive Charging System for EVs 3590
Zariff Meira Gomes, José Renes Pinheiro, Gilney Damm, Karim Kadem, Hassan Moussa

Dynamic Control of the Switching Behavior of SiC MOSFETs in Converter Operation 3599
Jochen Henn, Laurids Schmitz, Rik W. De Doncker

A Series Resonant Balancing Converter for Bipolar DC Grids on Ships 3607
Sachin Yadav, Zian Qin, Pavol Bauer

A V2G-Enabled Seven-Level Buck PFC Rectifier for EV Charging Application 3615
Anekant Jain, Ritika Agarwal, Krishna Kumar Gupta, Sanjay K. Jain

Experimental Demonstration of a 2.2kW Active-Clamp Converter for High-Current Wide-Voltage-Transfer Ratio Applications 3625
Philipp Rehlaender, Bastian Korthauer, Frank Schafmeister, Joachim Böcker

A Simplified Model for the Battery Ageing Potential Under Highly Rippled Load 3636
Tomáš Kacetl, Jan Kacetl, Nima Tashakor, Stefan Goetz

System Modeling and Design of a Hybrid Renewable Energy System for a Cable Network Head-End Station in Rural Area 3646
Tobias Schillinger, Thomas Schuhmann, Martin Eckart

Comparison of System-Level Availability in Industrial Grids 3655
G. Emmers, J. Driesen

Ageing Mitigation and Loss Control in Reconfigurable Batteries in Series-Level Setups 3665
Tomáš Kacetl, Jan Kacetl, Nima Tashakor, Stefan Goetz

Characterization of Conventional and Advanced Current Measurement Techniques Suitable for
WBG Semiconductor Devices 3676
Severin Klever, André Thönnessen, Rik W. De Doncker

Zero-Sequence Voltage Reduces DC-Link Capacitor Demand in Cascaded H-Bridge Converters for
Large-Scale Electrolyzers by 40% 3686
Roland Unruh, Frank Schafmeister, Joachim Böcker

Thermal Behavior Impact on the Electric Motor Shape Multi-Objective Optimization 3696
Aissam Riad Meddour, Anthony Babin, Nassim Rizoug, Christopher Vagg, Richard Burke,
Laid Degaa

Modelling Approaches of Power Systems Considering Grid-Connected Converters and Renewable
Generation Dynamics 3704
Jaume Girona-Badia, Vinícius Albernaz Lacerda, Eduardo Prieto-Araujo, Oriol Gomis-
Bellmunt, Stephan Kusche, Florian Pöschke, Horst Schulte

Efficiency and Lifetime Analysis of Several Airborne Wind Energy Electrical Drive Concepts 3711
Bakr Bagaber, Daniel Heide, Bernd Ponick, Axel Mertens

Design and Performance Analysis of Single-Phase Axial Flux Permanent Magnet Motor for
Coaxial Cascade 3722
Chu Wang, Xiaowei Hu, Xiaoya Wang, Weiwei Geng, Qiang Li, Jingning Hou

Comparison of Pulse Current Capability of Different Switches for Modular Multilevel Converter-
Based Arbitrary Wave Shape Generator Used for Dielectric Testing of High Voltage Grid Assets 3729
Dhanashree Ashok Ganeshpure, Ajeeth Phrassanna Soundararajan, Thiago Batista Soeiro,
Mohamad Ghaffarian Niasar, Peter Vaessen, Pavol Bauer

Accurate Modeling of IGBT-Based Converters in PLECS 3740
Anne Von Hoegen, Philipp Tillmann, Tetsuya Kojima, Rik W. De Doncker

Novel Analytical Method for Estimating the Junction-To-Top Thermal Resistance of Power
MOSFETs 3750
José Miguel Sanz-Alcaine, Francisco Jose Perez-Cebolla, Carlos Bernal-Ruiz, Asier Arruti,
Iosu Aizpuru

DC-Side Impedance for Handling Interoperability of Multi-Vendor Multi-Terminal HVDC
Systems 3757
Ashkan Nami, Adil Abdalrahman, Ying-Jiang Häfner, Malaya Kumar Sahu, Khirod Kumar
Nayak

Utilizing the Electroluminescence of SiC MOSFETs as Degradation Sensitive Optical Parameter 3766
Lukas A. Ruppert, Michael Laumen, Rik W. De Doncker

Characterization of GaN-On-AlN/SiC Transistors Towards Monolithic Integrability 3775
Nick Wieczorek, Xiaomeng Geng, Carsten Kuring, Oliver Hilt, Frank Brunner, Mihaela Wolf,
Joachim Würfl, Sibylle Dieckerhoff

Optimal Frequency for Dynamic Wireless Power Transfer 3786
Mincui Liang, Khalil El Khamlichi Drissi, Christophe Pasquier

A Wide-Input-Voltage-Range 50W Series-Capacitor Buck Converter with Ancillary Voltage Bus for Fast Transient Response in 48V PoL Applications... 3796
  Nameer Khan, James Xu, Gerard Villar Piqué, John Pigott, Henk Jan Bergveld, Alaa El Sherif, Olivier Trescases

Four-Level Boost Inverter Based on ANPC Topology with Switched-Capacitor Branch............................ 3804
  Robert Stala, Adam Penczek, Stanislaw Piróg, Aleksander Skala, Andrzej Mondzik, Zbigniew Waradzyn, Krishna Kumar Gupta, Pallavee Bhatnagar, Sanjay K. Jain, Kasinath Jena

Comparative Evaluation of Partially-Rated Energy Storage Integration Topologies for High Voltage Modular Multilevel Converters... 3813
  Zoe Blatsi, Sebastian Neira, Stephen Finney, Michael M. C. Merlin

Influence of Current Collapse Due to V_{ds} Bias Effect on GaN-HEMTs I_d-V_{ds} Characteristics in Saturation Region 3822
  Xuyang Lu, Arnaud Videt, Ke Li, Soroush Faramehr, Petar Igic, Nadir Idir

Deep-Learning Fault Detection and Classification on a UAV Propulsion System 3831
  Pierre-Yves Brulin, Fouad Khenfri, Nassim Rizoug

A Compact Solid State Transformer for Replacing Conventional Medium Power Transformer in Weight-Critical Applications.. 3838
  Leon Fauth, Felix Willer, Jens Friebe

Comparative Study of Single-Phase and Three-Phase DAB for EV Charging Application........................... 3846
  Nicola Blasuttigh, Hamzeh Beiranvand, Thiago Pereira, Marco Liserre

Dynamic Load Emulation for Automotive Power IC Robustness Validation 3855
  Alexander Ulbing, Daniel Kostynski, Markus Sievers

DAB Frequency Decoupling Control with Current Minimization 3862
  Simon Uicich, Jean-Yves Gauthier, Xuefang Lin-Shi, Bruno Allard, Arnaud Plat

Design and Performance Analysis of a Modified Proportional Multi-Resonant (PMR) Controller for Three-Phase Voltage-Source Inverters 3871
  Ahmad Ali Nazeri, Mahmoud Saeidi, Peter Zacharias

Proposition and Comparison of Several Solutions for High Induced Voltage Across Inactive Transmitting Coils in a Series-Series Compensation DIPT System 3883
  Wassim Kabbara, Tanguy Phulpin, Mohamed Bensetti, Antoine Caillierez, Serge Loudot, Daniel Sadarnac

Modeling and Measuring the Bearing Capacitance of Radially Loaded Bearings 3893
  Stefan Quabeck, Daniel C. Rodriguez, Rik W. De Doncker

Comprehensive Control of Matrix Converters in On-Board Electric Drive Applications............................ 3903
  Galina Mirzaeva

Power System Simulation Tool for Quick Benchmarking of Innovative MVDC Grids in E-Mobility Applications.. 3910
  Daniel Siemaszko, Philippe Noisette

An Artificial Intelligence Pipeline for Critical Equipment Thermal Conditioning System Design 3920
  Raik Orbay, Athanasios Tzanakis, Inko Marcaide, Jonas Löfgren, Torbjörn Thiringer, Thomas Bernichon

Aspects of Stability Issues of HVAC/HVDC Coupled Grids.. 3928
   Gianni Bakhos, Kosei Shinoda, Juan-Carlos Gonzalez-Torres, Abdelkrim Benchaib, Luigi Vanfretti, Seddik Bacha

Measurement of Coss-V Characteristic of the 1.7kV/900A SiC Power Module and Estimation of the Channel Current.. 3938
   Jacek Rabkowski, Fernando Gonzalez-Hernando, Mariusz Zdanowski, Irma Villar, Uxue Larrañaga

In-Slot Cooling of Electrical Machines using Traditional Techniques and Additive Manufacturing 3947
   Ahmed Hembel, Gokhan Cakal, Bulent Sarlioglu

Comparison of High-Power 2-Level and 3-Level Converters in Terms of Power Density, Costs and Performance... 3957
   Ludwig Schlegel, Wilfried Hofmann

Autonomous Characterization of Lithium-Ion Battery Model Parameters Utilizing a Mathematical Optimization Methodology 3966
   Hamzeh Beiranvand, Helge Krüger, Sandra Hansen, Marco Liserre, Christian Werlig, Andreas Würsig

SOC Governed Algorithm for an EV Cascaded H-Bridge Connected to a DC Charger 3975
   Giulia Tresca, Andrea Formentini, Filippo Gemma, Federico Lusardi, Riccardo Leuzzi, Pericle Zanchetta

Shaping the Transition from Si-Based Power Devices to SiC MOSFETs and GaN HEMTs 3984
   Gerald Deboy

Reinventing Batteries Through Nanotechnology 3986
   Yi Cui

Advancing GaN Power ICs: Efficiency, Reliability & Autonomy... 3987
   Dan Kinzer

Electrification Strategy of Volkswagen Group.. 3989
   Alexander Krick

Make it Fly — the Future of Sustainable Aviation.. 3991
   Tanja Neuland

The Instrumental but Extremely Challenging Role of Hydrogen Towards a Decarbonized Society 3992
   Stefan Linder

Short Circuit Behavior of Dual Three-Phase Permanent Magnet Synchronous Motors with Different Mutual Inductance in Electric Propulsion Application 3993
   Yinghui Yang, Georg Möhlenkamp

Hybrid Silicon-SiC Inverter – Combining the Best of Both Worlds 4003
   Hans-Günter Eckel, Felix Kayser, Pham Ha Trieu To

Robustness of SiC Trench MOSFETs 4004
   Christian Felgemacher

3D Predictive Fatigue Modeling of Power Modules 4005
   Ben Samples, Brandon Passmore

Heterogeneous Integration of Power Conversion using Power Supply on Chip and Power Supply in Package.. 4006
Cian Ó Mathúna, Seamus O'Driscoll

Driving Innovations for Power Electronics with Integratable and Sustainable Magnetics............................. 4008
Matt Wilkowski

Impact of Package Technology on the Switching Behavior of High-Voltage GaN FETs............................. 4011
Sebastian Klötzer

Impact of Power Electronics on Battery Operation 4012
Dirk Uwe Sauer

Trends in Power Electronics and Batteries for Electrified Vehicle Infrastructure.. 4013
Torsten Leifert

Impact of High Frequency Current Pulses on Battery Ageing 4014
Julia Kowal

Aircraft Electrification – System-Level Potentials for Aviation Decarbonization 4015
Kathrin Ebner, Antoine Habersetzer, Arne Seitz

About Power Electronics Challenges in Aviation 4016
Marco Bohllaender

Development of Electric Motors for Aircraft Applications.. 4017
Simon Wolfstädter

Powertrain Trends in Electric Trucks... 4018
Luciana C. Afonso

Modulation Strategy Impact of BEV Inverters on the Voltage Ripple and the High-Voltage Traction System Stability 4019
Cornelius Rettner

Zero Emission Trucks & Bodies 4020
Martin Glaser

Integrating Offshore Wind & Hydrogen - An Operator's View 4021
Florian Gremme

Status Quo and Future Prospects of Power Electronic Solutions for Electrolysis Plants 4022
Sven Schumann

Modular Power Supply System for Large Scale Water Electrolyzers... 4023
Ralf Juchem, Klaus Rigbers

Properties of a Lithium-Ion Battery as a Partner of Power Electronics... 4025
Alexander Blömeke, Katharina Lilith Quade, Dominik Jöst, Weihan Li, Florian Ringbeck, Dirk Uwe Sauer

Author Index

Dynamic Power Analysis of Inverter-Fed Drives based on the Switching Period of the Power Electronics

Alexander Stock
Hottinger Brüel & Kjær GmbH
Im Tiefen See 45
64293 Darmstadt, Germany
Email: alexander.stock@hbkworld.com
URL: https://www.hbkworld.com

Keywords

≪Measurements≫, ≪Variable speed drive≫, ≪Pulsed power converter≫, ≪Pulse Width Modulation (PWM)≫, ≪Real-time processing≫

Abstract

Maximizing efficiency is one of the key design criteria for inverter-fed drives. The basis of this optimization process is a reliable measurement of the efficiency. The efficiency determination is in turn based on the cyclic calculation of the active power quantities based the fundamental period of the electrical quantities (voltages and currents) which is defined in detail in the relevant standards. For this reason, periodic voltages and currents are assumed to result in a steady-state condition of the drive if the standards are interpreted accurately. In many modern, highly dynamic applications (automotive, robotics, etc.), however, steady-state conditions are rare during normal operation. The electrical quantities change continuously both in magnitude and frequency and are characterized by transient behavior. In this paper, a novel highly dynamic measurement method for the approximation of the active power and efficiency is presented with regard to these dynamic applications. This methodology still approximates the conventional fundamental cycle-based definitions in the steady state, but also delivers additional dynamic information during transient balancing processes. The dynamic power analysis uses the switching period of the power electronics as averaging interval. For this reason, with regard to real-time measurements, a robust online switching cycle detection is an essential requirement to be able to perform this power analysis on a real-time capable power analyzer.

Introduction

Many industrial applications, such as fan or pump drives, are mainly operated at steady-state conditions. On the other hand, the increase in performance in the area of control and information technology as well as the constantly increasing switching frequencies of new power semiconductors enable the realization of modern highly dynamic drive concepts. In this context, for example, the rapidly growing market of electromobility or applications in the field of robotics should be mentioned. In these latter applications, the drives are hardly operated in steady states. The operating points in such dynamic applications change very frequently and often continuously. Therefore, it is essential to be able to measure and evaluate the efficiency and losses of these drives in non-steady-state operation, both during the development process and in everyday use.

The loss or efficiency determination is regulated in relevant standards, see [1] just to mention one of several examples. The basis for the efficiency calculation is the active power, which represents the average energy transport per fundamental period related to this period length. It is calculated as the average of the instantaneous power over the fundamental cycle of the associated phase voltages and phase currents of the investigated drive, see standards [2, 3, 4]. If these standards are interpreted accurately, all definitions of active power refer to the period of the periodic voltages and currents. However, the voltages and currents of the previously mentioned highly dynamic drives in automotive or robotics are non-periodic signals, characterized by transient peaks and continuously varying frequency and amplitude. In practice, the definitions and procedures of the standards can also be applied to these dynamic signals. In this case the averaging interval may be determined by a zero crossing detection or phase locked loop (PLL) circuits. Since there are no periodic signals at all, the averaging interval determined in this way cannot represent the fundamental period.

This paper presents methods for determining novel power quantities that result from averaging over the switching period of the inverter instead of the fundamental period. These power quantities approximate the conventional definitions from the relevant standards during the steady state and combine them in a new power quantity, additionally including transient information from instantaneous quantities during balancing processes, see [5, 6, 7]. In this context, to mention an example, the dynamic active power is introduced as an approximation of the conventional active power during steady state. However, this dynamic active power additionally contains information of the instantaneous power during transient balancing processes. This method can be extended to several other averaged power quantities, fundamental quantities, RMS values, see [6, 7, 8]. To be able to perform this analysis on a real-time power analyzer, a robust online switching cycle detection is essential. This paper presents the methodology of the dynamic power analysis. Comparative measurements are used to illustrate the advantages of these dynamic definitions. Finally, it is shown how the measurement process of an efficiency mapping may be drastically accelerated by use of this novel analysis method.

Dynamic power analysis

Time domain approach

Fig. 1 shows a general abstracted equivalent circuit diagram of a n-phase system as it is typically used for power analysis according to [6, 7, 9, 10, 11].

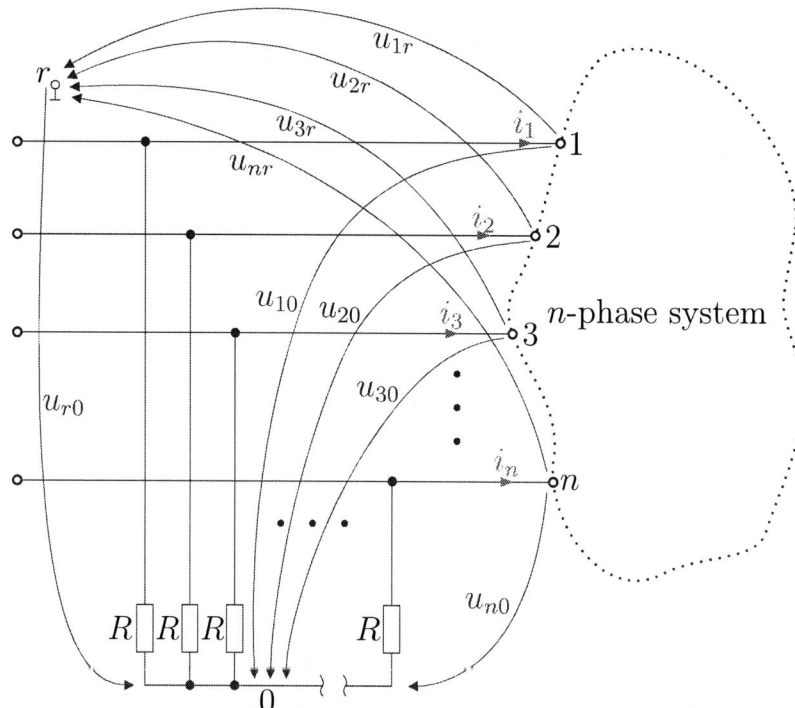

Fig. 1: Abstracted equivalent circuit diagram of an n-phase system as used for power analysis [6]

All conductor voltages and currents connected to the n-phase system and contributing to the power transfer must be considered in the power analysis [6, 10, 11] (but not the communication lines or protective earth conductors, etc.). The power analysis is completely defined by the terminal behavior of the measured system whereas the internal system topology is irrelevant for the power analysis of the total system. The voltages may be measured in theory against any arbitrary reference potential r resulting in $u_{1r}, u_{2r}, \ldots, u_{nr}$. In practice, of course, the reference potential should be galvanically connected to the measured circuit (e.g., ground potential, star point of an electric machine, artificial star point, negative DC link potential, etc.). Since the sum of the measured phase currents is zero due to Kirchhoff's law, the zero system voltage $u_{0r} = -u_{r0}$ can not carry a current and does therefore not contribute to the instantaneous power $p(t)$:

$$
\begin{aligned}
p(t) &= u_{1r} \cdot i_1 + u_{2r} \cdot i_1 + \cdots + u_{nr} \cdot i_n = (u_{10} + u_{0r}) \cdot i_1 + (u_{20} + u_{0r}) \cdot i_1 + \cdots + (u_{n0} + u_{0r}) \cdot i_n \\
&= u_{10} \cdot i_1 + u_{20} \cdot i_1 + \cdots + u_{n0} \cdot i_n + u_{0r} \cdot \underbrace{(i_1 + i_2 + \cdots + i_n)}_{=0} = u_{10} \cdot i_1 + u_{20} \cdot i_1 + \cdots + u_{n0} \cdot i_n \quad (1)
\end{aligned}
$$

However, the zero system voltage affects on the RMS values and therefore also on reactive and apparent power quantities. Since the result of the power analysis should be defined exclusively by the terminal behavior of the investigated system and therefore must not depend on the reference potential of the voltage measurement, the zero system voltage should always be subtracted first. Power analysis should be performed using the natural-zero-voltages $u_{10}, u_{20}, \ldots, u_{n0}$ only:

$$
u_{v0}(t) = u_{vr}(t) - u_{0r} \qquad \text{with} \qquad u_{0r} = \frac{1}{n} \cdot \sum_{\mu=1}^{n} u_{\mu r} \qquad \text{for} \qquad v \in \{1, 2, \ldots, n\} \tag{2}
$$

According to the relevant standards [2, 3, 4] and technical literature, e.g., [9, 10, 11], the active power P is defined as the time average of the instantaneous power $p(t)$ over the fundamental period T_{h1} of the associated periodic voltages u_{v0} and currents i_v:

$$
P = \overline{p(t)}\Big|_{T_{h1}} = \frac{1}{T_{h1}} \int_{T_{h1}} p(\tau) \, d\tau \tag{3}
$$

The basic concept of dynamic active power P_{dyn} is to shorten the averaging interval from the fundamental period to the switching period of the inverter T_s:

$$
P_{dyn} = \overline{p(t, T_s)}\Big|_{T_s} = \frac{1}{T_s} \int_{T_s} p(\tau, T_s) \, d\tau \tag{4}
$$

The switching cycle is derived from the measured inverter voltages. For typical pulse-width modulated (PWM) inverter output voltages, this can result in an increase of the averaging frequency by a factor of up to a few hundreds (e.g., $f_{h1} = 1/T_{h1} = 50\,\text{Hz}$, $f_s = 1/T_s = 8\,\text{kHz} \Rightarrow f_s/f_{h1} = 160$). Furthermore, assuming that typical PWM-generated inverter output voltages are applied to a symmetrical ohmic-inductive 3-phase load during steady state (e.g., a 3-phase machine), the equivalence of the conventional active power definition and the dynamic active power definition could be shown in [6, 7] for the theoretical limit value consideration of an infinitesimal switching period, i.e. for $T_s \to 0$ or $f_s \to \infty$. Based on this derivation, it is further shown in [6] that the dynamic active power coincides exactly with the active power for the theoretical case of infinite switching frequency during steady state, if the instantaneous power is constant for this limit value consideration:

$$\lim_{T_s \to 0} \left[P_{\mathrm{dyn}}\right] = P \quad \text{for} \quad \lim_{T_s \to 0} \left[p(t, T_s)\right] = p(t) = \mathrm{const} = p = P \tag{5}$$

For real applications with limited (finite) switching frequency, the higher the switching frequency compared to the fundamental frequency, the better the quality of the approximation of the conventional active power by the dynamic active power during the steady state. The relative error ΔP_{dyn} can be used as indicator for a maximum tolerable deviation:

$$P_{\mathrm{dyn}} \approx P \quad \text{for} \quad T_s \ll T_{h1} \quad \text{so that} \quad \left| \frac{\overline{p(t,T_s)}\Big|_{T_s} - P}{P} \right| \leq \Delta P_{\mathrm{dyn}} \tag{6}$$

Frequency domain consideration

The characteristics of the switching cycle-based active power can also be visualized in the frequency domain as function of the frequency f by applying a discrete Fourier transform (DFT) on sampled values of instantaneous power $p_k = p(kT_a)$ with the sample period T_a and $k \in \mathbb{N}$. The graphics of this subsection are based on simulated data. Fig. 2 shows, as an example, the amplitude spectrum $|\underline{\hat{p}}_\nu| = |\mathcal{DFT}\{p_k\}|$ of the total instantaneous power of a balanced ohmic-inductive 3-phase load operated with sinusoidal PWM (SPWM), as well as the amplitude response of a switching cycle-based averaging filter $|\underline{G}(\Omega)|$ with the sample frequency $f_a = 1/T_a$, the normalized angular frequency $\Omega = 2\pi \frac{f}{f_a}$, and assuming an integer number of samples per switching period $N = f_a/f_s$ with $N \in \mathbb{N}$, see [6]:

$$|\underline{G}(\Omega)| = \frac{1}{N} \cdot \left| \frac{\sin\left(\frac{1}{2}N\Omega\right)}{\sin\left(\frac{1}{2}\Omega\right)} \right| = \frac{f_s}{f_a} \cdot \left| \frac{\sin\left(\pi\frac{f}{f_s}\right)}{\sin\left(\pi\frac{f}{f_a}\right)} \right| = |\underline{G}(f)| \tag{7}$$

In this example, the inductance of the load is assumed to be sufficiently large resulting in purely sinusoidal load currents, whose switching frequency harmonics are therefore neglected. The amplitude spectrum of the instantaneous power is normalized to the product of the fundamental voltage amplitude $\hat{u}$, the amplitude of the sinusoudal current $\hat{i}$ and the cosine of the phase shift angle $\cos(\varphi)$ between the fundamental voltage and current waveform.

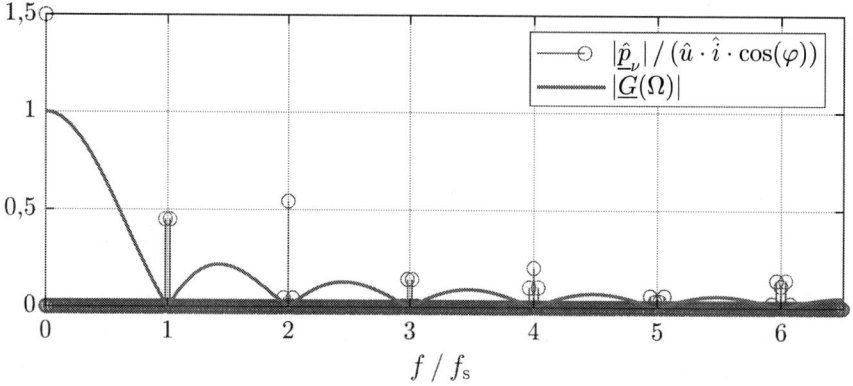

Fig. 2: Amplitude spectrum of the instantaneous power of a balanced ohmic-inductive 3-phase load fed by a SPWM-modulated inverter output voltage assuming a huge load inductance resulting in purely sinusoidal load currents [6]

Considering periodic voltages and currents, the active power P in (3), defined as fundamental cycle-based average of the instantaneous power $p(t)$, is thus represented by the DC-component of the instantaneous power. As it is well known from scientific literature, only corresponding harmonics of voltages and currents contribute to active power, resulting in $P = \frac{3}{2} \cdot \hat{u} \cdot \hat{i} \cdot \cos(\varphi)$, see [12], due to the purely sinusoidal currents. For this reason, as expected with regard to the normalization of the amplitude spectrum, $|\underline{\hat{p}}_0| = |\underline{\hat{p}}(f = 0 \cdot f_a)| = \frac{3}{2} = 1.5$, see Fig. 2.

Based on the definition in (4), the amplitude spectrum of the dynamic active power $|\underline{\hat{P}}_{\mathrm{dynv}}|$ can be calculated multiplying the amplitude spectrum of the instantaneous power $|\underline{\hat{p}}_v|$ by the amplitude response of the switching cycle-based average filter $|\underline{G}_v|$. For this, $|\underline{G}(\Omega)|$ is evaluated at discrete frequencies $\Omega = 2\pi \frac{v}{N-1}$ with $v \in \{0, 1, \dots, N-1\}$ leading to $= |\underline{G}(2\pi \frac{v}{N-1})| = |\underline{G}_v|$:

$$|\underline{\hat{P}}_{\mathrm{dynv}}| = |\underline{G}_v| \cdot |\underline{\hat{p}}_v| \tag{8}$$

As expected from Fig. 2, with regard to (7), $\lim_{f \to 0} |\underline{G}(f)| = 1$,see [6], whereas $|\underline{G}(f = kf_s)| = 0$, $k \in \mathbb{N}$. For this reason, $|\underline{\hat{P}}_{\mathrm{dyn0}}| = \frac{3}{2}$. For all $v \neq 0$, the harmonics $|\underline{\hat{P}}_{\mathrm{dynv}}|$ are strongly damped by at least a factor of approximately 100 compared to the DC-component, see Fig. 3:

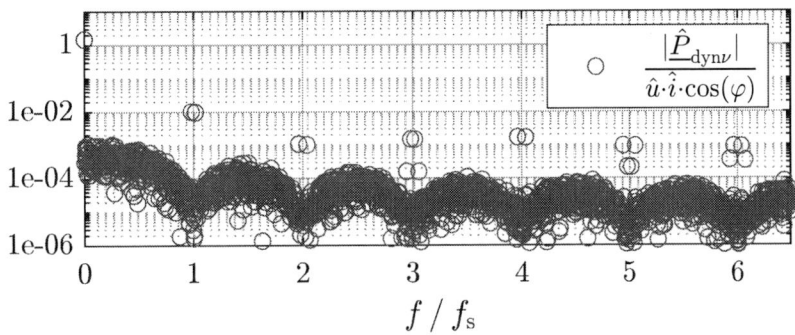

Fig. 3: Amplitude spectrum of dynamic active power normalized to its fundamental [6]

In addition to the DC component, integer multiples of the switching frequency and associated sidebands, i.e., integer multiples of the switching frequency plus/minus integer multiples of the fundamental frequency, occur in the amplitude spectrum of the instantaneous power in Fig. 2. It is obvious that the amplitudes of the sidebands decrease with increasing frequency gap from the switching frequency harmonics (related center frequencies). Therefore, when calculating the dynamic active power spectrum according to (8), these sidebands are strongly damped. Nevertheless, only the integer multiples of the switching frequency are completely eliminated by averaging. For this reason, the dynamic active power is in general not equal but a powerful approximation of the conventional active power during steady state.

If the switching frequency is increased while the fundamental frequency remains constant, the frequency gaps between the sidebands and the corresponding center frequencies (switching frequency harmonics) decrease with respect to the absolute value of the respective center frequencies. In this way, the harmonics in the sidebands are increasingly damped applying the switching cycle-based averaging filter. With regard to Fig. 2, continuing this argumentation graphically illustrates and reinforces the mathematical limit value consideration in (5). The switching frequency-based averaging for the theoretical case of an infinite switching frequency not only completely eliminates the integer multiples of the switching frequency, but also the associated sidebands. This is consistent with the mathematical time domain requirement (5): $\lim_{T_s \to 0} [p(t, T_s)] = \mathrm{const} \Rightarrow \lim_{T_s \to 0} [P_{\mathrm{dyn}}] = P$.

However, even at infinitely high switching frequency, the condition of constant instantaneous power must in general not always be fulfilled. In the following, an ohmic-inductive 2-pole system is considered instead of the 3-phase system. Assuming again a SPWM-generated inverter output voltage and a purely sinusoidal current due to sufficiently large inductance, the resulting amplitude spectrum of the instantaneous power is depicted in Fig. 4.

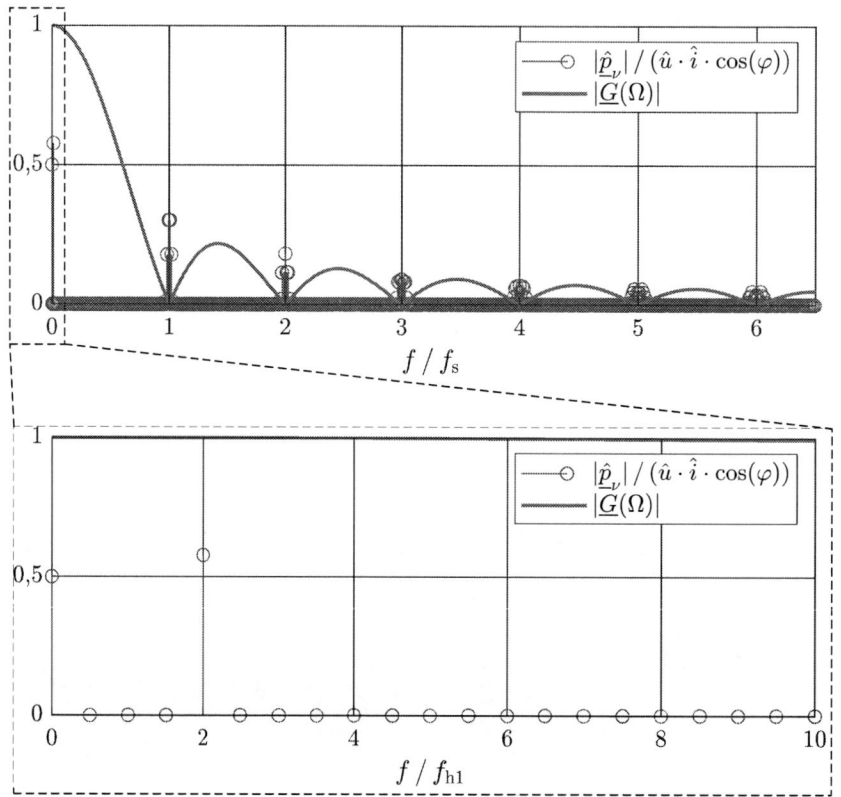

Fig. 4: Amplitude spectrum of the instantaneous power of an ohmic-inductive 2-pole load fed by a SPWM-modulated inverter output voltage assuming a huge load inductance resulting in purely sinusoidal load currents

It should be mentioned at first, that the zoomed bottom plot of Fig. 4 is normalized to the fundamental frequency f_{h1}, whereas the top plot is again normalized to the switching frequency in analogy to Fig. 2. Since only corresponding harmonics of voltages and currents contribute to active power, $P = \frac{1}{2} \cdot \hat{u} \cdot \hat{i} \cdot \cos(\varphi)$, the normalized DC component results in $|\underline{\hat{p}}_0| = \frac{1}{2} = 0.5$. However, in contrast to the 3-phase system, the amplitude spectrum of the instantaneous power in this case also contains the 2nd harmonic related to the fundamental $|\underline{\hat{p}}(f = 2f_{h1})| = \frac{\hat{u}\hat{i}}{2}$, see [12]. However, the 2nd harmonic is not eliminated by averaging over the switching cycle. On the contrary, it is propagated to $|\underline{\hat{P}}_{\text{dyn}\nu}|$ due to the factor $|\underline{G}(f = 2f_{h1})| \approx 1$. For this reason, the switching frequency-based averaging at the investigated two-pole system does not approximate the active power. It merely reflects the instantaneous power as it would occur using a purely sinusoidal supply voltage instead of a SPWM-modulated inverter.

Nevertheless, the dynamic power analysis based on the inverter's switching cycle is very suitable for PWM-operated inverter-fed multiphase systems ($n > 2$) as long as the switching frequency f_s is significantly higher than the fundamental frequency f_{h1}, typically at least $f_s \geq 10 f_{h1}$. The exact requirements furthermore depend on the applied PWM method and the individual load characteristics (e.g. current ripple or imbalances).

Comparative measurements

Startup process of a PMSM

In this section, comparative measurements of conventional fundamental cycle-based determination and dynamic switching cycle-based calculation methods are presented. In this way, the advantages of dynamic calculation can be illustrated.

First, a no-load start-up process of an double 2-level inverter-fed 3-phase permanent magnet synchronous machine (PMSM) is investigated. The system topology is shown in Fig. 5.

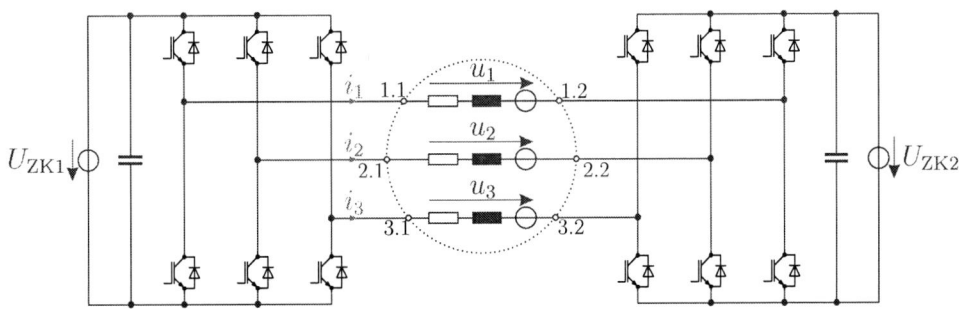

Fig. 5: Double 2-level inverter-fed PMSM[6]

The PMSM is accelerated from standstill to a constant steady state operating speed. During steady state, only the power loss must be supplied in order to control a constant speed. To accelerate the machine, torque generating currents must be injected by the inverter. The current in the first conductor i_1 is shown in the top plot of Fig. 6, normalized to the nominal current I_N of the PMSM. Based on the zero-crossings of this current, the corresponding fundamental period represented by the cycle signal $CycleDetect(i_1)$ can be calculated on the measurement system in real-time. For reasons of causality, applying the conventional active power calculation from (3), the first active power value can be calculated as average over the fundamental period at the earliest after the complete detection of this period, i.e. after its end. The bottom plot in Fig. 6 shows the instantaneous power p, the (conventional) active power P and the dynamic active power P_{dyn}, each normalized to the product of the averaged DC-link voltage of the inverter system $U_{ZK} = \frac{U_{ZK1}+U_{ZK2}}{2}$ and the nominal current I_N.

With regard to (1), the instantaneous power contains the switching frequency harmonics predominantly affected by the inverter output voltages. The instantaneous power does not visually reflect any useful information about the average energy consumption of the PMSM during this start-up process. As described above, the conventional active power is delayed by one fundamental period during real-time measurements due to causality. For this reason, on a power meter or data acquisition system, the active power result could be calculated and visualized earliest during the following fundamental period, while the next active power value is determined simultaneously. A comparison of p and P shows that this method of active power calculation is not useful for this transient acceleration process, especially with regard to the real-time calculation of active power. With a precise interpretation of the relevant standards [1, 2, 3], the active power during this balancing process is not defined anyway due to aperiodic voltages and currents. On the other hand, the switching cycle can be detected by evaluating an inverter output voltage. The dynamic active power calculation in (4) is therefore only delayed by one switching period during real-time measurements. Based on Fig. 6, it is clear that P_{dyn} highly dynamically reflects the information about the average energy transfer per switching period.

At first glance, it seems theoretically obvious to achieve a similar dynamic active power result by simply low-pass filtering the instantaneous power using a first order low pass filter or even higher order. However, there are reasonable arguments to perform the switching cycle-based averaging. For example, if a fundamental period contains N switching periods, a subsequent average over the N already averaged dynamic active power results delivers exactly the same result as directly applying the conventional active

Dynamic Power Analysis of Inverter-Fed Drives based on the Switching Period of the
Power Electronics

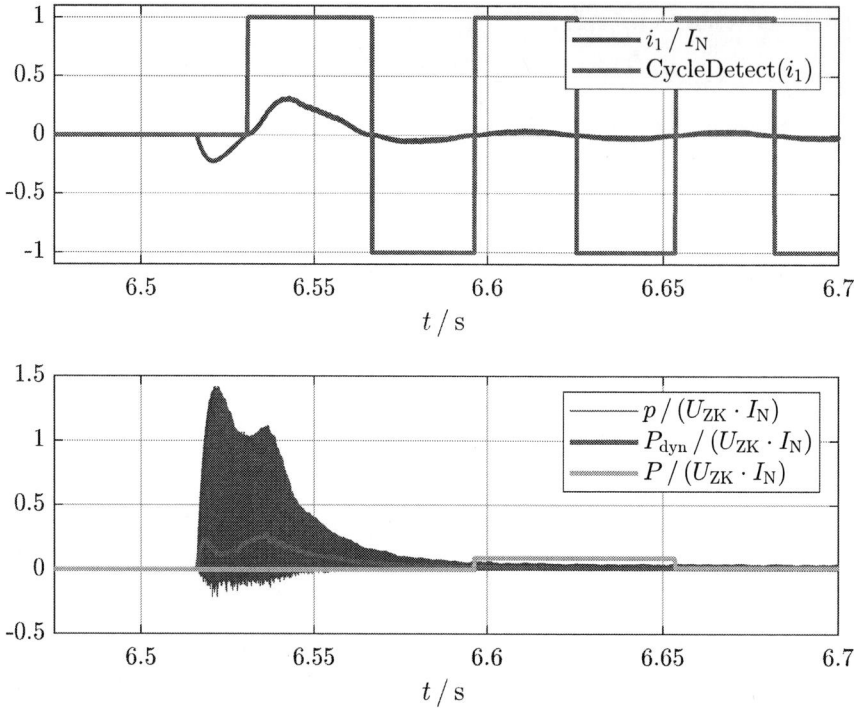

Fig. 6: Instantaneous power $p(t)$, active power P and novel dynamic active power approximation P_{dyn} during a balancing process between two different load operating points with constant speed; $f_{h1} < 30\,\text{Hz}$, $f_s \approx 16\,\text{kHz}$; [6]

power calculation as an average over the fundamental period. This ensures consistency in the active power data. In addition, knowledge of the fundamental frequency and switching frequency is required anyway for suitable low-pass filtering in order to determine a suitable cut-off frequency of the low-pass filter. This might also need to be adjusted during operation if, for example, the switching frequency changes.

Dynamic transition between different steady state operating points

In the measurement described below, the speed-controlled PMSM is mechanically connected to a torque-controlled load machine in order to be able to investigate different load operating points. Fig. 7 shows the power quantities p, P and P_{dyn}, normalized with regard to the active power P_{AP} related to the investigated motor operating point. The graphic shows two different load operating points including the dynamic transition in between. Initially, the PMSM is driven in generator mode, resulting in negative active power quantities. Then, a load torque step is generated with the torque-controlled load machine by changing both the sign and the magnitude of the torque. This leads to the subsequent motor steady state operation.

As with the previous example, the advantages of switching cycle-based dynamic active power are significant during the transient transition. While the conventional active power P increases stair-wise due to the comparatively long averaging interval of the fundamental period, the dynamic active power even contains the transient peaks and overshoots of the instantaneous power, caused by the inverter control. Furthermore, it can be seen that the two active power quantities show a very high degree of similarity at the respective steady-state operating points.

Conclusively, it can be stated that the dynamic active power P_{dyn} combines the information of the instantaneous power $p(t)$ during transient balancing processes with the information of the conventional active power P during steady state in a novel active power quantity.

Dynamic Power Analysis of Inverter-Fed Drives based on the Switching Period of the
Power Electronics

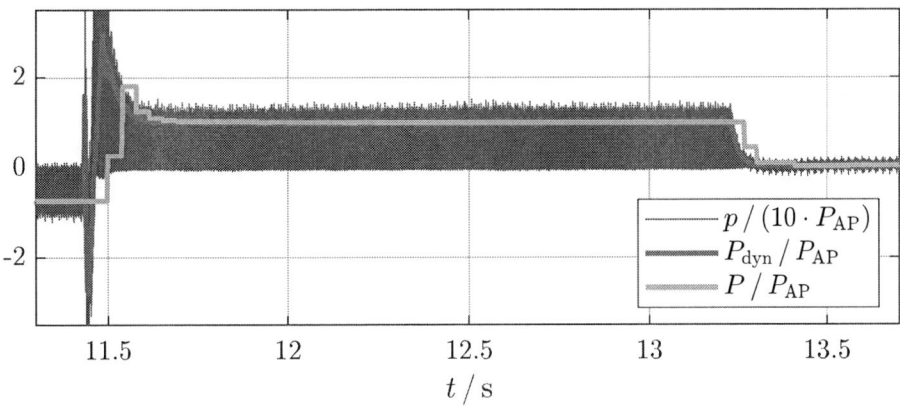

Fig. 7: Instantaneous power $p(t)$, active power P and novel dynamic active power approximation P_{dyn} during a balancing process between two different load operating points with constant speed; $f_{\mathrm{h1}} \approx 20\,\mathrm{Hz}$, $f_{\mathrm{s}} \approx 16\,\mathrm{kHz}$

Accelerated Efficiency Mapping

The commonly known procedure for creating an efficiency map is to measure the device under test (DUT) at different relevant steady-state operating points. In the case of an electric machine, all speed/torque-combinations of interest are approached sequentially and measured under steady state conditions. When a new operating point is invoked, the automation system typically waits until all balancing processes have been completed and the new steady state has been reached. Subsequently, a trigger signal is sent to the measurement system to acquire all relevant physical quantities such as voltages, currents, speed and torque. The electrical and mechanical active powers are then calculated from these measurands individually for each steady state operating point. This procedure can take a long time, especially if there are a large number of operating points to be measured.

If, on the other hand, dynamic power calculation is used, i.e., the active power quantities are determined by averaging over the switching cycle of the inverter, the requirement of a steady state conditions is obsolete. Thus, for each constant speed value, the entire relevant torque range can be passed through in a continuous linear manner. Fig. 8 compares efficiency maps generated using the conventional efficiency mapping process of the PMSM, see Fig. 8a, and one based on the dynamic method, see Fig. 8b. The two measurements were performed separately, but on the same machine from Fig. 5. In Fig. 8a, the stationary speed-torque operating points are marked by the red circles.

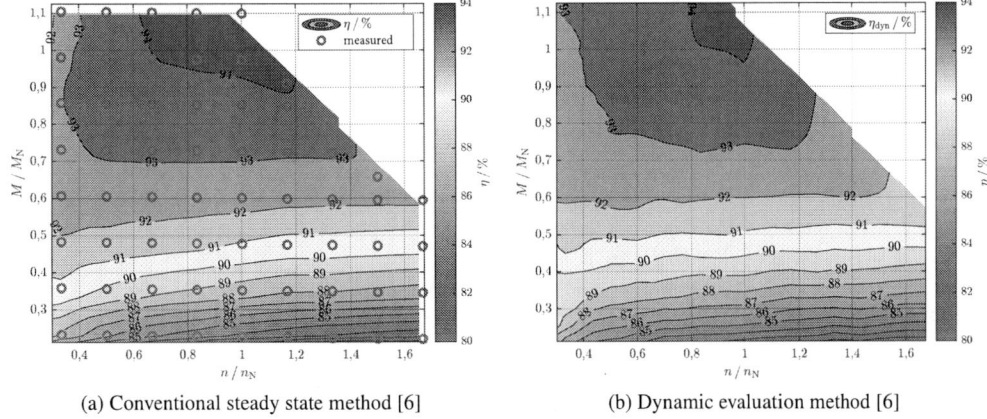

(a) Conventional steady state method [6] (b) Dynamic evaluation method [6]

Fig. 8: Comparison of a conventional steady state operating point based evaluation and a highly dynamic efficiency mapping method

The deviation of both measurements is less than 0.5 % in most relevant operating points. If a maximum of accuracy is required for the efficiency measurement, then the conventional evaluation during steady state operating points is recommended. However, if a pass-fail test is used to check whether a machine reaches a certain required threshold value of efficiency, the dynamic efficiency mapping is very well suited to generate a meaningful result with drastically reduced measurement duration. During the example measurement in Fig. 8, the required measuring time could be reduced to 10 % of the time needed for the conventional measurement. For reasons of comparability, the dynamic measurement was performed at stationary speeds whereas the torque was increased linearly. However, in general the speed could also be varied continuously. With regard to the automotive application, it is therefore possible to determine the efficiency map as an additional parameter in a Worldwide harmonized Light vehicles Test Procedure (WLTP) that is performed anyway. In doing so, an additional measurement process can be saved. For this reason the dynamic power analysis is very suitable for saving measurement time or even a complete measurement and thus resources and costs, respectively.

Conclusion

This paper presents a method for dynamic real-time power calculation for power meters, which is particularly well suited for highly dynamic applications such as they appear in automotive or robotics. Since the switching period of the power inverter represents the averaging interval of the cycle-based dynamic power quantities, reliable real-time detection of the switching frequency is essential. The presented methodology can be transferred to further cycle-based quantities, such as reactive or apparent power quantities, RMS values, power factor, efficiency, etc. It is very suitable to determine meaningful measurement results even during transient balancing processes in non-stationary operation. Furthermore, this paper presents how this measurement method can be used to perform accelerated efficiency mapping in order to save measuring time and resources in this way. The advantages of the presented method are substantiated by characteristic comparative measurements. Future investigations are concerned with the continuous optimization of real-time switching cycle detection so that it can be flexibly and reliably applied to numerous power inverter topologies.

References

[1] IEC/TS 60034-2-3: Rotating electrical machines – Part 2-3: Specific test methods for determining losses and efficiency of converter-fed AC induction motors. 11/2013.
[2] DIN 40110-1: Wechselstromgrößen Zweileiter-Stromkreise. 03/1994.
[3] DIN 40110-2: Wechselstromgrößen Mehrleiter-Stromkreise. 11/2002.
[4] IEEE Std 1459-2010: IEEE Standard Definitions for the Measurement of Electric Power Quantities Under Sinusoidal, Nonsinusoidal, Balanced, or Unbalanced Conditions. 19. 03. 2010.
[5] Stock, A.; Teigelkötter, J.; Kowalski, T.; Staudt, S.; Ackermans, P.; Lang, K.: Determination of active power on the basis of the switching frequency (Schaltfrequenzbasierte Wirkleistungsmessung). Patent WO 2018/228655 A1. HBM Netherlands B.V. 20. 12. 2018.
[6] Stock, A.: Messtechnische Analyse der Energieverluste von stromrichtergespeisten Antriebssystemen im nichtstationären Betrieb. Dissertation. University of the Federal Armed Forces Munich, 2021.
[7] Stock, A.; Teigelkötter, J.; Staudt, S.; Kowalski, T.: Highly dynamic analysis of active power and fundamental component approximation of inverter fed applications. In: IEEE 12th International Conference on Power Electronics and Drive Systems (PEDS) (Honolulu, USA). IEEE, 2018, pp. 291-296.
[8] Stock, A.; Teigelkötter, J.; Büdel, J.; Staudt, S.: Highly Dynamic Calculation of Power Quantities and Further Analysis of Inverter-Fed Machines. In: 20th European Conference on Power Electronics and Applications (EPE'18 ECCE Europe) (Riga, Latvia). IEEE, 2018.
[9] Depenbrock, M.: The FBD-method, a generally applicable tool for analyzing power relations. In: IEEE Transactions on Power Systems. Vol. 20. no. 2. IEEE, 1993, pp. 381-387.
[10] Staudt, V.: Ein Beitrag zu Leistungsbegriffen und Kompensationsverfahren für Mehrleitersysteme .Habilitation. Ruhr University Bochum, 2000.
[11] Staudt, V.: Fryze – Buchholz – Depenbrock: A time-domain power theory. In: International School on Nonsinusoidal Currents and Compensation 2008 (Łagów, Poland). IEEE, 2008, S. 1-12.
[12] Teigelkötter, J.: Energieeffiziente elektrische Antriebe: Grundlagen, Leistungselektronik, Betriebsverhalten und Regelung von Drehstrommotoren. Springer Vieweg. Wiesbaden, 2013.

Stability analysis in an inverter-dominant microgrid facing in-rush current of an induction machine

Nastaran Fazli, David Hammes, Sidney Gierschner, Hans-Günter Eckel
Institute of Electrical Power Engineering
University of Rostock
Rostock, GERMANY
Phone: +49 381 498-7115
Email: nastaran.fazli@uni-rostock.de

Acknowledgments

This paper was made within the framework of the research project "OffWiPP", which is supported by the Federal Ministry for Economic Affairs and Energy (03El4037A) on the basis of a decision by the German Bundestag.

Keywords

≪Grid-forming converters≫, ≪Renewable energy systems≫, ≪Induction motor≫, ≪Islanded operation≫, ≪Power system stability≫.

Abstract

In islanding operation with high shares of renewables, transient disturbances may put the system in danger of instability. In such a situation, the frequency-voltage profile of the remaining power system, does not only depend on the depth of the unbalance, which appears following a transient, but also to the percentage of renewables and their control topologies as well as the dominant load on the islanded area. Hence in this paper, the renewables are controlled to act as a grid-supporting current and voltage sources. Also the load characteristic is varying from dominant resistive, to dominant induction machine and converter- fed loads. The emergency measures for under- and over-frequency and voltage control is also considered the same in all simulation scenarios. The simulations are carried out in Matlab-Simulink via different scenarios to find out if the islanded power system and the generation units in that area are able to handle such a transient event considering different criteria via the emergency measures in the islanded power system.

Introduction

With continues growth of non-synchronous generation (NSG) units, following the trend of decarbonizing the globe, the share of renewable energy sources (RES) reached to an alarming level regarding power system security. In addition to the reduction of power system strength including mechanical inertia and short-circuit ratio with growing of NSGs, an investigation in 2013 shows that control interaction and instability of these units in higher shares is another issue which has to be considered [1].

On the other hand, the pressure on the installed equipment of the actual power system has increased the risk of events leading to system splits within continent Europe in recent years [2]. There are mainly two threats for an unintentional islanding event according to [3], first one refers to the the high unbalances after islanding, and the second one to the amount of inertia in the islanded area, that this work focus on the second part. Additionally, in the remaining power system after separation, stability does not only depend on the share of NSG to the conventional synchronous generator (SG) units but also to the load combination in this area [4].

Industrial loads for example with a high share of induction machines (IM) can have a different characteristic in response to the disturbances than residential load with more share of heating, lighting and converter-fed loads, as they don't share the frequency-supporting features of a directly-connected drive [5]. In addition to it, are the disturbances like turning on a huge IM or tripping transmission lines and etc, which affects frequency and voltage stability on the power system. This can drag the system towards instability border and cascading events especially if the system is separated from a bulky part of the power system and if SGs of that area are reaching their operational limits to support the power system with active power injection and reactive power supplying.

As mentioned in [2], the seven main concerns of power system operators, can nowadays be supported by grid-supporting voltage source (GSVS) control of converters. These types of control for a voltage source inverter (VSI) can emulate similar behavior to an SG by providing instantaneous reserve or inertia support and voltage control as much as their limitation on available energy, actual operating point and over current capability allows. By increasing share of RES, the task is to find out the border of how much state-of-the-art grid-supporting current source (GSCS)is required to be replaced by GSVS to ensure system stability. This depends on many criteria, namely, for which unbalances GSVS should be dimensioned, future speed of triggering under-frequency load shedding (UFLS) relays, if primary control is provided by RES and lastly to the load composition and taking into accounts transients which may occur during islanding of the power system.

Starting up of a huge IM is one of these transients that can drag an islanded power system to the border of instability. As during starting, the IMs draw a high current duo to their lower machine impedance in lower speed of the rotor if they are not compensated [6]. This will place the system especially with high shares of renewable to a alarming stage as the majority of the installed capacity of RES don't have an instantaneous reaction to such disturbances. However, the GSVSs will inherently sense these voltage and frequency variations and can participate in counteracting their effect.

Therefore, the effect of the inrush current of an IM during islanding is going to be investigated. In this paper, a similar power unbalance via hard start of an IM is going to be tested via different scenarios. The frequency–voltage stability in the islanded area will be evaluated using also the emergency measures like limited frequency sensitive mode – underfrequency (LFSM-U) and limited frequency sensitive mode – overfrequency (LFSM-O), voltage-var control and under-voltage load sheeding (UVLS). Among all the other participating factors like percentage of unbalances, how slow is the LFSM-O with limitation of conventional units and how fast are LSRs and etc, the focus in this paper is drawn to investigate the minimum required inertia with looking at the effect of the dominant load classes in this microgrid during the event.

Structure of the Power System

The power system structure is taken from the model of Kundur [7], as shown in Fig.1. The wind turbines (WT) of the available RES are a mixture of GSCSs and GSVSs.

RESs Control Strategy

In order to provide grid-supporting features and to fulfill grid requirements, the control of a branch of RESs in Fig.1 is realized via Grid-supporting current source (GSCS), which is an extension of the state-of the-art grid following current source converter. The GSCS is implemented in a power converter with voltage-oriented control, which controls the converter currents in the inner loop [8] and active and reactive powers in the slower outer loop. A Phase-locked loop (PLL) is used for grid synchronization and current limitations is applied on the controller current references to protect the converter semi-conductor switches. In the other branch of RES, a GSVS is applied using Rotating-voltage-vector control (RVVC) [12]. The active power loop

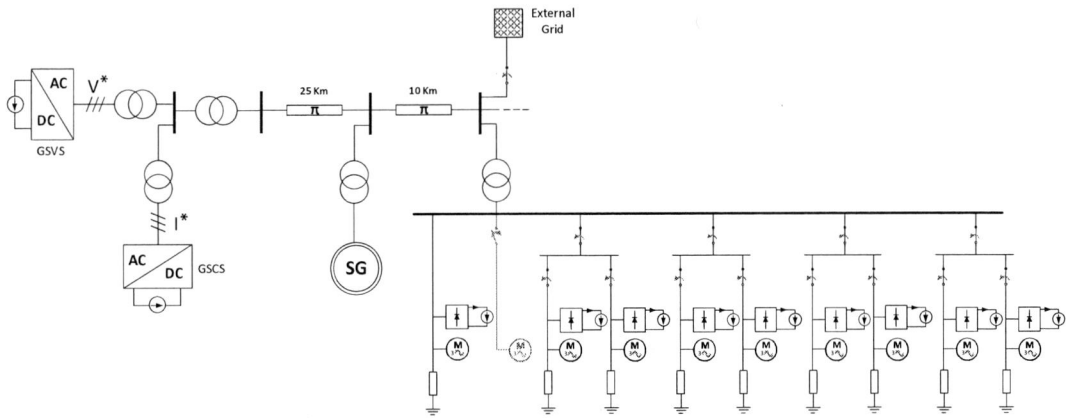

Fig. 1: Model of the simulated power system

uses the mechanical inertia constant in a similar way to a SG. The converter filter impedance is $x_f = 0.01$ pu and its inertia constant is H=5 s. Instantaneous reserve is realized via active power loop to support the grid frequency deviation and provide time for the containment reserve and load shedding relays (LSRs) to act. Primary control is also a choice of system operator and can be activated if WT is operating in sub-optimal curve [13]. A DC-link voltage controller is realized in the machine-side converter control to guarantee a steady DC voltage when grid frequency and voltage changes. This is done by varying the torque of permanent magnet synchronous generator of WT depending on the depth of frequency fall and choice of parameters in GSVS. Over-current limitation will be triggered when the currents go beyond 1.2 pu limit.

Synchronous Generator (SG)

An SG is used with an speed governor and active voltage regulator (AVR) using the SM AC2C excitation system in conformation with IEEE [14]. The transient impedance of SG is $x'_d = 0.1$ pu similar to GSVS filter and Mechanical inertia is H = 5 s similar to GSVS at nominal power.

Loads

Five main branches of load is realized in the simulated power system, as shown in Fig.1. Four branches, are each comprised of 12.5% of the total load power and the rest 50% is in the last branch. Each branch has a mixture of resistive loads (resembling heating and lighting loads), directly-connecting machines (IMs) and converter-connected drives (B6U). The proportion of each type of the load relative to the other types is verying in different simulation scenarios based on [5] to show a fraction of typical loads in residential, commercial or industrial area. According to the Fig. 2-4 of [5], a hypothesis of the load composition for three different load classes is considered as follows: in the class I, similar to the residential area that normally heating and cooking power consumption is high, a dominant share of resistive load 50%, 30% IM and 20% B6U is considered. The load Class II belongs to the industrial area, where majority of loads are motors. In this class, a share of 70% IM, 20% B6U and lastly 10% resistive loads is considered. And the load class III, is the situation when a high share of converter-connected drives exist. The loads in this class are 50% B6U, 30% IM, and 20% resistive load.

Emergency remedies including LFSM-O, LFSM-U, Voltage control, UFLS and UVLS

The LFSM-O is applied for both GSCS and GSVS converters with a time constant of 1.7 s to fulfill the grid requirements regarding over-frequency power reduction [9], [10]. According to

Table I: Simulated Scenarios

Load class I		Load Class II		Load Class III	
50% Resistive- 30% IM- 20% B6U		10% Resistive- 70% IM- 20% B6U		20% Resistive- 30% IM- 50% B6U	
Test 1	100% SG	Test 2	100% SG	Test 3	100% SG
Test 4	50% SG- 50% GSCS	Test 5	50% SG- 50% GSCS	Test 6	50% SG- 50% GSCS
Test 13	50% SG- 25% GSCS- 20%GSVS	Test 14	50% SG- 25% GSCS- 20%GSVS	Test 15	50% SG- 25% GSCS- 20%GSVS
Test 7	25% SG- 90% GSCS	Test 8	25% SG- 90% GSCS	Test 9	25% SG- 90% GSCS
Test 16	25% SG- 37.5% GSCS- 37.5%GSVS	Test 17	25% SG- 37.5% GSCS- 37.5%GSVS	Test 18	25% SG- 37.5% GSCS- 37.5%GSVS
Test 10	10% SG- 90% GSCS	Test 11	10% SG- 90% GSCS	Test 12	10% SG- 90% GSCS
Test 19	10% SG- 45% GSCS- 45%GSVS	Test 20	10% SG- 45% GSCS- 45%GSVS	Test 21	10% SG- 45% GSCS- 45%GSVS

these regulations, LFSM-O will be activated from 50.2 Hz with a droop slope of -20 pu to drop 50% of the generation when the frequency reaches to the threshold of tripping units (51.5 Hz). For SG, however, the LFSM-O is applied with a bigger time constant (8 s) due to limitations for conventional units that requires a slower reaction [11].

A voltage droop controller is also applied for both GSCS and GSVS according to grid connection standards [11]. Based on these regulations, the controller gain is chosen -2 pu and the time constant of the voltage controller's low-pass filter is chosen 10 ms. Based on these parameters, in 50% voltage drop, RESs will deliver 1 pu of overexcited reactive power within 30 ms. The dead-band for activation of the voltage controller is $\pm5\%$ of nominal voltage. For SG, a similar deadband is applied to the AVR. Also, the reactive power of SG is limited to 1 pu of the nominal apparent power by under excitation and over excitation limiting controllers.

The LFSM-U refers to the activation of the frequency containment reserves (FCR) when the system is in an emergency state in underfrequency events. This is only realized via SG here, as the RESs are designed to operate in maximum power point. The same limiattion of LFSM-O for power reduction is applied here for LSFM-U with SG. That SG can ramp the power with a rate limited to 20% min^{-1}.

The other emergency action is under-frequency load shedding (UFLS). These relays are located in four branches to be able to disconnect each time 12.5% of the total load at frequency thresholds of f= {49-48.8-48.6-48 Hz} as a last defense action for keeping stability in the power system. At each of these thresholds, the frequency has to stay below this value for four consecutive grid cycles until the tripping signal is sent and then there will be another 100 ms delay from the emitted signal until the relays opens up and disconnect the load. The Under voltage load shedding (UVLS) relays are also placed in eight sub-branches of the loads as a counter-measure action against voltage dips in an area. Load characteristic plays an import role in UVLS. For instance, voltage-dependent loads like resistive loads don't necessarily need a prompt reaction when under voltage happens, IMs, however, need a less delayed reaction from under voltage relays (around 1.5 s) [15]. Therefore a faster load shedding concept here is considered. Each UVLS relay drops 6% of the total load in the following time stamps [15] : -10% below nominal voltage with 1.5 s time delay, -10% below nominal voltage with 3 s time delay, -10% below nominal voltage with 5 s time delay.

Simulation Scenarios

The question of how much GSVSs is necessary to bring back the strength to the actual power system is being investigated in some optimistic and pessimistic studies [4] and [17]. In these concepts, grid following current sources are replaced by GSVSs to mitigate the concern of TSOs regarding system security. However, in most studies the influence of consumer side is not considered. In [16], this effect is partially being seen on the islanding event in Flensburg in 2019. Therefore, in this paper, first the line connecting the power system to the external grid is disconnected some ms before t= 0 s and then at t= 0 s, the starting process of a high power IM (with a nominal power equal to 4% of the total load), without any compensation is studied in different islanding scenarios with high RES and the three classes of loads. The results of some of these test scenarios are shown and compared in the following. The criteria for the analyzing test scenarios is that frequency and voltage stay inside normal operating band (f = [47.5 - 51.5 Hz] and V = ±10%), also using remedial measures. Another important criteria here is the Rate of Change of Frequency (ROCOF) and the number of needed load shedding (LS) before reaching to a new steady state value. Importance of ROCOF criteria is because the value of ROCOF is often used along with a threshold value for a fast UFLS. Also a high ROCOF can lead to malfunctioning of protection relays that are installed in medium voltage grid to detect islanding. According to [18], the setting for these relays are normally between 0.1 - 1.0 Hzs^{-1}. On the other hand, ROCOFs higher than -1 Hzs^{-1} not only can lead to malfunctioning of emergeny equipments in the power system, but also it might cause instability and lifetime threats for conventional generation units. Hence, in this work, ROCOF rate of -1 Hzs^{-1} is considered critical in choosing minimum required inertia to stand such a transient.

Comparing Load Classes I-III in the Base Scenario

Fig. 2 shows and compares the results of starting an IM at t = 0 s in different classes of loads with base scenario, where there is not any RESs participating in the generation. This is shown by T1 and 2 and 3 in Fig.2, (*T stands for the word test*). In all tests, voltage stability is assured thanks to fast voltage support of the SG. The voltage stays within ±10% Vn which is the allowed operation band. The best df/dt among these three tests belongs to T1 with highest percentage of resistive load. This is achieved because the voltage has droped with starting of IM, and as a result, resistive loads with quadratic relation with the supply voltage, have reduced their active power consumption. In this case, by reduction of ΔP, df/dt gets smoother and reaches a value of -0.38 Hzs^{-1} measured in a window of 500 ms. However, by recovery of the voltage, the ΔP also increases and ends up with activation of first step of LS after 7.2 s.

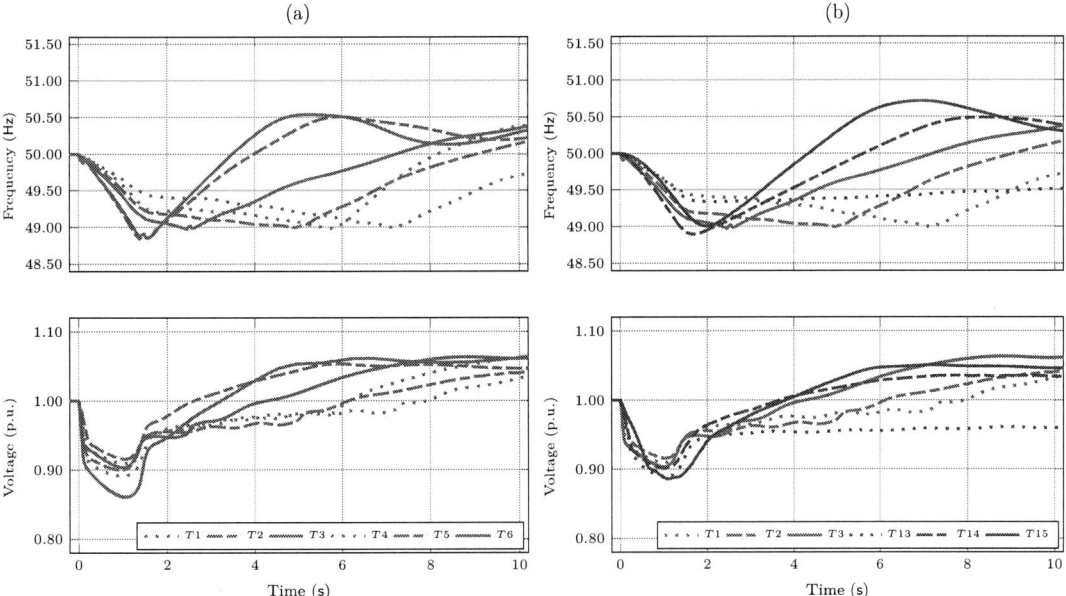

Fig. 2: Base tests with only SG are shown in gray. a) Tests with having SG+GSCS in Generation side are shown in light blue, b) Tests with having SG+ GSCS+ GSVS in generation side are shown in dark blue

In T2, with turning on process of IM, the voltage drops, however, not as low as the T1 and T3 due to high proportion of IM. The load of IM in the consumer side is a variable-torque load, whose mechanical torque is a function of square of rotational speed. Therefore, with reduction

Table II: Detail of test results in Fig.2

Tests	F min (Hz)	F max (Hz)	ROCOF (Hz/s)	Number of LS	LS time (s)	Vmin (pu)	Vmax (pu)
T1	49.00	50.40	0.38	once	7.20	0.91	1.06
T2	49.00	50.65	0.38	once	5.00	0.92	1.05
T3	49.00	50.70	0.38	once	2.47	0.90	1.07
T4	49.00	50.80	0.48	once	5.81	0.89	1.06
T13	49.35	49.60	0.40	none	-	0.89	0.96
T5	49.00	50.50	0.76	once	1.38	0.90	1.05
T14	49.00	50.49	0.75	once	1.57	0.90	1.03
T6	49.00	50.56	0.75	once	1.39	0.86	1.06
T15	49.00	50.70	0.64	once	1.67	0.88	1.05

of the voltage, the other IMs in the consumer side, have decreased their electrical torque as a function of the grid voltage and proportionally their rotational speed, and with that their power consumption has dropped. In this condition, the power factor of the load improves and this improves the voltage profile in the power system with load class II. The df/dt in T2 is slightly better than T3. It is also shown that by bringing back the voltage of T2 and T3, into the normal operating band, frequency starts deviating more in T3 comparing to T2 due to higher resistive loads and less self-regulating effect. Therefore, the fastest LS happens with T3 (after 2.47 s) and then it follows by T2 (after 5 s). In general, in all T1-3, the grid can keep the voltage stability and frequency stability after such a transient by only one step of load shedding. The details about test results is summarized in Table II.

In T4, 50% of the load power is generated from the RES with GSCS. Among T4-6, voltage stability is only maintained for T5 with IM dominant loads. The df/dt in all cases is worsen due to higher share of RESs in the generation side. Still T4 has the best df/dt as the voltage drops even lower than T1. The value of df/dt is slightly lower in T5 comparing to T6. Therefore, a faster load shedding happens with T5 after (1.38 s).

Comparing T4 versus T13 shows that both reach to a minimum voltage of below 0.9 pu. However, in contarst to T4 with GSCS, in T13 the effect of inertia support from GSVS is shown. As with almost similar voltage profile after the event, the ROCOF has improved and the frequency settles in a new equilibrium point without the need for load shedding in T13, in contrast to T4 that load shedding is happening after 5.81 s. Comparing T5 versus T14, shows an acceptable voltage support in both scenarios. In a similar way, df/dt is improved in T14 and this results in a delayed LS comparing to T5. In T15, voltage support from both SG and GSVS can result in a better voltage profile in contrast to voltage in T6. The oscillation in the voltage of T15 comes from over-current activation to protect GSVS from over-currents due to huge active and reactive power support. Again a better inertia support is seen with T15.

Comparing Generation Combinations in Load Class III

Fig. 3 shows the simulation results of another group of tests which have the same class of load (Class III) but their generation unit is changing in the percentage of SG and RESs with and without GSVS. As it is shown in Fig. 3(a), by increasing the share of RESs from T9 to T12, how much negative ROCOF decreases due to reduction of inertia in the islanded power system. The decrease of df/dt from -1.03 $Hz\,s^{-1}$ in T9 to -1.6 $Hz\,s^{-1}$ in T12 causes triggering of the first load shedding relay very fast after 0.68 s. In both tests (T9 and T12), 24% of the load has dropped in two steps of load shedding behind each other. When 50% of the RESs is replaced by GSVS in T18 and T21, the results shows that the ROCOF has improved. This is shown by Fig. 3(c) and (e) that facing the inrush current of the IM during islanding, causes a frequency deviation,

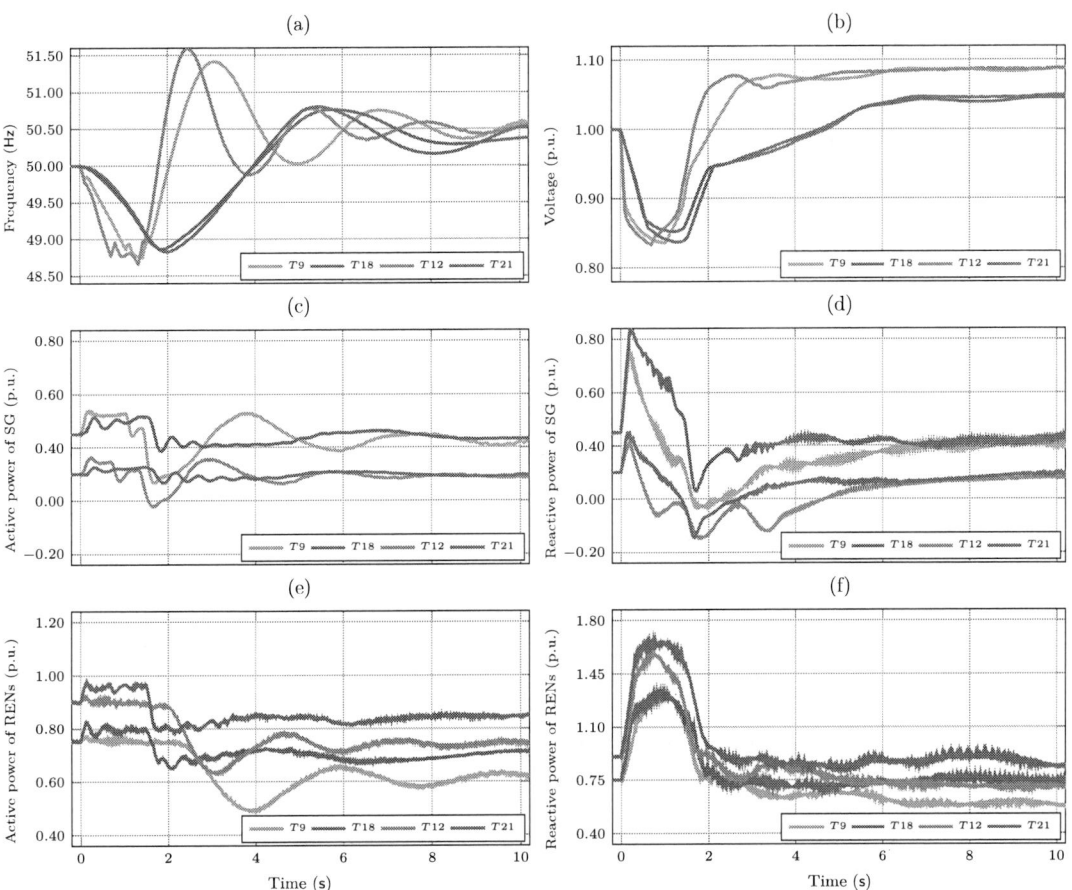

Fig. 3: a) frequency, b) voltage at the load bus, c) Active power of SG, d) reactive power of SG, e) active power of RESs including GSCS & GSVS, f) reactive power of RESs including GSCS & GSVS

due to ΔP in the islanded system. As a result, there is a deviation in the frequency that in T9 & T12, only SG is counteracting it (Fig.3(c)). Reducing the share of SG from 25% (T_G=2.5 s) to 10% (T_G=1 s), causes a steeper ROCOF once the IM is turning on. In contrast, in T18 with GSVS which has a 50% share of the RESs (T_G=6.2 s), a higher inertia can be provided via generation units comparing to T9. As it is shown in Fig.3(c), the SG doesn't compensate only for the frequency deviation, also RESs are participating in inertia power. In the similar manner is T21 (T_G=5.5 s), although the time constant has dropped comparing to T12, the frequency and voltage stability improved comparing to the similar cases with T9 and T12. The details about test results is summarized in Table III.

After deceleration phase, and activation of first step of load shedding, the frequency rises again into over-frequency region. As it is shown with T12, the frequency goes from normal frequency band to the band that normally generation units disconnect from the power system [3]. This is critical in this way that according to the data of historic events [16], separation of generation units can cause cascaded events and can end up with blackout. Therefore, in T12, the amount of inertia is insufficient in facing such transients. In other tests, the frequency stays in the allowed frequency band. In over-frequencies, LFSM-O of all RESs and SG are participating in curtailing their power. However, from the moment that df/dt becomes positive, the reaction of SG and GSVS is almost intrinsic due their voltage source nature. This reduces $+\Delta p$ in the system. On the other hand, for RESs with current source nature, the LFSM-O starts reducing the active

Table III: Detail of test results in Fig.3

Tests	F min (Hz)	Fmax (Hz)	ROCOF (Hz/s)	Number of LS	LS time (s)	Vmin (pu)	Vmax (pu)
T9	48.80	51.43	1.03	twice	1.04 - 1.43	0.83	1.08
T18	49.00	50.76	0.65	once	1.65	0.85	1.04
T12	48.80	51.60	1.60	twice	0.70 -1.33	0.83	1.08
T21	49.00	50.80	0.67	once	1.51	0.83	1.04

power only when the frequency passes 50.2 Hz which is the threshold value for activation of LFSM-O. This delayed reaction of GSCS comparing to SG and GSVS is shown in Fig.3(c) and (e) between T9 and T18. After 2nd load shedding, and with positive ROCOF, SG and GSVS both absorbed additional power in the system, however, GSCS power reduction starts more than 0.5 s later. Due to this slower reaction and dropping two branches of the load, the frequency rises the most in T12 (out of the allowed band). The over-frequencies are smaller due to max 12% load shedding in both T18 and T21 and also faster reaction from the generation units.

Looking at voltage profiles, the reduction in SG share has reduced the voltage from T9 to T12. This is due to inherent voltage source nature of the SG that reacts immediately to variation of voltage in its terminal. GSCS, however, with the droop controller reacts to the voltage variation once the voltage has passes the dead band of $\pm 5\%$ and with a time constant of 0.01 s. The RESs with GSVS on the other hand, have an intrinsic response to the voltage variation in the bus they are connecting to it. As the results show for the load class III, when the percentage of B6U is high in the load combination, with starting the inrush current of the IM, due to huge reactive current at the starting, the voltage drops hugely.

In this load type, a huge reactive current is drawn, when the voltage is dropping, in order to keep the active power constant. Hence, the reactive power consumption via the passive elements and filter of the B6U slightly increases. Converter-fed loads, in general, in case of the frequency variation don't participate in frequency-dependent active power reduction, in contrast to IMs, and in case of voltage fall, B6Us don't support voltage by reduction of reactive power consumption as well. Therefore, the voltage profile is the worst with converter-dominant loads.

Comparing Classes II-III in Various Generation Combinations

These series of simulation results are shown in Fig. 4. Comparing T9 and T12, shows that the negative ROCOF is slightly lower in Converter-dominant load, comparing to the IM dominant loads. So with both cases of T9 and T12, a faster load shedding is happening due to steeper ROCOF. In T12, the frequency goes out of allowed operation band and doesn't stay within stable borders. In all tests 9-12, the voltage is below normal operation band and after an under frequency load shedding and with voltage support from generation units, it is taken back to the allowed band without the need for UVLS. To compare these results with their equivalent tests (T17-21) when 50% of RES generation is provided via GSVSs, it is shown obviously that df/dt is improved in all tests of T17-21. Needing only one step of load shedding and a slight voltage improvement is due to support from GSVSs. The details about test results is summarized in Table IV. The difference between frequency profile between Fig. 4(a) and (b) shows that the ROCOF is steeper with Class II in contrast to Class III, when GSVSs have 50% of the RESs. The reason for this can be due to voltage supporting functionality of the GSVS RESs, that the voltage profile has improved in tests 17-21 comparing to the equivalent tests of 8 -12, with this difference that a very fast load shedding didn't drop the loads at very beginning similar to tests 8 12. Therefore, the results show that in T18 and T21 before any load shedding happens, the voltage profile goes so low that the remaining loads of the load class III, will reduce even further their power consumption (Resistive and IM loads). Therefore, in contrast to T17 and T20, the low voltage profile has helped the frequency deviation and T17 and T20, got a steeper df/dt

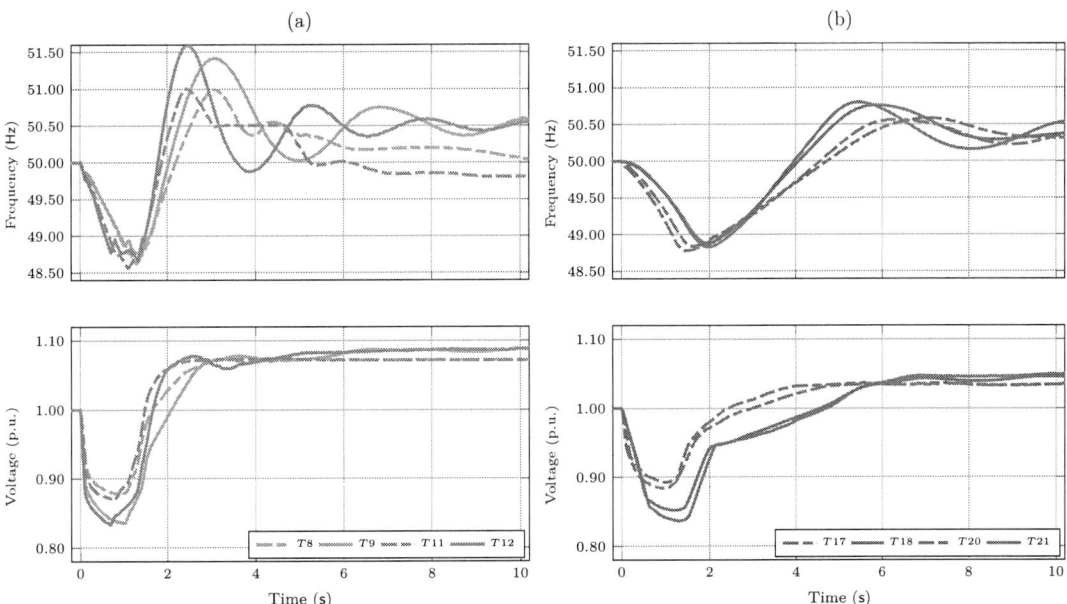

Fig. 4: a) Tests with having SG+GSCS in Generation side in light green and red, b) Tests with having SG+ GSCS+ GSVS in generation side in dark green and red

Table IV: Detail of test results in Fig.4

Tests	F min (Hz)	Fmax (Hz)	ROCOF (Hz/s)	Number of LS	LS time (s)	Vmin (pu)	Vmax (pu)
T8	48.80	50.98	1.04	twice	1.06 - 1.31	0.88	1.07
T17	49.00	50.58	0.76	once	1.50	0.89	1.03
T9	48.80	51.41	1.03	twice	1.04 −1.43	0.83	1.06
T18	49.00	50.76	0.65	once	1.65	0.85	1.04
T11	48.80	51.00	1.50	twice	0.8 − 1.11	0.87	1.07
T20	49.00	50.56	0.88	once	1.24	0.88	1.03
T12	48.80	51.60	1.60	twice	0.70 -1.33	0.83	1.08
T21	49.00	50.80	0.67	once	1.51	0.83	1.04

and a faster load shedding comparing to T18 and T21 with dominant B6U. Also we can see that ROCOF has improved from -1.04 $Hz s^{-1}$ in T8 to -0.76 $Hz s^{-1}$ in T17. The same trend exist between T9 and T18. In both T17 and T18 a ROCOF above -1 $Hz s^{-1}$ was resulted from deploying GSVSs in RESs. Also, the number of LS in Fig.4(a) doesn't increase from one step in contrast to Fig. 4(b) that in all tests two steps of load shedding is necessary and still the power system is not in the stable border, for instance in T12. Looking at the voltage profile at T9, shows that the voltage has dropped the most with load class III. As B6U will increase slightly the reactive power consumption due to its filter when the grid voltage is below rated value. This worsens the voltage profile comparing to load class II, because the IMs are feeding a variable torque load. In this case, IMs decreases the electromagnetic torque as a function of source voltage. This decelerates the IM and the active and reactive power consumption of the load is decreasing proportional to the square of the speed. Because of that, as it is shown, the voltage in T8 & T11 and T17 & T20 is higher than the ones in load class III.

Conclusion

Comparing different scenarios showed a grid dominant by the resistive load offers the best voltage-dependent active power reduction, when following the inrush current from IM, a huge reactive current is drawn which deteriorates the voltage profile in the load bus. If the main load is IM, depending on their connecting load, they could offer the best grid-supporting characteristics with self-regulation effect and voltage-dependent power consumption variation. Hence, the best voltage profile was seen with this load class in all test scenarios. And lastly, when converter-fed loads are dominant in the consumer side, it was shown that grid-friendly features of IMs have vanished. Instead, in order to keep the active power consumption constant, when the voltage has fallen, a converter-fed drive increased the current flow and with that, reactive power consumption was increased. This was shown by the voltage profile in all cases with dominant B6U in the consumer side.

Looking at the results for different share of non-synchronous generation units, showed that above 50% RESs, in all cases except load class I, the power system is not able to stay stable unless 50% of RESs are replaced by GSVS. Replacing GSCS with GSVS, in almost all cases, has improved either ROCOF, delayed, eliminated or reduced the number of consecutive load shedding and improved voltage profile. Even with 90% RES, the ROCOF stayed above -1 Hzs^{-1}, which is critical especially if the equipment for remedial actions in the power system are exhausted or old to deal with such unbalances in the islanded area. The results in this paper showed the necessity of considering the dominant load characteristic to hold up an islanding condition even facing transient events. These results in the small scale has shown another factor which plays a critical role in studying minimum required inertia in the power system.

References

[1] H. Urdal, R. Ierna, J. Zhu, C. Ivanov, A. Dahresobh, et al., "System strength considerations in a converter dominated power system," in 12th Wind Integration Workshop, London, England, 2013.

[2] ENTSO-E, "High Penetration of Power Electronic Interfaced Power Sources and the Potential Contribution of Grid Forming Converters," Technical Report, 2020.

[3] D. Duckwitz, "Power System Inertia Derivation of Requirements and Comparison of Inertia Emulation Methods for Converter-based Power Plants," Ph.D dissertation, Electrical Engineering and Computer Science of the University of Kassel, 2019.

[4] N. Fazli, S. Gierschner and H.-G. Eckel, "Evaluating frequency stability with consideration of load type in different share of renewables and emulated inertia in case of system split," 22nd European Conference on Power Electronics and Applications (EPE'20 ECCE Europe), 2020.

[5] "Modelling and Aggregation of Loads in Flexible Power Networks- Working Group C4.605- Report 566," 2014.

[6] P. J. Colleran and W. E. Rogers, "Controlled Starting of AC Induction Motors," in IEEE Transactions on Industry Applications, vol. IA-19, no. 6, pp. 1014-1018, Nov. 1983.

[7] Kundur, P."Power System Stability and Control", New York, USA: McGraw- Hill, Inc., 1994. 1776p. ISBN: 0-07-035958-X.

[8] R. Teodorescu, M. Liserre and P. Rodríguez. "Grid Converters for Photovoltaic and Wind Power Systems."(2011).

[9] VDE / FNN, "VDE-AR-N 4120:2015-01 Technical requirements for the connection and operation of customer installations to the high-voltage network (TCC High-Voltage)," 2015.

[10] European Commission, "Commission Regulation (EU) 2016/631 of April 2016 establishing a network code on requirements for grid connection of generators," 2016. [Online]. Available:

http://eur-lex.europa.eu/eli/reg/2016/ 631/oj.

[11] "VDE-AR-N 4120 (draft May 2017) Technical requirements for the connection and operation of customer installations to the high voltage network (TAR high voltage) (German edition)," 2017.

[12] N. Fazli, S. Gierschner, M. Gierschner, L. Cai and H. Eckel, "Rotating-Voltage-Vector Control for Wind Energy Plants Providing Possibility for Ancillary Services," 2018 20th European Conference on Power Electronics and Applications (EPE'18 ECCE Europe), 2018.

[13] Zhang, Z.-S.; Sun, Y.-Z.; Lin, J.; Li, G.-J.: "Coordinated frequency regulation by doubly fed induction generator-based wind power plants", IET Renewable Power Generation, 2012, 6, (1), p. 38-47, DOI: 10.1049/iet-rpg.2010.0208

[14] "IEEE Recommended Practice for Excitation System Models for Power System Stability Studies," in IEEE Std 421.5-2016.

[15] C. W. Taylor, "Concepts of undervoltage load shedding for voltage stability," in IEEE Transactions on Power Delivery, vol. 7, no. 2, pp. 480-488, April 1992.

[16] Thiesen, H, Jauch. C. "Determining the load inertia contribution from different power consumer groups," Energies 2020.

[17] R. Ierna, J. Zhu, H. Urdal, A. J. Roscoe, M. Yu, A. Dysko, and C. D.Booth, "Effects of VSM Convertor Control on Penetration Limits ofNon-Synchronous Generation in the GB Power System," in 15th WindIntegration Workshop, p. 8, Nov. 2016.

[18] H. Kazemi Kargar and J. Mirzaei, "New method for islanding detection of wind turbines," 2008 IEEE 2nd International Power and Energy Conference, 2008.

Self-Oscillating Capacitive Power Transfer with Multiple Receiver Capability and Coupling Path Adaption

Norbert Seliger
Technical University of Apllied Sciences Rosenheim
Hochschulstrasse 1
Rosenheim, Germany
Phone: +49 (08031) 805-2624
Fax: +49 (08031) 805-2702
Email: norbert.seliger@th-rosenheim.de
URL: http://www.th-rosenheim.de

Acknowledgments

The author would like to thank Dr. G. Deboy, R. Schmidt-Rudloff and Dr. M. Schlenk from Infineon Technologies AG for providing CoolGaN devices.

Keywords

≪Capacitive Coupling≫, ≪Half Bridge≫, ≪Wireless power transmission≫, ≪Gallium Nitride (GaN)≫, ≪HEMT≫, ≪Contactless energy transfer≫, ≪High frequency power converter≫.

Abstract

We present a capacitive power transfer system with self-adapting capability to multiple, variable load receivers. The proposed self-oscillating GaN-based half-bridge converter with load current feedback automatically adapts to the coupling path. We show design equations based on network theory and demonstrate experimentally more than 100W power transfer at 92% efficiency.

Introduction

Numerous research and development projects are currently devoted to the wireless power transfer (WPT) to portable electronic systems and electric vehicles [1]. A general challenge in conventional wireless power transmission is matching the transmission system to varying coupling paths and to changing loads (impedance matching approach), which theoretically limits the overall efficiency. On the other hand, high efficiency is achieved by energy transfer with sub-optimal power transfer, i.e. at unmatched loads in order to reduce the losses in the passive and active components (high-efficiency approach) [1].

Wireless energy transfer could be based on quasi-static electric fields (capacitive power transmission, CPT), on quasi-static magnetic fields (inductive power transmission, IPT [2] or on non-stationary electromagnetic fields [3]. Our study focuses on contactless power transfer via electric fields (capacitive power transmission, CPT), which is the preferred technique for bridging short distances [4].

If applications do not rely on regulations for constant frequency operation (which have been mainly issued for IPT systems), then improved system performance by self-tuning (self-oscillating) WPT approaches for voltage and power control [5, 6] have been proposed. Sun et al. [7] have published a self-oscillation resonant switching converter coping with multiple receiver loads and compensating changes in transfer spacing and coil misalignment, respectively. In such systems, however, sophisticated detection and control circuitry at the primary side is needed for self-adjustment.

A recent self-oscillation technique introduced as over-the-air positive wireless power transfer overcomes these limitations [8]. Liu et al. [9] have demonstrated a robust operation for a self-oscillating CPT system operating at variable loads. A multiple receiver operation for low power levels has been discussed in [10].

In this paper we follow the approach recently published by Seliger in [11] to design a self-adapting capacitive energy transfer for multiple receivers with the capability of robust operation at variable coupling paths. The next section briefly describes the converter system, followed by a section presenting first experimental results on a prototype demonstrator.

Self-oscillating Half-Bridge Converter

We report on a wireless energy transfer via the quasi-static electric field (capacitive power transmission, CPT). The structures for field coupling are good conductors of arbitrary shape acting as electrodes, with a dielectric material or air in between. In the following analysis, we represent these structures by lumped circuit elements (coupling capacitors).

The self-oscillating half-bridge converter driving a capacitively coupled load is depicted in Fig. 1. The wireless power system consists of n multiple receivers (load resistances $R_{load,n}$), each supplied with coupling capacitors $C_{p1,n}$ and $C_{p2,n}$, respectively. A supposed operation stage of the half-bridge converter

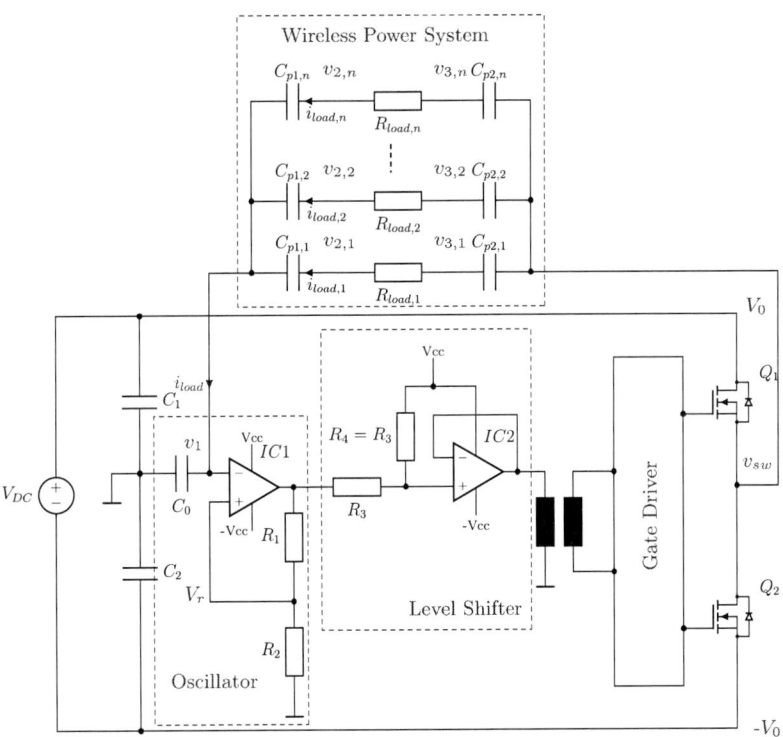

Fig. 1: Schematics of the self-oscillating half-bridge converter.

with transistor Q_1 as turned-on (and voltage $v_{sw}(t) = V_0$), gives a positive current i_{load} through the Wireless Power System in Fig. 1. The voltage $v_1(t)$ across capacitor C_0 rises until it reaches the reference voltage V_r, which is given by

$$V_r = \frac{R_2}{R_1 + R_2} V_{cc} = \beta V_{cc} \tag{1}$$

with the supply voltage $\pm V_{cc}$ of the operational amplifier $IC1$ forming the oscillator. Exceeding this reference voltage reverses the output voltage of $IC1$ to $-V_{cc}$ and changes the logic level of the level shifter circuit ($IC2$). The output voltage of this stage directly drives the gate signals of the half-bridge converter via a digital isolator circuit. Since the logical signal at the gate driver input has changed, transistor Q_1 is turned off and the low-side switch Q_2 is turned on. The switching node voltage $v_{sw}(t)$ changes from V_0 to $-V_0$ and impresses a negative load current i_{load} (reversed as shown in Fig. 1). Capacitor C_0 is discharged until it reaches the lower threshold voltage of $V_r = -\beta V_{cc}$, which completes one period of self-oscillation.

It should be noted that the load current i_{load} is the sum of the individual currents $i_{load,n}$ impressed on multiple receiver loads. Whenever there is a change in the loads, i.e. a change of an individual load $R_{load,n}$, or a change in the number of applied loads, n, the circuit will immediately react on that by the charging dynamics of C_0. Furthermore, variation in the coupling path capacitances $C_{p1,n}$ and $C_{p2,n}$ directly influences the charging/discharging current of C_0.

Multiple Load Operation

Referring to [12] and to Fig. 1, for a given load and coupling path arrangement, the switching node voltage is a rectangular waveform with 50% duty cycle and amplitude $V_0 = \frac{V_{DC}}{2}$. We write for each individual load current $i_{load,n}$ from network theory:

$$
\begin{aligned}
i_{load,n} &= C_{p1,n}\frac{d}{dt}\left[v_{2,n}(t)-v_1(t)\right] \\
&= \frac{v_{3,n}(t)-v_{2,n}(t)}{R_{load,n}} \\
&= C_{p2,n}\frac{d}{dt}\left[v_{sw}(t)-v_{3,n}(t)\right]
\end{aligned}
\tag{2}
$$

An analytical solution for the circuit waveforms can be found for the special case of N multiple receivers of the same load $R_{load,n} = R_{load}$, with coupling capacitances $C_{p1,n} = C_{p1}$ and $C_{p2,n} = C_{p2}$, respectively [10]. The total load current $i_{load} = Ni_{load,n}$ is expressed by Eq. (2):

$$
\begin{aligned}
i_{load} &= C_0\frac{dv_1(t)}{dt} \\
&= NC_{p1}\frac{d}{dt}\left[v_2(t)-v_1(t)\right] \\
&= N\frac{v_3(t)-v_2(t)}{R_{load}} \\
&= NC_{p2}\frac{d}{dt}\left[v_{sw}(t)-v_3(t)\right]
\end{aligned}
\tag{3}
$$

where the individual voltages are now $v_{2,n} = v_2$ and $v_{3,n} = v_3$. Solving for $v_1(t)$ during the charging of C_0 gives

$$
v_1(t) = \frac{1}{\gamma}\left[V_0 - (V_0 + \beta\gamma V_{cc})e^{-\frac{t}{\tau}}\right]
\tag{4}
$$

The capacitance ratio γ is given by $\gamma = 1 + \frac{C_0}{NC_{p1}} + \frac{C_0}{NC_{p2}}$. The time constant $\tau = \frac{C_0 R_{load}}{\gamma N}$ reflects the self-adapting behavior of the circuit. The switching period T is obtained as

$$
T = 2\tau\ln\left(\frac{V_0 + \beta\gamma V_{cc}}{V_0 - \beta\gamma V_{cc}}\right)
\tag{5}
$$

According to [12], the condition for start-up of the self-oscillation is derived from Eq. (4) and given by

$$
V_0 > \beta\gamma V_{cc}
\tag{6}
$$

From the analytical solution we find some remarkable properties of the self-oscillating CPT system: Changes in the number of receivers, N, or changes in the loads, R_{load}, or changes in the coupling paths (C_{p1}, C_{p2}) are compensated by adapting the switching period. We can further calculate the total average power at the loads, P_{load}, as

$$
P_{load} = N\frac{\tau}{T}\frac{(V_0 + \beta\gamma V_{cc})^2}{R_{load}}\left(1 - e^{-\frac{T}{\tau}}\right)
\tag{7}
$$

and compare with the load power for a direct wired connection, $P_0 = N \frac{V_0^2}{R_{load}}$. The resulting power ratio P_r is then

$$P_r = \frac{P_{load}}{P_0} = \frac{\tau}{T} \left(1 + \frac{\beta \gamma V_{cc}}{V_0}\right)^2 \left(1 - e^{-\frac{T}{\tau}}\right) \tag{8}$$

It should be emphasized that the power ratio is independent of the load resistance R_{load} [11] and weakly dependent on the capacitance ratio γ. It expresses the power transfer of the WPT system in comparison to a direct wired connection. Fig. 2a shows exemplary the variation of the capacitance ratio γ for the case of N equal loads with fixed C_0 and variable $C_{p1,n} = C_{p2,n} = C_p$. The related power ratio P_r is calculated in Fig. 2b for different resistance ratios β from Eq. (1) and for a switching amplitude of $V_0 = 100$ V. We find a power ratio close to 1 for a large variability in β and for $1 < \gamma < 10$, i.e. resulting in a high power transfer for multiple receiver operation ($N = 1 \dots 10$) and changes in the coupling capacitances C_p according to Fig. 2a.

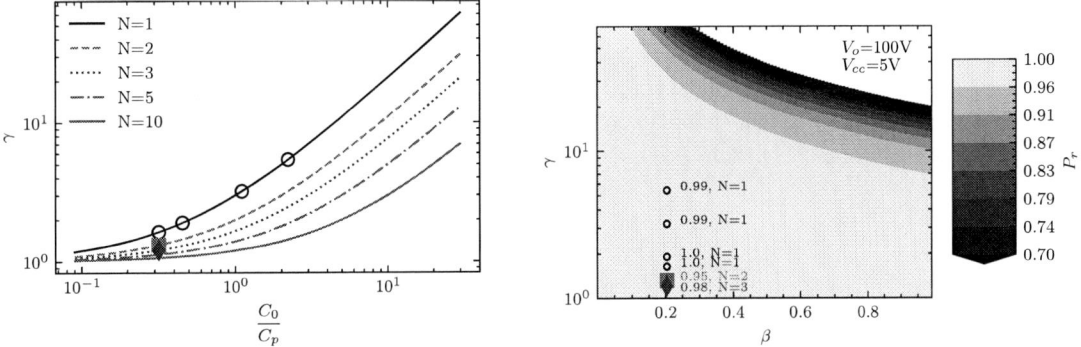

(a) Computed capacitance ratio γ for variable coupling capacitances ($C_{p1,n} = C_{p2,n} = C_p$) and N equal loads. Marked points: Experimental data.

(b) Computed power ratio P_r for variable coupling ratios γ and variable β. Marked points: Experimental results.

Fig. 2: Capacitance and power ratios for multiple load operation and variable coupling path.

For a given load scenario (N loads, load resistance R_{load}), the load power is controllable by the DC-link voltage V_{DC}, since $V_0 = \frac{V_{DC}}{2}$, and according to Eq. (7). For that purpose, a DC/DC converter (buck or boost converter) would be additionally required to connect a constant DC power supply with the variable DC link voltage V_{DC}.

Experimental Set-Up and Results

Prototype System

Fig. 3 shows the set-up of the prototype system. We have built a PCB for the self-oscillation circuit and for the half-bridge converter (operating with 650 V rated GaN-HEMTs). The loads (halogen lamps or power resistors mounted on a heat sink) are connected by coupling capacitors. For convenience, we use discrete capacitors of the same size $C_{p1,n} = C_{p2,n}$. Multiple receiver operation can be studied by changing the number of loads (in series with the capacitors) connected to the half-bridge inverter board. A high voltage power supply provides the DC link voltage V_{DC} and a second laboratory power source supplies the gate drive circuit and the oscillator board, respectively (cf. Fig. 3).

Alternatively, as has been shown in [11], the self-oscillating CPT transfer via parallel brass plates with a plastic sheet as a dielectric instead of using discrete components can be applied for a large range of capacitance values. Furthermore, utilizing the high permittivity of water enables self-oscillating CPT via metallic plates even in tap water [12].

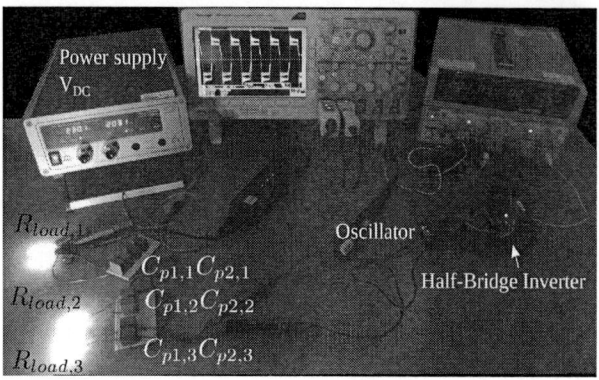

Fig. 3: Experimental set-up with three 42 W halogen lamps ($N = 3$ loads) and equal capacitances $C_{p1,n} = C_{p2,n}$.

Measurement results

In Fig. 4 we compare the analytical result for the load voltage $v_{load}(t)$ derived from the solution of $i_{load}(t)$ in Eq. (3) with a circuit simulation [13], and with the measurement on the prototype system. The simulated waveform nicely matches with the experimental signal. Since the analytical result is based on an ideal switching behavior, we expect an overestimation of the rise and fall times, and the switching period as well, for periods $T < 4\,\mu s$ [11].

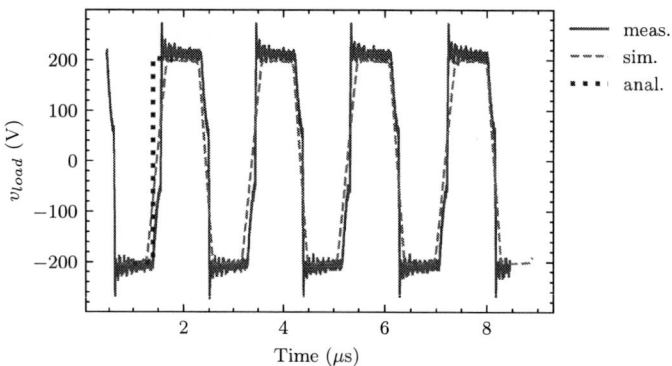

Fig. 4: Load voltage for a total load of three 42 W halogen lamps ($N = 3$) at a DC-bias voltage $V_{DC} = 2V_0 = 404\,\text{V}$. The switching period automatically adapts to $T = 1.9\,\mu s$. The total power P_{load} is 104 W.

From measurements of the root-mean-square values of the load currents and voltages, respectively, we determine the average power at the load, P_{load}. With the root-mean-square voltage measured at the switching node, $v_{sw}(t)$, we determine the maximum power to be transferable to the load for a direct wired connection, P_0. The experimental power ratios P_r are given in Fig. 2b, for the cases of multiple loads ($N = 1 \ldots 3$) and variation in coupling capacitances C_p. The related values for γ are indicated in Fig. 2a. Our experiments confirm the capability for high power transfer with multiple receivers and with changing coupling path conditions: The measured power ratio P_r is higher than 0.99 for a coupling ratio changing by a factor of 10 for a single load ($N = 1$), and slightly smaller ($0.95 \ldots 0.98$) for multiple loads.

With the input power P_{in} measured at the converter DC power supplies, we determine the efficiency η of the CPT system as:

$$\eta = \frac{P_{load}}{P_{in}} \tag{9}$$

For the operation data in Fig. 4 we get $\eta = 92\%$. We have determined the efficiency of the CPT system for a variety of parameters, i.e. the DC link supply voltage $V_{DC} = 2V_0$ and the load resistance R_{load}. Table I summarizes some of the experimental results. The half-bridge converter assures an efficiency over 90% for fundamental oscillation frequencies up to f_{osc}=600 kHz and supply voltages up to $V_{DC} = 400V$.

Table I: DC link voltage V_{DC}, R_{load} for $N = 1$, fundamental oscillation frequency $f_{osc} = 1/T$ and efficiency η (experimental results). The value of 291 kHz is computed for ideal switching (Eq.(5)).

V_{DC} (V)	$R_{load}(\Omega)$	f_{osc} (kHz)	η (%)
100	286	246 (291)	93
200	404	304	91
240	248	528	91
375	303	556	90

This prototype half-bridge converter can be operated up to 10 A at a maximum DC link voltage of $V_{DC} = 450V$. Therefore, power transfer up to kilowatts would be feasible.

Discussion

Self-oscillating capacitive power transfer has some major advantages as compared to conventional CPT systems. In our proposed converter, the load voltage and thus the load current are almost rectangular signals (cf. Fig. 4). The time-integrated load current results in a triangular voltage signal across the coupling capacitors, i.e. the energy transfer is due to a triangular electric field signal. Thus in a Fourier series representation, odd numbered harmonic signals for a specific oscillation period according to Eq. (5) can be identified, which are contributing with the corresponding harmonic amplitude to the overall transferred power.

Due to the non-resonant operation, there is no need for a compensation network on the transmitter as well as on the receiver side. Such compensation networks are adding complexity and costs to single frequency and multiple frequency resonant topologies [14], cause limited coupling distances (by network detuning due to reduced coupling capacitances), and deteriorate system efficiency [15].

By self-adapting to a change in the impedance of the coupling link, the self-oscillating CPT is robust against lateral misalignment in a Four Plate-horizontal or -vertical coupling path [16]. The system reacts analogous to a vertical displacement, i.e. it automatically responds by increasing or decreasing the oscillation period. The CPT answers to changes in the load R_{load} by adapting the time constant τ and period T in a comparable way. Adding multiple receivers to the wireless link has a similar effect.

Our analytical and experimental analysis reveals that, for a given coupling link, reducing the individual load, R_{load}, for a constant number of loads, N, reduces the period T. A corresponding lowering in T is found for a grow in N at constant individual loads R_{load}. Both operation conditions raise the switching losses in the converter resulting in a lower efficiency η. The drop in η at elevated DC link voltages in Table I, keeping the coupling link almost unchanged, manifest contributions of the oscillation frequency f_{osc}, the device voltage $V_{ds} = V_{DC}$ and the load current to the switching losses for the GaN-HEMTS [17].

The broadband energy transmission with variable switching period found here serves as a potential source for EMI disturbances. On the other hand, design strategies for the coupling structures in order to minimize electric field emission are already available [18].

Conclusion

A novel CPT system for multiple receiver operation and capability to self-adapting to the transfer path impedance has been developed and parametrically studied. The prototype system confirms the characteristics from network analysis, where a high power transfer and a high efficiency makes the self-adapting CPT system attractive for multiple load applications.

References

[1] Shu Yuen Ron Hui, Wenxing Zhong, and Chi Kwan Lee. A critical review of recent progress in mid-range wireless power transfer. *IEEE Transactions on Power Electronics*, 29(9):4500–4511, 2013.

[2] A. Alphones and Prasad Jayathurathnage. Review on wireless power transfer technology (invited paper). In *2017 IEEE Asia Pacific Microwave Conference (APMC)*, pages 326–329, November 2017.

[3] Bernd Strassner and Kai Chang. Microwave Power Transmission: Historical Milestones and System Components. *Proceedings of the IEEE*, 101(6):1379–1396, June 2013.

[4] Jiejian Dai and Daniel C Ludois. A survey of wireless power transfer and a critical comparison of inductive and capacitive coupling for small gap applications. *IEEE Transactions on Power Electronics*, 30(11):6017–6029, 2015.

[5] Alireza Namadmalan. Self-oscillating tuning loops for series resonant inductive power transfer systems. *IEEE Transactions on Power Electronics*, 31(10):7320–7327, 2015.

[6] Masood Moghaddami and Arif Sarwat. Self-tuning variable frequency controller for inductive electric vehicle charging with multiple power levels. *IEEE Transactions on Transportation electrification*, 3(2):488–495, 2016.

[7] Shubin Sun, Bo Zhang, Chao Rong, Xujian Shu, and Zhihao Wei. A Multi-receiver Wireless Power Transfer System Using Self-oscillating Source Composed of ZVS Full-bridge Inverter. *IEEE Transactions on Industrial Electronics*, 2021.

[8] Younes Ra'Di, Bhakti Chowkwale, Constantinos Valagiannopoulos, Fu Liu, Andrea Alù, Constantin R Simovski, and Sergei A Tretyakov. On-site wireless power generation. *IEEE Transactions on Antennas and Propagation*, 66(8):4260–4268, 2018.

[9] Fu Liu, Bhakti Chowkwale, and Sergei A Tretyakov. Self-oscillating capacitive wireless power transfer with robust operation. In *2018 IEEE International Symposium on Antennas and Propagation & USNC/URSI National Radio Science Meeting*, pages 2533–2534. IEEE, 2018.

[10] Fei Liu, Prasad Jayathurathnage, and Sergei A Tretyakov. Active metasurfaces as a platform for capacitive wireless power transfer supporting multiple receivers. In *2019 13th Int. Congress on Artificial Materials for Novel Wave Phenomena (Metamaterials)*, pages X–227. IEEE, 2019.

[11] Seliger N. Design of a Half-Bridge Converter for Self-Adapting Capacitive Wireless Power Transfer. *IEEE Conference on Energy Conversion 2021*, pages 1–6, 2021.

[12] Seliger N. A free oscillating Half-Bridge Converter for capacitively coupled Wireless Power Transfer. *2021 IEEE Forum on Research and Technologies for Society and Industry*, pages 1–6, 2021.

[13] Mike Engelhardt. Ltspice XVII, Analog Devices Corporation All rights reserved. *https://www.analog.com*, 1995-2021.

[14] Zhen Zhang, Xingyu Li, Hongliang Pang, Hasan Komurcugil, Zhenyan Liang, and Ralph Kennel. Multiple-Frequency Resonating Compensation for Multichannel Transmission of Wireless Power Transfer. *IEEE Transactions on Power Electronics*, 36(5):5169–5180, May 2021.

[15] Mehmet Zahid Erel, Kamil Cagatay Bayindir, Mehmet Timur Aydemir, Sanjay K. Chaudhary, and Josep M. Guerrero. A Comprehensive Review on Wireless Capacitive Power Transfer Technology: Fundamentals and Applications. *IEEE Access*, 10:3116–3143, 2022.

[16] Alberto Reatti, Luca Pugi, Fabio Corti, and Francesco Grasso. Effect of Misalignment in a Four Plates Capacitive Wireless Power Transfer System. In *2020 IEEE International Conference on Environment and Electrical Engineering and 2020 IEEE Industrial and Commercial Power Systems Europe (EEEIC/I&CPS Europe)*, pages 1–4. IEEE, 2020.

[17] Jacob Gareau, Ruoyu Hou, and Ali Emadi. Review of Loss Distribution, Analysis, and Measurement Techniques for GaN HEMTs. *IEEE Transactions on Power Electronics*, 35(7):7405–7418, July 2020.

[18] Hua Zhang, Fei Lu, Heath Hofmann, Weiguo Liu, and Chunting Chris Mi. Six-Plate Capacitive Coupler to Reduce Electric Field Emission in Large Air-Gap Capacitive Power Transfer. *IEEE Transactions on Power Electronics*, 33(1):665–675, January 2018.

An Electrically Driven Gas Compressor for Hydrogen Refueling Stations with Active Power Smoothing

Alfred Rufer

EPFL Ecole Polytechnique Fédérale de Lausanne

STI-DO-EPFL Station 11

CH1015 Lausanne, Switzerland

+4179 244 09 84

alfred.rufer@epfl.ch

Keywords

«Electrical drive», « Active power-decoupling circuit», « Capacitors», « Highly dynamic drive», « Force Control»

Abstract

A new electrically driven gas booster is described as an alternative to the classical air-driven gas boosters known for their poor energetic efficiency. These boosters are used in hydrogen refueling stations where the global energy account is of significancy. The proposed system uses a common crankshaft and two connecting rods to transform the rotational motion of the motor to the linear displacement of the original compressor pistons. The strongly fluctuating power of the compressor is smoothed by an active capacitive auxiliary storage device connected to the DC circuit of the power converter that feeds the electric motor.

The paper describes the typical behaviour of the compressor and of the coupling mechanism from the point of view of the needed forces, torques and power. The design of the auxiliary storage capacitor is given and the very specific waveforms are represented.

Introduction

Green Hydrogen is a promising energy vector in the sector of automotive vehicles especially in the context of reducing the emissions of green-house gases. Due to its low weight density, Hydrogen must be stored on-board in high pressure reservoirs. A simplified representation of a Green Hydrogen refueling station is given in Fig. 1. In such a system, the green fuel is generated from electric renewable source feeding a water electrolyser.

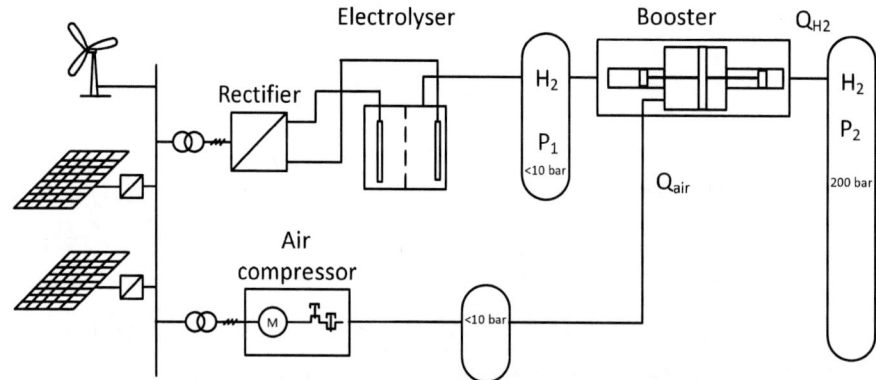

Fig. 1 Example of a Hydrogen refueling station for automotive vehicles.

Then, the green fuel is pressurized to a high-pressure level around 700 bar for personal vehicles or 350 bar for heavy transportation means. The pressurization machinery is realized conventionally by air driven gas boosters which are known for their simplicity, reliability and low costs [1].

However, the poor energetic efficiency of the used pneumatic motor in the gas booster addresses the question of the total efficiency of a complete filling station. In general, the main cause of the poor efficiency of pneumatic actuators resides in the fact that the air under pressure accumulated in the working chambers is simply released to the surrounding before initiating the return stroke. Fig. 2 shows a pressure-to-volume diagram of the thermodynamic change of state of the air in a pneumatic cylinder, where the constant pressure displacement work is represented with the finely hatched square surface W_2 between V_1 and V_2. This surface corresponds to the effectively produced mechanical work by a cylinder under constant pressure P_2. Left to the V_2 value of the volume, the surface under the decreasing curve (W_{2d}) corresponds to the recoverable expansion work corresponding to the lost energy when the exhaust valve is opened.

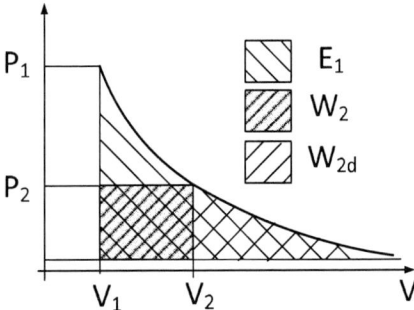

Fig. 2 Pressure-volume diagram of pressurized air work and expansion

A better use of the pressurized air of a cylinder has been proposed, using the principle of adding an expansion work component to the constant pressure displacement work, recovering so the internal energetic content of the intaken air [2-5]. In principle, the addition of an expansion chamber can increase the cylinder system efficiency from around 0.35 to 0.7 [4]. With the same goal to improve the efficiency, the principle of using the expansion energy has been applied to a pneumatic booster [6]. Another better adapted system is based on the use of an additional expansion cylinder mechanically coupled to the filling cylinder [7]. The resultant active force is in this case better adapted to the compression characteristic but needs a specific design to be able to provide a sufficient effort over the full length of the strokes.

In this paper, an electrically driven gas booster is proposed, where the mechanical force is provided by a classical electric motor. The coupling of the rotational motion of the motor to the linear displacement of the compression stages is realized on the base of a classical crankshaft and two connecting rods. The original compression cylinders are conserved and placed on both sides of the crankshaft. The strongly fluctuating compression force of the pistons at a low frequency results in a strongly modulated power of the electric motor. For smoothing the power taken from the electrical feeding grid, an active power compensation system using variable voltage capacitors is connected to the DC link of the frequency converter.

An electric motor instead of the common pneumatic actuator

Figure 3 shows on the top the classical air-driven gas booster with the complete feeding chain which comprises an electric motor driving an air compressor. The pressurized air reservoir serves not only as a simple power source for the air-drive of the compressor but plays additionally the role of power buffer for the strongly fluctuating power demand of the compression cylinders. The cascade of the partial efficiencies of the electric motor (87%), the air compressor (65%) and a classical pneumatic booster (35%) leads to a compression efficiency of around 20%.

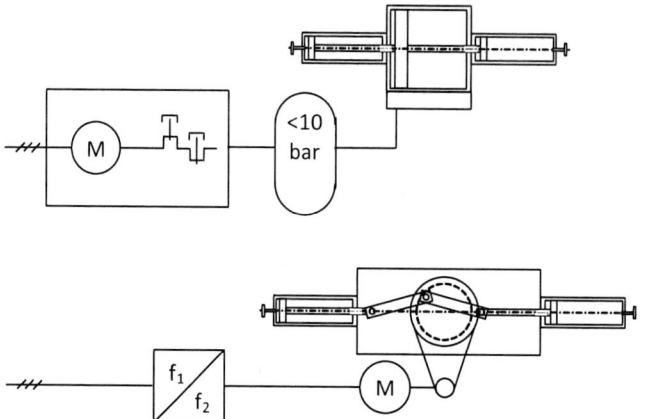

Fig. 3 Gas booster driven by a crankshaft and two connecting rods

In the lower part of the figure, the new proposed system is represented where the motion of the compression cylinders is provided by a crankshaft and two connecting rods driven by a variable frequency rotating field machine. The low operating frequency of the compression stages is determined by the temperature limit of the compression stages and needs to insert a mechanical reduction gear between motor and crankshaft. This new system greatly simplifies the supply chain of the compression stages by reducing the number of successive energy conversions and by eliminating the pneumatic compressor and actuator. The result is a highly improved energy efficiency. The partial efficiencies are estimated as 95% for the frequency converter, 92% for the motor and 97% for the gear, leading to a value of 84% for the driving equipment.

Regarding the strongly fluctuating power demand of the cylinders, the frequency converter is equipped with an active power compensation stage based on a capacitive storage circuit.

Coupling the rotational motion to the linear displacement of the pistons

The most known method for coupling a rotational motion to a linear displacement is given by the classical crankshaft system. In the proposed system, the alternating operation of a left and right compression cylinders is maintained. The crankpin is therefore coupled to two separated connecting rods (bottom of Fig. 3).

Mathematical description of the piston-crankshaft assembly

In Figure 4 a piston assembly is represented with the connecting rod and the crankshaft. The parameters are indicated as r, the radius of the crankshaft, l the length of the connecting rod, and Φ the angle of rotation of the crankshaft. The diameter of the piston, d and its position x are also indicated.

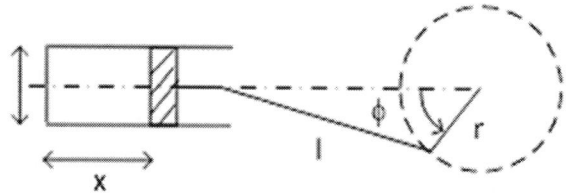

Fig. 4 Piston, crankshaft and connecting rod

The position of the piston is given through rel. (1)

$$x = r(1 - \cos\varphi) + \frac{\lambda}{2} r \sin^2\varphi \tag{1}$$

where the connecting rod ratio λ is used and is defined as

$$\lambda = \frac{r}{l} \tag{2}$$

The velocity of the piston is given by

$$v = \omega \cdot r \cdot \sin\varphi(1 + \lambda\cos\varphi) \tag{3}$$

In the simulation process, the torque developed by the motor is calculated through the indirect calculation of the power. If the force exerted on the piston is given by

$$F_p = p \cdot A \tag{4}$$

the mechanical power is defined by the product of the force by the piston's velocity:

$$\text{Pow} = F_p \cdot v \tag{5}$$

The torque is obtained by the division of the power by the angular velocity ω

$$M_{mot} = \text{Pow} / \omega \tag{6}$$

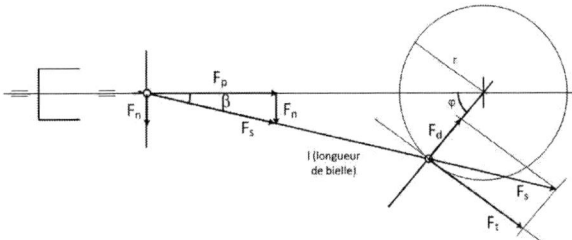

Fig. 5 Diagram of the forces for the calculation of the torque and reactions

A direct calculation of the torque is however possible with the extended model represented in Fig. 5 [8]. The force on the connecting rod F_s, the tangential force F_t and the reaction force F_n are also defined. The torque on the crankshaft is given by rel.7.

$$M_{mot} = F_t \cdot r \tag{7}$$

The tangent force F_t is given by (8)

$$F_t = F_s \cos(\pi / 2 - (\beta + \varphi)) \tag{8}$$

The force transmitted through the connecting rod F_s is calculated with the help of the piston force F_p and the angle beta (9)

$$F_s = \frac{F_p}{\cos\beta} \tag{9}$$

This angle is given by rel. (10)

$$\beta = arc\sin\left(\frac{r}{l} \cdot \sin\varphi\right) \tag{10}$$

The perpendicular reaction F_n is defined as per (11)

$$F_n = F_s \sin\beta = \frac{F_p}{\cos\beta} \sin\beta \tag{11}$$

Simulation results

In Figure 6, the dimensionless evolution of the volumes of the compression cylinders is represented. The blue curve is representing the evolution of the left cylinder while the red curve shows the behaviour of the right cylinder. The curves represent already the asymmetric variation of the volumes of the left and right compression cylinders due to properties of the used crankshaft and 180° shifted connecting rods. The velocity of the two compressing pistons is given in Fig.7.

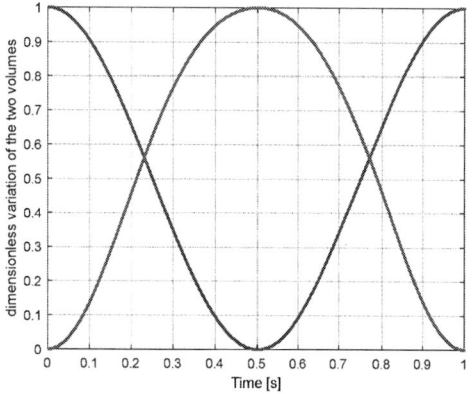

Fig. 6 Dimension-less variation of the volumes of the compression cylinders

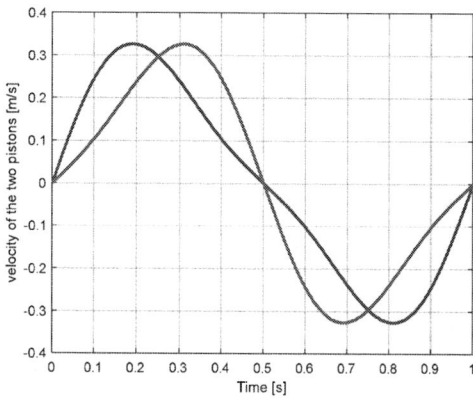

Fig. 7 Velocity of the pistons (m/s)

Figure 8 shows the evolution of the horizontal force of the left-side piston caused by the pressure in this cylinder. During the left-to-the-right stroke (0 to 0.5s), the diagram shows the force related to the pressure rise from the inlet pressure (15 bar) to the output pressure (160 bar). The elevation of the pressure is determined by a law of adiabatic evolution according rel. 15.

$$P_{gas} = P_{in\_gas} \left(\frac{V_{compr\_var}}{V_{comp\_max}} \right)^{\gamma} \quad \text{with } \gamma = 1.4 \tag{15}$$

During the second half-period (0.5s to 1s) which corresponds to the right-to-the-left stroke, the pressure level corresponds to the intake pressure (15 bar). The global force developed by the two pistons of the compression cylinders is represented in Fig. 9. One can see that the force components due to the intake pressure compensates partially the compression forces.

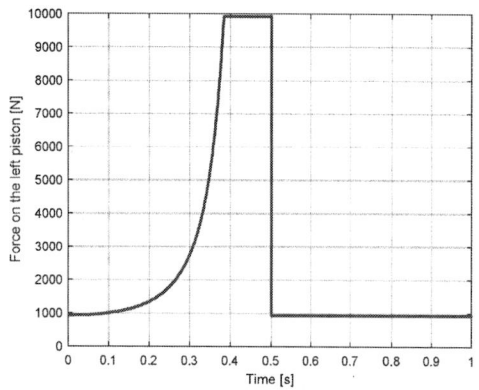

Fig. 8 Force developed by the left-side piston (N)

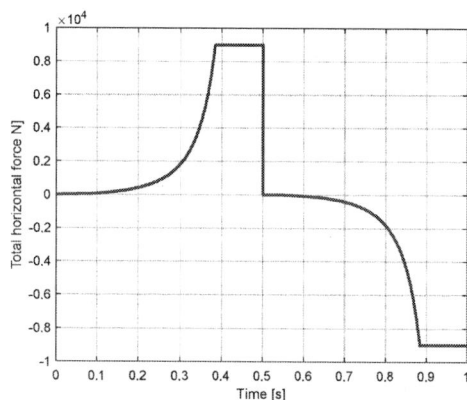

Fig. 9 Total force (horizontal) (N)

Figure 7 has shown the evolution of the piston's velocity. This information is further used to calculate the power developed by the compression cylinders (Fig. 10). This figure shows an interesting property of the system driven by a crankshaft and connecting rods, where the power (as well as the torque, see Fig.11) takes a negative value at the beginning of the compression phenomena. This effect is due to the asymmetrical evolution of the displacement and speed of the left and right cylinders. The value of the torque is obtained from the value of the powers according to relation 6.

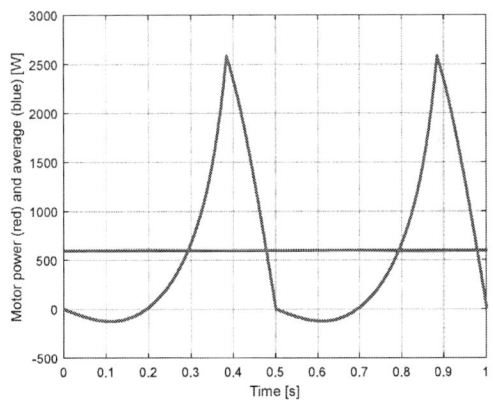

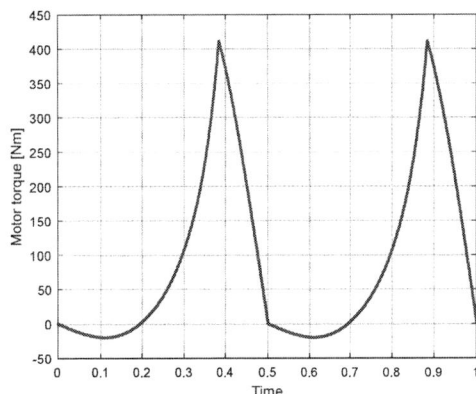

Fig. 10 Power received by the two pistons (W)   Fig. 11 Torque developed (crankshaft) (Nm)

The electric drive with an active power smoothing circuit

The waveforms of power and torque needed for the activation of the compression cylinders are represented in Fig. 10 and 11. These waveforms can be followed by a modern electric drive, for example a permanent magnet synchronous motor fed by a voltage source inverter. The power however should not be taken directly from the feeding grid where the power level is normally smooth. To obtain at the line side a fully smoothed power, a power compensating storage device is added and interconnected at the level of the intermediary DC circuit of the converter.

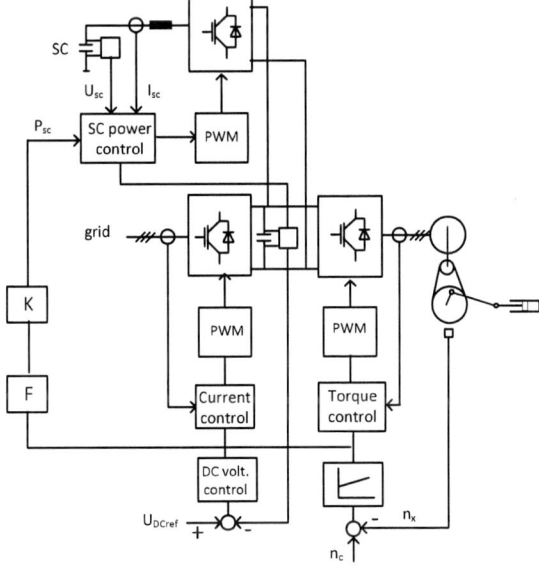

Fig. 12 Schematic diagram of the electric drive system with active power compensation circuit.

The active compensation device is composed of a capacitor which is periodically charged and discharged through a dedicated DC-DC conversion stage. The complete circuit of the drive fed by the voltage source converter and the active compensation circuit is represented in Fig. 12. The more detailed scheme of the power electronic circuits is given in Fig. 13.

The motor feeding converter is composed of two back-to-back connected voltage source converters. The two converters have at the motor-side a torque and speed controller, and at the line side a line current control with a superimposed DC-link voltage control (Fig. 12).

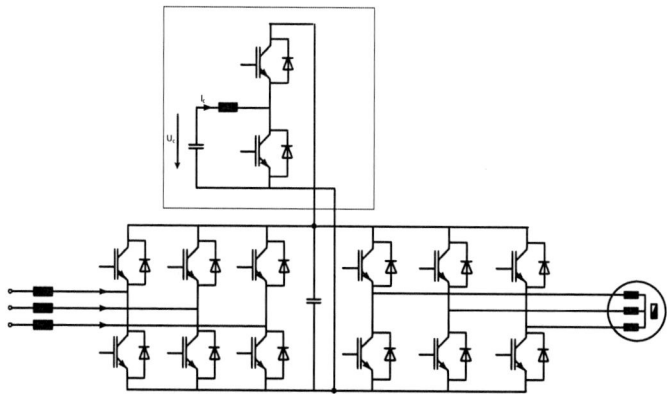

Fig. 13 Detailed scheme of the power electronic converter with active power compensation

According to the very specific waveform of the needed power and torque for the compression stages (Fig. 10 and 11), the driving motor current shows very strong variations. These variations and especially their corresponding low frequency can be seen in Fig. 14

The active power compensator is composed of a bidirectional step-down and step-up chopper circuit interfaced with the variable voltage storage capacitor and connected directly to the DC link of the frequency converter.

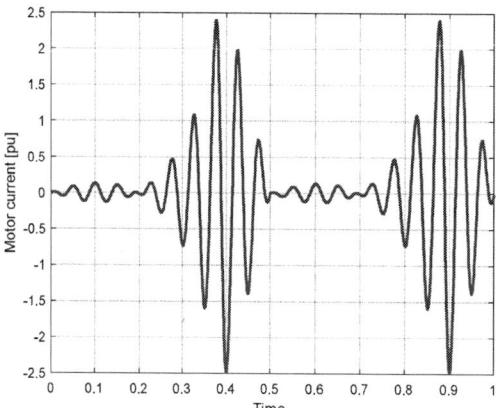

Fig. 14 Current in one phase of the motor (p.u)

The control of the active power compensator can be seen globally in Fig. 12. This control is based on the principle using a feed-forward power control signal obtained from the output of the speed controller at the motor side. Then, the control of the capacitor current is realized, taking in account the actual value of the capacitor voltage. The capacitor current reference is obtained by dividing the compensation power reference by the capacitor voltage. The detailed control circuit of the compensation capacitor is given in Fig. 15.

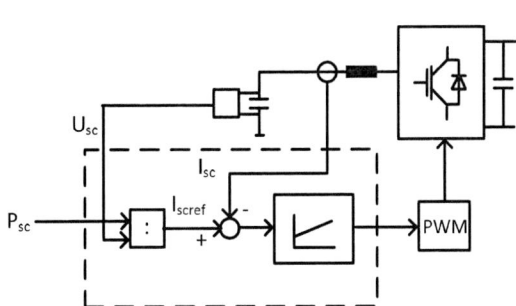

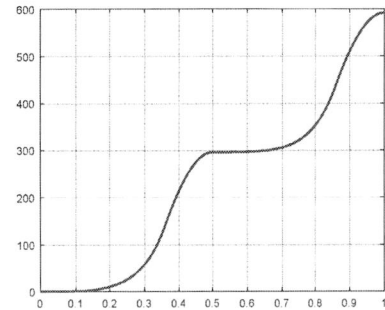

Fig. 15 Detailed scheme of the control of the compensation Capacitor

Fig. 16 Energy flow from the motor to the compression cylinders (J)

Design of the storage capacitor

The design of the storage capacitor is done on the base of the amount of energy to be stored and unstored in it. Fug. 16 illustrates the integral of the power, corresponding of the total energy flow.

During one cycle of the power which corresponds to one half of the revolution of the crankshaft, the energy variation is of 298 J. During this variation, the storage capacitor is charged and discharged in order to obtain a smooth average value at the level of the power exchange with the grid. The amounts of charged energy and of the discharged one are identic, and are equal to the half of the energy variation mentioned before

$$W_{Cch} = W_{Cdisch} = \frac{E_{var}}{2} = 298J / 2 = 149J \qquad (16)$$

The domain of use of the capacitor in terms of its discharge ratio [9] can be calculated. The design criterion of this element corresponds to the definition of its corresponding voltage variation. The capacitor is discharged from its 100% value down to a value of 50%, starting from a physical value of 600V.

The total energy content of the fully charged capacitor is

$$W_{Ctot} = \frac{1}{2}C \cdot U^2 \qquad (17)$$

And for a voltage variation from 100% to 50% the energy extracted is

$$W_{extr\_50} = \frac{1}{2}C \cdot (U^2 - \left(\frac{U}{2}\right)^2) = \frac{3}{4} \cdot \frac{1}{2}CU^2 \qquad (18)$$

Then, the value of the capacitor is defined from

$$149J = \frac{3}{4} \cdot \frac{1}{2}CU^2$$

$$C = 149J\frac{4}{3} \cdot \frac{2}{U^2} = 149J\frac{4}{3} \cdot \frac{2}{(600V)^2} = 0.00110F \qquad (19)$$

or 1100 μF

The waveform of the capacitor current of the active power compensation circuit is given in Fig. 17. The corresponding value of the capacitor voltage is represented in Fig. 18.

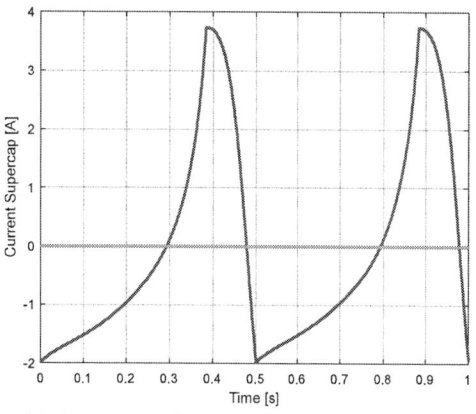

Fig. 17 Current exchanged with the compensator capacitor (A)

Fig. 18 Voltage of the storage capacitor (V)

Conclusions

An original electrically driven gas booster system is proposed as an alternative to the classical air-driven compressors where the energetic efficiency is problematic. A geared electric drive is proposed for delivering the strongly pulsating power demanded by the compression stages. Further, these power pulsations are compensated by an active storage circuit based on a controlled variable voltage capacitor. The paper has shown the characteristic waveform of the compressor forces, torque and pulsating power demand and proposed a design for the active power compensator. The power circuits are described together with their control functions.

References

[1] Ligen Y., Vrubel H., Arlettaz J., Girault H., Experimental correlations and integration of gas boosters in a hydrogen refueling station, International Journal of Hydrogen Energy, 45, (2020),16663-16671, Elsevier.

[2] Vito Gianfranco Truglia, High-Efficiency Engine Driven by Pressurized Air or Other Compressible Gases, Patent No US 9.677.400 B2, Jun. 13, 2017.

[3] Nègre Guy, Engine with an active mono-enery and/or bi-energy chamber with compressed air and/or additional energy and thermodynamic cycle thereof, US Patent US 7,469,527 B2, Dec. 30, 2008.

[4] Rufer A., A High efficiency pneumatic drive system using vane-type semi-rotary actuators, FACTA UNIVERSITATIS, Series: Electronics and Energetics, Vol. 34, Iss. 3, September 2021.

[5] Rufer A., Pneumatic cylinder assembly with enhanced energetic efficiency, PCT Patent application PCT/CH2021000002, May, 4th 2021.

[6] Yan Shi, Maolin Cai, Flow Characteristics of Expansion Energy Used Pneumatic Booster, Chinese Journal of Mechanical Engineering, Vol. 25, No. 5, 2012

[7] Rufer A., A pneumatic driven hydrogen compressor with reduced air consumption, Patent applications, 2021.

[8] George H Martin, Kinematics and Dynamics of machines, Mc Graw Hill Series in Mechanical Engineering, Waveland Press USA.

[9] Rufer A., Energy storage – Systems and components, CRC Press, Taylor and Francis group,2018

Unsymmetrical fault behavior of PLL based grid-connected converters

Philipp Hackl, Ziqian Zhang, Robert Schuerhuber
Institute of Electrical Power Systems, TU Graz, Austria
Tel.: +43 (0) 316 873 - 7567
E-Mail: philipp.hackl@tugraz.at
URL: https://www.tugraz.at/institute/iean

Keywords

«Grid-connected converter», «Stability analysis», «Transient analysis», «PLL», «Fault ride-through»

Abstract

The fault behavior of converters is essential for stable operation of converter-driven power grids. This paper introduces a stability criterion of unsymmetrical faults for phase-locked loop (PLL) based grid-connected converters. The focus lies on the stability of the PLL considering the grid conditions and the converter feed-in power during the fault. For this purpose, the approach is based on the phase portrait method where several situations are examined in theory. Finally, the stability criterion is validated with real-time system experiments.

Introduction

Due to the expansion of renewable energy sources, more and more converter-based systems are connected to the grid. This tends to reduce the short-circuit power of the grid and thus the stability of the converter is significantly affected negatively. A particularly critical case is the operation ride-through a short-circuit fault. In this case, the grid codes require the injection of reactive current into the grid to support the grid voltage and also for supplying sufficient current for protection relaying.

The converter control algorithm and the corresponding parameters dominate the stability of the converter-based generation. Thereby, a successful synchronization to the grid is achieved by different methods, which are reviewed in [1] without an analysis in the fault case. In most converters, a phase-locked loop (PLL) unit is used, which serves as base for the control algorithm and must remain stable during and after the fault to successfully ride-through a fault. In the literature numerous studies of PLL stability investigations concerning symmetrical faults exist, which are generally regarded as the most severe faults with the highest short-circuit current. In [2] a detailed design-oriented investigation of different PLL parameters with validations in experiments are examined, but only the transient stability of converters with equilibrium points are considered. Whereby [3] introduces a steady-state and transient stability criterion depending on the grid topology, grid voltage and injected converter current.

However, unsymmetrical faults occur far more frequently in power systems. In this case, a standard approach is to transform the investigated grid topology into an equivalent circuit in symmetrical component representation. This enables converters to fulfill the requirement of most grid codes e.g. [4] that require not only injection of positive sequence reactive power into the grid, but also negative sequence reactive power. In [5] the authors describe the coupling of the sequences in different grid faults and examine positive and negative sequence dominated instabilities. Although a steady-state stability criterion is defined, the theoretical approach investigating the transient stability is not specified in detail.

This paper introduces a steady-state and transient stability criterion of PLL based grid-connected converters under unsymmetrical faults. Thereby, it considers positive and negative sequence coupling with different current injection strategies. For this purpose, the synchronization method of the converter is explained, followed by a description of the influence of the grid topology. Subsequently, the relation of the power grid with the converter is introduced and the stability criterion is examined.

Finally, the stability criterion is validated with experiments using a real-time-system as described in [6]. Therefore, a typical single overhead line grid model with a converter and a Dy transformer, shown in Fig. 1, is used.

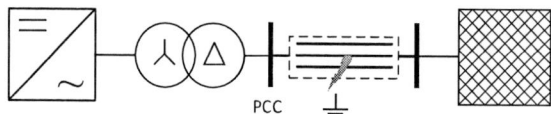

Fig. 1: Overview of investigated grid topology

In this paper the following conventions for notations are used.

X^+	positive sequence variable	ϑ_{index}	phase angle of complex variable
X^-	negative sequence variable	X_{index}	amplitude of complex variable
X^0	zero sequence variable	x_d	direct component of a variable
$\underline{X}_{index}$	complex variable	x_q	quadrature component of a variable

Modelling of converter control in relation to grid topology

The control strategies of most currently used converters require the terminal voltage as control quantity to feed-in the desired power. A phase-locked loop (PLL) is used for this purpose, whereby its stability is essential for the entire control algorithm. The importance of this reference phase angle of the PLL for the stability in symmetrical faults has already been discussed in detail in [3]. However, in the case of unsymmetrical faults, grid codes also require the injection of negative sequence reactive current into the grid voltage. Therefore, the negative sequence`s phase angle of the terminal voltage is also necessary for a successful fault-ride-through (FRT) control.

In order to enable the converter to operate stable in an unsymmetrical grid, there are various PLL strategies available [1]. A common used strategy is to control the positive and negative sequence current separately, which is shown in Fig. 2 for a single line-to-earth fault (red) and a line-to-line fault (green). Here it can be seen how the reference phase angles are obtained from the converter terminal voltage $\underline{U}$ through two separate PLLs. Each PLL controls the q-axis component of the voltages in a closed loop to get the reference phase angles.

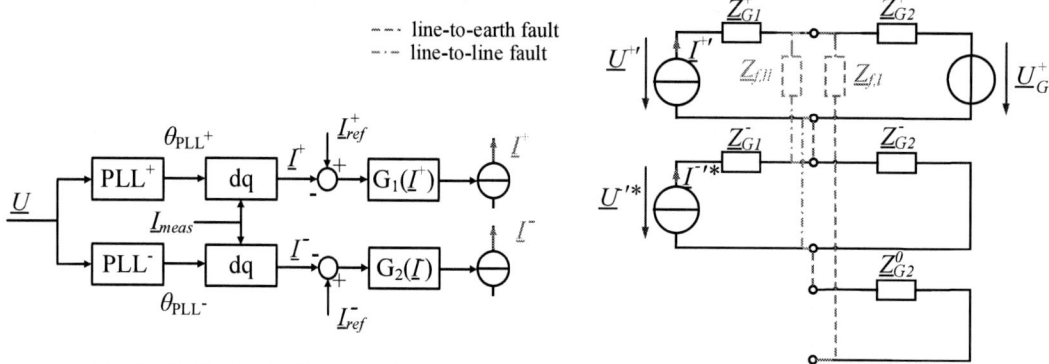

Fig. 2: (left) Block diagram of converter controller in positive and negative sequence
(right) Grid topology using symmetrical components

With these phase angles it is possible to decouple the converter current in a positive and negative sequence control algorithm. These currents can then be controlled through decoupled current controllers to the given reference currents. In this paper the research on the PLLs stability is mainly of interest, therefore the current control loops, G_1 and G_2 in Fig. 2, are assumed to be ideal. The converter is simplified as two controlled current sources, $\underline{I}^+, \underline{I}^-$, with reference currents $\underline{I}^+_{ref}, \underline{I}^-_{ref}$.

The following three factors mainly influence the voltage at the converter terminals: grid topology, fault type and fault location. Therefore, the grid topology from Fig. 1 is transformed into symmetrical components, shown in Fig. 2, to evaluate the resulting voltage during unsymmetrical faults at the converter terminals. These are connected in different configurations depending on the assumed unsymmetrical fault condition. The zero-sequence system of the converter voltage is decoupled via the delta-wye transformer.

In Fig. 2, the grid is represented by a Thevenin equivalent circuit, where Z_{G2} summarize the grid impedance and the impedance of the transmission line from the grid to the fault location. The converter is represented by an ideal current source with a series impedance Z_{G1} which include the impedances of the transformer and the transmission line from the converter to the fault location.

To calculate the converter terminal voltages, the grid voltage, grid impedance and also the converter output current must be considered. The positive and negative sequence is considered by the superposition principle. This results in the set of equations (1) on the high-voltage side, which are valid for all kinds of grid faults and topologies.

$$\underline{U}^{+'} = \underline{K}_1 \cdot \underline{U}_G^+ + \underline{Z}_{11} \cdot \underline{I}^{+'} + \underline{Z}_{12} \cdot \underline{I}^{-'*} \qquad (1a)$$
$$\underline{U}^{-'*} = \underline{K}_2 \cdot \underline{U}_G^+ + \underline{Z}_{21} \cdot \underline{I}^{+'} + \underline{Z}_{22} \cdot \underline{I}^{-'*} \qquad (1b)$$

The factors depend on the grid topology and the actual fault, whereby $\underline{K}_1, \underline{K}_2$ indicate the influence of the grid voltage on the positive/negative sequence voltage. $\underline{Z}_{11}, \underline{Z}_{22}$ define the impact of the positive/negative sequence current on the voltage of the same sequence, whereby $\underline{Z}_{12}, \underline{Z}_{21}$ are transfer impedance, coupling the positive and negative sequence systems. The variables $\underline{I}^{-'*}, \underline{U}^{-'*}$ are the conjugate complex values as explained in [7] [8].

To ensure proper operation of the converter, the measured converter terminal voltage must be converted to the dq reference frame. For this purpose, the determined phase angles of the PLLs are used.

$$u_d^+ + j \cdot u_q^+ = \underline{U}^+ \cdot e^{j \cdot \theta_{PLL^+}} \qquad (2a)$$
$$u_d^- + j \cdot u_q^- = \underline{U}^{-*} \cdot e^{-j \cdot \theta_{PLL^-}} \qquad (2b)$$

These expressions combined with (1) leads to the relation of the grid topology including the PLL, in (3). $\theta_{K_1}, \theta_{K_2}, \theta_{Z_{11}}, \theta_{Z_{22}}, \theta_{Z_{12}}, \theta_{Z_{21}}$ indicate the phase angle of the complex factors from (1), $\theta_{PLL^+}, \theta_{PLL^-}$ are the phase angles of the PLLs and $\theta_{I^+}, \theta_{I^-}$ are the phase angles of the positive and negative sequence converter currents.

$$u_d^+ + j \cdot u_q^+ = K_1 \cdot U_G^+ \cdot e^{j(\theta_{K_1} - \theta_{PLL^+})} + Z_{11} \cdot I^+ \cdot e^{j(\theta_{Z_{11}} + \theta_{I^+})} + Z_{12} \cdot I^- \cdot e^{j(\theta_{Z_{12}} - \theta_{1PLL} + \theta_{PLL^-} + \theta_{I^-})}$$
$$(3a)$$
$$u_d^- + j \cdot u_q^- = K_2 \cdot U_G^+ \cdot e^{j(\theta_{K_2} - \theta_{PLL^-})} + Z_{22} \cdot I^- \cdot e^{j(\theta_{Z_{22}} + \theta_{I^-})} + Z_{21} \cdot I^+ \cdot e^{j(\theta_{Z_{21}} + \theta_{PLL^+} - \theta_{PLL^-} + \theta_{I^+})}$$
$$(3b)$$

If the imaginary part of the voltage is shown separately and the time dependence of the PLLs are marked explicitly, the following expressions (4) are derived.

$$u_q^+\big(\theta_{PLL^+}(t)\big) = K_1 \cdot U_G^+ \cdot \sin\big(\theta_{K_1} - \theta_{PLL^+}(t)\big) + Z_{11} \cdot I^+ \cdot \sin(\theta_{Z_{11}} + \theta_{I^+}) + Z_{12} \cdot I^- \cdot$$
$$\sin(\theta_{Z_{12}} - \theta_{PLL^+} + \theta_{PLL^-} + \theta_{I^-}) \qquad (4a)$$
$$u_q^-\big(\theta_{PLL^-}(t)\big) = K_2 \cdot U_G^+ \cdot \sin\big(\theta_{K_2} - \theta_{PLL^-}(t)\big) + Z_{22} \cdot I^- \cdot \sin(\theta_{Z_{22}} + \theta_{I^-}) + Z_{21} \cdot I^+ \cdot$$
$$\sin(\theta_{Z_{21}} + \theta_{PLL^+} - \theta_{PLL^-} + \theta_{I^+}) \qquad (4b)$$

If the coupling between positive and negative sequence PLLs is neglected, which can be argued in grid conditions where $\underline{Z}_{12} \ll \underline{Z}_{11}$ and $\underline{Z}_{21} \ll \underline{Z}_{22}$ this results in (5).

$$u_q^+\big(\theta_{PLL^+}(t)\big) = K_1 \cdot U_G^+ \cdot \sin\big(\theta_{K_1} - \theta_{PLL^+}(t)\big) + Z_{11} \cdot I^+ \cdot \sin\big(\theta_{Z_{11}} + \theta_{I^+}\big) \tag{5a}$$

$$u_q^-\big(\theta_{PLL^-}(t)\big) = K_2 \cdot U_G^+ \cdot \sin\big(\theta_{K_2} - \theta_{PLL^-}(t)\big) + Z_{22} \cdot I^- \cdot \sin\big(\theta_{Z_{22}} + \theta_{I^-}\big) \tag{5b}$$

For a successful synchronization of the PLLs to the grid, the PLLs control the imaginary part of the terminal voltages to zero, explained in [3]. Therefore, a stable equilibrium point (SEP) of the PLLs can only be achieved, if the equation's (5) output is zero. This results to the analytical solution (6), if it is solved for the phase angles of the PLLs ($\theta_{PLL^+,sep}$, $\theta_{PLL^-,sep}$). Whereby the factor n (0, 1, 2, 3, …) indicates that there exist periodically solutions.

$$\theta_{PLL^+,sep} = \theta_{K_1} + \arcsin\left(\frac{Z_{11} \cdot I^+ \cdot \sin(\theta_{Z_{11}} + \theta_{I^+})}{K_1 \cdot U_G^+}\right) \pm n \cdot 2\pi \tag{6a}$$

$$\theta_{PLL^-,sep} = \theta_{K_2} + \arcsin\left(\frac{Z_{22} \cdot I^- \cdot \sin(\theta_{Z_{22}} + \theta_{I^-})}{K_2 \cdot U_G^+}\right) \pm n \cdot 2\pi \tag{6b}$$

To obtain a real number as phase angle the argument of the arcsin must be between -1 to 1. An equilibrium point can thus only be defined by the following conditions (7).

$$K_1 \cdot U_G^+ \geq Z_{11} \cdot I^+ \cdot \sin(\theta_{Z_{11}} + \theta_{I^+}) \tag{7a}$$

$$K_2 \cdot U_G^+ \geq Z_{22} \cdot I^- \cdot \sin(\theta_{Z_{22}} + \theta_{I^-}) \tag{7b}$$

In (3) and (4) it can be seen that the q-axes of the converter terminal voltages u_q are in a sinusoidal relation to the reference phase angles from the PLLs. For closer investigation (3) can be divided into three parts. The first parts, e.g. $K_1 \cdot U_G^+ \cdot \sin\big(\theta_{K_1} - \theta_{PLL^+}(t)\big)$, depend on the phase angle of the PLL and define the amplitudes regarding the grid voltage factors and the grid voltage itself. The second parts, e.g. $Z_{11} \cdot I^+ \cdot \sin\big(\theta_{Z_{22}} + \theta_{I^+}\big)$, are independent of the PLLs output. This leads to an offset of the function, which depends on the grid parameters and the injected currents of the converter. Finally, the third parts, e.g. $Z_{12} \cdot I^- \cdot \sin\big(\theta_{Z_{12}} - \theta_{PLL^+} + \theta_{PLL^-} + \theta_{I^-}\big)$, show the coupling of the positive and negative sequence grid factors, converter currents and PLL phase angles. This influences the phase shifts and also the amplitudes of the quadrature parts of the voltage u_q.

Fig. 3 shows the relation of the converter terminal voltage and phase angle of the positive sequence PLL for different grid conditions, depending on the aforementioned grid factors in (1). The phase angle of the PLL can only be stable at intersection points with the abscissa. These results for different grid conditions in either two equilibrium points (EP), one EP or no EP. If there are two EPs, a distinction must be made between stable equilibrium points (SEP) and unstable equilibrium points (USEP), as in [3].

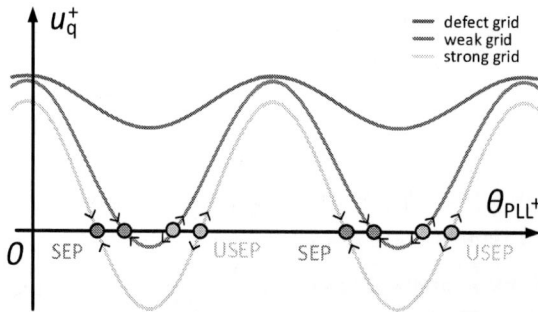

Fig. 3: Relation of converter terminal voltage and phase angle of PLL in positive sequence

Furthermore, Fig. 3 shows that a stronger grid (green curve) results in a small offset leading to a large negative u_q zone. On the other hand, a weak grid (purple curve) has a larger offset which leads to a smaller negative u_q zone. At certain grid conditions (red curve) it may happen that there is no intersection with the abscissa, whereby no SEP for the PLL exists. This leads to a criterion to determine if there is a SEP that can be reached by the PLL control. For example, the offset implies that the PLL control in a stronger grid has more chance to stabilize at the nearest SEP than for weaker grids. Due to the varying control structures and parameters of PLLs this leads to different dynamic behaviors even at same grid conditions. This can cause converter fail to operate, while others are able to withstand the same disturbance without any issues.

Steady-state and transient stability criterion during fault

This stability criterion shows under which conditions a PLL based converter can reach a stable equilibrium point (SEP) during an unsymmetrical fault. The applications are based on the values of table I with the topology shown in Fig. 1 during a line-to-line short circuit. The quadrature components of the terminal voltage in positive (left) and negative (right) sequence are shown in following figures (blue curves) according the equation (5). The figures also show the considering of the coupling of the PLLs (red curves) as mentioned in (4). The intersections of the curves with the abscissa are the possible equilibrium points, whereby the analytical solutions for the SEP by means of (6) are marked with a circle.

On the left side of Fig. 4, the sinusoidal relation of u_q^+ to the phase angle of the positive sequence PLL with an active power feed-in of 1 p.u. is shown. It can be seen, that during the fault because of the high active current injection and the weak grid there is a big offset. This reduces the negative u_q zone of the PLL and it is possible that PLLs with a high cut-off frequency can get unstable already. Because of the low current injection in the negative sequence, the coupling of the PLLs for u_q^+ is neglectable. For u_q^- there are safely reachable SEPs with and without the coupling of θ_{PLL^+}. Under these conditions it can be assumed that general well-tuned PLLs remain stable during the fault.

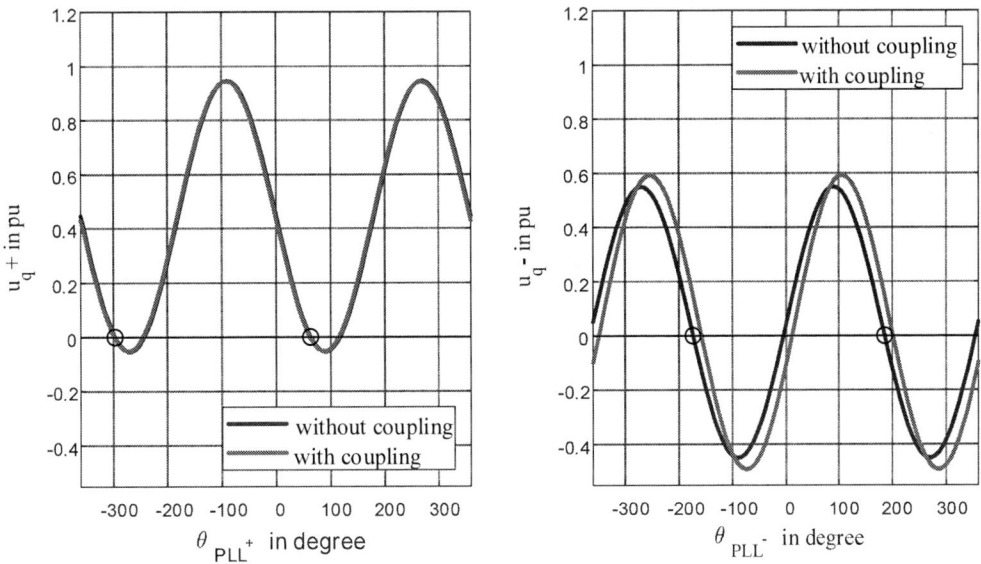

Fig. 4: Terminal voltage in relation to phase angle of PLLs with $I_d^+ = 1$ p.u. injection

With an injection of 1.2 p.u. as active power in the positive sequence, there is no longer an intersection of u_q^+ with the abscissa resulting in no SEP during the fault situation, seen in Fig. 5. The converter becomes unstable even if the negative sequence could reach a SEP.

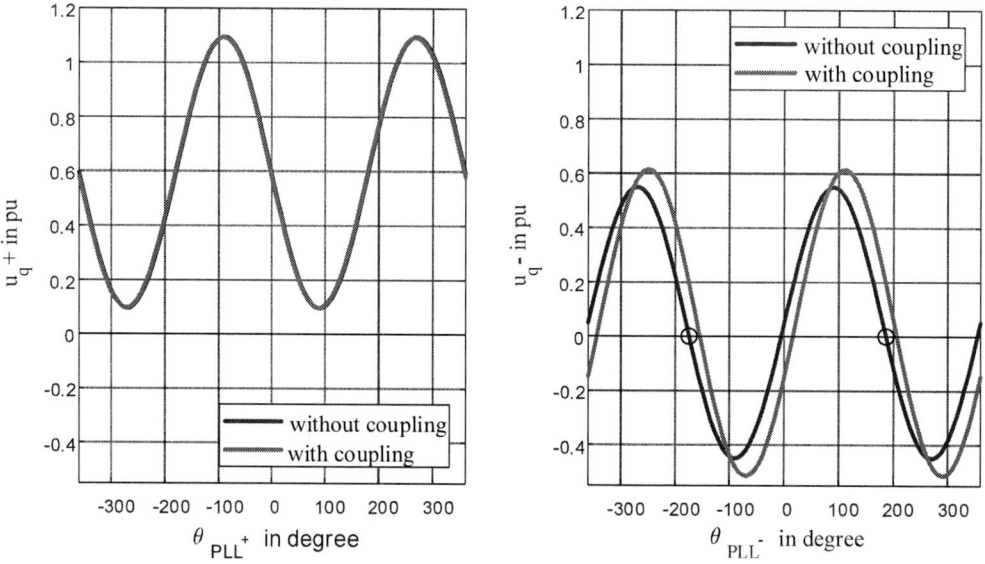

Fig. 5: Terminal voltage in relation to phase angle of PLLs with $I_d^+ = 1.2$ p.u. injection

Assuming the same active power feed-in of 1.2 p.u. in the positive sequence and an additional reactive power injection in the negative sequence, the coupling of the systems results in a SEP in the positive sequence, seen in Fig. 6. In case of a well-tuned PLL, the converter can continue to operate stable under this fault conditions. With less accurate PLLs, the positive sequence phase angle moves along this red line.

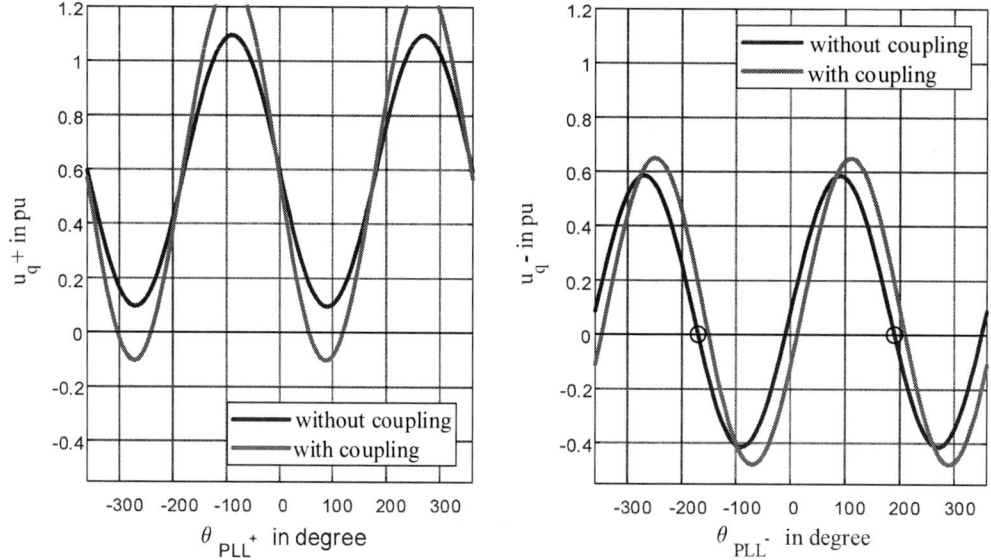

Fig. 6: Terminal voltage in relation to phase angle of PLLs with $I_d^+ = 1.2$ p.u. and $I_q^- = 1.2$ p.u. injection

The phase portrait method is an approach where the angular velocity (derivative of the phase angle over time) is plotted on the y-axis and the phase angle is plotted on the x-axis [9] [10], e.g. seen in Fig. 7. The black upper and lower curves limit the domain of attraction (DOA) in which any point, defined uniquely by its phase angle and angular velocity, moves towards a stable equilibrium point (SEP) in a reasonable time. Different DOAs are marked in different colors, each with a SEP inside.

If there is no intersection of u_q with the abscissa during the fault (no EP), then the phase angle of the PLL accelerates away from the pre-fault SEP along the red curve in Fig. 7. If the fault is cleared soon enough (e.g. 1), the SEP before the fault state can be reached quickly. If the fault is cleared later (e.g. 2), the next SEP is obtained, which takes longer. If the fault is cleared too late (in blue area, e.g. 3), no SEP can be reached and the converter becomes unstable. If there is a SEP during the fault, but it cannot be reached by the control the PLL accelerates and decelerates and circles around this SEP, as shown by the blue curve in Fig. 7. After the fault, the PLL converge very quickly to this SEP (e.g. 4 in Fig. 7) and the system remains stable.

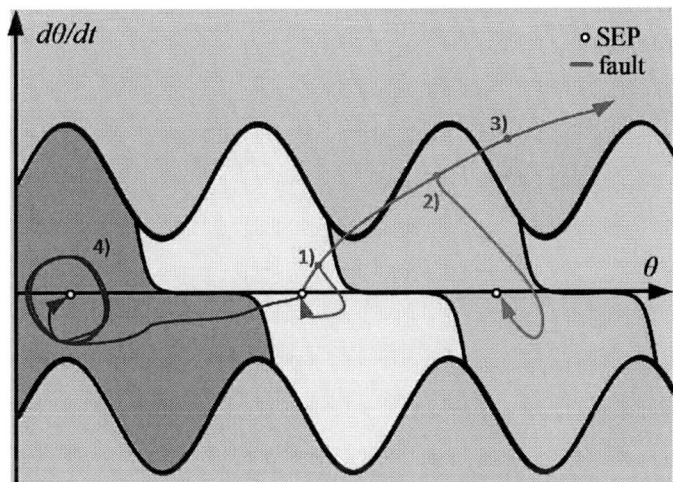

Fig. 7: Phase portrait method with stable regions (DOAs in different colors) and unstable region during fault and after fault clearing

Validation in Real-Time-System Experiments

To validate the stability criterion mentioned above, experiments are carried out. The grid topology (seen in Fig. 1) is emulated in a real-time simulation and the converter is based on the grid following control concept. Thereby a dual second order generalized integrator phase-locked loop (DSOGI-PLL) [11] is used to obtain the positive and negative sequence reference phase angles as base for the controlled current sources. The results show the line-to-line short-circuit between phase 2 and 3 with a fault duration of 0.4 s occurring at 0.2 s in the middle of the line. The transformer has a Dyg5 vector group and is assumed to be ideal. The low voltage side is solidly grounded and the high voltage side is operated slightly unbalanced (U_{L3} = 1.02 p.u.). The table I summarizes the most important parameters.

Table I: Parameters of real-time-system experiments

U_{HV} = 14 kV	U_{LV} = 690 V	$PLL^+_{cut\_off}$ = 20 Hz	fault time = 0.4 s
SCR = 1.5	P_{rated} = 10 MW	$PLL^-_{cut\_off}$ = 20 Hz	Z_{fault} = 0 Ω
XR_{ratio} = 7	f = 50 Hz	f_{RTS} = 10 kHz	Fault position = 50 %

For the first application, the PLL based converter continues to inject the nominal active power from the pre-fault state during the fault condition (seen in Fig. 8). In this case, the plot in the bottom left corner shows the transient response of the voltages u_q^+ and u_q^- in the fault case, although the PLL converges to a SEP. This corresponds to the steady-state results in Fig. 4. The phase portrait plot (right) shows how the trajectory from the pre-fault state travels towards a new SEP during the fault (red) and returns to the original SEP after the fault has been cleared. The negative sequence θ_{PLL^-} moves towards a new SEP during the fault and stays close at a new SEP after fault clearing.

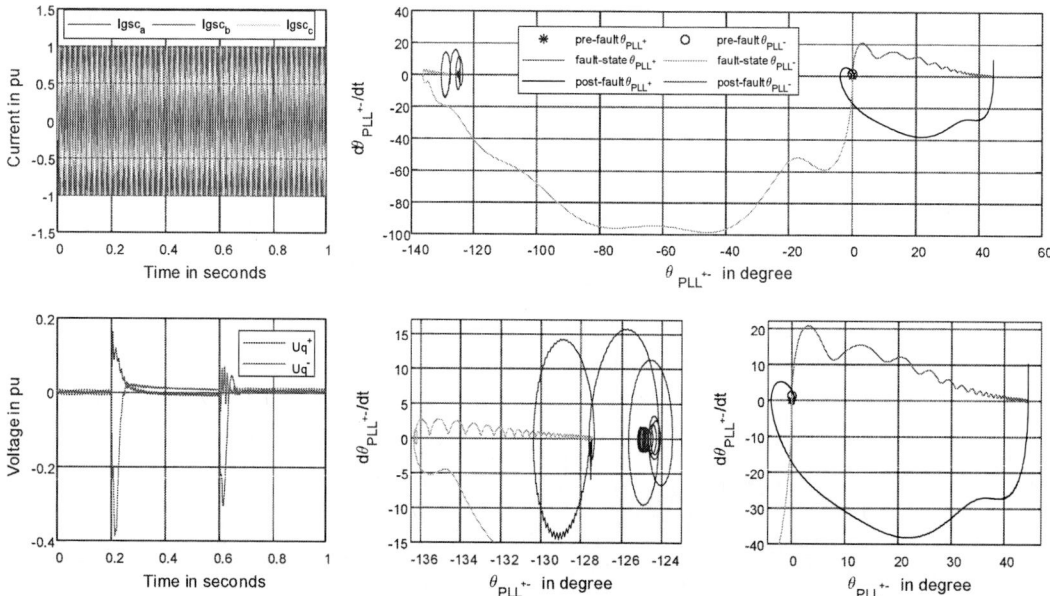

Fig. 8: Experimental validation of stability criterion with nominal active power injection during fault $(I_d^+ = 1 \text{ p.u.})$

Fig. 9 shows the current injection of 1 p.u. in pre- and post-fault state while during the fault the current feed-in is 1.2 p.u. as positive sequence active power. In this case, the increase in current can be seen in the upper left corner. Furthermore, in the fault case the voltage u_q^+ no longer intersects with the abscissa, whereby it is not possible to lead to a SEP under these conditions. This results to an acceleration of θ_{PLL^+} during the fault and consequently no SEP is reached after the fault clearing, in which case a SEP would be present again, which leads θ_{PLL^+} to become unstable. Due to the coupling of the PLLs, θ_{PLL^-} cannot reach its SEP during and after the fault, so it circles around a SEP.

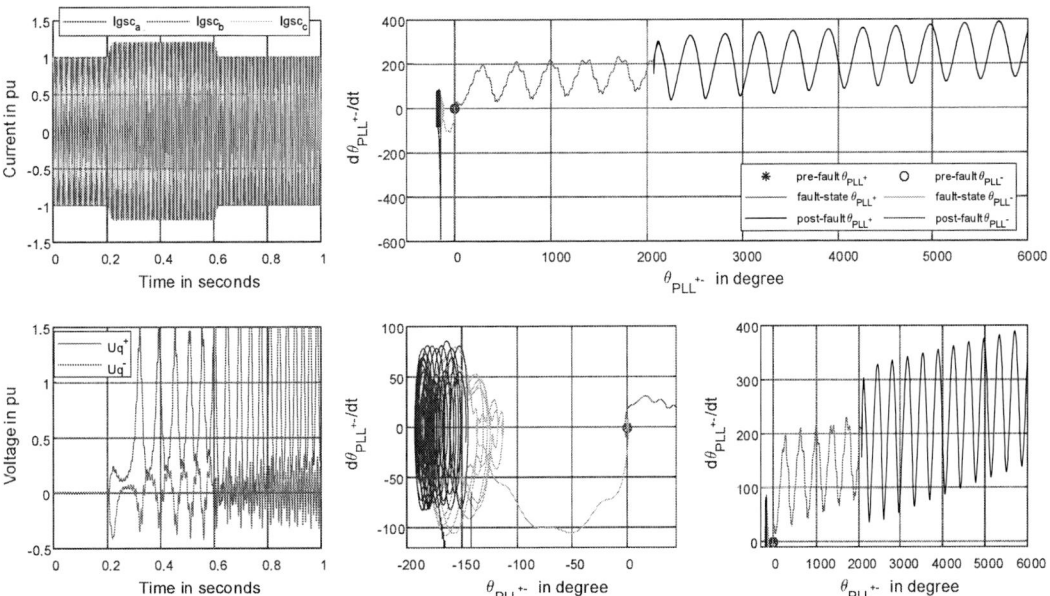

Fig. 9: Experimental validation of stability criterion with high active power injection during fault ($I_d^+ = 1.2$ p.u.)

The final application is that during the fault state, again as before, an increased positive sequence active power current, but also reactive power in the negative sequence is injected, seen in Fig. 10. This leads to the fact that, according to Fig. 6, the coupling of the PLLs results in a SEP through the reactive power injection. An optimal tuned PLL could thus also reach the SEP in the fault state. Unfortunately, the negative u_q zone for the selected parameters of the PLLs are to small to reach a SEP. However, the reactive power ensures that θ_{PLL^+} does not accelerate excessively and consequently circles around a SEP in positive and negative sequence. After the fault clearing, a SEP can thus be reached very quickly in both instances, whereby the θ_{PLL^+} also returns to the pre-fault state.

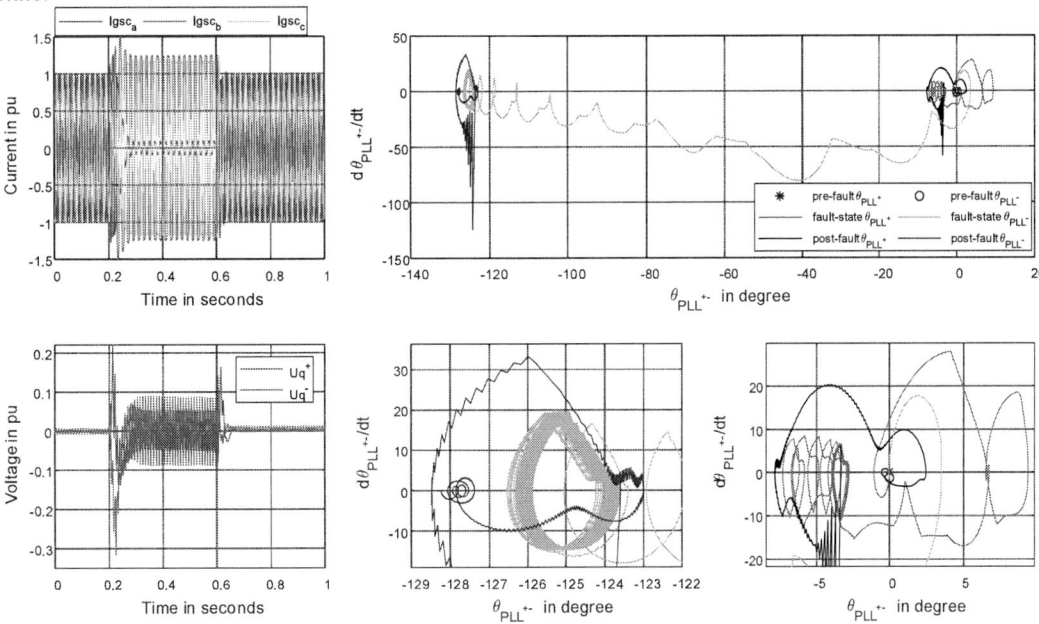

Fig. 10: Experimental validation of stability criterion with high active power and negative sequence reactive power injection during fault ($I_d^+ = 1.2$ p.u., $I_q^- = 1.2$ p.u.)

Conclusion

This paper investigates the fault behavior of PLL based grid-connected converters for unsymmetrical faults. The theoretical relation of stable equilibrium points (SEP) for the PLL is derived and leads to a stability criterion based on the grid conditions and converter feed-in power. It can be seen that with weaker grids it becomes more difficult for the PLL to stay stable during faults. Furthermore, the positive and negative sequence PLL are coupled through a grid factor and the converter feed-in current. On the one hand, the neglection of the coupling for low unsymmetrical grids results in a quite accurate analytical solution of the periodically SEP.

On the other hand, the coupling cannot be neglected during an unsymmetrical fault-state with a high injection of negative sequence current. For example, it is shown that a reactive current feed-in at the negative sequence can help to reach a SEP for the positive sequence PLL. In order to investigate the transient process of the pre-fault, fault and post-fault conditions accurately, the phase portrait method is been introduced. Subsequently, the stability criterium is evaluated for a line-to-line short-circuit through varying current injections. Finally, the stability criterion was validated with real-time-system experiments. It can be concluded that the approach is accurate to investigate the availability of SEPs for unsymmetrical grid faults. In future works, the behavior of different PLL algorithms and converter controllers will be investigated in more detail.

References

[1] N. Jaalam, N. A. Rahim, A. Bakar, C. Tan, and A. M. Haidar, "A comprehensive review of synchronization methods for grid-connected converters of renewable energy source," Renewable and Sustainable Energy Reviews, vol. 59, pp. 1471–1481, 2016, doi: 10.1016/j.rser.2016.01.066.

[2] H. Wu and X. Wang, "Design-Oriented Transient Stability Analysis of PLL-Synchronized Voltage-Source Converters," in IEEE Transactions on Power Electronics, vol. 35, no. 4, pp. 3573-3589, April 2020

[3] Z. Zhang, R. Schuerhuber, L. Fickert and K. Friedl, "Study of stability after low voltage ride-through caused by phase-locked loop of grid-side converter", in International Journal of Electrical Power & Energy Systems, Volume 129, 2021

[4] VDE/FNN, "Technical Requirements for the Connection and Operation of Customer Installations to the High Voltage Network (TAR High Voltage)", 2017

[5] X. He, C. He, S. Pan, H. Geng, and F. Liu, "Synchronization Instability of Inverter-Based Generation During Asymmetrical Grid Faults," IEEE Trans. Power Syst., vol. 37, no. 2, pp. 1018–1031, 2022, doi: 10.1109/TPWRS.2021.3098393.

[6] Z. Zhang, C. Lehmal, P. Hackl, and R. Schuerhuber, "Transient Stability Analysis and Post-Fault Restart Strategy for Current-Limited Grid-Forming Converter," Energies, vol. 15, no. 10, p. 3552, 2022, doi: 10.3390/en15103552.

[7] L. Harnefors, "Modeling of Three-Phase Dynamic Systems Using Complex Transfer Functions and Transfer Matrices," in IEEE Transactions on Industrial Electronics, vol. 54, no. 4, pp. 2239-2248, Aug. 2007

[8] B. Wang, S. Wang and J. Hu, "Dynamic Modeling of Asymmetrical-Faulted Grid by Decomposing Coupled Sequences via Complex Vector," in IEEE Journal of Emerging and Selected Topics in Power Electronics, vol. 9, no. 2, pp. 2452-2464, April 2021

[9] Z. Zhang, R. Schuerhuber, L. Fickert, F. Katrin, C. Guochu, and Z. Yongming, "Domain of Attraction's Estimation for Grid Connected Converters with Phase-Locked Loop," IEEE Trans. Power Syst., p. 1, 2021, doi: 10.1109/TPWRS.2021.3098960.

[10] H. Wu and X. Wang, "Transient Stability Impact of the Phase-Locked Loop on Grid-Connected Voltage Source Converters," in 2018 International Power Electronics Conference (IPEC-Niigata 2018 -ECCE Asia), Niigata, 52018, pp. 2673–2680.

[11] A. A. Nazib, D. G. Holmes, and B. P. McGrath, "Decoupled DSOGI-PLL for Improved Three Phase Grid Synchronisation," in 2018 International Power Electronics Conference (IPEC-Niigata 2018 -ECCE Asia), Niigata, 52018, pp. 3670–3677.

Stability Assessment and Optimization of MMC Energy Balancing for Drive Applications at Standstill Using an Averaging Approach

Qiuye Gui[1,2] and Hendrik Fehr[1] and Albrecht Gensior[1]

[1]Technische Universität Ilmenau, Germany; [2]Technische Universität Dresden, Germany

qiuye.gui@tu-dresden.de; hendrik.fehr@tu-ilmenau.de; albrecht.gensior@tu-ilmenau.de

Acknowledgments

This work was supported by *Deutsche Forschungsgemeinschaft*, DFG, grant GE 2502/5-1. The test bench for the experiments has been provided by Chair of Power Electronics, Technische Universität Dresden.

Keywords

≪Converter control≫, ≪Modular Multilevel Converters (MMC)≫, ≪Non-linear control≫, ≪Optimization≫

Abstract

A new controller design framework for periodic systems is presented and applied to the Modular Multi-level Converter energy balancing. It provides a rigorous stability analysis for the state-of-art averaging control and the approach using constants as reference for the instantaneous energies. An optimized controller with improved performance is derived and verified by simulation and experiment.

Introduction

This paper focuses on the energy balancing issue of Modular Multilevel Converters (MMCs) with half-bridge cells in dc operation, i.e. a Permanent Magnet Synchronous Machine (PMSM) is fed at standstill, see Fig. 1. The balancing task is challenging due to the strong coupling [1, 2] and the underactuation of the energy model [3, 4]. The state-of-art approach [5–7] regulates the average value of equivalent arm voltages or stored arm energies by identifying current and voltage harmonic components that influence the respective average power terms. However, it does not provide a state-space model for the average energy which is derived mathematically from the circuit model. What's more, although the average values of the energies are controlled, no stability w.r.t. the energy ripples is confirmed, as an instability is possible due to the strong coupling effect. On the other hand, dynamic phasor models [8] and harmonic state-space approaches [9] describe the dynamics of all energy harmonics, but increase the system order

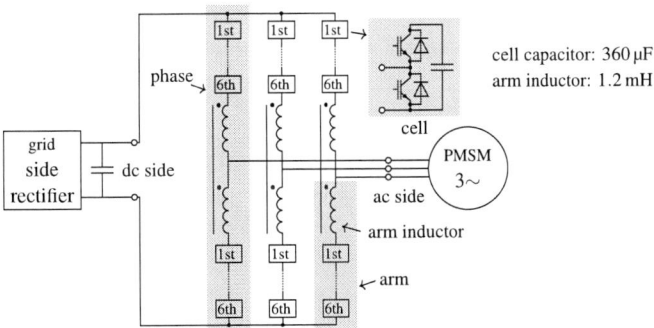

Fig. 1: Scheme of the application.

which makes them less attractive for model-based control design. To close the gap mentioned above, [3] provides an average energy model with mathematical background for a simple systematic model-based energy controller design. However, the model is based on a theorem from [10] which is limited to a finite time interval and thus cannot confirm stability of the proposed control system beyond that range. Moreover, the analysis in [3] focuses on the case which coincides with the average value control, and thus excludes approaches like [1] where a constant is used as reference for the instantaneous energies.

The present work provides a new controller design framework for periodic systems, and applies it to the energy controller design of an MMC. The framework extends the stability analysis in [3] to the whole time range $t \in [0, \infty)$. Compared with [9] which only provides a region of attraction in an uncertain small neighborhood of a stationary regime, it also provides a global stability assessment of the proposed control system. The energy reference can either be a stationary operating regime or a constant, which confirms the exponential convergence of average value controls like [5] and the stability of controls like [1] which use constants as reference for the instantaneous energies. Two controllers are derived: One with a simple choice of current harmonics and an optimized one leading to improved performance with two different common-mode voltage injections. Both results are verified by simulations and experiments.

A New Controller Design Framework Based on Averaging

In this section, the controller design problem for the state-space model[1]

$$\dot{x} = f_1 (t, x, u) \tag{1}$$

is considered, where the $n \times 1$ vector x and the $m \times 1$ vector u refer to the state and the input, respectively. Here, the task is to control the trajectory $t \mapsto x(t)$ tracking its reference $t \mapsto x_{\mathrm{ref}}(t)$ by means of u as the controller output with error dependency. With the error definition $x_{\mathrm{err}} = x - x_{\mathrm{ref}}$, (1) can be rewritten as

$$\dot{x}_{\mathrm{err}} = f_1 \left(t, x_{\mathrm{err}} + x_{\mathrm{ref}}(t), u(x_{\mathrm{err}})\right) - \dot{x}_{\mathrm{ref}}(t) =: f_2 \left(t, x_{\mathrm{err}}, u(x_{\mathrm{err}})\right). \tag{2}$$

If necessary, the integral of the error state can be easily introduced by means of regarding it as a part of the new error state. The next step should be the controller design, that is to specify an appropriate controller output $u(x_{\mathrm{err}})$ such that a desired dynamics or convergence of x_{err} can be achieved. However, if the function f_2 is nonlinear w.r.t. x_{err} or u, it may increase the controller design effort substantially. In the following, a new controller design framework is developed based on Theorem 10.4 in [11]. This framework deals with the case where the function f_2 in (2) is periodic in t, and allows a simpler controller design, that is, the time dependency of f_2 does not need to be considered during the controller design. First, the main results of Theorem 10.4 in [11] are recalled as follows:

Theorem 1 *Let $f\left(\tau, x_{\mathrm{err}}, u(x_{\mathrm{err}}), \varepsilon\right)$ be an n-dimensional function with the following properties:*

(P1) The function itself and its partial derivatives w.r.t. $(x_{\mathrm{err}}, \varepsilon)$ up to the second order are continuous and bounded for $(\tau, x_{\mathrm{err}}, \varepsilon) \in [0, \infty) \times \mathbb{D}_1 \times [0, \varepsilon_1]$, for some $\varepsilon_1 > 0$ and every compact set $\mathbb{D}_1 \subset \mathbb{D}_2$, where $\mathbb{D}_2 \subset \mathbb{R}^n$ is a domain containing the origin.

(P2) The function f is T-periodic in τ for some $T > 0$, i.e. $f\left(\tau + T, x_{\mathrm{err}}, u(x_{\mathrm{err}}), \varepsilon\right) = f\left(\tau, x_{\mathrm{err}}, u(x_{\mathrm{err}}), \varepsilon\right)$, and has the average $\overline{f}\left(x_{\mathrm{err}}, u(x_{\mathrm{err}})\right)$. It is obtained by averaging f w.r.t. τ at a vanishing ε and a τ-independent state x_{err}, i.e.

$$\overline{f}\left(x_{\mathrm{err}}, u(x_{\mathrm{err}})\right) = \frac{1}{T} \int_0^T f\left(\tau, x_{\mathrm{err}}, u(x_{\mathrm{err}}), 0\right) \mathrm{d}\tau. \tag{3}$$

(P3) The solution $\tau \mapsto x_{\mathrm{err}}(\tau)$ of the standard form of the original system

$$\frac{\mathrm{d}}{\mathrm{d}\tau} x_{\mathrm{err}} = \varepsilon f\left(\tau, x_{\mathrm{err}}, u(x_{\mathrm{err}}), \varepsilon\right) \tag{4}$$

---

[1] In this work, $\dot{x} := \mathrm{d}x/\mathrm{d}t$.

and the solution $\tau \mapsto \overline{\boldsymbol{x}}_{\mathrm{err}}(\tau)$ *of the standard form of the average system*

$$\frac{\mathrm{d}}{\mathrm{d}\tau}\overline{\boldsymbol{x}}_{\mathrm{err}} = \varepsilon \overline{\boldsymbol{f}}\left(\overline{\boldsymbol{x}}_{\mathrm{err}}, \boldsymbol{u}\left(\overline{\boldsymbol{x}}_{\mathrm{err}}\right)\right), \tag{5}$$

start at the same initial value, that is $\overline{\boldsymbol{x}}_{\mathrm{err}}(0) = \boldsymbol{x}_{\mathrm{err}}(0)$.

(P4) The origin $\overline{\boldsymbol{x}}_{\mathrm{err}} = \boldsymbol{0} \in \mathbb{D}_2$ *is an exponentially stable equilibrium point of (5) and* $\overline{\boldsymbol{x}}_{\mathrm{err}}(0) \in \mathbb{D}_3$ *where* $\mathbb{D}_3 \subset \mathbb{D}_2$ *is a compact subset of its region of attraction.*

Then there exists an $\varepsilon_2 > 0$ *such that for all* $0 < \varepsilon < \varepsilon_2$, $\|\boldsymbol{x}_{\mathrm{err}}(\tau) - \overline{\boldsymbol{x}}_{\mathrm{err}}(\tau)\| \leq c\varepsilon$ *(for some* $c > 0$*) is valid for all* $\tau \in [0,\infty)$, *and the origin of the original system (4) is exponentially stable if the following additional condition is satisfied:*

(P5) The function $\boldsymbol{f}(\tau, \boldsymbol{x}_{\mathrm{err}}, \boldsymbol{u}(\boldsymbol{x}_{\mathrm{err}}), \varepsilon)$ *satisfies* $\boldsymbol{f}(\tau, \boldsymbol{0}, \boldsymbol{u}(\boldsymbol{0}), \varepsilon) = \boldsymbol{0}$, *that is, the origin is an equilibrium point of the original system (4).*

Here, the state of the average system is represented by $\overline{\boldsymbol{x}}_{\mathrm{err}}$ in order to distinguish from $\boldsymbol{x}_{\mathrm{err}}$ for the original system. Under the required conditions, this theorem confirms a stability of the original system (4), that is, it's solution always lies in a neighborhood of the solution of the average system (5) which converges exponentially to $\boldsymbol{0}$. Therefore, an alternative method for the state-space model based control system design can be provided. Instead of considering a complicated original system, the controller design can be accomplished at a simplified average system, derived by (3). This method can be summarized as the flow chart shown in Fig. 2.

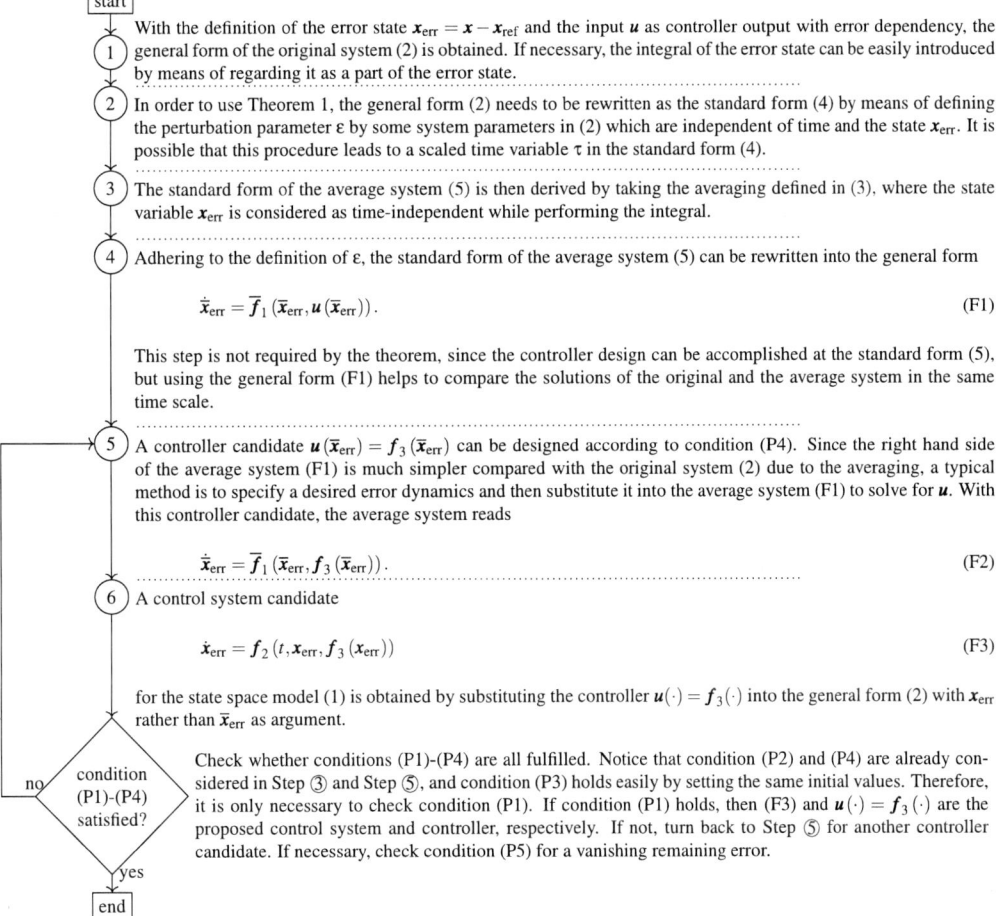

Fig. 2: Flow chart for the controller design framework based on Theorem 1.

Since the parameter ε is defined by some system parameters independent of time and state variable, it is expected that the solutions of a standard form and the corresponding general form are equivalent to each other in their individual time scales. Since the origin of the average system is exponentially stable, according to Theorem 1, with the designed controller and a sufficiently small ε, the solution $x_{\mathrm{err}}(t)$ of the proposed control system (F3) converges to a neighborhood of 0 as $t \to \infty$, and the remaining error vanishes if the origin is an equilibrium point of the proposed control system (F3). This stability is for every initial value $x_{\mathrm{err}}(0)$ lying in the region of attraction of the origin of the average system (F2).

Besides the conclusions, it is necessary to discuss the conditions required by Theorem 1. As for condition (P2), the function $f(\tau, x_{\mathrm{err}}, u(x_{\mathrm{err}}), \varepsilon)$ in (4) must be periodic in τ. Since the perturbation parameter ε is defined independent of t or x_{err}, a typical example is that the function $f_1(t, x_{\mathrm{err}} + x_{\mathrm{ref}}(t), u(x_{\mathrm{err}}))$ and the reference $x_{\mathrm{ref}}(t)$ in the system (2) are both periodic in t. The condition (P3) holds easily by setting the same initial values, while an appropriate controller design can ensure conditions (P1), (P4). This is expected to be accomplished without much effort since the controller design can be simplified after averaging. However, a challenge can be the choice of a sufficiently small ε. According to the analysis in [11], since ε_2 is generally difficult to calculate, a practical choice would be the definition of ε dependent on some parameters which can be changed by practical conditions or settings. In this way, it is possible that the parameter ε is able to approach a suitable value such that the proposed control system is stable.

Energy Model of Modular Multilevel Converters

In this text, a simplified energy model for MMCs as in [3] is considered:

$$\dot{e}_{s0} = v_{\mathrm{DC}} i_{s0} - \mathrm{Re}\left(\underline{v}_y^* \underline{i}\right) \qquad e_{s0} : \text{scaled total stored energy} \tag{6a}$$

$$\dot{e}_{d0} = -2 v_{y0} i_{s0} - \mathrm{Re}\left(\underline{i}_s^* \underline{v}_y\right) \qquad e_{d0} : \text{scaled total vertical energy difference} \tag{6b}$$

$$\dot{\underline{e}}_s = v_{\mathrm{DC}} \underline{i}_s - \underline{v}_y^* \underline{i}^* - 2\underline{i} v_{y0} \qquad \underline{e}_s : \text{complex energy sum} \tag{6c}$$

$$\dot{\underline{e}}_d = v_{\mathrm{DC}} \underline{i} - \underline{i}_s^* \underline{v}_y^* - 2\underline{i}_s v_{y0} - 2 i_{s0} \underline{v}_y \qquad \underline{e}_d : \text{complex energy difference.} \tag{6d}$$

The transformed energies can be summarized as $\boldsymbol{e} = \left(\begin{array}{cccccc} e_{s0} & e_{d0} & \mathrm{Re}(\underline{e}_s) & \mathrm{Im}(\underline{e}_s) & \mathrm{Re}(\underline{e}_d) & \mathrm{Im}(\underline{e}_d) \end{array} \right)^{\mathrm{T}}$. The currents i_{s0}, $\underline{i}_s$, and $\underline{i}$ denote the scaled dc current, the circulating current, and the complex ac-side current, respectively. The variables v_{DC}, v_{y0}, and $\underline{v}_y$ denote the dc-link voltage, the common-mode voltage, and the complex ac side voltage, respectively. Details of the variable definition and transformation can be found in [3]. Here, the current dynamics are neglected since they are much faster than the energy dynamics. Hence, in the dc operation, the ac-side current $\underline{i}$ and voltage $\underline{v}_y$ are considered to be constant. In order to realize the balancing task, harmonics w.r.t. a predetermined frequency ω_{cm} are injected in i_{s0}, $\underline{i}_s$, and v_{y0}. Therefore, the general form of injection reads

$$i_{s0} = I_{s0}^{[0]}(\boldsymbol{e}_{\mathrm{err}}, \boldsymbol{e}_{\mathrm{err},\mathrm{I}}) + \sum_{n \in \mathbb{Z}^+} \left(\underline{I}_{s0}^{[n]}(\boldsymbol{e}_{\mathrm{err}}, \boldsymbol{e}_{\mathrm{err},\mathrm{I}}) \, \mathrm{e}^{jn\omega_{\mathrm{cm}}t} + \underline{I}_{s0}^{[n]*}(\boldsymbol{e}_{\mathrm{err}}, \boldsymbol{e}_{\mathrm{err},\mathrm{I}}) \, \mathrm{e}^{-jn\omega_{\mathrm{cm}}t} \right) \tag{7a}$$

$$\underline{i}_s = \sum_{n \in \mathbb{Z}} \underline{I}_s^{[n]}(\boldsymbol{e}_{\mathrm{err}}, \boldsymbol{e}_{\mathrm{err},\mathrm{I}}) \, \mathrm{e}^{jn\omega_{\mathrm{cm}}t} \qquad \underline{i} = \underline{I}^{[0]} \qquad \underline{v}_y = \underline{V}_y^{[0]} \qquad v_{\mathrm{DC}} = V_{\mathrm{DC}}^{[0]} \tag{7b}$$

$$v_{y0} = \sum_{n \in \mathbb{N}_{\mathrm{cm}}} \left(\underline{V}_{y0}^{[n]} \mathrm{e}^{jn\omega_{\mathrm{cm}}t} + \underline{V}_{y0}^{[n]*} \mathrm{e}^{-jn\omega_{\mathrm{cm}}t} \right) \qquad \mathbb{N}_{\mathrm{cm}} : \text{made up of all harmonic orders of } v_{y0} \tag{7c}$$

where the coefficients $\underline{I}^{[0]}$, $\underline{V}_y^{[0]}$, $\underline{V}_{y0}^{[n]}$, and $V_{\mathrm{DC}}^{[0]}$ are constant, while the coefficients of i_{s0} and $\underline{i}_s$ might depend on the energy error $\boldsymbol{e}_{\mathrm{err}} = \boldsymbol{e} - \boldsymbol{e}_{\mathrm{ref}}$ and its integral $\boldsymbol{e}_{\mathrm{err},\mathrm{I}}$, obtained from $\dot{\boldsymbol{e}}_{\mathrm{err},\mathrm{I}} = \boldsymbol{e}_{\mathrm{err}}$, to facilitate the balancing task. This consideration covers the case where the coefficients are not constant due to the controller behavior. Two different common-mode voltage injections shown in Fig. 3 are considered. The coefficients for the red wave are given by (7c) with

$$n \in \mathbb{N}_{\mathrm{cm}} = \{1, 3\} \qquad \underline{V}_{y0}^{[1]} = 0.15 \cdot V_{\mathrm{DC}}^{[0]} \qquad \underline{V}_{y0}^{[3]} = -\left| \underline{V}_{y0}^{[1]} \right| / 6 \cdot \mathrm{e}^{j3 \arg\left(\underline{V}_{y0}^{[1]}\right)} \tag{8}$$

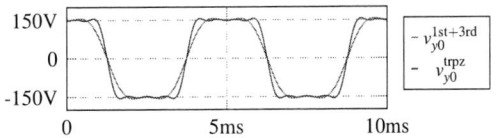

Fig. 3: Example of common-mode voltage: 1st+3rd harmonics $v_{y0}^{1st+3rd}$ (red, $\mathbb{N}_{cm} = \{1,3\}$, similar as [12] IV.B), approximated trapezoidal wave v_{y0}^{trpz} (blue, $\mathbb{N}_{cm} = \{1,3,5,7\}$).

where the third harmonic allows for an increased first harmonic $\underline{V}_{y0}^{[1]}$. The other wave in blue approximates a trapezoidal shape up to the 7th harmonic, given by (7c) with

$$n \in \mathbb{N}_{cm} = \{1,3,5,7\} \qquad \underline{V}_{y0}^{[n]} = \frac{V_{DC}^{[0]}}{4} \frac{\sin\left(n\frac{\pi}{2}\right)}{n\frac{\pi}{2}} \frac{\sin\left(n\frac{\pi}{10}\right)}{n\frac{\pi}{10}}. \tag{9}$$

In this text, the energy reference $\boldsymbol{e}_{ref}$, defined by $\dot{\boldsymbol{e}}_{ref} := \boldsymbol{p}_{ref}(\omega_{cm}t)$, is either considered as a constant $\boldsymbol{E}$ with the derivative $\boldsymbol{p}_{ref}(\omega_{cm}t) = \boldsymbol{0}$, or a stationary operating regime $\boldsymbol{e}_{rgm}$, where $\boldsymbol{p}_{ref}(\omega_{cm}t)$ is $2\pi/\omega_{cm}$-periodic without drift. The calculation of the regimes will be discussed after the controller design.

Controller Design Based on the Proposed Framework

Step ①: Substituting the injection (7) into the energy model (6) leads to the system

$$\dot{\boldsymbol{e}}_{err} = \boldsymbol{p}_{dc}(\boldsymbol{e}_{err}, \boldsymbol{e}_{err,I}) + \boldsymbol{p}_{ac}(\omega_{cm}t, \boldsymbol{e}_{err}, \boldsymbol{e}_{err,I}) - \boldsymbol{p}_{ref}(\omega_{cm}t) \tag{10a}$$

$$\dot{\boldsymbol{e}}_{err,I} = \boldsymbol{e}_{err} \tag{10b}$$

with $\boldsymbol{p}_{dc}(\cdot,\cdot) = \left(\begin{array}{cccccc} p_{dc,es0} & p_{dc,ed0} & \mathrm{Re}\left(\underline{p}_{dc,es}\right) & \mathrm{Im}\left(\underline{p}_{dc,es}\right) & \mathrm{Re}\left(\underline{p}_{dc,ed}\right) & \mathrm{Im}\left(\underline{p}_{dc,ed}\right) \end{array} \right)^{\mathrm{T}}$ and

$$p_{dc,es0} = V_{DC}^{[0]}I_{s0}^{[0]}(\boldsymbol{e}_{err}, \boldsymbol{e}_{err,I}) - \mathrm{Re}\left(\underline{V}_{y}^{[0]*}\underline{I}^{[0]}\right) \qquad \underline{p}_{dc,es} = V_{DC}^{[0]}\underline{I}_{s}^{[0]}(\boldsymbol{e}_{err}, \boldsymbol{e}_{err,I}) - \underline{V}_{y}^{[0]*}\underline{I}^{[0]*} \tag{11a}$$

$$p_{dc,ed0} = -\mathrm{Re}\left[4 \sum_{n \in \mathbb{N}_{cm}} \underline{V}_{y0}^{[n]*}\underline{I}_{s0}^{[n]}(\boldsymbol{e}_{err}, \boldsymbol{e}_{err,I}) + \underline{V}_{y}^{[0]*}\underline{I}_{s}^{[0]}(\boldsymbol{e}_{err}, \boldsymbol{e}_{err,I})\right] \tag{11b}$$

$$\underline{p}_{dc,ed} = V_{DC}^{[0]}\underline{I}^{[0]} - \underline{V}_{y}^{[0]*}\underline{I}_{s}^{[0]*}(\boldsymbol{e}_{err}, \boldsymbol{e}_{err,I}) - 2\underline{V}_{y}^{[0]}I_{s0}^{[0]}(\boldsymbol{e}_{err}, \boldsymbol{e}_{err,I})$$
$$- 2\sum_{n \in \mathbb{N}_{cm}}\left(\underline{V}_{y0}^{[n]}\underline{I}_{s}^{[-n]}(\boldsymbol{e}_{err}, \boldsymbol{e}_{err,I}) + \underline{V}_{y0}^{[n]*}\underline{I}_{s}^{[n]}(\boldsymbol{e}_{err}, \boldsymbol{e}_{err,I})\right). \tag{11c}$$

This corresponds to the general form (2), where $\boldsymbol{p}_{ac}$ and $\boldsymbol{p}_{dc}$ denote the power terms with and without explicit time dependency, respectively, while $\left(\begin{array}{cc} \boldsymbol{e}_{err}^{\mathrm{T}} & \boldsymbol{e}_{err,I}^{\mathrm{T}} \end{array} \right)^{\mathrm{T}}$ represents the state variable. Here, function $\boldsymbol{p}_{ac}$ is $2\pi/\omega_{cm}$-periodic in t, that is $\boldsymbol{p}_{ac}(\omega_{cm}t, \boldsymbol{e}_{err}, \boldsymbol{e}_{err,I}) = \boldsymbol{p}_{ac}(\omega_{cm}(t + 2\pi/\omega_{cm}), \boldsymbol{e}_{err}, \boldsymbol{e}_{err,I})$.

Step ②: The perturbation parameter can be defined as $\varepsilon = 1/\omega_{cm}$, which means that ε can be reduced by increasing the predetermined frequency ω_{cm}. By means of this definition, (10) can be rewritten as

$$\frac{\mathrm{d}}{\mathrm{d}\tau}\boldsymbol{e}_{err} = \varepsilon\left[\boldsymbol{p}_{dc}(\boldsymbol{e}_{err}, \boldsymbol{e}_{err,I}) + \boldsymbol{p}_{ac}(\tau, \boldsymbol{e}_{err}, \boldsymbol{e}_{err,I}) - \boldsymbol{p}_{ref}(\tau)\right] \qquad \frac{\mathrm{d}}{\mathrm{d}\tau}\boldsymbol{e}_{err,I} = \varepsilon\boldsymbol{e}_{err} \tag{12}$$

with the new time scale $\tau = \omega_{cm}t$. The term $\boldsymbol{p}_{ref}(\tau)$ is assumed to be explicitly independent of ε after the change of time scale, as will be confirmed in Step ⑥.

Step ③: As discussed at the end of the previous section, the term $\boldsymbol{p}_{ref}(\tau)$ is 2π-periodic in τ without drift. According to (3), the averaging of the right hand side of (12) reads

$$\frac{1}{2\pi}\int_0^{2\pi}\left[\boldsymbol{p}_{dc}(\boldsymbol{e}_{err}, \boldsymbol{e}_{err,I}) + \boldsymbol{p}_{ac}(\tau, \boldsymbol{e}_{err}, \boldsymbol{e}_{err,I}) - \boldsymbol{p}_{ref}(\tau)\right]\mathrm{d}\tau = \boldsymbol{p}_{dc}(\boldsymbol{e}_{err}, \boldsymbol{e}_{err,I}) \tag{13a}$$

$$\frac{1}{2\pi}\int_0^{2\pi} \boldsymbol{e}_{\mathrm{err}}\mathrm{d}\tau = \overline{\boldsymbol{e}}_{\mathrm{err}}.\tag{13b}$$

Notice that the states $\boldsymbol{e}_{\mathrm{err}}$ and $\boldsymbol{e}_{\mathrm{err,I}}$ are considered as time-independent while performing the integral. As a result, the average system

$$\frac{\mathrm{d}}{\mathrm{d}\tau}\overline{\boldsymbol{e}}_{\mathrm{err}} = \varepsilon \boldsymbol{p}_{\mathrm{dc}}\left(\overline{\boldsymbol{e}}_{\mathrm{err}}, \overline{\boldsymbol{e}}_{\mathrm{err,I}}\right) \qquad \frac{\mathrm{d}}{\mathrm{d}\tau}\overline{\boldsymbol{e}}_{\mathrm{err,I}} = \varepsilon\overline{\boldsymbol{e}}_{\mathrm{err}}\tag{14}$$

is obtained which corresponds to the standard form (5) in the framework. Here, the state is represented by $\left(\begin{array}{cc}\overline{\boldsymbol{e}}_{\mathrm{err}}^{\mathrm{T}} & \overline{\boldsymbol{e}}_{\mathrm{err,I}}^{\mathrm{T}}\end{array}\right)^{\mathrm{T}}$ in order to distinguish from $\left(\begin{array}{cc}\boldsymbol{e}_{\mathrm{err}}^{\mathrm{T}} & \boldsymbol{e}_{\mathrm{err,I}}^{\mathrm{T}}\end{array}\right)^{\mathrm{T}}$ for the original system.

Step ④: With the same definition $\varepsilon = 1/\omega_{\mathrm{cm}}$, system (14) can be rewritten as

$$\dot{\overline{\boldsymbol{e}}}_{\mathrm{err}} = \boldsymbol{p}_{\mathrm{dc}}\left(\overline{\boldsymbol{e}}_{\mathrm{err}}, \overline{\boldsymbol{e}}_{\mathrm{err,I}}\right) \qquad \dot{\overline{\boldsymbol{e}}}_{\mathrm{err,I}} = \overline{\boldsymbol{e}}_{\mathrm{err}}\tag{15}$$

which corresponds to the general form (F1) in the framework.

Step ⑤: For the system (15), the linear error dynamics

$$\dot{\overline{\boldsymbol{e}}}_{\mathrm{err}} = -k_{\mathrm{P,e}}\overline{\boldsymbol{e}}_{\mathrm{err}} - k_{\mathrm{I,e}}\overline{\boldsymbol{e}}_{\mathrm{err,I}} \qquad \dot{\overline{\boldsymbol{e}}}_{\mathrm{err,I}} = \overline{\boldsymbol{e}}_{\mathrm{err}}\tag{16}$$

is specified. This guarantees that the origin is globally exponentially stable if $k_{\mathrm{P,e}} > 0$, $k_{\mathrm{I,e}} > 0$. In this text, we choose $k_{\mathrm{I,e}} = k_{\mathrm{P,e}}^2/2$ in order to keep a good compromise between the rapidity and the overshoot of the solution of (16) [13]. Substituting the error dynamics (16) into the average system (15) leads to

$$-k_{\mathrm{P,e}}\overline{e}_{s0,\mathrm{err}} - k_{\mathrm{I,e}}\overline{e}_{s0,\mathrm{err,I}} = V_{\mathrm{DC}}^{[0]}I_{s0}^{[0]}\left(\overline{\boldsymbol{e}}_{\mathrm{err}}, \overline{\boldsymbol{e}}_{\mathrm{err,I}}\right) - \mathrm{Re}\left(\underline{V}_y^{[0]*}\underline{I}^{[0]}\right)\tag{17a}$$

$$-k_{\mathrm{P,e}}\overline{\underline{e}}_{s,\mathrm{err}} - k_{\mathrm{I,e}}\overline{\underline{e}}_{s,\mathrm{err,I}} = V_{\mathrm{DC}}^{[0]}\underline{I}_s^{[0]}\left(\overline{\boldsymbol{e}}_{\mathrm{err}}, \overline{\boldsymbol{e}}_{\mathrm{err,I}}\right) - \underline{V}_y^{[0]*}\underline{I}^{[0]*}\tag{17b}$$

$$-k_{\mathrm{P,e}}\overline{e}_{d0,\mathrm{err}} - k_{\mathrm{I,e}}\overline{e}_{d0,\mathrm{err,I}} = -\mathrm{Re}\left[4\sum_{n\in\mathbb{N}_{\mathrm{cm}}}\underline{V}_{y0}^{[n]*}\underline{I}_{s0}^{[n]}\left(\overline{\boldsymbol{e}}_{\mathrm{err}}, \overline{\boldsymbol{e}}_{\mathrm{err,I}}\right) + \underline{V}_y^{[0]*}\underline{I}_s^{[0]}\left(\overline{\boldsymbol{e}}_{\mathrm{err}}, \overline{\boldsymbol{e}}_{\mathrm{err,I}}\right)\right]\tag{17c}$$

$$-k_{\mathrm{P,e}}\overline{\underline{e}}_{d,\mathrm{err}} - k_{\mathrm{I,e}}\overline{\underline{e}}_{d,\mathrm{err,I}} = V_{\mathrm{DC}}^{[0]}\underline{I}^{[0]} - \underline{V}_y^{[0]*}\underline{I}_s^{[0]*}\left(\overline{\boldsymbol{e}}_{\mathrm{err}}, \overline{\boldsymbol{e}}_{\mathrm{err,I}}\right) - 2\underline{V}_y^{[0]}I_{s0}^{[0]}\left(\overline{\boldsymbol{e}}_{\mathrm{err}}, \overline{\boldsymbol{e}}_{\mathrm{err,I}}\right)$$
$$- 2\sum_{n\in\mathbb{N}_{\mathrm{cm}}}\left(\underline{V}_{y0}^{[n]}\underline{I}_s^{[-n]}\left(\overline{\boldsymbol{e}}_{\mathrm{err}}, \overline{\boldsymbol{e}}_{\mathrm{err,I}}\right) + \underline{V}_{y0}^{[n]*}\underline{I}_s^{[n]}\left(\overline{\boldsymbol{e}}_{\mathrm{err}}, \overline{\boldsymbol{e}}_{\mathrm{err,I}}\right)\right).\tag{17d}$$

Similar as the idea in [14, 15], the constant coefficients of the common-mode voltage avoid nonlinearities w.r.t. the candidates of the controller output in (17) such that the design effort can be reduced greatly. In the subsystem (17a) and (17b), only the coefficients $I_{s0}^{[0]}$ and $\underline{I}_s^{[0]}$ are available to accomplish the dynamics of $\overline{e}_{s0,\mathrm{err}}$ and $\overline{\underline{e}}_{s,\mathrm{err}}$, respectively. The corresponding solutions are

$$I_{s0}^{[0]}\left(\cdot,\cdot\right) = \left(-k_{\mathrm{P,e}}\overline{e}_{s0,\mathrm{err}} - k_{\mathrm{I,e}}\overline{e}_{s0,\mathrm{err,I}} + \mathrm{Re}\left(\underline{V}_y^{[0]*}\underline{I}^{[0]}\right)\right)/V_{\mathrm{DC}}^{[0]}\tag{18a}$$

$$\underline{I}_s^{[0]}\left(\cdot,\cdot\right) = \left(-k_{\mathrm{P,e}}\overline{\underline{e}}_{s,\mathrm{err}} - k_{\mathrm{I,e}}\overline{\underline{e}}_{s,\mathrm{err,I}} + \underline{V}_y^{[0]*}\underline{I}^{[0]*}\right)/V_{\mathrm{DC}}^{[0]}.\tag{18b}$$

However, from the subsystems (17c) and (17d), it is obvious that more than one current coefficient can be used for the dynamics of $\overline{e}_{d0,\mathrm{err}}$ and $\overline{\underline{e}}_{d,\mathrm{err}}$, respectively. In other words, the subsystem (17c) and (17d) is underdetermined and thus has infinite solutions. In the following, two solutions are derived, leading to two energy controllers:

• **simple choice**: Only one current coefficient is used for the dynamics of $\overline{e}_{d0,\mathrm{err}}$ or $\overline{\underline{e}}_{d,\mathrm{err}}$, respectively:

$$\underline{I}_{s0}^{[1]}\left(\cdot,\cdot\right) = \left(k_{\mathrm{P,e}}\overline{e}_{d0,\mathrm{err}} + k_{\mathrm{I,e}}\overline{e}_{d0,\mathrm{err,I}} - \mathrm{Re}\left(\underline{V}_y^{[0]*}\underline{I}_s^{[0]}\left(\cdot,\cdot\right)\right)\right)/\left(4\underline{V}_{y0}^{[1]*}\right)\tag{19a}$$

$$\underline{I}_s^{[1]}\left(\cdot,\cdot\right) = \left(k_{\mathrm{P,e}}\overline{\underline{e}}_{d,\mathrm{err}} + k_{\mathrm{I,e}}\overline{\underline{e}}_{d,\mathrm{err,I}} + V_{\mathrm{DC}}^{[0]}\underline{I}^{[0]} - \underline{V}_y^{[0]*}\underline{I}_s^{[0]*}\left(\cdot,\cdot\right) - 2\underline{V}_y^{[0]}I_{s0}^{[0]}\left(\cdot,\cdot\right)\right)/\left(2\underline{V}_{y0}^{[1]*}\right)\tag{19b}$$

Therefore, the whole controller is given by (18) and (19). It is very similar to the choice for lf-mode

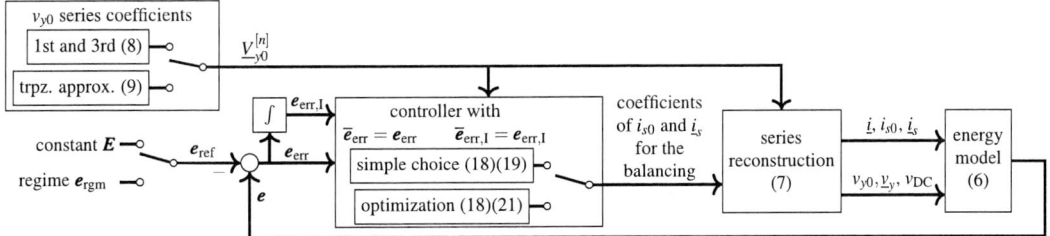

Fig. 4: Block diagram of the energy control system for the dc operation.

in [5] except for the usage of harmonics of order -1 in the complex circulating current. For this variant, the common-mode voltage is given by $v_{y0}^{\text{1st}+\text{3rd}}$ as the red wave in Fig. 3.

• **optimization** for arm current Root Mean Square (RMS): This variant is derived by a quadratic programming problem with linear constraints. The objective function is given by

$$f_{\text{obj}} = 4 \sum_{n \in \mathbb{N}_{\text{cm}}} \left| \underline{I}_{s0}^{[n]} (\cdot, \cdot) \right|^2 + \sum_{n \in \mathbb{N}_{\text{cm}}} \left(\left| \underline{I}_{s}^{[n]} (\cdot, \cdot) \right|^2 + \left| \underline{I}_{s}^{[-n]} (\cdot, \cdot) \right|^2 \right) \tag{20}$$

and it is subjected to (17c) and (17d). According to the variable definition given in [3], since the ac-side current $\underline{i} = \underline{I}^{[0]}$ is fixed by the load, and $I_{s0}^{[0]} (\cdot, \cdot)$ and $\underline{I}_{s}^{[0]} (\cdot, \cdot)$ are fixed by (17a) and (17b), respectively, the minimum of (20) leads to the minimal arm current RMS if the MMC enters a stationary regime where all current coefficients are constant. This optimization problem is convex, and the result leads to the optimized controller given by (18) and

$$\underline{I}_{s0}^{[n]} (\cdot, \cdot) = \underline{V}_{y0}^{[n]} \left(k_{\text{P,ed0}} \bar{e}_{d0,\text{err}} + k_{\text{I,ed0}} \bar{e}_{d0,\text{err,I}} - \text{Re} \left(\underline{V}_y^{[0]} \underline{I}_s^{[0]*} (\cdot, \cdot) \right) \right) / (4S) \tag{21a}$$

$$\underline{I}_{s}^{[-n]} (\cdot, \cdot) = \underline{V}_{y0}^{[n]*} \left(k_{\text{P,ed}} \bar{e}_{d,\text{err}} + k_{\text{I,ed}} \bar{e}_{d,\text{err,I}} + V_{\text{DC}} \underline{I}^{[0]} - \underline{V}_y^{[0]*} \underline{I}_s^{[0]*} (\cdot, \cdot) - 2\underline{V}_y^{[0]} I_{s0}^{[0]} (\cdot, \cdot) \right) / (4S) \tag{21b}$$

$$\underline{I}_{s}^{[n]} (\cdot, \cdot) = \underline{V}_{y0}^{[n]} \left(k_{\text{P,ed}} \bar{e}_{d,\text{err}} + k_{\text{I,ed}} \bar{e}_{d,\text{err,I}} + V_{\text{DC}} \underline{I}^{[0]} - \underline{V}_y^{[0]*} \underline{I}_s^{[0]*} (\cdot, \cdot) - 2\underline{V}_y^{[0]} I_{s0}^{[0]} (\cdot, \cdot) \right) / (4S) \tag{21c}$$

with $n \in \mathbb{N}_{\text{cm}}$ and $S = \sum_{m \in \mathbb{N}_{\text{cm}}} |\underline{V}_{y0}^{[m]}|^2$. This controller enables all feasible current harmonics for the subsystem (17c) and (17d). For this controller, the common-mode voltage can be given by either $v_{y0}^{\text{1st}+\text{3rd}}$ or v_{y0}^{trpz} in Fig. 3.

Step ⑥: Using the designed controller with the substitution $\bar{e}_{\text{err}} = e_{\text{err}}$ and $\bar{e}_{\text{err,I}} = e_{\text{err,I}}$, the energy control system in Fig. 4 is derived. Regarding the stationary operating regime e_{rgm}, it can be designed such that condition (P5) is fulfilled. In this case, the origin becomes an equilibrium point of the system which, by (P5), ensures a vanishing remaining energy error. This leads to $p_{\text{ref}} (\omega_{\text{cm}} t) = p_{\text{dc}} (\mathbf{0}, \mathbf{0}) + p_{\text{ac}} (\omega_{\text{cm}} t, \mathbf{0}, \mathbf{0})$ which is derived from (10) with $e_{\text{err}} = \mathbf{0}$ and $e_{\text{err,I}} = \mathbf{0}$. In other words, a feasible stationary operating regime can be obtained by integrating the energy model (6) analytically, where the corresponding current injection is determined by the energy controller with vanishing energy error, and the average value of the regime is set to the constant E. With the designed controller and the regime discussed above, it is obvious that all conditions in Theorem 1 hold, and the system in Fig. 4 is the proposed energy control system for dc operation.

According to Theorem 1, for a sufficiently large frequency ω_{cm} and with the same initial values, the error $\left(\begin{array}{cc} e_{\text{err}}^{\text{T}} (t) & e_{\text{err,I}}^{\text{T}} (t) \end{array} \right)^{\text{T}}$ of the proposed energy control system lies in a neighborhood of the solution $\left(\begin{array}{cc} \bar{e}_{\text{err}}^{\text{T}} (t) & \bar{e}_{\text{err,I}}^{\text{T}} (t) \end{array} \right)^{\text{T}}$ of the corresponding average system (16) which converges globally and exponentially to $\mathbf{0}$, and the remaining error vanishes when the energy reference is given by the regime e_{rgm}. This is confirmed by the simulations in Fig. 5. In other words, for a sufficiently large predetermined frequency ω_{cm}, the MMC energy $e(t)$ converges globally to a neighborhood of the constant reference E, or it tracks the calculated regime e_{rgm}.

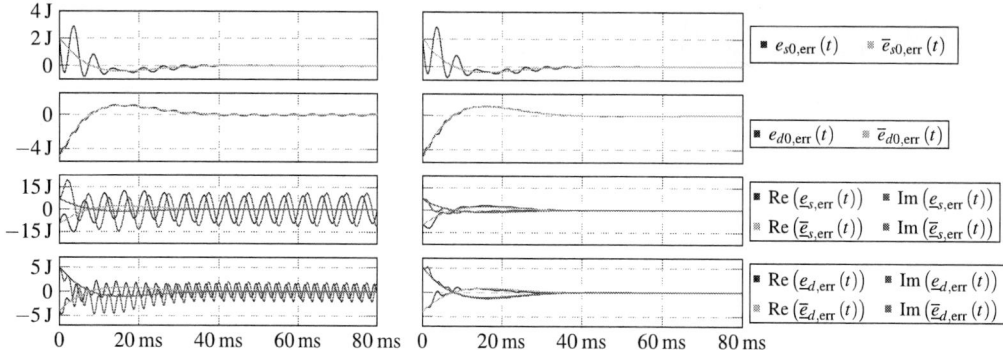

Fig. 5: Examples of error trajectory $t \mapsto \boldsymbol{e}_{\mathrm{err}}(t)$ of the proposed energy control system in Fig. 4 with the constant $\boldsymbol{E}$ (left column) and the stationary operating regime $\boldsymbol{e}_{\mathrm{rgm}}$ (right column) as the energy reference. The trajectory $t \mapsto \bar{\boldsymbol{e}}_{\mathrm{err}}(t)$ of the corresponding average system is determined by the designed error dynamics (16). The results are obtained by means of the simple choice (18), (19).

Measurement Results

For the implementation, the energy model in Fig. 4 is replaced by an MMC with current control and Pulse Width Modulation (PWM). The voltage of the dc-link capacitor is controlled to $600\,\mathrm{V}$ by an active grid side rectifier. The PWM frequency is $4884\,\mathrm{Hz}$ and the stator resistance of the PMSM is $0.59\,\Omega$. During the test, the PMSM is free from any mechanical load, so its rotor is oriented by the stator currents.

The proposed controllers are tested with a step of the average value of the energy reference from an imbalanced to a balanced value at $0\,\mathrm{ms}$. Fig. 6 shows the trajectories of (a) energy errors and (b) equivalent cell capacitor voltages for all six different variants indicated in the axis. In the simulations on the right, the energy error converges to zero whenever the regime $\boldsymbol{e}_{\mathrm{rgm}}$ is used as reference, while the errors do not vanish in case of the constant reference $\boldsymbol{E}$. These results coincide with the theoretical conclusions and show that the chosen value for the predetermined frequency ω_{cm} is large enough to ensure the stability of the proposed control system. The measurement of the energy error coincides well with the simulation, except for the small remaining errors in the measurement when the reference is given by the regime $\boldsymbol{e}_{\mathrm{rgm}}$. Fortunately, they do not result in a large mismatch between the measurement and the simulation of the cell capacitor voltage.

For a comprehensive comparison, some indexes reflecting the stationary and the transient performance are shown in Fig. 7: the peak-to-peak value $v_{C,\mathrm{pp}}$ of the equivalent cell capacitor voltages, the RMS value $I_{z,\mathrm{rms,sum}}$ of the arm currents for the stationary operation, and the Integral Time Absolute Error (ITAE) value $f_{\mathrm{ITAE}}(\boldsymbol{e}_z)$ of the difference between the arm energy $\boldsymbol{e}_z = \begin{pmatrix} e_{z1} & e_{z2} & e_{z3} & e_{z4} & e_{z5} & e_{z6} \end{pmatrix}^{\mathrm{T}}$ and its stationary solution. The results are plotted by varying controller parameter $k_{\mathrm{P,e}}$ and predetermined frequency ω_{cm} (settings see caption). The measured and simulated indexes are similar, except for $I_{z,\mathrm{rms,sum}}$ in Fig. 7b with the optimized controller, most likely because of an increased current control error due to the additional harmonics of higher order. The large reduction of $I_{z,\mathrm{rms,sum}}$ in the same figure is caused by the optimized controller and its objective function (20), which also reduces $v_{C,\mathrm{pp}}$, especially for the transient. Using the regime $\boldsymbol{e}_{\mathrm{rgm}}$ as reference systematically reduces $v_{C,\mathrm{pp}}$ and $I_{z,\mathrm{rms,sum}}$. However, the constant reference $\boldsymbol{E}$ is a noteworthy option, since it spares the costly regime determination. Compared with the common-mode voltage $v_{y0}^{\mathrm{1st+3rd}}$ (Fig. 3, red), the approximated trapezoidal wave v_{y0}^{trpz} (Fig. 3, blue) for the optimized controller leads to smaller $I_{z,\mathrm{rms,sum}}$ and $v_{C,\mathrm{pp}}$. On the whole, compared with the simple choice (18), (19), the optimization (18), (21) leads to a slight reduction of f_{ITAE}, except for the range $\omega_{\mathrm{cm}} > 2\pi \cdot 300\,\mathrm{rad/s}$. As expected, an increasing ω_{cm} leads to a smaller $v_{C,\mathrm{pp}}$. A larger controller parameter $k_{\mathrm{P,e}}$ decreases f_{ITAE} and $v_{C,\mathrm{pp}}$ during the transient, but increases $I_{z,\mathrm{rms,sum}}$ and $v_{C,\mathrm{pp}}$ slightly for the stationary operation.

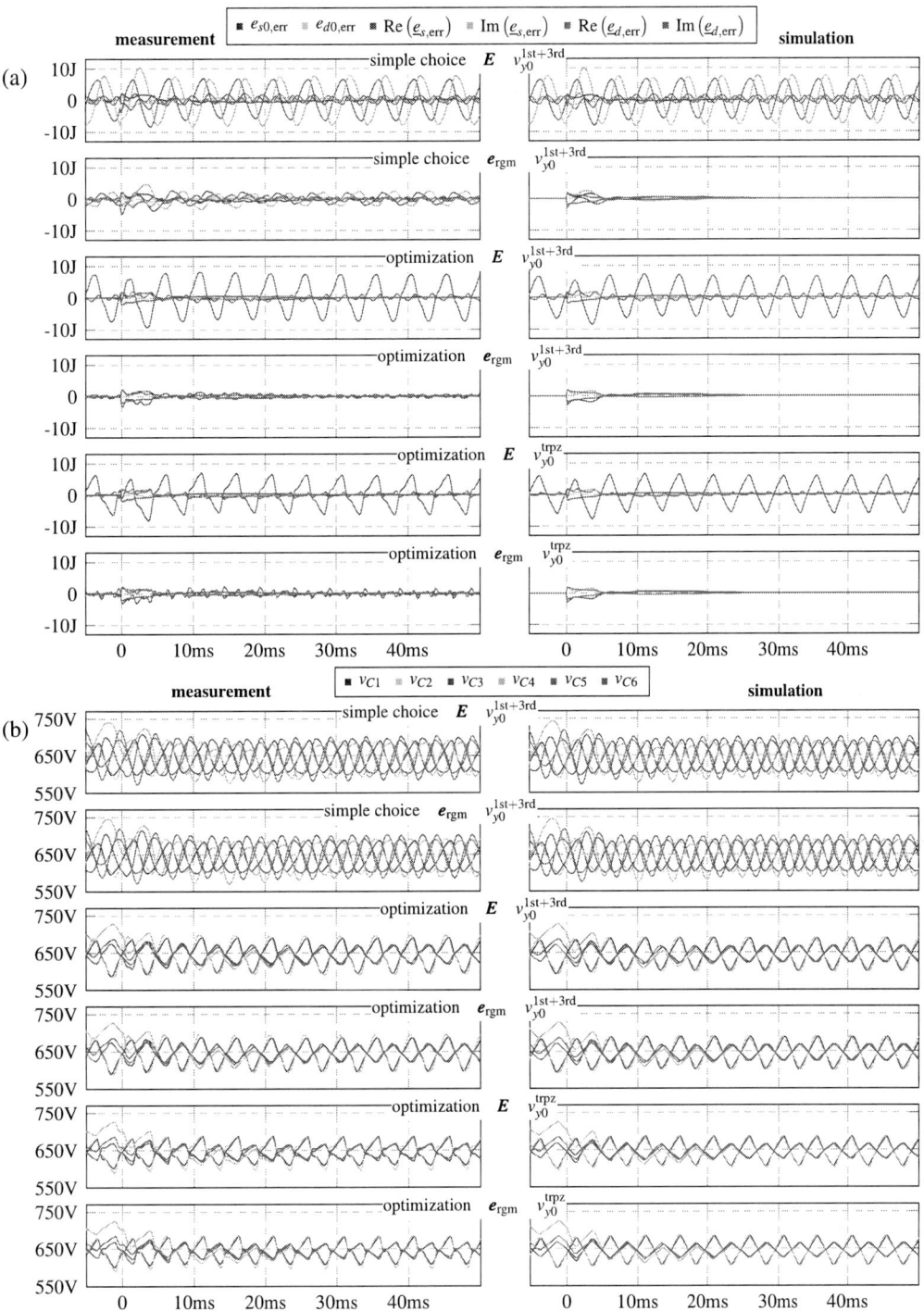

Fig. 6: Measurement and simulation results of the control system in Fig. 4: (a) energy errors (b) equivalent cell capacitor voltages. The average value of the energy reference steps at 0 ms, from an imbalanced value $(52\,\text{J}, -2\,\text{J}, -3\,\text{J}, 2\,\text{J}, 2\,\text{J}, -2\,\text{J})$ to a balanced value $(50\,\text{J}, 0, 0, 0, 0, 0)$. Settings: $\omega_{\text{cm}} = 203.5\,\text{rad/s}$, $k_{\text{P,e}} = 250\,\text{Hz}$, $k_{\text{I,e}} = k_{\text{P,e}}^2/2$, $\underline{i} = \underline{I}^{[0]} = 5\,\text{A}$.

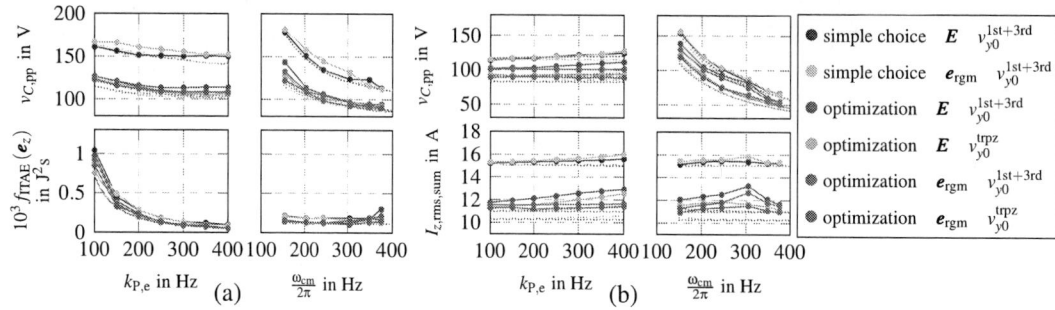

Fig. 7: Measurements (solid) and simulations (dotted) of the indexes reflecting (a) the transient and (b) the stationary performance of the proposed energy control system. Settings: $\omega_{cm} = 203.5\,\text{rad/s}$ for the 1st and 3rd column, $k_{P,e} = 250\,\text{Hz}$ for the 2nd and 4th column, $k_{I,e} = k_{P,e}^2/2$, $\underline{i} = \underline{I}^{[0]} = 5\,\text{A}$.

Conclusion

A new controller design framework for periodic systems is presented relying on the results in [11]. It is applied to the energy controller design of MMCs feeding a PMSM at standstill. This provides a rigorous stability analysis for the proposed energy control system, that is, an exponential convergence of the energy error in case of the calculated stationary operating regime e_{rgm} as the reference (this coincides with the more common approaches like in [5]), and a global stability in case of the constant energy references (this coincides with the less common approaches like in [1]). These theoretical conclusions are confirmed by simulations and measurements. Besides the controller with the simple choice, an optimized controller is derived by quadratic programming. The validity of the constant energy reference and the improved performance of the optimization w.r.t. the peak-to-peak value of the cell capacitor voltages and the arm current RMS is verified by measurement results.

References

[1] R. Lizana, et al. Control of arm capacitor voltages in modular multilevel converters. IEEE Trans. Power Electron., 31(2):1774–1784, 2016.

[2] H. Fehr, et al. Eigenvalue optimization of the energy-balancing feedback for modular multilevel converters. IEEE Trans. Power Electron., 34(11):11482–11495, 2019.

[3] A. Gensior, et al. Modeling and energy balancing control of modular multilevel converters using perturbation theory for quasi-periodic systems. IEEE Trans. Power Electron., 36(2):2201–2217, 2021.

[4] H. Bärnklau, et al. A model-based control scheme for modular multilevel converters. IEEE Trans. Ind. Electron., 60(12):5359–5375, 2013.

[5] J. Kolb, et al. Cascaded control system of the modular multilevel converter for feeding variable-speed drives. IEEE Trans. Power Electron., 30(1):349–357, 2015.

[6] A. E. Leon, et al. Energy balancing improvement of modular multilevel converters under unbalanced grid conditions. IEEE Trans. Power Electron., 32(8):6628–6637, 2017.

[7] M. Espinoza, et al. Modelling and control of the modular multilevel converter in back to back configuration for high power induction machine drives. In IECON 2016, pages 5046–5051, 2016.

[8] Ö. C. Sakinci, et al. Generalized dynamic phasor modeling of the mmc for small-signal stability analysis. IEEE Trans. Power Del., 34(3):991–1000, 2019.

[9] Y. Ma, et al. Stability analysis of modular multilevel converter based on harmonic state-space theory. IET Power Electronics, 12(15):3987–3997, 2019.

[10] F. Verhulst. Nonlinear Differential Equations and Dynamical Systems. Springer, 2000.

[11] H. K. Khalil. Nonlinear systems. Prentice Hall, Upper Saddle River, NJ, 3. ed. edition, 2002.

[12] M. Hagiwara, et al. Start-up and low-speed operation of an electric motor driven by a modular multilevel cascade inverter. IEEE Trans. Ind. Appl., 49(4):1556–1565, 2013.

[13] D. Graham, et al. The synthesis of "optimum" transient response: Criteria and standard forms. Transactions of the American Institute of Electrical Engineers, Part II: Applications and Industry, 72(5):273–288, 1953.

[14] Q. Gui, et al. Energy-balancing of a modular multilevel converter using an online trajectory planning algorithm. In EPE'20 ECCE Europe, 2020.

[15] Q. Gui, et al. Energy-balancing of a modular multilevel converter with pulsed dc load using an online trajectory planning algorithm. In EPE'21 ECCE Europe, 2021.

Turn-on Losses Optimization for Medium Power SiC MOSFET

Half-bridge Module

Pham Ha Trieu To, Felix Kayser, Hans-Günter Eckel
UNIVERSITY OF ROSTOCK
Albert-Einstein-Str. 2
18059 Rostock, Germany
Phone: +49 (0) 381 498 7135
Fax: +49 (0) 381 498 7102
Email: pham.to2@uni-rostock.de
URL: http://www.iee.uni-rostock.de

Keywords

<< Driver concepts >>, <<MOSFET>>, <<Silicon Carbide (SiC) >>, <<Switching losses>>

Abstract

This paper explains the mechanism of the parasitic turn-on (PTO) effect in a medium power SiC MOSFET half-bridge module and the relation between it and the reverse-recovery process of MOSFET's body diode. Based on that knowledge, a detail practical turn-on losses optimization process for medium power SiC MOSFET modules using PTO is presented. To quantify the stability of this method, some quantitative metrics are suggested to measure the critical values' sensitivity. The experimental measurements show that turn-on losses can be reduced 50% lower than conventional R_{gon} tuning method.

Introduction

The turn-on speed of a SiC MOSFET half-bridge module is limited by the overvoltage of its body diode during reverse-recovery which is the consequence of the commutation's loop stray inductance and the reverse-recovery current speed. In a conventional approach, a higher gate resistance is used to reduce the turn-on speed and hence protect the device from overvoltage [11]. The price for this protection is the higher turn-on losses on the MOSFET due to slower turn-on speed. This paper shows an effective process to achieve lower turn-on losses for a MOSFET half-bridge module which takes advantage of parasitic turn-on effect. The method was introduced in [1] for discrete chips with scale inductance but there was no explanation about the mechanism behind the method which is very important to understand the boundary conditions and also pros and cons of using parasitic turn-on effect.

Parasitic turn-on (PTO) is a complex effect which depends on many different parameters. To control PTO safely without destructing the module, it is necessary to have some metrics to quantify the risk of the module's destruction. There are two main destruction mechanisms: overvoltage of the MOSFET and the shoot-through which is presented by critical reverse-recovery losses. The overvoltage and the reverse-recovery energy are called critical parameters and their relationship with the PTO's control parameters are quantified by sensitivity factors. The optimization process is secured by selecting the lower sensitivity factor zones and proper safety margins for the overvoltage and threshold voltage. The experimental demonstration of the optimization process is also presented in this paper.

1. PTO in SiC MOSFET half-bridge modules

The double pulse test setup is shown in Fig. 1 with more detail of T31 parasitic elements. These capacitors C_{gd}, C_{gs}, C_{ds} are MOSFET T31's internal capacitors. L_{sin}, L_{sext} are the internal, external stray inductance of the source terminal of T31. L_{gin} and R_{gin} are T31's internal gate inductance and resistance. Lp is stray inductance between DC link voltage source and the MOSFET modules. According to [2], there is a magnetic coupling Mg in the SiC MOSFET module between T31's gate network circuit and $\frac{di_{ds}}{dt}$ of T32. The voltage induced by the magnetic coupling can be added or subtracted to T31's gate voltage depends on the internal structure of the device and is different from manufacturers to manufacturers. Double pulse test (DPT) is applied on the lower MOSFET T32, the upper MOSFET T31 is off at adjustable gate voltage v_{gsoff}. The voltage and current across T31 (v_r, i_r) were measured to calculate reverse-recovery energy E_{rr}. The voltage and current across T32 (v_{ds}, i_{ds}) were measured to calculate turn-on energy E_{on} of T32. The parasitic turn-on happens at T31 when its body diode is turned off by turning on T32. When the load current i_L is taken over by T32, the voltage across T31 is raising at speed $\frac{dv_r}{dt}$. The displacement current i_{gd} of C_{gd} then flows through the gate i_g and also C_{gs}, pulling up T31's gate voltage. T31's gate voltage at this time can be modeled by the equation (1):

$$v_{gsoff} = -(R_{goff} + R_{gin})i_g - (L_g + L_{gin})\frac{di_g}{dt} + v_{gs} - L_{sin}\frac{di_r}{dt} + M_g\frac{di_{ds}}{dt} \quad (1)$$

$$i_{gd} = i_g + i_{gs} \leftrightarrow i_g = C_{gd}\frac{dv_r}{dt} - (C_{gs} + C_{gd})\frac{dv_{gs}}{dt} = C_{gd}\frac{dv_r}{dt} - C_{iss}\frac{dv_{gs}}{dt} \quad (2)$$

In case of having low gate inductance and low internal gate resistance, equation (1) can be approximated:

$$v_{gs} = v_{gsoff} + R_{goff} \cdot \left(C_{gd}\frac{dv_r}{dt} - C_{iss}\frac{dv_{gs}}{dt}\right) - (M_g + L_{sin}) \cdot \frac{di_{ds}}{dt} \quad (3)$$

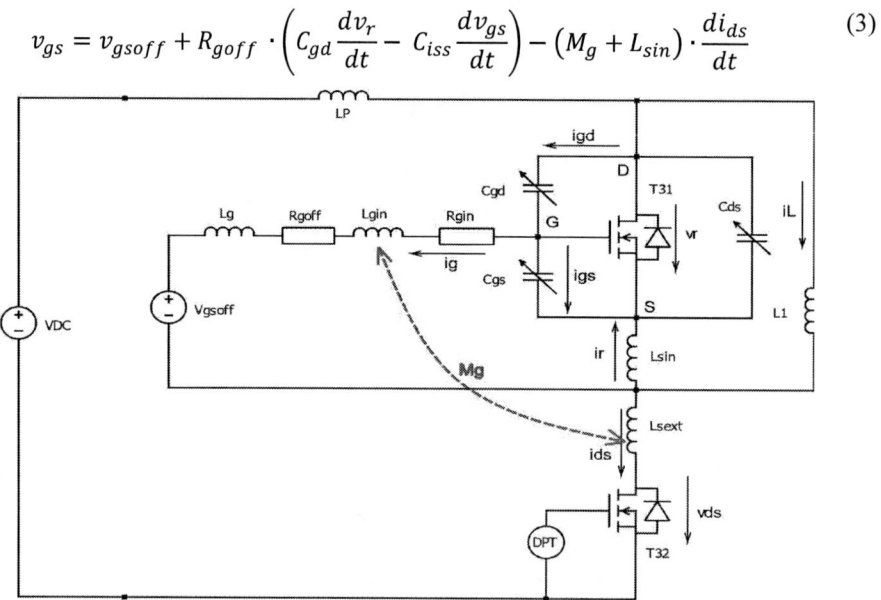

Fig 1: Medium power SiC MOSFET half-bridge double pule test set-up with detail T31 parasitic elements and magnetic coupling Mg between T31's gate circuit network and $\frac{di_{ds}}{dt}$ of T32

When the induced gate voltage of T31 is larger than the threshold voltage $v_{gs} > v_{th}$, PTO happens. Equation (3) is a combination of 3 components: v_{gsoff} is the static component, the voltage dynamic component: $R_{goff}\left(C_{gd}\frac{dv_r}{dt} - C_{iss}\frac{dv_{gs}}{dt}\right)$ and the current dynamic component: $-\left(M_g + L_{sin}\right)\frac{di_{ds}}{dt}$. If we look at the current dynamic part of equation (3), some interesting effect can be seen. When T32 is turned on, the current go through it always has $\frac{di_{ds}}{dt} > 0$. If the module has a positive magnetic coupling M_g, the current dynamic component of (3) is larger at faster turn-on speed (smaller R_{gon}) and hence reduce the parasitic turn on effect on T31. There is also the case when the module has a negative magnetic coupling [2], it will increase the T31's parasitic turn-on. The parasitic inductance on the source terminal L_{sin} also helps to reduce parasitic turn-on of T31. Hence the module without Kelvin connector has a positive effect on preventing parasitic turn-on. The voltage dynamic component depends on $R_{goff}, C_{gd}, C_{iss}$. Small R_{goff}, C_{gd} reduce the effect of dynamic voltage component. The static and current dynamic component are user controllable by changing v_{gsoff} and R_{gon}. The voltage dynamic component is a user uncontrollable part. This means the setup which has low R_{goff}, C_{gd} and high $C_{iss}, M_g + L_{sin}$ has more user control ability or user can control PTO more easily.

2. Reverse-recovery of SiC MOSFET's intrinsic body diode

There are three stages in the reverse-recovery of SiC MOSFET's body diode that can be seen in Fig 2. At the first stage, when T32 is turned on, the T31's body diode forward current starts decreasing until zero. At the second stage, the diode enters the reverse-biased state, its current continues flowing in reverse direction at the speed di_{rf}/dt because there are still a lot of carriers in the n- drift layer need to be swept out. When the reverse current reaches its peak value Irr, the depletion layer starts to form, the diode enters its third stage. When the depletion starts to expend, the drain-source voltage vr increases to the blocking voltage, the amount of carrier needing to be swept out of the depletion region also reduces and hence the reverse current also reduces at the speed di_{rr}/dt. In this stage, high dvr/dt can cause parasitic turn-on. Depending on the stray inductance of the external circuit (LP), the peak voltage over T31 at this stage is:

$$v_r = VDC - LP \cdot \frac{di_{ds}}{dt} = VDC + LP \cdot \frac{di_{rr}}{dt} \tag{4}$$

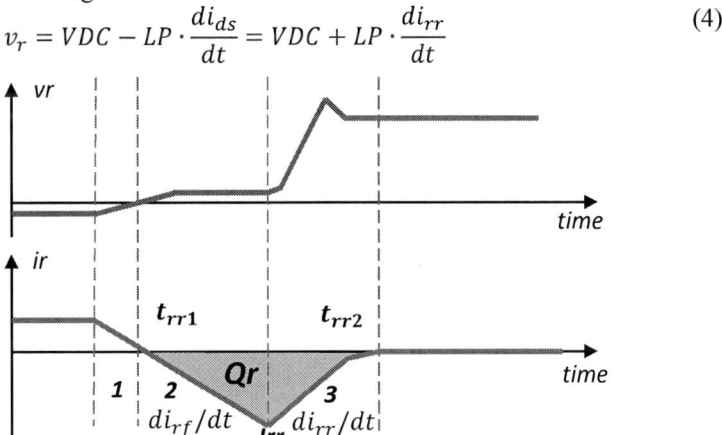

Fig 2: Reverse-recovery waveform of SiC MOSFET intrinsic body diode.

Apart from the overvoltage cause by di_{rr}/dt, T31 also encounters the oscillation from the combination of external stray inductance LP and the output capacitor C_{oss} of T31 (Fig 3.), the total voltage on T31 is:

$$v_r = VDC + LP \cdot \frac{di_{rr}}{dt} + \sqrt{\frac{LP}{C_{oss}}} \cdot I_L \cdot \sin(\omega t + \theta) \cdot e^{-\omega \tau t} \qquad (5)$$

With IL is the load current at the turn-on moment. $\omega \approx \frac{1}{\sqrt{LP.C_{oss}}}$ is the angle frequency of the oscillation, τ is the oscillation damping factor. R_P is the parasitic resistance of the external circuit. R_{bd} is the series resitance of the body diode in the drift region which can be seen in Fig 3.

$$\tau = \frac{R_P + R_{bd}}{2} \sqrt{\frac{C_{oss}}{LP}} \qquad (6)$$

At high junction temperature, the life time of the minority carriers in the drift region is longer [6,7,8] and hence the intrinsic carrier concentration in drift region is higher. That means the body diode's series resistance R_{bd} is lower at higher temperature [6]. Consequently, the turn-on oscillation is poorly damped and hence more overvoltage on the body diode has. For many medium power module applications, the modules are connected by a busbar which has a long commutation loop and a large parasitic inductance. The reverse-recovery oscillation becomes more severe in this scenario, especially at high current and high junction's temperature.

3. Parasitic turn-on and the reverse-recovery oscillation damping

From the damping factor of the reverse-recovery oscillation which is described in equation (6), to damp the oscillation, R_{bd} should be increased to have bigger damping factor. Since R_{bd} depends on the concentration of intrinsic carriers in the drift layer [6], R_{bd} can be increased by reducing the concentration of the intrinsic carriers in the drift layer. According to [9] this can be done by increasing the off-state gate voltage of T31. At higher gate voltage, the channel's resistance gets lower, a portion of charges in the drift region will be attracted to the channel area and hence temporarily reduces the charge concentration near by the body diode (Fig 3.). The same experiment result can be found in [10].

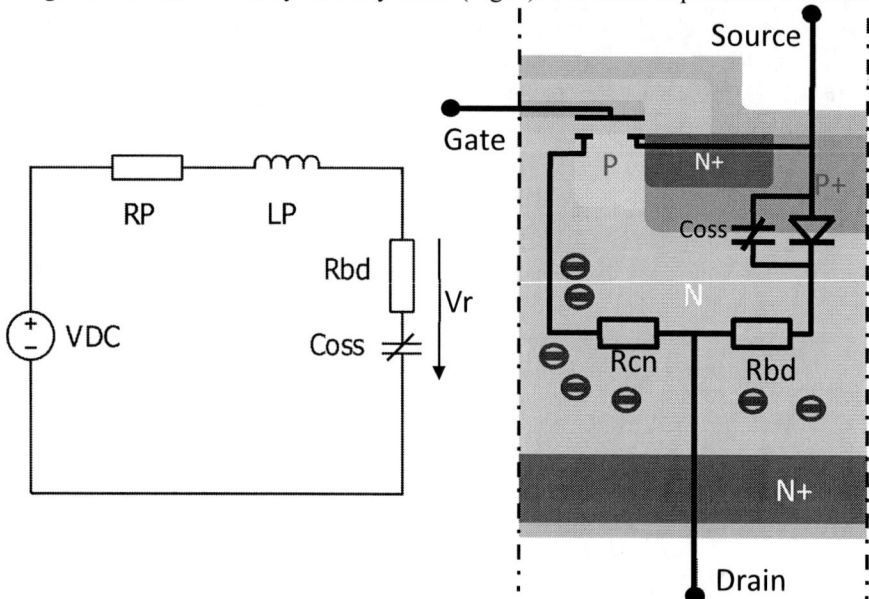

Fig 3: Cross section of power MOSFET with its intrinsic body diode, internal resistance and the oscillation circuit formed together with the external stray inductance LP.

In literature [9,10], the gate voltage is increased but not excess the threshold voltage. Hence, there were no parasitic turn-on. With parasitic turn-on, when the gate voltage excesses the threshold voltage for a short time only in the third stage of the reverse-recovery, the channel is partially opened and its resistance is decreased more than without PTO. This leads to further increase the body diode's series resistor R_{bd}, it consequently increases the damping factor and hence reduces further the overvoltage on T31. Too much parasitic turn-on (when the channel is completely opened or goes into the low ohmic region) can lead to huge reverse-recovery losses and may destroy the device. But if PTO is used in a controlled way such as adjusting the R_{bd} just enough to damp the oscillation to reduce the overvoltage, The R_{gon} of T31 can be lowered .This leads to a faster turn-on with much less turn-on losses and only slightly higher reverse-recovery losses. Fig 4. shows the voltage and current across the body diode during reverse-recovery process at 900VDC, 400A load current, 150°C junction temperature with different turn-off gate voltage v_{gsoff} = [-7, -3, -2, -1] V. It can be seen that during the third stage of the recovery process, when the depletion layer starts to form, the voltage vr fast increases at speed $\frac{dv_r}{dt}$ and starts oscillating. Because of high $\frac{dv_r}{dt}$, the gate voltage of T31 is parasitically turned on and hence reduces MOSFET T31's channel resistance. The charge carriers in drift region are now attracted to the area near to the channel and then temporarily reduce the charge concentration in the drift region near to the body diode. Because of less charge carrier concentration, the series resistor of the body diode is temporarily increased and damps the oscillation, the higher v_{gsoff} is, the better vr is damped. The overvoltage on T31 is further reduced when increase v_{gsoff}.

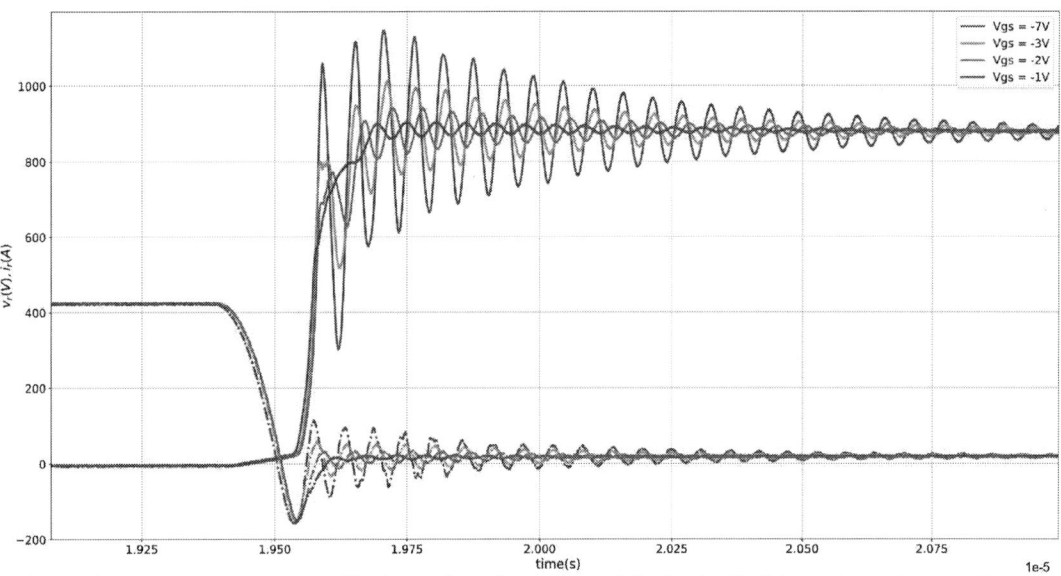

Fig 4: The reverse-recovery oscillation's damping effect of the body diode's series resistance R_{bd} at different v_{gsoff} which were measured at 150°C.

4. Turn-on losses optimization process

As mentioned in section 2 and 3, when there is a large external parasitic inductance and high load current, the oscillation amplitude becomes the major part of the overvoltage of the body diode during reverse-recovery. This oscillation can be damped by increasing the body diode's series resistance R_{bd} using the parasitic turn-on effect. In other cases, when there is a low external stray inductance and low load current, the parasitic turn-on has no positive effect on optimizing turn-on losses because the overvoltage is dominated by reverser-recovery current speed $\frac{di_{rr}}{dt}$. This is also the limitation for this optimization method.

Section 1 shows how the user can control the PTO by changing v_{gsoff} and R_{gon}. To increase the user's PTO control ability, R_{goff} should be small enough to have more independence from the voltage dynamic component. If the stray inductance LP is too large, active clamping or a decoupling capacitor may be used to further reduce R_{goff}.

The first step in the optimization process is finding the range of the controllable variable v_{gsoff} and R_{gon}. To do this, a double pulse test is set up. To find the controllable range of R_{gon}, first the v_{gsoff} should be set low enough to ensure that there is no parasitic turn-on. The user should keep the tolerance of the gate's oxide in mind while reducing v_{gsoff}. A conventional R_{gon} tuning process should be carried out to get the minimum value of $R_{gon} = Rn_{gon}$, which ensures the over voltage on T31 $Vrmax$ stay inside the safe operating area (SOA) of the MOSFET. The total turn-on energy and reverse-recovery energy $E_{summax} = E_{rr} + E_{on}$ should be recorded to have a threshold for the optimization process. The optimization process is only meaningful when the total turn-on losses is smaller than the conventional R_{gon} tuning. when using PTO to optimize losses, R_{gon} can be further reduced. So Rn_{gon} is the maximum value of R_{gon}. The minimum value R_{gonmin} is limited by the application's maximum turn-on speed. After this step, the user should have $R_{gon} \in [R_{gonmin}, Rn_{gon}]$. To find the v_{gsoff} control range, the gate's voltage of T31 should be measured and recorded during double pulse test with different v_{gsoff} voltages. The double pulse test in this case has to be done at the maximum junction temperature of the device since the overvoltage gets worst at high temperature. The lowest value $v_{gsoffmin}$ is the value, at which, the induced gate's voltage of T31 is equal to the threshold voltage of the MOSFET when turning on T32 at $R_{gon} = Rn_{gon}$. The maximum $v_{gsoffmax}$ should be closed to the threshold voltage of the MOSFET, but the user should have some safety margin for this value to avoid shoot-through. The control range of v_{gsoff} is now defined in the range $[v_{gsoffmin}, v_{gsoffmax}]$.

At the second step, the user should discretize the control parameters range into $R_{gon} \in [R_{gonmin}, R_{gon1}, R_{gon2}, \dots Rn_{gon}]$ and $v_{gsoff} \in [v_{gsoffmin}, v_{gsoff1}, v_{gsoff2}, \dots v_{gsoffmax}]$. The number of points depend on the user's choice. A large number of points may create more measurements but give more accurate result.

As the third step, double pulse testes in worst case scenario of the operation (maximum junction temperature, maximum DC link voltage and maximum load current) are made for each combination of v_{gsoff}, R_{gon} and the total turn-on energy and reverse-recovery energy $E_{sum} = E_{rr} + E_{on}$ should be recorded to compare with the threshold E_{summax} that has been measured in the first step.

At the last step, the parameter's sensitivity should be calculated to identify the stable zone of the method. The sensitivity factors are defined in the next section.

5. Parameters' sensitivity factors

During the turn-on process with parasitic turn-on effect, the overvoltage of T31 Vrmax and the reverse-recovery energy E_{rr} are the critical values, which are directly related to the module's destruction. These values should not be too sensitive with the control parameters. This means some small changes in the application's parameters, because of temperature or components' tolerance…, should not change much the critical values to assure a stable operation. To quantify this stability, four sensitivity factors are introduced:

$$S_{EV} = \frac{\partial E_{rr}}{\partial v_{gsoff}} \; for \; discrete \; value \; is \; \frac{\Delta E_{rr}}{\Delta v_{gsoff}} \qquad (7)$$

$$S_{ER} = \frac{\partial E_{rr}}{\partial R_{gon}} \; for \; discrete \; value \; is \; \frac{\Delta E_{rr}}{\Delta R_{gon}} \qquad (8)$$

$$S_{VV} = \frac{\partial V_{rmax}}{\partial v_{gsoff}} \; for \; discrete \; value \; is \; \frac{\Delta V_{rmax}}{\Delta v_{gsoff}} \qquad (9)$$

$$S_{VR} = \frac{\partial V_{rmax}}{\partial R_{gon}} \; for \; discrete \; value \; is \; \frac{\Delta V_{rmax}}{\Delta R_{gon}} \qquad (10)$$

S_{EV}, S_{ER} measure how sensitive the reverse-recovery energy E_{rr} to v_{gsoff} and R_{gon} respectively.

S_{VV}, S_{VR} measure how sensitive the body diode's overvoltage Vrmax to v_{gsoff} and R_{gon} respectively.

6. Experimental result

To demonstrate the optimization process, an experimental testbench is set up like in Fig 1. The device under test is a 1200V-375A, 62mm SiC MOSFET module. The module has no Kelvin connector, so there is a slop on vr's measurement during ir falling. The double pulse test setup (Fig 1.) has 26nH external inductance together with 20nH internal module's stray inductance. The worst case scenario is defined at 900VDC link voltage, 600A load current and 150°C junction temperature. All the differential voltage probes and Rogowski coil current probes are properly calibrated and de-skewed. At the first and second step of the optimization process, the conventional R_{gon} tuning process was done with $v_{gsoff} = -5V$ and having no parasitic turn-on involved. The smallest $Rn_{gon} = 4.7\ \Omega$, at maximum overvoltage on T31 is $Vrmax = 1175V$ (safety margin is 25V below 1200V), $R_{goff} = 2\Omega$. The total energy losses in this case was $E_{summax} = E_{rr} + E_{on} = 37$ mJ. R_{gon} range chosen for the optimization is $R_{gon} \in [0, 1, 2, 2.9]$. When turning on T32 with $R_{gon} = 4.7\ \Omega$ and adjusted $v_{gsoff} = -4.4V$, the parasitic turn-on gate voltage measured on T31 was equal to the MOSFET threshold voltage $v_{th} = 4.5V$. The $v_{gsoffmax}$ was chosen to -1V with a safety margin of 5.5 V below the threshold voltage v_{th}. v_{gsoff} was discretized into 8 points:
$v_{gsoff} \in [-4.4, -4, -3.5, -3, -2.5, -2, -1.5, -1](V)$.
At the third step, the double pulse measurements at worst case scenario for 32 combinations of v_{gsoff} and R_{gon} were made. The total energy $E_{sum} = E_{rr} + E_{on}$ and T31's overvoltage $Vrmax$ were recorded and compared to the conventional R_{gon} tuning. The result is displayed in Fig 5.

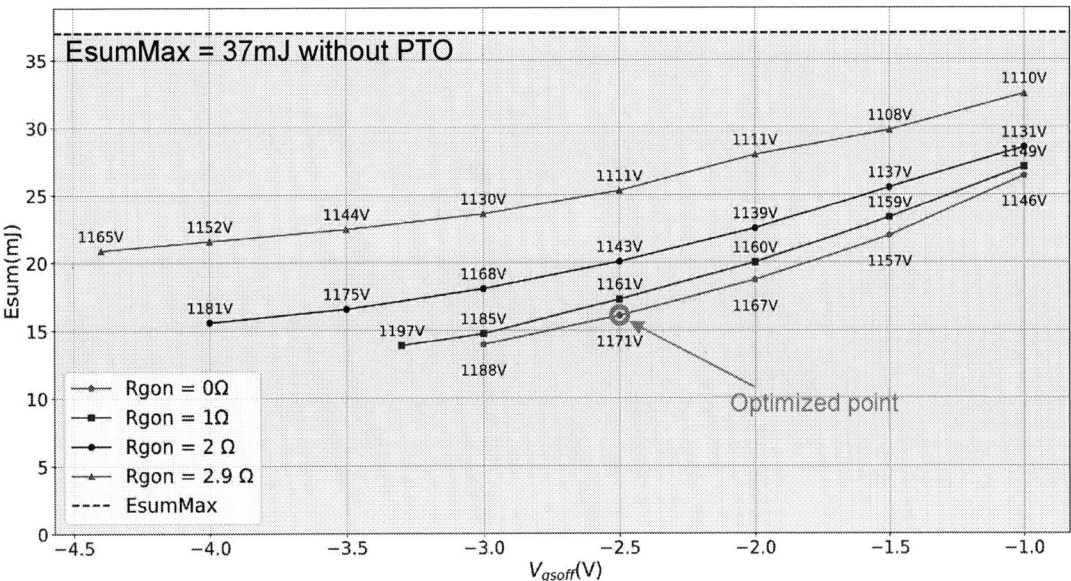

Fig 5: Total turn-on energy E_{sum} with different combination of v_{gsoff}, R_{gon}. The body diode's overvoltage Vrmax is marked at each measurement point. The green area depicts the total turn-on losses E_{sum} that can be further optimized compare to $E_{summax} = 37$ mJ conventional turn-on. The optimal turn-on energy $E_{sum} = 16\ mJ$ was found at $R_{gon} = 0\Omega$ and $v_{gsoff} = -2.5V$ where the overvoltage of T31 $Vrmax = 1171V$ is still below the safety margin.

At the last step, the discrete sensitivity factors $S_{EV}, S_{ER}, S_{VV}, S_{VR}$ are calculated. The example calculation for discrete S_{EV} is showed in table I. The same manner is applied for the other sensitive factors.

Table I: Example of calculating discrete sensitivity $S_{EV}(\frac{mJ}{V})$ factor.

$v_{gsoff}(V)$	Err(mJ)	$S_{EV}(mJ/V)$
-3	12.45	$S_{EV} = \dfrac{14.5 - 12.45}{-2.5 - (-3)} = 4.1$
-2.5	14.5	
-2	17.2	$S_{EV} = \dfrac{20.4 - 17.2}{-1.5 - (-2)} = 6.4$
-1.5	20.4	

The results of the calculation are displayed in Fig 6.,7.,8.and 9. It can be seen in Fig 6. That the reverse-recovery energy E_{rr} is less sensitive at lower v_{gsoff} values. At higher v_{gsoff}, the parasitic turn-on gets worse and hence increases the E_{rr} faster than at lower v_{gsoff}.

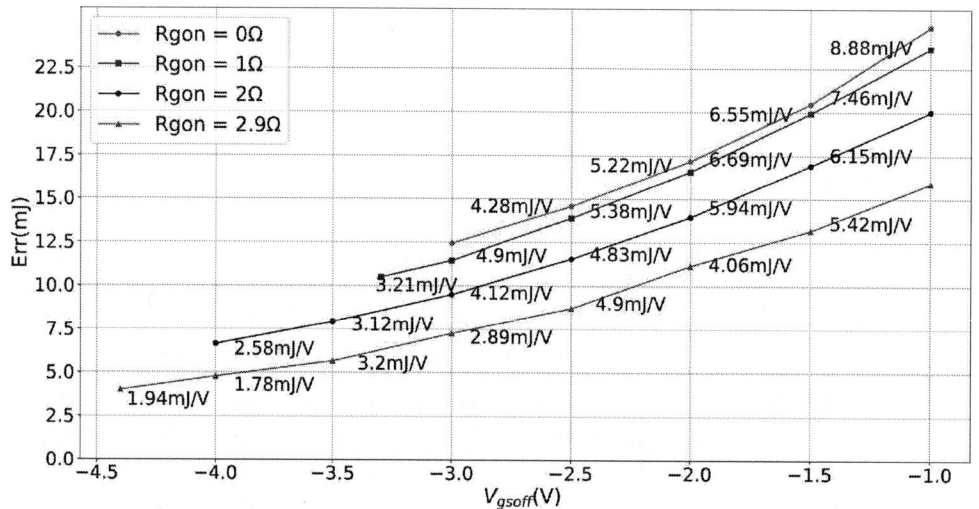

Fig 6: Reverse-recovery energy E_{rr} is less sensitive at lower v_{gsoff} range. The sensitivity factor S_{EV} is also marked on the graph.

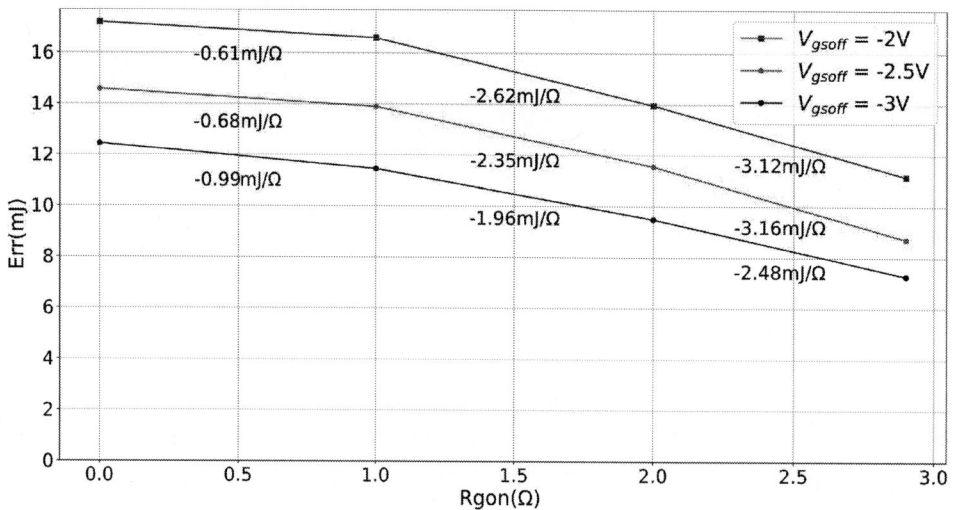

Fig 7: Reverse-recovery energy E_{rr} is less sensitive at lower R_{gon} range. The sensitivity factor S_{ER} mJ/Ω is marked on the graph.

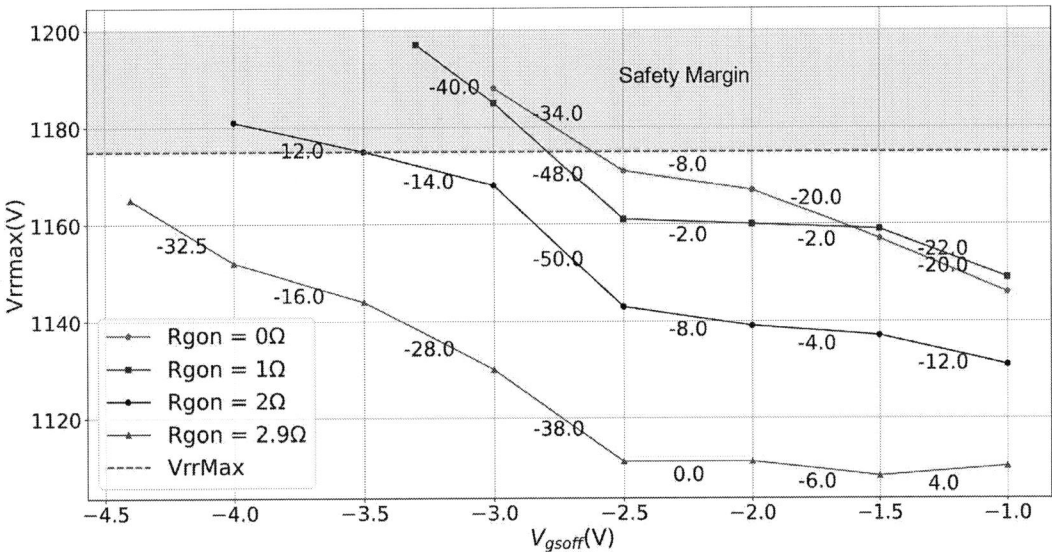

Fig 8: Body diode's overvoltage Vrmax is less sensitive at higher v_{gsoff} range. The sensitivity factor S_{VV} is also marked on the graph

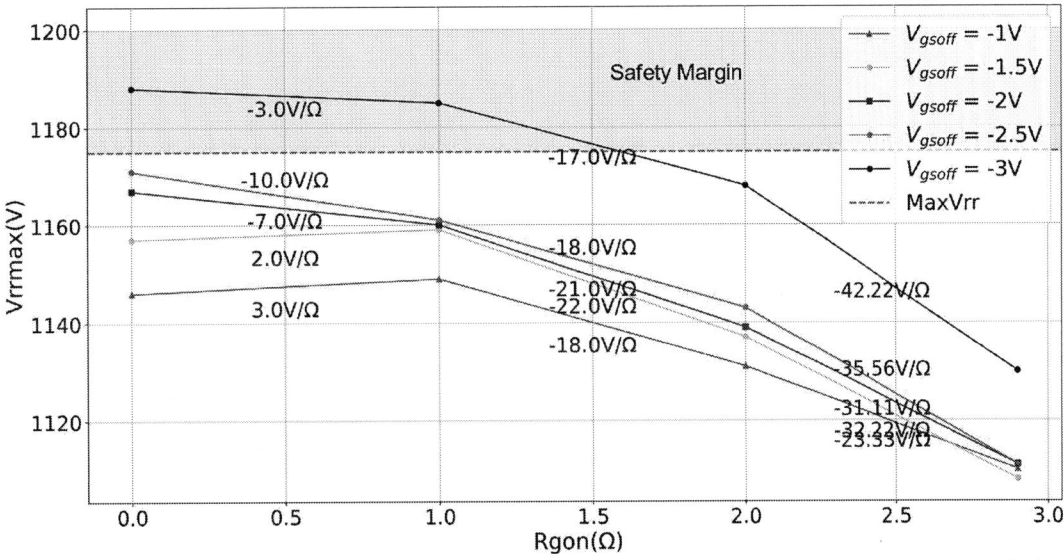

Fig 9: Body diode's overvoltage Vrmax is less sensitive to lower R_{gon} range. The sensitivity factor S_{VR} V/Ω is also marked on the graph.

In Fig 7, the reverse-recovery energy E_{rr} is less sensitive at lower R_{gon} range because in equation (3), when the dynamic current component has a positive $M_g + L_{sin}$, at high turn-on speed (lower R_{gon}), T31 has less parasitic turn-on so E_{rr} less sensitive at lower R_{gon} range. In Fig 8, the body diode's overvoltage $Vrmax$ is less sensitive at higher v_{gsoff} range, because the oscillation is suppressed by high resistance of R_{bd}. In Fig 9, the body diode's overvoltage $Vrmax$ is less sensitive to lower R_{gon} range because at the same v_{gsoff}, the R_{bd} value does not change much, the over voltage of the body diode is dominated by $LP \cdot \frac{di_{rr}}{dt}$ from equation (5). At low switching speed, a certain number of charges in the drift region are recombined before they are swept out [9]. This phenomenon makes the $\frac{di_{rr}}{dt}$ more sensitive at slow turn-on speed than at fast turn-on speed, hence the $Vrmax$ is less sensitive in the lower R_{gon} range (Fig 10.).

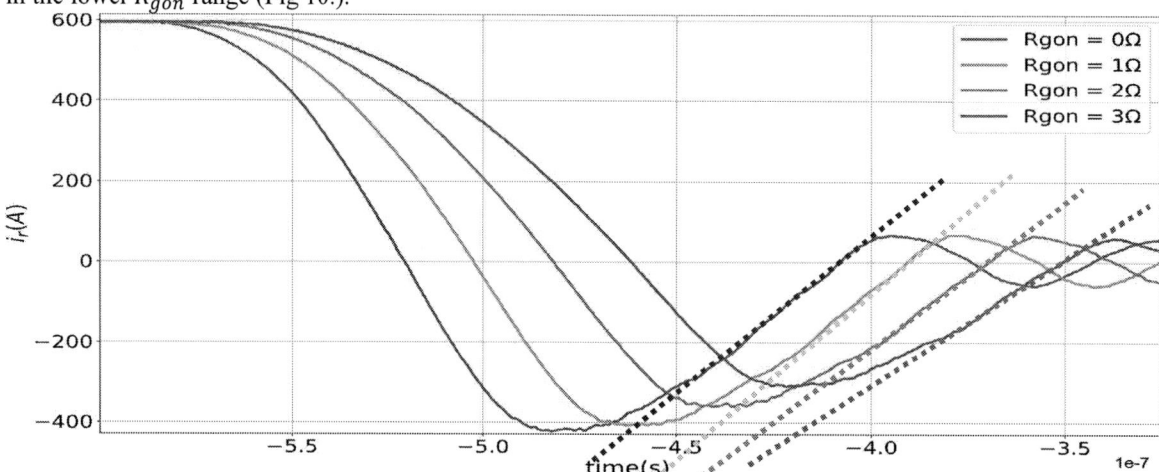

Fig 10: Reverse-recovery current speed $\frac{di_{rr}}{dt}$ is more sensitive at slow turn-on speed (red and green lines) than high turn-on speed (blue and orange lines) at the same v_{gsoff}

From the calculated sensitivities above there is a trade off between low and high v_{gsoff}. Low v_{gsoff} gives a stable E_{rr} but high-sensitive $Vrmax$. v_{gsoff} should be chosen with high priority to have a stable $Vrmax$ and less E_{rr}. After considering all the quantitative stability above, the lowest and stable turn-on losses can be optimized with the PTO method is 16mJ at $R_{gon} = 0\Omega$, $v_{gsoff} = -2.5V$, more than 50% lower than the case without parasitic turn-on effect. Fig 11. shows the low total turn-on losses and the reverse-recovery losses of the proposed optimization method versus the losses of a conventional R_{gon} tuning at different operating points.

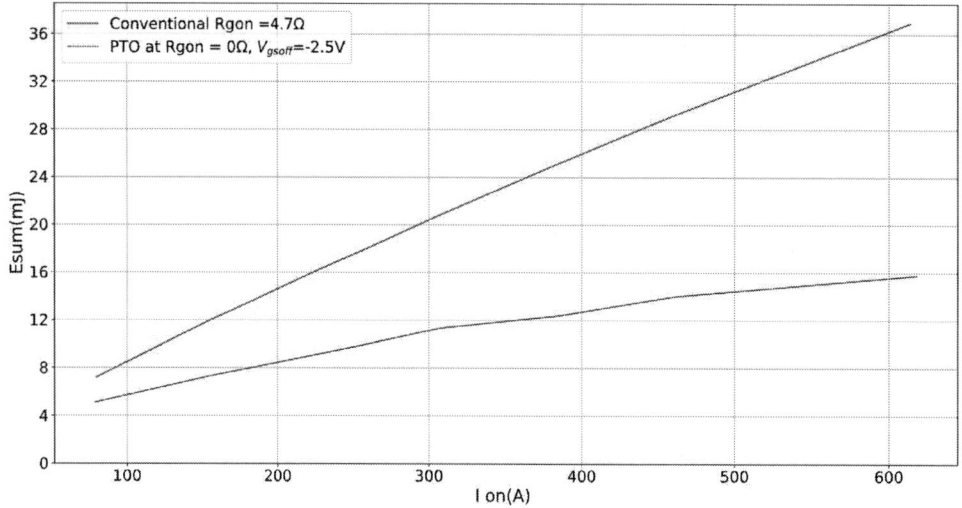

Fig 11. Comparison of total turn-on losses Esum = Err + Eon of the proposed method and the conventional R_{gon} tuning method at different operation points

Conclusion

The paper has shown the principle mechanism behind the parasitic turn-on effect and its relation with the reverse-recovery of the intrinsic body diode of SiC MOSFET in the half-bridge module. Also a step by step process to optimize turn-on losses using parasitic turn-on effect was introduced and demonstrated with experimental measurement. The results show, that the turn-on losses can be reduced by more than 50% compare to the conventional tuning method. The optimization point was selected at the less sensitive zone of the control parameters to ensure a safe turn-on.

References

[1] Patrick Hofstetter, Robert W. Maier, Mark-M. Bakran ," Parasitic Turn-On of SiC MOSFETs – Turning a Bug into a Feature", pp.6-12, PCIM Europe digital days 2020, 7 – 8 July 2020;

[2] Jorge Mari, Fabio Carastro, Max-Josef Kell, "Assessing the Presence of Parasitic Turn On in SiC Mosfet Power Modules", EPE'21 ECCE Europe, September 2021.

[3] Klaus Sobe1, Thomas Basler2, Blaz Klobucar," Characterization of the parasitic turn-on behavior of discrete CoolSiC™ MOSFETs", pp.54-60, PCIM Europe 2019, 7 – 9 May 2019, Nuremberg, Germany.

[4] A.März, T. Bertelshofer, M. Helsper, Mark-M. Bakran, Comparison of SiC MOSFET gate-drive concepts to suppress parasitic turn-on in low inductance power modules, EPE'17 ECCE Europe, pp.1-10.

[5] Mario Pulvirenti, Body Diode Reverse-recovery Effects on SiC MOSFET Half-Bridge Converters, 2020 IEEE Energy Conversion Congress and Exposition (ECCE), pp.2871-2877

[6] Kang Peng, Characterization and Modeling of SiC MOSFET Body Diode,2016 IEEE Applied Power Electronics Conference and Exposition (APEC), pp.2127-2135

[7] A. Bolotnikov "Utilization of SiC MOSFET body diode in hard switching applications," Materials Science Forum, vol.778-780, pp.947-950, Feb. 2014.

[8] O. Kordina, J.P. Bergman, C. Hallin, "The minority carrier lifetime of n-type 4H- and 6H-SiC epitaxial layers," Applied physics letters,1996.

[9] Paul Sochor,Understanding the Turn-off behavior of SiC MOSFET body diodes in fast witching applications,PCIM Europe digital days 2021, 3 – 7 May 2021, pp.290-297.

[10] Mario Pulvirenti, Body Diode Reverse-recovery Effects on SiC MOSFET Half-Bridge Converters, 2020 IEEE Energy Conversion Congress and Exposition (ECCE), pp.2871-2877.

[11] Paul Sochor, Commutation loop design for optimized switching behavior of CoolSiC MOSFETs using compact models, PCIM Europe digital days 2020, 7 – 8 July 2020, pp.1988-1995.

Oscillation Damping in a 500kW Hybrid Si/SiC Three-Level ANPC Inverter with Decoupling Capacitor

Pham Ha Trieu To, Hans-Günter Eckel
UNIVERSITY OF ROSTOCK
Albert-Einstein-Str. 2
18059 Rostock, Germany
Phone: +49 (0) 381 498 7135
Fax: +49 (0) 381 498 7102
Email: pham.to2@uni-rostock.de
URL: http://www.iee.uni-rostock.de

Keywords

«Voltage Source Inverter (VSI) », « Hybrid », « Converter control », « Silicon Carbide (SiC) »

Abstract

This paper presents a high-power hybrid Si/SiC 3L ANPC using only two SiC MOSFET. The high parasitic inductive commutation loop is decoupled by adding a decoupling capacitor near to the SiC MOSFET modules. Because of the decoupling capacitor, there are two main oscillation circuits are formed. These oscillations can lead to serious EMI issues and need to be damped. The damping schemes for the switching oscillation were experimental investigated and a special switching strategy was also developed to actively cut off the decoupling oscillation.

Introduction

The active neutral point clamped (ANPC) was famous for having flexible switching states and hence losses can be flexibly distributed among the power switches [4]. Silicon carbine (SiC) MOSFET has much lower switching losses compare to Si IGBT, but it is still not a good option for mass production because of its high cost. With 3 level output voltage and high switching frequency, the filter inductor's volume can be significantly reduced. In order to take advantages of both ANPC three level topology and SiC MOSFET at lower cost, many studies have proposed the hybrid ANPC (HANPC) topology, which uses both silicon IGBT and SiC MOSFET as power switches [5-12]. The main principle of the HANPC is using the flexible switching states to relocate the high switching frequency on SiC MOSFETs to reduce the switching losses. Those publications can be divided into two main streams: one is replacing 4 Si IGBT with 4 SiC MOSFET [5,6,10] (Fig 1.a), the other is replacing two Si IGBT with two SiC MOSFET [6,7,8,9,11,12] (Fig 1.b). Most of the publications concentrated on low power applications which can be done with discrete SiC chip on PCB or all switches are integrated in one module [11] which have low parasitic inductance commutation loop. For high power applications, the separated SiC MOSFET, IGBT modules are connected by busbar to form HANPC topology. In this scenario, the large stray inductance commutation loop becomes the major issue which increases the switching losses of the high power HANPC (Fig 1.b). In [5], a HANPC with four SiC MOSFET in megawatt scaled was introduced, which took the advantage of the small commutation loop (Fig.1a). On the other hand, four SiC MOSFET were used which again increased the cost of the inverter.

This paper proposes a high power HANPC which uses only two SiC MOSFET together with a decoupling capacitor (C_f) placed near to SiC MOSFET to decouple the high parasitic commutation loop (Fig.1c). This configuration can reduce the cost of the inverter and decouple the large inductive commutation loop from the SiC MOSFET module. During switching the SiC MOSFET, the oscillation circuit is created from the stray inductance of the commutation loop and the output capacitor (C_{oss}) of the MOSFET. Because of C_f, this oscillation circuit is divided into two loops. The low frequency (LF) oscillation loop or can also be called decoupling oscillation loop (Fig.2 blue dash line) and the high frequency (HF) oscillation loop (Fig.2 a green dash line). The theoretical model of the oscillation circuit was studied in [13]. These oscillations pose a threat to electromagnetic interference (EMI) compliance both in conducted emission and magnetic field radiated emission over regulatory frequency range from 150kHz to 30MHz [14,15]. The main challenge of this 2SiC hybrid ANPC concept is how to damp these oscillations. Ferrite core (FC) is proposed to damp the high frequency loop and high-power film capacitor C_f with external stainless-steel connections is proposed to damp both of the oscillations. A special switching scheme is also designed to actively cut off the low frequency oscillation path. The experimental 500kW 2SiC hybrid ANPC testbench is setup to verify the concept can be seen in Fig.8.

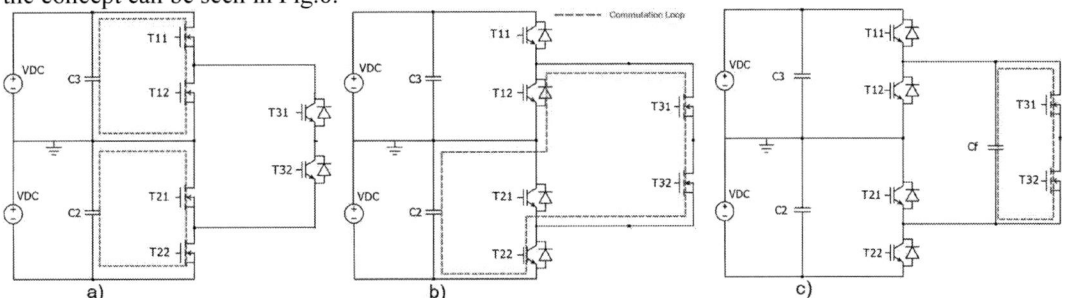

Fig. 1: HANPC topology, a) 4SiC with short commutation loop proposed in [5], b) 2SiC with long commutation loop, c) 2SiC with decoupling capacitor C_f is proposed in this paper.

1. Switching oscillation in HANPC

2SiC hybrid ANPC has 3 levels output: positive (P), zero (ZN, ZP) and negative(N) output voltage. The conventional switching combination is shown as the table I.a. When the switch is on, it is marked with 1 and off with 0. It can be seen that the IGBT switch (T11, T12, T21, T22) are switched at fundamental frequency and SiC MOSFET (T31, T32) are switched at carrier frequency when the switching states change between $P \leftrightarrow ZP$ or $N \leftrightarrow ZN$.

Table I. HANPC switching states

	a) Conventional switching states						b) Proposed switching states						
	T11	T12	T21	T22	T31	T32	T11	T12	T21	T22	T31	T32	Vout
P	1	0	1	0	1	0	1	0	0	0	1	0	VDC
ZP	1	0	1	0	0	1	0	0	1	0	0	1	0
ZN	0	1	0	1	1	0	0	1	0	0	1	0	0
N	0	1	0	1	0	1	0	0	0	1	0	1	-VDC

Let's take a look at the transition from $P \rightarrow ZP$ when the negative current IL direction is going into the inverter (Fig 2.a), the switches T11, T21 are on, T32 is off, T31 is turned from on (P) to off (deadtime) before turning on T32(ZP). With the assumption that DC link capacitors C2, C3 are very big and can be considered as voltage source VDC, the equivalent circuit in this case can be seen in Fig. 2b with L_{s1}, R_{s1} are stray inductance and resistance between DC link capacitor and decoupling capacitor C_f. L_{s2}, R_{s2} are stray inductance and resistance between C_f and T31's output capacitor C_{oss}, R_f is external resistance of C_f. According to [13], when $C_f \gg C_{oss}$ and $L_{s1} \gg L_{s2}$ (the conditions are normally archived with decoupling capacitor is put near to the MOSFET module), the loop's stray inductance will be fully decoupled and the oscillations can be studied separately for each individual loop. The oscillation voltage v_{ds} over T31 can be modeled with the equation:

$$v_{ds} = VDC + \widehat{V_L}.\sin(\omega_L t + \emptyset_L).e^{-\omega_L t.\tau_L} + \widehat{V_H}.\sin(\omega_H t + \emptyset_H).e^{-\omega_H t.\tau_H} \qquad (1)$$

With the angular frequency of LF and HF loop: $\omega_L \approx \frac{1}{\sqrt{L_{s1}.C_f}}$, $\omega_H \approx \frac{1}{\sqrt{L_{s2}.C_{oss}}}$,

the initial phase of LF, HFoscillations: $\emptyset_L, \emptyset_H$
oscillation amplitude of LF and HF loop:

$$\widehat{V_L} = \sqrt{\frac{L_{s1}}{C_f}}\, I_L, \; \widehat{V_H} = \sqrt{\frac{L_{s2}}{C_{oss}}}\, I_L \tag{1.a}$$

With IL is load current at the turn off moment
Damping factor of LF and HF loop:

$$\tau_L = \frac{R_{s1}+R_f}{2}\sqrt{\frac{C_f}{L_{s1}}}, \; \tau_H = \frac{R_{s2}+R_f}{2}\sqrt{\frac{C_{oss}}{L_{s2}}} \tag{1.b}$$

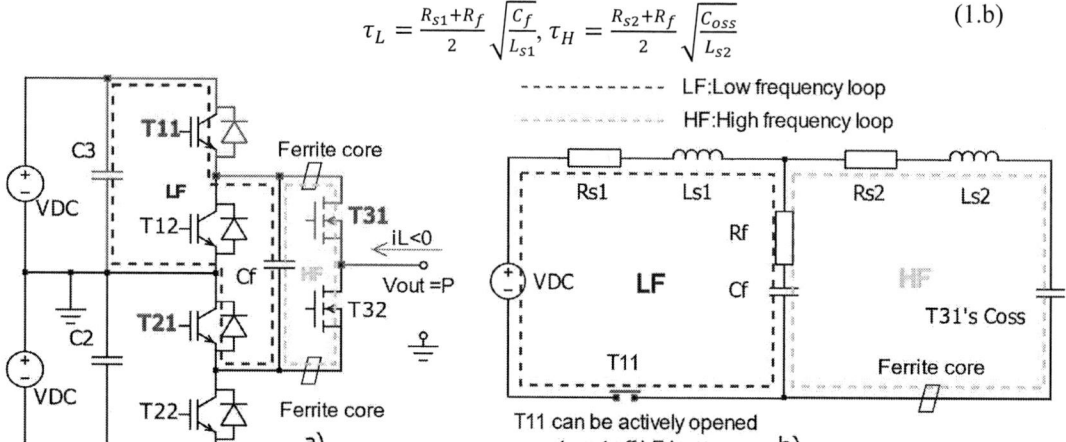

Fig. 2: a) Two oscillation circuits are formed by decoupling capacitor C_f, b) LF, HF's equivalent circuits, R_f is C_f's external resistance it can increase the damping ratio of both LF, HFloop. MnZn Ferrite core damp HF loop. T11 can be turned off to actively cut off the LF oscillation.

From equation (1.b), the resistance of the loop (R_{s1}, R_{s2}) can be increased in order to damp the oscillation for HF and LF loop. But increasing the resistance of the busbar or terminals will lead to high conduction losses due to DC current. By using capacitors which have external parasitic resistance R_f either than only the conventional film snubber capacitor, the resistance of both HF and LF loop can be increased without losses cause by DC current. The HF loop also can be damped effectively with ferrite core (FC) because the HF's frequency range of the switching is in the damping frequency range of the common ferrite cores on the market (from 1MHz to 50 MHz). Another way to damp the LF loop is actively cut off the oscillation loop by turning off one of the IGBT (for example T11 or T21). The details of these damping methods are explained in the next section of this paper.

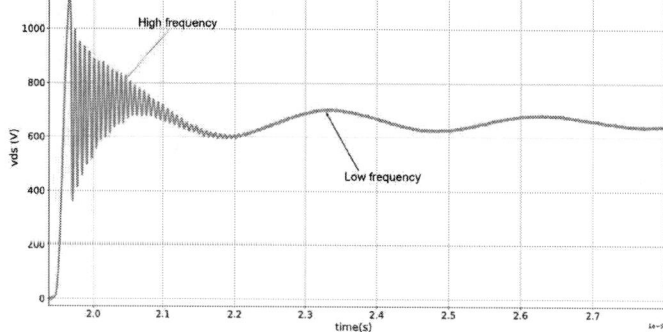

Fig. 3: Low and high frequency (LF, HF) oscillations on T31's voltage during turn-off

2. High Frequency oscillation damping

Damping switching oscillation with snubber RC circuit or ferrite cores are preferred than conventional reducing switching speed technique because of lower switching losses [15-18]. According to the snubber design guideline in [16], the RC snubber circuit values can be estimated in equations (2), (3): with C_{sn}, R_{sn}, PR_{sn} are snubber capacitor, snubber resistor, power losses on snubber resistor values.

IL_{max} is the maximum load current, Vds_{peak} is the peak voltage under safe operation area (SOA) of the SiC MOSFET module, f_{sw} is the switching frequency, t_r is the turn-on current rise time.

$$C_{sn} = \frac{(L_{s1} + L_{s2})IL_{max}^2}{\left(Vds_{peak} - VDC\right)^2} \tag{2}$$

$$R_{sn} = \frac{1}{6 \cdot C_{sn} \cdot f_{sw}} \tag{3}$$

$$PR_{sn} = \frac{1}{2} \cdot C_{sn} \cdot f_{sw} \cdot \left(V^2 ds_{peak} - VDC^2\right) + 1.125 \cdot f_{sw} \cdot \frac{(L_{s1} + L_{s2})^2 IL_{max}^2}{t_r \cdot R_{sn}} \tag{4}$$

In this 500kW 2SiC hybrid ANPC experimental setup, the stray inductance loop is quite large, around 115nH, $IL_{max} = 450$A, VDC = 750V, $Vds_{peak} = 1200\ V$, switching frequency $f_{sw} = 10\ kHz$. From equation (4), it can be seen that due to high load current, high switching speed and large parasitic inductance, the power losses on the snubber resistor is huge. This make the damping method using RC snubber is not practical in this scenario. In contract to RC snubber, some ferrite core materials can provide high equivalent resistance at high frequency range (>1MHz) and low equivalent resistance at low frequency range which can increase the damping ratio at high frequency and lower losses at low frequency range (< 1MHz) [19], those special properties of ferrite cores best fit for high power, high frequency oscillation damping. In the experiment, MnZn ferrite cores were placed on the high frequency loop of the oscillation circuit (Fig 2.a). Voltage vds across T31 was measured during turning off. The result shows in Fig 4.b, the high frequency oscillation was totally removed. Ferrite cores also introduced more stray inductance to the commutation loop that increased the over voltage more than the case without ferrite cores.

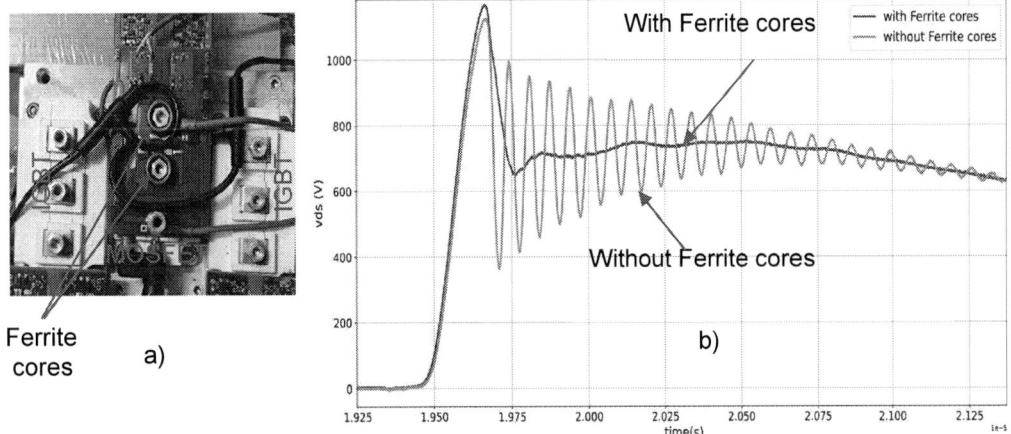

Fig. 4: High frequency (HF) damping effect of MnZn Ferrite cores. a) real implementation in 2SiC ANPC layout. b) voltage of T31 during turn-off with and without ferrite cores.

3. Low frequency oscillation active cut-off switching scheme

From Fig.2a, the oscillation circuit of LF makes a closed loop when both T11, T21 are on. The low frequency oscillation or the decoupling oscillation can be actively cut off by turning off either T11 or T21 if one of them does not carry the load current. In this paper, the switching states in table I.b is proposed based on that principle. The important thing is how to transit between the switching states safely with lowest switching losses and independent of the load current direction. One example for the transition from state P to ZP as describe in table II. is analyzed.

In the proposed switching scheme, the IGBTs are switched at the switching frequency but there are no switching losses because the switching is done at no load current. Moreover, the switching scheme is independent from the load current's direction because the current's location (upper or lower side of HANPC) is always defined when T31(upper side) or T32 (lower side) is on. These other transition $ZN \leftrightarrow N, N \leftrightarrow Z, ZN \leftrightarrow ZP$ use similar approach with the above example.

Table II: Transition from P to ZP

Step	Current direction IL < 0	Current direction IL > 0	Note
1			- State P: T11, T31: on. - Vout = VDC (P). - Load current is the red line of the circuit. The red color switches' names are the on state switches. When T31 is on, load current is always located on the upper side of ANPC with all its direction.
2			- T11, T31: on, T21 complemented to T11 is turned on, Vout = VDC. - LF loop is closed, LF starts oscillating - Load current is located on upper side of ANPC. T21(lower side of ANPC) is always turned on at no load current independent from IL's direction => no switching losses.
3			- Dead time: T11, T21: on, T31 is turned off - The oscillation in HF and LF are triggered when T31 is off. Load current is on T11 (upper side) if IL<0 and on T21 (lower side) if IL >0. - The switching state is independent from IL's direction.
4			-T11, T21: on, T32 is turned on -Vout = 0 (ZP) -When T32 is on, the load current is always relocated to the lower side of ANPC, no longer go through T11 with all IL directions. The switching state is independent from IL's direction.
5			- State ZP, T21, T32: on, T11 is turned off. - T11 is actively turned off to cut off the LF oscillation loop. - load current is located at lower side of ANPC. So T11 is always turned off at no load, no turn off losses. The switching state is independent from IL's direction.

The switching scheme is implemented in a FPGA and DSP controller board and do the experiment with the 2SiC hybrid ANPC. The result was measured at IL = 550A, VDC = 650V with ferrite cores the result showed that the low frequency oscillation or the decoupling oscillation is actively cut off at the time T11 is off when the inverter state transferred from P to ZP (Fig. 5).

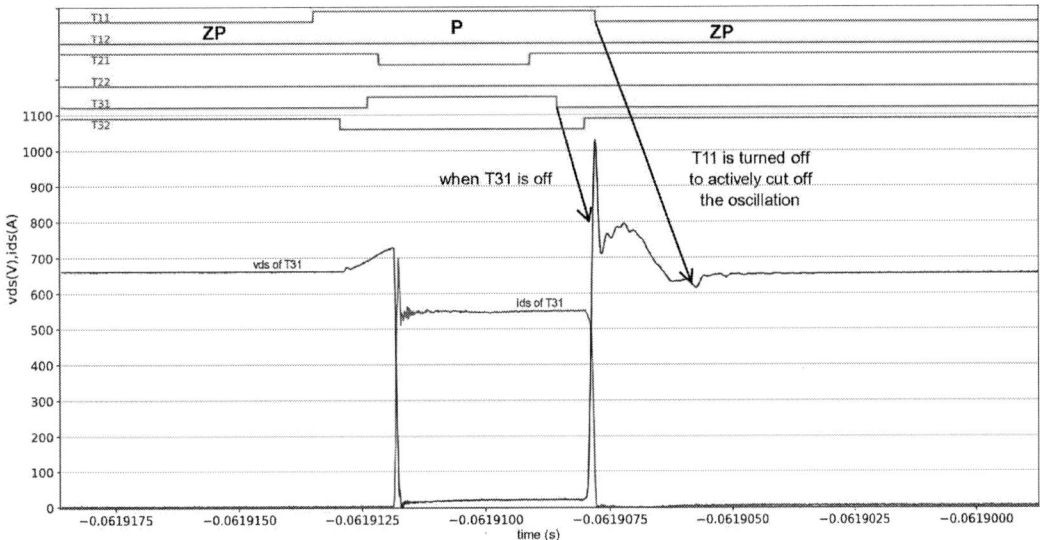

Fig 5: The decoupling oscillation (LF) on T31 is actively cut off during the transition from P to ZP.

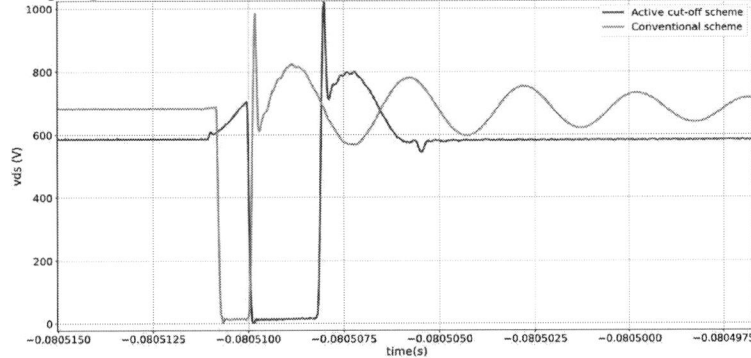

Fig 6: Compare conventional switching scheme and active cut-off scheme effect on the T31'svoltage oscillation when ANPC changed from state P to ZP (measured at different VDC)

Fig.6 shows the voltage of T31 during transition from state P to ZP with conventional switching scheme (measured at VDC = 750V) and with active cut-off scheme (measured at VDC = 600V). The decoupling oscillation or LF is cut off successfully right after the inverter reaching its new state ZP.

4. Decoupling capacitor

The decoupling capacitor C_f plays an important role in 2SiC hybrid ANPC topology. The value of C_f has to be large enough to decouple the large commutation loop and lower the oscillations' amplitude so that it is smaller than the overvoltage of the SiC MOSFET during turn-on and turn-off transient. During switching, there is current that charges and discharges C_f and causes losses on the internal capacitor's parasitic resistance (ESR) which can be heated up and destroy the capacitor. Those charging and discharging current are proportional to the capacitor value and hence set the upper limit value of the decoupling capacitor. The decoupling capacitor value can be estimated:
From equation (1), the maximum oscillation amplitude of the voltage across SiC MOSFET can be:

$$v_{dsmax} = VDC + I_L \left(\sqrt{\frac{L_{s1}}{C_f}} + \sqrt{\frac{L_{s2}}{C_{oss}}} \right) \qquad (5)$$

If we assume that the voltage on the decoupling capacitor C_f is constant and equal to VDC during the switching transient. The voltage across the SiC MOSFET during turn-off after fully decoupled is:

$$v_{ds} = VDC - L_{s2}\frac{di_{ds}}{dt} \tag{6}$$

The value of C_f should be chosen so the maximum oscillation amplitude is smaller than the overvoltage of SiC MOSFET after decoupling. From (5), (6)

$$C_f \geq \frac{I_L^2 L_{s1}}{\left(I_L \cdot \sqrt{\frac{L_{s2}}{C_{oss}}} + L_{s2} \cdot \frac{di_{ds}}{dt}\right)^2} \tag{7}$$

During turn-on, the overvoltage on SiC MOSFET in the half-bridge depends on the reverse recovery current speed $\frac{di_{rr}}{dt}$. For SiC MOSFET the reverse recovery current speed is significantly fast at high temperature [22]. The overvoltage on SiC MOSFET during turn-on can be calculated:

$$v_{ds} = VDC + L_{s2}\frac{di_{rr}}{dt} \tag{8}$$

In the same manner like turn-off case. The value of C_f for turn-on

$$C_f \geq \frac{I_L^2 L_{s1}}{\left(-I_L \cdot \sqrt{\frac{L_{s2}}{C_{oss}}} + L_{s2} \cdot \frac{di_{rr}}{dt}\right)^2} \tag{9}$$

To protect the SiC MOSFET module during switching, the value of the decoupling capacitor should be satisfied both turn-on and turn-off cases.

$$C_f \geq Max\left[\frac{I_L^2 L_{s1}}{\left(I_L \cdot \sqrt{\frac{L_{s2}}{C_{oss}}} + L_{s2} \cdot \frac{di_{ds}}{dt}\right)^2}, \frac{I_L^2 L_{s1}}{\left(-I_L \cdot \sqrt{\frac{L_{s2}}{C_{oss}}} + L_{s2} \cdot \frac{di_{rr}}{dt}\right)^2}\right] \tag{10}$$

According to [20] the RMS current goes through the decoupling capacitor can be estimated

$$I_{cRMS} = I_L\sqrt{f_{sw}\left(\frac{L_{s1}}{R_{s1} + ESR} + \frac{1}{4}C_f(R_{s1} + ESR)\right)} \tag{11}$$

The power losses on decoupling capacitor is

$$P_c = ESR \cdot I_{cRMS}^2 = ESR \cdot I_L^2 \cdot f_{sw} \cdot \left(\frac{L_{s1}}{R_{s1} + ESR} + \frac{1}{4}C_f(R_{s1} + ESR)\right) \tag{12}$$

With ESR is the internal parasitic resistance of decoupling capacitor C_f.

(10) shows the minimum value of the decoupling capacitor to protect the SiC MOSFET module from overvoltage. From (1.a), the larger C_f value is, the smaller amplitude of low frequency oscillation. But the upper limit value of C_f is limited by its internal parasitic resistance ESR can be seen from equation (12) and the cooling conditions.

From equation (1.b), there is another option to increase the damping factor of both low and high frequency loop is increasing the external resistor R_f of C_f. This external resistor can be physically installed by using stainless steels plates as can be seen in Fig.7. Because stainless steel has resistivity 40 times higher than copper, this material increases R_f value enough to damp the oscillation without increasing internal temperature of the capacitor.

Fig 7. shows a capacitor assembly with 1.3 µF high power film capacitor with ESR = 0.3 mΩ, max Irms = 850A, 900 Vrms rated voltage. The capacitor is connected to the SiC MOSFET module using 2 stainless steel plates. The voltage measured on T31 during turning off in case of using 1.5 µF normal film snubber capacitor and in case of using this special capacitor assembly. The stainless steel plates showed a clearly damping effect on both 2 oscillation circuits.

During continuous running the inverter at full load current 300Arms for two hours, the capacitor's temperature reached maximum only 28°C at 20°C ambient temperature.

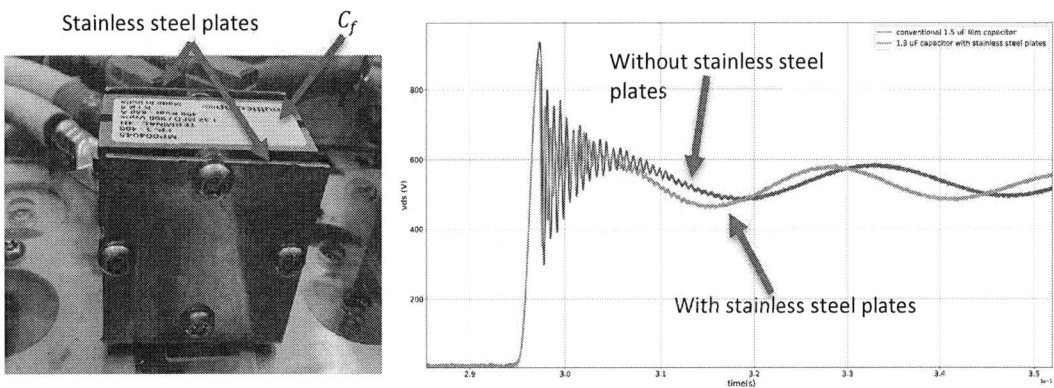

Fig 7: Decoupling capacitor with stainless steel plates assembly and the damping effect on both low and high frequency oscillation.

5. 500 kW 2SiC hybrid ANPC experiment.

The 500kW 2SiC ANPC (Fig. 8) was built to verify the damping concept. The inverter specification shows in table III.

Table III: 500kW 2SiC hybrid ANPC specification

Parameters	Value
Output power	520 kVA/500kW
DC link voltage	1500V
Output voltage	3Phases, 50Hz, VLL = 1000V
Output current	300Arms
Switching frequency	10kHz
Cooling method	Forced-air cooling

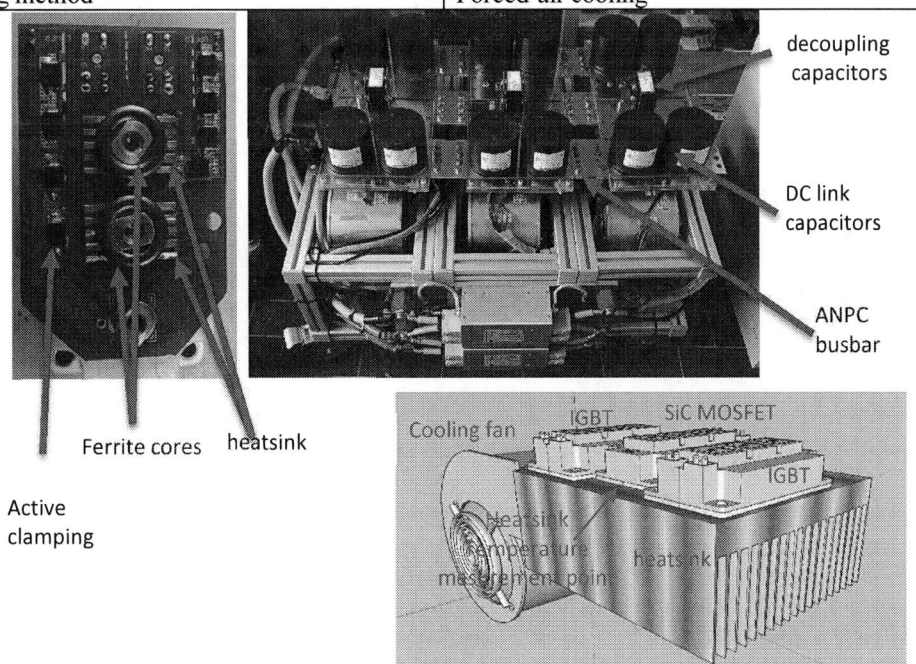

Fig 8: 500 kW 2SiC ANPC experiment inverter with forced-air cooling setup

To verify the damping effect of the damping scheme and check the temperature of critical components of the 2SiC ANPC during continuous running, one phase of the inverter was setup like in Fig 9. The inverter was run continuously for 2 hours at 305 Arms load current without any problem. The temperatures of the critical components were also marked on Fig 9.

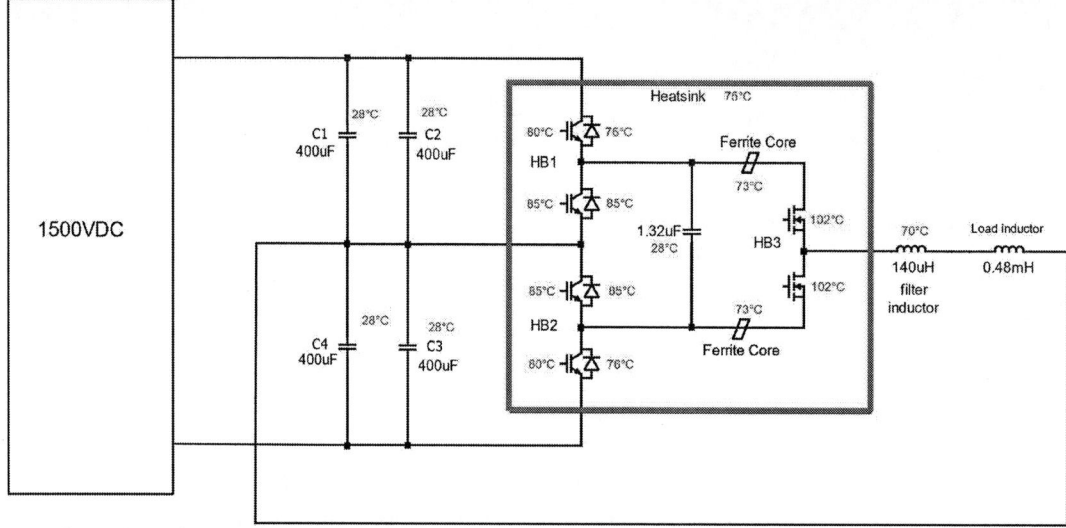

Fig 9: One phase 2SiC ANPC continuous test setup and its critical components' temperature at 305Arms for 2hours continuous running

The losses of all switches were simulated in Plecs at 305 Arms are shown in Table 4. And their junctions's temperature were also calculated in Table IV.

Table IV: Semiconductors' power losses and their junction temperature at 305 Arms load current.

	T11	T12	T21	T22	T31	T32	D11	D12	D21	D22	$\sum P$
Ploss(W)	92	183	183	92	228	228	14	102	102	14	1238
Tj(°C)	80	85	85	80	102	102	76	85	85	76	

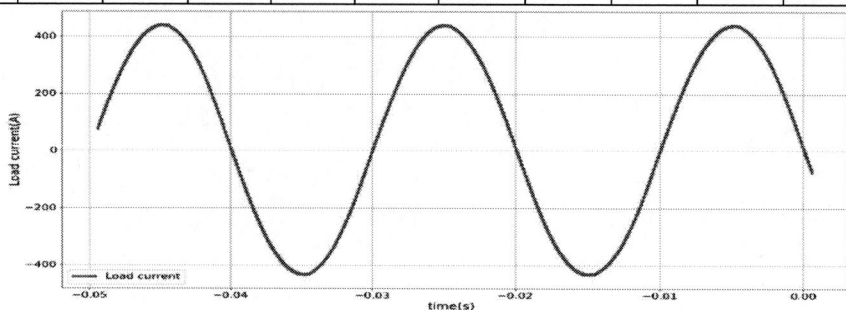

Fig 10: One phase load current waveform at 305 Arms

The experiment successfully archived the target designed at full load current 300Arms and had efficiency 99.25% (only semiconductors' losses) without overheating components. The switching oscillations are well damped by the special active cut-off switching scheme and ferrite cores. SiC MOSFET's junction temperature were still in the safe operating range.

Conclusions

In this paper, an experimental 500 kW 2SiC hybrid ANPC was presented and the switching oscillations of the SiC MOSFET module were successfully damped by using ferrite cores and a special active cut-off switching scheme. The paper also shows the method to choose the value of the decoupling capacitor and the possibility to damp the switching oscillations with stainless steel external resistance. The inverter was run at full load current for two hours without overheating its critical components. And the efficiency can reach 99.25% at rated current output.

References

[1] Christian R. Müller: hybrides ANPC-Modulkonzept für Solarwechselrichter: SiC-MOSFET trifft IGBT Elektronik Power Oktober 2019.

[2] Zhijian Feng: A High-Efficiency Three-Level ANPC Inverter Based on Hybrid SiC and Si Devices Energies 2020, 13, 1159; doi:10.3390/en13051159

[3] Chen, Q.; Wang, Q.; Li, G.; Ding, S. The Control of Unequal Power Losses Distribution in Three-Level Neutral-Point-Clamped VSC. In Proceedings of the 2012 15th International Conference on Electrical Machines and Systems (ICEMS), Sapporo, Japan, 21–24 October 2012.

[4] Bruckner, T.; Bemet, S. Loss Balancing in Three-Level Voltage Source Inverters applying Active NPC Switches. In Proceedings of the 2001 IEEE 32nd Annual Power Electronics Specialists Conference (IEEE Cat. No.01CH37230), Vancouver, BC, Canada, 17–21 June 2001.

[5] Zhang, D.; He, J.; Pan, D. A Megawatt-Scale Medium-Voltage High Efficiency High Power Density "SiC+Si" Hybrid Three-Level ANPC Inverter for Aircraft Hybrid-Electric Propulsion Systems. In Proceedings of the IEEE Energy Conversion Congress and Exposition (ECCE), Portland, OR, USA, 23–27 September 2018.

[6] Zhang, L.; Lou, X.; Li, C.; Wu, F.; Gu, Y. Evaluation of Different Si/SiC Hyrbid Three-Level Active NPC Inverters for High Power Density. *IEEE Trans. Power Electron.* 2019.

[7] Feng, Z.; Zhang, X.; Yu, S.; Wang, J. Loss Analysis and Measurement of ANPC Inverter Based on SiC & Si Hybrid Module. In Proceedings of the IEEE International Power Electronics and Application Conference and Exposition (PEAC), Shenzhen, China, 4–7 November 2018

[8] Zhijian Feng, Xing Zhang *, Jianing Wang and Shaolin Yu, A High-Efficiency Three-Level ANPC Inverter Based on Hybrid SiC and Si Devices, Energies2020,13, 1159; doi:10.3390/en13051159.

[9] Bong Hyun Kwon 1,2, Sang-Hun Kim 2 , Seok-Min Kim 2 and Kyo-Beum Lee 2, Fault Diagnosis of Open-Switch Failure in a Grid-Connected Three-Level Si/SiC Hybrid ANPC Inverter, Electronics 2020, 9, 399; doi:10.3390/electronics9030399

[10] Pan, D.; Zhang, D.; He, J.; Immer, C.; Dame, M. Control of MW-Scale High-Frequency "SiC+Si" Multilevel ANPC Inverter in Pump-Back Test for Aircraft Hybrid-Electric Propulsion Applications. *IEEE J. Emerg. Sel.Top. Power Electron.* 2020.

[11] Sahan, B.; Mueller, C.R.; Lenze, A.; Czichon, J.; Slawinski, M. Combining the benefits of SiC T-MOSFET and Si IGBT in a novel ANPC power module for highly compact 1500-V grid-tied inverters. In Proceedings of the PCIM Europe 2019, Nuremberg, Germany, 7–9 May 2019.

[12] Alireza Kouchaki; Giorgo Kapino; Morten Nymand Design of a High Frequency 3-Phase 3-Level Hybrid Active-NPC Inverter, 2018 20th European Conference on Power Electronics and Applications (EPE'18 ECCE Europe), pp.1-10

[13] Yoann Pascal; Denis Labrousse; Mickaël Petit; François Costa, Study of the Impedance of the Bypassing Network of a Switching Cell – Influence of the Positioning of the Decoupling Capacitors, 2019 IEEE International Workshop on Integrated Power Packaging (IWIPP) pp.120-124

[14] The IEC/CISPR 11, EN 55011 (Industrial, scientific and medical (ISM) radio-frequency equipment – Electromagnetic disturbance characteristics – Limits and methods of measurement)

[15] Tianjiao Liu, A Survey on Switching Oscillations in Power Converters, IEEE JOURNAL OF EMERGING AND SELECTED TOPICS IN POWER ELECTRONICS, VOL. 8, NO. 1, MARCH 2020, pp.893-908

[16] Rahul S. Chokhawala, Switching Voltage Transient Protection Schemes for High-Current IGBT Modules, IEEE TRANSACTIONS ON INDUSTRY APPLICATIONS, VOL. 33, NO. 6, NOVEMBER/DECEMBER 1997, pp.1601-1610

[17] Ivan Josifovic, Improving SiC JFET Switching Behavior Under Influence of Circuit Parasitics, IEEE TRANSACTIONS ON POWER ELECTRONICS, VOL. 27, NO. 8, AUGUST 2012, pp.3843-3853

[18] Tianjiao Liu, Experimental and Modeling Comparison of Different Damping Techniques to Suppress Switching Oscillations of SiC MOSFETs, 2018 IEEE Energy Conversion Congress and Exposition (ECCE), pp.7024-7031

[19] Adrian Suarez, Effectiveness Assessment of a Nanocrystalline Sleeve Ferrite Core Compared with Ceramic Cores for Reducing Conducted EMI, Electronics 2019, 8, 800

[20] Yi Zhang, Saed Sobhani, Rahul Chokhawala, Snubber Considerations for IGBT applications, International Rectifier Corporation Applications Engineering

[22] Kang Peng, Characterization and Modeling of SiC MOSFET Body Diode,2016 IEEE Applied Power Electronics Conference and Exposition (APEC), pp.2127-2135

Multi Busbar Sub-module Modular Multilevel STATCOM with Partially Rated Energy Storage Configured in Sub-stacks

Chuantong Hao, Wenhao Ma, Michael Merlin, Paul Judge, Stephen Finney
Institute of Energy System, School of Engineering, University of Edinburgh
Edinburgh EH9 3DW, Scotland, UK
Phone: +44 (0) 0736 75 86886
Email: Chuantong.Hao@ed.ac.uk

Acknowledgments

This work is supported by the joint scholarship of China Scholarship Council (CSC) and the University of Edinburgh.

Keywords

≪Modular Multilevel Converters (MMC)≫, ≪Multi-Busbar Submodule≫, ≪Mixed Modulation≫, ≪Static Synchronous Compensator (STATCOM)≫, ≪Partially Rated Energy Storage≫.

Abstract

This paper presents a modular multilevel STATCOM with partially rated energy storage configured in sub-stacks based on full bridge multi busbar Sub-module (SM). The soft-paralleling mechanism and doubled paths increase the current limit. The lower level controller of the proposed topology is detailed introduced and the performances are compared with a conventional single-busbar full bridge STATCOM controlled by classic SM sorting based low level controller. When providing reactive power compensation, power losses is reduced by 14.9%, and the maximum SM voltage deviation is reduced by 35.2%, while active power capability is further improved from power factor = 0.5 to power factor = 0.7 with the proposed control framework of the MBSM STATCOM.

Introduction

Static compensator (STATCOM) is commonly applied for supporting the stability of the grid by continuously absorbing or releasing reactive power in response to voltage variations [1]. Ancillary service such as inertial and frequency support is also provided when integrating STATCOM with Energy Storage (ES) Systems as active power can be extracted from or injected to ES [2]. Besides, Modular multilevel Converter (MMC) [3], [4] allows ES elements to be distributed in Sub-Modules (SM) so that the management of ES conditions can be achieved coping with MMC control algorithm.

Increasingly attention has been attracted for applications of the STATCOM with ES in distributed energy networks. Recent works have proved adapting MMC for ES will reduce harmonic distortion, switching frequency and power losses [5], [6]. Control algorithms that offer the flexibility to directly manipulate the active power components for state of charge (SoC) balancing of the batteries are presented in [7]. A delta-connected partially-rated ES STATCOM and the control structure are developed in [8]. The authors conclude that converters rated at 1 pu active power require 69% of the full bridge (FB) SMs to be integrated with ES (ES-SMs). The ES interface will be modelled as a controlled current source in simulation and analysis as its specific structure is not the focus of this paper.

Generally, additional circulating current is injected into the delta-configured loop in a STATCOM to enhance the voltage balancing of SMs and ES-SMs and increase the amount of power that can be extracted [9], [10]. However, the phase-leg current limit determines that the circulating current can not

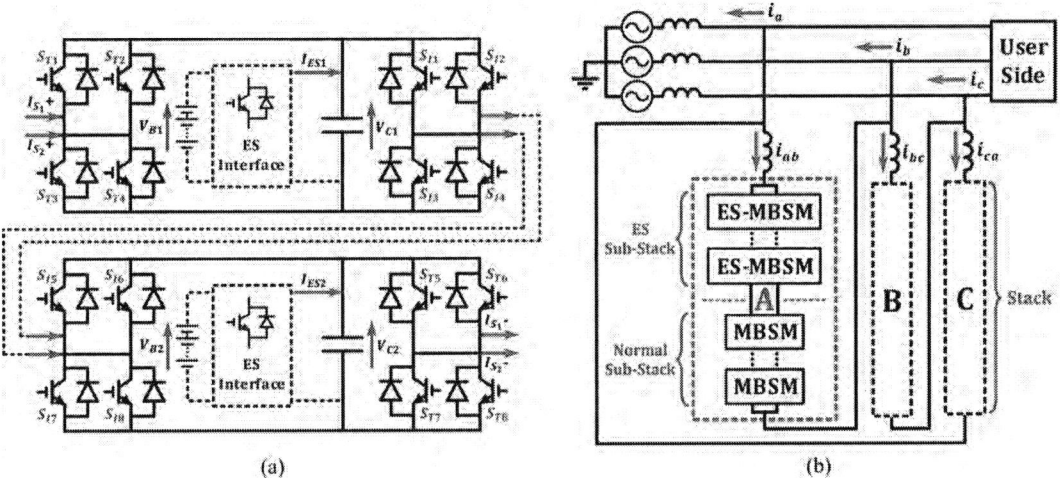

(a)   (b)

Fig. 1: The proposed topology: (a) ES-MBSMs depicting the connection manner of ES interface and two adjacent MBSMs, (b) The delta-configured MBSM PRES-STATCOM with an ES Sub-Stack and a normal Sub-Stack.

be too large, which prevents further reduction of the required ES-SMs fraction. Besides, the iterative loop algorithm calculates the gate signals at every controller step. Not only will the average switching frequency of semiconductors be large, but also the computational complexity increases sharply with the increase of the number of SMs [11].

This paper proposes a multi busbar sub-module (MBSM) based delta-configured partially-rated ES sub-stack STATCOM (PRES-STATCOM) and the corresponding mixed low level control framework. Capacitors in different MBSMs can be paralleled for energy balancing with soft-paralleling mechanism applied. The switching frequency, together with the maximum SM voltage deviation can be reduced to different extent depending the operation set-points. Since current is able to flow in two paths, MBSMs allow larger circulating current and consequently less fraction of ES MBSMs is required to release the same amount of active power compared with conventional FB based PRES-STATCOM.

The rest of the paper is organized as follows. Section 2 presents the converter topology and states of MBSMs. Section 3 introduces the low level control framework. Section 4 provides simulation results and analysis. Finally, the conclusion is presented in Section 5.

Converter Topology

Structure of MBSMs and the Proposed Converter

The structure of the ES-MBSMs together with the connection manners of the ES interface and two adjacent MBSMs are presented in Fig. 1 (a). Compared with the conventional FBSM, the MBSM consists of twice the number of interfaces, semiconductors and busbars [12] [13]. The capacitors in two adjacent MBSMs can thus be connected in parallel. As illustrated by the dotted lines, stack current is divided into two parts when flowing through MBSMs. The ES interface is connected to the SM capacitor, which could be directly-connected wires or a DC/DC converter. When the interface is disabled, the ES-MBSM becomes a capacitor-only MBSM that can only provide reactive power during normal operation.

Zero sequence circulating current results in delta-configured structure attracts more attentions than its star-configured counterpart. The circulating current amongst all three phases creates additional degrees of freedom for balancing capacitors voltage and managing SoC of the ESs. The MBSM based delta-configured PRES-STATCOM is presents in Fig. 1 (b). The stack in phase A is extended to show the detail of how different type of MBSMs are placed. ES-MBSMs and normal MBSMs are placed into two groups, forming an ES sub-stack and a normal sub-stack. The two sub-stacks can output voltage in opposite polarities so that SM voltage balancing is promoted. The MBSM STATCOM is different from the FBSM STATCOM in structure only inside the stacks. So the same higher level control method can be applied to both topology while their low level controller should be designed separately. The detailed low level controller will be introduced in the next section.

States of MBSMs

Conventional Sub-module Based States

MBSMs are able to operate in three modes as the same of conventional FBSMs when the same gate signals are assigned to the adjacent half-bridges. Therefore, controllers for FBSMs can also be applied to MBSMs with minor adjustments. Take the upper MBSM in Fig. 1 (a) as an example, the modes and the switching states of semiconductors are:

- Positive Voltage Mode (1),   when S_{T1}, S_{T2} and S_{I3}, S_{I4} are on while other switches stay off.
- Negative Voltage Mode (-1),   when S_{T3}, S_{T4} and S_{I1}, S_{I2} are on while other switches stay off.
- Zero Voltage Mode (0),   when S_{T1}, S_{T2} and S_{I1}, S_{I2} are on while other switches stay off.

Soft-paralleling Based States

Multi busbars can form two paths to connect the anodes and cathodes of the capacitors in two adjacent MBSMs. The capacitors can be paralleled and their voltage will be equal. The example in Fig. 1 (a) is applied again to illustrate. States of MBSMs are defined in two categories in terms of stack terminals ($S_{T1}, ..., S_{T8}$) and interconnections ($S_{I1}, ..., S_{I8}$). The modes of terminal and interconnections are presented in Fig. 2 (a) and Fig. 2 (b) respectively. The Soft-Parallel Modes are highlighted here to avoid inrush current caused by the directly paralleling of two capacitors with voltage difference. The directional conduction characteristics of anti parallel power diode limit the inrush current in the envelope formed by the stack current. Two semiconductors in the same half-bridge are blocked in soft-parallel mode, resulting in leakage current passing through the blocked semiconductors, Consequently, the soft-parallel mode can only be activated when the stack current is larger than the leakage current.

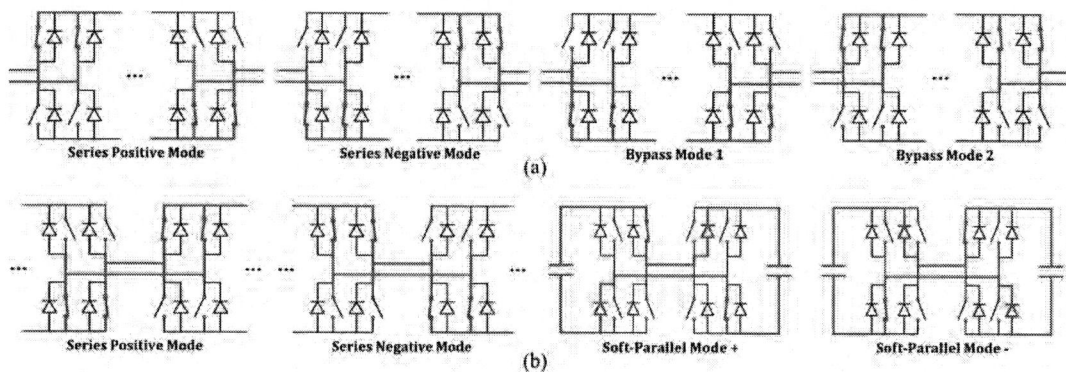

Fig. 2: MBSM states: (a) Terminal states, (b) Interconnection states.

Low Level Control System

Conventional Sub-module Based Controller

When MBSMs act as conventional FBSMs, the modes of two adjacent MBSMs can be opposite to accelerate the votlage balancing. Similar to the method proposed in [2] for PRS-MMC, the controller is an iteration algorithm which is powerful in terms of reducing the voltage differences of different capacitor at the cost of increasing computation complexity and switching frequency. The conventional sub-module based voltage balancing algorithm is summarized in **Algorithm 1**. The controller evaluates the preferential states of all MBSMs based on the SM voltages, stack current and voltage references in every controller step. The principle of preferential states is discharging the MBSM whose $V_{SM} > V_{aver}$

Algorithm 1 Conventional Sub-module based voltage balancing algorithm

Input: V_{stack}: Stack voltage reference; V_{SM}: MBSM capacitor voltages; I_{stack}: Stack current; N_{SM}: Number of MBSMs.

Output: Gate signals

  Calculate average SM voltage V_{aver}, calculate sorted index R by ranking $|V_{SM} - V_{aver}|$;

  Initial $V_{ref}(1) = V_{stack}$, $V_{avail}(1) = \sum V_{SM}$;

  for $i = 1 \rightarrow N_{SM}$ **do**

    Set $j = 1$, calculate preferential states array S_{pref}: if $(V_C(R(j)) - V_{aver}) \cdot I_{stack} \leq 0$, $S_{pref} = [1, 0, -1]$, else $S_{pref} = [-1, 0, 1]$;

      while $(j \leq 3) \wedge (S_{pref}(j)$ is **NOT** accepted based on (1)) **do**

        $j++$;

      end while

    $V_{ref}(i+1) = V_{ref}(i) - S_{pref}(R(i)) \cdot V_{SM}(R(i))$, $V_{avail}(i+1) = V_{avail}(i+1) - V_{SM}(R(i))$;

  end for

  Translate states to gate signals.

and charging the MBSM whose $V_{SM} \leq V_{aver}$ with stack current as long as the sum voltage of the rest MBSMs is higher than the remaining output voltage reference. The feasibility of the most preferential state is determined by equations (1).

$$Accept\ State = \begin{cases} 1, & if\ |V_{ref} - V_{SM}(R(j))| \leq V_{avail} - V_{SM}(R(j))/2 \\ 0, & if\ |V_{ref}| \leq V_{avail} - V_{SM}(R(j))/2 \\ -1, & if\ |V_{ref} + V_{SM}(R(j))| \leq V_{avail} - V_{SM}(R(j))/2 \end{cases} \tag{1}$$

Soft-paralleling Based Controller

Sub-stack Voltage References

Algorithm 2 Soft-paralleling based sub-stack voltage reference creation algorithm

Input: V_{stack}: Stack voltage reference; V_C: MBSM capacitor voltages; I_{stack}: Stack current; N_{ES}: Number of ES MBSMs; N_{Cap}: Number of Normal MBSMs;

Output: V_{ES}: ES sub-stack voltage reference; V_{Cap}: Normal sub-stack voltage reference; $Flag$: Flag indicating whether control the whole stack (= 1) or two sub-stacks (= 0);

1: Calculate average MBSM voltage V_{aver}, average ES-MBSM voltage V_{ESaver} and average normal MBSM voltage V_{Caver}, calculate the sign of stack current $sgn(I_{stack})$;

2: **if** $(sgn(I_{stack}) \cdot V_{stack}/V_{aver} \geq N_{Cap}) \vee (sgn(I_{stack}) \cdot V_{stack}/V_{aver} \leq -N_{ES}) \vee (V_{ESaver} < V_{Caver})$ **then**

3:    $Flag = 1$;

4: **else**

5:    $Flag = 0$;

6:    **if** $(sgn(I_{stack}) \cdot V_{stack}/V_{aver} \geq N_{Cap} - N_{ES})$ **then**

7:      $V_{ES} = V_{stack}/V_{aver} - sgn(I_{stack}) \cdot N_{Cap}$, $V_{Cap} = sgn(I_{stack}) \cdot N_{Cap}$;

8:    **else**

9:      $V_{ES} = -sgn(I_{stack}) \cdot N_{ES}$, $V_{Cap} = V_{stack}/V_{aver} + sgn(I_{stack}) \cdot N_{ES}$;

10:    **end if**

11: **end if**

The first step of the soft-paralleling based controller is calculating the voltage references of the whole stack or the sub-stacks. Voltage references for two sub-stacks will be generated when the required voltage output of the whole stack is not exceed the voltage capability of sub-stacks, as presented in **Algorithm 2**. As ES currents are all charged to SM capacitors in the ES sub-stack, their voltages are more likely to be higher than the rated value. Therefore, the ES sub-stack will be discharged by stack current while the normal sub-stack will be charged to reduce the average voltage differences.

Mixed Modulation Framework

Algorithm 3 Soft-paralleling based gate signals generation algorithm

---

Input: V_{ref}: Voltage reference of the stack (or sub-stack); V_{sm}: MBSM voltages of the stack (or sub-stack); *Carrier*: Carriers for all MBSMs in the stack (or sub-stack); N_{sm}: Number of MBSMs in the stack (or sub-stack); I_{stack}: Stack current;

Output: Gate signals

1: **for** $i = 1 \rightarrow N_{sm}$ **do**
2:    Assign $Carrier(i)$ to the corresponding terminal or interconnections;
3:    Compare $V_{ref}/\sum V_{sm}$, Get states: (1) $State(i) = 1$, if $N_{ref} > Carrier(i)$, (2) $State(i) = 0$, if $Carrier(i) \geq N_{ref} > -C(i)$, (3) $State(i) = -1$, if $N_{ref} \leq -Carrier(i)$;
4: **end for**
5: Translate states to gate signals.

The second step of the soft-paralleling based controller is illustrated in **Algorithm 3**. Overall, the principle is to generate gate signals with the voltage references and the carriers. If $Flag = 1$, the stack will be controlled as a whole part to track the voltage reference. If $Flag = 0$, the reference voltages of two sub-stacks will be applied to allow the sub-stacks generate different voltage outputs, as long as the sum voltage is equal to the total voltage reference.

Compared to **Algorithm 1**, the computational complexity of the soft-paralleling based controller is greatly reduced. Besides, the average semiconductor switching frequency is also limited by the carrier frequency. The phase disposition and shift carrier (PDSC) modulation framework is proposed here to enhance the voltage balancing of two sub-stacks while the carriers are assigned to the whole stack. PDSC is a mixed framework of phase disposition carrier modulation (PDC) [14] and phase shift carrier modulation (PSC) [15]. The carrier with the largest average value is assigned to the interconnection between two sub-stacks to make it operate longer time at soft-paralleling mode, as illustrated by the red line in Fig. 3 (c).

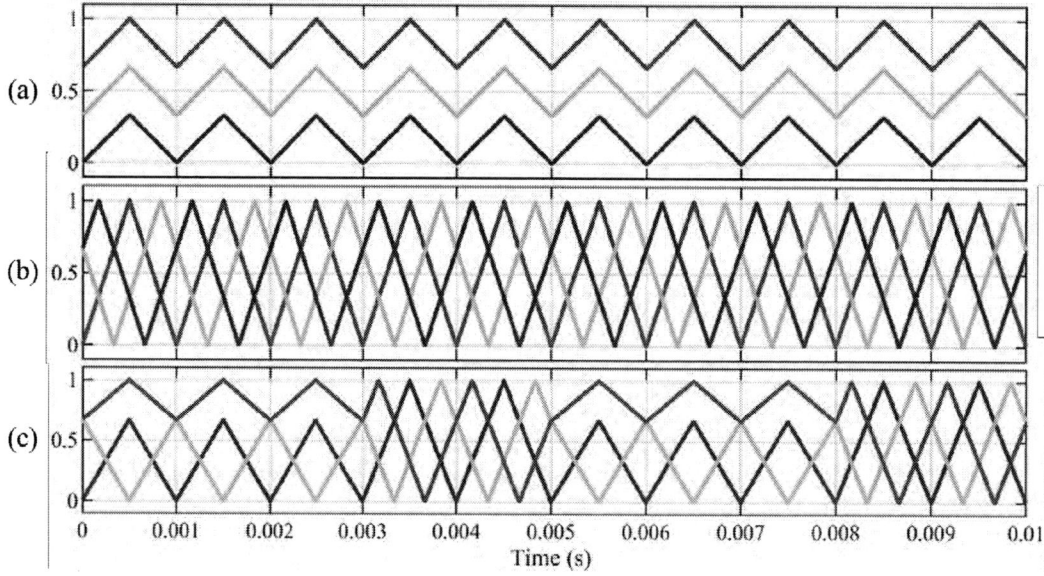

Fig. 3: Modulation frameworks: (a) Phase Disposition Carrier (PDC), (b) Phase Shift Carrier (PSC), (c) the proposed Phase Disposition and Shift Carrier (PDSC).

Controller Switching Criteria

The conventional SM based controller is more powerful in terms of forcing the capacitor voltages to converge however increases the computation complexity and switching frequency. The soft-paralleling mechanism can not function when the stack current is too low to activate the power diodes as interpreted before. A stack current threshold (I_{thres}) is set to select the proper controller mode. Whenever the absolute value of the stack current is larger than I_{thres}, the soft-paralleling based mode is selected, or the MBSM STATCOM will operate in the iteration based mode.

Simulation Results and Analysis

The performances of the proposed converter are verified by a MBSM STATCOM model built in MAT-LAB/ Simulink and the parameters are listed in Table I. The controller step is set as 5 μs and the PDSC frequency is 500 Hz. The losses curves of the IGBT module FZ1200R33HE3 produced by *Infineon* are applied for losses analysis.

Table I: Simulation Model Parameters

Parameters	Value	Parameters	Value
STATCOM nominal power	30 MVA	AC side line voltage (RMS)	18 kV
Nominal frequency	50 Hz	Simulation sampling time	5 μs
Branch inductance	0.1 pu	Phase inductance	0.1 pu
Nominal cell voltage	2000 V	MBSM capacitance	2.5 mF
Number of MBSMs per stack	15	Number of ES-MBSMs per stack	9

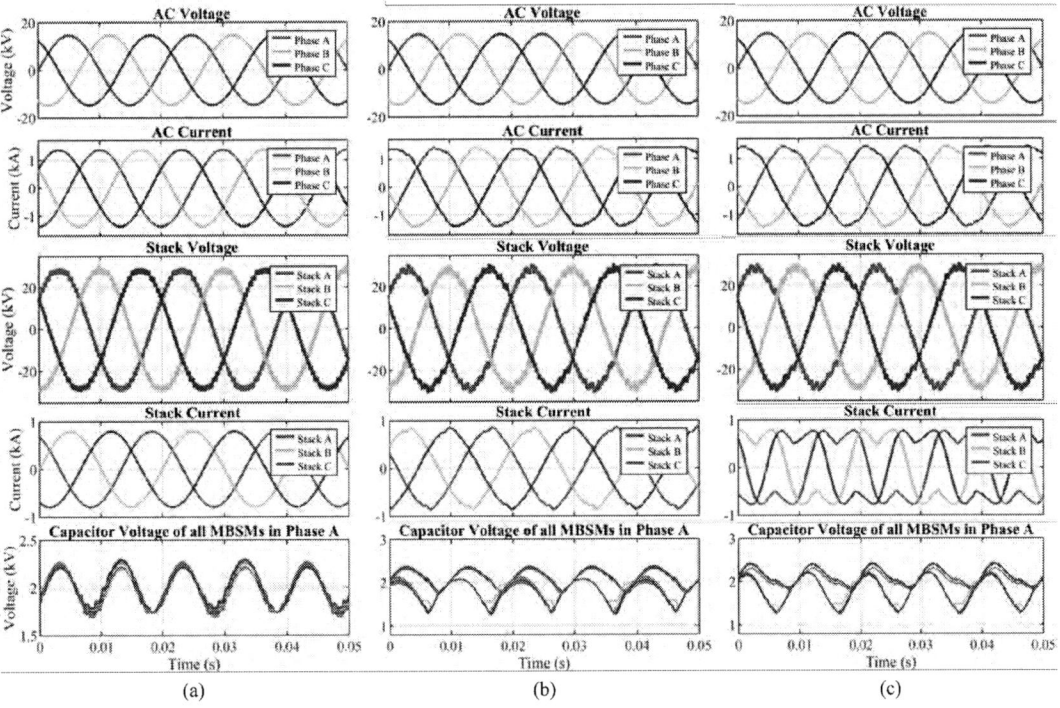

Fig. 4: Converter outputs when operating at: (a) $P_{ref} = 0$, $Q_{ref} = 1$ without circulating current, (b) $P_{ref} = 0.6$, $Q_{ref} = 0.8$ without circulating current, (c) $P_{ref} = 0.6$, $Q_{ref} = 0.8$ with constant amplitude (300A) third harmonic circulating current.

Fig. 4 illustrates the converter outputs at different set-points. When realising 100% reactive power, as is shown in Fig. 4 (a), the AC current and stack current are both of high quality with low THD at 0.1%. In addition, the maximum capacitor voltages deviation is approximately 9.1%. At the set-point of $P_{ref} = 0.6$ and $Q_{ref} = 0.8$ (Fig. 4 (b)), where the apparent power is equal to the former set-point, active power generated by ES-MBSMs results in voltage unbalance of different sub-stacks. The maximum SM voltage deviation increases to approximately 22.8% and consequently the output current has larger distortion. Fig. 4(c) presents the results when $P_{ref} = 0.6$ and $Q_{ref} = 0.8$ with constant amplitude (300A) third harmonic circulating current applied. The maximum voltage deviation is reduced to 19.4 %, verifying that additional circulating current is able to promote the SM voltage balancing.

The performances of the proposed MBSM STATCOM together with its low level controller are compared with a conventional FBSM STATCOM controlled by classic SM sorting low level controller [7]. The system parameters are the same as those listed in Table I. The power losses and maximum SM voltage deviation versus varying active powers with constant apparent power are illustrated in Fig. 5 and Fig. 6. The grey columns represent data when the converter is not stable. The MBSM STATCOM is able to release 0.7 pu active power while the conventional FBSM STATCOM becomes unstable when $Q_{ref} > 0.5$. The MBSM STATCOM performs better on conduction losses, switching losses and the maximum voltage deviation than its FBSM counterpart.

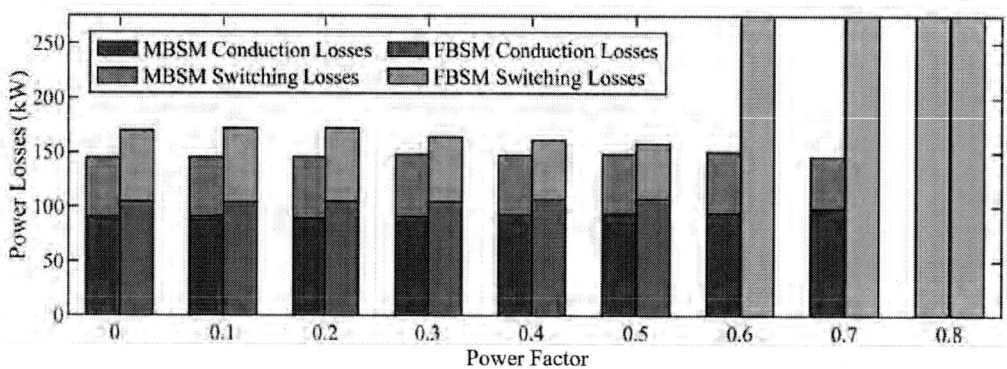

Fig. 5: Conduction losses and switching losses versus varying power factors with 1 p.u. apparent power. Grey columns represent data when the converter is not stable.

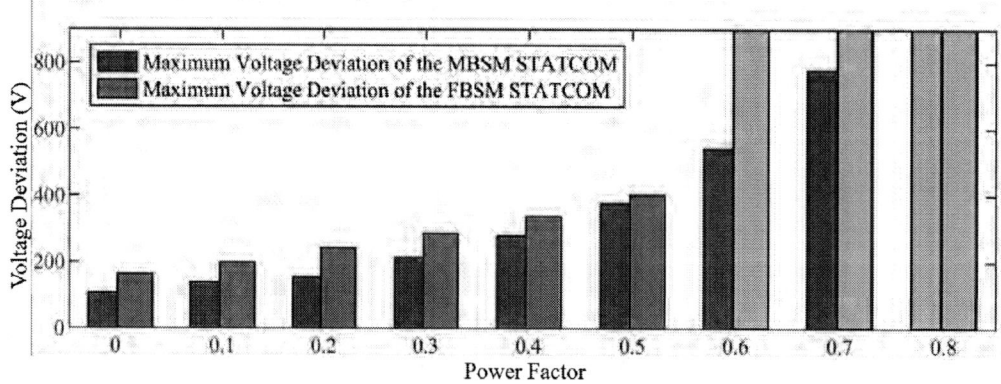

Fig. 6: Maximum SM voltage deviation versus varying power factors with 1 p.u. apparent power. Grey columns represent data when the converter is not stable.

Conclusion

This paper proposes a new type of partially rated energy storage STATCOM configured in ES-MBSM sub-stacks. The mixed control system and modulation framework are also presented. Compared with the conventional FB counterpart, the active power capability is increased from power factor = 0.5 to power factor = 0.7, while the power losses and the maximum SM voltage deviation are reduced by 14.9% and 35.2% respectively when operating at pure reactive power output. Moreover, the active power capability can be further improved with larger circulating current applied.

References

[1] J. A. Barrado, R. Grino and H. Valderrama-Blavi, "Power-Quality Improvement of a Stand-Alone Induction Generator Using a STATCOM With Battery Energy Storage System," in IEEE Transactions on Power Delivery, vol. 25, no. 4, pp. 2734-2741, Oct. 2010, doi: 10.1109/TPWRD.2010.2051565.

[2] P. D. Judge and T. C. Green, "Modular Multilevel Converter With Partially Rated Integrated Energy Storage Suitable for Frequency Support and Ancillary Service Provision," in IEEE Transactions on Power Delivery, vol. 34, no. 1, pp. 208-219, Feb. 2019, doi: 10.1109/TPWRD.2018.2874209.

[3] M. M. C. Merlin et al., "The Extended Overlap Alternate Arm Converter: A Voltage-Source Converter With DC Fault Ride-Through Capability and a Compact Design," in IEEE Transactions on Power Electronics, vol. 33, no. 5, pp. 3898-3910, May 2018, doi: 10.1109/TPEL.2017.2723948.

[4] M. Glinka and R. Marquardt, "A new AC/AC multilevel converter family," in IEEE Transactions on Industrial Electronics, vol. 52, no. 3, pp. 662-669, June 2005, doi: 10.1109/TIE.2005.843973.

[5] S. Yang, J. Fang, Y. Tang, H. Qiu, C. Dong and P. Wang, "Modular Multilevel Converter Synthetic Inertia-Based Frequency Support for Medium-Voltage Microgrids," in IEEE Transactions on Industrial Electronics, vol. 66, no. 11, pp. 8992-9002, Nov. 2019, doi: 10.1109/TIE.2018.2890491.

[6] L. Zhang, Y. Tang, S. Yang and F. Gao, "Decoupled Power Control for a Modular-Multilevel-Converter-Based Hybrid ACDC Grid Integrated With Hybrid Energy Storage," in IEEE Transactions on Industrial Electronics, vol. 66, no. 4, pp. 2926-2934, April 2019, doi: 10.1109/TIE.2018.2842795.

[7] M. Vasiladiotis and A. Rufer, "Analysis and Control of Modular Multilevel Converters With Integrated Battery Energy Storage," in IEEE Transactions on Power Electronics, vol. 30, no. 1, pp. 163-175, Jan. 2015, doi: 10.1109/TPEL.2014.2303297

[8] S. G. Mian, P. D. Judge, A. Junyent-Fer and T. C. Green, "A Delta-Connected Modular Multilevel STATCOM With Partially-Rated Energy Storage for Provision of Ancillary Services," in IEEE Transactions on Power Delivery, vol. 36, no. 5, pp. 2893-2903, Oct. 2021, doi: 10.1109/TPWRD.2020.3029312.

[9] Z. Li, R. Lizana, S. M. Lukic, A. V. Peterchev and S. M. Goetz, "Current Injection Methods for Ripple-Current Suppression in Delta-Configured Split-Battery Energy Storage," in IEEE Transactions on Power Electronics, vol. 34, no. 8, pp. 7411-7421, Aug. 2019, doi: 10.1109/TPEL.2018.2879613.

[10] W. Yang, Q. Song, S. Xu, H. Rao and W. Liu, "An MMC Topology Based on Unidirectional Current H-Bridge Submodule With Active Circulating Current Injection," in IEEE Transactions on Power Electronics, vol. 33, no. 5, pp. 3870-3883, May 2018, doi: 10.1109/TPEL.2017.2722011.

[11] A. Rashwan, M. A. Sayed, Y. A. Mobarak, G. Shabib and T. Senjyu, "Predictive Controller Based on Switching State Grouping for a Modular Multilevel Converter With Reduced Computational Time," in IEEE Transactions on Power Delivery, vol. 32, no. 5, pp. 2189-2198, Oct. 2017, doi: 10.1109/TPWRD.2016.2639529.

[12] S. M. Goetz, A. V. Peterchev and T. Weyh, "Modular Multilevel Converter With Series and Parallel Module Connectivity: Topology and Control," in IEEE Transactions on Power Electronics, vol. 30, no. 1, pp. 203-215, Jan. 2015, doi: 10.1109/TPEL.2014.2310225.

[13] Z. Li, R. Lizana F., Z. Yu, S. Sha, A. V. Peterchev and S. M. Goetz, "A Modular Multilevel Series/Parallel Converter for a Wide Frequency Range Operation," in IEEE Transactions on Power Electronics, vol. 34, no. 10, pp. 9854-9865, Oct. 2019, doi: 10.1109/TPEL.2019.2891052.

[14] A. M. Y. M. Ghias, J. Pou, G. J. Capella, V. G. Agelidis, R. P. Aguilera and T. Meynard, "Single-Carrier Phase-Disposition PWM Implementation for Multilevel Flying Capacitor Converters," in IEEE Transactions on Power Electronics, vol. 30, no. 10, pp. 5376-5380, Oct. 2015, doi: 10.1109/TPEL.2015.2427201.

[15] S. Bal, D. B. Yelaverthi, A. K. Rathore and D. Srinivasan, "Improved Modulation Strategy Using Dual Phase Shift Modulation for Active Commutated Current-Fed Dual Active Bridge," in IEEE Transactions on Power Electronics, vol. 33, no. 9, pp. 7359-7375, Sept. 2018, doi: 10.1109/TPEL.2017.2764917.

Three-Phase ZVS Inverter with Variable and Fixed Frequency Operation based on GaN Semiconductors

Benedikt Kohlhepp, Michael Lutsch and Thomas Dürbaum
Electromagnetic Fields, Friedrich-Alexander University Erlangen-Nürnberg (FAU)
Konrad-Zuse-Strasse 3/5
91052 Erlangen, Germany
Tel.: +49 (0)9131 85 28951
Fax: +49 (0)9131 85 27787
E-Mail: benedikt.kohlhepp@fau.de
URL: https://www.emf.tf.fau.de/

Keywords

«ZVS converters», «Voltage Source Inverter (VSI)», «Gallium Nitride (GaN)».

Abstract

As high efficiency is on everyone's lips, this paper studies an inverter using a ZVS modulation scheme promising high efficiency. It uses triangular current mode (TCM) to minimize the ripple while maintaining ZVS. In order to fulfill the required power density, high switching frequencies should be applied. Current threshold detection circuits in combination with this modulation scheme turn the inverter's legs into current sources. Since all phases of the inverter are tied together at the load's star point, there might be issues, if slight current threshold errors exist within the inverter. In order to demonstrate that the modulation scheme can handle situations with and without connection of the load's neutral point to the DC-link midpoint, both configurations are studied in a practical setup. The practical test setup gives satisfactory results with respect to the quality of the output waveforms using the THD of the currents.

Introduction

High efficiency and power density are typical demands in power electronics, which can only be achieved by moving the switching frequency up to several hundred kHz as thereby passives shrink in size. In order to enable that high switching frequencies, switching losses need to be eliminated, because these scale linearly with frequency [1]. Thus, lossless switching, especially zero voltage switching (ZVS), should be exploited [2]. In inverters, triangular current mode (TCM) realizes ZVS and current mode control [3] [4] [5] [6] [7]. Compared to conventional fixed frequency PWM operation, TCM has an increased current ripple, but eliminates switching losses [8]. From this modulation scheme, typically a variable switching frequency results, which has its highest value around the zero crossing of the sinusoidal current [9]. The switching frequency must be limited to a certain value capable for all components within the circuit. This is why, a modulation scheme with two operating modes, fixed (around zero crossing of the sinusoidal current) and variable (in all other regions) switching frequency, results [10].

In order to detect zero crossing of the high frequency inductor current during variable switching frequency operation, which is required to enable lossless switching, a current threshold detection circuit can be used [11]. Hence, the inverter acts like a current mode controlled one. In case all phases of the inverter operate in current mode control, three current sources feed the load. Non-idealities of the components and their tolerances can lead to the sum of the inductor currents (at the load's star point) to be non-zero. Thus, in applications without a neutral point connection at the load, this may results in problems.

Therefore, this paper first theoretically analyzes the modulation scheme and based on promising simulation results, an experimental setup is built. A first study at the practical setup uses a connection of the load's neutral point to the midpoint of the DC-link, which delivers smooth sinusoidal output waveforms as intended. Then, to prove the versatility of the modulation scheme using applications without neutral point connection, the potential of the neutral point is floating.

Inverter Topology and Operating Principles

Each leg of the two-level inverter studied (Fig. 1) comprises a representation of the load (modeling the fundamental frequency), a LC output filter, and a half-bridge. Furthermore, there is a circuit for the study of the star point connection of the load Y consisting of two capacitors C_M connected in series, paralleled to the DC-link.

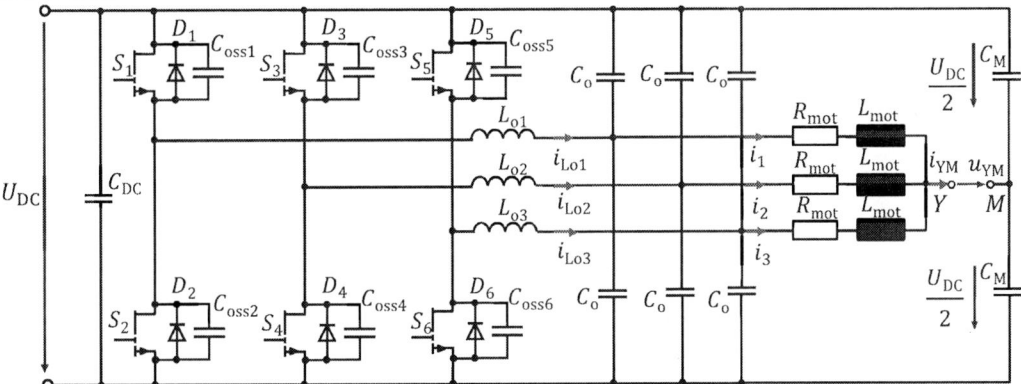

Fig. 1: Three-phase ZVS inverter with LC output filter and load

In order to study the modulation scheme, considering only one leg (half-bridge) of the inverter is sufficient. Thus, Fig. 2 (left) shows the circuit diagram including the modulator and current sensing circuit, all explanations refer to this half-bridge. The output side LC filter minimizes EMI emissions of the inverter. Properly designed, the inverter can achieve zero voltage switching (ZVS) by using a variable switching frequency modulation scheme in combination with the LC filter. Triangular current mode (TCM) allows applying ZVS during the entire sinusoidal period. Voltage sources at the input ($U_{DC}/2$) represent the DC-link capacitors. For explaining the behavior using a single phase representation of the inverter (Fig. 2 (left)), the load is connected to the midpoint of the DC-link capacitors. Using a three-phase inverter, the load is typically connected at the star point (see Fig. 1). The load model comprises an inductor and a resistor (Fig. 2 (right)) and corresponds to the terminal impedance typical of the grid or an electric machine. Simulations can be carried out just using the fundamental frequency to describe the load, as the LC output filter minimizes high frequency noise. Consequently, the resulting output current and component values can be gained by an AC analysis of the circuit for a given operating point of the inverter (output voltage, power and phase angle).

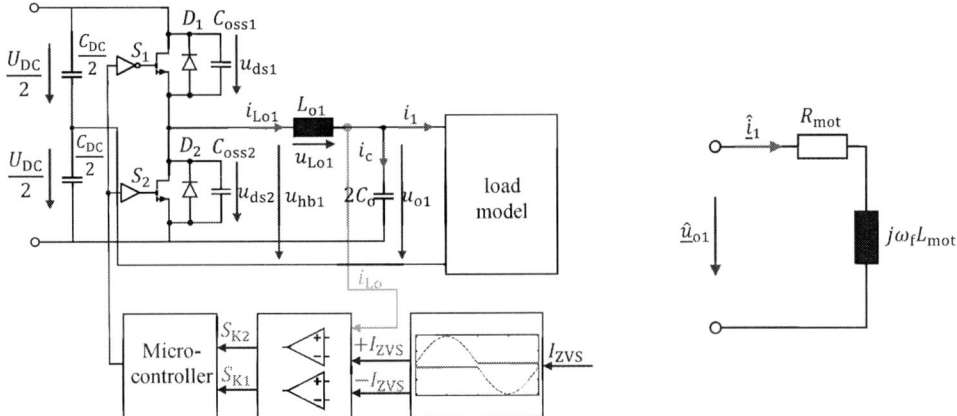

Fig. 2: Schematic of one half-bridge of the inverter including block diagram of the inductor current sensing circuit (left); single phase load model (right)

The inverter's output current is of sinusoidal shape with the fundamental period T_f and is given by:

$$i_1 = \hat{i}_1 \cdot \sin(\omega_f t) \qquad \text{where} \qquad \omega_f = 2\pi f_f = \frac{2\pi}{T_f} \qquad (1)$$

The corresponding output voltage generally is phase shifted to the current:

$$u_{o1} = \hat{u}_{o1} \cdot \sin(\omega_f t + \varphi_u) \qquad (2)$$

Modulation Scheme

The two aspects, generating a sinusoidal output voltage and reaching ZVS over the entire sinusoidal period, mentioned above require a suitable modulation strategy. In order to discharge the output capacitors (C_{oss}) of the lower and upper switch prior turn-on, which is required for ZVS, the inductor current needs to have a proper sign during each switching action. One switching action per switching period is typically uncritical with respect to ZVS[12]. During the positive half-wave the uncritical switching action is the one, at which the upper switch S_1 is turned off and, after the dead time has elapsed, the lower one turns on. The output current's sign and value certainly brings the voltage across the lower switch u_{ds2} to zero prior the lower switch turning on. In contrast, for reaching lossless turn-on (ZVS) for the other switching action, the modulation must provide a negative inductor current during this switching transition. The switching actions are vice versa during the negative half-wave. In order to reach the second goal (sinusoidal output current), the averaged inductor current over one switching period must coincide to the desired output current and the current phase angle of the sine to ensure high quality output waveforms. In case of fixed switching frequency modulation, a very high current ripple would be necessary to achieve ZVS over the entire sinusoidal period, which is why a variable switching frequency is used.

To reach the goals mentioned above, proper inductor current limits need to be set [13]. That's why a suitable current threshold detection circuitry is used. One approach sets the lower as well as the upper bound by comparators in conjunction with digital-analog converters [5] [6] [7] [14]. As these circuits are both costly and complex, another method, which combines a current threshold measurement and calculation to generate the switching pattern of the semiconductors, significantly minimizes cost [3] [11]. It uses only one comparator per half-wave and avoids digital-analog converters resulting in a hybrid modulation scheme.

The comparator generates the turn-off instant for the lower power semiconductor during the positive half-wave. Once the dead time t_{dead} has elapsed, the upper switch turns on featuring ZVS. The inductor current discharges the output capacitor of the upper semiconductor to reach ZVS during the dead time. A digital control circuit then derives the conduction time of the upper power semiconductor on the basis of theoretical equations. Thus, the digital controller delivers suitable gate signals to achieve ZVS and to provide sinusoidal output waveforms. The switching actions are vice versa during the negative half-wave of the sinusoidal period. Fig. 2 (left) gives the schematic of the converter and the block diagram of the inductor current detection circuit. As the modulation scheme is analogous for the negative half-wave, the explanations refer to the positive half-wave only.

In order to achieve high efficiency and exploit ZVS, the modulation scheme typically features a variable switching frequency. Assuming a resistive load, the lowest switching frequency occurs near the sinusoidal maximum and the highest is present around the zero crossing zone [9]. Especially if the modulation scheme is applied for the whole sinusoidal period, a wide frequency range results from the modulation scheme (see Fig. 3, dashed).

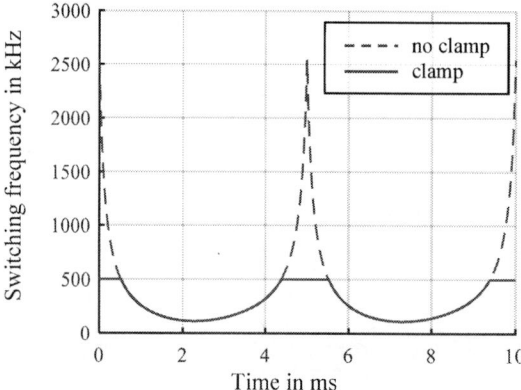

Fig. 3: Simulated switching frequency with and without clamping for a complete sinusoidal period

Since several components in power electronic circuits cannot handle very high switching frequencies, an upper limit for the switching frequency needs to be set. In addition, EMI standards give limits for the switching frequency in some applications to ensure to comply with the standards with a reasonable design of the circuit [10]. In this paper we use an upper limit for the switching frequency of 500 kHz, which is the lower limit for some automotive EMC standards. Another purpose of choosing 500 kHz as frequency boundary is to show the capabilities and limits of an implementation using a digital signal processor (DSP). Of course, some applications may favor lower frequency limits. As a result, the modulation scheme features fixed and variable switching frequency. Fig. 3 shows (solid) that fixed frequency operation occurs near the current's zero crossings, whereas variable switching frequency results in all other regions.

In order to ensure safe operation of the inverter without triggering the filter's resonance, its component values must be chosen properly. (8) shows, that the filter inductance value directly affects the resulting switching frequency. Furthermore, the inductance values as well as the filter capacitor's value impact the filter's resonance:

$$f_{res} = \frac{1}{2\pi\sqrt{2L_o C_o}}$$

(3)

Thus, L_o should be selected to set the resulting switching frequency range. Afterwards, the capacitor's values must be chosen to ensure the filter resonance lies significantly higher than the sinusoidal output frequency, as it should pass the low pass filter unaffected. Secondly, the lowest switching frequency occurring within the inverter should be significantly higher than the filter's resonance, to avoid unintended oscillations and proper filtering. Hence,

$$f_{s,min} \gg f_{res} \quad \text{and} \quad f_f \ll f_{res}.$$

(4)

To ensure ZVS over the whole sinusoidal period, the current at the switching action at which the lower switch turns off and the upper one turns on, must be chosen properly.

The absolute value of the current required to reach ZVS during variable switching frequency operation at the switching instant is:

$$I_{ZVS} = \frac{Q_{ZVS}}{t_{dead}} + \frac{1}{2} \cdot \frac{\frac{U_{DC}}{2} - u_{o1}\left(\frac{T_f}{2}\right)}{L_o} \cdot t_{dead}$$

(5)

For this equation, linear current slopes are assumed. Q_{ZVS} is the absolute value of the charge that needs to be provided by the inductor current during the dead time t_{dead} to reach lossless turn-on. $-I_{ZVS}$ must be present at the switching instant when the lower switch turns off, as (5) gives the absolute value. For a given dead time, the critical transition is present in case of the maximum current slope, which occurs during conduction of the upper switch. Equation (5) uses the maximum voltage across the inductor in this half-wave to ensure the ZVS condition is met over the entire half-wave. Assuming a resistive load, the maximum voltage is present near the zero crossing of the sinusoidal current at $T_f/2$.

The conduction time of the upper switch is given by:

$$t_{on,S1} = d \cdot T_s$$

(6)

The duty-cycle d can be derived from the voltages present in the inverter:

$$d = \frac{u_{o1}}{U_{DC}} + 0.5$$

(7)

To obtain the conduction time, also the switching period T_s must be known. If current slopes are supposed to be linear, (8) gives the switching period. The average inductor current $\overline{i_{Lo}}$ over one switching period should coincide the sinusoidal output current i_1 to achieve a high-quality output current.

$$T_s = \frac{2 \cdot (i_1 + I_{ZVS})}{\frac{\frac{U_{DC}}{2} - u_{o1}}{L_o} \cdot d}$$

(8)

If (8) delivers a switching frequency higher than the specified maximum frequency, it is clamped (fixed frequency operation). Then, (7) gives the duty cycle of the half-bridge. During fixed switching frequency operation, a higher current ripple results than required for lossless switching (ZVS) of the half-bridge switches.

Table 1 summarizes the inverter's parameters used for the theoretical study and in the practical setup. The intended application is to drive a 48V motor, but the modulation scheme itself can also be used for grid-tied inverters. Of course, other component values as well as another switching frequency range should be used for grid applications.

Table 1: Inverter operating parameters

Parameter	Value	Parameter	Value
L_o	2.3 µH	$T_{s,min}$	2 µs
Q_{ZVS}	50 nC	T_f	10 ms
U_{DC}	48 V	t_{dead}	50 ns
$\hat{u}_{o1}$	16.9 V	φ_u	13 °
$\hat{i}_1$	11 A		

Using the described modulation scheme and the parameters given in Table 1 a simulation generates the inductor currents (i_{Lo1}, i_{Lo2}, and i_{Lo3}) for the three-phase inverter (Fig. 1) shown in Fig. 4 (left). Near the zero crossings of the corresponding output currents (i_1, i_2, and i_3, given in Fig. 4 (right)), the inductor currents show an increased ripple compared to the waveforms only using TCM. This stems from fixed frequency operation. Furthermore, the inductor currents and output currents prove that the two goals set initially (ZVS and sinusoidal output waveforms) are reached.

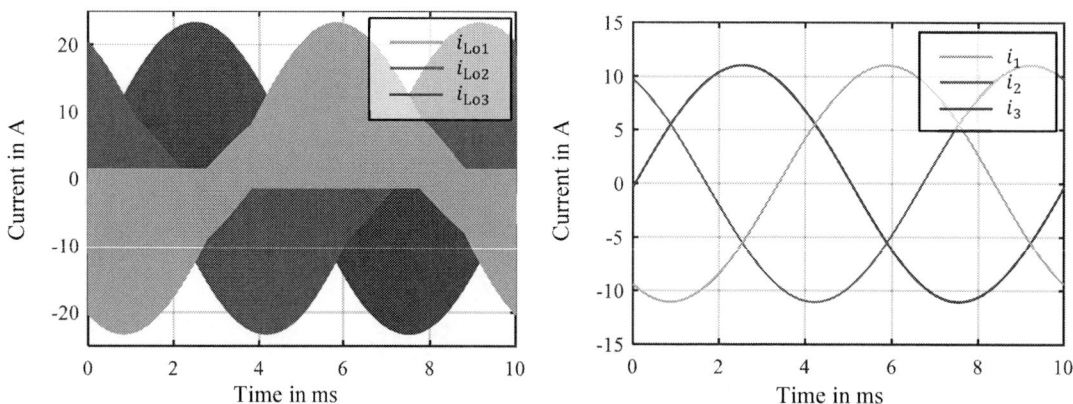

Fig. 4: Inductor current i_{Lo1}, i_{Lo2}, and i_{Lo3} gained by simulation (left); simulated output current i_1, i_2, and i_3 (right)

Analysis of Low-Frequency Behavior

To observe the low-frequency behavior e.g. for studying control algorithms of the inverter, the ripple caused by the switching actions of the half-bridge can be neglected. During fixed frequency operation of the inverter, the modulation scheme sets the averaged half-bridge voltage over one switching cycle $\bar{u}_{hb}$ which is of rectangular shape and is defined by its duty-cycle d and the DC-link voltage U_{DC}. Thus, a voltage source $\bar{u}_{hb}$ can be used to describe the modulation scheme's behavior and it behaves like a voltage mode control. Fig. 5 (left) gives the corresponding low frequency equivalent circuit. The half-bridge voltage's high frequency content is eliminated and their low frequency component passes the inverter's LC-filter and is present at the output of the inverter. From the output voltage and the load impedance Z_{mot} the inductor current $\bar{i}_{Lo}^*$ and output current $\bar{i}_1^*$ result. Note that 'derived variables' ($\bar{i}_{Lo}^*$ and $\bar{i}_1^*$) are indicated by an asterisk ('*'). This means that these variables result from the circuit topology and are not directly set by the modulation scheme. Therefore, in fixed frequency operation, the output voltage is set by the modulation scheme and the currents $\bar{i}_{Lo}^*$ and $\bar{i}_1^*$ result from the circuit topology.

In contrast, during variable switching frequency, a comparator triggers one switching action of the half-bridge, when the inductor current reaches a certain value (I_{ZVS}). As a consequence, the modulation scheme programs the averaged inductor current over one switching period. Thus, it acts like a current mode control. In the low frequency equivalent circuit (see Fig. 5 (right)), this behavior is considered by replacing the inductor and modulation scheme by a current source $\bar{i}_{Lo}$. During variable switching frequency operation, the averaged half-bridge voltage $\bar{u}_{hb}^*$ and the output voltage $\bar{u}_o^*$ result from the modulation scheme and form the derived values.

The derived values (variable with '*') match the intended ones (without '*') if an ideal converter is used, as the modulation scheme assumes an ideal behavior. In a practical setup however, the derived variables differ from the intended ones, as non-ideal behavior inevitably exists. This can be for example losses within the converter's components, load impedance deviations, the dead time of the half-bridge, inaccurate voltage measurements of the DC-link and output voltage, and deviations of the comparator thresholds from the ideal values. All these effects can lead to derived variables to differ from the intended ones.

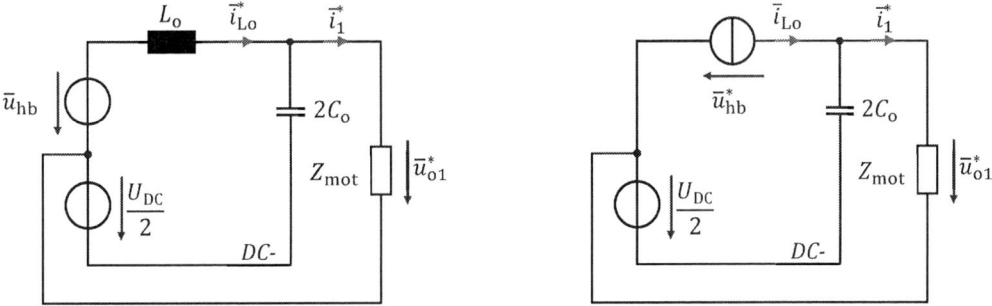

Fig. 5: Low frequency equivalent circuit for fixed switching frequency operating (left); low frequency equivalent circuit for variable switching frequency operation (right)

Experimental Setup

In order to show the applicability of the proposed modulation scheme in a real hardware, a three-phase prototype is built. The inverter uses GaN-HEMTs as half-bridge semiconductors and operates under the conditions mentioned in Table 1.

Study of the Inverter with Star Point Connection

In case the inverter's phases operate in variable switching frequency mode, they act like current mode controlled. As can be seen from Fig. 4 (left), there are time intervals, where all three phases of the inverter are in variable switching frequency operation. Thus, in an idealized inverter three current sources feed the load. As already stated in the previous analysis of the modulation scheme, in a practical setup, due to component tolerances and non-idealities, the averaged inductor currents may differ from the intended ones. Thus, the sum of all load currents at the load's neutral point (or star point) can be non-zero:

$$\sum_{k=1}^{3} i_k = i_1 + i_2 + i_3 = i_{\mathrm{YM}} \neq 0 \tag{9}$$

In Fig. 1, the connection Y to M is left open. But for the first study it is shorted to provide a return path for the residual current i_{YM} resulting from the non-ideal behavior of the modulation scheme and the inverter. With this configuration all phases of the inverter are decoupled, consequently operating independently [6] [15].

Fig. 6 (left) shows the inductor current of all three phases obtained by the practical test setup of a three-phase inverter with the load's star point connected to the midpoint of the DC-link. On the right-hand side of Fig. 6, the load currents as well as the residual current i_{YM} is depicted. Both measurements at the practical setup show that the waveforms look like the intended ones. It is worth noting that the load impedance directly couples the output current and voltage. Hence, studying only the current is sufficient. Fig. 7 shows a zoomed analysis of the star point current i_{YM}, which is less than 4 % of the output current. In order to evaluate the quality of the output waveforms, the total harmonic distortion (THD) is used for the output currents:

$$THD_k = \sqrt{\frac{\sum_{n=2}^{n_{OS}} \hat{i}_{k,\mathrm{FFT}}^2(n \cdot f_{\mathrm{f}})}{\hat{i}_{k,\mathrm{FFT}}^2(f_{\mathrm{f}})}} \text{ with } k = 1 \dots 3 \tag{10}$$

The THD delivers one number that represents the harmonic content of each current with respect to its fundamental component. n_{OS} represents the number of harmonics summed up for the calculation of the THD and is set to 500 within this study. $\hat{i}_{1,\mathrm{FFT}}$ is the fast Fourier transforms' (FFT) output for each inverter output current. The THD for the three output currents is given in Table 2 named as 'short' and is below 2 %, which is relatively low. These results show that the modulation method can be used, when the star point is connected to the midpoint of the DC-link.

Table 2: THD with and without star point connection

Current	Short	Floating
$\hat{i}_1$	1.68 %	1.04 %
$\hat{i}_2$	1.67 %	1.05 %
$\hat{i}_3$	1.57 %	0.92 %

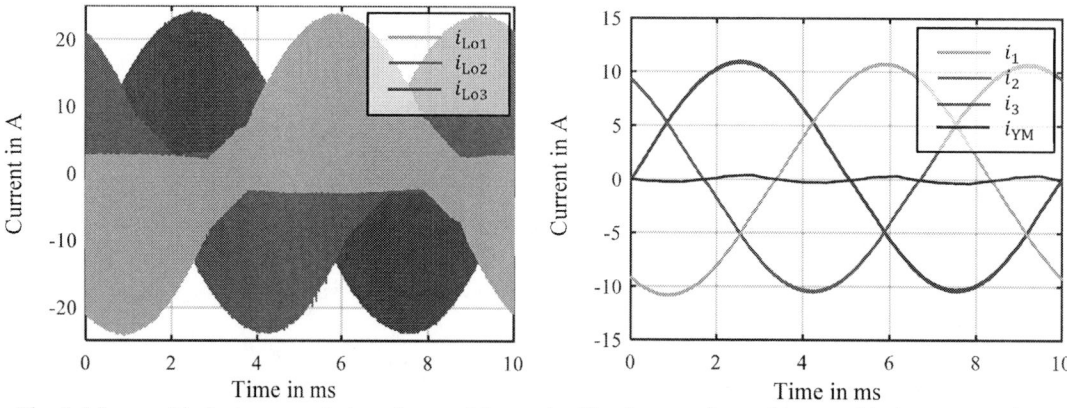

Fig. 6: Measured inductor currents i_{Lo1}, i_{Lo2}, and i_{Lo3} gained by the experimental setup with short connection between Y and M (left); measured output currents i_1, i_2, and i_3 and star point current i_{YM} (right)

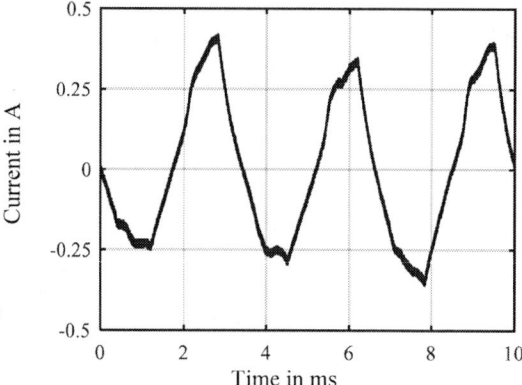

Fig. 7: Zoom of measured star point current i_{YM} with short connection between Y and M gained by the experimental setup

Study of the Inverter with Floating Star Point

Since in many applications the connection of load's star point to the DC-link is not present, the behavior of the inverter without connection will be investigated in the second part. Therefore, the connection between Y and M is not present anymore, which means that the sum of all load currents at the load's neutral point (or star point) is forced to be zero:

$$\sum_{k=1}^{3} i_k = i_1 + i_2 + i_3 = i_{YM} = 0 \tag{11}$$

Thus, the load's neutral point is floating and the voltage from Y to M u_{YM} can be studied. As already mentioned, due to component tolerances and non-ideal behavior of the modulation scheme and the inverter, the averaged inductor currents can differ from the intended ones. Thus, (11) is not satisfied anymore. If, in case of non-idealities, no return path from the load's neutral point is given, as the load does not feature a neutral point connection, the only alternative path for the currents, set by modulation i_{Lo1}, i_{Lo2}, and i_{Lo3} (see Fig. 5 (right)), are the filter capacitors. Consequently, this residual currents will distort the output voltage, as a current flowing through a capacitor will change its voltage. Of course, also in the ideal case, a current flows through the capacitors, since their voltage is ideally sine-shaped during a fundamental period. Assuming an ideal inverter, the capacitor's current can be obtained using:

$$i_{c,\sim} = 2 \cdot C_o \cdot \hat{u}_{o1} \cdot \omega_f \cdot \cos(\omega_f t + \varphi_u) \tag{12}$$

According to (12), this current is proportional to the output voltage u_{o1}, the filter capacitance C_o and the frequency of the sinusoidal current. The modulation scheme takes into account this current component which occurs inherently due to the LC filter. However, if non-idealities occur in the system, additional current may flow through the capacitors:

$$i_c = i_{c,\sim} + i_{c,res} \tag{13}$$

As the current in this case can differ from the sinusoidal shape, the output voltage can be distorted. In order to investigate the overall influence of this additional current on the output currents and voltages an experiment needs to be performed using the same setup with opened star point connection. Therefore, the inverter's output currents are shown in Fig. 8 (left) to prove that the modulation scheme can also handle this situation.

As the waveforms are of sinusoidal shape and do not differ from the intended ones, the additional current due to the non-idealities $i_{c,res}$ is negligible compared to the sinusoidal charging current $i_{c,\sim}$ resulting from the sinusoidal output voltage. Fig. 8 (right) shows a zoomed analysis of the voltage between Y and M. Again, the THD is evaluated. When the star point is floating, a THD of around 1% for all three currents results (see Table 2). These results demonstrate that the modulation scheme is capable of operating in applications without the load's star point connected to the DC-link.

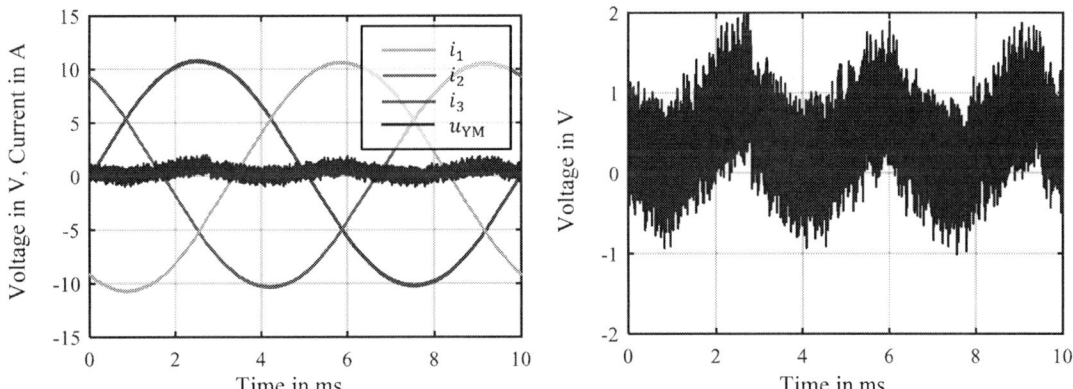

Fig. 8: Measured output currents i_1, i_2, and i_3 and star point voltage u_{YM} gained by the experimental setup with open between Y and M (left); zoom of measured star point voltage u_{YM} (right)

Conclusion

The paper relates to a hybrid modulation scheme for three-phase two-level inverters featuring two different operating modes. It comprises a combination of variable and fixed frequency to reach lossless switching for a whole sinusoidal period. After an analysis of the modulation strategy and its theoretical background, based on promising results gained by simulation, a prototype in real hardware is built. Since non-idealities and component tolerances can cause undesired behavior at the load's neutral point, this paper first conducts a study with the load's neutral point clamped to the midpoint of the DC-link. In order to use this modulation scheme also for applications without neutral point connection, a second investigation proves that this modulation scheme can also be applied at loads with a floating star point. The measurements conducted demonstrate that smooth sinusoidal waveforms can be obtained and the modulation scheme is suitable for this applications.

References

[1] B. Kohlhepp, D. Kübrich, and T. Dürbaum, "Switching Loss Measurement – A Thermal Approach Applied to GaN-Half-Bridge Configuration," in *2021 23st European Conference on Power Electronics and Applications (EPE '21 ECCE Europe)*, Sep. 2021, pp. 1–8.

[2] Quan Li and P. Wolfs, "A Review of the Single Phase Photovoltaic Module Integrated Converter Topologies With Three Different DC Link Configurations," *IEEE Trans. Power Electron.*, vol. 23, no. 3, pp. 1320–1333, May 2008, doi: 10.1109/TPEL.2008.920883.

[3] B. Kohlhepp, J. Heubeck, and T. Duerbaum, "Non-ideal Behavior of ZVS Inverter Comprising Variable and Fixed Frequency Operation: Analysis, Compensation, and Verification," *Power Electronic Devices and Components*, p. 100007, Apr. 2022, doi: 10.1016/j.pedc.2022.100007.

[4] L. Huber, B. T. Irving, and M. M. Jovanovic, "Closed-Loop Control Methods for Interleaved DCM/CCM Boundary Boost PFC Converters," in *2009 Twenty-Fourth Annual IEEE Applied Power Electronics Conference and Exposition*, Washington, DC, USA, Feb. 2009, pp. 991–997, doi: 10.1109/APEC.2009.4802783.

[5] B. Kohlhepp, T. Foerster, and T. Duerbaum, "High Frequency ZVS GaN-Inverter with Adaptive Dead Time," in *2021 56th International Universities Power Engineering Conference (UPEC)*, Middlesbrough, United Kingdom, Aug. 2021, pp. 1–6, doi: 10.1109/UPEC50034.2021.9548249.

[6] B. Kohlhepp and T. Duerbaum, "Novel DPWM Modulation Scheme for Three-Phase ZVS Inverters," in *2021 56th International Universities Power Engineering Conference (UPEC)*, Middlesbrough, United Kingdom, Aug. 2021, pp. 1–6, doi: 10.1109/UPEC50034.2021.9548210.

[7] B. Kohlhepp and T. Durbaum, "Modulation Strategy Comprising TCM with Frequency Limit and DPWM for Fast Switching GaN-Inverters," in *2021 IEEE 8th Workshop on Wide Bandgap Power Devices and Applications (WiPDA)*, Redondo Beach, CA, USA, Nov. 2021, pp. 205–210, doi: 10.1109/WiPDA49284.2021.9645127.

[8] B. Kohlhepp, D. Kübrich, M. Tannhäuser, and T. Dürbaum, "Modulation Method to Reduce Losses in Inverters with LC-Filters," in *2021 23st European Conference on Power Electronics and Applications (EPE '21 ECCE Europe)*, Sep. 2021, p. P.1-P.8.

[9] M. Haider *et al.*, "Novel ZVS S-TCM Modulation of Three-Phase AC/DC Converters," *IEEE Open J. Power Electron.*, vol. 1, pp. 529–543, 2020, doi: 10.1109/OJPEL.2020.3040036.

[10] Q. Huang and A. Q. Huang, "Variable Frequency Average Current Mode Control for ZVS Symmetrical Dual-Buck H-Bridge All-GaN Inverter," *IEEE J. Emerg. Sel. Topics Power Electron.*, vol. 8, no. 4, pp. 4416–4427, Dec. 2020, doi: 10.1109/JESTPE.2019.2940270.

[11] A. Amirahmadi *et al.*, "Hybrid ZVS BCM Current Controlled Three-Phase Microinverter," *IEEE Trans. Power Electron.*, vol. 29, no. 4, pp. 2124–2134, Apr. 2014, doi: 10.1109/TPEL.2013.2271302.

[12] B. Kohlhepp, D. Kübrich, M. Tannhäuser, and T. Dürbaum, "Adaptive dead time in high frequency GaN-Inverters with LC output filter," in *The 10th International Conference on Power Electronics, Machines and Drives (PEMD 2020)*, Online Conference, 2021, pp. 372–377, doi: 10.1049/icp.2021.0977.

[13] Q. Zhang, H. Hu, D. Zhang, X. Fang, Z. J. Shen, and I. Batarseh, "A Controlled-Type ZVS Technique Without Auxiliary Components for the Low Power DC/AC Inverter," *IEEE Trans. Power Electron.*, vol. 28, no. 7, pp. 3287–3296, Jul. 2013, doi: 10.1109/TPEL.2012.2225075.

[14] D. Zhang, Q. Zhang, H. Hu, A. Grishina, J. Shen, and I. Batarseh, "High efficiency current mode control for three-phase micro-inverters," in *2012 Twenty-Seventh Annual IEEE Applied Power Electronics Conference and Exposition (APEC)*, Orlando, FL, USA, Feb. 2012, pp. 892–897, doi: 10.1109/APEC.2012.6165924.

[15] B. Fan, Q. Wang, R. Burgos, A. Ismail, and D. Boroyevich, "Adaptive Hysteresis Current Based ZVS Modulation and Voltage Gain Compensation for High-Frequency Three-Phase Converters," *IEEE Trans. Power Electron.*, vol. 36, no. 1, pp. 1143–1156, Jan. 2021, doi: 10.1109/TPEL.2020.3002894.

Influences of Conductor Positions and Fast Rising Impulse Voltages on the Line-End Coil based on a Three-Phase High-Frequency Model

Ting Helmholdt-Zhu[1], Volker Grabs[2]

[1] Leibniz University Hannover
Welfengarten 1
30167 Hannover, Germany
helmholdt-zhu@stud.uni-hannover.de

[2] Lenze SE
Hans-Lenze-Straße 1
31855 Aerzen, Germany
volker.grabs@lenze.com

Acknowledgments

The work presented in this publication was supported by the research project UmSiChT, funded by the German Federal Ministry of Education and Research (BMBF). The responsibility for the content of this publication lies with the authors.

Keywords

≪Design optimization≫, ≪Electrical machine≫, ≪Genetic algorithm≫, ≪Pulse Width Modulation (PWM)≫, ≪Wide bandgap devices≫.

Abstract

Due to the steep-edged voltage impulses from the new generation of power electronics, the lifespan of stator insulation systems is significantly shortened. In this paper, a three-phase high-frequency (HF) model based on multi-conductor transmission line theories (MTLs) is developed to predict the electrical field distributions of each conductor in the line-end coils. In addition, the results of possible best and worst conductor spatial distributions in the stator slot from the one-phase HF model on the basis of genetic algorithms are also further investigated by the three-phase HF model. Finally, the theoretical results are validated through partial discharge measurements.

1. Introduction

One of the common failures in electrical machines are the turn-to-turn and turn-to-ground insulation breakdown [1, 2]. The recent trends of wide bandgap devices in power electronics, which switch faster than silicon devices and therefore cause higher inter-turn stress [3, 4], result in a more vulnerable dielectric strength between conductors and the conductor to ground in the line-end coils. Hence, in order to prevent an early malfunction, [5] introduces a one-phase HF model utilizing genetic algorithms (GAs) [6] to investigate the influences of stator conductor spatial positions and rise times of the steep-edged voltage impulses on the transient voltage distributions. However, under real operation conditions the electrical machines are usually supplied by three-phase impulse voltages from an inverter. Therefore, in this paper, the one-phase HF model is further developed to represent a three-phase machine. Comparisons between measured and simulated results through phase current and differential voltage (between phase V and the star point) are demonstrated to validate the three-phase model. Utilizing this validated model, the voltage distributions according to [5] are calculated and compared to the results of the one-phase model. Finally, the simulated outcomes are further validated via partial discharge measurements.

2. Three-Phase High-Frequency Model

The derivation of the three-phase model consists of three main steps [7]: voltage terminal condition, current terminal condition and star point terminal condition. Each step is served to eliminate redundant variables and information, so that the matrix equations contain only the unknown variables (node voltages of each conductor and three-phase input currents). The following sections start with the introduction of the one-phase model in detail and then further develop it into a three-phase electrical stator winding model.

2.1 One-Phase Model

A stator coil model needs to be separated into two distinct regions: the slot and the overhang (OV) regions due to different distributions of electrical and magnetic fields. Hence, the stator coil can be considered as five discontinuous sections [8], which are interconnected. In other words, there are no mutual coupling among these five sections [9]. Each section is implemented as a MTL [10] and analyzed through the 2n-port network. The admittance parameters (Y-parameters) are utilized to describe the network.

A one-phase model, for instance, is presented in Fig. 1, which has 2 coil groups and each coil group occupies two 3-turn coils. The mutual capacitance coupling in OV region between adjacent coils within the same coil group is considered. In order to simplify the HF model, some assumptions are made: the conductor spatial distribution for all slots are the same, thus, for different coils the admittance parameters (Y-parameters) are the same; the conductors are positioned parallel against each other and with uniform cross section. Taking this simplified model as an example, the one-phase model of coil group 1 is thus

$$
\begin{bmatrix}
I_{s1}^1 \\
I_{s2}^1 \\
-I_{r1}^1 \\
-I_{r2}^1 \\
\cdots \\
I_{s1}^2 \\
-I_{r1}^2 \\
\cdots \\
I_{s2}^2 \\
-I_{r2}^2 \\
\cdots \\
I_{s1}^3 \\
I_{s2}^3 \\
-I_{r1}^3 \\
-I_{r2}^3 \\
\cdots \\
I_{s1}^4 \\
-I_{r1}^4 \\
\cdots \\
I_{s2}^4 \\
-I_{r2}^4 \\
\cdots \\
I_{s1}^5 \\
I_{s2}^5 \\
-I_{r1}^5 \\
-I_{r2}^5
\end{bmatrix}_{60\times 1}
=
\begin{bmatrix}
\tfrac{1}{2}Y_{\mathrm{OV}} & & & & & \\
& Y_{\mathrm{Slot}} & & & & \\
& & Y_{\mathrm{Slot}} & & & \\
& & & Y_{\mathrm{OV}} & & \\
& & & & Y_{\mathrm{Slot}} & \\
& & & & & Y_{\mathrm{Slot}} \\
& & & & & & \tfrac{1}{2}Y_{\mathrm{OV}}
\end{bmatrix}_{60\times 60}
\begin{bmatrix}
U_{s1}^1 \\
U_{s2}^1 \\
U_{r1}^1 \\
U_{r2}^1 \\
\cdots \\
U_{s1}^2 \\
U_{r1}^2 \\
\cdots \\
U_{s2}^2 \\
U_{r2}^2 \\
\cdots \\
U_{s1}^3 \\
U_{s2}^3 \\
U_{r1}^3 \\
U_{r2}^3 \\
\cdots \\
U_{s1}^4 \\
U_{r1}^4 \\
\cdots \\
U_{s2}^4 \\
U_{r2}^4 \\
\cdots \\
U_{s1}^5 \\
U_{s2}^5 \\
U_{r1}^5 \\
U_{r2}^5
\end{bmatrix}_{60\times 1}
\tag{1}
$$

where

- $I_{si}^j, -I_{ri}^j, U_{si}^j$ and U_{ri}^j are ($N \times 1$) current and voltage vectors at sending as well as receiving ends of line. The parameter N indicates the number of turns per coil and in this case, it is equal to 3. The variable $i = 1, 2$ presents the number of coils of coil group 1, $j = 1, 2, ..., 5$, according to the five sections. For example, $\boldsymbol{I}_{s1}^1 = [I_{s1,1}^1, I_{s1,2}^1, I_{s1,3}^1]^T$, $\boldsymbol{I}_{r1}^1 = [I_{r1,1}^1, I_{r1,2}^1, I_{r1,3}^1]^T$ and $\boldsymbol{U}_{s1}^1 = [U_{s1,1}^1, U_{s1,2}^1, U_{s1,3}^1]^T$, $\boldsymbol{U}_{r1}^1 = [U_{r1,1}^1, U_{r1,2}^1, U_{r1,3}^1]^T$.

- $[\boldsymbol{Y}]$ is the admittance matrix of dimension $(10CN \times 10CN)$, and the number 10 comes from 2×5, as there are five sections and each section has the same number of sending and receiving variables, C is the number of coil per coil group and here it is equal to 2. Thus, the dimension of $[\boldsymbol{Y}]$ for this simplified example is (60×60). The definition and calculation of matrices $\boldsymbol{Y}_{OV}$, $\boldsymbol{Y}_{Slot}$ are demonstrated in [8].

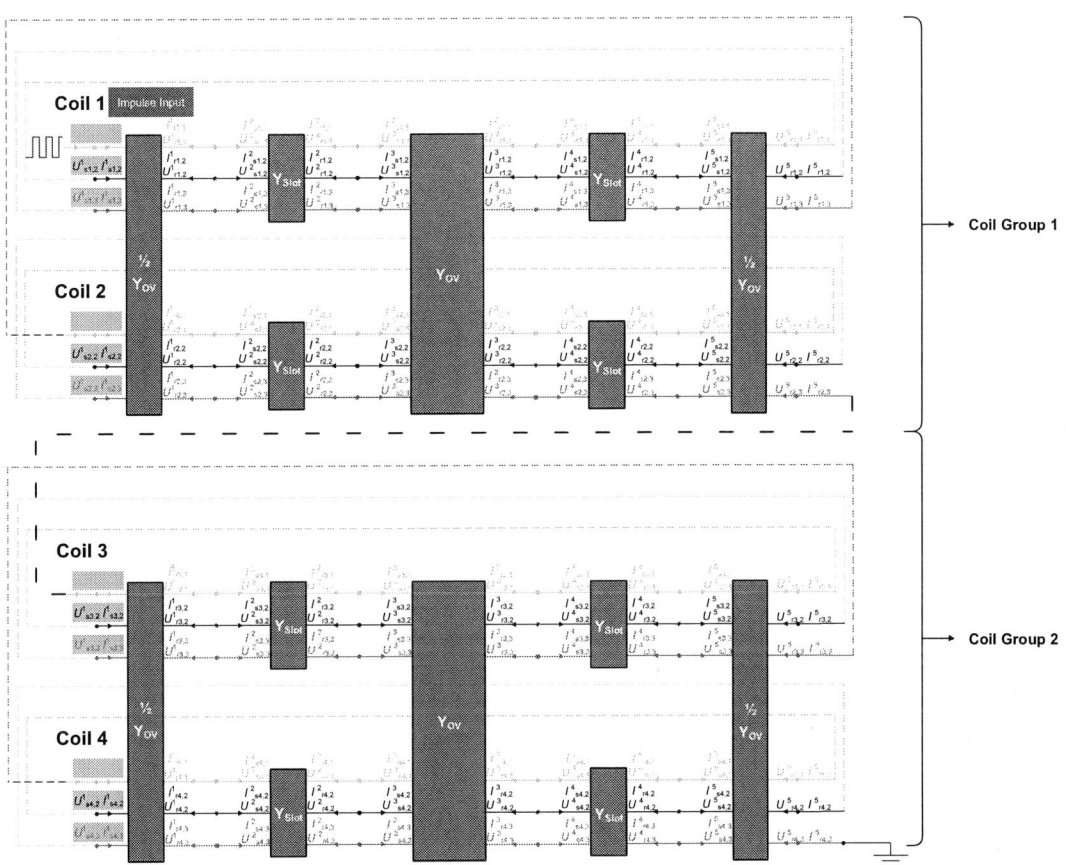

Fig. 1: Example of a one-phase model

Then, Eq. 1 can be summarized as

$$\left[\ \boldsymbol{I}_{C1}\ \right]_{60 \times 1} = [\boldsymbol{Y}]_{60 \times 60} \left[\ \boldsymbol{U}_{C1}\ \right]_{60 \times 1} \tag{2}$$

where C1 indicates the coil group 1. Hence, the result for the whole phase is

$$\begin{bmatrix} \boldsymbol{I}_{C1} \\ \boldsymbol{I}_{C2} \end{bmatrix} = \begin{bmatrix} [\boldsymbol{Y}]_{60 \times 60} & \\ & [\boldsymbol{Y}]_{60 \times 60} \end{bmatrix} \begin{bmatrix} \boldsymbol{U}_{C1} \\ \boldsymbol{U}_{C2} \end{bmatrix} \tag{3}$$

and C2 indicates the coil group 2, $\boldsymbol{I}_{C1} = [I_{s1}^1, I_{s2}^1, \cdots, -I_{r1}^5, -I_{r2}^5]^T$, $\boldsymbol{I}_{C2} = [I_{s3}^1, I_{s4}^1, \cdots, -I_{r3}^5, -I_{r4}^5]^T$ and $\boldsymbol{U}_{C1} = [U_{s1}^1, U_{s2}^1, \cdots, U_{r1}^5, U_{r2}^5]^T$, $\boldsymbol{U}_{C2} = [U_{s3}^1, U_{s4}^1, \cdots, U_{r3}^5, U_{r4}^5]^T$.

2.1.1 Voltage Terminal Condition

As mentioned, Eq. 3 can be simplified through terminal conditions. The first step is utilizing the voltage terminal condition. Figure 1 shows, that the receiving and sending node voltages between adjacent sections provide the same information, for instance, U_{r1}^1 and U_{s1}^2, U_{r1}^2 and U_{s1}^3, U_{r2}^1 and U_{s2}^2 etc., thus the receiving parameters can be considered as redundant information and be eliminated. The voltage vector $\begin{bmatrix} U_{C1} \\ U_{C2} \end{bmatrix}_{120 \times 1}$ is reduced to

$$
\begin{bmatrix} U_{C1,\text{New}} \\ \cdots \cdots \\ U_{C2,\text{New}} \\ \cdots \cdots \\ U_{\text{end}} \end{bmatrix}_{(60+1) \times 1}
=
\begin{bmatrix} U_{s1/2}^1 \\ U_{s1/2}^2 \\ U_{s1/2}^3 \\ U_{s1/2}^4 \\ U_{s1/2}^5 \\ \cdots \cdots \\ U_{s3/4}^1 \\ U_{s3/4}^2 \\ U_{s3/4}^3 \\ U_{s3/4}^4 \\ U_{s3/4}^5 \\ \cdots \cdots \\ U_{r4,3}^5 \end{bmatrix}_{(60+1) \times 1}
\tag{4}
$$

where U_{end} is the end terminal voltage. In this model it is equal to $U_{r4,3}^5$ (see Fig. 1). The corresponding transformation matrix is $[\boldsymbol{T}_U]$.

2.1.2 Current Terminal Condition

Similarly, the same approach is carried out for the current vector $\begin{bmatrix} \boldsymbol{I}_{C1} \\ \boldsymbol{I}_{C2} \end{bmatrix}_{120 \times 1}$ to cancel out the sending parameters. As the sending and receiving parameters are opposite in sign, after the addition the sending parameters are equal to 0 and the corresponding transformation matrix is $[\boldsymbol{T}_I]$. Afterwards, the current vector is further simplified by eliminating the receiving parameters, as they are redundant information. The final current vector is thus

$$
\begin{bmatrix} \boldsymbol{I}_{C1,\text{New}} \\ \cdots \cdots \\ \boldsymbol{I}_{C2,\text{New}} \\ \cdots \cdots \\ -\boldsymbol{I}_{\text{end}} \end{bmatrix}_{(60+1) \times 1}
=
\begin{bmatrix} I_{s1,1}^1 \\ \vdots \\ 0 \\ \cdots \cdots \\ 0 \\ \vdots \\ 0 \\ \cdots \cdots \\ -I_{r4,3}^5 \end{bmatrix}_{(60+1) \times 1}
\tag{5}
$$

The transformation matrix is $[\boldsymbol{T}_{I,\text{Eli}}]$.

2.2 Three-Phase Model

After the simplification of voltage and current terminal conditions, the expression of Eq. 3 is modified as

$$
\begin{bmatrix} \boldsymbol{I}_{C1,\text{New}} \\ \cdots \cdots \\ \boldsymbol{I}_{C2,\text{New}} \\ \cdots \cdots \\ -\boldsymbol{I}_{\text{end}} \end{bmatrix}_{(60+1) \times 1}
= [\boldsymbol{Y}_{\text{Mod}}]
\begin{bmatrix} \boldsymbol{U}_{C1,\text{New}} \\ \cdots \cdots \\ \boldsymbol{U}_{C2,\text{New}} \\ \cdots \cdots \\ \boldsymbol{U}_{\text{end}} \end{bmatrix}_{(60+1) \times 1}
\tag{6}
$$

where $[Y_{\text{Mod}}]$ is the modified admittance matrix

$$[Y_{\text{Mod}}]_{(60+1)\times(60+1)} = [T_{\text{I,Eli}}]_{(60+1)\times 120} \underbrace{\left([T_{\text{I}}]_{120\times 120}\left(\overbrace{[Y]_{120\times 120}[T_{\text{U}}]_{120\times(60+1)}}^{\text{Voltage Terminal Condition}}\right)\right)}_{\substack{\text{Current Terminal Condition} \\ \text{Current Terminal Condition}}} \qquad (7)$$

Based on Eq. 6 and the simplified one-phase model, a three-phase (U, V, W) stator winding is thus

$$\begin{bmatrix} I^1_{\text{sU1,1}} \\ 0 \\ \cdots\cdots \\ -I^5_{\text{rU4,3}} \\ \hline I^1_{\text{sV1,1}} \\ 0 \\ \cdots\cdots \\ -I^5_{\text{rV4,3}} \\ \hline I^1_{\text{sW1,1}} \\ 0 \\ \cdots\cdots \\ -I^5_{\text{rW4,3}} \end{bmatrix} = \begin{bmatrix} [Y_{\text{Mod}}] & & \\ & [Y_{\text{Mod}}] & \\ & & [Y_{\text{Mod}}] \end{bmatrix} \begin{bmatrix} U^1_{\text{sU1,1}} \\ \vdots \\ \cdots\cdots \\ U^5_{\text{rU4,3}} \\ \hline U^1_{\text{sV1,1}} \\ \vdots \\ \cdots\cdots \\ U^5_{\text{rV4,3}} \\ \hline U^1_{\text{sW1,1}} \\ \vdots \\ \cdots\cdots \\ U^5_{\text{rW4,3}} \end{bmatrix} \qquad (8)$$

In this paper, the end terminal condition is a star connection: $I^5_{\text{rU4,3}} + I^5_{\text{rV4,3}} + I^5_{\text{rW4,3}} = 0$ and $U^5_{\text{rU4,3}} = U^5_{\text{rV4,3}} = U^5_{\text{rW4,3}} = U_{\text{UN}}$, where U_{UN} is the star point voltage.

2.2.1 Star-Terminal Connection

Using $(I^5_{\text{rU4,3}} + I^5_{\text{rV4,3}} + I^5_{\text{rW4,3}} = 0)$ and $(U^5_{\text{rU4,3}} = U^5_{\text{rV4,3}} = U^5_{\text{rW4,3}} = U_{\text{UN}})$ to further eliminate redundant information of Eq. 8

$$\begin{bmatrix} I^1_{\text{sU1,1}} \\ 0 \\ \hline I^1_{\text{sV1,1}} \\ 0 \\ \hline I^1_{\text{sW1,1}} \\ 0 \\ \cdots\cdots \\ 0 \end{bmatrix} = [T_{\text{I,end}}] \begin{bmatrix} [Y_{\text{Mod}}] & & \\ & [Y_{\text{Mod}}] & \\ & & [Y_{\text{Mod}}] \end{bmatrix} [T_{\text{U,end}}] \begin{bmatrix} U^1_{\text{sU1,1}} \\ \vdots \\ \hline U^1_{\text{sV1,1}} \\ \vdots \\ \hline U^1_{\text{sW1,1}} \\ \vdots \\ \cdots\cdots \\ U_{\text{UN}} \end{bmatrix} \qquad (9)$$

where $[T_{\text{I,end}}]$ and $[T_{\text{U,end}}]$ are the corresponding transformation matrices and the new modified admittance matrix is expressed as $[Y'_{\text{Mod}}]$. In order to solve this matrix equation (Eq. 9), the unknown variables (node voltages and three-phase input currents) should be placed in a vector together through the transformation matrices $[\boldsymbol{T}_{\text{Mod,I}}]$ and $[\boldsymbol{T}_{\text{Mod,U}}]$

$$
\begin{bmatrix} I_{\text{sU1,1}}^1 \\ I_{\text{sV1,1}}^1 \\ I_{\text{sW1,1}}^1 \\ \cdots\cdots \\ \mathbf{0} \\ \cdots\cdots \\ 0 \end{bmatrix} = [\boldsymbol{T}_{\text{Mod,I}}][\boldsymbol{Y}_{\text{Mod}}'][\boldsymbol{T}_{\text{Mod,U}}] \begin{bmatrix} U_{\text{sU1,1}}^1 \\ U_{\text{sV1,1}}^1 \\ U_{\text{sW1,1}}^1 \\ \cdots\cdots \\ \vdots \\ \cdots\cdots \\ U_{\text{UN}} \end{bmatrix} \tag{10}
$$

The new admittance matrix is $[\boldsymbol{Y}_{\text{Mod}}'']$. In the voltage vector, the three input voltages ($U_{\text{sU1,1}}^1$, $U_{\text{sU1,1}}^1$ and $U_{\text{sW1,1}}^1$) are known, thus, the final step is the rearrangement of current and voltage vectors. The matrix $[\boldsymbol{Y}_{\text{Mod}}'']$ is, hence, partitioned into

$$
[\boldsymbol{Y}_{\text{Mod}}''] = \left[\begin{array}{c|c} \boldsymbol{Y}_{\text{Mod, 11}}'' & \boldsymbol{Y}_{\text{Mod, 12}}'' \\ \hline \boldsymbol{Y}_{\text{Mod, 21}}'' & \boldsymbol{Y}_{\text{Mod, 22}}'' \end{array} \right] \tag{11}
$$

Now the matrix equation (Eq. 10) is expressed as

$$
\begin{bmatrix} I_{\text{sU1,1}}^1 \\ I_{\text{sV1,1}}^1 \\ I_{\text{sW1,1}}^1 \\ \hline \mathbf{0} \\ 0 \end{bmatrix} = \left[\begin{array}{c|c} \boldsymbol{Y}_{\text{Mod, 11}}'' & \boldsymbol{Y}_{\text{Mod, 12}}'' \\ \hline \boldsymbol{Y}_{\text{Mod, 21}}'' & \boldsymbol{Y}_{\text{Mod, 22}}'' \end{array} \right] \begin{bmatrix} U_{\text{sU1,1}}^1 \\ U_{\text{sV1,1}}^1 \\ U_{\text{sW1,1}}^1 \\ \hline \vdots \\ U_{\text{UN}} \end{bmatrix} \tag{12}
$$

Moving all the unknown variables (node voltages) in the voltage vector to the left side of the equation

$$
\begin{bmatrix} I_{\text{sU1,1}}^1 \\ I_{\text{sV1,1}}^1 \\ I_{\text{sW1,1}}^1 \\ \vdots \\ U_{\text{UN}} \end{bmatrix} = \left[\begin{array}{c} \boldsymbol{Y}_{\text{Mod, 11}}'' - \boldsymbol{Y}_{\text{Mod, 12}}''(\boldsymbol{Y}_{\text{Mod, 22}}'')^{-1}\boldsymbol{Y}_{\text{Mod, 21}}'' \\ -(\boldsymbol{Y}_{\text{Mod, 22}}'')^{-1}\boldsymbol{Y}_{\text{Mod, 21}}'' \end{array} \right] \begin{bmatrix} U_{\text{sU1,1}}^1 \\ U_{\text{sV1,1}}^1 \\ U_{\text{sW1,1}}^1 \\ \mathbf{0} \\ 0 \end{bmatrix} \tag{13}
$$

As is shown, the left side of the matrix equation consists of all required unknown variables (input line currents and node voltages of each conductor in each section). At the right side, the modified admittance matrix is known based on the calculation of winding parameters of each conductor [8] and the voltage vector is also known, as the three-phase input voltages are obtained through measurement (illustrated in Section 3). The solving process is based on the fast Fourier transformation (FFT) and inverse-FFT, which is illustrated in [5][8].

3. Model Validation

The mathematical one-phase model is determined through two parts: Methods of determining conductor parameters and the structure of a single-coil model based on MTLs. The validations of these two parts are demonstrated in [8]. For a more complicated three-phase model, it is necessary to conduct further validation measurements.

3.1 Best Case I: Conductor Distribution

According to the results of [5], one of the optimized distributions is, that stator conductors start from the bottom of the slot and spread to the slot opening (for example, see Fig. 2(a)). It is named as 'Best Case I'. With the same distribution an electrical machine is constructed (see Fig. 2(b)(c)).

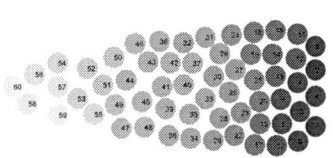

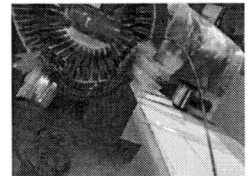

(a) Best Case I: Slot 1[5]   (b) Best Case I distribution for each coil (c) Separation of the slot in ten layers

Fig. 2: Best Case I electrical machine

Each coil are separated in six groups and each group contains ten conductors. Taking one line-end coil as an example, the first group is turn 1 to turn 10, then turn 11 to turn 20, turn 21 to turn 30, until the last group turn 51 to turn 60. Afterwards, conductors are positioned in the slot according to the group number and each group is separated through a thin film (see Fig. 2(b)(c)), so that conductors in different groups are not mixed to each other. Eventually, these thin films are removed and the electrical machine has a similar conductor distribution as the theoretical optimization results of Best Case I and can be utilized to validate the three-phase HF model according to Eq. 13.

3.2 Model Validation

First of all, the real three-phase impulse voltages ($U_{sU1,1}^1$, $U_{sV1,1}^1$ and $U_{sW1,1}^1$) with a 16 kHz switching frequency from a three-phase silicon carbide (SiC)-based inverter are measured and defined as input values for the three-phase HF model (see Fig. 3).

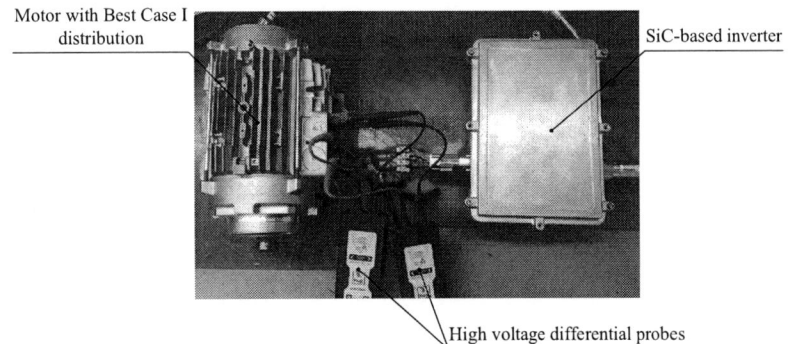

Fig. 3: Three-phase voltages measurement

Besides, the voltages are measured with two different rise times, which is realized through two gate resistances ($20\,\Omega$: $55\,\text{V/ns}$, $0.5\,\Omega$: $100\,\text{V/ns}$) of the SiC-based inverter.

3.2.1 Voltage Validation

The differential voltage between phase V (U_V) and the star point (U_{UN}) is measured as one of the comparison criterion (see Fig. 4). Using the results from $0.5\,\Omega$, it can be seen, that the simulated and measured differential voltages ($U_{V,\,UN}$) show a close agreement. The differential voltage ($U_{V,\,UN}$) oscillates during each switching operation. This is caused by the high-frequency parts contained in the switching edge, which excites the resonance frequency of the parasitic impedance from the conductors.

3.2.1 Current Validation

The phase current measurement is conducted through a current probe from Tektronix TCP0030, which has a bandwidth of 0-120 MHz. Figure 5 shows the measured and calculated current curves by $20\,\Omega$ and

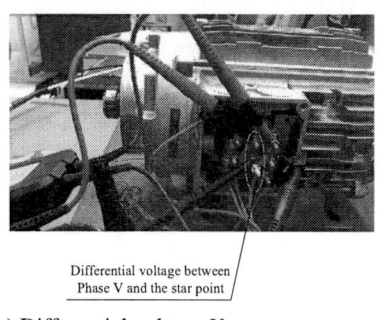

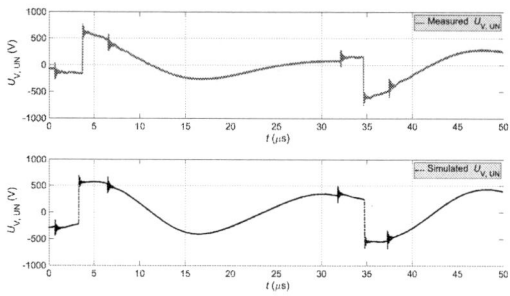

(a) Differential voltage $U_{V, UN}$ measurement

(b) Validation through differential voltage $U_{V, UN}$

Fig. 4: Voltage validation

0.5 Ω of phase U, for instance. In general, both simulated and measured values show the same trend: with decreased gate resistance, the peak amplitude of the current is increased.

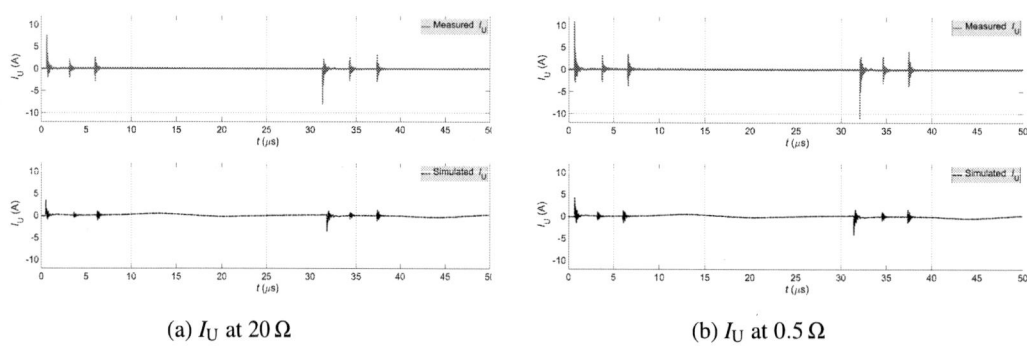

(a) I_U at 20 Ω

(b) I_U at 0.5 Ω

Fig. 5: Current validation

The variation of amplitudes between simulated and measured values is caused by following reasons: firstly, the limitation of the current probe due to the frequency bandwidth; secondly, the interruption from high-frequency noises of the SiC power electronics; thirdly, the exact values of the permittivity from stator insulation materials are unknown. Therefore, a precise prediction of the peak-amplitude is difficult to achieve in terms of both the limited measurement technique and the identifications of accurate material data. However, the purpose of this research is to compare and study the changes of different conductor distributions made on the transient voltage distribution in line-end coils. Therefore, the results can also be compared after a normalization. Using phase current I_U at 0.5 Ω as an example (see Fig. 6).

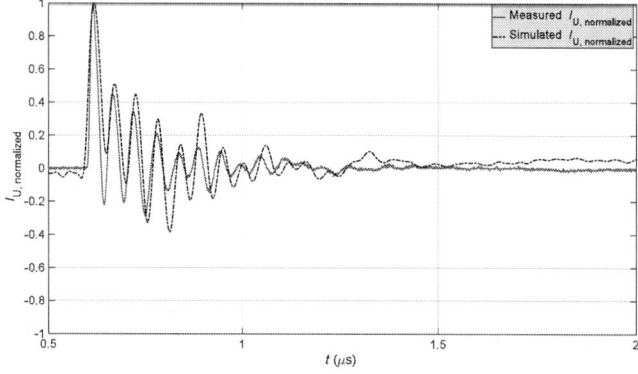

Fig. 6: Normalized phase current I_U

As demonstrated, the transient oscillations of measured phase current can be accurately predicted by the simulated results from the three-phase HF model in terms of the curve shape and the transient vibrations.

4. Results from Best and Worst Cases based on the Three-Phase HF Model

The results of field strength in line-end coils based on the three-phase model show the same advantages of utilizing the Best Case II distribution [5] (see Fig. 7(a)) (the other way around than Best Case I, see Fig. 7(a)). The Worst Case [5] indicates, that turn 1 and turn 60 (the last turn) of the line-end coil are placed directly next to each other (see Fig. 7(b)). It is noticeable, that the advantage of Best Case II distribution in terms of the maximum field strength (Fig. 7(b)(c)), which is already illustrated in [5] based on the one-phase HF model, is also validated through a three-phase HF model.

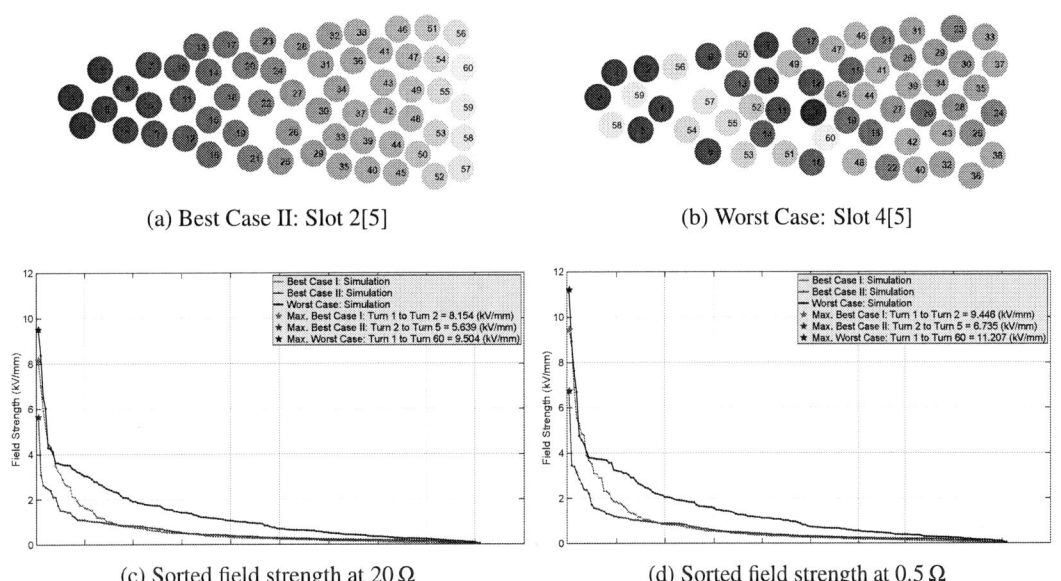

(a) Best Case II: Slot 2[5]　　　　　　　　　(b) Worst Case: Slot 4[5]

(c) Sorted field strength at 20 Ω　　　　　　(d) Sorted field strength at 0.5 Ω

Fig. 7: (a) and (b) illustrate conductor distributions of Best Case II and Worst Case. The color presents the turn number from turn 1 (dark blue) to turn 60 (light yellow); (c) and (d) show the sorted field strengths by 20 Ω and 0.5 Ω with Best Case I, II and Worst Case conductor distributions

5. Validation of Best and Worst Cases through Surge Voltage Tests

In order to validate the theoretical results of Best and Worst Cases distributions on the dielectric strength of the stator winding insulation system under impulse voltages, test stators with similar spatial arrangements are constructed (see Fig. 8).

Overall, there are three different types of test stators:
- Type 1: The conductors are distributed randomly, like the nowadays standard serial product.
- Type 2: The line-end coil based on the Worst Case distribution (the first and last turns placed directly next to each other), but with an extra sleeving of the first turn (Fig. 8(a)).
- Type 3: All coils are sorted based on the Best Case II distribution (Fig. 8(b)). In the meantime, with an exchange on the end of the star point, the spatial arrangement of conductors is then the other way around (from Best Case II to Best Case I). At the beginning, the conductors are separated in seven groups using kapton tape in the two OV regions (Fig. 8(b)). The first group (Group 1) contains turn 1 and turn 2. The rest groups have 3 turns each according to the sequence.

All the stators have the same winding topology with three parallel wires of each turn, one layer and three phases with double impregnation. Each version has two test stators, so, in total, there are 6 specimens. The measurement is conducted with the MTC2 surge voltage tester. The applied voltage starts from 2 kV and rises 250 V on each step, which contains 10 impulses. Each phase is tested for turn-turn insulation in

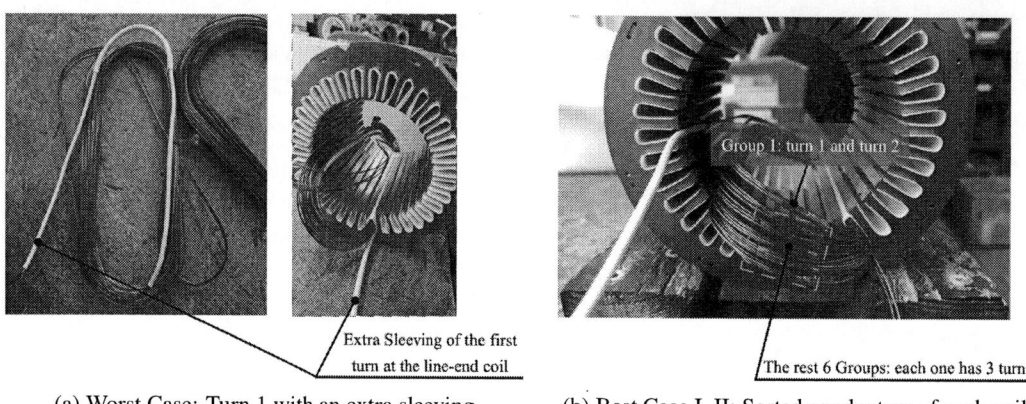

(a) Worst Case: Turn 1 with an extra sleeving   (b) Best Case I, II: Sorted conductors of each coil

Fig. 8: Best Case I, Best Case II and Worst Case distributions

terms of the repetitive partial discharge inception voltage (RPDIV) and each test is repeated three times (see Fig. 9).

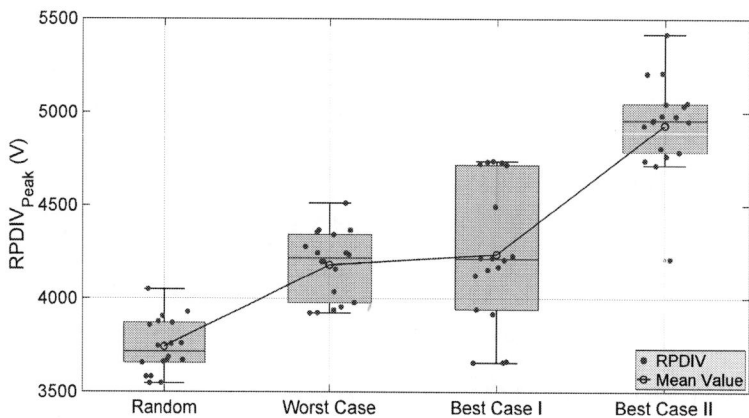

Fig. 9: RPDIV results of four different conductor distributions

As is shown, Best Case II represents an obvious higher inter-turn dielectric strength than Best Case I, Worst Case with an extra sleeving and Random distribution. Based on the mean value of Random distribution (3741.3 V), an enhancement of the RPDIV at Worst Case with an extra sleeving, Best Case I and Best Case II is of around 12 %, 13 % and 32 %, respectively.

5. Conclusion

In this paper, a three-phase HF model is introduced and validated through both voltage and current measurements. Besides, based on this model, the transient voltages along the first coil of each phase are calculated in terms of Best Case I, II and Worst Case distributions, which are the optimized results from one-phase HF model. The outcomes show again, that under Best Case II distribution the maximum value of electric field strength between nearby conductors is being the most restrained.

In addition, there are 6 test stators with different spatial conductor distributions being constructed to further validate the theoretical conclusions and the results show a high agreement: the RPDIV mean value of Best Case II stators are clearly the highest with an increase of 32 % compared to that of Random distributed stators. Furthermore, the advantage of utilizing an extra sleeving of turn 1 under worst case condition is not significantly demonstrated in the partial discharge measurement results. This partial discharge measurement should be further conducted with real SiC-inverters, so that the other effects (for example, the switching frequency, the peak voltage, etc.) can also be studied.

References

[1] D. E. Schump, "Testing to assure reliable operation of electric motors," 34th Annual Conference on Rural Electric Power, 1990, pp. B4/1-B4/6, doi: 10.1109/REPCON.1990.68527

[2] J. Geiman, "DC step-voltage and surge testing of motors", Maint. Technol., vol. 20, no. 3, pp. 3239, 2007

[3] A. Hoffmann and B. Ponick, "Method for the Prediction of the Potential Distribution in Electrical Machine Windings Under Pulse Voltage Stress," in IEEE Transactions on Energy Conversion, vol. 36, no. 2, pp. 1180-1187, June 2021, doi: 10.1109/TEC.2020.3026531

[4] Y. Xie, J. Zhang, F. Leonardi, A. R. Munoz, M. W. Degner and F. Liang, "Modeling and Verification of Electrical Stress in Inverter-Driven Electric Machine Windings," in IEEE Transactions on Industry Applications, vol. 55, no. 6, pp. 5818-5829, Nov.-Dec. 2019, doi: 10.1109/TIA.2019.2937068

[5] T. Helmholdt-Zhu and H. Borcherding, "Investigations on the Influences of Winding Positions and Rise Times on the Winding Isolation System within the Line-End Coil under Fast Rising Impulse Voltages," 2021 23rd European Conference on Power Electronics and Applications (EPE'21 ECCE Europe), 2021, pp. 1-10

[6] Sumathi, S. et al, "Evolutionary Intelligence," 1nd Ed., Springer, 2008

[7] J. L. Guardado and K. J. Cornick, "A computer model for calculating steep-fronted surge distribution in machine windings," in IEEE Transactions on Energy Conversion, vol. 4, no. 1, pp. 95-101, March 1989, doi: 10.1109/60.23156

[8] T. Helmholdt-Zhu, B. Knebusch and H. Borcherding, "High-Frequency Models for the Prediction of Transient Effects in Motor Windings under Fast Rising Impulse Voltages," PCIM Europe digital days 2020; International Exhibition and Conference for Power Electronics, Intelligent Motion, Renewable Energy and Energy Management, 2020, pp. 1-8

[9] B.S. Oyegoke, "Transient Voltage Distribution in Stator Winding of Electrical Machine Fed from a Frequency Converter," Finnish Academies of Technology, 2000

[10] C.R. Paul., "Analysis of Multiconductor Transmission Lines," 2nd. John Wiley & Sons, Inc., 2008

Simulation Tool for Optimization of Digital Active Gate Drive Sequence Using Genetic Algorithm

Hajime Takayama*, Shuhei Fukunaga†, and Takashi Hikihara*

* Department of Electrical Engineering,
Kyoto University
A1-412, Kyoto-university-Katsura, Nishikyo
Kyoto 615-8246, Japan
Phone: +81 75-383-2241
Email: h-takayama@dove.kuee.kyoto-u.ac.jp

† Division of Electrical, Electronic,
and Infocommunication Engineering,
Osaka University
Yamadaoka 2-1, Suita
Osaka 565-0871, Japan

Acknowledgments

This work was partially supported by the Program on Open Innovation Platform with Enterprises, Research Institute and Academia (OPERA) of Japan Science and Technology Agency, the Cross-ministerial Strategic Innovation Promotion Program (SIP), "Energy systems of an Internet of Energy (IoE) society"(Funding agency: JST), and JSPS KAKENHI Grant numbers 20H02151 and 21J22448.

Keywords

≪Digital control≫, ≪Hard switching≫, ≪Genetic algorithm≫, ≪Optimization method≫, ≪Silicon Carbide (SiC)≫, ≪Simulation≫, ≪Smart Gate Drivers≫.

Abstract

We demonstrate a simulation tool to optimize the operation of the digital active gate driver (DAGD) for SiC MOSFETs. The binary nature of DAGD's operating principle makes the genetic algorithm a preferable method. The optimization tool is developed by combining a Python-based program and SPICE simulation. Optimized solutions exhibit improved switching characteristics in wide operating conditions. The effect of conditions on obtained solutions is also analyzed.

Introduction

Active gate drive (AGD) has been gathering much attention in the hard switching of power devices. In particular, various types of AGDs for wide band-gap power devices have been studied because they face increased overshoots of the device voltage and current, which can lead to serious reliability problems such as electromagnetic interference and crosstalks [1, 2]. Various types of AGDs have been studied so far, such as gate-resistance-controlled methods [1], gate-voltage-controlled methods [3–7], and others [2,8]. The digitization of AGDs has also been attempted [4,5,8]. Unlike analog AGDs and other methods to suppress the overshoots, such as snubber circuits, they can adjust their operation without changing the circuit itself, which adds to the flexibility in their control strategy.

We particularly put the focus on the gate-voltage-controlled AGDs, which have been investigated in several research [3–7]. They directly designate a certain state of MOSFETs during the transient state of switching by setting particular gate-voltage level. This makes it easier to understand the relation between the control strategy and the device characteristics, which differs from device to device and, in addition, depends on the operating conditions.

For the pursuit of the potential of digitized AGDs, the digital control scheme to optimize their operation needs to be developed. Gate-voltage selection based on the modeling approach is investigated in [3].

However, they require an accurate parameter estimation or identification in advance, which is challenging for some applications. This work explores the potential of applying the genetic algorithm [9], one of the well-known metaheuristics, to obtain the optimized operation of AGD in a wide operating range. Since it takes a combinatorial-optimization approach, it can be applied to any device characteristics and operating conditions, even if they are unknown. A simulation tool using Python software and SPICE simulation is originally developed for this purpose. The obtained solutions are analyzed from the viewpoints of the device characteristics and the operating conditions.

GA-based operation optimization of digital active gate driver

This section introduces the digital active gate driver (DAGD) studied in [5] and the simulation tool using Python software and SPICE simulation. Due to the binary nature of the operation of DAGD, the application of the genetic algorithm is found preferable for optimizing its operation. The details of the algorithm and simulation settings are also described.

Configuration of 4-bit DAGD and simulated circuit

DAGD is designed from the architecture of the binary-weighted resistor digital-to-analog converter. The circuit configuration of 4-bit DAGD is shown in Fig. 1. The gate-source voltage (V_{GS}) waveform is shaped by the successive alteration of V_{GS} using a multi-bit gate signal sequence; V_{GS} is set to the value given by (1) every time the gate signals (b_0 to b_3) are changed at the clock rate of the outer controller such as FPGAs.

$$V_{GS} = \frac{\sum_{j=0}^{3} b_j 2^j}{\sum_{j=0}^{3} 2^j} V_{drv} \qquad (1)$$

The operation of DAGD is determined by means of this gate signal sequence, which proposes a great controllability of the V_{GS} waveform. The voltage-related characteristics of the MOSFET, such as the threshold voltage and the Miller plateau voltage, can be effectively taken into consideration in shaping the V_{GS} waveform. This advantage, however, requires proper control methods as the adjustment of the gate driver's operation is totally handled by the digital controller, not by adjusting the circuit itself.

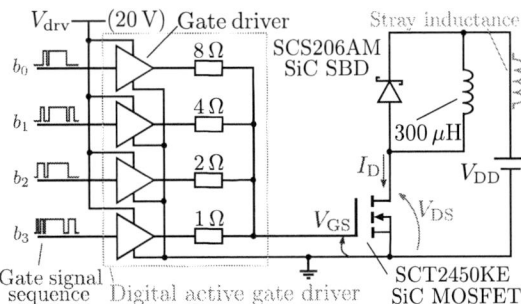

Fig. 1: Configuration of 4 bit DAGD. Double pulse testing circuit is assumed as the load.

The right part of Fig. 1 is an inductive-load test circuit known as the double-pulse testing, composed of a SiC MOSFET, a SiC Schottky barrier diode, a load inductor, and a power source (V_{DD}). It can set arbitrary operating conditions, a set of drain-source voltage (V_{DS}) and drain current (I_D), which needs to be taken into consideration in the AGD strategy. The test circuit also includes stray inductance of 80 nH at the wire from V_{DD} and 5 nH at each terminal of power devices.

For the model of the SiC MOSFET, we use the surface-potential-based model developed in [10]. Since it reflects the physical structure of the MOSFET using the surface equation, it can reproduce the accurate I–V and C–V characteristics even in the high-power region, where the actual switching operations in power converters take place. This helps the development of AGD strategy of DAGD as it is strongly related to these attributes.

Optimization tool based on GA

To find the optimum gate signal sequence, we here adopt a multi-objective genetic algorithm known as NSGA-II [9]. The optimization process is described in Fig. 2. The NSGA-II is implemented using a Python framework called DEAP [11], and the SPICE simulation is performed using SIMetrix software (Ver. 7.20j) [12].

Firstly, 60 genes are randomly selected to create an initial population. The genes are encoded into the gate signal sequences, which are the 4-bit time-series signal data saved in text-file format to be read by the netlist file. The Python program then sends a command to run the SIMetrix script, which simulates the double-pulse testing for all genes. After the simulation finishes, the program reads the waveform data to calculate the two fitness. Here, we use the switching loss and the amount of overshoot (peak of V_{DS} for turn-off and that of I_D for turn-on) as the two fitness, according to which the individuals are selected. It is noted that all the individuals pass the selection in the first loop as there are just 60 individuals in the population. The offspring population is generated from the current population through two-point crossover and one-bit-flip mutation, whose probabilities are set at 0.9 and 0.1, respectively. Their fitness are evaluated in the same way as described above. From the second loop, 60 individuals are selected from a set of the parent population and the offspring population, whose size is 120. By doing so, the Pareto front solutions are kept and updated. The number of generations is fixed at 30.

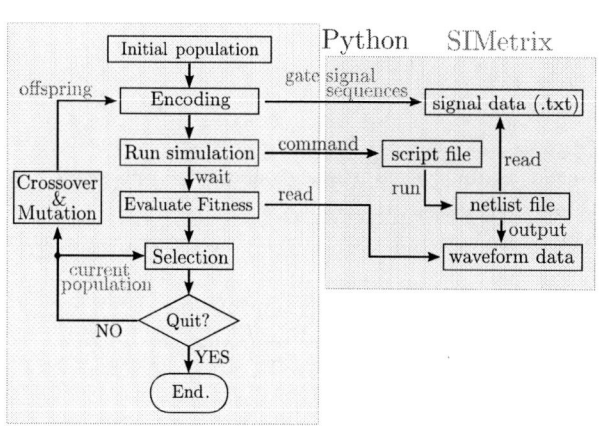

Fig. 2: Flowchart of optimization based on NSGA-II.

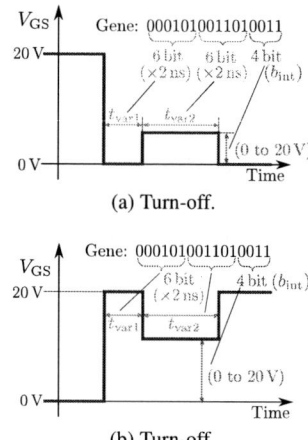

(a) Turn-off.

(b) Turn-off.

Fig. 3: Genotype of gate signal sequences.

Genotype of gate signal sequences

The multi-bit gate signal sequence of DAGD is naturally a binary sequence, which can be directly used as the genes and also can be easily converted to other genotypes. We adopt the genotype as shown in Fig. 3 to limit the degree of freedom into three, making the analysis simple and effective. The sequence for turn-off/on is described as follows (hereinafter, the prefix "0x" means the following number is hexadecimal): firstly, the gate signal is set at 0x0/0xF for a duration of t_{var1}; then, it is changed to b_{int} and kept for t_{var2}; finally, it is set at 0x0/0xF again to complete the turn-off/on operation. The first and following 6 bits of the gene represent t_{var1} and t_{var2}, respectively, while the last 4 bits represent b_{int}.

It is worthwhile to mention that this kind of V_{GS} waveform is sometimes referred to as the three-level (3-L) waveform in the research of voltage-controlled AGDs [3]. It is ideal for analyzing the relationship between a particular voltage level, which corresponds to b_{int}, and the resulting switching characteristics.

Verification of optimization tool

The proposed optimization tool is performed in simulation. Firstly, the tool is verified in a single operating condition, and the obtained solutions are analyzed in detail. A comparison with a standard gate driving and optimization results in other operating conditions are given afterward.

Optimization in a single operating condition

For the system verification, the operating condition is set at $(V_{DS}, I_D) = (120\,V, 4\,A)$. Figs. 4 (a) and (b) show the fitness maps of solutions in the final generation population at turn-off and turn-on, respectively. The selected values of b_{int} are also expressed by the color bar. These figures confirm that the Pareto-front solutions are successfully obtained by the optimization both at turn-off and turn-on. Here, we select some of the Pareto-front solutions marked with Roman numbers to see the transient behavior. The switching waveforms of these solutions are shown in Fig. 5. Solutions I and V, which are shown with triangles in Fig. 4, refer to the case without AGD, where $t_{var1} = t_{var2} = 0$ and $b_{int} = 0x0/0xF$ for turn-off/on.

In turn-off, 37 solutions out of 60 belong to the Pareto front. The solution I is located at the right edge of the front, meaning that it has the most significant surge voltage. Solutions with the value of b_{int} typically around 4 to 6 are scattered along the front. Solutions II and III have the same b_{int} but different t_{var1} and t_{var2}, which produces the difference in their fitness. It is an interesting result that several levels of V_{GS} can be used to obtain similar values of fitness.

In turn-on, 57 solutions belong to the Pareto front. Unlike at turn-off, the Pareto front can be divided into groups of solutions with the same value of b_{int} located closely to each other. For example, solutions around VII have $b_{int} = 0x9$, and solutions around VI have $b_{int} = 0x8$. The value of b_{int} increases as the solution locates further to the bottom-right.

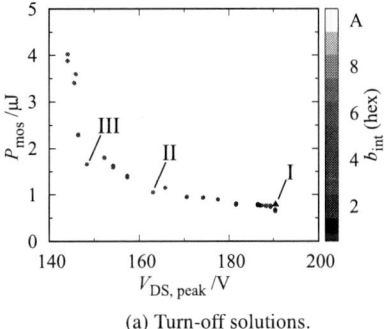

(a) Turn-off solutions.

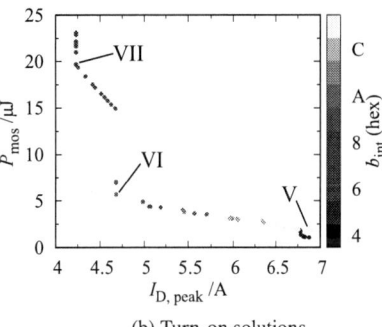

(b) Turn-on solutions.

Fig. 4: Fitness map of obtained solutions in final generation.

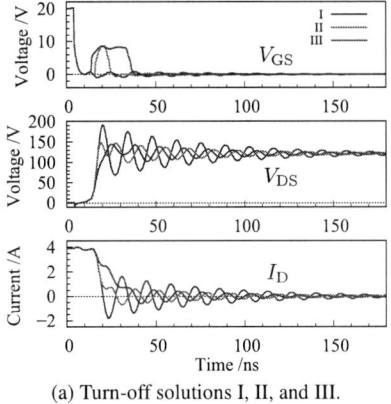

(a) Turn-off solutions I, II, and III.

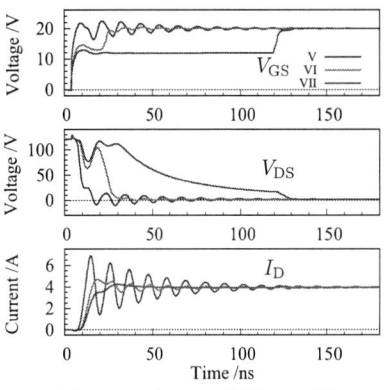

(b) Turn-on solutions V, VI, and VII.

Fig. 5: Selected switching waveforms of obtained solutions.

Comparison with gate-resistance-incremental method

The obtained Pareto front solutions are compared with those of standard gate-driving (SGD), where only 1-bit of the DAGD is used, and the gate resistance is set at $2^k\,\Omega$ $(k = 1, 2, 3\ldots)$. Figs. 6 (a) and (b) show the comparison at turn-off and turn-on, respectively, where SGD solutions are plotted with crosses. It is clear that the optimized DAGD solutions show a better trade-off between the two fitness in both cases. This is because the optimized solution of DAGD slows the switching only during the critical period, while in SGD the switching is stalled for the entire duration of the switching. These results show that DAGD outperforms SGD with a properly-optimized gate signal sequence, which can be selected depending on the design constraints.

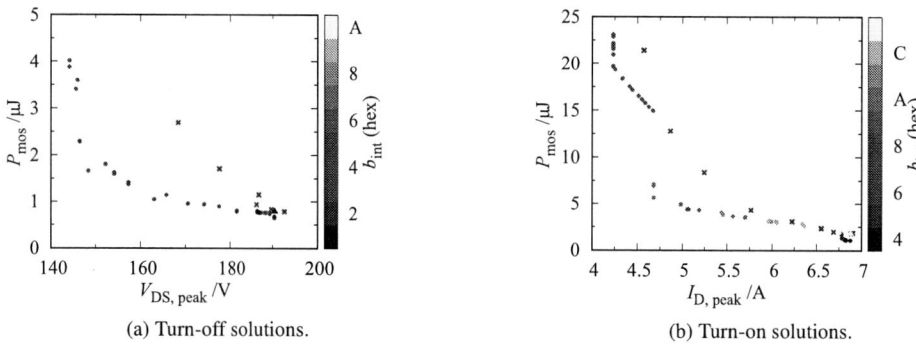

(a) Turn-off solutions.

(b) Turn-on solutions.

Fig. 6: Comparison of fitness between DAGD optimized by GA (dots) and SGD (crosses).

Optimization in different operating conditions

So far, the optimization tool of DAGD is verified in a single operating condition. We now apply it to a variety of operating conditions to see how the optimized solution changes accordingly. The operating conditions of $(V_{DS}, I_D) = (120\,\mathrm{V}, 8\,\mathrm{A})$, $(240\,\mathrm{V}, 4\,\mathrm{A})$, and $(240\,\mathrm{V}, 8\,\mathrm{A})$ are selected additionally. The focus is put on what value of b_{int} is selected, so compare the fitness maps of the optimized solutions.

Figure 7 shows the optimization results at turn-off in the three operating conditions. By comparing the three figures and also Fig. 4 (a), it is found that similar values of b_{int} of 0x4 to 0x6 are selected in every operating condition. This result is understood with the threshold voltage of MOSFETs, which does not change drastically when operating condition changes.

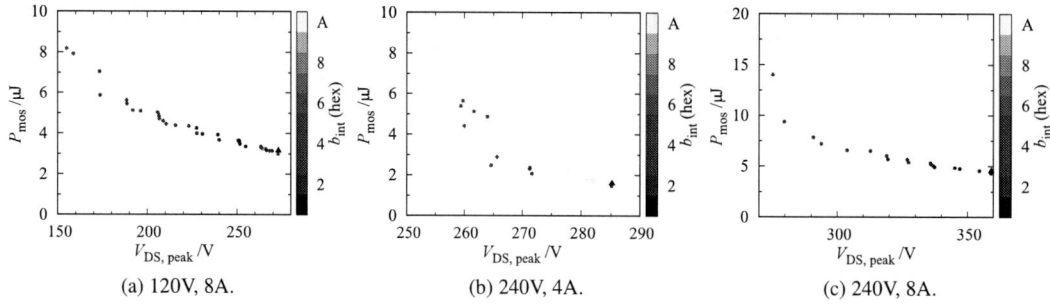

(a) 120V, 8A.

(b) 240V, 4A.

(c) 240V, 8A.

Fig. 7: Optimization results at turn-off in varied operating conditions.

Figure 8 shows the optimization results at turn-on. We clearly recognize a difference in the selection of b_{int}. For example, if we compare the solutions with the least $I_{D,peak}$, the selected bit is 0xB in Fig. 8 (a) and (c), while it is 0x9 in Fig. 4 (a) and Fig. 8 (b). This suggests that the effective value of b_{int} is higher for an operating condition with larger current. It complies with the fact that the Miller plateau voltage of the device increases as the operating current becomes larger.

These results confirm that the operating condition affects the AGD strategy mainly at turn-on. The proposed tool selects the appropriate gate signal sequences in every condition, dealing with such changes.

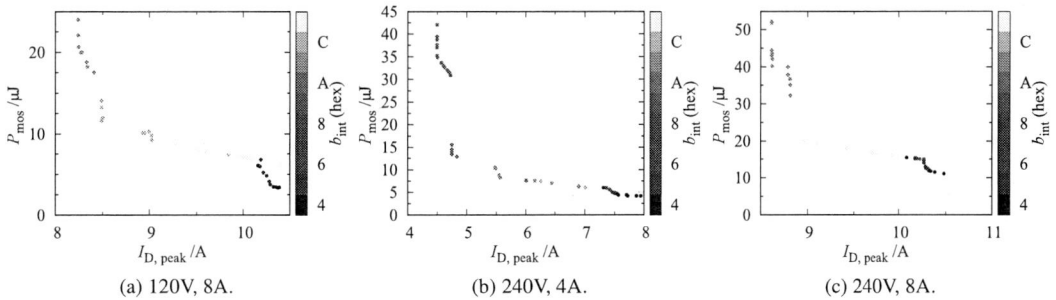

(a) 120V, 8A.　　　　　　　　(b) 240V, 4A.　　　　　　　　(c) 240V, 8A.

Fig. 8: Optimization results at turn-on in varied operating conditions.

Conclusion

This paper provided the Python-based simulation tool to optimize the gate signal sequence of DAGD using GA. The gate signal sequence is converted into a genotype with three variables to realize a simple and effective analysis of the resulting switching characteristics. It was shown that the Pareto-front solutions were successfully obtained, which outperformed the basic gate-driving method of increasing the gate resistance. It was also confirmed that the operating conditions play a critical role in the selection of effective V_{GS} level at turn-on in response to the shift of the Miller plateau voltage.

By utilizing the results obtained in simulation, the presented tool can be used for predicting the proper gate signal sequence in the experiment, according to the approved rate of overshoots and losses. Practical limitations, such as the time restrictions of the controllers and buffers, need to be taken into account. Also, the effect of the temperature and the resulting thermal drift are not considered in the simulation. A combination with a local search algorithm using closed-loop feedback will help optimize the operation furtherly according to such dispersion of device characteristics. The experimental verification of the optimization tool is now under development.

References

[1] A. P. Camacho, V. Sala, H. Ghorbani, and J. L. R. Martinez, "A Novel Active Gate Driver for Improving SiC MOSFET Switching Trajectory," *IEEE Trans. Ind. Electron.*, vol. 64, no. 11, pp. 9032–9042, 2017.

[2] Z. Zhang, J. Dix, F. F. Wang, B. J. Blalock, D. Costinett, and L. M. Tolbert, "Intelligent Gate Drive for Fast Switching and Crosstalk Suppression of SiC Devices," *IEEE Transactions on Power Electronics*, vol. 32, no. 12, pp. 9319–9332, 2017.

[3] S. Zhao, A. Dearien, Y. Wu, C. Farnell, A. U. Rashid, F. Luo, and H. A. Mantooth, "Adaptive Multi-Level Active Gate Drivers for SiC Power Devices," *IEEE Transactions on Power Electronics*, vol. 35, no. 2, pp. 1882–1898, 2020.

[4] D. J. Rogers and B. Murmann, "Digital active gate drives using sequential optimization," in *2016 IEEE Appl. Power Electron. Conf. and Expo. (APEC)*, ser. 2016 IEEE Appl. Power Electron. Conf. and Expo. (APEC), 2016, pp. 1650–1656.

[5] H. Takayama, T. Okuda, and T. Hikihara, "Digital active gate drive of SiC MOSFETs for controlling switching behavior—Preparation toward universal digitization of power switching," *International Journal of Circuit Theory and Applications*, no. July, p. cta.3136, 9 2021. [Online]. Available: https://onlinelibrary.wiley.com/doi/10.1002/cta.3136

[6] N. Idir, R. Bausière, and J. J. Franchaud, "Active Gate Voltage Control of Turn-on dv/dt and Turn-off dv/dt in Insulated Gate Transistors," *IEEE Trans. Power Electron.*, vol. 21, no. 4, pp. 849–855, 2006.

[7] P. D. Judge, R. Mathieson, and S. Finney, "A Six Level Gate-Driver Topology with 2 . 5 ns Resolution for Silicon Carbide MOSFET Active Gate Drive Development," in *12th Energy Conversion Congress and Exposition - ASIA (ECCE ASIA)*, 2021, pp. 2118–2123.

[8] K. Miyazaki, S. Abe, M. Tsukuda, I. Omura, K. Wada, M. Takamiya, and T. Sakurai, "General-Purpose Clocked Gate Driver IC With Programmable 63-Level Drivability to Optimize Overshoot and Energy Loss in Switching by a Simulated Annealing Algorithm," *IEEE Transactions on Industry Applications*, vol. 53, no. 3, pp. 2350–2357, 2017.

[9] K. Deb, A. Pratap, S. Agarwal, and T. Meyarivan, "A fast and elitist multiobjective genetic algorithm: NSGA-II," *IEEE Transactions on Evolutionary Computation*, vol. 6, no. 2, pp. 182–197, 2002.

[10] Y. Nakamura, M. Shintani, K. Oishi, T. Sato, and T. Hikihara, "A simulation model for SiC power MOSFET based on surface potential," in *International Conference on Simulation of Semiconductor Processes and Devices (SISPAD)*. Institute of Electrical and Electronics Engineers Inc., 10 2016, pp. 121–124.

[11] DEAP Project, "DEAP documentation," https://deap.readthedocs.io/en/master/, (accessed on 9th May 2022).

[12] SIMetrix Technologies Ltd., "SIMetrix Simulator," https://www.simetrix.co.uk, (accessed on 9th May 2022).

Analysis of balancing algorithms for Quasi-Two/Three-Level Single Phase Operation of a Flying Capacitor Converter

Stefan Mersche, Markus Bayer, Kai Rickert und Marc Hiller
Karlsruher Institut für Technologie (KIT) - Elektrotechnisches Institut
Kaiserstr.12
Karlsruhe, Germany
Phone: +49 (0) 721-608-42701
Email: stefan.mersche@kit.edu
URL: http://www.kit.edu

Keywords

≪Flying Capacitor converter≫, ≪Quasi-Two-Level≫, ≪Medium voltage converter≫, ≪silicon-carbide (SiC)≫, ≪Grid-connected converter≫.

Abstract

Today's standard medium voltage converters are operated with low switching frequencies and contain bulky passive components. One concept to change this is the Quasi-Two-Level operation (Q2O) of multilevel converters with fast switching semiconductors to minimize passive components. The Flying Capacitor Converter (FCC) with SiC semiconductors and operated with Q2O is thus a converter with minimized passive components. In this paper, a comparison of balancing algorithms for Quasi-Two-Level operation of a FCC is presented. The differences between the various methods are demonstrated by simulation and measurement results. Furthermore, a modulation principle with balancing for Quasi-Three-Level operation of a FCC is introduced. With the FCC, it is easy to upgrade from the Q2O to Quasi-Three-Level operation (Q3O) to take full advantage of a three-level output voltage.

Introduction

The generation of electrical energy is currently in a state of transition and will be massively expanded in the next few years due to global climate protection targets. The power of inverter-based generation plants for renewable energies is increasing and these can no longer be integrated exclusively into the low-voltage grids due to the increasing system power ratings. Thus, the grid applications for medium voltage converters will increase in relevance. Today's standard medium voltage converters are either based on the Modular Multilevel Converter (MMC) technology requiring a considerable amount of DC link capacitors or on various 3/5-level converter topologies which are operated at low switching frequencies and contain bulky line filters. Both concepts still lead to relatively high costs due to the high amount of passive components, which are major reasons for the slow spread of medium voltage power converters in grid applications. One promising concept is the Quasi-Two-Level operation (Q2O) of multilevel converters with fast switching semiconductors to minimize the passive components. The Flying Capacitor Converter with Quasi-Two-Level represents a promising variant of such a medium-voltage converter with a greatly reduced expenditure for the DC capacitors. The Quasi-Three-Level offers the possibility to further improve the quality of the output voltage. Compared to the Q2O, the capacitance of one DC capacitor per phase must be increased, but at the same time the filter elements on the AC side can be further reduced.

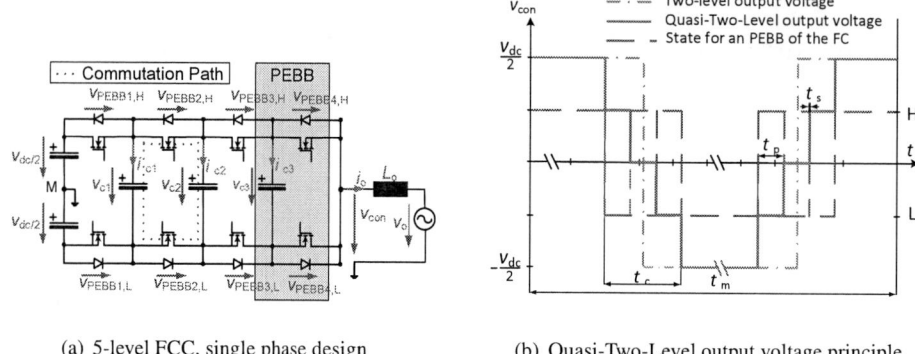

(a) 5-level FCC, single phase design   (b) Quasi-Two-Level output voltage principle

Fig. 1: 5-level flying capacitor converter, diagram of Quasi-Two-Level switching sequence

Fundamentals of the Flying Capacitor Converter

The concept of the Flying Capacitor Converter (FCC) was first introduced [1]. The multilevel voltage is generated by switching capacitors into the active current path. A n-level FCC ($n \in \mathbb{N}$) is made of $2 \cdot (n-1)$ power semiconductor switches and $(n-2)$ flying capacitors in addition to the DC link capacitor, an exemplary 5-level FCC is shown in fig. 1 (a). The commutation cells of the FCC consist of one high side and one low side semiconductor and the corresponding capacitor. To avoid a short circuit between the capacitors of two adjacent cells, only one transistor of each cell is switched on at the same time.

Each flying capacitor C_i has a different nominal voltage $v_{c,nom,i}$.

$$v_{c,nom,i} = v_{dc} \cdot \frac{n-1-i}{n-1} \quad i \in [1...(n-2)] \tag{1}$$

To simplify the construction of an FCC, the converter is built by using modular components i.e. Power Electronic Building Blocks (PEBB). Since the commutation cell is the repeating circuit element, the PEBB consists of one commutation cell without the second capacitor, as can be seen in fig. 1 (a). An n-level converter requires $(n-1)$ PEBBs. The required average forward blocking voltage of the semiconductors in a PEBB can be calculated with eq. (2).

$$v_{PEBB,nom} = \frac{v_{dc}}{n-1} \quad n \in \mathbb{N} \tag{2}$$

During operation, the blocking voltage across the semiconductors always deviates from the average blocking voltage in eq. (2), because the capacitor voltages are free floating and not clamped by external voltages sources. Hence, the voltage across the capacitors varies if the capacitors are switched into the output current path. This is the case for all output voltage levels except $+\frac{v_{dc}}{2}$ and $-\frac{v_{dc}}{2}$. In order to keep the blocking voltage of the semiconductors within a reasonable range , it is necessary to balance the capacitor voltages to keep their difference voltage within acceptable limits.

Depending on the operation mode, there are different ways to achieve this: Conventional multilevel operation with Alternative Phase Opposition Disposition (APOD) modulation or Phase Disposition (PD) works by exchanging carrier signals symmetrically for the corresponding semiconductor or with an active balancing algorithm [2].

For Q2O there are two types of balancing: passive or active. For the active method an additional voltage measurement is required. A method with passive balancing for a stationary operating point is presented in [3]. Furthermore, several active balancing algorithms are presented in literature [3, 4, 5]. For active balancing, there are essentially two degrees of freedom available in the modulation. In addition to the time period for which a capacitor is switched into the output current path, the current direction in the

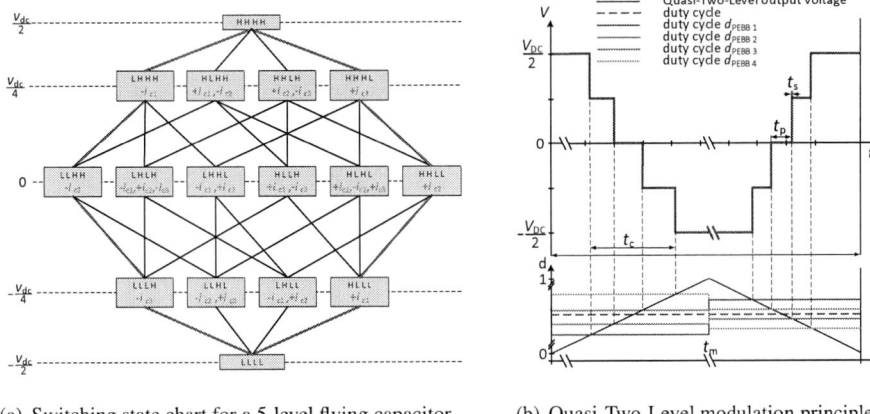

(a) Switching state chart for a 5-level flying capacitor   (b) Quasi-Two-Level modulation principle

Fig. 2: Quasi-Two-Level operation balancing and modulation

capacitor can also be varied due to the redundant switching states.

Quasi-Two-Level Operation

The basic idea of the Quasi-Two-Level modulation is to use the higher level multilevel topology in combination with a two-level modulation scheme. Hence, the applied control and modulation strategies are very similar to the simple ones used for a standard two-level converters. In addition, the advantages of multilevel converters in terms of output voltage waveform (reduced $\mathrm{d}v/\mathrm{d}t$, lower overvoltages) can be used even at high output voltages. The frequency spectrum of the AC output voltage is comparable to that of a 2-level converter. Figure 1 (b) shows a simplified output voltage (red line) of the Quasi-Two-Level modulation and for comparison the output voltage of a two-level converter (dashed green line). Each time a PEBB state (dashed blue line) is changed, a different multilevel voltage is generated. The quasi-two-level switching time t_C is typical below than 5% of the modulation time t_m and in the range of usual interlocking times. For better representation t_C was increased in Figure 1 (b) and Figure 2 (b). The time t_C is the sum of t_p of a switch event.

The Q2O of the FCC with an balancing algorithm was presented independently in [3, 4, 5]. The problem of balancing was fundamentally solved with two different methods. This principle of the modulation of the Quasi-Two-Level algorithm is illustrated in fig. 2 (b) by using the method described in [4]. A carrier-based modulation with an carrier for each PEBB is used to generate the gate signals. Each PEBB has different levels of duty cycle to switch with time delay so that the the Quasi-Two-Level output voltage is generated. The publications [3, 5] showed no significant differences by generation the gate-signals.

Algorithms of Balancing the Capacitor Voltages by Quasi-Two-Level Operation

The balancing methods published so far can be divided into two methods: The first methods [3, 5] has a fixed switching sequence and varies t_p, i.e. the time an intermediate voltage is active. The second methods, formulated and investigated in our earlier work in [4], varies the switching sequence while t_p fixed.

The algorithms of [5] only use the switching sequences marked in thick blue in fig. 2 (a). Depending on the capacitor voltage deviations, the time t_p is selected to be either minimal or maximal. The maximum of t_p depends on the capacitor capacitance and its permissible voltage range in operation. The minimum of t_p depends on the locking time of the used semiconductors, which is dependent on the switching times of the semiconductors ($t_p > t_s$). If at the next switching state the deviation of the actual capacitor voltage and the calculated mean value of the capacitor voltage $v_{c,\mathrm{nom},i}$ decreases, then this switching state is set with the maximum t_p, otherwise the switching state is set only for the minimum t_p.

Analysis of balancing algorithms for Quasi-Two/Three-Level Single Phase Operation of a Flying Capacitor Converter   MERSCHE Stefan Christoph

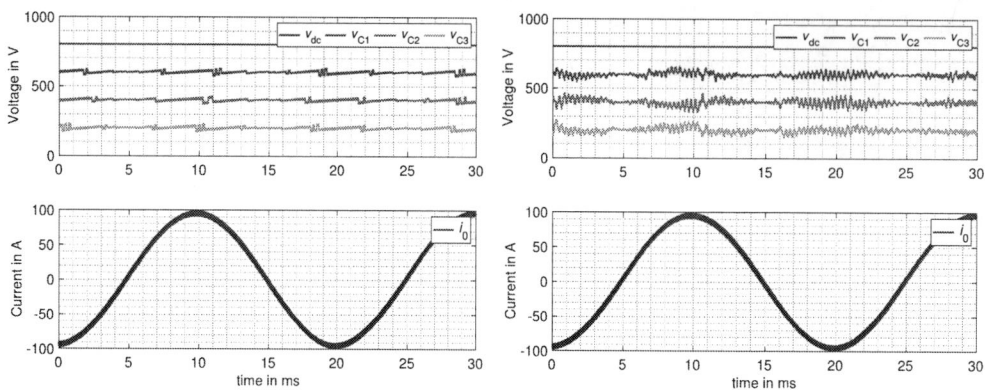

(a) Simulation results of the first methods [5] balancing algorithm - fixed switching sequences and different time t_p

(b) Simulation results of the second methods [4] balancing algorithm - variable switching sequences and fixed time t_p

Fig. 3: Comparison of capacitor voltage waveforms with the same simulation configuration

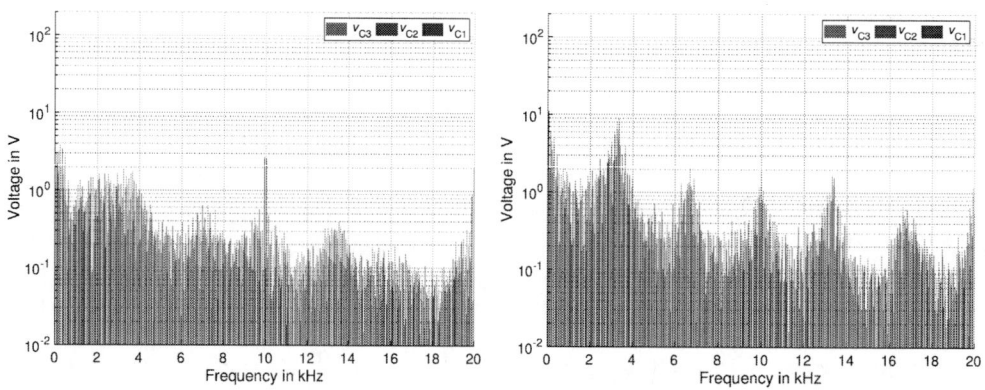

(a) Frequency spectrum of the first methods [5] balancing algorithm - fixed switching sequences and different time t_p

(b) Frequency spectrum of the second methods [4] balancing algorithm - variable switching sequences and fixed time t_p

Fig. 4: Comparison of the frequency spectrum of capacitor voltage waveforms

For the algorithms in the second methods, all possible switching sequences in fig. 2 (a) are used. An optimal path is selected depending on the deviation of the actual capacitor voltage and the calculated mean value of the capacitor voltage $v_{\mathrm{c,nom},i}$. Here, the allowable range of t_p is also limited by the above mentioned parameters, but a fixed value is used for this algorithm.

Comparison of the Balancing Algorithm for Quasi-Two-Level Operation with simulation Results

To compare the balancing principles, one algorithm from each methods was implemented in the same simulation environment. The setup is a 5-level FCC with capacitance of 1000nF, a switching frequency of 10 kHz and an inductance L_o of 1 mH in the output. At the output, a single-phase voltage source is connected to the center point M of the split DC link. The split DC link is clamped to $+\frac{v_{\mathrm{dc}}}{2}$ with a voltage source directly in the simulation. The hardware parameters of the components as well as the control and carrier-based modulation are identical. Therefore, the differences of the capacitor voltages are caused by the different algorithms. Figure 3 shows the various curves of the capacitor voltages and the output current. It can be seen that both algorithms keep the capacitor voltages stable within a tolerance band around their nominal set-point. The algorithm from the second methods [4] always has larger deviations in capacitor voltages than the algorithm from the first methods [5]. The algorithm from second methods produces a larger mean capacitor voltage deviation than the algorithm from first methods. In this comparison, the frequency spectrum of the capacitor voltages were also considered, whose absolute values are lower with algorithm from first methods than with algorithm from second methods. The frequency spectrum of the capacitor voltages of the respective algorithms are plotted in fig. 4. The high-frequency voltage deviation with algorithms of first methods are existing exclusively by the switching frequency. The variable switching sequence, on the other hand, does not show such a significant conetration at the switching frequency.

In summary, the results for the algorithm of the first methods are better than for the second methods. Moreover, the coding effort for the algorithm of the first methods is lower than for the one of the second methods. Thus, the first algorithm has significant advantages over the second in stationary operation. For dynamic changes and error cases, the second method has more degrees of freedom.

Quasi-Three-Level Operation

The basic idea of the Quasi-Three-Level modulation is to use the multilevel topology in combination with a three-level modulation scheme. The approach is very similar to that of the Quasi-Two-Level. For the passive components in the output filter, the Q3O makes a significant difference since the voltage deviations to a sine wave with the Q3O is significantly smaller. The Common Mode Voltage is just as smaller and thus a lower stress on the insulation. The modulation is comparable to a three-level modulation and the superimposed control is equal to a standard three-level control, only the balancing is fundamentally different. The Quasi-Three-Level Operation (Q3O) of FCC was introduced in [6].

The basic difference to the Q2O is that one capacitor C_i needs a larger capacitance. This capacitor is required for the zero-voltage output level of the Quasi-Three-Level operation and thus it is not only switched into the current path transiently for a short time t_p and should be able to carry the output current for a full modulation period t_m, depending on the modulation duty cycle. Considering fig. 2 (a) for the 5-level FCC, the possible zero-voltage switching combinations where only one capacitor carries the output current are the combinations with capacitor C_2. The corresponding capacitance of C_2 is calculated according to eq. (3). The design is equivalent to the dimensioning explained in [4] for Quasi-Two-Levels capacitors. In addition to the modulation period t_m, the capacitance depends on the maximum output current $\hat{i}_o$ and the acceptable capacitor voltage variation $\Delta v_{c,max}$ during operation. Usually $\Delta v_{c,max}$ is selected in the range of 10...15 % of $v_{PEBB,nom}$.

$$C_{C2} = \frac{t_m \cdot \hat{i}_o}{\Delta v_{c,max}} \tag{3}$$

A novel approach of Modulation and Balancing Principle of the Quasi-Three-Level Operation is described below.

Novel Modulation and Balancing Principle of the Quasi-Three-Level Operation

In fig. 5 (b), the principle of the Quasi-Three-Level modulation is illustrated. A multi-carrier method is used. Regarding the modulation, there are four states which are switched on for a longer time than t_p. These are the blue colored states in fig. 5 (a). The other states are used, as for the Quasi-Two-Level, only for a short time t_p.

For Quasi-Three-Level operation, the balancing of the capacitor voltages takes place in two stages. First, the capacitor C_2 is balanced. Thus, depending on the deviation of the current capacitor voltage from the calculated mean value, one of the states LLHH or HHLL is selected. This is done depending on the difference between the actual capacitor voltage and its nominal value. The state which reduces the deviation is chosen once per modulation period.

The second step is to balance the other capacitors. For this purpose, the states required for the next switching event are determined according to the balancing principle of [4]. In this principle, fixed time

Analysis of balancing algorithms for Quasi-Two/Three-Level Single Phase Operation of a Flying Capacitor Converter

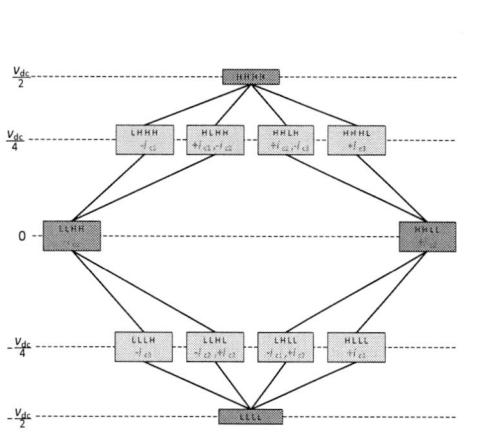

(a) Switching state chart for a 5-level flying capacitor for Quasi-Three-Level Operation

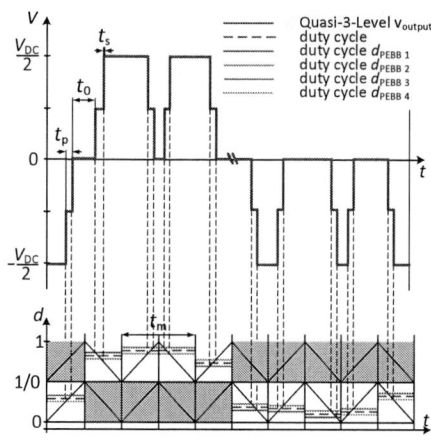

(b) Quasi-Three-Level modulation principle

Fig. 5: Quasi-Three-Level operation balancing and modulation

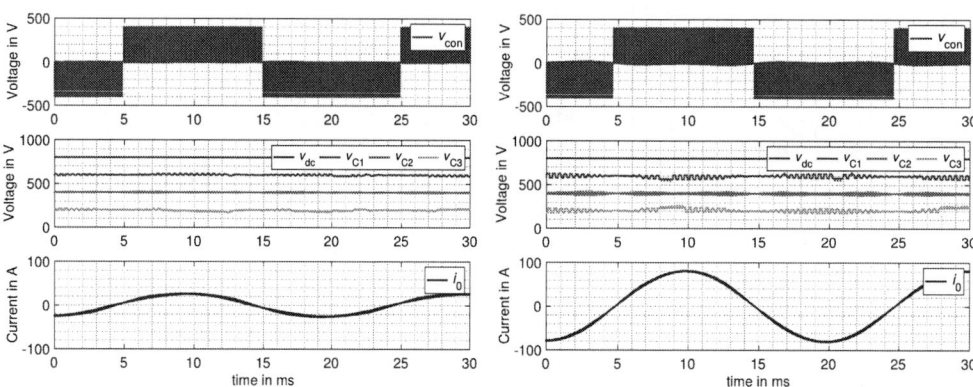

(a) Simulation results of the balancing algorithms for Quasi-Three-Level with an output current of 25 A

(b) Simulation results of the balancing algorithms for Quasi-Three-Level with an output current of 80 A

Fig. 6: Simulation results for Quasi-Three-Level with the balancing algorithms by various current value

periods t_p are used and balancing is done by choosing different states to reduce the deviation of capacitor voltages from their mean values. This is performed two times per modulation period.

The principle of the other methods of Quasi-Two-Level balancing would not be suitable here, because balancing is not possible with a fixed switching sequence and variable t_p. There is no charging and discharging of the capacitors realizable by shorter or longer t_p in the possible paths. This would require negative t_p, which it is not possible to realize.

Simulation Results of the Quasi-Three-Level Operation

The simulation setup is a 5-level FCC with capacitance of 1000nF for C_1 and C_3 - the capacitor C_2 with 0.08mF , a switching frequency of 10 kHz and an inductance L_o of 1 mH in the output. At the output, a single-phase voltage source is connected to the center point M of the split DC link. The split DC link is clamped to $+\frac{V_{dc}}{2}$ with a voltage source directly in the simulation. The simulation results of the Quasi-Three-Level are shown in fig. 6. To demonstrate the Quasi-Three-Level operation, the output voltage of the converter was plotted in addition to the capacitor voltage waveforms and the grid current. The balancing algorithm keeps the capacitor voltage within the permissible deviations and works stable. The algorithm including its implementation has not yet reached the same quality of the capacitor voltage

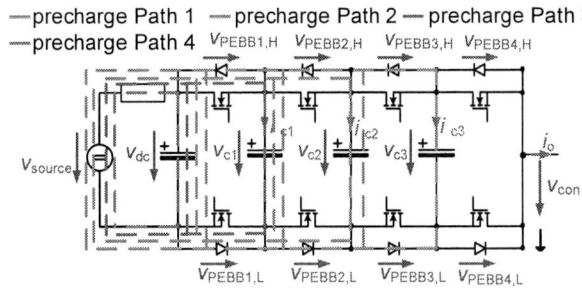
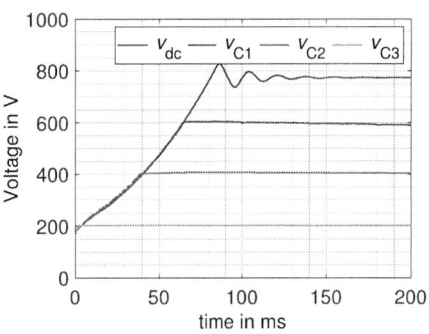

(a) Sketch of the Flying Capacitor with precharge paths of the capacitors

(b) Measurement of the precharging of the FCC with low voltage

Fig. 7: Start up for Operation by a single phase Flying Capacitor Converter

curve as the algoithms of the Quasi-Two-Level. The principle of selecting the path with the minimum deviation of the capacitor voltages leads to the result that a continuous deviation of the capacitor voltage is achieved in some operating ranges of a capacitor. These continuous deviations will be found in the capacitor voltages with always similar values of the output current. In the future, this effect should be examined in more detail.

Start up for Operation by a single Phase Flying Capacitor Converter

In order to operate the converter, the capacitors must be precharged so that the distribution of the DC link voltage results in the semiconductors after eq. (2). The precharging of the FCC with the flying capacitors and DC link capacitor is realized from the DC side.

Precharging works in such a way that in the first step all necessary semiconductors are switched on, so that all capacitors are connected in parallel. The current flow path is drawn in fig. 7 (a) as precharge path 1. Then, the voltage of the DC source is ramped up and the capacitors are charged with a constant current. When the voltage of the capacitors has reached the nominal value of capacitor C_3, the according semiconductors are switched off and the other capacitors are charged. The corresponding current flow is labeled as precharde path 2 in fig. 7 (a). The algorithm always switches off the respective semiconductors when the capacitor voltage has reached the setpoint of the respective capacitor. In the example of a 5-level FCC, the semiconductor is switched off a total of 3 times until the DC link capacitor is charged. In fig. 7 the precharge process is plotted from a measurement of the capacitor voltages. The overshooting of the DC link voltage is a result of the dynamics of the current and voltage regulation of the DC power supply. After the capacitors are charged, the balancing of the capacitor voltages and modulation of the output voltage as well as the regulation of the output side can start directly.

Measurements with using a single Phase Flying Capacitor Converter

In [7] the hardware of a Power Electronic Building Block for a FCC for Q2O was presented. The measurement results were generated with two PEBB's connected in series. The setup is a 5-level FCC with capacitance of 1000 nF, a switching frequency of 10 kHz and an inductance L_o of 1 mH in the output.

In the following, measurement results for the operating of the FCC with buck converter mode are presented. The buck converter operation was implemented with an LC filter and DC power supply at the output and a DC power supply to supply the DC link. In single-phase operation, buck converter mode is easier to implement than a single-phase voltage source in combination with a split DC link capacitor.

For the measurements, the voltage of each PEBB ($v_{PEBB,nom}$) was chosen to be 100 V. The voltage was selected low to increase the visibility of the voltage changes by the measurement. Therefore, the duty cycle in operation is in the range of 25 %, because the DC power supply at the output limits the output voltages up to 100 V and output current up to 100 A. .

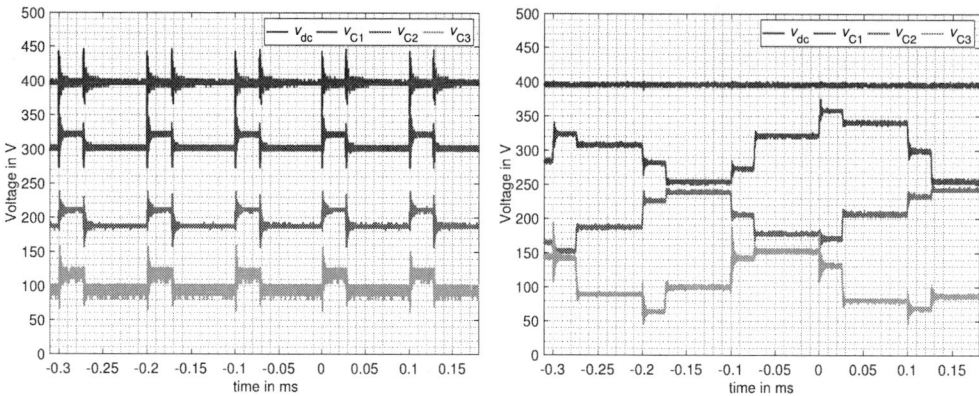

(a) The measured capacitor voltage of the first methods balancing algorithm - fixed switching sequences and different time t_p

(b) The measured capacitor voltage of the second methods balancing algorithm - variable switching sequences and fixed time t_p

Fig. 8: Measurement results by a single phase FCC for Q2O with buck converter operation

Comparison of the Balancing Algorithm for Quasi-Two-Level Operation with Measurements Results

In fig. 8 the capacitor voltage curves of the different balancing algorithms are shown. The measurement was executed with 75 A in each case, so that the capacitor voltage changes are similar. In fig. 8 (a) , t_p was varied in the range of 500 ns to 1000 ns. The switching sequence has been fixed as described in [5]. In fig. 8 (b) , t_p was fixed at 500 ns. The switching sequence has been selected according to [4].

It can be seen that the curve of the capacitor voltages of the measurement is similar to the simulation results. The difference is that in the measurement, a constant operating point with the same output current and duty cycle was used in comparison to the AC current in the simulation.

The different balancing algorithms result in different voltage deviations over time. The average voltage deviation for fig. 8 (a) is smaller than for fig. 8(b). The fixed switching sequence always charges each capacitor and discharges each in the same way. The magnitude of the voltage deviation depends on t_p and i_o.

In fig. 8 (b) the different switching sequences lead to voltage deviations, which need to be compensable by the algorithm. During the dimensioning of the capacitors and semiconductors of the prototype, a maximum permissible deviation of 80 V of the capacitor voltage was specified, which both of them fulfill.

What is noticeable in the measurement in fig. 8 (a) is that an oscillation is formed on the capacitor voltages due to the fixed switching sequence. This is not as noticeable with different switching sequences in fig. 8 (b). This is a disadvantageous characteristic, which has not been shown in the simulation results so far. Not all passive effects of the real structure are modeled in the simulation. The distributed leakage inductance in the commutation cells is implemented as a central element and the modeling of the semiconductors does not include a voltage-dependent output capacitance of the mosfets. Since in [5] the adjacent PEBBs are always switched one after the other, a resonance is formed in the structure. Therefore, an oscillation is formed between the Mosfet output capacitances and the distributed leakage inductance, which is transferred to the neighboring PEBBs when they also switch with a delay. The impact of this behavior is also visible in the overvoltages in the following chapter.

Comparison of the output voltage with different operation

The output voltage waveforms of the Quasi-Two-level and Two-level operation are illustrated in fig. 9,fig. 10 and fig. 11.

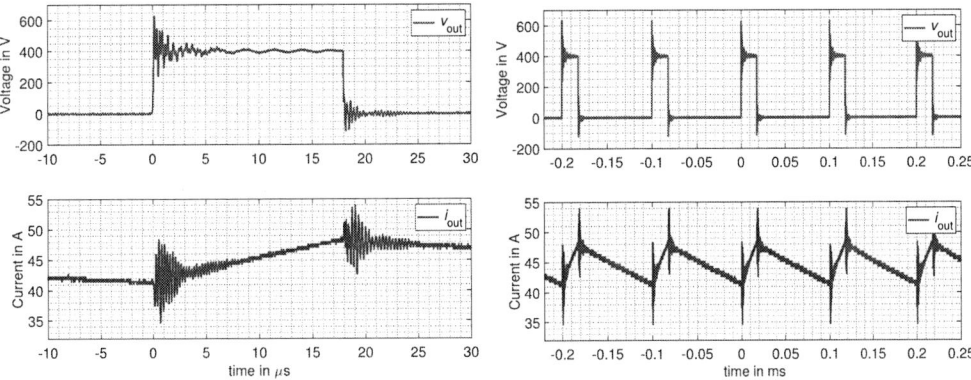

(a) The Two-Level Output voltage by buck converter operation with $t_\mathrm{p} = 0$

(b) An Overview of Two-Level Output voltage by buck converter operation with $t_\mathrm{p} = 0$

Fig. 9: Measurement results by a single phase Flying Capacitor Converter for Two-Level Operation

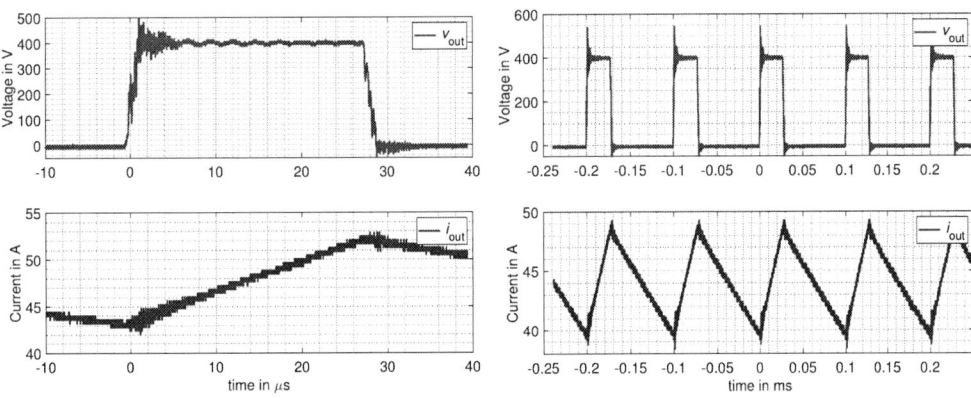

(a) The Quasi-Two-Level Output voltage by buck converter operation

(b) An Overview of Quasi-Two-Level Output voltage by buck converter operation

Fig. 10: Measurement results by a single phase FCC for Q2O with balancing the capacitor voltage with the first methods [5] balancing algorithm - fixed switching sequences and different time t_p

With the prototype, it is possible to switch all semiconductors synchronously ($t_\mathrm{p} = 0$) or with a time delay ($t_\mathrm{p} > 0$). In the graphs, the output voltage v_con is always measured against $-\frac{v_\mathrm{dc}}{2}$. There is always a plot with the switching events shown magnified and many events in a sequence.

In fig. 10 (a) and fig. 11 (a) the Quasi-Two-Level output voltage is shown and the three short-time switched multilevel voltage stages can be seen. In fig. 9 (a) the two-level output voltage is shown, which is generated by synchronously switching all semiconductors on the high or low side.

Depending on the choice of t_p and the balancing algorithm, the overvoltage at the output-voltage varies. In the setup, the overvoltage is the largest for the two-level operation. With the quasi-two level, the overvoltage is always smaller. However, it can be seen in fig. 11 (b), that depending on the individual switching frequencies, the overvoltage becomes smaller and larger. The overvoltage with the used switching sequences according to [5] are the switching sequences with the largest overvoltage. Considering the reduction of the overvoltage, the balancing principle with variable switching sequence with its greater complexity can outperform the principle with fixed switching sequence with a different weighting of the selection of successive switching states.

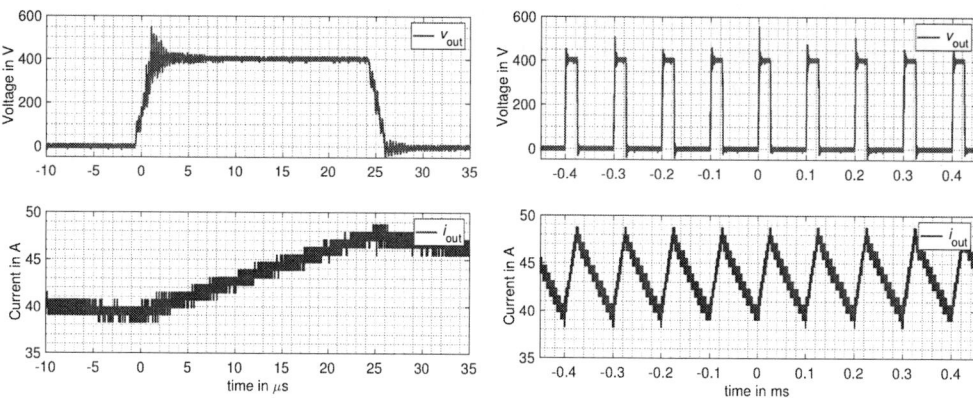

(a) The Quasi-Two-Level Output voltage by buck converter operation

(b) An Overview of Quasi-Two-Level Output voltage by buck converter operation

Fig. 11: Measurement results by a single phase FCC for Q2O with balancing the capacitor voltage with second methods [4] balancing algorithm - variable switching sequences and fixed time t_p

Conclusion

This paper presents a comparison and analysis of balancing principles for Q2O of the FCC. The algorithms with fixed switching sequence are easier to implement and generate smaller deviations of the capacitor voltage. Futhermore Simulation results and measurement results of Q2O were presented. The capacitor voltage deviations in simulation and measurements are comparable and thereby validate the function of the algorithms. With a fixed switching sequence, the measurements of the capacitor voltages have shown that an oscillation in the capacitor voltages is pronounced. The effects of the different balancing algorithms on the output voltage were analyzed with the measurement results of the Q2O. The overvoltage of the Quasi-Two-Level operation is always smaller than by Two-Level Operation, however, with the choice of the switching sequence this overvoltage can be additionally positively influenced. In addition, a Quasi-Three-Level operation of the FCC with a modulation principle and novel balancing algorithm is presented. For this purpose, simulation results of the balancing algorithm in Quasi-Three-Level operation are presented with a brief analysis of the resulting present. The Quasi-Three-Level operation can be achieved with only the increase the capacitance of one capacitor per phase. Pre-charge operation of capacitor voltages was explained and and verified experimentally by measurement results.

References

[1] T. Meynard and H. Foch, "Multilevel choppers for high voltage applications", EPE J., vol. 2, no. 1, pp. 45-50, Mar. 1992.

[2] G. Clos, L. Schindele, T. Franke, S. Gaertner, "Simple direct capacitor voltage balancing of a flying capacitor converter," EPE 2005 - Dresden, ISBN: 90-75815-08-5.

[3] P. Czyz, P. Papamanolis, T. Guillod, F. Krismer and J. W. Kolar, "New 40kV / 300kVA Quasi-2-Level Operated 5-Level Flying Capacitor SiC Super-Switch IPM," 2019 10th International Conference on Power Electronics and ECCE Asia (ICPE 2019 - ECCE Asia), Busan, Korea (South), 2019, pp. 813-820.

[4] S. Mersche, D. Bernet and M. Hiller, "Quasi-Two-Level Flying-Capacitor-Converter for Medium Voltage Grid Applications," 2019 IEEE Energy Conversion Congress and Exposition (ECCE), Baltimore, MD, USA, 2019, pp. 3666-3673, doi: 10.1109/ECCE.2019.8913201.

[5] S. Gierschner, Y. Hein, M. Gierschner, A. Sajid and H. Eckel, "Quasi-Two-Level Operation of a Five-Level Flying-Capacitor Converter," 2019 21st European Conference on Power Electronics and Applications (EPE '19 ECCE Europe), Genova, Italy, 2019, pp. P.1-P.9, doi: 10.23919/EPE.2019.8915445.

[6] M. Schweizer and T. B. Soeiro, "Heatsink-less Quasi 3-level flying capacitor inverter based on low voltage SMD MOSFETs," 2017 19th European Conference on Power Electronics and Applications (EPE'2017 ECCE Europe), Warsaw, 2017, pp. P1-P.10, doi: 10.23919/EPE17ECCEEurope.2017.8098916.

[7] S. Mersche, R. Schreier, P. Himmelmann and M. Hiller, "Medium Voltage Power Electronic Building Block for Quasi-two-level Operation of a Flying Capacitor Converter," 2021 23rd European Conference on Power Electronics and Applications (EPE'21 ECCE Europe), 2021, pp. P.1-P.10.

Instability in active balancing control of dc bus voltages in VSC converters interconnected via multi-winding transformers

Duro Basic, Sami Siala
General Electric, Power Conversion
18 Avenue de Québec, Villebon-Sur-Yvette, 91140 France
E-Mail: duro.basic@ge.com, sami.siala@ge.com
URL: https://www.gepowerconversion.com/

Keywords

«Parallel converters», « Pulse-Width-Modulated (PWM) converters », «Interleaving», «Grid connected converters»

Abstract

In high power applications, 3-phase ac/dc Voltage Source Converters (VSCs) are paralleled to increase the overall aggregated converter power rating, number of voltage steps, effective switching frequency and control bandwidth. For this, the interleaved Pulse Width Modulation (PWM) and current limiting converter paralleling inductors are utilized. Interphase Transformers (ITRs) or Multi-Winding Transformers (MWTs) are often adopted for the converter paralleling because they provide large inductances for the currents flowing among the converters while introduce relatively lower leakage inductances for the currents flowing from the converters to the power grid. However, large asymmetries in the inductances created by magnetic coupling in these transformers make it difficult to independently control active power flows in the converters. Thus, if the converter dc buses are not interconnected, instabilities in the active balancing control of the converter dc bus voltages are possible in certain operational conditions. In this paper MWT model and effects of magnetic coupling introduced by it on controllability of the power flows among the converters are presented. Operational condition leading to instability of the active balancing control is theoretically defined and validated in simulations.

Introduction

In high power applications, a number N of identical 3-phase ac/dc VSCs can be connected in parallel to increase power rating of the aggregated converter [1]-[4]. To increase the effective switching frequency, control bandwidth and reduce switching harmonics in the output voltage, the PWM carriers of the individual converters are mutually interleaved [5]. Thus, the converter paralleling requires utilization of coupling inductors to limit high frequency currents flowing among the converters which are produced by non-synchronous PWM switching. With the interleaved PWM, large paralleling inductors are needed to limit the current PWM switching harmonics [6]. To avoid large voltage drops across the paralleling inductors, ITRs can be used which provide large magnetizing inductance for the currents flowing among converters (cross currents) while introducing a negligible leakage inductance for the currents flowing from the converters to the power grid (cumulative currents) [4],[7]. Alternatively, the converters can be paralleled via MWTs as shown in Fig. 1, where the transformer leakage inductances provide the required paralleling inductances, grid-converter voltages matching and isolation [1] [2]. In such systems the MWT is normally designed to maximize the leakage inductances among the secondary windings [8] to minimize the PWM cross currents flowing among the secondary windings. In the converter systems with isolated or stacked dc busses (Fig. 1) the dc bus voltages of individual converters may diverge due to slight system imbalances. To ensure balanced dc bus voltages in all converters, the net active power absorption in each converter dc bus must be actively controlled, typically via control of the cross currents. However, unexpected instabilities in the dc bus voltage balancing control may be encountered in operatorial points characterised by low capacitive power factors. These instabilities have been previously reported in the STATCOM applications [9] with the converters paralleled via ITRs [4]. These instabilities have been linked in [9] to extremely large asymmetries of the inductances in the cumulative and cross current paths created by the converter paralleling via ITRs. In this paper it will be shown that such type of instabilities can also appear in the converter systems paralleled via MWTs.

The paper is organized as follows. Initially, modelling of MWTs using equivalent polygonal networks and coupled inductors is presented. Then the multi converter system model is derived using decompositions of the converter voltages and currents into so called cumulative components (defining power exchange with the power grid) and cross components (related to the cross currents and power flows among the converters). Based on such decompositions a multi-variable model is derived which describes cross-power flows among the converters in function of the cross currents. This multi-variable model considers the magnetic cross-coupling introduced by MWTs and gives an insight into the origin of the balancing control instability and allows quantification of the critical capacitive cumulative reactive current injection which initiates it. Finally, results of simulations of an example system are presented to validate the theoretical results and lustrate onset of the instability.

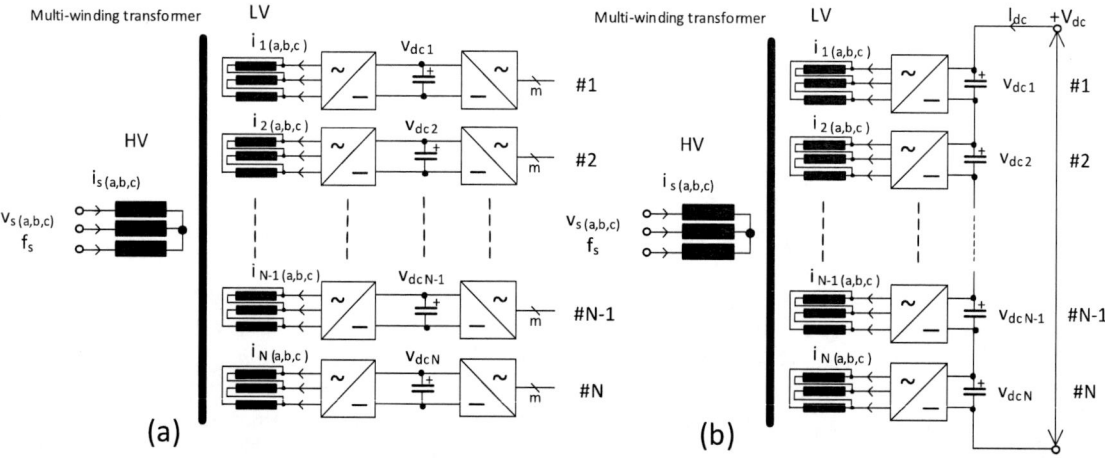

Fig. 1: Converter system based on N VSC converters interconnected using MWT with (a): individual back-back connected ac/dc/ac converters and (b): stacked ac/dc converters and common dc link.

Modelling of multi-winding transformers

The leakage impedances of MWTs with a primary (Higher Voltage, HV) and N secondary (Lower Voltage, LV) windings can be modelled by a polygonal network shown in Fig. 2 [10] [11]. The network parameters can be deduced from a matrix of short circuit voltages (shown in Table 1) which defines the short circuit voltages or reactances among all combinations of two windings in MWTs with N secondary windings [12]. This matrix can be obtained via calculations (in the transformer design phase) or in the short circuit tests (in the transformer validation phase). In Table 1 it is assumed that one winding (placed in a particular row) is energised while other winding (placed in a particular column) is short circuited. The voltage across the energised winding is increased until nominal current in that winding is reached. The short circuit voltage is then recorded in the table (at the intersection of the row /column corresponding to the position of the energised/short, circuited winding). The short circuit voltages in the matrix are typically expressed in per unit system where the base apparent power can be nominal apparent power of a single LV winding or power of the overall transformer (the base apparent power Sb should be clearly stated in the table). Dimension of the Table 1 depends on number of LV windings (N).

Table 1: Short circuit voltages $V_{SC\ LVi\text{-}HV}/\ V_{SC\ LVi\text{-}LVj}$ (p.u. impedances $X_{SC\ LVi\text{-}HV}/\ X_{SC\ LVi\text{-}LVj}$) which define magnetic couplings between all combinations of two windings in a MWT with N secondary windings.

Sb Base		Short-circuited Side						
		HV	LV1	LV2	LV3	LV4	LV5	LV6
Widing submited to rated current	HV							
	LV1	Vsc% $_{LV1\text{-}HV}$	N=1					
	LV2	Vsc% $_{LV2\text{-}HV}$	Vsc% $_{LV2\text{-}LV1}$	N=2				
	LV3	Vsc% $_{LV3\text{-}HV}$	Vsc% $_{LV3\text{-}LV1}$	Vsc% $_{LV3\text{-}LV2}$	N=3			
	LV4	Vsc% $_{LV4\text{-}HV}$	Vsc% $_{LV4\text{-}LV1}$	Vsc% $_{LV4\text{-}LV2}$	Vsc% $_{LV4\text{-}LV3}$	N=4		
	LV5	Vsc% $_{LV5\text{-}HV}$	Vsc% $_{LV5\text{-}LV1}$	Vsc% $_{LV5\text{-}LV2}$	Vsc% $_{LV5\text{-}LV3}$	Vsc% $_{LV5\text{-}LV4}$	N=5	
	LV6	Vsc% $_{LV6\text{-}HV}$	Vsc% $_{LV6\text{-}LV1}$	Vsc% $_{LV6\text{-}LV2}$	Vsc% $_{LV6\text{-}LV3}$	Vsc% $_{LV6\text{-}LV4}$	Vsc% $_{LV6\text{-}LV5}$	N=6

The Table 1 can be viewed as a generalisation of the standard short circuit voltage specification of two-winding transformers. For two winding transformers with single LV winding, there is only a single short circuit voltage value (Vsc% $_{LV1-HV}$). For 3 winding transformers with 2 LV secondaries, 3 short circuit voltages are needed to characterise all leakage reactances in the transformer. In a general case of a $N+1$ winding transformer with N LV windings, $N(N+1)/2$ short circuit voltages are needed.

The short circuit voltages in Table 1 are directly related to the short circuit impedances between all pairs of two winding ($X_{SC\ LVi-HV}$ or $X_{SC\ LVi-LVj}$). From them it is possible to identify all parameters of the polygonal network in Fig. 2, which interconnects all nodes/winding terminals. The winding resistances can be added to the inductance network to complete MWT model.

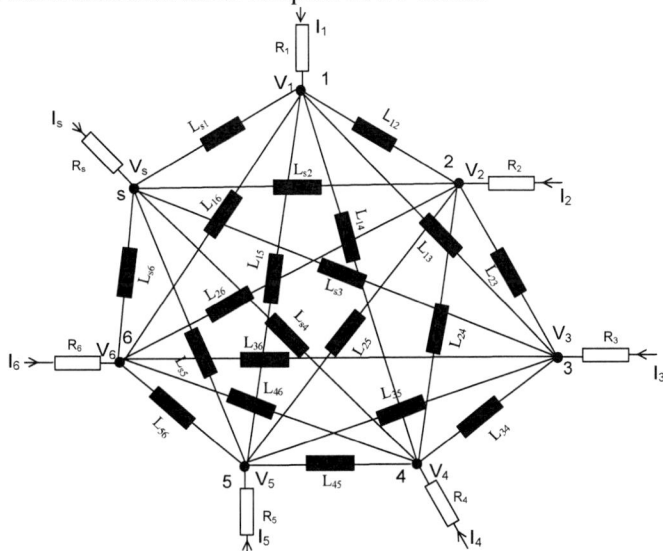

Fig. 2: Example of the equivalent polygonal scheme of a HV+6 LV MWT. All windings are referred to same voltage level (typically the secondary windings voltage level).

In the polygonal network Fig. 2 the winding terminal voltages and currents can be linked by the node equations written in the matrix form for the entire transformer model ($N+1$ nodes):

$$[I_t] = [Y_t]_{(N+1)\times(N+1)} [V_t] \tag{1}$$

In (1) $[I_t]$ is a vector containing all MWT windings currents, $[V_t]$ is a vector of all MWT windings voltages and $[Y_t]_{(N+1)\times(N+1)}$ is MWT admittance matrix. All reactances, voltages and currents are referred to same voltage (usually LV voltage) assuming equivalent star connection. In developed form, the MWT nodes matrix equation can be deduced by inspection from Fig. 2. It has the following general form:

$$
\begin{bmatrix} I_s \\ I_1 \\ I_2 \\ - \\ - \\ I_{N-1} \\ I_N \end{bmatrix}
=
\begin{bmatrix}
\dfrac{1}{jX_{ss}} & -\dfrac{1}{jX_{s1}} & -\dfrac{1}{jX_{s2}} & - & - & \dfrac{1}{jX_{s(N-1)}} & -\dfrac{1}{jX_{sN}} \\[2ex]
-\dfrac{1}{jX_{1s}} & \dfrac{1}{jX_{11}} & -\dfrac{1}{jX_{12}} & - & - & -\dfrac{1}{jX_{1(N-1)}} & -\dfrac{1}{jX_{1N}} \\[2ex]
-\dfrac{1}{jX_{2s}} & -\dfrac{1}{jX_{21}} & \dfrac{1}{jX_{22}} & - & - & -\dfrac{1}{jX_{2(N-1)}} & -\dfrac{1}{jX_{2N}} \\[2ex]
- & - & - & - & - & - & - \\[1ex]
- & - & - & - & - & - & - \\[1ex]
-\dfrac{1}{jX_{(N-1)s}} & \dfrac{1}{jX_{(N-1)1}} & -\dfrac{1}{jX_{(N-1)2}} & - & - & \dfrac{1}{jX_{(N-1)(N-1)}} & -\dfrac{1}{jX_{(N-1)N}} \\[2ex]
-\dfrac{1}{jX_{Ns}} & -\dfrac{1}{jX_{N1}} & -\dfrac{1}{jX_{N2}} & - & - & -\dfrac{1}{jX_{N(N-1)}} & \dfrac{1}{jX_{NN}}
\end{bmatrix}
\begin{bmatrix} V_s \\ V_1 \\ V_2 \\ - \\ - \\ V_{N-1} \\ V_N \end{bmatrix}
\tag{2}
$$

where X_{ij} for $i = s, 1..N$ and $j = s, 1..N$ are reactance defined as follows:

- for $i \neq j$, the matrix element is simply the admittance between two nodes:

$$Y_{ij} = -\frac{1}{jX_{ij}} \qquad \text{where} \qquad Y_{ji} = Y_{ij} \qquad X_{ij} = \omega_s L_{i,j} \qquad (3)$$

- The diagonal elements in the admittance matrix $(i = j)$ are a negative sum of all admittances connected to corresponding nodes:

$$\frac{1}{jX_{ss}} = \frac{1}{jX_{s1}} + \frac{1}{jX_{s2}} + . \qquad . + \frac{1}{jX_{s(N-1)}} + \frac{1}{jX_{sN}}$$

$$\frac{1}{jX_{11}} = \frac{1}{jX_{1s}} + \frac{1}{jX_{12}} + . \qquad . + \frac{1}{jX_{1(N-1)}} + \frac{1}{jX_{1N}}$$

$$--$$

$$\frac{1}{jX_{NN}} = \frac{1}{jX_{Ns}} + \frac{1}{jX_{N1}} + \frac{1}{jX_{N2}} + . \qquad . + \frac{1}{jX_{N(N-1)}} \qquad (4)$$

When only the leakage inductances and related couplings are considered with the magnetizing inductances neglected, the polygonal network is not connected to the ground reference node. Then, as only N winding currents are independent, it is possible to reduce dimension of the model by grounding the primary node (V_s=0). It leads to a reduced equation (5):

$$[I_{LV}] = [Y_t]_{N \times N} [V_{LV}] \qquad (5)$$

where the reduced admittance matrix with matrix dimensions $N \times N$ can be deduced from $[Y_t]_{(N+1) \times (N+1)}$ (by omitting the 1st column and 1st row in (2)):

$$[Y_t]_{N \times N} = \begin{bmatrix} \frac{1}{jX_{11}} & -\frac{1}{jX_{12}} & - & - & - & -\frac{1}{jX_{1(N-1)}} & -\frac{1}{jX_{1N}} \\ -\frac{1}{jX_{21}} & \frac{1}{jX_{22}} & - & - & - & -\frac{1}{jX_{2(N-1)}} & -\frac{1}{jX_{2N}} \\ - & - & - & - & - & - & - \\ - & - & - & - & - & - & - \\ - & - & - & - & - & - & - \\ -\frac{1}{jX_{(N-1)1}} & -\frac{1}{jX_{(N-1)2}} & - & - & - & \frac{1}{jX_{(N-1)(N-1)}} & -\frac{1}{jX_{(N-1)N}} \\ -\frac{1}{jX_{N1}} & -\frac{1}{jX_{N2}} & - & - & - & -\frac{1}{jX_{N(N-1)}} & -\frac{1}{jX_{NN}} \end{bmatrix} \qquad (6)$$

The matrix $[Y_t]_{N \times N}$ is invertible, and the leakage flux related reactance matrix $[X_\sigma]$ can be found:

$$[V_{LV}] = [X_\sigma][I_{LV}] \qquad \text{where} \qquad [X_\sigma] = [Y_t]_{N \times N}^{-1} \qquad (7)$$

$$\begin{bmatrix} V_1 \\ V_2 \\ - \\ - \\ V_{N-1} \\ V_N \end{bmatrix} = \begin{bmatrix} jX_{\sigma 11} & jX_{\sigma 12} & - & jX_{\sigma 1(N-1)} & jX_{\sigma 1N} \\ jX_{\sigma 21} & jX_{\sigma 22} & - & X_{\sigma 2(N-1)} & X_{\sigma 2N} \\ - & & & - & - \\ & & - & - & - \\ jX_{\sigma (N-1)1} & jX_{\sigma (N-1)2} & - & X_{\sigma (N-1)(N-1)} & X_{\sigma (N-1)N} \\ jX_{\sigma N1} & jX_{\sigma N2} & - & X_{\sigma N(N-1)} & X_{\sigma NN} \end{bmatrix} \begin{bmatrix} I_1 \\ I_2 \\ - \\ - \\ I_{N-1} \\ I_N \end{bmatrix} \qquad (8)$$

Introduction of the leakage reactance matrix (8) is important because its elements can be deduced from the short circuit voltage/impedance data provided in Table 1. The off-diagonal elements in $[X_\sigma]$ are found from Table 1 considering mutual coupling between LV windings in the short circuit tests:

$$X_{\sigma ij} = \frac{X_{sc\,LVi-HV} + X_{sc\,LVj-HV} - X_{sc\,LVi-LVj}}{2} \qquad i = 2...N \qquad j = 1...i-1 \qquad (9)$$

The diagonal elements are equal the short circuit impedance between the LV windings and primary:

$$X_{\sigma ii} = X_{sc\,LVi-HV} \qquad\qquad i = 1...N \qquad\qquad (10)$$

From the leakage reactance matrix $[X_\sigma]$ the leakage inductance matrix is $[L_\sigma] = 1/(j\omega_s)[X_\sigma]$. The $[L_\sigma]$ matrix is very important as it defines inductances seen between different LV windings in a MWTs. To minimize high frequency PWM cross currents, it is desirable to have as high as possible leakage inductances. Further, from the leakage reactance matrix $[X_\sigma]$ it is possible to reconstruct the full MWT model (2) which can be easily implemented in power electronics circuit simulators. For that the reduced admittance matrix $[Y_t]_{N\times N}$ is firstly found from (7):

$$[Y_t]_{N\times N} = [X_\sigma]^{-1} \qquad\qquad (11)$$

Then, from the admittance matrix $[Y_t]_{N\times N}$ the full MWT model admittance matrix $[Y_t]_{(N+1)\times(N+1)}$ can be constructed. The missing elements of the 1st column/row is obtained from the elements of the matrix $[Y_t]_{N\times N}$ (as sum of the row /column elements in the floating admittance matrix $[Y_t]_{(N+1)\times(N+1)}$ is zero):

$$\frac{1}{X_{1s}} = \frac{1}{X_{s1}} = -\left(Y_{t\,11} + Y_{t\,12} + \ldots + Y_{t\,1(N-1)} + Y_{t\,1N}\right)$$

$$\frac{1}{X_{2s}} = \frac{1}{X_{s2}} = -\left(Y_{t\,21} + Y_{t\,22} + \ldots + Y_{t\,2(N-1)} + Y_{t\,2N}\right)$$

$$\frac{1}{X_{Ns}} = \frac{1}{X_{sN}} = -\left(Y_{t\,N1} + Y_{t\,N2} + \ldots + Y_{t\,N(N-1)} + Y_{t\,NN}\right)$$

$$\frac{1}{X_{ss}} = -\left(\frac{1}{X_{s1}} + \frac{1}{X_{s2}} + \ldots + \frac{1}{X_{s(N-1)}} + \frac{1}{X_{sN}}\right) \qquad (12)$$

In this way the MWT admittance matrix $[Y_t]_{(N+1)\times(N+1)}$ (2) is completed. For simulation studies using the circuit simulation software (like PLECS) it is convenient to express the transformer model in the coupled reactance (inductance) form:

$$[V_t] = [Y_t]_{(N+1)\times(N+1)}^{-1} [I_t] \qquad\qquad (13)$$

However, it is not possible to find an inverse of the matrix $[Y_t]_{(N+1)\times(N+1)}$ because it is singular (its added row and column are found using linear combinations of other rows and columns). The problem can be resolved by attaching a shunt magnetizing reactance X_m at all or to just a single (let say primary winding) node. In the latter case, it leads to the following correction of the HV node admittance:

$$\frac{1}{X_{1s}} = \frac{1}{X_{1s}} + \frac{1}{X_m} \qquad\qquad (14)$$

With it, the matrix $[Y_t]_{(N+1)\times(N+1)}$ becomes invertible and the coupled inductor matrix can be finally found:

$$[L_t] = \frac{1}{j\omega_s}[Y_t]_{(N+1)\times(N+1)}^{-1} \qquad\qquad (15)$$

In the simulations with the actual voltage levels, it is possible to incorporate the transformer voltage transformation ratios defined by the ratios of the winding nominal (indexed by n) voltages and currents:

$$[T_v] = diag\left(\left[\frac{V_{sn}}{V_b}\ \frac{V_{1n}}{V_b}\ \frac{V_{2n}}{V_b}\ -\ -\ \frac{V_{(N-1)n}}{V_b}\ \frac{V_{Nn}}{V_b}\right]\right) \quad [T_i] = diag\left(\left[\frac{I_b}{I_{sn}}\ \frac{I_b}{I_{1n}}\ \frac{I_b}{I_{2n}}\ -\ -\ \frac{I_b}{I_{(N-1)n}}\ \frac{I_b}{I_{Nn}}\right]\right) \quad (16)$$

either via ideal transformers or directly via the coupled inductor matrix. In the direct approach, the final coupled inductor matrix of the MWT, which incorporates the transformation ratios, is:

$$[L_{MWT}] = [T_v][L_t][T_i] \qquad\qquad (17)$$

Modelling of multi converter systems interconnected via MWTs

In theoretical analysis of a multi converter systems created by paralleling of N converters it is useful decompose the converter (LV side) voltage and current vectors:

$$\left[v\right]=\left[v_1 \quad v_2 \quad v_3 \quad \cdots \quad v_{(N-1)} \quad v_N\right]^T \qquad \left[i\right]=\left[i_1 \quad i_2 \quad i_3 \quad \cdots \quad i_{(N-1)} \quad i_N\right]^T \qquad (18)$$

into the cumulative components defining interaction between the aggregated converter and power grid:

$$v_\Sigma = 1/N \sum_{i=1}^{N} v_i \qquad i_\Sigma = 1/N \sum_{i=1}^{N} i_i = i_s/N \qquad i=1{:}N \qquad (19)$$

and the cross or circulating components defining interactions among the converters:

$$\left[\Delta v\right]=\left[v\right]-v_\Sigma \qquad \left[\Delta i\right]=\left[i\right]-i_\Sigma \qquad i=1{:}N \qquad (20)$$

Considering that the diagonal elements in the admittance matrix (2) are negative sums of all admittances connected to the nodes, the first row in (2) links the converter cumulative voltage components and the grid voltage V_s via a simple scalar reactance (inductance) X_t (L_t) which represents the transformer short circuit impedance obtained in the test with all LV windings short circuited:

$$jX_t = \left(\frac{1}{jX_{s1}}+\frac{1}{jX_{s2}}+\cdots+\frac{1}{jX_{s(N-1)s}}+\frac{1}{jX_{sN}}\right)^{-1} \qquad L_t = X_t / \omega_s \qquad (21)$$

If the winding resistances are neglected, the following voltage equation describing the cumulative system is obtained:

$$v_\Sigma - v_s \approx L_t \frac{d}{dt}\left(Ni_\Sigma\right) \qquad \text{in steady state:} \qquad V_\Sigma - V_s \approx j\omega_s L_t NI_\Sigma \qquad (22)$$

The cross currents and voltages can be linked via the cross-coupling inductance matrix $[L_\sigma]$ (under assumption of symmetrical distribution of short circuit currents in LV windings in the short circuit test supplied from the HV side and with all LV windings shorted):

$$\left[\Delta v\right] \approx [L_\sigma]\frac{d}{dt}\left[\Delta i\right] \qquad \text{in steady state} \qquad [\Delta v] \approx j\omega_s[L_\sigma][\Delta i] = [X_\sigma][\Delta i] \qquad (23)$$

The cumulative mode (fundamental) current components define the active and reactive power flows between the aggregated converter and the grid. Thus, the active cumulative current component ($I_{\Sigma d}$, in phase with the grid voltage V_s) is used for control of the total active power flow and stored energy in the converter dc bus capacitors of all converters. The reactive cumulative current component ($I_{\Sigma q}$, in quadrature to the grid voltage V_s) defines the total reactive power injected into the grid. The grid power flow and cumulative current controls are identical to that in any single converter system and will be not discussed here. In steady state, if the winding resistances are neglected, the voltage equations of the cumulative sub-system written in the grid voltage vector oriented d,q frame are:

$$V_{\Sigma d} \approx V_s - \omega_s L_t NI_{\Sigma q} \qquad V_{\Sigma q} \approx \omega_s L_t NI_{\Sigma d} \qquad (24)$$

The cross currents can be used to trim the individual converter active power flows (balancing). The voltage and current components of the cross sub-system are linked via the leakage inductance matrix. In steady state, after neglecting the windings resistive voltage drops, the voltage equations describing the cross sub- system (in the grid voltage vector oriented d,q frame) are:

$$[\Delta v_d] \approx -\omega_s[L_\sigma][\Delta i_q] \qquad [\Delta v_q] \approx \omega_s[L_\sigma][\Delta i_d] \qquad (25)$$

For optimal system operation, deviations of the converter dc bus voltages from the average value:

$$\Delta v_{dc\,i} = v_{dc\,i} - v_{dc\,ave} \qquad i=1{:}N \qquad v_{dc\,ave} = 1/N \sum_{i=1}^{N} v_{dc\,i} \qquad (26)$$

should be minimized. Balance of the individual converter dc bus voltages can be actively controlled by redistributing the total power flow created by the cumulative active current component. This can be accomplished by controlling the fundamental comments of the cross currents and related cross-power flows by the dc bus balancing controller shown in Fig. 3. Based on differences of the converter dc bus voltages from their average value, the balancing controller sets the cross-power flow references $[\Delta p]^{Ref}$ which are further converted into the cross current references of each converter ($[\Delta i_d]^{Ref}$ and $[\Delta i_q]^{Ref}$).

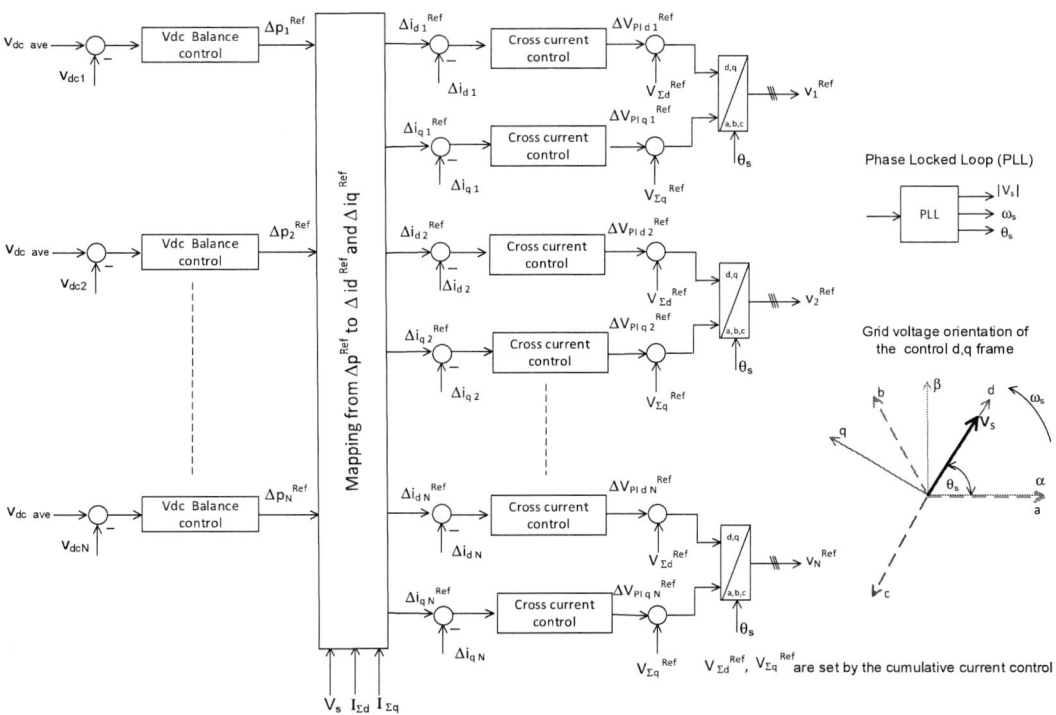

Fig. 3: The dc bus voltage balancing and cross current controls in the multi-converter system with the converters paralleled via the MWT.

Cross power flows and stability of converter dc bus voltage balance control

In this section it will be shown that differences between the effective inductances seen by the cumulative and cross currents, which is introduced by the MWT, may have a significant impact on the power flows among the converters. In general, the converter active power is defined by the scalar product of its voltage and current vectors. In the grid voltage oriented d,q frame (Fig. 3, $V_{\Sigma d} = V_s$, $V_{\Sigma q} = 0$), the converter active powers grouped into the power vector [p] are ([I] is $N \times N$ unity matrix):

$$[p] = -3/2[\Delta v]^T[\Delta i] = $$
$$-3/2\left\{ \left(V_{\Sigma d}[I] + [\Delta v_d]^T\right)\left(I_{\Sigma d}[I] + [\Delta i_d]\right) + \left(V_{\Sigma q}[I] + [\Delta v_q]^T\right)\left(I_{\Sigma q}[I] + [\Delta i_q]\right) \right\} \tag{27}$$

From (27) it is possible to separate the active power exchanges among the converters [Δp]:

$$[\Delta p] = -3/2\left\{ V_{\Sigma d}[\Delta i_d] + V_{\Sigma q}[\Delta i_q] + [\Delta v_d]I_{\Sigma d} + [\Delta v_q]I_{\Sigma q} \right\} \tag{28}$$

This equation for the active power flows produced by the cross currents (28) can be expanded by introduction of the voltage equations (25). The expanded form is ($V_{\Sigma d} = V_s$):

$$[\Delta p] = -3/2\left\{ V_s[\Delta i_d] - \omega_s\left(L_t N[I] - [L_\sigma]\right)\left(I_{\Sigma q}[\Delta i_d] - I_{\Sigma d}[\Delta i_q]\right) \right\} \tag{29}$$

Conventionally the active cross current components [Δi_d] are used to control the cross-power flows (power exchanges among converters) while the reactive components of the cross currents are kept at zero ([Δi_q] = 0). Then (29) is reduced to the following form:

$$[\Delta p] = -3/2\left\{ V_s[I] - \omega_s\left(L_t N[I] - [L_\sigma]\right)I_{\Sigma q} \right\}[\Delta i_d] \tag{30}$$

In this cross-power vector equation (30) there two terms can be distinguished:

- the first or principal term, proportional the grid voltage ($V_s[I]$) and the d axis cross currents.
- the second or parasitic term, defined by the cumulative mode reactive current and the transformer leakage inductance matrix (cross-coupling effect).

It should be emphasised that the second term is proportional to the cumulative mode reactive current. Thus, it can change its sign when character of the reactive current injection changes from the inductive (+sign) to capacitive (-sign). As the result, when the converters perform reactivate power injection, the cross-power flow among the converters is not defined by a simple scaling of $[\Delta i_d]$ by the factor $3/2V_s$ but by a matrix. Effectively in the $[\Delta p]^{Ref}$ to $[\Delta i_d]^{Ref}$ mapping we must consider effects of the parasitic cross coupling term (proportional to $I_{\Sigma q}$):

$$3/2\left\{ V_s[I] - \omega_s\left(L_t\, N[I] - [L_\sigma]\right) I_{\Sigma q} \right\} \qquad \text{where } [I] \text{ is a } N\text{x}N \text{ identity matrix} \qquad (31)$$

If effect of the second term (cross-coupling) can be neglected, the d axis cross current references can be derived from the cross-power references using the following simple mapping (with $[\Delta i_q]^{Ref} = 0$):

$$[\Delta i_d]^{Ref} = -\ [\Delta p]^{Ref} /(3/2 V_s) \qquad (32)$$

Otherwise, as far as the diagonal term in (31) $V_s[I]$ remains dominant, the cross-coupling in (31) does not cause power flow sign inversions and hence the active control of the dc bus voltage balance remains stable. Changes of the reactive current injection $I_{\Sigma q}$ may alter the control loop gains and dynamic performances but they will not lead to the control instabilities. To guarantee stability, the matrix $V_s[I] - \omega_s\left(L_t\, N[I] - [L_\sigma]\right) I_{q\Sigma}$ must be positive-definite or at least semi-definite. To fulfil that requirement the reactive (capacitive) cumulative current must satisfy the following criterion [9]:

$$I_{\Sigma q} \geq -V_s /\left(\omega_s\left(\lambda_{\max} - NL_t\right)\right) \quad (\lambda_{\max} \text{ is the largest eigenvalue of the cross-inductance matrix} [L_\sigma]) \quad (33)$$

If this (capacitive) current limit is violated the balancing control will become unstable due to power flow sign inversions in the eigen direction associated to $\lambda_{\max}$ [9]. By inspection of (29) it is possible to observe that the cross-coupling term in the cross-power equation is possible to eliminate when the converters are operating with nonzero power factors. For that the active and reactive components of the cross currents must be simultaneously controlled using the following map:

$$[\Delta i_q]^{Ref} = I_{\Sigma q} / I_{\Sigma d}\, [\Delta i_d]^{Ref}; \qquad I_{\Sigma q}[\Delta i_d] - I_{\Sigma d}[\Delta i_q] = 0; \qquad [\Delta i_d]^{Ref} = -[\Delta p]^{Ref} /(3/2 V_s) \qquad (34)$$

This option is obviously feasible when the $I_{\Sigma q} / I_{\Sigma d}$ ratio is low (operation with high power factors).

Example case

In the example case presented in this section a multi-converter system shown in Fig. 1b is considered. It is consisted of N=6 converters coupled with the 33kV/50Hz power grid via a 18 MVA MWT with 33kV HV/6 × 2.25kV LV windings. The transformer short circuit voltages are given in Table 2.

Table 2: MWT relative short circuit voltages/impedances (in % of single LV winding base).

Single LV Winding Power Base		Short-circuited Winding						
		HV	LV1	LV2	LV3	LV4	LV5	LV6
Energized Widing	HV							
	LV1	51,9%						
	LV2	30,5%	30,6%					
	LV3	19,7%	64,1%	30,6%				
	LV4	19,6%	96,0%	64,1%	30,6%			
	LV5	30,4%	127,4%	96,0%	64,1%	30,6%		
	LV6	51,6%	159,8%	127,9%	96,0%	64,1%	30,6%	

The short circuit voltages are used to construct the leakage inductance matrix $[L_\sigma]$ (L_{base}= 5.37×10⁻³ H):

$$[L_\sigma] = \frac{5.37\times 10^{-3}}{100}
\begin{bmatrix}
51.9\% & 25.9\% & 3.75\% & -12.3\% & -22.6\% & -28.9\% \\
25.9\% & 30.5\% & 9.8\% & -6.9\% & 17.6\% & -22.9\% \\
3.75\% & 9.8\% & 19.7\% & 4.4\% & -7.0\% & -12.4\% \\
-12.3\% & -6.9\% & 4.4\% & 19.6\% & 9.7\% & 3.5\% \\
-22.6\% & 17.6\% & -7.0\% & 9.7\% & 30.4\% & 25.7\% \\
-28.9\% & -22.9\% & -12.4\% & 3.5\% & 25.7\% & 51.6\%
\end{bmatrix} \qquad (35)$$

Once $[L_\sigma]$ is defined the total transformer leakage impedance for the cumulative system is found to be $NL_t \approx 18$ % (from eq. (12),(14) and (21)). Also, the eigen-values $[\lambda]$ of the matrix $[L_\sigma]$ are found:

$$[\lambda] = eig([L_\sigma]) = \frac{5.37 \times 10^{-3}}{100} diag\left([7.3\% \quad 9.6\% \quad 16.2\% \quad 18.0\% \quad 32.1\% \quad 120.6\%]\right) \tag{36}$$

The highest eigen-value $\lambda_{max} = max([\lambda])$ is approx. 120% of the single LV winding base (notice it is higher than 1 p.u.!). It defines the critical capacitive reactive current/power injection beyond which the dc bus voltage balancing instability is lost. From (33), and if the lowest value of the grid voltage of 80% of V_{sn} is assumed, it is possible to theoretically predict that the balancing control will become unstable when the capacitive reactive power/current injection exceeds the critical value of:

$$I_{\Sigma q} = -V_s / \left(\omega_s \left(\lambda_{max} - NL_t\right)\right) = -0.8\,p.u. / \left(1.2\,p.u. - 0.18\,p.u\right) = -0.784\,p.u. \,(604\,Arms / 854\,Apeak, Q = -11.3MVAr)$$

$$\tag{37}$$

This theoretical result, which predicts instability in the converter dc bus voltage balancing control when capitive current injection exceeds 11.3MVAr, has been confirmed in PLECS simulations using the MWT model created using the standard PLECS coupling inductor model which permeates are defined as explained by the matrix $[L_{MWT}]$ (17). The simulation results shown in Fig. 4 depicts evolutions of the total converter active and reactive powers, currents, and dc bus voltages when the converter total reactive (capacitive) current/power injection is increased from -5.3 MVAr to -13.3 MVAr in three steps. The active power flow has been maintained at constant, relatively low, value of 3.4 MW. It can be observed that, when the reactive power injection reaches -13.3MVAr (the theoretical limit), the converter individual dc buss voltages start diverging (while their average value remains stable), exactly as predicted by the theoretical result (37). Due to low active power transfer (low power factor), stabilization via compensation of the cross-coupling using $[\Delta i_q]$ has not been utilized ($[\Delta i_q]^{Ref}$ set to 0).

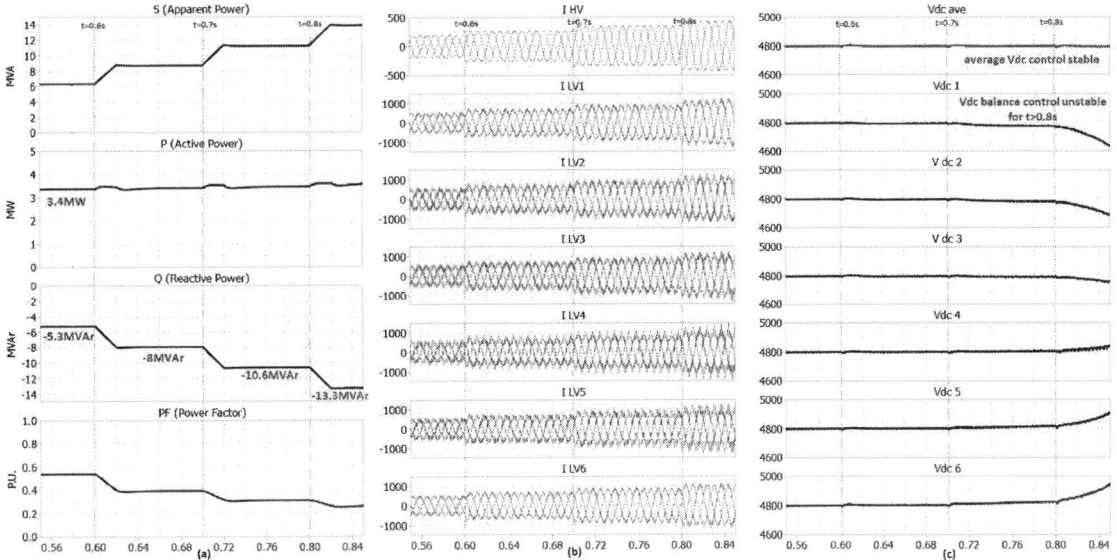

Fig. 4: Converter powers, currents and dc bus voltages. Divergence of the individual converter dc bus voltages starts when the reactive (capacitive) current injection exceeds the critical value of 0.784 p.u. (604Arms, 854Apeak).

Conclusions

High power converters can be constructed by using several VSC converters which are mutually paralleled via MWTs. If the converter dc busses are not mutually interconnected, balance of their dc bus voltages must be actively controlled. The active balancing control of the dc bus voltages is conventionally achieved by trimming the active power flows of the converters via control of active components of the cross currents at the fundamental frequency. However unexpected stability issues with the active dc voltage balance control may been encountered in the operational points characterized by low capacitive power factors and depressed grid voltages. This is particularly the case when MWTs are designed to provide high leakage reactance between the secondary windings (often exceeding 100%) to minimize the PWM switching current ripple.

In the theoretical analysis presented the origin of potential instabilities in the active balancing control has been explained. It has been linked to large differences in the effective inductances in the cumulative and cross currents paths which may is created by magnetically coupled MWT windings. While such high inductance asymmetry is desirable for minimization of the cross current ripple caused by the PWM interleaving and voltage drops in the cumulative current path, it also creates strong cross coupling terms proportional to the cumulative reactive current injection in the mapping of the cross currents into the cross-power flows. It has been theoretically shown that, if a critical level of the capacitive reactive current injection is exceeded, these cross-coupling terms may alter signs of the cross-power flows and thus destabilise the active dc bus voltage balancing control. This type of instability of the dc bus voltage balancing control has been confirmed by the simulation result presented.

The destabilizing cross coupling can be effectively eliminated, and the dc bus voltage balancing control stabilized, in operational points characterized by higher power factors. For that, it has been proposed to simultaneously control the active and reactive components of the cross currents. Due to the paper size limitations, evaluation of the performance of this compensation method was not included in this paper.

References

[1] H. Akagi, A. Nabae, S. Atoh, 'Control Strategy of Active Power Filters Using Multiple Voltage-Source PWM Converters, IEEE Transactions on Industry Applications, May/June 1986, Vol. IA-22, Issue 3, pp. 460-465.

[2] T. Konishi; S. Hase, A. Okui, S. Kinoshita, M. Sonetaka, 'Development of PWM converter with large capacity for electric railway substation', The Fifth International Conference on Power Electronics and Drive Systems, 2003. PEDS 2003, pp. 1264-1267.

[3] J. Chivite-Zabalza, M. Á. R. Vidal, Izurza-Moreno, G. Calvo, D. Madar, "A Large-Power Voltage Source Converter for FACTS Applications Combining Three-Level Neutral-Point-Clamped Power Electronic Building Blocks", IEEE Transactions on Industrial Electronics, Vol. 60, Issue 11, Nov. 2013, pp. 4759-4771.

[4] M. Morati, D. Girod, F. Terrien, V. Peron, P. Poure, S. Saadate, "Industrial 100-MVA EAF Voltage Flicker Mitigation Using VSC-Based STATCOM With Improved Performance", IEEE Transactions on Power Delivery, Vol. 31, No. 6, Dec. 2016, pp. 2494-2501.

[5] G.R. Walker, "Digitally-Implemented Naturally Sampled PWM Suitable for Multilevel Converter Control", IEEE Transactions on Power Electronics, Vol. 18, Issue 6, Nov. 2003, 1322-1329.

[6] D. Zhang; F. Wang; R. Burgos; R. Lai; D. Boroyevich, 'Impact of Interleaving on AC Passive Components of Paralleled Three-Phase Voltage-Source Converters', IEEE Transactions on Industry Applications, Vol. 46, Issue 3,,May-June 2010, pp. 1042 – 1054.

[7] I. G. Park, S. I. Kim, "Modelling and Analysis of Multi-Interphase Transformers for Connecting Power Converters in Parallel", 28th IEEE Power Electronics Specialists Conference. PESC97, 1997, pp. 1164-1170.

[8] J. Wang; A.F. Witulski; J. L. Vollin; T.K. Phelps; G.I. Cardwell, 'Derivation, calculation and measurement of parameters for a multi winding transformer electrical model', Fourteenth IEEE Annual Applied Power Electronics Conference and Exposition, APEC '99, 1991, Vol. 1, pp. 220-226.

[9] D. Basic; H. Baërd; S. Siala, 'Instability in Active Balancing Control of DC Bus Voltages in STATCOM Converters Paralleled via Interphase Transformers', IEEE Transactions on Power Delivery, Vol. 36, Issue 4, Aug. 2021, pp. 1992 – 2000.

[10] AESO, 'Transformer Modelling Guide', Revision 2, 2014, available on-line at https://www.aeso.ca/assets/linkfiles/4040.002-Rev02-Transformer-Modelling-Guide.pdf.

[11] J. El Hayek, 'Modeling aspects for power conversion locomotive transformers', 2010 IEEE Region 8 International Conference on Computational Technologies in Electrical and Electronics Engineering (SIBIRCON), 2010, pp. 513-517.

[12] M. Dudzik, A. Kobielski; J. Prusak, S. Drapik, 'Traction transformers, selected difficulties in object identification', International Symposium on Power Electronics Power Electronics, Electrical Drives, Automation and Motion, SPEEDAM, 20-22 June 2012, pp. 802-805.

Online Learning-based Islanding Detection Scheme for Grid-Connected Systems

Mohammed Ali Khan
Department of Electrical Power Engineering
Faculty of Electrical Engineering and Communication
Brno University of Technology
Brno, Czech Republic
Tel.: +420 77414 2485.
E-Mail: khan@vut.cz

V S Bharath. Kurukuru
Department of Electrical Engineering
Faculty of Engineering and Technology
Jamia Millia Islamia (A Central University)
New Delhi, India
Tel.: +91 95735 25724.
E-Mail: kvsb272@gmail.com

Rupam Singh
Institute for Intelligent System Technologies
Alpen-Adria-Universität Klagenfurt
Klagenfurt, Austria.
Tel.: +43 664 99826743.
E-Mail: Rupam.Singh@aau.at

Acknowledgment

This research work has been carried out in the Centre for Research and Utilization of Renewable Energy (CVVOZE). The authors gratefully acknowledge financial support from the Ministry of Education, Youth and Sports of the Czech Republic under BUT specific research program (project No. FEKT-S-20-6449).

Keywords

«Distributed generation », «Islanded operation», «Machine learning », and «Fault detection ».

Abstract

Data aggregation in smart grids is a key component for emergency responses during abnormalities in the grid. To efficiently utilize the aggregated data, and achieve fast identification of these abnormalities, this paper develops an online islanding detection approach. The development of the technique is realized with an online learning algorithm implemented using the large-scale support vector machine ($LaSVM$). The algorithm adopts a classification problem for islanding detection in grid-connected systems by considering a set of independent variables and unknown variables. The independent variables are related to the known islanding events in the grid-connected system, and the unknown variables are related to the dynamics of the grid operating in real-time. The proposed approach solves this problem by training the known and unknown variables and identifying new instances through sequential minimal optimization. The training and validation results provided indicate 99.8 % accuracy for islanding detection under standard operating conditions of the grid-connected system.

Introduction

Reliable, sustainable, and resilient electric power systems are essential for modern societies. These goals require the distribution and diversification of power sources, which could be facilitated by smart grids [1]. These smart grids enable bi-directional communication between the control units and the end-loads, contrary to the traditional utility grid that uses unidirectional power transmission. Generally, there are two operational modes for the distributed generation systems operating in the smart grid, i.e., islanded and grid-connected modes. The DGs transition between these modes is an example of time-sensitive events that require high priority in the queuing systems to reduce processing and detection time. Islanding may present a lot of adverse impacts on the power system as it may damage the DG by the influx of unregulated voltage with poor power quality and imbalance in frequency. To smoothen out the transition between both the mode of operation and fasted abnormality identification and islanding detection algorithm is necessary. Traditionally, the approaches utilized to detect these events are categorized as passive, active, and hybrid [2], and are mostly dependent on measuring multiple electrical characteristics.

Besides, the islanding detection can also be done either by remote or locally communication techniques [3]. Most remote techniques comprise the methods such as transfer trip, supervisory control, and data acquisition (SCADA), in which the parameters are transferred to a centralized control unit for operational decisions. Even though the method is highly reliable however the cost of communication is very high, and the cyber threats have also increased exponentially in recent years. Generally, the active and the passive islanding detection techniques are adopting local communication [4]. In the active islanding detection technique[5], a small perturb is added to the system parameter and observed for changes to determine the system operation. The active detection technique has an advantage such as a small non-detection zone and low cost of implementation, however, the introduction of perturb in signal may cause an increase in the noise and impact the overall harmonics in the system. Whereas in passive islanding detection [6], the different operating parameters of the system are monitored to identify any abnormality in the system. The implementation of the method is easy and very cost effective however the selection of threshold can be a concern and may lead to a false alert in case of a transient.

Further, to overcome these drawbacks islanding detection using artificial intelligence techniques has emerged based on certain machine learning algorithms [7]. Further, in [8] a Fuzzy Logic technique that is developed via the Decision Tree algorithm is proposed. The proposed technique was tested on data with and without noise, which performed perfectly with a 99.8 % of islanding detection rate. The authors in [9] extracted 62 features from voltage and current signals and trained the support vector machine classifier. In [10], the Neural network-based islanding detection technique is introduced to accurately classify the fault and prepare the system for disconnecting in case of fault.

All these techniques proved to be efficient when tested for single islanding scenarios. But for multiple or simultaneous islanding conditions, these techniques showed misclassification with a long classification time.

To overcome these elements, this paper develops a single parameter-based islanding classification approach using online learning methods. The proposed approach measures the voltage at the point of common coupling (PCC) and extracts its features to train with the large-scale support vector machines ($LaSVM$). The major contributions of this research are:

- Single measured parameter-based islanding classification approach for fast and accurate islanding detection.
- Online learning approach for accurate detection of multiple islanding events.

Further, the paper is organized as follows: In Section II, the system modeling is discussed with a focus on grid abnormalities and grid codes to overcome those abnormalities. In Section III, the information related to the classification problem along with the training algorithm is discussed in detail. In Section IV, the implementation of the developed algorithm is realized, and the corresponding results are discussed. Finally, the conclusion is provided in Section V.

System Modeling

Unintentional islanding may cause damage to the DGs and loads or even cause safety concerns to the maintenance personnel. Hence the system needs to be designed such that the fault is identified efficiently and the controller response to the fault clearance is adequate before the DGs are disconnected from the grid. An overview of the generalized grid-connected PV system is presented in Fig.1.

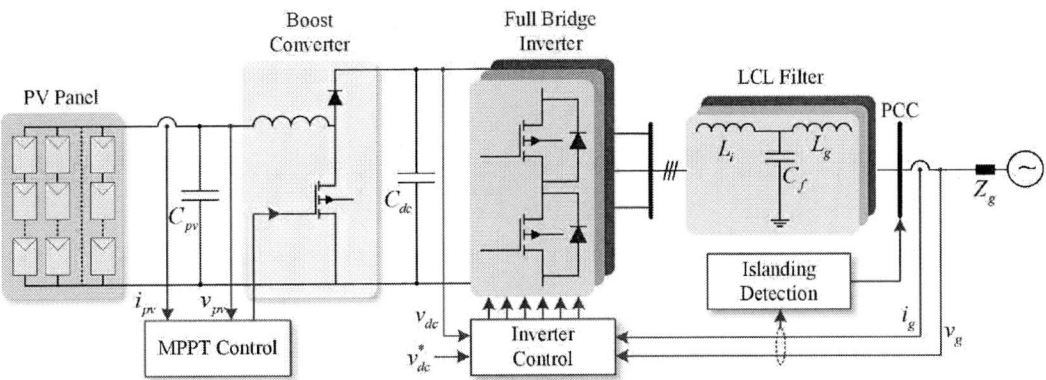

Fig. 1: Layout of three-phase grid-connected PV system. (v_{PV} and i_{PV} are the voltage and current measured at the PV output, v_{dc} is the DC link voltage and v_{dc}^* is the reference DC link voltage, C_{pv} is the DC link capacitance, l_g are grid side filter inductance, c_f is the filter capacitance, and v_g and i_g are the measured three-phase voltage and current at the point of common coupling)

System Controller Design

The control of the grid-connected inverter must be designed such that the power converter can support the utility in abnormal conditions and the system can operate independently of DGs if the fault persists after a certain duration as specified by the grid standards. The controller intends to regulate the inductor current (i_{L_g}) of the LCL filter by tuning the magnitude and the phase angle of the capacitor voltage (i_{C_f}). The fluctuation in the magnitude and phase angle of the of the i_{C_f} tend to determine the active and reactive response of the grid inductor (i_g) as illustrated in Fig. 2.

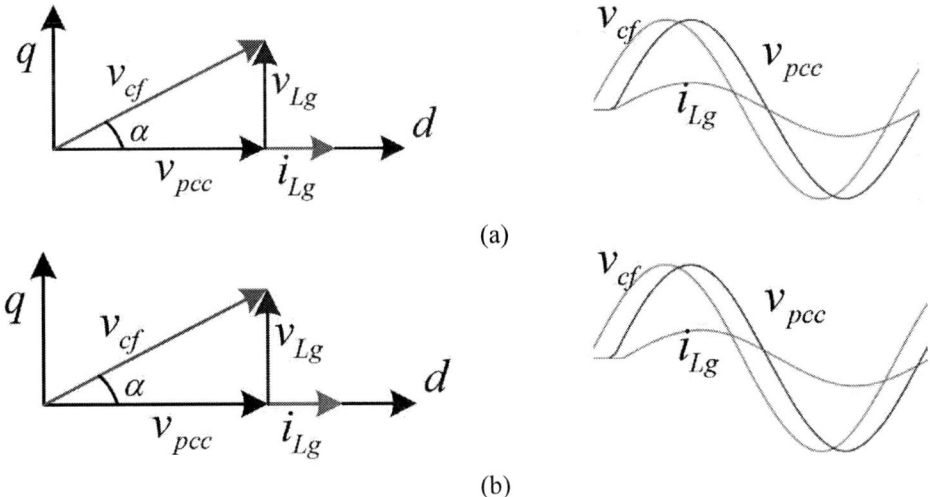

Fig. 2: Phasor representation of controller implementation. (a) Active power injection (b) Active and reactive power injection

It can be observed from Fig. 2, initially the i_{Lg} is present in the q-axis, hence resulting in active power injection, whereas the fluctuation in the magnitude and phase shift of the i_{Cf} corresponding to the v_g has resulted in a shift of v_{L_g} and further pushed i_{L_g} into d-axis with a phase shift of β. As a result, v_{L_g} has both d and q components are present and active as well as reactive power injection is taking place at such instant. The control scheme comprises two loops where the inner loop controls the voltage of the capacitor whereas the outer loop is responsible for the inductor current control. The detailed representation of the control scheme is presented in Fig .3.

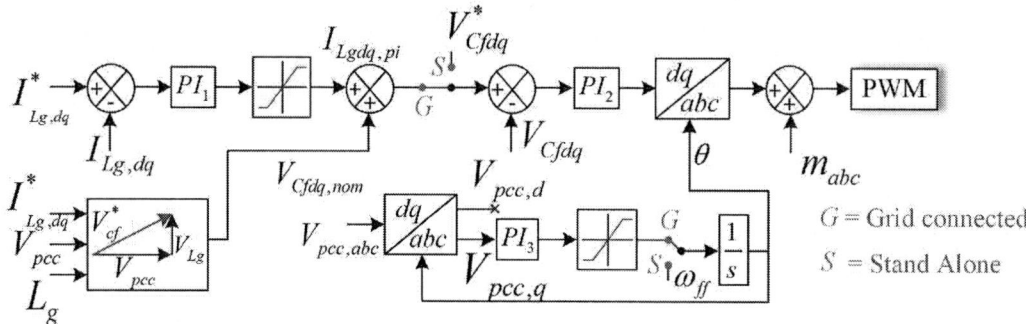

Fig. 3: Control block diagram

The nominal d and q components of capacitor voltage are represented as:

$$V_{Cfd,nom} = V_{pcc,d} + V_{Lg,d} = V_{pcc,d} + i^*_{Lg,q} \times \omega L_g \tag{1}$$

$$V_{Cfq,nom} = V_{Lg,q} = i^*_{Lg,d} \times \omega L_g \tag{2}$$

The capacitor voltage reference (V^*_{Cf}) can be achieved by adding the compensation value with the nominal value of the current controller as represented in (1) and (2). The phase-locked loop (PLL) and grid inductor controller are confined by the limiter to provide a stable voltage to the load during the fault clearance time. The limiter range is set up as per the grid standards (IEEE 1547) and can range as follow:

$$-0.1V_{g,peak} \leq I_{Lgdq,com} \leq 0.1V_{g,peak} \tag{3}$$

$$-0.01\omega_{nom} \leq \omega_{com} \leq 0.01\omega_{nom} \tag{4}$$

During the grid-connected mode of operation, the inverter tends to supply power to the load and the utilities, whereas during the clearance time, PLL and grid inductor control are saturated, and the capacitor voltage controls the inverter operation. In most cases, the Islanding detection algorithm only monitors the magnitude and frequency of voltage at PCC which may lead to false identification at times. Hence the new islanding detection algorithm aims to utilize the multiple features for more accurate and efficient fault identification.

Grid Parameters and Abnormalities

The Islanding approach provides additional protection to the DGs and protects the system from complete blackout. Islanding detection also aims at improving the operating condition of the power system and protecting the power electronics components from damage. Various grid standards have been formulated to regulate the operation of DGs with the utility. By specifying different integration requirements and various operating constraints, the distribution companies along with the manufacturers have come up with the grid codes for the smooth operation of the system. Further, these grid standards also protect the DG components during a grid abnormality as well as the life of line workers on maintenance duty. The multiple standards corresponding to the abnormality detection and DGs disconnection have been formulated by the different countries in association with the distribution companies in the region. The grid standards stipulate parameters for the grid forming/feeding and supporting aspects of the DG [11]. Few of the most used grid codes for the islanding detection are: UL1741 [12], IEEE 1547 [13], IEC 62116 [14], and IEEE 929 [15]. The codes are selected by the utilities and the government depending on the load that is to be supplied. Hence it is expected by the power system to follow one of the above grid standards and disconnect DGs from the utilities in case of system abnormalities.

The DGs while integrated with the utility is susceptible to faults that can damage the local load in case of unintentional islanding. A few of the most commonly occurring abnormalities in the grid-connected DG operation are discussed in the Table. 1.

Table I: Normal and abnormal operating conditions at the grid side

Class	Condition	Cause and Impact	State
NO	Normal Operation	The voltage at PCC (v_{pcc}) is within the limits specified by the grid standards.	Not Islanded
F1	Asymmetrical Faults	Voltage dips are generated at v_{pcc}.	Islanded
F2	Faults due to Impedance	Violation in power transferring capacity of the system.	Islanded
F3	Frequency Mismatch	The frequency exceeds the permissible limit.	Islanded
F4	Harmonics	Voltage transient is observed at v_{pcc}	Islanded
F5	Grid synchronization fault	A mismatch between distributed generation voltage and grid voltage.	Islanded

Classification Algorithm

Classification Problem: The classification problem estimates the value of an unknown class variable depending on the known values of one or more independent variables. For developing an islanding detection logic (IDL) for a grid-connected inverter, the independent variables deal with information related to the known islanding events, and the unknown variables are based on the dynamics of the grid operating in the real time. The sequence of elements in the independent variables is given as $\{\mathcal{F}, \mathcal{C}\}$ where $\mathcal{F}$ is an ordered list of known data for abnormalities affecting the grid voltage, and $\mathcal{C}$ corresponds to the labeled class for predicting a given hypothesis. Consider the projection of the d independent variables to a $\mathcal{D}$-dimensional feature space, the ordered list of known data for i^{th} attribute is termed as $\mathcal{F}_i$. For a given unknown hypothesis with the feature set $\mathcal{U}_t$, the output is estimated as:

$$\mathcal{C}_t = f(\mathcal{U}_t, \mathcal{D}, parameters), \tag{1}$$

where $\mathcal{C}_t$ is the class estimated for the features of the unknown hypothesis, and $parameters$ indicates the kernels used for transforming the input test data into higher dimensional feature space. Generally, these parameters are set during the monitoring process or learned by a classifier based on the nonlinearity of the known islanding data. To solve this classification problem, the independent variables are trained with support vector machines inspired by their online learning advantages.

Online Support Vector Machine: Generally, linear SVMs were used in the literature to achieve online learning with both linear and nonlinear data [16]. For linear data, the linear SVMs on the primal representation are identified to be efficient and scalable and their weight vectors can be efficiently computed without the need for external kernels. In the case of nonlinear data, this situation changes as the weight vectors cannot be expressed explicitly, and kernels are required for transforming the data into a higher dimensional feature space. This problem requires the frequent optimization of the duality principle [17] during the training process. To achieve this, the *LaSVM* is adapted which is a dual representation algorithm that employs optimization schemes to achieve higher accuracies quickly [18]. Generally, the *LaSVM* maintains a set of support vector indices $\mathcal{I}$ which correspond to α non-zero coefficients. Further, whenever a new instance is provided, the *LaSVM* performs two operations, Process, and Reprocess. The Process operation performs a sequential minimal optimization to develop a violation pair from the variables corresponding to the new instance and the variables from the current set of support vector indices $\mathcal{I}$. This operation adds new support vectors to the current support vector indices $\mathcal{I}$. Similarly, the Reprocess performs a sequential minimal optimization to develop a most violation pair from any two variables available in the current set of support vector indices $\mathcal{I}$. This operation eliminates the unnecessary support vectors to keep the current support vector indices $\mathcal{I}$ as small as possible. The Algorithm 1 presents the pseudo-code of *LaSVM*.

Algorithm 1: pseudo-code of *LaSVM*

Step 1:   Initialization

- Seed $\mathcal{S}$ with variables from each class $\mathcal{C}$.

- Set $\alpha \leftarrow 0$ and initiate the gradient g

Step 2:   Online Episode

- Predefine the number of episodes to repeat the process.

- Select feature set $\mathcal{F}_i$

- Compile and run Process($\mathcal{F}_i$)

- Run Reprocess

Step 3:   Finalizing

- Repeat the Reprocess in *Step 2* till the convergence criteria are achieved.

To achieve multi-class SVMs with the *LaSVM*, the *LaRank* is used as an extension. The operation of *LaRank* for dual objective optimization is given as

$$\text{maximize}_{\mathcal{b}} \sum_{i=1}^{l} \mathcal{b}_i^{\mathcal{C}_i} - \frac{1}{2} \sum_{i,j}^{l} \sum_{\mathcal{C}=1}^{d} \mathcal{b}_i^{\mathcal{C}} \mathcal{b}_j^{\mathcal{C}} k(\mathcal{F}_i, \mathcal{F}_j) \tag{2}$$

$$\text{subject to } \mathcal{b}_i^{\mathcal{C}_i} \leq \mathcal{R}\delta_{\mathcal{C},\mathcal{C}_i}, 1 \leq i \leq l, 1 \leq \mathcal{C} \leq d \tag{3}$$

$$\sum_{\mathcal{C}=1}^{d} \mathcal{b}_i^{\mathcal{C}} = 0, 1 \leq i \leq l \tag{4}$$

where $\mathcal{b}_i^{\mathcal{C}}$ is the coefficient of violation pair $(\mathcal{F}_i, \mathcal{C})$, $\mathcal{R}$ is the regularization parameter, and δ is the Kronecker symbol. Further, the *LaRank* modifies the support patterns and support vectors such that the Process and Reprocess operations are extended. In this modification, for a support vector $(\mathcal{F}_i, \mathcal{C})$ whose coefficients are $\mathcal{b}_i^{\mathcal{C}}$, $1 \leq i \leq l$, are non-zero, there exists some $\mathcal{C}$, $1 \leq \mathcal{C} \leq d$, for all the patterns of $\mathcal{F}_i$. The extended operations of Process and Reprocess are given as, Process New, Process Old, and Optimize. The Process New operation performs a sequential minimal optimization to develop a violation pair from two variables in a feasible direction corresponding to the new instance. Further, the Process Old operation performs a sequential minimal optimization to develop the most violation pair from the variables corresponding to a randomly selected support pattern. Lastly, the Optimize operation performs similar to the Process Old operation but developed the most violation pair from the support vectors associated with the support pattern. Moreover, the derivative of the dual objective optimization function in (2) concerning the coefficient $\mathcal{b}_i^{\mathcal{C}}$ gives:

$$g_i(\mathcal{C}) = \delta_{\mathcal{C}_i,\mathcal{C}} - \sum_j \mathcal{b}_j^{\mathcal{C}} k(\mathcal{F}_i, \mathcal{F}_j). \tag{5}$$

This identifies that, the gradient $g_i(\mathcal{C})$ is reliant on the coefficient of each class $\mathcal{C}$ individually. Further, the Process Old operation only works on d gradient computations due to the equality constraints with the added variables. To speed up these computations, the sparseness of the support vectors is exploited, resulting in the need for a new implementation for the effective computation of the gradients. To achieve this, the $LaRank$ only stores and updates the current support vector gradients. This improves the computation speed and minimizes the memory requirement.

Classifier Training

The proposed classifier is trained with the abnormalities simulated on a $15\ kW$ three-phase grid-connected PV system as shown in Fig. 1. As mentioned in Table. 1, one normal operating condition, and 5 abnormal operating conditions are simulated to develop the known islanding scenario dataset. The flow diagram of the online learning process along with the action of $LaSVM$ for islanding detection is given in Fig. 4. During a new instance in the system, the trained model adapts the process and reprocess approach through the sequential minimization optimization as discussed in Algorithm 1.

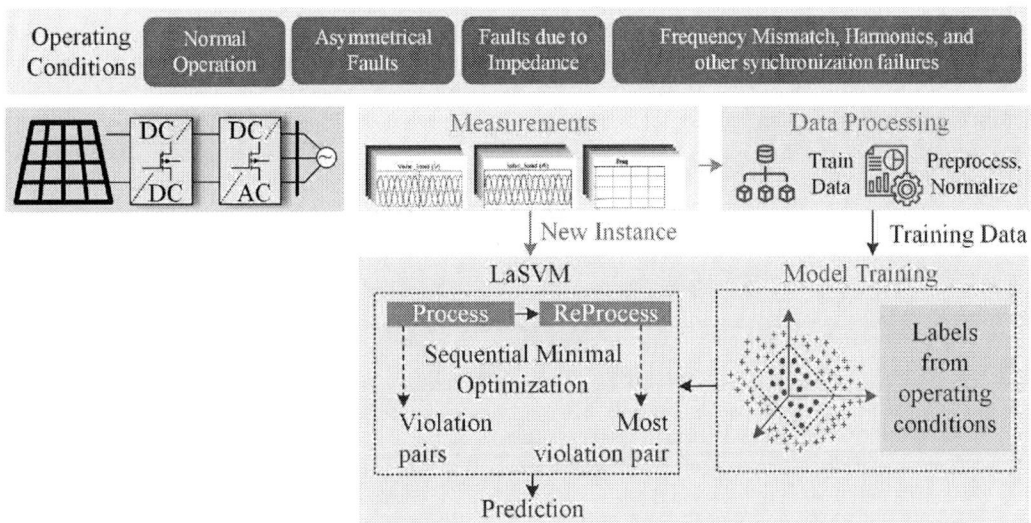

Fig 4. Flow diagram for online islanding detection in a grid-connected system

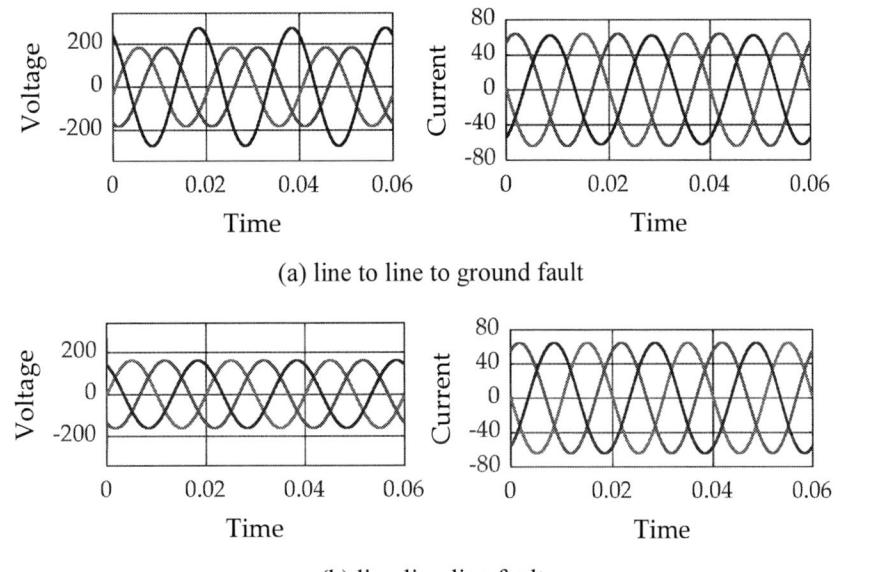

(a) line to line to ground fault

(b) line-line-line fault

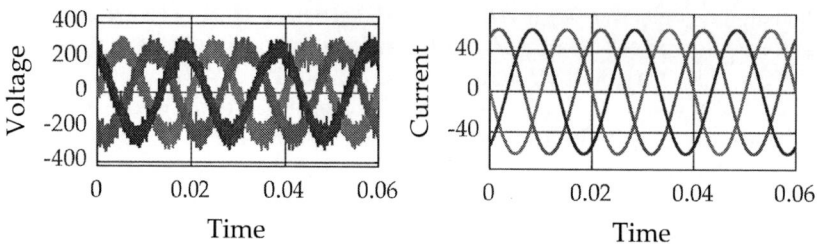

(c) Harmonics injected at the point of common coupling through grid side abnormalities

Fig 5. Sample plots of voltage and current measured during various grid abnormalities

A brief overview of the known variables used in the learning process is provided through the voltage and current measurements in Fig. 5. The data obtained from the preprocessing approach for different abnormalities are labeled with their corresponding classes to model the online SVM. The online SVM classifier is trained with C++ programming with the source code containing a C library for implementing the kernel cache, and the extended Process New, Process Old, and Optimize operations. The additional C++ programs *LaSVM* and *LaRank* are used to run the classifier training with the online SVM for multi-class classification. Initially, the known and unknown variable data is handled by splitting it into training and testing datasets. Each feature/value pair in the training data set has a target value separated by a space character which is used for making predictions with the unknown hypothesis in the testing condition. The corresponding data is trained with *LaSVM* and the corresponding results are shown in Fig. 6.

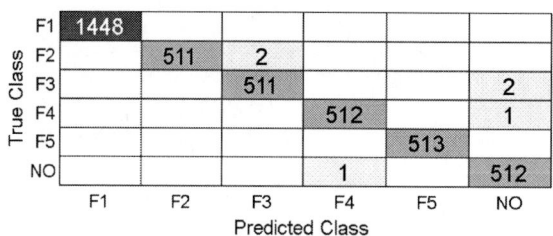

(a) Classification of samples

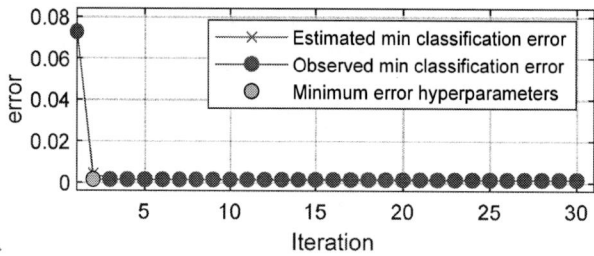

(b) Misclassification error

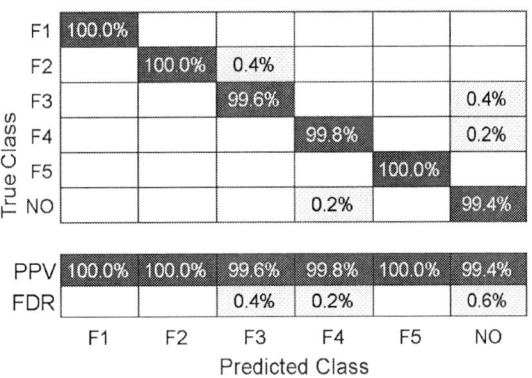

(c) Positive prediction and false detection rate

Fig. 6: *LaSVM* classifier training results

A total of 4013 samples are trained with 6 different classes. The online SVM formulates the violation pairs between the support variables with the processed data to form the support vectors and train the classifier according to the classes. The classifier is trained for 30 episodes. The classification samples in Fig. 6 (a) indicate 4 misclassified samples in the training data set. Further, the minimum classification error is plotted as shown in Fig. 6 (b) to identify the estimated and observed error during the training and testing process with the trained classifier. The minimum classification error is computed by the optimization process when considering all the sets of hyperparameter values tried so far, including the current iteration. This estimate is based on an upper confidence interval of the current classification error objective model observed after the best point hyperparameter [19], [20], and randomly performs the error checks by increasing the confidence limit after certain iterations. The best point hyperparameter indicates the region where the training process is optimized, and the minimum classification error indicates the episode where the classifier obtained maximum accuracy. The result in Fig. 6 (c) indicates the positive predictive value and false detection rate in % for the truly and falsely classified samples. The *LaSVM* is identified to have a training and validation accuracy of around 99.8 % with only 6 misclassified samples.

Conclusion

In this paper, an online learning algorithm is applied using the large-scale SVMs for approximative training of islanding scenarios in the grid-connected system. The training process is evaluated with a simulated three-phase grid-connected system. It adapted the *LaRank* paradigm to achieve fast approximation while classifying an untrained or new data set. Based on the results it is identified that the developed approach accurately trains and classifies different islanding scenarios for simultaneous abnormalities and achieved 99.8 % training and validation accuracy. The trained classifier has a detection speed of 0.4 *sec* for detecting known abnormalities. The results suggest that implementing online classifiers for islanding detection can provide a well-performing and robust way of obtaining approximation solutions with a good trade-off between accuracy and time.

References

[1] Y. Zhang, T. Huang, and E. F. Bompard, "Big data analytics in smart grids: a review," *Energy Informatics*, vol. 1, no. 1, p. 8, Dec. 2018, doi: 10.1186/s42162-018-0007-5.

[2] N. K., S. A. Siddiqui, and M. Fozdar, "Hybrid islanding detection method and priority-based load shedding for distribution networks in the presence of DG units," *IET Gener. Transm. Distrib.*, vol. 11, no. 3, pp. 586–595, 2017, doi: 10.1049/iet-gtd.2016.0437.

[3] B. Guha, R. J. Haddad, and Y. Kalaani, "Anti-islanding techniques for Inverter-based Distributed Generation systems - A survey," in *SoutheastCon 2015*, Apr. 2015, pp. 1–9, doi: 10.1109/SECON.2015.7133045.

[4]   M. A. Khan, A. Haque, V. S. B. Kurukuru, and M. Saad, "Islanding detection techniques for grid-connected photovoltaic systems-A review," *Renew. Sustain. Energy Rev.*, vol. 154, p. 111854, Feb. 2022, doi: 10.1016/j.rser.2021.111854.

[5]   M. Khodaparastan, H. Vahedi, F. Khazaeli, and H. Oraee, "A Novel Hybrid Islanding Detection Method for Inverter-Based DGs Using SFS and ROCOF," *IEEE Trans. Power Deliv.*, vol. 32, no. 5, pp. 2162–2170, Oct. 2017, doi: 10.1109/TPWRD.2015.2406577.

[6]   K. N. E. K. Ahmad, N. A. Rahim, J. Selvaraj, A. Rivai, and K. Chaniago, "An effective passive islanding detection method for PV single-phase grid-connected inverter," *Sol. Energy*, vol. 97, pp. 155–167, 2013.

[7]   O. N. Faqhruldin, E. F. El-Saadany, and H. H. Zeineldin, "A Universal Islanding Detection Technique for Distributed Generation Using Pattern Recognition," *IEEE Trans. Smart Grid*, vol. 5, no. 4, pp. 1985–1992, Jul. 2014, doi: 10.1109/TSG.2014.2302439.

[8]   S. R. Samantaray, K. El-Arroudi, G. Joos, and I. Kamwa, "A Fuzzy Rule-Based Approach for Islanding Detection in Distributed Generation," *IEEE Trans. Power Deliv.*, vol. 25, no. 3, pp. 1427–1433, Jul. 2010, doi: 10.1109/TPWRD.2010.2042625.

[9]   B. Matic-Cuka and M. Kezunovic, "Islanding detection for inverter-based distributed generation using support vector machine method," *IEEE Trans. Smart Grid*, vol. 5, no. 6, pp. 2676–2686, 2014, doi: 10.1109/TSG.2014.2338736.

[10]   M. A. Khan, V. S. Bharath Kurukuru, A. Haque, and S. Mekhilef, "Islanding Classification Mechanism for Grid-Connected Photovoltaic Systems," *IEEE J. Emerg. Sel. Top. Power Electron.*, vol. 9, no. 2, pp. 1966–1975, Apr. 2021.

[11]   IEEE, "IEEE STD 1547-2018," in *IEEE Standard for Interconnection and Interoperability of Distributed Energy Resources with Associated Electric Power Systems Interfaces*, 2018.

[12]   B. Bahrani, H. Karimi, and R. Iravani, "Nondetection zone assessment of an active islanding detection method and its experimental evaluation," *IEEE Trans. Power Deliv.*, vol. 26, no. 2, pp. 517–525, 2011, doi: 10.1109/TPWRD.2009.2036016.

[13]   IEEE, "IEEE 1547,2008," in *IEEE Standard for Interconnecting Distributed Resources with Electric Power Systems*, 2008.

[14]   International Electrotechnical Commission, "IEC 62116-2014," 2014.

[15]   "IEEE 929-2000 Systems, Recommended Practice for Utility Interconnected Photovoltaic (PV)," 2000.

[16]   J. Nalepa and M. Kawulok, "Selecting training sets for support vector machines: a review," *Artif. Intell. Rev.*, vol. 52, no. 2, pp. 857–900, Aug. 2019, doi: 10.1007/s10462-017-9611-1.

[17]   J. A. K. Suykens, C. Alzate, and K. Pelckmans, "Primal and dual model representations in kernel-based learning," *Stat. Surv.*, vol. 4, no. none, Jan. 2010, doi: 10.1214/09-SS052.

[18]   A. Bordes, S. Ertekin, J. Weston, and L. Bottou, "Fast kernel classifiers with online and active learning," *J. Mach. Learn. Res.*, vol. 6, pp. 1579–1619, 2005.

[19]   T. Bihl, J. Schoenbeck, D. Steeneck, and J. Jordan, "Easy and Efficient Hyperparameter Optimization to Address Some Artificial Intelligence 'ilities,'" 2020, doi: 10.24251/HICSS.2020.118.

[20]   J. Bergstra and Y. Bengio, "Random search for hyper-parameter optimization," *J. Mach. Learn. Res.*, vol. 13, pp. 281–305, 2012.

Difference in the design process of LCL filters for grid connected VSI when using SiC/GaN instead of Si semiconductors

Dennis Kampen, Lukas Fräger, Niklas Badenhop
BLOCK Transformatoren-Elektronik GmbH
Max-Planck-Str. 36-46
Verden, Germany
Tel.: +49 4231 678 0
E-Mail: dennis.kampen@block.eu
URL: http://www.block.eu
Arthur Mambetow
Future Energy – Institut für Energieforschung
Technische Hochschule Ostwestfalen-Lippe
Campusallee 12
Lemgo, Germany
URL: www.ife-owl.de

Keywords

«Active Front-End», «Filter Design Automation», «Passive filters», «LCL-type inverter», «Power quality»

Abstract

Although the design of LCL filters has been discussed extensively in recent years, the different requirements for the design process when using SiC/GaN semiconductors instead of classical Si semiconductors have not been presented in detail. Due to the higher switching frequency, new EMI limits in lower frequency range and other influencing factors, there are some differences, which must be taken into account in the design process. This enables a more resource-saving use of materials as well as a faster development time.

Introduction

LCL filters are used to reduce voltage and current harmonics of grid-connected inverters (VSI) to an acceptable level. A good design of the values of the three elements L_1, L_2 and C is crucial to keep the harmonics on the one hand and the costs, losses and size low enough on the other hand.

Classical LCL filter design process and evaluation when using SiC/GaN semiconductors

In this section, the classical design process of LCL filters in the literature is presented and discussed in terms of usability in SiC/GaN VSI.

Equivalent circuit assumed for the design

Typically, the equivalent circuit in Fig. 1 (left) is assumed when designing LCL filters [1]. The line impedance is neglected because the filter inductance values are significantly higher than the line impedances.

→When using SiC/GaN semiconductors, a much higher switching frequency is often used. As a result, the inductance values related to the inverter power are significantly lower and the line impedances should no longer be neglected in the filter design. In addition, the switching frequency with SiC/GaN mostly lays within EMI limits, especially since standards were extended down to 9kHz, i.e. [19].

Therefore, a line stabilization network (LISN, [18]) should be considered at least for designing the filter attenuation, Fig. 1 (right).

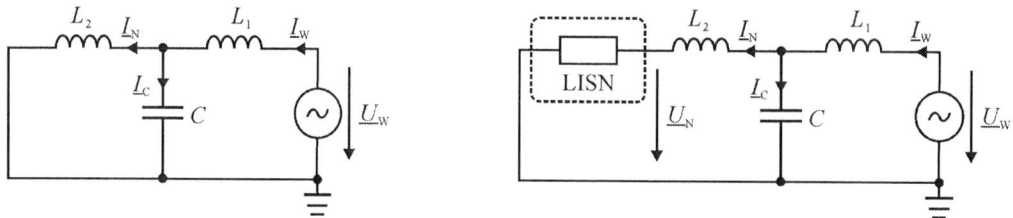

Fig. 1: Simplified single-phase equivalent circuit for an LCL filter design for frequencies other than the grid frequency most used in literature (left) and minimum required equivalent circuit (right) if switching frequency is in the frequency range of EMI-limits.
U_W = inverter output voltage, I_W = inverter output current, U_N = grid voltage, I_N = grid current.

Determination of a maximum total inductance

The maximum total inductance $L_1+L_2=L$ is defined by the permissible voltage drop across the filter at grid frequency ω_g. In [10] the maximal inductance is limited by the control dynamics. The inductance is often defined <5% [1] up to <10% [3] of the phase voltage of the network. The higher the inductance, the higher the required DC link voltage of the converter to be able to feed energy into the grid against the grid voltage.

$$L_{max} = \frac{(0,05 \ldots 0,15) \cdot U_N}{\omega_g \cdot I_N} \tag{1}$$

→Due to the often higher switching frequency when using SiC/GaN, the inductance required for the filter effect is so low that the maximum voltage drop is mostly not a design criterion anymore.

Determination of the minimum inverter side inductance L_1

In practice, the dominant approach is to define a maximum current ripple through the inverter-side inductance and calculate the inductance required to achieve it. The "x" depends on the converter topology, modulation method and modulation factor.

$$L_{1,min} = x \cdot \frac{U_{ZK}}{f_s \cdot \Delta I_{Pk-Pk}} \tag{2}$$

Where U_{ZK} = DC link voltage ΔI_{Pk-Pk} = peak-peak current ripple f_s = switching frequency

The current ripple is usually limited by the RMS value to max. 10% of the rated current RMS [1], [2], [4]. Only few papers suggest higher ripple values like 20% [5], [6], 30% [7] or even 40% [8].

The reason for limiting the current ripple to low values is, on the one hand, to limit the core losses as well as the additional copper losses of the inverter-side inductance L_1. This is very relevant for IGBT inverters, since the switching frequencies are so low that the inverter-side inductance is often designed with a core made of electrical steel and a solid copper conductor. Eddy current core and conductor skin- and proximity losses might become critical.

→ For SiC/GaN converters, however, the switching frequency is normally so high that the inductance on the converter side is designed with ferrite, powder or nanocrystalline core material and HF stranded wire. These inductors could carry significantly higher ripple currents.

Second, for at least older IGBT semiconductors, turn-off switching losses are higher than turn-on switching losses.

→ For SiC/GaN semiconductors turn-off losses are lower than turn-on losses. Thus, with increasing ripple, switching losses tend to increase for IGBTs while for SiC/GaN switching losses tend to decrease with increasing ripple [cmp. 9]. This can be seen, looking at the datasheet values of an older IGBT and a SiC MOSFET. Therefore, more ripple current can be accepted with SiC/GaN in comparison to IGBT.

Table I: Comparison of the relation between turn-on and turn-off energy of SIC MOSFET [11] and IGBT [12]

Type	Turn-on energy E_{on}	Turn-off energy E_{off}
IMZ120R045M1 CoolSiC 1200V SiC Trench MOSFET	280µJ	70µJ
IGW15T120 TrenchStop IGBT	1300µJ	1400µJ

Third, the ripple current has also an effect on the capacitor design, as it increases the RMS value of the capacitor current and therefore the size and costs of the capacitor.
→ Because the equivalent series resistance ESR of a foil capacitor is decreasing with frequency, higher ripple currents can be tolerated for high switching frequencies with SiC/GaN semiconductors in comparison to IGBT converters.

Determination of the grid side inductance L_2

For a given capacitance and total inductance L, the optimum filter effect can be achieved if $L_1=L_2$, since the resonance point is then at the lowest frequency and the filter therefore has higher attenuation at higher frequencies [3]. However, due to a high load on the converter and the inductance on the converter side caused by high ripple currents, a different ratio is often selected in practice. The ratio is often set to $a = 2$ according to the authors experience.

$$L_1 = a \cdot L_2 \tag{3}$$

→As shown above, when SiC/GaN semiconductors are used, higher ripple currents and thus a smaller inverter-side inductance L_1 can be used. This means that the optimum ratio $a = 1$ regarding the filter effect can be used in the design.

Checking the filter resonance point range

The resonance frequency f_r should be placed with a suitable distance between the grid frequency f_g and the switching frequency f_s. A usual placement of the resonance point in practice is [1], [3], [13]:

$$5 \ldots 10 \cdot f_g < f_r < \frac{f_s}{2}. \tag{4}$$

→This design rule is just as valid when using SiC/GaN semiconductors as it is for IGBTs, but with larger design space.

Choice of capacitor value

In [1], [2], [3], [4], [5] and [6], the capacitor C is first specified by the maximum permissible reactive power in order not to generate too high additional load for the inverter. Values between 5% and 15% [6] of the inverter rated power P_N are mentioned.

$$C_{max} = \frac{(0,05\dots0,15) \cdot P_N}{3 \cdot \omega_g \cdot U_N^2} \tag{5}$$

Where ω_g = fundamental grid frequency U_N = grid voltage

Especially for high power IGBT converters with correspondingly low switching frequency, this limit is very relevant and important for the design process.

→However, when using SiC/GaN semiconductors and a much higher switching frequency, the resonant frequency of the LCL filter is accordingly much higher. Therefore, the design criterion is irrelevant here, because a maximum limit of the capacitor is never reached in these applications.

Filter effect: Checking the attenuation of switching-frequency components

Typically, the gain of the inverter-side current ripple to the line-side current ripple is calculated as follows [1]:

$$\frac{\underline{I}_N(h)}{\underline{I}_W(h)} = \frac{1}{|1 + r \cdot (1 + \alpha \cdot x)|} \tag{6}$$

Where

$$\alpha = \frac{L_1 \cdot \omega_s^2}{\omega_g \cdot \frac{U_N^2}{P_N}} \qquad r = \frac{L_2}{L_1} \qquad x = 0,05\dots0,15 \ (\% \ of \ P_N) \tag{7}$$

Here, a damping of 20% [1] is often proposed. Hamonic limits, i.e. from IEEE519 are considered in some papers [1], [4] or [14] to set the attenuation.

→A fundamental problem with this approach is that limits of EMI standards >2kHz are not taken into account in the design. Today, there are EMI limits at least for the frequency range 9kHz-150kHz [19], [21] and up to 30MHz [20]. Examples are given in Fig.2.

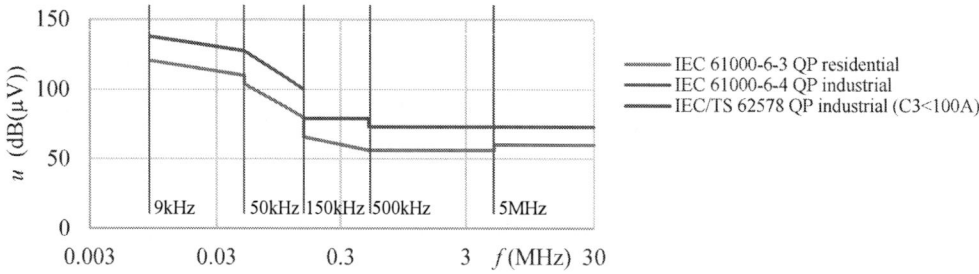

Fig. 2: Example limit values QP for interference voltages.

Typical switching frequencies of SiC/GaN usually lie exactly in this frequency range. Ignoring these limits in the design process, the filter values must be adjusted again and again in practice.

Here, it is now proposed to keep at least the switching frequency component of the grid voltage compared to the standard limit value at switching frequency $U_{N,\text{limit}}$. To calculate the corresponding max. grid ripple current, formula (8) can be used.

$$\underline{I}_{N,max}(\omega_s) = \frac{\underline{U}_{N,limit}(\omega_s)}{\underline{Z}_{LSIN}(\omega_s)} \tag{8}$$

The necessary current attenuation is calculated by

$$\frac{\underline{I}_{N,max}(\omega_s)}{\underline{I}_W(\omega_s)} = \frac{1}{X_C(\omega_s) \cdot (X_C(\omega_s) + X_{L2}(\omega_s) + Z_{LSIN}(\omega_s))} \tag{9}$$

The minimum required capacitor value can be found by resolving equation (9) to C.

→ Although this dimensioning of the capacitor value is much better than previous rules of thumb in the literature, it only provides a good starting value for the filter design. Especially due to the higher switching frequency when using SiC/GaN, parasitic effects of real components like inductors, capacitors, semiconductors, layout, cables etc. are all the more influential in the actual EMI emission. In addition, the component values of LCL filters and additional required components for higher frequency EMC compliance such as Y-capacitors and common-mode chokes are closer to each other, so they actually need to be considered together when designing LCL filters.

Control consideration and Damping

Passive or active damping for LCL filter is widely discussed, i.e. in [1],[3], [15], [16] or [17]. The current control closed loop might be unstable because of the filter resonance frequency. This is especially important for the design of LCL filters for high power IGBT inverters with low switching frequencies and therefore low filter resonance frequencies near to the bandwidth of the controller dynamics.

→If it is again assumed that the switching frequencies are significantly higher when using SiC/GaN instead of IGBT, the stability considerations concerning current control become less relevant. The filter resonance point lies in a significantly higher frequency range and thus at a far distance from the control bandwidth, if huge invests in control components should be avoided.

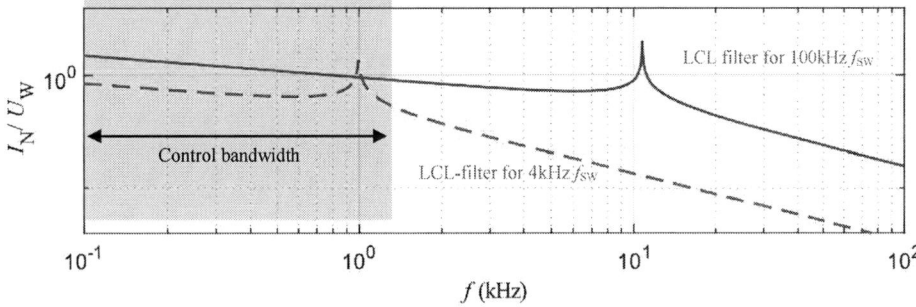

Fig. 3: Resonant frequency of the LCL filter for 4kHz and 100kHz switching frequency compared to the control bandwidth.

However, the range around the resonant frequency must be damped to prevent excessive excitation by the output voltage of the VSI or other voltage sources from the mains. But in contrast to the literature on the design of damping at low switching frequencies, at high switching frequencies with SiC/GaN, significant damping can already be accounted for and utilized by the parasitic properties of the inductors, see [22] and [13], as well as the cables, terminals and other system components. Only the still missing damping must then be supplemented by additional passive damping resistors.

Conclusion

In this paper, the classical design of LCL filters is critically discussed with respect to applicability to SiC/GaN inverters. It turns out that some steps of the classical design are no longer as relevant as for IGBT converters due to the usually higher switching frequency. There are also new design criteria. For example, the influence of the grid impedance due to smaller component values as well as new EMI limits in the frequency range down to 9kHz. The parasitic behavior of the components is very important and the design of the LCL filter should be made considering also the high frequency EMI-filter components. The evaluation shown here allows LCL filters for SiC/GaN converters to be designed with more background knowledge and gives opportunity for a faster design with less material invest.

References

[1] M. Liserre, F. Blaabjerg and S. Hansen, "Design and control of an LCL-filter-based three-phase active rectifier," in IEEE Transactions on Industry Applications, vol. 41, no. 5, pp. 1281-1291, Sept.-Oct. 2005

[2] A. Reznik, M. G. Simões, A. Al-Durra and S. M. Muyeen, " LCL Filter Design and Performance Analysis for Grid-Interconnected Systems," in IEEE Transactions on Industry Applications, vol. 50, no. 2, pp. 1225-1232, March-April 2014

[3] Y. Han et al., "Modeling and Stability Analysis of LCL -Type Grid-Connected Inverters: A Comprehensive Overview," in IEEE Access, vol. 7, pp. 114975-115001, 2019

[4] Araujo, Samuel & Engler, Alfred & Sahan, Benjamin & Antunes, F.L.M. LCL filter design for grid-connected NPC inverters in offshore wind turbines. Fraunhofer IWES. ICPE 2007

[5] [SIC] Y. Liu et al., "LCL Filter Design of a 50-kW 60-kHz SiC Inverter with Size and Thermal Considerations for Aerospace Applications," in IEEE Transactions on Industrial Electronics, vol. 64, no. 10, pp. 8321-8333, Oct. 2017, doi: 10.1109/TIE.2017

[6] T. C. Y. Wang, Zhihong Ye, Gautam Sinha and Xiaoming Yuan, "Output filter design for a grid-interconnected three-phase inverter," IEEE 34th Annual Conference on Power Electronics Specialist, 2003. PESC '03., 2003

[7] A. Kouchaki and M. Nymand, "Analytical Design of Passive LCL Filter for Three-Phase Two-Level Power Factor Correction Rectifiers," in IEEE Transactions on Power Electronics, vol. 33, no. 4, pp. 3012-3022, April 2018, doi: 10.1109/TPEL.2017.2705288.

[8] Y. Tang, W. Yao, P. C. Loh and F. Blaabjerg, "Design of LCL-filters with LCL resonance frequencies beyond the Nyquist frequency for grid-connected inverters," 2015 IEEE Energy Conversion Congress and Exposition (ECCE), 2015

[9] L. Fräger, S. Langfermann, M. Owzareck and J. Friebe, "An Analytic Inverter Loss Model for Design and Operation Space Optimization," 2021 23rd European Conference on Power Electronics and Applications (EPE'21 ECCE Europe), 2021

[10] F. Liccardo, P. Marino, C. Schiano and N. Visciano, "Active front-end design criteria," 10th International Conference on Harmonics and Quality of Power. Proceedings (Cat. No.02EX630), 2002

[11] Datasheet of IMZ120R045M1, INFINEON website, 10/2021, https://www.infineon.com/dgdl/Infineon-IMZ120R045M1-DataSheet-v02_06-EN.pdf?fileId=5546d46269bda8df0169de350d7b3a3e

[12] Datasheet of IGW15T120, INFINEON website, 10/2021 https://www.infineon.com/dgdl/Infineon-IGW15T120-DataSheet-v02_05-EN.pdf?fileId=db3a304412b407950112b4281d403d68

[13] Acharya S., Anurag A..: "Practical Design Considerations for MV LCL Filter Under High dv/dt Conditions Considering the Effects of Parasitic Elements", 9th IEEE International Symposium on Power Electronics for Distributed Generation Systems (PEDG), 2018

[14] S. Jayalath and M. Hanif, "Generalized LCL-Filter Design Algorithm for Grid-Connected Voltage-Source Inverter," in IEEE Transactions on Industrial Electronics, vol. 64, no. 3, pp. 1905-1915, March 2017

[15] R. Peña-Alzola, M. Liserre, F. Blaabjerg, M. Ordonez and Y. Yang, "LCL-Filter Design for Robust Active Damping in Grid-Connected Converters," in IEEE Transactions on Industrial Informatics, vol. 10, no. 4, pp. 2192-2203, Nov. 2014

[16] A Bento, R Luís, JF Silva, LCL filter design for a grid connected telecom station AC-DC converter using SiC devices, International Young Engineers Forum (YEF-ECE), 61-66, 2018

[17] K. H. Ahmed, S. J. Finney and B. W. Williams, "Passive Filter Design for Three-Phase Inverter Interfacing in Distributed Generation," 2007 Compatibility in Power Electronics, 2007

[18] CISPR 16-1-2:2014 Specification for radio disturbance and immunity measuring apparatus and methods - Part 1-2: Radio disturbance and immunity measuring apparatus - Coupling devices for conducted disturbance measurements

[19] IEC 61000-6-3:2020. Electromagnetic compatibility (EMC) - Part 6-3: Generic standards - Emission standard for equipment in residential environments

[20] IEC 61000-6-4:2018. Electromagnetic compatibility (EMC) - Part 6-4: Generic standards - Emission standard for industrial environments

[21] IEC/TS 62578. Power electronics systems and equipment - Operation conditions and characteristics of active infeed converter (AIC) applications including design recommendations for their emission values below 150 kHz.

[22] Kampen, D.: "Modeling inductors in frequency domain considering different flux densities for optimized control design in terms of efficiency and stability", PCIM, 2012

Analysis and design of a resonant DC/DC transformer in modular operation

Abraham López[1], Manuel Arias[1], Pablo F. Miaja[1] and Arturo Fernández[2]

[1]Universidad de Oviedo, Electronic Power Supply Systems Group (e-mail: lopezabraham@uniovi.es)
Edificio Departamental Oeste, Nº 3. Campus Universitario de Viesques. 33204 Gijón, Spain
[2]Electrical Power Management Section, European Space Agency (ESA), Noordwijk, Netherlands

Abstract: **Modular connection between DC/DC converters is commonly used for many applications, to adapt voltage and power ranges, and in order to achieve scalability. This work presents a modular connection between DC/DC modules, providing an accurate voltage and power sharing, in a reliable way. This option could be used in the intermediate bus architectures, to adapt voltage and power levels, achieving high efficiency. Another advantage of the modular connection is the possibility of standardization. This way, it is possible to reuse a whole system module, just only adapting its voltage level, using an intermediate stage, based on the connection of several DC/DC blocks. To validate this solution, several DC/DC resonant converts have been designed for an input and output voltage of 56 V and 28 V respectively, for a rated power of 200 W (per module) and for a switching frequency of 400 kHz. Therefore, by combining several of these designed modules, it is possible to work at higher voltages and powers in whole system.**

Keywords: *Resonant converter, isolated converter, standardization, modular converter, reliability.*

I. INTRODUCTION

Modular design of DC/DC converters has been widely used in industrial, military, or medical applications when high efficiency and power sharing are required This solution brings, also, the possibility to have flexible designs, just by combining several DC/DC blocks. In order to generate different voltage levels there are many options. The first one would be a Distributed Power Architecture (DPA), represented in Fig. 1. However, for higher powers, an Intermediate Bus Architecture (IBA) structure, as the one shown in Fig. 1b, has a better performance [2]. A regulated IBA structure (Fig. 1b) is normally composed by two stages. The first one (i.e. bus converter) takes a nominal input voltage (typically 48 V) and step it down to a range of 8V – 12 V [1]. The second stage is a point of load (PoL) converter that adapts the output voltage of the bus converter to the required voltage. This IBA structures have been used in industry, for many years, to distribute power with good performance. Other possibility is the unregulated IBA structure, as the one shown in Fig. 1c. This architecture is based on the use of an unregulated DC/DC transformer (DCX) that adapts the bus voltage level to the desired one. In general, it is more competitive than the regulated one due to its more optimized design of the passive components [2]. Nonetheless, due to its fixed gain, it must be specifically designed for given input and output voltages. IBA power scalability can be achieved by input-parallel, output-parallel connections, while voltage scalability can be achieved by series connections. In both cases, tolerances in components normally force the use of specific power-sharing techniques, being too complex or based on centralized controllers [3],[4].

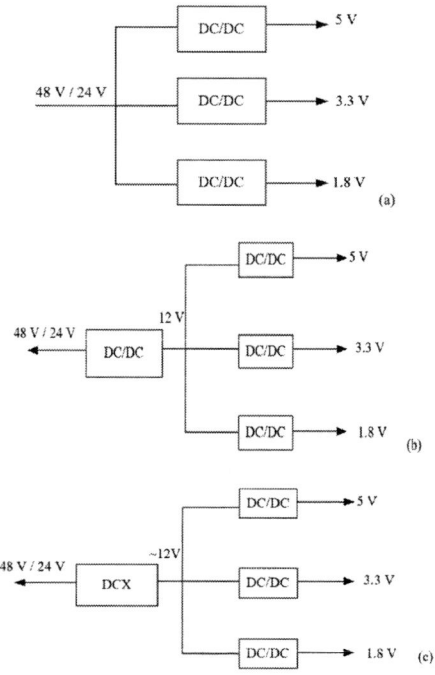

Fig. 1. DPAs for lower power applications. (a) conventional, (b) regulated IBA, (c) unregulated IBA

The aim of this work is analyzing a specific DCX modular topology for an unregulated IBA structure and how can be connected in ISOP (Input Series – Output Parallel) or IPOP (Input Parallel – Output Parallel) configurations for voltage and power scalability. In both cases good power sharing and input voltage sharing are achieved without requiring a complex control stage, being this its main advantage. The control stage only needs a common straight forward synchronization signal, so all the DCX modules will start each switching period at the same time. Thanks to this, the same topology can be easily used for different input voltage ranges by selecting the number of modules in ISOP connection (forming a string), equally sharing the input voltage, instead of redesigning the whole DCX converter. Similarly, power scalability is achieved by IPOP connection of these strings.

Traditional DCX architectures found in literature [5] are many times based on LLC resonant topologies because they can achieve high efficiencies due to their soft switching characteristics and because their high-power density. However, tolerances in their passive components can affect negatively the voltage and power sharing because their static gain is dependent on these passive component values. This forces the use of complex controls, centralized or master-slave structures, with their associated drawbacks, like being a weak point from the reliability point of view. The DCX topology used in this work presents a tolerance immunity because its static gain is only dependent on the turn ratio of the magnetic transformer.

This article is organized as follows. Section II describes DCX topology selected for this work. Section III describes the operation of different DCX modules connected. Section IV describes the robustness of this topology against tolerances in the resonant circuit. Experimental results are shown in Section V. Finally, the main conclusions to this work are shown in Section VI.

II. THE DC/DC TRANSFORMER TOPOLOGY (DCX)

The DCX topology (Fig. 2.a) has been introduced in [6] and consists in a resonant full bridge converter operating with fixed duty cycle and fixed switching frequency. In the secondary side, a center-tap rectifier is implemented. In each half switching period, the resonance takes place between the leakage inductance of the transformer (L_{lk_i}) and the output capacitor (C_O). If L_{lk_i}, Co and f_{sw} are conveniently selected, current will start and end at zero level at any load condition, leading to the waveforms shown in Fig. 2.b and ensuring ZCS in secondary diodes. If magnetizing current is conveniently adjusted, ZVS is reached on primary MOSFETs even at no load situation. All of this leads to a very high efficiency.

This topology presents a fixed static voltage gain only dependent on the turns ratio 'n' of the magnetic transformer (1):

$$V_o = V_{in} \cdot n \qquad (1)$$

This is the main advantage of the topology for this application in comparison to other options, such as DCXs based on the LLC resonant converter [5], whose static gain is dependent not only on the switching frequency, but also on the parameters of its resonant tank. This dependence forces the necessity of specific controls to ensure power and voltage sharing when tolerances are considered [3], [4]. On the other hand, the turns ratio of a transformer is a fixed value independent from tolerances or control signal variability. This makes the equalization of input voltages between series- or parallel-connected modules straightforward (this aspect will be fully analyzed in Section IV). This also makes possible to have a reduced set of predesigned transformers (with different turn ratios) as a straightforward way of changing the static gain of the module during its assembly process, without compromising standardization.

The resonant current and the output voltage can be expressed as:

$$v_o(t) = (V_{in} \cdot n) - I_o \cdot \sqrt{\frac{L_{lk_i}}{C_o}} \cdot sin(\omega_i \cdot t) \\ + (v_o(0) - V_{in} \cdot n) \cdot cos(\omega_i \cdot t) \qquad (2)$$

$$i_{lk\_i}(t) = I_o \cdot (1 - cos(\omega_i \cdot t)) + \frac{V_{in} \cdot n - v_O(0)}{\sqrt{\frac{L_{lk_i}}{C_o}}} \\ \cdot sin(\omega_i \cdot t) \qquad (3)$$

$$\omega_i = \frac{1}{\sqrt{L_{lk\_i} \cdot C_o}} \qquad (4)$$

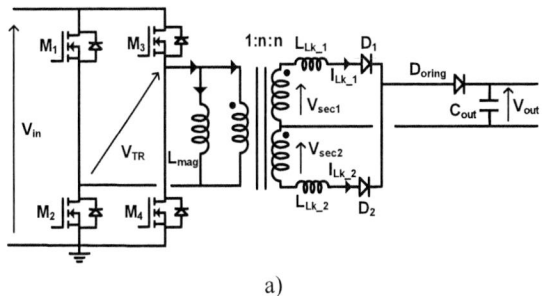

a)

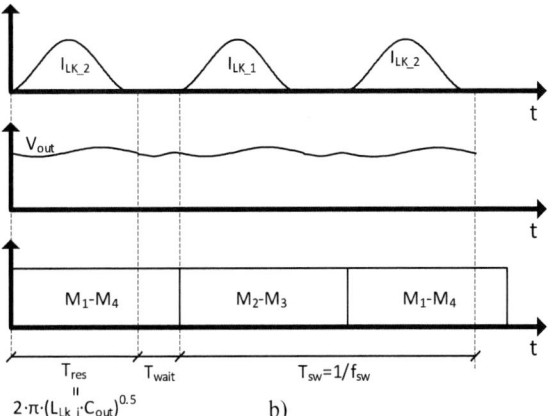

b)

Fig. 2. a) Schematic of the DCX module; b) Main waveforms

where $v_o(t)$ is the output voltage, V_{in} the input voltage, 'n' the turn ratio of the magnetic transformer, I_o is the output current, L_{lki} is the leakage inductance of the winding 'i' (as well as the resonant inductance), C_o is the output capacitance (which is also the resonant capacitance), $v_o(0)$ is the output voltage at the beginning of each resonance, $i_{lk\_i}(t)$ is the resonant current through leakage inductance 'i'. It should be considered that the time 't' restarts in zero for each resonant period (i.e. for each half switching period of the topology). The magnetizing inductance does not play any role in the resonant tank, as in the case of the LLC resonant converter [5]. This alleviates the design constraints of the transformer and makes easier the integration of the resonant inductance and the transformer in a single core. The magnetizing inductance is only tied to the condition of reaching ZVS under any load condition, so its value can be wisely adjusted.

The output capacitor is the resonant capacitor as well. This reduces the number of components, but it may also increase the output voltage ripple. In general, this is a drawback that makes this topology unsuitable for many applications. Nonetheless, in any application in which this converter in not supplying a final load directly, but a second stage converter, the high-frequency ripple of the DCX output capacitor will be filtered by the EMI filter of the second stage. This is the case of the IBA structure mention in Section I or any two-stage topology, in which EMI filters are mandatory. Moreover, the proposed DCX stage will presumably have a higher switching frequency than the second stage, so the filtering effect should be enhanced.

III. MODULAR OPERATION

Thanks to the simple static gain of the topology, which does not depend on the values of the resonant components, several DCX converters [7]-[10] can be connected in ISOP or IPOP reaching automatic input-voltage and power sharing without a complex central control, only sharing a simple clock signal (synchronization signal). The analysis assumes that all the DCX modules have equal values of L_{lk} for both windings ($L_{lk}=L_{lk1}=L_{lk2}$), and of L_{lk} and C_o for all modules. In Section IV, the effect of the tolerances in these values will be shown.

The key point of the proposed system is that the PoLs of a given IBA structure operate with a common input voltage. Therefore, there is no need to connect the output ports of the DCX modules in series to adapt the overall output voltage. Their outputs can be then designed for that predefined voltage and connected in parallel for power scalability. According to (1), this common output voltage to all the DCX modules leads to an equal input voltage level in all the modules, whether they are connected in series or in parallel. Hence, the input voltage is evenly shared by all the serialized modules without specific control. Once the input voltage range adaptation is achieved by selecting the appropriate number of serialized modules (m_s) and their static gain (G_v) through the implemented transformer, the resulting scheme can be parallelized 'm_p' times to reach power scalability (see Fig. 3).

For the mathematical analysis, the several DCX modules connected ('$m\_s \cdot m\_p$' modules) can be represented as in Fig. 4.a), where the square-pulse voltage sources represent the voltage at the secondary side of the magnetic transformers, with a given period with zero voltage applied to the transformer (dead times). Assuming a clock signal synchronizing all the DCX modules, the phase of all the square-pulse voltage sources is the same. As already explained, the input voltage of all the modules is equal, given equation (1) and the parallel connection of all the modules at their output. Consequently, the amplitude of those square-pulse voltage sources is equal as well. Hence, the voltage at the switching nodes (SN_{ij}) is equal for every module in every instant, so they are electrically equivalent, and the system can be represented as in Fig. 4.b), where all the resonant tanks are connected in parallel. Thus, the output voltage of this equivalent circuit is:

$$v_o(t) = (V_{in} \cdot n) - I'_o \cdot \sqrt{\frac{L_{lk}}{C_0 \cdot (m\_s \cdot m\_p)^2}}$$
$$\cdot sin(\omega't) + (V_C(0) - V_{in} \cdot n)$$
$$\cdot cos(\omega't) \tag{5}$$

being,

$$\omega' = \sqrt{\frac{1}{L_{lk} \cdot C_o}} = \omega \tag{6}$$

$$I'_o = I_o \cdot (m\_s \cdot m\_p) \tag{7}$$

The angular frequency is equal for a single resonant tank and for the equivalent one (see (4) and (6)). Replacing (6) and (7) in (5):

$$v_o(t) = (V_{in} \cdot n) - I_o \cdot \sqrt{\frac{L_{lk}}{C_o}}$$
$$\cdot sin(\omega \cdot t) + (V_{Co} - V_{in} \cdot n)$$
$$\cdot cos(\omega \cdot t) \tag{8}$$

The output voltage is the same for one stand-alone module or for several modules connected in any configuration at their input (see (2) and (5)). Therefore, the resonant currents can be calculated as (9):

$$i_{lk\_i}(t) = \frac{1}{L_{lk}} \cdot \int_0^{\frac{2\cdot\pi}{\omega}} [V_{in} \cdot n - v_o(t)] \cdot dt = I_o$$
$$\cdot (1 - cos(\omega \cdot t))$$
$$+ \frac{V_{in} \cdot n - v_0(0)}{\sqrt{\frac{L_{lk}}{C_o}}} \cdot sin(\omega \cdot t) \tag{9}$$

As can be seen from (3) and (9), all the resonant currents are equal and are not affected by the number of modules or by the array configuration (bear in mind the assumption of no tolerances in the resonant inductors). This means that, given the common output voltage, accurate power sharing between all the modules can be achieved without any kind of dedicated or complex sharing control, just with a clock signal that synchronizes all the primary full bridges. Delays or problems related to the transmission of this clock signal are minor, and they can be easily overcome.

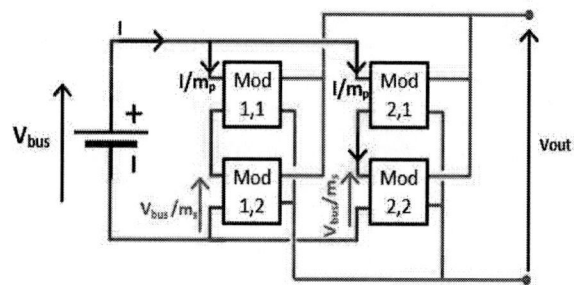

Fig. 3. Modular connection between DCX modules

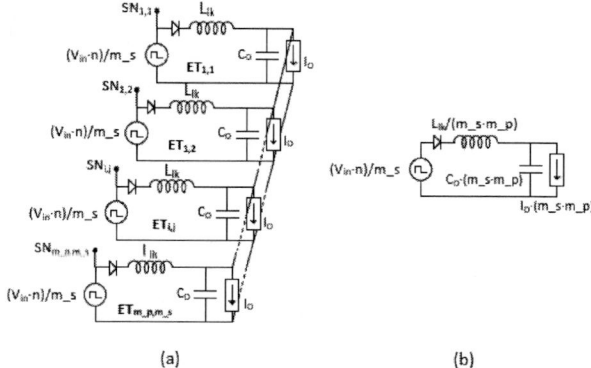

(a)   (b)

Fig. 4. a) Equivalent circuit for the DCX presented when resonant conditions are observed. b) simplified equivalent circuit derived from a)

IV. TOLERANCE INFLUENCE

Many DC/DC transformers use the LLC resonant topology in order to have high efficiency and power density [5]. Nonetheless, these LLC converters have a static gain dependent not only on the switching frequency but also on the parameters of its resonant tank (i.e. resonant capacitor, leakage inductance and magnetizing inductance). This dependence forces the necessity of specific and complex control stages to ensure power and voltage sharing when tolerances in the resonant elements are considered [3],[4]. On the other hand, the DCX topology used in this work has a fixed static voltage gain only dependent on the turn ratio of the magnetic transformer. This is a main advantage as the turns ratio of a magnetic transformer is a fixed value, robust against tolerances, aging, or control signal variability.

This ageing and tolerance analysis is valid not only for different values of L_{lk} between different modules, but also between leakage inductances of the same module (L_{lk1} and L_{lk2}).

Even when the resonant tanks of different modules are not exactly equal, if they fulfill the imposed conditions of starting and ending each resonant period with zero current, equation (1) is still valid, leading to the same square-pulse waveform in the input of each resonant tank, as previously explained using Fig. 4. Therefore, it is possible to establish the expressions for the equivalent inductance and capacitance as before, but considering the drift (due to aging and constructions tolerance) affecting each component as well:

$$\frac{1}{L_{eq}} = \frac{1}{L_{lk}} \cdot \sum_{i=0}^{m} \frac{1}{1 + dr_{L\_i}} \cdot \qquad (10)$$

$$C_{eq} = C_o \cdot \sum_{i=0}^{m} 1 + dr_{C\_i} \qquad (11)$$

where 'm' is the total number of modules, and dr_{L_i} and dr_{C_i} are the relative drift (positive or negative) of the inductor and the capacitor in module 'i'. Therefore:

$$v_o(t) = V_{in} \cdot n - I'_o \cdot \sqrt{\frac{L_{eq}}{C_{eq}}} \cdot sin(\omega' \cdot t)$$
$$+ (v_O(0) - V_{in} \cdot n) \cdot cos(\omega' \cdot t) \qquad (12)$$

$$\omega' = \sqrt{\frac{1}{L_{lk} \cdot C_o} \cdot \sum_{i=0}^{m} \frac{1}{1 + dr_{L\_i}} \cdot \frac{1}{\sum_{i=0}^{m} 1 + dr_{C\_i}}} \qquad (13)$$

$$I'_o = I_o \cdot m \qquad (14)$$

$$m = m\_s \cdot m\_p \qquad (15)$$

With (12) it is possible to obtain the current through any resonant tank:

$$i_{lk\_i}(t) = \frac{1}{L_{lk\_i}} \cdot \int_0^{\frac{2\cdot\pi}{\omega'}} [V_{in} \cdot n - v_o(t)] \cdot dt \qquad (16)$$

which yields:

$$i_{lk\_i}(t) = \frac{L_{eq}}{L_{lk\_i}} \cdot I_0 \cdot m \cdot (1 - cos\,\omega' \cdot t)$$
$$- \frac{Vc_0 - V_{in} \cdot n}{\omega' \cdot L_{lk_i}} \cdot sin\,\omega' \cdot t \qquad (17)$$

Considering these equations, it is clear that aging and tolerance does not compromise the operation of the DCX modules, if a common synchronization signal is used. As all the modules share the same input and output voltage levels, the resonant currents through the different inductors start at the same time and reach zero also at the same time (see (17)). This way the initial assumption, regarding current, is then verified and equation (2) is still valid.

Nonetheless, as can be derived form (17), depending on the module inductance value, its current peak value will change, leading to a power imbalance between modules. This power imbalance represents a drawback of the proposed topology and the prize to pay for the simple synchronization and the straightforward series/parallel connection of the modules at the input port. Nonetheless, as can be derived from (17), the capacitor value drift does not affect the current sharing between modules. Therefore, the power imbalance is considerably lower than in other DCX topologies in which the resonant capacitor is not parallelized between the modules (for instance, the LLC resonant converter [5]). In this example, value drifts of 10%, have been considered.

Fig. 5 shows a simulation of the resonant currents through the rectifier diodes (I_{lk}) for two modules in IPOP connection when LLC topology is used (Fig. 5 b) and when the proposed topology is used (Fig. 5 a). A center-tap rectifier has been used in both topologies. The control in both topologies is the same as well; a simple PWM signal synchronized among all the modules. In this simulation, ±10% tolerances in the resonant components have been considered for all the DCX modules. This way, it is possible to see how when the LLC topology is used without a specific control as in [3], [4], the tolerances lead to I_{lk} currents differing in amplitude and in phase. With the proposed DCX system, there is no phase difference between the resonant currents, (i.e. both starts and ends at the same point), and the amplitude differences are smaller than with the LLC architecture. This better performance of the DCX module proposed in this work is based on the equivalent resonant inductor (L_{lk}) for all the IPOP modules, being the parallel combination of all the L_{lki}, in the same way as happens with the resonant capacitor (C_O) (see Fig. 4 b). In the LLC topologies, only the equivalent magnetizing inductance is the parallel combination of the magnetizing inductance of each module.

Even when perfect synchronization can be achieved, the change in the common resonant frequency (see (13)) may lead to the loss of ZCS and, consequently, to a reduction in efficiency. Obviously, the switching period must be chosen so that enough time is given to the resonant current to reach zero. Hence, the switching period, and thus the period of the synchronization signal, must be equal to two times the resonant period in the worst case. This worst case is represented by having all the resonant elements of all the DCX modules affected by the maximum positive drift, thus achieving the longest resonant period. From (13):

$$T_{sw} = 2 \cdot \frac{2 \cdot \pi}{\omega'} = \frac{4 \cdot \pi}{\sqrt{\frac{1}{L_{lk} \cdot C_o} \cdot \frac{1}{1 + dr_{L\_i}} \cdot \frac{1}{(1 + dr_{C\_i})}}} \qquad (18)$$

In this case, the situation would be the one depicted in Fig. 2.b with T_{wait} equal to zero. This resonant period has to be adjusted at design stage and cannot be changed during operation, or a centralized and more complex control stage would be required.

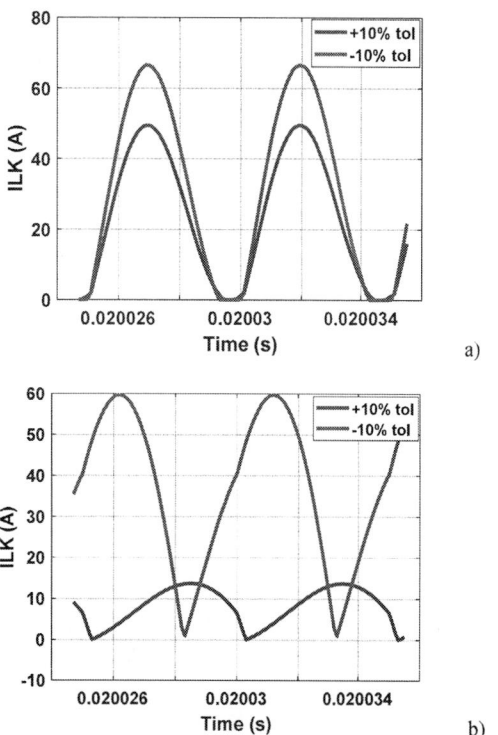

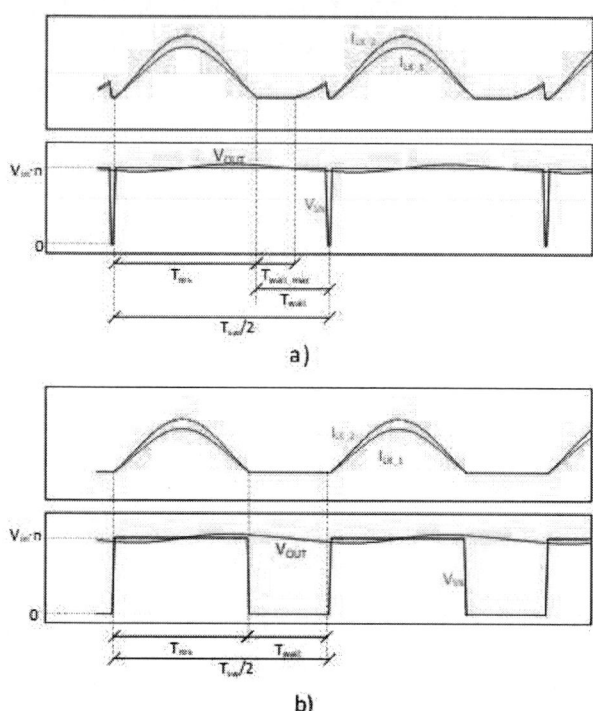

Fig. 5. Currents through the rectifier diodes in two modules in IPOP connection using a) DCX used; b) LLC topology

Fig. 6. a) Resonant current when T_{wait} is longer than T_{wait_max}; b) same situation but with the transformer short-circuited when the resonant current reaches zero until new switching period begins

In any other case, the resonant period will be shorter than half the maximum switching period, leading to a situation as the one depicted in Fig. 2.b with T_{wait} different from zero, where ZCS is still achieved but, during some part of the switching period there is no energy transference between primary and secondary of the DCX. During this time with zero resonant current, energy provided to the load comes exclusively from the output capacitor, leading to its voltage reduction. If this period is long enough, the situation will turn into the one shown in Fig. 6.a for two parallel modules. The output voltage V_{out} becomes lower than the voltage in the secondary side of the magnetic transformer ($V_{in} \cdot n$) during T_{wait}, forward biasing the rectifier diode and leading to a new resonance to take place. This resonance will be cut as soon as $T_{sw}/2$ is reached (a new half switching period started), losing ZCS in the diode. Moreover, the whole performance in the system is affected as the premise of starting and finishing each resonant period at zero current is not fulfilled. It is possible to calculate the output voltage evolution and obtain the maximum value for T_{wait}, denoted as T_{wait_max}, before the resonance is restarted inside the same switching half period (see Fig. 6a):

$$V_O(T_{res}) - \frac{I_o \cdot m \cdot T_{wait\_max}}{C_o \cdot \sum_{i=0}^{m} 1 + dr_{C_i}} = V_{in} \cdot n \quad (19)$$

$$T_{wait\_max} = [V_O(T_{res}) - V_{in} \cdot n] \cdot \frac{C_o \cdot \sum_{i=0}^{m} 1 + dr_{C_i}}{I_o \cdot m} \quad (20)$$

where $V_O(T_{res})$ is the output voltage right when the resonant current reaches zero.

In order to avoid restarting the resonance inside the same half period, the full bridge control scheme can be modified, and the primary side of the transformer can be short-circuited (by turning on MOSFETs M_1-M_3 or M_2-M_4) once the current through the diodes becomes zero. In this way, ZCS is still achieved but it

would be impossible to start the unwanted resonance unless the output voltage falls below zero. The short-circuit will be released when the next half period starts according to the synchronization signal. This strategy can be easily implemented with a pair of small-size magnetic cores, a very simple analog circuit and a dc-blocking capacitor in the power stage (if required). With this strategy, the situation depicted in Fig. 6.a becomes the one in Fig. 6.b. As can be seen, as soon as the resonant current becomes zero, the voltage in the secondary side of the transformer is clamped to zero until a new half period needs to be started. This enlarges the value of T_{wait_max} to:

$$T_{wait\_max} = V_O(T_{res}) \cdot \frac{C_o \cdot \sum_{i=0}^{m} 1 + dr_{C_i}}{I_o \cdot m} \quad (21)$$

This new value of T_{wait_max} ensures that the resonance is not restarted for any reasonable drift value which, as has been said, can be in the 10% range. This control variation is independent from the rest of the modules and does not compromise modularity or simplicity. Conditions for ZVS in the primary switches are dependent on the magnetizing current only (not on load current) and short-circuiting the primary side keeps the magnetizing current at the value it had when the short-circuit was applied. Therefore, ZVS and efficiency are not affected either.

To sum up, although the topology was already proposed in [5], the idea of connecting several of these DCX modules in series or parallel, its tolerance immunity analysis, and the method for not losing ZCS due to tolerances in the resonant tank components are considered as new contributions in this work.

V. EXPERIMENTAL RESULTS

This section presents experimental results using the designed DCX prototypes. On the first part, experimental results for one DCX module is shown. On the second part, several DCX modules are connected in Input-Parallel Output-Parallel (IPOP) and Input-Series Output-Parallel (ISOP) configurations.

A. Experimental results for a single DCX module

Four prototypes of DCX modules have been designed and built according to the schematic in Fig. 2. Its main characteristics are listed in Table I. A photograph of one of the prototypes is shown in Fig. 7, where MOSFETs are highlighted in green and diodes in red. Fig. 8 shows a detail of the V_{DS} and V_{GS} transition of one of the MOSFETs; it can be seen that ZVS is achieved. Fig. 9 shows the drain-source voltage in M_4 primary transistor (V_{DSM4}), the resonant currents through the rectifier diodes (I_{LK}), and the output voltage (V_{OUT}) behavior. It is possible to see that the output voltage ripple is not negligible in the DCX topology, however this ripple can be easily filtered considering the input filter of a DC/DC converter connected downwards. Fig. 10 shows the same waveforms as Fig. 9 but considering the voltage downwards this second-stage input filter. This way, it is possible to see how the voltage ripple is minimized without affecting the resonant current behavior. As can be seen, ZVS can be reached in primary switches and ZCS in secondary ones. Therefore, the efficiency of each module is very high, as shown in Fig. 11, reaching 98% at the rated power and around 97% at half the rated power. Different efficiencies shown in Fig. 11 are due to the two different versions of the prototypes designed.

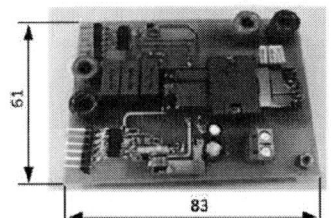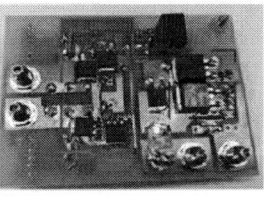

Fig. 7. Photograph of the prototype

Table I. Main specifications of the DCX designed

Input voltage (V_{in})	56 V
Output voltage (V_O)	28 V
Rated power (P_O)	200 W
Switching frequency (F_{SW})	400 kHz
Clock Source	ALTERA MAX 10M50DAF484C7G
Leakage inductance (L_K)	65 nH
Output capacitor (C_O)	0.3 µF
MOSFETs (M_1, M_2, M_3 y M_4)	PSMN063-150D
Rectifier diodes (D_1, D_2)	NRVBB60H100CTT4G
Magnetic core	EIR22/6/16
Magnetic material	N97
Drivers	IR2110 / SI8238BB
Turns	4:2

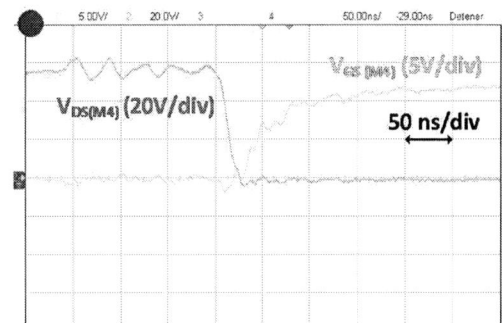

Fig. 8. ZVS achievement on primary switches

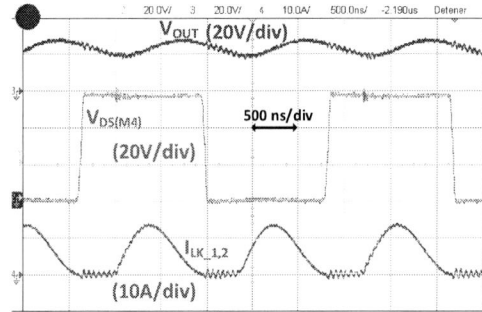

Fig. 9. V_{DS} (M_4), V_{OUT} and $I_{LK\_1,2}$ through the diodes

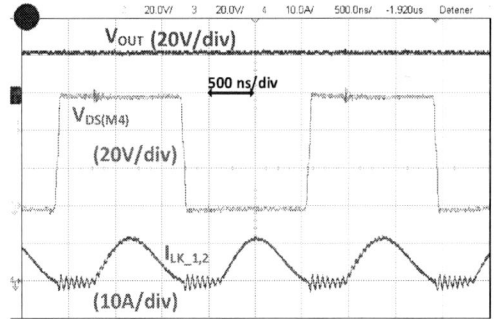

Fig. 10. V_{DS} (M_4), V_{OUT} and $I_{LK\_1,2}$ through the diodes, considering the DC/DC converter input filter

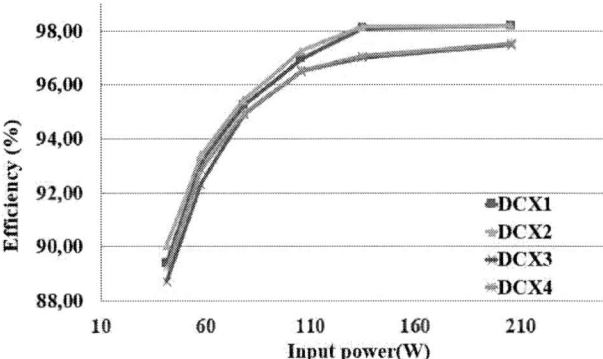

Fig. 11. Efficiency comparison between DCX modules

B. Experimental results for two DCX modules in IPOP configuration

Two DCX modules have been connected in IPOP configuration, increasing the rated power of the full system. Fig. 12 shows the output voltage (V_{OUT}) and the resonant currents through the diodes ($I_{LK\_1,2}$) for both DCX modules. In this case, variations in the leakage inductances have been artificially introduced in each module to increase the effect of the drifts and tolerances in the transformers. As can be seen, this does not affect the resonant behavior of the modules regarding ZCS. Both resonances finish at the same time (i.e. both present the same pulsation), but the amplitude of the resonant currents are different between both modules. As explained before, this is the price to pay for a simple control system based on a single clock signal. Nonetheless, it should be taken into account that this power deviation is relatively small, and it is not influenced by capacitor tolerance or aging. In this test, each DCX module is around the rated power (i.e. 200 W), making that the full power processed by the system will be 400 W (i.e. twice the rated power).

C. Experimental results using four DCX modules

By combining four modules using IPOP and ISOP configurations it is possible to build a system whose nominal input voltage is twice (i.e. 112 V) the rated voltage of one module (i.e. 56 V) while the output voltage is still equal to 28 V. At the same time, the rated power of the whole system (i.e. 800 W) is four times the rated power of a single DCX module (i.e. 200 W). Fig. 13 shows the combination of four modules using IPOP and ISOP configurations. Sub 1 and Sub 2 blocks represent the IPOP combination of a pair of modules, while both blocks are serialized at their inputs. Fig. 14 shows the input voltages in both subsystems (close to 56 V) and the output voltage (V_{OUT}) of the whole system (28 V). As can be seen, the input voltage is perfectly shared among the modules connected in series with the simple control proposed.

Fig. 15 represents the resonant currents through the rectifying diodes in the four modules (I_{DCX1}-I_{DCX4}) and Fig. 16 shows the average value of two of them. In this scenario, the leakage inductances were not artificially modified, as in Fig. 12, but kept the values resulting from the standard fabrication method (i.e. only affected by tolerances, but not aging). As can be seen, the resulting differences in the amplitudes of the resonant currents are small. This is mainly due to the high replicability of planar transformers whose windings are built with PCBs. As they are all connected in parallel at the output, power sharing is as accurate as current sharing.

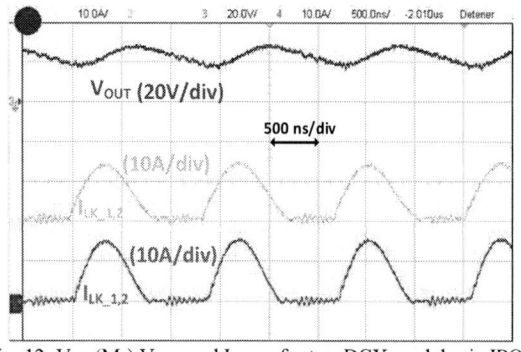

Fig. 12. V_{GS} (M_4) V_{OUT} and $I_{LK\_1,2}$ for two DCX modules in IPOP

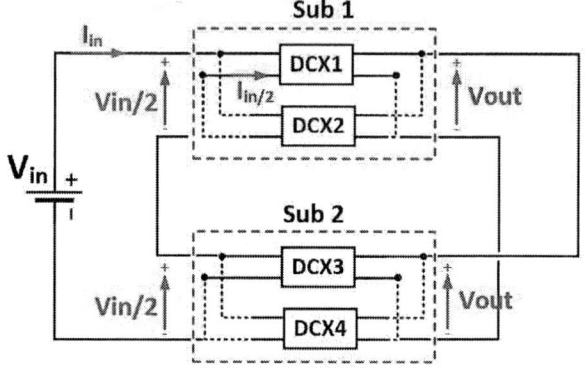

Fig. 13. Connection between four DCX modules

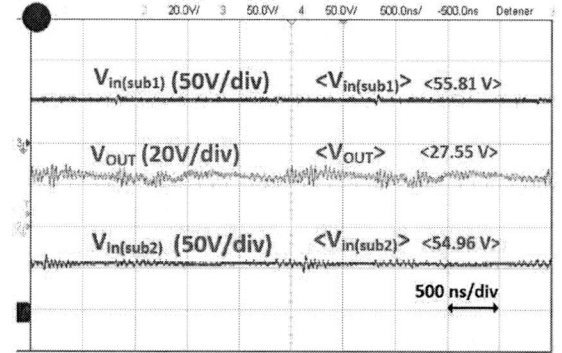

Fig. 14. Input voltages (V_{in}) for sub1 and sub2 with the whole system output voltage (V_{OUT})

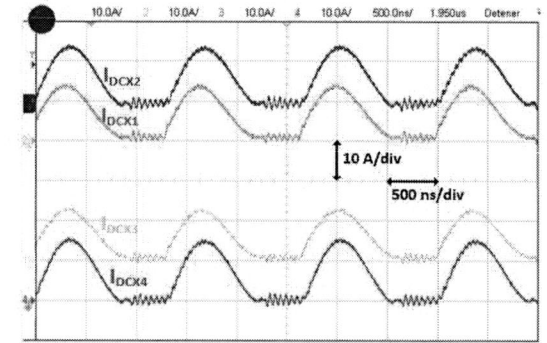

Fig. 15. Currents through the rectifier diodes in the four DCX modules. The shared power per module is nearly 200 W

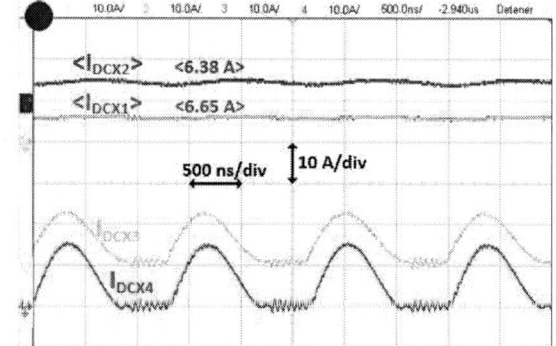

Fig. 16. Average current through DCX$_1$ and DCX$_2$ modules and resonant currents through the rectifier diodes in the DCX$_3$ and DCX$_4$ modules

VI. CONCLUSIONS

This work analyses the modular operation of a DCX topology used in distributed power architectures. It is based on an unregulated bus architecture, where the DCX adapts the intermediate bus voltage. It is a resonant topology in which high efficiencies can be achieved thanks to ZVS and ZCS at any load. Also, given the application, their outputs will be always connected in parallel, while their inputs can be connected in series or in parallel for both, voltage, and power scalability, respectively.

Its static gain is only dependent on its transformer turns ratio, which is a quite robust parameter against tolerances. Therefore, input voltage sharing is automatically achieved, with the only requirement of a simple synchronization signal. Regarding power sharing, it is low dependent on tolerances, so a good sharing can be achieved without a complex, or centralized control either.

The simplicity of this topology implies that the output capacitor is the resonant capacitor as well. The output voltage ripple is then not negligible. However, this high frequency ripple can be easily filtered without affecting the resonant behaviour and the modular operation of the proposed DCX modules, as shown in the experimental results presented in this work. These experimental results also show the optimum power and voltage sharing between modules, as well as the high efficiency they can achieve. These characteristics, along with a high efficiency and a correct power and voltage sharing between the DCX modules, make this solution especially suitable for intermediate bus architectures.

ACKNOWLEDGEMENTS

This work has been carried out by funding from the Spanish government through the PID2021-127707OB-C21 project, and the PRE2019-088425 grant. In the same way, this work has been supported by the Principality of Asturias and FICYT under the SV-PA-21-AYUD/2021/51931 project.

REFERENCES

[1] D. Reusch and F.C. Lee, "High frequency isolated bus converter with gallium nitride transistors and integrated transformer" in Energy Conversion Congress and Exposition (ECCE), 2012.

[2] W. Qin, X. Wu and J. Zhang, "A Family of DC Transformer (DCX) Topologies Based on New ZVZCS Cells with DC Resonant Capacitance," in IEEE Transactions on Power Electronics, vol. 32, no. 4, pp. 2822-2834, April 2017, doi: 10.1109/TPEL.2016.2572146.

[3] K. Yu, J. Du and H. Ma, "A novel current sharing method for multi-module LLC resonant converters," IECON 2017 - 43rd Annual Conference of the IEEE Industrial Electronics Society, Beijing.

[4] Piotr Czyz, Thomas Guillod, Florian Krismer, Jones Huber and Johann Kolar, "Design and experimental analysis of 166 kW medium-voltage medium frequency air-core transformer for 1:1- DCX applications", Published in: IEEE Journal of Emerging and Selected Topics in Power Electronics, February 2021.

[5] M. H. Ahmed, F. C. Lee, Q. Li and M. d. Rooij, "Design Optimization of Unregulated LLC Converter with Integrated Magnetics for Two-Stage 48V VRM," 2019 IEEE Energy Conversion Congress and Exposition (ECCE), Baltimore, MD, USA, 2019.

[6] Mohamed H. Ahmed, Fred C. Lee and Qiang Li. "Two-Stage 48-V VRM with intermediate bus voltage optimization for data centers", Published in: IEEE Journal of Emerging and Selected Topics in Power Electronics, February 2020, doi: 10.1109/JESTPE.2020.2976107

[7] Yuancheng Ren, Ming Xu, Julu Sun and F. C. Lee, "A family of high-power density unregulated bus converters," in *IEEE Transactions on Power Electronics*, vol. 20, no. 5, pp. 1045-1054, Sept. 2005. doi: 10.1109/TPEL.2005.854025.

[8] M. Barry, "Design issues in regulated and unregulated intermediate bus converters" in: Nineteenth Annual IEEE Applied Power Electronics Conference and Exposition, 2004. APEC '04. DOI: 10.1109/APEC.2004.1296045.

[9] Sumit Dutta, Sudhin Roy ,Subhashish Bhattacharya "A multi-terminal DC to DC converter topology with power accumulation from renewable energy sources with unregulated DC voltages" in 2013 Twenty-Eighth Annual IEEE Applied Power Electronics Conference and Exposition (APEC). DOI: 10.1109/APEC.2013.6520440

[10] Y. Chen, P. Wang, Y. Elasser and M. Chen, "LEGO-MIMO Architecture: A Universal Multi-Input Multi-Output (MIMO) Power Converter with Linear Extendable Group Operated (LEGO) Power Bricks," 2019 IEEE Energy Conversion Congress and Exposition (ECCE), Baltimore, MD, USA, 2019, pp. 5156-5163, doi: 10.1109/ECCE.2019.8912965.

Predictive Braking Algorithm for Soft Starter Driven Induction Motors

Hauke Nannen and Heiko Zatocil
OTH - Technical University of Applied Sciences
Amberg, Germany
h.nannen@oth-aw.de, h.zatocil@oth-aw.de

Gerd Griepentrog
Institute for Power Electronics and Control of Drives
Darmstadt, Germany
gerd.griepentrog@lea.tu-darmstadt.de

Keywords

≪Control of drive≫, ≪Thyristor≫, ≪Induction motor≫, ≪Industrial application≫

Abstract

Different algorithms exist for braking induction motors driven by soft starters. This paper investigates a model predictive algorithm for braking induction motors driven by soft starters in addition to state-of-the-art braking with grid phase rotation. Additionally, the impact of non-linear effects, e.g. saturation and the skin effect, on the predictive algorithm is discussed and evaluated based on simulations and measurements. Measurements of the different braking procedures are presented, compared and discussed. The measurements show that the total losses in the soft starter and the motor can be significantly reduced by the application of the predictive algorithm instead of the conventionally used phase angle control algorithm.

1 Introduction

Direct-on-line induction motors are the common solution for industrial applications with a fixed rotor speed. Besides acceleration and operation of induction motors, the braking of motors can be required. For example, braking is often necessary in applications with high inertia such as circular saws,

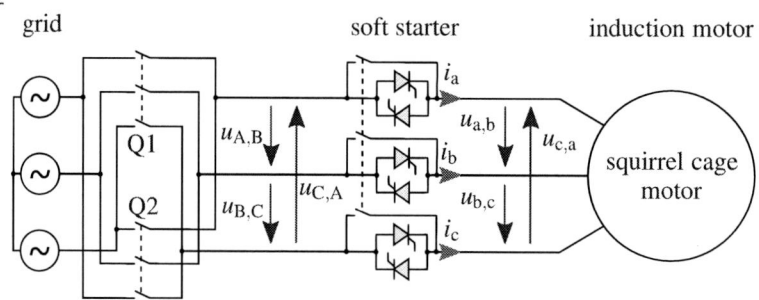

Fig. 1: Soft starter with contactors for inversion of grid rotation direction

mills and centrifuges. Without a braking function, the motor would take a lot of time to get to a standstill. Obviously, all these applications can be equipped with mechanical brakes, but this would lead to additional investment, maintenance and a larger construction. Due to these disadvantages, braking features have been discussed in the past [16]. The most prominent braking strategies are braking with DC currents [2, 3, 4, 7], braking with phase rotation [16] and braking using additional capacitors [10, 17]. All these approaches need additional equipment such as electric contactors and rectifiers. When a soft starter is used in the application, several strategies for braking induction motors are possible. Soft starters often use quasi-DC braking or improved braking by a combination of phase rotation and phase angle control.

Special approaches such as discrete frequency control (DFC) can also be used [6]. Depending on the approach, additional equipment (i.e. contactors) is necessary. Combinations of different algorithms for each time section of the braking procedure often yield the best performance. Nevertheless, all braking strategies have one thing in common: they heat up the motor during braking.

This paper complements the known strategies for braking with a model predictive algorithm, because the major part of the braking power is converted into heat within the electrical machine. In [13] this algorithm approach was used to accelerate a motor. Using this approach leads to a significant loss reduction during start-up, as compared with phase-angle-controlled soft starters. The present paper investigates, whether braking with the predictive algorithm also leads to this loss reduction. Furthermore, the investigations also include a configuration where no additional equipment for braking is necessary.

In the first section of this paper, the state-of-the-art braking procedures for soft starters are presented. The next section presents the prediction algorithm and its modification for braking. Afterwards, the measurements made demonstrate the performance of this approach.

2 State-of-the-Art Braking Strategies

This section presents the state-of-the-art braking strategies for soft starter driven induction motors. As the other approaches are much more common in industrial soft starters, DFC will not be taken into account in the comparison. One way to brake an induction motor is to interchange two phases of the grid. This changes the rotation direction of the grid and the main field in the stator of the motor. Hence, motor slip s changes from $s \approx 0$ to $s \approx 2$. When the motor's speed reaches zero, it is disconnected from the grid. Mostly, the interchange of motor phases is done by contactors or by a set of at least five thyristor pairs.

Using a soft starter can improve the braking. Figure 1 shows a typical configuration of contactors and a soft starter. The contactors Q1 and Q2 are used for changing the rotation direction. After the reconnection with flipped phases, the soft starter performs a phase angle control. This leads to reduced braking currents.

3 Predictive Algorithm

Publications in the recent years have established a predictive algorithm approach for soft starters driving different types of motors [11, 12, 13, 14, 19]. During start-up, these algorithms reduce the losses in the motor and soft starter significantly. The main idea behind these algorithms is to predict the motor torque and current curves for possible firing combinations and to assess their usefulness in an iterative process. The next sections show the theoretical background of these algorithms and adoption for braking.

Motor Model

A motor model is necessary for motor current and torque prediction. A detailed derivation can be found in [13]. Basis for the prediction is a standard induction motor model:

$$L_1 = L_{1\sigma} + L_h \quad , \quad L_2 = L_{2\sigma} + L_h \quad , \quad \sigma = 1 - \frac{L_h^2}{L_1 \cdot L_2} \tag{1}$$

$$\frac{d\vec{I}_1^S}{dt} = \frac{1}{\sigma \cdot L_1} \cdot \vec{U}_1^S - \frac{R_1 \cdot L_2^2 - R_2 \cdot L_h^2}{\sigma \cdot L_1 \cdot L_2^2} \cdot \vec{I}_1^S + \frac{R_2 \cdot L_h}{\sigma \cdot L_1 \cdot L_2^2} \cdot \vec{\Psi}_2^S - j \cdot \frac{\Omega_L \cdot L_h}{\sigma \cdot L_1 \cdot L_2} \cdot \vec{\Psi}_2^S \tag{2}$$

$$\frac{d\vec{\Psi}_2^S}{dt} = \frac{R_2 \cdot L_h}{L_2} \cdot \vec{I}_1^S - \frac{R_2}{L_2} \cdot \vec{\Psi}_2^S + j \cdot \Omega_L \cdot \vec{\Psi}_2^S \tag{3}$$

$$M_M = \frac{3}{2} \cdot p \cdot \frac{L_h}{L_2} \cdot \vec{\Psi}_2^S \times \vec{I}_1^S \tag{4}$$

The model consisting of Equations (2) – (4) is used for prediction. Before the prediction can be started, the model has to be initialised with the actual system state. The initialisation of the current state $\vec{I}_1^S$ can be done with the measured currents. As the rotor flux linkage cannot be measured directly with the soft starter configuration, a flux model is needed for rotor flux estimation. Therefore, Equation (3) is implemented to calculate the actual rotor flux linkage with the measured electric rotor speed Ω_L and stator currents $\vec{I}_1^S$.

For solving Equations (2) and (3), a prediction for grid voltage $\vec{U}_1^S$ and electric rotor speed Ω_L is also necessary. Therefore, the actual grid voltage can be rotated with the grid frequency f_{grid} and the prediction sample time t_{step}. The motor speed can be assumed as being constant due to the high system inertia J_{total}.

Together with the parameters, the initial and the estimated values, the prediction can be done. Now, the model is used to predict the current and torque over the prediction horizon t_{pred}. Prediction is done for all four possible firing opportunities (A & B, B & C, C & A, A & B & C) by solving the equations with an Explicit Euler method using a fixed step with t_{step}. This calculation will be repeated cyclically with new actual values within a defined step with t_{cycle}. A more detailed explanation can be found in [13].

Decision Criteria

After the motor model-based prediction of the variation over time for the current $\vec{I}_1^S$ and the torque M_M for each possible firing opportunity, it is necessary to decide whether the specific firing is useful or not. For braking the motor, it has to obviously generate a negative amount of torque, but some additional aspects must also be taken into account, as discussed in this section.

To brake the motor, the firing needs to generate a minimum amount of negative torque $M_{p,avgmin}$ over the prediction horizon t_{pred}. The absolute torque value $|M_M|$ is also limited, to prevent the application from damage:

$$\text{mean}\left(M_M\left(t = 0...t_{pred}\right)\right) < M_{p,avgmin} \quad , \quad \left|M_M\left(t = 0...t_{pred}\right)\right| > M_{p,max} \tag{5}$$

In addition to the torque requirements, it is useful to limit the maximum allowed peak current $i_{p,max}$, as well as the minimum allowed conduction time $t_{p,mc}$. This makes the system more robust, and firings with a small amount of torque are suppressed. Finally, a criterion for managing the rotor flux linkage is necessary. Without a significant rotor flux linkage amplitude $|\vec{\Psi}_2^S|$, no torque can be created afterwards. This necessitates the setting of a limit on the rotor flux $\Psi_{p,min}$, which must be still available at the end of each firing:

$$\left|\vec{I}_1^S\left(t = 0...t_{pred}\right)\right| < i_{p,max} \quad , \quad t_{pred} > t_{p,mc} \quad , \quad \left|\vec{\Psi}_2^S\left(t = t_{pred}\right)\right| > \Psi_{p,min} \tag{6}$$

When one of the predicted firing opportunities fulfills the requirements of Equations (5) and (6), the respective firing will be initiated on the real hardware. This creates a firing pattern that is more or less equal to the predicted one and considers the rotor flux. The main difference is that all other known braking strategies are based on fixed pulse patterns. Thus, in the predictive algorithm, firing is only applied to the system when a requested torque is produced, and the resulting rotor flux linkage at the end is within a limitation.

4 Impact of Non-Linear Effects

The motor model, which is described by Equations (1) – (4), deals with constant motor parameters. In reality, non-linear effects such as the saturation of the magnetic material, skin effect in the rotor bars and resistance variations due to temperature variations of the stator and/or rotor winding may occur. As these effects are not covered by the motor model used by the predictive algorithm, the predicted variations of stator current and torque might vary from the actual ones. The impact of these effects is examined in the next sections.

Saturation

Generally, two saturation effects can be observed in induction motors: the saturation of the main flux paths, resulting in a decreasing main inductance, and the saturation of the leakage flux paths, which leads to varying leakage inductances.

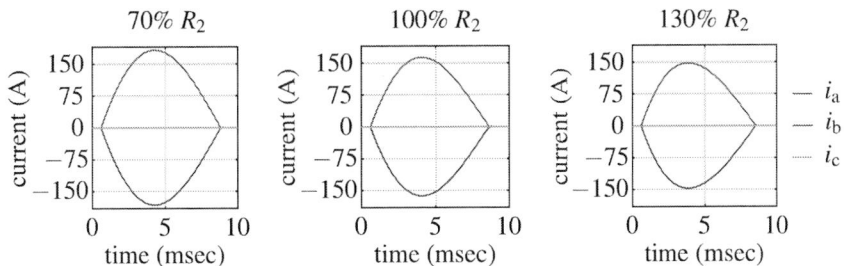

Fig. 2: Simulated variation of rotor resistance at $n = 1050\,\mathrm{rpm}$ with no magnetisation

Induction motors used in industrial applications typically have a rotor time constant higher than $100\,\mathrm{ms}$.

$$\tau_{\mathrm{rot}} = \frac{L_2}{R_2} \tag{7}$$

In contrast to that, a single stator current pulse only lasts for a few $10\,\mathrm{ms}$. Thus, this short current pulse has only a small impact on the main magnetisation level of the motor. Additionally, the rotor flux level $|\vec{\Psi}_2^S|$ is controlled to a defined level and will not exceed a certain reasonable level. Consequently, the saturation of the main flux can be neglected during the prediction of the motor currents and torque. For motors with distinct saturation phenomena during a single pulse, the resulting behaviour of the main inductance can be characterised and integrated into the motor model, according to [9].

As shown in [8], an analytical description of the saturation of the leakage paths is quite impossible. The resulting saturation behaviour severely depends on the geometry of the motor and a lot of other attributes. Some parts of the leakage paths saturate (e.g. the face side leakage paths), while some do not. In [15] one component of leakage saturation effects is described analytically, but not the whole resulting effect.

Skin Effect

During the operation of a soft starter, the skin effect in the rotor cage can be generated in two ways. First, it occurs during steady state operation with high slip values. As shown later, while applying the predictive algorithm, the stator currents during the start-up of the motor have a non-sinusoidal form. Thus, during this operation mode as well, skin effects occur in the rotor, as the non-sinusoidal currents also consist of high frequency components. In both cases, the skin effect leads to an increase in rotor resistance R_2 as the rotor currents are pushed to the upper part of the rotor slots. For the same reason, the rotor leakage inductance $L_{2\sigma}$ decreases simultaneously.

Figure 2 shows the simulation results for a single stator current pulse for different values of R_2 while all other motor parameters were kept constant. Obviously, even huge variations in rotor resistance have no severe impact on the magnitude and variation in stator currents. Nevertheless, the skin effect influences the produced torque as well as the resulting stator currents. If this impact cannot be neglected, it can be integrated in the motor model [1, 5, 18].

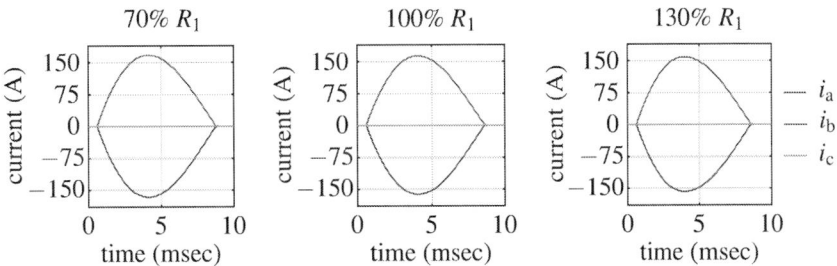

Fig. 3: Simulated variation of stator resistance at $n = 1050\,\mathrm{rpm}$ with no magnetisation

Temperature

In addition to saturation and the skin effect, temperature variations also lead to time-varying motor parameters, especially stator and rotor resistance. The impact of a rotor resistance variation on the resulting stator current pulse is already shown in Figure 2. Figure 3 shows the resulting stator current pulse for three different values of the stator resistance. As expected, the current shapes differ, but the differences are minimal.

Resulting Prediction Error

As shown above, all three non-linear phenomena affect the produced torque and the stator currents. Figure 4 shows the predicted and the real motor currents over a few current pulses during a motor start-up on the test bench. Obviously, the differences between the predicted and the measured motor currents are small and in a reasonable range. The general current shapes and magnitudes are predicted with good precision.

In industrial applications the current and the speed are controlled by dedicated controllers. Additionally, these controllers compensate the impact of possible prediction errors so that these errors do not disturb the resulting system behaviour.

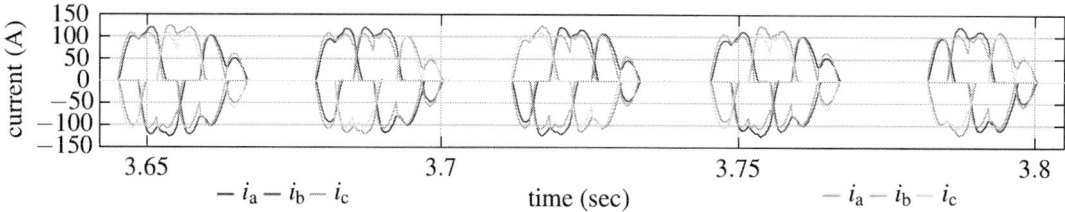

Fig. 4: Exemplary comparison of the predicted current (lighter colour) and the measured current (full colour)

5 Experimental Results

For the comparison of the state-of-the-art and the predictive braking approach with a focus on the losses in the motor and the soft starter, it is necessary to define, measure and calculate the dissipated energy.

The two important values for a loss comparison are the dissipated energy in the motor E_{mot} and that within the soft starter E_{thy} during braking. They limit the number of starts and stops per hour. The overall electrical energy consumption E_{elec} during braking is measured with a power analyser between the motor and the soft starter. Together with the energy stored in the rotating mass E_{mech} and the dissipated energy due to friction E_{fric}, the dissipated energy in the motor E_{mot} can be calculated:

$$E_{mot} = E_{elec} + \Delta E_{mech} - E_{fric} \qquad (8)$$

ΔE_{mech} represents the difference in mechanical energy stored in the rotating mass. The mechanical energy in the rotating mass can be calculated with the system inertia J_{total} and the current rotor speed n as follows:

$$E_{mech} = 1/2 \cdot J_{total} \cdot (2\pi \cdot n)^2 \qquad (9)$$

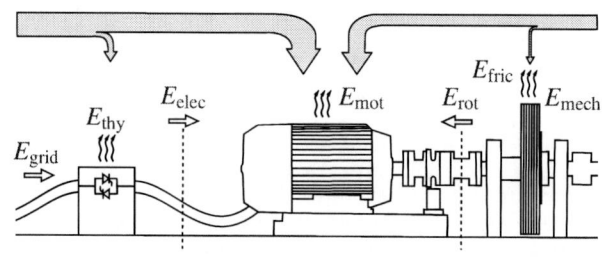

Fig. 5: Energy flow during braking

When the motor is stopped, all the energy of the rotating mass is transformed by the motor into electrical energy or heat. The dissipated energy within the soft starter E_{thy}, can be calculated based on the measured currents and a simple loss model with forward voltage U_f and dynamic resistance R_{on}:

$$E_{thy} = \int_{t_{start}}^{t_{stop}} U_f \cdot (|i_a| + |i_b| + |i_c|) + R_{on} \cdot (i_a^2 + i_b^2 + i_c^2) dt \qquad (10)$$

When the first current pulse for braking is done, the braking begins at t_{start} and ends at t_{stop}, when the motor is stopped.

Figure 5 shows the energy flow in the system during braking. E_{grid} is defined as energy consumption from grid, and E_{rot} is the amount of energy that is transferred from the rotating mass into the motor. To measure the dissipated energy due to friction, the motor was coasted to a standstill. The measurements showed that the mechanic losses E_{fric} are neglectable on this test bench.

Experimental Setup

The main parts of the used test setup are shown in Figure 5. Central elements are a standard industrial induction motor for direct line-start, contactors for phase rotation and a soft starter. They are connected as shown in Figure 1. The induction motor is connected to a speed sensor and a mechanical unit. With this mechanical unit additional inertia can be added to the rotor shaft using disks. With these disks, the system inertia J_{total} can be adjusted. All the shown data are measured with a high-precision power analyser LMG671 from ZES ZIMMER. The algorithms for controlling the thyristors are implemented on a dSPACE rapid controller prototyping system (RCP-system).

Regarding the braking time, the parameters of the algorithms were chosen in such a way that the total braking times for both algorithms could be compared. As the resulting braking time for the predictive algorithm depends on its parameters, a general comparison with an explicit result is difficult.

Phase Angle Control Algorithm

In Figure 6 a braking with phase rotation and phase angle control is shown. The procedure starts with a motor, which is connected to the grid via bypass and runs in steady state. First, contactor Q1 and the bypass contactor are opened, and then the motor runs for a short time without grid connection. Afterwards, contactor Q2 is closed, and the soft starter starts with phase angle control in the opposite direction, until the rotor is stopped. Figure 6 shows the current shape for phase angle control, which is well known, and continuous speed reduction.

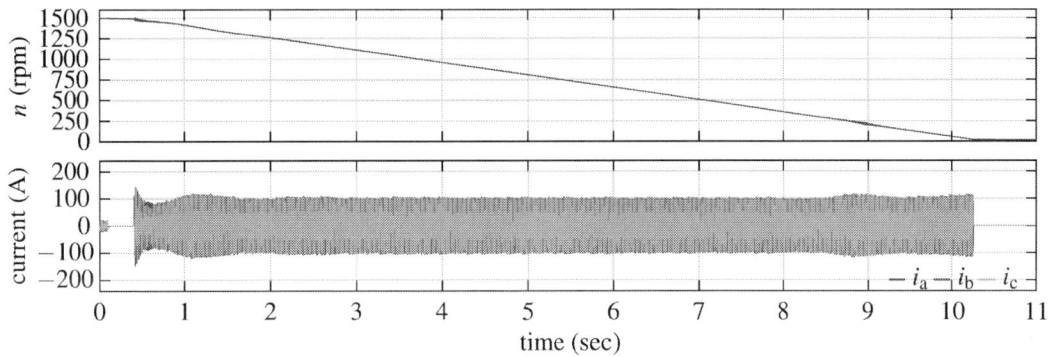

Fig. 6: Braking with phase rotation and phase angle control for current reduction

Predictive Algorithm

The predictive approach does not use any additional contactors, only a pure soft starter. Thus, the soft starter can only create braking torque by connecting the stator windings to a grid that voltage space vector rotates in the same direction as the rotor flux linkage.

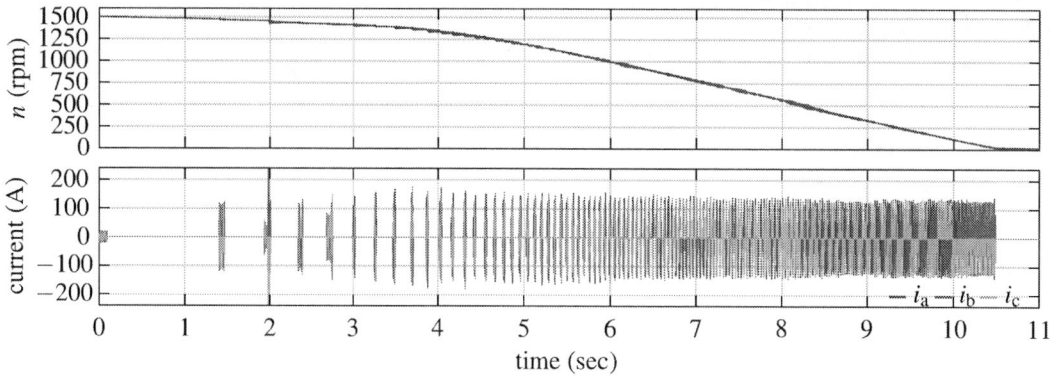

Fig. 7: Braking with predictive algorithm and without phase change

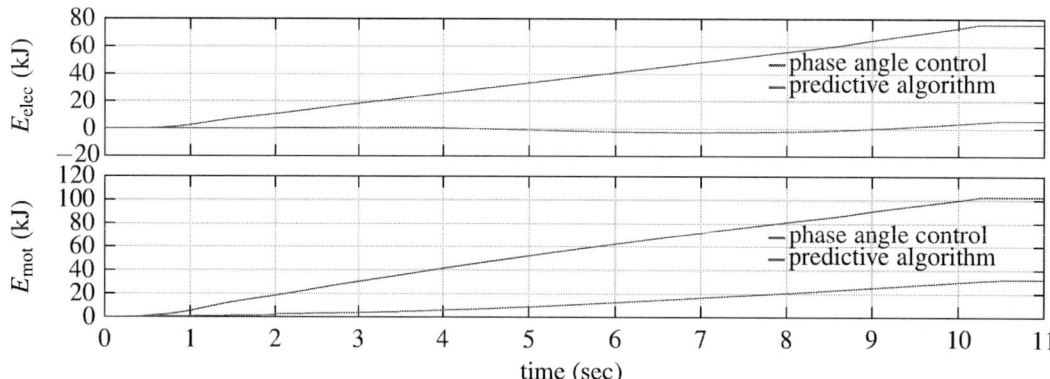

Fig. 8: Comparison of the accumulated energy flow during braking

Figure 7 shows that the braking consists of two time phases. In the first part, the algorithm leads to a kind of current burst mode. The current flows in short intervals, which are divided by time spent without current. This time decreases with every period until the second phase of braking starts, which features continuous braking. In the end of the second phase, the rotor is stopped.

In the first phase, the braking torque is less than in the second phase. This is the consequence of the large time periods without current in the first phase. The reason is the equal direction and nearly equal rotation speed of the grid and the motor, which leads to fewer opportunities for negative torque creation.

Comparison

Previous investigations have shown that the losses during start-up can be lowered by using a predictive algorithm for firing pulse generation [13]. This is investigated for braking.
In Figure 8, the energy flow during braking is shown. In the upper diagram, the electrical energy consumed from the power supply is shown.

The diagram also indicates that the phase angle control algorithm needs electrical energy from the net during the whole braking process. Although negative torque is generated by the motor, the converted energy is not transferred from the rotor to the stator but transformed to ohmic losses directly in the rotor. In contrast the predictive algorithm needs, in sum, nearly no electrical energy from the power supply during braking. This is because the motor works in real generator mode while using the predictive algorithm and, as a consequence, converts mechanical energy to electrical energy in the rotor and transforms this energy from the rotor to the stator winding. This transformed energy is used in the stator to, e.g. supply the stator losses.

The lower diagram in Figure 8 shows the overall motor losses during braking. The comparison between the two control approaches shows that the application of the predictive algorithm can help tremendously decrease the total motor losses, as the algorithm does not utilise a continuous current flow but a discrete one, which lowers the stator losses.

Figure 9 shows the calculated dissipated energy (Equation (8) and (10)) in the motor and the soft starter obtained from the measurements shown in Figures 6 and 7. Motor losses decrease by around 65 % when using the predictive approach. Thus the thermal stress for the motor reduces significantly.

Comparing the predictive algorithm approaches with the phase rotation braking, the losses in the thyristors can be reduced by 50 %, depending on which values are compared. Due to this significant loss reduction, the same soft starter can stop even larger motors or can perform more starts and stops per hour.

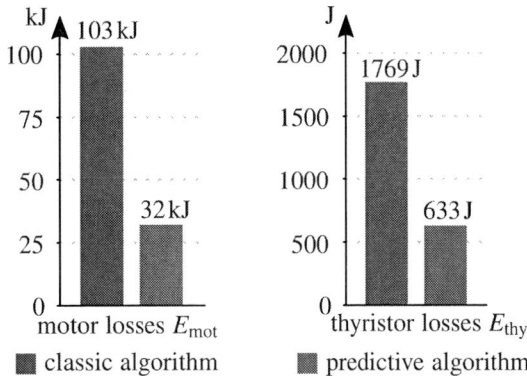

Fig. 9: Thyristor and motor losses during braking

6   Conclusion

This paper investigates different braking procedures for soft starter-driven induction motors. In addition to opposite phase rotation braking, the predictive approach is presented. The measurement results achieved on the test bench demonstrate that the losses in the motor can be lowered significantly. With these approaches, it is possible to stop the motor much more often without damaging it because of overtemperature. The presented measurements show that braking with soft starters is also possible without any additional contactors for phase reversal. This can significantly reduce the system costs for industrial realisations. The losses in the soft starter are shown to be reduced significantly by using the predictive approach. This enables more starts and stops per hour with the same soft starter, or the same soft starter can be used to stop even larger motors.

References

[1] M. Benecke, R. Doebbelin, G. Griepentrog, and A. Lindemann. *Skin Effect in Squirrel Cage Rotor Bars and Its Consideration in Simulation of Non-steady-state Operation of Induction Machines: Progress in electromagnetics research symposium March 20-23, 2011 Marrakesh, Morocco ; proceedings.* The Electromagnetics Academy, Cambridge, Mass., 2011.

[2] O. I. Butler. Stopping time and energy loss or a-c motors with d-c braking. *Transactions of the American Institute of Electrical Engineers. Part III: Power Apparatus and Systems*, 76(3):285–290, 1957.

[3] O. I. Butler and M. N. Abdel-Hamid. D-c dynamic braking of double-cage induction motors. *Transactions of the American Institute of Electrical Engineers. Part III: Power Apparatus and Systems*, 77(3):1035–1038, 1958.

[4] C. F. Evert. Dynamic braking of squirrel-cage induction motors [includes discussion]. *Transactions of the American Institute of Electrical Engineers. Part III: Power Apparatus and Systems*, 73(2):242–244, 1954.

[5] G. Huth. Description of transient deep bar effect with numerical defiend equivalent circuits (in german). *Archiv für Elektrotechnik*, 70(1):31–37, 1987.

[6] M. Laabidi, B. Rebhi, F. Kourda, M. Elleuch, and L. Ghodbani. Braking of induction motor with the technique of discrete frequency control. In *2010 7th International Multi- Conference on Systems, Signals and Devices*, pages 1–6, 2010.

[7] W. A. Lapierre and N. Metaxas. D-c dynamic braking of induction motors. *Electrical Engineering*, 72(9):785, 1953.

[8] T. A. Lipo and A. Consoli. Modeling and simulation of induction motors with saturable leakage reactances. *IEEE Transactions on Industry Applications*, IA-20(1):180–189, 1984.

[9] G. Müller, K. Vogt, and B. Ponick. *Calculation of Electrical Machines (in German)*, volume 2 of *Electrical Machines*. Wiley-VCH, Weinheim, 6. edition, 2009.

[10] S. S. Murthy, G. J. Berg, C. S. Jha, and A. K. Tandon. A novel method of multistage dynamic braking of three-phase induction motors. *IEEE Transactions on Industry Applications*, IA-20(2):328–334, 1984.

[11] H. Nannen and H. Zatocil. Initial rotor position determination of a soft starter driven synchronous motor. In *PCIM Europe 2017; International Exhibition and Conference for Power Electronics, Intelligent Motion, Renewable Energy and Energy Management*, pages 1–6, 2017.

[12] H. Nannen and H. Zatocil. Sensorless start-up of soft starter driven line-start pmsm based on back emf measurement. In *2017 IEEE 26th International Symposium on Industrial Electronics (ISIE)*, pages 354–361, 2017.

[13] H. Nannen, H. Zatocil, and G. Griepentrog. Novel predictive start-up algorithm for soft starter driven induction motors. In *IECON 2020 The 46th Annual Conference of the IEEE Industrial Electronics Society*, pages 3071–3078, 2020.

[14] H. Nannen, H. Zatocil, and G. Griepentrog. Predictive firing algorithm for soft starter driven induction motors. *IEEE Transactions on Industrial Electronics (Early Access)*, 2022.

[15] H. M. Norman. Induction motor locked saturation curves. *Electrical Engineering*, 53(4):536–541, 1934.

[16] H. C. Specht. Electric braking of induction motors. *Proceedings of the American Institute of Electrical Engineers*, 31(5):583–596, 1912.

[17] A. K. Tandon, S. S. Murthy, and B. P. Singh. Experimental studies on a novel braking system for induction motors. *IEEE Transactions on Industry Applications*, IA-20(5):1238–1243, 1984.

[18] P. Vas. *Sensorless Vector and Direct Torque Control*, volume 42 of *Oxford Science Publications*. Oxford [u.a.], reprinted. edition, 2003.

[19] H. Zatocil and H. Nannen. Sensorless start-up of soft starter driven ie4 motors. In *2017 19th European Conference on Power Electronics and Applications (EPE'17 ECCE Europe)*, pages 1–9, 2017.

Ambient Electromagnetic Energy Harvesting Circuit using Rectennas Manufactured with Stereolithography Resin

Xuan Viet Linh Nguyen, Tony Gerges, Jacques Verdier, Philippe Lombard, Michel Cabrera, Bruno Allard
Univ Lyon, INSA Lyon, Université Lyon1, Ecole Centrale de Lyon, CNRS, Ampère, UMR5505,
69621 Villeurbanne, France
xuan.nguyen@insa-lyon.fr

Jean-Marc Duchamp, Philippe Benech
Univ. Grenoble-Alpes, CNRS, Grenoble INP, G2Elab, 21 avenue des martyrs, CS 90624,
38031 Grenoble, France

Acknowledgements

Authors would like to appreciate Murata Integrated Components for their helps for our project (IPCEI INCA Nano 2022).

Keywords

«Electromagnetic Energy Harvester», «Elastic/Plastic deformation», «Energy management system», «Additive Manufacturing»

Abstract

The paper discusses exploratory results to validate main performances of an ambient electromagnetic field energy harvester implemented on a limited surface of non-dedicated polymer object fabricated using 3D Molded Interconnect Devices (3D MID) technology or Plastronics. The prototype is experimented with discrete devices (-7.5 dBm minimum total input RF power, 14.75% global efficiency).

Introduction

Nowadays, Internet of Things (IoT) devices become more and more popular in our life with many practical applications (sensors, smartwatches, smartphones, etc.). Nevertheless, the massive development of these devices leads a dramatic issue of requiring a huge amount of batteries which always contain many harmful components to the environment and to human's health. Therefore, many researches about replacing the batteries by the cleaner energy, such as vibrations, light, pressure, temperature [1] have been studied. Since the number of wireless communication devices increases rapidly, the ambient electromagnetic power density improves consequently. Developing the self-powered devices by harvesting the Radio Frequency (RF) energy becomes an interesting solution for replacing the batteries [2]–[5]. As the shell of devices are used mostly to protect the components inside and rarely for electronic purposes, it is convenient to implement the electrical functions, the RF components, the antennas, etc. on the surface of the 3D plastic packages.

We report here a system to harvest ambient electromagnetic energy, implemented on the surface skin of a 3D plastic object with the purpose of feeding a vehicle sensor for instance. The system consists in a RF-to-DC rectifier used to convert electromagnetic (EM) waves to DC energy and a Power Management Circuit (PMC) used to boost and regulate the rectified voltage of rectifier. The plastic object is fabricated using 3D MID methods: Stereolithography – an additive manufacturing method and electroless metallization.

The system is studied firstly with discrete components for verification purpose prior to the design of an integrated PMC with a companion capacitive Interposer to aggregate the various capacitors required in the circuit operation. A magnetic device is considered in the PMC that is too complex to integrate yet. The number of components to assemble on the object skin to create the function should be small as brazing a component on the object surface is challenging.

This paper is organized as follows. The RF-to-DC rectifier is presented in the first section. The method of fabricating, metallizing and choosing the components for the circuit will be described shortly in this section as well. Afterwards, a study about the electrical relation between the rectifier and the PMC to optimize the total efficiency of the system will be mentioned. Then, the principle of the PMC as well as the measurement results of the system in discrete form will be studied in the third section. Finally, a conclusion and perspectives will be given.

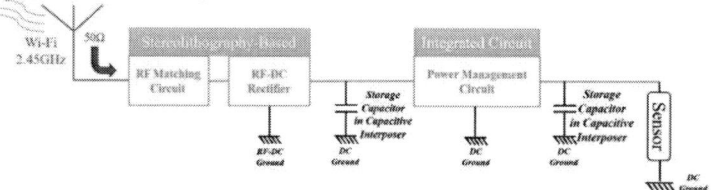

Fig. 1: Schematic of the targeted system

Stereolithography-based RF-to-DC rectifier

Fabrication of 3D printed and metallized rectifier

The polymer substrates are 3D-printed with a Stereolithography (SLA) printer Form 3 from Formlabs (Fig. 2). The high temperature resin FLHTAM02 from Formlabs is selected thanks to its good thermal resistance (around 238°C) for future brazing operation. After printing, the parts are cleaned in isopropanol to remove resin residuals and then UV cured with thermal post-processing to enhance their mechanical strength.

The electronic circuit and the antennas are fabricated by metallizing the substrates by copper Electroless Deposition (ELD) (method detailed in [6]). The thickness of the deposited copper is measured with X-ray fluorescence (Bowman (Serial B-XRF). The average measured value is (7 ± 1) µm. The electroless copper conductivity is measured with four-point-probes used with S-302 stand (Signatone) and attached to a Keithley (2450 SourceMeter). The measurements are carried out over 10 samples. The measured conductivity is around $(45 \pm 4.5) \times 10^6$ S/m.

RF-to-DC rectifier design

RF-to-DC rectifier plays a crucial role of converting the electromagnetic waves into DC power in the concept of a radio frequency energy harvester (RFEH) circuit. The proposed rectifier is a combination in parallel of four series topology rectifiers connected serially to RF chokes, which are used to prevent the load from RF signals, with one common storage capacitor (C_{common}) at the output of the circuit. The schematic and the realized circuit are illustrated in Fig. 2. The rectifying diode is the low-barrier Schottky diode SMS7630-079LF from Skyworks Solutions Inc. selected due to its low threshold voltage. In each branch of rectifier, to match the circuit to 50 Ω and to ensure the closed loop of DC current in the branch, a shorted circuit stub is employed. The load is chosen as 1 kΩ where the rectifier achieves its Maximum Power Point Tracking (MPPT) condition. C_{common} is set as 1 µF to optimize the operation of the PMC [7]. The storage capacitors in the circuit are the Surface Mounted Device (SMD) components of Murata used in Power Applications. Discrete SMD components are selected in the first step. Once the circuit is validated, they can be replaced by a capacitive interposer fabricated by Murata Integrated Passives to reduce the size as well as to improve the performances of the circuit. The circuit is simulated and optimized with the aid of Advanced Design System (ADS) Momentum for an input power of -20 dBm at the frequency of 2.45 GHz at each branch.

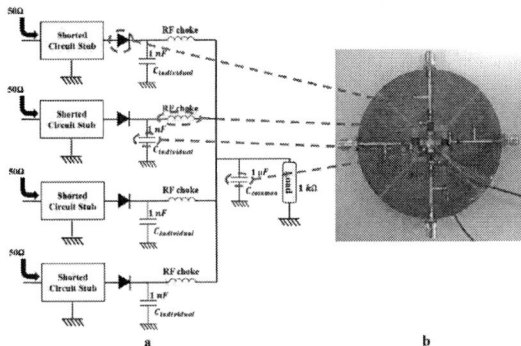

Fig. 2: Stereolithography-based RF-to-DC rectifier (a. Schematic of designed rectifier; b. Realized circuit)

Measuring the internal impedance of rectifier

Method of identifying the internal impedance

The MPPT condition of the rectifier is obtained when its output load is identical to the internal impedance of the circuit. This internal impedance is defined as the junction resistance, also known as Zero Bias Resistance (ZBR), of the rectifying diode at high frequencies [8]. Identifying the value of this resistance is necessary in order to optimize the efficiency of the system.

The relation between the incident power (P_i) at the input of the rectifier, the reflected loss (P_r) from the impedance matching condition and the transmitted power (P_t) towards the rectifying diode is given as

$$P_i = P_r + P_t \ (1)$$

Additionally, the RF-to-DC conversion efficiency of the circuit (η_{RF-DC}) can be expressed as

$$\eta_{RF-DC} = \frac{V_{out}^2}{R_L P_i} \ (2)$$

where, V_{out} is the rectified voltage at the load (V), R_L is the load (Ω) and P_i is the incident power at the input of the rectifier (W).

It is worthy to notice that, when a well-designed matching circuit is applied at a pre-defined input power and frequency, P_r can be neglected. Then, the following approximation is led

$$P_i = P_t \ (3)$$

The Eq. (2) is then written as

$$\eta_{RF-DC} = \frac{V_{out}^2}{R_L P_t} \ (4)$$

The internal impedance of the rectifier can be pointed out from η_{RF-DC} in variation of the load. However, when the load varies, P_r will not stay constant since the matching circuit is served for only one value of load, frequency and input power. P_t is not identical consequently. This leads the impacts to the accuracy of measured results. Ideally, for each value of load, frequency and input power, a corresponding matching circuit should be used. Literally, this approach requires a very complicated manipulation. Here, a method of determining the internal impedance of the rectifier using only one rectifier circuit for any measurement is considered. The method is based on changing P_i and measuring P_r so that P_t determined from Eq. (1) reaches the desired value of input power where the circuit is well-matched to 50 Ω and getting η_{RF-DC} for each value of load. For our measurements, we set P_t at different values, i.e. -30 dBm, -20 dBm, -10 dBm and 0 dBm for the frequency of 2.45 GHz. The measurement schematic and experimental setup are illustrated in Fig. 3.

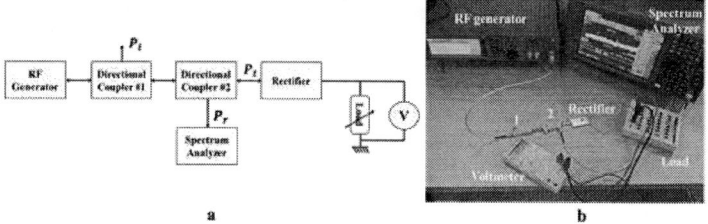

Fig. 3: The proposed method of identifying the internal impedance of the rectifier (a. Measurement schematic; b. Experimental setup (1 – Directional Coupler #1 used to measure the incident power P_i, 2 – Directional Coupler #2 used to measure the reflected loss P_r)).

The measurements are carried out on a mono-diode series topology rectifier with the Schottky diode SMS7630-079LF shown in Fig. 4.

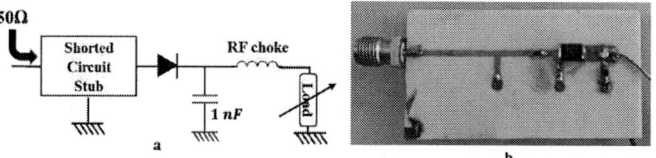

Fig. 4: The measured mono-diode series topology (a. Schematic of the circuit; b. Realized rectifier)

The simulated and measured results are plotted in Fig. 5. The internal impedance of the mono-series diode rectifier varies as functions of the input power of the circuit. The values of this impedance and η_{RF-DC} are respectively 5 kΩ (3.5 %) at -30 dBm, 3 kΩ (15.7 %) at -20 dBm, 2.5 kΩ (34.4 %) at -10 dBm and 2 kΩ (45.1%) at 0 dBm. In fact, the measured rectified voltage is sensitive to the accuracy of the source power level. 0.1 dB of shifting of the source power impact up to 2% the measured rectified voltage.

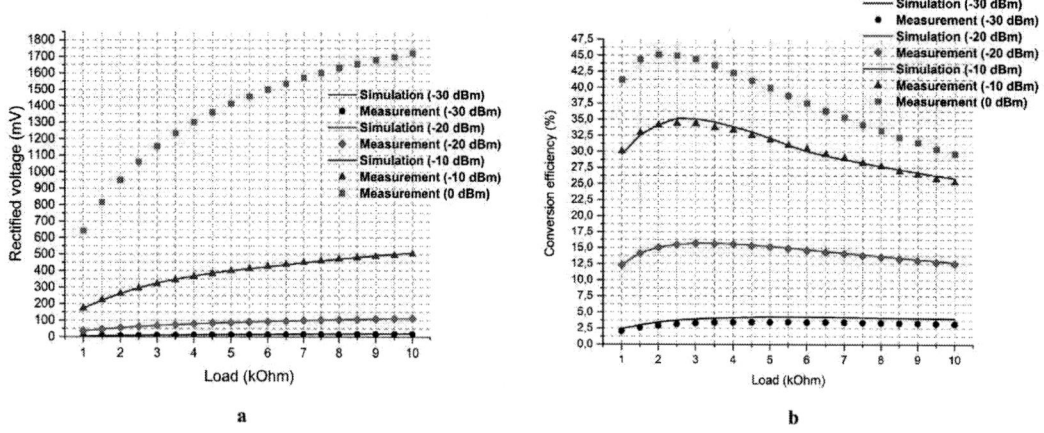

Fig. 5: The measured results of mono-diode series rectifier used to identify the internal impedance (a. Measured rectified voltage; b. Corresponding conversion efficiency)

In [5], this value varies as function of the input power level, which are 2.6 kΩ at 0 dBm, 4 kΩ at -20 dBm and 7.2 kΩ at -35 dBm. However, the lossy impacts from the matching circuit are not concerned in the measurements. In [8], authors show that the ZBR of this rectifying diode is around 5.1 kΩ at low current across the diode. In [9], a study of obtaining the optimal load of the rectifier at different temperature (from -20 °C to 30 °C) is carried out. At the ambient temperature (around 23 °C), ZBR of SMS7630 diode is from 6 kΩ to 7 kΩ at different input power level, which are from -40 dBm to -20 dBm. Comparing to our state-of-art, our measured results are very close to them. The difference between the results may come from the deployed method and/or the variation of the measured diode models.

Internal impedance of 4-port rectifier

The desired circuit is a RF-to-DC rectifier of four inputs illustrated in Fig. 2. The internal impedance of the circuit should be determined to maximize its conversion efficiency. The topology of each branch is mono-series diode, the Thevenin equivalent schematic of our 4-port rectifier in term of DC circuit is presented in Fig. 6. Since, the branches are linked in parallel, with the hypothesis is that if the input power at each input is lower than -30 dBm, it will be considered as zero, the equivalent internal impedance of the circuit ($Z_{circuit}$) can be calculated ideally as

$$\frac{1}{Z_{circuit}} = \sum_{i=1}^{4} 1/Z_{int\ imp_i} \ (5)$$

where $Z_{int\ imp_i}$ is the internal impedance of branch "i" at a defined input power at 2.45 GHz (Ω).

In fact, in situation of energy harvesting, the input power of each branch is not nor constant nor identical (from -30 dBm to 0 dBm), $Z_{circuit}$ will not fixed at a certain value consequently. Here, our proposed circuit is preferred to have ability of working in all possibilities. The value of $Z_{circuit}$ of all possibilities are calculated and plotted in Fig. 6. According to Fig. 6, the values of $Z_{circuit}$ are gathered mostly around the value of 1 kΩ. This value is therefore chosen as the value of the load.

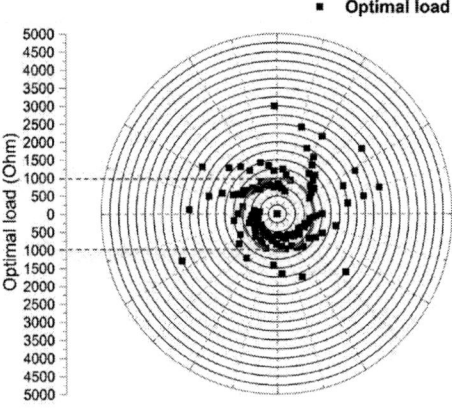

Fig. 6: The calculated optimal load of the 4-port rectifier

Discrete Power Management Circuit

The idea of our PMC was introduced in [10]. The circuit in Fig. 7 consists of a self-oscillating Armstrong converter for cold start-up and a Flyback converter in Discontinuous-Conduction Mode (DCM) for continuous operation. A low-consumption oscillator TS3002 of Silicon Labs is used to generate the switching frequency for the Flyback converter and two Under-Voltage-Lock-Out (UVLO) circuits discriminate the circuit operation. A Flyback converter in DCM mode operates the Maximum Power Point Tracking (MPPT) conditions and features a better efficiency than the Armstrong's one. It should be noticed that the Flyback converter supplies the oscillator and a natural feedback effect is obtained to track MPPT conditions. UVLO#1 is used to keep the Flyback converter operation during the variation of load. UVLO#2 has the role of regulating the output voltage.

A 2D circuit is fabricated with discrete components available on the market.

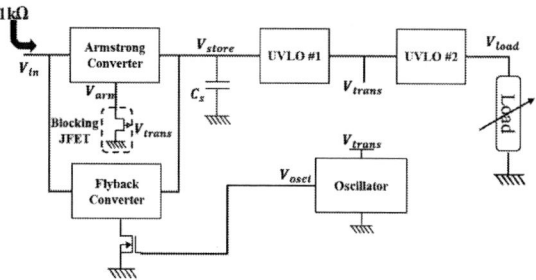

Fig. 7: Block diagram of the Power Management Circuit

The circuit has the following operation phases: Cold start, Transition from Amstrong to Flyback, Normal operation and Deactivation.

Cold start phase: When the voltage across the storage capacitor C_{s1} reaches to a minimum starting voltage [10], the Armstrong converter oscillates and boosts the voltage and stores the energy in another storage capacitor C_{s2}. The ideal minimum starting voltage (V_{start_min}) can be estimated as [10]

$$V_{start\_min} \approx (2I_{DSS}R_S + |V_p|)/m \tag{1}$$

Where, R_s is the impedance of source of circuit (Ω), I_{DSS} and V_p are Zero-Gate Voltage Drain Current (A) and Threshold Voltage (V) respectively of JFET used in Armstrong and m is the amplification ratio of transformer used in Armstrong. The selected JFET is J201 ($I_{DSS} = 0.6\ mA$ and $V_p = -0.6\ V$) and the transformer is WE750311681 from Würth Elektronik ($m = 10$). Thus, the estimated minimum voltage is 0.18 V.

Transition phase: the Armstrong converter keeps working until the voltage across C_{s2} reaches the closed threshold voltage ($V_{closed1}$) of UVLO#1 (1.2 V). Once UVLO#1 is closed, C_{s2} will feed the oscillator (V_{trans}) to start the Flyback converter. Simultaneously, V_{trans} will feed also the Gate of the blocking JFET (J177) to deactivate Armstrong converter.

Normal operation phase: the Flyback converter is self-supplied as long as the voltage V_{trans} is high enough to keep the oscillator working. The voltage across C_{s2} continues to increase until it reaches the closed threshold voltage ($V_{closed2}$) of UVLO#2 (1.57 V). When UVLO#2 is closed, C_{s2} feeds the load. When the load supply voltage (V_{load}) decreases below the opened threshold voltage ($V_{opened2}$) of UVLO#2 (1.4 V), UVLO#2 disconnects the load, and C_{s2} will start to re-charge.

In this phase of operation, the input impedance of the circuit (R_{input}) is determined as [7]

$$R_{input} = 2L_1 f_{sw}/D^2 \tag{2}$$

Where, L_1 is the primary inductance of the transformer (H), f_{sw} and D are respectively the switching frequency (Hz) and the duty cycle (50%) generated by the oscillator. A LPD6235-205 transformer from Coilcraft is selected due to its high primary inductance (2 mH). As presented previously, in order to achieve the optimal efficiency, the input impedance of the circuit should be fixed at 1 kΩ, the MPPT load of the rectifier. Therefore, f_{sw} and D of the oscillator are set as 65 kHz and 50%.

Deactivation phase: When V_{store} is lower than the opened threshold voltage ($V_{opened1}$) of UVLO#1 because of the degradation of the RF input voltage for example, UVLO#1 will disconnect the Flyback convertor. The Armstrong converter will be reactivated as long as the input voltage of circuit is higher than 0.26 V.

Measurement results

The PMC is tested with the rectifier in Fig. 2 and the total input power of -7.5 dBm of the rectifier at the frequency of 2.45 GHz. The measured input voltage of PMC is 0.264 V, where the circuit starts to operate. The electrical behavior of the PMC is presented in Fig. 8.

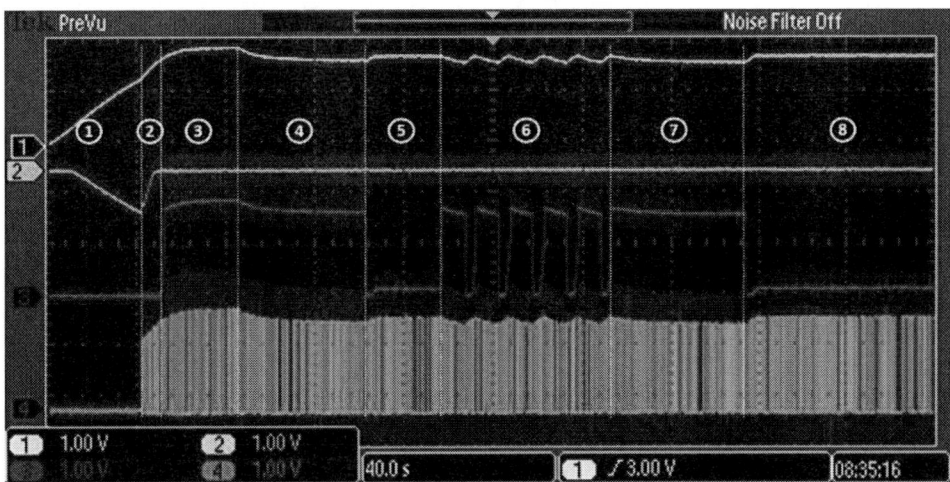

Fig. 8: The measured results of discrete Power Management Circuit (Yellow: V_{store} ; Blue: V_{arm} ; Purple: V_{load} ; Green: V_{osci})

In Fig. 8, the yellow curve presents the variation of voltage of storage capacitor C_{s2} ; the blue curve shows if the Armstrong circuit activated or deactivated; the purple curve is the voltage of the load; the green curve is the signal generated by the oscillator of the Flyback converter.

1) First, the load at the PMC's output is 1 MΩ. The circuit is currently in "cold start" phase. The storage voltage V_{store} across the storage capacitor C_s is pulled up due to the Armstrong converter. During the process, this voltage is not high enough to close UVLO#1 ($V_{closed1}$ < 1.2 V), thus, the Flyback converter is not activated yet and V_{load} stays at zero.

2) When V_{store} reaches $V_{closed1}$, the Flyback converter is activated and the Armstrong converter is turned off. However, V_{store} is still lower than the threshold voltage $V_{closed2}$ (1.57 V) of UVLO#2. Therefore, there is no supply of the load. The circuit is in "transition" phase.

3) The Flyback converter increases V_{store}. When V_{store} is higher than $V_{closed2}$, the load is supplied (V_{load} is identical to V_{store}). It should be noticed that the load is 1 MΩ. The transmission of energy from C_s to the load occurs immediately. The circuit is in "normal operation" phase.

4) The load is reduced to 100 kΩ. V_{load} is reduced to 1.62 V, but still higher than the threshold voltage $V_{opened2}$ (1.4 V) of UVLO#2. The PMC input power is sufficiently high to cover the circuit losses and the load consumption.

5) The load is then set to 10 kΩ. V_{load} drops until it is lower than $V_{opened2}$. UVLO#2 is then opened, the Flyback converter increases until V_{store} is higher than $V_{closed2}$. UVLO#2 is closed again. Since, the load is low, its consumption and the losses in the circuit increase [7], [10], this explains the slight output voltage. The Flyback converter keeps in operation as UVLO#1 is closed.

6) The load is fixed at 80 kΩ, i.e. the limit value of load in "normal operation" phase. We have a discontinuous supply of the load. This load consumes more energy, thus, C_s discharges rapidly.

7) The load is set to 100 kΩ with a similar operation as in case 4.

8) The load is set to 10 kΩ with a similar operation as in case 5.

The output voltage is 1.62 V for a load of 100 kΩ with an input voltage of 0.264 V in *"normal operation"* phase. The PMC boost ratio is then 6.14. The global efficiency of the system (η_{global}) in *"normal operation"* phase and for 100 kΩ at 0.264V input voltage is 14.75%.

The amplified voltage is around 1.62 V for a load of 100 kΩ with an input of 0.264 V. Hence, the amplified ratio of PMC is 6.14. The global efficiency of system (η_{global}) can be calculated as

$$\eta_{global} = V_{load}^2/(R_{load} \times P_{RF\ total}) = 1.62^2/(100k \times 10^{-3.75}) = 14.75\%$$

Conclusion

The discrete component circuit validates the operation of the rectifier and the PMC. The global efficiency is 14.75 % for a total power level of -7.5 dBm of the rectifier.

The performances' improvement of the circuit at the MPPT load depends on the capability of the rectifier with respect to the input power level as shown in Fig. 5 for one PMC. The other key element is the antenna gain. In addition, the efficiency of the Flyback used in the PMC is tributary on the electrical properties and the dimensions of the coupled inductance of the transformer [7].

The price of the energy harvesting circuit is a secondary issue compared to life-cycle assessment (LCA) of the device. Assembling of Armstrong and Flyback components is a good candidate for the energy harvesting applications [10] even though it requires a large number of components that contribute to a high level of gray energy. With the ambient EM energy available in the air [4], [11], during the life duration of the payload circuit, the harvested energy might be at least equal to its equivalent gray energy.

References

[1] S. Sudevalayam and P. Kulkarni, "Energy Harvesting Sensor Nodes: Survey and Implications," *IEEE Commun. Surv. Tutor.*, vol. 13, no. 3, pp. 443–461, 2011, doi: 10.1109/SURV.2011.060710.00094.

[2] E. Vandelle, D. H. N. Bui, T.-P. Vuong, G. Ardila, K. Wu, and S. Hemour, "Harvesting Ambient RF Energy Efficiently With Optimal Angular Coverage," *IEEE Trans. Antennas Propag.*, vol. 67, no. 3, pp. 1862–1873, Mar. 2019, doi: 10.1109/TAP.2018.2888957.

[3] "Design and experiments of a dual-band rectenna for ambient RF energy harvesting in urban environments - Khemar - 2018 - IET Microwaves, Antennas & Propagation - Wiley Online Library." https://ietresearch.onlinelibrary.wiley.com/doi/10.1049/iet-map.2016.1040 (accessed May 28, 2022).

[4] U. Muncuk, K. Alemdar, J. D. Sarode, and K. R. Chowdhury, "Multiband Ambient RF Energy Harvesting Circuit Design for Enabling Batteryless Sensors and IoT," *IEEE Internet Things J.*, vol. 5, no. 4, pp. 2700–2714, Aug. 2018, doi: 10.1109/JIOT.2018.2813162.

[5] S.-E. Adami *et al.*, "A Flexible 2.45-GHz Power Harvesting Wristband With Net System Output From −24.3 dBm of RF Power," *IEEE Trans. Microw. Theory Tech.*, vol. 66, no. 1, pp. 380–395, Jan. 2018, doi: 10.1109/TMTT.2017.2700299.

[6] T. Gerges *et al.*, "Investigation of 3D printed polymer-based heat dissipator for GaN transistors," in *2021 23rd European Conference on Power Electronics and Applications (EPE'21 ECCE Europe)*, Sep. 2021, p. P.1-P.9.

[7] A. Capitaine, "Récupération d'énergie à partir de piles à combustible microbiennes benthiques," Theses, Université de Lyon, 2017. Accessed: May 28, 2022. [Online]. Available: https://tel.archives-ouvertes.fr/tel-02090785

[8] S. Hemour *et al.*, "Towards Low-Power High-Efficiency RF and Microwave Energy Harvesting," *IEEE Trans. Microw. Theory Tech.*, vol. 62, no. 4, pp. 965–976, Apr. 2014, doi: 10.1109/TMTT.2014.2305134.

[9] X. Gu, L. Guo, S. Hemour, and K. Wu, "Optimum Temperatures for Enhanced Power Conversion Efficiency (PCE) of Zero-Bias Diode-Based Rectifiers," *IEEE Trans. Microw. Theory Tech.*, vol. 68, no. 9, pp. 4040–4053, Sep. 2020, doi: 10.1109/TMTT.2020.2992024.

[10] S.-E. Adami, "Optimisation de la récupération d'énergie dans les applications de rectenna," phdthesis, Ecole Centrale de Lyon, 2013. Accessed: May 28, 2022. [Online]. Available: https://tel.archives-ouvertes.fr/tel-00967525

[11] H. Takhedmit, "Ambient RF power harvesting: Application to remote supply of a batteryless temperature sensor," in *2016 IEEE International Smart Cities Conference (ISC2)*, Sep. 2016, pp. 1–4. doi: 10.1109/ISC2.2016.7580800.

Boost/Buck-boost Based Grid Connected Solar PV Micro-inverter with Reduced Number of Switches and Having Power Decoupling Capability

Arup Ratan Paul[*], Arghyadip Bhattacharya[†] and Kishore Chatterjee[‡]

IIT Bombay, Mumbai, India

Email: aruprpaul@ee.iitb.ac.in[*], arghyaiitb@ee.iitb.ac.in[†], kishore@ee.iitb.ac.in[‡]

Keywords

≪DC-AC converter≫, ≪Grid-connected inverter≫, ≪Renewable energy systems≫, ≪Transformerless PV inverter≫, ≪Micro-inverter≫ ≪Power Decoupling≫.

Abstract

A boost/buck-boost based transformer-less micro-inverter suitable for interfacing a 35 V, 220 W PV module to a single phase 220 V ac grid is proposed in this paper. The intermediate capacitor between the the boost stage and the buck-boost stage helps in achieving power decoupling between the dc side and the ac side. Since the inverter is endowed with inherent power decoupling feature, it is more reliable compared to the inverters utilizing high value electrolytic capacitors. The micro-inverter utilizes six switches, two of them operating at high frequency, two at line frequency and the rest two switches operate at high frequency either during the positive half cycle or in the negative half cycle. Both the boost stage and buck-boost stage is operated in discontinuous conduction mode for all possible operating conditions in order to achieve higher voltage gain, and negligible turn on losses of the high frequency switches. The direct connection between the negative terminal of the PV module and the grid neutral ensures the leakage current flow to be zero. The analyses of the proposed micro-inverter is carried out. The viability if the proposed scheme is validated by detailed simulation studies in the MATLAB/Simulink platform.

Introduction

The renewable energy resources are becoming more popular in the whole world as they are environment-friendly and inexhaustible in nature. In countries like India, where the sunlight is abundant throughout the year, solar power generation is a viable alternative to the conventional fossil fuel based electricity generation. In order to integrate the solar photovoltaic (PV) systems to the existing ac grids the extracted power must be processed through an inverter. In case of grid connected rooftop solar PV inverters, micro-inverters are becoming more popular over central or string inverters as they offer modularity, plug and play feature and for their capability of extracting maximum available power from each PV module [1]. However due to smaller power rating and higher voltage gain requirement, the efficiency is poor compared to central or string inverters. In order to improve the efficiency of a micro-inverter, less number of power conversion stages are desired [2,3]. Hence transformer-less single stage topologies are taken into consideration.

In most of the countries the distribution voltage level is 220-230 V. Interfacing a 35 V PV module to a 220 V ac grid without using a transformer is a major challenge. Further in case of a transformer-less micro-inverter topology, the leakage current issue also needs to be considered [4,5]. The micro-inverter topologies reported in [6–16] are doubly grounded, i.e, the negative terminal of the PV module is directly connected to the grid neutral. Hence the leakage current flow is zero in the aforementioned micro-inverters. However, the micro-inverters reported in [6–12,14,15] are not suitable for interfacing a 35 V module to a 220 V ac grid as they offer lower voltage gain and suitable for interfacing to a 110 V ac grid. The micro-inverters reported in [13,16,17] are suitable for interfacing a 35 V PV module to 220 V ac grid. However, the leakage current issue has not been addressed in [17].

In case of single phase inverters, there is a mismatch in the instantaneous power between the dc side and the ac side, which leads to significant second order harmonic ripple in the dc bus voltage. However, such ripple in the voltage across PV terminals decreases the maximum power point tracking (MPPT) efficiency [18, 19]. The simplest and cheapest solution to this issue is to connect a high value capacitor across the PV module such that the voltage ripple across it becomes low. If the voltage ripple has to be less the 5%, the capacitance needs to be connected across a 35 V, 220 W PV module is more than 13 mF as per the design guidelines provided in [20]. However, the usage of electrolytic capacitor is not a reliable solution as their lifetime is very poor compared to a PV module. High value thin film capacitor is also not a practical solution for a micro-inverter as it would increase the volume and cost of the inverter significantly. It is to be noted that, the micro-inverters reported in [6, 8, 9, 11–15, 17] do not have the power decoupling capability and hence a large value capacitor needs to be connected across the PV terminals. A viable solution to this issue is to connect an additional active power decoupling circuit (APDC) across the PV module, wherein the value of capacitance required is significantly less and hence thin film capacitors can be utilized. A number of APDC are reported in [20–22]. However, APDC required two additional high frequency (HF) switches and an inductor, which add losses to the system. Hence attempts have been made to achieve the power decoupling inherently in the main micro-inverter circuit [7, 10, 16].

Among all the aforementioned micro-inverters, only the topology reported in [16] has all of the following features:

- It is suitable for interfacing a 35 V PV module to a 220 V ac grid.
- It is doubly grounded, and hence the leakage current flow is zero.
- It is having the inherent power decoupling capability.

However that micro-inverter utilizes seven switches. Further it is operated in continuous conduction mode (CCM) for all possible operating conditions. Hence the duty ratio of the boost stage is as high as 0.85. Also the volume of the inductors are significantly high. In order to overcome all the aforementioned drawbacks of micro-inverter reported in [16], a micro-inverter having only six switches are proposed in this paper. The micro-inverter is having inherent power decoupling capability and suitable for interfacing a 35 V PV module to a 220 V ac grid. The leakage current issue is also been addressed. The proposed micro-inverter is operated in discontinuous conduction mode (DCM) in order to achieve (i) same voltage gain at lower duty ratio, (ii) reduction in volume of the inductors and (iii) negligible turn on losses of the HF switches. A detailed simulation studies on the proposed micro-inverter is carried out to confirm its viability.

Principle of the Proposed Topology

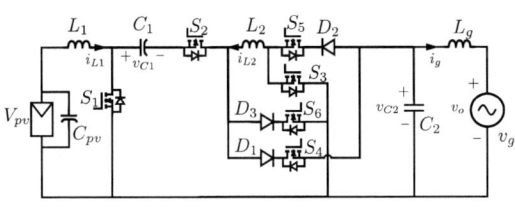

Fig. 1: Schematic diagram of the proposed micro-inverter

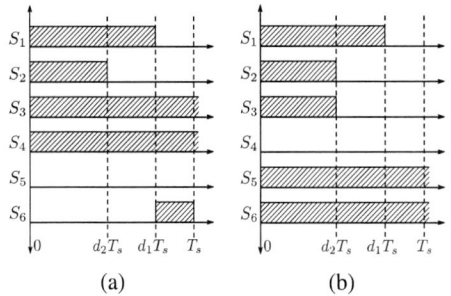

Fig. 2: Switching pulses: (a) positive half cycle and (b) negative half cycle

The schematic diagram of the proposed micro-inverter topology is shown in Fig. 1. This inverter topology is also realized by integrating a boost stage followed by a buck-boost stage as is the case in [16]. However, the topological configuration is modified and hence the number of switches is reduced to six. The intermediate capacitor, C_1 acts as a coupling element between the boost stage and the buck-boost stage, and it is also used for power decoupling. The duty ratio of the boost stage, d_1 is used to control the current drawn from the PV module and maintain the converter operating at MPP. The duty ratio of the buck-boost stage, d_2 is used to maintain the voltage across C_1, and accordingly the power is fed to the grid. The gating pulses are given to the switches as shown in Fig. 2. It is to be noted that the switches, S_1 and S_2 operate HF, S_4 and S_5 ant line frequency (LF), and the remaining two switches, S_3 and S_6 operates at HF for either in positive half cycle (PHC) or in negative half cycle (NHC) of the grid voltage. The switching time period, T_s can be divided into three intervals. In case of this configuration of the inverter, the inductor, L_2 has to be designed in such a way that the current through it has to be become zero before $d_1 T_s$ for all operating conditions. The current through L_1 is also discontinuous.

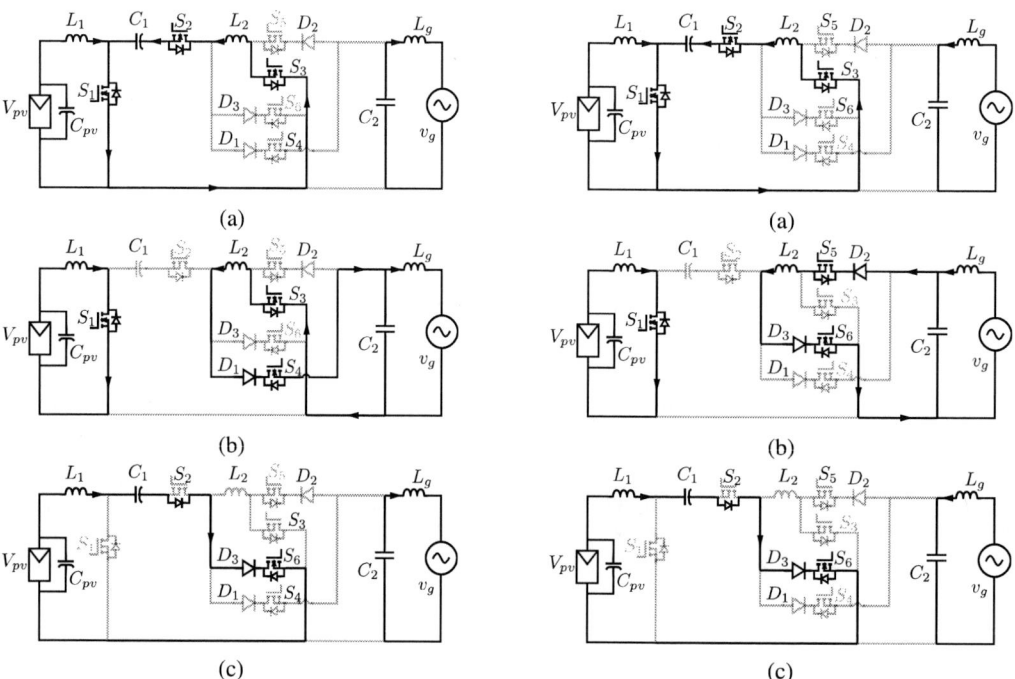

Fig. 3: Modes of operation for PHC: (a) Mode-I, (b) Mode-II, and (c) Mode-III

Fig. 4: Modes of operation for NHC: (a) Mode-I, (b) Mode-II, and (c) Mode-III

Positive Half Cycle

The modes of operation in PHC are shown in Fig. 3. The switches, S_3 and S_4 are turned on for the whole half cycle whereas S_5 is kept off.

Mode-I $(0 \leq t < d_2 T_s)$: The switches, S_1 and S_2 are turned on while the S_6 is kept off. The currents through L_1 and L_2 start rising from zero. Both the capacitors, C_1 and C_2 gets discharged in this mode. The D_1 blocks the current through S_4.

Mode-II $(d_2 T_s \leq t < d_1 T_s)$: In this interval, S_2 is turned off, while the states of the other switches remain the same as in mode-I. The current, i_{L1} flowing through L_1 continues to rise while the current, i_{L2} flowing through L_2 starts decreasing. Before the end of this mode i_{L2} becomes zero. The capacitor C_1 is in floating condition. The inductor L_2 supplies the power to the load.

Mode-III $(d_1 T_s \leq t < T_s)$: At the start of this interval, S_1 is turned off and S_6 is turned on. The current, i_{L2} remains zero in this mode. The current, i_{L1} starts decreasing and becomes zero before or at the end of this mode depending on the loading condition. The capacitor C_1 gets charged in this interval. It is to be noted that, the body diode of S_2 conducts in this interval and hence no gate pulse is given to S_2.

Negative Half Cycle

The modes of operation in NHC are shown in Fig. 4. The switches, S_5 and S_6 are turned on for the whole half cycle whereas S_4 is kept off.

Mode-I ($0 \leq t < d_2 T_s$): The switches, S_1, S_2 and S_3 are turned on. The currents through L_1 and L_2 start rising from zero. Both the capacitors, C_1 and C_2 gets discharged in this mode. The D_1 blocks the current through S_4.

Mode-II ($d_2 T_s \leq t < d_1 T_s$): In this interval, S_2 is turned off, while the states of the other switches remain the same as in mode-I. The current, i_{L1} flowing through L_1 continues to rise while the current, i_{L2} flowing through L_2 starts decreasing. Before the end of this mode i_{L2} becomes zero. The capacitor C_1 is in floating condition. The inductor L_2 supplies the power to the load.

Mode-III ($d_1 T_s \leq t < T_s$): At the start of this interval, S_1 is turned off and S_6 is turned on. The current, i_{L2} remains zero in this mode. The current, i_{L1} starts decreasing and becomes zero before or at the end of this mode depending on the loading condition. The capacitor C_1 gets charged in this interval. It is to be noted that, the body diode of S_2 conducts in this interval and hence no gate pulse is given to S_2.

Combined Operation

The governing equations of the inductors, L_1 and L_2 for different modes are provided in Table I. It may be noted that, all these equations are valid in both PHC and NHC.

Table I: Governing equations of the inductors, L_1 and L_2

Mode-I	Mode-II	Mode-III	
$0 < t < d_1 T_s$		$d_1 T_s < t < (1-\Delta_1)T_s$	$(1-\Delta_1)T_s < t < T_s$
$0 < t < d_2 T_s$	$d_2 T_s < t < (1-\Delta_2)T_s$	$(1-\Delta_2)T_s < t < T_s$	
$L_1 \dfrac{di_{L1}}{dt} = V_{pv}$ (1)		$L_1 \dfrac{di_{L1}}{dt} = v_{C1}$ (2)	$L_1 \dfrac{di_{L1}}{dt} = 0$ (3)
$L_2 \dfrac{di_{L2}}{dt} = v_{C1}$ (4)	$L_2 \dfrac{di_{L2}}{dt} = \lvert v_g \rvert$ (5)	$L_2 \dfrac{di_{L2}}{dt} = 0$ (6)	

$\Delta_1 T_s$: interval for which i_{L1} remains zero $\qquad$ $\Delta_2 T_s$: interval for which i_{L2} remains zero

Applying volt-second balance across L_1 and L_2 respectively, the voltage gain of the boost stage and the buck-boost stage are obtained as,

$$\frac{v_{C1}}{V_{pv}} = \frac{1-\Delta_1}{1-d_1-\Delta_1} \quad \text{and} \quad \frac{v_g}{v_{C1}} = \frac{\text{sgn}(v_g)d_2}{1-d_2-\Delta_2}. \tag{7}$$

Therefore, the voltage gain of the inverter can be obtained as,

$$\frac{v_g}{V_{pv}} = \frac{\text{sgn}(v_g)(1-\Delta_1)d_2}{(1-d_1-\Delta_1)(1-d_2-\Delta_2)}. \tag{8}$$

However, the output voltage is decided by the grid voltage whereas the input voltage is decided the MPP voltage at steady state. Further, the average voltage across C_1 is also maintained to a predefined value. Therefore the voltage gain is remains independent of the loading conditions. Since both the stages are operated in DCM, the duty ratios are related to the transferred power. Hence d_1 and d_2 need to be estimated to feed the desired power. The duty ratios of the boost stage and the buck-boost can be respectively estimated as,

$$d_{1_0} = \sqrt{\frac{2I_{L1}L_1(v_{C1}-V_{pv})}{v_{C1}V_{pv}T_s}} \quad \text{and} \quad d_{2_0} = \frac{2}{v_{C1}}\sqrt{\frac{V_g I_g L_2}{T_s}}\lvert \sin \omega t \rvert. \tag{9}$$

The current fed to the grid is controlled indirectly by controlling i_{L2}. Therefore, the reference, i_{L2}^* has to be estimated from I_{gm}^* and it is given by,

$$i_{L2}^* = I_{gm}^*\left(\lvert \sin \omega t \rvert + \frac{V_{gm}}{v_{C1}}\sin^2 \omega t\right). \tag{10}$$

Configuration of Control

The main objective of the controller is to feed sinusoidal current with appropriate magnitude to the grid while operating at MPP. The configuration of control is shown in Fig. 5. The boost stage and the buck-boost stage of the proposed micro-inverter is separated by the intermediate capacitor, C_1. The voltage across C_1 is controlled by a controller having slower dynamics compared the current controller. Hence the two stages can be controlled individually.

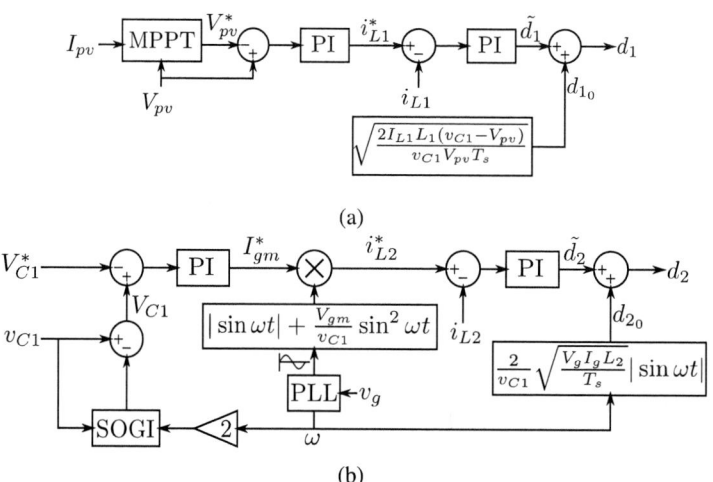

Fig. 5: Block diagram of the control configuration: (a) boost stage and (b) buck-boost stage

The controller dedicated for the boost stage is shown in Fig. 5(a), which can also be termed as a input controller. An incremental conductance based MPPT controller is utilized, which generates the reference voltage, V_{pv}^*. The error between the reference and the measured V_{pv} is passed through a PI based voltage controller to generate the reference current, i_{L1}^*. The duty ratio, d_1 is estimated using the expression provided in (9), which is the feed forward component of d_1. The measured i_{L1} is compared with i_{L1}^* and the generated error is passed through a PI controller the generate the feedback component, $\tilde{d}_1$. Finally, $\tilde{d}_1$ is added with the feed forward component to obtain the actual duty ratio d_1.

The controller for the buck-boost stage is depicted in Fig. 5(b), which can also be termed as a output controller. The power drawn from the PV module at the boost stage needs to be fed to the grid in order to maintain V_{C1} to a predefined reference, where V_{C1} is the average voltage across C_1. A second order generalized integrator (SOGI) filter tuned at twice the grid frequency is utilized to filter out the the second order harmonics in v_{C1}. The PI controller, which maintains V_{C1} at V_{C1}^*, generates the reference peak output current I_{gm}^*. The reference current, i_{L2}^* is obtained from I_{gm}^* using (10). The reference sinusoidal wave having unit magnitude is generated from the grid voltage using the single phase PLL. The feed forward component of d_2 is estimated using the expression provided in (9). The feedback component of d_2 is $\tilde{d}_2$ is generated by a PI controller which maintains i_{L2} to its reference, i_{L2}^*. At the end, the feedback component is added with the feed forward component to obtain the actual duty ratio of the buck-boost stage, d_2.

Design Guidelines

Input Inductor, L_1

The input inductor, L_1 needs to be designed in such a way, so that the current through it remains discontinuous for all possible operating conditions. Hence the limiting condition for L_1 is obtained as,

$$L_1 \le \frac{V_{pv}T_s(V_{C1} - V_{pv})}{2I_{pv}V_{C1}}. \tag{11}$$

Output Inductor, L_2

Since the current, i_{L2} must become zero before the end of mode-II, the limiting condition for L_2 can be obtained as,

$$L_2 \leq \frac{(V_{C1} - V_{PV})V_{C1}}{V_{PV}(V_{gm} + V_{C1})} \left(\frac{I_{L1}}{I_{gm}}\right) L_1.$$

(12)

Intermediate Capacitor, C_1

The intermediate capacitor, C_1 acts as a coupling between the dc side and the ac side by allowing second order harmonic ripple. Therefore

$$C_1 = \frac{P}{2\pi f_g V_{C1} \Delta v_{C1}},$$

(13)

where P is the power being transferred, f_g is the frequency of the grid voltage, and Δv_{C1} is the maximum allowable voltage ripple across C_1.

Output Capacitor, C_2

The voltage ripple, Δv_{C2} is maximum at $\omega t = \pi/2$. Therefore

$$C_2 = \frac{I_{gm} d_2 T_s}{V_{gm} \frac{\Delta v_{C2}}{V_{C2}}},$$

(14)

where the value of d_2 can be calculated from (9).

Simulated Performance of the Proposed Micro-inverter

Detailed simulation studies of the proposed micro-inverter is carried out on MATLAB/Simulink platform. The parameters chosen from the simulation are depicted in Table II.

Table II: Parameters for Simulation

System Parameters		Inverter Parameters	
Parameters	Values	Parameters	Values
V_{mpp} and V_{oc} at STC	34.7 V and 44 V	L_1	38 μH
I_{mpp} and I_{sc} at STC	6.35 A and 6.6 A	C_1	100 μF
MPP power, P_{mpp} at STC	220.34 W	L_2	200 μH
Grid voltage, V_g	220 V	C_2	0.47 μF
Grid frequency, f_g	50 Hz	L_g	1 mH
Switching frequency, f_s	50 kHz	C_{pv}	20 μF

V_{oc}: Open circuit voltage of the PV module V_{mpp}: Voltage of the PV module at MPP
I_{sc}: Short circuit current of the PV module I_{mpp}: Current from the PV module at MPP
STC: Standard test condition (Solar insolation=1 kW/m^2, Temp.=25°C)

The solar irradiance is varied over time as depicted in Fig. 6. The power drawn from the PV module is also shown in this figure. It can be inferred from the figure that MPP is tracked properly with variation of solar irradiance over time. The waveforms of the voltage across PV module and the current drawn from it are shown in Fig. 7. In Fig. 8, the grid voltage and the current fed to the grid are shown. From Fig. 7 and Fig. 8, it can be inferred that the micro-inverter is capable of interfacing a 35 V PV module to a 220 V ac grid. THD of the grid current is observed to be 4.09% while negotiating the solar irradiance of 1 kW/m^2.

The voltage across the capacitor, C_1 is depicted in Fig. 9. The average voltage is maintained at 250 V, and second order harmonic ripple is 27 V peak to peak when the irradiance is 1 kW/m^2. It can be observed from Fig. 10 that i_{L2} becomes zero before i_{L1} starts decreasing. The plots of duty ratios, d_1 and d_2 for

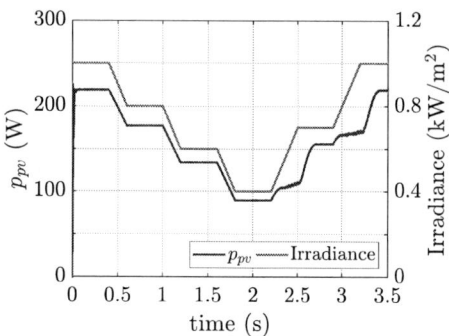

Fig. 6: Waveform of solar irradiance and the power drawn from the PV module

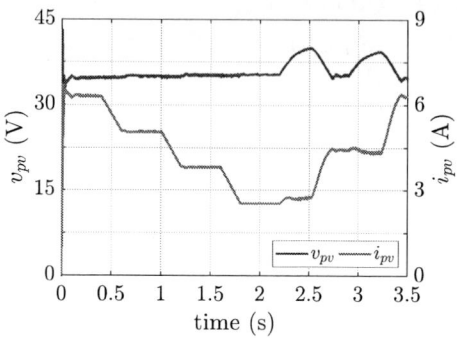

Fig. 7: PV module voltage and its current waveform

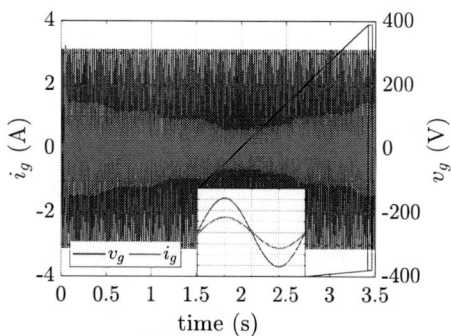

Fig. 8: Grid side voltage and current waveform, and magnified plot over 20 ms

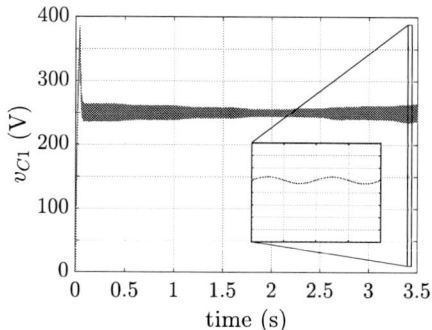

Fig. 9: Voltage waveform across C_1, and magnified plot over 20 ms

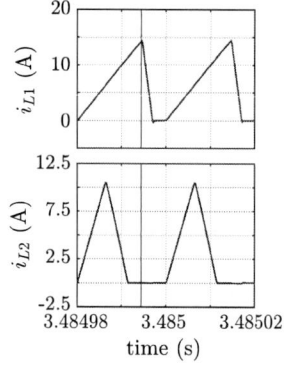

Fig. 10: Waveform of i_{L1} and i_{L2} magnified over 40 μs near $\omega t = \pi/2$

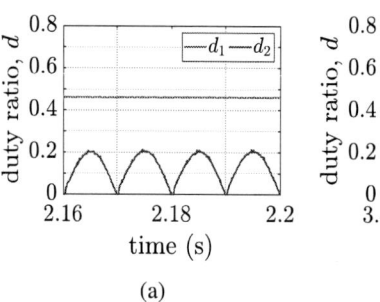

(a)   (b)

Fig. 11: Waveform of duty ratio over 40 ms for (a) 400 W/m^2, and (b) 1 kW/m^2

different solar irradiances are shown in Fig. 11. At rated power, the value of d_1 is 0.73, which is less compared to that reported in [16]. The waveform of d_2 is a rectified sine wave. Depending on the solar irradiance level, the duty ratios change as the inverter is operated in DCM.

The efficiency of the proposed micro-inverter at STC is estimated as 95%. The estimated maximum efficiency of the inverter is 95.8% when the irradiance level is 800 W/m^2.

Conclusion

A transformer-less solar PV micro-inverter topology is proposed in this paper. Due to the chosen modified operation, one switch is reduced compared to that in [16]. The volume of the inductor is also reduced significantly as the inverter is operated in DCM. Further, DCM operation ensures negligible turn on losses of the switches. This micro-inverter is suitable for interfacing a 35 V PV module to a 220 V grid and is having the feature of power decoupling. The negative terminal of the PV module is connected to the grid neutral, which ensures negligible leakage current flow.

References

[1] A. Bidram, A. Davoudi, and R. S. Balog, "Control and Circuit Techniques to Mitigate Partial Shading Effects in Photovoltaic Arrays," *IEEE Journal of Photovoltaics*, vol. 2, no. 4, pp. 532–546, 2012.

[2] D. Meneses, F. Blaabjerg, O. Garcia, and J. A. Cobos, "Review and Comparison of Step-up Transformerless Topologies for Photovoltaic AC-module Application," *IEEE Transactions on Power Electronics*, vol. 28, no. 6, pp. 2649–2663, 2013.

[3] K. Alluhaybi, I. Batarseh, and H. Hu, "Comprehensive Review and Comparison of Single-Phase Grid-Tied Photovoltaic Microinverters," *IEEE Journal of Emerging and Selected Topics in Power Electronics*, vol. 8, no. 2, pp. 1310–1329, 2019.

[4] E. Gubia, P. Sanchis, A. Ursua, J. Lopez, and L. Marroyo, "Ground Currents in Single-Phase Transformerless Photovoltaic Systems," *Progress in photovoltaics: research and applications*, vol. 15, no. 7, pp. 629–650, 2007.

[5] W. Li, Y. Gu, H. Luo, W. Cui, X. He, and C. Xia, "Topology Review and Derivation Methodology of Single-Phase Transformerless Photovoltaic Inverters for Leakage Current Suppression," *IEEE Transactions on Industrial Electronics*, vol. 62, no. 7, pp. 4537–4551, July 2015.

[6] H. Patel and V. Agarwal, "A Single-Stage Single-Phase Transformer-Less Doubly Grounded Grid-Connected PV Interface," *IEEE Transactions on Energy Conversion*, vol. 24, no. 1, pp. 93–101, March 2009.

[7] A. Jamatia, V. Gautam, and P. Sensarma, "Single Phase Buck-Boost Derived PV Micro-inverter with Power Decoupling Capability," in *Power Electronics, Drives and Energy Systems (PEDES), 2016 IEEE International Conference on*. IEEE, 2016, pp. 1–6.

[8] V. Gautam and P. Sensarma, "Design of Ćuk-Derived Transformerless Common-Grounded PV Microinverter in CCM," *IEEE Transactions on Industrial Electronics*, vol. 64, no. 8, pp. 6245–6254, 2017.

[9] A. Kumar and P. Sensarma, "A Four-switch Single-stage Single-phase Buck–Boost Inverter," *IEEE Transactions on Power Electronics*, vol. 32, no. 7, pp. 5282–5292, 2016.

[10] A. Jamatia, V. Gautam, and P. Sensarma, "Power Decoupling for Single-Phase PV System UsingĆuk Derived Microinverter," *IEEE Transactions on Industry Applications*, vol. 54, no. 4, pp. 3586–3595, July 2018.

[11] A. Kumar and P. Sensarma, "New Switching Strategy for Single-mode Operation of a Single-stage Buck–Boost Inverter," *IEEE Transactions on Power Electronics*, vol. 33, no. 7, pp. 5927–5936, 2017.

[12] M. Rajeev and V. Agarwal, "Analysis and Control of a Novel Transformer-Less Microinverter for PV-Grid Interface," *IEEE Journal of Photovoltaics*, 2018.

[13] A. Bhattacharya, A. R. Paul, and K. Chatterjee, "A Single Phase Single Stage SEPIC-ĆUK Based Non-Isolated High Gain and Efficient Micro-Inverter," in *2019 IEEE 46th Photovoltaic Specialists Conference (PVSC)*. IEEE, 2019, pp. 0708–0715.

[14] A. R. Paul, A. Bhattacharya, and K. Chatterjee, "A Novel SEPIC-Ćuk Based High Gain Solar Micro-Inverter for Integration to Grid," in *2019 National Power Electronics Conference (NPEC)*. IEEE, 2019, pp. 1–5.

[15] A. Sarikhani, M. M. Takantape, and M. Hamzeh, "A Transformerless Common-Ground Three-Switch Single-Phase Inverter for Photovoltaic Systems," *IEEE Transactions on Power Electronics*, vol. 35, no. 9, pp. 8902–8909, 2020.

[16] A. R. Paul, A. Bhattacharya, and K. Chatterjee, "A Novel Single Phase Grid connected Transformer-less Solar Micro-inverter Topology with Power Decoupling Capability," in *2020 IEEE International Conference on Power Electronics, Drives and Energy Systems (PEDES)*. IEEE, 2020, pp. 1–6.

[17] F. Zhang, Y. Xie, Y. Hu, G. Chen, and X. Wang, "A Hybrid Boost–Flyback/Flyback Microinverter for Photovoltaic Applications," *IEEE Transactions on Industrial Electronics*, vol. 67, no. 1, pp. 308–318, 2019.

[18] C. R. Sullivan, J. J. Awerbuch, and A. M. Latham, "Decrease in Photovoltaic Power Output from Ripple: Simple General Calculation and the Effect of Partial Shading," *IEEE Transactions on Power Electronics*, vol. 28, no. 2, pp. 740–747, 2012.

[19] S. B. Kjaer, J. K. Pedersen, and F. Blaabjerg, "A Review of Single-phase Grid-connected Inverters for Photovoltaic Modules," *IEEE Transactions on Industry Applications*, vol. 41, no. 5, pp. 1292–1306, 2005.

[20] H. Hu, S. Harb, N. Kutkut, I. Batarseh, and Z. J. Shen, "A Review of Power Decoupling Techniques for Microinverters with Three Different Decoupling Capacitor Locations in PV Systems," *IEEE Transactions on Power Electronics*, vol. 28, no. 6, pp. 2711–2726, 2013.

[21] A. Kyritsis, N. Papanikolaou, and E. Tatakis, "A Novel Parallel Active Filter for Current Pulsation Smoothing on Single Stage Grid-connected AC-PV Modules," in *Power Electronics and Applications, 2007 European Conference on*. IEEE, 2007, pp. 1–10.

[22] Y. Sun, Y. Liu, M. Su, W. Xiong, and J. Yang, "Review of Active Power Decoupling Topologies in Single-Phase Systems," *IEEE Transactions on Power Electronics*, vol. 31, no. 7, pp. 4778–4794, 2015.

Operation and Selection of Multilevel Power Converters for Doubly Fed Induction Generator-based Wind Turbines

Kapil Jha[1], Joseph Banda[2], Hridya I[1], Arvind Tiwari[3]
[1]GE Research, Bangalore, KA, India
[2]Norwegian University Of Science And Technology, Norway
[3]GE Research, Niskayuna, NY, USA
E-Mail: kapil.jha@ge.com

Keywords

DFIG, Wind Energy, Multi-level Converter, ANPC Converter, NPC Converter

Abstract

Operation of Doubly fed induction generator (DFIG) based wind turbine with 3-Level back-to-back converter has been discussed. Converter operation and loss analysis on a 3.2 MW-60Hz turbine shows that classical NPC converter is suitable for line side converter, while Active-NPC (ANPC) converter is optimal for rotor side converter realization.

Introduction

Conventionally, 2-level voltage source converters are used in building back-back power electronic system for DFIG based turbines [1] due to their cost, reliability, and performance. To reduce overall levelized cost of electricity, wind turbine manufacturers are continuously targeting for higher output power from individual turbine. As the turbine power increase, the voltage and/or current rating of the back-back converter system need to increase as well. Higher current in electrical system leads to higher losses, hence, at certain power level it becomes imperative to use converters which can produce higher voltage reliably. In this paper, operation of multi-level converter with DFIG has been evaluated.

Several multilevel converter topologies have been proposed in literature to generate higher output voltage, e.g. NPC and Active-NPC converter, flying capacitor based multilevel converter, neutral point piloted converter, cascaded H-bridge multilevel inverter, etc. [2]. Neutral point clamped (NPC) and Active-NPC (ANPC) converters, are widely used in industry due their performance and reliability. In this paper, operation of NPC and ANPC converters with DFIG based wind turbine have been critically evaluated and converter selection has been suggested accordingly.

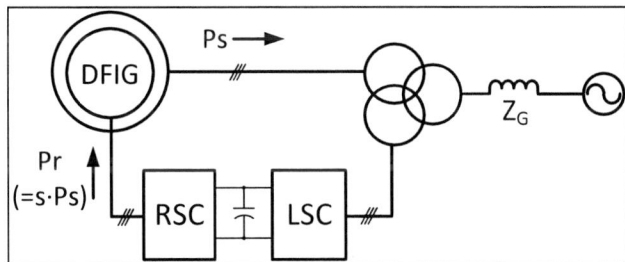

Fig. 1: Electrical single line schematic of DFIG based wind turbine

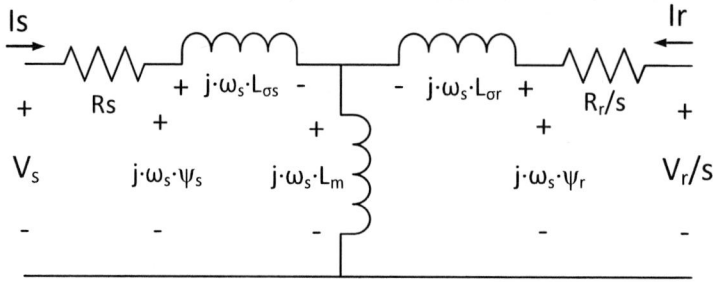

Fig. 2: One phase Steady-state equivalent circuit of doubly fed induction generator referred to stator

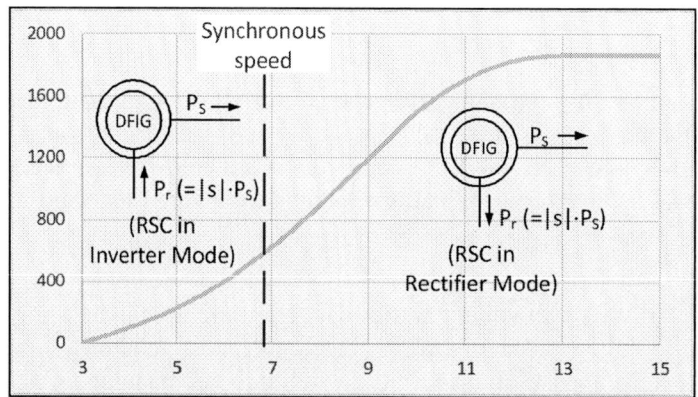

Fig. 3: Sub-synchronous and super-synchronous operation and direction of power flow in DFIG

Overview of DFIG based wind turbine system and operating regions

A simple single line electrical architecture of a doubly fed induction generator (DFIG) based wind turbine with key components is shown in Fig 1. Here, DFIG stator is connected directly to the grid while rotor is coupled to a back-back bidirectional power electronic converter system, viz., the rotor side converter and line side converter. Purpose of the rotor side converter is to generate appropriate rotor voltage at its terminals to generate desired real and reactive power from DFIG. The line side converter controls the shared dc-link of back-back converter system.

One phase equivalent circuit of DFIG in stator reference frame is shown in Fig. 2 [3]. Here, V_s and V_r are stator and rotor voltages, I_s and I_r are stator and rotor currents, respectively. R_s and R_r are stator and rotor resistances, $L_{\sigma s}$ and $L_{\sigma r}$ are stator and rotor leakage inductances,

Table 1: Steady-state expressions of DFIG voltages, currents, and fluxes in stator reference frame

$\bar{V}_s - \bar{I}_s \cdot R_s = j \cdot \omega_s \cdot \bar{\Psi}_s$ (1)	$\bar{V}_r - \bar{I}_r \cdot R_s = j \cdot s \cdot \omega_s \cdot \bar{\Psi}_r$ (4)	$P_r = s \cdot P_s$ (7)
$\bar{\Psi}_s = \bar{I}_s \cdot L_s + \bar{I}_R \cdot L_m$ (2)	$L_s = L_{\sigma s} + L_m$ (5)	$P_{mech} = P_s + P_r$ (8)
$\bar{\Psi}_R = \bar{I}_s \cdot L_m + \bar{I}_R \cdot L_r$ (3)	$L_r = L_{\sigma r} + L_m$ (6)	

Table 2: DFIG voltage, current and flux phasors in under-excited, UPF, and over-excited conditions at sub-synchronous and super-synchronous operations

	under-excited (Qs < 0)	UPF (Qs = 0)	over-excited (Qs > 0)
Sub-Synchronous Operation (s > 0)			
Super-Synchronous Operation (s < 0)			

L_s and L_r are stator and rotor inductances, respectively. ω_s is grid voltage angular speed, s is slip of operation, and L_m is the magnetizing inductance. Ψ_s and Ψ_r are stator and rotor fluxes, respectively.

Steady state expressions of voltages and fluxes are given in (1) – (8) for DFIG machine in stator referred reference frame in Table 1. Using these equations, steady-state magnitudes of fluxes, power, and currents of the machine can be determined at any given machine parameters, speed, and voltages.

It is an inherent property of DFIG that partial power of stator flows through the rotor (7), as shown in Fig. 3 [4]. The magnetizing current for the DFIG is provided by RSC. In sub-synchronous operation, the RSC operates in inverter mode while in super-synchronous operation it acts as a rectifier at various power factors. Region specific grid code requires a wind turbine to support rated reactive power across all operating points [5]. The phasor plots at under-excited, unity power factor (UPF), and over-excited condition for DFIG at sub-synchronous as well as super-synchronous speed have been plotted in Table 2. Here, Qs is the reactive power output from DFIG stator. In sub-synchronous operation, both V_r and I_r are in positive half of phasor plot suggesting inverter mode operation for RSC. While, in super-synchronous operation, V_r and I_r are in opposite half of each other indicating rectifier mode of operation for RSC. Hence, RSC operates both at leading as well as lagging power factor. In contrast, LSC usually operates at fixed power factor, depending upon speed of the DFIG. In sub-synchronous operation LSC operates in rectifier mode at -1 power factor, while, it operates in inverter mode at UPF at super-synchronous speed.

Selection of NPC vs ANPC converter for DFIG

In schematic of Fig. 4(a), one phase leg of neutral point clamped converter is shown. Here, T1-T6 are power semiconductor devices. In NPC converter, T1-T4 are active switches (IGBT/IGCT/Power MOSFET), with antiparallel diode, while T5 and T6 are power diodes. In an ANPC converter, all six switches (T1-T6) are active switches with antiparallel diodes, and this configuration provide additional option to distribute the losses using controls [6-7].

Table 3: System parameters for simulation validation

Attribute	Value
P_{mech} (MW)	3.2
Qs (MVar)	1.0
f (Hz)	60
Poles	6
ω (RPM)	1500
s	-0.25
IGBT module with diode	FZ3600R17HP4
Dc-link (V)	~2200
f_{sw}	2 kHz

1. Modulation and Switching

The switching scheme of active power semiconductor switches is mentioned in Table 3 (a) and (b) for NPC and ANPC converters. The sine-triangle modulation scheme has been used to generate the switching pattern, where the modulation signal of low frequency is compared with higher frequency triangular carrier signal to generate switching output. Here m is the modulation index and S is the switching state. S is logic high at an instant when m is higher than the carrier triangular signal and vice versa. $\bar{s}$ is complementary signal of S, as shown in Fig. 4(b).

2. Commutation path

The commutation path for NPC as well as ANPC converters have been evaluated to analyze the operation under non-UPF condition. It is important to note that both in NPC as well as ANPC converter, the voltages and current through the device T1 and T4 will remain same at similar operating condition. Only difference in current and power distribution will occur in devices T2, T3, T5, and T6. The commutation path has been plotted at an instant of $m>0$ in rectifier mode, as shown in Fig. 5. In Fig. 5(a), when S is logic high, the current flows into the capacitors via diodes of T1 and T2 and when S is logic low, current flows into the neutral mid-point via T3 and T6. However, in ANPC converter both in positive and zero state, the current flows through it as shown in Fig. 5(b). Hence, in NPC converter, significant switching losses occur in T2 along with conduction losses, however, in ANPC converter, no switching loss would occur in device T2. The conduction losses will be identical in NPC and ANPC converters in rectifier mode operation. However, in rectifier operation, switching losses will be lower in ANPC converter as compared to NPC converter. In ANPC converter, switching losses will occur in devices T1 and T5 in positive half of modulation ($m>0$), however, in NPC converter switching losses will occur in devices T1, T2, T3, and T6. Hence, total losses in ANPC converter will also be lower as compared to NPC in rectifier mode of operation.

Table 4: Switching pattern for (a) NPC and (b) ANPC converter

(a)					(b)						
NPC	s1	s2	s3	s4	ANPC	s1	s2	s3	s4	s5	s6
$m>0$	S	1	$\bar{S}$	0	$m>0$	S	1	0	0	$\bar{S}$	S
$m<0$	0	$\bar{S}$	1	S	$m<0$	0	0	1	S	S	$\bar{S}$

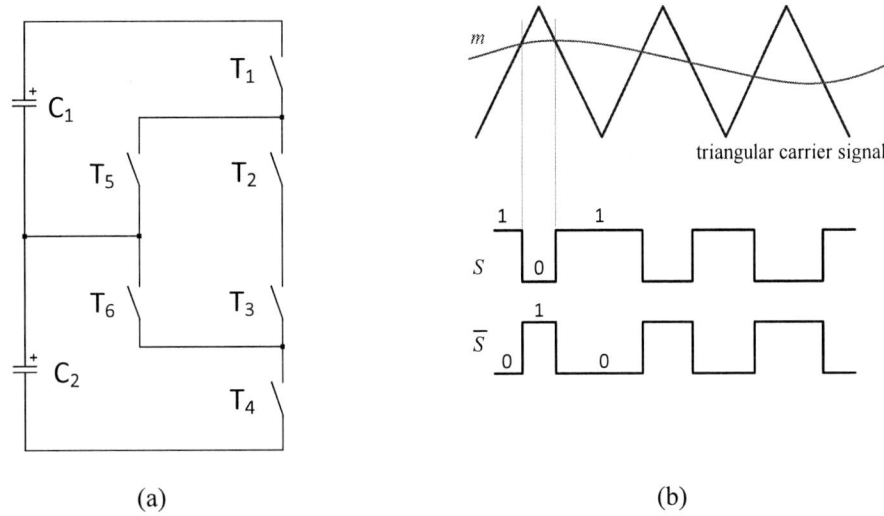

(a)   (b)

Fig. 4: (a) Generic schematic of neutral point clamped converters; (b) Synthesis of switching pulses

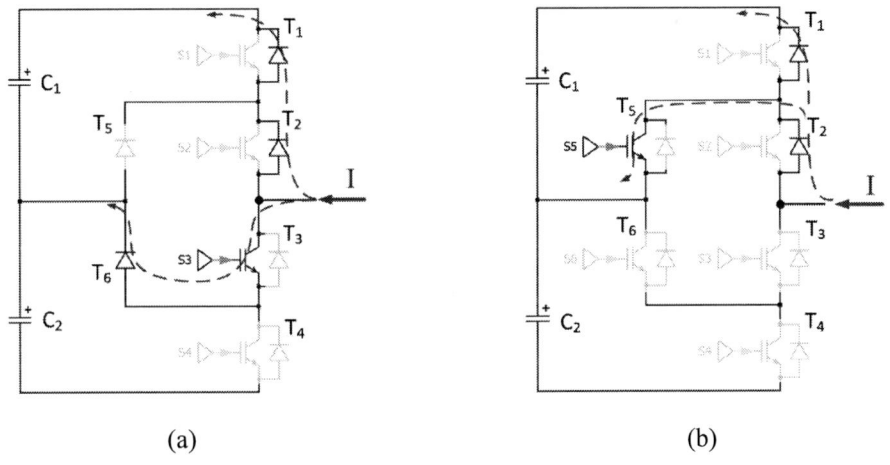

(a)   (b)

Fig. 5: Current flow direction in (a) NPC and (b) ANPC converter in rectifier mode of operation during active and zero states

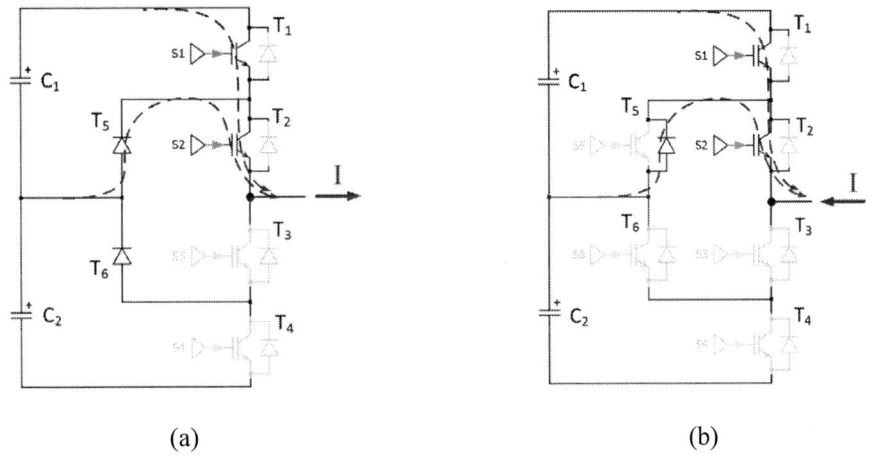

(a)   (b)

Fig. 6: Current flow direction in (a) NPC and (b) ANPC converter in inverter mode of operation during active and zero states

3. Loss Analysis and Converter Selection

The RSC process maximum power at rated operation of turbine in super-synchronous region and both RSC and LSC converter ratings and thermal requirements are designed for the same. Since both NPC and ANPC converters are identical except for switches T5 and T6, losses in switches T1 and T4 will remain the same for both converters at same operating conditions. However, the loss distribution will be different in switches T2, T3, T5, and T6.

Specifications of the rated operation are mentioned in Table 4. Here, P_{mech} is DFIG output power in megawatts, it is combination of P_s and P_r as given by (8). Qs is the DFIG stator reactive power output in MVar. Grid frequency (f) is 60 Hz, and DFIG synchronous speed is 1200 RPM. At rated operation, machine is rotated at 1500 RPM at -0.25 slip. The dc-link is maintained at 2.2 kV for both NPC as well as ANPC converters. RSC switching frequency is 2 kHz.

Comparison of loss distribution in NPC and ANPC converters has been done for RSC, as shown in Fig. 7. Here, phase modulation (m), output current (I_r), conduction losses (P_{cond}), switching losses (P_{sw}), and total losses (P_{Tot}) have been plotted in devices T2 and T5 for both NPC and ANPC converters, as shown in Fig. 7(a) and Fig. 7(b), respectively. It is important to recall that in device T2 in NPC and ANPC converters represents IGBT+anti-parallel diode, however, device T5 in NPC is diode while device T5 in ANPC converter is IGBT+anti-parallel diode, as shown in Fig. 5. Losses in Fig. 7 are combination of both IGBT and diode, wherever applicable.

As shown in Fig. 7, combined total losses in NPC converter of devices T2 and T5 (P_{Tot_T2} + P_{Tot_T5}) are more than 20% higher as compared to combined losses of devices T2 and T5 in ANPC converter, due to higher switching losses in NPC converter. As discussed previously, device T2 in RSC realized using NPC converter will experience both conduction and switching losses, at rated operation, as shown in Fig. 7(a). Here, device T2 bears approx. 75% of combined total losses of devices T2 and T5. The cooling requirements for this converter needs to be designed as per total losses in device T2.

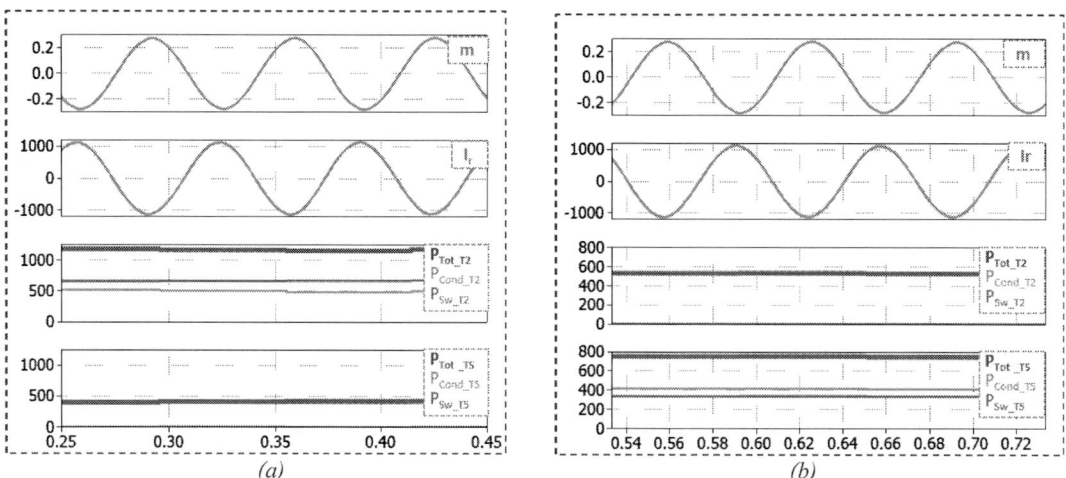

Fig. 7: At Rated Power condition, total transistor losses at T2 and T5 - in (a) NPC and (b) ANPC based 3-level converter

Table 5: Comparison of ANPC vs NPC converter for RSC realization

Total Losses	ANPC converter has 20% lower losses as compared to NPC converter
Max device losses	35% less in ANPC converter as compared to NPC converter
Loss distribution (T2:T5)	75:25 in NPC vs 41:59 in ANPC

However, in case of ANPC converter, as shown in Fig. 7(b), using the switching pattern described in Table 3(b), T2 device only observe conduction losses. Here, device T2 bears approx. 41% of combined total losses of devices T2 and T5. In this way, both T2 and T5 devices in ANPC converter are significantly less-stressed as compared to NPC converter. Therefore, ANPC converter cooling requirement would be much lesser as compared to NPC converter, also, efficiency of ANPC converter would also be better. Maximum thermal losses in a device in ANPC converter are approximately 35% lower than maximum thermal losses in NPC converter, which significantly reduces cooling requirements in ANPC converter. same has been summarized in Table 5. Hence, ANPC converter is better suited for RSC realization.

LSC in DFIG operates in inverter mode at rated condition which usually operates at UPF. Both NPC and ANPC converters will observe same loss distribution in LSC and same total losses as well, as shown in Fig. 6. Hence, NPC is more suitable to realize LSC as it requires lower active switches and cost saving is possible as NPC requires lower number of control and gate drive cards as compared to ANPC converter.

Conclusion

In this paper, operation of DFIG and variant of neutral point clamped multi-level converters have been discussed. Based upon the thermal and loss analysis, classic neutral point clamped converter appears more suitable for realization of line side converter in DFIG turbine while active-NPC converter is optimal for realization of rotor side converter. Simulation validation of the results with a 3.2 MW-60Hz DFIG turbine at rated operation show that maximum losses in individual device can be reduced by 35% in RSC with ANPC converter as compared to NPC converter, and efficiency of system also increases. NPC converter has been chosen for LSC as both loss distribution and efficiency in ANPC and NPC converter are found to be identical at rated power operation of DFIG.

References

[1] J. M. Carrasco, L. G. Franquelo, J. T. Bialasiewicz, E. Galvan, R. C. Portillo Guisado, M. A. M. Prats, J. I. Leon, N. Moreno-Alfonso, "Power-electronic systems for the grid integration of renewable energy sources: a survey", IEEE Trans. on Ind. Electron., vol. 53, no. 4, pp. 1002-1016, June 2006.

[2] S. S. Fazel, S. Bernet, D. Krug, and K. Jalili, "Design and Comparison of 4-kV Neutral-Point-Clamped, Flying-Capacitor, and Series-Connected H-Bridge Multilevel Converters," IEEE Trans. on Ind. Electronics, vol. 43, no. 4, pp. 1032-1040, July 2007.

[3] G. Abad, J. López, M. Rodríguez, L. Marroyo, G. Iwanski, Doubly Fed Induction Machine: Modeling and Control for Wind Energy Generation Applications, Wiley-IEEE Press, 2011.

[4] "GE's 1.87-87," GE Renewable Energy. [Online]. Available: https://www.ge.com/content/dam/ge-renew-new/downloads/brochures/wind-onshore-turbine-1.85-87-gea30627d-r1.pdf

[5] E. H. Enrique, "Generation Capability Curves for Wind Farms," IEEE Conf. on Technologies for Sustainability (SusTech), pp. 103-106, July 2014.

[6] T. Brückner, S. Bernet, and H. Güldner, "The Active NPC Converter and Its Loss-Balancing Control," IEEE Trans. on Ind. Electronics, vol. 52, no. 3, pp. 855-868, June 2005.

[7] Y. Deng, J. Li, K. H. Shin, et. al, "Improved modulation scheme for loss balancing of three-level active NPC converters," IEEE Trans. Power Electron., vol. 32, no. 4, pp. 2521–2532, Apr. 2017.

A detailed View on the Trapezoidal Operation for MMC Type Braking Chopper in Medium Voltage Application

Patrick Hofstetter, Viktor Hofmann, Dennis Karwatzki
Siemens AG, Large Drives Applications
Vogelweiherstraße 1-15
90441 Nürnberg, Germany
+49 152 38939179
Email: Patrick.Hofstetter@siemens.com
URL: http://www.siemens.com

Keywords

≪Modular Multilevel Converters (MMC)≫, ≪Control of drive≫, ≪Industrial application≫, ≪Reliability≫, ≪Braking Chopper≫

Abstract

A detailed view on the trapezoidal operation of the modular multilevel converter (MMC) type braking chopper is given. The influence of different limitations on the possible operational range is derived and an optimization algorithm is suggested. Finally, the analysis is verified on an exemplary medium voltage application and validated by simulation.

Introduction

Braking choppers are typically applied in drive systems having only two-quadrant (2Q) operation to provide motor braking capability. Furthermore, they can be used in converter systems for protection against high voltages, which may occur when a load feeds back energy into the converter. The easiest way to realize a braking chopper is using a switch in series with a braking resistor, see Fig. 1 (I).

By applying this circuit to the DC-link, electrical energy can simply be converted into heat in the braking resistor R_{BR} by closing the switch. A Modular Multilevel Converter (MMC) is shown on the left side of Fig. 1 and is often equipped with half bridge submodules (a). As the MMC is typically used in high and medium voltage applications, the standard braking chopper is usually not applied, as this would result in large quantities of high blocking semiconductors connected in series to realize the switch. The typical static and dynamic voltage imbalances between these semiconductors would have to be minimized by selecting IGBTs with low parameter deviations such as e.g. leakage current or threshold voltage. Additionally, passive snubbers or active methods are probably needed. In medium voltage applications a high blocking SiC MOSFET [1] may be applicable in future to replace the series connection. Another problem of a conventional braking chopper at an MMC is the influence of the arm inductances L_{arm}, which induce overvoltages at turn-off events during the braking chopper operation. Therefore, in literature braking chopper topologies based on modular cells were presented and analysed for MMC applications. One way is to use the submodules with distributed braking resistors $R_{BR,dis}$ such as (b) directly in the MMC instead of the common half bridge submodules (a). The submodules with resistors can alternatively be equipped into an external braking chopper arm (II) with or without an external braking resistor R_{BR}. The advantage of submodules with integrated braking resistor is the possibility to easily discharge the submodule capacitor and thus to simply achieve energy balance in the cells. The disadvantages of the resistor submodules are typically higher costs due to higher hardware efforts and the necessity to cool the distributed resistors. These kind of braking chopper topologies are for example being analyzed

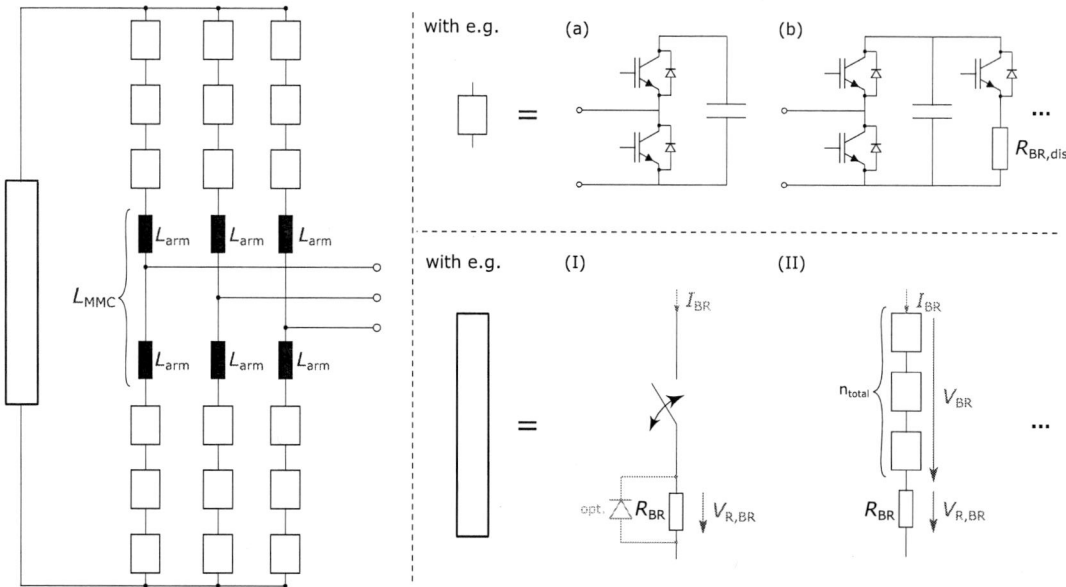

Fig. 1: MMC Converter and different types of half bridge submodules as well as braking chopper topologies such as (a) half bridge submodule, (b) half bridge submodule with distributed chopper resistor $R_{BR,dis}$, (I) Classical braking chopper or (II) MMC type braking chopper

in [2, 3, 4, 5, 6, 7]. The main advantage of using common half bridge submodules (a) in both MMC and external braking chopper arm (II) is the reduction in component variety. Consequently, complexity in production lines and costumer service efforts are minimized. Accordingly, only the submodule type (a) for both the MMC and the MMC type braking chopper (II) is analyzed in this paper. An equivalent circuit diagram with the voltage definitions is given in Fig. 2.

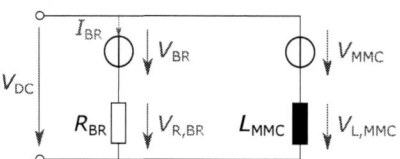

Fig. 2: Equivalent circuit diagram corresponding to the MMC of Fig. 1 with MMC type braking chopper (II), which is equipped with bridge modules (a)

In the braking chopper path, the voltages consist of the modulated voltage of the braking chopper cells V_{BR} and the voltage drop across the braking resistor $V_{R,BR}$. The MMC is represented by the modulated DC voltage V_{MMC} and the voltage drop across the effective DC side arm inductance $V_{L,MMC}$. There are two possible ways to operate the chopper to achieve energy balance in the cells. On the one hand, you can use a continuous operation, where the energy balancing is done by circulating currents, as shown in [7, 8]. On the other hand, there is the possibility to drive a trapezoidal braking current, which is shown in [2, 6, 9]. The focus of this paper is on the latter operation. In contrast to previous publications a detailed view on the limitations and the maximum possible operation ranges is shown.

In order to simplify following explanations, the possible switching states of the half bridge submodules is defined in Table. I.

Table I: Submodule switching states of the half bridge submodules from Fig. 1(a)

switching state	upper IGBT	lower IGBT	voltage
S_0	off	on	0
S_1	on	off	$V_{C,SM}$
S_2	off	off	0 @ neg. I $V_{C,SM}$ @ pos. I

Operation principle

When the current demand for the braking choppper $I^*_{BR,RMS}$ is greater than zero, an active braking chopper operation is performed. The main part of the power is dissipated in the resistor during t_{on}, when all n_{total} submodules of the chopper are in S_0 ($V_{BR} = 0\,V$). This is schematically illustrated in the operation principle in Fig. 3.

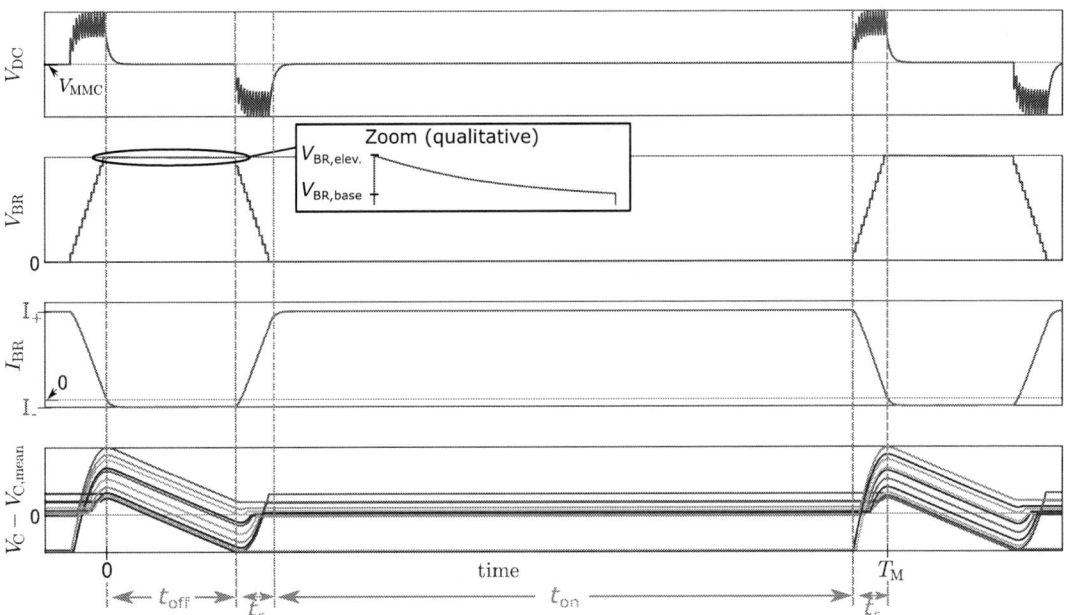

Fig. 3: Trapezoidal operation principle of the braking chopper during active operation based on simulation for $n_{total} = 12$

During the voltage ramps (with t_r), the capacitor voltages of the cells will be charged, when $V_{BR} < V_{DC}$. When all the modules are in S_1, the braking chopper cells are discharged, as long as $V_{BR} > V_{DC}$. In order to keep the cell voltages below their maximum voltage rating and to ensure an energy balance, there is a minimum time $t_{off,min}$ for this chopper state. As the capacitor voltages of the braking chopper submodules must stay balanced, the switching order should be done as explained in [9]. When the cells switch from S_1 to S_0, the order is from highest to lowest V_C. In contrast to this, when the cells switch from S_0 to S_1, the order is from lowest to highest.

When the braking chopper is not needed for energy dissipation it should be in an idle mode, which is explained below. When the current demand for the braking chopper is zero ($I^*_{RMS} = 0\,A$), the braking chopper enters the idle mode, where it should waste as few energy as possible. There is the possibility to simply keep all the cells in S_1. Essentially, the braking chopper path would correspond to a series connection of the submodule capacitors and the braking resistor. Considering an operation of an MMC with a diode rectifier, the rectifier's voltage ripple would cause relevant currents and losses in the braking resistor. Another solution to achieve low losses in idle mode is to set all the submodules into S_2. This way the submodule capacitors can only be charged by V_{DC} but not discharged by it, but there would still be the problem of voltage balancing between the cells. MMC submodules are typically equipped with balancing and discharge resistors in parallel to the IGBTs and capacitor, respectively. These resistors already help to achieve a better balance in capacitor voltages. Additionally, it is proposed to keep the submodule with the highest capacitor voltage in S_0. This can for example be checked every modulation cycle T_M, so that the appropriate submodule can slowly be discharged by the discharge and balancing resistors, whereas the other modules are charged to V_{DC}. The proposed switching states during active operation and idle mode are summarized in Fig. 4.

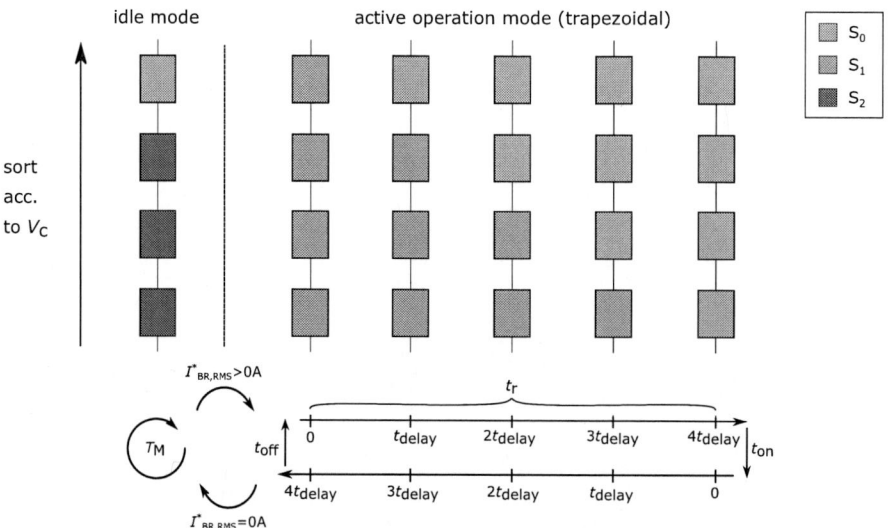

Fig. 4: Switching states during active and idle mode for $n_{total} = 4$ submodules in the braking chopper

Design Optimization

Without the need of cell energy balance, the maximum dissipated power $P_{R,BR,max}$ at a specific V_{DC} would only be limited by the maximum admissible RMS current I_{nom} of the submodules, by choosing $R_{BR} = V_{DC}/I_{nom}$. In the following, the determination of $P_{R,BR,max}$ and the corresponding R_{BR} at a specific V_{DC} will be derived under consideration of energy balancing and thus the presence of a minimum chopper off-state $t_{off,min}$ to discharge the braking chopper cells. The influence of these additional limitations will then be derived. Finally, the possible operating range for an exemplary medium voltage application will be shown and the results will be validated by simulation.

Determination of the maximum dissipation power

The power which is dissipated in one modulation period T_M can simply be controlled by the duty cycle and is calculated by

$$P_{R,BR} = I_{BR,RMS}^2 \cdot R_{BR}. \tag{1}$$

To determine $P_{R,BR,max}$, the maximum RMS current of the trapezoidal current profile (compare Fig. 3) is estimated by

$$I_{BR,RMS,max} = \sqrt{ f_M \cdot \left[\frac{2t_r \left(I_+^3 - I_-^3 \right)}{3 \left(I_+ - I_- \right)} + I_+^2 \cdot \underbrace{\left(T_M - t_{off,min} - 2t_r \right)}_{t_{on,max}} + I_-^2 \cdot t_{off,min} \right] }, \tag{2}$$

where I_+ and I_- describe the maximum positive and negative braking chopper currents, respectively and f_M the modulation frequency. In order to determine $t_{off,min}$, the discharged energy during this time has to be equal to the charged energy during the ramps. During t_{on} the braking chopper submodule capacitors are only insignificantly discharged by the discharge and balancing resistors. Therefore, it has no effect on the determination of $t_{off,min}$. The voltage levels of the braking chopper before and after the charging of the ramps are called $V_{BR,base}$ and $V_{BR,elev.}$, respectively and can be seen in the "zoom" of Fig. 3. The initial sum of submodule capacitor voltages at step zero ($i = 0$) is given by

$$V_{C,total,0} = V_{BR,base} = n_{total} \cdot V_{C,nom} \tag{3}$$

with the nominal voltage of the braking chopper submodules $V_{C,nom}$. The voltage after the ramps $V_{BR,elev.}$ can be calculated by evaluating the sum of the submodule capacitor voltages $V_{C,total,i+1}$ for each single step i of the consecutive switching of the cells from $n = n_{total} - 1$ active submodules to $n = 1$ and back to $n = n_{total} - 1$. Depending on the sum of voltages of the activated submodules, the capacitors are charged or discharged. If the voltage of the active submodules is higher than V_{DC}, they discharge and the voltage after one step with t_{delay} is calculated by

$$V_{C,total,i+1} = V_{DC} + (V_{C,total,i} - V_{DC}) \cdot e^{-\frac{t_{delay}}{R_{BR} \cdot C/n}} \tag{4}$$

with the submodule capacitance of C. Most of the time, the sum of voltages of the active cells is lower than V_{DC} during the ramps. Hence, the submodule capacitors are charged. In this case the calculation of the voltage after one step can be performed by

$$V_{C,total,i+1} = V_{DC} \cdot \left(1 - e^{-\frac{t_0 + t_{delay}}{R_{BR} \cdot C/n}} \right) \tag{5}$$

with

$$t_0 = -\ln \left(1 - \frac{V_{C,total,i}}{V_{DC}} \right) \cdot R_{BR} \cdot C/n. \tag{6}$$

The last step's $V_{C,total,i+1}$ corresponds to $V_{BR,elev.}$, which should discharge exactly to the base level $V_{BR,base}$ during $t_{off,min}$, compare „zoom" in Fig. 3. The value of $t_{off,min}$ can therefore be calculated by

$$t_{off,min} = -\ln \left(\frac{V_{BR,base} - V_{DC}}{V_{BR,elev.} - V_{DC}} \right) \cdot R_{BR} \cdot C/n_{total}. \tag{7}$$

Choice of submodule capacitance

Along with the value of the braking resistor, the submodule capacitance C may be a variable, which can typically be chosen freely, but influences $P_{R,BR,max}$. On the one hand lower capacitance values seem to lead to a smaller $t_{off,min}$ according to (7), as the cell capacitor discharge is faster due to a smaller time constant. On the other hand $V_{BR,elev}$ of the equation will also be higher for lower capacitances. Mainly, this is influenced by the steps calculated in (5) with (6). In principal, a lower C and thus a smaller time constant leads to a faster charge but also a faster discharge of the capacitors, which finally cancels each other out when considering $P_{R,BR,max}$. Consequently, less submodule capacitors can be used compared to the MMC submodules, which saves costs. But one has to keep in mind that a lower capacitance results in a higher submodule voltage during the charging process, which must not exceed the over voltage limit of the submodule.

Influence of limitations on the operating area

For a given system, the possible operating area of the braking chopper can be described by $I_{\text{BR,RMS,max}}$ and the specific resistance

$$R_{\text{spec}} = \frac{R_{\text{BR}}}{V_{\text{DC}}}. \tag{8}$$

Without further limitations, the maximum current is given by $V_{\text{DC}}/R_{\text{BR}}$, compare blue curve in Fig. 5.

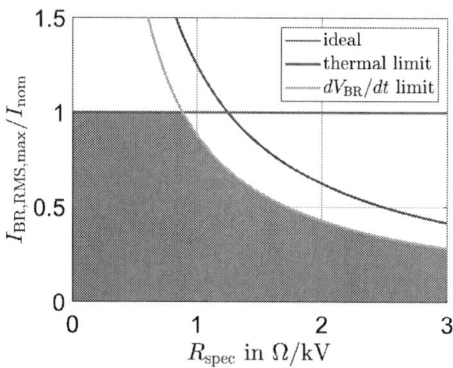

Fig. 5: Possible operating area for the trapezoidal braking chopper operation

This would be the case, if all submodules would be switched at the same time ($t_{\text{delay}} = 0$ & $t_{\text{r}} = 0$) and if there wouldn't be the need of a minimum off time of the braking chopper. As explained in the introduction, switching the submodules at once leads to high overvoltages at the DC link due to the arm inductances. This is simply described by

$$V_{\text{DC}}(t) = V_{\text{MMC}} + V_{\text{C,total}} \cdot e^{-\frac{R_{\text{BR}}}{L_{\text{MMC}}} \cdot t}. \tag{9}$$

Due to the consecutive switching, this changes to

$$V_{\text{DC}}(t) = V_{\text{MMC}} + \sum_{n=0}^{n_{\text{total}}-1} H(t - n \cdot t_{\text{delay}}) \cdot V_{\text{C,n}} \cdot e^{-\frac{R_{\text{BR}}}{L_{\text{MMC}}} \cdot (t - n \cdot t_{\text{delay}})} \tag{10}$$

with the Heaviside function $H(t)$. An example was already shown in the first plot of Fig. 3. The consecutive switching is basically used to reduce $\frac{dV_{\text{BR}}}{dt}$ and thus the $\frac{dI_{\text{BR}}}{dt}$ consequently lowering overvoltages on V_{DC}. The current limitation due to $t_{\text{off,min}}$ can therefore be interpreted as a $\frac{dV_{\text{BR}}}{dt}$ limitation. According to the calculation with the equations (2)-(7), this reduces the theoretical maximum RMS current to the values given by the yellow curve in Fig. 5. This gets worse for higher modulation frequencies. But the lower f_{M}, the higher is the voltage ripple of the MMC cell capacitors. Therefore, a minimum modulation frequency is needed to ensure a stable drive operation.

The thermal limits are additional factors limiting the maximum possible RMS current, compare the red curve in Fig. 5. To keep it simple in this paper, it considers a thermal limit which is given by the mechanical components of the submodule. The resulting possible operating area is illustrated in green. In order to maximize the possible power of the braking chopper, the resistor value and current should be as high as possible, compare (1). This is given by the intercept point of the two limiting curves. This point can easily be determined by iteration:

As a reasonable starting value for R_{BR}, $V_{\text{DC}}/I_{\text{max}}$ is used with the maximum absolute peak current rating of the submodules I_{max}. $I_{\text{BR,RMS,max}}$ is determined according to the calculation with the equations (2)-(7). R_{BR} is then increased until

$$I_{\text{BR,RMS,max}} = I_{\text{nom}} \tag{11}$$

with the nominal current of the submodules I_nom. This way, the algorithm follows the yellow curve of Fig. 5 from a reasonable starting point to the intercept with the thermal limit. Finally, $P_\text{R,BR,max}$ is calculated by (1).

Example of a medium voltage application and validation by simulation

The MMC type braking chopper should be used in medium voltage application. A number of MMCs operating at different modulated DC voltages V_MMC should be equipped with braking choppers. The specifications of one half bridge submodule are given in Table II.

Table II: Specifications of a half bridge submodule based on a medium voltage application

	value	description
$V_\text{C,nom}$	1kV	nominal cell voltage
I_nom	1kA	nominal RMS current
I_max	2kA	absolute maximum peak current rating
t_delay	10µs	time between consecutive switching steps
C_mod	2mF	capacitor value of the submodule

The chopper modulation frequency is set to $f_\text{M} = 600\,\text{Hz}$ to meet the MMC cell voltage ripple requirements. The arm inductance of the MMC is presumed to be $L_\text{MMC} = 100\,\mu\text{H}$. The braking chopper voltage rise/drop of about

$$\frac{dV_\text{BR}}{dt} = \frac{V_\text{C,nom}}{t_\text{delay}} = 0.1\,\text{kV}/\mu\text{s} \tag{12}$$

is assumed to be low enough to not develop critical $\frac{dI_\text{BR}}{dt}$ and thus voltages in this inductance. For various V_MMC, the results with different numbers of submodules (SM) are shown in Fig. 6a).

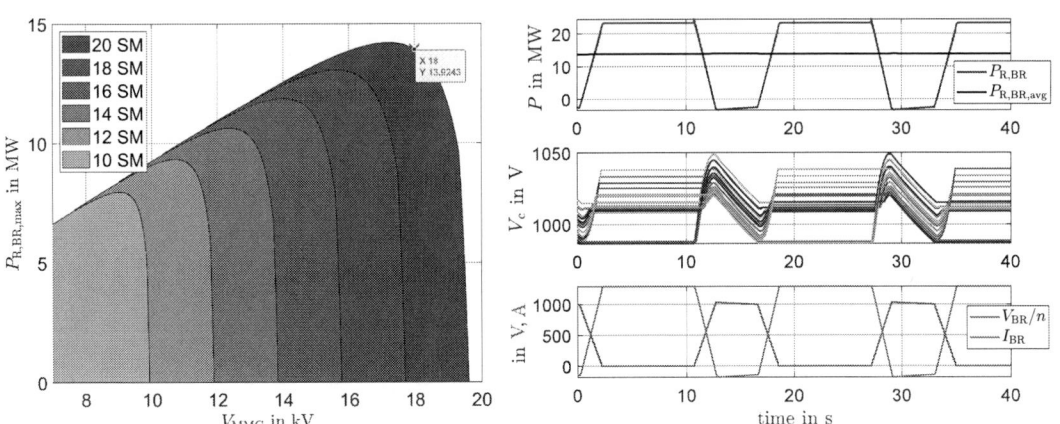

(a) Possible operating range for the MMC braking chopper

(b) Simulation result for $n = 20$ at $V_\text{MMC} = 18\,\text{kV}$ at the maximum duty cycle

Fig. 6: Calculation and simulation results for the considered application

The individual plots can be optically divided into two parts. On the left side of the maximum in Fig. 6(a), the maximum power is essentially limited by I_nom of the submodules. The right side shows the effect of higher $t_\text{off,min}$. To validate the calculation, different operation points were simulated successfully. For example, Fig. 6(b) shows the operation point with $n = 20$ submodules at $V_\text{MMC} = 18\,\text{kV}$ with an optimized $R_\text{BR} = 13.94\,\Omega$. On the one hand, it can be seen that the average power of 13.81 MW matches very well with the calculation of 13.92 MW. On the other hand, the cell voltages and thus the energy of the braking chopper cells does not increase over time and matches with $V_{C,nom}$ of about $1\,kV$. When a lower duty cycle is needed, the cell voltage level would be lower as the discharge phase during t_{off} would be longer.

Conclusion

The trapezoidal operation of the MMC type braking chopper was analyzed in detail. First, the operation principle was explained and the determination of the maximum braking chopper power was analyzed. With this, the influence of the consecutive switching and thermal limits were shown and an optimization algorithm was derived. Based on these calculations, the possible operating range was derived for an exemplary medium voltage application and finally validated by simulation results.

References

[1] K. Vechalapu, S. Bhattacharya, E. Van Brunt, S. Ryu, D. Grider, and J. Palmour, "Comparative evaluation of 15-kv sic mosfet and 15-kv sic igbt for medium-voltage converter under the same dv/dt conditions," *IEEE Journal of Emerging and Selected Topics in Power Electronics*, vol. 5, no. 1, pp. 469–489, 2017.

[2] B. Xu, C. Gao, J. Zhang, J. Yang, B. Xia, and Z. He, "A novel dc chopper topology for vsc-based offshore wind farm connection," *IEEE Transactions on Power Electronics*, vol. 36, no. 3, pp. 3017–3027, 2021.

[3] V. Hussennether, J. Rittiger, A. Barth, D. Worthington, G. Dell'Anna, M. Rapetti, B. Hühnerbein, and M. Siebert, "Projects borwin2 and helwin1 – large scale multilevel voltage-sourced converter technology for bundling of offshore windpower," in *CIGRÉ Session - B4-306*, pp. 1–11, 2012.

[4] Y. Okazaki, S. Shioda, and H. Akagi, "Performance of a distributed dynamic brake for an induction motor fed by a modular multilevel dscc inverter," *IEEE Transactions on Power Electronics*, vol. 33, no. 6, pp. 4796–4806, 2018.

[5] M. Wang, Y. Hu, W. Zhao, Y. Wang, and G. Chen, "Application of modular multilevel converter in medium voltage high power permanent magnet synchronous generator wind energy conversion systems," *IET Renewable Power Generation*, vol. 10, no. 6, pp. 824–833, 2016.

[6] J. Maneiro, S. Tennakoon, C. Barker, and F. Hassan, "Energy diverting converter topologies for hvdc transmission systems," in *2013 15th European Conference on Power Electronics and Applications (EPE)*, pp. 1–10, 2013.

[7] A. Birkel, A. Schön, and M. Bakran, "Analysis and semiconductor based comparison of energy diverting converter topologies for hvdc transmission systems," in *2015 17th European Conference on Power Electronics and Applications (EPE'15 ECCE-Europe)*, pp. 1–10, 2015.

[8] V. Hofmann and P. Hofstetter, "Design and modulation optimization of an mmc based braking chopper," in *2022 24th European Conference on Power Electronics and Applications (EPE'22 ECCE Europe)*, pp. 1–7, 2022.

[9] S. Schoening, P. K. Steimer, and J. W. Kolar, "Braking chopper solutions for modular multilevel converters," in *Proceedings of the 2011 14th European Conference on Power Electronics and Applications*, pp. 1–10, 2011.

Influence of operating frequency on high-power medium-voltage medium-frequency transformers

Thomas B. Gradinger[1], Ralph M. Burkart, and Marko Mogorovic

Hitachi Energy Switzerland Ltd.
URL: http://www.hitachienergy.com
[1] Segelhofstrasse 1A
5405 Baden-Dättwil, Switzerland
Tel.: +41 (0)79 150 59 69
E-Mail: thomas.gradinger@hitachienergy.com

Keywords

«Solid-state transformer», «medium voltage», «transformer», «design optimization».

Abstract

This paper provides a fundamental insight into the scaling laws of medium-frequency transformers (MFTs). Economic MFTs are a key enabler for solid-state transformers and viable MVDC solutions. An understanding of the influences on the MFT cost structure, and of the limitations to the reduction of size, mass, and cost, is therefore key. Simplified scaling laws are developed that suggest a potential of an increase in switching frequency to improve the MFT performance metrics. They suggest that for fixed MFT geometry, an increase in frequency can lead to lower losses under constrained inductances. The scaling laws also show that a core-size reduction at a given frequency must always be accompanied by increasing core losses. In practical MFT design, size reduction is only observed up to a certain "scaling saturation" frequency. In the studied examples featuring nanocrystalline cores, it is located near 7 kHz. Insight into the scaling behavior is provided by the analysis of the constraint functions used in the MFT optimization. A constrained loss ratio may limit the designs at all frequencies and inhibit core-size reduction. The "scaling saturation" frequency corresponds to a transition between saturation limit and thermal limit of the core, and can be shifted upward by improved core cooling. This enables an attractive compromise between cost savings and loss increase, and shows the importance of sufficient core cooling.

Introduction

Nowadays, the electric distribution grids are facing a proliferation of modern high-power electric applications ranging from several MW to several hundred MW, such as renewables [1], railways [2], e-mobility [3], and data centers [4, 5]. Consequently, the solutions for conversion from high-power MVAC to controlled LVAC or LVDC, and thus the solid-state transformer (SST) concept, are more and more under the focus of the industry [6, 7]. Moreover, the SST is an enabler of the emerging MVDC grids, with the high promise of enabling a significantly more efficient transmission, and an opportunity to increase the capacity of existing MVAC infrastructure by refurbishing and converting it into MVDC [8]. The most popular SST topology is a modular solution consisting of multiple identical cells in a so-called, input-series-output-parallel (ISOP) connection (or vice versa, IPOS, depending on the power-flow direction). The MFT is a key component of each SST cell, providing both input-output voltage matching and galvanic insulation as one of the mandatory features of the SST, as required by standards for interconnection of MV and LV [9].

If we are to analyze the economic viability of an MVDC system based on the total cost of ownership (TCO), this value will depend on the cost of all components, including the SST. Hence, highly efficient and economic solutions for the SST and therefore the MFT, as a significant contributor to the SST cost and losses, are a key. As the cost to be optimized is not CapEx (capital expenditure), but TCO including the cost of losses, optimization of investment cost must be done under constraints of efficiency.

An important choice in the context of cost optimization is that of the operating frequency of the MFT. It is known to have a strong influence on the MFT and can be chosen freely, up to some efficiency limit of the converters connected to the MFT. With the advent of SiC, the available frequency range has extended even higher, allowing significant freedom of choice that can be used for SST and MFT optimization [10–12]. In their simplest form, the equations governing transformer design suggest a size reduction with frequency [13] according to

$$A_c A_w \propto \frac{S}{J f_{sw} \hat{B}},$$
(1)

where S and J are apparent power and rms current density, respectively, f_{sw} is the switching frequency, $\hat{B}$ is the magnetic peak-flux density, and A_c and A_w are the cross-sections of the core and windings, respectively. It has been shown, however, that there are limits to the size reduction. In [10], it was demonstrated that this limit is due to cooling, and that there is an optimum frequency depending on the cooling performance. While the basic findings in this study still appear to be valid, the present study focuses on several important extensions and modifications. Most importantly, these include

- the choice of higher power per MFT, which is important to avoid uneconomically high insulation effort for MV applications;
- the choice of nanocrystalline core material, which covers best the considered range of f_{sw} of 1 to 15 kHz;
- the consistent consideration of the inductance requirements coming from the converters;
- a detailed and realistic insulation model;
- detailed analysis of the design limits imposed via constraint functions, including temperatures and efficiency, and including the roles of the individual components (core, windings); and
- an analysis of cost, rather than volume and mass, only.

As winding conductor, litz wire is considered in the present study. Next to foil, this is one of the dominant conductor types for medium-frequency windings. The focus on litz wire does not correspond to a loss of generality, because both conductor types are adaptable in sub-conductor size (strand diameter or foil thickness), which is the essential feature in the context of the present study. For the insulation, dry casting, with separately cast LV and HV windings, is assumed.

MFT optimization tool

As a basis for the present study, a MATLAB tool was used for MFT design and optimization. This tool – called *MFT optimization tool* for brevity – is described in [14]. For convenience, a summary of this description is provided here. The core of the tool is a function that evaluates an individual MFT design, based on a technical requirements specification (TRS) and a vector x of *primary variables* that fully define the MFT design. $x = \{x_d, x_c\}$ consists of a vector x_d of discrete variables and a vector x_c of continuous variables. Examples of discrete variables are the number of turns N_t of a winding, the strand diameter d_s of litz wire, and the core material. Examples of continuous variables are core-window width Δx_c, core-window height Δz_c, and current density in a winding conductor. x is also called the *design space*, and a particular point within it, i.e., a value of x, is called a *design point*. The MFT optimization tool is versatile and covers different core materials, winding-conductor types, and insulation types. The variables in x and the length of x may change depending on the type of MFT design.

Optimization is done with respect to a cost function u, for which a minimum is sought; and under constraints, expressed by constraint functions K_i. u is a weighted rms of cost (C), mass (m), boxed volume (V_{box}), and loss ratio, i.e., losses per real power (λ):

$$u \equiv \sqrt{w_c (C/C_{opt})^2 + w_m (m/m_{opt})^2 + w_V (V_{box}/V_{box,opt})^2 + w_\lambda (\lambda/\lambda_{opt})^2},$$
(2)

where the w_z are weighting factors ($z = C, m, V_{box}$, or λ), and z_{opt} is the value of z at the optimum. The constraint functions are used to enforce upper temperature limits of the materials used as well as an upper limit on the loss ratio, and to avoid core saturation. They can also be used to enforce upper and lower limits on magnetizing and leakage inductance. A valid solution has $K_i \leq 0 \; \forall \; i$, while $K_i = 0$ indicates an active constraint. The optimization is based on sweeping x, using gridded interpolants for u and the K_i. The MFT optimization tool is mostly based on simplified physical models that are quick to evaluate, and that describe the electromagnetic and thermal behavior of the MFT and enable the design of the electrical insulation. Typically, analytical expressions [15] are used, equipped with "engineering coefficients". The insulation model is based on realistic assumptions and includes consideration of the tests required by international standards [16].

Scaling laws

As will be seen, the MFT cost function is strongly influenced by the constraints on the inductances. From the point of view of the converters connected to the MFT, two inductances are key: the magnetizing inductance L_m and the leakage inductance L_σ. Both L_m and L_σ need to assume a certain value, or lie within a certain range, to facilitate converter design. This holds both for LLC resonant converters and for DABs (dual-active bridges). In the following, the constraints for L_m and L_σ are expressed as constraints for L_m and L_m/L_σ. Key geometric parameters of the MFT, that are used in the scaling laws, are identified in Fig. 1. It is noted that the considered MFT has both the primary and secondary winding distributed among both core legs. Each leg is surrounded by a primary-winding portion on the inside, and a secondary-winding portion on the outside. The portions of a winding can be series or parallel connected.

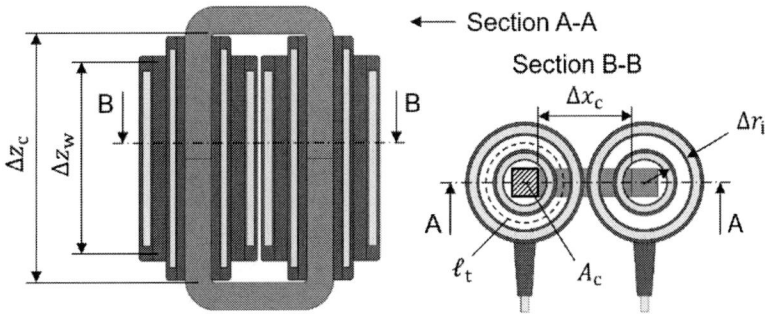

Fig. 1: Key geometric parameters of MFT: mean winding height, Δz_w; core-window height, Δz_c; core-window width, Δx_c; mean turn length, ℓ_t; mean radial distance between windings due to insulation, Δr_i; core cross-section, A_c.

Assuming that the needed mean turn length is mainly influenced by the core cross-section, the leakage inductance scales as follows:

$$L_\sigma \propto \frac{N_t^2 \ell_t \Delta r_i}{\Delta z_w} \propto \frac{N_t^2 A_c^{1/2} \Delta r_i}{\Delta z_w}. \tag{3}$$

Δr_i results from the TRS via the insulation requirements, while A_c (via core-leg width) and N_t are part of the design space. Δz_w results from Δz_c after subtracting the insulation distances and can therefore also be considered to be part of the design space.

With R_m denoting the magnetizing reluctance, L_m scales as follows:

$$L_m \propto \frac{N_t^2}{R_m} \propto \frac{N_t^2 \mu_0 A_c}{\frac{\ell_c}{\mu_{r,c}} + \ell_g} \propto \frac{N_t^2 A_c}{\ell_g}. \tag{4}$$

Here, ℓ_c and ℓ_g denote the length of the magnetic path in the core and the air-gap width, respectively; and $\mu_{r,c}$ is the relative permeability of the core. The assumption is made that the reluctance is dominated by the air gap, which is a good design principle that ensures robustness and the possibility for fine-tuning L_m, if needed. After selecting a design point, ℓ_g is adapted to obtain the required L_m.

The saturation constraint for the core is

$$B_{sat} \geq \hat{B} = \frac{\hat{\Phi}}{A_c} \propto \frac{N_t \hat{I}_m}{R_m A_c} \propto \frac{N_t \hat{I}_m}{\ell_g}, \tag{5}$$

where $\hat{\Phi}$ and $\hat{I}_m$ are the peaks of the magnetic flux and the magnetizing current, respectively. The amount of magnetizing current is relevant for achieving soft switching in the semiconductors.

For the ratio of inductances, one gets

$$\frac{L_{\mathrm{m}}}{L_{\sigma}} \propto \frac{A_{\mathrm{c}}\Delta z_{\mathrm{w}}}{\ell_{\mathrm{g}} A_{\mathrm{c}}^{1/2}\Delta r_{\mathrm{i}}} \propto \frac{A_{\mathrm{c}}^{1/2}\Delta z_{\mathrm{w}}}{\ell_{\mathrm{g}}}. \tag{6}$$

From this scaling law, one notes that a high value of $L_{\mathrm{m}}/L_{\sigma}$, which corresponds to good magnetic coupling between primary and secondary, tends to yield a tall MFT with a large core cross-section and a small air gap.

To develop an understanding of the MFT's scaling with the switching frequency f_{sw}, next, the loss scaling is studied under the assumption that both L_{m} and L_{σ} scale with $1/f_{\mathrm{sw}}$. This assumption is reasonable from the point of view of converter requirements to achieve frequency invariant current shapes with constant rms and corner values. As a starting point, an optimized MFT design at a reference frequency f_{sw}^{*} is taken. Then, for $f_{\mathrm{sw}} > f_{\mathrm{sw}}^{*}$, it is assumed that the MFT design does not change, except for N_{t}. According to (3) and (4), $L_{\{\mathrm{m},\sigma\}} \propto N_{\mathrm{t}}^{2}$, such that

$$N_{\mathrm{t}} \propto f_{\mathrm{sw}}^{-1/2}. \tag{7}$$

For the core losses P_{c}, the Steinmetz equation is adopted since it facilitates simple scaling laws. Using (5) and (7),

$$P_{\mathrm{c}} \propto f_{\mathrm{sw}}^{\alpha}\hat{B}^{\beta} \propto f_{\mathrm{sw}}^{\alpha} N_{\mathrm{t}}^{\beta} \propto f_{\mathrm{sw}}^{\alpha-\beta/2}. \tag{8}$$

The Steinmetz parameters α and β depend on the core material. In the present study, the focus is on nanocrystalline material, which is the one that covers best the analyzed frequency range of 1 to 15 kHz. For Vitroperm 500 F, we measured $\alpha = 1.14$ and $\beta = 1.98$, such that

$$P_{\mathrm{c}} \propto f_{\mathrm{sw}}^{0.15}. \tag{9}$$

The winding losses P_{w} depend on the DC resistance R_{dc} and on the AC-to-DC resistance ratio F_{R}, which is frequency dependent:

$$P_{\mathrm{w}} \propto R_{\mathrm{dc}} F_{R}(f_{\mathrm{sw}}). \tag{10}$$

With the overall cross-section of the winding staying the same, the conductor cross-section can grow if there are fewer turns, such that

$$R_{\mathrm{dc}} \propto \frac{N_{\mathrm{t}}^{2}\ell_{\mathrm{t}}}{\Delta z_{\mathrm{w}} F_{\mathrm{Cu}}} \propto \frac{N_{\mathrm{t}}^{2} A_{\mathrm{c}}^{1/2}}{\Delta z_{\mathrm{w}} F_{\mathrm{Cu}}}, \tag{11}$$

where F_{Cu} is the copper filling ratio, i.e., the ratio of copper to total litz-wire cross-section. Using (7) and noting that A_{c} and Δz_{w} remain constant,

$$R_{\mathrm{dc}} \propto \frac{1}{f_{\mathrm{sw}} F_{\mathrm{Cu}}}. \tag{12}$$

It is assumed that F_{R} can be kept approximately constant when raising f_{sw} by reducing the litz-wire's strand diameter d_{s} such that its ratio to the skin depth δ stays constant. For the frequency range of interest, it suffices to consider d_{s} in the range of 0.1 to 0.5 mm (AWG, American Wire Gauge, 38 to 24), which is well feasible and commercially available. Since $\delta \propto f_{\mathrm{sw}}^{-1/2}$, one gets $d_{\mathrm{s}} \propto f_{\mathrm{sw}}^{-1/2}$. Reducing d_{s} generally leads to a reduction of F_{Cu}, which can be assumed to scale like

$$F_{\mathrm{Cu}} \propto d_{s}^{\gamma}. \tag{13}$$

In the range of d_{s} of 0.1 to 0.5 mm, $\gamma = 0.4$ is a reasonable assumption. Using (12), one obtains $R_{\mathrm{dc}} \propto f_{\mathrm{sw}}^{-0.8}$, such that also

$$P_{\mathrm{w}} \propto f_{\mathrm{sw}}^{-0.8}. \tag{14}$$

This suggests significant loss-reduction potential for the windings by increasing the frequency. A potential limiting factor of the reduction of the total MFT losses is the growth of P_{c} according to (9). However, since the exponent of 0.15 of P_{c} is small, the growth of P_{c} should not dominate the reduction in P_{w} up to rather high frequencies.

As an alternative to profiting from reduced losses, it should be possible to reduce the MFT in size and, consequently, cost. To this end, the core and conductor cross-sections are reduced, increasing the losses, either just a bit, or until they are back to those at f_{sw}^{*}. In view of this option, it is interesting to discuss what happens to the losses if the core shrinks. For this discussion, it is assumed that f_{sw} stays constant, and that also L_{m} and L_{σ} have fixed values as imposed by the TRS. The core volume can shrink by a reduction of either A_c or ℓ_c, including a combination of the two. First, the effect of a reduction in A_c is analyzed. Under the assumption of a preserved inductance ratio, one finds from (6) that

$$\ell_g \propto A_c^{1/2}. \tag{15}$$

This implies a drop in ℓ_g. Keeping L_m constant means, according to (4), that

$$N_t \propto \frac{\ell_g^{1/2}}{A_c^{1/2}} \propto A_c^{-1/4}, \tag{16}$$

resulting in an increase in N_t. Together with (11), this means that the winding losses remain constant:

$$P_w \propto N_t^2 A_c^{1/2} = \text{const.} \tag{17}$$

To determine the scaling of the core losses, it is noted that $\ell_c \propto \Delta z_c + \Delta x_c$. If it is assumed that the drivers of Δz_c and Δx_c are Δz_w and Δr_i, respectively, then

$$\ell_c \propto 1 + \frac{\Delta r_i}{\Delta z_w}. \tag{18}$$

With the core volume $V_c \propto A_c \ell_c$, and based on (5), one obtains

$$P_c \propto V_c \hat{B}^\beta \propto A_c \left(1 + \frac{\Delta r_i}{\Delta z_w}\right)\left(\frac{N_t \hat{I}_m}{\ell_g}\right)^2, \tag{19}$$

and, using (15) and (16),

$$P_c \propto A_c^{-1/2}. \tag{20}$$

The effect of a reduction in ℓ_c on the losses is analyzed assuming that this reduction is driven by a decrease in Δz_w. It is then noted from (6), that

$$\ell_g \propto \Delta z_w, \tag{21}$$

implying a reduction in air-gap width. From (3),

$$N_t \propto \Delta z_w^{1/2}, \tag{22}$$

one notes that the number of turns decreases. Using again (11), inserting (22), and keeping A_c constant shows that the winding losses are again constant:

$$P_w \propto \frac{N_t^2}{\Delta z_w} = \text{const.} \tag{23}$$

For the core losses, one can start from (19), fix A_c, and use (22) and (23):

$$P_c \propto \left(1 + \frac{\Delta r_i}{\Delta z_w}\right)\left(\frac{N_t \hat{I}_m}{\ell_g}\right)^2 \propto \left(1 + \frac{\Delta r_i}{\Delta z_w}\right)\frac{1}{\Delta z_w}. \tag{24}$$

Scaling laws (20) and (24) demonstrate that no matter by which mechanism the core shrinks, the size reduction is accompanied by an increase in core losses, while the winding losses remain constant. Importantly, one can conclude that *if* the core is thermally limited, further size reduction will not be possible. The admissible temperatures would not only be exceeded because of the rise in P_c, but also because of the reduction in core surface and, hence, cooling surface.

Results from MFT optimization tool

Cases considered

To test the scaling laws that were developed under simplifying assumptions, parameter studies were made using designs obtained with the MFT optimization tool. Four cases were studied, which shared the TRS described in Tab. 1. They are typical for MFTs used in the cells of SSTs of ISOP topology, rated for a power in the MW range. Nanocrystalline core material (Vitroperm 500) and copper litz wire was used in accordance with the assumptions in the scaling laws. The MFT has an O-core as depicted in Fig. 1. The weighting factors w_c, w_m, w_V, and w_λ in Eq. (2) were set to 0.8, 0, 0.2, and 0, respectively. This puts a strong emphasis on cost, while also considering the volume a little to avoid weird designs that strongly grow in volume for a marginal cost benefit. No weight was attributed to λ, since the efficiency is constrained directly, as described below. The frequencies studied are in the range of 1 to 15 kHz. The ratio L_m/L_σ was initially assumed to be constant to preserve the ratio between magnetizing and load current. This constraint is not a strict necessity and will be relaxed below.

Property	Value
real power	500 kW
cell voltage on HV side	750 V
HV-to-LV turns ratio	2
$2\pi f_{sw}L_m$	39.8 Ω (case 1 to 3)
L_m/L_σ	80 (case 1 to 3)
insulation	dry-cast
DC insulation voltage	25 kV
equivalent "highest voltage for equipment" U_m [16]	17.5 kV rms
1 min. applied AC test voltage [16]	38 kV rms
lightning-impulse test voltage [16]	95 kV
cooling	air forced
ambient temperature	35 °C

Tab. 1: Common MFT TRS of the cases studied with the MFT optimization tool.

Case 1 – constrained efficiency

In a first case, the efficiency was constrained by limiting the loss ratio to $\lambda_{lim} = 0.6$ %. In Fig. 2, the MFT performance metrics V_{box}, C, m, and u are shown as a function of f_{sw}. It can be seen that in the range of 1 to about 7 kHz, the MFT benefits from the increase in frequency, with all the metrics decreasing. This corresponds to the expectations based on the developed scaling laws. However, the decrease levels off at $f_{sw,sat} \cong 7$ kHz, and no further improvement can be observed when continuing to increase f_{sw}. It is concluded that $f_{sw,sat}$ is a "scaling saturation frequency", at which the scaling laws obtained under the simplifying assumptions made cease to apply.

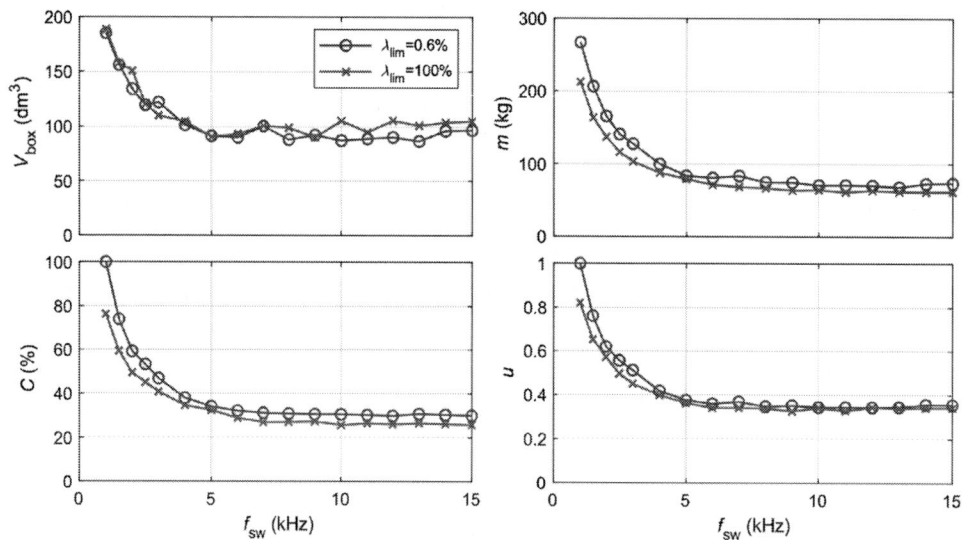

Fig. 2: Cases 1 and 2, MFT performance metrics.

To gain insight into this behavior, it is necessary to look at further MFT properties as a function of frequency. At this point, it is noted that the curves over f_{sw} are often not smooth, but exhibit jumps, or kinks. This has to do with the primary variables and properties of the MFT that are of integer type, such as N_t, but also d_s or the depth of nanocrystalline cores (that equals the ribbon width), because suppliers standardize their products to discrete sizes. When, at some transition frequency, the optimum jumps from one discrete value of a property to the next, a corresponding jump in all properties results.

In Fig. 3, winding properties are summarized. Above $f_{sw,sat}$, there is only a weak trend of N_t to decrease with f_{sw}. In contrast, below $f_{sw,sat}$, the behavior resembles that of scaling law (7) derived under the assumption of a fixed MFT geometry. The DC resistance of the HV winding shows a gradual decrease with f_{sw}. The decrease of the AC resistance is a little weaker in comparison as a result of an increase in F_R. The loss data are shown in Fig. 4. λ equals λ_{lim} over the whole frequency range, indicating an active constraint. At low f_{sw}, about 90 % of the losses are created in the windings, and only 10 % in the core.

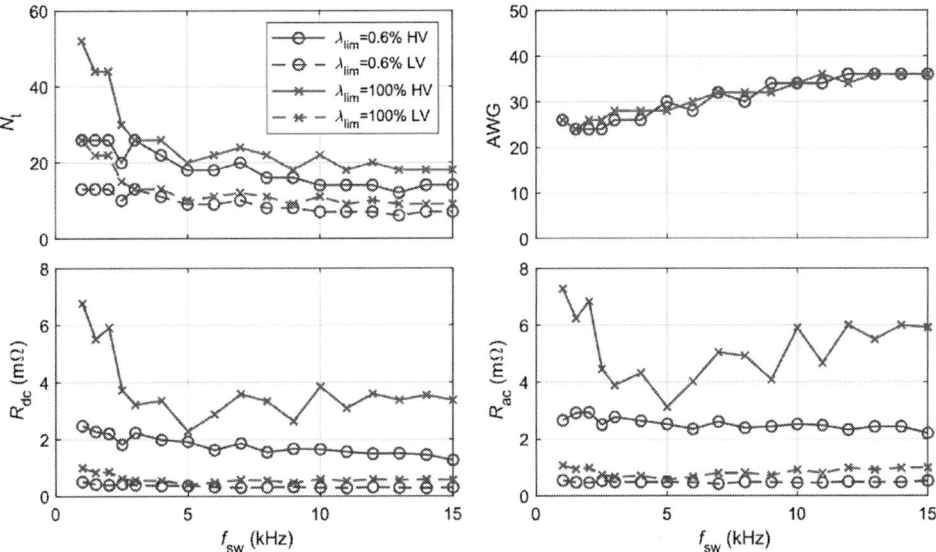

Fig. 3: Cases 1 and 2, winding properties.

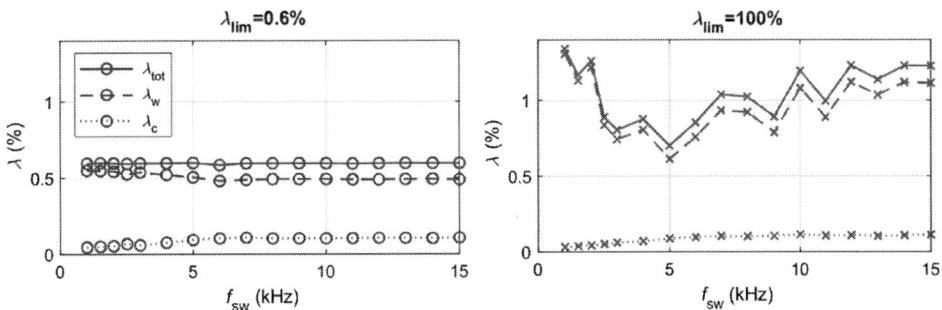

Fig. 4: Cases 1 (left) and 2 (right), loss ratio.

To understand why no benefit can be obtained by increasing f_{sw} beyond $f_{sw,sat}$, it is instructive to look at the constraint functions that are plotted in Fig. 5. Up to about $f_{sw,sat}$, the core is saturated, as indicated by the corresponding active constraint. From $f_{sw,sat}$ on, the core is not saturated any more, but thermally limited. Throughout the entire frequency range studied, the constraint on λ is active, while the windings are generally not thermally limited. Referring to the scaling laws derived, the behavior below $f_{sw,sat}$ can be thought of as the result of two steps: (i) increase of f_{sw} for fixed size, reducing the losses; and (ii) using the created loss margin to shrink the MFT according to scaling laws (17), (20), (23), and (24). These steps are useful for the understanding, while the *MFT optimization tool* finds the optimum directly by sweeping x. Importantly, below $f_{sw,sat}$, the other active constraint, the one on core saturation, scales favorably with frequency, allowing the core to shrink on increasing f_{sw}. The core-temperature constraint, active above $f_{sw,sat}$ does not scale favorably with frequency, preventing further size reduction. The behavior is confirmed in Fig. 6, where the core cost – a direct indicator of the core volume – is shown. Another important observation in Fig. 6 is that the cost of the core is at most of the frequencies at least twice that of the windings (LV and HV winding together), and thus dominating the cost of the MFT. This shows the key role of the core in any approach to reduce MFT cost. Another constraint, which is plotted in Fig. 5, is that of the minimum air gap needed for robustness and fine-tuning of L_m. As can be seen, the constraint is always far from active, confirming that the assumptions made in the derivation of scaling law (4) are valid.

A general remark is added regarding the interpretation of the constraint functions. It is important to remember the discrete nature of the physical model of the MFT, as discussed above. Therefore, it may happened that a constraint function cannot reach exactly zero, and a negative value close to zero indicates an active constraint. The next integer value of a design variable would lead to a positive constraint-function value and a violation of the constraint.

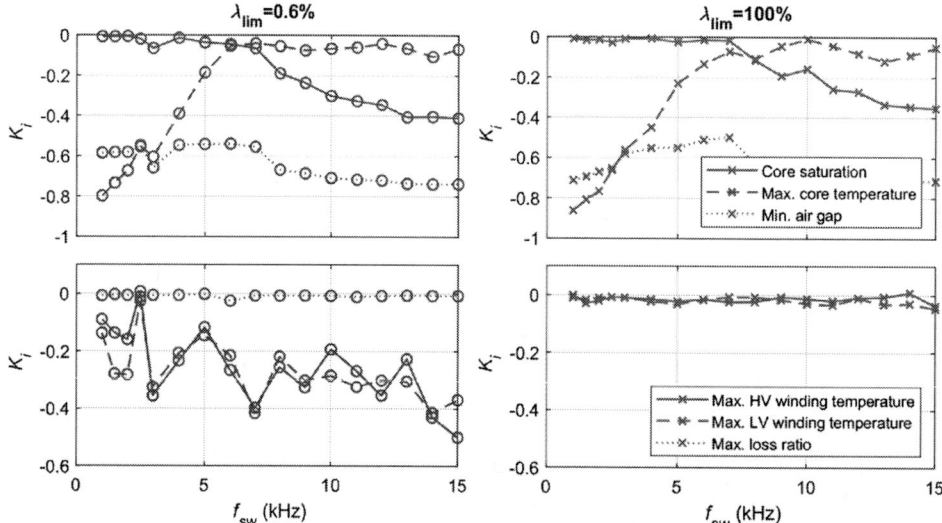

Fig. 5: Cases 1 (left) and 2 (right), constraint functions.

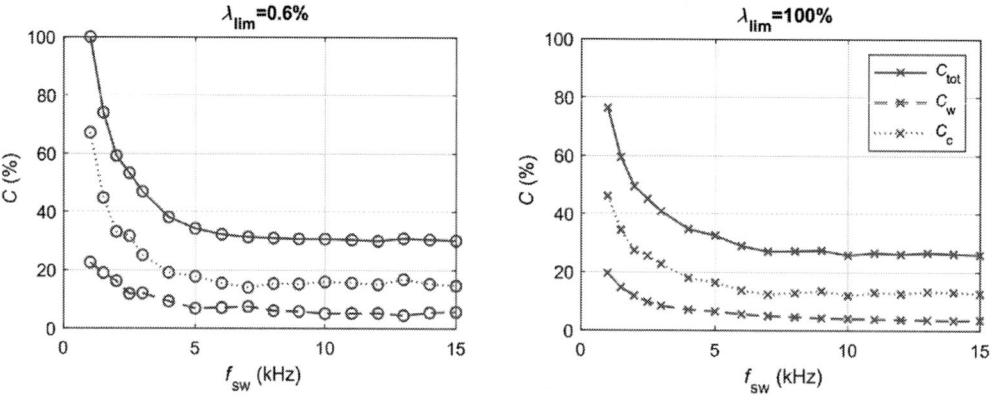

Fig. 6: Cases 1 (left) and 2 (right), MFT costs and mass.

Case 2 – unconstrained efficiency

In case 2, the efficiency constraint was relaxed to investigate its influence on the MFT performance metrics. In Fig. 2 it can be seen that, while the size increases a little, there is a slight decrease in mass and cost, resulting also in a minor decrease in u. Overall, there is a parallel shift of curves rather than a more pronounced scaling with f_{sw}. In particular, $f_{sw,sat}$ remains at about 7 kHz.

As can be seen in Fig. 3, compared to case 1 there is a noticeable increase in N_t at low frequencies, which enables the core to shrink and leads to a significant increase in losses. It is concluded that the loss-ratio constraint has been the limiting factor preventing this in case 1. Above $f_{sw,sat}$, N_t also increases compared to case 1, but less than at low frequencies. The result is the same, i.e., a reduction in core size (as confirmed in Fig. 6) at the expense of higher losses. The new limiting constraint is the winding temperature. From $f_{sw,sat}$ on, the core temperature is also at or near the limit, as the corresponding constraint function is close to zero. Hence, it can be concluded that in case 2 the MFT is entirely thermally limited. Further size and cost reduction of the MFT can only be enabled through more aggressive cooling.

Case 3 – unconstrained efficiency, better core cooling

Regarding the thermal limitation found in case 2, the key role of the core is noted again, and its dominant contribution to cost. Therefore, in a third case, the influence of cooling efficiency was studied. Cases 1 and 2 had a heat-transfer coefficient (HTC) between core surface and air of 36 $W/(m^2K)$, corresponding to realistic forced air cooling with an air speed of several m/s. More aggressive cooling with the HTC doubled to 72 $W/(m^2K)$ shifts $f_{sw,sat}$ up from 7 to about 16 kHz. The winding losses, and thereby the total losses, are slightly reduced (with λ dropping by up to 0.2 %) thanks to smaller core cross-section. This behavior is remarkable, since often, better cooling is only an enabler of *higher* losses.

Case 4 – variation of constraints on inductances

A fourth case was included to study the influence of the inductance constraints, again for unconstrained loss ratio. The results are summarized in Fig. 7. With L_m still constrained, the blue and green curves show the effect of a deviation of L_m/L_σ from the baseline value of 80. The grey curves are for relaxed L_m/L_σ constraint, and the black ones for additionally relaxed L_m. It is observed that $L_m/L_\sigma = 80$ is close to the natural value of L_m/L_σ, while the natural value of L_m is higher than the originally prescribed value of 39.8 $\Omega/(2\pi f_{sw})$. Interestingly, once the constraint on L_m/L_σ has been relaxed, the additional relaxation of L_m does not lead to any change. For L_m/L_σ below its natural value, the cost function increases slightly below $f_{sw,sat}$, while for L_m/L_σ above its natural value, there is a more pronounced increase of the cost function above $f_{sw,sat}$. Overall, it can be concluded that there is significant flexibility in the choice of the inductances without a major drawback in terms of cost function, while the value of $f_{sw,sat}$ remains largely unchanged.

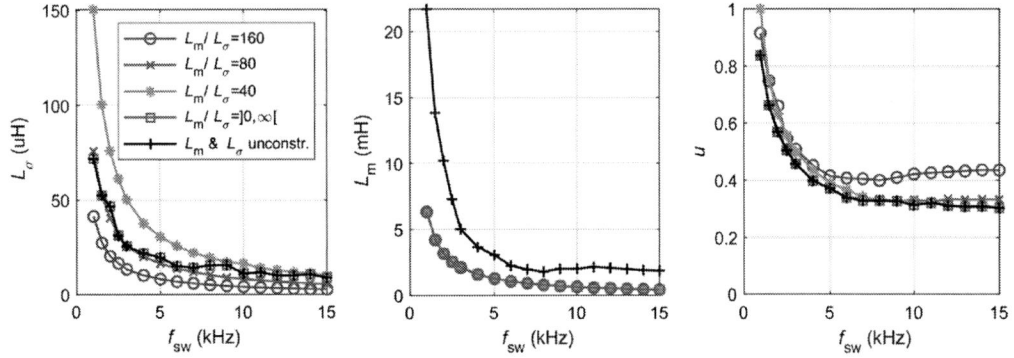

Fig. 7: Case 4, inductances and cost function.

Conclusion

Economic MFTs are a key enabler for economic SSTs and MVDC solutions. An understanding of the influences on the MFT cost structure, and of the limitations to the reduction of size, mass, and cost, is therefore key. Simplified scaling laws suggest a potential of an increase in switching frequency to improve the MFT performance metrics. They suggest that for fixed MFT geometry, an increase in frequency can lead to lower losses under constrained inductances. Alternatively, the size can be reduced while reducing the losses less or keeping them constant. The scaling laws also show that a core-size reduction at a given frequency must always be accompanied by increasing core losses.

In practical MFT design, size reduction is only observed up to a certain "scaling saturation" frequency. The analysis of the constraint functions used in the MFT optimization provides insight into this behavior. In case of the loss ratio constrained to ≤ 0.6 %, this constraint limits the designs at all frequencies. In particular, core-size reduction is not possible, even though the core's contribution to the total losses is modest. Relaxing the loss constraint leads to moderately smaller designs, while increasing the losses significantly, but does not shift the scaling saturation frequency. The designs are thermally limited in this case. Improving the core cooling shows the key role of the core in MFT design, in particular in terms of cost contribution. An upward shift in the scaling saturation frequency can be achieved, resulting in cost savings without a penalty of increased losses. The results hence clearly indicate that too weak core cooling should be avoided in MFTs using nanocrystalline cores.

Overall, the results suggest that, for the considered MFT technology, there is little motivation to increase the switching frequency of the converter beyond about 7 kHz, which is the scaling saturation limit. Higher frequencies, however, may still be interesting from the point of view of acoustic noise. Switching at 10 kHz, for example, allows to safely move the noise out of the audible range, in particular with the core's magnetostriction emitting noise at twice the switching frequency.

References

[1] World's first 5 MW DC converter, https://impulse.schaffner.com/de/power-magnetics-5-mw-gleichstromwandler, 29 October 2021.

[2] C. Zhao, D. Dujic, A. Mester, J. K. Steinke, M. Weiss, S. Lewdeni-Schmid, T. Chaudhuri, and P. Stefanutti, "Power electronic traction transformer – medium voltage prototype", IEEE Trans. on Industrial Electronics, vol. 61, no. 7, pp. 3257–3268, July 2014.

[3] D. Dujic, "Electric vehicles charging – an ultrafast overview", keynote at PCIM Asia, Int. Exhibition and Conf. for Power Electronics, Intelligent Motion, Renewable Energy and Energy Management, Shanghai, China, 26–28 June 2019.

[4] X. Wang, "Optimal DC power distribution system design for data center with efficiency improvement", PhD thesis, University of Wisconsin-Milwaukee, August 2014.

[5] L. Wang, F. Zhang, J. A. Aroca, A. V. Vasilakos, K. Zheng, C. Hou, D. Li, and Z. Liu, "GreenDCN: a general framework for achieving energy efficiency in data center networks", IEEE Journal on Selected Areas in Communications, vol. 32, no. 1, pp. 4-15, January 2014.

[6] M. K. Das, C. Capell, D. E. Grider, S. Leslie, J. Ostop, R. Raju, M. Schutten, J. Nasadoski, and A. Hefner, "10 kV, 120 A SiC half H-bridge power MOSFET modules suitable for high frequency, medium voltage applications", IEEE Energy Conversion Congress and Exposition, pp. 2689-2692, September 2011.

[7] D. Wang, J. Tian, C. Mao, J. Lu, Y. Duan, J. Qiu, and H. Cai, "A 10-kV/400-V 500-kVA electronic power transformer", IEEE Trans. on Industrial Electronics, vol. 63, no. 11, pp. 6653–6663, November 2016.

[8] J. Yu, K. Smith, M. Urizarbarrena, N. MacLeod, R. Bryans, and A. Moon, "Initial designs for the ANGLE DC project; converting existing AC cable and overhead line into DC operation", 13th IET Int. Conf. on AC and DC Power Transmission (ACDC), pp. 1-6, 2017.

[9] M. Mogorovic, "Modeling and design optimization of medium frequency transformers for medium-voltage high-power converters," PhD thesis, EPFL Lausanne, Switzerland, 2019.

[10] U. Drofenik, "A 150 kW medium frequency transformer optimized for maximum power density", 7th Int. Conf. on Integrated Power Electronics Systems (CIPS), Nuremberg, Germany, 6–8 March 2012.

[11] M. Mogorovic and D. Dujic, "Sensitivity analysis of medium frequency transformer design", Int. Power Electronics Conference (IPEC-Niigata – ECCE Asia), pp. 2170-2175, 2018.

[12] M. Mogorovic and D. Dujic, "Sensitivity analysis of medium-frequency transformer designs for solid-state transformers", IEEE Trans. on Power Electronics, vol. 34, no. 9, pp. 8356-8367, September 2019.

[13] N. Mohan, T. M. Undeland, and W. P. Robbins, "Power electronics: converters, applications, and design", 3rd ed., John Wiley & Sons, 2002.

[14] T. Gradinger and M. Mogorovic, "Foil-winding design for medium-frequency medium-voltage transformers", 23rd European Conf. on Power Electronics and Applications (EPE ECCE Europe), online, 6–10 September 2021.

[15] A. Van den Bossche and V. C. Valchev, "Inductors and transformers for power electronics", Taylor & Francis, March 2005.

[16] IEC 60076-3, 2nd ed., "Power transformers – Part 3: Insulation levels, dielectric tests and external clearances in air", March 2000.

[17] M. Mogorovic and D. Dujic, "Analysis of the effectiveness of the series inductor integration into the MFT for SST Applications", PCIM Europe digital days, Intern. Exhibition and Conf. for Power Electronics, Intelligent Motion, Renewable Energy and Energy Management, pp. 1-7, 2020.

Output Power Characteristics of Isolated Secondary-Resonant SAB DC-DC Converter for Output Voltage Variation

Shota Yamashita, Kohei Budo, Takaharu Takeshita
Dept. of Electrical and Mechanical Engineering, Graduate School of Engineering
Nagoya Institute of Technology
Gokiso, Showa, Nagoya
Aichi, 466-8555 Japan
Tel.: +81 / (52) 735 − 5441
Fax: +81 / (52) 735 − 5432
E-Mail: take@nitech.ac.jp
URL: http://motion.web.nitech.ac.jp/

Acknowledgements

A part of this work was supported by JSPS KAKENHI Grant Number JP21H01310.

Keywords

«DC-DC converter», «Isolated converter», «High frequency power converter», «Soft switching», «Dual Active Bridge (DAB)»

Abstract

The Secondary-Resonant Single Active Bridge (SR-SAB) converter is proposed as a high power, low-cost, compact, and high-efficiency DC-DC converter. The SR-SAB converter is composed of the H-bridge circuit, the high-frequency transformer, and the diode rectifier circuit with resonant capacitors connected in parallel. When primary H-bridge circuit switches, the resonant capacitors cause LC resonance. Due to LC resonance, the secondary diode rectifier circuit switching is delayed and the secondary voltage is lags behind the primary voltage like the DAB converter. While LC resonance occurs, large voltage is applied to leakage inductance and secondary current increase significantly. Therefore, the SR-SAB converter achieves high total power factor like DAB converter with simple SAB converter circuit. This paper presents the output power characteristics of the SR-SAB converter for the output voltage variation. The effectiveness of output power characteristics for output voltage variation of the SR-SAB converter is verified by experiments.

Introduction

The electric vehicle (EV) is attracting public attentions to solve environmental problems. The compact and high-efficiency isolated DC-DC converter is needed for EV's battery charger [1]. The bidirectional isolated Dual-Active-Bridge (DAB) DC-DC converter has been proposed. Fig.1 shows the DAB converter. Fig.2 shows the conventional SAB converter. Fig.3 shows the SR-SAB converter. The DAB converter is composed of H-bridges on the primary and secondary side and high-frequency transformer [2], [3]. Fig.4 shows waveforms of the DAB converter, the conventional SAB converter and the SR-SAB converter. Square waves of the primary voltage v_1 and secondary voltage v_2 are generated by controlling H-bridge circuits. The DAB converter transforms the power using phase difference between the primary voltage v_1 and the secondary voltage v_2. While the secondary voltage v_2 lags behind the primary voltage v_1, the leakage inductance voltage v_L, the voltage applied to the leakage inductance, is difference of the primary voltage v_1 and the secondary voltage v_2 (= sum of the input DC voltage V_{in} and the output DC voltage V_{out}), the secondary current i_2 increases significantly. Therefore, the DAB converter achieves high total power factor and the DAB converter can transform power even if the input DC voltage V_{in} is lower than the output DC voltage V_{out}. By controlling the phase between the primary voltage v_1 and the secondary voltage v_2, the DAB converter achieves output power control and bidirectional power transformation [4], [5]. The simple and low-cost isolated DC-DC converter compared with DAB converter is needed for unidirectional power conversion such as battery charger.

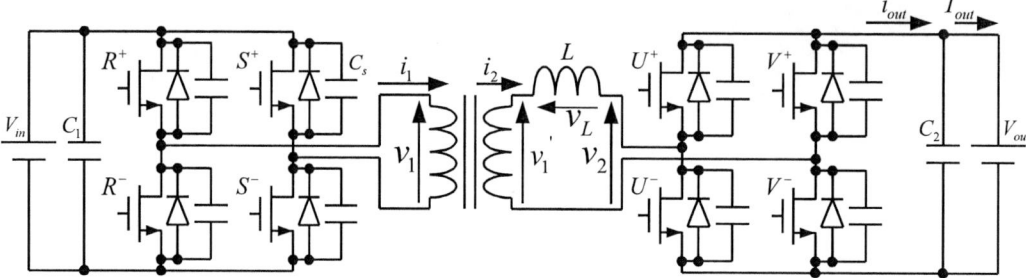

Fig. 1 : Conventional DAB converter circuit configuration

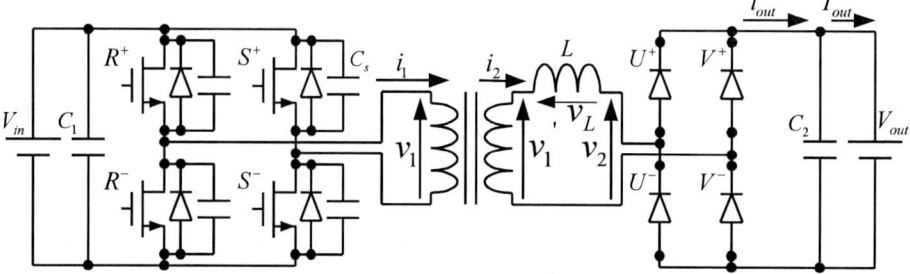

Fig. 2 : Conventional SAB converter circuit configuration

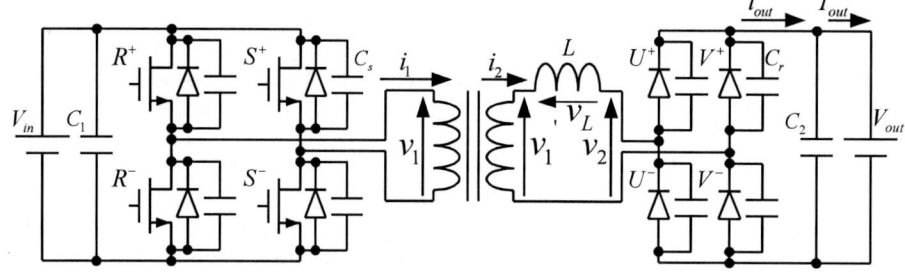

Fig. 3 : Proposed SR-SAB DC-DC converter circuit configuration

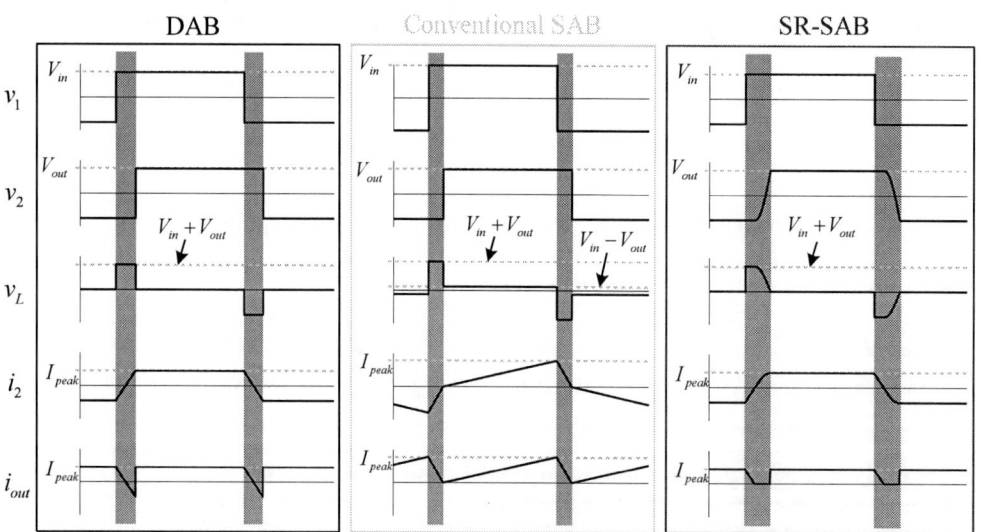

Fig. 4 : Waveforms of the DAB converter, the conventional SAB converter and the SR-SAB converter

The unidirectional isolated Single-Active-Bridge (SAB) DC-DC converter has been proposed. The SAB converter is composed of the H-bridge circuit, high-frequency transformer and the diode rectifier circuit [6],[7],[8]. The SAB converter is a low-cost and compact unidirectional isolated DC-DC converter. Square wave of the primary voltage v_1 is generated by controlling H-bridge circuit. The secondary voltage v_2 depends on on/off state of the secondary diodes. The phase difference between the primary voltage v_1 and the secondary voltage v_2 is short and the secondary current i_2 dose not increase sufficiently. Therefore, total power factor of the conventional SAB converter is low compared with the DAB converter.

The authors have proposed a high power, low-cost, compact, and high-efficiency isolated Secondary-Resonant SAB (SR-SAB) DC-DC converter [9][10][11]. The SR-SAB converter is composed of the H-bridge circuit on the primary side and the diode rectifier circuit with resonant capacitors connected in parallel on the secondary side. Because the resonant capacitors delay turn-on of diodes of the secondary rectifier circuit, the secondary voltage v_2 lags behind the primary voltage v_1. By the delay of the primary voltage v_1, the voltage applied to leakage inductor v_L is difference of the primary voltage v_1 and the secondary voltage v_2 and the secondary current i_2 increases significantly like the DAB converter. Therefore, the SR-SAB converter achieves high total power factor, like the DAB converter, with simple circuit configuration like the conventional SAB converter.

This paper presents the analysis and the output power characteristics of the SR-SAB converter for the output DC voltage V_{out} variation. The output power of the proposed SR-SAB converter can be expressed by using only four-points in the secondary current i_2. First, the secondary current i_2 waveform of the SR-SAB converter is derived by considering the behaver of the secondary side. Next, the output power characteristics for output DC voltage V_{out} variation of the SR-SAB converter is derived by using secondary i_2 current waveform. Finally, the effectiveness of the output power characteristics is verified by comparing the experimental results with theoretical characteristics.

Overview of the SR-SAB Converter

Waveform of the SR-SAB and the conventional SAB converter

Table 1 shows condition of the waveform of the waveforms of the conventional SAB converter and the SR-SAB converter. In the conventional SAB converter, the primary voltage v_1 with the amplitude as the input DC voltage V_{in} is generated by controlling four switches R^+, R^-, S^+, and S^-. The secondary voltage v_2 is a square wave with the amplitude as the output DC voltage V_{out}. The secondary voltage v_2 is lags behind the primary voltage v_1. When secondary current i_2 reaches zero, the diodes of the secondary side turns-on, and the secondary voltage v_2 switches and the phase delay period shown in red in the Fig.4 (the period between switching of the primary voltage v_1 and switching of the secondary voltage v_2) finishes. The leakage inductance voltage v_L is difference of the secondary voltage v_2 minus the primary voltage v_1. Sum of the input DC voltage V_{in} and the output DC voltage V_{out} is applied to the leakage inductance voltage v_L during the phase delay period and difference of input DC voltage V_{in} minus the output DC voltage V_{out} is applied during the other periods. The secondary current i_2 increases by the leakage inductance voltage v_L. The secondary current i_2 increases significantly during the phase delay period and increases slowly during the other period. Because the phase delay period is short, the secondary current i_2 dose not increase sufficiently. Therefore, total power factor of the conventional SAB converter is low. The output current i_{out} is the rectified secondary current i_2.

The condition of waveforms in Fig.4 is shown in Table 1. In the SR-SAB converter, the primary voltage v_1 with the amplitude as the input DC voltage V_{in} is generated by controlling four switches R^+, R^-, S^+, and S^-. The secondary voltage v_2 is a square wave with sign wave-like rising edge with the amplitude as the output DC voltage V_{out}. The secondary voltage v_2 is lags behind the primary voltage v_1. When secondary current i_2 reaches zero, resonant capacitors C_r start discharging by the secondary LC resonance between leakage inductor L and resonant capacitors C_r of secondary side. While the secondary LC resonance occurs, resonance capacitors C_r have charge, therefore, the diodes of the secondary side are off state and the secondary voltage v_2 rises with sign wave-like rising edge. When resonance capacitors C_r finish discharging, the secondary LC resonance finishes and the phase delay period shown in red in the Fig.4 finishes. This means that the phase delay period is extended by the secondary LC resonance. The leakage inductance voltage v_L is difference of the secondary voltage v_2 minus the primary voltage v_1. During the phase delay period, large voltage is applied to the leakage inductance voltage v_L and during the other period, difference of input DC voltage V_{in} minus the output DC voltage V_{out} is applied. The secondary current i_2 increases by the leakage inductance voltage v_L. The secondary current i_2 increases significantly during the phase delay period. Because the phase delay period is extended by the secondary LC resonance, the secondary current i_2 increase sufficiently. Therefore, total power factor of the SR-SAB converter is high. During the secondary LC resonance period, the secondary current i_2 flow only resonant capacitors C_r and transformer, therefore, the output current i_{out} is zero during the secondary LC resonance period and the output current i_{out} is the rectified secondary current i_2 during the other period.

Table 1 shows parameters of the SR-SAB converter and the conventional SAB converter. The peak current I_{peak} is the peak current the secondary current i_2. The peak current I_{peak} of the SR-SAB converter is lower than the conventional SAB converter. Total power factor of the SR-SAB converter is higher than the conventional SAB converter.

Characteristics of the SR-SAB for output voltage variation

Fig.5 shows the operating waveforms of the SR-SAB converter under $V_{in} = V_{out}$ condition as black line, under $V_{in} > V_{out}$ condition as red dashed line, under $V_{in} < V_{out}$ condition as blue dashed line, restrictively.

Table 1 : Parameters of the SR-SAB converter and the conventional SAB converter

	SR-SAB converter	Conventional SAB converter
Input DC voltage V_{in}	265 V	345 V
Output DC voltage V_{out}	265 V	265 V
Output power P_{out}	2.5 kW	2.5 kW
Leakage inductance L	92 uH	92 uH
Resonant capacitor C_r	43 nF	-
Frequency of transformer f	20 kHz	20 kHz
Peak current I_{peak}	11.5 A	19.2 A
Total power factor	0.92	0.67

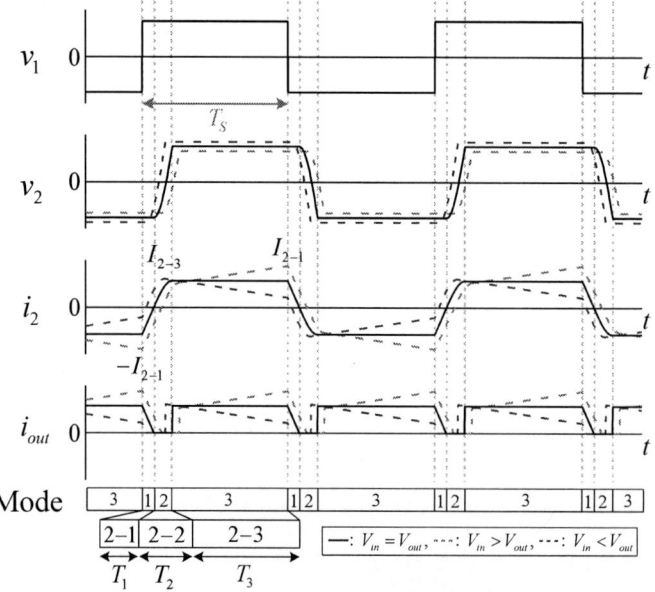

Fig. 5 : Operating waveform of the SR-SAB converter

The operation of the SR-SAB converter is divided into three operation modes. The Mode 2-1 starts when the primary voltage v_1 switches. During the duration T_1 of the Mode 2-1, sum of the input DC voltage and output DC voltage is applied to the leakage inductance voltage v_L and the secondary current i_2 increases significantly. The initial value I_{2-1} is the initial value of the Mode2-1 of the secondary current i_2. The Mode 2-1 finishes when secondary current i_2 reaches zero, and the Mode 2-2 starts. During the duration T_2 of the Mode 2-2, the secondary LC resonance occurs and the diodes of secondary side are off state. Therefore, large voltage is applied to the leakage inductance voltage v_L and the secondary current i_2 increases significantly. The Mode 2-2 finishes when secondary voltage v_2 reaches the output voltage V_{out}, and the Mode 2-3 starts. During the duration T_3 of Mode 2-3, diodes of secondary side are on state. The leakage inductance voltage v_L is the input DC voltage minus the output DC voltage V_{out}, therefore, increase or decrease in the secondary current i_2 depends on the relationship between the input DC voltage V_{in} and the output DC voltage V_{out}. The initial value I_{2-3} is the initial value of the Mode 2-3 of the secondary current i_2. The Mode 2-3 finishes when the primary voltage v_1 switches again.

The red dushed waveforms in Fig.5 shows the waveforms of SR-SAB converter under $V_{in} > V_{out}$ condition. In the Mode 2-3, secondary current i_2 increases and the initial value I_{2-1} is larger than $V_{in} = V_{out}$ condition. Because the initial value I_{2-1} is larger, the duration T_1, period of time until secondary current i_2 reaches zero, is longer than $V_{in} = V_{out}$ condition. The duration T_2 is short and the initial value I_{2-3} is small because duration of the secondary LC resonance depends on the output DC voltage V_{out}.

The blue dushed waveforms in Fig.5 shows the waveforms of SR-SAB converter under $V_{in} < V_{out}$ condition. In the Mode 2-3, secondary current i_2 decreases and the initial value I_{2-1} is smaller than $V_{in} = V_{out}$ condition. Because the initial value I_{2-1} is smaller, the duration T_1 is shorter than $V_{in} = V_{out}$ condition. The duration T_2 is long and the initial value I_{2-3} is large because duration of the secondary LC resonance depends on the output DC voltage V_{out}.

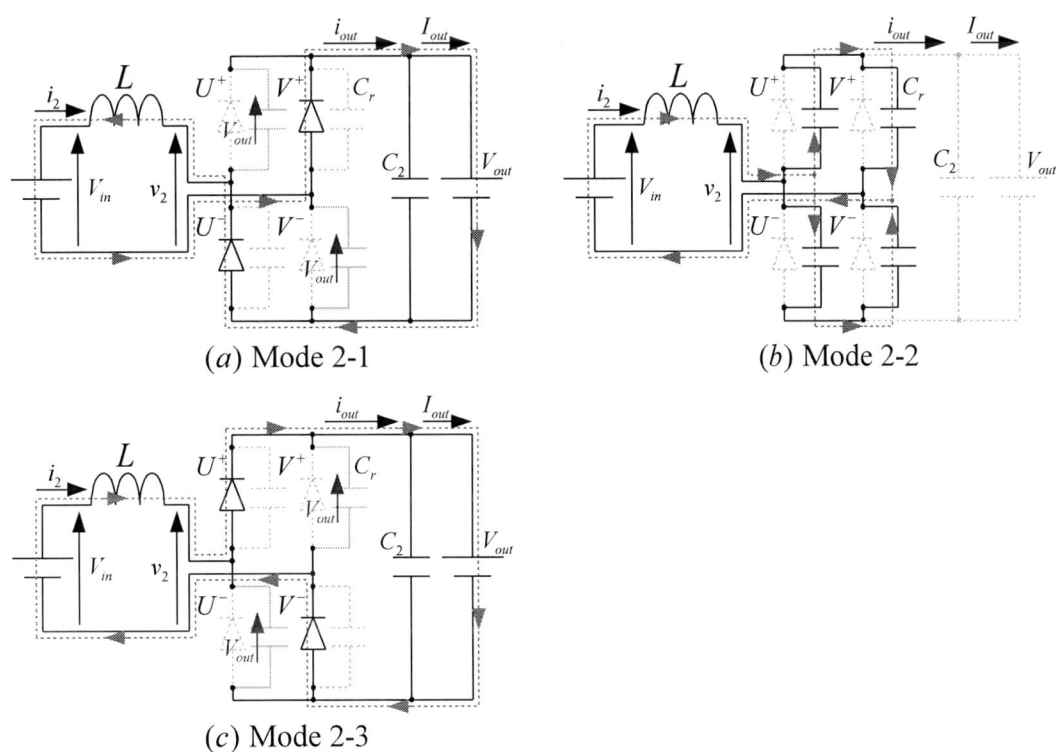

(a) Mode 2-1   (b) Mode 2-2

(c) Mode 2-3

Fig. 6 : The secondary circuit operation

Operation Theory of Secondary-Converter

This chapter explains the operation theory of the secondary side to obtain the output power P_{out} expressed by the secondary current i_2. Fig.6 shows the secondary connection diagrams of the Mode 2-1, the Mode 2-2, and the Mode 2-3, respectively in the Fig.5 based on the conduction pattern of secondary diodes when the primary switches R^+ and S^- are on-state. The primary voltage v_1 is used as the DC voltage source because the primary switches R^+ and S^- are on-state. In the Mode 2-1, the secondary current i_2 increases toward zero because the primary voltage $v_1 = V_{in}$ and the diodes U^- and V^+ are on-state. When the secondary current i_2 reaches zero, all diodes on the secondary side are off-state, then the Mode 2-2 begins. In the Mode 2-2, the secondary LC resonance between leakage inductor L and resonant capacitors C_r occurs. All diodes of secondary side keep off-state. When the capacitor voltage connected to diode U^+ and V^- are zero-voltage, diode U^+ and V^- are on-state, then the Mode 2-3 begins. In the Mode 2-3, the secondary diodes U^+ and V^- are on-state. When the primary switches R^- and S^+ are on-state, the secondary side operates with the secondary current flowing in the reverse direction of the diagram.

Fig.6 (a) shows the secondary circuit and the flow of secondary current i_2 in the Mode 2-1 with red lines. In this paper, time $t = 0$ is defined as the timing of the primary voltage v_1 switched from $v_1 = -V_{in}$ to $v_1 = V_{in}$. The secondary current i_2 of the Mode 2-1 is obtained from voltage-equation of Fig.6(a) by using the initial value $i_2(0) = -I_{2-1}$ as follows:

$$i_2(t) = \frac{V_{in} + V_{out}}{L} t - I_{2-1} \quad (0 \leq t \leq T_1) \tag{1}$$

$$i_{out}(t) = |i_2(t)| \quad (0 \leq t \leq T_1) \tag{2}$$

The duration T_1 of the Mode 2-1 is obtained from $i_2(T_1) = 0$ as follows:

$$T_1 = \frac{L}{V_{in} + V_{out}} I_{2-1} \tag{3}$$

Fig.6(b) shows the secondary circuit and the flow of the secondary current i_2 during the Mode 2-2 with the red line. The resonant capacitor C_r and the transformer leakage inductance L cause the secondary LC resonance. The secondary current i_2 flows through the resonant capacitor C_r as shown in Fig.6(b). Because the secondary current i_2 increases by phase difference created by the secondary LC resonance. The current flows only through the resonant capacitor, and the output current i_{out} is zero. The secondary current i_2 of the Mode2-2 is obtained from voltage equation of Fig.6(b) by using the initial value $i_2(T_1) = 0$ and the duration T_1 of the Mode 2-1 as follows:

$$i_2(t) = \sqrt{LC_r}(V_{in} + V_{out}) \sin \frac{1}{\sqrt{LC_r}}(t - T_1) \quad (T_1 \leq t \leq T_1 + T_2) \tag{4}$$

$$i_{out}(t) = 0 \quad (T_1 \leq t \leq T_1 + T_2) \tag{5}$$

The secondary LC resonance in secondary circuit finishes when the voltage of the resonance capacitor C_r of U^- and V^+ reaches zero and diode of U^- and V^+ turn on. Therefore, the duration T_2 of the Mode 2-2 is obtained as follows:

$$T_2 = \sqrt{LC_r} \cos^{-1}\left(\frac{V_{in} - V_{out}}{V_{in} + V_{out}}\right) = A \cos^{-1}\frac{\beta}{2 - \beta} \tag{6}$$

$$I_{2-3} = i_2(T_2) = 2\frac{A}{L}V_{in}\sqrt{1 - \beta} \tag{7}$$

where $A = \sqrt{LC_r}$ and $\beta = (V_{in} - V_{out})/V_{in}$.

Fig.6(c) shows the secondary circuit and the flow of the secondary current i_2 during the Mode 2-3 with red lines. The secondary current i_2 is obtained from voltage-equation of Fig.6(c) by using the initial value $i_2(T_1 + T_2) = I_{2-3}$ and the duration T_2 of the Mode 2-2 as follows:

$$i_2(t) = \frac{V_{in} - V_{out}}{L}(t - T_1 - T_2) + I_{2-3} \quad (T_1 + T_2 \le t \le T_1 + T_2 + T_3) \tag{8}$$

$$i_{out}(t) = |i_2(t)| \quad (T_1 + T_2 \le t \le T_1 + T_2 + T_3) \tag{9}$$

Characteristics Output Power

The output power P_{out} is obtained using the four-parameters I_{2-1}, I_{2-3}, T_1, and T_3 as follow:

$$P_{out} = \frac{1}{T_S}\int V_{out}\, i_{out}\, dt = \frac{V_{out}}{2T_S}\{(I_{2-1} + I_{2-3})T_3 + I_{2-1}T_1\}, \tag{10}$$

Where

$$T_S = T_1 + T_2 + T_3 \tag{11}$$

This chapter explains the derivation method of four parameters I_{2-1}, I_{2-3}, T_1, and T_3 of the output power P_{out} in (10) to obtain the output power P_{out} characteristics for the output DC voltage V_{out} variation $\beta(= (V_{in} - V_{out})/V_{in})$ which means the relation between input DC voltage V_{in} and output DC voltage V_{out}. Fig.7 shows the four parameters I_{2-1}, I_{2-3}, T_1, and T_3 on the secondary current i_2 in the half period T_S.

Because the secondary current i_2 cyclically changes, equations are obtained from $i_2(T_S) = -i_2(0)$ as follow:

$$i_2(T_S) = \frac{V_{in} - V_{out}}{L}(T_S - T_1 - T_2) + I_{2-3} = I_{2-1} \tag{12}$$

Parameters T_1, T_3, and I_{2-1} are obtained from equations (3), (6), (7), (11), and (12)

$$T_1 = \frac{1}{2}\beta(T_S - T_2) + A\sqrt{1 - \beta} \tag{13}$$

$$T_3 = \frac{1}{2}(2 - \beta)(T_S - T_2) - A\sqrt{1 - \beta} \tag{14}$$

$$I_{2-1} = \frac{V_{in}}{2L}(-\beta^2 + 2\beta)(T_S - T_2) + \frac{V_{in}}{L}A(2 - \beta)\sqrt{1 - \beta} \tag{15}$$

Because β is close to zero, following approximation can be used.

$$(-\beta + 1)^{\frac{1}{2}} \approx 1 - \frac{1}{2}\beta \tag{16}$$

$$\cos^{-1}\frac{\beta}{2 - \beta} \approx \left(\frac{\pi}{2} - \frac{1}{2}\beta\right) \tag{17}$$

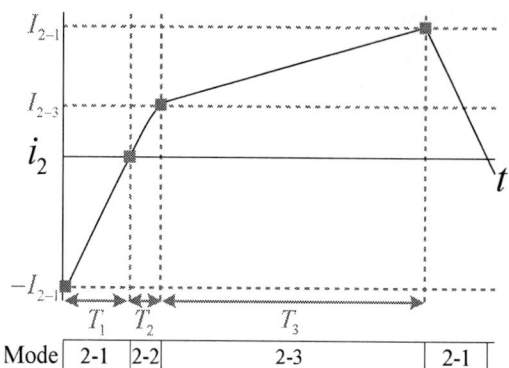

Fig. 7 : Four points of secondary current i_2

Parameters are simplified by applying approximation (16) and (17) to (6), (7), (13), (14), and (15) as follow:

$$T_1 \approx \frac{1}{4} A \beta^2 + \left\{ \frac{1}{2} T_S - \left(\frac{\pi}{2} + \frac{1}{2} \right) A \right\} \beta + A, \tag{18}$$

$$T_2 \approx A \left(\frac{\pi}{2} - \frac{1}{2} \beta \right) \tag{19}$$

$$T_3 \approx -\frac{1}{4} A \beta^2 + \left\{ -\frac{1}{2} T_S + \left(\frac{\pi}{2} + 1 \right) A \right\} \beta + T_S - \left(\frac{\pi}{2} + 1 \right) A, \tag{20}$$

$$I_{2-1} \approx \frac{V_{in}}{L} \left[-\frac{1}{4} A \beta^3 + \left\{ -\frac{1}{2} T_S + \left(\frac{\pi}{4} + 1 \right) A \right\} \beta^2 + \left\{ T_S - \left(\frac{\pi}{2} + 2 \right) A \right\} \beta + 2A \right] \tag{21}$$

$$I_{2-3} \approx 2 \frac{A}{L} V_{in} \left(1 - \frac{1}{2} \beta \right) \tag{22}$$

The output power P_{out} is derived substituting (18), (19), (20), (21), and (22) for (10) and expressed by circuit parameters V_{in}, V_{out}, L, C_r and switching frequency $(1/2T_S)$. The output power P_{out} is derived as simplified equation as follow:

$$
\begin{aligned}
P_{out} = \frac{V_{in}^2}{2 T_S L} \Big[&\frac{1}{8} A^2 \beta^5 + \left\{ -\frac{1}{8} A^2 + \frac{1}{2} A T_S - \left(\frac{\pi}{4} + \frac{3}{4} \right) A^2 \right\} \beta^4 \\
&+ \left\{ \frac{1}{2} T_S^2 - \left(\frac{\pi}{2} + \frac{5}{2} \right) A T_S + \left(\frac{1}{8} \pi^2 + \frac{3}{2} \pi + \frac{13}{4} \right) A^2 \right\} \beta^3 \\
&+ \left\{ -\frac{3}{2} T_S^2 + \left(\frac{3}{2} \pi + 6 \right) A T_S - \left(\frac{3}{8} \pi^2 + \frac{15}{4} \pi + \frac{13}{2} \right) A^2 \right\} \beta^2 \\
&+ \left\{ T_S^2 - (\pi + 8) A T_S + \left(\frac{1}{4} \pi^2 + \frac{9}{2} \pi + 6 \right) A^2 \right\} \beta \\
&+ 4 A T_S - (2\pi + 2) A^2 \Big]
\end{aligned}
\tag{23}
$$

Table 2 : Experimental conditions

Input DC voltage V_{in}	265 V
Leakage inductor L	92 μH
Resonant capacitor C_r	43 nF
Frequency of transformer $1/(2T_S)$	20 kHz
DC capacitor C_1, C_2	1500 μF

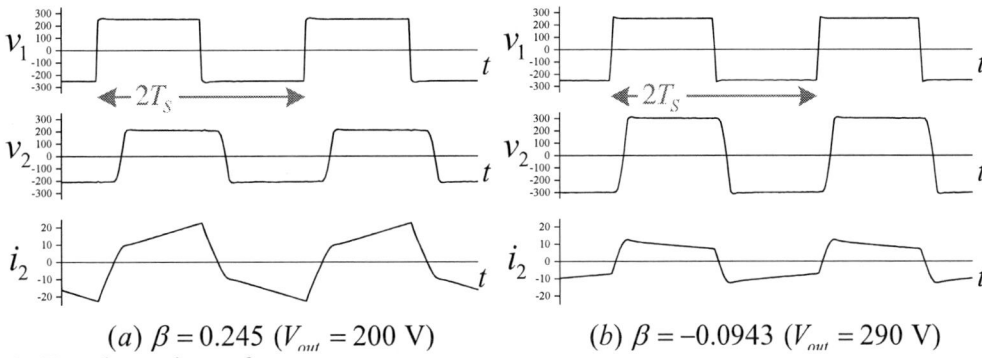

(a) $\beta = 0.245$ ($V_{out} = 200$ V)   (b) $\beta = -0.0943$ ($V_{out} = 290$ V)

Fig. 8 : Experimental waveforms

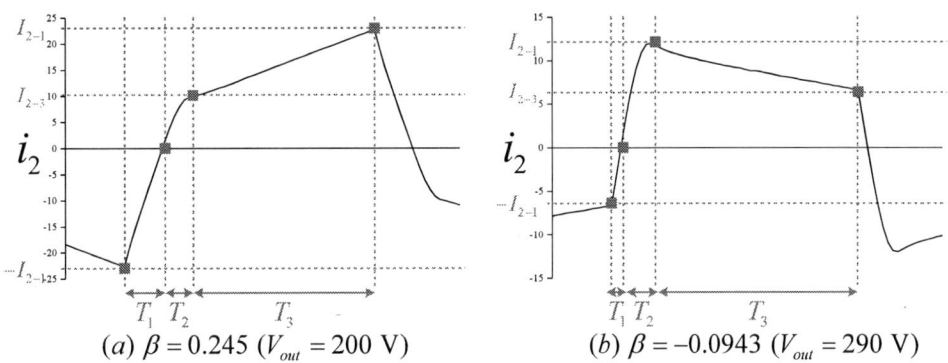

(a) $\beta = 0.245$ ($V_{out} = 200$ V)   (b) $\beta = -0.0943$ ($V_{out} = 290$ V)

Fig. 9 : Experimental secondary current i_2 waveform for one period

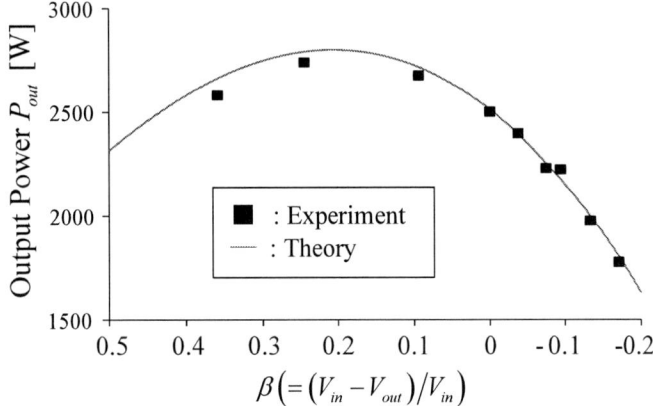

Fig. 10 : Experimental result of the output power with respect to β

Experimental Results

Table 2 shows experimental conditions. Leakage inductor L and resonant capacitor C_r are designed to provide 2.5 kW output power under the output DC voltage $V_{out} =$ DC voltage V_{in} condition.

Fig.8 shows the experimental waveforms under $\beta = 0.245$ ($V_{out} = 200$ V) and $\beta = -0.0943$ ($V_{out} = 290$ V) condition. The effective value of the secondary current i_2 is high under the input DC voltage V_{in} > output DC voltage V_{out} condition. The effective value of the secondary current i_2 is low under the input DC voltage V_{in} < output DC voltage V_{out} condition.

Fig.9 shows the experimental secondary current i_2 waveform for one period as black line and the theorical secondary current parameter values in equations (18), (19), (20), (21) and (22) as red square symbols. The experimental secondary current i_2 waveforms are agreement with the theorical secondary current i_2 parameter values.

Fig.10 shows the experimental result of the output power P_{out} as black square symbols and the theorical output power P_{out} in equation (23) as a red line. The experimental output power P_{out} characteristics are agreement with theorical output power P_{out}.

Conclusion

The output power characteristics of the SR-SAB converter for the output DC voltage variation obtained as simplified equation using four parameters of secondary current. The effectiveness of the output power characteristics of the SR-SAB converter is verified by experiments.

References

[1] Yu Fang, Songyin Cao, Yong Xie and P. Wheeler, "Study on bidirectional-charger for electric vehicle applied to power dispatching in smart grid," 2016 IEEE 8th International Power Electronics and Motion Control Conference (IPEMC-ECCE Asia), 2016, pp. 2709-2713, doi: 10.1109/IPEMC.2016.7512726.

[2] B. Zhao, Q. Song, W. Liu and Y. Sun, "Overview of Dual-Active-Bridge Isolated Bidirectional DC–DC Converter for High-Frequency-Link Power-Conversion System," in IEEE Transactions on Power Electronics, vol. 29, no. 8, pp. 4091-4106, Aug. 2014, doi: 10.1109/TPEL.2013.2289913.

[3] M. H. Kheraluwala, R. W. Gascoigne, D. M. Divan and E. D. Baumann, "Performance characterization of a high-power dual active bridge dc-to- dc converter", IEEE Trans. Ind. Appl., vol. 28, no. 6, pp. 1294-1301, 1992.

[4] S. Inoue and H. Akagi, "A Bi-Directional DC/DC Converter for an Energy Storage System," APEC 07 - Twenty-Second Annual IEEE Applied Power Electronics Conference and Exposition, 2007, pp. 761-767, doi: 10.1109/APEX.2007.357601.

[5] R. Huang and S. K. Mazumder, "A Soft-Switching Scheme for an Isolated DC/DC Converter With Pulsating DC Output for a Three-Phase High-Frequency-Link PWM Converter," in IEEE Transactions on Power Electronics, vol. 24, no. 10, pp. 2276-2288, Oct. 2009, doi: 10.1109/TPEL.2009.2022755.

[6]G. D. Demetriades and H. P. Nee, "Characterisation of the Soft-switched Single-Active Bridge Topology Employing a Novel Control Scheme for High-power DC-DC Applications," 2005 IEEE 36th Power Electronics Specialists Conference, 2005, pp. 1947-1951, doi: 10.1109/PESC.2005.1581898.

[7] K. Park and Z. Chen, "Analysis and design of a parallel-connected single active bridge DC-DC converter for high-power wind farm applications," 2013 15th European Conference on Power Electronics and Applications (EPE), 2013, pp. 1-10, doi: 10.1109/EPE.2013.6631854.

[8] Ryo Haneda, Hirofumi Akagi and Kenji Hukuda, "Output Voltage Regulation of a Unidirectional Isolated DC-DC Converter Used as an Auxiliary Power Supply for Electric Commuter Trains", IEEJ Trans. IA, vol. 137, no. 5, pp. 406-413, 2017.

[9] C. A. Tuan, H. Naoki and T. Takeshita, "Unidirectional Isolated High-Frequency-Link DC-DC Converter Using Soft-Switching Technique," 2019 IEEE 4th International Future Energy Electronics Conference (IFEEC), 2019, pp. 1-7

[10]Cao Anh Tuan, Takaharu T., "Analysis of Unidirectional Secondary Resonant Single Active Bridge DC-DC converter", Energies 2021, 14, 6349, 2022, p14

[11] Cao Anh Tuan, Takaharu Takeshita, "Output Power Characteristics of Unidirectional Secondary-Resonant Single-Active-Bridge DC-DC Converter using Pulse Width Control", IEEJ Journal of Industry Applications, Volume 11, Issue 2, pp. 359-368, 2022.

Hardware and Control Design of a High Precision Modular Power Converter based on GaN Technology for Particle Accelerator Magnets

Thomas Margreiter[1,2,*], Ivan De Cesaris[1], Maurizio Incurvati[2], Sebastien Pelletier[1], Martin Schiestl[2], Ronald Stärz[2]

[1]EBG MedAustron GmbH, Marie-Curie-Straße 5, 2700 Wiener Neustadt, Austria
[2]MCI Management Center Innsbruck, Maximillianstraße 2, 6020 Innsbruck, Austria
*Corresponding author: thomas.margreiter@medaustron.at

Keywords

≪Particle accelerators≫, ≪Gallium nitride (GaN)≫, ≪Interleaved converters≫, ≪Converter control≫, ≪Hardware design≫.

Abstract

Particle accelerators are exploited in cancer treatment. Due to advanced performance requirements in this field, an industry standard modular power converter by means of hardware prototyping is developed. Moreover, a sophisticated controller is implemented. Measurements proof its suitability, acting as an ideal voltage source for accelerator magnets, requiring fast transients.

Introduction

One application for particle accelerators is their utilization in cancer treatment by means of ion beams. Carbon ions have a drastically increased effectiveness and physical impact compared to similar treatment techniques [1]. This method is far less invasive for healthy tissue surrounding tumour cells compared to others. The cause can be explained by the physical concept called Bragg peak. Particle therapy has proven effectiveness for treating tumours close to radiation sensitive organs, since the maximum dose can be focused to the closest vicinity of the tumour itself [1]. In order to deflect the mentioned beams precisely towards their target and to keep them on their trajectory within the particle accelerator, it is necessary to supply various magnets on the path between extractor and human body with precise current waveforms. Since their magnetic fields are proportional to the current flowing through the magnets, in terms of amplitude and frequency, controlling these currents means controlling the fields and thus the beam direction. To fulfil the aforementioned requirements, tight specifications on reference speed and control accuracy are needed. Therefore, a novel modular power converter that is capable of this performance including a sophisticated control strategy has to be developed.

System Overview and Specifications

Important parameters are summarized in Table I. Typical current waveforms for driving so called scanning magnets are illustrated in Fig. 1(a) and (b). Those magnets act as a final beam aiming device in this case. The characteristic waveforms are also applicable in the reverse direction. As is shown by Fig. 1(b) a current ramp can be decomposed into individual steps. One step size can be calculated with

$$\Delta I_{max} = \frac{V_{out_{max}}}{L_{m_{max}}} \Delta T_{max} = \frac{200\,\text{V}}{22\,\text{mH}} 300\,\mu\text{s} = 2.7\,\text{A} \tag{1}$$

and is therefore set to 3 A for experimentation purposes. The corresponding reference steps and hence the overall waveforms are generated by means of trajectory theory [2]. A fifth order polynomial is utilized providing sufficient smoothness for the control algorithm. A step is initiated by a new setpoint input and is followed by delay time T_{dmax}. Subsequently, the voltage has its ramp up phase during T_{Uramp} and the current increases within the time T_{Iramp}. The overall cycle period T_{cycle} is determined by means of

$$T_{cycle} = 4 \frac{I_{nom}}{\Delta I_{max}} (T_s + T_{flat})$$

(2)

that results in 20 ms, given the parameters from Table I and the previously calculated current step size. The given ripple requirement from Table I is not defined by the user but chosen in such a range because it is assumed to be the state of the art for these class of converters. This links to the stated voltage ripple in worst case conditions.

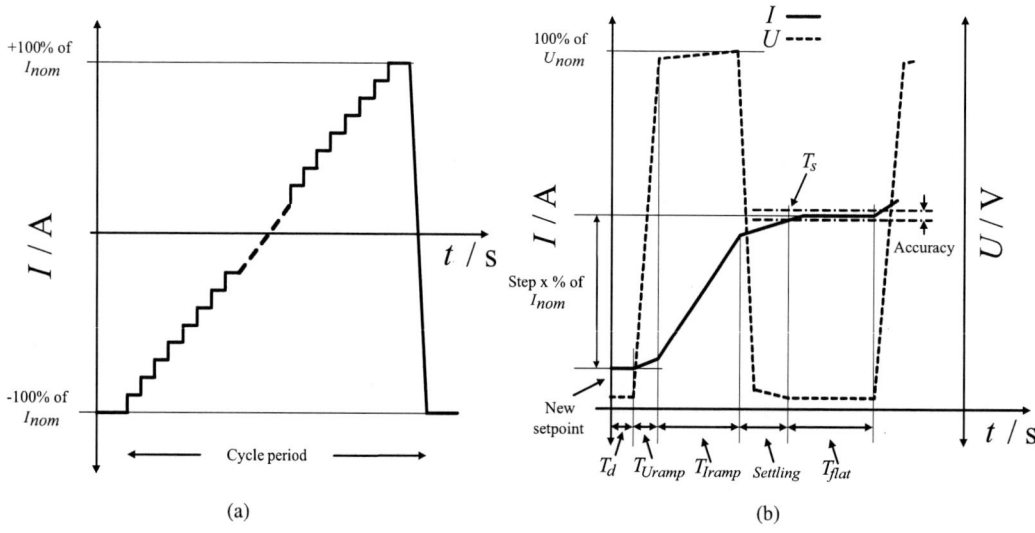

(a)   (b)

Fig. 1: (a) Current waveforms with stepped ramp current ranging from $-I_{nom}$ to $+I_{nom}$; (b) Close up to the vicinity of one current step with corresponding voltage waveforms [3].

Table I: Parameters and specifications for the modular power converter [3]. Summarized are settling time T_s, flattop time T_{flat}, switching frequency f_{sw}, nominal output current I_{nom}, DC-Link voltage U_{DC} as well as load inductivity L_m and resistance R_m and output quantity ripple metrics at worst case conditions (L_m = 340 µH, R_m = 125 mΩ) and 200 kHz.

Parameter	Value
Cycle period	20 ms
T_s	300 µs
T_{flat}	600 µs
I_{nom}	±15 A
U_{DC}	200 V
GaN Model	IGOT60R070D1
f_{sw}	100 kHz
L_m	340 µH up to 22 mH
R_m	125 mΩ up to 250 mΩ
U_{pp} ripple	200 ppm
I_{pp} ripple	10 ppm

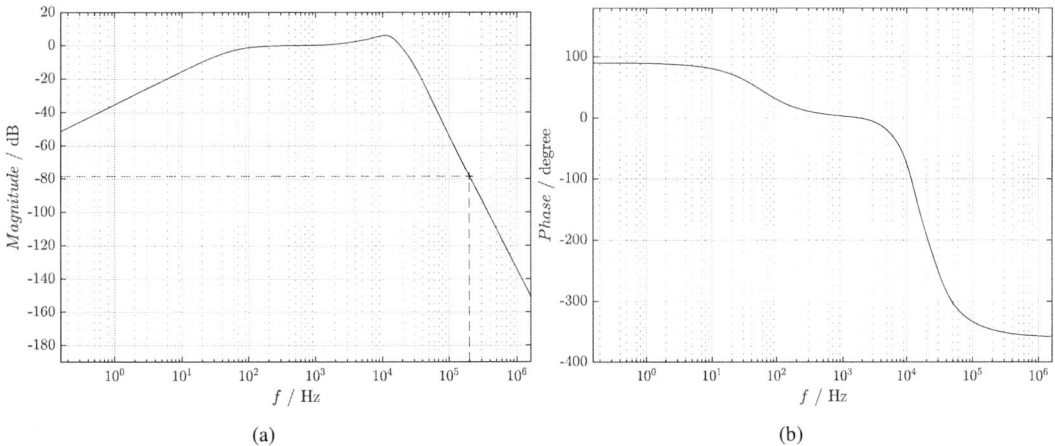

(a)   (b)

Fig. 2: Bode plots for the Bessel output filter cascaded to a minimum load model with corresponding amplitude plot (a) and phase plot (b) of the voltage transfer function. The 200 kHz point is marked with a + and dashed lines in the magnitude plot.

Fig. 4(a) provides an overview of the designed and assembled power converter. From left to right first the DC-Link with bulk capacitors of 1320 μF are visible. The next section integrates the designated connectors for the GaN4Q's. Furthermore, in the middle area the utilized XMC4700 is visible and the remaining area is a section designated for auxiliaries. The board integrates gallium nitride (GaN) components as switches for its H-bridge. Moreover, it has already proven high performance capabilities in initial testing [4].

Although a thorough comparison between different technologies to implement the power electronics stage is outside the scope of this paper, it is worth to analyze shortly the motivations that lead to the choice of the GaN technology. Given the combination of high transient voltage and high bandwidth, the components that fit best are the wide-bandgap (WBG) devices so called SiC and GaN. SiC allow higher power densities when higher currents and DC-link voltages are needed as they are typically 1200 V devices. GaN FETs are rated up to 600 V and at the present current capabilities and packaging solutions are shown to be optimal at power levels from some kW to some tens of kW [5]. Optimal selection of the device is however also related to the application. In the present proposal one of the advantages considered in selecting the GaN devices was the absence of the body-diode in the structure of the GaN. This leads to further reduced losses as there is no reverse recovery charge and improved EMI as well as contributes to reducing additional harmonics that may interfere with the performance of the LCLC filter. Switching energy is in principle also favourable for the GaN versus the SiC devices. This may allow in future designs to push the switching frequency to higher values, further reducing the overall voltage/current ripple and improving the control bandwidth capability. In the proposed prototype the switching frequency has been limited to 100 kHz to cope with off-the shelf characteristics of power chokes used in the filter and with specifications constraints.

A candidate for a power supply that is presumably capable of fulfilling the aforementioned requirements is the so called GaN4Q board as depicted in Fig. 5(a). In order to gain the ability for higher output powers and simultaneously reduce the output ripple to a reasonable amount, two of these boards are interleaved and placed within an industry standard Euro-crate 19" rack. A Bode plot for the utilised fourth order Bessel filters on these boards in combination with the minimum output load as characterised in Table I is depicted in Fig. 2(a) and (b). The two degrees of freedom design approach for the filter itself takes into consideration the worst case current ripple through the first inductor (L_1) stage and the desired cut-off frequency [6]. A value of $\pm 25\%$ of the maximum RMS output current of one GaN4Q board is assumed for that purpose. The filter transfer function can be expressed as

$$G(s) = \frac{k_1 \cdot s + 1}{k_5 \cdot s^5 + k_4 \cdot s^4 + k_3 \cdot s^3 + k_2 \cdot s^2 + k_1 \cdot s + 1} \tag{3a}$$

$$\text{with} \quad k_1 = R_D C_D, \tag{3b}$$

$$k_2 = L_1 C_1 + (L_1 + L_2)(C_2 + C_D), \tag{3c}$$

$$k_3 = L_1 C_1 R_D C_D + (L_1 + L_2) C_2 R_D C_D, \tag{3d}$$

$$k_4 = (C_2 + C_D) L_1 L_2 C_1, \tag{3e}$$

$$\text{and} \quad k_5 = L_1 L_2 C_1 C_2 R_D C_D. \tag{3f}$$

The corresponding topology of the filter stages is depicted in Fig. 3(a). Note: The filter components are split up in order to avoid short circuit paths for certain switch configurations. Hence, the given nomenclature adds a proceeding capital letter (A or B) to distinguish between the two GaN4Q boards and a succeeding number to distinguish between the individual branches within each filter structure. The harmonics injected due to the steep transients of the current trajectories steps has to be dissipated in the filter damping resistors R_D. Therefore, a significant limiting factor for possible minimum rise and fall times (and subsequently also the maximum closed loop bandwidth) is the maximum power that R_D is able to dissipate. In the case of the presented design and with the chosen resistors the dissipated power is not allowed to exceed 12 W per board. Fig. 4(b) shows an overview for the design of the modular power converter. The control of the individual GaN4Q's is conducted by means of an XMC4700 from Infineon which handles the voltage measurements for a closed loop control. As it is shown later in more detail, the integrated ADC of the XMC acts somewhat as a bottleneck among other effects. This can be traced back to its highest possible Bit-resolution of 12 Bit and considering a maximum measurable power supply output voltage of ± 200 V. Limit cycling effects for the closed loop are therefore investigated.

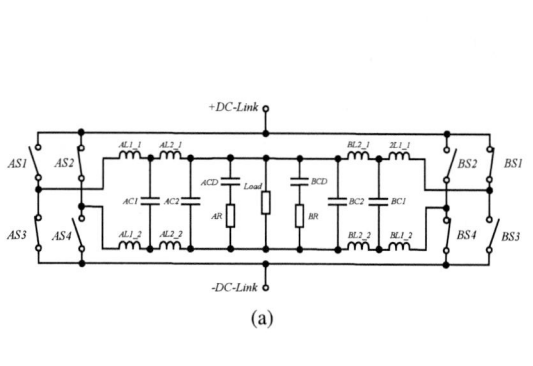

(a)

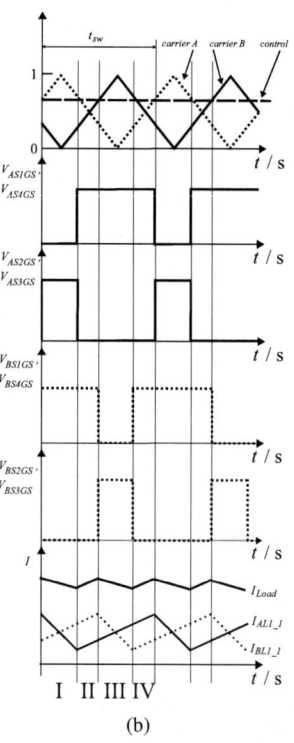

(b)

Fig. 3: (a) Two GaN4Q topologies interleaved in switching state I. (b) Simplified switches and the individual Bessel filters and the corresponding typical waveforms for the utilized modulation strategy. The switching states are numbered from I to IV. Dead times are not depicted.

Hardware Design and Interleaved Modulation

The utilized topology of the two H-bridges is illustrated in Fig. 3(a). The corresponding waveforms for the individual switches are presented in Fig. 3(b). As it can be seen, the governing property of the interleaved bipolar switching strategy are two carrier signals, phase shifted by 180°. Another major benefit of the implemented modulation can be seen in the load current. Double the switching frequency appears on the output of the power supply. Therefore, the filter structure can be designed smaller for the

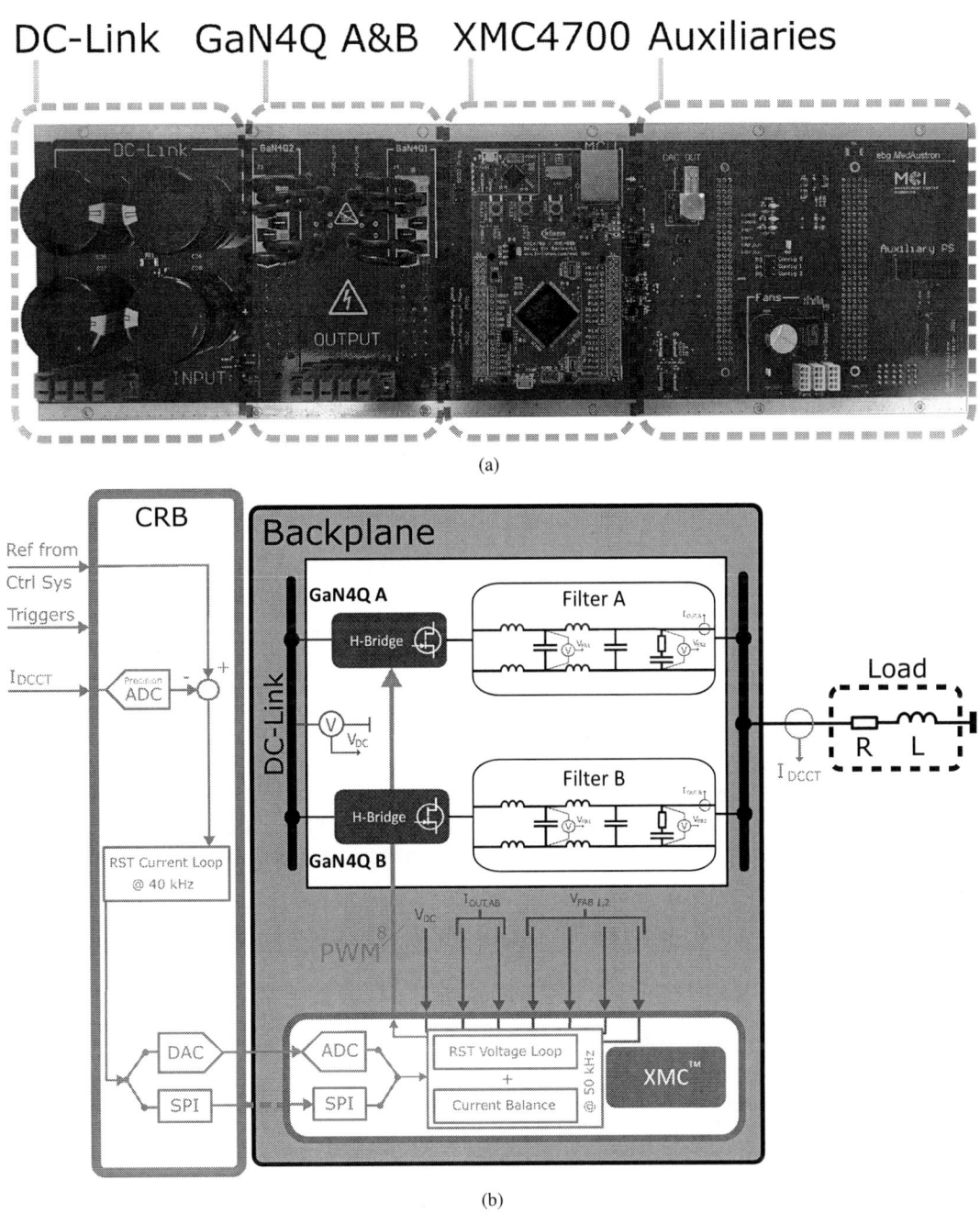

Fig. 4: (a) Backplane PCB with attached XMC4700 and section explanation overview; (b) Schematic power supply overview with backplane, two GaN4Q boards, DC-Link, CRB and corresponding interfaces.

(a)   (b)

Fig. 5: (a) Assembled GaN4Q prototype [4]; (b) Back view on the subrack with connected GaN4Q boards.

same ripple criterion since the unwanted frequencies are further pushed into the stop band of the utilized Bessel filters. A back view on the subrack is depicted in Fig. 5(b) that complies with the IEC 60297 standard.

RST-controller Methodology

The RST-controller is a feed-forward and feedback control structure as illustrated in Fig. 6(a). It consists of R, S and T polynomials to be synthesised by means of pole-zero cancellation and utilizing Diophantine equations [4, 7, 8]. The reason for using this type of controller is the relative heuristic approach for its design as is shown in this section. Another major reason are the current step waveforms as mentioned earlier. These can be seen as trajectories and therefore a feedforward controller is a tailored solution in this case [2]. The RST-controllers closed loop characteristic polynomial A_{cl} can be specified with

$$A_{cl} = A_m A_o, \tag{4}$$

where A_m is usually chosen as a second order transfer function and A_o is a so called observer polynomial that is introduced in order to account for uncertainties in the plant model G, that in turn is determined by means of system identification. The imposed closed loop dynamics for A_m are a damping factor ζ_m of 1 and a bandwidth f_m of 5 kHz. For A_o, the observer damping ζ_o is 1.2 and an observer frequency f_o of 5.5 kHz is chosen. All transfer functions including the one from the plant are given in digital signal processing (DSP) format (i.e., negative powers of z) and can be factorized as

$$G(z^{-1}) - \frac{B^- B^+}{A^- A^+}, \tag{5}$$

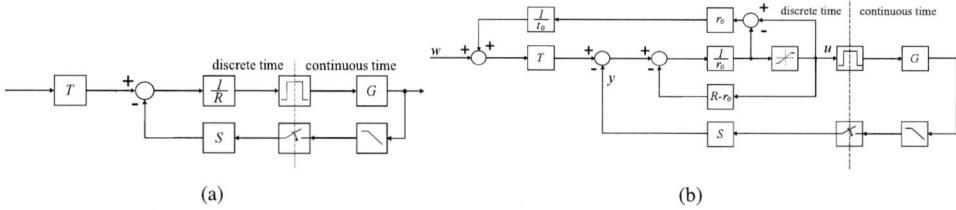

(a)   (b)

Fig. 6: (a) RST-control loop with a process model and sampling from continuous- to discrete-time and reconstruction. (b) Actual RST-controller implementation with command correction and control saturation.

where B^- and A^- contain all unstable zeros and poles that lie outside or on the border of the unit circle that have to be compensated. In contrast, B^+ and A^+ contain the poles and zeros that are stable, hence to be cancelled for the design process. The system identification by means of step data from the actual hardware results in a 9th-order transfer function that delivers the best results for controller design and subsequent utilisation in the actual hardware. That transfer function includes the interleaved GaN4Q boards, the filters, and the load. The load itself can be modelled as a first order filter according to the values for L_m and R_m presented in Table I. Controller constraints in terms of number of integrators are set [9]. In order to fulfil the steady state criterion, at least one integrator is included in the closed loop. The design process ends up with so called primary and auxiliary Diophantine equations and can be stated as

$$A^- R_1 R' + B^- S_1 S' = A_m A_o, \tag{6a}$$

$$T = B'_m A_o. \tag{6b}$$

Subsequently, the linear system of equations is solved for unknown R' and S' polynomials with (6a). R and S can now be computed. T is determined by (6b). The minimum order of the polynomials is determined by the order of the previously mentioned plant transfer function. Evidently, the resulting controller performance, especially for the actual system, is highly dependent on the identified plant model and the imposed uncertainties as well as the chosen observer polynomial. The actual control structure as it is implemented in software is illustrated in Fig. 6(b). For the present system only a voltage control loop is considered. Therefore, the developed power converter on its own can be seen as a voltage source where the current control can be implemented externally.

Measurement Analysis

Fig. 7(a) and 7(b) depict a step response to 50 V for an ohmic and ohmic-inductive load. The desired rise time of 300 μs can be reached and the maximum bandwidth can be stated as 5 kHz. The modular power converter driving desired current waveforms as shown in Fig. 1(a) through an ohmic-inductive load by means of a setpoint sequence are depicted in Fig. 8(b). The output voltages in that case are shown in Fig. 8(a). Furthermore, closed loop results are investigated for a minimum load with $L_m = 340$ μH, $R_m = 125$ mΩ and an added assumed connections resistance of 0.5 Ω. Hence, the measurement under these conditions for a desired step to 1.5 V is illustrated in Fig. 9(a) with the corresponding current in Fig. 9(b). Evidently, a strong ripple content is present in both measurements. It should be mentioned that the

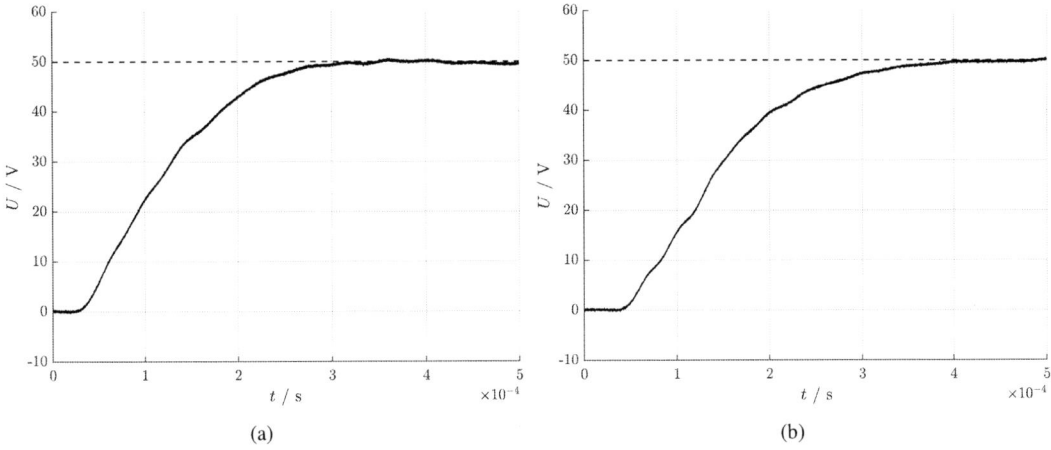

(a)   (b)

Fig. 7: (a) Closed loop voltage step response for an ohmic ($R_m = 6.1$ Ω) and (b) ohmic-inductive load ($L_m = 340$ μH, $R_m = 6.2$ Ω) with two GaN4Q boards (solid) to a reference of 50 V (dashed) and operated in interleaved mode. A moving average is used for the measured signals that corresponds to a cut-off frequency of 22 MHz.

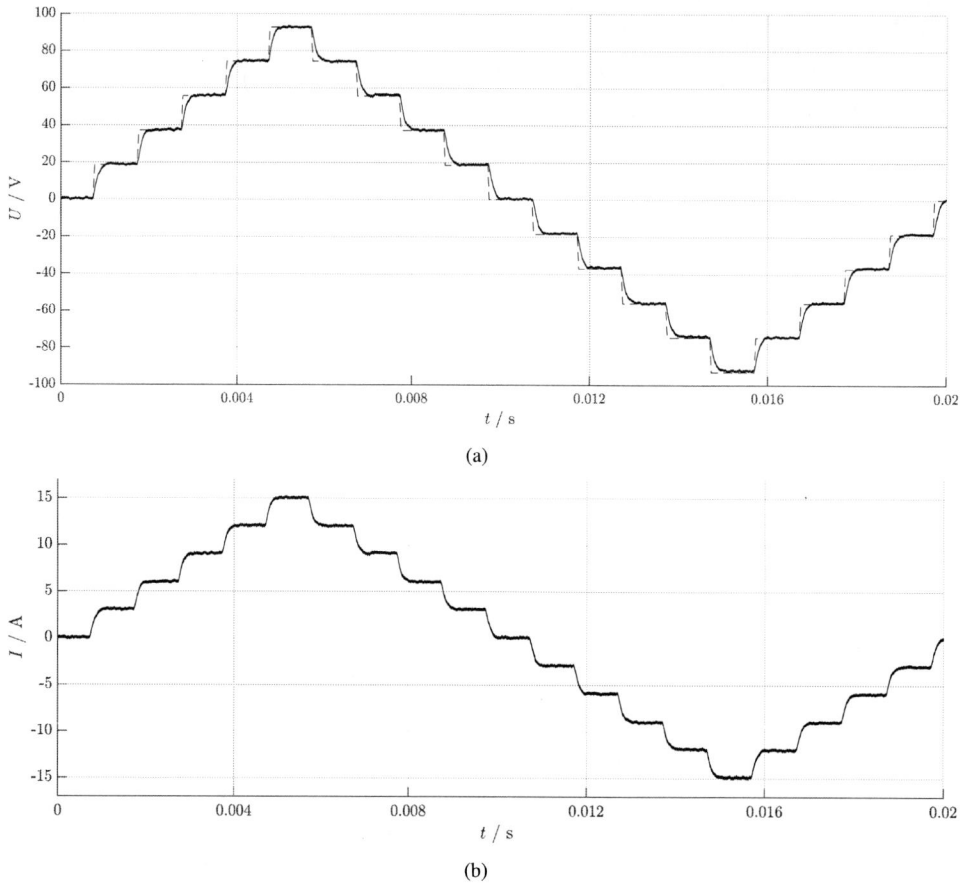

(a)

(b)

Fig. 8: (a) Staircase output voltage and (b) current waveform at 50 Hz for an ohmic-inductive load with two GaN4Q boards in interleaved mode. The output voltage setpoint waveform (dashed) is superimposed to the measured voltage waveform (solid). A moving average is used for the measured signals that corresponds to a cut-off frequency of 22 MHz.

amplitude in the current ripple would be significantly lower for an actual scanning magnet setup with a drastically increased inductance. This ripple is further investigated with a desired step to 0.5 V. A close up to a result for this condition is visible in Fig. 10. Investigating the utilized 12 Bit ADC gives a maximum resolution of 97 mV for the full-scale measurable voltage of 400 V. This leads to the assumption of limit cycle as a cause for the present unwanted contents in the signal. Further investigations revealed the PWM generation as the actual bottleneck. The utilized XMC4700 provides a clock frequency of 144 MHz and with the center aligned counter gives a resolution of 9 Bit for a desired PWM frequency of 100 kHz and an actual clock period of 16 ns. This Bit-resolution is in good accordance with the amplitude of the spikes visible in Fig. 10.

Conclusion

The capability of the modular power converter to be able to operate under trajectory tracking mode is confirmed with tests under various loads. For a first prototype, the power supply has shown capabilities that make it suitable of operating magnets with good static and dynamic responses. From the closed loop measurement analysis in Fig. 9(a),(b) and Fig. 10, two major points of improvement can be determined for the prototype and its controller. First, a dedicated 16 Bit SAR ADC shall be used. Furthermore, a microcontroller with a high resolution PWM option has to be implemented.

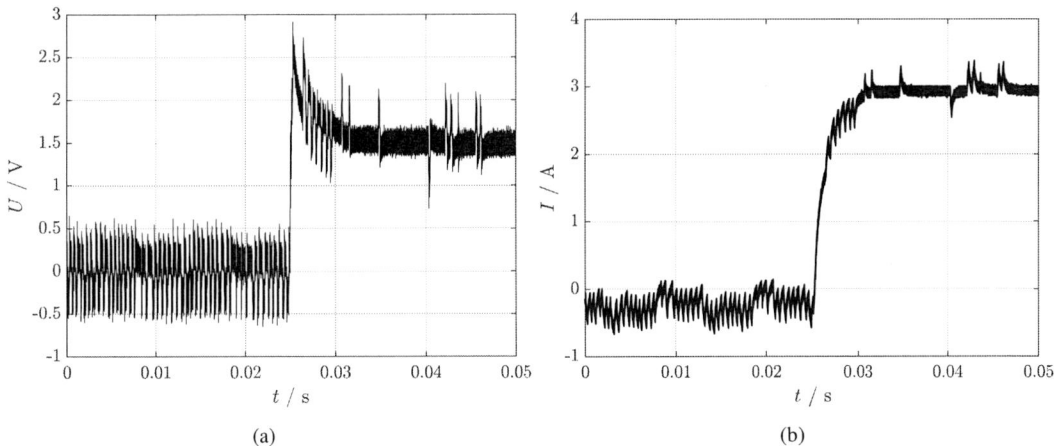

(a)   (b)

Fig. 9: (a) Voltage and (b) current step response for a minimum ohmic-inductive load condition with L_m = 340 µH, R_m = 125 mΩ and an assumed connection resistance of 0.5 Ω. Desired voltage closed loop parameters are a bandwidth of 1 kHz and damping of 0.7.

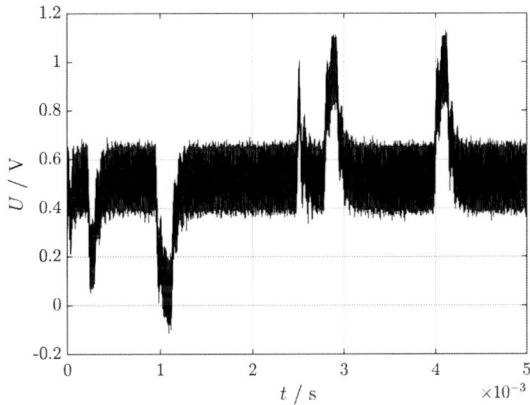

Fig. 10: Limit cycle voltage ripple close up.

References

[1] Linz U.: Ion Beam Therapy, Biological and Medical Physics - Biomedical Engineering, Springer, 2012

[2] Biagiotta L., and Melchiorri C.: Trajectory Planning for Automatic Machines and Robots, Springer, 2008

[3] Teixeira Alen R. and De Cesaris I.: Technical Specification Power Converter Family F, 2019EC010_POF-A_1906051_RTE, unpublished, EBG MedAustron GmbH, 2019

[4] Incurvati M., Margreiter T., Riedler T., and Stärz R.: High Precision Four Quadrant Converter with GaN Technology, IPAC, 2021

[5] Masoud Beheshti: Wide-bandgap semiconductors: Performance and benefits of GaN versus SiC, https://www.ti.com/lit/an/slyt801/slyt801.pdf?ts=1653423500177

[6] Künzi R.: Passive Power Filters, CERN proceedings of the CAS-CERN School, 2014, pp. 265-289

[7] Åström K., and Wittenmark B.: Computer-Controlled Systems: Theory and Design, Prentice Hall, 1990

[8] Godoy E., and Ostertag E.: RST-Controller Design: A Rational Teaching Method based on Two Diophantine Equations, IFAC, 2006

[9] Veenstra M., Beuret A., De Cesaris I., and Fraboulet P.: High-Performance Digital Control of Magnet Power Supplies. unpublished, EBG MedAustron GmbH, 2011

Battery cycler to generate open li-ion cell aging data and models

Matthias Luh, Thomas Blank
Institute for Data Processing and Electronics (IPE), Karlsruhe Institute of Technology (KIT)
Hermann-von-Helmholtz-Platz 1
76344 Eggenstein-Leopoldshafen, Germany
Tel.: +49 / (0) 721 608-29167
E-Mail: Matthias.Luh@kit.edu, Thomas.Blank@kit.edu
URL: https://www.ipe.kit.edu/english/

Acknowledgements

This work is funded by the German Research Foundation (DFG) as part of the Research Training Group 2153 "Energy Status Data – Informatics Methods for its Collection, Analysis and Exploitation". We thank Dr. Fabian Jeschull and his colleagues from the Institute for Applied Materials (IAM) at KIT for performing the comparative electrochemical impedance spectroscopy and interpreting its results.

Keywords

Batteries, Lifetime, Power cycling, Battery impedance measurement, Gallium Nitride (GaN)

Abstract

Battery degradation is relevant for the lifetime, cost, and life cycle analysis of electric vehicles and stationary storages. Publicly available, reusable battery aging data is scarce and aging experiments are time-consuming and expensive. This paper starts with an overview of existing battery aging data and models. We then present our battery cycler hardware, which we intend to use to generate open battery degradation data and models. After analyzing the switching behavior, efficiency, and control behavior of the hardware, this paper gives an overview of the generated data and the user interface of the battery cycler. Finally, we provide an outlook on the further development of the project.

Introduction

Batteries play an increasingly important role in the CO2 reduction targets introduced to mitigate climate change: They are used in electric vehicles (BEV, HEV, FCEV) and stationary storage systems (home, industrial, and large-scale grid energy storage systems). Battery aging is an important aspect to consider when developing or using cells, battery systems, and applications that utilize these batteries, regardless of the cell chemistry (e.g., different kinds of lithium, sodium or lead-acid cells). The degradation of battery cells limits lifetime and, therefore, the overall cost of the application. It also affects the environmental balance and the carbon footprint of the product using the battery.

However, battery degradation is complex and dependent on many aspects. The aging of a lithium-ion cell (which is probably the most important and widespread cell type at present and in the near future) can be further differentiated into calendar aging and cyclic aging. Cycling aging occurs when the cell is charged or discharged (i.e., cycled). It is dependent on the depth of discharge (or, more specifically, the maximum and minimum voltage or State of Charge (SoC) when charging and discharging the cell), the temperature of the cell, the charging / discharging rate (i.e., the current), and the number of cycles. Calendar aging occurs at all times, even when the cell is not used, and is therefore dependent on time, but also the temperature and the voltage/SoC at which the cell is idling [1], [2]. A review of the different electrochemical aging mechanisms can be found in [3] and [4].

Modeling cell aging by considering each electrochemical effect (e.g., the temperature-dependent electrolyte decomposition) might help improve cell chemistries. However, it is too complex and afflicted with many unknown factors for product designers that select a battery cell for their application, analyze the impact of different modes of operation on the selected battery cell, or run an aging model in the product's battery management system (BMS). Instead, heuristic mathematical models can be used for these purposes. However, cell manufacturers usually provide little information on cell aging and the influence of the parameters mentioned before. To acquire such models, an (accelerated) aging measurement of cells cycling or resting under different operating conditions can be performed. The resulting data set can then be used to fit mathematical models (e.g., the "NREL" [5] or more sophisticated models [6]) or use artificial intelligence to generate a cell aging model. However, this process is both time-consuming and very expensive because not only many cells but also cell cycler equipment with many channels are required for the cell aging measurements and deep expertise in cell aging is beneficial. Large EV manufacturers or suppliers have the means and also a substantial financial interest to run those experiments (for example, to define the battery warranty conditions). However, these companies are not interested in publishing these measurement results or only publish results that are not sufficient to derive a reusable cell model [7]. The publications with the best insight into aging results that we found are:

- A comparison of a 240 mAh "LiNi$_{0.5}$Mn$_{0.3}$Co$_{0.2}$O$_2$ / artificial graphite (NMC532/AG)" pouch cell ("million-mile" battery) [8] with a commercial, cylindrical 2.05 Ah Sanyo UR18650E cell (previously published in [9])
- A comprehensive aging experiment using commercial, cylindrical Sanyo UR18650W cells with 1.5 Ah capacity and a "LiNi$_{1/3}$Co$_{1/3}$Mn$_{1/3}$ + LiMn$_2$O$_4$ composite cathode" [6], [10] as well as a similar study from the same research group using commercial, cylindrical A123 26650 LiFePO$_4$ cells with 2.2 Ah capacity [11]

While the authors provide impressive results and include many graphs showing the capacity fade of the cells under different operating conditions, no raw data of the measurements are available. It is not well possible and subject to many uncertainties to use the graphs and information given in the paper to deduce a cell aging model that can be used in other investigations, e.g., to determine the capacity fade of an EV battery under different operating conditions in simulations. Due to an absence of comprehensive, public, reusable battery degradation data and models, very simple aging models are often used in research and small and medium enterprises when analyzing the aging of cells in applications. For example, some scientific authors that analyze the influence of electric vehicle charging or Vehicle-to-Grid (V2G) assume that there is a constant number of charging and discharging cycles and, optionally, a maximum lifespan of the battery in years [12]. Others assign linear and quadratic terms to energy- and power-related battery aging costs [13]. Therefore, depending on the battery model and aging assumptions used, statements about the economic viability of V2G range from "the potential profit […] was outweighed by the cost of battery degradation" [14] to "no increased" or even "decreased aging" due to V2G [15]. Uncertainty regarding battery degradation may even prevent applications like Vehicle-to-Grid from becoming widespread and gaining acceptance, despite its enormous potential macroeconomic benefit.

We aim to produce a comprehensive cell aging dataset and model that is accessible to the public and can be reused in other investigations. For this purpose, we would like to select a representative cell that is comparable in its cell chemistry and electrical properties to cells typically used in electric cars. Part of these cells shall be exposed predominantly to cyclic aging, the other only to calendar aging. Additional cells could be exposed to cycles typical for the application, e.g., WLTP driving profiles or representative V2G patterns. The cells shall age under different parameters. For cyclically aged cells, the temperature (T_{cyc}) as well as the current rate ($C_{cyc\pm}$) and voltage limits for charging/discharging ($V_{chg/dischg}$) shall be varied. For calendar-aged cells, the temperature (T_{cal}) and the voltage (V_{cal}) at which the cell rests shall be varied. For reasons of redundancy and a better statistical significance, even when using cells of the same batch, it is recommended to age several cells (N_{red}) at the same parameter set. Using a relatively small set of parameters, the total number of cells quickly becomes very large, e.g., for $T_{cyc}=T_{cal}=4$, $C_{cyc\pm}=4$, $V_{chg/dischg}=3$, $V_{cal}=4$, $N_{red}=3$, the total number of cells is:

$$N = N_{red} \cdot \left(T_{cyc} \cdot C_{cyc\pm} \cdot V_{chg/dischg} + T_{cal} \cdot V_{cal} \right) = 192 \qquad (1)$$

As commercial cyclers for a large number of cells quickly become very expensive, we aim to develop our own battery cycler hardware. A prototype of this hardware and the first results we gathered while developing the system are presented in this paper. The battery degradation data collected with the hardware will be subject to another publication after the aging measurements ran long enough to draw robust conclusions about aging.

Experimental setup

We developed a battery cycler prototype (shown in Fig. 1) that can charge and discharge four individual battery cells with up to 4.5 V and 7.5 A, control their temperature using different pools with a liquid cooling/heating fluid, capture the cell voltages, currents, and temperatures, and store the acquired measurement data in an easily accessible, reusable format both locally and online. In the final setup, we plan to split up the functions of the board (cycling & measurement, temperature control) into separate boards and allow each cycler board to control up to 12 cells, which is more practical and cost-effective.

The block diagram in Fig. 2 gives a schematic overview of the testbench. The prototype board, including the cycling and measurement hardware ("slave board"), is shown on the left side. It is controlled by an Infineon Aurix TriCore TC375 processor with software we developed using the TASKING TriCore Software Development Toolset.

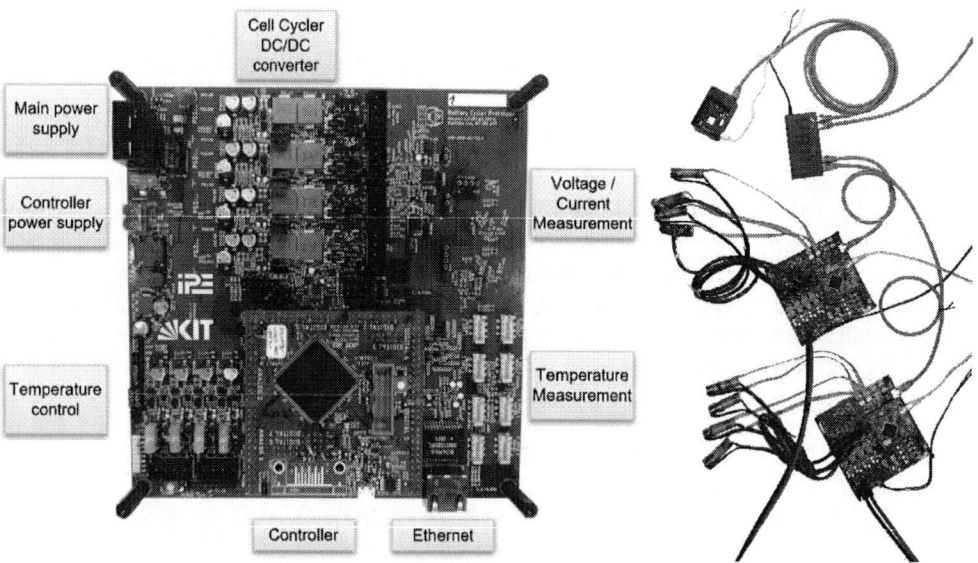

Fig. 1: Battery Cell Cycler Prototype Board (left) and exemplary setup with two cycler boards (right)

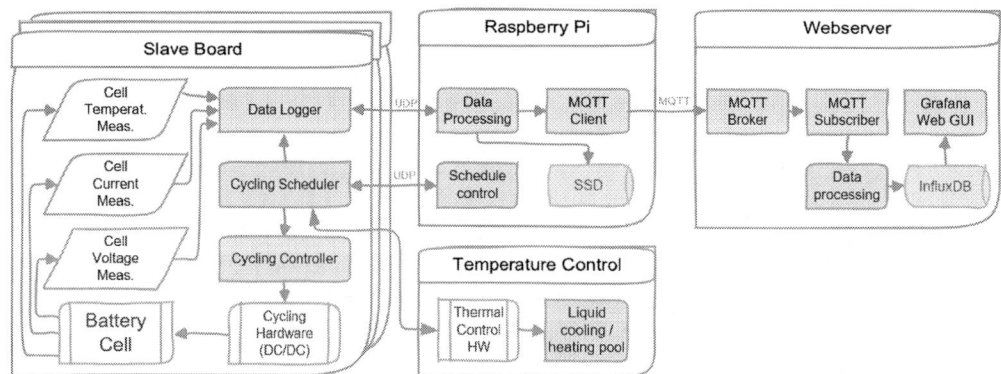

Fig. 2: Cycler testbench hardware/software components and communication structure

For each cell, an individual non-isolated DC/DC converter comprising a half-bridge is used, which adjusts the common input voltage to the voltage required to charge and discharge the cell. Gallium nitride (GaN) and silicon (Si) MOSFETs (EPC2045 from EPC Co. and RQ3E100BN from Rohm) are compared on the prototype board to examine the most suitable type for the final board. A two-stage LC-LC filter filters the output of the DC/DC converter to obtain a smooth cell voltage and current. Fig. 3 shows a simplified schematic of the converter.

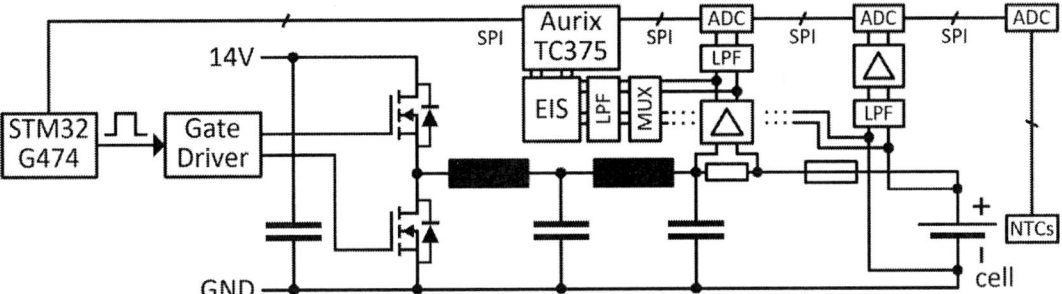

Fig. 3: Simplified schematic of the DC/DC converter with two-stage LC-filter and cell measurements

While the Aurix processor is very powerful for processing and communication, its minimum PWM timer resolution is only 10 ns. Using a switching frequency of 500 kHz, this corresponds to a 0.5 % duty cycle resolution. If an input voltage of 14 V is selected for the DC/DC converter, the open circuit output voltage could be adjusted approximately in 70 mV steps, which is too coarse for the battery cycler. Instead, the gate drivers are controlled by the high-resolution timer module (HRTIM) of an additional STM32G474 controller. The HRTIM can generate up to 12 PWM signals with a duty cycle resolution of 183.8 ps or 0.00919 %, resulting in converter output steps of approximately 0.13 mV. Using two processors per board makes the design more expensive but also safer since both processors must allow the converter to operate and each can stop cycling in case of an error or a timeout.

The cell voltages are acquired with a resolution of ca. 69 µV in a range from 0 to 4.5 V using a differential voltage measurement circuit. A four-point measurement for separate power flow and voltage measurement contacts with four individual cables and tabs welded to the two poles of the cylindrical cells is used. A shunt and a current-sense amplifier with fast overcurrent detection are used to measure a cell current of up to ±11.25 A with a resolution of ca. 343 µA. Shunts with higher resistances can be used to increase accuracy in case lower cell currents are sufficient for cell cycling. There are two temperature measurement circuits per cell using NTCs connected to the casing as well as the negative pole contact of the cylindrical cells.

While the cycling controller handles the combined constant current and voltage (CC/CV) control, the cycling scheduler is responsible for the higher-level schedule. It determines when the cells are cycled, initiates check-ups in which the remaining capacity and aged impedance of the cells are measured every few days to weeks, issues a shutdown due to faults, and, if possible, handles an automatic restart.

A data logger module captures every relevant measurement and state every two seconds, stores the data on an SD card on the cycler board, and sends this data set to a Raspberry Pi via Ethernet. Multiple slave boards are connected to a Raspberry Pi. In the final setup, independent slave and Raspberry Pi boards shall be used to perform redundant measurements of similar parameter sets. On the Raspberry, the data set is stored on a local SSD and forwarded to a web server via MQTT. The web server unpacks the MQTT message and stores the individual measurements into an InfluxDB database. A Grafana web interface can be conveniently used to access the database and visualize real-time or previous measurements in a web browser remotely using customizable dashboards.

We included an option to measure the complex, frequency-dependent cell impedance using electrochemical impedance spectroscopy (EIS). Instead of an analog circuit that discharges the cell with a small sinusoidal current, the STM controller outputs a PWM pattern that excites the cell with a sinusoidal current using the DC/DC converter. Analog multiplexers connect the low-pass filtered current and differential voltage measurement signals of one cell at a time to the EIS unit. The amplitude and phase characteristics of the voltage and current measurement low-pass filters are matched to the same -3 dB cut-off frequency of approximately 33 kHz to reduce their effect on the EIS measurement while still significantly damping switching noise. Op-amps are used to hold the maximum / minimum voltage and current values and amplify their difference for an ADC. Schmitt triggers allow the processor to measure their phase angle to calculate the complex impedance.

Preliminary results

The battery degradation measurements have not started yet and will be performed on a successor board. We plan to publish the collected raw data along with post-processed measurements and a deduced aging model of the observed cell after the cells aged on the final testbench for at least ½ to 1 year and conclusions about aging behavior can be drawn. Nevertheless, in this publication, we will present first measurement results using the prototype board, including measurements of the DC/DC converter switching and filtering behavior comparing GaN and Si MOSFETs, the behavior of the CC/CV controller, and the electrochemical impedance spectroscopy.

Switching Behavior

Fig. 4 shows the switching behavior and output signals of the DC/DC converter used for cycling (charging/discharging) the cell. In the example, the cell is discharged with 5 A and has a terminal voltage of approximately 3.5 V. Three configurations of the half-bridge are compared: One using RQ3E100BN silicon MOSFETs (left column), one with EPC2045 GaN MOSFETs using the same gate resistor values (center column) and a half-bridge using the same GaN MOSFETs but with optimized gate resistor values and a snubber circuit at the output of the half-bridge (2.7 nF, 2.2 Ω).

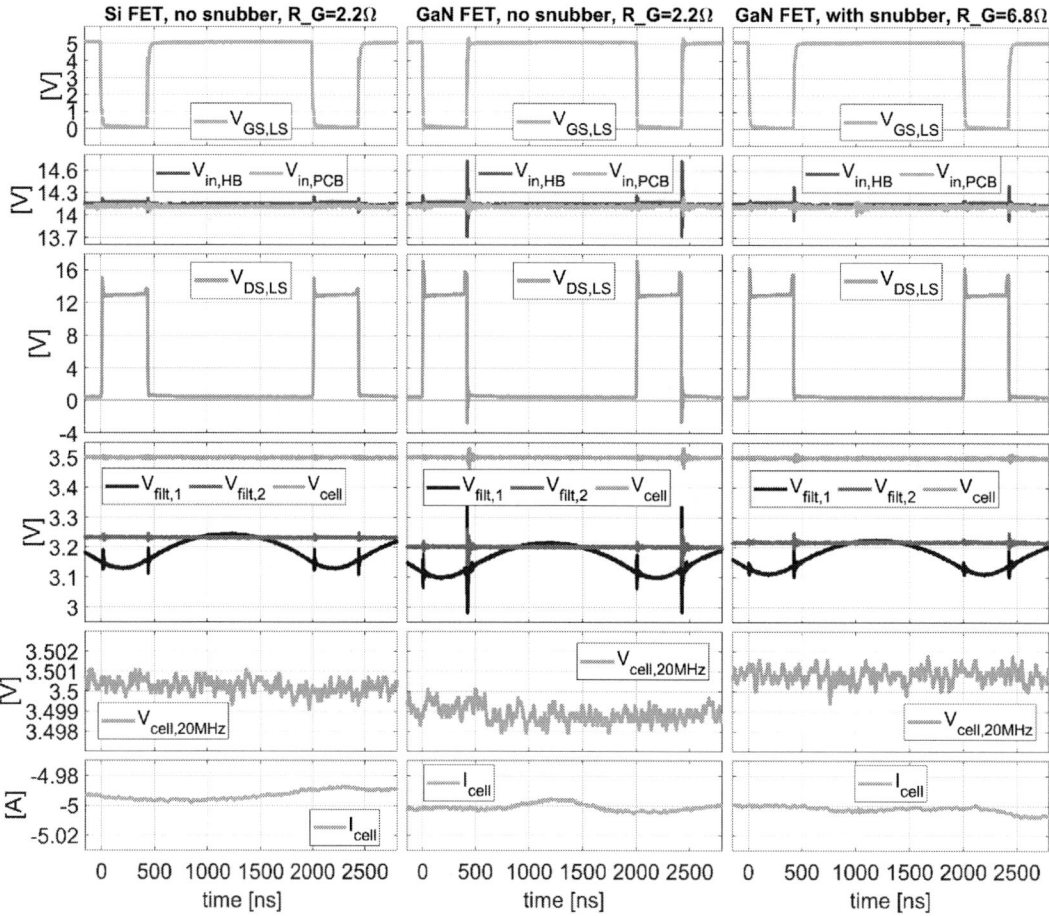

Fig. 4: Switching behavior and output signals of the cell cycling DC/DC converter

The signals were captured with a Tektronix MSO58 oscilloscope with Tektronix TPP0500B voltage probes and a Pico Technology TA018 current probe. The upper row shows the gate-source voltage of the low-side FET (turquoise), followed by the DC link voltage right in front of the examined half-bridge (red) and at the input of the battery cycler PCB (yellow), the drain-source voltage of the low-side FET (orange), the voltage at the output of the DC/DC filter behind the first LC filter (dark blue) as well as behind the second LC filter (pink) and right at the battery cell (cyan). The bottom row shows the battery cell current (green). The battery cell voltage signal is additionally recorded in a 20 MHz bandwidth-limited measurement. The comparison of the two cell voltage signals shows that most of the noise lies in the frequency range above 20 MHz.

The switching behavior is compared in more detail in Fig. 5. It can be seen in both figures that the drain-source voltage V_{DS} is significantly overswinging when turning off and on the GaN MOSFET, while the Si MOSFET only generates relatively small ringing. The oscillations are damped by the LC filters but still influence the battery cell voltage. A higher gate resistor value ($R_{G,ext}$ = 6.8 Ω instead of 2.2 Ω) was chosen for the GaN half-bridge to limit the drain-source voltage slew rates, as can be seen in Fig. 5 and Table 1. While the higher gate resistance and snubber in the gallium nitride half-bridge reduced oscillations, the silicon half-bridge still produced a less noisy cell voltage, especially during turn-on operations. In addition, the reverse voltage drop across the body diode of the silicon FET is smaller, which results in lower conduction losses during switching transitions.

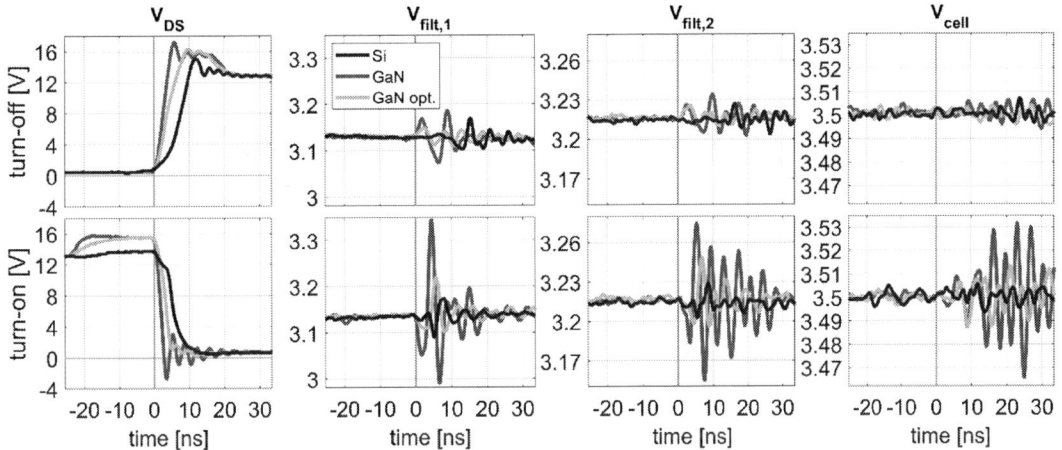

Fig. 5: Detailed comparison of switching behavior and output signals for turn-off (top row) and turn-on (bottom row) of the low-side silicon MOSFET with R_G = 2.2 Ω (dark blue), the GaN MOSFET with R_G = 2.2 Ω without snubber (red) and the GaN MOSFET with R_G = 6.8 Ω and with snubber (yellow)

Table 1: V_{DS} slew rates of the low-side MOSFET in different half-bridge configurations

Half-bridge configuration:	Si, no snubber	GaN, no snubber		GaN, with snubber	
	R_G = 2.2 Ω	R_G = 2.2 Ω	R_G = 6.8 Ω	R_G = 2.2 Ω	R_G = 6.8 Ω
turn-off	+1.9 V/ns	+3.8 V/ns	+3.5 V/ns	+3.0 V/ns	+2.6 V/ns
turn-on	−4.0 V/ns	−6.9 V/ns	−5.8 V/ns	−6.5 V/ns	−4.8 V/ns

Efficiency

The DC/DC converter was optimized for a dynamic, highly accurate low-noise operation at moderate efficiency. A comparison of the efficiencies of different half-bridge configurations at various output voltages for an output current of +3 A (charging) and -5 A (discharging) is shown in Fig. 6. The efficiency was measured by a Hioki PW3390 power analyzer using high accuracy Hioki CT6862-05 current sensors using a range of 15 V and 5 A. The voltages were measured at the input power supply plug and the cell output plug, so all losses on the PCB (including converter, filters, current measurement shunts, fuses, relay, PCB traces, and plugs) are included, but the cell and power supply cable losses are not.

As is typical for a buck converter, the efficiency at constant input voltage (here: 14 V) decreases with decreasing output voltage and increasing output current. The efficiency is about 1.5 % higher when no snubber is used. The efficiencies of Si and GaN FETs without snubber are relatively similar even though the typical $R_{DS,on}$ datasheet value of the silicon FET (11.0 mΩ) is about twice as high as the one of the GaN FET (5.6 mΩ). However, conduction losses in the converter are dominated by the filter and the cell fuse (ca. 35 mΩ).

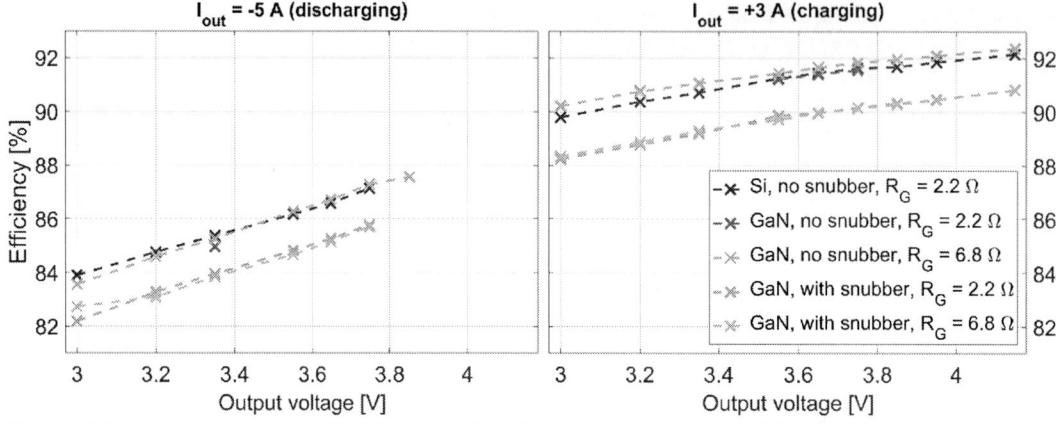

Fig. 6: Efficiency comparison of different half-bridge topologies at $V_{in} = 14$ V

Controller performance

The CC/CV controller is capable of a seamless transition from constant current to constant voltage. It considers the input voltage, the estimated open-circuit voltage of the cells, the cell resistance measured in the last check-up, and a fixed combined resistance for cables, PCB, filter, and MOSFETs to reach the desired output current quickly. The control algorithm runs on the Aurix processor with a frequency of 10 kHz for each cell. It calculates the desired PWM duty cycle and sends it to the STM controller via SPI. A running counter value and a CRC8 are used to detect faults in the transmission. The STM controller applies the duty cycle to the PWM running at a frequency of 500 kHz as soon as it receives and decodes the message. One PWM signal per cell is sent to an On Semi NCP81151B synchronous buck gate driver, which is connected to the two MOSFETs of the half-bridge. The PWM signals for the cells are shifted in phase to minimize the overall input current and voltage ripple.

Turn-on and turn-off procedures with an initial cell voltage of 3.5 V and a setpoint of 3.7 V / 3.0 A (charging) as well as an initial cell voltage of 3.7 V and a setpoint of 3.5 V / -5.0 A (discharging) are shown in Fig. 7. The controller reaches the desired output current after 1-3.5 ms for the first time and stabilizes the current after approximately 4-5 ms. Between 100 and 200 ms after turn-on, the current ripple measured with the oscilloscope settles down to less than ±67 mA, the voltage ripple after the first LC-filter to ± 70 mV, and the ripple after the second cell to less than ±3 mV (including measurement noise). In steady-state, the 20 MHz bandwidth limited signal of the cell voltage has a ripple of less than ±1 mV and the cell current noise is below ±15 mA. The voltage and current signals used for data acquisition and control, as well as for the EIS circuit, are low-pass filtered to further reduce switching-induced ripple and other noise.

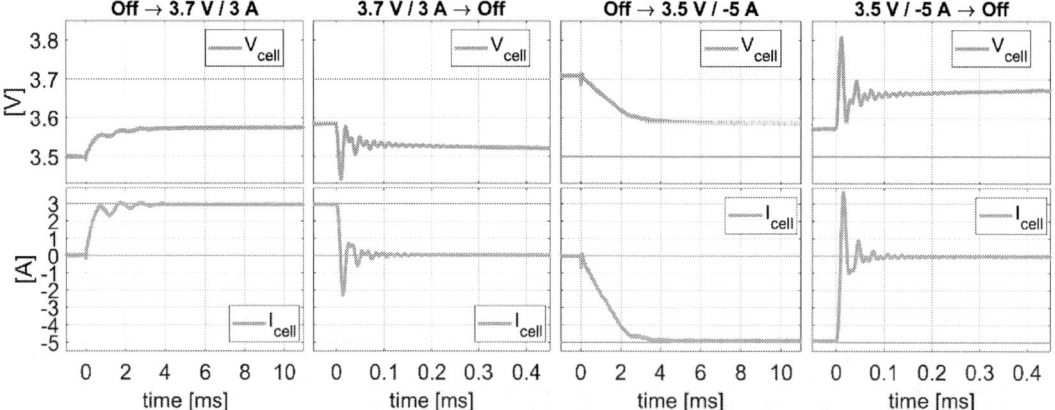

Fig. 7: Cell voltage (cyan) and current (green) when turning on the controller with a charging current of 3 A (leftmost column) and when turning off afterwards (second column) as well as when turning on with a discharging current of -5 A (third column) and turning off afterwards (rightmost column)

In Fig. 8, the same controller parameters were applied as before, but the battery cell voltage before turn-on is closer to the desired voltage limit. Hence, the controller operates in CV mode. The cell voltage limit is not violated, but the controller needs approximately 10-20 ms to reach the voltage setpoint.

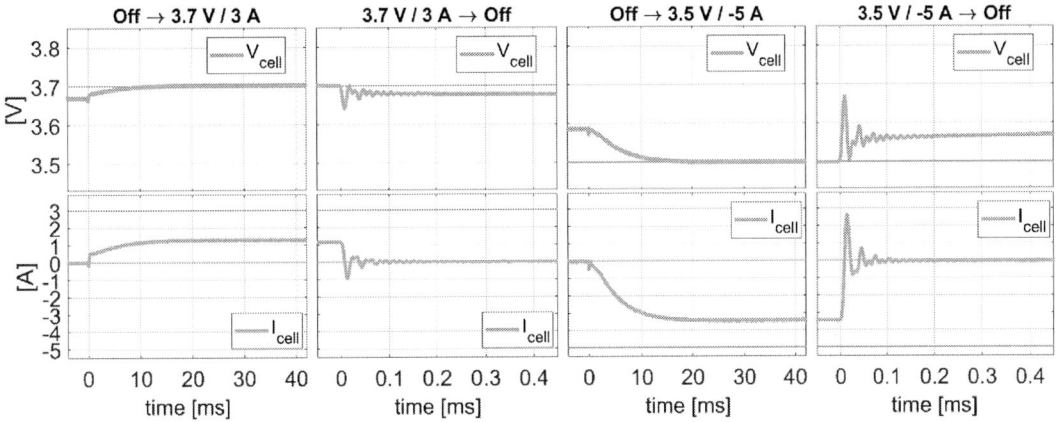

Fig. 8: Same signals and controller settings as in Fig. 7, but this time the controller runs into the current limit and smoothly transitions into constant voltage (CV) mode

Accuracy

We compared the cycler's voltage and current measurements with commercial high-accuracy devices for a few exemplary measurement points. A more comprehensive range of accuracy measurements across the whole temperature, voltage, and current range and all cell channels was not performed.

When the battery cycler was inactive, one indicated cell voltage from the cycler was 3.4231 V, while we measured 3.4246 V with a Keithley 2470 source meter in the off state, a deviation of 1.5 mV or 0.044 %. In two other instances, the cell voltage was overestimated by 0.6 mV and 1.2 mV or 0.017 % and 0.033 %, respectively. When the cycler was discharging a cell with a set current of -5 A, the cycler indicated an average output current of -5.0000 A, while the Hioki PW3390 power analyzer using CT6862-05 current sensors measured -4.9952 A, a deviation of 4.8 mA or 0.096 %. When charging with a set current of +3 A, the cycler indicated an average current of +3.0000 A, while the power analyzer measured +2.9988 A, a deviation of 1.2 mA or 0.04 %.

Electrochemical Impedance Spectroscopy

We performed an Electrochemical Impedance Spectroscopy (EIS) for an LG INR18650HG2 cell with our battery cycler and a commercial BioLogic VSP at different temperatures (2°C, 12°C, 22°C, 37°C) and open-circuit voltages (3.5, 3.8 V). The resulting impedance curves are shown in Fig. 9. In general, the curves match relatively well, indicating that the EIS functionality implemented on the cycler works.

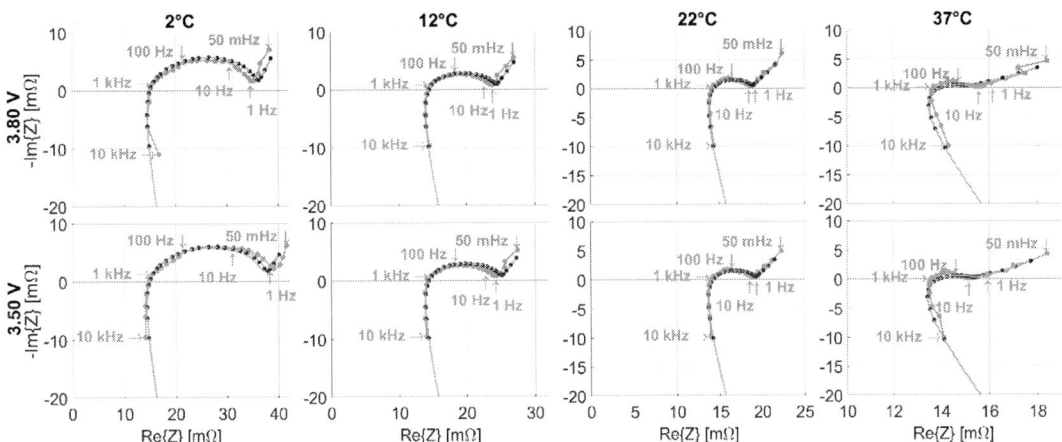

Fig. 9: EIS comparison using our battery cycler (orange) and a commercial BioLogic VSP (blue)

However, significant deviations occurred at high and very low frequencies (above 5 kHz and below 200 mHz). The average absolute amplitude error for all data points in the figure is 1.54 %, with a standard deviation of 1.46 % and a maximum error of +13.17%. The average absolute phase error is 1.03°, with a standard deviation of 0.95° and a maximum error of 4.30°.

Although the EIS results suggest the cycler is unsuitable for high-precision measurements, it is conceivable to use it for qualitative analyses, e.g., to observe the change in impedance over the cell's life.

Data acquisition and visualization

As described before, collected data can be visualized online using the open-source tool Grafana. The data is ready to be displayed approximately 2 seconds after being collected. An example dashboard showing data of a single cell captured with the prototype is shown in Fig. 10: In the left column, the cell current (orange), voltage (dark blue), and an estimation of the open-circuit voltage (light blue) as well as the power (red) is shown. The center column shows the cell temperature (green), the estimated state of charge (yellow), as well as the added charge (light purple) and energy (dark purple) since the charging or discharging began. Important recent measurements and states are summarized in the right column.

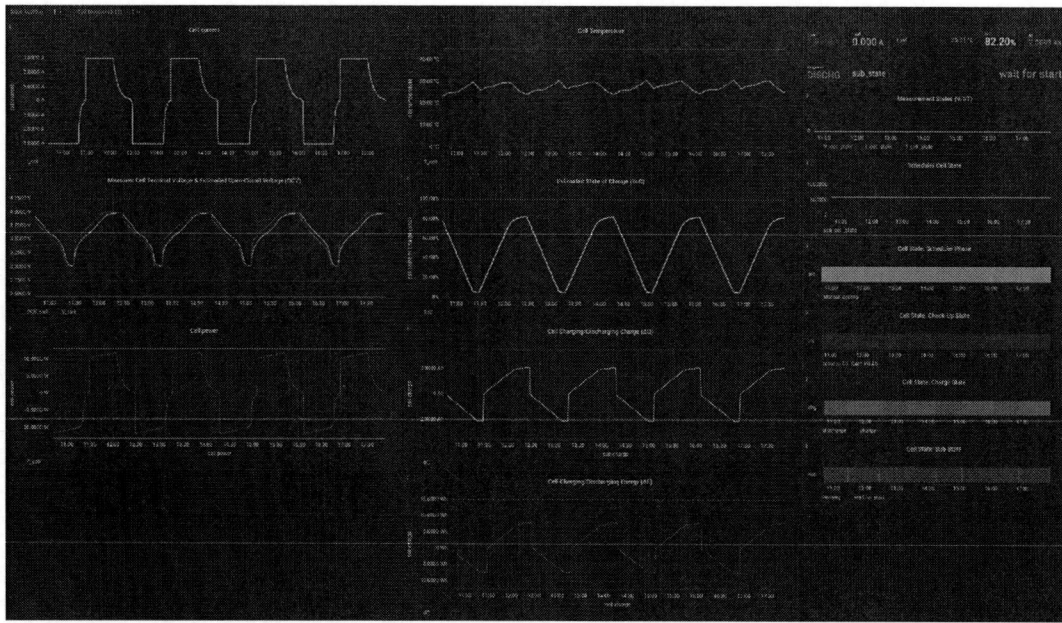

Fig. 10: Visualization of cell measurements in a web browser using Grafana.

Additional dashboards exist for the thermal management measurements and states as well as for the EIS and long-term battery aging results. On the latter, the charge and energy extracted and charged per cycle, as well as the coulomb and energy efficiency, are shown. In addition, the total charge and energy extracted over the life of the cell and the number of cycles can be displayed.

Conclusion

Battery aging is a critical aspect when estimating lifetime and cost or conducting a life cycle analysis of EVs and stationary storage systems. A lack of freely available, easily reusable cell aging data and models imposes many challenges and uncertainties when investigating these technologies and their use cases. Published cell aging measurements that are to some extent suitable for generating a specific cell aging model were summarized at the beginning of this paper.

We presented the prototype of a new battery cycler that we intend to use to cost-effectively perform comprehensive cell degradation experiments on li-ion battery cells to publish aging data and models that can be reused in other studies and for product development.

We compared the switching performance and efficiency of GaN and Si MOSFETs for the DC/DC converter to charge and discharge the cells. The silicon FETs cause fewer oscillations in the output voltage and are slightly more efficient and easier to handle in our application than the BGA GaN FETs. Therefore, we prefer the Si MOSFETs for our final design. We presented the performance of the CC/CV controller, which is capable of dynamically and accurately controlling the cell current and voltage within less than 20 ms. Afterwards, we give an insight into the electrochemical impedance spectroscopy that we have implemented on the cycler and show that it matches relatively well with that of a commercial device. Finally, we presented the online data visualization tool which can be used to monitor all cells.

Outlook

While the prototype was used to analyze different hardware and software components of the cycler, we are currently developing a more practical, cost-optimized version that can reliably and efficiently cycle more cells fully automated. Last but not least, we still have to carry out the actual experiment in which cells will be operated over a period of several months to years and analyze the measurements to generate models for degradation, which can be used in calculations and simulations. Once a model for a specific cell is derived, evaluating it with realistic profiles of different applications is desirable. It is also conceivable to analyze if the model can easily be adjusted to cells with other capacities, form factors, or chemistries using a much smaller set of reference measurements combined with the original model.

References

[1]   Xu B., Oudalov A., Ulbig A., et al.: Modeling of Lithium-Ion Battery Degradation for Cell Life Assessment, IEEE Transactions on Smart Grid, vol. 9, no. 2, pp. 1131–1140, 2018, DOI 10.1109/TSG.2016.2578950

[2]   Keil P.: Aging of Lithium-Ion Batteries in Electric Vehicles, Technische Universität München, 2017

[3]   Vetter J., Novák P., Wagner M.R., et al.: Ageing mechanisms in lithium-ion batteries, Journal of Power Sources, vol. 147, no. 1, pp. 269–281, 2005, DOI 10.1016/j.jpowsour.2005.01.006

[4]   Alipour M., Ziebert C., Conte F.V., et al.: A Review on Temperature-Dependent Electrochemical Properties, Aging, and Performance of Lithium-Ion Cells, Batteries, vol. 6, no. 3, 2020, DOI 10.3390/batteries6030035

[5]   Smith K., Warleywine M., Wood E., et al.: Comparison of Plug-In Hybrid Electric Vehicle Battery Life Across Geographies and Drive-Cycles, SAE Technical Papers, 2012, DOI 10.4271/2012-01-0666

[6]   Purewal J., Wang J., Graetz J., et al.: Degradation of lithium ion batteries employing graphite negatives and nickel–cobalt–manganese oxide + spinel manganese oxide positives: Part 2, chemical–mechanical degradation model, Journal of Power Sources, vol. 272, pp. 1154–1161, 2014, DOI 10.1016/j.jpowsour.2014.07.028

[7]   Belaid S., Mingant R., Petit M., et al.: Strategies to Extend the Lifespan of Automotive Batteries through Battery Modeling and System Simulation: The MOBICUS Project, in 2017 IEEE Vehicle Power and Propulsion Conference (VPPC), 2017, DOI 10.1109/VPPC.2017.8330949

[8]   Harlow J.E., Ma X., Li J., et al.: A Wide Range of Testing Results on an Excellent Lithium-Ion Cell Chemistry to be used as Benchmarks for New Battery Technologies, J. Electrochem. Soc., vol. 166, no. 13, pp. A3031–A3044, 2019, DOI 10.1149/2.0981913jes

[9]   Ecker M., Nieto N., Käbitz S., et al.: Calendar and cycle life study of Li(NiMnCo)O2-based 18650 lithium-ion batteries, Journal of Power Sources, vol. 248, pp. 839–851, 2014, DOI 10.1016/j.jpowsour.2013.09.143

[10] Wang J., Purewal J., Liu P., et al.: Degradation of lithium ion batteries employing graphite negatives and nickel–cobalt–manganese oxide + spinel manganese oxide positives: Part 1, aging mechanisms and life estimation, Journal of Power Sources, vol. 269, pp. 937–948, 2014, DOI 10.1016/j.jpowsour.2014.07.030

[11] Wang J., Liu P., Hicks-Garner J., et al.: Cycle-life model for graphite-LiFePO4 cells, Journal of Power Sources, vol. 196, no. 8, pp. 3942–3948, 2011, DOI 10.1016/j.jpowsour.2010.11.134

[12] Lehtola T. and Zahedi A.: Cost of EV battery wear due to vehicle to grid application, in 2015 Australasian Universities Power Engineering Conference (AUPEC), 2015, DOI 10.1109/AUPEC.2015.7324824

[13] Schuller A., Dietz B., Flath C.M., et al.: Charging Strategies for Battery Electric Vehicles: Economic Benchmark and V2G Potential, IEEE Transactions on Power Systems, vol. 29, no. 5, pp. 2014–2022, 2014, DOI 10.1109/TPWRS.2014.2301024

[14] Hoke A., Brissette A., Maksimović D., et al.: Electric vehicle charge optimization including effects of lithium-ion battery degradation, in 2011 IEEE Vehicle Power and Propulsion Conference, 2011, DOI 10.1109/VPPC.2011.6043046

[15] Rosekeit M., Lunz B., Sauer D.U., et al.: Bidirektionales Ladegerät für Elektrofahrzeuge als Energiespeicher im Smart Grid - Bi-directional Charger for Electric vehicles as Energy Storage in the Smart Grid, 2012

Function Blocks of a Highly-Integrated All-in-GaN Power IC for DC-DC Conversion

Michael Basler, Richard Reiner, Stefan Moench, Patrick Waltereit, and Rüdiger Quay
Fraunhofer Institute for Applied Solid State Physics IAF
Tullastraße 72
Freiburg, Germany
Tel.: +49 761 5159 - 196
Fax: +49 761 5159 - 71196
E-Mail: michael.basler@iaf.fraunhofer.de
URL: www.iaf.fraunhofer.de

Acknowledgements

This work was supported by the Fraunhofer Internal Programs under Grant No. PREPARE 840 229.

Keywords

«Emerging technology», «Gallium Nitride (GaN)», «Monolithic power integration», «Power integrated circuit», «DC-DC converter».

Abstract

GaN-on-Si technology is on the advance for the use in power ICs thanks to wide bandgap performance combined with a lateral structure and a low-cost carrier substrate. A common GaN power IC platform with several active and passive devices, as well as analog and digital circuits is presented. This platform is used to integrate periphery function blocks of power electronics circuits, such as driving, sensing, protection, and control. In detail, an efficient GaN-based gate driver with integrated bootstrap capacitances is realized, which has a current consumption of only 2.2 mA at a supply voltage of 5 V and a switching frequency of 1 MHz. Furthermore, a GaN-based voltage mode control is described based on a PWM generator with error amplifier and verified with measurements. Finally, the circuit design of an all-in-GaN power IC for DC-DC conversion with half-bridge, driver, level shifter, dead time and voltage mode control is presented. With small additional chip area, further function blocks can be integrated to the power device(s) to realize a modern highly-efficient and highly-functional GaN-based conversion component for the next generation of power electronics.

Introduction

Gallium nitride (GaN) technology is increasingly used in power integrated circuits (ICs) due to its superior physical properties combined with its lateral structure. These characteristics enable the rapid development of a new generation of power electronics, progressively reducing the size, losses and costs of these systems and their applications. In this way, first GaN power ICs on low-cost Silicon (Si) substrates have been successfully launched on the market [1–3]. Usually, Si-based BCD (Bipolar-CMOS-DMOS) technologies are used for power or power management ICs. However, the next generation of power IC platforms require more area-efficient power devices. This offers the GaN-on-Si technology with a similar range of active and passive devices to give designers the greatest possible flexibility. GaN power ICs consisting of power devices with additional periphery or function blocks such as gate driver, sensing, protection circuity and even control can lead to an increased performance.

The development of GaN power ICs was initiated over a decade ago by academia and then driven and accelerated by industry, with GaN power ICs recently being made commercially available. Especially the companies Efficient Power Conversion EPC, Navitas Semiconductor, and Innoscience as well as start-ups like Wise-Integration, GaNPower International, and Cambridge GaN Devices CGD offer GaN

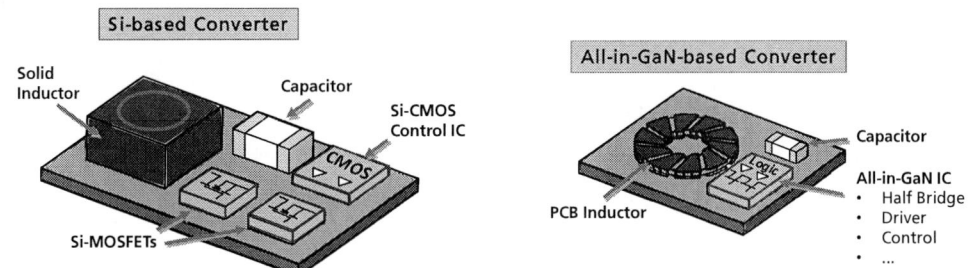

Fig. 1: Low-voltage DC-DC converter: (Left) State of the art with Si power MOSFETs, discrete Si-based driver and control and (right) vision of the future with an all-in-GaN-based converter with GaN power IC, which integrated power devices, driver and control.

ICs with several function blocks such as gate driver, sensing and protection circuits [1]. In particular, EPC's ePower™ Stage includes a half-bridge with driver, level shifter, logic and under-voltage lockout (UVLO) [3]. This GaN IC should be suitable for 48 V systems in battery-powered and motor applications, in the automotive or industrial sector, and in IT infrastructures. The interest in the introduction of 48 V system is the reduction of power consumption and cabling volume compared to 12 V solutions. Usually low-voltage DC-DC converters or point-of-load (PoL) converter consist of two Si-based power MOSFETs configured to a half-bridge with drivers and control in CMOS technology as well as a solid power inductor, shown in Fig. 1. Recent DC-DC converters with higher power density resulting from higher switching frequencies mostly use GaN-based power semiconductors as a single device or as a half-bridge [4]. The next step for even more compact systems with higher functionality is the integration of the control into the GaN IC, in addition to the power devices and drivers, as an all-in-GaN-based converter, shown in Fig. 1.

This work presents two function blocks, a gate driver with integrated bootstrap capacitors, and a GaN-based voltage mode control of a monolithic all-in-GaN power IC for DC-DC conversions up to 48 V. The used GaN technology is described and the design of the two function blocks as well as the all-in-GaN power IC is presented. Measurements of the functional parts are shown. The all-in-GaN IC shows a higher level of integration into GaN technology with higher functionality and power density in the future.

GaN Power IC Platform

The epitaxial structures of the GaN-on-Si technology starts with a conductive Si substrate. A multilayer GaN buffer and an AlGaN barrier are grown above. The AlGaN/GaN heterojunction establishes a highly conductive two-dimensional electron gas (2DEG) with a sheet resistance of ~600-800 $\Omega/\square$, which is used as drift and depletion zone for the high-electron-mobility Transistors (HEMT) and other devices. A p-GaN layer caps the devices and is structured into the p-GaN gate. The p-GaN layer shifts the threshold voltage of the standard Schottky gate into positive values. Four metallization with corresponding passivations are used for ohmic contacts, interconnections and field plates. Fig. 2 shows a simplified cross-section of the GaN power IC platform with several active and passive devices.

The active devices are enhancement-/depletion-mode (e-/d-mode) HEMTs, Schottky barrier diodes (SBDs), and lateral field-effect rectifiers (LFERs) [5] for low-voltage (LV) or high-voltage (HV) applications from 12 – 650 V. To increase the operating voltage, the depletion zone must be extended, which in turn increases the drift zone and the on-resistance. The field plate allows the depletion zone to be reduced while maintaining the same breakdown voltage. As an example, the gate-width scaled on-resistance result in 6.5 $\Omega\times$mm for a 12 V-class and 15.1 $\Omega\times$mm for a 650 V-class e-mode HEMT. Layout structures are also being investigated that enable area-efficient power devices for low-voltage applications [6].

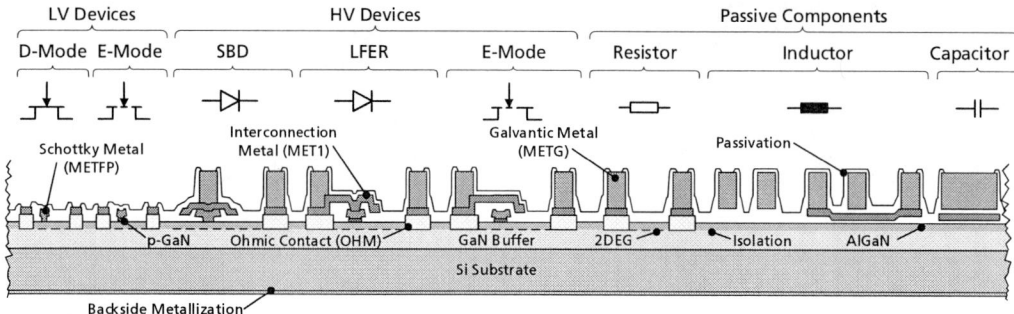

Fig. 2: Simplified cross-section of the GaN-on-Si technology as GaN power IC platform. In the platform, there are several active devices such as e-/d-mode HEMTs, Schottky barrier diodes (SBDs), and lateral field-effect rectifiers (LFERs) as low-voltage (LV) or high-voltage (HV) devices. In addition, there are passive components such as resistors, inductors, and capacitors.

Passive components can also be integrated into the technology. Resistors can be realized by the four metal layers or the 2DEG. On-chip capacitors can be realized as metal-insulator-metal (MIM) structures or can be extended by stacking resulting in a capacitance density of $0.22 - 1.16$ fF/µm^2. The p-GaN gate capacitors offer the counterpart to MOS capacitors from Si-based IC technologies. This capacitor has a high capacitance density of 1.19 fF/µm^2, but it is highly non-linear with a low voltage range from the threshold voltage up to ~6.5 V. In addition, on-chip spiral inductors are designed with inductances of <50 nH. The use as power inductors was investigated in [7] with associated thermal considerations.

The work uses a digital and analog library previously created by the authors, which includes area-efficient standard cells for layout design, as well as schematic symbol and simulation models based on measurement data, which allows a complete design flow. The technology is limited to the direct-coupled FET logic (DCFL) due to the lack of p-type devices. Unlike CMOS logic, DCFL has large static currents, limiting the amount of GaN logic that can be practically integrated. The digital library (NOT, NOR, NAND, AND, OR, RS flip-flop, Schmitt-Trigger) have a miniaturized standard cell with a height of 50 µm, supply voltage of 5 V, max. static currents of ≤0.3 mA, and high noise margins. The analog building blocks include differential amplifiers and comparators with e-/d-mode differential pair for sensing, protection circuits or analog controls. In addition, voltage references with and without temperature compensation, current mirrors and linear regulators are also included in the analog library [8].

With these devices and building blocks in this GaN power IC platform, power electronic circuits with power devices and additional function blocks such as gate driver, protection and sensing circuits and even control can be realized and monolithically integrated.

GaN-Based Gate Driver

GaN power devices with integrated gate drivers allow to increase the switching frequency by reducing the gate loop inductance. Furthermore, critical ringing and overshoot can be reduced. There are different gate driver concepts: first approaches started with a push-pull stage [9], over approaches with additional logic inverters or NOT gates [10] up to drivers with boot-strapping concepts [10, 11] and even with dV/dt control [12, 13], temperature compensation [14], and over-current protection (OCP) [15–18]. All these gate drivers have the same push-pull output stage (consisting of two e-mode transistors in series). The push-pull stage has the disadvantage, if the pull-up transistor is turned-on with the supply voltage V_{DD}, that the output swing is reduced by the threshold voltage $V_{TH,E}$. There are three possibilities to avoid the output voltage drop and to ensure a rail-to-rail output: First, V_{DD} can be increased, which in turn causes higher gate stress and additionally degradation as well as positive V_{DD}-shift of the p-GaN gates [10]. Second, the output voltage is pulled-up to V_{DD} with a bootstrap circuit with one supply voltage [10, 11]. Last possibility, the pull-up transistor is either turned-on with $2 \times V_{DD}$ with an additionally generated second supply voltages [15], e.g. also realized by bootstrapping.

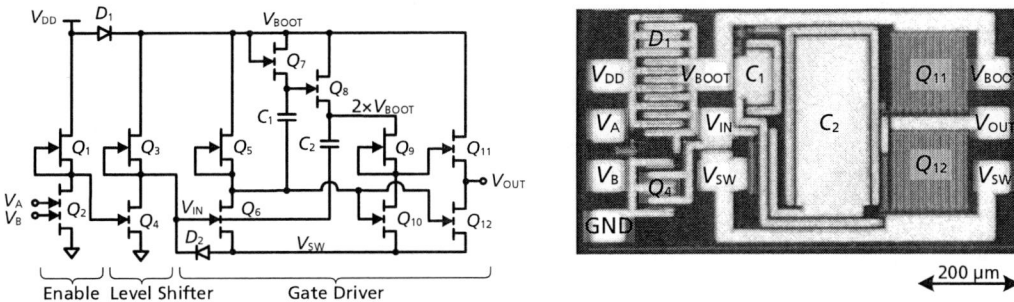

Fig. 3: (Left) Schematic and (right) chip photo of a bootstrap-based gate driver additionally with level shifter, bootstrap diode for bootstrap supply for high-side gate driver and input logic for enabling.

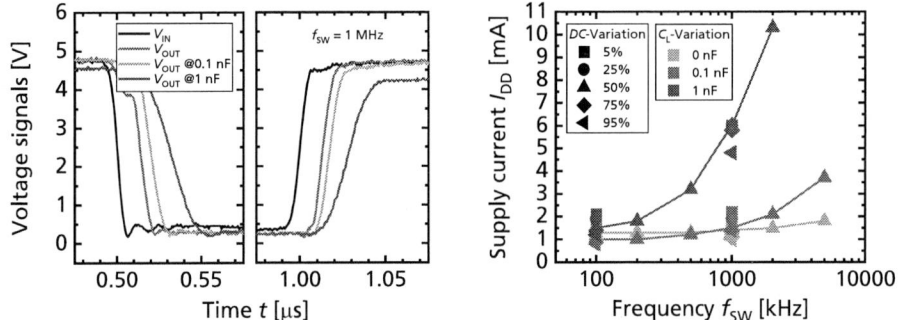

Fig. 4: (Left) Falling and rising edge of the gate driver with f_{SW} = 1 MHz, V_{BOOT} = 5 V and three different load capacitors of C_L = 0/0.1/1 nF. (Right) Power supply current I_{DD} as function of the frequency f_{SW} with different duty-cycles and load capacitors.

In the GaN power IC platform of this work, a gate driver has been implemented according to [15]. Fig. 3 shows the schematic and layout of the bootstrap-based gate driver additionally with a level shifter, bootstrap diode for high-side gate driver and an input logic for enabling. The circuit is realized on a small chip area of 865×455 µm² with pads. The transistors of the gate driver are all e-mode HEMTs of the 12 V-class. Here Q_5 and Q_6 form a logic inverter, Q_7, Q_8 with C_1 and C_2 a bootstrap circuit and Q_9-Q_{12} a single input inverted buffer. This bootstrap circuit provides a voltage of $2 \times V_{DD}$ instead of $2 \times V_{DD}$-V_{T0} for the not gate of the single inverted input buffer, whereby V_{DD} is here equal to V_{BOOT} and V_{T0} is the turn-on voltage in case of only a bootstrap diode. The stacked MIM capacitor C_1 and C_2 have a capacitance of approx. 5 pF and 50 pF. The gate widths of the push-pull stage are 3 mm and of the two NOT gate 10/100 µm for pull-up/down device. The gate driver can be used for high-side (HS) and low-side (LS) and has been designed and optimized for low power-consumption.

Fig. 4 shows the measured falling and rising edge of the input V_{IN} and output V_{OUT} of the fabricated gate driver with f_{SW} = 1 MHz, V_{BOOT} = 5 V and three different load capacitors of C_L = 0/0.1/1 nF. Without load capacitor, the propagation delay for on $t_{D,ON}$ and off $t_{D,OFF}$ is about 13 ns each, resulting mainly from the multi-stage approach. These delays increase with larger load capacitance: 17.32 ns and 19.71 ns for 0.1 nF and 28.55 ns and 30.65 ns for 1 nF. Fig. 4 also shows the power supply current I_{DD} as function of the frequency f_{SW} with different duty-cycles DC and load capacitors. The supply current increases with increasing frequency and load capacitor. However, as DC increases, the current decreases, because more energy is shorted at the gate. The max. switching frequency of the gate driver is >5 MHz. The typical input capacitance of the main GaN power transistors of the half-bridge is about 100 pF. At a switching frequency of 1 MHz, the max. supply current with 0.1 nF is 2.2 mA, which is efficient. The quiescent supply current is about 1.6 mA, mainly from the static currents of the logic circuit which is realized without a CMOS technology. In comparison, the EPC2152 ePower™ Stage with GaN half-bridge driver, logic and UVLO has a quiescent current of 22 mA and a current consumption of 29 mA at a switching frequency of 1 MHz [19]. Due to the high quiescent currents in GaN compared to CMOS drivers, an external series cut-off circuit is proposed in [20]. Compared with CMOS drivers, the quiescent currents are increased, but the current consumption with higher frequencies is smaller with GaN drivers [21, 22].

GaN-Based Voltage Mode Control

The voltage mode control (VMC) is the most widespread control method of a synchronous buck converter due to its simplicity. The output voltage of the power stage is compared with a reference voltage V_{REF} and the duty-cycle is adjusted based on the error. There are already publications on GaN-based control, such as voltage mode control with duty-cycle switcher [16] or the cycle-by-cycle peak current control [17]. Also PWM generators based on a comparator and sawtooth generator have been published [23–25]. For closed loop control, a compensation network for the error amplifier is used to ensure stable operation of the synchronous buck converter.

In the GaN power IC platform of this work, a voltage mode control has been implemented according to [24] with error amplifier, which are shown in Fig. 5. The sawtooth generator in turn consists of a hysteresis comparator with two external resistors R_1, R_{FB} for the feedback network and a charging unit (NOT gate) with external sawtooth capacitor C_{SAW}. The error amplifier as well as comparator also for the sawtooth generator is a two-stage amplifier consisting of a differential amplifier with source follower, which were already presented in [8]. The circuit is realized on a not fully filled test IC with chip area of 1×1 mm^2 and pads for higher flexibility. The transistors and diodes are all devices of the 12 V-class.

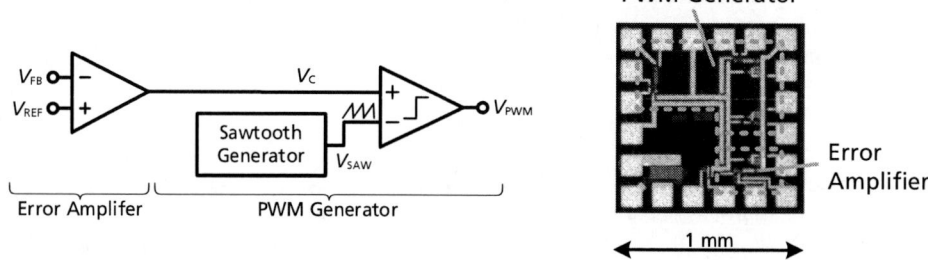

Fig. 5: (Left) Schematic and (right) chip photo of a voltage mode control based on an error amplifier and PWM generator consisting of a sawtooth generator and comparator.

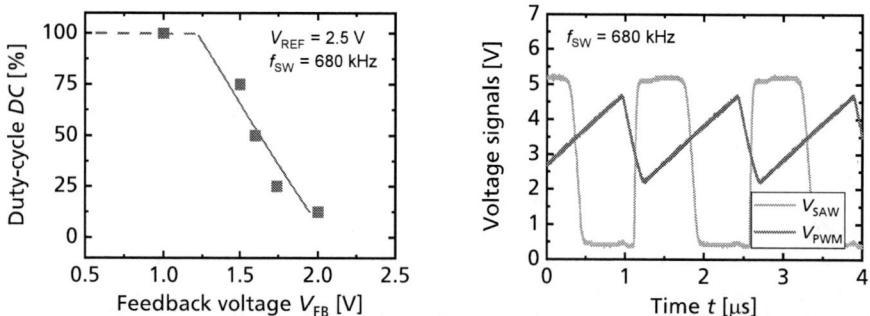

Fig. 6: (Left) Duty cycle DC of the PWM generator as function of the feedback voltage V_{FB} of the error amplifier at five measured points and linear fit over the $DC < 100\%$. (Right) Sawtooth and PWM time signals with duty-cycle of 50%.

The sawtooth generator has the following external components: $R_{FB} = 100$ kΩ, $R_1 = 15$ kΩ, $C_{SAW} = 470$ pF. These components can change the sawtooth frequency f_{SW} as well as the amplitude of the sawtooth as analyzed in [23–25]. Depending on the values, the amplitude of the sawtooth increases up to a maximum of 6 V with increased sawtooth capacitance and covers sawtooth frequencies of about 0.5-1.5 MHz. Fig. 6 shows the measured and resulting duty-cycle DC of the PWM generator as function of the feedback (input) voltage V_{FB} and $V_{REF} = 2.5$ V at five measured points and a linear fit over the $DC < 100\%$. The measured frequency $f_{SW} = 680$ kHz results from the R_{FB}, R_1, C_{SAW} dimensioning. The min. DC is limited by the falling edge of the sawtooth and is 12% in this operating point. Fig. 6 also shows the PWM and sawtooth time signals at $DC = 50\%$ and $f_{SW} = 680$ kHz. The PWM generator and the error amplifier require a supply voltage of 12 V. The current consumption is ~10 mA ±2 mA depending on the operation point. The power consumption of this circuit can be further reduced by smaller dimensioning of the resistors, or using other means to increase the resistance (for lower static consumption) of the logic devices.

All-in-GaN Power IC for DC-DC Conversion

The all-in-GaN power IC integrates a symmetrical half-bridge, driver for HS and LS switch, bootstrap diode, level shifter, dead time and voltage mode control. This GaN IC was designed for a synchronous buck converter with DC-DC conversion up to 48 V. The chip is realized on an area of 3×2.5 mm² with the presented GaN power IC platform and is shown in Fig. 7.

The half bridge consists of HS and LS HEMT with a gate width of 105 mm, a channel length of 4.7 μm and comb layout. The half-bridge is suitable up to 48 V with breakdown voltage over 80 V. The HS gate driver is shown in Fig. 8. Both gate drivers are based on a DCLF inverter Q_1/Q_2, a bootstrapped DCFL inverter Q_3/Q_4, D_2, C_1 and a push-pull stage Q_5/Q_6 (different to the previously presented GaN-based gate driver). The push-pull stage has the disadvantage, if the pull-up transistor is turned-on with the supply voltage V_{DD}, that in the high-level output voltage has a voltage drop of the threshold voltage. A bootstrap circuit of the previous DCFL inverter consisting of diode D_2 and capacitor C_1 avoids this voltage drop,

Fig. 7: GaN Power IC for a synchronous buck converter consisting of a half-bridge with driver, level shifter, dead time control, and voltage mode control. (Left) Schematic and (right) chip photo.

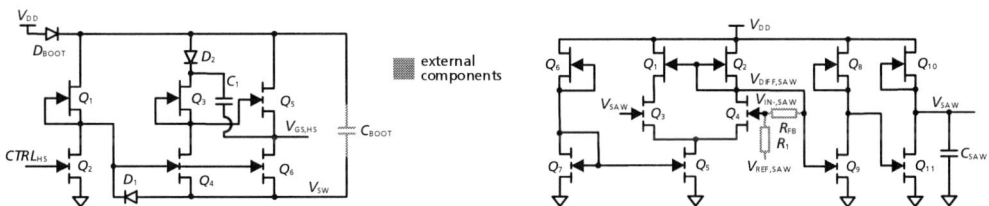

Fig. 8: Schematic of (left) HS gate driver with bootstrap voltage supply and (right) sawtooth generator consisting of a hysteresis comparator and a charging unit.

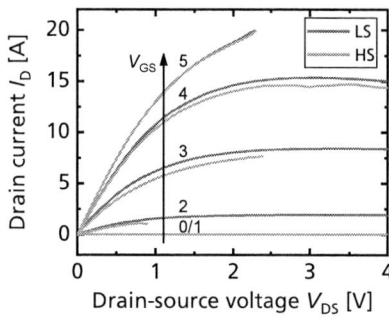 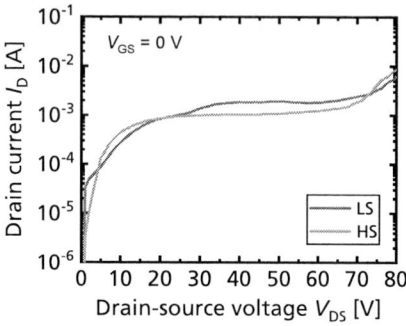

Fig. 9: Symmetrical half-bridge consisting HS/LS HEMT with comb layout: (Left) Output characteristic with different gate-source voltages from 0 V to 5 V and (Right) breakdown characteristic at $V_{GS} = 0$ V.

as described before. These capacitors C_1 in the HS and LS gate driver are also integrated in the GaN IC, such that no external capacitors are required for the gate drivers. In addition, compared to the layout in Fig. 3, the capacitors are further area-optimized by realizing them as stacked MIM capacitors under a connection pad, which increases the area efficiency and reduces the external number of components. The first DCFL inverter is used in the HS gate driver for the level shifting of the signal. In addition, this level shifter has a protection diode D_1 to limit the input of the gate driver in off-state. This gate driver approach based on three stages has the advantage of a non-inverting single input, unipolar supply, and rail-to-rail output. For a bootstrap supply of the HS gate driver, an additional diode D_{BOOT} was integrated. However, the bootstrap capacitor C_{BOOT} must be wired externally due to the high capacitance value, shown in Fig. 7. The dead time is realized by logic gates and integrated MIM capacitors. The voltage mode control is based on a PWM generator consisting of comparator and sawtooth generator and an error amplifier with external type III compensation network. The sawtooth comparator in turn consists of a hysteresis comparator and a charging unit, shown in Fig. 8 and similar to the previously presented GaN-based voltage mode control. The feedback network of the hysteresis comparator R_{FB} and R_1 is wired externally, but the capacitor for the sawtooth comparator C_{SAW} is again integrated. The hysteresis band or amplitude and output voltages of the sawtooth generator can adjusted by the resistance ratio R_{FB}/R_1 or R_1/R_{FB} and the reference voltage $V_{REF,SAW}$ of the used comparator. All periphery of the GaN IC has a supply voltage of 5 V.

The IC has already been processed and statically characterized. Fig. 9 shows experimental on-wafer pre-characterization of the symmetrical half-bridge. The output characteristic shows different curves with a gate-source voltage from 0 V to 5 V. The on-resistance is 68 mΩ at $V_{DS} = 0.5$ V and $V_{GS} = 5$ V. The drain current I_D in blocking mode with $V_{GS} = 0$ V is <2 mA at $V_{DS} = 50$ V. Depending on the thermal management and the output voltage, this IC can provide an output power >120 W at an output voltage of 24 V depending on the output current.

The integrated GaN synchronous buck converter demonstrates a higher level of integration into the GaN technology with higher functionality resulting in low system cost or bill of materials (BOM). At the same time, the losses can be further reduced, especially the losses of the driver and the control, and a higher power density of the system can be realized. The area of the driver, sensing, and control is only 28.6% while the power devices (in this case the half-bridge) is 71.4% of the chip area. This is a small additional area to the power devices for further functionality and higher compactness. Furthermore, the area of the control has several pads for higher flexibility, which has a negative effect on the area utilization. However, the IC can also be used without the control or in boost converter configuration. Thus, the power electronic designer has a higher flexibility and can use the individual function blocks independently.

Conclusion

This work presents function blocks of a highly-integrated all-in-GaN power IC for DC-DC conversion. Therefore, a GaN power IC platform is shown with several active devices such as e-/d-mode HEMT, SBD, LFER for low and high voltages and passive components such as resistors, capacitors and inductors. In addition, digital and analog libraries are described. With the help of this GaN power IC platform two function blocks are designed. On the one hand a highly-integrated gate driver with a bootstrapping approach and on the other hand a voltage mode control. The gate driver is characterized by its low power consumption, e.g. 1.6 mA quiescent supply current and 2.2 mA at a switching frequency of 1 MHz and a load capacitance of 0.1 nF. The voltage mode control integrates of a PWM generator consisting of sawtooth generator and comparator, and error amplifier. Exemplary measurements are shown at a switching frequency of 680 kHz. Finally, a circuit design of an all-in-GaN power IC consisting of symmetrical half-bridge, half-bridge gate driver, level shifter, dead time control and voltage mode control is described. IV measurements of the half-bridge are shown as a first verification. Thus, this highly-integrated GaN IC shows an example of a modern highly-functional and highly-efficient GaN-based power conversion component for the next generation of power electronics resulting in low system cost or bill of materials.

References

[1]   Yole, "Power GaN 2021: Epitaxy, Devices, Applications & Technology Trends: Market and Technology Report 2021," 2021. [Online]. Available: www.yole.fr

[2]   A. Lidow, "The Path Forward for GaN Power Devices," in *2020 IEEE Workshop on Wide Bandgap Power Devices and Applications in Asia (WiPDA Asia)*, 2020, pp. 1–3.

[3]   D. Kinzer, "Monolithic GaN Power IC Technology Drives Wide Bandgap Adoption," in *2020 IEEE International Electron Devices Meeting (IEDM)*, 2020, 27.5.1-27.5.4.

[4]   M. Basler *et al.*, "High-Power Density DC-DC Converters Using Highly-Integrated Half-Bridge GaN ICs," in *PCIM Europe digital days 2021; International Exhibition and Conference for Power Electronics, Intelligent Motion, Renewable Energy and Energy Management*, 2021, pp. 1–8.

[5]   M. Basler *et al.*, "Large-Area Lateral AlGaN/GaN-on-Si Field-Effect Rectifier with Low Turn-On Voltage," *IEEE Electron Device Letters*, p. 1, 2020, doi: 10.1109/LED.2020.2994656.

[6]   R. Reiner *et al.*, "Design of Low-Resistance and Area-Efficient GaN-HEMTs for Low-Voltage Power Applications," in *PCIM Europe digital days 2021; International Exhibition and Conference for Power Electronics, Intelligent Motion, Renewable Energy and Energy Management*, 2021, pp. 1–8.

[7]   M. Basler *et al.*, "Monolithic Integration of Inductive Components in a GaN-on-Si Technology," in *CIPS 2020; 11th International Conference on Integrated Power Electronics Systems*, 2020, pp. 1–6.

[8]   M. Basler *et al.*, "Building Blocks for GaN Power Integration," *IEEE Access*, vol. 9, pp. 163122–163137, 2021, doi: 10.1109/ACCESS.2021.3132667.

[9]   S. Ujita *et al.*, "A compact GaN-based DC-DC converter IC with high-speed gate drivers enabling high efficiencies," in *2014 IEEE 26th International Symposium on Power Semiconductor Devices & IC's (ISPSD)*, 2014, pp. 51–54.

[10]   G. Tang *et al.*, "High-speed, high-reliability GaN power device with integrated gate driver," in *2018 IEEE 30th International Symposium on Power Semiconductor Devices and ICs (ISPSD)*, 2018, pp. 76–79.

[11]   S. Moench *et al.*, "A 600V p-GaN Gate HEMT with Intrinsic Freewheeling Schottky-Diode in a GaN Power IC with Bootstrapped Driver and Sensors," in *2020 32nd International Symposium on Power Semiconductor Devices and ICs (ISPSD)*, 2020, pp. 254–257.

[12]   H. -Y. Chen *et al.*, "A Domino Bootstrapping 12V GaN Driver for Driving an On-Chip 650V eGaN Power Switch for 96% High Efficiency," in *2020 IEEE Symposium on VLSI Circuits*, 2020, pp. 1–2.

[13]   W. L. Jiang *et al.*, "Monolithic Integration of a 5-MHz GaN Half-Bridge in a 200-V GaN-on-SOI Process: Programmable dv/dt Control and Floating High-Voltage Level-Shifter," in *2021 IEEE Applied Power Electronics Conference and Exposition (APEC)*, 2021, pp. 728–734.

[14]   Y. -Y. Kao *et al.*, "Fully Integrated GaN-on-Silicon Gate Driver and GaN Switch With Temperature-Compensated Fast Turn-on Technique for Achieving Switching Frequency of 50 MHz and Slew Rate of 118.3 V/Ns," *IEEE Journal of Solid-State Circuits*, p. 1, 2021, doi: 10.1109/JSSC.2021.3103875.

[15]   H. Xu, G. Tang, J. Wei, Z. Zheng, and K. J. Chen, "Monolithic Integration of Gate Driver and Protection Modules With <italic>P</italic>-GaN Gate Power HEMTs," *IEEE Transactions on Industrial Electronics*, vol. 69, no. 7, pp. 6784–6793, 2022, doi: 10.1109/TIE.2021.3102387.

[16] R. Sun, Y. C. Liang, Y. Yeo, C. Zhao, W. Chen, and B. Zhang, "Development of GaN Power IC Platform and All GaN DC-DC Buck Converter IC," in *2019 31st International Symposium on Power Semiconductor Devices and ICs (ISPSD)*, 2019, pp. 271–274.

[17] M. Kaufmann and B. Wicht, "A Monolithic GaN-IC With Integrated Control Loop for 400-V Offline Buck Operation Achieving 95.6% Peak Efficiency," *IEEE Journal of Solid-State Circuits*, p. 1, 2020, doi: 10.1109/JSSC.2020.3018404.

[18] W. L. Jiang *et al.,* "An Integrated GaN Overcurrent Protection Circuit for Power HEMTs Using SenseHEMT," *IEEE Transactions on Power Electronics*, vol. 37, no. 8, pp. 9314–9324, 2022, doi: 10.1109/TPEL.2022.3158655.

[19] Efficient Power Conversion, *EPC2152 80 V, 15 A ePower™ Stage.* [Online]. Available: https://epc-co.com /epc/Portals/0/epc/documents/datasheets/EPC2152_datasheet.pdf (accessed: Jan. 31 2022).

[20] Navitas Semiconductor, *GaNFast™ Power IC NV6113 Datasheet.* [Online]. Available: https:// navitassemi.com/wp-content/uploads/2022/06/NV6113-Datasheet-FINAL-05-19-22.pdf (accessed: Jun. 10 2022).

[21] Renesas, *EL7202, EL7212, EL7222 High Speed, Dual Channel Power MOSFET Drivers Datasheet.* [Online]. Available: https://www.renesas.com/eu/en/document/dst/el7202-el7212-el7222-datasheet (accessed: Jun. 10 2022).

[22] Texas Instruments, *UCC2751x Single-Channel, High-Speed, Low-Side Gate Driver (With 4-A Peak Source and 4-A Peak Sink) Datasheet.* [Online]. Available: https://www.ti.com/lit/ds/symlink/ucc27516.pdf?ts= 1654847904989&ref_url=https%253A%252F%252Fwww.ti.com%252Fproduct%252FUCC27516 (accessed: Jun. 10 2022).

[23] A. Li *et al.,* "Monolithic Comparator and Sawtooth Generator of AlGaN/GaN MIS-HEMTs With Threshold Voltage Modulation for High-Temperature Applications," *IEEE Transactions on Electron Devices*, vol. 68, no. 6, pp. 2673–2679, 2021, doi: 10.1109/TED.2021.3075425.

[24] H. Wang, A. M. H. Kwan, Q. Jiang, and K. J. Chen, "A GaN Pulse Width Modulation Integrated Circuit for GaN Power Converters," *IEEE Transactions on Electron Devices*, vol. 62, no. 4, pp. 1143–1149, 2015, doi: 10.1109/TED.2015.2396649.

[25] Y. Shen, Z. Li, A. Li, and W. Liu, "Monolithically Integrated PWM Circuit Based on AlGaN/GaN MIS-HMETs for All-GaN Smart Power System," in *2021 IEEE 14th International Conference on ASIC (ASICON)*, 2021, pp. 1–4.

Comparison of Redundancy Requirements for Modular Multilevel Converter Considering Manufacturer Reliability Inputs and Mission Profile

Diego Velazco[1,2], Guy Clerc[1,2], Emmanuel Boutleux[1,2], François Wallart[1]

[1] SUPERGRID INSTITUTE SAS
23 rue Cyprian
Villeurbanne, France
Tel.: +33 / (0) – 4 28 01 23 23
E-Mail: Diego.VELAZCO@supergrid-institute.com
URL: https://www.supergrid-institute.com/

[2] Univ. Lyon, Univ. Claude Bernard Lyon1, INSA Lyon, Ecole Centrale Lyon, AMPERE
UMR CNRS 5005

Acknowledgements

This work was supported by a grant overseen by the French National Research Agency (ANR) as part of the "Investissements d'Avenir" Program (ANE-ITE-002-01).

Keywords

«Modular Multilevel converters (MMC)», «HVDC», «Mission profile», «Lifetime», «Reliability».

Abstract

During the design phase, target reliability values of components – including semi-conductors - allow the computation of converter level reliability and redundancy requirements. This work proposes a more accurate method based on manufacturer lifetime models and mission profile evaluation. The new method is applied on a Modular Multilevel Converter.

I - Introduction

The growing needs of power transmission systems are pushing towards the rapid development of HVDC technologies. Moreover, HVDC has been recognized as a major contender for performing long-distance bulk-power transmission [1], [2]. Among HVDC technologies, voltage source converters (VSC-HVDC) are considered the most suitable to interact with existing AC grids [3]. Nowadays, the most promising VSC-HVDC topology is the Modular Multilevel Converter (MMC) [4], [5].

The MMC for HVDC applications is a three-phase converter as seen in Fig. 1 (a). Each phase comprises an upper and a lower arm. Each arm is composed of a stack of submodules (SMs) and an arm inductor. The SMs are connected in series in order to withstand the high voltage levels inherent to HVDC systems. Multiple SM configurations are proposed in the literature, but the two most popular are the half-bridge (HB-SM) and the full-bridge (FB-SM) configurations. Today, in practical applications, the HB-SM is the most used due to cost considerations. To guarantee the proper functioning of the converter, a large number of SMs must be installed in each arm. In the case of HVDC applications the number of HB-SMs is in the range of some hundreds per arm.

A HB-SM consists of a storage capacitor C, two IGBT T1 and T2 and two antiparallel connected diodes D1 and D2 as depicted in Fig. 1 (b). The different semiconductors allow the insertion, or the bypass of the capacitor and they will be active based on the command signals and the direction of the current.

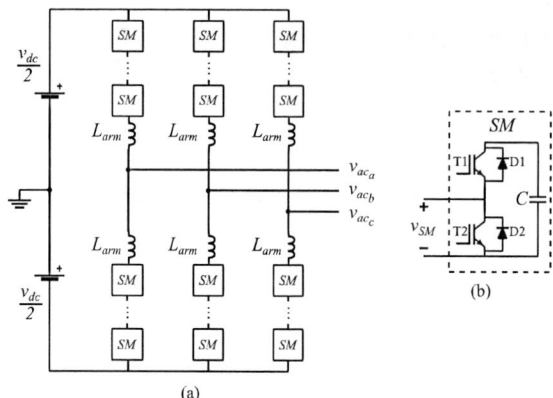

Fig. 1: (a) MMC topology, (b) Half-bridge submodule

Given the criticality of energy outages and its potential repercussions, special care should be taken for assessing the reliability of HVDC technologies such as the MMC. According to the ENTSOE, improved reliability (e.g. decreasing 1-2 trips per year) and improved availability (e.g. decreasing 1-2 outages days per year) can bring significant cost savings to TSOs and / or society [6].

Having such demanding reliability constraints, MMC arms are equipped with additional SMs for providing the redundancy required for attaining the availability requirements of an HVDC system. This work intends to shed a light on the estimation of the number of redundant SMs. In order to do so, two case studies will be compared: In the first case, the redundancy computation will use out-of-the-shelf reliability indicators from the manufacturer. In the second case, the redundancy will be computed using the mission profile of the converter.

In order to be able to consider the mission profile of the converter and obtain the number of redundant semiconductors, the thermal loading of power devices will be determined thanks to the methodology for lifetime estimation developed in [7].

This work will present the profile-based lifetime estimation methodology in Section II. Then Section III will depict the methodology for performing reliability predictions of the different elements of a SM. Section IV presents the methodology for translating the reliability calculations of a SM into redundancy requirements. Section V discusses the reliability predictions and redundancy estimations and gives a comparison on the two case studies. Finally, section VI gives the conclusions on this research.

II – Methodology for profile-based lifetime estimation

The methodology employed for considering the mission profile for the lifetime estimation of the power semiconductors [7] can be summarized in Fig. 2. It consists of 5 different stages: the modeling of the converter, the calculation of the losses of the semiconductor devices, the calculation of the thermal loading, the use of the rain flow counting algorithm [8] for organizing the thermal profiles and finally, the use of the lifetime models provided by the manufacturer as well as the Miners rule [9] for estimating the lifetime consumption.

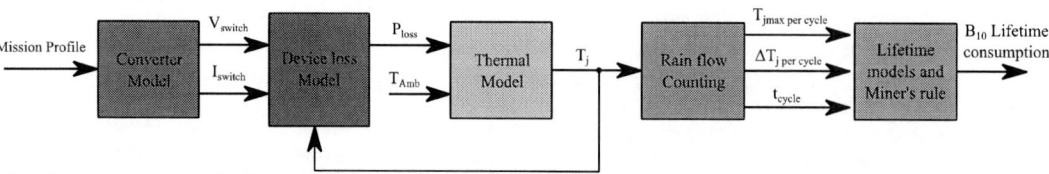

Fig. 2: Flowchart of the lifetime estimation methodology

An important step in the lifetime estimation methodology is the assessment of the temperature profile of the different semiconductors. This stage is performed considering two different timescales as seen in

[7]. The thermal profiles obtained for the studied mission profile, which corresponds to an offshore wind farm (OWF), can be seen in Fig. 3. Then the thermal profiles are organized thanks to the rain flow counting algorithm as seen in the left portion of Fig. 4.

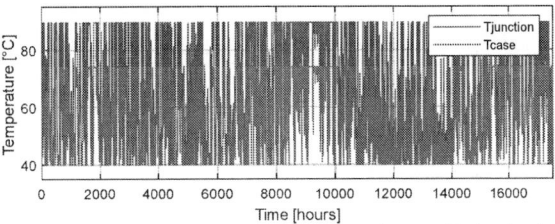

Fig. 3: Thermal profile of T2 under the evaluated period for the studied mission profile

One of the hallmarks presented in this methodology is the introduction of the model-based extrapolation method [7] for obtaining the B10 number of cycles to failure. The extrapolation method is necessary to be able to use the manufacturer lifetime models [10] for estimating the lifetime consumption in the evaluated period. The extrapolation method is constructed thanks to the Norris-Landzberg lifetime model [11], yielded by expression (1) and which results in the extrapolation grid shown in the central portion of Fig. 4.

$$N_{f1} = N_{f2} \left(\frac{\Delta T_2}{\Delta T_1}\right)^{\alpha} \left(\frac{t_2}{t_1}\right)^{\beta} exp\left(\gamma \left(\frac{1}{T_1} - \frac{1}{T_2}\right)\right) \tag{1}$$

In (1), Nf_1 represents the number of cycles to failure for condition 1 (resp. for Nf_2), ΔT_1 the temperature swing for condition 1 (resp. for ΔT_2), t_1 the cycle duration for condition 1 (resp. for t_2) and T_1 the temperature reference for condition 1 (resp. for T_2). Coefficients α, β and γ can vary according to the evaluated conditions. The grid shown in the central portion of Fig. 4 allows for the calculation of the new coefficients α, β and γ for each cycling condition. These coefficients are computed for each mesh with least-squares algorithm considering the vertices of the mesh.

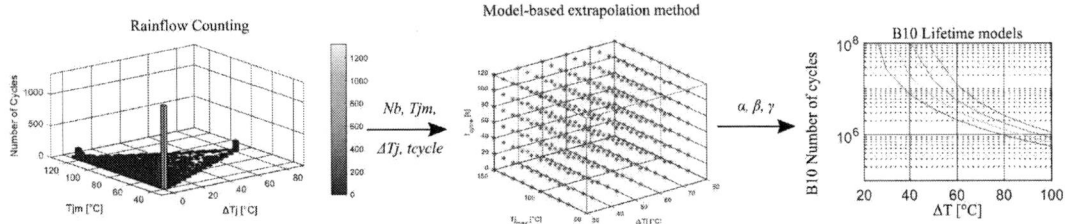

Fig. 4: Grid for model-based extrapolation method

The manufacturer provides different curves for performing the lifetime estimation of the different elements of the semiconductor devices such as the chip solder, the baseplate solder, and the bond wires [10]. The yearly B10 lifetime consumption, $CL_{i,k}$ for the kth element (chip solder, dbc solder or bondwire) and i^{th} thermal cycle identified is obtained with (2). In (2), nb_i is the number of cycles for a given thermal condition and $Nf_{i,k}$ is the number of cycles to failure for the same condition. Moreover, the expected time LF_k for a 10% failure rate of the kth element of the semiconductors can be calculated with (3). In (3), N_{cyc} is the total number of thermal cycles in the evaluated period. Some of the lifetime consumption results for the evaluated profile can be seen in Fig. 5.

$$CL_{i,k} = \frac{nb_i}{N_{fi,k}}, k \in \{\text{chip solder, dbc solder, bondwire}\} \tag{2}$$

$$LF_k = \frac{1}{\sum_{i=1}^{N_{cyc}} CL_{i,k}}, k \in \{\text{chip solder, dbc solder, bondwire}\} \tag{3}$$

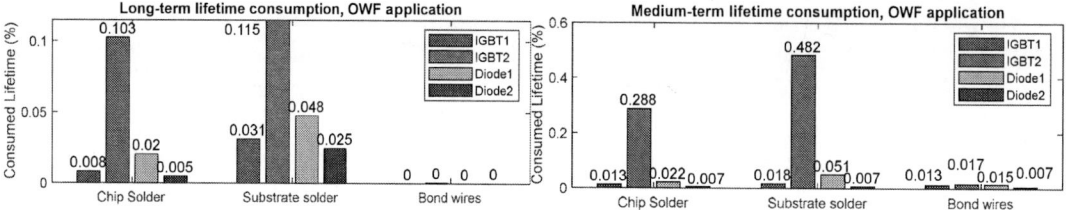

Fig. 5: Lifetime consumption results (a) Long-term timescale (b) Medium-term timescale

III – Methodology for reliability predictions

In this work the reliability predictions are performed considering two different case studies: manufacturer off-the-shelf reliability indicators and mission profile-based reliability estimations.

For the first case, the inputs considered are the failure rates of the different elements that compose a SM, such as the power semiconductors, the storage capacitor and the electronic board that drives the SM as done in [12]. The failure rates of these components, expressed in FIT are depicted in Table I.

Table I: FIT values for the SM elements [12]

Component	FIT
IGBT	100
Capacitor	300
Control Board	1200

Having the failure rates λ, of the elements of a SM the calculation of the failure rate of a SM λ_{SM}, can be calculated with (4) by adding the failure rates of the different elements of the SM and subsequently the reliability of a SM $R_{SM}(t)$ can be computed employing (5).

$$\lambda_{SM} = 2 \cdot \lambda_{IGBT} + \lambda_{capacitor} + \lambda_{board} \tag{4}$$

$$R_{SM}(t) = e^{-(\lambda_{SM} \cdot t)} \tag{5}$$

It is also worth mentioning that for the case in which only off-the-shelf reliability indicators are employed, the reliability function is obtained employing an exponential distribution. This method does not consider wear-out failures. The resulting reliability distribution for this case can be seen in Fig. 6.

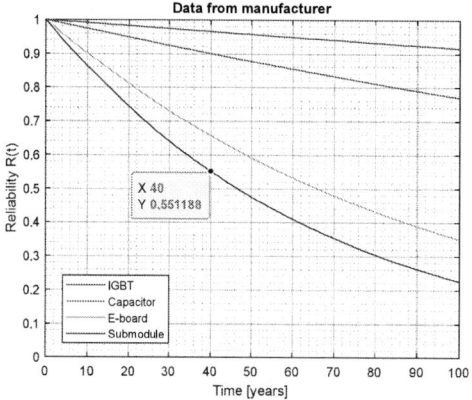

Fig. 6: Reliability of a SM and its composing elements considering off-the-shelf reliability indicators

The second case considers the mission profile experienced by the converter and employs the results obtained in section II as inputs. The profile evaluated in this work corresponds to a 1 GW converter installed in an offshore wind farm. After having calculated the lifetime consumptions of the elements of the semiconductors in the evaluated period and under the evaluated profile, the lifetime for a 10% failure rate is calculated by using the data from [10]. This result can be translated into a device level reliability function as depicted in Fig. 7.

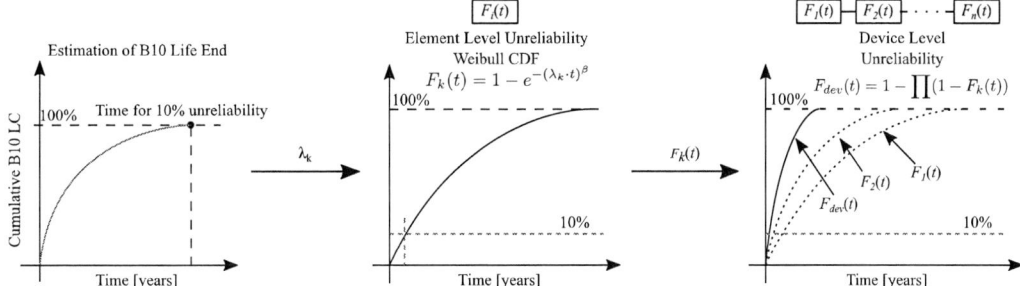

Fig. 7: Methodology for calculating device level reliability indicators

As the manufacturer B10 lifetime models [10] are fitted using a two parameter Weibull distribution (6), (7) will allow the calculation of the failure rate describing the cumulative density function CDF, for the elements of the semiconductors. In (7), the CDF $F_k(LF_k) = 0.1$. It is worth noting that different shape factors β were evaluated for the semiconductor devices in this study, as this parameter represents the variance of the instantaneous failure distribution of the studied elements. Having calculated the individual CDF for the elements of the semiconductors, the unreliability for the semiconductor devices $F_{dev}(t)$ can be obtained with (8).

$$F_k(t) = 1 - e^{-(\lambda_k \cdot t)^{\beta}}, k \in \{chip\ solder, dbc\ solder, bondwire\} \tag{6}$$

$$\lambda_k = -\frac{[ln(1-F_k(LF_k))]^{1/\beta}}{LF_k}, k \in \{chip\ solder, dbc\ solder, bondwire\} \tag{7}$$

$$F_{dev}(t) = 1 - \prod\left(1 - F_k(t)\right), dev \in \{T1, D1, T2, D2\} \tag{8}$$

Once the unreliability function of all the semiconductors is obtained, the unreliability of the SM can be obtained with (9). In (9), m represents all the different elements of the SM such as the semiconductor devices T1, D1, T2 and D2, as well as the capacitor and electronic board. The failure distribution of the capacitor and the electronic board are calculated with an exponential distribution, as done in the previous case study (off-the-shelf reliability indicator). With the failure distribution of a SM, its reliability function can be deduced with (10). Fig. 8 depicts the reliability functions of the elements of a SM. The results in Fig. 8 were obtained with $\beta = 2.5$.

$$F_{SM}(t) = 1 - \prod\left(1 - F_m(t)\right), m \in \{T1, D1, T2, D2, Capacitor, Board\} \tag{9}$$

$$R_{SM}(t) = 1 - F_{SM}(t) \tag{10}$$

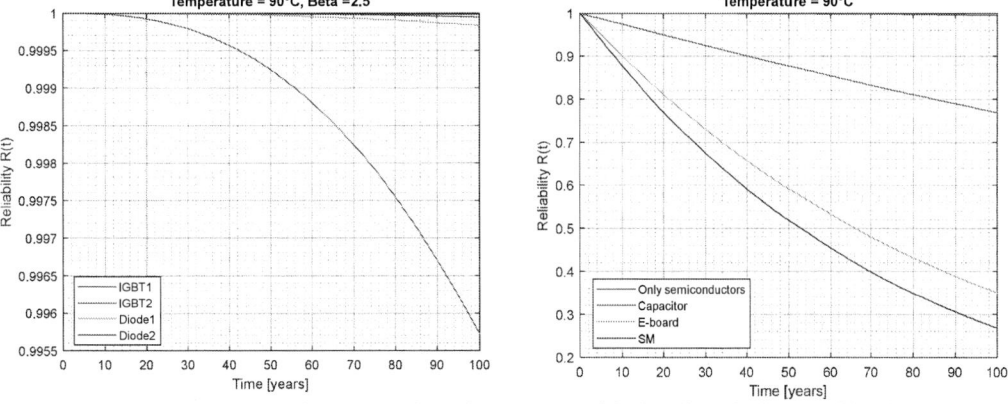

Fig. 8: Reliability of a SM and its composing elements considering the mission profile (β=2.5)

In Fig. 9, the resulting reliability functions of the SMs for the two case studies can be appreciated. The profile-based case study considered different shape factors β for the reliability computations. The values considered are 1, 1.5, 2, 2.5, 5 and 10 as depicted in Fig. 9.

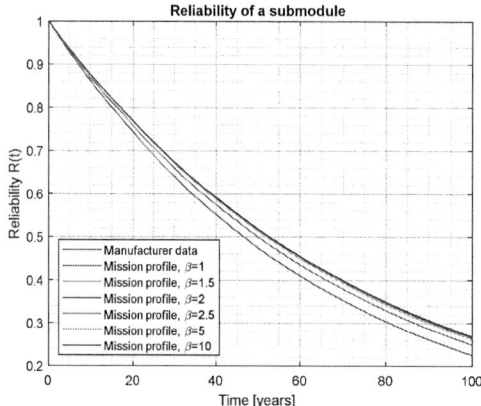

Fig. 9: Reliability functions of a SM for the studied cases

It can be seen from Fig. 9, that manufacturer off-the-shelf data yields the more pessimistic reliability prediction when compared to the profile-based reliability estimation. The results considering the mission profile are notably dependent on the thermal conditions chosen for the study. The lifetime models employed depend greatly on the thermal profile experienced by the semiconductors and more particularly on the variation of the experienced thermal cycles ΔT.

For the purpose of this study and in order to obtain realistic results, a maximum junction temperature of 90°C was considered for the semiconductor devices as seen on Fig. 3. This requirement was then used for dimensioning the thermal characteristics of the heatsink associated to the semiconductors. If a higher maximum junction temperature would have been considered, the profile-based reliability calculations would yield significantly lower and worse results.

Additionally, it can be said from Fig. 9 that, given the thermal constraint imposed to the semiconductors, the shape factor β in the range [1.5 10] doesn't have a great influence on the final reliability computation of the SM. Moreover, since the reliability function for the capacitor and the control board are the same in all the configurations, the final SM reliability estimation is mostly dependent on those elements, as they have considerably higher failure rates than the semiconductors.

IV – Methodology for redundancy estimations

Knowing the reliability distribution of a SM allows the calculation of the reliability function of an MMC arm. In order to make a realistic evaluation of the arm reliability a Markov chain was developed for this purpose as depicted in Fig. 10. Along with the Markov chain, Monte Carlo simulations were performed for assessing multiple converter lifetimes. For the purpose of the simulations a target mission time of 40 years was considered.

Consequently, different redundancy levels were tested. For each redundancy level, 1000 simulations were carried out. The minimum number of SMs to be installed in an arm should guarantee a reliability of 99.5 % by the end of the target mission time of the converter. Maintenance is also considered for the simulations. At each maintenance period, if the number of failed SMs is smaller than the number of redundant SMs, all the failed ones are replaced. However, if at any point throughout the simulations the amount of failed SMs is bigger than the number of redundant SMs, then the converter has failed that test. A maintenance intervention every two years was considered for the Monte Carlo simulations. The arm reliability is approximated to the number of arm survivals for a given redundancy level.

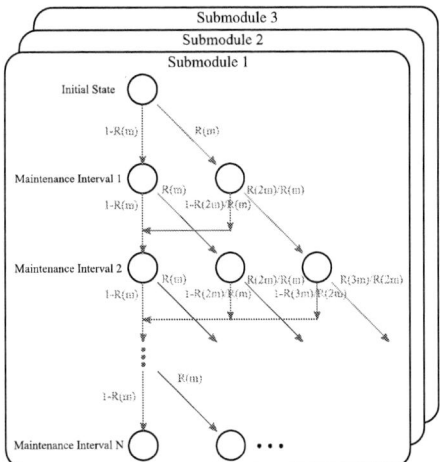

Fig. 10: Markov chain for calculating the availability of the converter for a target mission time

This methodology allowed for the calculation of the redundancy requirements of the MMC for the two case studies. These results can be seen in Fig. 11.

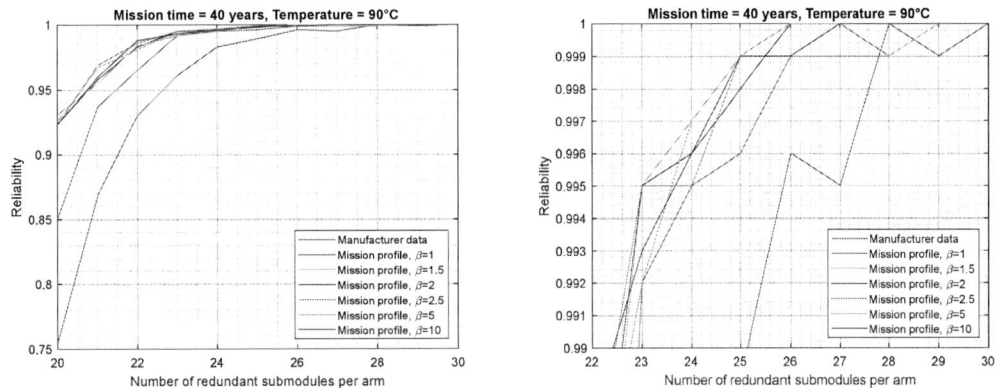

Fig. 11: Reliability of an MMC arm vs number of redundant SMs

As mentioned previously in this section, the chosen criteria for determining the number of redundant SMs required for an MMC arm is to guarantee a reliability of 99.5% throughout its operating life. Thus, the number of redundant SMs required for each of the evaluated cases can be seen in Table II.

Table II: Number of redundant SMs for a 40-year target mission time and 99.5% reliability

Case Study	Manufac-turer data	Mission profile, β=1	Mission profile, β=1.5	Mission profile, β=2	Mission profile, β=2.5	Mission profile, β=5	Mission profile, β=10
Redundant SMs	26	24	23	24	23	24	24

V – Discussion on the results of the reliability predictions and redundancy estimations

The number of redundant SMs reflects the reliability functions calculated in section III. The results depicted in Table II and Fig. 11 indicate that using the manufacturer reliability indicators yields the worst-case scenario for redundancy dimensioning. As mentioned in section III, the thermal constraints imposed to the converter have a significant impact on the redundancy requirements obtained in the profile-based redundancy calculations.

The results seen in Table II show that the shape factor β used for calculating the reliability distributions of the semiconductors has almost no impact on the final redundancy estimations. This is mainly due to the fact that the failure rates of the storage capacitor and the control board of the SM are much higher than the failure rates of the semiconductors in both case studies.

In a Weibull distribution, the shape factor β allows the modeling of the wear-out of the studied element when β>1. The reason why there is not a noticeable loss of reliability in spite of the different β evaluated, is that the acceleration in the reliability decline happens much after the target mission time of the converter. An example concerning the reliability of the chip solder of IGBT 2 can be seen in Fig. 12.

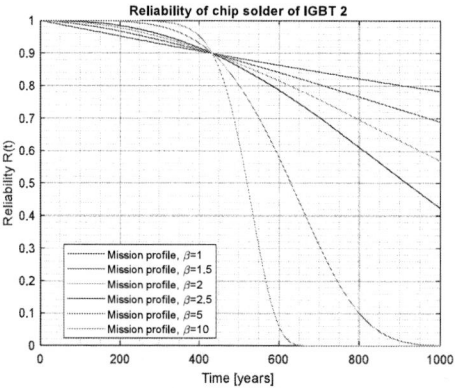

Fig. 12: Reliability of chip solder of IGBT 2 for different β

The reliability functions seen on Fig. 12 are obtained considering a fixed time span for a loss of reliability of 10%, which corresponds to the B10 definition in [10]. The fixed time span for the loss of reliability is the result of (3) in the lifetime estimation methodology described in section II.

Another noticeable result from this research is the estimation of the number of failed SMs at each maintenance period. This result allows the calculation of the approximated amount of SMs required for a proper functioning of an MMC arm throughout all the target mission time of the converter (40 years). These calculations were possible thanks to the Monte Carlo simulations and the Markov chain depicted in Fig. 10. In Fig. 13(a), the number of failed SMs for the profile-based redundancy calculations can be seen. The results correspond to a shape factor β = 2.5. Consequently, in Fig. 13(b), the number of SMs employed during the operating life of the converter arm can be seen.

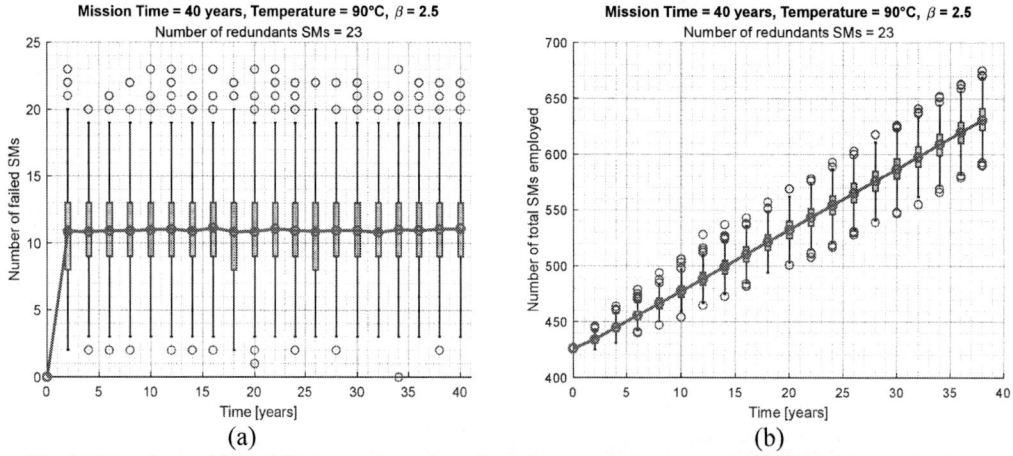

Fig. 13: (a) Number of failed SMs per arm at each maintenance intervention (1000 Monte Carlo tests), (b) Number of total SMs employed during the operating life of the converter arm

In Fig. 13(a), it can be seen that the maximum number of failed SMs at a given maintenance intervention is equal to the number of redundant SMs installed. If the number of failed SMs would have been higher, the MMC arm wouldn't have survived and the whole converter would have stopped working. The average number of failed SMs per maintenance period is 11, but it can have a large variance as evidenced on Fig. 13(a).

It can be seen from Fig. 13(b), that the total number of SMs employed throughout the operating life of the converter is indeed much higher than the minimum number of SMs required for the arm functioning. The studied MMC arm requires only 400 SMs to operate, it requires 23 redundant SMs to guarantee a reliability of at least 99.5% by the end of its target mission time and in total, it will require 631 SMs in average. This means that the converter arm requires more than 200 additional SMs than the ones with which it started working, which by no means is a negligeable amount and can have a great impact on the OPEX of the converter. Moreover, an MMC has six arms, so the total number of SMs employed in the whole converter throughout its target mission time can ascend to 3800 SMs in average, a very large amount when compared to the 2400 SMs, required for MMC operation. A comparison between the total number of SMs employed for all the cases can be seen in Fig. 14.

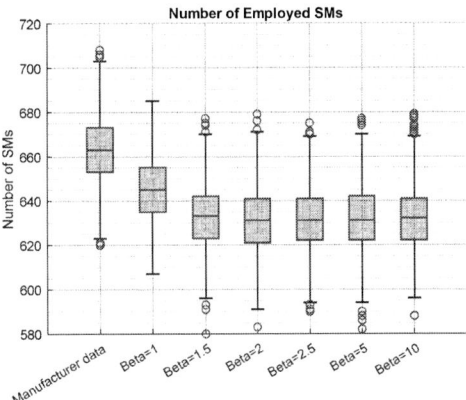

Fig. 14: Comparison between the number of employed SMs

As mentioned previously, the shape factor β employed for the profile-based redundancy calculations does not affect notably the number of redundant SMs (Fig. 11). However, when $\beta=1$ a slight impact on the total number of SMs employed can be seen in Fig. 14. The average number of employed SMs when $\beta=1$ is 645, compared to 631 for the other profile-based redundancy estimations where $\beta>1$. It is worth reminding that when $\beta=1$, the Weibull distribution behaves as an exponential distribution. As it was shown in Fig. 12, with $\beta>1$ an acceleration in the decline of the reliability distribution is expected. However, this decline happens much after the target mission time of the converter, that means that before this decline the reliability distributions that have a higher shape factor β will have higher reliability values, as it was seen in Fig. 12.

Concerning the total number of the employed SMs that is determined by the manufacturer off-the-shelf reliability indicators, it can be clearly seen in Fig. 14 that this input yields the worse results. The exponential distribution was used to model the reliability function in this case. As previously seen on section III, the reliability function obtained with the manufacturer inputs is the more pessimistic and the number of total employed SMs reflect this condition. The average number of SMs employed with this information ascends to 663, which is bigger than the number of employed SMs coming from the profile-based methodology (645 for $\beta=1$ and 631 for $\beta>1$).

VI – Conclusion

In this research, the methodology for lifetime estimation and reliability predictions were presented. Additionally, the methodology for performing redundancy computations for an MMC arm was introduced. Two case studies were treated in this paper, one uses off-the-shelf reliability indicators and the other uses the mission profile experienced by the converter. The profile-based results were strongly influenced by the thermal constraints imposed to the converter.

There is a direct correlation between the reliability function of the SM and the number of redundant SMs to be installed to attain a target reliability by the end of the operating life of the converter. Moreover, this correlation is also reflected on the total number of SMs to be employed throughout the operating life of the converter. It was seen that, in terms of the number of redundant SMs required, the more pessimistic results came from the manufacturer off-the-shelf reliability indicators, whereas the profile-based calculations offered similar results for all the evaluated shape factors β.

When analyzing the total number of SMs employed throughout the operating life of the MMC, the methodology based on manufacturer reliability inputs yielded the poorest results as expected from the reliability predictions. The results obtained with the profile-based methodology show two different behaviors. When β=1 the number of employed SMs is larger than when β>1. This is mainly due to the fact that the shape factor influences on the reliability decline after the operating life of the converter.

Finally, it can be said that the study of the number of failed and employed SMs opens up many perspectives for future research. The authors have already started working on some, such as the choice of alternative maintenance strategies concerning the replacement of healthier SMs in order to optimize the number of redundant SMs required or the total number of SMs employed throughout the active life of the MMC.

The methodologies for lifetime estimation and reliability predictions developed in this work, can be generalized to other converter topologies.

References

[1]   M. P. Bahrman and B. K. Johnson, "The ABCs of HVDC transmission technologies," *IEEE Power Energy Mag.*, vol. 5, no. 2, pp. 32–44, Mar. 2007, doi: 10.1109/MPAE.2007.329194.

[2]   J. D. Páez, D. Frey, J. Maneiro, S. Bacha, and P. Dworakowski, "Overview of DC–DC Converters Dedicated to HVdc Grids," *IEEE Trans. Power Deliv.*, vol. 34, no. 1, pp. 119–128, Feb. 2019, doi: 10.1109/TPWRD.2018.2846408.

[3]   J. C. Gonzalez-Torres, J. Mermet-Guyennet, S. Silvani, and A. Benchaib, "Power system stability enhancement via VSC-HVDC control using remote signals: Application on the Nordic 44-bus test system," in *15th IET International Conference on AC and DC Power Transmission (ACDC 2019)*, 2019, pp. 1–6.

[4]   R. Marquardt, A. Lesnicar, and J. Hildinger, "Modulares Stromrichterkonzept für Netzkupplungsanwendungen bei hohen Spannungen," *ETG-Conf.*, 2002.

[5]   S. Allebrod, R. Hamerski, and R. Marquardt, "New transformerless, scalable Modular Multilevel Converters for HVDC-transmission," in *2008 IEEE Power Electronics Specialists Conference*, Jun. 2008, pp. 174–179. doi: 10.1109/PESC.2008.4591920.

[6]   ENTSO-E, "Improving HVDC System Reliability." ENTSO-E, Nov. 2018. [Online]. Available: https://eepublicdownloads.entsoe.eu/clean-documents/Publications/Position%20papers%20and%20reports/entsoe_pp_HVDC_181205_web.pdf

[7]   D. Velazco, G. Clerc, E. Boutleux, F. Wallart, and L. Chédot, "IGBT Lifetime Estimation in a Modular Multilevel Converter for bidirectional point-to-point HVDC application," in *2020 22nd European Conference on Power Electronics and Applications (EPE'20 ECCE Europe)*, Sep. 2020, pp. 1–10. doi: 10.23919/EPE20ECCEEurope43536.2020.9215880.

[8]   M. Matsuichi and T. Endo, "Fatigue of metals subjected to varying stress," 1968.

[9]   Miner, M.A., "Cumulative Damage in Fatigue," *J. Appl. Mech.*, no. 12, pp. A159–A164, 1945.

[10]   N. Kaminski, "Load-cycling capability of HiPak IGBT modules. APPLICATION NOTE 5SYA 2043-04." ABB Group, Feb. 04, 2014.

[11]   K. C. Norris and A. H. Landzberg, "Reliability of Controlled Collapse Interconnections," *IBM J. Res. Dev.*, vol. 13, no. 3, pp. 266–271, May 1969, doi: 10.1147/rd.133.0266.

[12]   J. Wylie, M. C. Merlin, and T. C. Green, "Analysis of the effects from constant random and wear-out failures of sub-modules within a modular multi-level converter with varying maintenance periods," in *2017 19th European Conference on Power Electronics and Applications (EPE'17 ECCE Europe)*, Sep. 2017, p. P.1-P.10. doi: 10.23919/EPE17ECCEEurope.2017.8099246.

Impact of Insulation and Cooling on Performance due to Reliability-Oriented Design of Electrical Machines

Lucas Vincent Hanisch, Jonas Franzki, Markus Henke
Institute for Electrical Machines, Traction and Drives, TU Braunschweig
Hans-Sommer-Straße 66
Braunschweig, Germany
Phone: +49 (0) 531-3913906
Email: l-v.hanisch@tu-braunschweig.de
URL: http://www.tu-braunschweig.de/imab

Keywords

≪Permanent magnet motor≫, ≪Insulation≫, ≪Thermal model≫, ≪Cooling≫, ≪Reliability≫

Abstract

In this paper, a multiphysics model of a permanent magnet synchronous motor (PMSM) is created to investigate the influence of the cooling and insulation system on the temperature distribution and performance of the machine. In thermal computer fluid dynamic (CFD) studies, the influence of the coolant flow, the winding head impregnation and the insulation paper on the temperature distribution of the machine is investigated. Finally, the influence of thermal effects on the steady-state limit torque of the machine is evaluated. It is systematically shown that insulation materials with high thermal conductance allow torque increases of about 3.3%. Torque can be increased by about 1.9% per reduced 0.1 mm insulation thickness of the slot liner. Alternatively, the temperature impact on insulation can be reduced at constant torque leading to improved insulation lifetime and reliability.

Introduction

Increasing power densities of electrical machines for advanced applications in electromobility lead to an intensified focus on the electrical insulation system as a critical component. The reason for this is that with increased voltage and power density, the percentage of insulation failures in the stator winding of electrical machines has increased from 36% to 66% [1]. The damage of insulation is the consequence of erosion processes that accelerate with increased temperature and lead to winding short circuits [2]. To prevent these damage mechanisms and to take them into account in the design of electrical machines, an early focus on insulation stresses and reliability leading to a reliability-oriented design is becoming increasingly important [3].

Therefore, many studies investigate the lifetime and reliability of windings and insulation systems as a function of temperature. In [4], the partial discharge inception voltage is used to assess the condition of enamel-insulated copper wires and their behavior under different thermal boundary conditions is investigated. In [5], similar results were obtained for slot liners.

The objective of the research presented in this paper is to investigate the thermal effects and interactions of different insulation systems and cooling mechanisms, and to determine the impact on machine performance.

For this purpose, a PMSM is investigated with a multiphysics model. Since the power loss calculation is very important for the thermal investigations, it is evaluated in more detail in 2D and 3D simulations. In the thermal analysis, the slot is homogenized according to [6] and modeled as a composite of conductor, insulating varnish, and impregnating resin. This model is analyzed in CFD studies to evaluate different thermal effects on steady state performance.

Electromagnetic modeling of the PMSM

In this chapter, the electrical modeling of the PMSM is explained. The correct meshing of the model is explicitly discussed in order to be able to perform precise loss calculations in an acceptable time and reduce numerical noise [7].

Comparison of 2D and 3D model

When calculating losses from 2D models, one should consider that end effects such as leakage fluxes at the axial ends of the machine are neglected and that iron and magnetic losses tend to be underestimated [7]. For the machine under consideration, the 2D and 3D meshes including the data on the meshes can be seen in Fig. 1 and Table I. The 3D meshing has 1.19 million elements, almost 17 times as many as the 2D meshing. The 3D mesh resolution was limited by time and RAM resources. Edge length and time step were chosen in a way that a harmonic stepping according to the circumferential speed was achieved. Table I shows, that in 2D simulation a much higher resolution can be achieved at lower computation time.

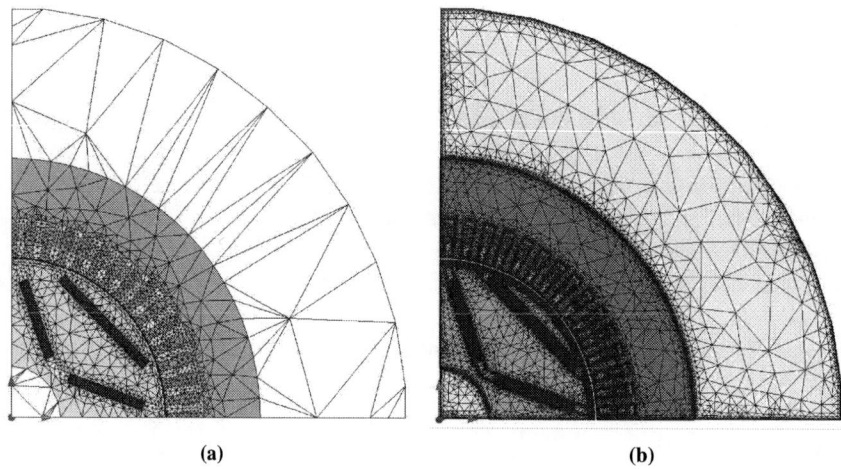

(a)   (b)

Fig. 1: Comparison of 3D mesh (a) and 2D mesh (b)

Table I: Quantitative comparison of the two meshes

Attribute	Value
3D simulation	
Number of circumferential segments	180
Time step	16.667 µs
Max. edge length - surfaces	2.72969495 mm
Max. edge length - magnets	1.5 mm
Element number	1.19 Mio.
2D simulation	
Number of circumferential segments	600
Time step	10 µs
Max. edge length	0.40945424251787 mm
Element number	70,823

Loss calculation

Fig. 2a and Fig. 2b show the results of loss calculations at 220 Nm and 5000 rpm. It confirms the result of [8] that the power losses calculated by the simulation settle only after the first electrical period (6 ms). Up to this point, especially the hysteresis losses have not yet reached a steady state, but double from 3 to 6 ms. It is therefore reasonable to use the power losses from 6 ms for the evaluation.

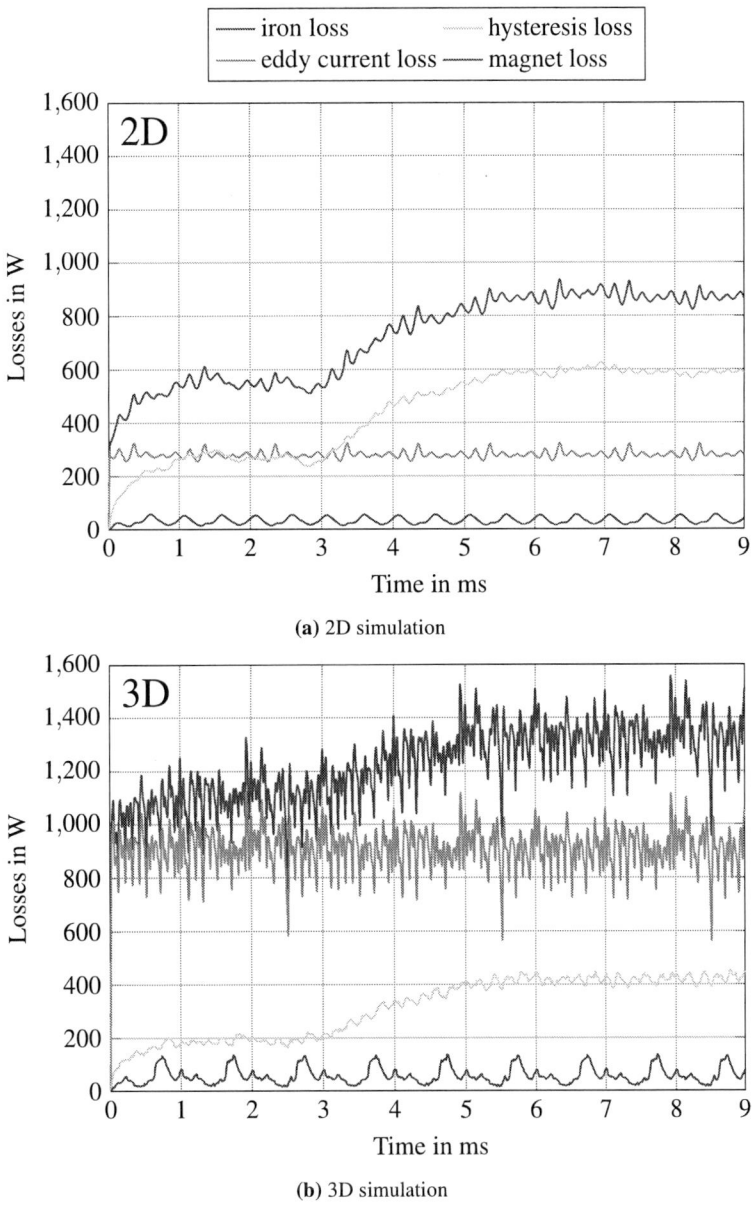

(a) 2D simulation

(b) 3D simulation

Fig. 2: Losses plotted against simulation time, M = 220 Nm, n = 5000 rpm

Beyond the plots, there are also significant differences between the 3D and 2D simulations. For example, in the 3D simulation, the eddy current losses reach significantly higher values than the hysteresis losses, whereas this is observed in reverse in the 2D simulation. Moreover, in the 3D simulation, the eddy current and iron losses are subject to high noise.

The losses of the 2D simulation are only higher than those of the 3D simulation in the case of hysteresis losses. On average, the losses calculated via 2D simulation are about 20% lower than those of the 3D simulation. The difference could be caused by the consideration of the third dimension, which is pointed out in particular by Lundmark et al. in [7]. However, they also show that too low temporal and spatial resolution can strongly distort the results. This is suspected due to the discussed strong noise behavior in 3D simulation. They additionally point out that for a very high resolution of the 2D simulation, a better match with a high resolution 3D simulation can be achieved. The resolution of the 2D simulation with 1,200 time steps per revolution is already at a very high level [7]. For the reasons listed, from now on the 2D simulation results will be used for the thermal calculations.

Thermal modeling

To investigate dynamic aspects, the first step is to analyse the continuous operation of the electric machine at the nominal point (current density of 8 A/mm^2 for 220 Nm at 5,000 rpm). Electromagnetically, it can be assumed that the power losses are constant in time. This is permissible because the thermal time constant is much larger than the electromagnetic one. A meander cooling jacket is used for the simulated machine and shown in Fig. 3. The loops are implemented in such a way that a cooling channel runs axially every 30° and is connected to the next axial cooling channel by a meander loop above each winding head. In this way, one axial channel cools five slots and the circumference can be reduced to just these five slots. When choosing a meander cooling jacket, a challenge remains: Per meander, the coolant also heats up, so that in the last meander the coolant is warmer than in the first. This is accompanied by a lower temperature difference to the stator in the last meander and thus also a poorer cooling effect. This can be taken into account by either

- a meander loop is solved and the fluid dynamic and thermal results of the output are defined as a new input, so that step by step all meander loops are solved and the full representation of the machine is created, or
- the engine is not thermally modeled and consequently only the power loss is directly impressed on the cooling jacket, or
- an increased input temperature of the coolant for the individual meander loop is assumed, so that approximately the average cooling temperature of the fluid is reflected, whereby a certain result blurring is accepted

so that a circumferential reduction is still possible.

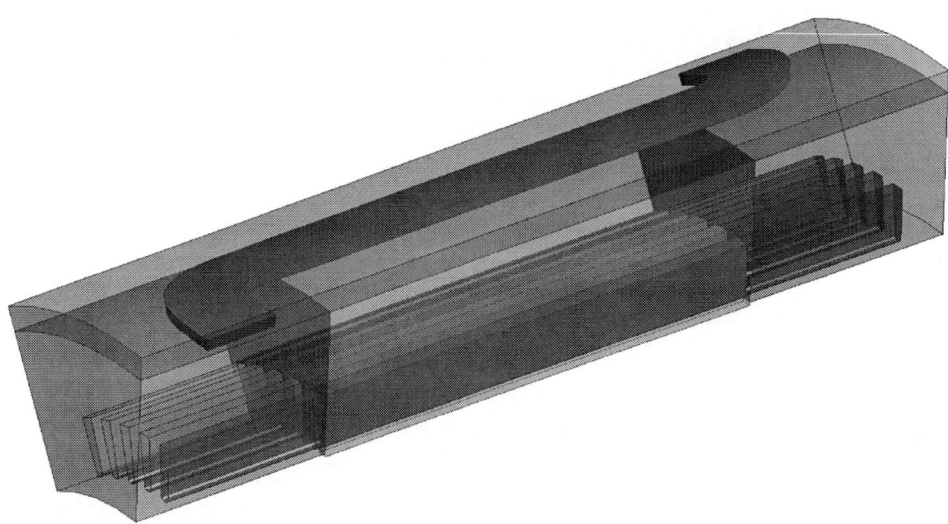

Fig. 3: Modeled section of the stator with meander-shaped cooling jacket

Investigation of thermal effects and insulation

The influence of the coolant flow on the temperature is to be investigated first. Fig. 4 shows on the left the maximum temperature of different machine components above the coolant flow for two operating points with current densities of 14 A/mm^2 (T_{max} = 180°C are reached) and 8 A/mm^2. It can be seen that, especially in the range below 3 l/min, an increase in flow rate can still reduce the temperature considerably. Above 6 l/min, the temperatures are hardly lowered. An exclusive consideration of the volume flow is not sufficient for an assessment of the quality of the cooling system, because it only describes the benefit achieved by a higher flow velocity, but not the required effort. To take the effort into account, the Stanton number is used as a quality criterion. This puts the heat transfer in relation to the required flow. In Fig. 4, it can be seen on the right that the Stanton number drops for increasing coolant flows, especially above 6 l/min. It can be seen that increasing the coolant flow increases the heat transfer coefficient, but the velocity of the fluid increases to a greater extent.

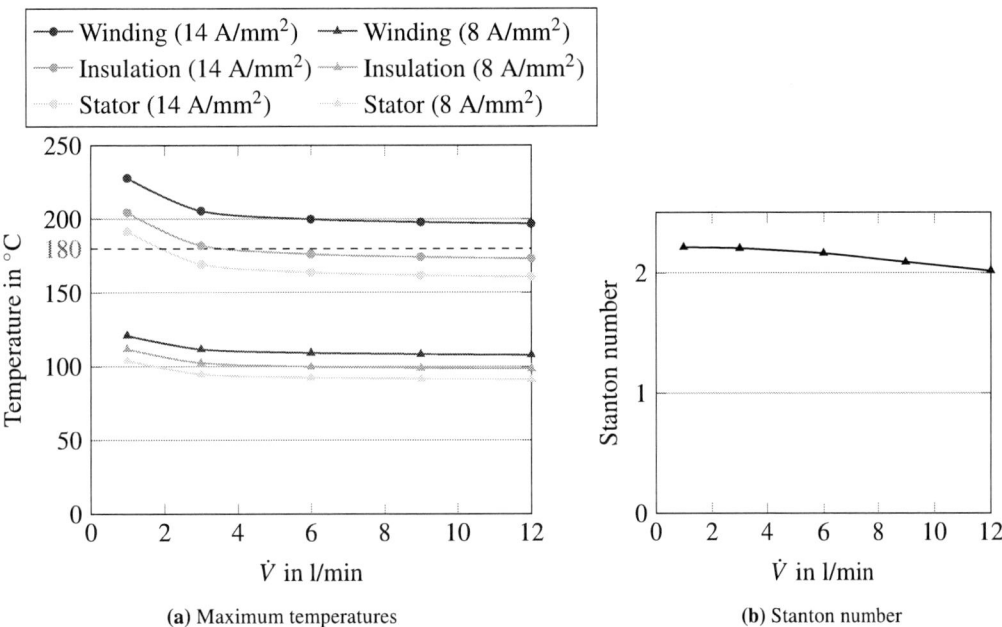

(a) Maximum temperatures

(b) Stanton number

Fig. 4: Maximum temperatures and Stanton number plotted against coolant flow rate

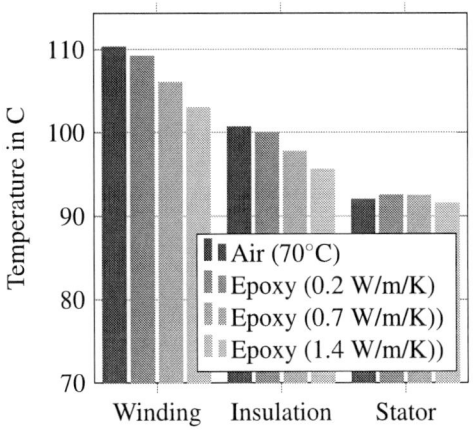

Fig. 5: Maximum temperatures for different potting materials

In addition to cooling, the influence of the insulation system on the thermal performance of the machine is investigated. For this purpose, Fig. 5 evaluates the temperature with different epoxy-based insulating resins. The results show that an epoxy impregnation causes a temperature reduction in the winding and insulation compared with air, irrespective of the thermal conductivity. On the stator, the epoxy impregnation initially has a temperature-increasing effect, since an additional heat input is formed from the winding head via the epoxy into the stator. Depending on the thermal conductivity, epoxy impregnation can reduce the temperature in the slot insulation by up to 5 degrees. In addition to the electrical properties studied here, thermally optimized insulation resins may differ in mechanical or chemical properties. This must be taken into account in the processing of the insulation resins, which may result in the revision of manufacturing steps.

Finally, the thickness of the slot liner was varied and the effects on temperature were studied. In Fig. 6, a linear relationship between temperature and insulation thickness can be observed. This observation is consistent with equation for heat transfer ($\Delta T = -\frac{\dot{Q}}{A\lambda}\delta$), which states that the temperature difference ΔT increases linearly with insulation thickness δ. The average slope - i.e. the temperature increase that occurs per layer increase of the insulation by 0.1 mm - for the maximum temperature is approx. $1,44\ \frac{K}{0,1\ mm}$ for the winding and at $1,52\ \frac{K}{0,1\ mm}$ for the insulation. The average temperature of the insulation shows a lower dependence on the layer thickness ($0,75\ \frac{K}{0,1\ mm}$). The temperature of the stator remains almost unaffected.

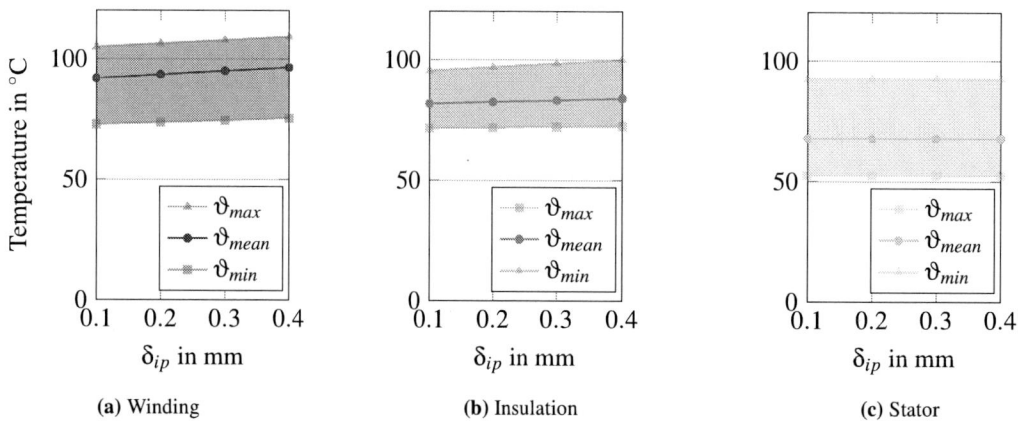

Fig. 6: Temperatures plotted against the thickness of the insulation paper δ_{ip}

Influences on the limit torque of the PMSM

Finally, the influence of the slot liner on the torque density of the PMSM is investigated. For this purpose, on the one hand the layer thickness of the slot liner is reduced from 0.4mm to 0.1mm. On the other hand, two different slot liners with a thermal conductivity of 0.2 W/mK (IP1) and 0.38 W/mK (IP2) are compared.

The evaluation of Table II shows that a slot liner 2 with higher thermal conductivity already results in a torque increase of 3.3% at the same thickness compared to slot liner 1 with 0.4 mm thickness. Reducing the layer thickness by 75% to 0.1 mm would result in a torque increase of 5.8%, i.e. a torque increase of about 1.9% per 0.1 mm reduction in insulation thickness. A combination of both measures (higher conductivity and 25% thickness) would finally result in a maximum torque increase of 7.6%. It should be noted that the performance increase due to thinner insulation papers is opposed by a reduced robustness against electrical breakdown and partial discharge. In addition, high thermal conductive insulation papers are more expensive than ordinary insulation papers and are therefore avoided in applications with increased focus on costs.

Table II: Results of investigated torque/IP dependency

Insulation material (thickness)	Current density A/mm^2	Torque Nm	Torque increase	
			Nm	%
IP1 (0,4 mm)	14,2	428	–	–
IP2 (0,4 mm)	14,6	442	14	3,3
IP1 (0,1 mm)	14,9	453	25	5,8
IP2 (0,1 mm)	15,1	460	32	7,6

Conclusion and Outlook

It was shown that an increased coolant flow significantly reduces the temperature of the PMSM only up to a certain volume flow and that the increased cooling effort can be taken into account with the Stanton number. The temperature reduction in the winding and insulation as a result of different epoxy resins and the layer thickness of the slot liner was shown. This can be exploited in a reliability-oriented machine design, so that either the performance of the machine can be increased with the same geometry, or the machine dimensions can be reduced while maintaining the same reliability. However, this also increases cost and weight, which is why it should be subject to a trade-off analysis. At the maximum, the heat dissipation of the PMSM could be optimized by using a more conductive and thinner insulation paper to such an extent that a higher current density enables a torque increase of 7.6% at the same temperature.

In future investigations, the simulation results are to be verified on reference machines or motorettes with different insulation materials. In addition to investigations of the thermal characteristics or torque of the electrical machine, the reliability of the various insulation systems can be assessed by partial discharge measurements.

References

[1] J. He, C. Somogyi, A. Strandt and N. A. O. Demerdash, "Diagnosis of stator winding short-circuit faults in an interior permanent magnet synchronous machine," 2014 IEEE Energy Conversion Congress and Exposition (ECCE), 2014, pp. 3125-3130, doi: 10.1109/ECCE.2014.6953825.

[2] Pietrzak P, Wolkiewicz M. Comparison of Selected Methods for the Stator Winding Condition Monitoring of a PMSM Using the Stator Phase Currents. Energies. 2021; 14(6):1630. https://doi.org/10.3390/en14061630

[3] M. Galea, P. Giangrande, V. Madonna and G. Buticchi, "Reliability-Oriented Design of Electrical Machines: The Design Process for Machines' Insulation Systems MUST Evolve," in IEEE Industrial Electronics Magazine, vol. 14, no. 1, pp. 20-28, March 2020, doi: 10.1109/MIE.2019.2947688.

[4] F. Pauli, N. Driendl and K. Hameyer, "Study on Temperature Dependence of Partial Discharge in Low Voltage Traction Drives," 2019 IEEE Workshop on Electrical Machines Design, Control and Diagnosis (WEMDCD), 2019, pp. 209-214, doi: 10.1109/WEMDCD.2019.8887790.

[5] K. Bae et al., "Current State and Development Trends of Insulation Systems in BEV Traction Motors Steered by Electric Powertrain Innovation," PCIM Europe digital days 2021; International Exhibition and Conference for Power Electronics, Intelligent Motion, Renewable Energy and Energy Management, 2021, pp. 1-8.

[6] L. Idoughi, X. Mininger, F. Bouillault, L. Bernard and E. Hoang, "Thermal Model With Winding Homogenization and FIT Discretization for Stator Slot," in IEEE Transactions on Magnetics, vol. 47, no. 12, pp. 4822-4826, Dec. 2011, doi: 10.1109/TMAG.2011.2159013.

[7] S. T. Lundmark and P. R. Fard, "Two-Dimensional and Three-Dimensional Core and Magnet Loss Modeling in a Radial Flux and a Transverse Flux PM Traction Motor," in IEEE Transactions on Industry Applications, vol. 53, no. 3, pp. 2028-2039, May-June 2017, doi: 10.1109/TIA.2017.2671416.

[8] S. T. Lundmark, A. Acquaviva and A. Bergqvist, "Coupled 3-D Thermal and Electromagnetic Modelling of a Liquid-cooled Transverse Flux Traction Motor," 2018 XIII International Conference on Electrical Machines (ICEM), 2018, pp. 2640-2646, doi: 10.1109/ICELMACH.2018.8506835.

Long Switching Horizon Model Predictive Controller for High-Speed Integrated Modular Motor Drives

Martin Schiestl, Maurizio Incurvati, Ronald Stärz and Markus Schmid
EMERGING APPLICATIONS LAB, MCI MANAGEMENT CENTER INNSBRUCK
Maximilianstraße 2
6020 Innsbruck, Austria
Phone: +43 512 2070 - 3936
Email: martin.schiestl@mci.edu, maurizio.incurvati@mci.edu, ronald.staerz@mci.edu
URL: https://www.mci.edu/en/

Acknowledgments

The authors would like to thank Mr. Davide Bagnara for the interesting discussions and helpful advise as well as Infineon Technologies AG for the provided part samples.

Keywords

≪Axial Machines≫, ≪Control of drive≫, ≪High power density systems≫, ≪High-speed drive≫, ≪Model Predictive Control≫, ≪Multiphase drive≫, ≪Wide bandgap devices≫

Abstract

High-speed motor drives are becoming more common in industrial applications. As they are characterised by low inductance, high switching frequencies of the inverter are required, therefore limiting the control cycle time. To cope with these issues, a direct model predictive controller with long switching horizon is proposed in this paper.

Introduction

The concept of Model Predictive Control (MPC) is known since some decades but due to high computational effort of the algorithm, the application to power electronics and motor drives is rather new. The development of fast and powerful microprocessors over the last decades have made it possible to use MPC in power electronics and drives. Furthermore new concepts like Integrated Modular Motor Drives (IMMD) would benefit from the use of MPC if this control methodology could be used also in high-speed applications [1]. Non-linear systems are also ideal candidates for MPC [2] as any non-linearity can be introduced in the model used for the algorithm. These represents a considerable advantage over other control methods that are based on linearised or piece-wise linear models. Among the different MPC algorithms applied to electrical drives the following are to be addressed. Direct torque control (DTC) [3], [4], [5] with very good torque performance and short response times but inherent torque and flux ripples [6]. Model predictive torque control with a finite control set (FCS-MPTC) [7], [8] improving traditional DTC at the cost of a higher calculation effort. To reduce the torque and flux ripple the dead-beat direct torque control [9], [10] was proposed, resulting however in higher calculation effort and effectiveness relying on a precise model of the system. Direct current control (DCC)[11] and trajectory based control [12] are to be mentioned. In 2018, Walz et. al [13], proposed a hysteresis based multiple prediction step system (MPC with bounds) with a short prediction horizon and on a low switching frequency. In 2015, Xie et al. [14] proposed a dead-beat-direct torque and flux control (DB-DTFC) to reduce the calculation effort with a prediction horizon of $N_p = 1$. The control method proposed in this paper is a Direct MPC and DB applied to a high-speed core-less axial flux integrated modular motor drive CAF-PMM-IMMD

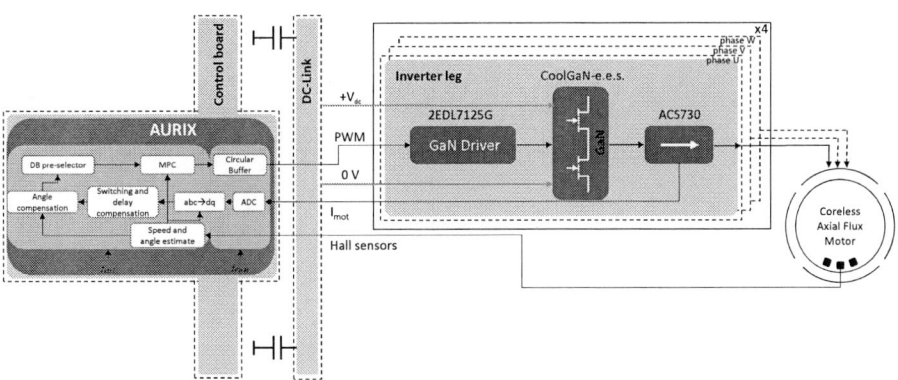

Fig. 1: Block diagram of controller and CAF-PMM-IMMD system.

[1] with low inductance (51 μH). A long switching horizon of length $N_s = 10$ is used to determine the trajectory of the stator currents. DB is used to reduce the set of voltage vectors. Delay compensation and receding horizon policy are implemented. The switching horizon is synchronous to the PWM frequency $f_s = 100$ kHz, while the prediction horizon is synchronous to the measurements' sampling time and control cycle at $f_p = 10$ kHz. In the paper it will be shown by simulations and experimental results that the long switching horizon MPC (LSH-DMPC) is a feasible and robust control method resulting also in low computational burden. Furthermore the proposed algorithm is implemented on a custom hardware therefore showing the feasibility for being applied in real industrial drives. The block diagram of the CAF-PMM-IMMD where the LSH-DMPC is implemented is shown in Fig.1.

This paper is organised as follows: at first the discrete time model of a PMSM and general theoretical framework of LSH-DMPC is introduced; the next section illustrates the outcome of the simulations; following the description of the software implementation and experimental results; finishing up with the conclusions.

Long Switching Horizon DMPC

The discrete model of the core-less axial flux motor is provided with

$$\hat{i}_s(k+1) = \mathbb{A}i_s(k) + \mathbb{B}u_s(k) + \mathbb{C} \quad \text{with} \quad i_s(k) = \begin{bmatrix} i_d \\ i_q \end{bmatrix}, \ u_s(k) = \begin{bmatrix} u_d \\ u_q \end{bmatrix} \tag{1}$$

$$\mathbb{A} = \begin{bmatrix} 1 - \frac{R_s T_s}{L_d} & \frac{L_q T_s \omega_e(k)}{L_d} \\ -\frac{L_q T_s \omega_e(k)}{L_d} & 1 - \frac{R_s T_s}{L_q} \end{bmatrix}, \quad \mathbb{B} = \begin{bmatrix} \frac{T_s}{L_d} & 0 \\ 0 & \frac{T_s}{L_q} \end{bmatrix}, \quad \mathbb{C} = \begin{bmatrix} 0 \\ \frac{-\psi_{PM} T_s \omega_e(k)}{L_q} \end{bmatrix}. \tag{2}$$

Note that k is the $k-th$ switching period and $T_p = N_s T_s$ is defined as the control time period.
The core concept of the proposed algorithm is to generate an optimal switching sequence to be applied (at PWM frequency) during the control cycle that is executed at a lower speed thus providing sufficient time to execute the high number of calculations comprised in the MPC. The ratio between control cycle frequency and PWM switching frequency is constant and set to $f_P/f_s = 10\text{kHz}/100\text{kHz} = 1/10$.

Fig. 2a shows the sampling of the currents and of actuation of optimal voltage vectors. Fig. 2b depicts in detail the process of selection of the optimal voltage vectors based on DB controller + MPC approach.

The control problem for the LSH-DMPC can be formulated as follows:

$$g(i) = \|\mathbf{y}^*(i) - \mathbf{y}(i)\|^2 + \lambda_u \|\Delta \mathbf{u}(i)\|^2, \quad \text{with} \quad \forall i = k + N_s, \ldots, k + 2N_s - 1 \tag{3}$$

subject to

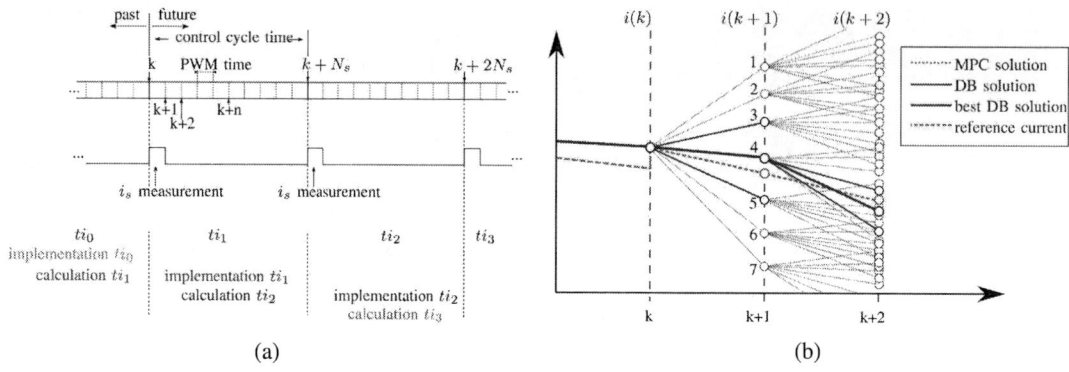

(a)   (b)

Fig. 2: (a) Time scheduling of LSH-MPC with sampling at control period. (b) Prediction of the current trajectory during the switching horizon using DB to pre-select the voltage vectors.

$$\mathbf{y}(k+N_s) = \hat{\mathbf{y}}(k+N_s). \tag{4}$$

It is defined *current time interval* $k,\ldots,k+N_s-1$ the time period during which the switching sequence calculated in the preceding control period is applied. It is defined *subsequent time interval* $k+N_s,\ldots,k+2N_s-1$ the time period for which the switching sequence is going to be calculated. Note that, as shown in Equ. 4, the controlled output $\mathbf{y}(k+N_s)$ at the beginning of the subsequent control time interval is assumed to be equal to the extrapolated value $\hat{\mathbf{y}}(k+N_s)$. In this way the computation delay is compensated.

The switching sequence and optimisation problem are defined as:

$$\mathbf{U}(i) = \left[\mathbf{u}^T(i)\,\mathbf{u}^T(i+1)\,\ldots\,\mathbf{u}^T(i+N_s-1)\right] \tag{5}$$

$$\begin{cases} \hat{\mathbf{x}}(m+1) = \mathbb{A}\hat{\mathbf{x}}(m) + \mathbb{B}\mathbf{u}_{opt}(m) \\ \hat{\mathbf{y}}(m+1) = \mathbb{C}\hat{\mathbf{x}}(m+1) \\ \quad\hat{\mathbf{x}}(k) = \mathbf{x}_{\mathbf{meas}}(k), \quad \forall m = k,\ldots,k+N_s-1 \end{cases} \qquad \begin{cases} \mathbf{x}(i+1) = \mathbb{A}\mathbf{x}(i) + \mathbb{B}\mathbf{u}(i) \\ \mathbf{y}(i+1) = \mathbb{C}\mathbf{x}(i+1) \\ \quad \forall i = k+N_s,\ldots,k+2N_s-1 \end{cases} \tag{6}$$

where $\mathbf{U}(i)$ is the sequence of inverter switch positions of the subsequent time interval, the m-index is for the current time interval, the i-index is for the subsequent time interval. For the switching effort:

$$\begin{cases} \Delta\mathbf{u}(i) = \mathbf{u}(i) - \mathbf{u}(i-1) \\ \mathbf{U}(i) \in \mathbb{U}_{DB} \subset \mathbb{U} = 7 \text{ Voltage Vectors} \\ \mathbb{U} \in \{1+j0, \frac{1}{2}+j\frac{\sqrt{3}}{2}, -\frac{1}{2}+j\frac{\sqrt{3}}{2}, -1+j0, -\frac{1}{2}-j\frac{\sqrt{3}}{2}, \frac{1}{2}-j\frac{\sqrt{3}}{2}, 0\} \end{cases} \tag{7}$$

In Equ. 6 the optimized switching sequence $\mathbf{u}_{opt}(m)$ is used to extrapolate the trajectories of the states and of the output the final value of which will be used as initial value to calculate the new optimised switching sequence. Equ. 6 left, shows that the measured states are used as initial values for the current time interval therefore applying the receding horizon policy. According to Equ. 7, the selected switching states are a subset of the possible 7 voltage vectors in a three-phase two-level inverter and are pre-selected by the dead-beat control (DB subscript). It should be noted that the algorithm is implemented in dq–frame while the selection of voltage vectors by the DB control is carried out in $\alpha\beta$–frame.

The DB controller estimates the 3 adjacent voltage vector that are applied to Equ. 1 according to:

$$u_s(k) = \mathbb{R}i_s(k) + \omega_e(k)\mathbb{G}i_s(k) + \frac{\mathbb{H}}{T_s}(i_s^*(k) - i_s(k)) + \mathbb{F}\omega_e(k)\psi_{PM} \tag{8a}$$

$$\mathbb{R} = \begin{bmatrix} R_s & 0 \\ 0 & R_s \end{bmatrix}, \quad \mathbb{G} = \begin{bmatrix} 0 & -L_q \\ L_d & 0 \end{bmatrix}, \mathbb{H} = \begin{bmatrix} L_d & 0 \\ 0 & L_q \end{bmatrix}, \mathbb{F} = \begin{bmatrix} 0 \\ 1 \end{bmatrix} \tag{8b}$$

A compensation of the rotor rotation during the control cycle is implemented by:

$$\hat{\theta}_e(k) = \theta_e + \omega_e T_p + \omega_e(k-1)T_s; \quad \forall k \in (k, k+N_s-1) \tag{9}$$

where θ_e is the electrical angle measured at the beginning of each control period k, $\omega_e T_p$ adds the compensation due to the control period, $\omega_e(k-1)T_s$ adds the compensation due to each PWM cycle. The flowchart in Fig. 3b summarises the described control algorithm.

Simulations

The proposed system has been simulated into MATLAB/Simulink. Whereas power electronics, motor, ADCs and the model predictive controller are either modelled or implemented in C-pseudo code. The implemented controller has been tested under several operating points. The following are the main building blocks of the simulation: Three-phase inverter: implemented through the Simscape/Specialized Power Systems toolbox. Switches and body diodes are represented by their basic characteristics such as forward voltage drop and differential resistance, Modulator: triangular carrier modulator with discretized steps, frequency $f_s = 100\,\text{kHz}$. Although not strictly needed for MPC implementation, the carrier modulator approach allows an easy comparison between a standard PI controller in the dq-frame and MPC. The switch on/off state is selected by simply setting the reference either to the maximum or minimum value of the carrier, Motor: model of the CAF-PMM including rotor inertia and bearing friction. Simulations can be also run at an ideal constant fundamental frequency in order to be able to test the performance of the control algorithm with short simulation time. General drive parameters are given in Tab. Ia, Rotor angle: transitions of hall-sensors have been implemented in order to test the algorithm for the angle estimate used in the real implementation. Ideal electrical angle can be also generated, ADCs: quantization as well as timing of the ADCs as implemented in the hardware. In particular the synchronous sampling has been implemented: the ADC of first phase is triggered at the peak of the carrier and the second phase is measured 100 ns afterwards, MPC: model predictive controller has been implemented into C-pseudo code using a triggered MATLAB function. The code is executed at $f_p = 10\,\text{kHz}$. This allowed to easily export the MPC to the microcontroller with only minor changes.

In Fig. 5 the three-phase currents are shown and zoomed in order to show the control period of 100 µs, the PWM carrier and the states of the top switch of each inverter leg showing therefore the commutations selected by the MPC. Furthermore an insight is provided in the output values of the ADCs at PWM frequency as well as their value as seen by the controller. Fig. 3a shows the current waveform for three different current levels. As can be seen by direct inspection of the figure, the tracking of the i_q reference (i_d was set to zero) is nearly perfect as well as the ripple remains bounded between approximately $\pm 2\,\text{A}$.

Although a thorough comparison between MPC and conventional PI-dq control is outside the scope of the present paper, some simulations have been run in order to highlight the superior performance of the proposed controller. The comparison has been done generating a large current step response from zero to maximum current at different electrical frequencies. The PI-dq control bandwidth has been set to 200 Hz for stability reasons and the pole-zero cancellation technique has been applied. As can be seen in Fig. 8a, the MPC generates an almost instantaneous response at all frequencies. On the contrary the PI-dq

Table I: (a) Drive parameters. (b) Parameters for Robustness Simulations.

(a)

# of pole pairs	5
Rated voltage / current	24 V / 15 A$_{rms}$
Phase resistance	12 mΩ
d/q axis inductance	4 µH + 47 µH
Flux linkage	1 mWb

(b)

	Val 1	Val 2	Val 3
Electrical frequency (Hz)	100	1250	
Flux constant (% var. of nom.)	-20	0	+20
Resistance (% var. of nom.)	-50	0	+50
Inductance (% var. of nom.)	-50	0	+50
Reference current (A)	5	15	

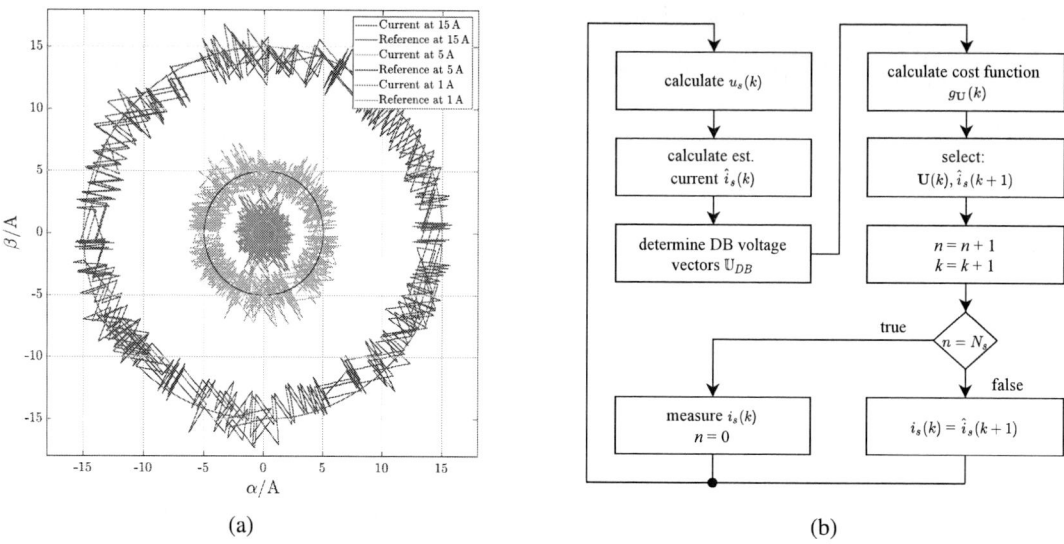

(a)   (b)

Fig. 3: (a) Simulated currents in the αβ−frame at three different operating points and electrical frequency of 300 Hz, (b) Flowchart of the control algorithm.

control shows not only a slower response but also increased overshoot as long as the electrical frequency increases. The reason for this behavior has been identified in the non compensated control and sampling delays that ultimately leads to an unstable system for the PI-dq control at maximum electrical frequency of 1250Hz or 15.000rpm. The delay compensation is on the other hand intrinsic in the adopted MPC algorithm.

Robustness Analysis

In order to analyse the robustness and stability of the proposed algorithm, 162 simulations have been run varying the parameters shown in Tab. Ib. The results are displayed in Fig. 4a. As a figure of merit the RMS error between the reference current and the motor current in the αβ−frame has been chosen, according to

$$e_{\alpha\beta,rms} = \sqrt{\frac{1}{N_s} \sum_{n=1}^{N_s} (i_{\alpha,ref} - i_\alpha)^2 + (i_{\beta,ref} - i_\beta)^2} \tag{10}$$

where N_s is the number of samples considered in the simulation. Note that five cycles at the fundamental frequency have been assumed to calculate the RMS error. The plots show contour lines of the RMS error as a percentage of the operating motor current generated by varying the five parameters shown in Tab. Ib. Note that the parameters are changed only in the MPC algorithm and not in the model of the motor assumed in the simulation. The RMS error contains both the information related to the ripple content as well as to the error in tracking the reference circle (in the αβ−frame, compare Fig. 3a). Horizontal axis represents the MPC model motor resistance. Vertical axis represents the MPC model motor inductance. In each plot the operating point at nominal R, L is located in the centre. It can be noted that in general, at lower motor electrical frequencies, the RMS error is relatively insensitive to even large variations of motor parameters R, L, Ψ and operating current I. On the other side, at maximum frequency 1250 Hz, the RMS error is more sensitive to under estimates of the motor resistance and quite sensitive to changes of the motor flux constant Ψ. Finally it should be noted that RMS error levels of around 25% at 5 A and 10% at 15 A show that the RMS error contains only the ripple and the tracking of the reference current is practically perfect.

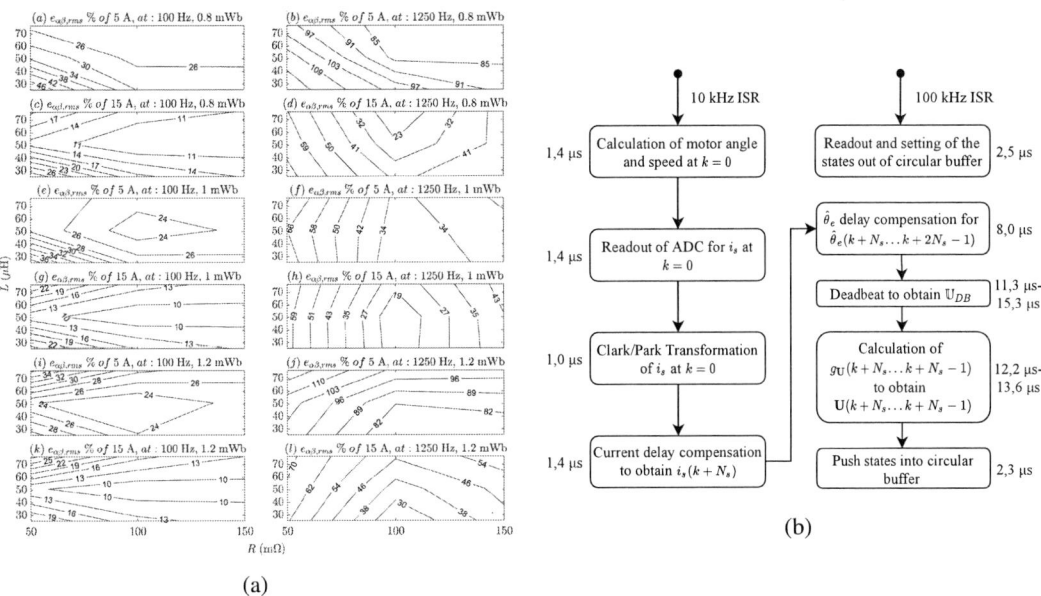

(a)

(b)

Fig. 4: (a) RMS error of the current calculated in the αβ−frame. Parameters: R phase resistance of the model, L phase inductance of the model, f_{mot} electrical frequency, Ψ motor flux constant, I_{rms} phase current, (b) Flowchart of the implemented algorithm with the corresponding execution times. The control ISR (10 kHz) and the PWM ISR (100 kHz) share the available 100 µs per control cycle.

Software implementation

The proposed algorithm is implemented on a SAK-TC233LP-32F200N AC which is an automotive grade TriCore™ v1.6E 32-bit MCU with a base frequency of 200 MHz. The algorithm is utilising two interrupt service routines (ISR) whereas one is triggered at the PWM frequency of the inverter and one at control frequency. Both ISR share a common circular buffer which stores the calculated on-times for the inverter. The calculated on-times are taken out of the circular buffer and set accordingly in switching PWM triggered ISR. Hence this ISR has the highest priority and the control ISR is interrupted on a regular basis.

The control ISR starts by calculating the motor angle and speed out of the hall sensor signals. Latter is done by estimating the angular velocity based on the $\Delta\theta$ to the last time step. This velocity is averaged with the ones of earlier timesteps and used in combination of the hall sensor signals to calculate the actual motor angle. In the next step, the ADC of the phase current measurements are read out. Hereby two of them are obtained and the third calculated.

With the calculated motor angle as well as with the measured currents the Clark/Park transformation is done which serves as starting point for the current trajectory extrapolation to compensate the calculation delay. Later estimation, see Listing 1, is based on the motor model, the angular velocity, the current estimation one time step earlier as well as the chosen voltage vector active for the particular time step which has been calculated the control cycle prior. The last current estimation, which is for the time step $k + N_s$, is then used for the calculation of the first voltage vector.

Listing 1: Current trajectory extrapolation for the time steps $k = 1$ up to $k = N_s$.

```
for (int k = 0 ; k < 10; k++){
    MPC->id_est[k+1] = MPC->A_d*MPC->id_est[k] + B_d_times_omega_el_rad*MPC->iq_est[k] + MPC->C_d*MPC->DB.
        d_best_store[k];
    MPC->iq_est[k+1] = MPC->A_q*MPC->iq_est[k] + -B_q_times_omega_el_rad*MPC->id_est[k] + MPC->C_q*MPC->DB.
        q_best_store[k] + Psi_C_q_times_omega_el_rad;
}
```

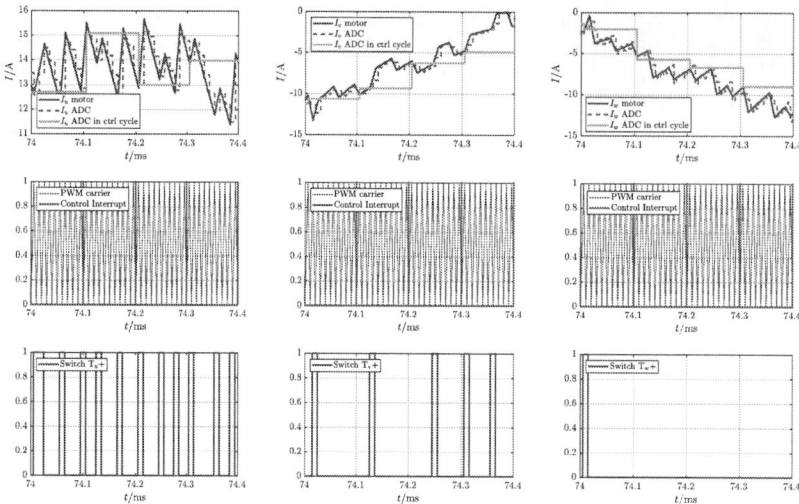

Fig. 5: Waveforms at 15 A, 300 Hz. First row: phase currents showing also ADC output. Second row: PWM carrier and Control Interrupt. Third row: State of top switch of each leg.

Following the current trajectory extrapolation, the estimated electrical angle $\hat{\theta}_e$ (Equ. 9) for the time steps $k+N_s$ to $k+N_s-1$ are calculated, based on the measured motor angle θ_e and the current angular velocity ω_e. These angles are then used to perform the Park transform to convert from the $\alpha\beta$-frame to the dq-frame to then select the correct sector, see listing 2. The motor current for the subsequent time interval is then calculated for each switching cycle basing on the motor model and the pre-selected voltage vectors $\mathbb{U}_{DB}$.

Listing 2: Pre-calculations for dead-beat selection.

```
for (int k = 0 ; k < 10; k++){
    MPC->ud[k] = MPC->R_s *MPC->id[k] - Lq_times_omega_el_rad*MPC->iq[k] + MPC->inv_C_d*(i->d_ramp-MPC->id[k]);
    MPC->uq[k] = MPC->R_s *MPC->iq[k] + Ld_times_omega_el_rad*MPC->id[k] + MPC->inv_C_q*(i->q_ramp - MPC->iq[k])
        - Psi_times_omega_el_rad;
    MPC->ualpha[k] = MPC->cosf_theta_est[k] * MPC->ud[k] - MPC->sinf_theta_est[k]*MPC->uq[k];
    MPC->ubeta[k] = MPC->cosf_theta_est[k] * MPC->uq[k] + MPC->sinf_theta_est[k]*MPC->ud[k];
    MPC->DB[k] = atan2f(MPC->ubeta[k], MPC->ualpha[k]);

        ...
}
```

The reference current is subtracted by the individual currents whereas this difference is squared hence building the cost function. Hereby the minimal result is then chosen as optimised voltage vector for the current time step, see listing 3. The process is then repeated along the switching horizon.

After the selection of the best voltage vectors, the corresponding states are pushed into the circular buffer. The states stored in the circular buffer are then implemented by the PWM ISR as shown in Fig. 4b.

Listing 3: Minimisation of the cost function to obtain **U**.

```
for (int vv = 1; vv!=3; ++vv){
    MPC->DB_d[k][vv] = MPC->cosf_theta_est[k]*MPC->a[k][vv] + MPC->sinf_theta_est[k]*MPC->b[k][vv];
    MPC->DB_q[k][vv] = MPC->cosf_theta_est[k]*MPC->b[k][vv] - MPC->sinf_theta_est[k]*MPC->a[k][vv];
    MPC->i_d[k+1][vv] = i_d_temp + MPC->C_d*MPC->DB_d[k][vv];
    MPC->i_q[k+1][vv] = i_q_temp + MPC->C_q*MPC->DB_q[k][vv];
    MPC->imax = 1e6;
    if ((MPC->i_d[k+1][vv]*MPC->i_d[k+1][vv] + MPC->i_q[k+1][vv]*MPC->i_q[k+1][vv]) < 100)
        MPC->imax = 0;
```

```
    ...
    g = sqr_isd_vv + sqr_isq_vv + MPC->imax;

    if(g < min){

        ind = vv;

        min = g;

    }

}
```

Execution Time

In Fig. 4b one can find next to each routine the corresponding execution time. The complete algorithm takes 64.0 µs to 69.4 µs to complete. Whereas the first four blocks and the last one of the control loop are not computational expensive due to the consideration of only one timestep. The voltage vector calculation in combination with the cost function and the theta estimation on the other hand are time consuming. Latter is due to several factors. Starting for the theta estimation with consists of three floating point operations, 11 summations, 10 floating sine and 10 floating cosine operations. Hence each time step that needs to be calculated accounts for roughly 800 ns.

The calculations of the voltage vectors are due to the nested loops more resource demanding. As for every time step three voltage vectors need to be assessed and several operations need to be done 30 times per control loop cycle. This leads to a total of 406 floating point, 400 summations and 10 arc tangent operations. Every time step can therefore be accounted with 2.35 µs - 2.89 µs. The total CPU load of the algorithm is between 64%-69%.

As the native PWM frequency of the motor is 200 kHz, the frequency has had to be reduced as the computational effort is increased by a vast amount. If latter speed wants to be restored one has several options available. Although the algorithm leaves some room for improvement, the targeted frequency increase can only be achieved if tradeoffs are made. These can be changing the cost function to an absolute approach instead of a squared sum or a linearisation of angle and current estimation. Another variant can be the change to a faster µC like the SAK-TC333LP-32F300F AA with 300 MHz or one with a higher core count like the SAK-TC264DC-40F200W BC.

Experimental Results

Several electrical test are carried out on the motor. All tests and measurements done are with no mechanical load on the shaft. The measurement setup can be seen in Fig. 6. All values used in the following

Fig. 6: Measurement setup used. On the right side the mounted motor is depicted. The individual phases of the motor and the individual half-bridges are brought out with cables of the lid. Hereby the additional phase inductance of 47 µH is connected as seen at the bottom of the picture. Two current probes of the type TCPA300 are used to measure the phase current which are connected to a PicoScope 3405D MSO.

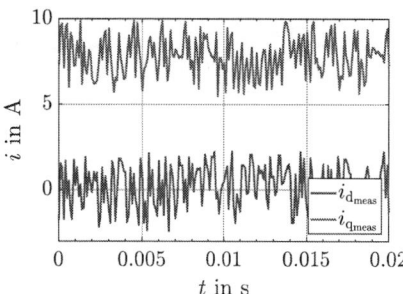 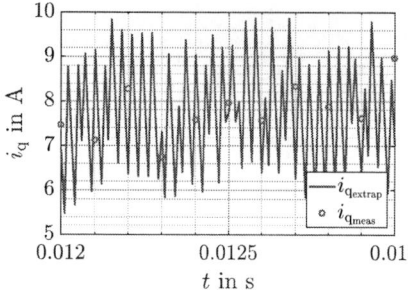

Fig. 7: Measured i_q and i_d. A longer timespan is depicted left and a zoomed in illustration including the extrapolated current on the right.

plots are read out of the µC in order to better understand what latter is seeing. The depicted current probes are only used to double checking the ADC measurements.

Current Extrapolation

The current extrapolation discussed in section and is vital for the following calculation of the voltage vector. In Fig. 4b the structure of the algorithm is shown. The measured currents are obtained in each control ISR by the ADCs while it's value at the end of the current switching horizon is extrapolated and used for the computation of the predicted current that occurs at control ISR (10 kHz); note however that the predicted currents need to be computed for each switching cycle (100 kHz). As it can be seen in Fig. 7 the ripple current for $i_{q_{meas}}$ and $i_{d_{meas}}$ is in the same order of magnitude as in Fig. 3a. In Fig. 7 on the right a portion of the graph is shown, where one can see the good agreement of $i_{q_{meas}}$ and $i_{q_{extrap}}$. A higher inductance motor hence a lower current ripple would improve the implemented current extrapolation.

Voltage Vector Calculation

As stated in section and , the extrapolated currents are used to calculate back into the $\alpha\beta$-frame were the voltage vector is chosen. According to latter the resulting current is computed from which the next voltage vector can be chosen. In Tab. II this process is shown for one control cycle. In the first two columns θ_{meas} and the resulting angle of the conversion to the $\alpha\beta$-frame θ_{DB} is stated. It has to be noted that θ_{meas} and θ_{DB} are two different angles. In particular, θ_{DB} is related to the sector considered in the calculations.

θ_{DB} is determining $\mathbb{U}_{DB}$, hence the two possible voltage vectors additional to the zero vector. In the columns 3-8 the resulting i_d and i_q out of the three possible voltage vector is shown. From latter the difference between the reference current to the computed is squared. Both are summed up and the lowest

Table II: Depiction of the process a voltage vector is chosen.

θ/rad		i_d/A			i_q/A			State		
DB	meas	$\mathbb{U}_{DB0}$	$\mathbb{U}_{DB1}$	$\mathbb{U}_{DB2}$	$\mathbb{U}_{DB0}$	$\mathbb{U}_{DB1}$	$\mathbb{U}_{DB2}$	u	v	w
-0.76	-1.34	-2.3	1.18	-1.48	3.5	4.56	7.04	0	1	0
0.98	-1.30	1.34	2.31	-1.21	3.45	6.96	6.04	0	0	1
-1.26	-1.26	-0.93	1.51	2.63	5	2.31	5.77	1	1	1
-0.25	-1.22	-0.7	2.88	0.56	3.97	4.59	7.39	1	1	1
0.15	-1.18	-0.52	0.88	-2.72	2.96	6.32	5.85	0	1	1
2.17	-1.13	1.13	-2.5	-0.41	5.18	4.87	1.89	1	1	1
1.43	-1.09	1.32	-0.64	-2.31	4.07	7.13	3.9	1	1	1
1.14	-1.05	1.46	-0.37	-2.18	2.97	6.11	2.95	0	0	1
-1.3	-1.01	-0.11	1.59	3.53	5.03	1.82	4.89	1	1	1
0.71	-0.97	0.11	2.17	-1.45	3.97	6.96	7.25	1	1	1

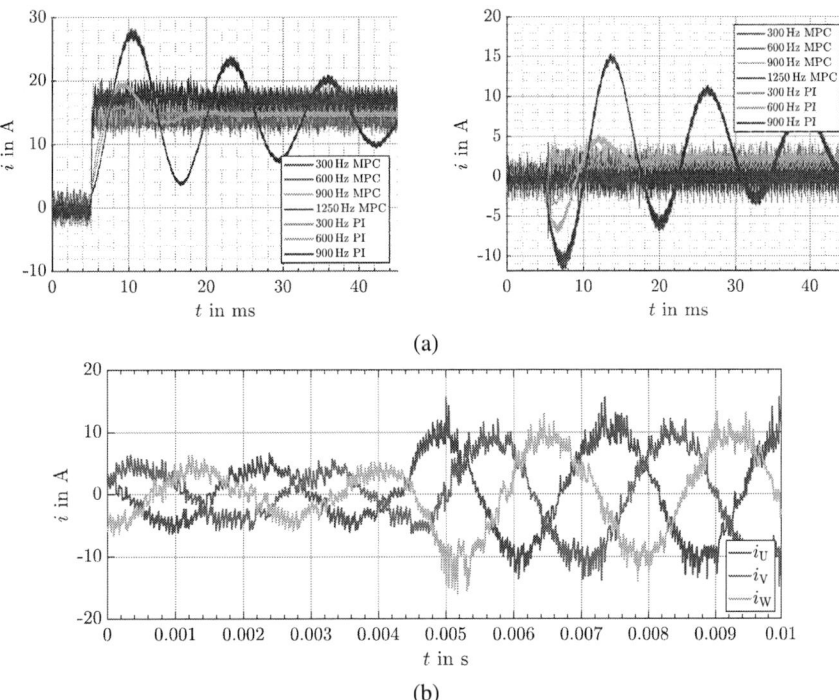

Fig. 8: (a) Step response comparison of PI-dq and MPC showing i_q (left) and i_d (right). (b) Measurement of the phase currents of a step of $i_{q,ref}$ from 4 A to 8 A.

result is leading to the best switching state to track the reference. These values are indicated by the background colour green in Tab. II.

Dynamic Behaviour and Phase-currents

In order to better compare the measurements to the simulation at an $i_{q,ref}$ step from 4 A to 8 A the phase currents are measured, see Fig. 8b. The motor reacts near instant to the reference change and shows no sign of overshoot or oscillations. It has to be stated, that the implemented model has to have a strong agreement to the actual used hardware. This has to be ensured as the differential part in Equ. 8a is inherently very sensitive to high ripple as well as to noise. In combination with a deviation of model to real setup the MPC will show a reduced performance. The influence of noise can be seen in the phase currents in Fig. 8b especially when approaching higher reference currents.

Conclusion

In this paper, the feasibility of a long switching horizon MPC for low sampling to carrier ratios f_p/f_s is proven. Simulations and measurements show the effectiveness of the proposed LSH-DMPC. A robustness analysis will be carried out proving stability and performance of the LSH-DMPC. Experimental investigation describes the implementation on a custom designed control and power electronics investigating in detail execution time of the different sections of the algorithm as well as other implementation issues. The proposed system shows therefore the feasibility of being implemented in an actual industrial application. In future work, investigation on parameters' sensitivity, ADC optimisation and multiple converter control will be carried out.

References

[1] M. Schiestl, F. Marcolini, M. Incurvati, F. G. Capponi, R. Stärz, F. Caricchi, A. S. Rodríguez, and L. Wild, "Development of a high power density drive system for unmanned aerial vehicles," IEEE Transactions on Power Electronics, vol. 36, no. 3, pp. 3159–3171, 2021.

[2] J. Rodriguez, P. Cortes, Predictive Control of Power Converters and Electrical Drives. John Wiley & Sons, 2012.

[3] I. Ludtke, "Direct torque control of induction motors," in IEE Colloquium on Vector Control and Direct Torque Control of Induction Motors. IEE, 1995.

[4] T. Geyer, "Computationally efficient model predictive direct torque control," IEEE Transactions on Power Electronics, vol. 26, no. 10, pp. 2804–2816, oct 2011.

[5] G. Tobias, "Generalized model predictive direct torque control: Long prediction horizons and minimization of switching losses," Proceedings of the 48h IEEE Conference on Decision and Control (CDC) held jointly with 2009 28th Chinese Control Conference, pp. 6799–6804, 2009.

[6] P. Vas, Sensorless Vector and Direct Torque Control. Oxford University Press, 1998. [Online]. Available: https://www.ebook.de/de/product/12940375/peter_vas_sensorless_vector_and_direct_torque_control.html

[7] M. Preindl and S. Bolognani, "Model predictive direct torque control with finite control set for PMSM drive systems, part 1: Maximum torque per ampere operation," IEEE Transactions on Industrial Informatics, vol. 9, no. 4, pp. 1912–1921, nov 2013.

[8] J. Rodriguez, M. P. Kazmierkowski, J. R. Espinoza, P. Zanchetta, H. Abu-Rub, H. A. Young, and C. A. Rojas, "State of the art of finite control set model predictive control in power electronics," IEEE Transactions on Industrial Informatics, vol. 9, no. 2, pp. 1003–1016, may 2013.

[9] J. S. Lee, R. D. Lorenz, and M. A. Valenzuela, "Time-optimal and loss-minimizing deadbeat-direct torque and flux control for interior permanent-magnet synchronous machines," IEEE Transactions on Industry Applications, vol. 50, no. 3, pp. 1880–1890, may 2014.

[10] B. Kenny and R. Lorenz, "Stator- and rotor-flux-based deadbeat direct torque control of induction machines," IEEE Transactions on Industry Applications, vol. 39, no. 4, pp. 1093–1101, jul 2003.

[11] J. C. R. Martinez, R. M. Kennel, and T. Geyer, "Model predictive direct current control," 2010 IEEE International Conference on Industrial Technology, pp. 1808–1813, 2010.

[12] S. Bolognani, S. Bolognani, L. Peretti, and M. Zigliotto, "Design and implementation of model predictive control for electrical motor drives," IEEE Transactions on Industrial Electronics, vol. 56, no. 6, pp. 1925–1936, jun 2009.

[13] S. Walz, R. Lazar, and M. Liserre, "Multi-step model predictive control for a high-speed medium-power pmsm," 2018 IEEE Energy Conversion Congress and Exposition (ECCE), pp. 5040–5046, sep 2018.

[14] W. Xie, X. Wang, F. Wang, W. Xu, R. M. Kennel, D. Gerling, and R. D. Lorenz, "Finite-control-set model predictive torque control with a deadbeat solution for pmsm drives," IEEE Transactions on Industrial Electronics, vol. 62, no. 9, pp. 5402–5410, sep 2015.

Standalone Power Management System for Flexible Piezo Electric Nano Generators (PENG) Based on the Co-Polymer P(VDF:TrFE)

Alexander Wölk[*], Mahmoud Shousha[*], Shashank Shekhawat Singh[*], Martin Haug[*], Lorandt Fölkel[*], Michael Brooks[*], Asier Alvarez[¥], Andreas Petritz[¥], Philipp Schäffner[¥], Jonas Groten[¥], Andreas Tschepp[¥], Barbara Stadlober[¥]

[*]Würth Elektronik eiSos Group, Germany
[¥]Joanneum Research, Graz, Austria
E-Mail: mahmoud.shousha@we-online.de
URL: Würth Elektronik eiSos Group

Acknowledgements

The authors would like to acknowledge the funding from the European Union through the H2020, SYMPHONY project (https://www.symphony-energy.eu/). The project SYMPHONY receives funding from the European Union's Horizon 2020 research and innovation programme under Grant Agreement No. 862095.

Keywords

«Piezo actuators», «Maximum power point tracking», «Power management», «Energy harvester».

Abstract

This paper presents the design and manufacturing of a PENG based on flexible PVDF. The power management system (PMS) used to convert its ac voltage into a dc one, providing MPPT is presented. The system is tested to cover two use cases, namely energy storage in a battery and in a standalone configuration. The PMS is able to produce adjustable output voltages ranging from 0.8V to 1.5V.
With a regulated 1.5V the PMS is capable of producing up to 900nJ of energy from bending single deformation of the harvester. The PMS can harvest as low as 1.4µW from the PENG. A low power MPPT system, consisting of a single comparator is shown. In addition, insights about selecting optimal input capacitance of the PMS under different use cases is given in this paper. The MPPT technique improved the harvested output power by up to 44%.

Introduction

The Internet of Things (IoT) is the talk of the town, in which self-sustainable autonomous sensors and battery systems have gained popularity for everyday life. Energy harvesting (EH) technology can play a major role in realizing these standalone sensors and battery systems [1]-[3]. Energy harvesting has been attracting attention for a long time now, due to its ability to power small sensor nodes without the need of a battery or direct power connection [4]. Among EH concepts, piezoelectric nanogenerators (PENG) have the advantage of a simple design and high integration density. Usually such generators are made of piezoelectric ceramics like PZT, but these are brittle and due to their stiffness possess a high resonance frequency [5]. Polymer-based PENGs, in contrast, are a flexible alternative for harvesting at lower frequencies and outputs close to 1 mW have been shown [8]. In addition, they are lead free [6] and can be realized with low costs in a large area with techniques like roll-to-roll (R2R) or screen printing [7]. Energy harvesting from these polymer-based PENGs could be one of the potential sources for low-power devices, because of its sheer advantages in the form of flexibility, good stability, and its ease of handling and shaping [9]. A solution to make these low-power devices autonomous is by extending battery-operating time by storing the energy harvested from the PENGs along with achieving maximum power point tracking (MPPT) for higher efficiency. A major application that could benefit by extending the battery operating time is embedding a wearable energy harvesting device in everyday activities of athletes. These wearable energy harvesters can be implemented in shoes [10, 11], bags [12, 13], and clothing [14-16].

In this work, a piezoelectric energy harvesting system has been developed, by translating mechanical energy into electrical energy and rectifying the electrical energy into a regulated dc output voltage using a power management system (PMS). An important aspect of this application is that the higher power loads require both an EH and a battery, whereas lower power loads can be sufficiently powered by an EH alone. This requires a PMS that can handle both, battery-powered and battery-less systems. Hence, two application specific cases have been discussed in this paper to cover a wide range of energy harvesting applications, along with considering the overall size of the PMS to make it cost effective. A series of experiments are performed to characterize these harvester prototypes, proving that the harvester can serve as a wearable power supply for low power applications. Maximum power point tracking results in tuning the design to operate at the point where the EH produces its maximum energy and has its highest efficiency [17]. Hence, the MPPT feature has been implemented in the PMS to maximize the harvested energy. To combat the high power consumption of typical micro controller based MPPT implementations, a design using a simple circuit is utilized [21]-[25].

Flexible Transducer Fabrication and Characterization

The flexible ferroelectric transducers were fabricated in a capacitor-like structure on a 125 μm thin PET substrate as shown in the schematic in Fig. 1. The transducer structure was fully screen-printed and comprises a layer of the ferroelectric co-polymer P(VDF:TrFE)80:20 sandwiched between two layers of PEDOT:PSS electrodes as displayed in Fig. 1. Due to its piezo- and pyroelectrical characteristics, P(VDF:TrFE) can detect changes in stress, strain and even temperature. P(VDF:TrFE)-based transducers can convert a part of the mechanical/thermal input energy into electrical energy and this transducer property can be exploited for energy harvesting devices that transduce waste mechanical to electrical energy. The energy output of a flexible transducer (PENG) during periodic manual bending is shown in Fig. 1b. By varying the external resistance, the maximum output power P_{out} reached 25 μW for an optimum load resistance R_L = 6.9 MΩ, corresponding to a peak power density of 14 mW/cm3.

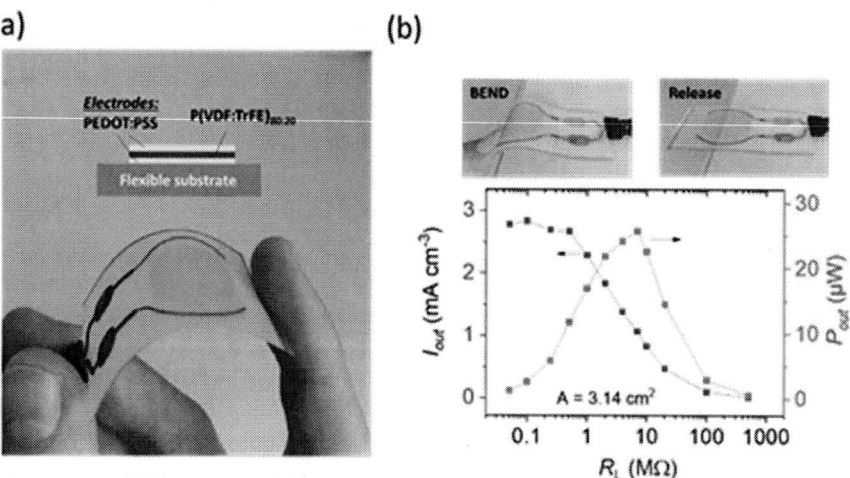

Fig. 1: Flexible P(VDF:TrFE)-based transducers. a) Photograph and schematics. b) Energy output by periodic manual bending: The current density of a random sample, I_{out}, as well as the corresponding output power, P_{out}, are plotted as a function of load resistance R_L.

Power Management System Practical Implementation

The PMS is developed based on the following block shown in Fig. 2 for a standalone and battery storage system. The energy source used for realizing the design is a polymer-based PENG which works on mechanical vibration and strain energy. Piezoelectric generators can generate high output voltages [18], up to a few volts. However, due to their small current characteristics, only in Microamps, they produce relatively low output power [18]-[20]. Hence, low leakage diodes and a very low quiescent current buck converter are utilized, resulting in a lower ambient current consumption of the harvester system.

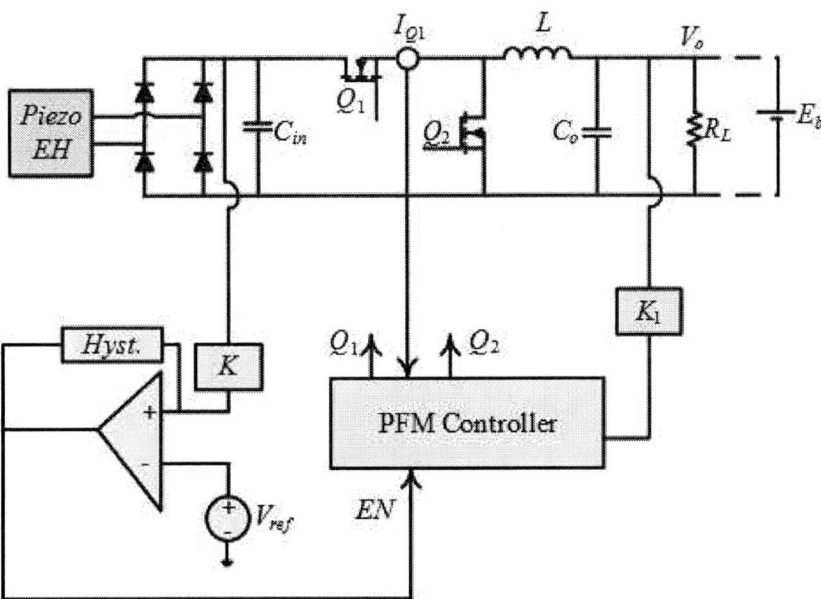

Fig. 2: Block diagram of the PMS.

The PMS system is composed of a PENG, a full-bridge rectifier, MPPT circuit with voltage divider (K) and comparator, a dc-dc buck converter with pulse-frequency modulation (PFM) control and a battery/load. The Piezo EH is modeled as an equivalent circuit consisting of a sinusoidal current source, a capacitor, C_p, and a large resistor, R_p, [26, 27]. The PENG's sinusoidal and random vibrational outputs are rectified through the full-bridge rectifier by converting the ac output into a dc one. Four low leakage diodes are used to realize the full-bridge rectifier.

As mentioned, the converter is realized with a pulse-frequency modulation (PFM) controller resulting in higher efficiency at light loads. PFM mode allows the buck converter to vary the switching frequency according to the load, ensuring that the inductor current always remains positive. This results in lower conduction losses compared to the conventional continuous conduction mode of operation (CCM), as well as provides the advantage of reduced switching losses due to the switching frequency decreasing with the load current. The rectified voltage from the PENG will charge the input capacitor C_{in} of the dc-dc converter up to ideally the open circuit voltage of the PENG, V_{oc}, if the converter is disabled. The energy harvested accumulates on the input capacitor C_{in}, until it is transferred by the buck converter to the output capacitor, C_o. Once the voltage across the input capacitor exceeds the predetermined reference, the buck converter is enabled and a regulated output voltage can be achieved. Hence, a high voltage and low-current output from the energy harvester is converted to an adjustable low-voltage (0.8V-1.5V) output for the standalone usage/or battery storage system. Low C_{in} and C_o capacitances are selected accordingly to start the buck converter with minimum current consumption, although resulting in a ripple-regulated output voltage, which is desired for a correct PFM and MPPT operation.

Maximum power transfer of a PENG occurs when the electric load matches the internal impedance of the energy harvester or, when the voltage across the electric load is half of the rectified open-circuit voltage, V_{oc}. Unlike more power-hungry solutions, the implementation of MPPT described in this paper has been achieved with only a voltage divider and a comparator. Through the electrical characteristics of the PENG shown above, it was determined that after rectification (1V drop across each diode), maximum power can be extracted at around 3V. Hence, the inverting input of the comparator is set to 0.2V (V_{ref}) and the non-inverting input of the comparator is fed through a voltage divider gain (K), comprising of two series-connected resistors. As the rectified voltage charges C_{in}, the divider output voltage also increases accordingly. Once the divider voltage goes above the reference voltage, the comparator provides an enable signal, resulting in the turn-on of the buck converter. C_{in} starts discharging by transferring energy to C_o, hence, the divider voltage drops below V_{ref}. Then, the EN signal gets driven low from the comparator, resulting in the converter shutting down. C_{in} charges back to its maximum value, maintaining the optimum voltage from the PENG.

Optimization of the Input Capacitor

In order to find the optimal input capacitor value for the harvesting circuit, several measurements were taken and evaluated. Since the buck converter is basically disconnected until the enabling voltage is reached, the load of the harvester circuit in this phase can be modeled as the rectifying stage plus the input capacitor. To find the local MPP of this circuit, the harvester was excited continuously using a shaker and the output power was calculated. Figure 3 shows the output power of the harvester for different input capacitance values with varying loads.

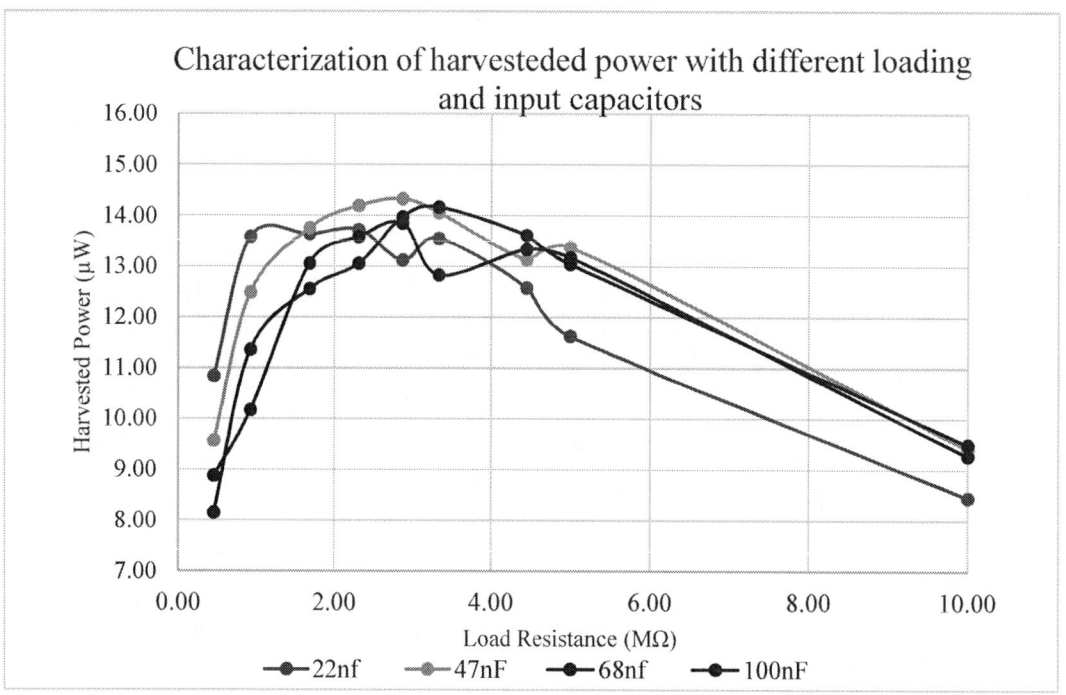

Fig. 3: Output power of the harvester for different input capacitance values with varying loads.

It can be noted that the output power varies against loading of the harvester with a slight variation due to the input capacitance value. This observation is important to consider when designing the PMS to ensure operation near the MPP, regardless of the input capacitance value. Hence, it is possible to track the MPP of the harvester, if one can track and change the input resistance of the buck converter as seen by the harvester. That happens by controlling the input voltage of the buck converter.

From Fig. 3, it can be interpreted that the input capacitance value has a negligible effect on the system design. However, if we consider energy as the judgement criterion to select the input capacitor, the picture will be different. That is due to the fact that a smaller capacitor will charge faster, improving the system performance during high frequency excitation of the harvester, such as the case where the harvester is mounted to a motor or wind turbine blades, similar to the Fig. 3 test conditions. Such use cases with continuous excitation will prefer a smaller input capacitance value as the input voltage does not drop quickly because of the nature of the use case. For other use cases where there is only a single excitation event, such as a smart floor, a bigger input capacitor will be more beneficial as it will hold the charge for longer time periods and results in more energy output. Another figure showing the output power vs input capacitor value could not be obtained due to the lack of a systematic single excitation source.

Maximum Power Point Tracking

There are several well-known methods to track the MPP of a system, such as perturb and observe, incremental conductance or the constant voltage method [28]. The problem with these methods is that they all need to measure the instantaneous voltage or current levels. Additionally a microcontroller is needed to perform calculations using the measured values to ensure MPPT. But these continuous measurements additional to the microcontroller need a lot of energy, which has to be provided, in case of a standalone version, by the system. It is questionable if the improvement through these calculations outweigh the additional energy consumption in a standalone system. Thus a standalone energy harvesting system stands to benefit if the MPPT implementation only relies on low energy components.

It was shown in the previous paragraph that the local maximum power point is highly dependent on the load resistance seen by the harvester (the input resistance of the buck converter), while the input capacitor has hardly any effect on that point. This means that the system should continue charging the input capacitor until that capacitor's voltage reaches the ideal level in relation to the input impedance of the converter and the converter's load resistance. It is possible to increase the output power of the harvester by varying the enabling voltage level of the converter. This is done using a low power comparator, which enables the converter at a certain input voltage level to increase the energy output of the converter compared to the system without the MPPT. The comparator was tuned to generate the enabling signal for the converter, when the MPP voltage is reached. Also, a hysteresis for the comparator was added to ensure the enable signal stays high as long as the input voltage is high enough for the converter to work. To verify this implementation, several measurements were taken using different PCBs. The input and output capacitors were varied to show how these values affected the energy output.

Experimental Results

To demonstrate the previously described concepts, different prototypes were built and tested. High impedance probes (>10MΩ) were used to collect all the data due to the low power nature of the application. The PMS was powered only by a PENG to realize a standalone use case. To prove the effectiveness of the PMS, all the measurements were collected with a single deformation event, not through continuous excitation which inherently generates more energy. The output voltage waveform of the PMS was processed using the CSV measurement provided by the oscilloscope. These data points were used to calculate the RMS voltage level of the output signal and then the output power and energy. This was done at different output voltage levels and different input and output capacitance values. To demonstrate the effectiveness of the presented MPPT concept, the MPPT block was disabled and enabled and the harvested output energy was compared.

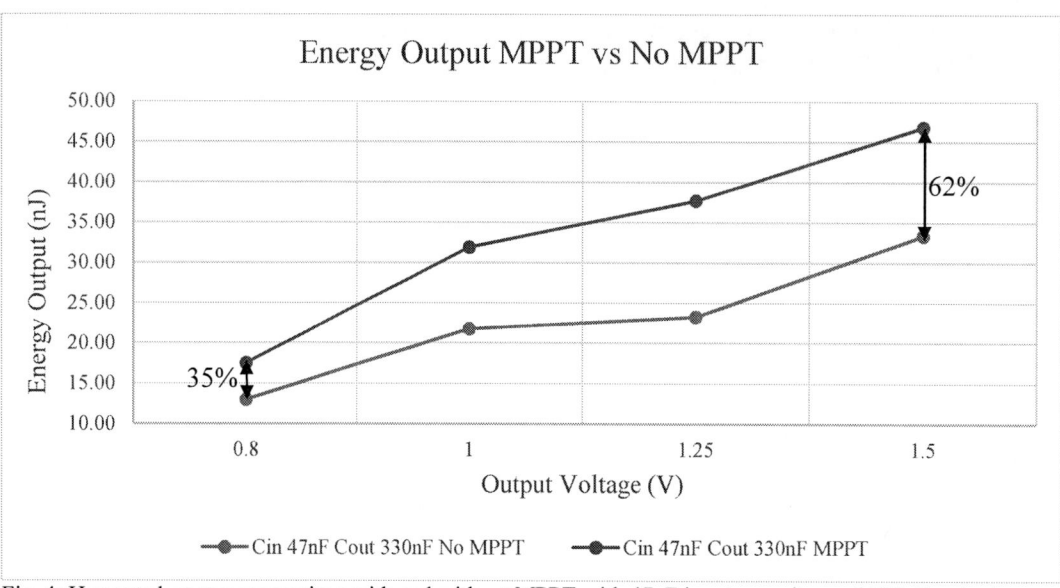

Fig. 4: Harvested energy comparison with and without MPPT with 47nF input capacitor and 330nF output capacitor.

Figure 4 shows a comparison between the harvested energy with and without the MPPT block using a 47nF input capacitor and a 330nF output capacitor. The MPPT technique presented in this work results in an improvement of the harvested energy between 35-62% with a single deformation event.

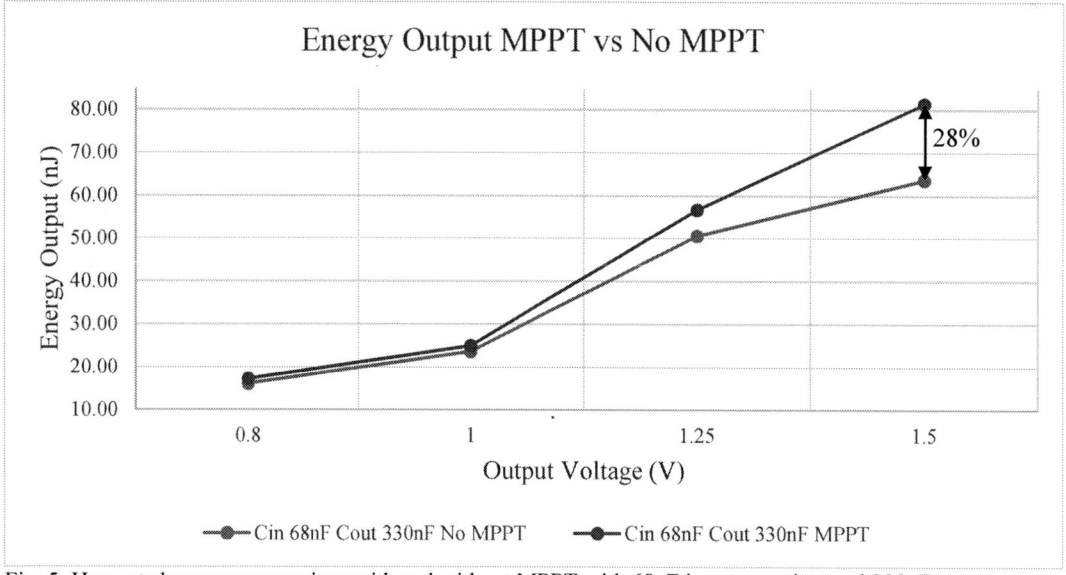

Fig. 5: Harvested energy comparison with and without MPPT with 68nF input capacitor and 330nF output capacitor.

The same test was repeated with a 68nF input capacitor and a 330nF output capacitor. The PMS was able to harvest more energy as explained earlier. The MPPT even improved the harvested energy by up to 28% at 1.5V. The general improvement of the output energy makes sense since the larger input capacitor can store more charges and needs longer to charge up than the smaller input capacitor. The size of the input capacitor does have limitations. To further increase the output energy, a 100nF input capacitor was used, but the circuit would not function properly because the input voltage of the comparator could not reach a level, where it would enable the converter. This showed, that 68nF is the optimal input capacitance value for the PMS evaluated for these measurements.

In order to optimize the output capacitor value, the output capacitor was changed between 100nF and 470nF and the previous test was repeated. The conclusion was that small output capacitance values results in a larger overshoots and had negative impact on the PFM stability while larger output capacitance values slow down the output voltage, reducing the useful time where the load can benefit from the output voltage value. Therefore, 330nF was used as an optimal capacitance value for the presented PMS.

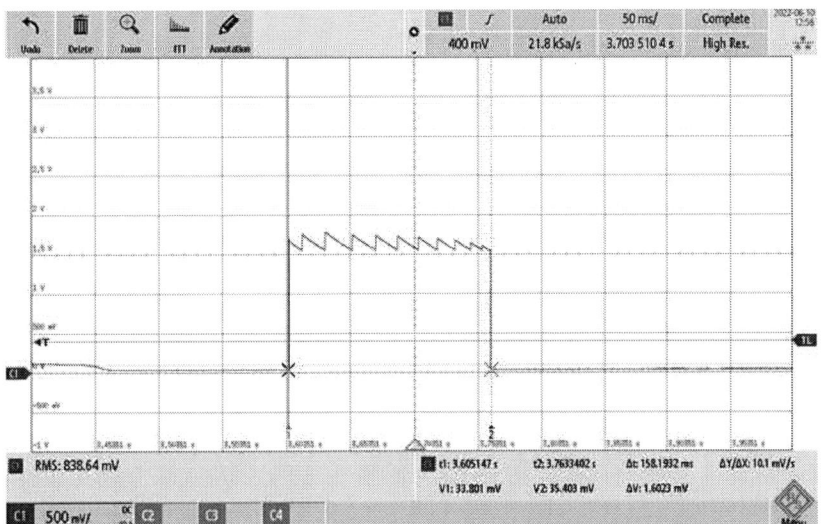

Fig. 6: Output voltage behavior of the standalone use case, with a nominal value of 1.5V.

Figure 6 shows the regulated output voltage with the optimal input and output capacitor values at 1.5V with a single deformation. It can be noted that the PFM controller regulated the output voltage at the valley point and the output capacitor keeps the output voltage ripples at an acceptable amplitude. The voltage waveform is regulated for 160mSec which is enough for sensing and transmission in many IoT systems [29].

The harvested energy was also measured versus the output load resistance and a comparison between the harvested energy with and without MPPT is shown in Fig. 7. It is worth mentioning that these results were collected with the optimal input and output capacitor values, as determined before. It is clearly visible that the system with MPPT improves the energy output of the converter while maintaining almost exactly the same output power. This can be achieved due to the fact that the PENG produces the most energy when the load applied to it matches the internal impedance of the harvester. By manipulating the input voltage of the converter and enabling it at a later point one can artificially generate a different load resistance seen by the harvester to optimize the energy conversion. This lead to an improvement of up to 44% compared to the system without the MPPT.

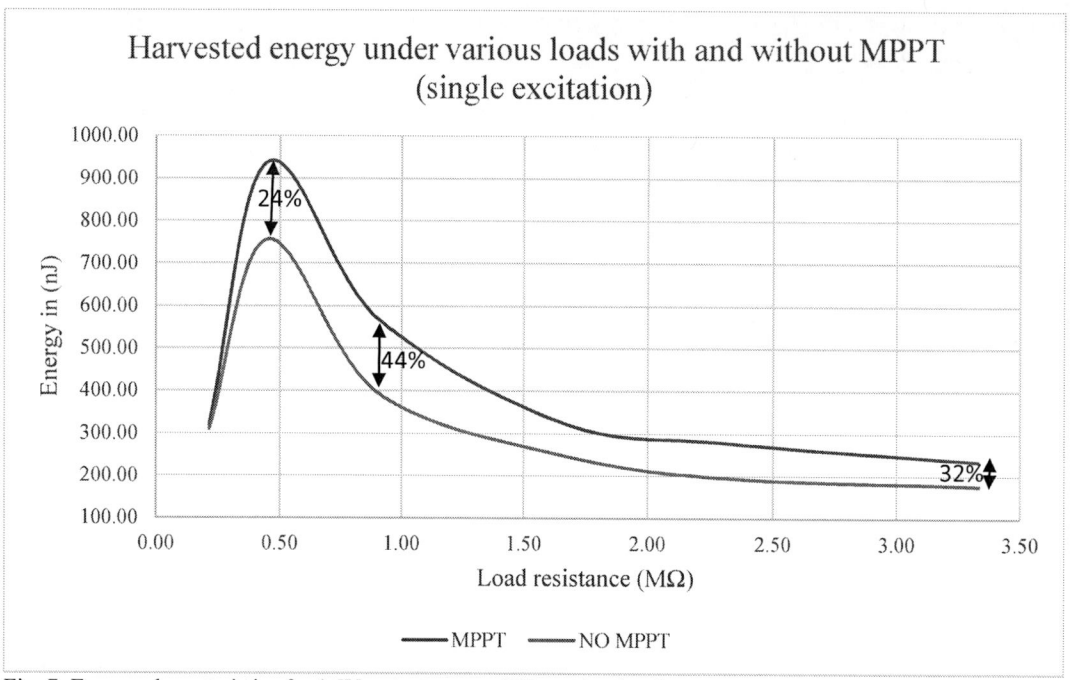

Fig. 7: Energy characteristics for 1.5V output versus different load resistances.

Figure 8 shows how the converter is activated after a short startup phase, when the input voltage reaches 3V (the predetermined MPPT value).The input voltage to the comparator is measured through the voltage divider, and when the 0.2V reference is met, the converter is enabled. The input voltage of the converter decreases as the converter is transferring energy to the output.

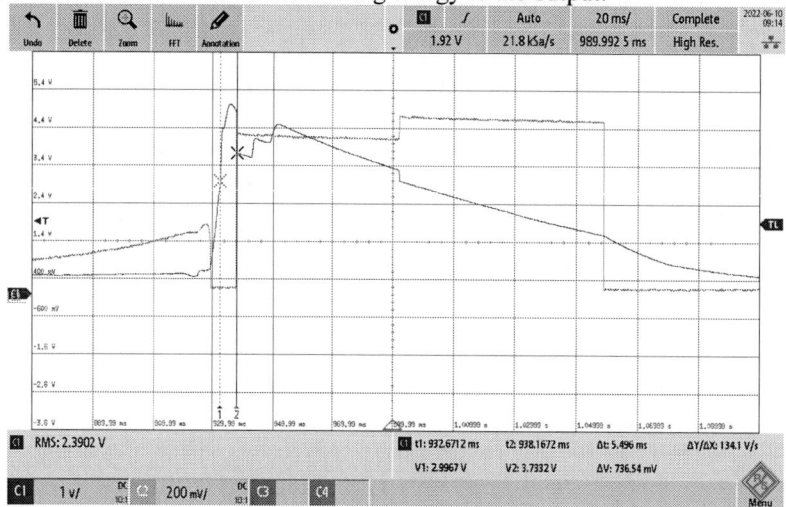

Fig. 8: Input voltage signal (blue), regulated output voltage (pink).

The PENG was also developed to capture, accumulate and store energy. The prototype is capable of efficiently managing harvested energy to charge the battery powered system. A low-power SMD battery (TDK- Ceracharge) is used for the design, requiring an input voltage between 1.5V-1.6V which can be delivered by the regulated output of the buck converter. The battery together with a blocking diode, to prevent self-discharge, were connected in parallel to the power management system (PMS). Fig. 9 shows a measurement of the voltage across a resistor connected in series with the battery, to show the behavior of the charging current entering the battery.

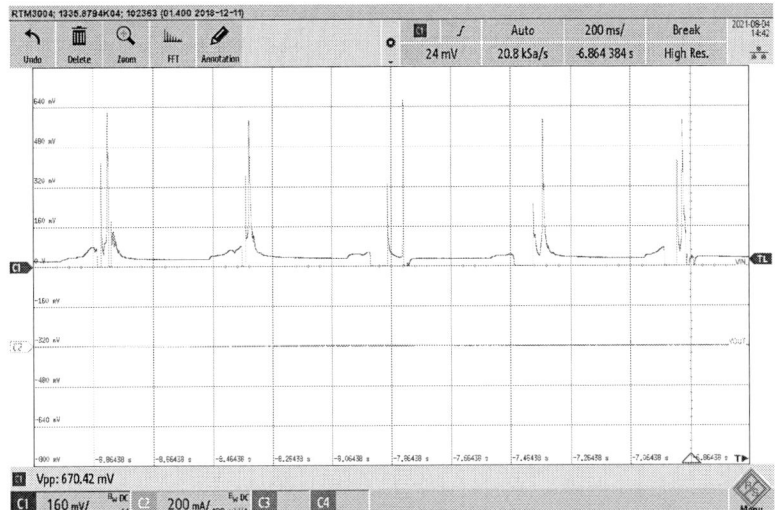

Fig. 9: Energy stored in a battery.

Conclusions

In this paper, we have presented an energy harvesting system that is capable of delivering a peak output energy of up to 900nJ with a single deformation. The presented MPPT solution improves the harvested output energy by up to 44% while providing a regulated output voltage. The operation of the PMS covered two different use cases, both as a standalone application and with energy storage. Insights about selection of the optimal input and output capacitances were shown and implemented in this paper.

References

[1] Kortuem G., Kawsar F., Sundramoorthy V., Fitton D. Smart Objects as Building Blocks for the Internet of Things. IEEE Internet Comput. 2010;14:44–51. doi: 10.1109/MIC.2009.143.

[2] D. Dinulovic, M. Shousha, M. Albatool, T. Zafar, J. Bickel,H. Ngo, and M. Haug, "Dual-Rotor Electromagnetic-Based Energy Harvesting System for Smart Home Applications," in *IEEE Transactions on Magnetics*, vol. 57, no. 2, pp. 1-5, Feb. 2021, Art no. 8001005, doi: 10.1109/TMAG.2020.3014065.

[3] Kamalinejad P., Mahapatra C., Sheng Z., Mirabbasi S., Leung V.C.M., Guan Y.L. Wireless Energy Harvesting for the Internet of Things. IEEE Commun. Mag. 2015;53:102–108. doi: 10.1109/MCOM.2015.7120024.

[4] M. Safaei, H.A. Sodano, S.R. Anton, "A review of energy harvesting using piezoelectric materials: State-of-the-art a decade later (2008-2018)". Smart Mater. Struct. 2019, 28, doi:10.1088/1361-665X/ab36e4.

[5] S. Mishra, L. Unnikrishnan, S.K. Nayak, S. Mohanty, "Advances in Piezoelectric Polymer Composites for Energy Harvesting Applications: A Systematic Review". Macromol. Mater. Eng. 2019, 304, 1–25, doi:10.1002/mame.201800463.

[6] B. Stadlober, M. Zirkl, M. Irimia-Vladu, "Route towards sustainable smart sensors: Ferroelectric polyvinylidene fluoride-based materials and their integration in flexible electronics". Chem. Soc. Rev. 2019, 48, 1787–1825, doi:10.1039/c8cs00928g.

[7] S. Khan, L Lorenzelli, R.S. Dahiya, "Technologies for printing sensors and electronics over large flexible substrates: A review." IEEE Sens. J. 2015, 15, 3164–3185, doi:10.1109/JSEN.2014.2375203.

[8] N. Godard, L. Allirol, A. Latour,S. Glinsek, M. Gérard, J. Polesel, F. Domingues Dos Santos, E. Defay, "1-mW Vibration Energy Harvester Based on a Cantilever with Printed Polymer Multilayers." Cell Reports Phys. Sci. 2020, 1, doi:10.1016/j.xcrp.2020.100068

[9] Starner, T.; Paradiso, J.A. Human generated power for mobile electronics. Low Power Electron. Des. 2004, 1–35.

[10] Kymissis, J.; Kendall, C.; Paradiso, J.; Gershenfeld, N. October Parasitic power harvesting in shoes. Proceedings of the Second International Symposium on Wearable Computers, Digest of Papers, Pittsburgh, PA, USA, USA, 19–20 October 1998; pp. 132–139.

[11] Ishida, K.; Huang, T.C.; Honda, K.; Shinozuka, Y.; Fuketa, H.; Yokota, T.; Sakurai, T. Insole pedometer with piezoelectric energy harvester and 2v organic circuits. IEEE J Solid State Circuits 2013, 48, 255–264.

[12] Granstrom, J.; Feenstra, J.; Sodano, H.A.; Farinholt, K. Energy harvesting from a backpack instrumented with piezoelectric shoulder straps. Smart Mater. Struct. 2007, 16, 1810–1820.

[13] Feenstra, J.; Granstrom, J.; Sodano, H. Energy harvesting through a backpack employing a mechanically amplified piezoelectric stack. Mech. Syst. Signal Process. 2008, 22, 721–734.

[14] Swallow, L.M.; Luo, J.K.; Siores, E.; Patel, I.; Dodds, D. A piezoelectric fibre composite based energy harvesting device for potential wearable applications. Smart Mater. Struct. 2008, 17, 025017.

[15] Ramsay, M.J.; Clark, W.W. June Piezoelectric energy harvesting for bio-MEMS applications. Proceedings of the SPIE's 8th Annual International Symposium on Smart Structures and Materials, International Society for Optics and Photonics, Newport Beach, CA, USA, 4–8 March 2001; pp. 429–438.

[16] Yang, B.; Kwang-Seok, Y. Piezoelectric shell structures as wearable energy harvesters for effective power generation at low-frequency movement. Sens. Actuators A Phys. 2012, 188, 427–433.

[17] M. Shousha, D. Dinulovic, M. Haug, T. Petrovic and A. Mahgoub, "A Power Management System for Electromagnetic Energy Harvesters in Battery/Batteryless Applications," in IEEE J. Emerg. Sel. Topics Power Electron. Vol., no. , pp. 1-15, 2019.

[18] Z. Qin, H. Talleb, S. Yan, X. Xu, and Z. Ren, "Application of PGD on parametric modeling of a piezoelectric energy harvester," IEEE Trans. Mag., vol. 52, no. 11, pp. 1–11, Nov. 2016.

[19] A. Khaligh, P. Zeng, and C. Zheng, "Kinetic energy harvesting using piezoelectric and electromagnetic technologies-state of the art," IEEE Trans. Ind. Electron., vol. 57, no. 3, pp. 850–860, Mar. 2010.

[20] F. Khameneifar, S. Arzanpour, and M. Moallem, "A piezoelectric energy harvester for rotary motion applications: Design and experiments," IEEE/ASME Trans. Mechtron., vol. 18, no. 5, pp. 1527–1534, Oct. 2013.

[21] Z. J. Chew and M. Zhu: "Adaptive maximum power point finding using direct VOC/2 tracking method with microwatt power consumption for energy harvesting," IEEE Trans. Power Electron. 33 (2018) 8164 (DOI: 10.1109/TPEL.2017.2774102).

[22] D. H. Jung, et al.: "Thermal and solar energy harvesting boost converter with time-multiplexing MPPT algorithm," IEICE Electron. Express 13 (2016) 20160287 (DOI: 10.1587/elex.13. 20160287).

[23] S. Stanzione, et al.: "A high voltage self-biased integrated DC-DC buck converter with fully analog MPPT algorithm for electrostatic energy harvesters," IEEE J. Solid-State Circuits 48 (2013) 3002 (DOI: 10.1109/JSSC.2013.2283152).

[24] Y. K. Teh, et al.: "Design of transformer-based boost converter for high internal resistance energy harvesting sources with 21mV selfstartup voltage and 74% power efficiency," IEEE J. Solid-State Circuits 49 (2014) 2694 (DOI: 10.1109/JSSC.2014.2354645).

[25] S. Lee and J. Jeong: "An off-chip input capacitor-less boost converter with fast MPPT for energy harvesting," IEICE Electron. Express 11 (2014) 20140385 (DOI: 10.1587/elex.11.20140385).

[26] Y. D. Ye, et al.: "A self-powered zero-quiescent-current active rectifier for piezoelectric energy harvesting," IEICE Electron. Express 15 (2018) 20180739 (DOI: 10.1587/elex.15.20180739)

[27] S. Roundy, et al.: "Improving power output for vibration-based energy scavengers," IEEE Pervasive Comput. 4 (2005) 28 (DOI:10.1109/MPRV.2005.14).

[28] M. Tung, D. Aiguo, P. Hu, D. Nirmal, and K. Nair.: "Evaluation of Micro Controller Based Maximum Power Point Tracking Methods Using dSPACE Platform." In Proc. Australian University Power Engineering Conference, 2006.

[29] M. Shousha, D. Dinulovic, M. Brooks and M. Haug.: "A miniaturized cost effective shared inductor based energy management system for ultra-low-voltage electromagnetic energy harvesters in battery powered applications," 2018 IEEE Applied Power Electronics Conference and Exposition (APEC), 2018, pp. 703-707.

Analysis and Estimation of Neutral-Point Voltage Balancing Ability of an Optimized Balancing Algorithm for Grid Connected Active-NPC converter

Joseph Banda[1], Kapil Jha[2], Hridya Ittamveettil[2],
Arvind Kumar Tiwari[3], Fernando Ramirez[3]
[1]NTNU, Norway
[2]General Electric, India
[3]General Electric, United States
E-Mail: kapil.jha@ge.com

Keywords

Neutral point clamped inverter, Capacitor voltage balancing, Grid-connected converter, multi-level converters, Neutral current ripple

Abstract

Neutral-point (NP) voltage balancing is a well-known challenge associated with Neutral Point Clamped (NPC) multi-level voltage source converters. Most of the literature on multilevel NPC converters discusses on multiple balancing techniques that are used to nullify dc or ac unbalance and claims the adaptability of these techniques over wide range of modulation indices and power factors. This paper details out the analysis for estimating the maximum balancing ability of an optimized neutral point balancing technique applied on a grid connected 3-Level Active-NPC (ANPC) converter with supporting simulation and experimental results. Also, a simplified analysis to study the impact of grid harmonics on the neutral point potential is presented and validated with the help of simulation results.

Introduction

Multilevel converters are widely used for medium-voltage high power applications regarding the integration of renewable energy sources like wind and solar. The major advantages they offer include increased voltage capability thereby reducing the overall system losses, reduced total harmonic distortion, lower Common Mode Voltages (CMV) [1]. Because of its simplicity, reliability and higher manufacturing readiness level, NPC converter is one of the well accepted 3-level topologies by the industry. The additional third level is realized by clamping the pole voltage to the dc-link mid-point. The mid-point is achieved by splitting the dc-link into two halves using series connected capacitors. In such converters, it becomes essential to control the voltage of each series connected capacitor, or mid-point voltage, also known as neutral-point (NP) voltage.

Various balancing techniques have been proposed in the literature to control the neutral point voltage of NPC converter over a wide range of power factors and modulation indices [2-11]. In [2], a comprehensive study around NP balancing controls with mathematical derivations for various operating conditions is discussed. A new balancing strategy is proposed in [3] to modify bridge currents for better performance under no load or light load conditions. The bigger challenge of balancing the neutral point voltage occurs at higher modulation indices as there will be no margins to modify the modulation indices. Such a case is discussed in [4] and the introduced method is power flow direction independent. Closed loop small signal transfer function is presented in [5] to design the NP balancing proportional-integral (PI) regulator. This technique aims to reduce the injected common mode voltage for neutral point balancing, thereby limit the size of the dc link capacitance. In [6], an attempt to utilize 3 level NPC for active filtering is proposed and a 6th harmonic zero sequence component is utilized for balancing the neutral point voltage. As 6th being the lowest even and triplen harmonic that produces the non-zero neutral-point current and doesn't reflect in phase currents of the converters. Some attempts were made to optimize the balancing algorithms based on the chosen pulse width modulation (PWM) technique [7]. Theoretical expressions for the maximum zero sequence voltage are derived using interpolation methods in [8]. Inherent neutral point balancing ability of the converter is comprehensively studied in [9] and dependency of switching frequency of the converter is highlighted and methodologies were suggested for improving this inherent balancing ability. In [10], zero sequence voltage is utilized for balancing scheme to reduce both dc balancing and ac oscillations at the neutral point. With a little compromise on the output voltage harmonic distortion, new strategy is proposed in [11] for enhanced balancing ability. To summarize, any balancing algorithm injects either dc, or specific harmonics into modulation signal or modify bridge currents for getting nonzero average mid-point current. These techniques differ in terms of their ability to do NP balancing w.r.t the power factor range, dependency on load currents, and amount of CMV injected, etc. In this paper, the analysis of optimized NP balancing algorithm mentioned in [12,13] has been performed to estimate the true maximum balancing ability. The impact of even harmonic sequence in grid voltage on NP potential has also been analyzed. The challenge of NP voltage balancing is identical for NPC and ANPC converters. The additional advantage with ANPC is the flexibility to choose the path for the mid-point current during zero pole voltage level thereby achieving uniform loss distribution across the devices. In this paper, ANPC converter has been chosen for the validation in simulation and hardware. Theoretical analysis of balancing algorithm has been performed at two critical modulation indices, and validation of analysis is done using simulation and hardware results.

Grid connected Active-NPC (ANPC) converter

Fig. 1 shows a single line and single-phase schematic of grid connected ANPC converter with an output LC filter. The switching pulses for the devices S_1–S_6 for A-phase are based on the logic in Table-I, where M_a is the phase-A modulation signal and S is the switching state. S is logic high (represented as 1) when M_a is greater than the carrier signal and logic low (represented as 0) vice versa. $\bar{S}$ is complement of S. The gate pulses for the devices are derived from the control logic shown in Fig. 2 using third harmonic injected PWM with 2 level shifted carriers. The converter is operating as Active Front End (AFE) rectifier, where in the d-axis controls the total dc-link voltage (V_{dc}) and q-axis-the reactive power flow (I_q). Feedback quantities are the 3-phase grid voltages (V_{grid}) and inductor currents (I_{abc}). Inputs to the phase-locked-loop (PLL) block are grid voltage V_{grid} and estimated angle θ_{pll}, PLL aligns the grid voltage space vector along the d-axis [14-15]. The modulation signals $M_{(abc)}^*$ coming out of dq control is getting added with V_{cmv} coming out of the dc-link balancing control block as in Fig. 2. A detailed analysis of the dc-link NP balancing control is done in the following section.

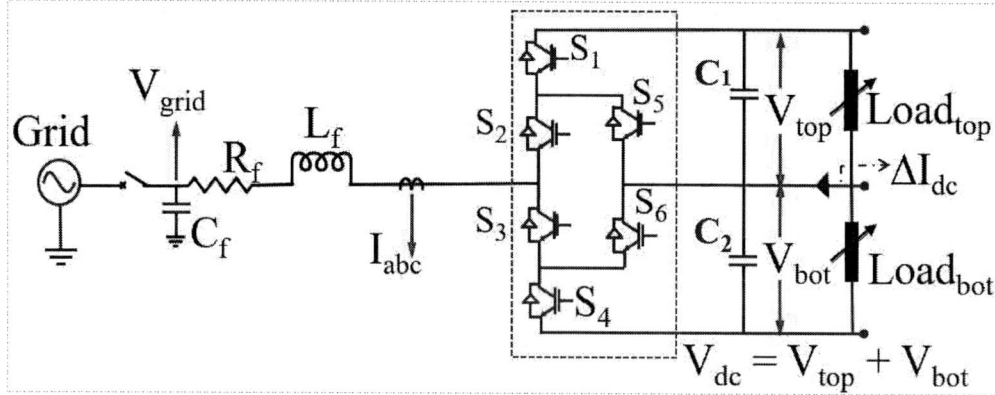

Fig. 1: System configuration of grid connected ANPC converter with grid voltage-oriented control

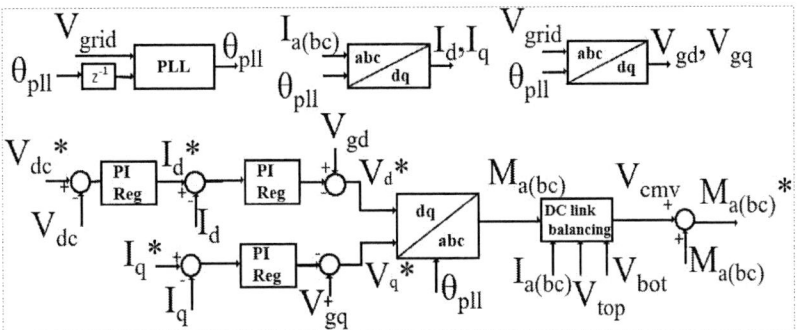

Fig. 2: *dq* control blocks for AFE rectifier operation of ANPC converter

Table I: Switching logic of ANPC converter

Condition	S1	S2	S3	S4	S5	S6
$M_a \geq 0$	S	1	0	0	$\bar{S}$	S
$M_a < 0$	0	0	1	S	S	$\bar{S}$

NP balancing and midpoint current in ANPC converter

3-Level ANPC converter uses active devices (IGBT/IGCT/MOSFET) with anti-parallel diodes that creates a path for the current to flow to the mid-point when the pole voltage is clamped to the mid-point of dc-link. Ideally, the sum of these instantaneous currents from all the 3 phase legs will result in zero mid-point current when averaged over a modulation cycle. But practically, there could be slight mismatch in the components of the converter, system dynamics, line voltage distortions from the grid resulting in the distorted modulation indices which produces a non-zero average mid-point current that could drift the NP voltage [16]. This could make the entire dc-link voltage appears across one capacitor and zero voltage across the other. Hence, while implementing any NP balancing algorithm, it becomes crucial to establish theoretical relationship of mid-point current with modulation signal, line current, power factor and grid harmonics to compensate the non-zero averaged mid-point current by injecting suitable CMV without distorting the converter line to line voltage and currents.

$$\sum I_{mp}(k)_{k=1,2,3} = -(|m_a|i_a + |m_b|i_b + |m_c|i_c) \tag{1}$$

$$I_{mp} = -3 * \Sigma i * m * sign(m) \tag{2}$$

$$i = I_0 + \sum_{h=1,2,3...}^{\infty}(I_h \sin(h\theta + \varphi_h)) \tag{3}$$

$$m = M_0 + \sum_{v=1,2,3,..}^{\infty}(M_v \sin(v\theta + \varphi_v)) \tag{4}$$

$$sign(m) = \frac{4}{\pi}\sum_{n=1,3,5}^{\infty}\frac{1}{n}(\sin(n\theta + n\theta_1)) \tag{5}$$

$$I_{mp(avg\_e)} = (M_1 * I_{h\_e} * \frac{2}{\pi}) * C * \sin(\varphi_h) \tag{6}$$

The generalized method for the estimation of mid-point current is given by eqns. (1)-(5) for a 3-phase 3-level ANPC converter with harmonics in modulation signal and line current. where, I_{mp} is the mid-point current, m_a, m_b, m_c are the modulation signals, i_a, i_b, i_c are the line currents, i, m, $sign(m)$ are the distorted phase current, modulation signal, and signum function of modulation, respectively. Here, I_0, M_0, I_h, M_v are the dc and ac harmonic components in line current and modulation, respectively.

Impact of grid harmonics on neutral point balancing

In eqns. (1)-(4), I_h, M_v represent the peak of the harmonic component in line current and modulation signal, respectively. These two components are modified to obtain dc bias in the mid-point current over a fundamental cycle, as given in (6), where I_{h_e} is the peak of even harmonic component in current and C is algebraic constant. If I_{h_e} is introduced by the grid voltage distortion, as shown in Fig. 3(a), it can create NP imbalance, when there is no active control of NP potential. As given in (6), even harmonics can result in a non-zero average mid-point current under certain conditions. In Fig. 3(b), 5% of 2nd harmonic is added in positive sequence to the line current of 100 A peak at unity modulation index, and the resultant mid-point current does not have any dc-bias and there is no drift in ΔV_{dc}. Here, phase angle (φ_h) is 90°. However, in Fig. 3(c), 5% of 2nd harmonic when applied in negative sequence drifts the NP voltage. On the contrary, 4th harmonic current in positive sequence will create drift in ΔV_{dc}, however, 4th harmonic in negative sequence does not create drift in ΔV_{dc}.

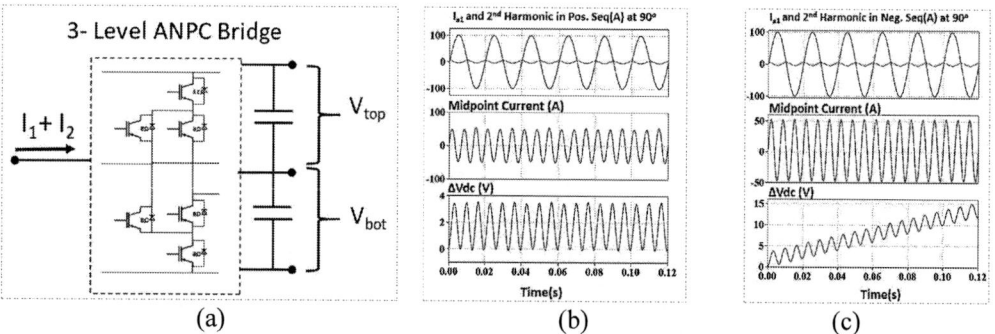

Fig. 3: (a) 3L-ANPC converter with 2nd harmonic in line current, fundamental line current, 5% of 2nd harmonic current at 90° phase, mid-point current and ΔV_{dc} in positive sequence in (b), and negative sequence in (c), respectively.

Optimal Neutral point balancing algorithm

Optimal NP balancing algorithm compensates for the dc unbalance by CMV injection [12-13,17-18]. It is superior to other balancing techniques as it does not inject CMV aggressively and prevents overmodulation operation across all power factors and modulation indices. As shown in Fig. 4, the difference in the top and bottom capacitance voltages (V_{top} & V_{bot}) has been calculated and same has been passed through Low-Pass Filter (LPF) and a Proportional-Integrator (PI) regulator. Output of PI controller (*DCBalRegOut*) acts like a dynamic gain whose value depends on the magnitude of voltage imbalance and sign decides the direction for generating the compensating mid-point current. To improve the dc bus utilization, third harmonic injection has been done in modulation signals (M_{a_cmd}, etc.) [10]. The modified modulation commands ($M_{a\_cmd\_th}$, etc.) are passed through a min-mid-max classifier to obtain instantaneous maximum and minimum values.

The limit calculator module computes the maximum, minimum and mid common mode offset voltages that can be added to modulation commands, at a particular instant. It is the key benefit of this algorithm as instantaneous offset computation is done dynamically considering the available margin in modulation commands. Using three instantaneous computed offset voltages (CMV_{Max}, etc.) and instantaneous line currents (I_{a_phase}, etc.), the NP current calculator block (I_{NP}) generate three NP currents ($I_{\_CMVMax}$, etc.) by the addition of these offset voltages (CMV_{Max}, etc.) to modulation indices using eqn. (1). Dynamic CMV calculator compares the three currents and computes the product of PI regulator output (*DCBalRegOut*) with the offset voltage that produces maximum positive or maximum negative neutral point current based on the sign of regulator output (*DCBalRegOut*). The final output from the NP algorithm (CMV^*) is added to modulation signals to generate final modulation command ($M_{a\_cmd}^*$, etc.). These modulation commands are used to generate PWM signals using sine-triangle switching technique [18]. Therefore, as described above, optimum NP balancing algorithm prevents overmodulation and ensure maximum balancing capability at an operating point. In next section, maximum balancing ability of this NP balancing scheme has been analyzed.

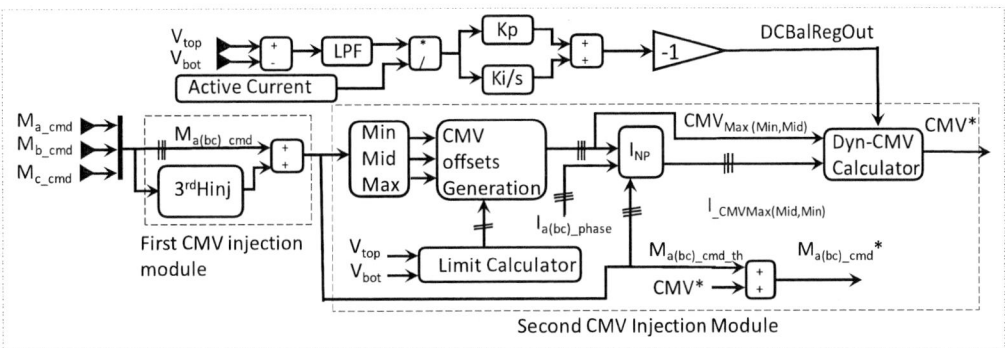

Fig. 4: Control structure of optimal NP balancing with CMV injection.

Dependency of maximum balancing ability on modulation index and power factor

The balancing ability is defined as the ratio of maximum average mid-point current (with CMV injection) to the peak of line current. For optimized balancing algorithm mentioned in the above section, the measure of computed balancing ability with modulation index and power factor angle ($\emptyset = 0°$, 30°, 60°, and 90°) is shown in Fig. 5. The balancing ability is highest in unity power factor at modulation index of 0.5 and it decreases as the power factor decrease. The balancing ability is minimum at unity modulation index and zero power factor operation. In Fig. 5, two cases plotted on the dotted line, viz., case-A and case-B have been chosen for further analysis. The power factor angle of these cases is 62°. case-A has lowest balancing ability of 0.042 per unit at unity modulation index and case-B has highest balancing ability of 0.43 per unit at modulation index of 0.5. Fig. 6(a) and 6(b) shows simulated outputs of three CMV offsets (CMV_{Max}, etc.), computed mid-point currents (I_{CMVMax}, etc.) with these offsets added to modulation commands, and the instantaneous maximum NP current (I_{MP_out}) from the converter with optimized balancing technique for case-A and case-B, respectively.

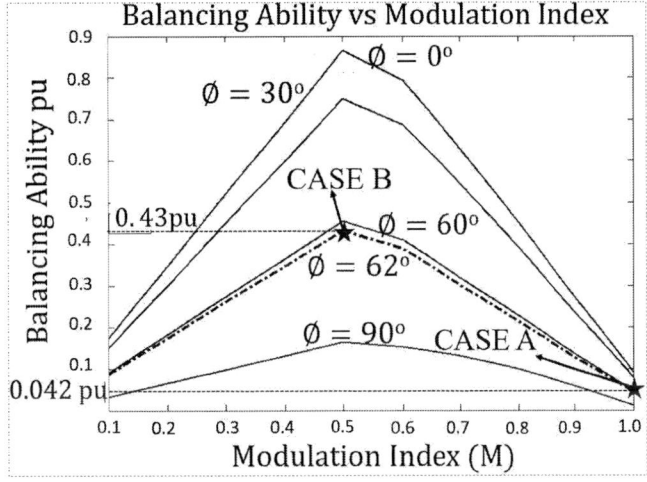

Fig. 5: Maximum balancing capability vs modulation index

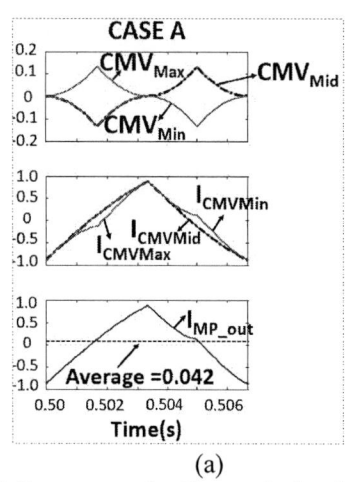
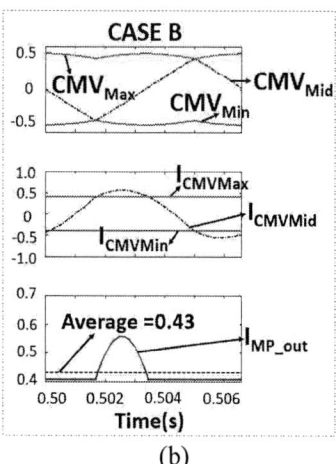

(a)   (b)

Fig. 6: (a) Common mode offsets, calculated mid-point currents from offsets, maximum mid-point current for Case-A, (b) for Case-B.

Experimental validation

Experimental validation of discussed NP balancing algorithm and analysis was performed on a 3-phase, 10 kVA, grid connected 3-level ANPC converter. Experimental results were obtained for case-A and case-B at operating conditions given in Table-II. The ANPC converter has been operated in rectifier mode using resistors as a load. The top and the bottom capacitors are connected to separate resistors. Unbalance in capacitor voltages is created by choosing appropriate resistor values.

Table II: Operating conditions for experimental validation

Case	Modulation index	Power factor angle	$I_{abc\_peak} = sqrt(2) *I_{01}$ (A)	V_{dctop} (V)	V_{dcbot} (V)	ΔI_{dc} (mA)	I_{mid_max} (mA)
A	1.0	61.6°	3.78	104	96	170	158
B	0.5	61.6°	3.78	104.2	95.7	66	1625

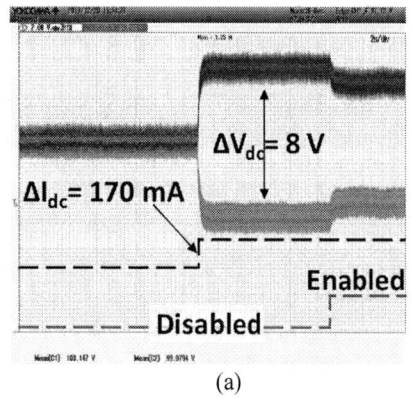
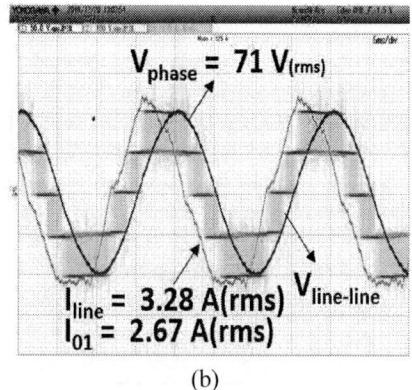

(a)   (b)

Fig. 7: Experimental results showing dc balancing ability: (a) top and bottom capacitor voltage and (b) grid current, grid phase voltage, and converter voltage (L-L) for case-A

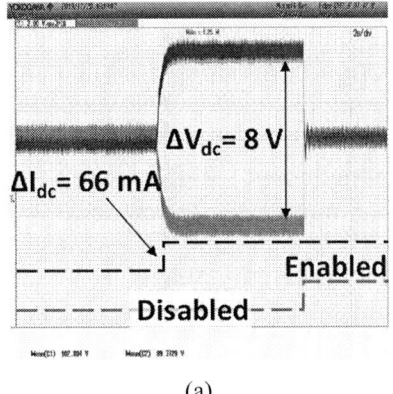 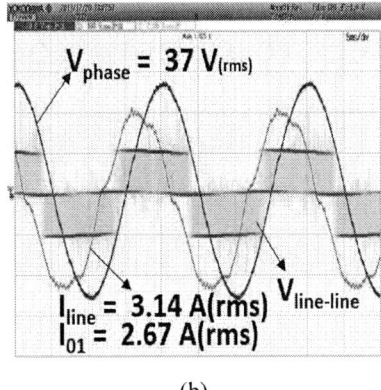

(a)   (b)

Fig. 8: Experimental results showing dc balancing ability: (a) top and bottom capacitor voltage and (b) grid current, grid phase voltage, and converter voltage (L-L) for case-B

In case A, the converter is operating at unity modulation index and power factor angle of approx. 62°. The amplitude of peak fundamental line current is 3.78 A (= 2.67*1.414). Due to unbalance in load resistors across capacitors, a NP current (ΔI_{dc}) of approx. 170 mA flows through dc-link mid-point and creates a dc unbalance (ΔV_{dc}) of 8 V across top and bottom capacitors, as shown in Fig. 7 (a). The computed maximum balancing ability of the converter in this condition is 158 mA (0.042*3.78), which is lower than the unbalance current. Hence, after enabling the algorithm, the unbalance in capacitor voltages reduces but it does not cancel out completely. Fig. 7(b) shows the grid voltage, grid current, and converter line to line voltage for case-A.

In case B, the converter is operating at a modulation index of 0.5 and the NP current (ΔI_{dc}) due to unbalance in load resistor creates 66 mA. At this operating condition, the maximum balancing ability is 1625mA (0.43*3.78) (given in Fig. 5), which is higher than the unbalance current. Therefore, the capacitor voltages get balanced as soon as the balancing algorithm is enabled, as in Fig. 8(a), confirming the analysis. Fig. 8(b) shows the grid voltage, grid current and converter line to line voltages for case-B respectively. In Fig.8(b), grid voltage (phase rms) is reduced to 37 V compared to 71 V in Fig. 7(b). This is done to obtain a modulation index of 0.5 in Case B from 1.0 in Case A without changing the DC link voltage.

Conclusion

In this paper, theory, operation, and analysis of optimized balancing algorithm applied to a grid connected ANPC converter has been discussed. Theoretical expressions for the computation of mid-point current, dependency on the CMV and margins in modulation indices to avoid overmodulation have been presented. The impact of even order sequence of grid harmonics on the NP voltage has been provided. The balancing ability of 3-level ANPC converter with optimized algorithm at two selected operating points has been analyzed and validated using simulation and experimental results.

References

[1] J. Rodriguez, Jih-Sheng Lai, and Fang Zheng Peng, "Multilevel inverters: a survey of topologies, controls, and applications," in *IEEE Trans. on Ind. Electron.*, vol. 49, no. 4, pp. 724-738, Aug. 2002

[2] N. Celanovic and D. Boroyevich, "A comprehensive study of neutral-point voltage balancing problem in three-level neutral-point-clamped voltage source PWM inverters," in *IEEE Transactions on Power Electronics*, vol. 15, no. 2, pp. 242-249, March 2000.

[3] M. Marchesoni, P. Segarich, and E. Soressi, "A new control strategy for neutral-point-clamped active rectifiers," *IEEE Trans. Ind. Electron.*, vol. 52, no. 2, pp. 462–470, Apr. 2005.

[4] R. M. Tallam, R. Naik and T. A. Nondahl, "A carrier-based PWM scheme for neutral-point voltage balancing in three-level inverters," in *IEEE Transactions on Industry Applications*, vol. 41, no. 6, pp. 1734-1743, Nov.-Dec. 2005.

[5] A. Bendre, G. Venkataramanan, D. Rosene and V. Srinivasan, "Modeling and design of a neutral-point voltage regulator for a three-level diode-clamped inverter using multiple-carrier modulation," in *IEEE Transactions on Industrial Electronics*, vol. 53, no. 3, pp. 718-726, June 2006.

[6] H. Akagi and T. Hatada, "Voltage balancing control for a three-level diode-clamped converter in a medium-voltage transformer less hybrid active filter," *IEEE Trans. Power Electron.*, vol. 24, no. 3, pp. 571–579, Mar. 2009.

[7] J. Zaragoza, J. Pou, S. Ceballos, E. Robles, C. Jaen and M. Corbalan, "Voltage-Balance Compensator for a Carrier-Based Modulation in the Neutral-Point-Clamped Converter," in *IEEE Transactions on Industrial Electronics*, vol. 56, no. 2, pp. 305-314, Feb. 2009.

[8] C. Wang and Y. Li, "Analysis and Calculation of Zero-Sequence Voltage Considering Neutral-Point Potential Balancing in Three-Level NPC Converters," in *IEEE Transactions on Industrial Electronics*, vol. 57, no. 7, pp. 2262-2271, July 2010.

[9] J. Shen, S. Schröder, R. Rösner and S. El-Barbari, "A Comprehensive Study of Neutral-Point Self-Balancing Effect in Neutral-Point-Clamped Three-Level Inverters," in *IEEE Transactions on Power Electronics*, vol. 26, no. 11, pp. 3084-3095, Nov. 2011.

[10] J. Pou, J. Zaragoza, S. Ceballos, M. Saeedifard and D. Boroyevich, "A Carrier-Based PWM Strategy with Zero-Sequence Voltage Injection for a Three-Level Neutral-Point-Clamped Converter," in *IEEE Transactions on Power Electronics*, vol. 27, no. 2, pp. 642-651, Feb. 2012.

[11] U. -M. Choi, J. -S. Lee and K. -B. Lee, "New Modulation Strategy to Balance the Neutral-Point Voltage for Three-Level Neutral-Clamped Inverter Systems," in *IEEE Transactions on Energy Conversion*, vol. 29, no. 1, pp. 91-100, March 2014.

[12] J. Shen, S. Schroder, B. Duro and R. Roesner, "A Neutral-Point Balancing Controller for a Three-Level Inverter with Full Power-Factor Range and Low Distortion," in *IEEE Tran. on Ind. Appl.*, vol. 49, no. 1, pp. 138-148, Jan.-Feb. 2013.

[13] J. Shen, S. Schroder, R. Roesner, "DC-link voltage balancing system and method for multilevel converters," U.S. Patent 8441820B2, 2011.

[14] J. S. Siva Prasad, T. Bhavar, R. Ghosh, G. Narayanan, "Vector control of three phase AC/DC frontend converter," *Sadhana* 33(5): 591–613.

[15] Banda, J.K., Jain, A.K. "Single-current-sensor-based active front-end-converter-fed four quadrants induction motor drive," Sadhana 42, 1275–1283 (2017).

[16] Josep Pou, "Modulation and control of three-phase PWM multilevel converters," PhD Thesis, Universitat Politècnica de Catalunya, 2002.

[17] Qin Lei, Jie Shen, Stefan Schroeder "System and method for unified common mode voltage injection," U.S. Patent 9755545B2, 2014.

[18] J. W. Kimball and M. Zawodniok, "Reducing Common-Mode Voltage in Three-Phase Sine-Triangle PWM With Interleaved Carriers," in *IEEE Trans. on Power Electronics*, vol. 26, no. 8, pp. 2229-2236, Aug. 2011.

A Direct Model Predictive Control Strategy of Back-to-Back Modular Multilevel Converters Using Arm Energy Estimation

Akseli Hakkila[*], Antonios Antonopoulos[†], Petros Karamanakos[*]

[*]Faculty of Information Technology and Communication Sciences, Tampere University,
33101 Tampere, Finland
Email: akseli.hakkila@ieee.org, p.karamanakos@ieee.org

[†]School of Electrical and Computer Engineering, National Technical University of Athens,
15780 Zografou, Greece
Email: antoniosantonopoulos@mail.ntua.gr

Keywords

≪Model Predictive Control≫, ≪Modular Multilevel Converters (MMC)≫, ≪Optimal control≫, ≪Multi-objective optimization≫.

Abstract

This paper presents a model predictive control (MPC) algorithm for modular multilevel converters (MMCs). To meet the control objectives of phase current reference tracking and circulating current minimization, the proposed control scheme calculates the optimal number of submodules (SMs) to be inserted in each arm. In doing so, favorable steady-state and dynamic performance is achieved. Moreover, by estimating—instead of measuring—the arm energies in the predictive stage of the control loop, the proposed control scheme results in self-stabilizing open-loop arm energy balancing, while avoiding potential stability issues. Furthermore, to reduce the computational complexity of the MPC algorithm, the optimization problem is simplified by controlling each phase separately and assuming that the SM capacitors are balanced within an arm. To ensure that this assumption is always satisfied, a subsequent capacitor voltage balancing algorithm is designed to select the individual SMs that are switched on and off. The performance of the proposed control strategy is validated with simulations for a high voltage dc system (HVDC) that consists of two MMCs with 20 SMs per arm in a back-to-back configuration.

Introduction

The modular multilevel converter (MMC) [1] is an excellent candidate for high-voltage applications, such as high voltage dc (HVDC) transmission systems [2], since its scalability allows operation at different voltage levels with low power losses and harmonic distortions. Control of MMCs, however, is a nontrivial task since multiple control objectives need to be simultaneously met. Specifically, output current reference tracking, circulating current elimination, and capacitor voltage balancing need to be achieved during both steady-state and transient operating conditions.

To achieve the above control goals, conventional control techniques are commonly employed that rely on linear control theory. Such methods decompose the multiple-input multiple-output (MIMO) control problem into several single-input single-output (SISO) control loops that are arranged in a cascaded manner and utilize proportional-integral (PI) controllers. Using cascaded control loops, however, works well in steady state, but it limits the transient performance since the system dynamics are not fully decoupled [3].

The aforementioned control issues, however, can be effectively tackled by modern control methods, such as model predictive control (MPC) [3]. MPC, especially in its version as direct controller (i.e., when a

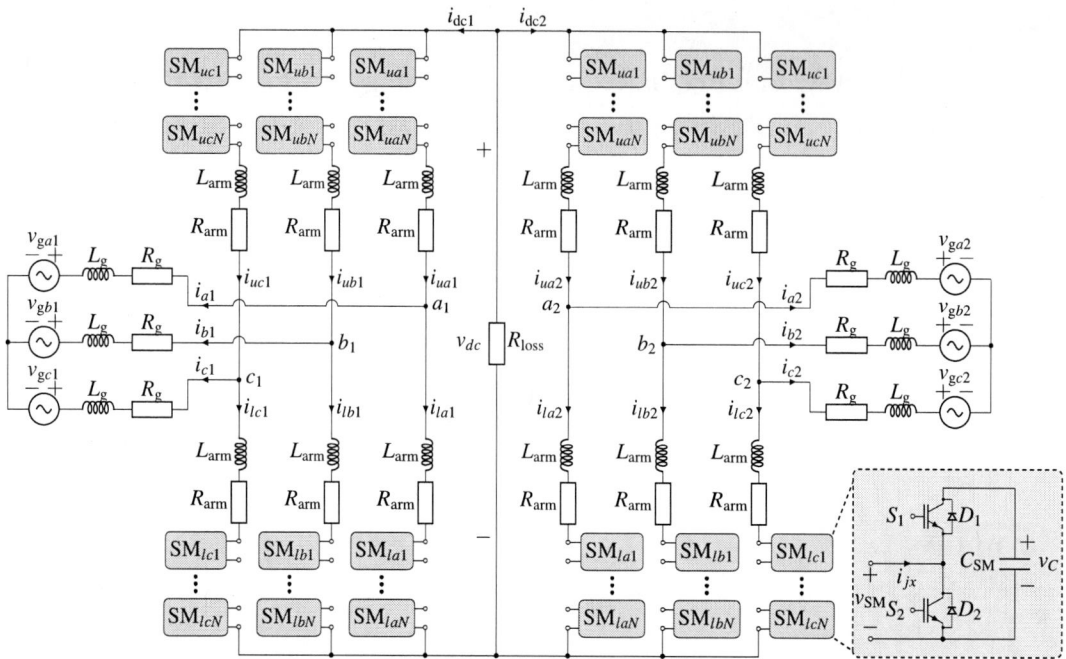

Fig. 1: Circuit diagram of a back-to-back connected MMC-HVDC system.

dedicated modulator is not used to generate the switching signals) can handle MIMO, nonlinear systems, such as MMCs, in a single control loop and address the multiple—and often conflicting—objectives in one computational stage [4]. Nevertheless, the lack of a modulator in direct MPC implies that the underlying optimization problem is an integer program (IP), which can be computationally intractable as its size increases [5]. This is exactly the case with MMCs, since the number of candidate solutions increases exponentially with the number of possible switch positions. Given that the latter is high due to the big number of used submodules (SMs), it can be understood that use of MPC for MMCs is not an easy task.

Given the above, this paper proposes an MPC algorithm for back-to-back MMCs used in HVDC systems that achieves superior performance by successfully addressing all the relevant control objectives. The MPC problem is simplified by controlling the arm energies based on estimates (instead of measurements) of the SM capacitor voltages. Note that, even though this technique has been shown to be effective with conventional control methods [6], when used in conjunction with MPC lead to increased circulating currents and high current total demand distortion (TDD) as it did not account for the arm energy ripple [7]. Moreover, to tackle the problem of the pronounced computational demands of direct MPC, techniques are implemented in this work that keep the computational complexity of the optimization problem modest. The efficacy of the proposed control scheme is verified with simulations acquired based on MMCs with 20 SMs per arm.

System Modeling

The system under consideration is the back-to-back connected MMCs shown in Fig. 1. Both MMCs consist of three phase legs, each of which is divided into two arms, namely the upper and lower arm. Each arm has N series-connected SMs, and an arm inductor L_{arm}—with internal resistance R_{arm}—to limit fault and circulating currents. The half-bridge SM, as seen in Fig. 1, consists of two pairs of active semiconductor switches with freewheeling diodes and a capacitor C_{SM}, which is ideally maintained at a voltage of v_{dc}/N, where v_{dc} is the dc-link voltage. For the back-to-back HVDC system, resistor R_{loss} is added in parallel to the MMC units to model the switching losses of the converters [2].

The MMC can be controlled by inserting and bypassing SMs into the arms. An inserted SM produces a voltage of C_{SM} at its terminal, while a bypassed SM has a terminal voltage of zero. Therefore, each of the upper and lower arms can produce $N + 1$ different voltage levels. Given this, there are two ways for the MMC to produce the desired output phase voltage, i.e., either by controlling both arms simultaneously such that N SMs are always inserted, or by controlling the SMs of each arm independently from each other. In the former case, the MMC can produce $N + 1$ voltage levels at its ac terminals. In the latter, there are $2N + 1$ available voltage levels resulting in lower distortion in the output current. Therefore, independent control of the arms is preferable, even though it can lead to a more complicated controller design, or increased computational burden, as in the case of MPC.

Besides the advantages stemming from the independent control of the arms, also controlling each phase independently employs the controller with another degree of freedom, as explained in the following section. Hence, in this section, the single-phase model of the MMC is considered. As the arm voltages depend on the number of inserted SMs, assuming that all capacitors are balanced within an arm, i.e., they all have they same voltage, means that knowledge of which specific SMs are inserted to produce the desired voltage is not required. Therefore, the modeling—and subsequent control—of the MMC can be greatly simplified by considering the number of inserted SMs n_{jx} in each arm $j \in \{u, l\}$ of each phase $x \in \{a, b, c\}$, instead of the switching state of each SM.

Considering the above assumption, the arm voltage v_{jx} produced by the SMs is given by

$$v_{jx} = \frac{n_{jx}}{N} v_{jx}^{\Sigma}, \tag{1}$$

where v_{jx}^{Σ} is the sum of all capacitor voltages in the arm. Assuming that all SMs have the same capacitance C_{SM}, the dynamics of the arm capacitor voltage sums can be described by

$$\frac{dv_{jx}^{\Sigma}}{dt} = \frac{n_{jx}}{C_{SM}} i_{jx}, \tag{2}$$

where i_{jx} is the arm current.

By applying Kirchhoff's current law to the ac terminals, the phase current can be given as

$$i_x = i_{ux} - i_{lx}. \tag{3}$$

Using Kirchhoff's voltage law to the upper and lower arms, the dynamics of the phase legs are governed by

$$L_g \frac{di_x}{dt} + L_{arm} \frac{di_{ux}}{dt} = \frac{1}{2} v_{dc} - R_g i_x - R_{arm} i_{ux} - v_{ux} - v_{gx}, \tag{4a}$$

$$L_g \frac{di_x}{dt} - L_{arm} \frac{di_{lx}}{dt} = -\frac{1}{2} v_{dc} - R_g i_x + R_{arm} i_{lx} + v_{lx} - v_{gx}. \tag{4b}$$

Adding (4a) and (4b), and by using (1) and (3), the dynamics of the phase current are described by

$$(2L_g + L_{arm}) \frac{di_x}{dt} = \frac{n_{lx} v_{lx}^{\Sigma} - n_{ux} v_{ux}^{\Sigma}}{N} - (R_{arm} + 2R_g) i_x - 2v_{gx}. \tag{5}$$

The upper and lower arm currents consist of a common- and a differential-mode component, i.e.,

$$i_{ux} = i_x/2 + i_{comm,x} \tag{6a}$$
$$i_{lx} = -i_x/2 + i_{comm,x}, \tag{6b}$$

where $i_{comm,x}$ is the common-mode current which flows through both arms and it is defined as

$$i_{comm,x} = \frac{i_{ux} + i_{lx}}{2} = \frac{i_{dc}}{3} + i_{zx}, \tag{7}$$

where i_{zx} is the circulating current. As can be observed in (7), in steady-state operation, the common-mode current includes a third of the dc-link current, as the latter is assumed to be equally divided among all phases. In addition, a circulating current component may exist in the common-mode current which is caused by the instantaneous voltage difference between the dc side and the phase leg. Subtracting (4b) from (4a), and with the help of (1) and (7), the common-mode current dynamics are given by

$$2L_{\mathrm{arm}}\frac{di_{\mathrm{comm},x}}{dt} = v_{\mathrm{dc}} - \frac{n_{lx}v_{lx}^{\Sigma} + n_{ux}v_{ux}^{\Sigma}}{N} - 2R_{\mathrm{arm}}i_{\mathrm{comm},x}. \tag{8}$$

Following, by inserting (6) into (2), the capacitor voltage sum dynamics can be given as a function of the phase and common-mode currents, i.e.,

$$\frac{dv_{ux}^{\Sigma}}{dt} = \frac{n_{ux}}{2C_{\mathrm{SM}}}i_x + \frac{n_{ux}}{C_{\mathrm{SM}}}i_{\mathrm{comm},x}, \tag{9a}$$

$$\frac{dv_{lx}^{\Sigma}}{dt} = -\frac{n_{lx}}{2C_{\mathrm{SM}}}i_x + \frac{n_{lx}}{C_{\mathrm{SM}}}i_{\mathrm{comm},x}. \tag{9b}$$

With the equations derived above that fully describe the dynamics of the MMC, i.e., (5), (8), and (9), the single-phase MMC model can be written in a state-space representation. To this aim, the state vector is chosen as $x = [i_x \; i_{\mathrm{comm},x} \; v_{ux}^{\Sigma} \; v_{lx}^{\Sigma}]^T$, while the input vector is $u = [n_{ux} \; n_{lx}]^T$. As can be seen from the equations the system is nonlinear. Specifically, the model of the MMC is described by the bilinear state-space model as [8]

$$\frac{dx(t)}{dt} = \underbrace{\begin{bmatrix} \frac{-R_{\mathrm{arm}}+2R_{\mathrm{g}}}{L_{\mathrm{arm}}+2L_{\mathrm{g}}} & 0 & 0 & 0 \\ 0 & \frac{R_{\mathrm{arm}}}{L_{\mathrm{arm}}} & 0 & 0 \\ 0 & 0 & 0 & 0 \\ 0 & 0 & 0 & 0 \end{bmatrix}}_{F} x(t) + \underbrace{\begin{bmatrix} 0 & 0 & -\frac{1}{N(L_{\mathrm{arm}}+2L_{\mathrm{g}})} & 0 \\ 0 & 0 & -\frac{1}{2NL_{\mathrm{arm}}} & 0 \\ \frac{1}{2C_{\mathrm{SM}}} & \frac{1}{C_{\mathrm{SM}}} & 0 & 0 \\ 0 & 0 & 0 & 0 \end{bmatrix}}_{G_1} u_1(t)x(t)$$

$$+ \underbrace{\begin{bmatrix} 0 & 0 & 0 & \frac{1}{N(L_{\mathrm{arm}}+2L_{\mathrm{g}})} \\ 0 & 0 & 0 & -\frac{1}{2NL_{\mathrm{arm}}} \\ 0 & 0 & 0 & 0 \\ -\frac{1}{2C_{\mathrm{SM}}} & \frac{1}{C_{\mathrm{SM}}} & 0 & 0 \end{bmatrix}}_{G_2} u_2(t)x(t) + \underbrace{\begin{bmatrix} -\frac{2v_{\mathrm{g}}(t)}{L_{\mathrm{arm}}+2L_{\mathrm{g}}} \\ \frac{v_{\mathrm{dc}}(t)}{2L_{\mathrm{arm}}} \\ 0 \\ 0 \end{bmatrix}}_{g(t)}, \tag{10}$$

where F is the system matrix, G_1 and G_2 are the input matrices, $g(t)$ is an offset vector, and u_1 and u_2 are the first and second elements of u, respectively.

Finally, for control purposes, the continuous-time state-space model is discretized by using the forward Euler method. This yields

$$x(k+1) = Ax(k) + B_1 u_1(k)x(k) + B_2 u_2(k)x(k) + b(k), \tag{11}$$

with $A = I + FT_s$, $B_1 = G_1 T_s$, $B_2 = G_2 T_s$ and $b = g T_s$, where T_s is the system sampling interval and I is the identity matrix of the same dimension as F.

Control Algorithm

The mathematical model of the MMCs derived in the previous section serves as the prediction model for the proposed MPC-based algorithm. In the sequel of this section, the derivation of the control scheme in question is provided.

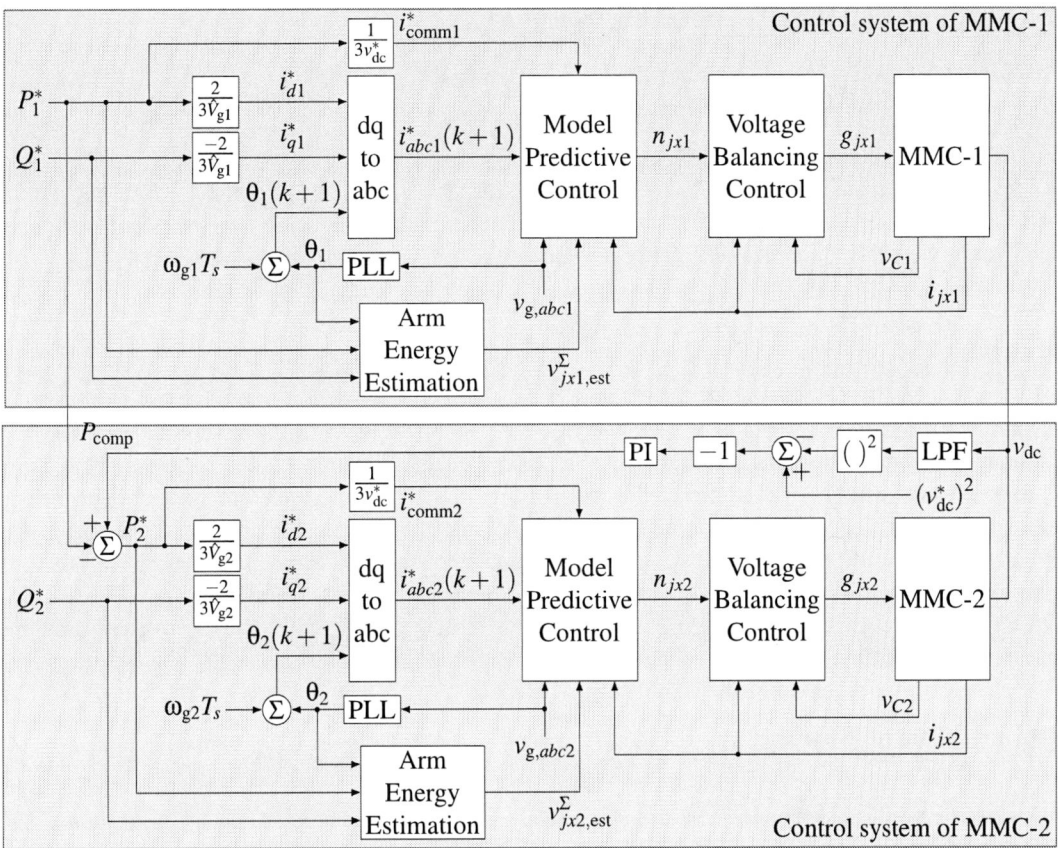

Fig. 2: Complete block diagram of the proposed MPC-based algorithm of the MMC-HVDC system.

Control Structure

The proposed control structure for the MMC-HVDC is shown in Fig. 2. As can be seen, a hierarchical control architecture is adopted which consists of the following three parts:

1. At the upper layer, a PI-based controller controls the dc-link voltage by manipulating the active power reference of MMC-2. In doing so, the controller accounts for the power losses of both MMC units.

2. At the middle layer, as discussed later, the proposed MPC algorithm directly computes the number of the to-be-inserted SMs into each arm to achieve phase current reference tracking and circulating current minimization.

3. At the lower layer, an SM capacitor voltage balancing mechanism selects which individual SMs are inserted/bypassed such that the SM capacitor voltages are balanced. In doing so, the gating signals for the SM switches are generated.

Arm Energy Estimation

Controlling the arm energies, i.e., v_{jx}^{Σ}, can be challenging since the circulating current is used to transfer active power between the converter arms. During steady-state operation the circulating current needs to be minimized. However, during transients, when imbalances in the arm energies occur, a small circulating current is required to redistribute the energy stored in the arms in order to balance them. Typically, this issue is solved by adding an extra component to the common-mode current using PI-based controllers to regulate the total energy and the energy difference between the upper and lower arms [9]. In this paper, however, the open-loop arm energy balancing method proposed in [6] is adopted and utilized in conjunction with MPC.

To this aim, the j arm energy is estimated according to [6]

$$W_{Cj}^{\Sigma} = W_C^{\Sigma *} \pm \frac{\hat{V}_g i_{\text{comm}}^* \cos(\theta)}{\omega_g} \mp \frac{\left(\frac{v_{\text{dc}}^*}{2} - R_{\text{arm}} i_{\text{comm}}^*\right) \hat{I}_x \cos(\theta - \phi)}{2\omega_g} + \frac{\hat{V}_g \hat{I}_x \sin(2\theta - \phi)}{8\omega_g}, \tag{12}$$

where $\hat{V}_g$ and $\hat{I}_x$ are the grid peak voltage and current, respectively, ω_g the grid angular frequency, θ the grid angle, ϕ the power angle, and $W_C^{\Sigma *}$ the average arm energy that is equal to the energy corresponding to the dc-link voltage reference. With (12), the sum of the arm capacitor voltages is given by

$$v_j^{\Sigma} = \sqrt{\frac{2NW_{Cj}^{\Sigma}}{C_{\text{SM}}}}. \tag{13}$$

Using the estimated arm energies works well in steady state and provides a self-stabilizing effect [6]. During transients, when there is an imbalance in the arm energies, a fundamental frequency circulating current appears, which helps to balance the arm energies [10, 11]. This results in a desirable open-loop arm energy balancing, implying that the MPC algorithm does not have to actively address the balancing problem. As a result, the controller design is simplified since the MPC objective function can account only for the grid and circulating currents tracking problem. It is worth mentioning, however, that, as shown in Fig. 2, even though the MPC part of the control structure does not require information about the capacitor voltages, the subsequent voltage balancing controller does. This means that the measurements of the capacitor voltages cannot be completely avoided.

Optimal Control Strategy

Given the open-loop arm balancing startegy described above, the main control objective of the MPC algorithm is to regulate the grid currents along their sinusoidal references so that grid standards, e.g., the IEEE 519 standard [12], are met. Moreover, the circulating currents that may exist in each phase leg need to be minimized to avoid power losses in the phase legs, increase of the switching devices current rating, and pronounced voltage ripples in the SM capacitors [2, 13]. These control tasks need to be met while keeping the average device switching frequency low so that the switching power losses are minimized. Finally, it is noteworthy that the MPC algorithm is not responsible for the SM capacitor voltage balancing, since this task is undertaken by the subsequent balancing controller, as previously mentioned.

With the aforementioned control goals, the objective function for the MPC problem is formulated as

$$J(k) = (\boldsymbol{y}^*(k+1) - \boldsymbol{y}(k+1))^T \boldsymbol{Q} (\boldsymbol{y}^*(k+1) - \boldsymbol{y}(k+1)) + \lambda_u \|\Delta \boldsymbol{u}(k)\|_1, \tag{14}$$

where $\boldsymbol{y} = \begin{bmatrix} i_x & i_{\text{comm},x} \end{bmatrix}^T$ and $\boldsymbol{y}^*$ are the output and corresponding reference vectors, respectively. Moreover, $\Delta \boldsymbol{u}(k) = \boldsymbol{u}(k) - \boldsymbol{u}(k-1)$ indicates the control effort, defined as the difference in the SM modules between two consecutive time steps. Note that $\boldsymbol{Q} = \text{diag}(\lambda_x, \lambda_{\text{comm}})$, with $\lambda_x, \lambda_{\text{comm}} > 0$, and $\lambda_u > 0$ are weighting factors that assign different penalization priorities between the output reference tracking and switching effort terms, respectively. These weighting factors are commonly chosen by examining the Pareto optimal fronts. Finally, it is worth mentioning that function (14) considers the common-mode—rather than the circulating—current since in the single-phase model of the MMC the circulating current cannot be properly predicted due to the coupling of the phases. This, however, does not compromise the controller performance, as the circulating current i_{zx} is effectively eliminated, as shown later.

Considering the big number N of SMs per arm, minimizing (14) in real time for all possible SM combinations can lead to a computationally intractable problem. To keep the computational complexity of the proposed MPC algorithm at bay, the following simplifications are adopted:

1. As already discussed, instead of solving for the optimal switching state of the MMC i.e., switch positions, the direct MPC algorithm computes the optimal number of inserted SMs per arm. This reduces the number of possible solutions from 2^{6N} to $(N+1)^6$.

```
 1: Calculate $i_x$, $i_{comm}$ and $v_j^\Sigma$ with (3), (7) and (13)
 2: $\boldsymbol{x}(k) \leftarrow \begin{bmatrix} i_x & i_{comm} & v_u^\Sigma & v_l^\Sigma \end{bmatrix}^T$
 3: $J_{opt} \leftarrow \infty$
 4: for $\Delta n_u = -\Delta n_{max}, \ldots, \Delta n_{max}$ do
 5:     $n_u \leftarrow n_u(k-1) + \Delta n_u$
 6:     if $n_u < 0$ or $n_u > N$ then
 7:         continue
 8:     end if
 9:     for $\Delta n_l = -\Delta n_{max}, \ldots, \Delta n_{max}$ do
10:         $n_l \leftarrow n_l(k-1) + \Delta n_l$
11:         if $n_l < 0$ or $n_l > N$ then
12:             continue
13:         end if
14:         Calculate $J$ with (14)
15:         if $J < J_{opt}$ then
16:             $J_{opt} \leftarrow J$
17:             $n_{u,opt} \leftarrow n_u$
18:             $n_{l,opt} \leftarrow n_l$
19:         end if
20:     end for
21: end for
22: return $n_{u,opt}$, $n_{l,opt}$
```

(a) MPC algorithm.

```
 1: Divide SMs into groups of inserted and bypassed SMs.
 2: Sort the groups by capacitor voltages.
 3: $\Delta n_{jx} \leftarrow n_{jx,opt}(k) - n_{jx,opt}(k-1)$
 4: if $\Delta n_{jx} > 0$ then
 5:     if $i_{jx} > 0$ then
 6:         Insert $\Delta n_{jx}$ SMs with the lowest voltages
 7:     else
 8:         Insert $\Delta n_{jx}$ SMs with the highest voltages
 9:     end if
10: else if $\Delta n_{jx} < 0$ then
11:     if $i_{jx} > 0$ then
12:         Bypass $\Delta n_{jx}$ SMs with the highest voltages
13:     else
14:         Bypass $\Delta n_{jx}$ SMs with the lowest voltages
15:     end if
16: else
17:     Apply the last-applied gating signals
18: end if
```

(b) Voltage balancing algorithm.

Fig. 3: Pseudocode of the MPC and voltage balancing algorithms.

2. As mentioned, the direct MPC algorithm controls each phase independently. This reduces the size of the optimization variable from 6 to 2, and, consequently, the possible solutions from $(N+1)^6$ to $(N+1)^2$ per phase.

3. Similar to, e.g., [14], hard constraints $\|\Delta\boldsymbol{u}(k)\|_\infty \leq \Delta n_{max}$ are added to limit the change in the inserted SMs to Δn_{max} per arm. This limits the possible solutions to $(2\Delta n_{max} + 1)^2$ per phase. As a result, the complexity of the controller is independent of the SM number N, which renders the proposed control scheme feasible for MMCs with any number of SMs. This, however, comes at a cost of a somewhat deteriorated transient performance.

Given the above, the optimization problem underlying direct MPC takes the form

$$
\begin{aligned}
\underset{\boldsymbol{u}(k)}{\text{minimize}} \quad & J(k) \\
\text{subject to} \quad & \boldsymbol{u}(k) \in \{0, \ldots, N\}^2 \\
& \|\Delta\boldsymbol{u}(k)\|_\infty \leq \Delta n_{max},
\end{aligned}
\tag{15}
$$

which is solved by enumerating all possible $(2\Delta n_{max} + 1)^2$ solutions for each phase as shown in the pseudocode of the MPC algorithm provided in Fig. 3a. Finally, it is worth pointing out that the offset vector $\boldsymbol{g}$ in (10) as well as the arm energy estimates (12) depend on the time-varying dc-link voltage. Because the MMC-HVDC system does not have separate dc capacitors, and the independent control of the upper and lower arms gives rise to a time-varying $n_{ux} + n_{lx}$ in the phase legs, the dc-link voltage experiences a high frequency ripple. To avoid the performance deterioration of the MPC-based algorithm, the dc-link voltage reference—instead of its instantaneous value—is used in (10) and (12). Such a simplification, however, can be justified by the fact that an outer-loop is used to regulate the actual dc-link voltage along it reference, meaning that the dc-link voltage is successfully controlled, and its average value is equal to its desired one.

Finally, once the optimal number of SMs per arm is computed, the voltage balancing controller is activated [15]. Note that this controller operates such that not only the SM capacitor balancing is achieved, but also that excessive and unnecessary switching is avoided. In doing so, it is ensured that the device switching frequency remains low. The pseudocode of the balancing algorithm is shown in Fig. 3b.

Table I: HVDC system parameters

Parameter	Symbol	SI Value	Per unit value
Dc-link voltage	V_{dc}	40 kV	1
Rated rms line-to-line voltage	V_R	20 kV	0.5
Rated apparent power	S_R	30 MVA	1
Grid side resistance	R_g	0.05 Ω	0.0015
Grid side inductance	L_g	5 mH	0.0481
Arm resistance	R_{arm}	0.1 Ω	0.0031
Arm inductance	L_{arm}	3 mH	0.0289
Submodule capacitance	C_{SM}	6 mF	61.5624
Grid-1 angular frequency	ω_{g1}	$2\pi 50$ rad s^{-1}	1
Grid-2 angular frequency	ω_{g2}	$2\pi 60$ rad s^{-1}	1.2
SMs per arm	N	20	

Performance Assessment

To assess the performance of the proposed control scheme under both steady-state and transient operating conditions, simulations are carried out for an MMC-HVDC system with 20 SMs per arm. The system parameters are shown in Table I. The weighting factors are heuristically chosen as $\lambda_x = 1$, $\lambda_{comm} = 0.35$ and $\lambda_u = 9 \times 10^{-5}$, resulting in an average device switching frequency of 140 Hz. Moreover, the sampling interval is chosen as $T_s = 100\,\mu s$, while $\Delta n_{max} = 1$, and the MPC prediction horizon length is $N_p = 1$. All results are shown in the per unit (p.u.) system.

The steady-state performance of the controller is shown in Fig. 4a and Fig. 4b for MMC-1 and MMC-2, respectively. The phase currents accurately track their references, with the associated TDD being as low as 0.8 % for both systems. As a result, the produced harmonics are well below the limits imposed by the IEEE 519 grid standard [12], see the last row in Fig. 4. Furthermore, the circulating currents are effectively eliminated with their rms value being 0.0084 p.u. Finally, the arm energy estimates, shown with dashed lines, are close to the actual arm energies and the arm total capacitor voltage ripple is around 5 % of the nominal value.

To test the transient behavior of the proposed MPC scheme, the active power reference is halved at $t = 40$ ms. The corresponding results are shown in Fig. 5a and Fig. 5b for MMC-1 and MMC-2, respectively. Owing to the direct nature of the controller, the phase currents reach their new desired values within only 2 ms. A small spike is observed in the circulating current. Nevertheless, it disappears in less than 2 ms, as the circulating current settles back to zero. Moreover, after the power reference step-change, a low-amplitude fundamental harmonic appears in the circulating currents. This enables the balancing of the capacitor voltages, as verified by the presented results. Finally, the capacitor voltage sums are well balanced even though they are not addressed in the objective function.

To provide further insight into the workings of the proposed MPC algorithm, Fig. 5c shows the active power of MMC-1 for different levels of Δn_{max}. Therein, the drawbacks of limiting Δn_{max} can be seen. By limiting Δn_{max} the available voltage to be inserted within a sampling interval is limited, thus causing slower transients. Moreover, as shown in Fig. 5a and Fig. 5b, due to the magnetization of the arm and filter inductors and the small prediction horizon, the settling time is further increased. Larger values of Δn_{max} can achieve faster transients, as shown in Fig. 5c.

Conclusion

This paper proposed a direct MPC scheme to control the grid and circulating currents of back-to-back MMCs. To do so, the number of SMs that need to be inserted is computed by solving a constrained optimization problem in real time. Moreover, an estimation scheme is implemented to provide the arm energies of the MMCs. By using these estimates in the prediction model, open-loop arm energy balancing

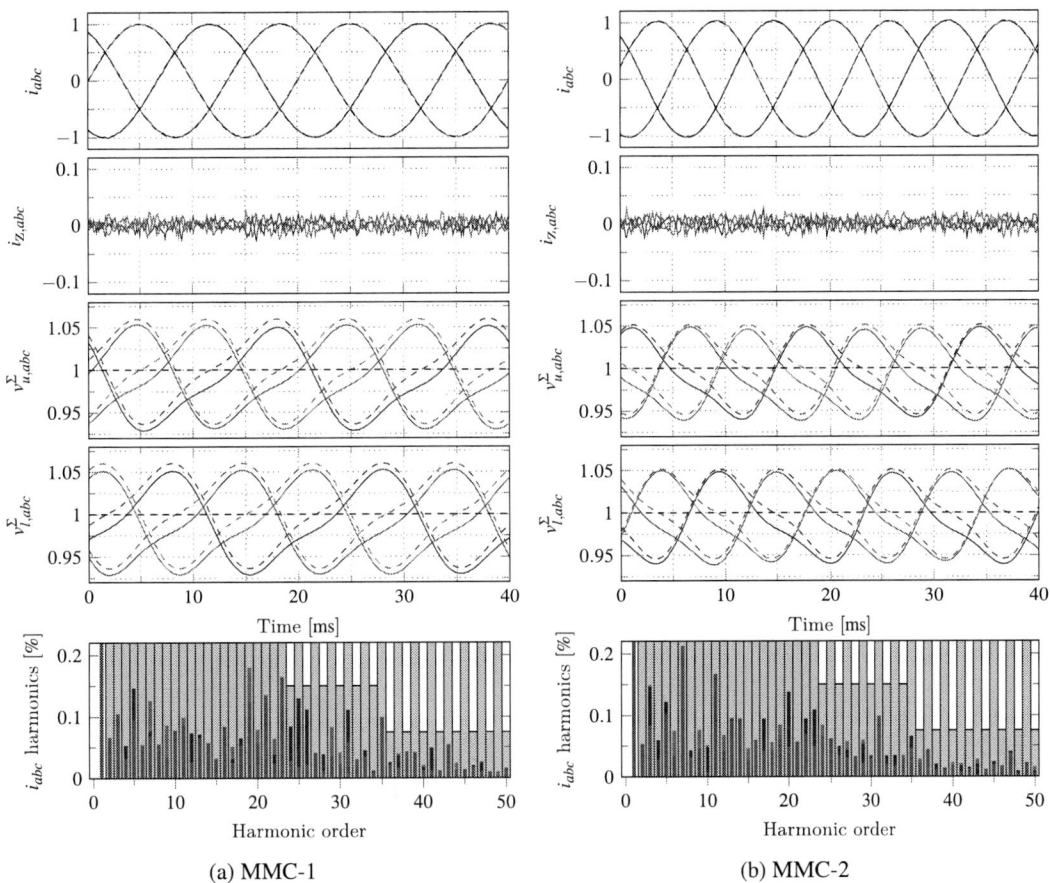

(a) MMC-1   (b) MMC-2

Fig. 4: Steady-state performance for operation under nominal conditions. Top to bottom: grid currents, circulating currents, (measured and estimated) arm total capacitor voltages, grid current harmonic spectra along with the IEEE 519 limits.

is ensured, while potential stability problems are avoided. Consequently, as verified by the presented results, favorable steady-state and transient performance is achieved.

References

[1] A. Lesnicar and R. Marquardt, "An innovative modular multilevel converter topology suitable for a wide power range," in *Proc. IEEE Power Tech. Conf.*, (Bologna, Italy), pp. 1–6, Jun. 2003.

[2] M. Saeedifard and R. Iravani, "Dynamic performance of a modular multilevel back-to-back HVDC system," *IEEE Trans. Power Del.*, vol. 25, pp. 2903–2912, Oct. 2010.

[3] T. Geyer, *Model predictive control of high power converters and industrial drives*. Hoboken, NJ, USA: Wiley, 2016.

[4] P. Karamanakos, E. Liegmann, T. Geyer, and R. Kennel, "Model predictive control of power electronic systems: Methods, results, and challenges," *IEEE Open J. Ind. Appl.*, vol. 1, pp. 95–114, 2020.

[5] P. Karamanakos and T. Geyer, "Guidelines for the design of finite control set model predictive controllers," *IEEE Trans. Power Electron.*, vol. 35, pp. 7434–7450, Jul. 2020.

[6] L. Angquist, A. Antonopoulos, D. Siemaszko, K. Ilves, M. Vasiladiotis, and H.-P. Nee, "Open-loop control of modular multilevel converters using estimation of stored energy," *IEEE Trans. Ind. Appl.*, vol. 47, pp. 2516–2524, Nov./Dec. 2011.

[7] F. Zhang, W. Li, and G. Joós, "A voltage-level-based model predictive control of modular multilevel converter," *IEEE Trans. Ind. Electron.*, vol. 63, pp. 5301–5312, Aug. 2016.

[8] V. Verdult, M. Verhaegen, and V. Verdult, "Bilinear state space systems for nonlinear dynamical modelling," *Theory in Biosc.*, vol. 119, no. 1, pp. 1–9, 2000.

[9] A. Antonopoulos, L. Angquist, and H.-P. Nee, "On dynamics and voltage control of the modular multilevel converter," in *Proc. Eur. Power Electron. Conf.*, (Barcelona, Spain), pp. 1–10, Sep. 2009.

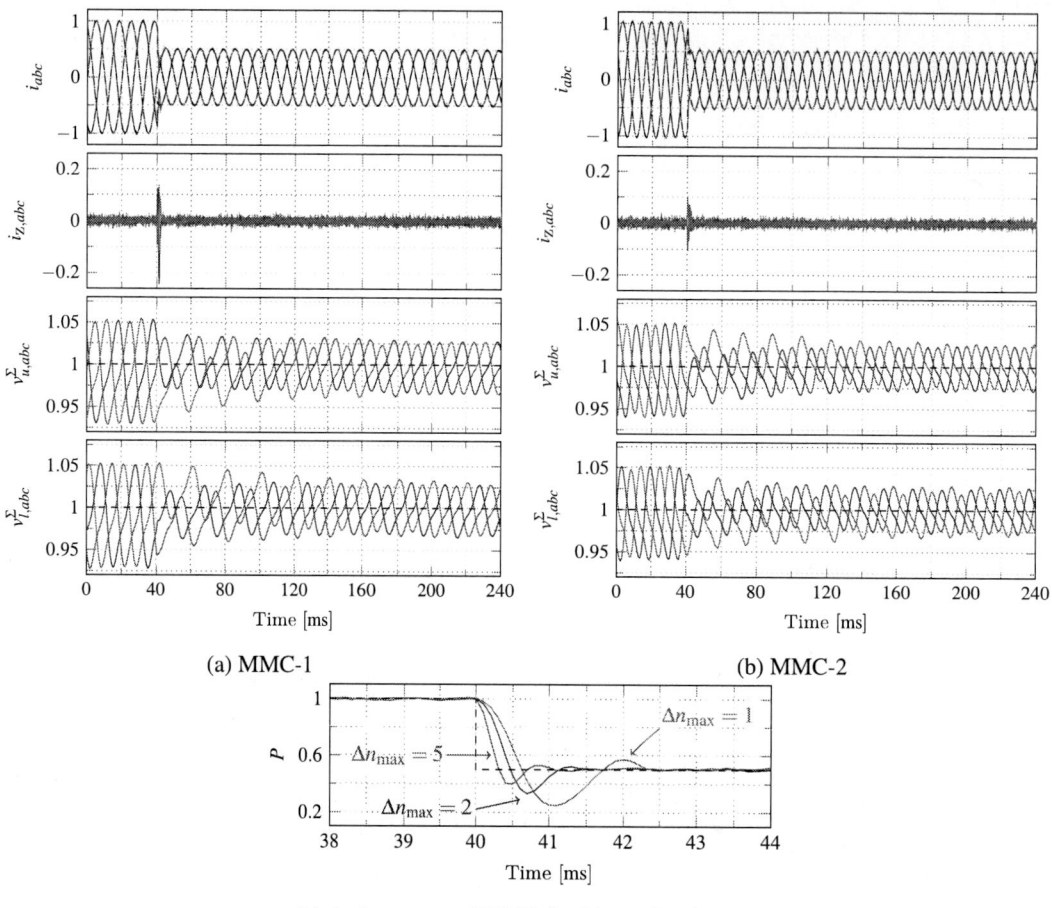

(a) MMC-1   (b) MMC-2

(c) Active power of MMC-1 with varying Δn_{max}.

Fig. 5: Transient performance for operation under an active power reference step. Top to bottom: grid currents, circulating currents, measured arm total capacitor voltages.

[10] A. Antonopoulos, L. Ängquist, L. Harnefors, K. Ilves, and H.-P. Nee, "Global asymptotic stability of modular multilevel converters," *IEEE Trans. Ind. Electron.*, vol. 61, pp. 603–612, Feb. 2014.

[11] L. Harnefors, A. Antonopoulos, S. Norrga, L. Angquist, and H.-P. Nee, "Dynamic analysis of modular multilevel converters," *IEEE Trans. Ind. Electron.*, vol. 60, pp. 2526–2537, Jul. 2013.

[12] IEEE Std 519-2014 (Revision of IEEE Std 519-1992), "IEEE recommended practices and requirements for harmonic control in electrical power systems," Jun. 2014.

[13] B. Bahrani, S. Debnath, and M. Saeedifard, "Circulating current suppression of the modular multilevel converter in a double-frequency rotating reference frame," *IEEE Trans. Power Electron.*, vol. 31, pp. 783–792, Jan. 2016.

[14] M. Vatani, B. Bahrani, M. Saeedifard, and M. Hovd, "Indirect finite control set model predictive control of modular multilevel converters," *IEEE Trans. Smart Grid*, vol. 6, pp. 1520–1529, May 2015.

[15] Q. Tu, Z. Xu, and L. Xu, "Reduced switching-frequency modulation and circulating current suppression for modular multilevel converters," *IEEE Trans. Power Del.*, vol. 26, pp. 2009–2017, Jul. 2011.

Study on Commutation Loop Inductance and Current Distribution to DC-link Capacitors in a GaN Half-bridge

Benedikt Kohlhepp, Samuel Faber, Jeremias Kaiser and Thomas Dürbaum
Electromagnetic Fields, Friedrich-Alexander University Erlangen-Nürnberg (FAU)
Konrad-Zuse-Strasse 3/5
91052 Erlangen, Germany
Tel.: +49 (0)9131 85 28951
Fax: +49 (0)9131 85 27787
E-Mail: benedikt.kohlhepp@fau.de
URL: https://www.emf.tf.fau.de/

Keywords

«Parasitic inductance», «Gallium Nitride (GaN)», «Half-bridge».

Abstract

Modern wide bandgap power semiconductors allow for increased efficiency compared to conventional semiconductors by faster switching transients. In order to harness the full performance from these components, parasitic inductances must be minimized. FEM simulations can deliver the parasitic inductances and thus, can be used to optimize the PCB layout and the placement of semiconductors and DC-link capacitors. Also cost constraints apply to most of the power electronic circuits. FEM simulations can facilitate the determination of cost optimized designs by reducing the amount of DC-link capacitors. Therefore, the current distribution incorporating the DC-link capacitors is required. An analysis of a GaN half-bridge shows the potential to omit one poorly utilized MLCC capacitor.

Introduction

Today's demands regarding power electronic circuits, high efficiency and power density as well as low cost call for advanced circuit topologies and modern power electronic components [1]. Thus, wide bandgap (WBG) semiconductors are inevitably required to increase the efficiency [2]. In order to estimate the efficiency, accurate loss predictions are necessary for each component involved [3] [4]. While conduction loss calculation of power semiconductor devices is relatively easy [2], switching loss estimation is quite challenging especially for WBG semiconductors [5] [6] [7]. As WBG devices can feature extremely fast switching transitions, circuit parasitics stemming from the device's package, the DC-link capacitor's interconnection, and the printed circuit board (PCB) layout become more prominent. For a profitable application of WBG semiconductors, circuit parasitics have to be reduced in order to fully exploit their advantages with regard to fast switching. Current GaN power semiconductors feature extremely low inductive packages as these are ball grid array (BGA) packages, also called chip scale package [8]. Due to these optimized packages, the resulting switching transients mainly depend on the inductances induced by the PCB interconnection and DC-link capacitors [9]. During the switching transitions, the commutation loop inductance causes overvoltages [10]. Furthermore, the semiconductor's output capacitance in conjunction with the commutation loop's inductance induces oscillations in the power loop. Overvoltages as well as the oscillations can lead to increased switching losses and deteriorate the electromagnetic interference (EMI) behavior [11].

Low cost means using the right (optimized) components, which are driven to their limits. Thus, in case several DC-link capacitors connected in parallel are used, the current distribution to the individual capacitor is of relevance. This allows an evaluation of the component utilization. Therefore, the current distribution to the capacitors can be used to find the optimal placement of the capacitors. In the best case scenario, this can lead to savings in components and hence, to lower cost.

In order to gain the required information regarding the commutation loop inductance, different measurement methods like vector network analyzers (VNA) [12], time domain reflectometry (TDR) [13] or impedance measurements [14] can be applied. Another possibility is to use numerical simulations [9] [15] exploiting e.g. FEM, which also allow to study the current distribution within the PCB and components. Therefore, this paper first introduces the half-bridge circuit and the modeling of all components involved within the simulation. Then, the commutation loop inductance is studied for different models of the DC-link capacitors. After that, the current distribution to the DC-link capacitors is studied depending on the frequency and also for different models of the DC-link capacitors. The paper also shows the current density on the PCB's copper layers.

Half-Bridge Configuration

As a basic building block for power electronic converters, the half-bridge circuit is used for the study of the parasitic inductances of the PCB and the current distribution. Fig. 1 (left) depicts the schematic of the half-bridge with both semiconductors T_{HS} and T_{LS}, the accompanying gate circuits and DC-link capacitor C_{DC}. The gate circuits comprise separated connections for turn-on and –off with $R_{g,HS,On}$, $R_{g,LS,On}$ and $R_{g,HS,Off}$, $R_{g,LS,Off}$. The schematic includes the boot-strap C_{Boot} as well as the bypass capacitor C_{Bypass}. Fig. 2 (left) shows the section of the half-bridge of the PCB layout which is derived from the schematic as top and side view. As can be seen from Fig. 2 (left), the PCB is made up from a 4-layer stackup with a copper thickness of 70 µm. The power semiconductors feature a low inductive BGA package and are 100 V GaN-HEMTs. The gate-driver also comes in a BGA package, allowing low inductive interconnections.

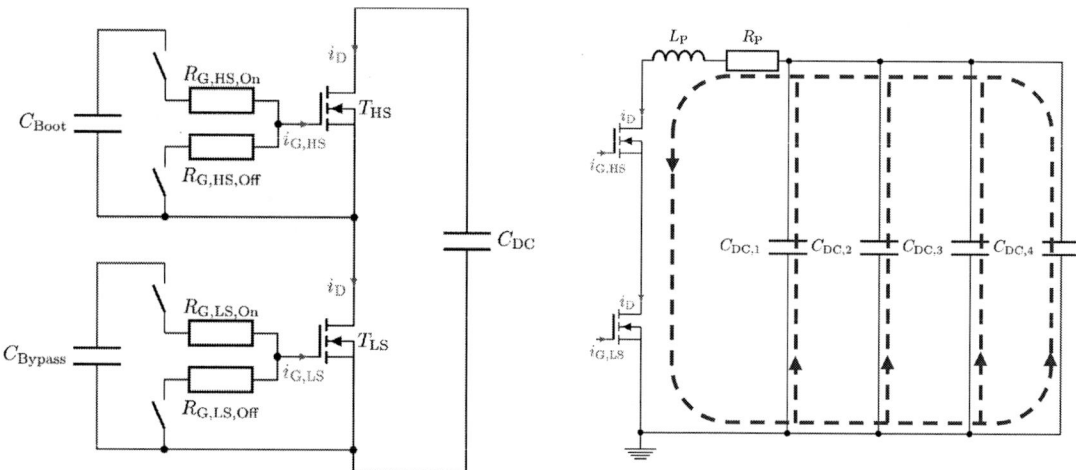

Fig. 1: Schematic of the half-bridge circuit and both gate circuits (left); equivalent circuit of the half-bridge (right)

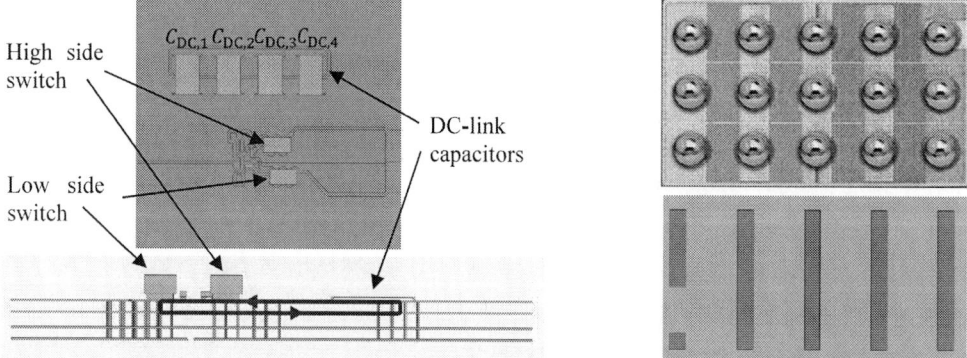

Fig. 2: Layout of the half-bridge circuit and both gate circuits; top view (upper) and side view (lower) (left); solder bump view of the studied device *[8]* and simulation model for the device (right)

All components are mounted on the top layer of the PCB as SMD devices. The DC-link capacitors $C_{DC,1}$, $C_{DC,2}$, $C_{DC,3}$, $C_{DC,4}$ (see Fig. 2 (left)), are placed directly next to the half-bridge semiconductors. The half-bridge's commutation loop starts at the lower pads (positive terminals) of the DC-link capacitors and connects to the drain of the upper switch on the top layer. A copper polygon on the top layer links both half-bridge semiconductors and forms the half-bridge midpoint. Filled vias within the source pads of the lower switch bring the commutation loop to the first inner layer (distance 360 µm). In this layer, the commutation loop passes back to the DC-link capacitors again. The ground pads of the DC-link capacitors also feature filled vias, and thus, close the commutation loop. As only the first two layers are used for commutation loop design, low commutation loop inductance can be expected [16]. On the top layer the gate loops are also routed. These do not have discrete (external) gate resistors, which ensures small gate loops. In general, for each loop inductances can be defined. Furthermore, mutual couplings must be considered, which represent the interaction between the loops. In some literature, this fact is neglected, but for an accurate study, it is vital [15]. In this paper, only the overall commutation loop inductance L_P shall be studied. The resulting equivalent circuit shows Fig. 1 (right). In addition to the commutation inductance, [9] also examines the inductance of the gate circuits and the mutual coupling.

Component Models

Suitable models for all components are crucial for gaining the circuit parasitics of the PCB. The DC-link capacitors consist of four SMD ceramic capacitors in a 1210 case and are located relatively close to the semiconductors. As their impedance decreases with higher frequencies, they are replaced by a short in simulation. In order to study the impact of the capacitor's shape, four different models are used. The first one is a thin copper yoke with a thickness of 100 µm and a distance of 100 µm above the PCB surface, which can be seen in Fig. 3 (a). Secondly, a copper yoke is used which has the same dimensions with respect to the pads as the MLCC capacitors (see Fig. 3 (b)). The height is chosen so that the thickness of the conductive material is in the range of twice the skin depth at 100 MHz. An approximation of the skin depth δ gives:

$$\delta = \sqrt{\frac{2}{2\pi f \mu \kappa}} \tag{1}$$

Thin copper yoke (a)   Copper yoke with dimensions of pads (b)   Solid block of copper (c) or tin (d)

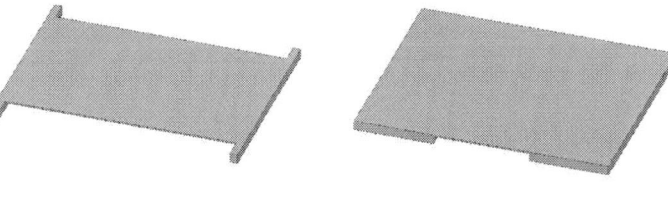 

Fig. 3: Different models for the MLCC DC-link capacitors

Equation (1) uses the permeability μ and the electrical conductivity κ [17]. Thus, Fig. 3 (b) represents a rough estimation of an "averaged" current path through the capacitors at high frequencies. Configuration (c) and (d) in Fig. 3 are solid blocks with the dimensions of the MLCC capacitors. There is a groove on the bottom side of the blocks, which forms two contact surfaces (pads) of the capacitor. These have the same size as the pads of the capacitors. (c) is a copper block ($58 \cdot 10^6$ 1/Ωm). Obviously, modeling the MLCC capacitors by solid conductive blocks is not realistic. As the current distribution depends on the electrical conductivity of the materials, two different materials with different electrical conductivity should be studied. The cause for this behavior is the skin as well as the proximity effect, which impact the current distribution within conductive materials at high frequencies. This is why tin with an electrical conductivity of $8.7 \cdot 10^6$ 1/Ωm is studied as well. It has a significantly lower electrical conductivity compared to copper. Furthermore, accurate models for the GaN devices are necessary. In on-state the semiconductors can be modeled by copper blocks with identical dimensions. These devices feature an extremely low inductance as they come in a ball grid array (BGA) package. The solder bumps (balls) must be considered as well, as their size is not negligible compared to the packages. The drain's and source's balls are column wise connected together to cuboids. The bump for the gate is also cubic. Fig. 2 shows on the right side the solder bump view of the device as well as the solder bump representation (grey) within the simulation model (green).

Study on Commutation Loop Inductance

This section studies the commutation loop inductance of the GaN-half-bridge circuit. For the analysis, ANSYS Maxwell3D eddy current is applied, which uses the quasi-stationary approximation of the Maxwell equations to solve field problems [18]. This analysis neglects displacement currents and thus wave propagation. It allows to obtain the current distribution within conductive material by using the finite element method (FEM). From the current distribution, the DC and AC inductance for the frequencies of interest can be obtained.

Fig. 4 (left) shows the commutation loop's inductance depending on the frequency from 10 kHz up to 100 MHz for the four configurations (a) to (d) given in Fig. 3. All curves drop with rising frequency, as for low frequencies, the inductance comprises the inner as well as the outer inductance. The inner inductance results from the energy stored within the conductor. As a consequence of the skin effect within the conductors (PCB traces and planes) the current is superseded to its surface with increasing frequency. For 10 kHz, the skin depth, according to (1), is much higher than the PCB traces' thickness (70 µm). The PCB interconnections are made of copper with $\kappa = 58 \cdot 10^6$ 1/Ωm. Thus, the skin effect's impact is relatively small for low frequency. At 1 MHz, the copper thickness and skin depth according to (1) are in the same range and for high frequencies (100 MHz), it is well below the copper thickness. As a consequence, with rising frequency, the inner inductance vanishes.

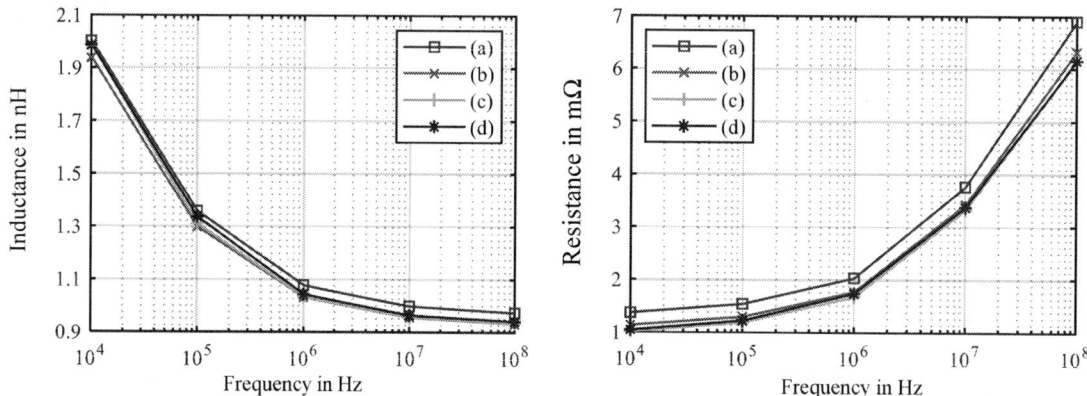

Fig. 4: Comparison of the commutation loop's self inductance depending on the frequency (left); comparison of the commutation loop's resistance depending on the frequency (right)

A second effect also impacts the resulting inductance. With rising frequency, the current distribution within the loop changes. This results in a current distribution that minimizes the inductance. Thus, the self-inductance typically drops with rising frequency. For low frequency, configuration (a) has the highest inductance for all frequencies studied, as the contact surface of the model (a) is only a narrow web. Furthermore, (a), (c), and (d) exhibit higher inductance compared to (b), as the current path is longer. For high frequencies, the inductance of (b), (c), and (d) are approximately the same, since the current is restricted to the lower part of the solid conductive blocks of (c) and (d) to minimize inductance. Thus, the current paths of this three models are nearly the same. All in all, the deviation in inductance is less than 50 pH. This study shows that the modeling of the capacitors has only a negligible effect on the inductance at high frequencies. Of course, the layer thickness of the copper also impacts the resulting inductance. For the intended application, 70µm copper thickness is chosen, as relatively high currents occur. Furthermore, [19] confirms the results gained by the study conducted here. [19] reports a slightly lower inductance, but due to manufacturing reasons, the commutation loop studied here has larger dimensions resulting in higher inductance. Hence, other reasons may impact the selection of the capacitor model used. This could be, for example, the computational effort (simulation time) or the memory consumption. Solid blocks (c) and (d) typically require more memory because the entire volume must be finer discretized by FEM. Fig. 4 (right) shows the commutation loop's resistance depending on the frequency for the four configurations of Fig. 3.

Studying the resistance shows, that (a)'s resistance is slightly higher for the complete frequency range under investigation. This results from the longer current path and the thin conductive structure of the yoke (a). All curves increase with rising frequency, which stems from the skin and proximity effect.

Study on Current Distribution of Capacitors

This section studies the current distribution to the MLCC capacitors. Firstly, configuration (c) (solid copper blocks), is used to model the capacitors. In order to ensure accurate simulation results, the mesh should be fine at the capacitor's cross sections and the commutation loop. Fig. 5 depicts the current density on the capacitor's cross section (in the middle between both pads) at 10 kHz (upper) and 10 MHz (lower). Using a frequency of 10 kHz shows that the current nearly fully utilizes the cross section of the capacitors. An increased current density is visible near the first copper layer of the PCB (lower part of the capacitors). However, if a significantly higher frequency is set, the current is displaced to the lower part of the surface. This results in a higher current density in the lower part than at the top. This result gained by simulation corresponds to the expectations, since the current flows at high frequencies in such a way that the loop area, i.e. the inductance, is low. It can also be seen from this figure that capacitors 3 and 4 carry a higher current than capacitors 1 and 2. This applies regardless of the frequency investigated.

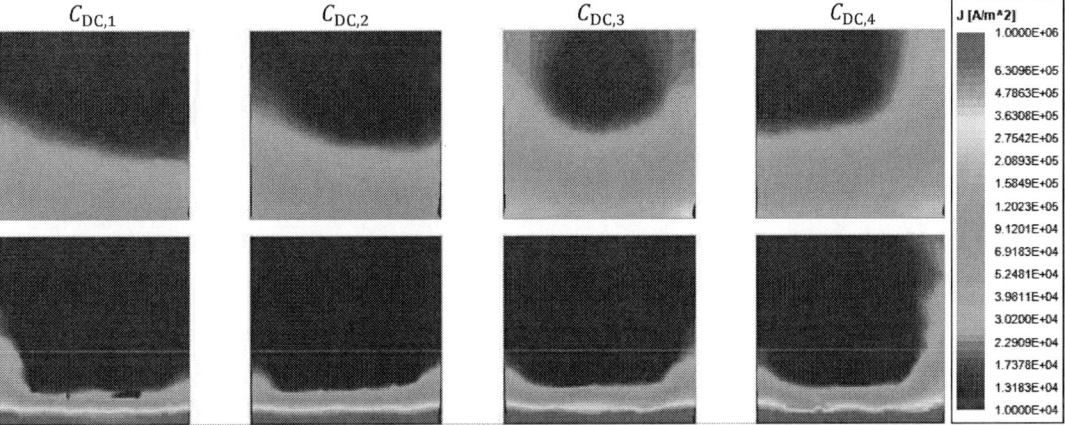

Fig. 5: Current density (logarithmic) on the DC-link capacitor's cross section for the four capacitors at 10 kHz (upper); current density on the DC-link capacitor's cross section for the four capacitors at 1 MHz (lower)

For a quantitative assessment of the current distribution along the capacitors, the current density must be integrated over the cross section of each individual capacitor. Therefore, the normal component $\hat{J}_y$ of the current density is integrated:

$$\hat{i} = \left| \iint_A \hat{\vec{J}} \cdot d\vec{A} \right| = \left| \iint_A \hat{J}_y dA \right| \tag{2}$$

A is the cross section of the individual capacitor. These calculations are directly carried out using the calculation tool provided by ANSYS Maxwell3D. Fig. 6(left) shows the current distribution of the four MLCC capacitors using configuration (c) in the range of 10 kHz up to 100 MHz. Regardless of the frequency, capacitor 1 takes less than 16 % and capacitor 2 conducts approximately 20 % of the overall current. Studying the current shares of capacitor 3 and 4 shows that these components together carry roughly 2/3 of the overall current regardless of the frequency studied.

In order to better understand the current distribution of the capacitors, the current density on the surface of the PCB can also be analyzed. Fig. 7 in the upper part shows the current density at the PCB's top layer for 10 kHz, 1 MHz, and 100 MHz for configuration (c) (solid copper blocks). In order to see the return path of the commutation loop, the first inner layer must be displayed as well. The lower part of Fig. 7 depicts the first inner layer also for solid copper blocks. The MLCC DC-link capacitors are indicated by dashed red rectangles in all six plots. All field distributions of Fig. 7 show a top view to the PCB. These current density distributions can then be used to draw conclusions about the current paths on the PCB. It can be seen that, regardless of the frequency, a very high current density occurs at the finger-like structure of the power semiconductors (top layer), which is due to the fact that the conductive surface is small there. At a frequency of 100 MHz, the current flows mainly at the edges of the finger-like structure of the half-bridge switches. Thus it seems that the total current in the loop is smaller than at lower frequencies. For all simulations a current of 1 A is used. Studying the top layer at 10 kHz confirms that capacitor 3 takes the highest portion of the overall current, as the current density is highest at its terminals compared to all other capacitors. Furthermore, from the top layer at 10 kHz it can be seen that the PCB area between the upper half-bridge switch and capacitor 3 has the highest current density (it occurs in light blue and green). For higher frequencies, the current shifts more and more to capacitor 4 (see Fig. 6 (left)), which can also be verified by the field plots of PCB's top layer for 1 MHz and 100 MHz. For 100 MHz, the field plot is a little bit blurred, as the mesh is not fine enough to plot the current density. Nevertheless, the variation in results for a more detailed mesh will not give better results, but significantly enhances computational effort. When looking at the current density within the first inner layer of the PCB (see Fig. 7 (lower)), there is an area looking like a 'hole', which is a recess in the first inner layer of the board, as there are vias connecting the top layer to the bottom of the board. These are mainly needed for heat dissipation (thermal vias) ensuring sufficient cooling of the upper half-bridge switch during operation. As this "conductive material free" zone directly lies between the upper half-bridge switch and capacitor 3, this is one reason, why capacitor 4 takes the largest portion of the overall current.

Secondly, the current distribution on the capacitors will also be studied for the different configurations from Fig. 3. This study can give an indication of whether the current distribution on the PCB in the simulation changes with different models of the capacitors. If it remains approximately the same with different MLCC capacitor models, it can be concluded that the modeling of the capacitors has only a minor influence on the current distribution resulting in the simulation. Thus, Fig. 6 (right) depicts the current distribution to the capacitors at 1 MHz for the different capacitor models (a) to (d) given in Fig. 3. In principle, the distribution remains constant regardless of the model used for the capacitors. Models (b) to (d) in particular show almost the same results. For (a) there is a little deviation from the others visible. This model consists of only one thin copper bar, thus introducing more inductance compared to the other models. This can already be seen from the evaluation of the commutation inductance in Fig. 4 (left). For this reason, the current is more evenly distributed among the four capacitors, since their connection to the half-bridge switches concerning the inductance is now no longer dominated by their position on the board, but rather by the capacitor model.

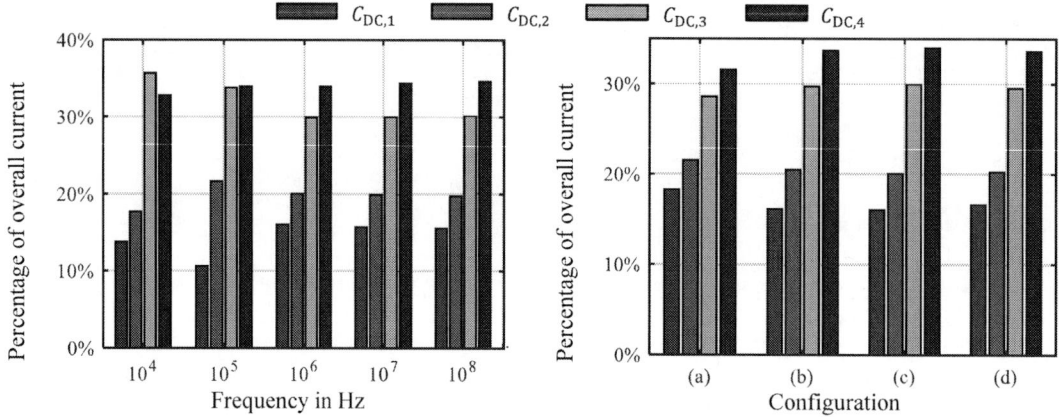

Fig. 6: Current distribution for the DC-link capacitors depending on the frequency for configuration (c) (left); comparison of different configurations at 1 MHz (right)

Discussion

From the study, one can conclude that the model used for the capacitors has only a minor impact on the resulting commutation loop inductance as well as on the current distribution to the paralleled DC-link capacitors within the studied PCB layout.

On the basis of this study, it can now be considered how the capacitors should ideally be placed. Since capacitor 1 only takes a very small proportion of the total current, this component may be omitted. This measure reduces the cost of the capacitors by 25 %. However, it should be noted that it must be verified whether the total capacitance of the parallel connection of 3 capacitors is still sufficient for the application.

In addition to the capacitance value itself, the capacitors must also be capable of handling the ripple current induced by the half-bridge operation.

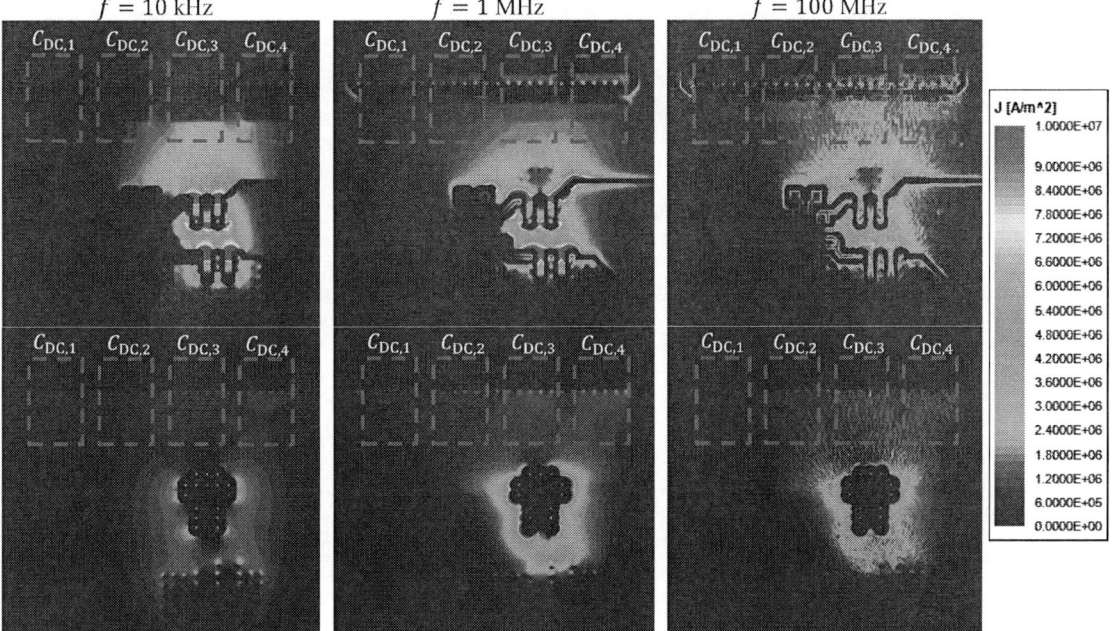

Fig. 7: Current density within the top layer (top view) of the PCB for different frequencies (upper); current density within the first inner layer (top view) of the PCB for different frequencies (lower)

Conclusion

In order to gain full performance from fast switching semiconductors like GaN, parasitic elements within the commutation cell must be minimized. As these inductances are in the sub-nH range, measuring them is quite difficult. Thus, FEM simulations of the PCB is the only way to obtain this information. In a GaN half-bridge circuit, the parasitic inductances are studied for different models of the MLCC DC-link capacitors. The commutation loop's inductance, which is one of the relevant inductances concerning the switching behavior, shows only a small impact (930 pH to 970 pH) depending on the capacitor's model. In order to build a cost effective design, the current distribution to the DC-link capacitors must also be investigated if several are connected in parallel. The analysis gives recommendations for further actions concerning component savings.

References

[1] B. Kohlhepp, J. Heubeck, and T. Duerbaum, "Non-ideal Behavior of ZVS Inverter Comprising Variable and Fixed Frequency Operation: Analysis, Compensation, and Verification," *Power Electronic Devices and Components*, p. 100007, Apr. 2022, doi: 10.1016/j.pedc.2022.100007.

[2] B. Kohlhepp, D. Kübrich, M. Tannhäuser, A. Hoffmann, and T. Dürbaum, "Test Setup for Dynamic ON-State Resistance Measurement of High- and Low-Voltage GaN-HEMTs Under Hard and Soft Switching Operations," *IEEE Transactions on Instrumentation and Measurement*, vol. 69, no. 10, pp. 7740–7751, Oct. 2020, doi: 10.1109/TIM.2020.2985186.

[3] B. Kohlhepp, M. Barwig, and T. Dürbaum, "Efficiency Comparison of an Asymmetrical Half-Bridge PWM Converter with Schottky and Synchronous Rectification," in *2018 53rd International Universities Power Engineering Conference (UPEC)*, Sep. 2018, pp. 1–6, doi: 10.1109/UPEC.2018.8542033.

[4] B. Kohlhepp, D. Kuebrich, S. Peller, and T. Duerbaum, "Test setup using DC metering approach for loss measurements of inductive components – principle, characterization, validation and application," *IET power electron.*, pp. 1–15, Jun. 2021, doi: 10.1049/pel2.12155.

[5] B. Kohlhepp, D. Kübrich, R. Schwanninger, and T. Dürbaum, "Switching Loss Estimation of GaN-HEMTs by Thermal Measurement Procedure," in *PCIM Europe digital days 2021; International Exhibition and Conference for Power Electronics, Intelligent Motion, Renewable Energy and Energy Management*, May 2021, pp. 1–8.

[6] B. Kohlhepp, D. Kübrich, and T. Dürbaum, "Switching Loss Measurement – A Thermal Approach Applied to GaN-Half-Bridge Configuration," in *2021 23st European Conference on Power Electronics and Applications (EPE '21 ECCE Europe)*, Sep. 2021, pp. 1–8.

[7] B. Kohlhepp, D. Kuebrich, and T. Duerbaum, "Experimental Study of the Coss-Losses Occurring During ZVS Transitions – Emphasis on Low and High Voltage GaN-HEMTs," in *CIPS 2020; 11th International Conference on Integrated Power Electronics Systems*, Mar. 2020, pp. 1–6.

[8] Efficient Power Conversion Corporation, "Datasheet EPC2045 – Enhancement Mode Power Transistor." El Segundo, 2018.

[9] B. Kohlhepp, S. Faber, J. Kaiser, and T. Dürbaum, "Extraction of Parasitic Elements of a Printed Circuit Board applied to a GaN Half-Bridge," in *PCIM Europe digital days 2022; International Exhibition and Conference for Power Electronics, Intelligent Motion, Renewable Energy and Energy Management*, May 2022, pp. 1–8.

[10] Efficient Power Conversion Corporation, "Application Note - How to Design an eGaN FET-Based Power Stage with an Optimal Layout." El Segundo, 2018.

[11] Y. Shen, J. Jiang, Y. Xiong, Y. Deng, X. He, and Z. Zeng, "Parasitic Inductance Effects on the Switching Loss Measurement of Power Semiconductor Devices," in *2006 IEEE International Symposium on Industrial Electronics*, Montreal, Que., Jul. 2006, pp. 847–852, doi: 10.1109/ISIE.2006.295745.

[12] T. Liu, T. T. Y. Wong, and Z. J. Shen, "A New Characterization Technique for Extracting Parasitic Inductances of SiC Power MOSFETs in Discrete and Module Packages Based on Two-Port S-Parameters Measurement," *IEEE Trans. Power Electron.*, vol. 33, no. 11, pp. 9819–9833, Nov. 2018, doi: 10.1109/TPEL.2017.2789240.

[13] S. Hashino and S. Toshihisa, "Separation measurement of parasitic impedance on a power electronics circuit board using TDR," in *2010 IEEE Energy Conversion Congress and Exposition*, Atlanta, GA, Sep. 2010, pp. 2700–2705, doi: 10.1109/ECCE.2010.5618046.

[14] R. S. Krishna Moorthy *et al.*, "Estimation, Minimization, and Validation of Commutation Loop Inductance for a 135-kW SiC EV Traction Inverter," *IEEE J. Emerg. Sel. Topics Power Electron.*, vol. 8, no. 1, pp. 286–297, Mar. 2020, doi: 10.1109/JESTPE.2019.2952884.

[15] X. Geng, K. Kuring, M. Wolf, O. Hilt, J. Würfl, and S. Diekerhoff, "Study on the Optimization of the Common Source Inductance for GaN Transistors," in *2021 23st European Conference on Power Electronics and Applications (EPE '21 ECCE Europe)*, Sep. 2021, p. P.1-P.10, doi: to be published.

[16] Efficient Power Conversion Corporation, "White paper - Optimizing PCB Layout." El Segundo, 2019.

[17] M. Albach, "Induktivitäten in der Leistungselektronik." Wiesbaden: Springer Fachmedien, 2017.

[18] ANSYS Inc., "Maxwell3D Help." Canonsburg, 2020.

[19] D. Reusch and J. Strydom, "Understanding the Effect of PCB Layout on Circuit Performance in a High-Frequency Gallium-Nitride-Based Point of Load Converter," *IEEE Trans. Power Electron.*, vol. 29, no. 4, pp. 2008–2015, Apr. 2014, doi: 10.1109/TPEL.2013.2266103.

Cooperative Control of Online Impedance Spectroscopy Monitoring Method and Maximum Power Point Tracking Method for Photovoltaic Panels

Xin Wang[*], Zhixue Zheng[*], Michel Aillerie[*], Alexandre De Bernardinis, Jean-Paul Sawicki
LABORATOIRE MATÉRIAUX OPTIQUES, PHOTONIQUE ET SYSTÈMES (LMOPS),
UNIVERSITÉ DE LORRAINE & CENTRALESUPÉLEC
Metz, France
* Corresponding authors: xin.wang@univ-lorraine.fr, zhixue.zheng@univ-lorraine.fr,
michel.aillerie@univ-lorraine.fr,
alexandre.de-bernardinis@univ-lorraine.fr,
jean-paul.sawicki@univ-lorraine.fr

Marie-Cécile Péra, Daniel Hissel
FCLAB, CENTRE NATIONAL DE LA RECHERCHE SCIENTIFIQUE (CNRS),
FRANCHE-COMTÉ ÉLECTRONIQUE MÉCANIQUE THERMIQUE ET OPTIQUE-
SCIENCES ET TECHNOLOGIES (FEMTO-ST) INSTITUTE, UNIVERSITÉ
BOURGOGNE FRANCHE-COMTÉ,
Belfort, France
marie-cecile.pera@univ-fcomte.fr,
daniel.hissel@univ-fcomte.fr

Acknowledgements

Authors gratefully thank the Chinese Scholarship Council (CSC) and French national research agency ANR in the framework of the EREMITE project (ANR-19-CE05-0008-01) for their financial support to the achievement of the work.

Keywords

«Photovoltaics», « Impedance measurement», «Control methods for electrical systems», «Solar Field»

Abstract

Health monitoring is essential for the photovoltaic (PV) panels to improve the system's efficiency and stability. Existing health monitoring methods are applied either offline or need additional equipment. Impedance spectroscopy (IS) methods can characterize online the internal processes of electrochemical systems with corresponding equivalent circuit model. However, it is rarely used for PV panels. This paper focuses on the online implementation of IS based on existing power converter for real-time health monitoring of PV panels. The injection region and type of perturbation signal for PV panels during IS measuring are analyzed. Based on PV panel's equivalent circuit model, a theoretical analysis of the influence of temperature and irradiance on internal parameters is made. Besides, a cooperative control method of online IS method and maximum power point tracking (MPPT) method is proposed. Even during the IS measuring, the quasi-maximum output power of PV panels can be maintained. The feasibility of the proposed method is verified by both simulation and experiment under different operating conditions.

Introduction

Due to the depletion of conventional energy sources and environmental impacts, the role of renewable sources for energy generation such as solar energy has become a prior choice nowadays [1]. As for other energetic sources and systems, health monitoring is essential for photovoltaic (PV) panels to detect the occurring faults and the degradation states, and to further improve the operational efficiency. The commonly used I-V curves of PV panels can provide fundamental electrical information in the full operating voltage range. Based on these curves, the state of PV panels can be monitored and several typical faults such as partial shading, short circuit and natural degradation can be diagnosed [2]. However, during the voltage sweep which is necessary to obtain information about the health state of the panel, the output voltage varies from open circuit to short circuit. Hence, in micro-grid applications, the PV panels can no longer assure (quasi-) maximum output power during the tests [3]-[5]. Another diagnostic category is based on imaging techniques [6], such as infrared imaging, lock in thermography (LIT), and electroluminescence imaging. However, these techniques are either limited by their high cost or offline applications, which are not suitable for online small power-scale applications.

Impedance spectroscopy (IS) is a widely used diagnostic tool in electrochemical systems such as batteries and fuel cells in combination with their AC equivalent models [7]-[9]. The basic principle is to impose a small perturbation signal (current or voltage) over the operation point of tested system and measure the corresponding response signal (voltage or current). If a current perturbation is carried out, the mode is called "galvanostatic". Otherwise, if a voltage perturbation is applied, the mode is called "potentiostatic". The impedance information is then extracted and calculated to characterize the system. Compared with I-V curves and other detection tools, IS method has more advantages for health monitoring of PV panels as elements of a microgrid due to the following reasons:

- IS method is non-destructive. It doesn't require the disconnection of PV panels from the microgrid during the operation, while assures a quasi-maximum output power of PV panels. Thus, the full operational functionality of the system can be maintained.
- IS method is capable of capturing the dynamics of the PV panel operation and characterizing the internal processes. When combined with an AC equivalent circuit model, the internal representative parameters of PV panels can be extracted and used for online health monitoring.
- IS method can be easily implemented for embedded applications. There is no need of additional equipment (e.g., electrochemical workstation) when based on existing power converters.

In the literature, it has been verified that IS method can describe accurately the internal states of PV panels by comparing with results obtained from frequency response analyzers and commonly used I-V curves [10]-[17]. Based on a commercial electrochemical workstation Zahner IM6, internal parameters of a single PV cell were tested in [10]. Due to the limitation of measuring current (± 3 A) and voltage range (± 4 V), this method is only applicable for a single PV cell. A converter-based IS method was implemented and verified in [12] [13]. However, the IS measurements were either made at open-circuit voltages or were achieved by open-loop regulation of duty cycle of the converter. In summary, the online application of IS method for PV panels' health monitoring is still in its infancy.

This work carried out in the framework of the EREMITE project funded by the Chinese Scholarship Council (CSC) and French Research Agency (ANR), aims to implement the IS method based on an existing DC-DC converter. The integration of the IS function into the converter controller can achieve a smooth transition between two modes, i.e., traditional maximum power point tracking (MPPT) mode and IS mode. Consequently, an additional electrochemical workstation is no longer required. The efficiency and reliability of the PV panels can be optimized in an autonomous way. In section II, both static and dynamic models of PV panel are briefly introduced. The AC equivalent circuit model and the online implementation of IS method are emphasized in section III. The feasibility of the proposed methodology has been verified by simulation results based on MATLAB/Simulink. In Section IV, experiments are further performed to study the influences of operating temperature and irradiance levels on the IS measurements. Conclusions are drawn in the final section.

PV panel characteristics

PV panels are made of several types of semiconductors using different manufacturing processes. In general, a PV cell can be essentially considered as a *p-n* junction. Under the radiation of light, carries move between different layers, which leads to current flow into the external circuit. The output current is affected by temperature and irradiance that affect the number of charge carriers and the moving speeds.

In general, an equivalent static model is applied to characterize the performance of PV panels, as shown in Fig. 1 (a). R_s and R_{sh} represent the series and shunt resistances, respectively. V is the output voltage and I is output current. D is a Schottky diode, which allows to describe the influence of temperature and irradiance on the output current. However, the static model can only react with the resistance information (i.e., R_s and R_{sh}) of PV panel under static conditions. To characterize the dynamic performance of PV panels, an equivalent dynamic model is necessary. In this paper, a dynamic model in [19] was referenced, while the reverse-bias conduction mode isn't considered in our case. Hence, the equivalent dynamic model is further simplified as the circuit shown in Fig. 1 (b).

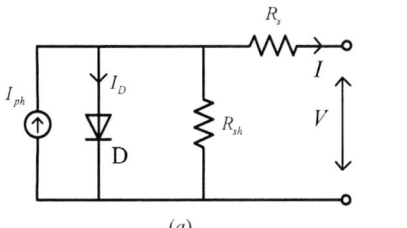 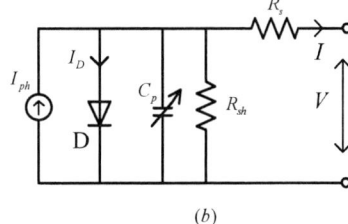

| (a) | (b) |

Fig. 1 (a) Equivalent static model; (b) Equivalent dynamic model

Compared with equivalent static model, a variable capacitor C_p is added in the dynamic model. And it can be expressed as follows:

$$C_p = \frac{q\tau I_D}{akT} \qquad (1)$$

Where a is the ideality factor ($1 < a < 2$), q is the absolute value of electron's charge ($q = 1.6*10^{-19}$ C), k is the Boltzmann's constant ($k = 1.35*10^{-23}$ J/K), T is the Kelvin temperature of the panel and τ represents the lifetime of carriers inside of PV panel.

Based on the Kirchhoff's law, the output voltage and current can be expressed follows:

$$\begin{cases} I = I_{ph} - I_D - I_{C_p} - I_{R_{sh}} \\ V = V_D + I \cdot R_s \end{cases} \qquad (2)$$

Where I_{ph}, I_{cp} and I_{Rsh} represent photovoltaic current, the current of C_p and R_{sh}, respectively. V_D is the direct voltage of the diode. And the equivalent resistance of the diode can be written as:

$$R_D = V_D / I_D \qquad (3)$$

To achieve a high energy generation efficiency, an MPPT method for PV systems is necessary. There are numerous MPPT methods proposed in the existing literature. In this paper, a traditional perturbation and observation (P&O) method is considered [18]. Since this work focuses majorly on the online IS method, more details on the MPPT methods will not be given herein.

Online Impedance Spectroscopy Method

For small and medium power distributed systems, PV modules are commonly equipped with an optimizer with MPPT function and a power converter. This section introduces an online implementation method to obtain the IS measurements based on its connected converter.

AC equivalent circuit of PV panels

An AC equivalent circuit is required for interpreting the IS measurements. Based on the equivalent dynamic model a corresponding AC equivalent circuit of PV panel is drawn in Fig. 2 (a). The model describes a Nyquist plot with a single time constant of R_p in parallel with a capacitor C_p, which is shifted from the origin by a series resistance R_s. R_p represents the associated recombination resistance of R_{sh} and R_D in Fig. 1 (b).

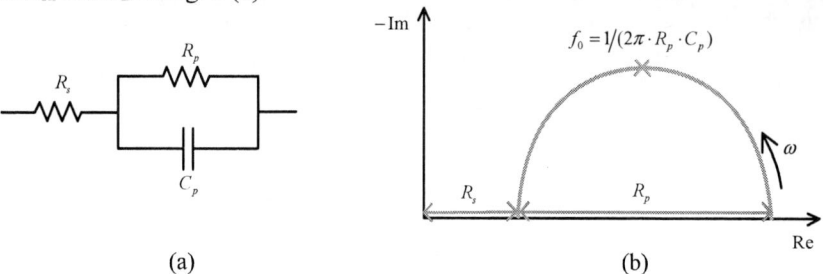

(a)   (b)

Fig. 2 (a) AC equivalent circuit model connected with a boost converter; (b) An ideal Nyquist curve with several representative parameters

By imposing a small voltage/current perturbation over the operating point of the tested system and measuring the corresponding current/voltage response, the Nyquist curve of PV panel can be obtained, which contains the impedance information over a large frequency range. As shown in Fig. 2 (b), the parameters of PV panels can be further extracted based on the Nyquist curve, where f_0 is the frequency of the top point of the curve and ω represents the angular frequency. The internal states of PV panel can thus be monitored by real-time IS measurements.

Online control method

The system controller setup is shown in Fig. 3 (a), including MPPT and IS method. In this setup, an AC current perturbation signal (i.e., the galvanostatic mode) is injected to achieve the IS measuring. The detail of the control loop for tracking the AC perturbation signal is shown in Fig. 3 (b), where $i_{\_ref\_ac}$ represents the reference the perturbation signal, I_0 is the DC component of output current, $i_{\_ref}$ is the sum of $i_{\_ref\_ac}$ and I_0, K_{PWM} is the gain of the converter's regulation, the equivalent impedance in s region is written as $sL+R_L$, and G(s) is the transfer function of controller used in Fig. 3 (a). The transfer function of the control block in Fig. 3 (b) can be expressed as follows:

$$\varphi(s)=\frac{I_{PV}(s)}{I_{\_ref}(s)}=\frac{G(s)\cdot K_{PWM}/(sL+R_L)}{1+G(s)\cdot K_{PWM}/(sL+R_L)}$$

(4)

In our system, a PI controller (G(s) = K_p+K_i/s) is used, whose appropriate parameters can be obtained based on the Bode diagram of the transfer function. In this control block, considering the frequency of the AC perturbation, the values of K_p and K_i are set to 400 and 100 respectively. Under this set of parameters, the controller frequency bandwidth can reach 10 kHz, as shown in Fig. 3 (c).

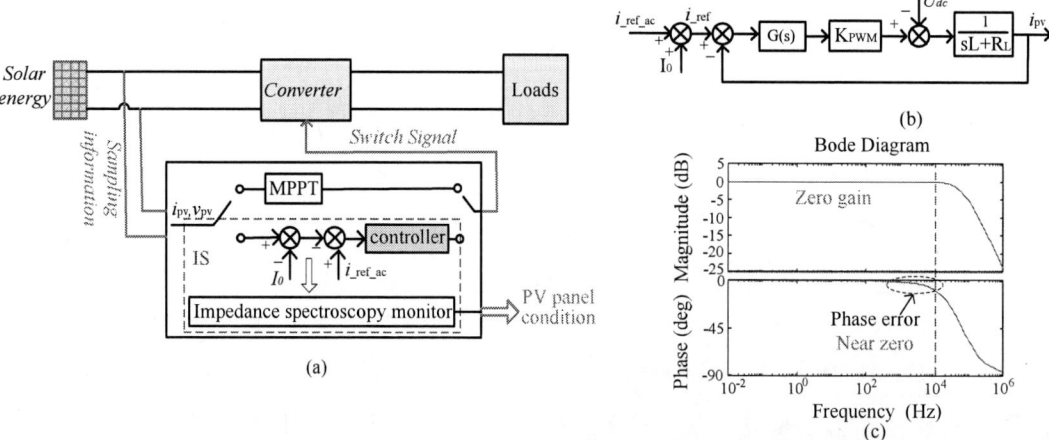

Fig. 3 (a) PV system with integrated IS method; (b) Control loop of IS online measurement

Simulation results

Based on the AC equivalent circuit in Fig. 2 (a) and the control method in Fig. 3, the cooperative control of online IS monitoring method and MPPT method is verified in MATLAB/Simulink. And the simulated results are shown in Fig. 4. The value of the DC bus voltage is set to 60 V and the switching frequency is equal to 10 kHz.

In Fig. 4 (a), based on the changing rate of I_{pv}, I-V curve is divided into one non-linear region (II) and two quasi-linear regions (I and III). According to the validity condition of IS measurements in [20], the injecting region must be a linear or quasi-linear region, in order to fulfill the linearity condition. Hence, the two quasi-linear regions (I and III) can be chosen initially for AC signal injection.

In region I, taking P_{qi} as an injection point, it can be observed that a significant voltage perturbation signal in the potentiostatic mode is required in order to produce a measurable current response signal. In the galvanostatic mode, a sufficient AC current perturbation signal is difficult to reach, since the current variation range is relatively narrow in this region.

In region III, taking P_{qv} as an injecting point, it can be observed that due to the relatively large voltage variation range, it would be more flexible to implement either the potentiostatic mode or the galvanostatic mode. Concrete analysis of the injection mode will be further made based on measured I-V curves in section IV.

In Fig. 4 (b), PV panel operates in MPPT mode at the beginning; then the scanning is started to achieve IS measuring. Once the scanning is completed, the PV panel continues to operate in MPPT mode. As can be seen in the responses, during the whole scanning process, PV panels are always connected with the microgrid bus and maintain an output power. Compared with MPPT process, the output power decreases little during IS measuring process (less than 5% of maximum P_{out}). Hence, IS method can assure a quasi-maximum PV output power during health monitoring.

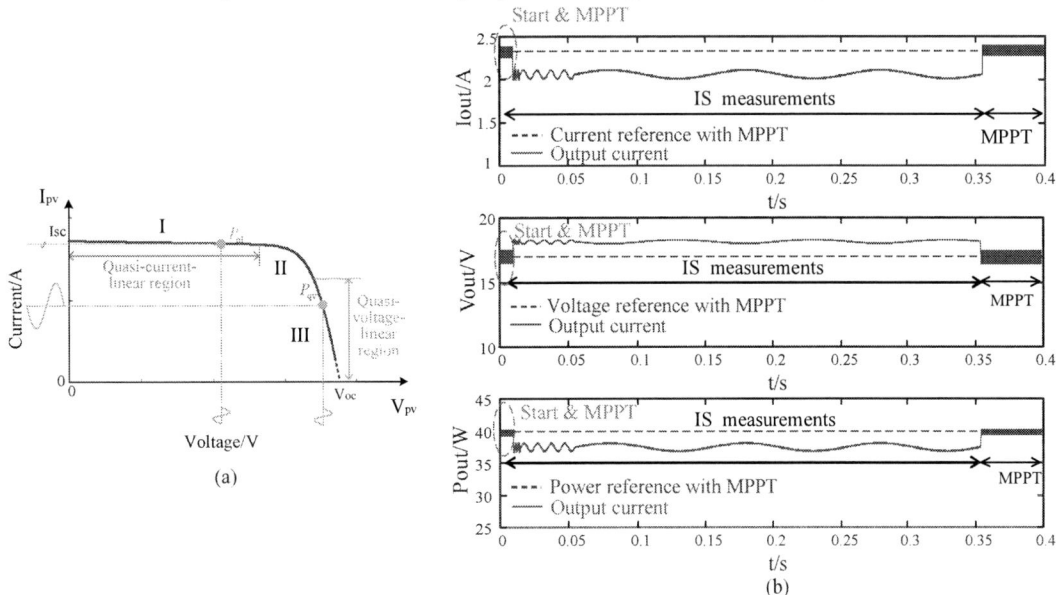

Fig. 4 (a) Region defined on I-V curve; (b) Simulation waveforms of the PV panel output current, voltage and power operating under the proposed control method

Experimental validation

Based on the proposed control method, experiments are performed to test the influence of temperature and irradiance to a PV panel in the laboratory. The experimental platform is shown in Fig. 5. A homemade magnetically coupled boost (MCB) converter was utilized herein, due to its high step-up ratio (up to 10) and high efficiency (> 92%) [21]. The output current is acquired by a high-accuracy and wide bandwidth current transducer (IT 65-S ULTRASTAB. The operating temperature is measured by a thermal camera (FILR E8) as shown in Fig. 5(b) and an average surface temperature is taken for information. The irradiance level is measured by a solar power meter PYR 1307.

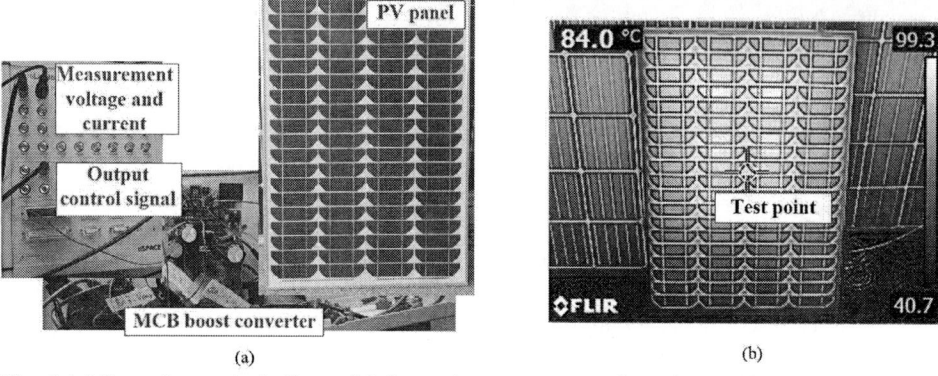

(a)   (b)

Fig. 5 (a) Experimental platform; (b) Operating temperature based on FLIS E8

A set of I-V curves under different operating temperatures with the same irradiance are shown in Fig. 6 (a). And a group of I-V curves under different irradiances with the same operating temperature are shown in Fig. 6 (b). If the potentiostatic mode is chosen, a common voltage injection point should be found for I-V curves at different operating temperatures. As can be observed in Fig. 6 (a), the PV panel output voltage is highly sensitive to changes of operating temperature especially in the quasi-voltage-linear region (region III). When the operating temperature changes significantly, it will be difficult to find a common injection point to pass through the region III of all the I-V curves. Taking $P_{qv}(V_{pv} = 17\text{ V})$ as an injection point, it can be observed that P_{qv} is in the nonlinear region II of I-V curve at 55 °C (at P_{MPP}), while it is out of the voltage range of the I-V curve at 84 °C. Meanwhile, for the galvanotactic mode, there is no such limitation. A common current injection point can be easily found while assures its location in region III, e.g., P_{1T}, P_{2T}, P_{3T} in Fig. 6 (a).

Under the influence of irradiance, the short circuit current (I_{sc}) is highly sensitive to the changes of irradiance, while the PV panel output voltage in region III keeps basically coincident among different curves, as shown in Fig. 6 (b). In this case, it would be easier to find a common current injection point. Therefore, the galvanostatic mode may be preferred. But if a common voltage injection point can be found, the potentiostatic mode can also be chosen, e.g., P_{1G}, P_{2G}, P_{3G} in Fig. 6 (b).

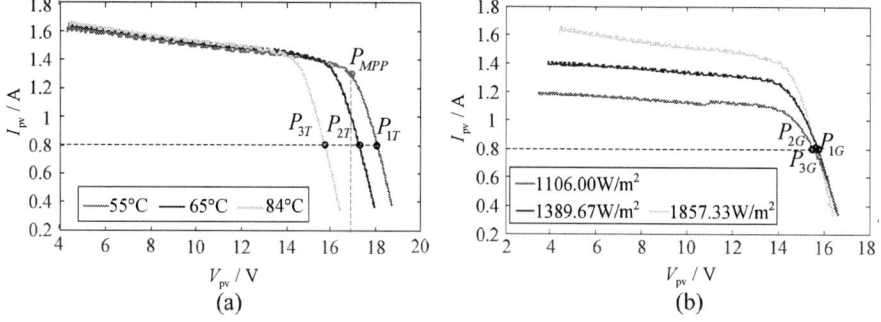

(a)   (b)

Fig. 6 Experimental injecting points: (a) inject current-perturbation signal; (b) inject voltage-perturbation signal

In accordance with proposed control method, Nyquist curves and bode diagrams of different operating temperatures and irradiances are obtained and shown in Fig. 7 and Fig. 9, respectively. A free EIS analyzer is utilized to extract the parameters to AC equivalent circuit, shown as R_s, R_p and C_p in Table I and Table II. And the changes of all parameters under different operating condition are shown in Fig.

8 and Fig. 10. In addition, the total resistance information under static condition is given for comparison, by calculating the slope at the injection point of each I-V curve, shown as R_{IV} in Table I and Table II. In principle, there exists an equivalence between R_{IV} and the total equivalent resistance (R_s+R_p) of IS measurements when the injection signal frequency is very low (thus can be regarded as quasi-static condition).

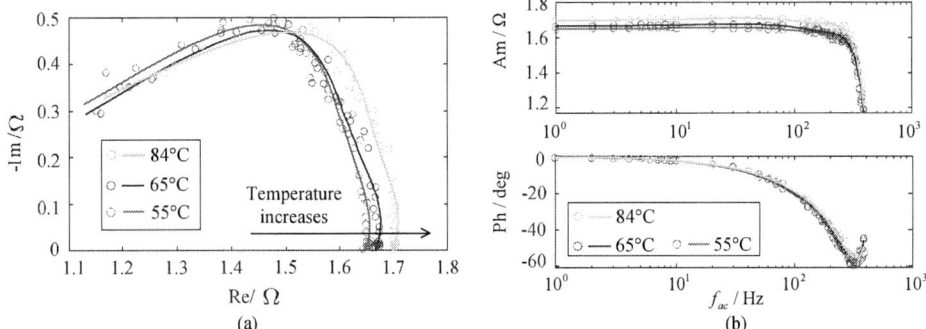

Fig. 7 Experimental results of different operating temperatures: (a) Nyquist curves; (b) Bode diagrams

As the operating temperature or irradiance change, a good consistency between the total equivalent resistance (R_s+R_p) and R_{IV} can be observed, as shown in Fig. 8 (a) and Fig. 10 (a). The error between the two resistances is less than 7%, which shows a good coherence between I-V curves and IS measurements at low frequencies. In parallel with the total equivalent resistance, IS measurement can equally provide equivalent capacitance information and separated resistance information of R_s and R_p, which is more powerful than I-V curves for health monitoring. As shown in Fig. 8 (b), R_p decreases with the increase of operating temperature, while R_s increases. When the irradiance increases, both R_p and R_s decrease as shown in Fig. 10 (b).

Table I: Internal parameters of PV panel at different operating temperatures

Temperature T (°C)	Equivalent Resistance in series R_s (Ω)	Equivalent Resistance in parallel R_p (Ω)	Total Equivalent Resistance R_s+R_p (Ω)	Equivalent Capacitance in parallel C_p (uF)	Total Equivalent Resistance on I-V curves R_{IV} (Ω)
55	0.76	0.92	1.68	397.98	1.69
65	0.82	0.88	1.70	438.43	1.71
84	0.87	0.86	1.73	398.15	1.73

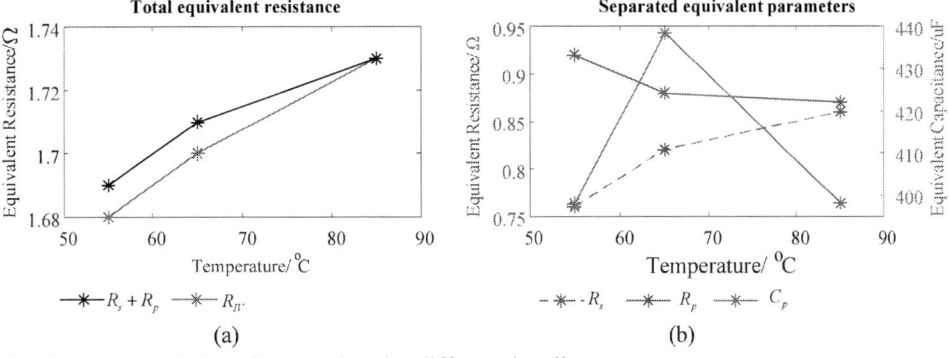

Fig. 8 Parameters' changing trend under different irradiances

When the operating temperature increases, the short circuit current (I_{sc}) is lightly affected while the output voltages of different injecting points (V_{P1T}, V_{P2T}, V_{P3T}) decrease significantly, as shown in Fig. 6 (a). Based on equation (2), the value of I_D is nearly constant while the value of V_D decreases as the operating temperature increases. Hence, the value of R_D will decrease. Since R_p is recombined by R_D and R_{sh}, its value will decrease the same as shown in Fig. 8 (b).

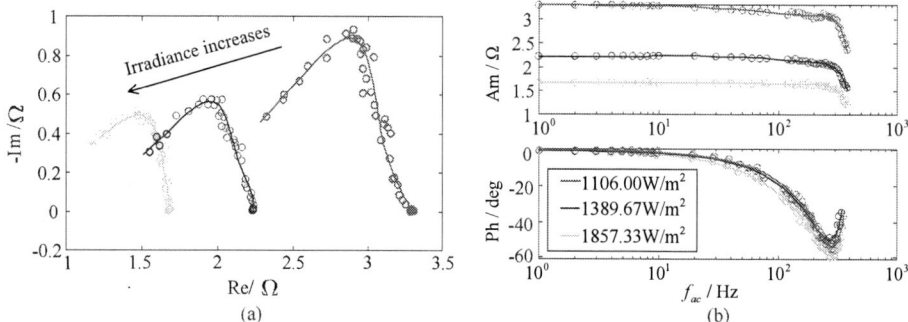

Fig. 9 Experimental results of different irradiances: (a) Nyquist curves; (b) Bode diagrams

When the irradiance increases, I_{sc} increases obviously as shown in Fig. 6 (b). In this condition, the output voltage of different injecting points (V_{P1G}, V_{P2G}, V_{P3G}) can be seen as a similar value. On the basis of equation (2), the value of I_D increases while the value of V_D is constant as the irradiance increases. R_p is composed of R_D and R_{sh} in parallel. The decrease of R_D will cause the value of R_p decreases, as shown in Fig. 10 (b).

Table II: Internal parameters of PV panel under different irradiances

Irradiance G (W/m^2)	Equivalent Resistance in series R_s (Ω)	Equivalent Resistance in parallel R_p (Ω)	Total Equivalent Resistance R_s+R_p (Ω)	Equivalent Capacitance in parallel C_p (uF)	Total Equivalent Resistance on I-V curves R_{IV} (Ω)
1857.33	0.75	0.96	1.71	373.91	1.83
1389.67	1.27	0.98	2.25	464.78	2.23
1106.00	1.73	1.56	3.29	253.26	3.21

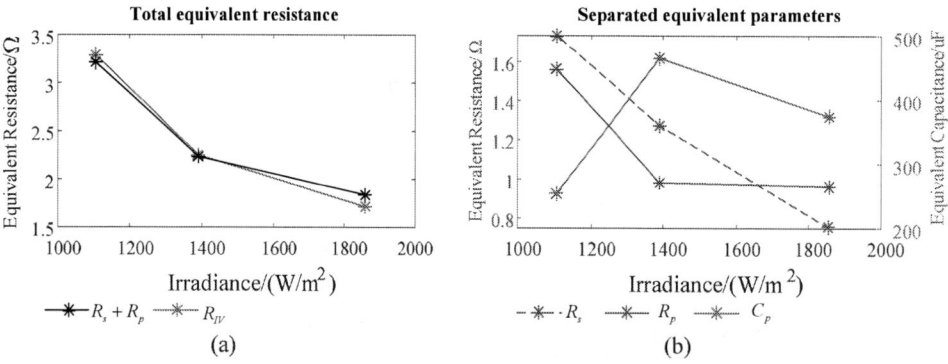

Fig. 10 Parameters' changing trend under different irradiances

According to the experimental results, it can be concluded that: when the irradiance changes, the values of R_p and R_s have the same changing trend; when the operating temperature changes, the values of R_p and R_s have the opposite changing trend. Based on the information of internal parameters obtained from IS measurement, the different operating conditions (i.e., different operating temperatures and irradiances herein) of PV panel can be identified.

Conclusion

In this paper, the online health monitoring of PV panels is emphasized. A collaborative control method between the MPPT and the IS measuring modes is proposed. Real-time measurements of IS based on DC-DC converter can be realized without the requirement of additional special equipment. Based on the AC equivalent circuit model, the internal parameters of PV panel under different operating temperatures and irradiances can thus be extracted and further utilized for online health monitoring. In the next step, IS measurements under different fault scenarios will be further focused on. If used for fault detection, different frequency segments relative to different types of faults can be divided to reduce unnecessary frequency measurements. Meanwhile, it can be noticed that the DC-DC converter design (e.g., the semiconductor material, its topology, and parameters) will affect the bandwidth of system control loop. And its influence to IS measurement will be additionally researched in the next step.

References

[1] V. Narayanan, S. Kewat, and B. Singh, "Solar PV-BES Based Microgrid System with Multifunctional VSC," IEEE Transactions on Industry Applications, vol. 56, no. 3, pp. 2957–2967, 2020

[2] Z. Zhang, M. Ma, H. Wang, H. Wang, W. Ma, and X. Zhang, "A fault diagnosis method for photovoltaic module current mismatch based on numerical analysis and statistics," Solar Energy, vol. 225, no. July, pp. 221–236, 2021

[3] A. Eskandari, J. Milimonfared, and M. Aghaei, "Line-line fault detection and classification for photovoltaic systems using ensemble learning model based on I-V characteristics," Solar Energy, vol. 211, no. July, pp. 354–365, 2020

[4] F. You and J. Benedikt, "An optimized-load-impedance calculation and mining method based on I-V Curves: Using broadband class-E power amplifier as example," IEEE Transactions on Industrial Electronics, vol. 66, no. 7, pp. 5254–5263, 2019

[5] B. Zbib and H. Al Sheikh, "Fault Detection and Diagnosis of Photovoltaic Systems through I-V Curve Analysis," 2nd International Conference on Electrical, Communication and Computer Engineering, ICECCE 2020, no. June, pp. 12–13, 2020

[6] K. Abdulmawjood, S. S. Refaat, and W. G. Morsi, "Detection and prediction of faults in photovoltaic arrays: A review," Proceedings - 2018 IEEE 12th International Conference on Compatibility, Power Electronics and Power Engineering, CPE-POWERENG 2018, pp. 1–8, 2018

[7] M. Crescentini et al., "Online EIS and Diagnostics on Lithium-Ion Batteries by Means of Low-Power Integrated Sensing and Parametric Modeling," IEEE Transactions on Instrumentation and Measurement, vol. 70, 2021

[8] H. Wang, A. Gaillard, and D. Hissel, "A review of DC/DC converter-based electrochemical impedance spectroscopy for fuel cell electric vehicles," Renewable Energy, vol. 141, pp. 124–138, 2019

[9] A Narjiss, D. Depernet, D. Candusso, F. Gustin, and D. Hissel, "On-line diagnosis of a PEM fuel cell through the PWM converter," Proceedings of FDFC 2008, pp. 734–739, 2008.

[10] D. Depernet, J. Vernier, P.A. Gril. Caractérisation de panneaux photovoltaïques par mesure d'impédance. Symposium de Génie Electrique, Université de Lorraine [UL], Nancy, France, 2018.

[11] D. Chenvidhya, K. Kirtikara, and C. Jivacate, "A new characterization method for solar cell dynamic impedance," Solar Energy Materials and Solar Cells, vol. 80, no. 4, pp. 459–464, 2003

[12] O. I. Olayiwola and P. S. Barendse, "Power Electronic Implementation of Electrochemical Impedance Spectroscopy on Photovoltaic Modules," ECCE 2020 - IEEE Energy Conversion Congress and Exposition, pp. 3654–3661, 2020

[13] M. A. Varnosfaderani and D. Strickland, "Online Electrochemical Impedance Spectroscopy (EIS) estimation of a solar panel," Vacuum, vol. 139, pp. 185–195, 2017

[14] S. Osawa, T. Nakano, S. Matsumoto, N. Katayama, Y. Saka, and H. Sato, "Fault diagnosis of photovoltaic modules using AC impedance spectroscopy," 2016 IEEE International Conference on Renewable Energy Research and Applications, ICRERA 2016, vol. 5, pp. 210–215, 2017

[15] M. Lohrasbi, P. Pattanapanishsawat, M. Isenberg, and S. S. C. Chuang, "Degradation study of dye-sensitized solar cells by electrochemical impedance and FTIR spectroscopy," 2013 IEEE Energytech, Energytech 2013, pp. 13–16, 2013

[16] M. I. Oprea, S. V. Spataru1, D. Sera, P. B. Poulsen, S. Thorsteinsson, R. Basu, A. R. Andersen, K. H.B. Frederiksen., "Detection of potential induced degradation in c-Si PV panels using electrical impedance spectroscopy," Conference Record of the IEEE Photovoltaic Specialists Conference, vol. 2016-November, no. 2, pp. 1575–1579, 2016

[17] O. I. Olayiwola and P. S. Barendse, "Characterization of Silicon-Based Photovoltaic Cells Using Broadband Impedance Spectroscopy," IEEE Transactions on Industry Applications, vol. 54, no. 6, pp. 6309–6319, 2018

[18] T. Esram and P. L. Chapman, "Comparison of photovoltaic array maximum power point tracking techniques," IEEE Transactions on Energy Conversion, vol. 22, no. 2, pp. 439–449, 2007.

[19] K. A. Kim, C. Xu, L. Jin, and P. T. Krein, "A dynamic photovoltaic model incorporating capacitive and reverse-bias characteristics," IEEE Journal of Photovoltaics, vol. 3, no. 4, pp. 1334–1341, 2013.

[20] X.-Z. Yuan, S. Chaojie, W. Haijiang, Z. Jiujun, Electrochemical Impedance Spectroscopy in PEM Fuel Cells: Fundamentals and Applications (Springer, London, 2010), pp. 95–136

[21] P. Petit, M. Aillerie, J. P. Sawicki, and J. P. Charles, "High efficiency DC-DC converters including a performed recovering leakage energy switch," Energy Procedia, vol. 36, pp. 642–649, 2013.

Benefits of switching from Si to SiC modules with further converter optimization

Antxon Arrizabalaga[1], Mikel Mazuela[1], Iosu Aizpuru[1], June Urkizu[1], Jon Aztiria[1]

[1] MONDRAGON UNIBERTSITATEA
Fundazioa eraikuntza; Jauregi Bailara, z.g, 20120
Hernani, Spain
Tel.: +34 / +(34) – 943794700.
E-Mail: aarrizabalaga@mondragon.edu
URL: https://www.mondragon.edu/

Keywords

«Wide bandgap devices», «DC-AC», «Switching losses», «Optimization», «Passive filters»

Abstract

SiC semiconductors have better characteristics than Si, improving power electronics converters performances. A prototype that can switch semiconductor technology without changing any other part of the system is built and tested, showing the efficiency improvements achieved with SiC. Finally, a theoretical system level converter optimization is done applying the experimental results.

Introduction

With the increased use of power electronics, the technical requirements of converters are also increasing [1], [2]. Higher power density, meaning less volume and less weight, as well as higher efficiency are needed. With higher integration of renewable energies, and switching devices connected to the grid, the waveform quality of inverters is also a concern nowadays [3]. Increasing the switching frequency can help in the filtering objective, without the need to increase the filter, which is one of the bulkiest parts of grid connected inverters [4].

In this aspect, state of the art silicon (Si) semiconductors have a natural technical limit [1], which directly affect to the converter performance, limiting the ability to reach the requisites of several applications. Silicon carbide (SiC) high power modules break the boundaries of Si, enabling higher power densities, with reduced size, weight and higher efficiency at converter level [5].

Fig. 1 shows the characteristics of SiC normalized to Si. As it is seen, SiC has a higher breakdown electric field, which allows to withstand higher voltage with a thinner drift layer, Fig: 2 (a), reducing the specific on resistance, and breaking Si boundaries, shown in Fig: 2 (b). In addition, the higher electron saturation velocity allows to switch from one switching state to other faster than Si [6], [7]. The faster switching means less switching losses, as it will be seen later.

When comparing Si and SiC 1700 V high power half bridge modules, the first difference is the transistor technology. For medium-high voltage and power levels, insulated gate bipolar transistors (IGBT) are used with Si, while the higher breaking voltage of SiC allows to form metal oxide-semiconductor field-effect transistors (MOSFET). These transistors do not suffer from the direct voltage drop that IGBTs suffer, showing better conduction performance in low currents. The second main difference occurs in the switching, in which MOSFETs do not have a tail current in the turn off. In addition, the faster switching of the SiC with respect to Si makes the switching of SiC MOSFETs more efficient.

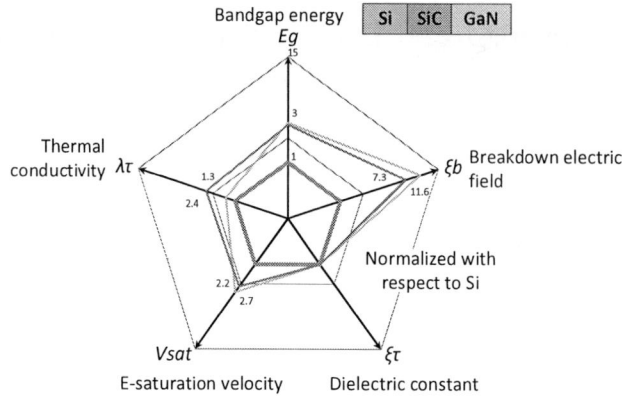

Fig. 1: Key normalized electrical and thermal characteristics of Si , GaN and SiC, [8].

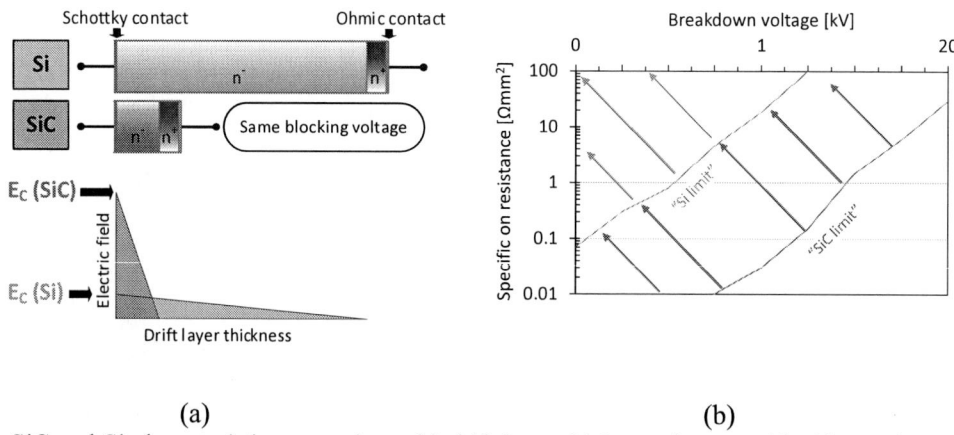

(a)   (b)

Fig: 2: SiC and Si characteristic comparison, (a) drift layer thickness for same blocking voltage, (b) specific on-resistance vs breakdown voltage, [6].

The previously mentioned better performance leads to an improved efficiency of the converter. This can lead to smaller cooling systems (CS), reducing the volume of the converter [9], [10], so it can help to meet the most restrictive requirements in several applications [11]–[13]. Renewable energy can be one of those applications, reducing the volume of the photovoltaic (PV) inverter for example, being able to install it in the rooftops of residential buildings [14]. Another industrial application for the improvements SiC brings to power electronics is the electric vehicle (EV) [5]. A smaller layout of the converter can allow new locations for inverters in the electric cars, and higher power density will allow to have more power transmitted to the mechanical shaft, improving the overall performance of the EVs [5].

This paper presents an experimental verification of the efficiency improvements achieved by switching from Si to SiC technology. The experimental platform has been designed to replace Si and SiC modules in the same system, in order to get a fair comparison of the device performance. However, an optimized converter is required to maximize the benefits of SiC modules.

Experimental platform

This section presents the designed prototype. The objective is to analyze the effect of replacing the state of the art Si technology with high power SiC modules. The same power stage, with the same components are used, replacing only the semiconductors. This approach isolates the impact of the technology in the efficiency of the converter. To make a fair comparison, the tests are performed in a controlled environment using passive loads. This allows to test both technologies under the same operation points and conditions.

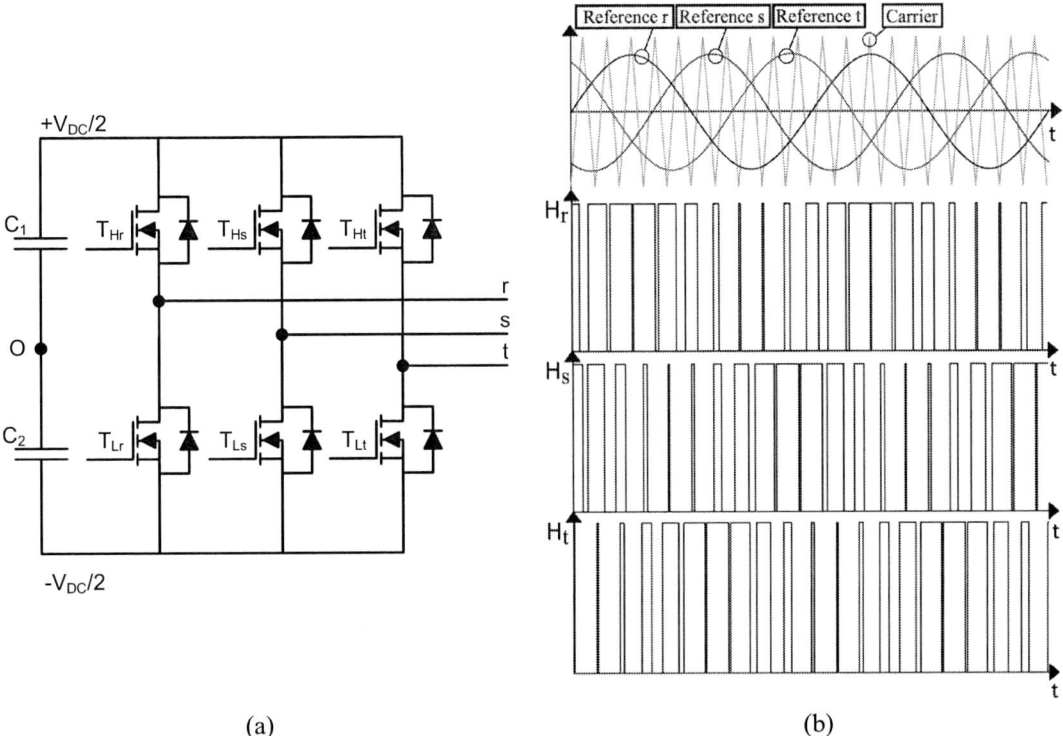

(a)   (b)

Fig. 3: Three phase two level converter, (a) topology scheme and (b) PWM modulation diagram with gate driving signals for each phase.

The 3 phase 2 level (2L) topology is selected, and shown in Fig. 3 (a). Fig. 3 (b) shows the driving PWM modulation. The reference of each phase is 120° out of phase. High and low switches are complementary, and represented with H and L in Fig. 3. Only the driving signal of the high switch is shown per phase.

Converter design

The key components of the converter design are the semiconductor modules. Si and SiC modules must have the same package, in order to be replaceable. The universal 62 mm module is selected. The main characteristics of the selected modules are shown in Fig. 4, together with the pinout. Fig. 5 shows the designed converter. By taking out the DC link capacitors and the drivers, all fixed to the busbar, the lower layer of the converter is accessible. The design ensures the module replacement is done in a fast and safe way, allowing to test both Si and SiC technologies efficiently.

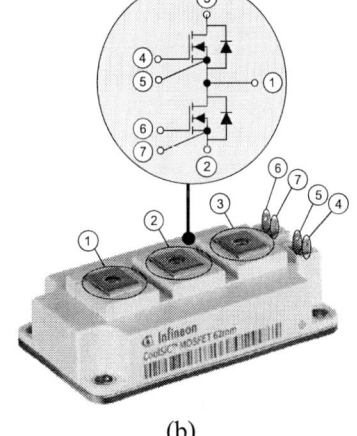

Technology	Si IGBT	SiC MOSFET
Reference	FF300R17KE4	FF8MR20KM1*
I_C-I_D	300 A	185 A
V_{CE}-V_{DS}	1700 V	2000 V
*Under development, preliminary samples.		

(a)   (b)

Fig. 4: Selected 62 mm modules, (a) Characteristics, (b) package pinout.

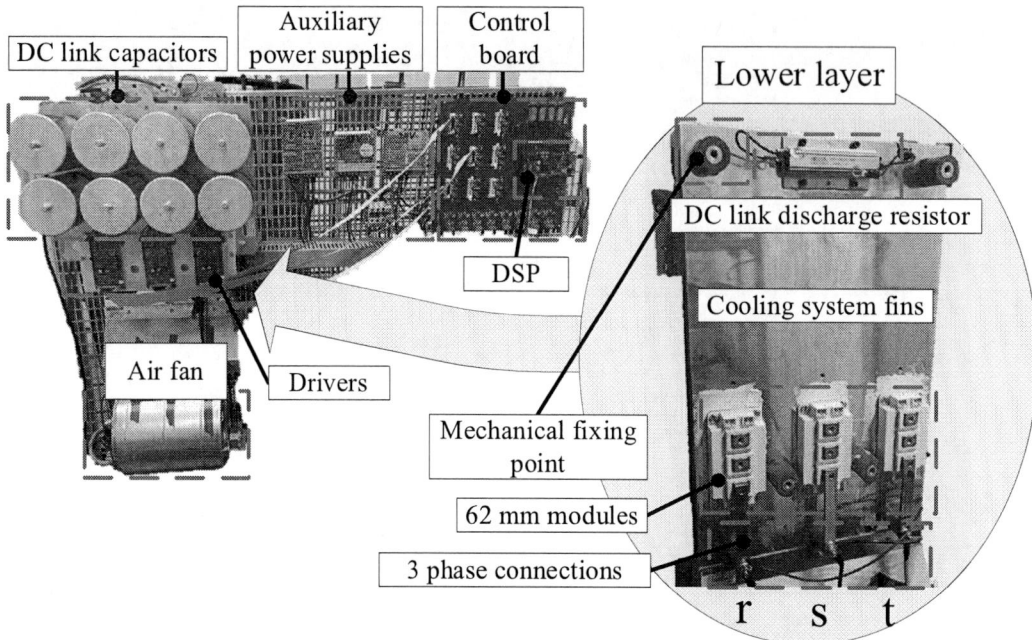

Fig. 5: The designed converter, with the replaceable 62 mm modules in the lower layer.

Table I: List of the components used in the power stage.

Component	Reference	Qty	Detail
Drivers	2ASC-17A1HP	3	Compatible for Si and SiC.
DC link capacitors	947D102K901CJRSN	8	900 V and 1 mF each.
Discharge resistor	HS150 18K	1	18 kΩ and 150 W.
Busbar	Own design	1	Custom made.
Cooling system	SC-C300P	1	Radial 35 m³/min forced airflow.
DSP	Delfino C2000	1	High resolution PWM.
Control board	Own design	1	8 analog measurements. 9 conditioned driving signals. 4 external relays for security circuit breakers.
DC power supply	Yaskawa 4A0100	1	Regenerative power converter with 150 A DC current.

Table I shows the components used to build the common power stage. The same drivers can work with Si and SiC, with a maximum voltage of 2000 V in the DC link. The DC link capacitors are set in 4 parallel branches of 2 capacitors in series. With this configuration, high voltage capability of the converter is ensured, as well as having high capacitance in the DC link. A discharge resistor is installed as a security measure, to ensure the capacitors are discharged when the converter is turned off. The custom made busbar connects all the before mentioned components, giving them mechanical stability. If the busbar is removed, all the attached components are also removed, making the lower layer of the converter easily accessible.

The control board is an own design, interfacing between the digital signal processor (DSP) and the drivers. It also allows to perform 8 different measurements and drives the security circuit breakers with 4 relays. The control is implemented in MatLab-Simulink and deployed to the DSP. Finally, a regenerative power converter, also supplying the power, sets the DC voltage.

The designed power converter has a rated power of 125 kVA. However, the grid in the testing facility can only provide less than 50 kVA. Testing the converter at higher power in another facility is a future objective.

Test results

The tests are performed keeping the same fixed DC link voltage and modulation index m_a. The active power is increased from 5 to 40 kW, reaching the limit of the testing facility. The load inductors are changed in every testing point, to ensure the power factor PF is kept between the desired margins. The fundamental frequency f_0 of the synthesized sinewave is fixed in order to get the necessary impedance with the available load inductors. The switching frequency is varied from 2.5 to 10 kHz, to evaluate its effect in both technologies. Table II shows the testing conditions and operation points. The same points are repeated for both Si and SiC modules, in order to make a fair comparison.

The converter is run in the operation conditions previously described. The obtained line voltage and phase currents are shown in Fig. 8 for the SiC modules at full power. Only two phases are measured because it is a balanced system. The peak current value in this test is 110 A, corresponding to 40 kW.

Efficiency tests

First, both technologies are tested in all the power range at 2.5 kHz switching frequency, Fig. 6 (a). Even if the low switching frequency favors Si modules, the converter with SiC has better efficiency in all the power range. The efficiency difference is the highest at low power, where switching losses dominate over the conduction losses. At high power, the efficiency gap is smaller, due to the better conduction characteristics of the Si IGBTs at high current. However, the Si converter does not reach 98 %, while the SiC converter goes over 99 %.

Fig. 6 (b) presents the efficiency of both technologies at maximum power (40 kW), but different switching frequencies. It is observed how the SiC converter keeps high efficiency, over 98.8 % even at 10 kHz, while its Si counterpart drops below 94.5 %. This result shows the possibility to increase the switching frequency with SiC power modules and still keep high efficiency. The switching frequency increment allows improving other components of the system, such as the filters.

Table II: Test operation points for both Si and SiC modules.

Parameter	Value	Unit
DC voltage	630	V
Modulation index	0.9	/
Power factor	0.75-0.85	/
Fundamental frequency	250	Hz
Switching frequency	2.5-10	kHz
Active power	5-40	kW

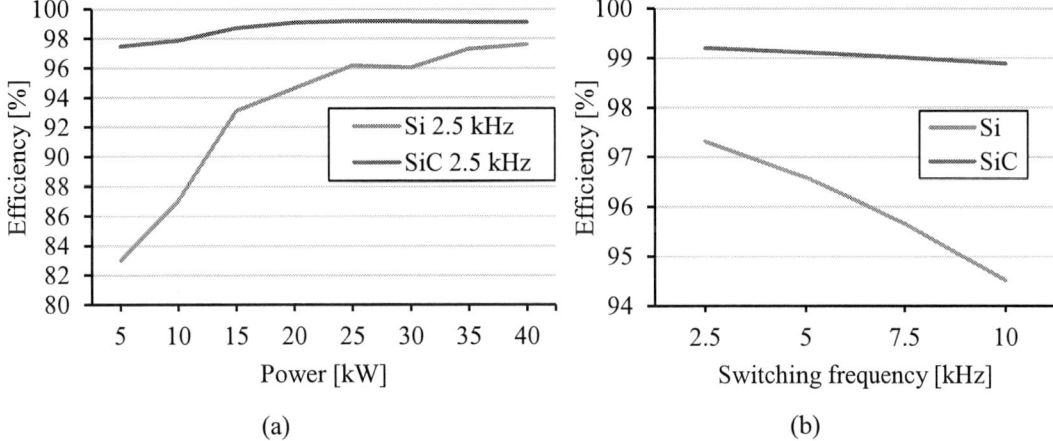

(a)   (b)

Fig. 6: Efficiency test results for both Si and SiC technologies, (a) at 2.5 kHz switching frequency and all the power range, (b) at maximum power and different switching frequencies.

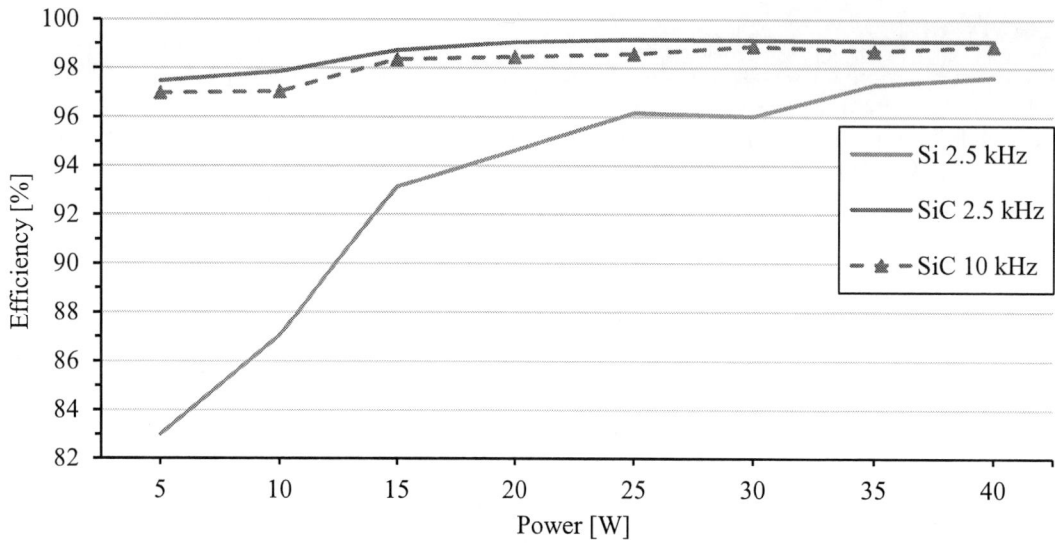

Fig. 7: Efficiency test results for both Si and SiC technologies in all the power range, including SiC at 10 kHz switching frequency.

Fig. 7 includes the efficiency of the SiC technology at 10 kHz switching frequency in all the power range. Although the increased switching frequency penalizes efficiency, the evolution is similar to the 2.5 kHz system, and still achieves high values. The worst point is at low power, but the efficiency is kept over 95 %. The highest value is 98.88 %, only losing 0.22 % with respect to the system at 2.5 kHz.

The output waveforms of the converter with SiC modules are shown in Fig. 8. The waveforms in Fig. 8 (a) are obtained at low switching frequency, 2.5 kHz. The ones in Fig. 8 (b) are obtained at 10 kHz switching frequency. The better output waveform quality is visible in the high switching frequency case, both for the line voltage and phase current. With an improved output waveform such as the one in Fig. 8 (b), the filtering needs are reduced. This directly impacts the total volume of the converter, as it is studied in the next section.

Converter volume optimization

The cooling system used for the tests in the previous section is oversized for security reasons. Thus, a theoretical converter volume optimization can be performed, based on the efficiency test results obtained in the previous section. The cooling systems and the required filters volumes are compared for Si and SiC technologies. In the case of the SiC, the increase in the switching frequency is also considered. The analysis shows the possible converter volume reduction with the use of SiC technology.

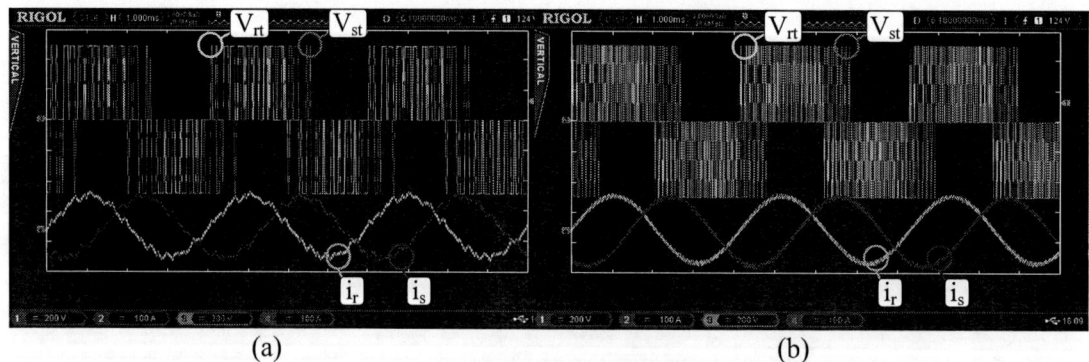

Fig. 8: Line voltage and phase current of the SiC converters at 40 kW, (a) at 2.5 kHz switching frequency, (b) at 10 kHz.

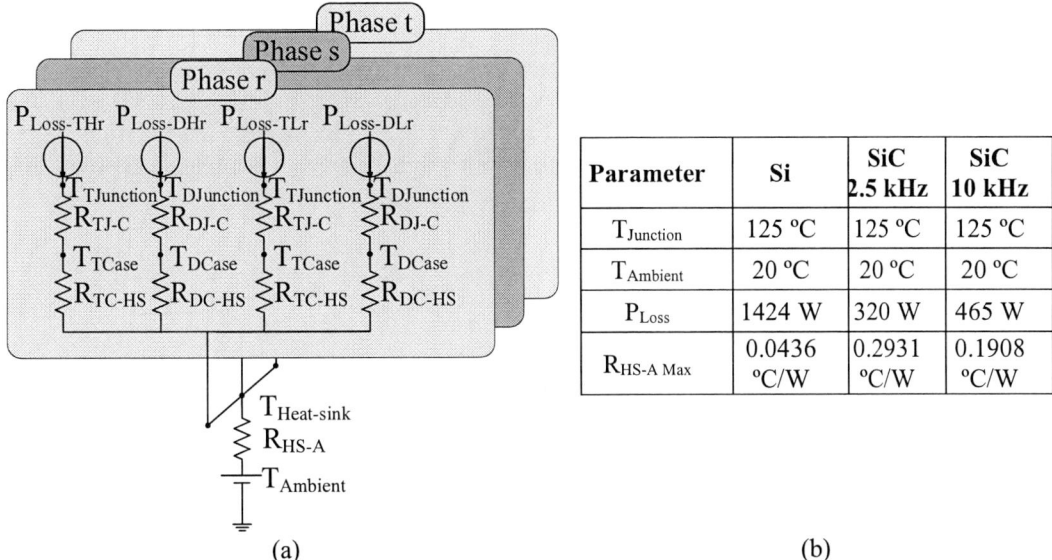

Parameter	Si	SiC 2.5 kHz	SiC 10 kHz
$T_{Junction}$	125 °C	125 °C	125 °C
$T_{Ambient}$	20 °C	20 °C	20 °C
P_{Loss}	1424 W	320 W	465 W
$R_{HS-A\ Max}$	0.0436 °C/W	0.2931 °C/W	0.1908 °C/W

| (a) | (b) |

Fig. 9: Cooling system analysis, (a) equivalent thermal circuit, (b) most relevant parameters.

Cooling system calculation

Fig. 9 shows the analysis of the cooling system. First, the equivalent thermal circuit is drawn, Fig. 9 (a), paralleling the modules in phase r, s and t. Each module has two semiconductors, high H and low L, and each semiconductor a transistor T and a diode D. R_{J-C} and R_{C-HS} represent the thermal resistance from the junction to the case and from the case to the heat-sink respectively. These parameters are found in the datasheets of the modules. Maximum junction temperature and ambient temperature are shown in Fig. 9 (b). The power losses are the maximum registered for each configuration, in the tests shown in Fig. 7, and presented in Fig. 9 (b). As it is not possible to measure if the losses are generated in the diode or in the transistor, the worst case scenario is analyzed. This is when all the losses are generated in the path with the highest R_{th}. The maximum allowable thermal resistance of the cooling system R_{HS-A} is calculated for each case.

Next, commercial cooling systems are plotted in Fig. 10, depending on their volume and thermal resistance. For high thermal resistances, axial air fans are enough, while for low thermal resistances, high volume is required together with radial air fans. A trend approximating the volume of the cooling system depending on the thermal resistance is calculated (1), and shown with a black dashed line in Fig. 10.

$$Vol_{CS} = 0.3622 \cdot R_{th}^{-0.988} \tag{1}$$

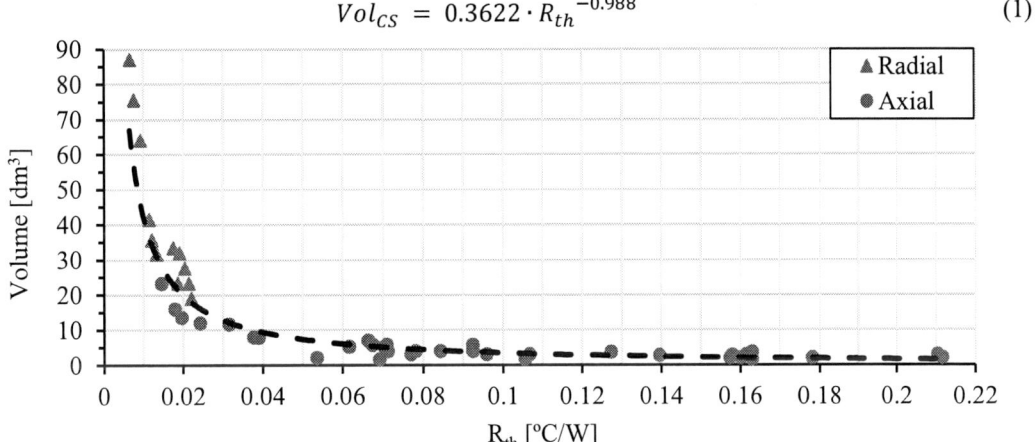

Fig. 10: Commercial cooling systems depending on their thermal resistance and volume, with the calculated potential trend in black.

Filter volume calculation

An LC line filter is usually added in the output of the converter to shape the output current. The inductance and capacitance values in the filter are calculated using expressions (2) and (3) presented in [15] and [16] respectively. V_{dc} is 630 V , ΔI_{out} is defined as 10 % of the I_{out} and Att_{req}, which refers to the required attenuation of the filter, is set to 0.01 in order to have enough damping in the switching frequency [17]. m refers to the converter topology level, 2 in this case.

$$L_f = \frac{V_{dc}}{6(m-1) \cdot \Delta I_{out} \cdot f_{sw}} \tag{2}$$

$$C_f = \frac{1}{(2\pi \cdot f_{sw})^2 \cdot L_f \cdot Att_{req}} \tag{3}$$

To calculate the volume of the inductor, the area product A_p technique proposed in [18] is used. Equation (4) uses the factor k_L to relate the area product and inductor volume. As this factor is dependent on the switching frequency, a polynomial approximation is performed to calculate k_L in [19], and shown in (5). In the case of the capacitor, [16] identifies a relation between volume, the rated voltage and capacitance for every technology. To estimate the required volume, commercial foil capacitors are analyzed, considering their volume and rated voltage, Fig. 11 (a). It is identified that the volume is linearly dependent on the rated capacitance, for a same rated voltage. 800 V series is selected, and the linear expression shown in (6), represented with a black dashed line in Fig. 11(a), is used to estimate the required capacitors' volume in the filter, introducing the capacitance value C_f in micro farads (µF). Fig. 11 (b) shows the volume of the required LC filter for each switching frequency. (7) shows the whole filter volume calculation. After adding the volumes of the capacitor and the inductor, the total volume is multiplied by three in order to consider the three phases of the converter.

$$Vol_{L_f} = k_L \cdot A_p^{\frac{3}{4}} \tag{4}$$

$$k_L = 2.676 \times 10^{-5} \cdot f_{sw} + 19.71 \tag{5}$$

$$Vol_{C_f} = 0.0016 \cdot C_f + 0.0062 \tag{6}$$

$$Vol_{Filter} = 3 \cdot (Vol_{C_f} + Vol_{L_f}) \tag{7}$$

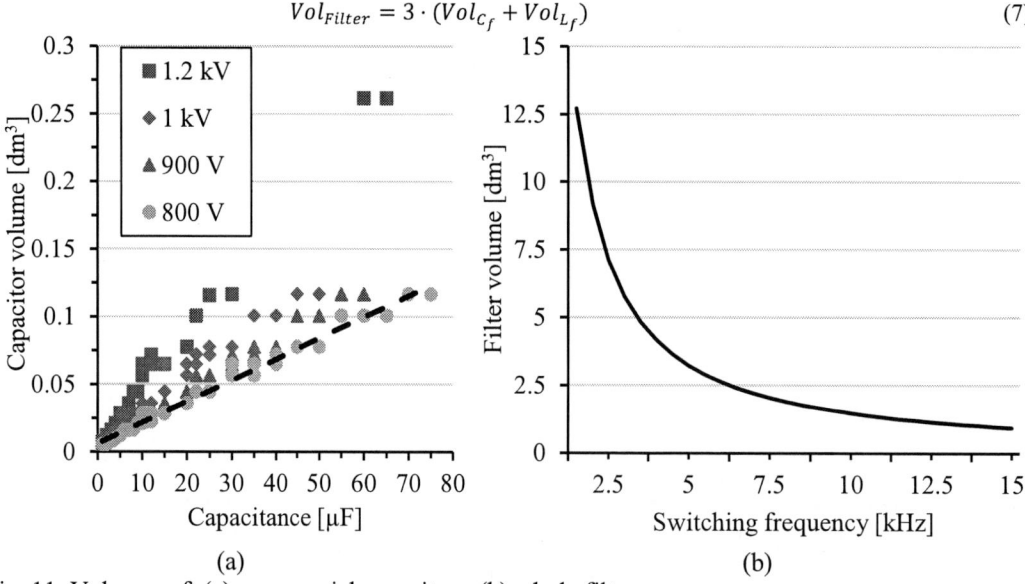

Fig. 11: Volumes of, (a) commercial capacitors, (b) whole filter.

Table III: optimized converter characteristics, with the use of SiC.

	Si at 2.5 kHz	SiC at 2.5 kHz	SiC at 10 kHz	Unit
Experimental efficiency at nominal power	97.60	99.19	98.88	%
Required CS volume	8.00	1.21	1.86	dm^3
Required CS volume reduction	/	84.87	76.75	%
Required filter volume	7.10	7.10	1.47	dm^3
Required filter volume reduction	/	/	79.29	%
Combined volume	15.1	8.31	3.33	dm^3
Combined volume reduction	/	44.96	77.94	%

Combined volume analysis

Finally, the cooling system and the filter volume are added for each case, to calculate the combined volume of the converter. Table III shows the results, being SiC at 10 kHz the solution that would provide the best volume reduction. The better efficiency of the SiC devices allow to reduce the cooling system compared to the Si converter. Finally, and due to the good switching performance of SiC, increasing the switching frequency does not increase the losses considerably, keeping the cooling system volume low.

If the switching frequency is increased in the SiC converter, the filter can also be reduced, and the combined volume reduction is maximum. The efficiency of the optimized SiC converter is still as high as 98.88 %.

Conclusion

A functional inverter is presented in this work, having the capability to change the semiconductor technology from Si IGBTs to SiC MOSFETs while keeping the rest of the system unchanged, see Fig. 5. This provides the possibility to compare the semiconductor technologies in a fair way. The results in efficiency are shown, improving 1.59 % at nominal power, Fig. 6 (a). In addition, the evolution of the efficiency is analyzed when the switching frequency is increased, Fig. 6 (b). Taking advantage of the better switching performance of the SiC MOSFETs, it is observed that the switching frequency can be increased from 2.5 kHz to 10 kHz and still improve the efficiency of the state of the art Si converter 1.28 %.

The whole system should be optimized in order to fully use the SiC semiconductor capabilities. This optimization is shown in Table III, achieving 77.94 % combined volume reduction with SiC MOSFETs switching at 10 kHz, while the efficiency at nominal power is still improved, as mentioned before. In addition, reducing the filter 79.29 % is found valuable, because the cost and weight of the filter are a significant part of the whole converters.

The advantages of migrating from Si IGBTs to SiC MOSFETs are experimentally proved in a general purpose inverter. A theoretical converter optimization is performed, achieving valuable improvements at system level; that can be applied to meet the restrictive requirements some applications are demanding nowadays. The improved system could easily benefit electric vehicles, with its reduced volume and weight, as well as domestic PV systems, by allowing to install inverters attached to solar panels in rooftops, for example.

References

[1]   M. F. Yaakub, M. A. M. Radzi, F. H. M. Noh, and M. Azri, "Silicon carbide power device characteristics, applications and challenges: An overview," Int. J. Power Electron. Drive Syst., vol. 11, no. 4, pp. 2194–2202, 2020.

[2]   H. Zhang and L. M. Tolbert, "Efficiency Impact of Silicon Carbide Power Electronics for Modern Wind Turbine Full Scale Frequency Converter," IEEE Trans. Ind. Electron., vol. 58, no. 1, 2011.

[3]   F. Blaabjerg and K. Ma, "Wind energy systems," Proc. IEEE, vol. 105, no. 11, pp. 2116–2131, 2017.

[4]   J. W. Kolar et al., "Impact of Magnetics on Power Electronics Converter Performance State-of-the-Art and Future Prospects Magnetics Committee," Zurich, 2017.

[5]   A. Matallana et al., "Power module electronics in HEV/EV applications: New trends in wide-bandgap semiconductor technologies and design aspects," Renew. Sustain. Energy Rev., vol. 113, no. June, p. 109264, 2019.

[6]   T. Kimoto, "Material science and device physics in SiC technology for high-voltage power devices," Jpn. J. Appl. Phys., vol. 54, no. 4, 2015.

[7]   D. Garrido, "Impacto de los semiconductores de banda ancha prohibida en el diseño de convertidores de potencia," Mondragon Unibertsitatea, 2019.

[8]   B. W. Williams, Principles of Power electronics. Glasgow: Barry W. Williams, 2006.

[9]   H. Zhang, L. M. Tolbert, and B. Ozpineci, "Impact of SiC devices on hybrid electric and plug-in hybrid electric vehicles," IEEE Trans. Ind. Appl., vol. 47, no. 2, pp. 912–921, 2011.

[10]   E. Gurpinar, "Wide-Bandgap Semiconductor Based Power Converters for Renewable Energy Systems," Univ. Nottingham, 2017.

[11]   A. Castellazzi, E. Gurpinar, Z. Wang, A. S. Hussein, and P. G. Fernandez, "Impact of wide-bandgap technology on renewable energy and smart-grid power conversion applications including storage," Energies, vol. 12, no. 23, pp. 1–14, 2019.

[12]   A. S. Abdelrahman, Z. Erdem, Y. Attia, and M. Z. Youssef, "Wide Bandgap Devices in Electric Vehicle Converters: A Performance Survey," Can. J. Electr. Comput. Eng., vol. 41, no. 1, pp. 45–54, Dec. 2018.

[13]   C. Sintamarean, E. Eni, F. Blaabjerg, R. Teodorescu, and H. Wang, "Wide-band gap devices in PV systems - Opportunities and challenges," in 2014 International Power Electronics Conference, IPEC-Hiroshima - ECCE Asia 2014, 2014, pp. 1912–1919.

[14]   A. Singh et al., "Development and Validation of a SiC Based 50 kW Grid-Connected PV Inverter," 2018 IEEE Energy Convers. Congr. Expo. ECCE 2018, no. September, pp. 6165–6172, 2018.

[15]   M. Mazuela, "Análisis y desarrollo de una novedosa topología de convertidor multinivel para aplicaciones de media tensión y alta potencia," Mondragon Unibertsitatea, 2015.

[16]   J. W. Kolar et al., "PWM converter power density barriers," Fourth Power Convers. Conf. PCC-NAGOYA 2007 - Conf. Proc., no. May, 2007.

[17]   A. Anthon, Z. Zhang, M. A. E. Andersen, D. G. Holmes, B. McGrath, and C. A. Teixeira, "The benefits of SiC mosfets in a T-type inverter for grid-tie applications," IEEE Trans. Power Electron., vol. 32, no. 4, pp. 2808–2821, 2017.

[18]   W. G. Hurley and W. H. Wölfle, "Transformers and inductors for power electronics: theory, design and applications". Wiley-Blackwell, 2013.

[19]   E. Gurpinar and A. Castellazzi, "Single-Phase T-Type Inverter Performance Benchmark Using Si IGBTs, SiC MOSFETs, and GaN HEMTs," IEEE Trans. Power Electron., vol. 31, no. 10, pp. 7148–7160, Oct. 2016.

On the reduction of output capacitance in two-level three phase PFC boost rectifier for pulsating loads.

Tania C. Cano[1], Douglas Pedroso[1], Alberto Rodríguez[2], Ignacio Castro[1], Diego G. Lamar[2].

[1]*Collins Aerospace Applied Research & Technology Ireland*
4th Floor, Penrose Business Centre, Penrose Wharf, Cork, Ireland

[2]*University of Oviedo*
Edificio Torres Quevedo (Departamental Oeste), bloque 3
Campus Universitario, 33204 Gijón, España (Spain)

E-Mail: Tania.CuestaCano@collins.com
URL: www.collinsaerospace.com/

Acknowledgements

This work has been funding by the European program (H2020) in the context of EASIER project. And has been supported by the Principality of Asturias and FICYT under project SV-PA-21-AYUD/2021/51931 and by the Spanish Government under project PID2021-127707OB-C21.

Keywords

«AC-DC converter», «Capacitors», «Converter control», «Pulse current charge/discharge», «Volume reduction».

Abstract

This work explores the dc-link capacitance reduction of a traditional three phase rectifier. A review of different methods for reducing the current across the dc-link capacitor, and consequently its size, is presented in this paper. In this work, the achievable capacitance reduction is explored by the action of a dual-loop control. A two-level power factor correction six-switch voltage source rectifier feeding a high demanding pulse load is analyzed. As a baseline, the output capacitor is designed from an energy storage perspective to achieve a specified maximum voltage ripple. The control design is performed in the frequency domain to get the best disturbance rejection under certain requirements and implementation constraints with the help of system time response evaluation. This work proposes a theoretical analysis that can be applied for the pursuit of converter weight reduction or other figures of merit such as volume or cost. The conclusions achieved with that theoretical study provide capacitance reduction ratios for the two-level power factor correction six-switch voltage source rectifier when implementing a classic dual-loop control algorithm.

Introduction

In power electronics, the use of capacitors acting as decouplers between supply and load is widespread. In that case, the capacitor must be designed to supply the load power demand during a certain time. Some examples can be found in dc-dc converters, Modular Multilevel Converters (MMC) or single-phase ac-dc converters [1]. In some of these cases, the capacitor is designed from an energy perspective, resulting in high capacitive values and consequently bulky capacitors. Furthermore, in some applications of three phase converters, the output capacitor has also to be sized for energy storage target. For instance, in Back-to-Back (BTB) converters, the robustness of the dc link is crucial and the capacitor should compensate any mismatch between the rectifier and the inverter [2].

The new paradigm of the electric mobility requires improvements in the specific power density of power converters. Transportation electrification is demanding higher power density power converters so the new electric solutions can substitute the original systems without increasing the total weight and volume of the propulsion system, which could derive into implementation problems and extra cost [2]–[4]. Reactive components are constantly under study in order to reduce their volume and weight because they are the highest contribution to power density of power converters. In the case of capacitor size reduction, an effective method is based on the minimization of current through the capacitor. The current through the capacitor is the sum of the current due to the instantaneous power imbalance between the power converter and the load, and the current harmonic content.

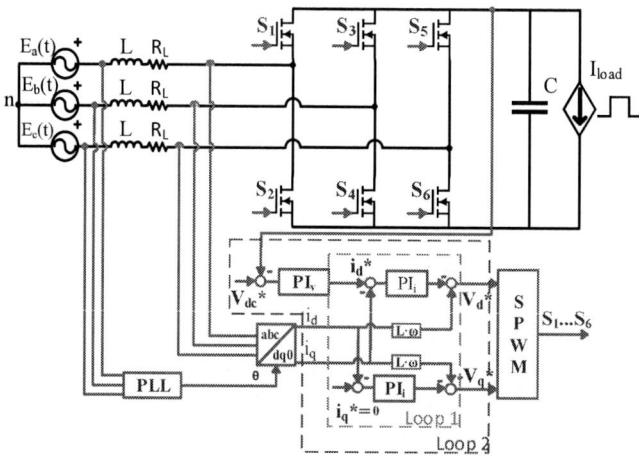

Fig. 1. Application scheme

Several methods to reduce either current type can be classified in three categories [5]: inverter topology modification [6], power balance control and innovative modulation schemes [7]. The power balance control method proposes different control schemes to improve the controller response speed, some of them can be found in [8], [9]. However, these control schemes add complexity to the control implementation. In [10], the BTB minimum capacitance is defined for different switching frequencies based on the linear relation between the inductance and the capacitance when the capacitor is purely selected and designed from an energy perspective. The main contribution of this work is a methodology to reduce the output capacitor by analysing the dynamics limit of a classic dual-loop control scheme, see Fig. 1, guaranteeing using the voltage ripple as a constraint.

The case under study consists of a two-level power factor correction six-switch voltage source rectifier (PFCVSR) supplying a pulsating high current load. The load of the converter can be considered as a pulsating current source connected to the output capacitor with a long period (i.e., on the order of milliseconds) when compared to the switching period of the power converter. In this work, the capacitance reduction is attained with the design of the control for the rectifier. The control design process includes the digital time delay introduced by digital implementation (Td). The analysis trades-off the control effort (i.e., control loop required bandwidth and phase margin), the switching frequency, the output voltage ripple and the required capacitance, yielding as a result a mapping of the minimum capacitive values required to achieve the desired performance.

The switching frequency range under study is limited for practical reasons (Table I). Although input filter design is not studied in this paper, the minimum switching frequency is fixed over 40 times the mains frequency not to penalize its design [11]. Although current power devices technology allows to manage high voltage (i.e. hundreds of volts) and medium current level (i.e. tens of amps) up to MHz range, switching frequency has been set 10 times lower to work in a more conservative range.

PFC rectifier control design

Table I shows the data of the system under study. It should be noted that the methodology procedure can be applied to other system specifications if the system is correctly modelled. For example, this is potentially useful for adapting the converter power rating, but if higher frequencies are considered parasitic elements may start playing an important role. The PFCVSR control design presented in this work is based on d-q synchronous reference frame converter average model assuming a sinusoidal PWM modulation [12]. A classic dual-loop control structure is used, where the outer control loop is responsible for maintaining a constant output capacitor voltage and the inner control loop is designed to control the boost inductor currents. Feedforward compensation is also implemented in the current control loop to eliminate the inherent d-q phases cross coupling, see Fig. 1. Therefore, the current control loop dynamics can be reduced to the input boost inductor, which can be simply modelled as a first order plant [13]. The stability requirements can be stablished in terms of phase margin (PM) by studying the control loop in the frequency domain. The minimum acceptable PM is set to 45° in both control loops. Mathematical expressions are derived from both current and voltage open loop gain (OL) to define the PI controller parameters as function of the phase margin (PM) and the crossover frequency (ω_c). At ω_c conditions (1) and (2) are imposed.

$$|OL(\omega_c)| = 1 \qquad (1)$$

$$\angle OL(\omega_c) = -180^0 + PM \quad (2)$$

Table I. Specifications of the case under test.

Variable	Value
Ac frequency	400-800 Hz
Nominal dc bus voltage (V_n)	430 V
Maximum voltage variation	10%
Minimum switching frequency	32 kHz
Maximum switching frequency	100 kHz
Pulse load amplitude (I_{load})	150 A
Pulse load period (T)	2 ms
Modulation index (M_p)	0.8962
T_d	T_{sw}

a) Current control loop design.

The current control loop, control loop 1 in Fig. 1, generates the PWM modulation index required to reproduce a sinusoidal input current in phase with the input voltage and with the required amplitude. The current OL expression is given by:

$$OL_i(s) = Kp_1 \frac{Ti_1 \cdot s + 1}{Ti_1 \cdot s} \cdot \frac{1}{T_d \cdot s + 1} \cdot \frac{1}{L \cdot s + R_L} \quad (3)$$

Although more complex and detailed approaches can be used to model the effect of T_d [14], the approach detailed in [13] is used to model it as a first-order system helping on reducing the complexity of the analysis.

Substituting the conditions defined by (1) and (2) in (3), the next mathematical expressions are yielded,

$$Ki_1 = \frac{1}{Ti_1} = \frac{\omega_c}{\tan(-90° + PM + \tan^{-1}(T_d \cdot \omega_c) + \tan^{-1}\left(\frac{L}{R_L} \cdot \omega_c\right)\right)} \quad (4)$$

$$Kp_1 = \frac{Ti_i \cdot \omega_c}{\sqrt{(Ti_i \cdot \omega_c)^2 + 1}} \cdot \sqrt{(T_d \cdot \omega_c)^2 + 1}. R_L . \sqrt{(^L/_{R_L} \cdot \omega_c)^2 + 1} \qquad (5)$$

$$PI_i = Kp_1 \cdot \frac{s + Ki_1}{s} \qquad (6)$$

, they determined the current controller parameters: Kp_1 that is the PI proportional gain and Ki_1, the PI integral action. They are defined in function of the design constraints, such as PM, ω_c, the inductor time constant defined by the ratio of L, the inductance value, and R_L, which is its series resistance, and T_d. Ti_1 represents the PI control time constant and it is the inverse of Ki_1 and the PI current controller is defined by a PI series form (PI_i). Note that the T_d and the inductor time constant ($^L/_{R_L}$) degrade the loop stability which will be compensated by the PI integral action.

Note that control loop stability can be studied by evaluating (4). The theoretical maximum for the crossover frequency ω_{clim}, can be obtained by equating Ti_1 to zero. T_d can be easily estimated by taking into account the sampling and updating rating [15]. It is normally expressed in terms of sampling periods, which in this case is related with the switching period. For example, T_d can be 1 or 0.5 sampling cycles. Evaluating (4) for $Ti_1 > 0$ the minimum $^L/_{R_L}$ for different combinations of ω_c, switching frequency and PM are obtained, Fig.2. These limits indicate the dynamic constraint of the inductor design for a certain PM and ω_c. That is, there are $^L/_{R_L}$ that will turn the system unstable and control compensation is not possible. Those limits are used to stablished constraints to the capacitance reduction study meaning that the attained solution will always be a feasible solution from a stability perspective.

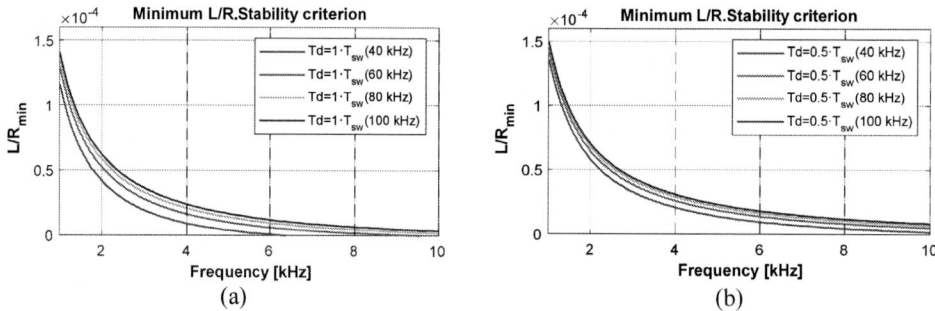

Fig. 2 Current control loop stability minimum boundaries when the PM is 45° and updating occurs at: (a) sampling time, (b) half the sampling time.

b) Voltage control loop.

The control loop 2, voltage control loop, in Fig. 1 is designed to regulate a constant voltage level. Load changes are seen as voltage disturbances by the control loop, so the design is focused on disturbance rejection. The effect of the control loop 1 dynamics has been approximated by a first order transfer function defined by this time constant response (T_{CL}). This can be done if the inner control loop, in this case control loop 1, dynamics are significantly higher than the outer control loop, the control loop 2 in this work. Voltage OL expression is given by,

$$OL_V(s) = Kp_2 \frac{Ti_2 \cdot s + 1}{Ti_2 \cdot s} \cdot \frac{1}{T_{CL} \cdot s + 1} \cdot \frac{3}{4} \cdot M_d \cdot \frac{ESR \cdot C \cdot s + 1}{C \cdot s} \qquad (7)$$

Substituting the conditions (1) and (2) in (7) the following mathematical expressions are derived,

$$Ki_2 = \frac{1}{Ti_2} = \frac{\omega_c}{tan(PM + tan^{-1}(T_{CL} \cdot \omega_c) - tan^{-1}(C \cdot ESR. \omega_c))} \qquad (8)$$

$$Kp_2 = \frac{Ti_2 \cdot \omega_c}{\sqrt{(Ti_2 \cdot \omega_c)^2 + 1}} \cdot \frac{\sqrt{(T_{CL} \cdot \omega_c)^2 + 1} \cdot C \cdot \omega_c}{\sqrt{(C \cdot ESR \cdot \omega_c)^2 + 1}} \cdot \frac{4}{3 \cdot M_d}. \quad (9)$$

$$PI_v = Kp_2 \cdot \frac{s + Ki_2}{s} \quad (10)$$

, they determined the current controller parameters: Kp_2 that is the PI proportional action and Ki_2, the PI integral action. The voltage controller is defined by a PI series (PI_v). As can be seen the OL gain depends on the modulation index (M_d), which decreases when load current increases. The voltage control loop is designed for the worst case, i.e. for the peak pulse amplitude. If the equivalent series resistance (ESR) of the output capacitor is expected to be big enough to change the design system plant [16], [17], it should be taken into account for the voltage control loop design.

In terms of zeros and poles, ESR adds a zero in the left s-plane side and in terms of frequency response it modifies OL gain and phase. From a time response perspective, it can be translated into slower response and initial voltage peak for high values of ESR. For simplicity, in the study presented in this work ESR effect has been neglected considering the expectation of very low ESR value. Nonetheless, it is important to take it into account otherwise.

PI controllers design criteria.

Fig. 3 illustrates the different design steps followed to determine the capacitance reduction. The solution with the best performance is selected based on (4)-(10) analysis. Best performance selection is evaluated with a performance factor (σ) that is defined in terms of the main close loop (CL(s)) time step response characteristics: the rise time (T_{rise}), the settling time ($T_{settling}$) and the overshoot percentage ($Overshoot(\%)$). For simplicity, in this work σ has been defined as the multiplication of all these characteristics:

$$\sigma = T_{settling} \cdot T_{rise} \cdot Overshoot(\%) \quad (11)$$

The best design is selected based on the minimum σ value, which is assumed to give a good trade off among the three parameters. In terms of control loop 1 the system response studied is focused on reference tracking improvement and in the case of the control loop 2 the system response studied is focused on the disturbance rejection improvement. Both control loop responses are analyzed independently under the same performance factor, although control loop 1 is studied first because it limits the theoretical maximum crossover frequency of control loop 2. In both cases, from a set of scenarios defined under constraints of minimum PM, a pre-defined switching frequency and a certain L/R_L, the case with the lowest σ is chosen as the best design.

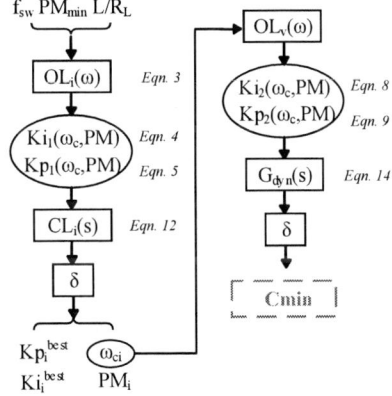

Fig. 3.Minimum capacitance calculation flowchart.

To perform the minimum capacitance identification, the equations that defined the loop control design has been made widespread. The control loop 1 and control loop 2 transfer function expressions that determines the reference tracking and disturbance rejection respectively have been rearranged to more general expressions,

$$\frac{id(s)}{idref(s)} = \frac{Kp_1' \cdot \frac{Ti_1 \cdot s + 1}{Ti_1 \cdot s} \cdot \frac{1}{Td \cdot s + 1} \cdot \frac{1}{L/_{R_L} \cdot s + 1}}{1 + Kp_1' \cdot \frac{Ti_1 \cdot s + 1}{Ti_1 \cdot s} \cdot \frac{1}{Td \cdot s + 1} \cdot \frac{1}{L/_{R_L} \cdot s + 1}} \tag{12}$$

$$\frac{Vo(s)}{io(s)} = \frac{1}{C} \cdot \frac{s \cdot (T_{CL} \cdot s + 1)}{T_{CL} \cdot s^3 + s^2 + Kp_2' \cdot s + \frac{Kp_2'}{Ti_v}} = \frac{1}{C} \cdot G_{dyn} \tag{13}$$

Where,

$$Kp_1' = \frac{Kp_1}{R_L} \tag{14}$$

$$Kp_2' = Kp_2 \cdot \frac{3 \cdot Md}{4 \cdot C} \tag{15}$$

It can be deduced from (14) that the value of the capacitor does not affect the dynamics (i.e., in terms of poles and zeros position). It only affects to the gain of the transfer function Therefore, analyzing G_{dyn} in terms of performance factor σ, the best disturbance rejection dynamics can be identified independently of the capacitance value and the operation point. The best disturbance rejection response that can be expected under f$_{sw}$, PM and $L/_{R_L}$ ratio constraints can be obtained for a normalized case where i_{load} and the nominal voltage are set to unity. This yields a normalized voltage variation, ΔV_{pu}.

From ΔV_{pu}, the minimum capacitance for any voltage variation ($\Delta V_{required}$, expressed as an absolute value) and any certain current level demand (i_{load}) can be deduced with,

$$C_{min} = \Delta V_{pu} * \frac{i_{load}}{\Delta V_{required}} \tag{16}$$

Capacitance reduction evaluation

The main contribution of this work is the study of the limits of the output capacitance reduction by the action of the classic dual control loop. The capacitance reduction ratio in terms of the base case can be calculated. The base case has been set as energy storage capacitance design to limit a maximum voltage variation for a high current pulse demand of a duration of 2 ms. As mentioned in the introduction section, the obtained data can be also used for proving the effectiveness of more complex control algorithms proposed for capacitance reduction.

Table I specifications and an $L/_{R_L}$ ratio equal to $0.5 \cdot 10^3$ (i.e., as matter of example and based on a preliminary inductor design) have been used for the analysis presented in this paper. Fig. 4.a shows the results for an example in which i_{load} and $\Delta V_{required}$ are set to 1. The results shown in Fig. 4.b can be easily obtained from the case with i_{load} and $\Delta V_{required}$ set to 1, using (16).

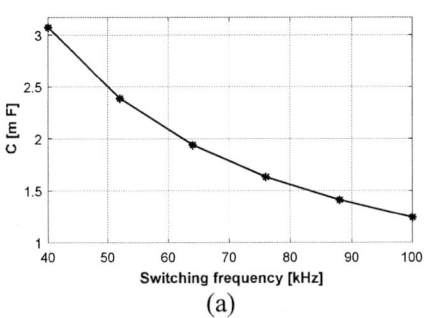 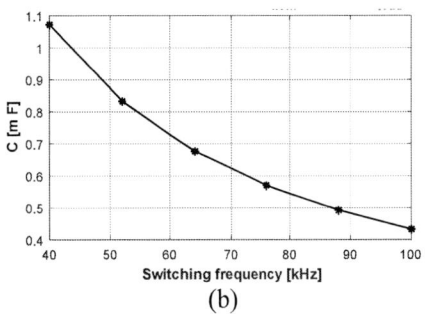

| (a) | (b) |

Fig. 4. Capacitance reduction study results. (a)Case where i_{load} and $\Delta V_{required}$ are set to 1 (b) Results particularization for a given operating point (Table I).

To validate the theoretical study results, a simulation has been performed. The simulation has been developed in MATLAB/Simulink software and a good approximation to the real control implementation in a DSP platform was achieved. Simulation and analytical results are compared with good agreement, Fig. 6.

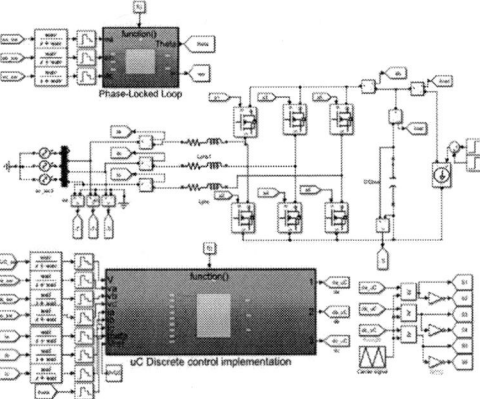

Fig. 5. MATLAB/Simulink discrete system simulation with C-code based control loop implementation.

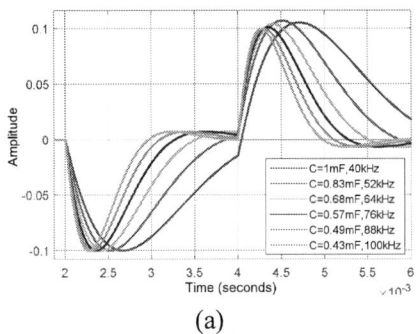 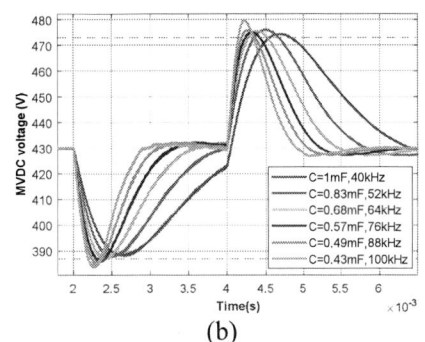

| (a) | (b) |

Fig. 6.Step response evaluation with the minimum capacitance calculated. Red dashed lines indicate the allowable voltage variation band (a) Theoretical time response, based on (7). (b) PFCVR switching model simulation implemented in MATLAB/Simulink.

Fig. 6.a shows the voltage variation of an equivalent linear system when a current of a unit pulse is demanded and Fig. 6.b displays the voltage variation of the switching model simulation of Fig. 5 for a 150 A pulse current. Finally, Table II illustrates the capacitor reduction ratios for different switching frequencies when comparing with a purely based energy storage capacitor (C_{base}) under the specifications described in Table I.

Table II. Reduction ratio estimation.

f_{sw} [kHz]	C_{min} [mF] @430V,150A	Reduction ratio [%] (C_{base}= 6.7mF)
40	1.08	83.8
52	0.83	87.6
64	0.68	89.8
76	0.57	91.4
88	0.49	92.7
100	0.43	93.5

Conclusion

Optimization tools to help increase the power density of power converters are a need for the electric mobility market. In this work a methodology to estimate and reduce the size of output capacitor has been shown achieving reduction ratios as high as 93.5%.

The aim of this work is to present a useful methodology for power rectifier weight and volume optimization by identifying the reduction ratio of the output capacitor size by means of traditional control technique. In the case of study presented in this work, reduction ratios higher than 80% were achieved by improving the control dynamics of a three-phase ac-dc converter. As it can be seen in Fig. 4, the capacitance reduction is not linearly related with the switching frequency. The generated data is intended to be useful for evaluating more complex control algorithms focus on the capacitor reduction of PFCVR and voltage source converters in general. Additionally, a combination of these results with the study of the system losses can give a trade-off between losses and power density.

Finally, the proposed methodology for a PFCVR has been demonstrated in MATLAB/Simulink with a discrete model. This model is key in the future work to predict the operation of the prototype to be implemented.

References

[1] R. W. Erickson y D. Maksimovic, *Fundamentals of Power Electronics*, 3.ª ed. Springer International Publishing, 2020. doi: 10.1007/978-3-030-43881-4.
[2] R. Lai, «Analysis and Design for a High Power Density Three-Phase AC Converter Using SiC Devices», dic. 2008, Consulted: 18th of December 2020. [Online]. Available in: https://vtechworks.lib.vt.edu/handle/10919/30155
[3] R. Wang, «High Power Density and High Temperature Converter Design for Transportation Applications», jun. 2012, Consulted: 18th of December 2020. [Online]. Available in: https://vtechworks.lib.vt.edu/handle/10919/28264
[4] C. Brando, «High-density multilevel power converters for use in renewable and transportation applications.», University of Illinois at Urbana-Champaign, 2019. [Online]. Available in: http://hdl.handle.net/2142/106246
[5] H. Ye y A. Emadi, «An interleaving scheme to reduce DC-link current harmonics of dual traction inverters in hybrid electric vehicles», *2014 IEEE Applied Power Electronics Conference and Exposition - APEC 2014*, March. 2014, pp. 3205-3211. doi: 10.1109/APEC.2014.6803764.
[6] Gui Jia Su, «Electrical motor/generator drive apparatus and method.», US20110074326A1, 2011

[7]   D. Zhang, F. Wang, R. Burgos, R. Lai, y D. Boroyevich, «DC-Link Ripple Current Reduction for Paralleled Three-Phase Voltage-Source Converters With Interleaving», *IEEE Trans. Power Electron.*, vol. 26, n.º 6, pp. 1741-1753, June. 2011, doi: 10.1109/TPEL.2010.2082002.

[8]   L. Malesani, L. Rossetto, P. Tenti, y P. Tomasin, «AC/DC/AC PWM converter with reduced energy storage in the DC link», *IEEE Trans. Ind. Appl.*, vol. 31, n.º 2, pp. 287-292, March. 1995, doi: 10.1109/28.370275.

[9]   W. Lee y S. Sul, «DC-Link Voltage Stabilization for Reduced DC-Link Capacitor Inverter», *IEEE Trans. Ind. Appl.*, vol. 50, n.º 1, pp. 404-414, January. 2014, doi: 10.1109/TIA.2013.2268733.

[10]   J. W. Kolar *et al.*, «PWM Converter Power Density Barriers», en *2007 Power Conversion Conference - Nagoya*, April. 2007, p. P-9-P-29. doi: 10.1109/PCCON.2007.372914.

[11]   «RTCA DO-160 - Environmental Conditions and Test Procedures for Airborne Equipment | Engineering360». https://standards.globalspec.com/std/9894152/RTCA%20DO-160.

[12]   D. Boroyevich, R. Burgos, I. Cvetkovic, y B. Wen, «Modeling and Control of Three-Phase High-Power High-Frequency Converters», p. 133, 2018.

[13]   V. Blasko y V. Kaura, «A new mathematical model and control of a three-phase AC-DC voltage source converter», *IEEE Trans. Power Electron.*, vol. 12, n.º 1, pp. 116-123, January. 1997, doi: 10.1109/63.554176.

[14]   L. Xueyan y Y. Zheng, «Comparison of time delay processing methods in control system», en *2015 4th International Conference on Computer Science and Network Technology (ICCSNT)*, December. 2015, vol. 01, pp. 1502-1505. doi: 10.1109/ICCSNT.2015.7491014.

[15]   N. Hoffmann, F. W. Fuchs, y J. Dannehl, «Models and effects of different updating and sampling concepts to the control of grid-connected PWM converters — A study based on discrete time domain analysis», en *Proceedings of the 2011 14th European Conference on Power Electronics and Applications*, August. 2011, pp. 1-10.

[16]   E. D. Hagh, P. Mohammadalizadeh, y E. Babaei, «Effects of ESR on stability and frequency response of Cuk converter by using signal flow graph method», en *The 6th Power Electronics, Drive Systems Technologies Conference (PEDSTC2015)*, February. 2015, pp. 83-88. doi: 10.1109/PEDSTC.2015.7093254.

[17]   J. Leyva-Ramos, M. G. Ortiz-Lopez, y L. H. Diaz-Saldierna, «The effect of ESR of the capacitors on modeling of a quadratic boost converter», en *2008 11th Workshop on Control and Modeling for Power Electronics*, August. 2008, pp. 1-5. doi: 10.1109/COMPEL.2008.4634679.

Cognitive Insights into Metaheuristic Digital Twin based Health Monitoring of DC-DC Converters

Abdul Basit Mirza[1], Kushan Choksi[1], Sama Salehi Vala[1], Krishna Moorthy Radha[2], Madhu Sudhan Chinthavali[2] and Fang Luo[1]

[1]Department of Electrical and Computer Engineering, Stony Brook University, USA

[2]Oak Ridge National Lab (ORNL), USA

abdulbasit.mirza@stonybrook.edu

Acknowledgements

This work was supported by Oak Ridge National Laboratory (ORNL) funded through the Department of Energy (DOE) - Office of Electricity's (OE), Transformer Resilience and Advanced Components (TRAC) program led by the program manager Andre Pereira. The authors would also like to acknowledge the National Science Foundation (NSF Award No. 1846917) for lending financial support for this work.

Keywords

Digital Twin, Genetic Algorithm (GA), Health Monitoring, Metaheuristic Optimization, Particle Swarm Optimization (PSO), Sensitivity Analysis.

Abstract

Reliability of components has always been a major concern to the performance and stability of DC-DC converters. After long-term operation, these passive components and switching devices start to degrade and become weak to withstand normal electrical and thermal stresses. An insightful digital interface to the physical layer known as Digital Twin (DT) can be a sustainable solution for ensuring reliability. This paper extends the DT concept to component level health monitoring in DC-DC converters. The proposed concept is noninvasive and does not require additional sensors. The working principle is to minimize the weighted least squared error between the digital twin output and the measured data of state variables through metaheuristic optimization. An application for Two-Phase Interleaved Boost Converter with reverse coupled inductor is considered and Hardware-in-the-loop (HIL) platform is used for sensitivity analysis for component degradation. Further, the optimization problem is solved using the following two popular metaheuristic optimization methods: Particle Swarm Optimization (PSO) and Genetic Algorithm (GA). Further, the performance of both methods for 20 executions in terms of computational time; convergence rate and dispersion are compared. It is evident from the results that GA outperforms PSO with 50 % less execution time and better accuracy > 95 %.

Introduction

In recent years, the use of DC-DC converters in various applications such as industrial, transportation and renewable energy has increased significantly. This poses a key concern for reliability of the converter as it determines the stability of the whole system. DC-DC converters comprise power semiconductors switches and energy storing elements such as inductor and capacitor. These components degrade over time due to environmental strains and switching action. [1]-[2]. For instance, the capacitance drops 10-20% [3] and R_{DSon} of the MOSFET increases 5-20% [4] due to aging.

Several methods have been proposed in literatures for health monitoring. Mostly, these methods rely on inserting extra sensors in the converter to sense a voltage or current signal primarily related to the health monitoring parameter [5], [6] or using averaged model with reduced sensitivity [7]. Further, the associated complexity such are additional DSPs and signal lines makes the whole design complex and vulnerable to errors.

Recently, the Digital Twin (DT) concept has been extended for health monitoring of DC-DC converters. The DT virtually replicates converter, which can be used to estimate the component values by matching the response of digital twin to that of the actual converter. A case for estimating parameters for buck converter using DT through average least square error minimization objective function f_{obj} through Particle Swarm Optimization (PSO) is proposed [8]. Although the proposed method is effective, it requires large memory for data acquisition and the accuracy for ESRs prediction is low.

This paper proposes an improved DT model for health monitoring of DC-DC converters. The proposed method is non-invasive and does not require need for additional sensors. It is based on minimizing the weighted least-squared error between response of the DT and the measured data over the health parameters, making it a multi-objective optimization problem. The efficacy of the proposed method is demonstrated on a higher order two-phase interleaved boost converter with coupled inductor with more parameters to estimate. For sensitivity analysis, Hardware-in-the-loop (HIL) testing is used for simulating component degradation. The optimization problem is solved using the following two metaheuristic methods: Particle Swarm Optimization (PSO) and Genetic Algorithm (GA), along with performance comparison. Finally, health monitoring is performed on a hardware prototype and results are compared.

Digital Twin Modeling for Two-Phase Interleaved Boost Converter

Fig. 1 shows the schematic of a two-phase interleaved boost converter with reverse coupled inductor. iL_1, iL_2 and v_c represents the inductor currents and the output capacitor voltage. R_{L1}, R_{L2}, R_{DSon1}, R_{DSon2} and R_C are the parasitics resistances of the coupled inductor, switches, and output capacitor. Health monitoring involves estimating the following seven parameters: L, C, R_{L1}, R_{L2}, R_{DSon1}, R_{DSon2} and R_C. The coupling factor K_C of the converter can be assumed to be constant as it primarily depends on the physical layout of the winding, which does not change over time.

The proposed health monitoring approach is based on minimizing the least-squared error between response of the DT and the measured data. This requires DT to model both the transient and steady-state responses. The transient response is sensitive to the values of the passive components such as L, C. To cater this, the actual state space model $\dot{x} = Ax + B$, is considered instead of the averaged state space model. The state space $\dot{x} = Ax + B$ with state variables iL_1, iL_2 and v_c and output voltage v_{out} for four possible switching states S_1-D_2, D_1-D_2, S_2-D_1 and S_1-S_2 are given in (1)-(4) respectively. The quantity d represents the instantaneous duty cycle of the converter. The differential equations in state-space representation in (1)-(4) are in continuous time domain. However, the data from the converter is in discrete time, sampled at sampling frequency f_s. This requires discretization of the differential equations with adequate accuracy. Among, the popular discretization methods such as 4th Order Runge-Kutta (RK), Trapezoidal, Backward and Forward Euler, RK is chosen as it provides better accuracy. Further, although RK's computational time is larger than the other methods, it is not a concern as degradation is not time sensitive. The discretized expressions using 4th Order RK with time step T_s, for iL_1, iL_2 and v_c are given in (5)-(7). The parameters k_{a1}-k_{a4}, k_{b1}-k_{b4} and k_{c1}-k_{c4} are implicit auxiliary variables of RK method. From v_c^{n+1}, v_{out}^{n+1} can be computed by using the respective expression for each switching state in (1)-(4).

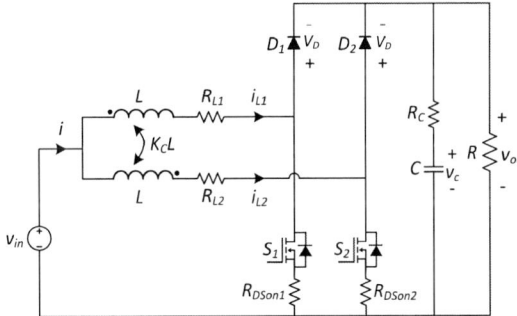

Fig. 1. Two-phase interleaved boost converter.

$$\begin{cases} A = \begin{bmatrix} -L(R_{L1} + R_{DSon1}) & -K_C L R_{L2} + \frac{-K_C L R R_C}{R_C + R} & \frac{-K_C L R}{R_C + R} \\ -K_C L(R_{L1} + R_{DSon1}) & -L R_{L2} - \frac{L R R_C}{R_C + R} & -\frac{L R}{R_C + R} \\ 0 & \frac{R}{C(R + R_C)} & -\frac{1}{C(R_C + R)} \end{bmatrix} \\ B = \begin{bmatrix} -K_C L V_D + L(1 + K_c) v_{in} \\ -L V_D + L(1 + K_c) v_{in} \\ 0 \end{bmatrix} \\ v_{out} = \frac{R R_C}{R_C + R} i_{L2} + \frac{R v_c}{R_C + R} \end{cases} \tag{1}$$

$$\begin{cases} A = \begin{bmatrix} -L R_{L1} & -K_C L R_{L2} & 0 \\ -K_C L R_{L1} & -L R_{L2} & 0 \\ \frac{R}{C(R + R_C)} & \frac{R}{C(R + R_C)} & -\frac{1}{C(R_C + R)} \end{bmatrix} \\ B = \begin{bmatrix} L(1 + K_c)(-dv_{in}/(1 - d) + V_D) \\ L(1 + K_c)(-dv_{in}/(1 - d) + V_D) \\ 0 \end{bmatrix} \\ v_{out} = \frac{R R_C}{R_C + R} (i_{L1} + i_{L2}) + \frac{R v_c}{R_C + R} \end{cases} \tag{2}$$

$$\begin{cases} A = \begin{bmatrix} -L R_{L1} - \frac{X R R_C}{R_C + R} & -K_C L(R_{L2} + R_{DSon2}) & \frac{-L R}{R_C + R} \\ -K_C L R_{L1} + \frac{Y R R_C}{R_C + R} & -L(R_{L2} + R_{DSon2}) & \frac{K_C L R}{R_C + R} \\ \frac{R}{C(R + R_C)} & 0 & -\frac{1}{C(R_C + R)} \end{bmatrix} \\ B = \begin{bmatrix} -L V_D + L(1 + K_c) v_{in} \\ -K_C L V_D + L(1 + K_c) v_{in} \\ 0 \end{bmatrix} \\ v_{out} = \frac{R R_C}{R_C + R} i_{L1} + \frac{R v_c}{R_C + R} \end{cases} \tag{3}$$

$$\begin{cases} A = \begin{bmatrix} -L(R_{L1} + R_{DSon1}) & -K_C L(R_{L1} + R_{DSon1}) & 0 \\ -K_C L(R_{L1} + R_{DSon1}) & -L(R_{L1} + R_{DSon1}) & 0 \\ 0 & 0 & -\frac{1}{C(R_C + R)} \end{bmatrix} \\ B = \begin{bmatrix} L(1 + K_c) v_{in} \\ L(1 + K_c) v_{in} \\ 0 \end{bmatrix} \\ v_{out} = \frac{R v_c}{R_C + R} \end{cases} \tag{4}$$

$$iL_1^{n+1} = iL_1^n + \frac{T_s}{6}(k_{a1} + 2k_{a2} + 2k_{a3} + k_{a4}) \tag{5}$$

$$iL_2^{n+1} = iL_2^n + \frac{T_s}{6}(k_{b1} + 2k_{b2} + 2k_{b3} + k_{b4}) \tag{6}$$

$$v_c^{n+1} = v_c^n + \frac{T_s}{6}(k_{c1} + 2k_{c2} + 2_{c3} + k_{c4}) \tag{7}$$

Health Monitoring Through Metaheuristic Optimization

Health monitoring can be performed by matching the response of DT with that of the actual converter by minimizing the sum of squared error for N data points, like multivariate regression, between the measured and the DT values for iL_1, iL_2 and v_{out}. This becomes a multi-objective optimization problem with the following objective functions:

$$f_{obj1}(x) = \sum_{k=1}^{N} \left(i_{L1,m}^k - i_{L1,d}^k \right)^2 \tag{8}$$

$$f_{obj2}(x) = \sum_{k=1}^{N} \left(i_{L2,m}^k - i_{L2,d}^k\right)^2 \tag{9}$$

$$f_{obj3}(x) = \sum_{k=1}^{N} \left(v_{out,m}^k - v_{out,d}^k\right)^2 \tag{10}$$

defined on $x = (L, C, R_{L1}, R_{L2}, R_{DSon1}, R_{DSon2}, R_C)$, with inequality constraints only. The inequality constraints cater the component tolerances as well as degradation. The subscripts m and d corresponds to measured and DT values. The multi-objective optimization problem can be converted into single objective f_{obj} by using weighted sum method [10], given by (11).

$$
\begin{aligned}
&\text{minimize} && f_{obj} = \alpha f_{obj1} + \beta f_{obj2} + \gamma f_{obj3} \\
&(L, C, R_{L1}, R_{L2}, R_{DSon1}, R_{DSon2}, R_C) && L_{min} \leq L \leq L_{max1} \\
& && C_{min} \leq C \leq C_{max} \\
& && R_{L1\_min} \leq R_{L1} \leq R_{L1\_max} \\
& && R_{L2\_min} \leq R_{L2} \leq R_{L2\_max} \\
& && R_{DSon1\_min} \leq R_{DSon1} \leq R_{DSon1\_max} \\
& && R_{DSon2\_min} \leq R_{DSon2} \leq R_{DSon2\_max} \\
& && R_{C\_min} \leq R_C \leq R_{C\_max}
\end{aligned} \tag{11}
$$

α, β and γ represents the weights for the objective functions. The optimization problem is complex to be solved through derivative-based methods such as geometric programming or convex optimization; the response of the state variables is based on piecewise differential equations. The resultant solutions of the differential equations are continuous at the boundary but not differentiable. As a result, meta-heuristic-based optimization methods are best suited for this problem. Two popular metaheuristic methods: PSO [11] and GA [12] are selected to solve the optimization problem in (11). The algorithms have been proven to be effective in solving highly non-linear optimization problems in the power electronics domain [13]-[14]. Compared with deterministic methods, these methods are derivative free and search for optimal solution through concept of biological evolution and collaborative behavior of biological populations. PSO is inspired by the ability of bird flocks to adapt to new environment through information sharing. GA on the other hand is inspired by Charles Darwin's theory of natural evolution to arrive to an optimum solution through concept of chromosomal crossover and mutation. Although the searching process for both methods lead to longer computational time, it is not critical for health monitoring as degradation happens slowly.

Health Monitoring Validation

To validate the performance of the proposed DT, a 60 V to 100 V 2.5 kW two-phase interleaved boost converter, switching at f = 50 kHz, is simulated in Typhoon HIL 402 with gating signals generated from Texas Instrument LAUNCHXL-F28379D DSP (Fig. 2). HIL platform is chosen as it gives flexibility in varying the component values dynamically. The data for 5 switching cycles, as opposed to 80 cycles in [8], is sampled and passed to PSO and GA based optimization program for health monitoring with $\alpha = \beta = 2\gamma = 1/length(data)$ in (11). $length(data)$ refers to the number of data points used for computation, which in turn depends on the sampling frequency and the number of switching cycles considered. For instance, with 1 MHz sampling, the weight $1/length(data)$ comes out to be 0.01. The lower and upper bounds of the inequality constraints are set to ± 30 % of their nominal value. Table I tabulates the nominal values of the health monitoring parameters. The coupling factor K_C is fixed to 0.8 throughout the analysis, as coupling factor depends on the orientations and placement of windings on the core, which do not change with degradation.

The learning factors r_a and r_b for particle best P_{Best} and global best G_{Best} value for computation of particle new position (pos^{i+1}) for PSO in (12) are both set to 2.05. Further, the inertia weight ω for velocity V is initialized with a large value and then reduced dynamically in each iteration to ensure better convergence [11].

$$pos^{i+1} = \omega V^i + 2r_a(P_{Best} - P^i) + 2r_b(G_{Best} - G^i) \qquad (12)$$

Table I: Converter Parameters

Parameter	Nominal Value
L	100 µH
C	47 µF
R_{L1}	50 mΩ
R_{L2}	50 mΩ
R_{DSon1}	10 mΩ
R_{DSon2}	10 mΩ
R_C	3 mΩ

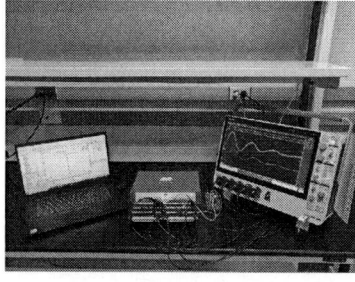

Fig. 2. HIL validation test setup.

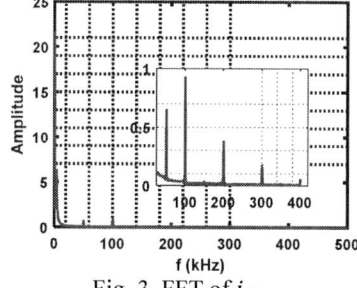

Fig. 3. FFT of i_{L1}.

For GA, roulette wheel selection method with Boltzmann probability distribution is chosen [12]. For performance comparison, the swarm size in PSO and number of chromosomes in GA are both set to 200; the number of iterations during a single execution is set 80. Moreover, the sampling frequency has a huge influence on the convergence and correct estimation of health parameters. The channel currents i_{L1} and i_{L2} for the boost converter in Fig. 1 contain fourth order harmonic of the switching frequency f (Fig. 3). To avoid aliasing, sampling frequency f_s is set to $1\ MHz$ ($2.5 \times 4f$). The effect of aliasing is shown in Fig. 4 for i_{L1}, where the DT is not able to follow the measured data if $f_s < 2 \times 4f$.

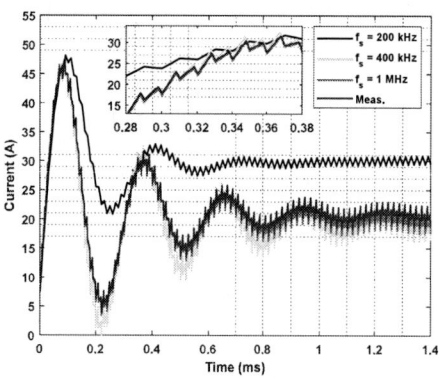

Fig. 4. Impact of sampling frequency f_s.

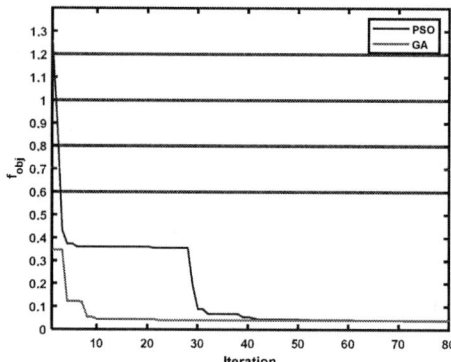

Fig. 5. PSO and GA f_{obj} descent comparison.

Fig. 5 compares the descending process of f_{obj} using PSO and GA. The f_{obj} value after first iteration using GA is 3.7 times lower than PSO. Also, GA minimizes f_{obj} in a smaller number of iterations. The average execution times observed for PSO and GA are 25.5 s and 13.2 s respectively. Both algorithms can minimize the f_{obj} to 0.005. The results for nominal (Table I) and sensitivity analysis are compared and summarized in Fig. 6 for both optimization methods. The horizontal line represents the average value for 20 independent executions. It is evident from the results that both PSO and GA are able to estimate L, C within 3 % tolerance. However, the estimated values for ESRs ($R_{L1}, R_{L2}, R_{DSon1}, R_{DSon2}, R_c$), using PSO have more dispersion around the set/average value compared with GA. Table II compares the standard deviations σ_{PSO} and σ_{GA}, observed for the estimated results of ESRs in Fig 6. Based on the results, it can be concluded that GA is best suited for the DT-based health monitoring with better accuracy.

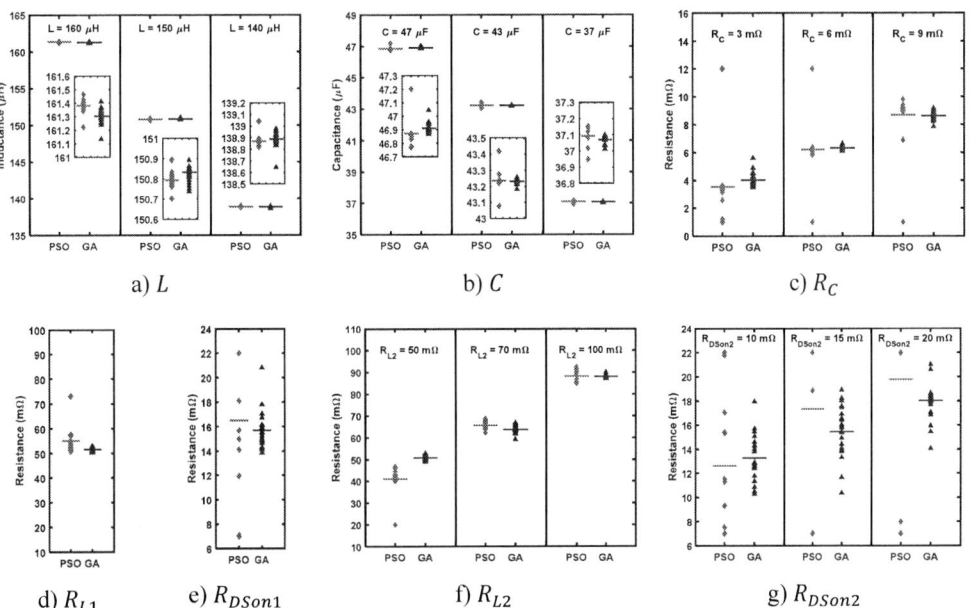

a) L b) C c) R_C

d) R_{L1} e) R_{DSon1} f) R_{L2} g) R_{DSon2}

Fig. 6. Estimated parameters using PSO and GA for nominal and perturbed cases.

Table II: Estimated ESRs Standard Deviation Comparison

Parameter	R_{L1}	R_{L2}			R_{DSon1}	R_{DSon2}			R_C		
Case (mΩ)	50	50	70	100	10	10	15	20	3	6	9
σ_{PSO} (mΩ)	4.9	5.3	2.3	2.7	6.3	6.2	6.8	5.2	2.1	1.7	1.8
σ_{GA} (mΩ)	0.8	1.1	1.8	0.6	1.5	1.9	2.1	1.7	0.5	0.2	0.3

Hardware Prototype Validation

A 20 to 36 V low-voltage hardware prototype (Fig. 7), switching at 50 kHz, is built, and tested to validate the effectiveness of the proposed concept. The switches are realized using two CREE KIT-CRD-8FF65P modules, based on CREE C3M0060065J, 650 V 65 mΩ MOSFETs. Based on the results from the previous section, an optimization is run using GA, owing to its better accuracy. The nominal measured values and the estimated values of the health parameters are tabulated in Table III. The estimated values are in line with the nominal values, justifying the efficacy of the method. The maximum percentage error observed is 2 % for L and C and 7 % for the ESRs.

Table III: Nominal and estimated parameters

Parameter	Nominal Value	Estimated Value
L	662.3 µH	657.3 µH
C	120.0 µF	124.1 µF
R_{L1}	169.4 mΩ	182.7 mΩ
R_{L2}	142.8 mΩ	126.3 mΩ
R_{DSon1}	65.0 mΩ	72.3 mΩ
R_{DSon2}	65.0 mΩ	67.7 mΩ
R_C	9.2 mΩ	11.5 mΩ

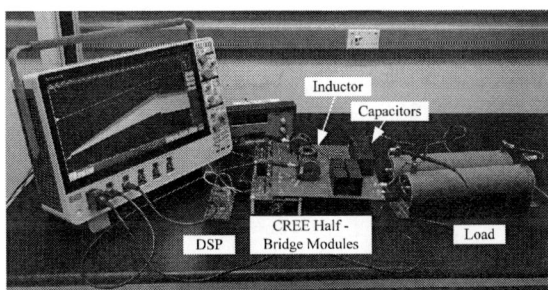

Fig. 7. Test setup.

Conclusion

Health monitoring of DC-DC converter components is essential for ensuring peak efficiency and reliability in the long run. An improved metaheuristic DT concept is proposed for determining the health parameters of passive components and the power devices of complex DC-DC converters. The proposed concept is validated using PSO and GA optimization techniques. Based on the results, GA has better accuracy with less dispersion and executes faster than PSO, while utilizing less memory resources.

References

[1] H. Givi, E. Farjah and T. Ghanbari, "A Comprehensive Monitoring System for Online Fault Diagnosis and Aging Detection of Non-Isolated DC–DC Converters' Components," in *IEEE Transactions on Power Electronics*, vol. 34, no. 7, pp. 6858-6875, July 2019.

[2] A. Marquez *et al.*, "Power Devices Aging Equalization of Interleaved DC–DC Boost Converters via Power Routing," in *IEEE Journal of Emerging and Selected Topics in Industrial Electronics*, vol. 1, no. 1, pp. 91-101, July 2020.

[3] H. Soliman, H. Wang and F. Blaabjerg, "A Review of the Condition Monitoring of Capacitors in Power Electronic Converters," in *IEEE Transactions on Industry Applications*, vol. 52, no. 6, pp. 4976-4989, Nov.-Dec. 2016.

[4] S. Dusmez, H. Duran and B. Akin, "Remaining Useful Lifetime Estimation for Thermally Stressed Power MOSFETs Based on on-State Resistance Variation," in *IEEE Transactions on Industry Applications*, vol. 52, no. 3, pp. 2554-2563, May-June 2016.

[5] S. Bęczkowski, P. Ghimre, A. R. de Vega, S. Munk-Nielsen, B. Rannestad and P. Thøgersen, "Online Vce measurement method for wear-out monitoring of high power IGBT modules," *2013 15th European Conference on Power Electronics and Applications (EPE)*, 2013.

[6] Z. Wang, B. Tian, W. Qiao and L. Qu, "Real-Time Aging Monitoring for IGBT Modules Using Case Temperature," in *IEEE Transactions on Industrial Electronics*, vol. 63, no. 2, pp. 1168-1178, Feb. 2016.

[7] B. X. Li and K. S. Low, "Low Sampling Rate Online Parameters Monitoring of DC–DC Converters for Predictive-Maintenance Using Biogeography-Based Optimization," in *IEEE Transactions on Power Electronics*, vol. 31, no. 4, pp. 2870-2879, April 2016.

[8] Y. Peng, S. Zhao and H. Wang, "A Digital Twin Based Estimation Method for Health Indicators of DC–DC Converters," in *IEEE Transactions on Power Electronics*, vol. 36, no. 2, pp. 2105-2118, Feb. 2021.

[9] C. Zheng, H. Ma, J. Lai and L. Zhang, "Design Considerations to Reduce Gap Variation and Misalignment Effects for the Inductive Power Transfer System," in *IEEE Transactions on Power Electronics*, vol. 30, no. 11, pp. 6108-6119, Nov. 2015.

[10] Boyd, S., 2021. *Convex Optimization.* [online] Web.stanford.edu. Available at: https://web.stanford.edu/~boyd/cvxbook/bv_cvxbook.pdf.

[11] Clerc, M., 2006. *Particle Swarm Optimization.* 1st ed. John Wiley & Sons, Ltd.

[12] Haupt, R. and Haupt, S., 2004. *Practical genetic algorithms.* Hoboken, N.J.: Wiley-Interscience.

[13] S. E. De León-Aldaco, H. Calleja and J. Aguayo Alquicira, "Metaheuristic Optimization Methods Applied to Power Converters: A Review," in *IEEE Transactions on Power Electronics*, vol. 30, no. 12, pp. 6791-6803, Dec. 2015.

[14] M. S. Mohammed and R. A. Vural, "Evolutionary Design Automation of High Efficiency Series Resonant Converter for Photovoltaic Systems," in *IEEE Transactions on Power Electronics*, vol. 35, no. 11, pp. 11332-11343, Nov. 2020.

A Three-Phase Isolated Secondary-Resonant Single-Active-Bridge DC-DC Converter with a Delta-Star Connected Transformer

Atsushi Nishio, Kohei Budo, Mai Van Tuan and Takaharu Takeshita
Dept. of Electrical and Mechanical Engineering, Graduate School of Engineering,
Nagoya Institute of Technology
Gokiso, Showa
Nagoya, 466-8555 Japan
Phone: +81 (52) 735-5441
Fax: +81 (52) 735-5432
Email: take@nitech.ac.jp
URL: http://motion.web.nitech.ac.jp/

Acknowledgments

A part of this work was supported by JSPS KAKENHI Grant Number JP21H01310.

Keywords

≪Power converters for EV≫, ≪Converter circuit≫, ≪DC-DC converters≫, ≪Soft switching≫, ≪Three-phase system≫

Abstract

This paper presents a three-phase isolated secondary-resonant single-active-bridge DC-DC converter with a delta-star connected transformer. The total power factor of the transformer can be improved by using the LC resonance compared with a conventional converter. The effectiveness of the proposed circuit is verified by experiments using a 2.5 kW 265 V, 15 kHz laboratory prototype.

Introduction

Lately, various problems such as global warming and the energy crisis have happened. The spread of electric vehicles (EVs) is one of the solutions to solve these problems. Recently, many kinds of isolated DC-DC converters have been researched and developed for applications of battery chargers [1]-[8]. There are two types of isolated DC-DC converters: single-phase circuits and three-phase circuits. A three-phase isolated DC-DC converters are suitable for a high-power conversion compared with a single-phase one [9]. By these reasons, the demand for a three-phase isolated DC-DC converter increases.

Various topologies of three-phase isolated DC-DC converters has been proposed. The single-active-bridge (SAB) DC-DC converter [5],[9],[10] for an unidirectional power converter has been proposed. It has a three-phase inverter on the primary side and a diode rectifier circuit on the secondary side. Because the secondary side of the transformer is composed of a diode rectifier circuit, there are few numbers of necessary parts, and this circuit can be downsized. Furthermore, it can reduce a switching loss by using soft-switching. A three-phase isolated DC-DC converter has four types of transformer connections, star-star connection, delta-delta connection, delta-star connection, and star-delta connection. The SAB can achieve high-efficiency and high power density by using star-delta or delta-star connection of the transformer [11]. Because the secondary circuit of the SAB is composed of diodes, the value of the total power factor of the transformer becomes low. Therefore, a loss becomes large.

This paper presents a three-phase isolated secondary-resonant single-active-bridge (SR-SAB) DC-DC converter. The proposed circuit is more suitable for high power applications than single-phase SR-SAB

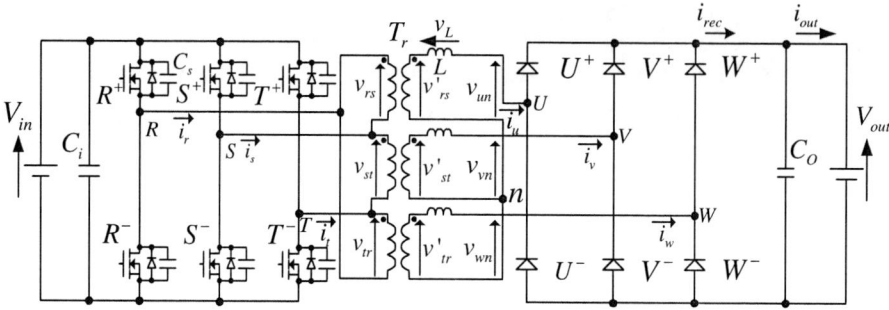

(a) Conventional three-phase isolated SAB DC-DC converter with a delta-star connected transformer.

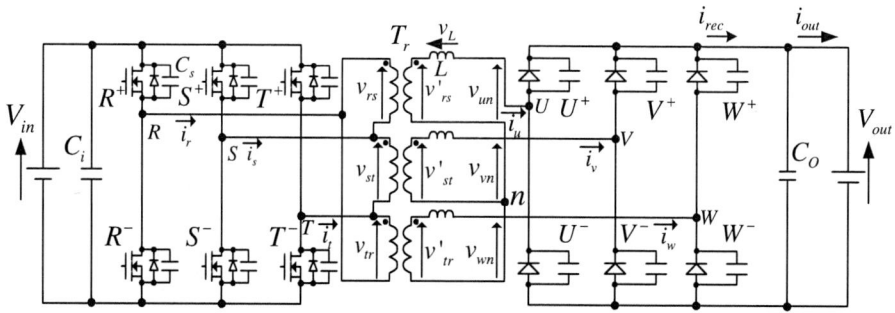

(b) Proposed three-phase isolated SR-SAB DC-DC converter with a delta-star connected transformer.

Fig. 1: Isolated three-phase DC-DC converter.

[3],[4] because it uses a three-phase transformer. The primary side of the transformer is composed of a three-phase inverter, the secondary side of the transformer consists of a diode rectifier circuit composed of six diodes with the resonant capacitor in parallel. The SR-SAB can achieve soft-switching. The SR-SAB can also raise the value of the total power factor of the transformer because the SR-SAB can smoothly change a transformer current by using the LC resonance between the leakage inductor of the transformer and the resonant capacitor. The theory of a three-phase isolated SR-SAB DC-DC converter with a delta-star connected transformer is derived, and the effectiveness of the theory is verified by experiments.

Overview of Three-Phase Isolated DC-DC Converter

Circuit Configuration of a Conventional SAB Converter

Fig. 1(a) shows a conventional three-phase isolated SAB DC-DC converter with a delta-star connected transformer. The SAB is an isolated DC-DC converter. The primary circuit of the high frequency transformer T_r is composed of a three-phase inverter composed of six switches $R^+ - T^-$ with the soft-switching capacitors C_s in parallel. The secondary circuit is composed of a three-phase diode rectifier circuit composed of six diodes $U^+ - W^-$. When the input DC voltage V_{in} and the output DC voltage V_{out} are equal, transformer current i_u does not flow. Therefore, the SAB cannot send electric power to the load V_{out}. The SAB needs the higher input voltage V_{in} than the output voltage V_{out}, and then the value of the total power factor of the transformer becomes low.

Circuit Configuration of a Proposed SR-SAB Converter

Fig. 1(b) shows a proposed three-phase isolated SR-SAB DC-DC converter with a delta-star connected transformer. The primary circuit of the high frequency transformer T_r is the same configuration as the SAB in Fig. 1(a). The difference from the conventional circuit is the configuration of the secondary circuit. The secondary circuit is composed of a three-phase diode rectifier circuit composed of six diodes

$U^+ - W^-$ with the resonant capacitor C_r in parallel. To make V_{in} and V_{out} equal, the turn ratio of the transformer is set to 2:1. The proposed circuit can reduce switching loss by achieving soft-switching in the primary side. The proposed circuit can reduce a recovery loss on the diode because the recovery current does not flow in the secondary diode-bridge by the LC resonance. Since LC resonance occurs at the leakage inductance L of the high-frequency transformer T_r and the capacitor C_r, electric power can be sent even in situations when input and output voltages are equal. The total power factor of the transformer is also improved.

Advantages of a Proposed SR-SAB

Fig. 2 shows the switching patterns of the primary phases R, S and T and the voltage and current waveforms of the high frequency transformer. The waveforms were generated under the same output power, 3.0 kW. The simulation conditions of the waveforms are shown in Table I. The blue line shows waveforms of a conventional SAB converter, the red line shows waveforms of a proposed SR-SAB converter. The square voltage waveform v'_{rs} with voltage level 0, $\pm V_{in}/2$ is generated by using the switches of phases R and S are switched every 120 degrees and their duty ratios are 50 %. The primary voltage $v'_{rs}(= v_{rs}/2)$ is the secondary equivalent voltage of the primary voltage taking into account that the transformer turn ratio is 2:1. The secondary voltage v_{un} is the line voltage on the secondary side of the transformer. The voltage v_L is the voltage difference between the primary voltage v'_{rs} and the secondary voltage v_{un}, and the transformer current i_u flows when this voltage is applied to the transformer. The durations T_0 shown in Fig. 2 is the period when LC resonance is occurred in the phase R.

The waveforms of the SAB are shown in Fig. 2 as blue lines. The output DC voltage V_{out} is 265 V, and the input DC voltage V_{in} is raised to 360 V to obtain the output power P_{out}, 3.0 kW. The primary voltage v'_{rs} is square voltage waveform with voltage level 0, $\pm V_{in}/2$. The secondary voltage v_{un} is square voltage waveform with voltage level 0, $\pm V_{out}/3$, $\pm 2V_{out}/3$. The transformer current i_u becomes trapezoidal waveform.

The waveforms of the SR-SAB under the condition $V_{in} = V_{out}$ are shown in Fig. 2 with red lines. The primary voltage v'_{rs} is the square voltage waveform with voltage level 0, $\pm V_{in}/2$. During the period of T_0 shown in the Fig. 2, LC resonance is occurred and the secondary voltage v_{un} varies sinusoidally. In the period T_0, the reactor voltage v_L becomes large because the voltage difference between the primary and secondary voltages increases, and the transformer current i_u increases sinusoidally. The transformer current i_u of the SR-SAB becomes a near sinusoidal waveform.

The SR-SAB can make a phase difference be-

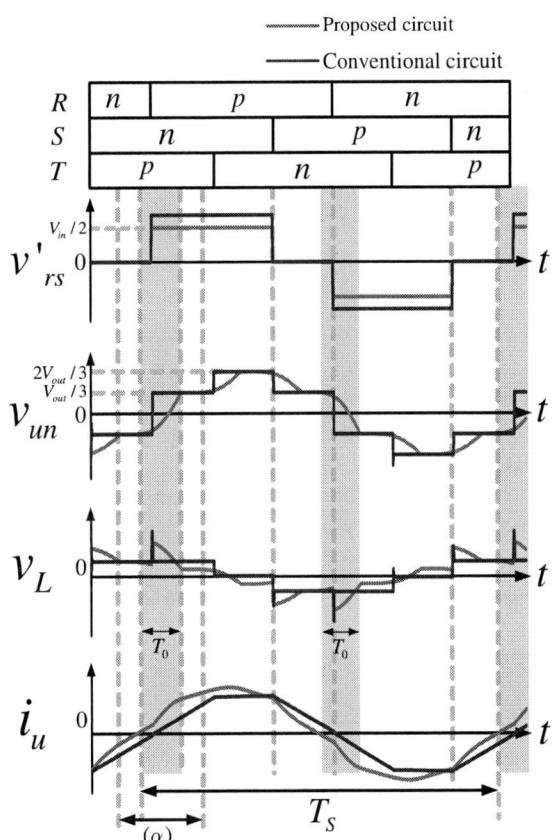

Fig. 2: Theoretical waveforms.

Table I: Simulation conditions.

Parameter	SAB	SR-SAB
Output power P_{out}	3.0 kW	
Output voltage V_{out}	265 V	
Input voltage V_{in}	360 V	265 V
DC capacitors C_i, C_o	1500 μF	
Resonant capacitor C_r	–	80 nF
Leakage inductance L	93 μH	
Frequency of transformer f_s	15 kHz	
Total Power Factor	0.78	0.92

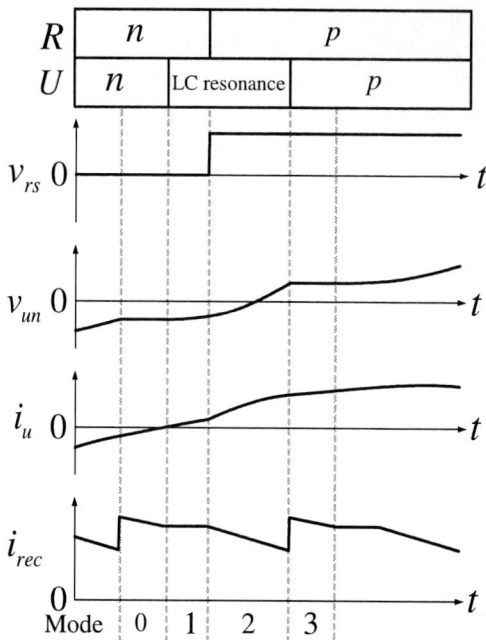

Fig. 3: Magnification waveforms of proposed circuit.

tween the primary voltage v'_{rs} and the secondary voltage v_{un} by using LC resonance. The waveform value of the reactor voltage v_L of the SR-SAB is higher than that of the SAB, especially during the LC resonance term. The transformer current i_u is increased significantly during the LC resonance term when the phase difference between the primary voltage v'_{rs} and the secondary voltage v_{un} becomes particularly large.

The total power factor of the transformer is obtained as (1) by using the value of the output power P_{out}, the effective value of the primary voltage V_{rsrms} and the effective value of the primary transformer current I_{rrms}. The total power factor under the simulation conditions is calculated to be 0.78 for the SAB and 0.92 for the SR-SAB. This is why the SR-SAB is better suited for high power conversion than the SAB.

$$TPF = \frac{P_{out}}{3 \times V_{rsrms} \times I_{rrms}} \qquad (1)$$

Switching Operation of the Proposed Circuit

The magnified waveforms of the theoretical waveforms in Fig. 2 are shown in Fig. 3. The period of (α) is divided into four operation modes Mode 0–Mode 3. The switching pattern of the primary phase R and the conduction pattern of the secondary phase U are also shown. The current i_{rec} is generated by being rectified the transformer current i_u, i_v and i_w. The switches of phase R are switched when shifting from Mode 1 to Mode 2. The diode of the phase U turns off when the operation mode changes from Mode 0 to Mode 1, then LC resonance occurs in Modes 1 and 2, and conduction resumes when Mode 3 begins. Thus, the switching timing of the switch on the primary side and the conduction timing of the diode on the secondary side are offset by caused LC resonance in the secondary side circuit. The voltage difference between the primary voltage v_{rs} and the secondary voltage v_{un} becomes larger in Mode 1 and Mode 2 by the secondary voltage v_{un} increased gradually by LC resonance, to avoid switching timing of the diode on the secondary side at the same timing as the switching of the switch on the primary side. In this way, the transformer current i_u is increased significantly. This is the reason that power can be converted even when the input DC voltage V_{in} and output DC voltage V_{out} are equal or when the input DC voltage V_{in} is lower. These characteristics indicate that the proposed the SR-SAB is more suitable for high power conversion than conventional the SAB.

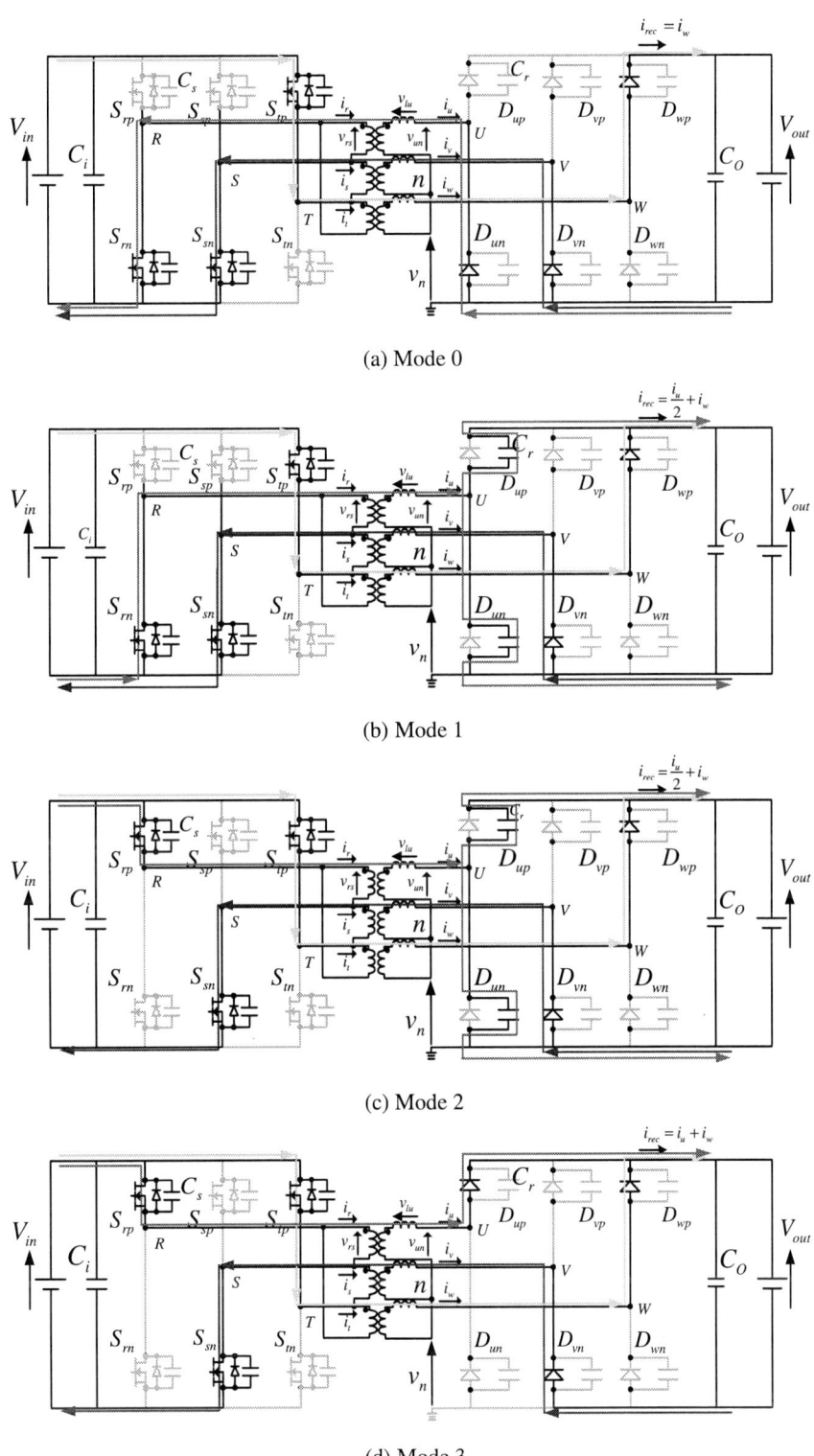

(a) Mode 0

(b) Mode 1

(c) Mode 2

(d) Mode 3

Fig. 4: Connection diagram of the proposed SR-SAB.

Theory of Secondary Converter

Fig. 4 shows the commutation modes from the diode U^- to U^+ on the secondary diode rectifier circuit. In Mode 0, the diode U^- is conducting, and after the resonance period of Mode 1 and Mode 2, the diode U^+ is conducting in Mode 3. Fig. 3 shows the magnification of the theoretical waveforms in Fig. 2. The diode U^- becomes on-state during Mode 0, and diodes U^- and U^+ become off-state during Mode 1, and the diode U^+ becomes on-state during Mode 3. In Mode 0, the primary voltage v'_{rs} is zero. As the diode U^- is on-state, the secondary voltage v_{un} is $-V_{out}/3$, and the transformer current i_u increases by the difference of the primary voltage v'_{rs} and the secondary voltage v_{un}. When the transformer current i_u becomes zero, the circuit becomes to Mode 1. In Mode 1, the diode U^- becomes off-state, and the LC resonance between the leakage inductor L and two capacitors C_r connected parallel to diodes U^+ and U^- occurs. The transformer current i_u increases from zero, and the secondary voltage v_{un} is the sinusoidal waveform in Mode 1. When the on-state switch of three-phase inverter of the primary side is switched from R^- to R^+ during Mode 1, the primary voltage v'_{rs} becomes $V_{out}/2$. Mode 1 is finished then. In Mode 2, the transformer current i_u continues increasing, and the secondary voltage v_{un} also increases in the sinusoidal waveform by the LC resonance like Mode 1. When the secondary voltage v_{un} becomes $V_{out}/3$, the diode U^+ becomes on-state, the circuit becomes to Mode 3. In Mode 3, the primary voltage v'_{rs} is $V_{out}/2$ and the secondary voltage v_{un} is $V_{out}/3$, and therefore the transformer current i_u increases. The following voltage equation of the transformer in Fig. 4 is obtained:

$$\begin{bmatrix} v'_{rs} \\ v'_{st} \\ v'_{tr} \end{bmatrix} = \frac{L}{3}\frac{d}{dt}\begin{bmatrix} i_u \\ i_v \\ i_w \end{bmatrix} + \begin{bmatrix} v_u \\ v_v \\ v_w \end{bmatrix} - \begin{bmatrix} v_n \\ v_n \\ v_n \end{bmatrix} \tag{2}$$

The voltage v_n is obtained in (3) by using the relations $i_u + i_v + i_w = 0$ and $v'_{rs} + v'_{st} + v'_{tr} = 0$ in (2):

$$v_n = \frac{1}{3}(v_u + v_v + v_w) \tag{3}$$

Mode 1

Fig. 4(b) shows the connection diagram of Mode 1. Under the condition that the time when the transformer current i_u becomes zero is defined as $t = 0$, and the time when the primary-side of the circuit switches during LC resonance is determined as T_1, the period from $t = 0 - T_1$ is Mode 1. When the transformer current i_u becomes zero, the diodes U^+ and U^- are turned off, and the current flows to the reactor L and the capacitors C_r equally. Therefore, the capacitor of the diode U^+ is discharged and the capacitor of the diode U^- is charged by half of the transformer current i_u. The voltage waveform and the current waveform becomes sinusoidal waveform by LC resonance. The voltage of the phase U v_u is obtained as follows:

$$v_u = \frac{1}{2C_r}\int i_u dt \tag{4}$$

The neutral point potential v_n is calculated in (5):

$$v_n = \frac{1}{3}V_{out} + \frac{1}{6C_r}\int i_u dt \tag{5}$$

The secondary interline voltage of the transformer in Fig. 4(b) is obtained from (4) and (5):

$$v_{un} = v_u - v_n = -\frac{1}{3}V_{out} + \frac{1}{6C_r}\int i_u dt \tag{6}$$

The voltage equation of the phase U of Mode 1 is obtained as follows:

$$\frac{1}{3}V_{out} = \frac{1}{3}L\frac{di_u}{dt} + \frac{1}{3C_r}\int i_u dt \qquad (7)$$

The transformer current i_u in Fig. 4(b) is calculated from (7):

$$i_u(t) = V_{out}\sqrt{\frac{C_r}{L}}\sin\frac{t}{\sqrt{LC_r}} \qquad (8)$$

The resonant frequency f_0 is obtained:

$$f_0 = \frac{1}{2\pi\sqrt{LC_r}} \qquad (9)$$

The transformer current in the phase where the LC resonance occurs becomes a sinusoidal waveform with the resonant frequency f_0. Substituting the transformer current i_u in (8) into (4), the voltage v_u is calculated as follows:

$$v_u(t) = \frac{1}{2}V_{out}\left(1 - \cos\frac{t}{\sqrt{LC_r}}\right) \qquad (10)$$

The secondary interline voltage of the transformer v_{un} in Fig. 4(b) is obtained from (7):

$$v_{un} = v_u - v_n = -\frac{1}{3}V_{out}\cos\frac{t}{\sqrt{LC_r}} \qquad (11)$$

The secondary interline voltage of the transformer v_{un} becomes sinusoidal waveform during LC resonance. Since $v_v = 0$ and $v_w = V_{out}$ in Mode 1, and at $t = 0$, $i_u = 0$, $i_v = -I_0$ and $i_w = I_0$, so the transformer currents i_v and i_w in Fig. 4(b) are obtained by (2):

$$i_v(t) = -I_0 - \frac{1}{2}i_u(t) \qquad (12)$$

$$i_w(t) = I_0 - \frac{1}{2}i_u(t) \qquad (13)$$

While the transformer current i_u increases, both transformer currents i_v and i_w decrease. The output current i_{rec} in Fig. 4(b) is obtained as follows:

$$i_{rec}(t) = \frac{1}{2}i_u(t) + i_w(t) = I_0 \qquad (14)$$

From (14), the output current i_{rec} becomes a constant value.

Mode 2

Fig. 4(c) shows the connection diagram of Mode 1. Mode 2 continues from the time the primary circuit switches during the LC resonance of Mode 1 to the time the diode U^+ conducts. The voltage and current waveform becomes sinusoidal by the LC resonant as the same as Mode 1. The transformer current i_u increases more significantly because the resonant current in Mode 2 is added to the resonance current in Mode 1. When the value of v_u equals to V_{out}, Mode 2 is finished. The voltage equation of phase U in Mode 2 is obtained as follows:

$$\frac{1}{6}V_{out} = \frac{L}{3}\frac{di_u}{dt} + \frac{1}{3C_r}\int i_u dt \tag{15}$$

The transformer current i_u in Fig. 4(c) is calculated by (15) as follows:

$$i_u(t) = i_{u1}(t) + i_{u2}(t) \tag{16}$$

$$i_{u1}(t) = V_{out}\sqrt{\frac{C_r}{L}}\sin\frac{t}{\sqrt{LC_r}} \tag{17}$$

$$i_{u2}(t) = \frac{3}{2}V_{out}\sqrt{\frac{C_r}{L}}\sin\frac{t-T_1}{\sqrt{LC_r}} \tag{18}$$

The first term $i_{u1}(t)$ on the right-hand side of (17) is the resonant current flows in Mode 1, and the second term $i_{u2}(t)$ is the resonant current which is appeared in Mode 2. The frequency of the resonant current $i_{u2}(t)$ is f_0. Since the resonant current $i_{u2}(t)$ is 1.5 times larger in amplitude than $i_{u1}(t)$, the resonant current $i_u(t)$ increases significantly in Mode 2 compared to Mode 1. The voltage v_u of the midpoint voltage of phase U and the interline voltage of the secondary side v_{un} in Fig. 4(c) are obtained by the following equations:

$$v_u = v_u(T_1) + \frac{1}{2C_r}\int i_u dt = V_{out}\left(\frac{5}{4} - \frac{1}{2}\cos\frac{t}{\sqrt{LC_r}} - \frac{3}{4}\cos\frac{t-T_1}{\sqrt{LC_r}}\right) \tag{19}$$

$$v_{un} = v_u - v_n = -\frac{1}{3}V_{out}\cos\frac{t}{\sqrt{LC_r}} + \frac{1}{2}V_{out}\left(1 - \cos\frac{t-T_1}{\sqrt{LC_r}}\right) \tag{20}$$

The voltage v_u and v_{un} increase more significantly than Mode 1. When the phase U voltage v_u satisfies the rerlation $v_u = V_{out}$, the diode U^+ conducts and Mode 2 finishes. The following equation is obtained from the relation of $v_u(T_1 + T_2) = V_{out}$, the duration of Mode 2 is T_2 :

$$2\cos\frac{T_1+T_2}{\sqrt{LC_r}} + 3\cos\frac{T_2}{\sqrt{LC_r}} = 1 \tag{21}$$

The transformer current i_v and i_w in Fig. 4(c) is obtained from the (2) as follows:

$$i_v(t) = -I_0 + \frac{3V_{out}}{4L}(t - T_1) - \frac{1}{2}i_u(t) \tag{22}$$

$$i_w(t) = I_0 - \frac{3V_{out}}{4L}(t - T_1) - \frac{1}{2}i_u(t) \tag{23}$$

The transformer current i_v and i_w vary with the half amplitude of the resonant current i_u. The output current i_{rec} in Fig. 4(c) is obtained:

$$i_{rec}(t) = \frac{i_u}{2} + i_w = I_0 - \frac{3V_{out}}{4L}(t - T_1) \tag{24}$$

The output current i_{rec} in Mode 2 decreases at a constant rate with a slope $3V_{out}/4L$.

Mode 3

Fig. 4(d) shows the connection diagram of Mode 1. In Mode 3, the transformer currents are $i_u > 0$, $i_v < 0$, $i_w > 0$, so diodes U^+, V^-, W^+ are on state. The voltage v_u between the diodes U^+ and U^- becomes $v_u = V_{out}$. In the same way, the voltages v_v and v_w are $v_v = 0$ and $v_w = V_{out}$. From (4), the neutral point potential v_n is obtained as $v_n = 2V_{out}/3$. From the voltage equation (2), the transformer currents $i_u - i_w$

are obtained as follows by using the initial current values $i_u(T_1 + T_2)$, $i_v(T_1 + T_2)$, $i_w(T_1 + T_2)$:

$$
\begin{cases}
i_u(t) = i_u(T_1 + T_2) + \dfrac{V_{out}}{2L}(t - T_1 - T_2) \\[4mm]
i_v(t) = i_v(T_1 + T_2) + \dfrac{V_{out}}{2L}(t - T_1 - T_2) \\[4mm]
i_w(t) = i_w(T_1 + T_2) - \dfrac{V_{out}}{L}(t - T_1 - T_2)
\end{cases}
\tag{25}
$$

From (25), the transformer currents i_u and i_v in Fig. 4(d) increase at a constant slope $V_{out}/2L$ and the transformer current i_w decreases at a constant slope V_{out}/L. Mode 3 is finished when the transformer current i_w becomes zero. The following relation stands for the initial current of Mode 1 I_0:

$$
I_0 = i_u(T_1 + T_2 + T_3) = -i_v(T_1 + T_2 + T_3)
\tag{26}
$$

The initial current value of Mode 1 I_0 is obtained as follows:

$$
I_0 = \frac{3V_{out}}{2L}(T_2 + T_3)
\tag{27}
$$

Using the equations $T_1 + T_2 + T_3 = T_s/6$ and $T_s = 1/f_s$, the initial value I_0 is obtained as follows:

$$
I_0 = \frac{3V_{out}}{2L}\left(\frac{1}{6f_s} - T_1\right)
\tag{28}
$$

The output current i_{rec} in Fig. 4(c) is calculated by $i_{rec} = i_u + i_w$ as follows:

$$
\begin{aligned}
i_{rec} &= i_u(T_1 + T_2) + i_w(T_1 + T_2) - \frac{V_{out}}{2L}(t - T_1 - T_2) \\[2mm]
&= \frac{3V_{out}}{2L}(T_2 + T_3) - \frac{3V_{out}}{4L}T_2 + \frac{1}{2}V_{out}\sqrt{\frac{C_r}{L}}\sin\frac{T_1 + T_2}{\sqrt{LC_r}} \\[2mm]
&\quad + \frac{3}{4}V_{out}\sqrt{\frac{C_r}{L}}\sin\frac{T_2}{\sqrt{LC_r}} - \frac{V_{out}}{2L}(t - T_1 - T_2)
\end{aligned}
\tag{29}
$$

In Mode 3, the output current i_{rec} in Mode 3 decreases with a constant slope $V_{out}/2L$.

Experimental Results

Table II shows the experimental conditions. Fig. 5 shows the experimental waveforms under the condition of Table II. The waveforms of the primary voltage v_{rs} and the secondary voltage v_{un} are shown in Fig. 5. The waveform of the transformer current i_u is shown in Fig. 5.

The theoretical waveforms are obtained by experiments. The input power P_{in} and the output power P_{out} are measured by power analyzer WT1600, YOKOGAWA. The input power is 2.6 kW and the output power is 2.5 kW. The measured efficiency of the circuit is 95.3 %.

Table II: Experimental conditions.

Parameter	Value
Output power P_{out}	2.5 kW
Input voltage V_{in}, Output voltage V_{out}	265 V
DC capacitors C_i, C_o	1500 μF
Resonant capacitor C_r	80 nF
Leakage inductance L	93 μH
Turn ratio of the transformer	2:1
Frequency of transformer f_s	15 kHz

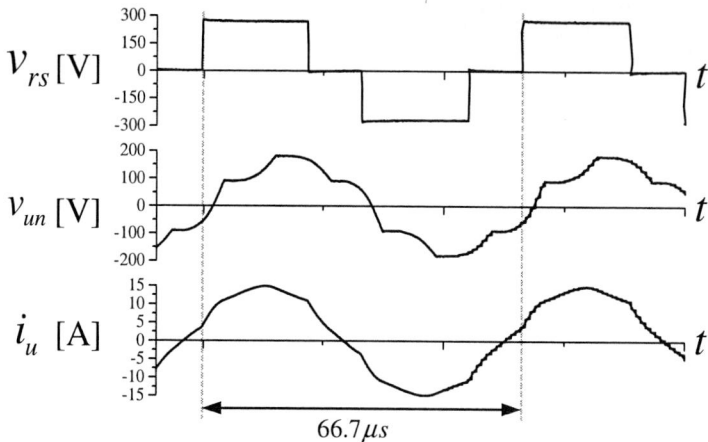

Fig. 5: Experimental waveforms. ($V_{in} = V_{out} = 265$ V, $f_s = 15$ kHz)

Conclusions

This paper presents the three-phase isolated SR-SAB DC-DC converter with a delta-star connected transformer. The advantages of the SR-SAB are clarified by comparing with the conventional SAB. The proposed SR-SAB converter can achieve high-efficiency compared with the conventional SAB by improving the value of the total power factor, and reducing the recovery loss of the diodes by the LC resonance between the leakage inductor and the capacitor of the diode rectifier circuit. The effectiveness of the theory is verified by experiments.

References

[1] R. W. A. A. De Doncker, D. M. Divan and M. H. Kheraluwala, "A three-phase soft-switched high-power-density DC/DC converter for high-power applications," *in IEEE Transactions on Industry Applications*, vol. 27, no. 1, pp. 63-73, Jan.-Feb. 1991

[2] C. Meyer and R. W. De Doncker, "Design of a Three-Phase Series Resonant Converter for Offshore DC Grids," *2007 IEEE Industry Applications Annual Meeting*, 2007, pp. 216-223

[3] Tuan, C.A.; Takeshita,T. "Analysis of Unidirectional Secondary Resonant Single Active Bridge DC-DC Converter," *Energies 2021*, 14, 6349, 2022, p.14

[4] Cao Anh Tuan, Takaharu Takeshita, "Output Power Characteristics of Unidirectional Secondary-Resonant Single-Active-Bridge DC-DC Converter using Pulse Width Control," *IEEJ Journal of Industry Applications*, Volume 11, Issue 2, 2022, pp. 359-368

[5] Y. Sang, A. Junyent-Ferré and T. C. Green, "Operational Principles of Three-Phase Single Active Bridge DC/DC Converters Under Duty Cycle Control," *in IEEE Transactions on Power Electronics*, vol. 35, no. 8, pp. 8737-8750, Aug. 2020

[6] Yeh Ting, S. de Haan and J. A. Ferreira, "A DC-DC Full-Bridge Hybrid Series Resonant Converter enabling constant switching frequency across wide load range," *Proceedings of The 7th International Power Electronics and Motion Control Conference*, 2012, pp. 1143-1150

[7] M. Stieneker and R. W. De Doncker, "Dual-active bridge dc-dc converter systems for medium-voltage DC distribution grids," *2015 IEEE 13th Brazilian Power Electronics Conference and 1st Southern Power Electronics Conference (COBEP/SPEC)*, 2015, pp. 1-6

[8] J. Jacobs, A. Averberg and R. De Doncker, "Multi-Phase Series Resonant DC-to-DC Converters: Stationary Investigations," *2005 IEEE 36th Power Electronics Specialists Conference*, 2005

[9] C. Sommer, A. Mertens, I. Larrazabal and I. Kortazar, "Analytical investigation of the three-phase single active bridge for offshore applications," *2016 18th European Conference on Power Electronics and Applications (EPE'16 ECCE Europe)*, 2016, pp. 1-10

[10] Y. Ting, S. de Haan and J. A. Ferreira, "The partial-resonant single active bridge DC-DC converter for conduction losses reduction in the single active bridge," *2013 IEEE ECCE Asia Downunder*, 2013, pp. 987-993

[11] J. Jacobs, M. Thommes and R. De Doncker, "A transformer comparison for three-phase single active bridges," *2005 European Conference on Power Electronics and Applications*, 2005, pp. 10 pp.-P.10

A Novel Concept to Optimize Core Loss in Planar Magnetic Based on an Unbalanced-Flux- Approach

Sobhi Barg[1], Kent Bertilsson[1], Grover Torrico[2]
[1]Mid Sweden University
[2]Huawei R&D Sweden
Sundsvall, Sweden
Sobhi.barg@miun.se, kent.bertilsson@miun.se, Grover.Torrico@huawei.com

Acknowledgements

The authors gratefully acknowledge the Swedish Energy Agency and Vinnova for the financial support of this work.

Keywords

« unbalanced-flux », « balanced-flux », « core loss », « magnetic power factor », « safe operating area».

Abstract

This paper presents a new method to design planar magnetics. Unlike existing magnetics which have a balanced-flux distribution, the   proposed method is based on the principle of unbalanced-flux distribution. The Steinmetz model, derived for this design principle, shows that the unbalanced-flux method reduces the core loss by more than 50%. The core loss reduction brings several benefits to planar magnetics such as: high magnetic power factor, better thermal performance and larger safe operating area (SOA). The proposed method is experimentally evaluated and compared with the balanced-flux method. The obtained results confirmed the advantages of the unbalanced-flux method found from the theoretical study. The core loss is decreased by more than 50%, the magnetic power factor is increased by 73% and the SOA is much larger.

Introduction

In recent years, the use of planar magnetics in switching mode power supplies (SMPS) has witnessed a sharp increase thanks to their advantages compared to high profile magnetics. These advantages are centered on two main features: higher thermal performance and lower manufacturing cost in mass production [1]. The emerging of the Wide-band-Gap devises and the natural soft switching of the resonant converters such as LLC and CLLC converters enable to push the switching frequency up to MHz range, which becomes as an indispensable solution to increase the power density and the efficiency of planar magnetics. Certainly, the magnetic components is still considered as the main obstacle to achieve high power density and efficiency for SMPS. To address the matter, several techniques such as matrix transformers, integrated magnetics, fractional turn, flux cancellation, interleaved windings, single and multiphase circuits have been tested at MHz switching frequency. Numerous studies that have been carried out argued that the power density and the efficiency has improved because of increasing the switching frequency to MHz range [2-5]. Undoubtedly, the overall power density and efficiency of the converter has increased. Nevertheless, despite this improvement, the existing works lacks the scientific evidences which prove that the choice of the MHz is what contributes to the increase of the power density and the efficiency of planar magnetics from one side, and that this achievement is not possible at few hundreds frequency range from another side. To briefly assess the MHz solution, we can consider the 1kW-380/12 V LLC converter, which is widely assessed in academic and industrial research. The volume of the realized magnetics in this circuit varies between 11 and 14 cm^3 (Tab. I). According to [6], the same power could be carried out with an ETD39 (13 cm^3) core at 300 kHz in a full-bridge and half-bridge converters. With the assumption that the transformer design for LLC and full-bridge converters is principally the same, several questions prop up to the surface about the efficacy of the MHz switching frequency. The evidence lies in the fact that the increase in the power density and the efficiency was achieved thanks to GaN features (low volume, high efficiency) and not to the magnetic devices. This can be understood by calculating the magnetics-to-converter volume ratio (MCVR) for both kinds of design: the few 100s kHz design and the MHz design. The common estimated MCVR for the few 100s kHz design varies from 30% to 50%, but, it ranges between 50% and 75% for the MHz design (see Tab. I, Fig. 1 and Fig. 2) [2-5].

Fig. 1. Planar magnetics size for 1 kW-380/12V LLC converter [2].

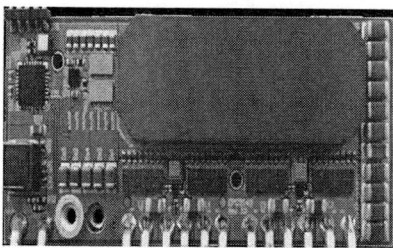

Fig. 2.Magnetics size in EPC 9149 kit [5].

TABLE. I KEY DATA OF SOME BUILT LLC CONVERTERS AT 1 MHZ

References	[2]	[3]	[4]
Converter power [W]	800	1000	1000
Frequency [MHz]	1	1	1
Magnetics box volume [cm^3]	11	12.44	14.2
Converter Power density [W/cm^3]	54.9	37.6	-
Magnetics density [W/cm^3]	72.72	80.38	70.4
MCVR	0.75	0.47	-

To sum up, increasing the switching frequency is generally beneficial to reduce the core volume, but at the same time, it leads to decrease the efficiency and the power capability as it will be shown in section II. The power capability can be characterized by the magnetic power factor (MPF), defined by $MPF = f\,B_m^2$. The MPF characterizes the maximum power capability of a magnetic core under thermal constraint for a given f and B_m. plotting the MPF of a given core in the (B, f) space with consideration of the thermal and saturation constraints determines its safe operating area (SOA).

The designer needs to have the necessary knowledge about the frequency effect on the core loss and the MPF of the magnetic core in order to optimize his design.

One fundamental principle of the existing magnetics is they have a balanced-flux distribution within the full core length. From our perspectives, this is the main obstacle that limits to improve the magnetics performance, which may ultimately lead to increase the core loss. Definitely, core loss is a detrimental factor to the efficiency and the power density of the magnetic components.

Unlike the balanced-flux approach and the existing design methods, a new method based on the principle of unbalanced-flux distribution is developed. The suggested method helps to reduce significantly the core loss, which in turns increases the power density, the efficiency, the MPF and the thermal performance of the magnetic core.

This paper is structured as follows: section II analyzes the limitations of the MHz approach and presents a new tool to optimize the (f, B) operating region of a given magnetic core. The unbalanced-flux method is developed in section III. The experimental part and the verification with the theoretical results is presented in section IV. Finally, the paper's major contributions are dealt with in the conclusion.

Principle and analysis of the Unbalanced-flux approach

The principle of the unbalanced-flux design is to generate unequal magnetic flux density between both: the central and the outer parts of the magnetic core. The central part is the central leg around which, the coil is wound to generate the required magnetic flux density. The outer part includes all remaining sections to close the magnetic path. In the balanced-flux design, the magnetic flux density has an approximate uniform density within the full core. One single condition for such design is that the cross section of the outer part is half the one of the central part (Fig. 3). This condition is not applied in the unbalanced-flux method. An example of the unbalanced-flux approach, where the outer and the central cross section are equal, is shown in Fig. 4.

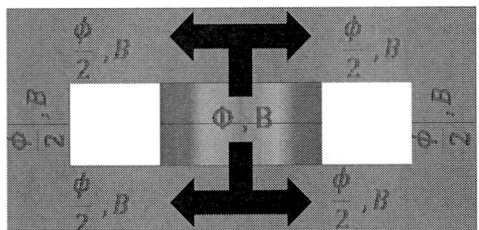

Fig. 3. Φ and B within central and outer parts for a balanced-flux design ($A_o=A_c/2$).

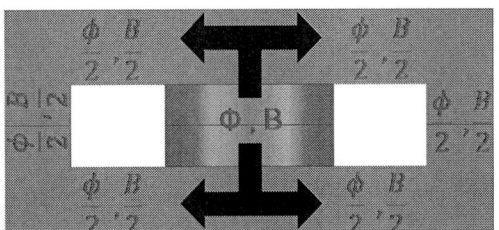

Fig. 4. Φ and B within central and outer parts for unbalanced-flux design, case ($A_o=A_c$).

One important characteristic of planar magnetics that has a significant influence on the proposed method, is the volume ratio between the outer-part volume and the total volume. This ratio is denoted as δ and expressed as follows:

$$\delta = \frac{V_{co}}{V_{c\_b}} \tag{1}$$

Where V_{c_b} and V_{co} are the total core volume and the outer-part volume respectively of a balanced-flux-based magnetic core. δ depends on the core dimensions. The expression of δ for a typical ETD shape planar core as function of the core dimensions is given by (2) and its variation with respect to the ratio between the window-width and the window-height is shown in Fig. 5. Typical planar cores usually have f/h higher than 2 and Fig.5 clearly shows that δ is higher than 80%.

$$\delta = \frac{V_{co}}{V_{c\_b}} = \frac{h+2f+d}{2\left[h+f+\frac{d}{2}\right]} \tag{2}$$

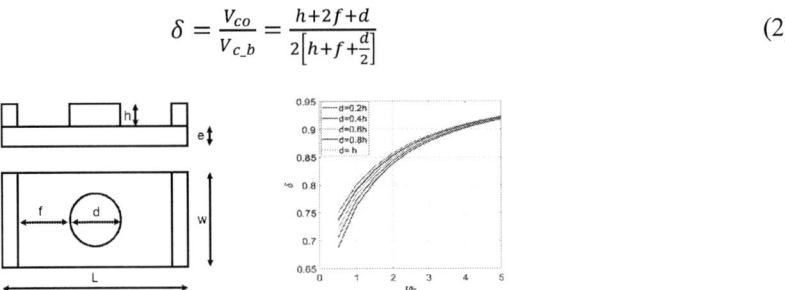

Fig. 5. Core dimensions and variation of δ with respect of f/h and for different d/h.

The Steinmetz equation is a common model to calculate the core loss [7-8].

$$P_{c\_b} = k\, f^{\alpha} B_m{}^{\beta}\, V_{c\_b} \tag{3}$$

B_m is the peak magnetic flux density, (k, α, β) are the Steinmetz parameters. This model could not be applied to the unbalanced-flux design method because the magnetic flux density is not uniform. Thus, it is required to use the superposition technique through which the core loss is calculated for each part and then sum up all the loss terms [9]. Consequently, the total core loss can be expressed as follows:

$$P_{c\_unb} = P_{cc} + P_{co} = k\,f^\alpha B_m{}^\beta\,\delta\,V_{c\_b} + k\,f^\alpha B_{mo}{}^\beta\,V_{o\_unb} \tag{4}$$

where P_{c_unb}, P_{cc}, P_{co} are the core loss of: the full core, the central part and the outer part respectively.

The relationship between the magnetic flux density and the cross section of the central and outer parts is given as follows:

$$\emptyset_{mo} = \frac{1}{2}\emptyset_{mc} \tag{5}$$

$$\frac{B_{mo}}{B_m} = \frac{1}{2}\frac{A_c}{A_o} \tag{6}$$

In the unbalanced-flux design, the volume of the outer part increases by a factor of $\left(\frac{2A_o}{A_c}\right)$ compared to the balanced-flux design. Using (1), V_{o_unb} can be written by the following equation:

$$V_{o\_unb} = \delta\left(\frac{2A_o}{A_c}\right)V_{c\_b} \tag{7}$$

Substituting (1), (6) and (7) in (4), we get the core loss of the unbalanced-flux design:

$$P_{c\_unb} = k\,f^\alpha B_m{}^\beta\left[\left(\left(\frac{A_c}{2\,A_o}\right)^{\beta-1} - 1\right)\delta + 1\right]V_{c\_b} \tag{8}$$

The core loss ratio (R_L) between the unbalanced and the balanced flux designs:

$$R_L = \frac{P_{c\_unb}}{P_{c\_b}} = \left(\left(\frac{A_c}{2\,A_o}\right)^{\beta-1} - 1\right)\delta + 1 \tag{9}$$

Fig.6 shows the evolution of R_L with respect to A_c/A_o for different cases of β. As it can be seen, the core loss can be substantially decreased by reducing A_c/A_o. Obviously, the case ($A_c/A_o = 2$) corresponds to the balanced-flux design where R_L is unit. For $A_c = A_o$, the core loss can be reduced by about 50%. R_L can reach up to 30% for A_c/A_o equals to 0.5. It can also be seen that magnetic materials with high β achieve high R_L. This means, if we consider two balanced-flux magnetic cores (core (X) and core (Y)) of two different materials having respectively β_1 and β_2, where β_1 is bigger than β_2 and let's suppose core (1) has higher core loss. Then, core (1) could have lower loss than core (2) in the unbalanced-flux design as shown in equation (10).

$$R_{Lx} = \frac{P_{c\_unb}(A)}{P_{c\_unb}(B)} = C_b\,\frac{\left(\left(\frac{A_c}{2\,A_o}\right)^{\beta_1-1} - 1\right)\delta + 1}{\left(\left(\frac{A_c}{2\,A_o}\right)^{\beta_2-1} - 1\right)\delta + 1} \tag{10}$$

Where $C_b = \frac{P_{c\_b}(A)}{P_{c\_b}(B)}$.

Fig.7 shows the loss ratio R_{Lx} (10) between core (X) and core (Y) for the case (β_1=2.3 and β_2=2) in the unbalanced-flux design. As an example, for C_b=1.1, core (Y) starts to have lower core loss than core (X) in the unbalanced-flux design when A_c/A_o is lower than 1.3.

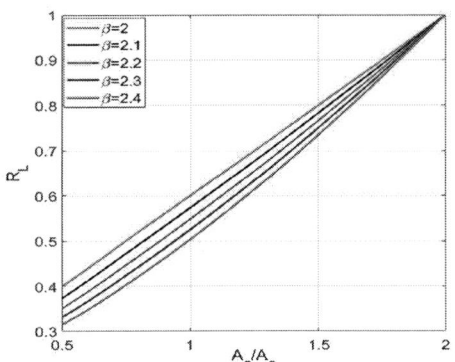
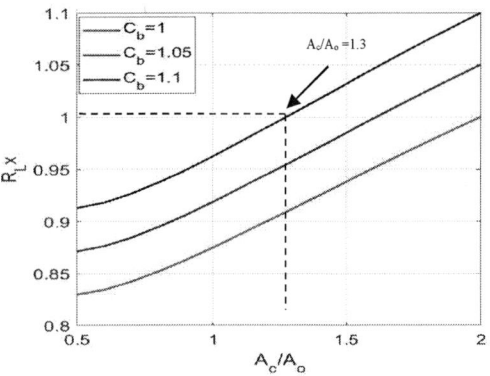

Fig. 6. Variation of R_L with respect to Ac/Ao for different β, δ=0.8.

Fig. 7. Variation of R_{Lx} with respect to Ac/Ao for different C_b, δ=0.8, $β_1$=2.3 and $β_2$=2.

One disadvantage of the unbalanced flux design is the increase of the core volume. The volume-increase-factor (R_V) is expressed by (11) and its variation as function of A_c/A_o is depicted in Fig.8. For instance, it is about 80% for equal A_c and A_o.

$$R_V = \frac{V_{c\_unb}}{V_{c\_b}} = \left(\left(\frac{2A_o}{A_c}\right) - 1\right)\delta + 1 \tag{11}$$

While the increase in the volume is detrimental to the power density, it is advantageous to the core loss density which is beneficial to the thermal performance. It is written as follow:

$$P_{c\_u} = \frac{P_{c\_unb}}{V_{c\_unb}} = k\,f^\alpha\,B_m{}^\beta\,\frac{\left[\left(\left(\frac{A_c}{2A_o}\right)^{\beta-1} - 1\right)\delta + 1\right]}{\left(\left(\frac{2A_o}{A_c}\right) - 1\right)\delta + 1} \tag{12}$$

The core loss density ratio between the balanced and the unbalanced flux designs (R_{cu}) is given by:

$$R_{cu} = \frac{P_{c\_u}(unbalanced)}{P_{c\_u}(balanced)} = \frac{\left[\left(\left(\frac{A_c}{2A_r}\right)^{\beta-1} - 1\right)\delta + 1\right]}{\left(\left(\frac{2A_r}{A_c}\right) - 1\right)\delta + 1} \tag{13}$$

Despite the increase in the core volume, the core loss density has significantly decreased as given by (13) and Fig. 9. As an example, R_{cu} decreases by about 70% for A_c/A_o equals to 1. The core loss density reduction has a huge effect on improving the thermal performance of the magnetic core. Eventually, it enables to extend the (SOA) in comparison to the balanced flux design.

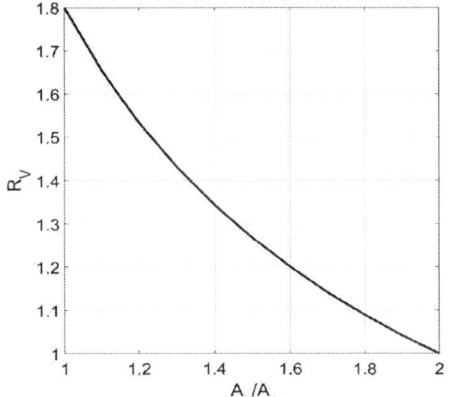
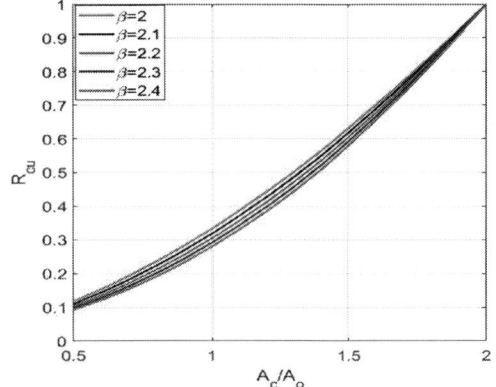

Fig. 8. Variation of R_V with respect to Ac/Ao, δ=0.8.

Fig. 9. Variation of R_{cu} R_{cu} with respect to Ac/Ao for different β, δ=0.8.

The proposed method enables to significantly reduce the core loss in comparison to the balanced-flux design. The reduction factor depends of the ratio between the cross section of the central and the outer parts. It is about 50% when both cross sections are designed equally. The loss reduction also depends on β. magnetic materials with high β have more capability to reduce the core loss. Despite the volume increase in the proposed design, the core loss density has significantly decreased. It brings several important advantages such as high power capability, better thermal performance and wider SOA. These advantages will be experimentally evaluated in the following section.

Experimental verification and discussion

In this section, the evaluation of the proposed method is established. To verify the proposed magnetics design, two unbalanced-flux magnetic cores are built (core2 and core3) and their performances are compared to a balanced-flux magnetic core (core1) (Fig. 10). The details of the tested cores dimensions are given in Tab. II and its magnetic material is DMR96 [9]. To achieve a reasonable comparison of the proposed method, we have designed core2 with nearly equal volume to core1 while core3 has twice the volume of core1. The shape of Core1 was chosen in a way it has better thermal performance compared to core2 and core3 as explained in the following remarks:

•Core1 and core 2 have approximately same volume, however core1 has larger base surface and slightly lower height.

•Core3 is a two integrated transformers. It has double the core volume to core1 but its base surface is only 1.7 time the base surface of core1.

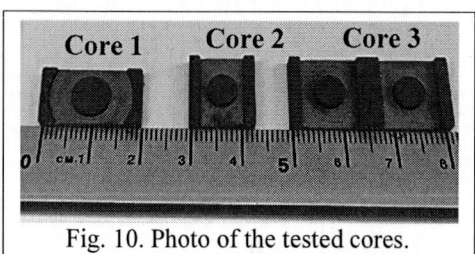

Fig. 10. Photo of the tested cores.

TABLE. II GEOMETRY CHARACTERISTICS OF THE TESTED MAGNETIC CORES

	Core 1		Core 2	Core 3
Flux	balanced		unbalanced	
Volume(mm^3)	1284		1114	2254
A_c (mm^2)	47.8		28.3	2*28.3
A_o (mm^2)	24.75		28.3	2*28.3
δ	0.9		0.97	0.97
Base area (mm^2)	226.5		192.7	384

The comparison criteria of this study are: core loss, temperature rise, the MPF and the SOA. For each core, we have applied the same magnetic flux density at a specific frequency and we measure its core loss and temperature using a fast calorimetric measurement technique. This measurement technique was used to measure switching devices and magnetic core loss and the reader is referred to [10-12] for more details about its principle. In the following, we discuss the comparison results:

Core loss and temperature evaluation

• *Core1 VS Core2*: though core2 has nearly same volume as core1, its core loss is much smaller as shown in Fig.11. It can also be noticed that core2 has also much lower temperature. Fig.12 (a) shows that the core loss density of the balanced flux design is 2 to 3.5 times the core loss density of the unbalanced-flux design over the flux density range [0.1-0.25] T.

Concerning the thermal behavior, it is clear that the core2 has achieved much better thermal performance than core1. For instance, in case (f) for B=0.15 T, the registered temperature for core2 is 43.2°C, however it peaks at 78.2°C for core1 (Fig.11). It should be clearly noticed that the main reason, for the low temperature registered for core2, is the low core loss and certainly is not because of having lower thermal resistance. This can be synthesized by comparing the temperature of both cores at equal core loss. For instance, we can take the case (0.15T, 650 kHz) for core2 and the case (0.1T, 650 kHz) for core1, in which both cores have approximately the same core loss and nearly same temperature.

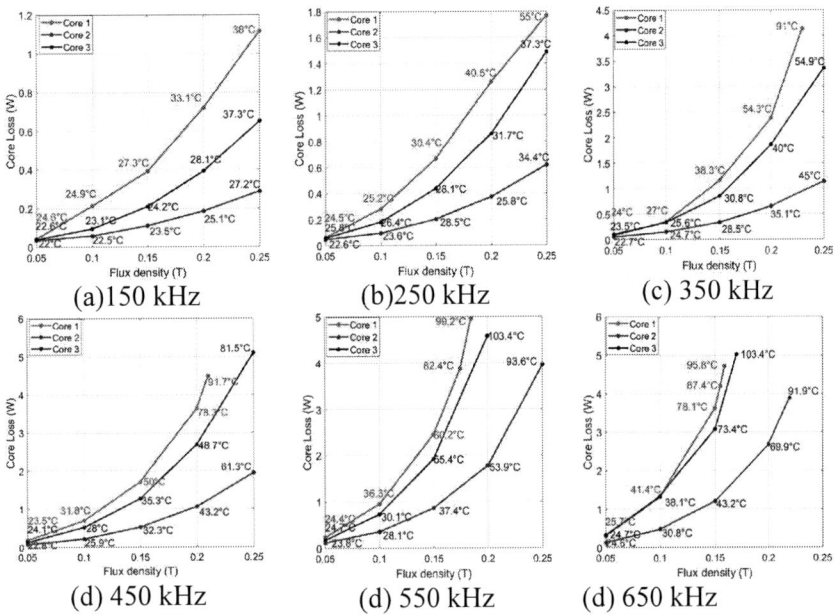

Fig. 11. Measured core loss and temperature of the tested cores.

• *Core1 VS Core3*: in general, core3 has achieved lower core loss and lower temperature rise despite its volume is twice the volume of core1. Fig.12 (b) shows that the core loss density ratio between the balanced and the unbalanced flux designs varies between 2 and 3.5 times in the interval [0.1-0.25] T. Three exceptions are registered for 0.1T at 650, 150 and 350 kHz, which could be due to some measurement errors. It can also be seen that core has achieved lower temperature than core1 at same core loss. As an example, at (0.1 T, 650 kHz), the core loss are same, but the temperature for core1 and core3 are 41.4 °C and 38.1 °C respectively. This means that core3 has lower effective thermal resistance than core1 despite that core1 has a higher outer surface-to-volume ratio.

In the previous comparison cases, the registered core loss density ratio, at 0.05 T, is smaller than 2. This is due to the core loss behavior at low temperature and low flux density. We can also notice that the loss density ratio depends on B and f. This aspect is very important and will be furtherly investigated in future works.

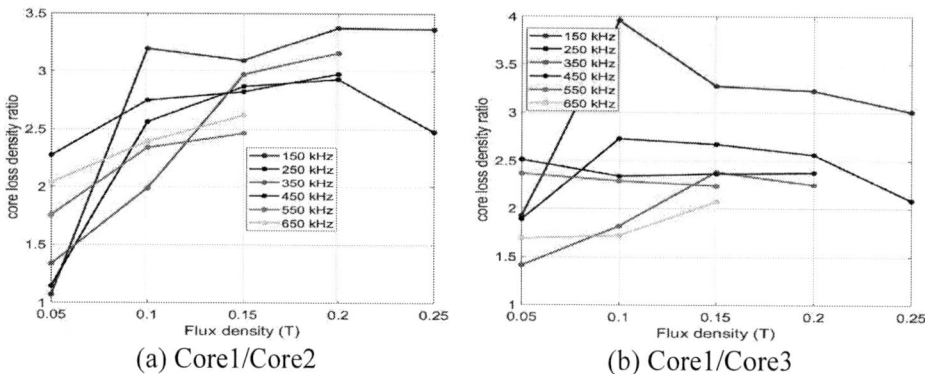

(a) Core1/Core2   (b) Core1/Core3

Fig. 12. Core loss density ratio between balanced and unbalanced flux designs

MPF and SOA evaluation

The core loss for core 1, core2 and core3 are measured over the frequency range [0.15-0.65] MHz and the corresponding MPF and SOA are given in Fig.13 and Fig.14. The obtained results show that the unbalanced-flux method enables to increase significantly the MPF. The MPF_{max} of core2 and core3 peaked at 34.4 $MHz.B^2$ and 28.1 $MHz.B^2$ respectively compared to 19.8 $MHz.B^2$ for core1. This means that core2 has 73% more power capability than core1 despite it has lower base surface and slightly lower volume. Additionally, despite the previous advantages accounted for core1, core2 has achieved much better performance such as, lower core loss, higher MPF and larger SOA. Fig.13 shows the limits of the SOA for core2 and core1. Similarly, Fig.14 shows the SOA for core3 to core1. As it can be seen, the SOA has been considerably increased which gives the designer more freedom to select the suitable combination of (B, f) in order to meet with the design requirements. Higher MPF and larger SOA definitely enable to reduce the number of turns which in turns can bring down the winding loss.

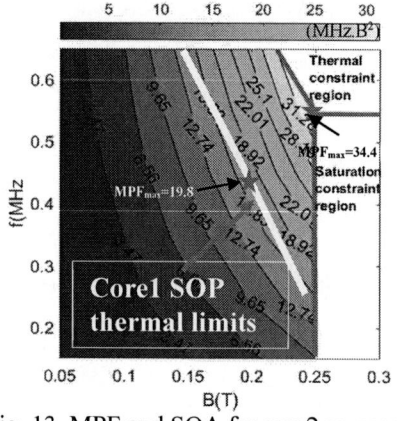
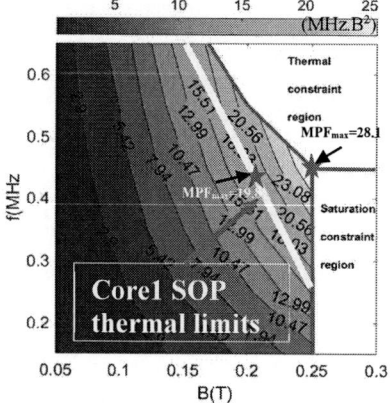

Fig. 13. MPF and SOA for core2 vs core1.   Fig. 14. MPF and SOA for core3 vs core1.

Conclusion

This paper has introduced a new design method based on the principle of the unbalanced-flux distribution in order to improve the power density and the efficiency of planar magnetics. Although these objectives are not dealt with in this work, the benefits offered by the proposed method such as low core loss, better thermal performance, high magnetic power factor (MPF) and larger safe operating area can only help to boost the efficiency and the power density. The theoretical and the experimental results, performed on the tested cores, have shown that core loss can be minimized by more than 50%, which in turn helps to improve the MPF by 73% and to enlarge significantly the SOP and the thermal performance. This study has opened a new direction in the design of magnetic components, however, more results and investigations need to be addressed. It will concern the optimization approach, the thermal modeling and more importantly its applications on the SMPS and how it affects the efficiency and the power density compared to the literature.

References

[1] Z. Ouyang, M. A. E. Andersen, "Overview of Planar Magnetic Technology—Fundamental Properties" *IEEE Transactions On Power Electronics, Vol. 29, No. 9, September 2014.*

[2] C. Fei, F. C. Lee and Q. Li, "High-Efficiency High-Power-Density LLC Converter With an Integrated Planar Matrix Transformer for High-Output Current Applications" *IEEE Transactions On Industrial Electronics, Vol. 64, No. 11, November 2017.*

[3] Y. Liu, K. Chen, C. Chen, Y. Syu, *IEEE*, Guan-Wei Lin, Katherine A. Kim, and Huang-Jen Chiu, "Quarter-Turn Transformer Design and Optimization for High Power Density 1-MHz *LLC* Resonant Converter," *IEEE Transactions On Industrial Electronics, Vol. 67, No. 2, February 2020.*

[4] M. K. Ranjram, and D. J. Perreault, "A 380-12 V, 1-kW, 1-MHz Converter Using a Miniaturized Split-Phase, Fractional-Turn Planar Transformer" *IEEE Transactions On Power Electronics, Vol. 37, No. 2, February 2022.*

[5] "EPC9149KIT-36-60 V Input, 9 - 15 V Output,83 A Output Fixed Conversion Ratio 1 kW LLC,1/8th Brick size Module" available on: *https://epc-co.com/epc/Products/DemoBoards/EPC9149.aspx*

[6] A. I. Pressmann, K. L. Billings, T. Morey, "Switchmode power supply design," *McGraw-Hill 3rd edition, 2009.*

[7] S. Barg, M. Hanen, K. Ammous, A. Ammous, "An Improved Empirical Formulation for Magnetic Core Losses Estimation Under Nonsinusoidal Induction," *IEEE Transactions On Power Electronics, Vol. 32, No. 3, March 2017*

[8] C. P. Steinmetz, "On the law of hysteresis," *AIEE Trans.*, vol. 9, pp. 3–64, 1892. *Reprinted under the title "A Steinmetz contribution to the ac power   evolution," Introduction by J. E. Brittain. Proc. IEEE,vol. 72, no. 2, pp. 196–221, 1984.*

[9] DMR96 datasheet, available: http://www.chinadmegc.com/

[10] S.Barg, F. Alam, K. Bertilsson, "Modeling of the Geometry Effect on the Core Loss and Verification with a Measurement Technique Based on the Seebeck Effect and FEA", Accepted in *the 45th Annual Conference of the IEEE Industrial Electronics Society, IECON 2019.*

[11] S. Barg, K. Bertilsson, "Core Loss Modeling and Calculation for Trapezoidal Magnetic Flux Density Waveform," *IEEE Transactions On Industrial Electronics-early access paper.*

[12] S. Barg, K. Bertilsson, "Core Loss Modeling and Calculation for Trapezoidal Magnetic Flux Density Waveform," *IEEE Open Journal of Power Electronics, Vol. 2, December 2021.*

Model Reduction using Singular Perturbation Methods for a Microgrid Application

Lasse Gnärig[1], Albrecht Gensior[2], Saioa Burutxaga Laza[3], Miguel Carrasco[4]
Carsten Reincke-Collon[5]
[1]Technische Universität Dresden, Germany, [2]Technische Universität Ilmenau, Germany,
[3]RWE Battery Solutions GmbH, Germany, [4]eMIS Deutschland GmbH, Germany
[5]Aggreko Deutschland GmbH, Germany
[1]lasse.gnaerig@tu-dresden.de, [2]albrecht.gensior@tu-ilmenau.de,
[3]saioa.burutxaga@rwe.com, [4]m.carrasco@emis-deutschland.com,
[5]carsten.reincke-collon@aggreko.com

Keywords

≪Grid Application≫ ≪Modelling≫ ≪Perturbation Theory≫ ≪Model Reduction≫

Abstract

The paper deals with the modeling of an island grid for the purpose of simulation. Since the behavior of the system is affected by phenomena taking influence in different time-scales, a model reduction scheme is proposed that leads to simpler models which reduce the computational burden in simulations. Compromises in the modeling procedure are addressed by investigating simulation studies.

Introduction

In contrast to conventional power systems, microgrids mainly fed by inverters have a comparably low physical inertia. This is why studies of stability analysis are of great interest. However, simulating the behavior of such grids may become challenging because the state dimension of the model grows quickly with the size of the network. Furthermore, since such models represent multiple processes associated with different time scales, the step size of the solver is limited by the fast phenomena. Both of these properties increase the computational effort for a simulation of the system [1–5]. In order to overcome this problem, reduced order models can be introduced. The approach proposed here uses singular perturbation theory as an appropriate tool for models with multiple time scales to obtain reduced models of lower order. By applying such methods, a subsystem associated with a short time scale is identified and decoupled from the residual system that evolves on a larger time scale [6–9]. By choosing different sets of variables considered to describe the fast dynamics, multiple models of descending order can be obtained. For the investigation of certain phenomena it is then possible to pick an appropriate model from this set of models [1].

A systematic model reduction procedure for large-signal microgrid models was presented in [1] and also applied in [2–4]. When applying singular perturbation methods, determining the states associated with a short time scale is a non-trivial task. In [2] and [3] this was done by examining the eigenvalues of the state matrix of the linearized microgrid model, whereas in [1] and [4] perturbation parameters were evaluated.

This paper provides a detailed dynamic model of a small microgrid consisting of two grid-forming inverters feeding a passive load. Furthermore, an optional connection to a stiff main grid is considered. The structure of the full order model is adopted from [10]. In order to obtain a detailed model, in contrast to [1], the dynamics of the transmission network and the load are taken into account. A model reduction procedure is introduced which does not require linearization nor the calculation of eigenvalues. Instead, the perturbation parameters are used, derived from the normalized model equations. Normalization ensures that

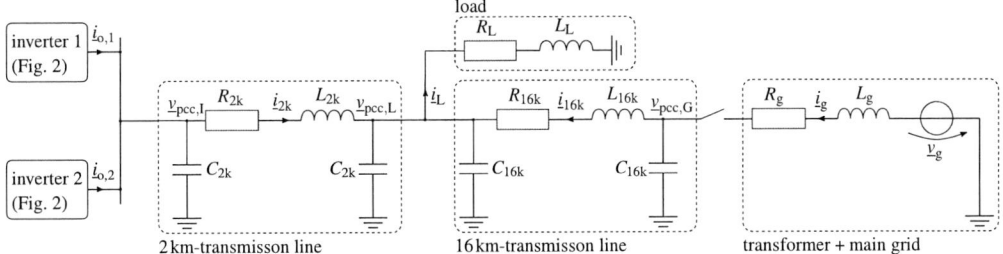

Fig. 1: Configuration of the grid application under investigation. All currents and voltages refer to the common reference frame DQ set by inverter 1.

the magnitudes of the state variables, which can defer greatly, do not complicate the finding of the fast states and ensure the comparability of the dimensionless perturbation parameters. Moreover, a comparison is given between a model obtained by the usual approach of neglecting transmission lines with a model found by the approach taken here.

The paper is organized as follows: First, the model of the microgrid is introduced using the modeling procedure presented in [10]. Subsequently, the obtained full order model is normalized and the state variables are separated, according to the time scale they are associated to. Depending on this choice, multiple reduced order models are introduced. The models are evaluated by comparing simulation results. A conclusion is given at the end.

System Model

The considered configuration of the grid is depicted in Fig. 1. It can be operated in two modes. In grid-connected mode, there exists a connection to a stiff main grid and the two inverters support the grid. In islanded mode, this connection is cut and the grid is formed by the two inverters. In the following, the system model is introduced step-by-step for the inverters, the transmission lines, the load and the connection to the main grid. For the subsequent steps of model reduction, it is important that the variables are denoted in a common coordinate system. Here, a rotating coordinate system is used and three-phase quantities x_k, $k = 1, 2, 3$ are represented as complex variables as

$$\underline{x} = \frac{2}{3} e^{-j\theta} \left(1 \quad -\frac{1}{2} + j\frac{\sqrt{3}}{2} \quad -\frac{1}{2} - j\frac{\sqrt{3}}{2} \right) \left(x_1 \quad x_2 \quad x_3 \right)^{\mathrm{T}} \tag{1}$$

where θ is the grid angle of the first inverter modeled by

$$\dot{\theta} = \omega_1. \tag{2}$$

The first converter is taken as a reference because the mains with its fixed grid frequency is only available in grid-connected mode, see also [10] and [1].

In a later step, singular perturbation methods are applied for model reduction. In general, this is possible if certain parameters are much larger than others which allows then to split the system behavior into a 'fast' and a 'slow' one. For such a comparison, it is necessary to normalize the model which leads to dimensionless variables. This ensures that the magnitudes of the state variables—which may differ greatly—do not complicate the identification of the 'fast' variables. Consequently, all state variables x_i are normalized to their maximum expectable value $\hat{X}_i$ with $i = 1, 2, \ldots, n$ where n denotes the order of the system. The corresponding normalized state variable is then given by $\tilde{x}_i = x_i / \hat{X}_i$. For that, the apparent power delivered by an inverter is normalized to its nominal power $\hat{S} = 625\,\mathrm{kVA}$, and all voltages contained in the state equations are normalized to $\hat{V} = 490\,\mathrm{V}$ which is the phase to ground amplitude of the nominal grid voltage. Consequently, all currents of an inverter model according to Fig 2 are normalized to $\hat{I} = 3\hat{S}/(2\hat{V})$, whereas all currents that are to be assigned to the lines, the load or the main grid according to Fig. 1 are normalized to $2\hat{I}$. This is justified by the fact that the microgrid is fed by two inverters. With this, the parameter $\hat{R} = \hat{U}/\hat{I}$ denotes a resistance. Furthermore, each angular frequency is normalized to

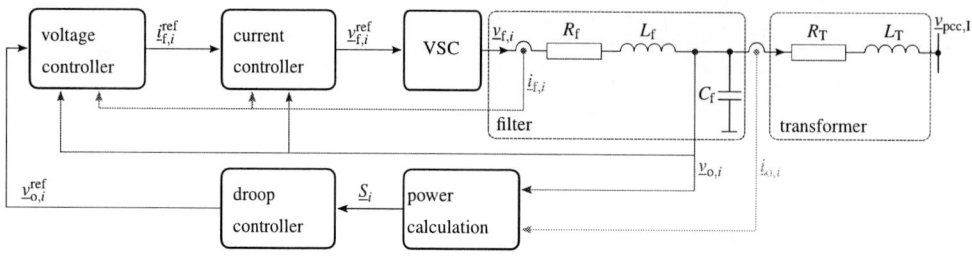

Fig. 2: Configuration of the grid-forming inverter $i = 1, 2$.

the nominal grid frequency ω_{nom}. The time t is normalized to a large time constant given by the cut-off frequency ω_c of the droop controller which will be introduced later. Hence, the relation of the time scales is captured by $\tilde{t} = \omega_c t$. For reasons of compactness, these preparatory steps are already included in the modeling which proceeds next. Note that the parameters on the left-hand-side of the following system equations, renamed as ε with an appropriate index, prepare the model for the model reduction procedure.

Inverter Model

The control scheme and schematic associated with the i-th, $i \in \{1, 2\}$, inverter is depicted in Fig. 2. It is adapted from [10] and is also used in [1]. A superordinated droop controller captures the power fed into the network. Based on this information, the droop controller determines the reference values for the frequency and the amplitude of the voltage $\underline{v}_{\mathrm{o},i}$. In order to obtain the reference value $\underline{v}_{\mathrm{f},i}^{\mathrm{ref}}$ for the voltage outputted by the voltage source converter (VSC), each inverter is equipped with a voltage controller and a subordinated current controller. The output voltage of the VSC is filtered by an LC-Filter, where its output voltage $\underline{v}_{\mathrm{o},i}$ is adapted to the voltage level of the transmission network by a transformer.

Using the common coordinates defined by the first inverter, the dynamics of the i-th inverter are modelled by

$$\frac{\mathrm{d}\underline{\tilde{S}}_i}{\mathrm{d}\tilde{t}} = \varepsilon_S \frac{\mathrm{d}\underline{\tilde{S}}_i}{\mathrm{d}\tilde{t}} = -\underline{\tilde{S}}_i + \underline{\tilde{v}}_{\mathrm{o},i}\underline{\tilde{i}}_{\mathrm{o},i}^* \tag{3a}$$

$$\frac{\omega_c}{\pi n_P \hat{S}}\frac{\mathrm{d}\delta_i}{\mathrm{d}\tilde{t}} = \varepsilon_\delta \frac{\mathrm{d}\delta_i}{\mathrm{d}\tilde{t}} = \mathrm{Re}\left(\underline{\tilde{S}}_1\right) - \mathrm{Re}\left(\underline{\tilde{S}}_i\right) \tag{3b}$$

$$\tilde{\omega}_i = 1 - \frac{2\pi n_P \hat{S}}{\omega_{\mathrm{nom}}}\mathrm{Re}\left(\underline{\tilde{S}}_i\right) \tag{3c}$$

$$\underline{\tilde{v}}_{\mathrm{o},i}^{\mathrm{ref}} = \left(1 - \frac{n_Q \hat{S}}{\hat{V}}\mathrm{Im}\left(\underline{\tilde{S}}_i\right)\right) \mathrm{e}^{\mathrm{j}\delta_i} \tag{3d}$$

$$\frac{\omega_c}{\omega_{\mathrm{nom}}}\frac{\mathrm{d}\underline{\tilde{\gamma}}_{1,i}}{\mathrm{d}\tilde{t}} = \varepsilon_{\gamma_1} \frac{\mathrm{d}\underline{\tilde{\gamma}}_{1,i}}{\mathrm{d}\tilde{t}} = -\left(\frac{2\omega_{\mathrm{cf}}}{\omega_{\mathrm{nom}}} + \mathrm{j}\tilde{\omega}_1\right)\underline{\tilde{\gamma}}_{1,i} - \tilde{\omega}_i^2 \underline{\tilde{\gamma}}_{2,i} + \frac{2k_1 \hat{V}\omega_{\mathrm{cf}}}{\hat{\Gamma}_1 \omega_{\mathrm{nom}}}\left(\underline{\tilde{v}}_{\mathrm{o},i}^{\mathrm{ref}} - \underline{\tilde{v}}_{\mathrm{o},i}\right) \tag{3e}$$

$$\frac{\omega_c}{\omega_{\mathrm{nom}}}\frac{\mathrm{d}\underline{\tilde{\gamma}}_{2,i}}{\mathrm{d}\tilde{t}} = \varepsilon_{\gamma_2} \frac{\mathrm{d}\underline{\tilde{\gamma}}_{2,i}}{\mathrm{d}\tilde{t}} = \underline{\tilde{\gamma}}_{1,i} - \mathrm{j}\tilde{\omega}_1 \underline{\tilde{\gamma}}_{2,i} \tag{3f}$$

$$\underline{\tilde{i}}_{\mathrm{f},i}^{\mathrm{ref}} = k_{\mathrm{p,v}}\hat{R}\left(\underline{\tilde{v}}_{\mathrm{o},i}^{\mathrm{ref}} - \underline{\tilde{v}}_{\mathrm{o},i}\right) + \frac{\hat{\Gamma}_1}{\hat{I}}\left(1 + \tilde{\omega}_i^2 L_{\mathrm{f}}C_{\mathrm{f}}\right)\underline{\tilde{\gamma}}_{1,i} - \underline{\tilde{i}}_{\mathrm{f},i} \tag{3g}$$

$$\underline{\tilde{v}}_{\mathrm{f},i}^{\mathrm{ref}} = \underline{\tilde{v}}_{\mathrm{f},i} = \frac{k_{\mathrm{p,i}}}{\hat{R}}\left(\underline{\tilde{i}}_{\mathrm{f},i}^{\mathrm{ref}} - \underline{\tilde{i}}_{\mathrm{f},i}\right) + \underline{\tilde{v}}_{\mathrm{o},i} \tag{3h}$$

$$\omega_c \frac{L_{\mathrm{f}}}{\hat{R}}\frac{\mathrm{d}\underline{\tilde{i}}_{\mathrm{f},i}}{\mathrm{d}\tilde{t}} = \varepsilon_{i_{\mathrm{f}}} \frac{\mathrm{d}\underline{\tilde{i}}_{\mathrm{f},i}}{\mathrm{d}\tilde{t}} = \underline{\tilde{v}}_{\mathrm{f},i} - \underline{\tilde{v}}_{\mathrm{o},i} - \frac{1}{\hat{R}}\left(R_{\mathrm{f}} + \mathrm{j}L_{\mathrm{f}}\omega_{\mathrm{nom}}\tilde{\omega}_1\right)\underline{\tilde{i}}_{\mathrm{f},i} \tag{3i}$$

$$\omega_c C_{\mathrm{f}}\hat{R}\frac{\mathrm{d}\underline{\tilde{v}}_{\mathrm{o},i}}{\mathrm{d}\tilde{t}} = \varepsilon_{v_{\mathrm{o}}} \frac{\mathrm{d}\underline{\tilde{v}}_{\mathrm{o},i}}{\mathrm{d}\tilde{t}} = \underline{\tilde{i}}_{\mathrm{f},i} - \underline{\tilde{i}}_{\mathrm{o},i} - \mathrm{j}\hat{R}C_{\mathrm{f}}\omega_{\mathrm{nom}}\tilde{\omega}_1\underline{\tilde{v}}_{\mathrm{o},i} \tag{3j}$$

$$\omega_c \frac{L_{\mathrm{T}}}{\hat{R}}\frac{\mathrm{d}\underline{\tilde{i}}_{\mathrm{o},i}}{\mathrm{d}\tilde{t}} = \varepsilon_{i_{\mathrm{o}}} \frac{\mathrm{d}\underline{\tilde{i}}_{\mathrm{o},i}}{\mathrm{d}\tilde{t}} = \underline{\tilde{v}}_{\mathrm{o},i} - \underline{\tilde{v}}_{\mathrm{pcc,I}} - \frac{1}{\hat{R}}\left(R_{\mathrm{T}} + \mathrm{j}L_{\mathrm{T}}\omega_{\mathrm{nom}}\tilde{\omega}_1\right)\underline{\tilde{i}}_{\mathrm{o},i} \tag{3k}$$

The submodel (3) can be subdivided as follows:

- Equations (3a)-(3d) and constitute the droop controller. The apparent power is filtered by a first order low-pass filter with the cut-off frequency ω_c which yields $\underline{\tilde{S}}_i$. With this information, the reference values $\tilde{\omega}_i$ and $\tilde{v}_{o,i}^{\text{ref}}$ for the output voltage and the frequency, respectively, are determined. These references are governed by the frequency-active power droop law (3c) and the voltage-reactive power droop law (3d), where n_P and n_Q are the droop parameters [10]. Since the voltage-reactive power droop law according to [10] defines only the magnitude but not the phase angle of the reference voltage $\tilde{v}_{o,i}^{\text{ref}}$, the difference angle δ_i must be taken into account, when referring to the common coordinate system. This angle, whose derivative is given by (3b) represents the load distribution among the i-th and the first inverter and results from the active power droop control of the individual inverters that causes different inverter frequencies $\tilde{\omega}_i$ [10]. Since the angle δ_i is dimensionless and expected to vary within a interval $[-\pi/2, \pi/2]$, the angle is not normalized. According to (3b), $\delta_1 = 0$ applies at all times.
- The cascaded voltage controller with inner current loop is constituted by (3e)-(3h). According to (3e)-(3g), the voltage controller, whose dynamics are captured by (3e) and (3f), is implemented as so-called proportional resonant controller with the proportional gain $k_{p,v}$, the resonant gain k_I and the frequency ω_{cf}. The reference current $i_{f,i}^{\text{ref}}$ is propagated to the subordinated current controller (3h) with gain $k_{p,i}$. In order to obtain dimensionless state variables that represent the voltage controller, the states $\tilde{\gamma}_{1,i}$ and $\tilde{\gamma}_{2,i}$ are normalized to $\hat{\Gamma}_1 = 2\hat{I}/(1 + \omega_{\text{nom}}^2 L_f C_f)$ and $\hat{\Gamma}_2 = \hat{\Gamma}_1/\omega_{\text{nom}}$, respectively. Note that the operation of the VSC is assumed to be ideal, hence $\underline{v}_{f,i}^{\text{ref}} = \underline{v}_{f,i}$ applies.
- Equations (3i) and (3j) describe the dynamics of the LC-filter. The filter features an inductance L_f and a capacitance C_f, respectively. Furthermore, the filter is damped by a resistance R_f.
- The grid connecting transformer is modeled according to (3k) by its short circuit inductance L_T and resistance R_T.

Load Model

The load is considered to be a passive RL-load. In the common reference frame it can be modeled as

$$2\omega_c \frac{L_L}{\hat{R}} \frac{d\tilde{i}_L}{d\tilde{t}} = \varepsilon_{i_L} \frac{d\tilde{i}_L}{d\tilde{t}} = \tilde{v}_{\text{pcc,L}} - \frac{2}{\hat{R}}\left(R_L + j\omega_{\text{nom}} L_L \tilde{\omega}_1\right)\tilde{i}_L \tag{4}$$

where $\underline{v}_{\text{pcc,L}}$ is the voltage of the network node that is connected to the load and R_L and L_L are the parameters of the load.

Model of the Connection to the Main Grid

The local grid is optionally connected to a main grid by a transformer. Considering the grid frequency as ω_{nom} and its voltage amplitude as v_{nom}, the connecting transformer can be modelled as

$$\frac{\omega_c}{\pi n_P \hat{S}} \frac{d\delta_g}{d\tilde{t}} = \varepsilon_\delta \frac{d\delta_g}{d\tilde{t}} = \underline{\tilde{S}}_1 + \underline{\tilde{S}}_1^* \tag{5a}$$

$$2\omega_c \frac{L_g}{\hat{R}} \frac{d\tilde{i}_g}{d\tilde{t}} = \varepsilon_{i_g} \frac{d\tilde{i}_g}{d\tilde{t}} = e^{j\delta_g} - \tilde{v}_{\text{pcc,G}} - \frac{2}{\hat{R}}\left(R_g + j\omega_{\text{nom}}\tilde{\omega}_1 L_g\right)\tilde{i}_g \tag{5b}$$

with its parameters R_g and L_g. Due to the droop control, the frequency of the the grid-forming inverters may differ from the the main grid frequency ω_{nom}. This is why, a difference angle δ_g between the main grid voltage and the voltage of the first inverter must be considered. As shown in (5b), this angle has to be taken into account when referring to the common reference frame.

Network Model

According to Fig. 1, the dynamics of the network are captured by

$$\frac{1}{2}\omega_c \hat{R} C_{2k} \frac{d\tilde{v}_{\text{pcc,I}}}{d\tilde{t}} = \varepsilon_{v_{\text{pcc,L}}} \frac{d\tilde{v}_{\text{pcc,I}}}{d\tilde{t}} = \frac{1}{2}\left(\tilde{i}_{o,1} + \tilde{i}_{o,2}\right) - \tilde{i}_{2k} - \frac{j}{2}\omega_{\text{nom}}\hat{R}C_{2k}\tilde{\omega}_1\tilde{v}_{\text{pcc,I}} \tag{6a}$$

$$2\omega_c \frac{L_{2k}}{\hat{R}} \frac{d\tilde{i}_{2k}}{d\tilde{t}} = \varepsilon_{i_{2k}} \frac{d\tilde{i}_{2k}}{d\tilde{t}} = \tilde{v}_{\text{pcc,I}} - \tilde{v}_{\text{pcc,L}} - \frac{2}{\hat{R}}\left(R_{2k} + j\omega_{\text{nom}}L_{2k}\tilde{\omega}_1\right)\tilde{i}_{2k} \tag{6b}$$

$$\frac{1}{2}\omega_{\mathrm{c}}\hat{R}\left(C_{2\mathrm{k}}+C_{16\mathrm{k}}\right)\frac{\mathrm{d}\tilde{\underline{v}}_{\mathrm{pcc,L}}}{\mathrm{d}\tilde{t}}=\varepsilon_{v_{\mathrm{pcc,L}}}\frac{\mathrm{d}\tilde{\underline{v}}_{\mathrm{pcc,L}}}{\mathrm{d}\tilde{t}}=\tilde{\underline{i}}_{2\mathrm{k}}+\tilde{\underline{i}}_{16\mathrm{k}}-\tilde{\underline{i}}_{\mathrm{L}}-\frac{\mathrm{j}}{2}\omega_{\mathrm{nom}}\hat{R}\left(C_{2\mathrm{k}}+C_{16\mathrm{k}}\right)\tilde{\omega}_{1}\tilde{\underline{v}}_{\mathrm{pcc,L}} \tag{6c}$$

$$2\omega_{\mathrm{c}}\frac{L_{16\mathrm{k}}}{\hat{R}}\frac{\mathrm{d}\tilde{\underline{i}}_{16\mathrm{k}}}{\mathrm{d}\tilde{t}}=\varepsilon_{i_{16\mathrm{k}}}\frac{\mathrm{d}\tilde{\underline{i}}_{16\mathrm{k}}}{\mathrm{d}\tilde{t}}=\tilde{\underline{v}}_{\mathrm{pcc,G}}-\tilde{\underline{v}}_{\mathrm{pcc,L}}-\frac{2}{\hat{R}}\left(R_{16\mathrm{k}}+\mathrm{j}\omega_{\mathrm{nom}}L_{16\mathrm{k}}\tilde{\omega}_{1}\right)\tilde{\underline{i}}_{16\mathrm{k}} \tag{6d}$$

$$\frac{1}{2}\omega_{\mathrm{c}}\hat{R}C_{16\mathrm{k}}\frac{\mathrm{d}\tilde{\underline{v}}_{\mathrm{pcc,G}}}{\mathrm{d}\tilde{t}}=\varepsilon_{v_{\mathrm{pcc,G}}}\frac{\mathrm{d}\tilde{\underline{v}}_{\mathrm{pcc,G}}}{\mathrm{d}\tilde{t}}=\tilde{\underline{i}}_{\mathrm{g}}-\tilde{\underline{i}}_{16\mathrm{k}}-\frac{\mathrm{j}}{2}\omega_{\mathrm{nom}}\hat{R}C_{16\mathrm{k}}\tilde{\omega}_{1}\tilde{\underline{v}}_{\mathrm{pcc,G}}. \tag{6e}$$

where all quantities are denoted in the common coordinate system defined by the first inverter.

Model Reduction

Singular Perturbation Theory

The full-order model (3)-(6) is considered here to describe the behavior of the system in the most detailed way. From practical experience, it is well known that some of its variables change 'slowly' in time while others change 'fast'. For some applications, the 'fast' behavior is outside the focus of investigation. Furthermore, due to the quite high number $n = 41$ of states, a simulation might be computationally expensive. This motivates to work with reduced models instead that approximate the behavior of the full-order model. Such a model reduction can be obtained by applying singular perturbation theory. This has been done for electrical machines [11], power electronic converters [12, 13], and also for island grids [1–4].

An introduction to singular perturbation theory can be found in [6, 8, 9]. According to these references, the standard form for singular perturbation model reduction reads

$$\dot{x} = f(t,x,z,\varepsilon) \tag{7a}$$
$$\varepsilon\dot{z} = g(t,x,z,\varepsilon). \tag{7b}$$

with the 'fast states' $z = (z_1,\ldots,z_m)^{\mathrm{T}}$, the 'slow states' $x = (x_1,\ldots,x_{n-m})^{\mathrm{T}}$, and the perturbation parameter ε. The subsystem (7b) is referred to as boundary-layer system. Setting the perturbation parameter ε to zero and solving

$$0 = g(t,x,z,0) \tag{8}$$

for z yields the quasi steady state solution $z^{\mathrm{qss}} = \phi(t,x)$ for the fast states. The reduced model is then given by

$$\dot{x} = f(t,x,\phi). \tag{9}$$

To find out whether the fast variables z converge to their steady state solution z^{qss}, a stability condition must be satisfied that guarantees that z will remain close to its equilibrium z^{qss}, while the slow states x are slowly varying [6, 8, 9]. This stability property is met if the equilibrium z^{qss} of the boundary-layer system (7b) is exponentially stable. According to [6, 8, 9], the stability property is satisfied if the Jacobian $\partial g/\partial z$ at $\varepsilon = 0$ is nonsingular for all $(x,z) \in D$ and the real parts of its eigenvalues are negative for all $(x,z) \in D$, where D is a region of interest for the present analysis. Hence, the stability property can be expressed as

$$\det\left(\frac{\partial g}{\partial z}\right)\bigg|_{\varepsilon=0} \neq 0, \qquad \forall (x,z) \in D \tag{10a}$$

$$\mathrm{Re}\left[\lambda\left\{\frac{\partial g}{\partial z}\right\}\right]\bigg|_{\varepsilon=0} < -c < 0, \qquad \forall (x,z) \in D, \tag{10b}$$

in which $\lambda\{\partial g/\partial z\}$ is an eigenvalue of the Jacobian matrix, and c is a fixed positive number.

The full-order model (3)-(6) can be brought into the standard form (7) by applying the following procedure: In a preparatory step, the model (3)-(6) is rewritten in a rearranged order as

$$\varepsilon \circ \dot{\zeta} = \psi(t,\zeta,\varepsilon), \tag{11}$$

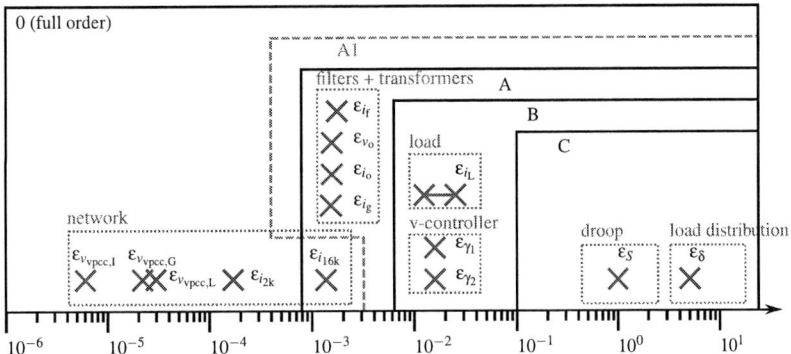

Fig. 3: Comparison of perturbation parameters

with the state vector $\boldsymbol{\zeta} = (\zeta_1, \ldots, \zeta_n)^{\mathrm{T}}$, $\zeta_i \in \{\tilde{\delta}_1, \ldots, \tilde{\underline{v}}_{\mathrm{pcc,G}}\}$ and the corresponding perturbation parameters $\boldsymbol{\varepsilon} = (\varepsilon_1, \ldots, \varepsilon_n)^{\mathrm{T}}$ with $\varepsilon_i \in \{\varepsilon_\delta, \ldots, \varepsilon_{v_{\mathrm{pcc,G}}}\}$. The vector $\boldsymbol{\psi} = (\psi_1, \ldots, \psi_n)^{\mathrm{T}}$ of functions contains the corresponding functions on the right-hand-side. The rearrangement is done such that $\varepsilon_{i+1} \leq \varepsilon_i$. Next, choose $\varepsilon \in (\varepsilon_n, \varepsilon_1)$ in a way that the parameters ε_i are split in two non-empty groups: the group with $\varepsilon_i \leq \varepsilon$ and the group $\varepsilon_i > \varepsilon$. Rewrite the lines of (11) as

$$\dot{\zeta}_i = \frac{1}{\varepsilon_i}\psi_i(t, \boldsymbol{\zeta}, \boldsymbol{\varepsilon}) \quad \text{for } \varepsilon_i > \varepsilon, i = 1, 2, \ldots, n - m \tag{12}$$

$$\varepsilon\dot{\zeta}_i = \frac{\varepsilon}{\varepsilon_i}\psi_i(t, \boldsymbol{\zeta}, \boldsymbol{\varepsilon}) \quad \text{for } \varepsilon_i \leq \varepsilon, i = n - m + 1, n - m + 2, \ldots, n. \tag{13}$$

By subsequently setting

$$\boldsymbol{x} = (\zeta_1, \ldots, \zeta_{n-m})^{\mathrm{T}}$$

$$\boldsymbol{z} = (\zeta_{n-m+1}, \ldots, \zeta_n)^{\mathrm{T}}$$

$$\boldsymbol{f}(t, \boldsymbol{x}, \boldsymbol{z}, \boldsymbol{\varepsilon}) = \left(\frac{1}{\varepsilon_1}\psi_1(t, \boldsymbol{\zeta}, \boldsymbol{\varepsilon}), \ldots, \frac{1}{\varepsilon_{n-m}}\psi_{n-m}(t, \boldsymbol{\zeta}, \boldsymbol{\varepsilon}) \right)^{\mathrm{T}}$$

$$\boldsymbol{g}(t, \boldsymbol{x}, \boldsymbol{z}, \boldsymbol{\varepsilon}) = \left(\frac{\varepsilon}{\varepsilon_{n-m+1}}\psi_{n-m+1}(t, \boldsymbol{\zeta}, \boldsymbol{\varepsilon}), \ldots, \frac{\varepsilon}{\varepsilon_n}\psi_n(t, \boldsymbol{\zeta}, \boldsymbol{\varepsilon}) \right)^{\mathrm{T}},$$

the model is obtained in the standard form (7).

Reduced-Order Models

The values of the elements of $\boldsymbol{\varepsilon}$, i.e. the perturbation parameters, are shown in Fig. 3. It is possible to identify groups by value and by topic and it is interesting to see that these two kinds of grouping may lead to different results. The basic idea is now to derive models of reduced order in a stepwise fashion. Therefore, according to a grouping scheme, some equations are considered to constitute the fast subsystem. Grouping by value is equivalent to choosing ε in order to obtain the form (12), (13). The quasi steady-state solution of the fast subsystem is substituted into the slow subsystem, see (8) and (9). This gives a reduced order model. The same procedure can be repeated and leads to models with an even lower order of the system.

For the particular system investigated here, this process is shown in Fig. 4. Since it is common practice to group by topic and to neglect the dynamics of the connecting network [1], Model A1 is derived from Model 0. However, grouping by value leads to Model A, where the dynamics of the long transmission line is associated with the slow subsystem because $\varepsilon_{i_{16k}}$ is much larger than the other parameters associated with the network. Models B and C are subsequently derived from Model A. As a result, a series of models of decreasing order is obtained.

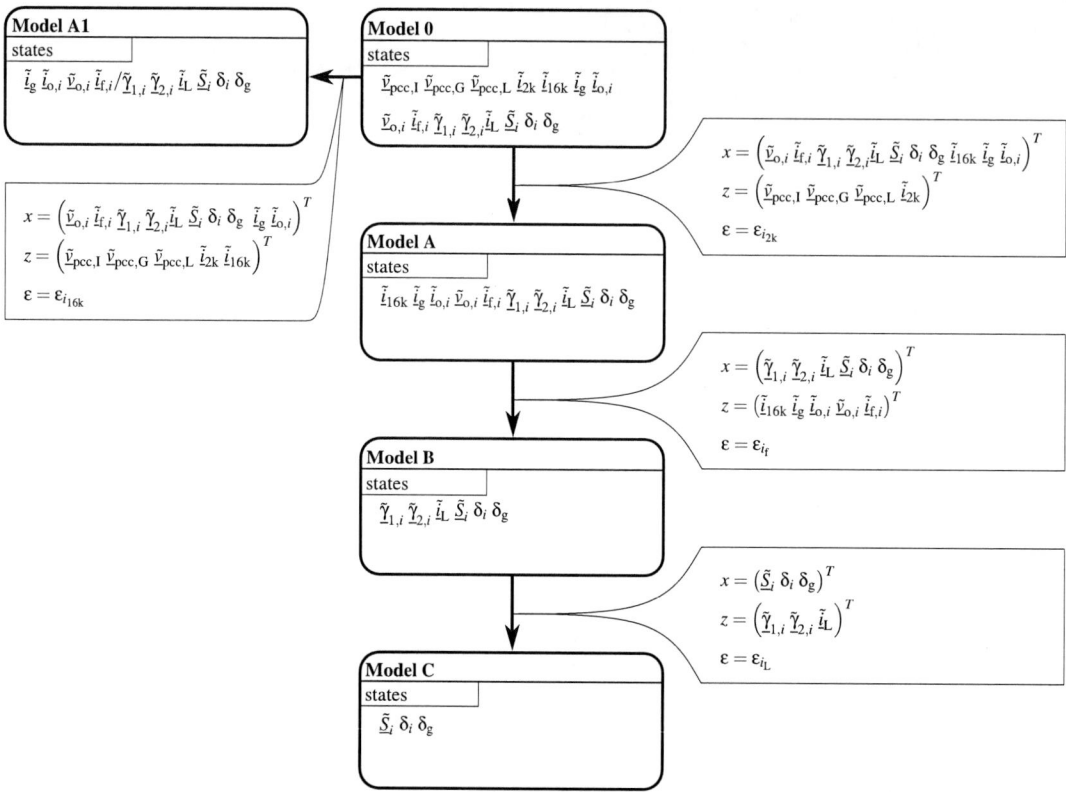

Fig. 4: Successive model reduction procedure

Numerical Case Studies

The effect on the modeling accuracy due to model reduction is investigated by comparing simulation results of the reduced-order models with the full-order model. The parameters of the system are provided by Table I. The system behavior under the influence of a sudden load step is considered. In a first series of experiments, the system is simulated in islanded mode. The corresponding results are depicted in Figs. 5 and 6. In a second experiment, the heuristic approach of model reduction leading to Model A1 is compared to Model A. For this purpose, the sytem operates in grid connected mode and the results are given in Fig. 7. From these results, the following conclusions can be drawn:

- Model A is particularly suitable for examining effects evolving in the interval $t \in [0, 0.02\,\text{s}]$ which are caused by the dynamics of the filter and the transformers. Resonance effects as a result of the network dynamics are not captured by Model A.

Table I: Parameters of the system

component	values
filter	$L_f = 200\,\mu\text{H}, C_f = 540\,\mu\text{F}, R_f = 2\,\text{m}\Omega$
inverter transformer	$L_T = 177\,\mu\text{H}, R_T = 15.4\,\text{m}\Omega$
2 km-line	$L_{2k} = 9.6\,\mu\text{H}, C_{2k} = 4.17\,\mu\text{F}, R_{2k} = 5.8\,\text{m}\Omega$
16 km-line	$L_{16k} = 77\,\mu\text{H}, C_{16k} = 16.7\,\mu\text{F}, R_{16k} = 23\,\text{m}\Omega$
mains transformer	$L_g = 89.5\,\mu\text{H}, R_g = 6.2\,\text{m}\Omega$
nominal voltage and frequency	$v_{\text{nom}} = \frac{\sqrt{2}}{\sqrt{3}}600\,\text{V}, f_{\text{nom}} = 50\,\text{Hz}$
droop controller	$n_P = 0.5\,\mu\text{Hz}\,\text{W}^{-1}, n_Q = 33.3\,\mu\text{Vvar}^{-1}, \omega_c = 5\,\text{s}^{-1}$
voltage and current controller	$k_{p,v} = 0.6\,\text{S}, k_I = 6\,\text{S}, \omega_{cf} = 10\,\text{s}^{-1}, k_{p,i} = 0.9\,\Omega$

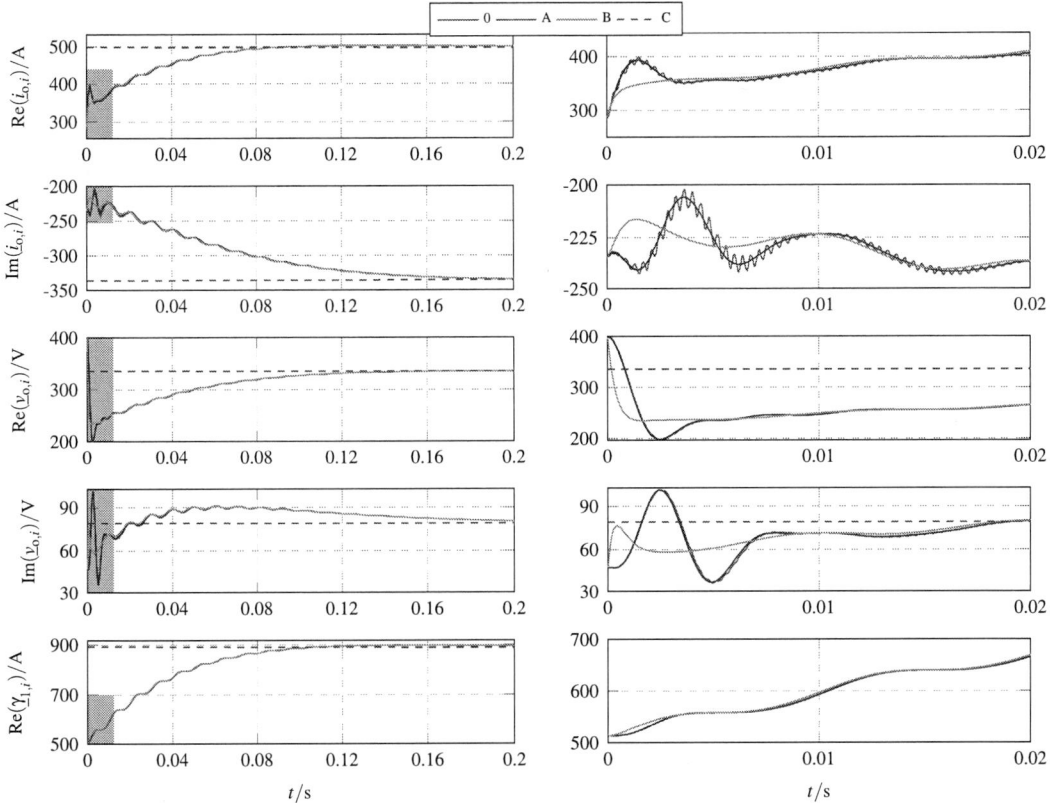

Fig. 5: Simulation results in islanded mode. A load step at $t = 0$ is considered. The results are plotted on a long time scale (left) and on a short time scale (right). The output current $\underline{i}_{o,i}$ of the i-th inverter, the voltage $\underline{v}_{o,i}$ across the filter capacitor of the i-th inverter, and a state $\underline{\gamma}_{1,i}$ associated with the voltage controller of the i-th inverter are displayed

- Model B delivers sufficiently accurate results for investigating processes that take place in the interval $t \in [0, 0.2\,\mathrm{s}]$. Therefore, Model B is suitable to examine effects due to the voltage controller dynamics.
- As shown in Fig. 6, when considering processes taking place in $t \in [0, 0.2\,\mathrm{s}]$, Model C yields poor results. Therefore, Model C is not suitable for investigating effects caused by the low-pass filters of the droop controller. However, Model C can be used for long time simulations of the microgrid, where the power sharing among the inverters is of interest.
- As shown in Fig. 7, Model A delivers significantly better results compared to Model A1. This is observed even though Model A1 has only one complex state less than Model A. This highlights the importance of the separation according to the corresponding perturbation parameters. Thus, it can

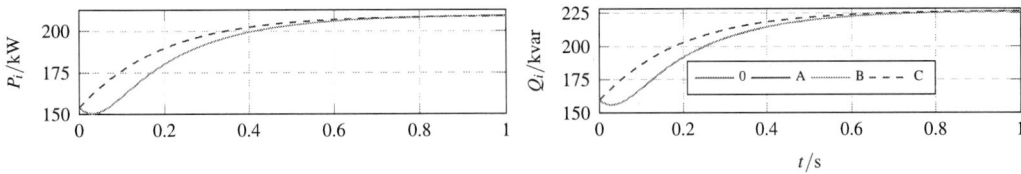

Fig. 6: Simulation results in islanded mode. A load step at $t = 0$ is considered. The low-pass filtered active and reactive Power P_i and Q_i delivered by the i-th inverter are displayed.

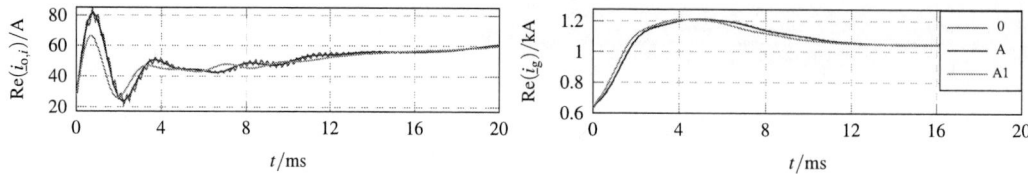

Fig. 7: The system is simulated in grid-connected mode. A load step at $t = 0$ is considered. The simulation results for the full order Model 0, the reduced Model A, and the reduced Model A1, where the dynamic network model has been replaced by an algebraic model are compared. The real values of the i-th inverter output current $\underline{i}_{o,i}$ and the current $\underline{i}_g$ delivered by the main grid are displayed.

Table II: Comparison of the models in terms of computational effort

	number of states	solver steps
Model 0	41	3489
Model A	33	360
Model A1	31	353
Model B	17	336
Model C	7	57

be concluded that the heuristic approach of neglecting the dynamic behavior of the transmission network, which is common in the modeling of network applications [1], can lead to inaccurate results. The models are compared in terms of computational effort by investigating the number of solver steps. The obtained results are shown in Table II for a time horizon of $t = 2\,\mathrm{s}$ using a stiff solver (ode23tb).

Conclusion

Singular perturbation methods were successfully applied for model reduction in a microgrid application. The method proposed here allows to separate the full-order dynamics into a fast and a slow subsystem. The splitting can be done according to the respective modeling purpose. As a result, multiple reduced models of descending order were obtained that reduce the computational effort for simulations. The method does not require calculation of eigenvalues nor empirical knowledge of the system. Moreover, it was shown that the heuristic approach of neglecting the dynamic behavior of the transmission network, which is common in the modeling of network applications, can lead to inaccurate results which can be avoided by using the method.

References

[1] L. Luo and S. V. Dhople. Spatiotemporal model reduction of inverter-based islanded microgrids. *IEEE Transactions on Energy Conversion*, 29(4):823–832, 2014.

[2] K. Kodra, Ningfan Z., and Z. Gajić. Model order reduction of an islanded microgrid using singular perturbations. In *2016 American Control Conference (ACC)*, pages 3650–3655, 2016.

[3] Y. Peng, et al. Reduced order modeling method of inverter-based microgrid for stability analysis. In *2017 IEEE Applied Power Electronics Conference and Exposition (APEC)*, pages 3470–3474, 2017.

[4] V. Mariani, F. Vasca, and J. M. Guerrero. Analysis of droop controlled parallel inverters in islanded microgrids. In *2014 IEEE International Energy Conference (ENERGYCON)*, pages 1304–1309, 2014.

[5] M. Rasheduzzaman, J. A. Mueller, and J. W. Kimball. Reduced-order small-signal model of microgrid systems. *IEEE Transactions on Sustainable Energy*, 6(4):1292–1305, 2015.

[6] P.V. Kokotovic, R.E. O'Malley, and P. Sannuti. Singular perturbations and order reduction in control theory — an overview. *Automatica*, 12(2):123–132, 1976.

[7] G. Peponides, P. Kokotovic, and J. Chow. Singular perturbations and time scales in nonlinear models of power systems. *IEEE Transactions on Circuits and Systems*, 29(11):758–767, 1982.

[8] H. K. Khalil. *Nonlinear Systems*. Pearson Education. Prentice Hall, 2002.

[9] P. Kokotovic, H. K. Khalil, and J. O'Reilly. *Singular Perturbation Methods in Control: Analysis and Design*. Classics in Applied Mathematics. Society for Industrial and Applied Mathematics (SIAM), 1999.

[10] N. Pogaku, M. Prodanovic, and T. C. Green. Modeling, analysis and testing of autonomous operation of an inverter-based microgrid. *IEEE Transactions on Power Electronics*, 22(2):613–625, 2007.

[11] Z. Sorchini and Ph. T. Krein. Formal derivation of direct torque control for induction machines. *IEEE Trans. Power Electron.*, 21(5):1428–1436, Sep. 2006.

[12] J. W. Kimball and Ph. T. Krein. Singular perturbation theory for DC-DC converters and application to PFC converters. *IEEE Trans. Power Electron.*, 23(6):2970–2981, Dec. 2008.

[13] A. Gensior and H. Fehr. Modeling and energy balancing control of modular multilevel converters using perturbation theory for quasi-periodic systems. *IEEE Trans. Power Electron.*, 36(2):2201–2217, 2021.

Drive Level Parameter Identification of an Induction Motor

Andreas Bünte   Alex Hald   Andreas Kirsch

University of Applied Sciences
Interaktion 1
33619 Bielefeld, Germany
Phone: +49 (521) 106-0
Email: andreas.buente@fh-bielefeld.de
URL: http://www.fh-bielefeld.de

Keywords

≪Induction motor≫, ≪Estimation technique≫, ≪System identification≫, ≪Magnetic saturation≫.

Abstract

A method to identify the electrical parameters of an inverter-fed induction motor is presented, which determines the saturation characteristic of the mutual inductance and the other parameters separately. Current displacement effects are taken into account. An advantage of this scheme is the insensitivity to voltage errors of the inverter. To achieve this, the identification of the constant parameters is based on the fact that the inverter primarily influences only the real part of the impedance. For the identification of the saturation characteristic, the impedance difference for 2 different but closely spaced frequencies is evaluated for different magnetic operating points.

Introduction

Models of the induction motor used for drive control include at least four parameters (stator resistance R_S, rotor resistance R_R, mutual inductance L_m and leakage inductances $L_{S\sigma} = L_{R\sigma}$) (cf. Fig 1 a)). The parameters R_S, R_R and $L_{R\sigma,S\sigma}$ are assumed to be constant and the number of pole pairs is assumed to be known. For the setting of the magnetic operating point, which is determined by the flux-forming component, the stator current or when the machine is used in a wide field weakening range, the saturation behavior of the mutual inductivity $L_m = f(i_\mu)$ should be known. The knowledge of the iron loss resistance R_{fe} is not necessary but can be useful if, for example, an optimum operating point has to be set in view of losses in the partial load range. Due to the importance for the control of induction motors, there are focused research activities to determine these parameters, e.g. [2], [9] and [13]. The various methods can be classified according to whether they determine the parameters during operation (online) or whether during the commissioning of the drive (offline).

Online methods have the advantage that they can track variable parameters, e.g. the temperature dependent resistances. However, the quality of the results depends on the operating points and the dynamics with which the drive system is operated. Offline methods can be differentiated according to the area of application, whereby the justifiable effort varies:

- *Manufacturer Level Identification:* The comparatively greatest effort can be made in the manufacturing plant of the motor or drive system. In particular, measuring devices are available for the electrical quantities, for speed and for torque. The motor can be stimulated and mechanically loaded as desired, but this requires time-consuming mechanical integration into the test rig. This identification can be performed by specialized personnel.
- *Drive Level Identification:* In the simplest case, parameter determination is carried out by the means of the drive system alone, i.e. stimulation, measurement and evaluation are carried out with

the inverter and the associated control. For safety reasons, a rotation of the shaft is not allowed. The need for specialized personnel is not desired.

This paper focuses on drive level offline methods. Manufacturer level measurements, especially no-load and short-circuit measurements are used to verify the measurement results [5]. At standstill, the motor is supplied with small stator voltage and requires only small amounts of electrical power and energy for stimulation. However, for the determination of the saturation characteristic of the main inductance, the stator current must be of the level of the rated current magnitude.

Models of the Induction Machine

If the iron losses (R_{fe}) are not taken into account, the equivalent circuit 1 a) shows the electrical parameters of the induction machine required for the control. The equivalent circuit has five components, but the corresponding admittance is given by the frequency response

$$\underline{Y}_2(j\omega) = \frac{1 + j\omega b_1}{a_0 + j\omega a_1 + (j\omega)^2 a_2} \tag{1}$$

and has only four coefficients b_1, a_0, a_1 and a_2, which depend on the electrical parameters. Therefore, an additional assumption regarding the leakage inductances may be made without loss of the generality. Three variants in particular are common here:

- $L_{S\sigma} = L_{R\sigma}$: The assumption of equal leakage inductances is mostly used.
- $L_{S\sigma} = 0$: This assumption leads to the so-called Γ-equivalent circuit, which is often used for stator flux-oriented control methods.
- $L_{R\sigma} = 0$: This assumption leads to the so-called inverse-Γ-equivalent circuit.

If the iron losses are taken into account, the equivalent circuit has three independent energy stores; a 3rd order system with the frequency response

$$\underline{Y}_3(j\omega) = \frac{1 + j\omega b_1 + (j\omega)^2 b_2}{a_0 + j\omega a_1 + (j\omega)^2 a_2 + (j\omega)^3 a_3} \tag{2}$$

is obtained. The number of parameters is now in equilibrium and an assumption about the distribution of leakage inductances is not mathematically exact but is often made as an approximation.

Fig. 1 b) and c) take into account the current displacement or deep bar effects [3], [8]. For better clarity,

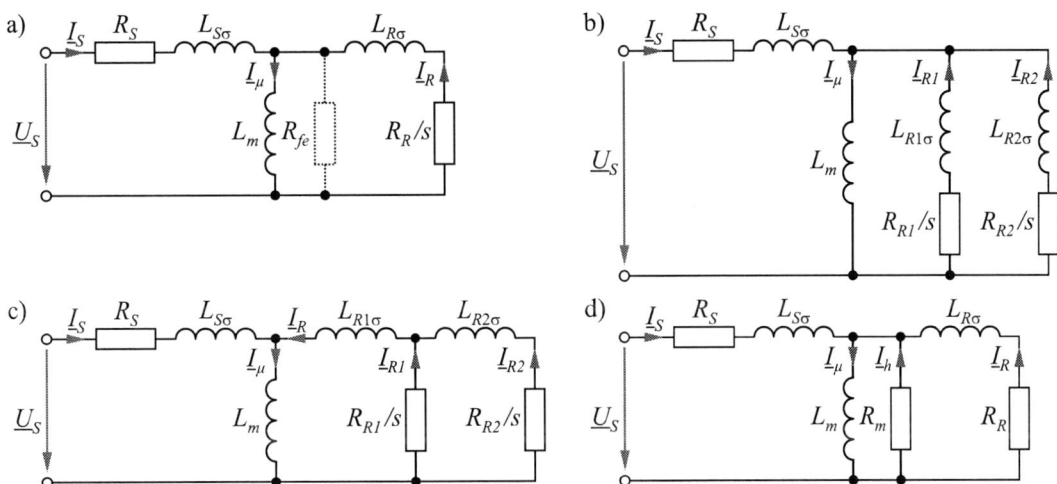

Fig. 1: Steady-state equivalent circuit of an induction motor a) with and without iron losses b) with current displacement, without iron losses, c) alternative form with current displacement, without iron losses, d) at standstill with one energized axis with combined iron losses and current displacement effects.

the same notation is used in both diagrams. With the exception of the stator resistance R_S, different parameters result for the different equivalent circuits. Both equivalent circuits can be described by the frequency response (2). The frequency response has six degrees of freedom while the equivalent circuits 1 b) and c) have seven. Analogous to the procedure for the inverse-Γ-equivalent circuit, the leakage inductance $L_{R1\sigma}$ in Fig. 1 d) can be assumed to be zero. As a consequence, R_{R1}/s is parallel to the mutual inductance L_m and, if the iron losses are also modeled, parallel to R_{fe}. By this, a separate determination of the iron losses and the current displacement effects is only possible if the slip s can be varied. If a rotation is not permitted, the slip is equal to 1 and corresponding resistances are combined ($R_m = R_{R1} \| R_{fe}$). The equivalent circuit Fig. 1 d) is obtained and since $R_{R1} \ll R_{fe}$ is valid the current displacement effects are dominant for R_m. With the exception of the Table II, space vector quantities in the amplitude invariant form are used in the following [7].

Since the motor must not produce any torque during identification, the rotor and stator flux vector must have the same direction. This behavior can be forced by energizing the motor only on a single axis. In the following the α-axis is regarded as the energized axis, $u_{S\beta} = 0$ and $i_{S\beta} = 0$ are valid. A standstill frequency response test (SSFR) is performed [6].

In this paper, the models Fig. 1 a) and d) are used. Therefore, the calculation of the electrical parameters from the frequency responses is given in Table I for both variants.

Table I: Calculation of the electrical parameters

Modell Fig. 1 a)	Modell Fig. 1 d)	
$R_S = a_0$	$R_S = a_0$	$L_{S\sigma} = \dfrac{a_3}{b_2}$
$R_R = \dfrac{a_1}{b_1} - a_0$	$R_h = \dfrac{a_2 b_2 - a_3 b_1}{b_2^2}$	$L_m = a_1 - \dfrac{a_3}{b_2}$
$L_m = \sqrt{\dfrac{(R_R b_1)(a_1 b_1 - a_2 - a_0 b_1^2)}{b_1}}$	$R_R = \dfrac{L_m^2 R_h}{(a_1 b_1 - a_2) R_h - L_m^2}$	$L_{R\sigma} = \dfrac{b_2 L_m R_h^2}{(a_1 b_1 - a_2) R_h - L_m^2}$
$L_{R,S\sigma} = R_R b_1 - L_m$		

According to [2], the single-axis supply differs from the operation with rotating fields with respect to the magnetization behavior. With rotating fields in the steady state (e. g. no-load test), the length of the space vector $|i_\mu|$ and also the flux $|\psi_m| = |L_m(i_m u) \cdot i_m|$ is constant. With sinusoidal currents in the single-axis test, the entire saturation characteristic is run through. Depending on the measuring method, a conversion of the saturation characteristics is therefore necessary.

Experimental setup

The rated data of the induction motor used are listed in Table II. To simplify the measurements and ensure a good accuracy, the machine was fed with an extra-low voltage inverter. This inverter is equipped with Vishay SIR622DP MosFETs, which are driven by Diodes Incorporated DGD0504FN-7 gate drivers. The electronics were supplied directly from a stabilized voltage source with $U_{dc} = 70$ V. The switching frequency is $f_S = 10$ kHz. LEM HO10-P/SP33 were used for the current measurement.

Since the machine was only stimulated with low frequencies or only with small amplitudes, the electronics do not reach their control limits despite the low dc link voltage. For this reason, the measurement methods described in this paper can be transferred to systems with mains voltage.

Measurement of the Constant Parameters

In [2], to determine the parameters of the model Fig. 1 a) without R_{fe}, the frequency response is measured by cross-correlation with sinusoidal currents and a single axis test. This method is suitable for the drive level.

	Table II: Name plate data of motor		Table III: Electrical parameter of motor,				
Parameter	Symbol	Value	Mod. Fig. 1 a)			Mod. Fig. 1 d)	
Manufacturer	WEG Group		Par.	Lock. Rot.	SSFR	Par.	SSFR
Type	12463324			No-Load			
Rated frequency	f_N	50 Hz	R_S	2.52 Ω	2.91 Ω	R_S	2.90 Ω
Rated power	P_N	3.0 kW	R_R	1.65 Ω	1.29 Ω	R_m	2.53 Ω
Rated voltage	U_N	400 V	L_m	272 mH	47.37 mH	R_R	1.21 Ω
Rated current	I_N	5.9 A	$L_{R,S\sigma}$	8.83 mH	13.42 mH	L_m	43.63 mH
Rated speed	n_N	2880 min^{-1}	R_{fe}	1570 Ω	–	$L_{S\sigma}$	20.65 mH
Rated power factor	λ_N	0.86				$L_{R\sigma}$	10.97 mH

The voltage source inverter has some parasitic effects, that causes differences $\Delta u_S = u_S - u_S^*$ between reference and actual voltage. The voltage change is especially large when current zero crossings occur. To avoid this, a DC component is superimposed on the current, e. g. $i_{S\alpha} = \bar{i} + \hat{i}\sin\omega t = I_N/\sqrt{2} + \frac{\sqrt{2}}{10}I_N\sin\omega t$.

The remaining voltage error influences the parameter identification if voltage detection is not used. One error is caused by the switching lag times which are necessary to avoid short-circuits at the inverter. This results in a difference between the reference duty cycle and the real duty cycle dependent on the turn-on-delay, the turn-off-delay and the signs of the phase currents. This error depends on the phase currents and the junction temperature. The same is true for the forward losses of the IGBTs and diodes of the inverter. If the temperature dependence is neglected, the voltage error uniquely depends on the current and corresponding describing function $\Delta U_S(j\omega)/I_S(j\omega)$ is real, as shown in [2]. Furthermore, if the current amplitude and offset is constant for different frequencies, the real part of the corresponding describing function is real and constant. By this, the inverter error voltage affects the stator resistance and the determination of $L_{S\sigma}$ and R_R are not influenced by the inverter. The determination of the saturation characteristic is not possible with constant current amplitude and offset and is done in a second step.

The proposed measurement was reproduced for a motor according to Table II. The frequencies were varied in the range $0.1\,\mathrm{Hz} \leq f \leq 1000\,\mathrm{Hz}$. The measurement was performed on the heated motor. Then the measurement data were approximated by the 2nd order model (Eq. (1)) and the 3rd order model (Eq. (2)). Fig. 2 shows the results and Table III the parameters of the motor according to Table I. The following effects can be observed:

- Since the conductance were approximated, and its magnitudes become small for high frequencies, the model uncertainties have an effect mainly at high frequencies.
- The inverter has a significant influence on the stator resistance.
- Higher frequencies yield to model uncertainties, which can be reduced with the 3rd order model. This model does not completely eliminate the model uncertainties; deviations can be observed from about 100 Hz. For an improvement, further increasing the order is possible. Retiere [10] proposes the use of a noninteger order model.
- The calculated mutual inductances differ strongly from the no-load test and are unreliable. The deviations can be explained: With regard to the magnetic behavior, the excitation selected for this measurement leads to the passage of minor loops. Their slopes differ significantly from those of the major loops determined in the no-load test [1].
- The rotor resistance obtained with the locked-rotor test is 21.8 % higher than the value obtained with the SSFR test. The short circuit test was performed at 50 Hz. In the SSFR measurement, smaller frequencies are weighted more heavily, at which the current displacement effects are smaller. The plausibility could be shown with a locked-rotor test with a smaller feed frequency. With the 3rd model, R_R is further decreased.
- The value for R_m obtained with the 3rd order model is in the order of magnitude of the stator resistance. Obviously, it models not only the iron losses but primarily current displacement effects.
- The estimated values of the leakage inductances are significantly larger with the SSFR test than

the value of the locked-rotor test. The data of the locked-rotor test were determined with rated current. A comparative measurement with the mean value of the SSFR test resulted in a higher value, i.e. the leakage inductance for the motor under test shows significant saturation influences, which makes the comparability of the data difficult.

The frequency response measurement was performed in a range of $0.1\,\text{Hz} \le f \le 1000\,\text{Hz}$ with the aim to show the model inaccuracies. These high frequencies are not necessary for parameter identification. Estimation with the 3rd model provides stable values when measured up to about 200 Hz, and for the 2nd model measurements up to about 80 Hz are sufficient. Furthermore, the number of measurements can be limited to approximately $n_f = 10$.

For the field-oriented control of the induction motor, the 2nd order model is usually used, since the iron losses do not significantly affect the dynamics and the current displacement effects are not important due to the small slip frequencies. However, by comparing the values of both models, conclusions can be drawn as to how pronounced the current displacement effects are. Further optimization of the drive control may be possible as a result [11], [12]. The iron losses and the current displacement effects cannot be separated at standstill. If the exact knowledge of the iron losses and the current displacement effects becomes necessary, the identification should be made on manufacturer level or with suitable online procedures.

Measurement of the Mutual Inductance

Zero torque during the measurement can be met with the single-axis test. For the determination of the main inductance, two difficulties are associated with the single-axis test:

- At standstill, the determination of the saturation characteristic is difficult. Since $s = 1$ applies, the frequency must not be too high, otherwise the small rotor impedance will short-circuit the comparatively high-impedance mutual inductance. Due to this, the motor is fed only with small voltage. If the excitation is done with the inverter and a voltage measurement is not available the

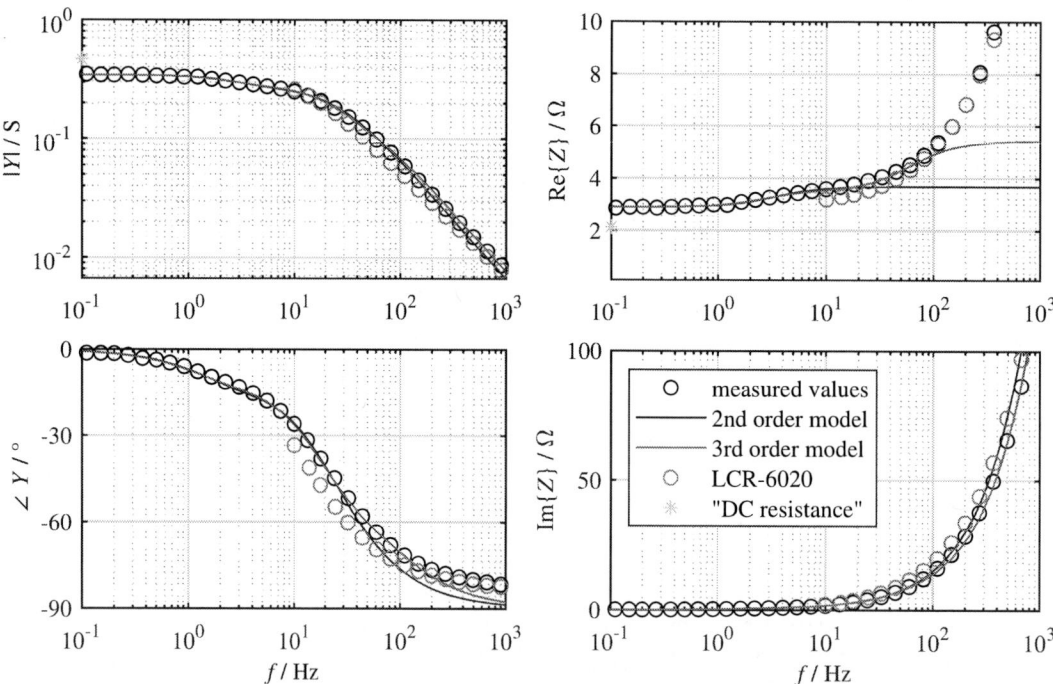

Fig. 2: Measured conductance and impedance (hot induction motor) and fitted models. Verification (DC and $f \ge 10\,\text{Hz}$) is done with an LCR meter (GW Instek) and a cold induction motor. For the sake of visualization, the DC value is shown at 0.1 Hz.

results are seriously distorted by the fault voltage caused by the inverter.
- The magnetic use of the motor in single-axis operation differs from the use in rotating field operation. A conversion of the respective characteristic curves becomes necessary.
- An evaluation of the real part alone is not useful, because the stator resistance must be known exactly (cf. (7)), which is not possible in view of the converter effects.

Comparison of Magnetizing Curves

For the operating point adjustment of the asynchronous motor, the saturation characteristic determined with the no-load test is used. In the no-load test, the magnetizing current can be set with variable stator voltage and thus a characteristic curve $L_m(i_\mu)$ can be determined. During a no-load measurement, the motor operates with a rotating flux whose amplitude is constant. The same applies to the magnetizing current $|i_\mu| = \sqrt{(i_{S\alpha} + i_{R\alpha})^2 + (i_{S\beta} + i_{R\beta})^2}$, which depends on the stator and rotor currents. This magnetizing current determines the instantaneous magnetic operating point.

In the case of single-axis supply, the magnitude of the magnetizing current is not constant and depends on the test signal. In the following, sinusoidal test signals without offset are considered. When an offset is used, from a magnetic point of view a minor-loop is passed through. The properties of this minor-loop do not allow a simple conclusion to be made about the main loop. If only the fundamental component is evaluated during the single-axis test, a fundamental inductance $L_{fm}(\widehat{I}_\mu)$ is determined. For this inductance, the entire covered range of the magnetization characteristic is relevant. An approximate conversion of the saturation characteristic is introduced in [2] and used in the following.

It could be assumed that the function $L_m(i_\mu)$ is even. If the saturation curves could be fitted by the polynomials

$$L_m(i_\mu) = \sum_{n=0}^{N} a_n i_\mu^{\,n} \tag{3}$$

$$L_{fm}\left(\widehat{I}_\mu\right) = \sum_{n=0}^{N} a_{fn} \widehat{I}_\mu^{\,n}, \tag{4}$$

the coefficients satisfy the equation

$$a_{fn} = k_n \cdot a_n \tag{5}$$

with

$$k_n = \begin{cases} 2 \prod_{l=0}^{n/2} \frac{2l+1}{2l+2} & \text{if } n \text{ even} \\ \frac{4}{\pi} \prod_{l=1}^{(n+1)/2} \frac{2l}{2l+1} & \text{else} \end{cases} \tag{6}$$

Identification of the inductance with the imaginary part

In [2] the authors suggested to evaluate only the imaginary parts of the impedance (Fig. 1 a) with $L_{S\sigma} = L_{R\sigma} = L_\sigma$ and $L_m = f(\widehat{i}_\mu)$)

$$\underline{Z}_2(L_m, \omega) = \frac{1}{\underline{Y}_2(\omega)} = R_S + \frac{R_R(\omega L_m)^2}{R_R^2 + \omega^2 (L_m + L_\sigma)^2} + \mathrm{j}\left(\omega(L_m + L_\sigma) - \frac{\omega^3 L_m^2 (L_m + L_\sigma)}{R_R^2 + \omega^2 (L_m + L_\sigma)^2}\right) \tag{7}$$

measured at different current levels for the determination of the saturation characteristic, since the imaginary parts are largely not influenced by the inverter. The measurements were performed at low frequencies, ensuring good sensitivity of the imaginary parts from the main inductance. For these frequencies, the 2nd order model is sufficiently accurate. For one current level the mutual inductance is given with:

$$L_m\left(\widehat{I_\mu}\right) = \frac{R_R^2 - \omega^2 L_\sigma^2 - \sqrt{\left(R_R^2 - \omega^2 L_\sigma^2\right)^2 + 4R_R^2 \operatorname{Im}\{\underline{Z}_2\}\left(2\omega L_\sigma - \operatorname{Im}\{\underline{Z}_2\}\right)}}{2\omega \operatorname{Im}\{\underline{Z}_2\} - 4\omega^2 L_\sigma} - L_\sigma \tag{8}$$

Here the magnetizing current depends on the stator current and the frequency.

$$\widehat{I_\mu} = \widehat{I_{Sd}}\sqrt{\frac{R_R^2 + \omega^2 L_\sigma^2}{R_R^2 + \omega^2\left(L_m + L_\sigma\right)^2}} \tag{9}$$

According to [2], a suitable measurement frequency can be determined with the relative sensitivities

$$E_x(\omega) = \frac{\partial \operatorname{Im}\{\underline{Z}_2\}}{\partial x} \cdot \frac{x}{\operatorname{Im}\{\underline{Z}_2\}} \tag{10}$$

of the evaluated imaginary part. The imaginary part should depend as much as possible on the inductance and as little as possible on the remaining parameters. The results are shown in Fig. 3. The measurement frequency should be to the left of the intersection of the sensitivities E_{Lm} and E_{RR}. Measurement errors in leakage inductance only have an effect at significantly higher frequencies and are less critical. For the following investigations $f = 0.2\,\mathrm{Hz}$ was used.

Very good results were obtained for various motors using this method. However, for motors with larger power ratings, two problems can be observed:

- Due to the large rotor time constants and the associated small measurement frequencies, the total measurement time becomes very large.
- Residual errors of the inverter also affect the imaginary part of the measured impedance. Due to the small impedances of high motor powers, the inductances are therefore estimated too small.

Very good results were obtained for various motors using this method.

Identification of the inductance with double frequency measurement

In this paper, a new approach to determine the main inductance by the difference $\Delta\underline{Z}(L_m) = \underline{Z}(L_m, \omega_2) - \underline{Z}(L_m, \omega_1)$ is presented. This double frequency measurement results in interesting features:

- The result is independent of R_S. Measurement errors in this quantity have no effect.
- If the stator current amplitude $\widehat{i_S}$ is the same for both frequencies and the inverter errors are only weakly frequency-dependent, the inverter errors are also suppressed. This applies to both the real and imaginary parts.
- Two measurements are required for one point of the saturation curve. The total measurement time becomes longer.
- If the stator current amplitude $\widehat{i_S}$ is the same for both frequencies, the magnetizing current will be different for the two frequencies.
- The choice of frequencies $\omega_{1,2}$ represent degrees of freedom.

The determination of the frequencies is critical. Therefore, this selection is considered below. When choosing the frequencies, the following must be taken into account:

- The frequencies should not differ too much, otherwise the magnetizing currents will differ significantly and thus different operating points of the characteristic curve will be measured for both frequencies.
- The frequency difference should not be too small, otherwise $\Delta\underline{Z}_{meas}$ cannot be measured robustly.
- The frequencies must not be too high, otherwise the rotor circuit becomes dominant.
- The frequencies should not be too small, otherwise the total measurement time becomes unattractively large.

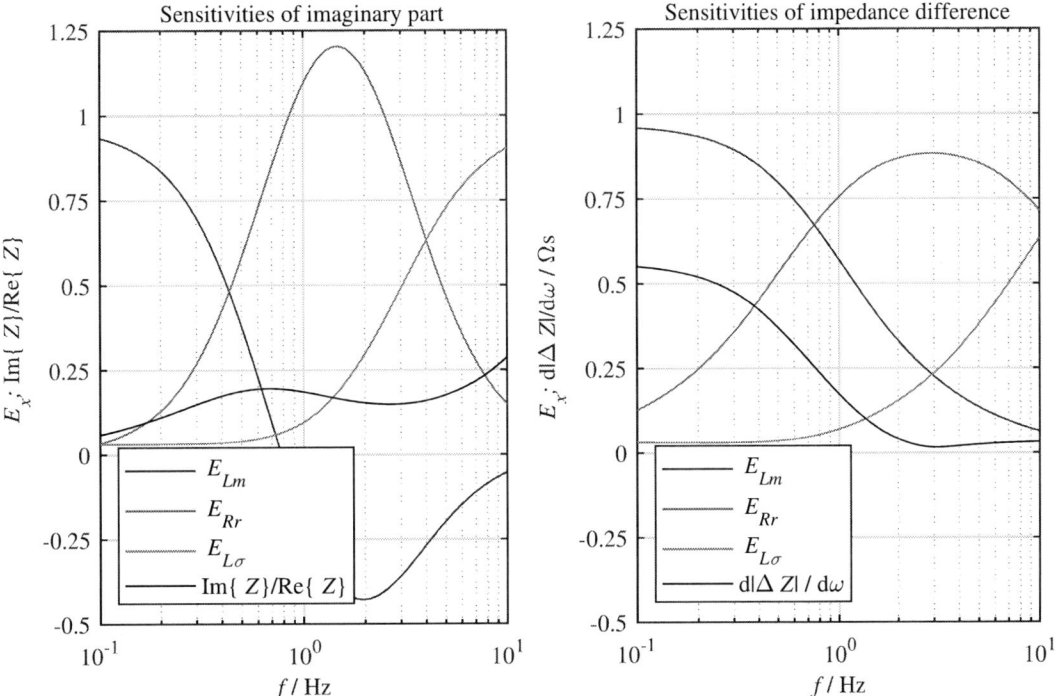

Fig. 3: Relative sensitivities of $\mathrm{Im}\{\underline{Z}_2\}$ (left) and of $|\underline{Z}_2(\omega_2) - \underline{Z}_2(\omega_1)|$ (right). For further consideration of the obtainable accuracies, the ratio of the imaginary and real parts as well as the derivation according to the frequency are additionally shown.

Also for this measurement the absolute frequencies can be determined by means of the relative sensitivities. These are determined with the equation

$$E_x(\omega) = \frac{\partial |\Delta\underline{Z}(L_m)|}{\partial x} \cdot \frac{x}{|\Delta\underline{Z}(L_m)|} \tag{11}$$

and shown in Fig. 3 (right). In this case, measurement errors in leakage inductance only have an effect at significantly higher frequencies and are less critical. The measurement frequencies should be to the left of the intersection of the sensitivities E_{Lm} and E_{RR}. For this method, the intersection point is at higher frequencies than for the determination via the imaginary part. By this, the disadvantage of a double measurement can be compensated by the choice of a higher frequency. Proven choice in simulations with different motors is:

$$\omega_1 = (0.2\ldots0.4) \cdot \frac{R_R}{L_m} \tag{12}$$

For a comparability of the results $\omega_1 = 2\pi \cdot 0.2 \text{ s}^{-1}$ is chosen also here. The remaining degree of freedom is the frequency difference $\Delta\omega = \omega_2 - \omega_1$. Measurements on the test rig show that the quality of the results depends strongly on this difference frequency. The practical results in this paper are obtained for $\omega_2 = 1.1 \cdot \omega_1$.

With the measured difference $\Delta\underline{Z}_{meas}$, the inductance L_m can be determined by solving the equation $0 = \underline{Z}_2(L_m, \omega_2) - \underline{Z}_2(L_m, \omega_1) - \Delta\underline{Z}_{meas}$. However, due to measurement errors and differences in magnetizing current, this equation is not expected to be simultaneously satisfiable for the real and imaginary parts at the same time. Instead, the cost functional

$$J(L_m) = |\underline{Z}_2(L_m, \omega_2) - \underline{Z}_2(L_m, \omega_1) - \Delta\underline{Z}_{meas}|^2 \tag{13}$$

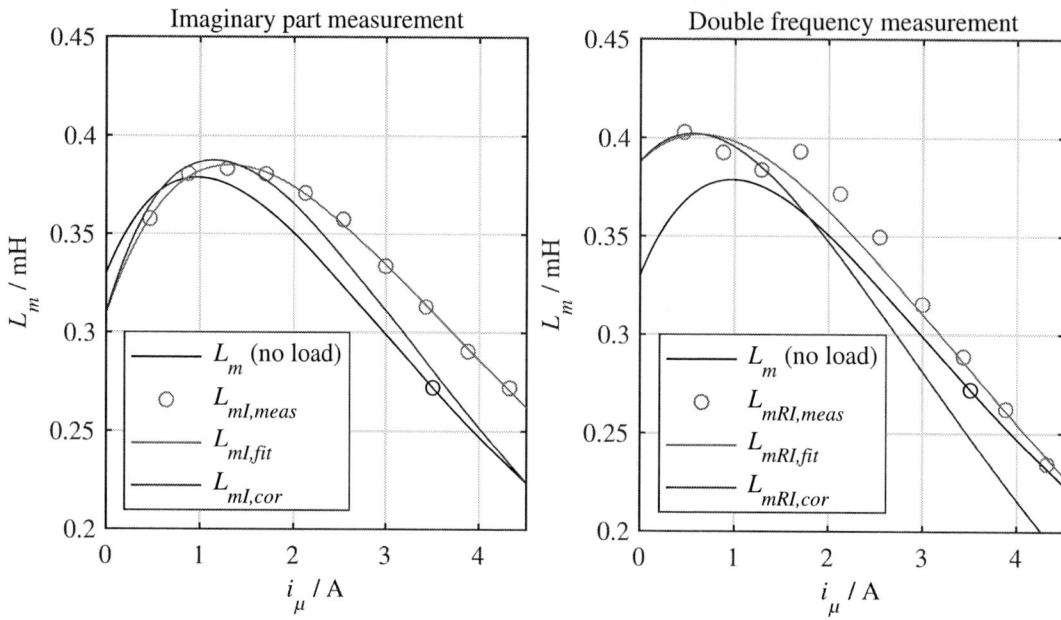

Fig. 4: Magnetizing curves of the one-frequency test (left) and the double frequencies test (right). L_m is the result of the no-load test, $L_{mI,mRI}$ are the results of the different tests and $L_{m,cor}$ are the results of the transformation according to (3)–(6). The rated magnetizing point is marked.

is minimized using numerical optimization. This cost functional describes the distance between the measured difference and the calculated difference in the complex plane and the minimum provides an estimation for L_m.

With regard to the suppression of the influence of the voltage errors of the inverter, the stator current amplitude should be kept constant. Then, the magnetizing currents differ depending on the two measuring frequencies (cf. (9)). With the choice $\omega_1 = 2\pi \cdot 0.2\,\text{s}^{-1}$; $\omega_2 = 1.1 \cdot \omega_1$, the difference for the magnetizing current for the motor under test is less than 1.5 % and negligible. In the following, the average value of the magnetizing currents at both frequencies is assigned to the calculated inductance.

Practical results of mutual inductance identification

Fig. 4 shows the measurement results for the motor according to Table II. The no-load characteristic curve describes the instantaneous values and can be considered as the reference. For both methods, the fundamental inductance was measured for 10 different stator current amplitudes and the measuring points were fitted with a polynomial. Since the characteristics of the fundamental inductance correspond to the single-axis supply, they must be converted to the no-load characteristic using equations (3)–(6). $L_{mX,cor}$ are the final results.

The following can be noted:
- In view of a drive level identification a measurement of the saturation characteristic is possible with both methods.
- For this setup, the results are better with the evaluation of the imaginary part alone. For the rated magnetizing current, there is a deviation of 3.3 % for the mutual inductance in this case and a deviation of -9.1 % for the double-frequency measurement.
- The measurement via the imaginary part for small currents seems to be more robust than the differential evaluation. This behavior could not be improved by increasing the frequency spacing $\Delta\omega$.
- The measurement at one frequency takes about 10 s. With large stator currents, the winding temperature can increase noticeably and influence the resistance. Thus, the sequence of measurements has an influence on the results.

Conclusion

An identification scheme was presented that allows parameter identification of an induction motor at standstill. In a first step the values of the resistances and the leakage inductances are determined by a frequency response test for both, a 2nd order and 3rd order model. The 3rd order model considers current displacement effects. These effects can be modeled with a combined resistor R_m which includes also iron losses. The 3rd model can be used to represent the system behavior up to much higher frequencies. Whether this is useful depends on the control method used.

In a second step, the saturation-dependent main inductance was determined. For this purpose, a new method was presented in which the differential impedance of two frequencies is evaluated. Inverter influences can thus be partially suppressed. The principle function of the presented method could be confirmed with measurements. In comparison with a known method, in which only the imaginary part of the impedance is evaluated, the robustness does not yet seem to be optimal. Further research is needed to improve this behavior.

References

[1] Bünte, A. and Grotstollen, H.: Parameter Identification of an Inverter–Fed Induction Motor at Standstill with a Correlation Method, 5. Eur. Conf. on Power Electronics and Applications (EPE), vol. 5, Brighton: 1993, pp. 97–102.

[2] Bünte, A. and Grotstollen, H.: Offline Parameter Identification of an Inverter–Fed Induction Motor at Standstill, 6. Eur. Conf. on Power Electronics and Applications (EPE), vol. 3, Sevilla: 1995, pp. 492–496.

[3] Corcoles, F.; Pedra, J.; Salichs, M. and Sainz, L.: Analysis of the induction machine parameter identification, IEEE Transactions on Energy Conversion, vol. 17, no. 2, 2002, pp. 183–190.

[4] Grotstollen, H. and Wiesing, J.: Torque capability and control of a saturated induction motor over a wide range of flux weakening, IEEE Transactions on Industrial Electronics, vol. 42, no. 4, 1995, pp. 374–381.

[5] IEEE: IEEE Standard Test Procedure for Polyphase Induction Motors and Generators, IEEE Std 112, 2017.

[6] IEEE: IEEE Guide for Test Procedures for Synchronous Machines, IEEE Std 115, 2009.

[7] Leonhard, W.: 30 Years Space Vectors, 20 Years Field Orientation, 10 Years Digital Signal Processing with Controlled AC-Drives, a Review, EPE Journal, vol. 1, 1991, pp. 13...20, 89...102.

[8] Monjo, L.; Kojooyan-Jafari, H.; Corcoles, F. and Pedra, J.: Squirrel-Cage Induction Motor Parameter Estimation Using a Variable Frequency Test, IEEE Transactions on Energy Conversion, vol. 30, no. 2, 2015, pp. 550–557.

[9] Odhano, S. A.: Self-Commissioning of AC Motor Drives, PhD thesis, Politecnico di Torino, Turin, 2014.

[10] Retiere, N. M., Ivanes, M. S.: An Introduction to Electric Machine Modeling by Systems of Noninteger order. Application to double-cage induction machine, IEEE Transactions on Energy Conversion, vol. 14, no. 4, 1999, pp. 1026–1032.

[11] Seok, J. K., Sul, S. K.: Pseudorotor-Flux-Oriented Control of an Induction Machine for Deep-Bar-Effect Compensation, IEEE Transactions on Industry Applications, vol. 34, no. 3, 1998, pp. 429–434.

[12] Schubert, M.; Koschik, S.; De Doncker, R. W.: Fast Optimal Efficiency Flux Control for Induction Motor Drives in Electric Vehicles Considering Core Losses, Main Flux Saturation and Rotor Deep Bar Effect. Application to double-cage induction machine, Annual IEEE Applied Power Electronics Conference and Exposition (APEC), 2013, pp. 811–816.

[13] Toliyat, H. A.; Levi, E. and Raina, M.: A review of RFO induction motor parameter estimation techniques, IEEE Transactions on Energy Conversion, vol. 18, no. 2, 2003, pp. 271–283.

Impedance Stability of Single-Phase *LCL* Grid-Connected Voltage Source Inverters with Wideband Gap Devices Under Different Control Approaches

Ramy Ali, Terence O'Donnell
UNIVERSITY COLLEGE DUBLIN
Dublin, Ireland
E-Mail: ramy.ali@ucdconnect.ie, terence.odonnell@ucd.ie

Acknowledgements

This work has emanated from research conducted with the financial support of Science Foundation Ireland under Grant No. SFI/16/IA/4496. The opinions, findings, and conclusions or recommendations expressed in this material are those of the authors and do not necessarily reflect the views of the Science Foundation Ireland.

Keywords

« Wide bandgap devices (WBG) », « Voltage Source Inverter (VSI) », « *LCL*-type inverter », « Impedance model », « Stability analysis ».

Abstract

This paper discusses the impedance stability of the grid connected VSI that is built using wide bandgap (WBG), SiC, or GaN semiconductors devices. The use of WBG devices have the benefits of higher switching frequency, lower losses, and smaller size. The impedance stability of the WBG-based VSI is studied under both grid side and inverter side current control. The impedance model of the VSI is developed and validated in Matlab/Simulink. Using this model, the variation of the VSI output impedance at increased switching frequencies is studied. The impedance stability is examined for the case where the VSI is connected at different points in a distribution network. The CIGRE European benchmark LV network is used. The analysis reveals that the impedance stability of the grid-side current-controlled WBG-based VSI is improved at higher switching frequencies. Simulation results are provided for verifying the theoretical analysis.

Introduction

Recently, with the motivation of improving efficiency and reducing the footprint of the voltage source inverter (VSI), the wide bandgap (WBG) semiconductors materials such as SiC, GaN have been adopted in the grid connected VSI. These devices have the potential to offer a smaller size, higher switching frequency, better reliability, and superior efficiency than the conventional silicon-based switches [1]. Although clearly of advantage in the VSI, it is of interest to also ask if this move to higher switching frequencies might have stability or power quality implications for the power system, especially in the context of high inverter-based resources (IBR) penetration levels.

Typically, the different stability concerns of the grid connected VSI have been investigated using one of two approaches: the state-space method in the time domain or the impedance-based method in the frequency domain. The state-space approach depends on the identification of the eigenvalues of the system which requires a detailed description of the system parameters. In contrast, the impedance-based method is a powerful approach as it evaluates the stability of a given converter based on its terminal characteristics. It essentially characterizes the VSI as a current source in parallel with an impedance, and the grid it connects to as voltage source in series with an impedance. The VSI-grid system is stable if ratio of the grid impedance Z_g to the inverter output impedance Z_o satisfies the Nyquist stability criterion (NSC) [2].

The grid connected VSI is typically equipped with an *LCL* filter to reduce the PWM switching harmonics. The output impedance Z_o of the grid connected VSI is a frequency dependent impedance as various control loops govern its properties [3]. As different control loops are effective in different frequency ranges, the output impedance characteristic varies across a wide frequency range. For instance, slower control loops, such as the dc link/power control loop, in addition to the phase locked loop (PLL), shape the low frequency behavior of the output impedance. The current control loop, on the other hand, has a larger bandwidth and dominates the medium-to-high frequency area of the output impedance. The *LCL* filter parameters determine the high frequency characteristic of the VSI output impedance [4]. On the other side, the grid impedance Z_g is also a frequency-dependent impedance and characterized by the impedance of long transmission lines/cables, transformers, and loads. Therefore, the interaction between both frequency-dependent impedances might occur and lead to harmonic resonances and instability problems over a wide range of frequencies from few Hertz (Hz) to several kilohertz (kHz).

Deploying WBG devices in the VSI results in a substantially higher switching frequency of operation, and the *LCL* filter parameters and the current controller bandwidth will be affected as they are dependent on the switching frequency. The switching frequency of the VSI has a significant impact on the crossover frequency (f_c) and thus the bandwidth of the current controller as it is usually designed to be in the range of $[f_{sw}/40, f_{sw}/5]$ [5]. As previously stated, the current controller bandwidth affects the characteristics of the VSI output impedance, therefore the current controller is a major control loop that could produce impedance stability issues in the VSI-grid system in the medium to high frequency range.

For the current control loop, the control target of the *LCL* grid-connected VSI can be the inverter-side current feedback (ICF) or the grid-side current feedback (GCF) [6]. Both choices have pros and cons. For instance, the ICF offers a more cost-effective option because the current sensor on the inverter-side of the VSI may be used for both overcurrent protection and control [7]. The ICF, on the other hand, is hampered by its inability to control the power factor on the grid [8]. Even though the GCF has the advantage of directly managing the power injected into the grid, ensuring the power factor of the grid, it requires an additional current sensor, which adds to the total cost of the system.

This work investigates the impact of the use of WBG devices and the consequent increase in switching frequency on the impedance stability of the VSI under both inverter side and grid side current control approaches. Although other works have studied the relative stability of both the ICF and GCF control approaches using the root locus method [9], these studies were conducted for the relativity lower switching frequencies of conventional silicon switches [4] and did not consider the impedance stability. This work investigates whether the impedance stability issues are likely to be more or less of an issue as VSI switching frequencies increase.

Modeling of VSI System

The configuration of a single-phase grid connected VSI, and its control system is depicted in Fig. 1. The VSI is composed of a full bridge DC-AC inverter that is supplied from a dc voltage source U_{dc}, which should be always higher than the peak ac voltage to generate the inverter output voltage, u_i. The VSI is connected to an *LCL* filter to damp the PWM switching harmonics, where L_1 is the inverter-side inductor, C_f is the filter capacitor, R_d is the passive damping resistor, and L_2 is the grid-side inductor. Generally, the inverter-side inductor L_1 is designed to achieve

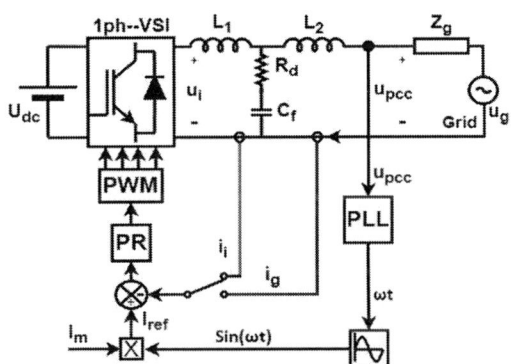

Fig. 1: The system configuration of a single-phase grid connected VSI with an *LCL* filter employing PR controller

maximum allowable output current ripple of 20-30% of the rated output current [10]. The filter capacitor C_f is chosen as a compromise between high-frequency harmonic attenuation and reactive power consumption at the fundamental frequency. For low- and medium-power VSI, the permitted reactive power of the capacitor is typically in the range of 2% to 15% of the rated power [11]. The damping resistance is usually designed to be around one-third of the capacitive reactance at the LCL resonance frequency [12]. Lastly, the grid-side inductor L_2 is set to comply with the grid code on VSI harmonic emission. It is usually selected to achieve a maximum harmonic attenuation rate of 20%. The value of L_2 is a trade-off between the filter cost and the total harmonic distortion (THD) of the grid current [13].

The low voltage (LV) network at the point of common coupling (PCC) in Fig. 1 is represented by an ideal voltage source u_g in series with a grid impedance Z_g, which depends on the type of grid, e.g., cable or overhead lines, urban or rural, etc. The PCC voltage is fed to the PLL which captures the voltage phase angle to generate the sinusoidal reference current I_{ref} of the current control loop. In this work, the injected grid current is controlled in the stationary reference frame ($\alpha\beta$ frame), so a proportional-resonance (PR) controller is employed which shows a more robust performance [14]. It is worth mentioning that in this work, a passive damping for the intrinsic LCL filter resonance [15] is used for the sake of simplicity. To analyze the stability of the VSI-grid system using the impedance-based stability approach, a linearized, small-signal, model of the VSI should be developed. Furthermore, the grid impedance of the LV network where the VSI is connected should also be evaluated to assess the VSI-grid system stability.

Small Signal Impedance Model of the VSI

The impedance stability study is performed using the impedance model of the LCL grid connected VSI, which is based on inverter-side current feedback (ICF) and grid-side current feedback (GCF). This model is mathematically created and validated using a Matlab/Simulink-based test system. Fig. 2 shows the control block diagram for both ICF and GCF.

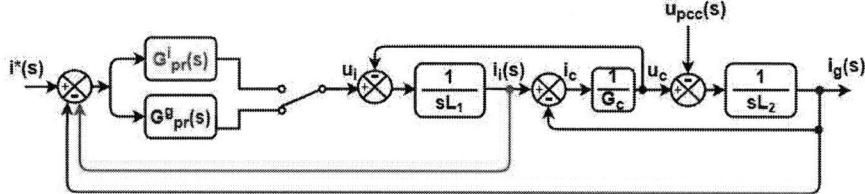

Fig. 2: Control block of LCL grid connected VSI with ICF (Red) and GCF (Blue).

To distinguish between the ICF and GCF controllers, G^i_{pr} is used for ICF, and G^g_{pr} for the GCF. In Fig. 2, G_{pr} represents the transfer function of the PR controller which can be written as in (1), where k_p is the proportional gain, and k_r is the resonant integral gain.

$$G_{pr} = k_p + k_r \frac{2\omega_{co}s}{s^2 + 2\omega_{co}s + \omega_g^2} \tag{1}$$

Furthermore, the instability caused by infinite gain at the nominal grid frequency ω_g can be avoided by choosing a cut-off frequency, ω_{co} that is substantially lower than the nominal grid frequency ($\omega_{co} \ll \omega_g$) [16]. To simplify the study, the parasitic resistances of the LCL filter's inductors are omitted, which also corresponds to the worst working circumstances of the filter. It is also assumed that any impact of the self-resonant frequency (SRF) of the inductor can be neglected in the frequency range of interest. Typically, the SRF of the inductor occurs in the MHz to GHz range [17] and is lower for higher inductance, lower current devices which is not the case here.

The current controller gains (k_p, k_r) can be fine-tuned using the PID tuner app in Matlab/Simulink, or they can be estimated mathematically with some efforts for fine-tuning. Because the crossover frequency

(f_c) is lower than the resonance frequency of the *LCL* filter, the influence of the *LCL* filter capacitor can be ignored when calculating the open loop gain at frequencies less than or equal to f_c, and the transfer function of the PR controller can be approximated to k_p at higher or equal to f_c [18]. Here the design of the GCF controller gains is given as an example to illustrate the design approach. The open loop gain of the GCF can be expressed as:

$$T_{GCF}(s) = G_{pr}^g \cdot \frac{1}{s^2 L_1 L_2 G_c + (L_1 + L_2)s} \tag{2}$$

The amplitude of $T_{GCF}(s)$ at f_c is unity, so the $|T_{GCF}(j\omega_c)|$ can be depicted as:

$$|T_{GCF}(j\omega_c)| = \left| \frac{k_p}{j\omega_c(L_1 + L_2)} \right| = 1,$$

$$k_p = \omega_c(L_1 + L_2) \tag{3}$$

Furthermore, the resonant gain (k_r) can be estimated as in [19]:

$$k_r = \frac{k_p \cdot \omega_c}{10} \tag{4}$$

where, $\omega_c = 2\pi f_c$ and the f_c is usually selected to be one-tenth of the switching frequency f_{sw} ($f_c = \frac{1}{10} f_{sw}$). It is worth highlighting that k_p, k_r values generated from (4) and (5) are estimates that may need to be fine-tuned.

Further, the equivalent transfer function of the filter capacitor branch G_c can be expressed in (5)

$$G_c = \frac{s C_f}{1 + s C_f R_d} \tag{5}$$

The inverter output impedance Z_o is measured at the PCC in Fig. 1 for both control choices, ICF and GCF, for the sake of comparability. Therefore, Z_o can be calculated by ignoring the input reference signal in Fig. 2 and calculating the ratio of $u_{pcc}(s)$ and $i_g(s)$ as $i_i(s)$ is the feedback signal (the red arrow).

The expression of the inverter output impedance Z_o^i for the ICF can be calculated as:

$$Z_o^i = \frac{u_{pcc}(s)}{-i_g(s)} = \frac{s^2 L_1 L_2 G_c + s L_2 G_c G_{pr}^i + s(L_1 + L_2) + G_{pr}^i}{s L_1 G_c + G_c G_{pr}^i + 1} \tag{6}$$

The VSI with GCF controller is analyzed using the same setup in Fig. 1 by using $i_g(s)$ as a feedback signal (the blue arrow). According to the control block diagram shown in Fig. 2, the expression of the output impedance of the VSI-based GCF, Z_o^g can be expressed as:

$$Z_o^g = \frac{u_{pcc}(s)}{-i_g(s)} = \frac{s^2 L_1 L_2 G_c + s(L_1 + L_2) + G_{pr}^g}{s L_1 G_c + 1} \tag{7}$$

To assess the impedance stability of the *LCL* grid connected VSI for the higher switching frequencies associated with the use of wide bandgap semiconductors, the rated data of the VSI, *LCL* filter parameters and current controller gains at different switching frequencies as well as the grid impedances at the point of connection to the grid are listed in Table I. It is worth mentioning that the current controllers have been designed to have a cross-over frequency of one-tenth of the switching frequency for ICF and GCF.

Table I: VSI rated data, *LCL* filter parameters, current controller gains, and the grid impedances

Parameter	Symbol	Switching Frequency			
		10 kHz	20 kHz	50 kHz	100 kHz
Rated Power	P_o	5 kW			
DC link Voltage	U_{dc}	400 V			
Grid voltage	U_g	230 V			
Grid frequency	f_g	50 Hz			
Inverter-side Inductor	L_1	2.6 mH	1.3 mH	0.52 mH	0.26 mH
Grid-side Inductor	L_2	0.65 mH	0.33 mH	0.13 mH	0.065 mH
Filter Capacitor	C_f	5 µf	5 µf	5 µf	5 µf
Damping resistor	R_d	3.5 Ω	3.5 Ω	3.5 Ω	3.5 Ω
PR controller gains of ICF	$k_p^i, k_r^i, \omega_{co}$	21, 2310, 0.5	21.7, 4781, 0.5	23, 8988, 0.5	20, 1.9e4, 0.5
PR controller gains of GCF	$k_p^g, k_r^g, \omega_{co}$	18.4, 2017, 0.5	16.4, 3616, 0.5	13, 7224, 0.5	15, 1.6e4, 0.5
Z_g at Bus 1	R_g^{B1}, L_g^{B1}	0.06 Ω, 0.35 mH			
Z_g at Bus 2	R_g^{B2}, L_g^{B2}	0.13 Ω, 0.76 mH			

Furthermore, a test system constructed in Matlab/Simulink is used to validate the analytical formula for the small-signal output impedance of both ICF and GCF based VSI. The test system includes a disturbance voltage source in series with the grid voltage, as well as the entire switching model of the grid connected VSI. The disturbance voltage has an amplitude of 8% of the rated voltage and is swept across a wide frequency range of 5 Hz to 10 kHz with a resolution of 5 Hz. A Fourier analysis is used to collect both the terminal voltage and injected current at each frequency of disturbance to determine the impedance at that frequency. The impedance acquired in this approach is then compared to that obtained using the analytical formulas in (6) and (7). The magnitude and phase graphs that result from the validation of ICF and GCF, respectively, are shown in Fig. 3 and Fig. 4. This validation considers the filter parameters and current controller gains at a switching frequency of 10 kHz. The magnitude and phase graphs are comparable with the theoretical estimates except in the low frequency area around 50 Hz. This is because removing the fundamental frequency component's effect from the test system in Matlab/Simulink is challenging, especially when the current value at 50 Hz is high.

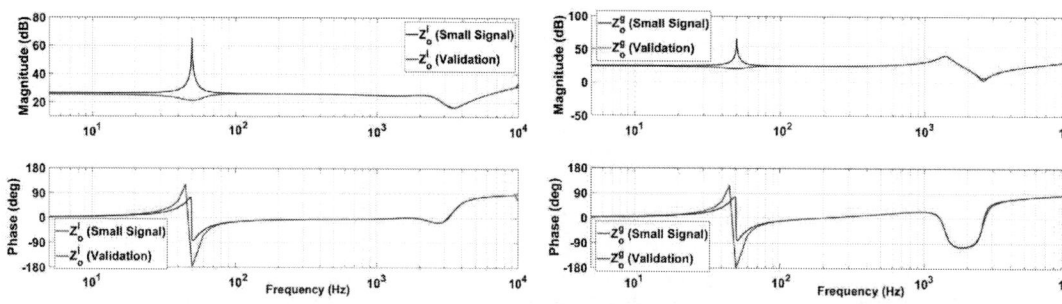

Fig. 3: Small-signal and validation of Z_o^i at f_{sw}=10 kHz

Fig. 4: Small-signal and validation of Z_o^g at f_{sw}=10 kHz

Output Impedance of WBG based VSI

The WBG-based VSI operates at a substantially greater switching frequency ($f_{sw} \geq 50$ kHz) than the traditional silicon-based inverter, which operates at a switching frequency of up to 10 kHz [19]. To analyze the impedance stability of the VSI-grid system, the VSI output impedance, for both ICF and GCF, will be examined over various switching frequencies that are typically used in silicon-based and WBG-based VSI. Fig. 5 shows the frequency response of VSI output impedance, Z_o^i at various switching frequencies. The phase plot of the ICF-based VSI never crosses the -90° phase, indicating that it is stable at different switching frequencies. The phase angle, on the other hand, approaches a deeper lag as the switching frequency rises, implying that the impedance stability of the grid connected VSI suffers as the switching frequency rises. In contrast, as the switching frequency increases, Z_o^g gains more damping because the phase plot at the magnitude dip moves further away from the -90° limit, as shown in Fig. 6.

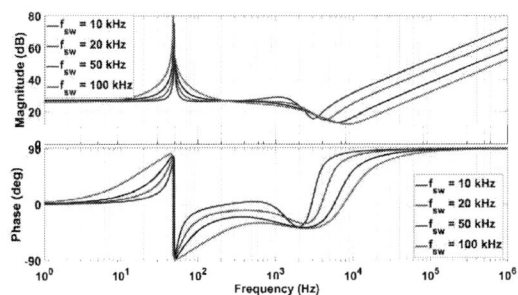

Fig. 5: Bode plots of Z_o^i at different f_{sw}

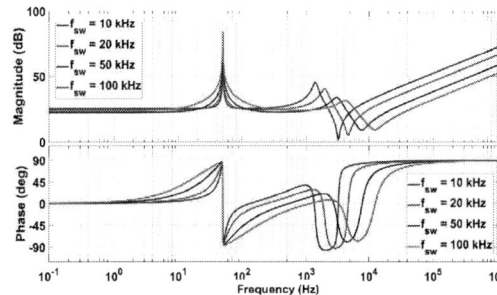

Fig. 6: Bode plots of Z_o^g at different f_{sw}

Study Cases of Impedance Stability and Power Quality Issues of the ICF and GCF based VSI in the LV Distribution Grid

Because the LV distribution network is mostly a radial system, the grid to which the VSI is connected has varying grid impedances, Z_g, depending on the distance between the connection point and the MV transformer. As a result, the VSI output impedance may interact with the multiple grid impedances, causing the VSI-grid system to become unstable. The grid impedance is determined in this study using the CIGRE European benchmark LV network [20], [21] shown in Fig. 7, where the impedance at the closest, bus B1, and the farthest, bus B2, points to the MV transformer are evaluated and listed in Table I.

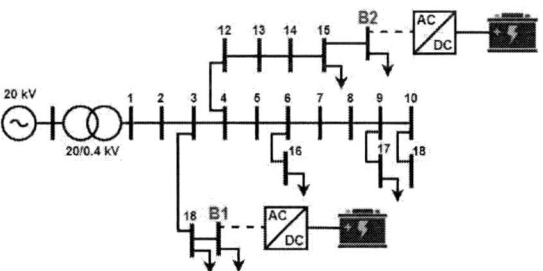

Fig. 7: The single line diagram of the CIGRE LV distribution feeder.

The ICF-based VSI is still more stable than the GCF, but its phase plot has less damping at high frequencies. On the other hand, the GCF-based VSI acquires better impedance stability at higher switching frequencies, as seen in the preceding analysis. Therefore, the case studies in this section will only look at the GCF-based VSI to demonstrate its improved impedance stability at higher switching frequencies. Furthermore, the highest grid impedance is found at the farthest point from the MV transformer, at bus B2, which reflects the worst-case scenario that could jeopardize the stability of the VSI-grid system. As a result, this is the only case considered in this study.

As mentioned earlier, the Nyquist stability criterion (NSC) can be employed to examine the stability of VSI-grid system based on the ratio of the grid impedance Z_g to the inverter output impedance Z_o. However, the NSC is a relative measure for the VSI-grid system's stability. Hence, the frequency response, Bode plot, of both the VSI output impedance and the grid impedance can be used to obtain an absolute norm of the VSI-grid system stability. When assessing the impedance stability of the VSI-grid system, two crucial points in the Bode plots of both impedances should be observed [22]:

1. The intersection points of the inverter output impedance Z_o and the grid impedance Z_g magnitude curves, which represent the zero-dB crossing points (resonance frequency) as both impedances have the same magnitude.

2. The phase margin angle φ_{PM} at the intersection point of Z_o and Z_g which can be calculated as $\varphi_{PM} = 180 - [\varphi_{Z_g} - \varphi_{Z_o}]$, where φ_{Z_o} and φ_{Z_g} are the phase angle of the inverter output impedance and the grid impedance respectively. A positive phase margin angle ensures the stability of the VSI-grid system [23].

The interaction between the VSI and the grid is analyzed using the VSI output impedance Z_o and the grid impedance Z_g, as shown in Fig. 8. The Bode plot in Fig. 8 predicts the instability of the injected grid current I_g of the VSI-based GCF with f_{sw}= 10 kHz because the phase margin angle at the resonance frequency of 2440 Hz is negative, -2.8°. As a result, as shown in Fig. 9(a), the injected grid current I_g is substantially distorted, and the harmonic spectrum of I_g in Fig. 9(b) is dominated by the resonance frequency at zero-dB crossing of Z_o^g and Z_g. The grid current I_g of the GCF based

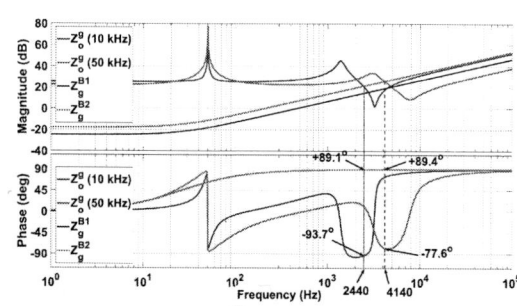

Fig. 8: Bode plot of Z_o^g at different f_{sw} and connected to Z_g at bus B1 and bus B2.

VSI with f_{sw}= 50 kHz, on the other hand, remains stable because the phase margin angle at the resonance frequency of 4140 Hz is positive, +13° as seen in the Bode plot in Fig. 8. The simulation findings in Fig. 10(a) and the harmonic spectrum of I_g in Fig. 10(b) both support this conclusion.

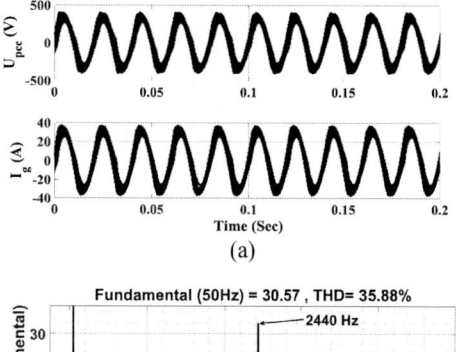

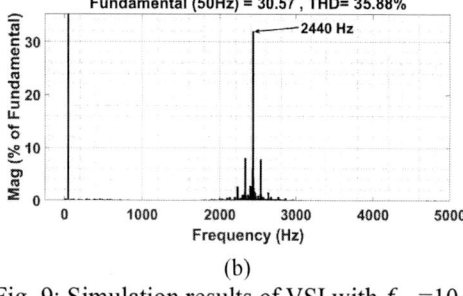

(a)

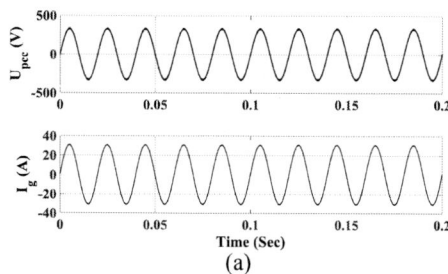

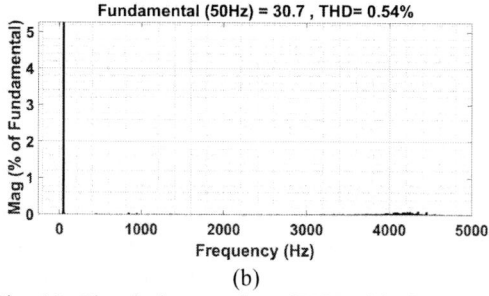

(a)

(b)

Fig. 9: Simulation results of VSI with f_{sw}=10 kHz connected to bus B2. (a) U_{pcc} and I_g waveforms (b) Harmonic spectrum of I_g.

(b)

Fig. 10: Simulation results of VSI with f_{sw}=50 kHz connected to bus B2. (a) U_{pcc} and I_g waveforms (b) Harmonic spectrum of I_g

Further case study is depicted in Fig. 11, where the VSI operates at f_{sw}= 20 kHz and f_{sw}= 100 kHz. The output impedance of the grid connected VSI, Z_o^g and the grid impedance, Z_g are used to test the system stability. According to the Bode plot in Fig. 11, the grid tied VSI is unstable when works at f_{sw}= 20 kHz and connected at bus B2, because the phase margin angle is negative, -0.4° at the resonance frequency of 3070 Hz.

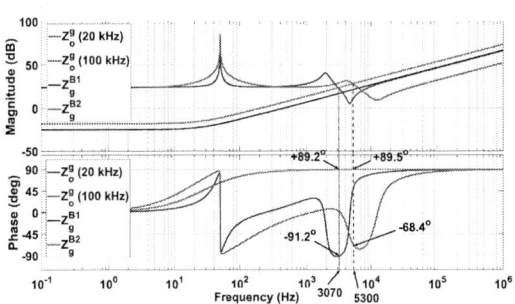

Fig. 11: Frequency response of Z_o^g at various f_{sw} and the VSI is connected at bus B1 and bus B2.

Therefore, as shown in Fig. 12, the injected grid current I_g of the VSI-based GCF with f_{sw}= 20 kHz is unstable. Further, as depicted in Fig. 12(b), the harmonic spectrum is dominated by the resonance frequency. In contrast, when the switching frequency f_{sw} is raised to 100 kHz, the stability of the grid connected VSI improves significantly because the phase margin angle is positive, +22.1° at the resonance frequency of 5300 Hz. As a result, the VSI-grid system has enough damping to attenuate the resonance frequency while maintaining the stability. The injected grid current I_g is perfect as shown in Fig. 13(a), and the harmonic spectrum of I_g in Fig. 13(b) has an acceptable power quality level.

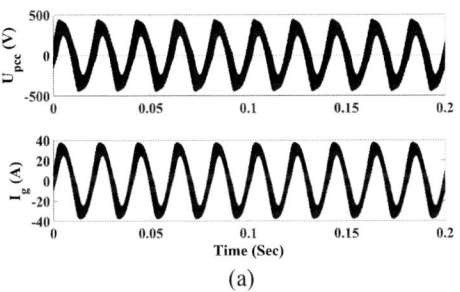

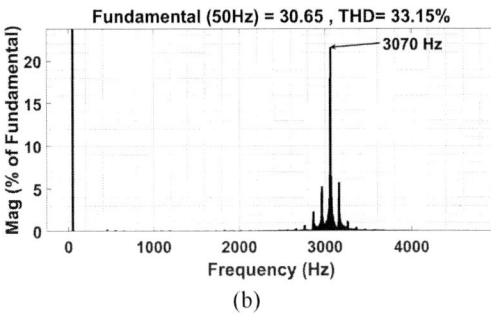

Fig. 12.: Simulation results of VSI with f_{sw}=20 kHz connected to bus B2. (a) U_{pcc} and I_g waveforms (b) Harmonic spectrum of I_g.

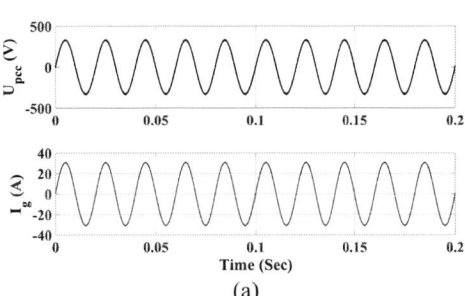

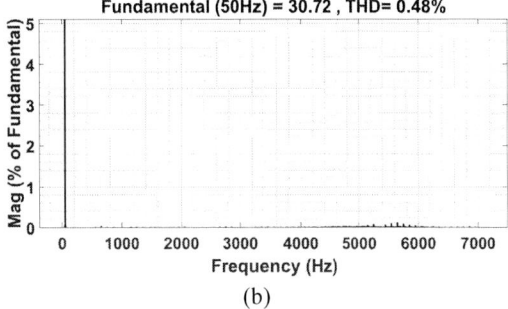

Fig. 13.: Simulation results of VSI with f_{sw}=100 kHz connected to bus B2. (a) U_{pcc} and I_g waveforms (b) Harmonic spectrum of I_g

Conclusion

The impedance stability of the WBG-based grid-connected VSI, which operates at higher switching frequencies, is explored at various points within the LV network. In Matlab/Simulink, the analytical expression of the VSI output impedance under both inverter-side and grid-side current control approaches is developed by using the small signal model of the single phase *LCL* grid connected inverter and the impedance expressions are validated by utilizing a test system that is constructed in Matlab/Simulink. The analysis and diverse study situations reveal that the impedance stability of the VSI-grid system is improved for GCF-based VSI at higher switching frequencies. The VSI-grid system obtains higher damping at higher switching frequencies, which ensures the stability of the grid connected inverter while also enhancing the quality of the injected grid current.

References

[1]   Singh S.: Recent Advancements in Wide Band Semiconductors (SiC and GaN) Technology for Future Devices, Silicon 2021, pp. 1–8.

[2]   Amin M.: Small-Signal Stability Assessment of Power Electronics Based Power Systems: A Discussion of Impedance-and Eigenvalue-Based Methods, IEEE Transactions on Industry Applications, vol. 53, no. 5, pp. 1–17, Sep. 2017.

[3]   Ali R.: Parameters Influencing Harmonic Stability for Single-phase Inverter in the Low Voltage Distribution Network, Proc. 2021 IEEE PES Innov. Smart Grid Technol. Eur. Smart Grids Towar. a Carbon-Free Futur. ISGT Eur. 2021, pp. 1–5, 2021.

[4]   Yu Y.: Modeling and analysis of resonance in LCL-type grid-connected inverters under different control schemes," *Energies*, vol. 10, no. 1, 2017.

[5]   Bao C.: Step-by-step controller design for LCL-Type Grid-Connected inverter with capacitor-current-feedback active-damping," IEEE Trans. Power Electron., vol. 29, no. 3, pp. 1239–1253, 2014.

[6]   Zheng C.: The Interaction Stability Analysis of a Multi-Inverter System Containing Different Types of Inverters, Energies 2018, vol. 11, no. 9.

[7]   Pan D.: Analysis and Design of Current Control Schemes for LCL-Type Grid-Connected Inverter Based on a General Mathematical Model, IEEE Transactions on Power Electronics, vol. 32, no. 6, pp.–16, Jun. 2017.

[8]   Zhou X.: A novel inverter-side current control method of LCL-Filtered inverters based on high-pass-filtered capacitor voltage feedforward, IEEE Access, vol. 8, pp. 1-11, 2020.

[9]   Dannehl J.: Limitations of voltage-oriented PI current control of grid-connected PWM rectifiers with LCL filters, IEEE Transactions on Industrial Electronics, vol. 56, no. 2, pp. 1-9, 2009.

[10]   Han Y.: Modeling and stability analysis of LCL-type grid-connected inverters: A comprehensive overview, IEEE Access, vol. 7, pp. 114975–115001, 2019.

[11]   Liserre M.: Design and control of an LCL-filter-based three-phase active rectifier, IEEE Trans. Ind. Appl., vol. 41, no. 5, pp. 1281–1291, 2005.

[12]   Beres R. N.: A Review of Passive Power Filters for Three-Phase Grid-Connected Voltage-Source Converters," IEEE J. Emerg. Sel. Top. Power Electron., vol. 4, no. 1, pp. 54–69, 2016.

[13]   Wang X.: Passivity-based design of passive damping for LCL-filtered voltage source converters, 2015 IEEE Energy Convers. Congr. Expo. ECCE 2015, pp. 3718–3725, 2015

[14]   Yang Z.: Stability Investigation of Three-Phase Grid-Tied PV Inverter Systems Using Impedance Models," IEEE J. Emerg. Sel. Top. Power Electron., vol. 6777, no. c, pp. 1–13, 2020.

[15]   Gharanikhajeh K.: A Harmonic Mitigation Technique for Multi-Parallel Grid-Connected Inverters in Distribution Networks," IEEE Trans. Power Deliv., pp. 1–1, Oct. 2021.

[16]   Bianchi N.: Active power filter control using neural network technologies, IEE Proceedings-Electric Power Appl., vol. 150, no. 2, pp. 139–145, 2003.

[17]   Berlingard Q.:RF performances at cryogenic temperature of inductors integrated in a FDSOI technology," Solid. State. Electron., p. 108285, 2022.

[18]   Zhou S.:"An Improved Design of Current Controller for LCL-Type Grid-Connected Converter to Reduce Negative Effect of PLL in Weak Grid," IEEE J. Emerg. Sel. Top. Power Electron., vol. 6, no. 2, pp. 648–663, 2018.

[19]   Yao W.: Design and Analysis of Robust Active Damping for LCL Filters Using Digital Notch Filters, IEEE Trans. Power Electron., vol. 32, no. 3, pp. 2360–2375, 2017

[20]   Barsali S.: Benchmark systems for network integration of renewable and distributed energy resources; 2014.

[21]   Jafarian M.: Grid Impedance Characterization To Provide a Robust Phase-Locked Loop Design for PV Systems," Proc. 2021 IEEE PES, ISGT Eur. 2021

[22]   Sowa I.: Impedance-based analysis of harmonic resonances in HVDC connected offshore wind power plants," Electr. Power Syst. Res., vol. 166, no. April 2018, pp. 61–72, 2019.

[23]   Jia L.: An Adaptive Active Damper for Improving the Stability of Grid-Connected Inverters under Weak Grid," IEEE Trans. Power Electron., vol. 33, no. 11, pp. 9561–9574, 2018.

Design and Modulation Optimization of an MMC Based Braking Chopper

Viktor Hofmann, Patrick Hofstetter
SIEMENS AG
Vogelweiherstr. 1-15
Nuremberg, Germany
Tel.: +49 / 162 – 691 6655
E-Mail: viktor.hofmann@siemens.com
URL: http://www.siemens.com

Keywords

«Modular Multilevel Converters (MMC)», «Braking Chopper», «High voltage power converters», «Medium voltage», «Industrial application».

Abstract

This paper investigates MMC based braking chopper configurations with a centralized resistor. It analyzes, optimizes and compares different cell configurations and modulation methods. A detailed description of the limiting factors and boundary conditions as well as the available degrees of freedom are provided and the theoretical derivations are validated by simulation.

Introduction

The Modular Multilevel Converter (MMC) [1] is a well-suited converter topology for medium- and high-voltage applications, which provides a lot of features such as high output voltage quality, efficiency and easy scalability [2] – [4]. In medium-voltage applications, it is often used as an adjustable-speed motor drive for large-capacity fans or compressors [5], [6]. In high-voltage direct current (HVDC) applications, the MMC can be utilized e.g. to connect offshore wind parks to the onshore grid [7], [8].

In both application fields, the systems often require the capability of dynamic braking. In case of a variable speed drive, the dynamic braking is primarily used for emergency purposes. Furthermore, it is more cost effective than regenerative braking [9] or applied when there is no possibility of regenerative braking [10]. In general, this is the case when a diode rectifier is used as the front-end inverter (Fig. 1(a)). In HVDC applications, the system configuration typically consists of two back-to-back VSC terminals (Fig. 1(b)). These common system configurations are illustrated in Fig. 1. For the discussed connection of wind parks, it is necessary that the system applies to the grid code requirements. The AC grid voltage may drop for several periods and disable the capability of regenerative braking. It is essential to dissipate the generated wind power during this time to avoid DC overvoltage or even major system damages.

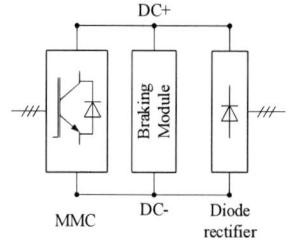

(a) MMC with diode front-end

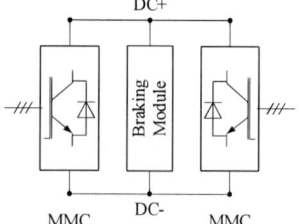

(b) Back-to-back MMC

Fig. 1: Typical system configurations of a braking application

The braking module is usually installed on the DC side between the terminals DC+ and DC-. There are several chopper topologies, which provide different advantages and disadvantages. The most common types are shown in Fig. 2. The classic way to realize a braking chopper, is the series connection of an IGBT and a braking resistor with a parallel diode (Fig. 2(a)). For higher system voltages, the IGBTs and diodes have to be designed as several series connected semiconductors. The power dissipation is performed by switching all semiconductors simultaneously in a two-level PWM operation, where the full DC-link voltage is applied across the chopper resistor. This topology offers a very simple design and control. The drawbacks are the very high dv/dt and di/dt as well as a bad EMI performance.

To reduce dv/dt and di/dt, modular chopper topologies were proposed, which offer a strictly modular design and good scalability [11, 12]. The main difference is the implementation of the chopper resistor. One possibility is to use a series connection of common MMC submodules (SM), such as half bridges (HB) and full bridges (FB), in series to a braking resistor (Fig. 2(b)). In this case, there is a single lumped resistor, which can be mounted outdoor and supplied by an air cooling.

Another solution is to distribute the braking resistor into the SMs. Typically, this option needs a water cooling and usually, the resistors cannot be mounted separately from the cell. It requires a special cell design and is challenging in diverting the high amount of generated heat. A possible design of such an SM is illustrated in Fig. 2(c). There are a lot of different cell designs. For example, it is possible to realize the switches 1 and 2 as passive components or the switch 3 as a single IGBT without an antiparallel diode to reduce the semiconductor effort. Furthermore, investigations were performed to combine the distributed resistor SM with a lumped resistor to reduce the SM cooling effort and increase the system performance [12].

The various topologies have different design criteria and operation modes. The general advantages and disadvantages were analyzed in [13]. This paper focuses on the topology shown in Fig. 2(b). A sinusoidal modulation method and an investigation on aspects of semiconductor utilization was presented in [14]. This paper focuses on the optimum design for different operation points. It considers the usage of only half bridges (HB) and full bridges (FB) within the braking chopper as well as a combination of both submodules (HB + FB) within a MMC braking chopper. Furthermore, it investigates the usage of different modulation methods, such as a superimposed higher frequency sinusoidal or a trapezoidal modulation. The derived optima will be validated by simulations.

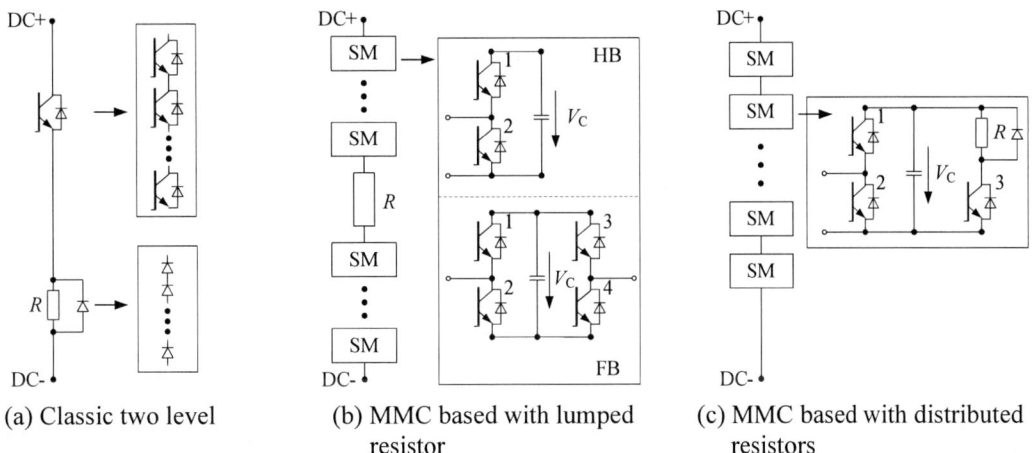

(a) Classic two level

(b) MMC based with lumped resistor

(c) MMC based with distributed resistors

Fig. 2: DC chopper topologies

Functional Principle

The considered MMC based braking chopper topology is illustrated in Fig. 2(b) and the functional principle will be explained by means of the equivalent circuit diagram shown in Fig. 3. The series connection of submodules can be seen as an adjustable voltage source. v_{Br} describes the modulated voltage of the braking chopper submodules and $v_{Arm,xp/n}$ represents the modulated arm voltages of the positive (p) and negative (n) MMC arm, respectively, in a phase x. L_{Arm} is the arm inductance and R_{Arm} summarizes all resistive parts of an MMC arm.

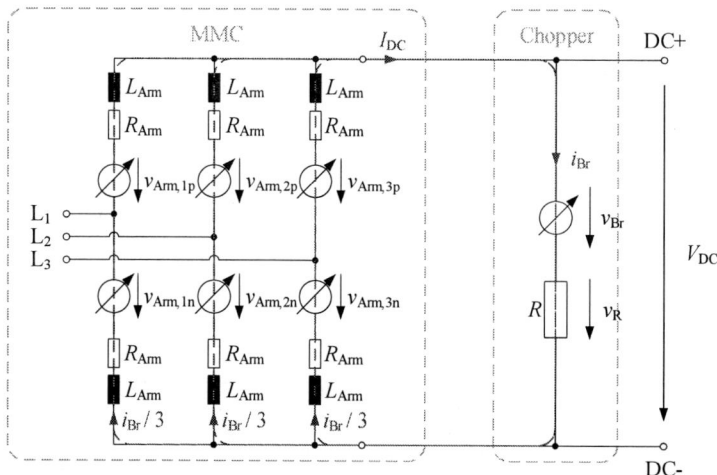

Fig. 3: Equivalent circuit diagram of MMC based braking chopper

In idle mode, the braking chopper modulates the DC voltage $v_{Br} = V_{DC}$ and there is no current flow through the braking resistor ($i_{Br} = 0$). Due to zero-current switching, the losses are neglectable during this operation.

During active chopper operation, the dissipation power is controlled by the DC current component $i_{Br,DC}$ of the chopper current i_{Br}. It is assumed that the total power which results from the MMC is dissipated by the braking resistor. Consequently, the DC component of the chopper current is equal to the MMC DC current I_{DC} and the dissipated power P_{Br} is given by:

$$P_{Br} = V_{DC} \cdot i_{Br,DC} = R \cdot i_{Br,rms}^2 \tag{1}$$

$$i_{Br,DC} = I_{DC} \tag{2}$$

Furthermore, it is necessary to ensure that the net energy flow to the cell capacitors $\overline{e_{Br}}$ is zero:

$$\overline{e_{Br}} = \int_0^{2\pi}(v_{Br} \cdot i_{Br})\mathrm{d}\omega t = 0 \tag{3}$$

Therefore, an AC current is modulated, which circulates between the MMC and the braking chopper. In this paper, three different AC current profiles are considered: a sinusoidal current (1H), a sinusoidal current that is additionally superimposed with a third harmonic (3H) and a trapezoidal current (TR). Consequently, the AC current component is given by:

$$i_{Br,AC,1H} = \hat{I}_{AC} \cdot \sin(\omega t) \tag{4}$$

$$i_{Br,AC,3H} = \hat{I}_{AC} \cdot (\sin(\omega t) + 1/6\sin(3\omega t)) \tag{5}$$

$$
i_{\mathrm{Br,AC,TR}} =
\begin{cases}
\dfrac{dv/dt}{R \cdot 2\pi f} \cdot \omega t & 0 \le \omega t < \tau \\[2mm]
\hat{I}_{\mathrm{AC}} & \tau \le \omega t < \pi - \tau \\[2mm]
-\dfrac{dv/dt}{R \cdot 2\pi f} \cdot (\omega t - \pi) \quad \text{if} \quad & \pi - \tau \le \omega t < \pi + \tau \quad , \quad \text{with } \tau = \dfrac{\hat{I}_{\mathrm{AC}} \cdot R \cdot 2\pi f}{dv/dt} \qquad (6) \\[2mm]
-\hat{I}_{\mathrm{AC}} & \pi + \tau \le \omega t < 2\pi - \tau \\[2mm]
\dfrac{dv/dt}{R \cdot 2\pi f} \cdot (\omega t - 2\pi) & 2\pi - \tau \le \omega t < 2\pi
\end{cases}
$$

$\hat{I}_{\mathrm{AC}}$ describes the amplitude of the modulated current, dv/dt the permissible voltage rise and f the modulated fundamental frequency. The resulting braking chopper current is finally given by:

$$
i_{\mathrm{Br}} = i_{\mathrm{Br,DC}} + i_{\mathrm{Br,AC}} \tag{7}
$$

The voltage drop across the resistor is given by Ohm's law and the modulated voltage of the braking chopper v_{Br} has to ensure a constant DC voltage between the terminals DC+ and DC− at any time:

$$
v_{\mathrm{Br}} = V_{\mathrm{DC}} - R \cdot i_{\mathrm{Br}} \tag{8}
$$

These equations clearly describe the braking chopper operation. The energy requirement of the chopper cells is fulfilled by adjusting the modulated AC current component. For a given system configuration and operation point, the AC amplitude of the modulated current is calculated by solving eq. (3) for the different modulation methods:

$$
\hat{I}_{\mathrm{AC,1H}} = \sqrt{2 \left(\frac{V_{\mathrm{DC}} \cdot I_{\mathrm{DC}}}{R} - I_{\mathrm{DC}}^2 \right)} \tag{9}
$$

$$
\hat{I}_{\mathrm{AC,3H}} = \sqrt{\frac{72}{37} \left(\frac{V_{\mathrm{DC}} \cdot I_{\mathrm{DC}}}{R} - I_{\mathrm{DC}}^2 \right)} \tag{10}
$$

$$
\hat{I}_{\mathrm{AC,TR}} = \frac{dv/dt}{8Rf} \cdot \left\{ 1 - 2\cos\left[\frac{1}{3} \left(\pi + \tan^{-1} \left(\frac{8f\sqrt{I_{\mathrm{DC}} \cdot R(V_{\mathrm{DC}} - I_{\mathrm{DC}} \cdot R)(48(R \cdot f \cdot I_{DC})^2 - 48 V_{\mathrm{DC}} \cdot I_{\mathrm{DC}} \cdot R \cdot f^2 + (dv/dt)^2)}}{96(R \cdot f \cdot I_{DC})^2 - 96 V_{\mathrm{DC}} \cdot I_{\mathrm{DC}} \cdot R \cdot f^2 + (dv/dt)^2} \right) \right) \right] \right\} \tag{11}
$$

Exemplary voltage and current profiles are illustrated in Fig. 4 for different modulation methods.

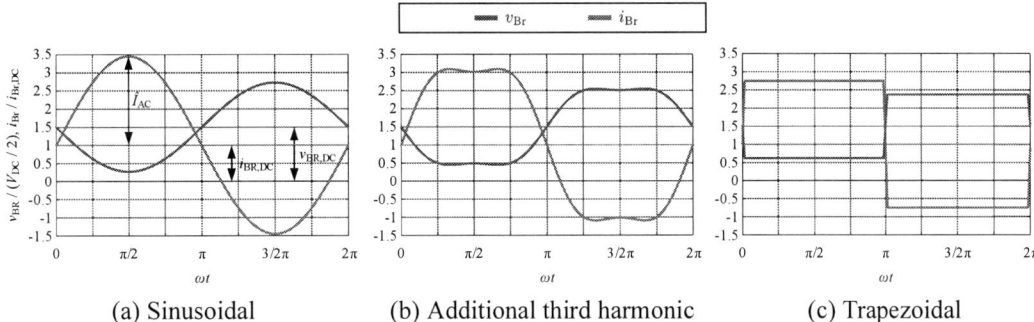

(a) Sinusoidal   (b) Additional third harmonic   (c) Trapezoidal

Fig. 4: Exemplary modulated fundamental voltage (blue) and current profiles (red) of the braking chopper for different modulation methods

Design optimization

In this section a design optimization is performed for different modulation methods and cell configurations of the braking chopper. Several limitations are discussed and the optimum point of operation is derived.

Theoretical operation area

An important issue is the power dissipation capability of the braking chopper, which can be described by the maximum admissible DC current. As the braking chopper can easily be scaled for higher voltage ratings by connecting several submodules in series, the performed analysis is not limited to any certain voltage level. There are several degrees of freedom and different limitations, which must be considered. The basic limitations depend on the used cell types as well as on the applied modulation method. For a given system, the possible operation area of the braking chopper can be described by the DC current and the specific resistance R_{Spec}:

$$R_{Spec} = \frac{R}{V_{DC}} \tag{12}$$

The various limitations will be discussed in the following. First, an MMC braking chopper is considered, which is only assembled with HB submodules and sinusoidal modulated. Since only a positive or zero voltage level can be modulated with half bridge cells, this topology is limited to a modulated chopper voltage that is always greater zero. The green hatched area in Fig. 5 shows the possible operation area of this topology, where I_{Nom} describes the nominal current rating of the applied semiconductors and is assumed to be $I_{Nom} = 1400A$.

The admissible operation area can be expanded, if at least a portion of full bridges is installed in the braking chopper. In this mixed cell configuration, it is important to pay attention to an energy balanced operation, because the HB can only be discharged during a negative cell current. In dependency of the operation point, a certain amount of full bridges must be available to ensure a stable operation [15]. The possible operation area of this mixed cell configuration (HB+FB) is limited by the red line in Fig. 5. At this point, the chopper current is always positive and the HB cannot be discharged anymore.

It is possible to shift both before mentioned limitations by using other modulation methods. An additional third voltage harmonic or the trapezoidal modulation expands the possible operation area of the HB chopper due to the lower absolute voltage level of the modulated chopper voltage (cf. Fig. 4 and eq. (1)-(11)). In contrast, the possible operation range of the mixed cell chopper configuration decreases, because of the reduced negative chopper current. The trapezoidal modulation leads to the same possible operation area of the HB and mixed cell braking chopper.

To enable an operation in the black hatched area, it is necessary to use bipolar submodules such as full bridges. FB are not limited by the before mentioned restrictions.

The black line marks the absolute limit of the considered operation principle for all braking chopper configurations. At this point, the DC current component of the braking chopper $i_{Br,DC}$ causes a voltage drop across the braking resistor that is equal to V_{DC}. Consequently, no active voltage modulation must be performed by the braking chopper at this point and the total power is already dissipated completely by the DC component. Higher power ratings would require the modulation of a negative DC chopper voltage. This would lead to a non-constant DC terminal voltage V_{DC}, when it is attempted to ensure the energy balance in the chopper cells.

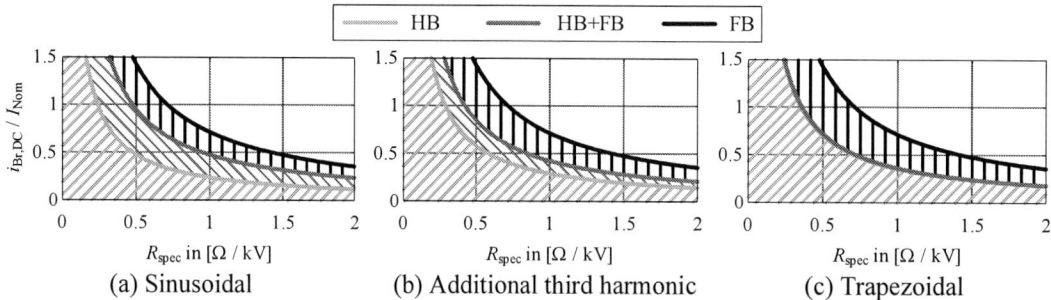

Fig. 5: Possible operation area of different cell configurations and different modulation methods

Partial load operation

Another important aspect is the partial load operation. The chopper must be able to operate at all partial load conditions. Therefore, a partial load factor a is introduced:

$$a = \frac{i_{Br,DC}}{I_{DC}} \tag{13}$$

$a = 1$ describes a full load operation as it was discussed before and $a = 0$ represents the idle mode. The maximum voltage level, which has to be modulated by the chopper arm, increases at lower partial load factors a (cf. eq. (1) – (11)). Fig. 6 illustrates the maximum modulated arm chopper voltage $V_{Br,Max}$ for different modulation methods in dependence of the partial load factor a. The maximum modulated voltage level is an important design criterion and determines the minimum needed amount of cells. The maximum modulated voltage level can be reduced by using a trapezoidal modulation. This reduces the hardware effort and consequently the total costs. This illustration is irrespective of the used cell type and also valid for the different current limits of the appropriate cell types according to Fig. 5. Increasing the partial load factor from $a = 0$ to $a = 1$ (Fig. 6) is equivalent to increasing the DC current from zero to the possible limit of the appropriate modulation method and cell configuration for a fixed resistance as it is depicted in Fig. 5.

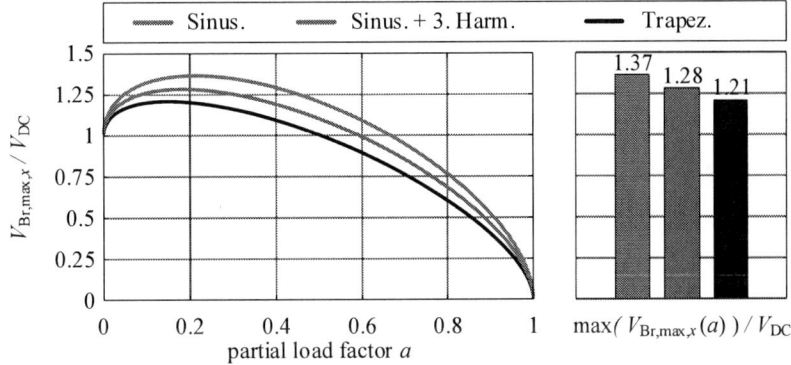

Fig. 6: Maximum modulated voltage level of a braking chopper in dependence of the partial load factor a.

Beside the total amount of cells, the cell capacitor sizing is a further design criterion. It is determined by the energy fluctuation e_{Br} and the resulting energy ripple ΔE_{Br}, respectively:

$$e_{Br} = \int_0^{t'} (v_{Br} \cdot i_{Br}) d\omega t \tag{14}$$

$$\Delta E_{Br} = \max(e_{Br}) - \min(e_{Br}) \tag{15}$$

The dimensioning of the cell capacitors is also affected by the desired tolerance band of the submodules K_T. The energy fluctuation corresponds to a permissible cell voltage range. Consequently, the relationship between the submodule capacitance C_m, the energy ripple and the permissible voltage fluctuation as well as the nominal cell voltage $V_{C,0}$ is specified by:

$$\frac{\Delta E_{Br}}{N_{Br}} = 2K_T C_m V_{C,0}^2 \tag{16}$$

Fig. 7 illustrates the relative energy ripple for different cell types and modulation methods. The values are related to the modulation frequency f_M and the maximum power P_{Max}. This power is reached for a specific chopper configuration at full load operation $a = 1$ (cf. Fig. 5). Higher power ratings and lower modulation frequencies increase the absolute energy ripple accordingly.

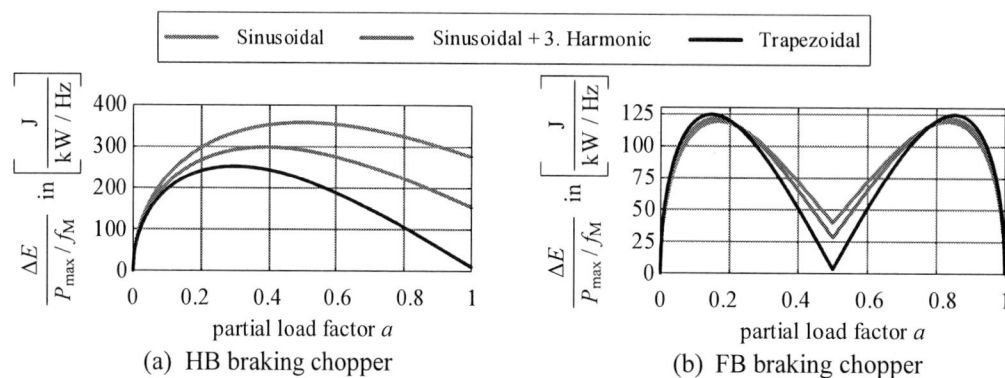

(a) HB braking chopper   (b) FB braking chopper

Fig 7: Energy ripple in partial load operation for different cell types and modulation methods

The FB chopper configuration (Fig. 7(b)) offers the lowest energy ripple at a partial load factor $a = 0.5$. This operation is marked by the red line in Fig. 5(c), that also represents the absolute operation limit of the trapezoidal modulated HB and mixed cell configuration. At this point, the chopper voltage and current profiles modulate alternately the minimum values within a fundamental modulation period. A low energy ripple is also established at no load and full load operation (the full load operation is represented by the black line in Fig. 5). All modulation methods provide the same maximum power for a specific chopper configuration (cf. Fig. 5). The maximum energy ripple is approximately in the same range for all modulation methods.

For HB chopper configurations (Fig. 7(a)), the trapezoidal modulation method is beneficial in comparison to the sine-based variants. It provides the lowest relative energy ripple and consequently the highest utilization of the installed capacitors. Note, that the discussed modulation methods result in different maximum power values (cf. Fig. 5), when a HB chopper is considered.

Thermal admissible current

Another limitation is the thermal admissible current of the semiconductors. The thermal load is evaluated by calculating the semiconductor losses of the cells. This is performed by an average loss model according to [16], which offers a high accuracy for steady state operation. Afterwards, the junction temperature can be determined by means of a thermal model. Fig. 8 shows the resulting thermal limitation of a HB and a FB braking chopper for different modulation methods. The trapezoidal modulation increases the power dissipation capability significantly in comparison to a sinusoidal modulation (note, that the sinusoidal modulation method has its maximum at $R_{\text{spec}} = 0.42 \ \Omega/\text{kV}$ and $i_{\text{Br,DC}}/I_{\text{Nom}} = 0.56$). For higher ratings of the specific resistor R_{spec}, the chopper operation is not limited by thermal issues.

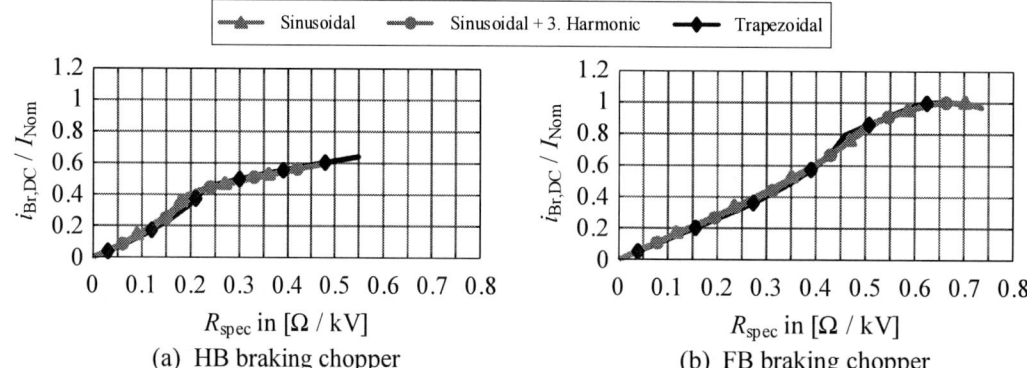

(a) HB braking chopper   (b) FB braking chopper

Fig. 8: Thermal limitation of the HB and FB braking chopper for different modulation methods

Resulting operation area

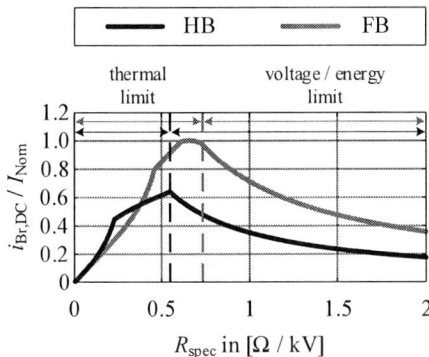

Fig 9: Resulting admissible operation area for a trapezoidal modulation method and different braking chopper configurations.

The analysis shows, that the trapezoidal modulation method is beneficial in several concerns. The half bridge cell type should be used for common power ratings and provides lower hardware costs. By assembling the braking chopper with full bridge submodules, it is possible to expand the possible operation area and realize higher power ratings. Fig. 9 shows an overview of the resulting admissible operation range for a trapezoidal modulation method and different braking chopper topologies. HB and FB chopper configurations are thermally limited for lower specific resistor values (dashed lines in Fig. 9). At higher resistor values, voltage and energy limitations delimit the braking chopper operation.

The most relevant design criteria were analyzed up to now. However, there are further issues in a real system configuration, such as a maximum admissible peak current of the semiconductors or the admissible rms current of passive and mechanical components (e.g. bus bars), that won't be discussed in this paper.

Simulation results

To verify the functionality of the proposed modulation method and validate the analysis, full scale simulation studies are performed. The investigated system is a grid connected back-to-back MMC configuration and the equivalent circuit diagram is illustrated in Fig. 10. The system design is based on typical medium voltage applications. Both grid sides operate at a $f_G = 50$ Hz grid frequency and AC phase-to-phase voltages of $v_{sLI/II,x} = 11$kV. The MMC DC voltage is set to $V_{DC} = 18$kV.

A HB chopper configuration is applied with $N_{Br} = 20$ chopper cells. An exemplary design will be explained in the following. To ensure a sufficient voltage availability during partial load operation (cf. Fig. 6), the nominal cell voltage is set to $V_{C,0} = 1.1$kV. A $C_m = 5$ mF submodule capacitance is chosen. In this way, the braking chopper and all MMC arms are equipped with the same amount of cells and also with the same cell type. By evaluating eq. (16), the chopper submodule capacitance can be further reduced.

According to Fig. 9, the highest chopper power is achieved at a specific resistor $R_{spec} = 0.549\,\Omega/\mathrm{kV}$. This leads to an absolute resistance $R_{Br} = 9.882\,\Omega$ with a maximum admissible DC current $i_{Br,DC} = 898$ A and the maximum power $P_{max} = 16.16$ MW, respectively. Various operating scenarios are performed to verify different operation modes. The simulated operation sequence is illustrated in Table I. Fig. 11 shows the simulation results.

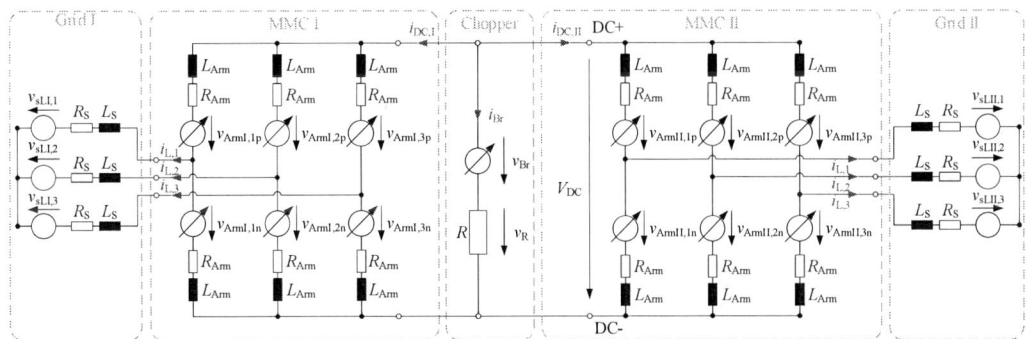

Fig. 10: Equivalent circuit diagram of the simulated system.

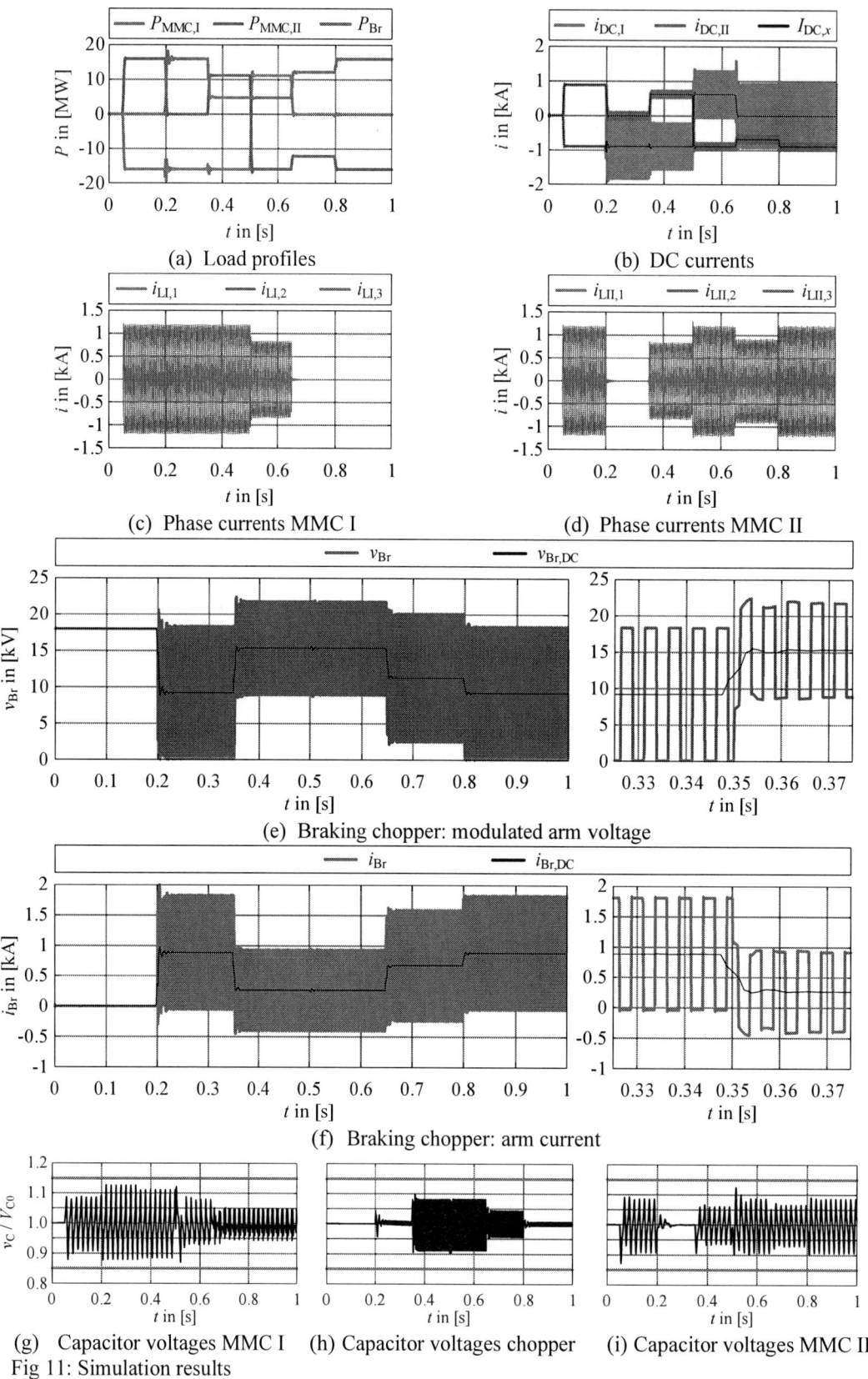

(a) Load profiles

(b) DC currents

(c) Phase currents MMC I

(d) Phase currents MMC II

(e) Braking chopper: modulated arm voltage

(f) Braking chopper: arm current

(g) Capacitor voltages MMC I

(h) Capacitor voltages chopper

(i) Capacitor voltages MMC II

Fig 11: Simulation results

Table I: Simulated operation sequence

	$P_{\text{MMC,I}}$ [MW]	P_{BR} [MW]	$P_{\text{MMC,II}}$ [MW]
$t_1 = 0.05\text{s}$	-16	0	16
$t_2 = 0.20\text{s}$	-16	16	0
$t_3 = 0.35\text{s}$	-16	4.8	11.2
$t_4 = 0.50\text{s}$	11.2	4.8	-16
$t_5 = 0.65\text{s}$	0	12.2	-12.2
$t_6 = 0.80\text{s}$	0	16	-16

The simulation starts in an idle mode. MMC I is voltage controlled and MMC II operates in current control mode. At $t = 50\text{ms}$ a first load step is performed to $P_{\text{MMC,I}} = -16\,\text{MW}$ and $P_{\text{MMC,II}} = 16\,\text{MW}$. This represents a total power transfer from grid I to grid II. The Chopper is activated at $t = 200\text{ms}$ and dissipates the total power. Consequently, no energy is transferred to MMC II leading to zero DC and AC currents (Fig. 11(b) and Fig. 11(d)). The chopper operates with a modulation frequency $f_{\text{M}} = 200\,\text{Hz}$. Due to the operation mode, the circulating chopper current is established between the braking chopper and MMC I. This leads to an increase of the capacitor voltages in this converter (Fig. 11(g)).

At $t = 350\text{ms}$ the dissipated chopper power is reduced to $P_{\text{Br}} = 4.8\,\text{MW}$ and the remaining energy is transferred to MMC II. This operation point results in the biggest cell voltage ripple of the braking chopper (cf. Fig. 7(a)). At this point, the partial load factor is $a = 0.3$ and the energy ripple is $\Delta E_{\text{Br}} = 20.4\,\text{kJ}$, according to Fig. 7(a). Evaluating eq. (16), the considered system configuration and operation point leads to a voltage tolerance band $K_{\text{T}} = 0.084$. That means, the voltage variation is $\pm 8.4\%$ in relation to the nominal capacitor voltage. That matches well with the simulation (Fig. 11(h)).

The load flow direction is reversed from MMC II to MMC I at $t = 500\text{ms}$. At $t = 650\text{ms}$ a further load step is performed to $P_{\text{MMC,II}} = -12.2\,\text{MW}$ and the total power is dissipated in the braking chopper. This represents a partial load factor $a = 0.76$. According to Fig. 7(a), this operation point halves the energy ripple, what is also in accordance to the simulation results.

Finally, the power rating is increased to $P_{\text{MMC,II}} = -16\,\text{MW}$ at $t = 800\text{ms}$ and the chopper operates at nearly full load. The energy ripple is drastically reduced at this operation point. It can be seen that all requested points of operation are in a steady state mode in an appropriate time and the tolerance band is not exceeded, even during transient load steps. This shows the high flexibility of the considered operation method.

Conclusion

This paper investigates MMC based chopper topologies with a centralized resistor. It considers different cell configurations, such as half bridges, full bridges or a mix of both cell types. Furthermore, different modulation methods are analyzed. This paper presents a detailed description of the most relevant limiting factors and boundary conditions as well as the available degrees of freedom. Design and modulation optimizations are performed and it is shown that the trapezoidal modulation method is beneficial in comparison to sine-based modulations in many concerns. The HB chopper configuration should be used for common power ratings and provides an attractive solution with regard to system costs. The theoretical derivations are validated by a full-scale simulation model showing good correlation between the theoretical calculation and the generic system operation.

References

[1] R. Marquardt, "Modular Multilevel Converter: An universal concept for HVDC-Networks and DC-Bus-applications," International Power Electronics Conference, pp. 502-507, 2010.

[2] M. Perez, S.Bernet, J. Rodriguez, S. Lizana and R. Kouro, "Circuit topologies, modelling, control schemes and applications of modular multilevel converters," IEEE Trans. Power Electron., vol. 30, no. 1, pp. 4-17, 2015.

[3] M. Malinowski, K. Gopakumar, J. Rodriguez and M. Pérez, "A survey on Cascaded Multilevel Inverters," IEEE Trans. Ind. Electron., vol. 57, no. 7, pp. 2197-2206, 2010.

[4] M. Glinka and R. Marquardt, "A New AC/AC Multilevel Converter Family," IEEE Trans. Ind. Electron., vol. 52, no. 3, pp. 662-669, 2005.

[5] M. Hiller, D. Krug, R. Sommer, S. Rohner, "A new highly modular medium voltage converter topology for industrial drive applications," IEEE EPE'09 ECCE Europe, pp. 1-10, 2009.

[6] A. Antonopoulos et al., "Modular multilevel converter AC motor drives with constant torque from zero to nominal speed," IEEE Trans. Ind. App., vol. 50, no. 3, pp. 1982-1993, 2014.

[7] Y. Sun et al., "Predictive Power Control of Modular Multilevel Converter for Wind Energy Integration via HVDC," IEEE PEDG 2020, pp. 334-339, 2020.

[8] R. Mourouvin et al., "AC/DC Dynamic Interactions of MMC-HVDC in Grid-Forming for Wind-Farm Intergration in AC Systems," IEEE EPE'20 ECCE Europe, pp. 1-9, 2020.

[9] Y. Okazaki, S. Shioda and H. Akagi, "Performance of a Distributed Dynamic Brake for an Induction Motor Fed by a Modular Multilevel DSCC Inverter," IEEE Trans. Power Electron., vol. 33, no. 6, pp. 4796-4806, 2018.

[10] M. Hagiwara, K. Nishimura and H. Akagi, "A medium-voltage motor drive with a modular multilevel PWM converter," IEEE Trans. Power Electron., vol. 25, no. 7, pp. 1786-1799, 2010.

[11] P. Hofstetter, V. Hofmann and D. Karwatzki, "A detailed View on the Trapezoidal Operation for MMC Type Braking Chopper in Medium Voltage Applications," IEEE EPE' 22 ECCE Europe, p. 1-8, 2022.

[12] B. Xu et al., "A Novel DC Chopper Topology for VSC-Based Offshore Wind Farm Connection," IEEE Trans. Pow. Electron., vol. 36, no. 3, 2021.

[13] J. Maneiro; S. Tennakoon; C. Barker and F. Hassan, "Energy diverting converter topologies for HVDC transmission systems," IEEE EPE'13 ECCE Europe, pp.1-10, 2013.

[14] A. Birkel; A. Schön and M.-M. Bakran, " Analysis and semiconductor based comparison of energy diverting converter topologies for HVDC transmission systems," IEEE EPE'17 ECCE Europe, pp. 1-10. 2017.

[15] V. Hofmann and M.-M. Bakran, "An Optimized Hybrid-MMC for HVDC," Proc. PCIM Europe, pp. 1-8, 2016.

[16] A. Schön, A. Birkel and M.-M. Bakran, "Modulation and Losses of Modular Multilevel Converters for HVDC Applications," Proc. PCIM Europe, 2014.

Modeling the arrangement of drill holes for orthogonal biasing in controllable inductors for power electronic converters

Jonas Pfeiffer[1], Christoph Drexler[1], Pierre Küster[2], Peter Zacharias[2]
and Michael Schmidhuber[1]

[1]SUMIDA COMPONENTS &
MODULES GMBH
Advance R&D Europe
Dr. Hans-Vogt-Platz 1
94130 Obernzell, Germany
Tel.: +49 / (0)8591 - 937 552
E-Mail: mschmidhuber@eu.sumida.com
URL: https://www.sumida.com

[2]UNIVERSITY OF KASSEL
Department of Electrical Power Engineering
Wilhelmshöher Allee 71
34121 Kassel, Germany
Tel.: +49 / (0)561 - 804 6344
Fax: +49 / (0)561 - 804 6521
E-Mail: peter.zacharias@uni-kassel.de
URL: https://www.uni-kassel.de/eecs/evs

Acknowledgements

The authors thank Dr. Eckard Specht from the Otto-von-Guericke-University Magdeburg in Magdeburg, Germany, and his great website packomania.com which was a significant basis for this paper.

Keywords

«Device modelling», «Device optimization», «Magnetic device», «Hardware (not only Software)», «Passive component»

Abstract

Magnetic devices are essential components of power electronic converters. Unfortunately, they usually are also the bulkiest and heaviest components in the converter system. Controllable magnetic devices using active premagnetization are a promising approach regarding further volume and weight reduction. One method of active premagnetization is orthogonal biasing. To be able to use this method, drill holes in the ferrite core material are necessary. Because of the material's fragility and brittleness, manufacturing ferrite cores is a difficult and expensive process. For this reason, an effective arrangement of the drill holes is critical.

In this paper a basic model of drill hole arrangement based on congruent circle packing is presented for circular and square core cross sections. 2D-FEM-simulations and measurement results are made for comparison and verification. The deviations between calculation and measurement results are discussed and opportunities for model improvement are given.

Introduction

The reduction of volume, weight and costs is an essential goal in the development process of power electronic converters. By constantly improving wide band gap technologies, such as SiC and GaN semiconductor power devices, the switching frequency of novel converter systems is further increasing. A higher switching frequency can be an option to reduce the volume and thereby the weight and costs of the magnetic devices. However, even with the usage of higher switching frequencies, magnetic devices remain the bottleneck regarding a significant volume and weight reduction of the converter system.

According to [1], a promising approach regarding further volume and weight reduction are controllable magnetic devices. If an additional winding (auxiliary winding) is wound on or introduced within the core material the auxiliary current impacts the magnetic core's saturation in the closer vicinity of the auxiliary winding. Thereby, the characteristic of the magnetic device can be affected actively, which enables an additional degree of freedom regarding the device's design process.

As it is shown in [2], there are three different methods to design controllable magnetic devices. Their different designs, impacts and effects on the magnetic component are also explained in detail in the above-mentioned reference. One of the presented methods is orthogonal biasing which is the method investigated in this paper.

To realize this method, drill holes through the ferrite core are needed to introduce the auxiliary winding. However, manufacturing ferrite is a complex process because of the material's fragility and brittleness. Special tools like an Ultrasonic/Sonic Driller/Corer (USDC) are needed to guarantee precise machining and avoid damage to the core material. The use of these tools is very expensive, especially if many small drill holes are required. Furthermore, the core's effective magnetic cross section becomes smaller the more holes are drilled into the ferrite material. Assuming an unchanged main current through the magnetic device, a reduced cross section area leads to a higher flux density and thus to higher core losses.

For these reasons, the question arises as to the minimum number of drill holes as well as their arrangement for a given auxiliary current to achieve a maximum saturation effect of a defined core section. In this paper, a basic model for the arrangement of drill holes for circular and square magnetic cross sections is presented, discussed and verified by simulation as well as measurement results.

Systematization of orthogonal biasing

Orthogonal biasing is one of three different methods of active premagnetization. It can be seen as a special variant of mixed biasing, which is also called "Virtual-Air-Gap" (VAG)-principle. The VAG-principle is investigated and discussed in [3].

If an auxiliary current flows through the auxiliary winding, a small area of the ferrite material around the drill hole gets saturated. Thereby, the local permeability in this area decreases which has an impact of the magnetic core's overall reluctance. The higher the auxiliary current, the higher the core's overall reluctance. The impedance value of the magnetic device thus becomes controllable.

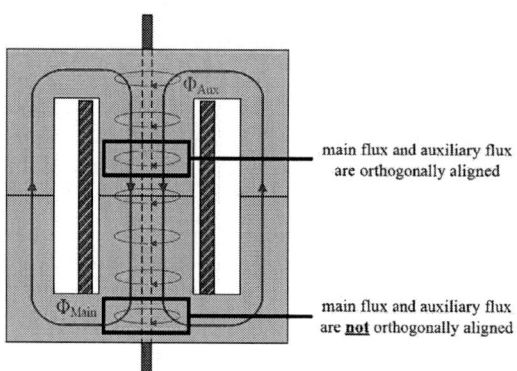

Fig. 1: Orthogonal biasing of an E-core (center leg). Main flux (red) and auxiliary flux (blue) are orthogonally aligned in the center leg but not in the yokes of the core

If a magnetic core section is orthogonally biased, main and auxiliary magnetic fluxes are orthogonally aligned. In theory, according to the superposition principle, both fluxes should not affect each other. Thereby, mathematical modeling becomes significant simpler. However, practical investigations show influences between main and auxiliary windings, e.g. due to the core geometry as the following example illustrates (see Fig. 1): If a center leg of an E-core should be affected by orthogonal biasing, the easiest way of machining is to drill a hole through the middle of the center leg and the yoke and introduce an auxiliary winding into the hole. Within the center leg itself, main and auxiliary magnetic flux are orthogonally aligned. However, within the yoke of the E-core the direction of the main flux changes because it closes in the outer legs. The auxiliary flux on the other hand remains in circles around the auxiliary winding. As a result, main and auxiliary magnetic flux are not orthogonally aligned in the yoke, which has an impact on the magnetic devices' characteristic.

Preliminary considerations of modeling and measurement prototypes

An exclusively orthogonal biased core is difficult to design due to the implementation of the auxiliary winding which must be introduced into and brought out of the core section. As mentioned in the previous section, drilling through the center leg of a ferrite core, e.g. an E-core as it is shown in Fig. 2 a), leads to a section within the yokes, where main and auxiliary magnetic fluxes are not orthogonally aligned because of the main flux's change of direction (see Fig. 1). This will lead to non-orthogonal biasing effects and with that a non-negligible impact on the magnetic device's characteristic in high auxiliary current ranges with negative effects on the model's accuracy.

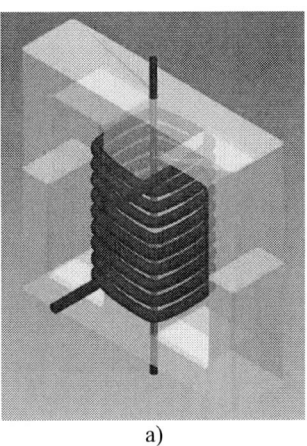

a)　　　　　　　　　　　　　　　　　　　b)

Fig. 2: Possible variants of the E 80/38/20 core for simulation and measurement verification of the model; light blue: ferrite core; red: main winding; dark blue: auxiliary winding

To avoid the described impact in the yoke of the E-Core, an alternative implementation of the auxiliary winding was investigated, as it is shown in Fig. 2 b). The leg of a core, e.g. the center leg of an E-core, is cut out an manufactured with a drill hole and a groove in which the auxiliary winding is placed. Thereby, no drilling through the yoke is needed. However, two small air gaps between center leg and yokes remain due to the cutting.

Both variants were simulated using the 3D FEM simulation software JMAG. SUMIDA Fi324 was used as ferrite material. For the variant shwon in Fig. 2 b) two air gap lengths between the center leg and the yokes of 0.1 mm each were assumed due to cutting off the centerleg from the yokes and glue it again.
The 3D FEM simulation results of the alternative variant in Fig. 2 b) show a negligible impact within the yokes. However, a non-negligible impact in the center leg around the outgoing of the auxiliary winding is ascertainable (see Fig. 3). Furthermore, the air gaps introduced into the core by cutting the center leg weaken the impact of orthogonal biasing on the magnetic device. These effects are also associated with higher manufacturing effort of the auxiliary winding's implementation than drilling a hole through the yoke and the center leg. For these reasons, it was decided to use the implementation variant shown in Fig. 2 a) for simulation and measurement verification, being aware of the described additional impact of the core section within the yokes which will lead to deviations of the model's calculations from the measurement results, especially if the auxiliary current increases. The manufactured cores for measurement verification are shown in Fig. 4.

Further considerations regarding general conditions were made to limit the number of the investigated shapes of a core's magnetic cross section. According to [4], the most frequent aspect ratio of an E-core's center leg is 1:1, which is a square. Moreover, there are numerous core geometries with circular (center) legs. Thus, the model will focus on circular and square cross sections.

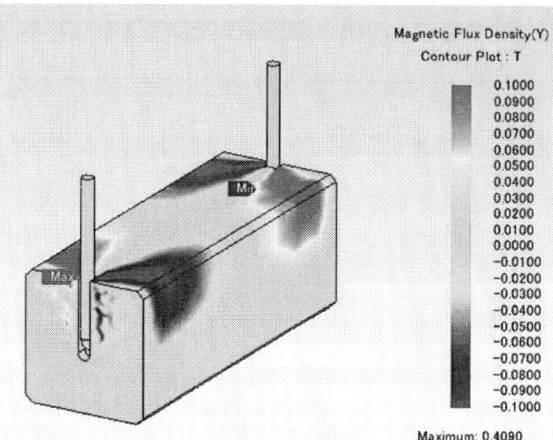

Fig. 3: Magnetic flux density (y-portion) around the outgoing of the auxiliary winding in the center leg of the E-core variant shown in Fig. 2 b) (main winding, yokes and outer legs are hidden)

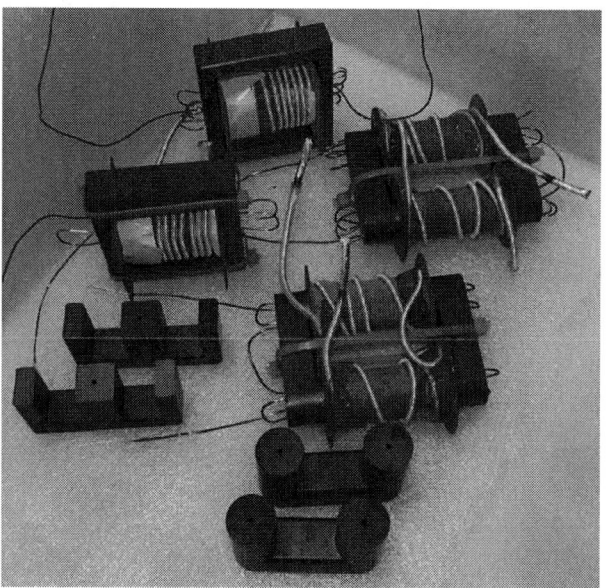

Fig. 4: Magnetic cores (URR 70/50/22 and E 80/38/20) with drill holes for measurement verification

Modeling the arrangement of drill holes for circular cross sections

It is assumed that the auxiliary magnetic flux caused by the auxiliary current, which is used for orthogonal biasing, flows in perfect circles around the auxiliary winding. The auxiliary magnetic flux density at the outside of the auxiliary winding is anti-proportional to the radius and decreases with $\frac{1}{r}$. Thus, the arrangement of the drill holes can be attributed to packing of congruent circles which is a well investigated mathematical field. Because of that, the optimal arrangement is well known and in parts mathematically proven. The best ways to pack two up to 20 congruent circles in a circular cross section is shown in [5].

To simplify the decreasing magnetic flux density, the material around the drill hole is divided in three different areas: saturated, "half-saturated" and unsaturated. By increasing the auxiliary current, the area will increase until the maximum radius of the circle is reached. The areas and their increase are shown in Fig. 5.

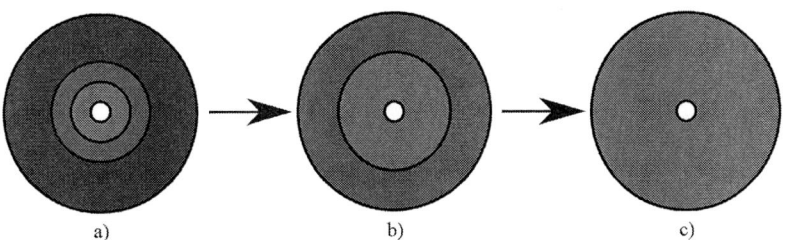

a)   b)   c)

Fig. 5: Simple saturation model of the material around the drill hole of a circular cross section. The different areas increasing according to the increasing auxiliary current;
red: saturated area; green: "half-saturated" area; blue: unsaturated area

It is obvious that the best packing density in a circular cross section is achieved by drilling a single hole in the middle of the magnetic core's leg. However, to achieve a maximum biasing effect it has to be ensured that the maximum auxiliary current is high enough the saturate the complete cross section. If the auxiliary current is limited, more than one drill hole could be needed. As already mentioned, drilling holes in ferrite material is a very expensive. Furthermore, a reduction of the magnetic core's cross section goes along with each additional drilling hole. For these reasons it seems reasonable to drill as few holes as possible to an upper limit of ten drill holes. According to [5], circular cross sections with one, four and seven drill holes have the highest packing density. Using seven drill holes, a maximum packing density of 77,78% can be achieved. A higher maximum packing density is only achieved if 19 drill holes are manufactured (80,32%).
A pair of URR 70/50/22 cores is used for modeling. Originally an URR 70/65/22 should be used but its legs had to be shortened due to the maximum drilling depth of the USDC tool. The holes were drilled in both legs.
To model the differential inductance of the core under the impact of orthogonal biasing the magnetic equivalent circuit of the core is calculated. Due to the increasing auxiliary current the circular effective area of the cross sections of both legs is decreasing until both legs are completely saturated, as it is shown in Fig. 5 c). In case of more than a single drill hole maximum saturation is achieved when the borders of the circles touch each other as it is shown in [5]. Thereby it is assumed that only the legs but not the yoke are affected by the orthogonal biasing because a strict orthogonal alignment of main and auxiliary flux is assumed.

Verification through measurements for circular cross sections

To verify the calculation model presented in the previous section, small signal measurements are executed using a Keysight 4980A Precision LCR Meter. Several URR 70/50/22 cores made of Fi324 (SUMIDA) were manufactured with one, four and seven drill holes according to the arrangements presented in [5]. The differential inductance value of the magnetic device in dependence of the auxiliary current is measured.
Because of deviations of the initial inductance values for $I_{Aux} = 0$ A between calculation model and measurement results the relative values of the device's inductance are used. The relative deviation between calculated and measured values are shown in Fig. 6.

The results in Fig. 6 show deviations smaller than 10% for auxiliary currents up to $I_{Aux} = 1$ A. For higher auxiliary currents, the deviations increase significantly. The main reason for this is assumed to be the impact of the yoke where main flux and auxiliary flux are not orthogonally aligned, as already discussed in the sections "systematization of orthogonal biasing" and "preliminary considerations of modeling and measurement prototypes". The more material is saturated through a high auxiliary current and/or a high number of drill holes, the higher the impact of the yoke section. Most deviations are negative which means that saturation in measurement results is higher than calculated.
Furthermore, 2D FEM simulations using FEMM 4.2 show that the saturation around the drill hole is not a circle if more than one drill hole is used (see Fig. 7). This also results in deviations between the calculation model and the measurement results. The reason for this is the vectorial superposition of the individual fields at the points of contact.

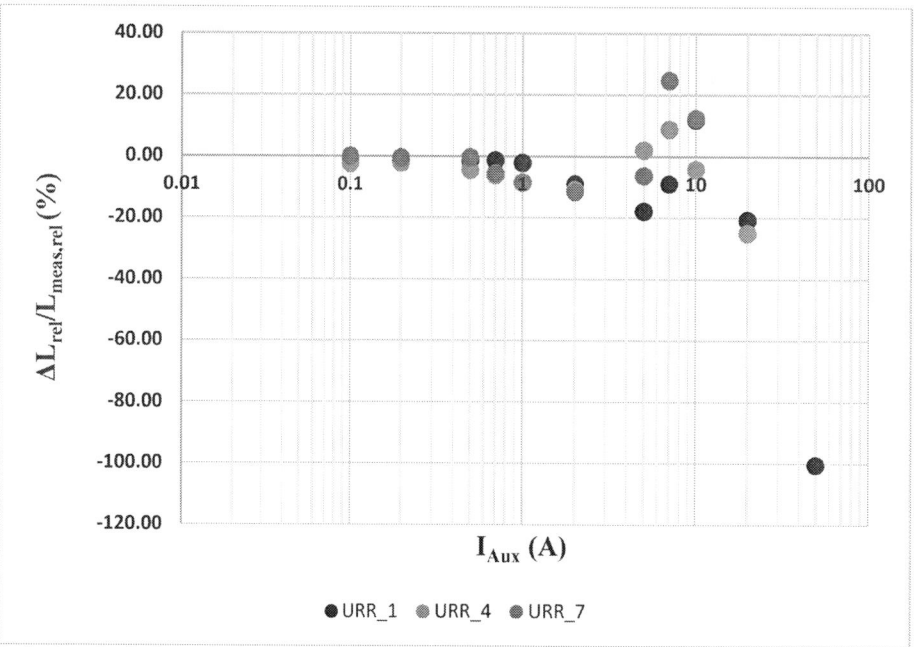

Fig. 6: Relative deviation of relative inductance value between calculation model and measurement results for an URR 70/50/22 core with one, four and seven drill holes within each leg

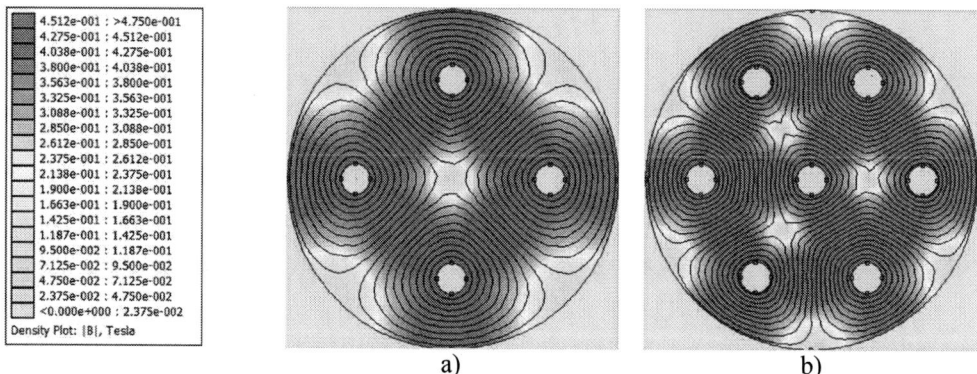

a)   b)

Fig. 7: Simulated magnetic flux density distribution of an URR 70/50/22 core's leg with four and seven drill holes ($I_{Aux} = 5\ A$)

Modeling the arrangement of drill holes for square cross sections

In a first step, the calculation model described in the section "modeling the arrangement of drill holes for circular cross sections" and shown in Fig. 5 is adapted to square cross sections as it is shown in Fig. 8, following the approach of circle packing.

The best ways to pack one up to 20 respectively 27 congruent circles in a square cross section are shown in [6] and [7]. According to the given references, the highest maximum packing densities using less than ten drill holes are achieved manufacturing one, four or nine holes (all 78,54%) which are thus the arrangements of investigation. A higher maximum packing density is achieved if 30 drill holes are manufactured (79,2%).

A pair of two E 80/38/20 cores is used for modeling. The holes are drilled through the center leg of the core. The calculation of the magnetic equivalent circuit is equivalent to the URR core in the previous mentioned section.

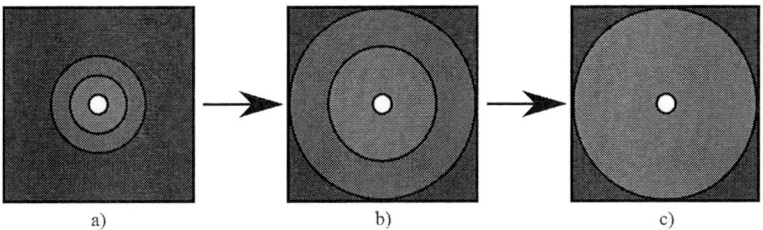

Fig. 8: Simple saturation model of the material around the drill hole of a square cross section. The different areas increasing according to the increasing auxiliary current;
red: saturated area; green: "half-saturated" area; blue: unsaturated area

The cross section is also simulated using 2D FEM software FEMM 4.2. As Fig. *9* shows, a circular expansion of the magnetic flux density is only present with small auxiliary currents. If the auxiliary current further increases a star-shaped density distribution is formed which finally leads into a cross-shaped density distribution. The reason for that is, that the distance from the middle of the drill hole to the middle of the outer edge of the square is shorter than the distance from the middle of the drill hole to the corner of the square. As a result, more cross-sectional area is available in the diagonal for the auxiliary magnetic flux which leads to a lower flux density in this area.

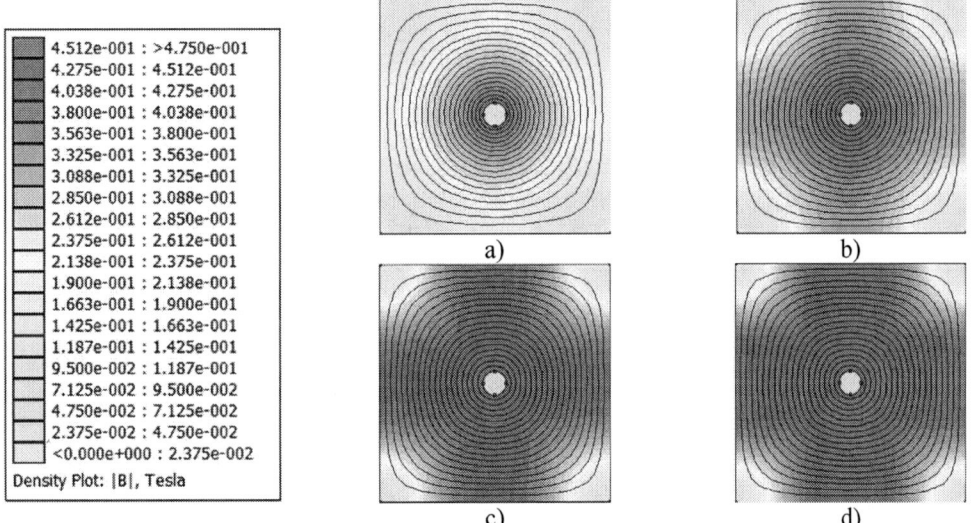

Fig. 9: Simulated magnetic flux density distribution of an E 80/38/20 core's center leg with a single drill hole; a): $I_{Aux} = 1\ A$; b): $I_{Aux} = 2\ A$; c): $I_{Aux} = 5\ A$; d): $I_{Aux} = 10\ A$

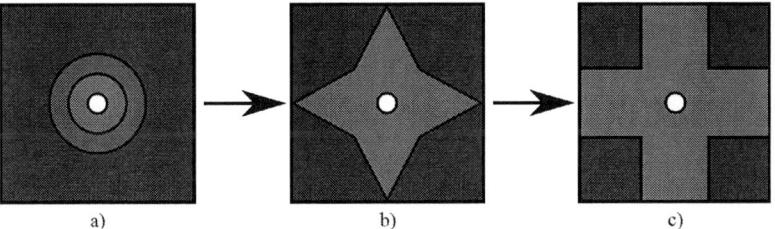

Fig. 10: Improved saturation model of the material around the drill hole of a square cross section. The different areas increasing according to the increasing auxiliary current;
red: saturated area; green: "half-saturated" area; blue: unsaturated area

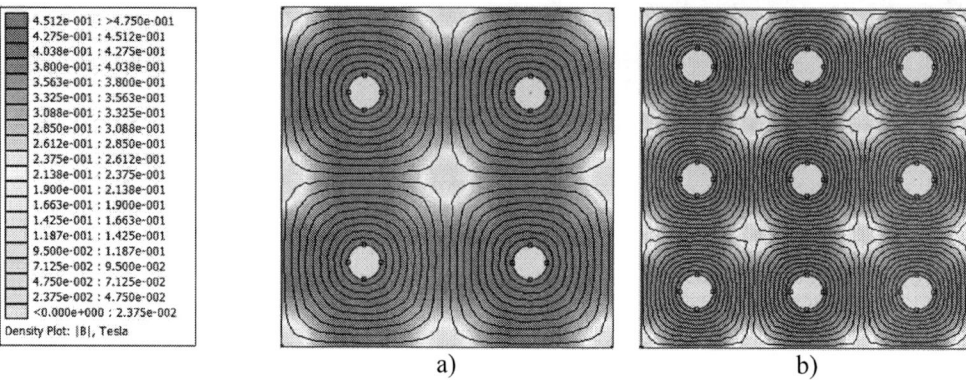

a)   b)

Fig. 11: Simulated magnetic flux density distribution of an E 80/38/20 core's center leg with four and nine drill holes ($I_{Aux} = 5\ A$)

According to the simulations results the calculation model is adjusted as it is shown in Fig. 10. A significant advantage of the investigated square cross sections is that the calculation model is easily transferable from a single drill hole to four or nine drill holes just by decreasing the square's edge length (see Fig. 11). Despite the different flux distribution, the arrangement of the drill holes according to circle packing remains unchanged.

Verification through measurements for square cross sections

Again, several E 80/38/20 cores made of Fi324 (SUMIDA) with one, four and nine drill holes according to [6] and [7] are manufactured and measured in small signal measurement using a Keysight 4980A Precision LCR Meter. The relative deviation between calculated and measured values are shown in Fig. 12. The circular-shaped markings represent the simple circular model (see Fig. 8) whereas the cross-shaped markings represent the improved model (see Fig. 10).

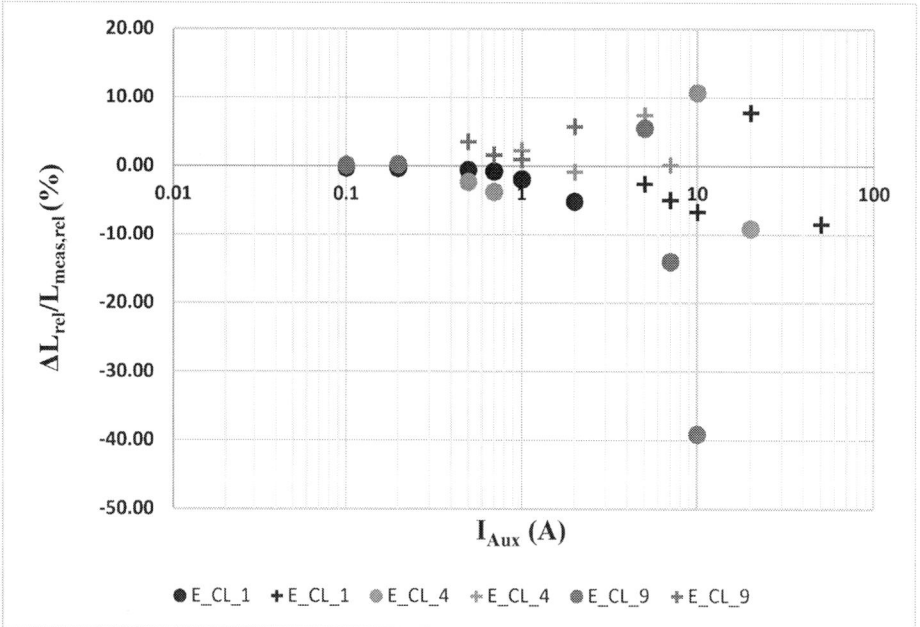

Fig. 12: Relative deviation of relative inductance value between calculation model and measurement results for an E 80/38/20 core with one, four and nine drill holes within the center leg (CL); circular-shaped markings: simple model; cross-shaped markings: improved model

Discussion and opportunities for improvement

The presented calculation model using a magnetic equivalent circuit is very simple and should be seen as a first approach regarding this topic. Still, most of the relative deviations between calculation model and measurement results are smaller than 20% for circular cross sections and smaller than 10% for square cross sections. However, there are plenty of opportunities for model improvement.

Just a few parameters used in the calculation model base on measurements of the used URR and E cores whereas most of the parameters is taken from the datasheets. These datasheet parameters are usually determined by measuring small toroidal cores which can lead to deviation if large core geometries are used as in this paper. Initial measurements of the core geometries that should be used in the application to gain more precise parameters for calculation will generate a higher accuracy of the model. However, the measurement of the cores is accompanied by an increased effort regarding time and available equipment.

Another opportunity for improvement is the division of the cross section. In the presented model the section is divided into an unsaturated, a "half-saturated" and a saturated area in dependence of the magnetic flux density. To achieve a higher accuracy of the calculation model, the division should be finer by increasing the number of different areas. Another approach could be to generate a function of the auxiliary magnetic flux as it is presented in [8]. Using that approach the magnetic flux density can be calculated very detailed in every point of the cross section. However, a computer-aided evaluation should be programmed for this approach.

The improved model for square cross sections is easily transferable from cores with a single drill hole to cores with more than one hole. Regarding circular cross sections a transferability to an increased number of drill holes is not trivial. Further investigations have to be executed to generate an improved calculation model with higher accuracy.

The most significant factor for deviations between calculation model and measurements is the impact of the yoke. Especially if a magnetic core with many drill holes is used, the saturation of the yoke as a large impact even by using medium auxiliary currents. Further investigations regarding the implementation of the auxiliary winding as they are presented in the section "preliminary considerations of modeling and measurement prototypes" should be made and evaluated in terms of their technical manufacturability. A more promising approach is to include the impact of the yoke in the calculation model.

It should be mentioned that the deviation shown in this paper might strongly depend on the selected operation point. Especially if high auxiliary currents are used to premagnetize the core what leads to a strong impact of the yoke, the materials hysteresis curve deforms as it is shown in [2]. Further investigations with regard to the impact of the operation point are useful.

Conclusion

Magnetic devices tend to be the bottleneck regarding volume and weight reduction in power converter systems. Even the constant improvement auf wide band gap semiconductor power devices and the associated increase of the switching frequency can only provide limited relief.

A promising approach regarding further volume und weight reduction are controllable magnetic devices using active premagnetization. One method of active premagnetization is orthogonal biasing. To be able to use this method drill holes within the magnetic core material are needed to introduce an auxiliary winding. Especially ferrite is a very fragile and brittle core material making its manufacturing complex and expensive. For this reason, the minimum number of drill holes as well as their arrangement for a given auxiliary current to achieve a maximum saturation effect of a defined core cross section is essential. In this paper, a basic model for the arrangement of drill holes for circular and square magnetic cross sections based on congruent circle packing was presented.

An exclusively orthogonal biased core is difficult to design because of the implementation of the auxiliary winding. Within the yoke of a magnetic core, magnetic main flux and magnetic auxiliary flux are not orthogonally aligned what leads to a significant impact of the yoke section regarding the controllability of the magnetic device. Two variants of a possible implementation of the auxiliary winding were simulated using 3D FEM software. Because of a significant larger manufacturing effort in connection with a non-negligible impact in the center leg around the outgoing of the auxiliary winding, it was decided to use the stanadard variant, being well aware of the yoke's negative impact on the models accuracy.

Calculation models for circular and square cross sections based on congruent circle packing were developed. Circle packing is a well investigated field in mathematical research which is why the optimal arrangement is well known and in parts mathematically proven. To calculate the inductance value of the magnetic device under the impact of orthogonal baising a magnetic equivalent circuit model was used. A leg of an URR 70/50/22 core was used for circular cross sections while the center leg of an E 80/38/20 core was used for square cros sections. The magnetic flux density distirbution was simulated using a 2D FEM software. Its results were compared to the calculation model. Because of a cross-shaped distribution of the magnetic flux density in square cross sections, the related calculation model was improved.

Both mentioned cores were manufactured with different number of drill holes and their inductance value was measured in small signal measurements.
The relative deviations of the relative differential inductance values of calculation and measurements show a good fitting (smaller than 10% deviation) for small auxilairy currents up to $I_{Aux} = 1$ A for circular cross sections respectively $I_{Aux} = 5$ A for squre cross sections. Most values are smaller than 20% deviation for circular cross sections and smaller than 10% deviation for square cross section. For larger auxiliary currents the impact of the yoke decreases the model's accuracy significantly as expected.

Because of the calculation models simpleness it can be seen as a first approach regarding the search for the most effective drill hole arrangement. However, its accuracy should be improved and further investigations regarding general arrangement of drill holes should be executed.

References

[1] Dennis Eichhorst, Jonas Pfeiffer and Peter Zacharias, "Weight reduction of DC/DC converters using controllable inductors", PCIM Europe, Nuremberg, Germany, 2019. Part of ISBN: 978-3-8007-4938-6

[2] Jonas Pfeiffer, Pierre Küster, Ilka E. M. Schulz, Jens Friebe and Peter Zacharias, "Review of flux interaction of differently aligned magnetic fields in inductors and transformers", IEEE Access, Vol. 9, pp. 2357-2381, 2020, DOI: 10.1109/ACCESS.2020.3047156

[3] Dale S. L. Dolan, "Modelling and performance evaluation of the virtual air gap variable reactor", Ph.D. dissertation, Grad. Elect. And Comp. Eng., Univ. Toronto, ON, Canada, 2009. URI: http://hdl.handle.net/1807/17754

[4] SUMIDA Components & Modules GmbH, "Components, Modules and Cores", Product Catalog, Obernzell, Bavaria, Germany, 2010

[5] R. L. Graham, B. D. Lubachevsky, K. J. Nurmela and P. R. J. Östergard, "Dense packings of congruent circles in a circle", Discrete Mathematics, Vol. 181, Issue: 1-3, pp. 139-154, Amsterdam, Netherlands, 1998, DOI: 10.1016/S0012-365X(97)00050-2

[6] Ronald Peikert, "Dichteste Packung von gleichen Kreisen in einem Quadrat", Elemente der Mathematik 49.1, pp. 16-26, Birkhäuser Verlag, Basel, Switzerland, 1994, PURL: http://resolver.sub.uni-goettingen.de/purl?PPN378850199_0049

[7] M. Goldberg, "The packing of equal circles in a square", Mathematics Magazine, Vol. 43:1, pp. 20-30, 2018, DOI: 10.1080/0025570X.1970.11975991

[8] Peter Zacharias, "Magnetic Components", 1st Edition, Springer Nature, Wiesbaden, Germany, 2022, ISBN: 978-3-658-37205-7

A Sectorized FCS-MPC Transformerless SST For Power Transmission Application

Gabriel Gaburro Bacheti[1], Renner Sartório Camargo[2], Emilio José Bueno[3],
Marco Liserre[4], Lucas Frizera Encarnação[5]

[1,5]Department of Electrical Engineering, Federal University of Espírito Santo (UFES),
Av. Fernando Ferrari, 514, Vitória 29075-910, Brazil
[2]Department of Control and Automation Engineering, Federal Institute of Espírito Santo
(IFES), Rod. ES010, Serra 29173-087, Brazil
[3]Department of Electronics, Alcalá University (UAH), Plaza San Diego, 28801 Madrid, Spain
[4]Chair of Power Electronics, Christian-Albrechts-Universitat, Kaiserstr. 2, Kiel, Germany
Emails: gabriel.bacheti@edu.ufes.br, rscamargo@ifes.edu.br, emilio.bueno@uah.es,
ml@tf.uni-kiel.de, lucas.encarnacao@ufes.br

Funding

This research was cofunding by Community of Madrid through the Pluriannual Agreement with the University of Alcalá in its line of stimulus to the research of young doctors, within the framework of the V PRICIT (CM/JIN/2021-019), National Research Council – CNPq (grant numbers 409024/2021-0 and 311848/2021-4) and Espírito Santo Research and Innovation Support Foundation – FAPES (grant numbers 514/2021 and 415/2021).

Keywords

≪Multi-level converters≫, ≪Model predictive control≫, ≪High voltage power converters≫, ≪Solid-State Transformer≫, ≪Optimization algorithm≫.

Abstract

This work proposes a sectorized optimization with Graph Theory applied in cascaded multilevel transformerless SST. The sectorized optimization can reduce the FCS-MPC combinational explosion due to the multiple cascaded H-bridges cells, while the proposed control is capable to avoid the prohibitive states maintaining the converter controllability and power quality.

Introduction

The insertion of high-voltage and high-power industrial equipment [1], the rise of renewable sources transforming the world's energy matrix, and even the use of low-voltage low-power non-linear loads, tend to increase the distribution and transmission systems complexity [2]. Therefore, problems related to overload feeders, bidirectional power flow, voltage variation, and harmonic content have become an issue for distribution and transmission systems. Consequently, the electric power systems' behavior is changing, being before passive and static, and now becoming more active and dynamic [3]. In this context, the Solid-State Transformer (SST) comes up as an interesting alternative to the Line Frequency Transformer (LFT) since it can perform ancillary services, providing a dynamic control for distribution and transmission systems. The SST is lighter and less bulky, provides harmonic filtering ability, supports voltage variations, and presents controlled bidirectional active and reactive power flow [3]. Therefore, SSTs can be considered smart transformers, and are the key point for a change in electric power systems [4] [5].

To operate with a high-power demand, the SST topology should be able to synthesize high-voltage levels, which is impossible with conventional 2-level converters. Thus, several studies were proposed to overcome this issue, developing specific converters to meet the growing demand for high-power applications [1]. Multilevel converters have an inherent potential to disrupt the technological barrier for power semiconductors, mainly because of the advantages they present over conventional converters, such as synthesizing higher voltage levels using lower power devices, reducing voltage stress and switching frequency, presenting reduced harmonic content, and less common-mode noise generation [1]. The literature presents the Dual Active Bridge (DAB), a well-established module, which can constitute multilevel converters, and demonstrates the ease of control and a great use for application in microgrids [4]. However, this structure presents an increase in volume, weight, cost, and construction complexity of the SST by increasing the number of semiconductors used in conversion stages [2]. So, a possible alternative with the absence of the High Frequency Transformer (HFT) element and reduction of power loss, shows up as an interesting open field to be developed, which has already started to be studied by the authors [2] using only the Cascaded H-bridge modules (CHB) without galvanic isolation. This structure can be presented in back-to-back (B2B) configuration for more complex applications, in which the converter should be able to operate in all four quadrants, as for high-power applications, considering that the bidirectional power flow is indispensable [2].

In CHB-B2B, voltage levels can be established by the number of H-bridges modules used, and the type of association made, such as input-parallel output-parallel (IPOP), input-series output-parallel (ISOP), input-series output-series (ISOS), or input-parallel output-series (IPOS) [4]. All these configurations present a reduced components number when compared to other classic multilevel topologies [2]. However, as a disadvantage, because the proposed configuration does not present galvanic isolation provided by the HFT, several short-circuit switching states make it impossible for driving the CHB-B2B with classical control and PWM modulation. Therefore, a Finite Control Set Model-based Predictive Control (FCS-MPC) with a Graph Theory approach was proposed by the authors to define the possible switching matrix, eliminating the prohibitive states and, at the same time, ensuring the SST controllability and power quality. Graph Theory is a mathematical tool that facilitates the development of almost intuitive algorithmic rules [6], by the definition of structures called graphs, formed by vertex and arcs concepts. The correspondence between these elements can be related to various problems from different areas, providing the desired solutions. However, Graph Theory implementation in power electronics converters and systems is still in the exploring stage and an overview of this research topic is summarized in [7], approaching general application, milestones and benefits.

Nonetheless, due to computational effort issues, the proposed solution was limited to four H-bridges modules achieving only a 5-level high-voltage (HV) and two 3-level medium-voltage (MV) in an ISOP CHB-B2B configuration [2]. As a static switch can have two possible states, the converter switching possibilities will increase exponentially. Therefore, the more H-bridges modules in CHB structure are used, and consequently, the greater number of levels an SST has, the more challenging the implementation of FCS-MPC techniques will be, due to processing limitations, as the sampling period tends to be short [8]. Thus, to solve this problem, different approaches have been proposed. In [9], an equivalent cost function is used with only the possible output voltage vectors, reducing the prediction model dimension which implies that the computational burden is lower. The optimization stage is performed by an exhaustive searching algorithm. In [10], a modified cost function is implemented with a sphere decoding algorithm to optimize a problem in a new space unconstrained solution. In [11], a hierarchical method is used, splitting the optimization into two stages. The first stage analyzes only the possible output voltage vectors while the second stage uses the redundant stages. Alternatively, in this paperwork, the authors propose a different methodology to get around an exhausting search inside the FCS-MPC sample space. Although the FCS-MPC principles are maintained, the multilevel converter is sectorized, dividing it into smaller parts which significantly reduces the switching matrix to be evaluated. Thus, the computational cost of executing the algorithm is considerably reduced. Therefore, a 19-level 110 kV : 30 kV SST with reduced components and without HFT galvanic isolation, with multiple output windings composed of 18 H-bridges modules in ISOP back-to-back configuration is implemented to demonstrate the proposed strategy applicability. The aforementioned FCS-MPC with Grapy Theory approach developed by the authors is also used in this work. Simulation results prove the proposed sectorized optimization FSC-MPC with Graph Theory effectiveness.

This paperwork is organized as follows: the proposed topology and Graph Theory analysis are provided first. The FCS-MPC equations are then presented, along with the proposed sectorized optimization. The findings of MatLab/Simulink are then displayed, followed by the conclusion.

Proposed SST Topology and Graph Theory analysis

Currently, SSTs are mainly used at the distribution voltage level, which varies from 2.3 kV to 35 kV [4]. For transmission voltage levels, the H-bridges modules increment must be performed to achieve high-voltage and high-power applications. The Union for the Coordination of the Transmission of Electricity (UCTE) consists of the largest synchronously operated electrical grid. Thereby, this system is very important for the European and world economy, and its constant improvement is necessary. Some Distribution System Operators (DSOs) associated with UCTE, which partly operate in Europe high-voltage grids, work with 30 kV to 110 kV 50 Hz transmission lines. Considering the SSTs advantages and aiming to expand its application to high-voltage and high-power systems, this work presents a new topology to be applied in the UCTE grid. Fig. 1a presents the proposed SST topology, which uses a 9 H-bridges modules series connection in the HV stage, and 3 disassociated series connections with 3 H-bridges modules in the MV stage. In this arrangement, a naturally 3:1 transformation ratio becomes possible to be obtained. As the secondary MV side has multiple windings, the SST structure could be sectorized in three different parts, as highlighted in Fig. 1b. This sectorized structure should be used in the Graph Theory analysis and also in the proposed optimization process to reduce de computational effort. Fig. 1c shows the proposed simplified scheme in 3-phase way, although this paperwork is addressed only the results for one phase.

The CHB-B2B may generate internal short-circuits according to the switching state. The conditions that can lead to a short-circuit must be defined, which are: i) the terminals of each capacitor become shorted, and ii) the opposite terminals of a group of capacitors become connected, which implies a ring with series capacitor connections. A connected and undirected graph from Graph Theory can be modeled considering the electrical nodes as vertices, and electrical switches as edges [12]. Thus, by finding simple paths that interconnect vertices, that represent both capacitor terminals, it is possible to visualize part of the prohibitive states. The other part is obtained by joining disjoint paths that interconnect different pairs of opposite terminals for any number of capacitors. Fig. 1b presents the sectorized structure composed of 12 legs, providing 212 (4,096) combinations. When applying Graph Theory to this topology, it is observed that only 576 of these switching states are useful, with the other combinations causing short-circuit states. It is important to emphasize that, with only 14% of the possible states, the sectorized part of the SST manages to synthesize all 7-level voltages, showing the applicability of the proposed topology and the usefulness of Graph Theory in mapping prohibitive states.

FCS-MPC and sectorized optimization

The FCS-MPC has attracted attention in the power electronics research community, for being a simple and intuitive way to control the converters, besides its ability to handle multivariable goals, showing good controllability and fast dynamic response, and capacity to incorporate straightforwardly nonlinearities and constraints into the control law [8], as the proposed 19-level CHB-B2B prohibitive switching states. For this converter, the HV and MV stages have together 72 switches, presenting 2^{36} (68,719,476,736) possible switching states, due to the leg complementary switches. Even with Graph Theory application aiming to reduce this amount with the short-circuit states removal becomes infeasible, since the digital FCS-MPC implementation have to perform mathematical calculations for each of these states, and currently, there is no availability of hardware to meet these demand requirements. The solution adopted to allow the system's control is to sectorize the converter into three parts, as seen in Fig. 1b. Therefore, with a smaller composition, each sector presents 2^{12} (4,096) possible switching states, but within these, only

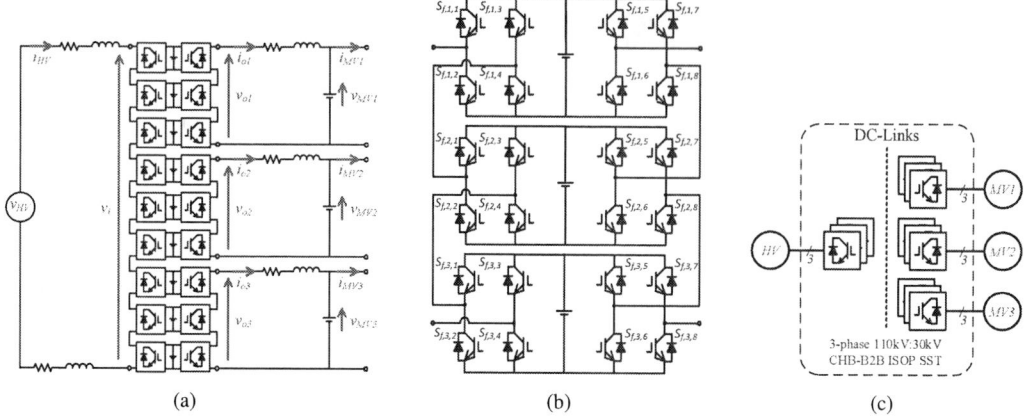

(a)   (b)   (c)

Fig. 1: New topology schematic. a) 19-level proposed SST topology, b) 7-level sectorized structure and c) 3-phase overview.

576 do not represent short-circuits. Thus, the new converter's switching matrix consists of the combination of the reduced switching matrices (S_s) obtained by Graph Theory, and thus, the converter control can be performed semi-independently for each sector, due to the MV control of each feeder haven't relations among themselves. However, the HV input current control is unique and a strategy that relates the three switching matrices adopted must be implemented. Therefore, the sampling space is drastically reduced, with the prediction of 3x576 (1,728) switching states, which represents a 99.99% reduction (1,728 / 68,719,476,736), making the control system practicable.

The FCS-MPC strategy implementation requires the dynamics equations for the DC-links, HV, and MV stages. The state of the H-bridge cell, with $S_{f,n,j}$ representing the jth S switch in the nth H-bridge cell of the fth feeder is represented by $d_{HVf,n}$ and $d_{MVf,n}$, for HV and MV stages respectively (see Fig. 1b), and are obtained by equations (1) and (2). Therefore, $d_{HVf,n}$ and $d_{MVf,n}$ can assume three logical values (1, 0, 1), according to the switches' state defined by different rows in S_s.

$$d_{HV,f,n} = S_{f,n,1}S_{f,n,4} - S_{f,n,2}S_{f,n,3} \tag{1}$$

$$d_{MV,f,n} = S_{f,n,5}S_{f,n,8} - S_{f,n,6}S_{f,n,7} \tag{2}$$

Analyzing Fig. 1a, the equation for nth DC-link voltage level of the fth feeder ($v_{dcf,n}$) can be obtained in (3), considering i_{of} the L filter current for the fth feeder and assuming that the values of the capacitors for each DC link (C_{dc}) are equal for a generalized equation. For MV predictive model, the controlled variables are the fth feeder's MV output (v_{MVf}) and the control loop is composed of the load connection point and LC output filter. Considering v_{of} as the fth feeder synthesized output voltage as described in (4), l_i the inductive components of inverters' LC filters, with r_i modeling its electrical losses, the following relationship (5) for i_{of} is found.

$$\frac{dv_{dcf,n}}{dt} = \frac{1}{C_{dc}} \left(d_{HVf,n}i_{HV} - d_{MVf,n}i_{of} \right) \tag{3}$$

$$v_{of} = \sum_{n=1}^{3} d_{MVf,n}v_{dcf,n} \tag{4}$$

$$\frac{di_{o_f}}{dt} = \frac{1}{l_i}\left(v_{of} - r_i i_{of} - v_{MVf}\right) \tag{5}$$

Relying on the dynamics of capacitive components in inverters' *LC* filters (C_i), the relationships of the control variables are presented in (6). In the HV stage, the controlled variable is its current (i_{HV}) and the control loop is formed by the grid voltage (v_{HV}) and the *L* filter, made up of r_r and l_r elements, the first one modeling electrical losses of the inductor. Using the same procedures as above and considering v_i as the HV synthesized output voltage, calculated through the summation in equation (7), the i_{HV} the relation is obtained and shown in (8).

$$\frac{dv_{MV_f}}{dt} = \frac{1}{C_i}\left(i_{o_f} - i_{MV_f}\right) \tag{6}$$

$$v_i = \sum_{f=1}^{3} \sum_{n=1}^{3} d_{HV_{f,n}} v_{dc_{f,n}} \tag{7}$$

$$\frac{di_{HV}}{dt} = \frac{1}{l_r}\left(v_{HV} - r_r i_{HV} - v_i\right) \tag{8}$$

The discrete equations necessary for the FCS-MPC application are obtained using the Euler numerical integration method in equations (3), (5), (6), and (8), in which [k] and [k+1] are the present and future instants respectively. To adapt equation (7) for the sectorized control strategy, the v_i term, which has 576 values possibilities according to switching, is replaced by N, a vector containing, in ascending order of values, the possible levels synthesized by the converter for nominal values of the DC-links. The discrete equations (9), (10), (11) and (12) infer different values for different switching. For example, an iteration with a variable i traversing all rows of S_s results in varying predicted values.

$$v_{dc_{f,n}}[k+1] = v_{dc_{f,n}}[k] + \frac{T_s}{C_{dc}}\left(d_{HV_{f,n}}[k]\,i_{HV}[k] - d_{MV_{f,n}} i_{o_f}\right) \tag{9}$$

$$i_{o_f}[k+1] = i_{o_f}[k] + \frac{T_s}{l_i}\left(v_{o_f}[k] - r_i i_{o_f}[k] - v_{MV_f}[k]\right) \tag{10}$$

$$v_{MV_f}[k+1] = v_{MV_f}[k] + \frac{T_s}{C_i}\left(i_{o_f}[k+1] - i_{MV_f}[k]\right) \tag{11}$$

$$i_{HV}[k+1] = i_{HV}[k] + \frac{T_s}{l_r}\left(N - r_r i_{HVk} - v_{HV}[k]\right) \tag{12}$$

The next step is to define the reference signals for the control variables, which for each feeder (v_{MV}^*) and DC-link (v_{dc}^*) consist of their respective nominal operating ratings (Table I). The HV current reference (i_{HV}^*) is obtained from PQ theory [13] together with a Second-Order Generalized Integrator (SOGI) [14]. Thus, preliminary cost functions (g_{pf}) can be defined for each feeder (13), considering only the DC-link and inverter stages, and calculating average DC-links voltages (14) for the proposed sectorized methodology. Weights are defined to designate which control object should be prioritized: v_{MVf} voltage synthesis (W_{MV}), regulation of DC-links (W_{dc}) or balance of DC-links (W_{bl}).

$$g_{pf} = W_{dc} \sum_{n=1}^{3} \left|v_{dc}^* - v_{dcf,n}[k+1]\right| + W_{bl} \sum_{n=1}^{3} \left|v_{av_f} - v_{dcf,n}[k+1]\right| + W_{MV}\left|v_{MV}^* - v_{MVf}[k+1]\right| \tag{13}$$

$$v_{avf} = \frac{1}{3} \sum_{n=1}^{3} v_{dcf,n}[k+1] \tag{14}$$

The most significant control challenge is to control the system's input current described in (12) as it is dependent on the voltage levels presented in DC-links ($v_{dcf,n}$) and their respective possible series inter-connections forming v_i (see Fig. 1a). Typically, the FCS-MPC methodology would use the full converter switching matrix, leading to a computational explosion. However, as mentioned, the sectorized strategy uses a reduced switching matrix obtained by Graph Theory, in which a direct application cannot achieve one of the control objectives, which consists of regulating the i_{HV} current. Therefore, the principles proposed for the solution of this issue are to initially treat the voltage variables of the DC-links as controlled on nominal ratings, and the definition of an auxiliary's matrix set (M). Each auxiliary matrix represents one specific converter voltage level, and consequently, one N array row (totalizing 19 matrices), with its row's elements containing ascending ordered different logical states possibilities for HV stage ($d_{HVf,n}$ [k]: -1, 0, 1). They correspond too into switched H-bridges voltages levels ($-v_{dc}$, 0, v_{dc}), which, together, form the voltage level that each matrix is related. Thus, the greater the voltage modulus related to each of the matrices, smaller is the number of combination possibilities for logical switching. For example, the matrix associated to the minimum voltage level $-9v_{dc}$ has only one combination. For such, all H-bridges modules must switch $-v_{dc}$, corresponding to a final v_i equal to $-9v_{dc}$ (summation of $-9v_{dc}$ matrix rows).

$$-9v_{dc} \; matrix = \begin{bmatrix} -1 & -1 & -1 & -1 & -1 & -1 & -1 & -1 & -1 \end{bmatrix}$$

The proposed sectorized control strategy initially performs the same procedures as the FCS-MPC to control i_{HV}. However, instead of considering the sectorized switching matrix and, consequently, different switched voltages for each H-bridge module, it considers the voltage levels that can be synthesized by all of them (N elements) and selects the optimal voltage among the possible 19-levels. This can be done through the predictions of i_{HV}[k+1] using (12) and the minimization of another cost function (15), and therefore, it is defined which auxiliary matrix (M_{opt}) is used for a given instant of time. Meanwhile, each auxiliary matrix itself does not specify exactly the voltage level that each H-bridge module must switch; this is accomplished by analyzing the voltage level that each DC-link has at a given instant of time, the power flow (current direction) to which the modules are subjected, in addition to the others control objectives, as described in (13), since the rectifier v_i voltage corresponds to the selected optimal voltage.

$$g_r = |i_{HV}^* - i_{HV}[k+1]| \tag{15}$$

Analyzing equation (9), it is verified that the contribution of the HV side depends on the logic state of the H-Bridges and on the signal that the measured current i_{HV} have at a given instant of time. Being this positive, the capacitors must be charged if the logic state of the H-bridges ($d_{HV_{f,n}}$) corresponds to 1, discharged if this value corresponds to -1, and maintain their voltage levels for a logical value equal to 0. Contrary results should occur for the condition of $i_{HV} < 0$. Thus, a principle is defined that capacitors with a lower voltage level, below the nominal ratings, at a given instant of time, have charging priority over the others. Those that have a higher voltage level, above nominal ratings, must have priority to be discharged (provide the necessary power flow to the system's feeders). Based on these ideas, and once M_{opt} has been selected, each row must be reordered logically: The direction of i_{HV} current is analyzed. If it is positive, the DC-links are sorted ascending according to their voltage levels. Otherwise, the voltage levels of each module are sorted in descending. Relating each column of M_{opt} to a specific DC-link, each row is reordered according to the ordering of the DC-links, determining which capacitors needs to be charged and discharged, keeping the balance between them, thus forming an ordered matrix denominate OM. For example, considering the $-v_{dc}$ matrix, if at a certain point in time, $v_{dc2,1} < v_{dc3,1} < v_{dc1,2} < v_{dc1,1} < v_{dc3,2} < v_{dc3,3} < v_{dc1,3} < v_{dc2,3} < v_{dc2,2}$ and $i_{HV} > 0$, the ordered $-v_{dc}$ matrix consists as shown below.

	$v_{dc2,1}$	$v_{dc3,1}$	$v_{dc1,2}$	$v_{dc1,1}$	$v_{dc3,2}$	$v_{dc3,3}$	$v_{dc1,3}$	$v_{dc2,3}$	$v_{dc2,2}$
	-1	-1	-1	-1	-1	1	1	1	1
	-1	-1	-1	-1	0	0	1	1	1
$M_{opt}(-v_{dc}) =$	-1	-1	-1	0	0	0	0	1	1
	-1	-1	0	0	0	0	0	0	1
	-1	0	0	0	0	0	0	0	0

$\downarrow$

	$v_{dc1,1}$	$v_{dc1,2}$	$v_{dc1,3}$	$v_{dc2,1}$	$v_{dc2,2}$	$v_{dc2,3}$	$v_{dc3,1}$	$v_{dc3,2}$	$v_{dc3,3}$
	-1	-1	1	-1	1	1	-1	-1	1
	-1	-1	1	-1	1	1	-1	0	0
$OM(-v_{dc}) =$	0	-1	0	-1	1	1	-1	0	0
	0	0	0	-1	1	0	-1	0	0
	0	0	0	-1	0	0	0	0	0
	$OM_{1,1}$	$OM_{1,2}$	$OM_{1,3}$	$OM_{2,1}$	$OM_{2,2}$	$OM_{2,3}$	$OM_{3,1}$	$OM_{3,2}$	$OM_{3,3}$

$-1 \rightarrow charge\ capacitors;\ 0 \rightarrow mantaining\ voltage\ level;\ 1 \rightarrow discharge\ capacitors$

Finally, the proposed algorithm must perform iterations that go through all rows of OM. At each iteration, new global cost functions g_{gf}, defined in (16), are calculated for each one of the feeders, through preliminary calculated cost functions g_{pf} and the comparison of each OM row's elements ($OM_{f,n}$) with a voltage level determined by the sectorized switching matrix (S_s) of each converter's sectors ($d_{HV_{f,n}}v_{dc_{f,n}}$). The W_{HV} weight is defined to assign i_{HV} control priority. At each iteration, is realized a minimization of the global cost functions g_{gf}, and a final cost function g_g is obtained through the summation of g_{gf} elements considering different feeders. Finally, after all iterations, the last minimization is performed in g_g to select which OM line is optimal for the switching. Therefore, it is possible to determine the optimal switching for each sector (S_{sopt_f}). Fig. 2 presents the overview operation of the sectorized FCS-MPC methodology applied to the proposed 19-level CHB-B2B, and Fig. 3 highlights the SST FCS-MPC algorithm.

$$g_{gf} = g_{pf} + W_{HV} \sum_{n=1}^{3} \left| OM_{f,n} v_{dc_{f,n}} - d_{HV_{f,n}} v_{dc_{f,n}} \right| \tag{16}$$

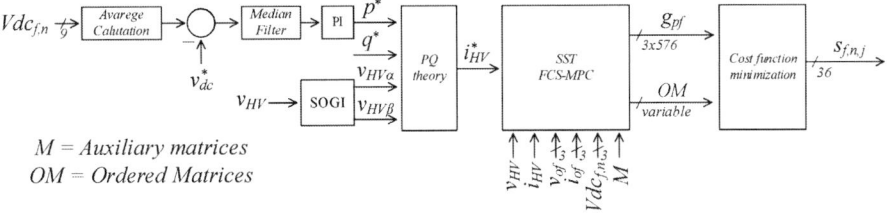

$M = Auxiliary\ matrices$
$OM = Ordered\ Matrices$

Fig. 2: Control diagram overview.

Simulation results

The steady-state and transient performance of the proposed 19-level ISOS CHB-B2B converter with the FCS-MPC optimization algorithm are verified by simulation, through Matlab/Simulink platform, and the parameters used for the circuit are described in Table I. The steady-state load scenario parameters are described in Table II, in which an apparent power demand of 10 MVA was defined for each one of the feeders, with an active power imbalance of 20% between the first and second feeders. Different load types are used for each feeder, representing high-power industrial equipment and non-linear loads, which brings this MV system closer to practical applications.

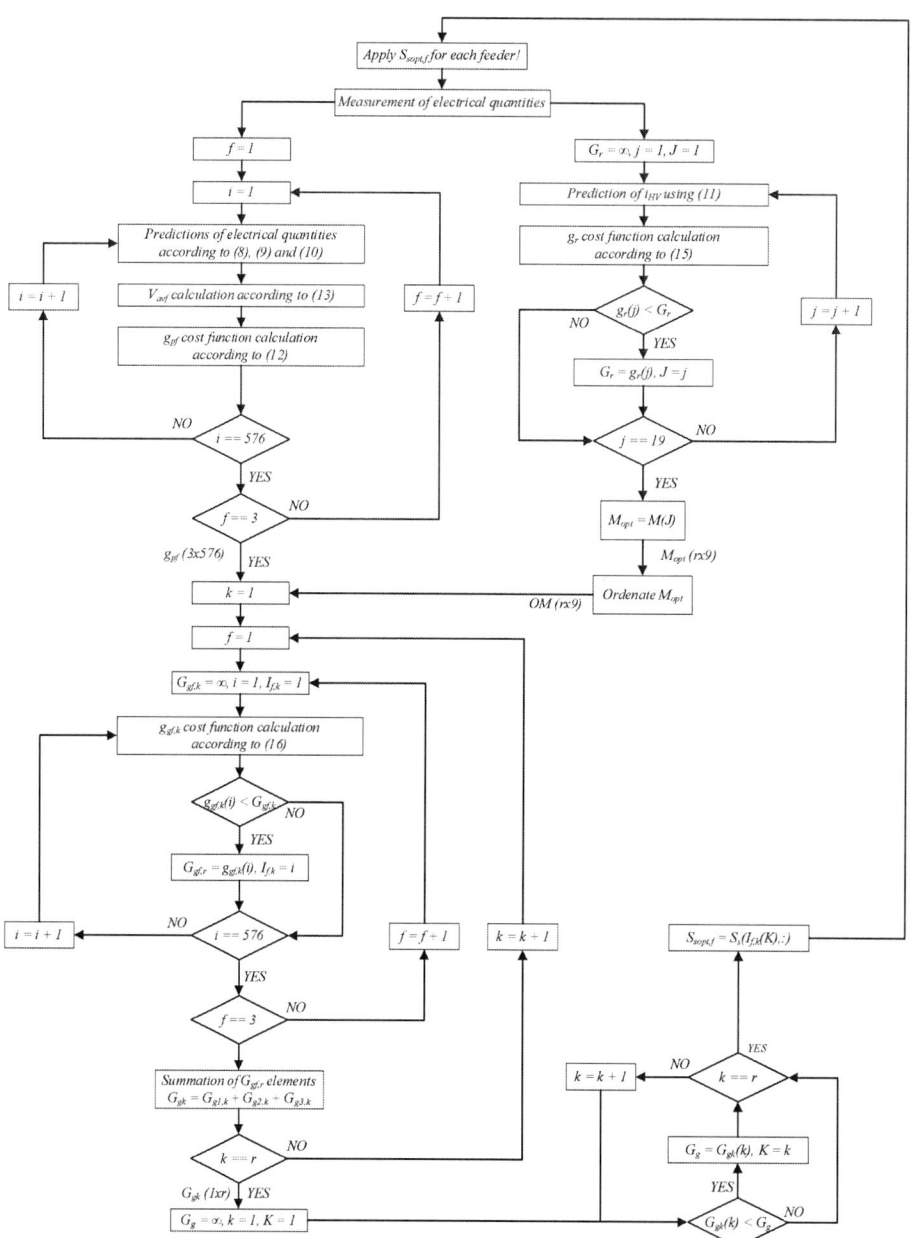

Fig. 3: SST FCS-MPC algorithm.

In Fig. 6, it can be observed the steady-state response of the FCS-MPC controlled variables: HV current (i_{HV}), DC-link voltages ($v_{dcf,n}$) and MV feeders (v_{MVf}). The voltage synthesized by the HV converter (v_i) and its reference calculated by the control system (N_{sel}) are also highlighted. As it can be seen in Fig. 4a, the sectorized FCS-MPC technique can synthesize a sinusoidal rectifier current that properly tracks the reference (i^*_{HV}), allowing the DC links regulation with a high power factor. It can also be noted in Fig. 4b that for synthesizing this current, the 19 possible voltage levels on the HV stage were achieved, which will guarantee a lower harmonic distortion (THD (i_{HV}) = 2.11%). Analyzing Fig. 4c, it is verified that all DC-link voltages are properly regulated with a maximum 0,54% oscillation at the nominal value, even for different feeders' load types, essential for the correct control functioning. Finally, is observed in Fig. 4d that all inverter output voltages can track the sinusoidal (v^*_{MV}) reference, even for non-linear load conditions (i_{MV3} current), ensuring a 110 kV:30 kV transformer voltage rating, as expected. Table

III shows the total harmonic distortion for the output voltages and currents, the first having an adequate harmonic content, reflecting in energy quality, since it is independent of the load. Therefore, THD results corroborate that the proposed 19-level ISOS CHB-B2B converter meets the IEEE Std 519™ required limits, namely a maximum of 8% THD for the feeders' output voltage and 5% for grid input current [15].

In Fig. 7, it can be observed the transient response of the FCS-MPC controlled variables. Initially, the specified loads expressed in Table II are operating in steady-state regime with the inductive load operating with 50% less reactive power. In 1.5s, the reactive power demand of the inductive load is increased by 50%. Later, at 1.55s, the resistive load is decreased by 20%. Fig. 5a shows that even with a load disturbance, the system's control is able to synthesize i_{HV} properly. Fig. 5b presents that the converter continues to switch the expected 19 voltage levels, while Fig. 5c shows that the DC-links continued to be balanced and controlled. Finally, Fig. 5d demonstrates that all voltage outputs from the feeders are still properly synthesized, in addition to changes in load currents according to the variation of the power demanded.

Table I: Simulation Parameters.

Parameter	Symbol	Value
HV voltage	v_{HV}	$110\sqrt{2}/\sqrt{3}\,kV$
MV voltage	$v_{MV,f}$	$30\sqrt{2}/\sqrt{3}\,kV$
DC-link nominal voltage	V_{dcn}	$10\,kV$
DC-link capacitance	C_{dcn}	$5\,mF$
Rectifier inductance	l_r	$100\,mH$
Rectifier resistance	r_r	$0.001\,m\Omega$
Inverter inductance	l_i	$7.5\,mH$
Inverter resistance	r_i	$1.5\,m\Omega$
Inverter capacitance	C_i	$3.54\,\mu F$
Grid frequency	f_{HV}	$50\,Hz$
The median filter window length	M	200
PI controller proportional gain	K_p	1
PI controller integral gain	K_i	50
Time Step	T_s	$50\,\mu s$
Input current weight	W_{HV}	1
Output voltage weight	W_{MV}	500
DC-link voltage weight	W_{dc}	10
Balance weight of DC-links	W_{bl}	5

Table II: Load Parameters.

Load type	Feeder	Values
Resistive	1	$90\,\Omega$
Inductive	2	$112.5\,\Omega\,/\,477,46\,mH$
Nonlinear	3	$2\,mH\;(AC)\,\mid\,70\,\Omega\,/\,1\,H\;(DC)$

Table III: THD Analysis.

Feeder	THD (v_o)	THD (i_o)
1	1.36%	1.36%
2	1.34%	1.08%
3	1.67%	42.86%

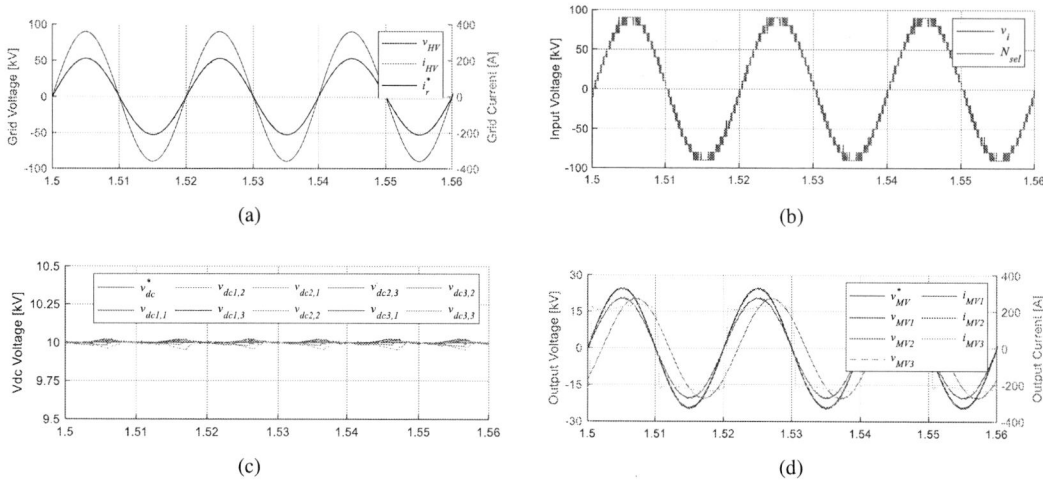

Fig. 4: Steady-state results. a) HV measurements b) HV converter stage, c) DC-links, and d) MV measurements.

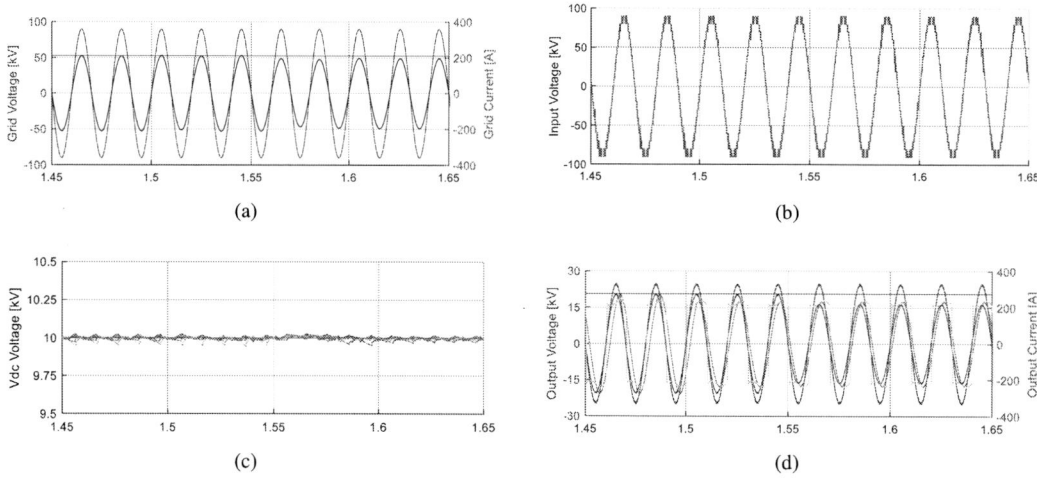

Fig. 5: Transient results. a) HV measurements b) HV converter stage, c) DC-links, and d) MV measurements.

Conclusion

Based on the steady-state results, the proposed sectorized FCS-MPC methodology for the 19-level CHB-B2B converter ensures a reduced computational effort, avoiding possible internal short-circuits, balancing DC-links voltage, and enabling HV current control and multiple windings MV output with low harmonic distortion, meeting the IEEE Std 519™ requirements, even in presence of non-linear loads. The proposed control algorithm also showed a good response to load disturbances, keeping all control variables regulated for active and reactive power demand variations.

References

[1] J. Rodriguez, Jih-Sheng Lai and Fang Zheng Peng, "Multilevel inverters: a survey of topologies, controls, and applications," in IEEE Transactions on Industrial Electronics, vol. 49, no. 4, pp. 724-738, Aug. 2002, doi: 10.1109/TIE.2002.801052.

[2] B. F. de Andrade Campos, R. S. Camargo, E. J. B. Peña and L. F. Encarnação, "Single-Phase AC/AC Multilevel H-Bridge Transformerless Converter with SST Functionalities," 2021 IEEE 15th International Conference on Compatibility, Power Electronics and Power Engineering (CPE-POWERENG), 2021, pp. 1-8, doi: 10.1109/CPE-POWERENG50821.2021.9501201..

[3] M. A. Hannan et al., "State of the Art of Solid-State Transformers: Advanced Topologies, Implementation Issues, Recent Progress and Improvements," in IEEE Access, vol. 8, pp. 19113-19132, 2020, doi: 10.1109/ACCESS.2020.2967345.

[4] X. She, R. Burgos, G. Wang, F. Wang and A. Q. Huang, "Review of solid state transformer in the distribution system: From components to field application," 2012 IEEE Energy Conversion Congress and Exposition (ECCE), 2012, pp. 4077-4084, doi: 10.1109/ECCE.2012.6342269.

[5] M. Liserre, G. Buticchi, M. Andresen, G. De Carne, L. F. Costa and Z. -X. Zou, "The Smart Transformer: Impact on the Electric Grid and Technology Challenges," in IEEE Industrial Electronics Magazine, vol. 10, no. 2, pp. 46-58, June 2016, doi: 10.1109/MIE.2016.2551418.

[6] H. Taha, Operations research an introduction, Boston: Pearson, 2016.

[7] Y. Li, J. Kuprat, Y. Li and M. Liserre, "Graph-Theory-Based Derivation, Modeling and Control of Power Converter Systems," in IEEE Journal of Emerging and Selected Topics in Power Electronics, doi: 10.1109/JESTPE.2022.3143437.

[8] S. Vazquez, J. Rodriguez, M. Rivera, L. G. Franquelo and M. Norambuena, "Model Predictive Control for Power Converters and Drives: Advances and Trends," in IEEE Transactions on Industrial Electronics, vol. 64, no. 2, pp. 935-947, Feb. 2017, doi: 10.1109/TIE.2016.2625238.

[9] C. Xia, T. Liu, T. Shi and Z. Song, "A Simplified Finite-Control-Set Model-Predictive Control for Power Converters," in IEEE Transactions on Industrial Informatics, vol. 10, no. 2, pp. 991-1002, May 2014, doi: 10.1109/TII.2013.2284558.

[10] T. Geyer and D. E. Quevedo, "Multistep Finite Control Set Model Predictive Control for Power Electronics," in IEEE Transactions on Power Electronics, vol. 29, no. 12, pp. 6836-6846, Dec. 2014, doi: 10.1109/TPEL.2014.2306939.

[11] J. Moon, J. Gwon, J. Park, D. Kang and J. Kim, "Model Predictive Control With a Reduced Number of Considered States in a Modular Multilevel Converter for HVDC System," in IEEE Transactions on Power Delivery, vol. 30, no. 2, pp. 608-617, April 2015, doi: 10.1109/TPWRD.2014.2303172.

[12] G. G. Bacheti, R. S. Camargo, T. S. Amorim, I. Yahyaoui, and L. F. Encarnação, "Model-Based Predictive Control with Graph Theory Approach Applied to Multilevel Back-to-Back Cascaded H-Bridge Converters," Electronics, vol. 11, no. 11, p. 1711, May 2022, doi: 10.3390/electronics11111711.

[13] H. Akagi, Instantaneous power theory and applications to power conditioning, Hoboken, N.J. Piscataway, N.J: Wiley IEEE Press, 2007.

[14] P. Rodríguez, R. Teodorescu, I. Candela, A. V. Timbus, M. Liserre and F. Blaabjerg, "New positive-sequence voltage detector for grid synchronization of power converters under faulty grid conditions," 2006 37th IEEE Power Electronics Specialists Conference, 2006, pp. 1-7, doi: 10.1109/pesc.2006.1712059.

[15] "IEEE Recommended Practice and Requirements for Harmonic Control in Electric Power Systems," in IEEE Std 519-2014 (Revision of IEEE Std 519-1992) , vol., no., pp.1-29, 11 June 2014, doi: 10.1109/IEEESTD.2014.6826459.

Inductance Estimation for Square-Shaped Multilayer Planar Windings

Theofilos Papadopoulos, Antonios Antonopoulos
National Technical University of Athens
School of Electrical and Computer Engineering
Zografou, Greece
E-Mail: teopap@mail.ntua.gr

Keywords

«Planar Magnetics», «Planar Transformer», «High-Frequency Transformer», «High Power Density Systems»

Abstract

In this paper the inductance of square-shaped multilayer planar windings (MLPW) for power applications is investigated. Three well-known equations Wheeler's, Rosa's, and Monomial, which are suitable for single-layer (L1) planar windings, are extended to properly appertain to multilayer architectures. The proposed procedure is verified for two-layer (L2) windings, and the extension to three-layers (L3) or more is discussed, for different values of the geometric parameters. Furthermore, the dependance of the coupling factor k with respect to the distance between the layers is investigated for L2. Finally, an experimental verification is carried out, for a number of selected windings.

Introduction

The continuous effort towards miniaturization and high-efficiency systems is bringing passive elements to the foreground, as they are taking up a large portion of the overall volume in state-of-the-art electronic converters. Within this scope, planar windings (PWs) are considered, due to their low profile and suitability for high-frequency applications. PW-based magnetic components have accurately predetermined values of inductance and capacitance, and can be printed directly on a printed circuit board (PCB) [1], lowering the cost of the manufacturing process. Their inherent characteristic of well-known inductance and capacitance values can be utilized in resonant converters (including series-resonant, LLC, CLLC etc.) [2], [3], even in combination with common-mode noise rejection [4], where the exact knowledge of the resonant tank values is of paramount importance for an efficient soft-switching control scheme, over a wide range of operating conditions.

Several equations have been proposed in literature [5], for the estimation of the total inductance of PWs, mainly focusing on square-shaped, single-layer designs of relatively small dimensions [6]. Generalizing the shape or the dimensions is not an easy task, and methods deducing very complex equations using electromagnetic field analysis have been suggested [7], [8]. Another approach is to adapt well-known and relatively simple equations for inductance estimation of PWs, namely Wheeler's (WH) [9], [10], Rosa's (RS) [11], and the Monomial (MN) [10], to different shapes. It has been shown [12] that extending from relatively small, square-shaped to considerably larger, rectangle-shaped windings is possible, with very accurate results.

This study aims for a different extension of the three aforementioned equations to MLPWs, namely for two, three and four layers. The limitations of this extension are discussed, aiming to provide a better understanding of the effect different design factors impose on the characteristics of an MLPW.

Analysis of Planar Winding Design Parameters

The fundamental equations for inductance estimation of PWs under consideration are

$$L_{\text{Wheeler}} = 1.17\mu_0 \text{N}^2 \frac{D+d}{1+2.75\dfrac{D-d}{D+d}}, \tag{1}$$

$$L_{\text{Rosa}} = 0.3175\mu_0 \text{N}^2 (D+d) \left[\ln\left(2.07\frac{D+d}{D-d} \right) + 0.18\left(\frac{D-d}{D+d} \right) + 0.13\left(\frac{D-d}{D+d} \right)^2 \right], \tag{2}$$

$$L_{\text{Monomial}} = 1.62 \cdot 10^{-12} N^{1.78} \left(10^6 D \right)^{-1.21} \left(10^6 w \right)^{-0.147} \left(10^6 \frac{D+d}{2} \right)^{2.4} \left(10^6 s \right)^{-0.03}, \tag{3}$$

where D and d are the outer and inner-side lengths, respectively, w is the width of the copper trace, s is the spacing between two adjacent traces and N is the total number of turns, as presented in Fig. 1. Three extra parameters are necessary to be defined for the extension to multiple layers, namely the number of turns per layer N_L, the number of layers N_L, where $N = N_T N_L$, and the distance between two consecutive layers O, as presented in Fig. 2a.

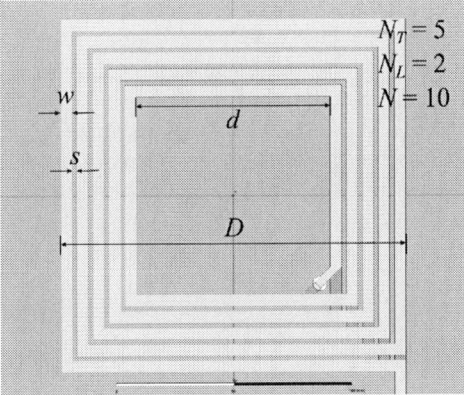

Fig. 1: Two-Layer Planar Winding (L2PW) of $N = N_T N_L = 5\times2$ turns and inner connection, with each parameter D, w, s, d presented.

Several observations can be made about the behavior of the inductance value for L1PWs. The term $(D + d)/2$, which is the average diameter of the square copper ring, is present in all three equations in such a way that the inductance is proportional to it. Eq. (2) also contains the filling factor $(D - d)/(D + d)$ term in the numerator, which is always less or equal to one, multiplied by terms also less than one, which reduces its weight on the final result. This means that large outer and inner sides lead to large inductance values for a given number of turns N. However, for a given voltage and current, w and s are limited by the electrical ratings, and there is no significant room for adjustments. Consequently, the physical dimensions of the board and the number of turns N determine dimensions of the outer and the inner sides, respectively.

To overcome this design limitation and increase the inductance without stretching the physical dimensions significantly, MLPWs are introduced, as a way to increase d for a given D (through shifting some of the turns to different layers), without compromising the nominal power of the winding or changing the total number of turns. Nevertheless, the effect of this action on the estimation of inductance should be addressed in more detail, considering also that a coupling factor is now introduced between the multiple layers. This coupling factor provides an insight on the behavior of the windings in case they are utilized as a high-frequency transformer.

Two indicative schematics of an L2 and an L4PW are presented in Fig. 2, with the distance between the layers exaggerated. The independent design parameters are D, w, s, N_T, N_L, and O. While N_L can be considered indirectly by substituting $N = N_T N_L$ into the aforementioned equations, O is not taken into account, although it is expected to be inversely correlated to the coupling factor, and hence, inversely correlated to the total inductance. The inner-side length d is a parameter dependent on the rest of the dimensional variables, and can be expressed as a function of the other parameters, as in

$$d = D - 2\frac{N}{N_L}(w+s) + 2s = D - 2N_T(w+s) + 2s .$$

(4)

For a winding capable of transferring a few tens of kW, the applied voltage is typically up to hundreds of V and the current up to tens of A. According to the IPC-2221 standard, to safely handle these values, the traces should be designed with w from 3 to 5 mm and s from 0.1 to 2 mm, depending on the position of the trace (internal or external PCB layer) and the type of the insulation mask. In this investigation, D is selected to vary from 100 to 210 mm and the number of turns per layer is set to $N_T = 5$, while d can be calculated from (4). It should be noted that when s is one order of magnitude less that w, the term $2 N_T (w + s)$ is much greater from the term $2s$, hence d depends strongly on D and the product $N_T (w + s)$. For fixed values of the outer side length and the number of turns, the sum $(w + s)$ determines the inner side, and therefore it strongly influences the total inductance.

All simulations are conducted with Maxwell3D of ANSYS, for an analysis region of 700x700x700 mm³ and a meshing of around 50 thousand tetrahedras. The excitation current is sinusoidal with 1 A amplitude and a frequency of 100 kHz. The current and frequency are reported for the sake of completeness, since they mainly affect a potential use of a ferrite core in this setup, which this beyond the scope of this study. Furthermore, it was confirmed by simulations that the connection placement between the layers does not affect in any significant manner the total inductance, and can be done either from the inner or the outer side of the winding. Hence, the inter-layer connectors are placed in such a way to minimize their length: at the inner side for L2, at the inner and then outer side for L3, etc.

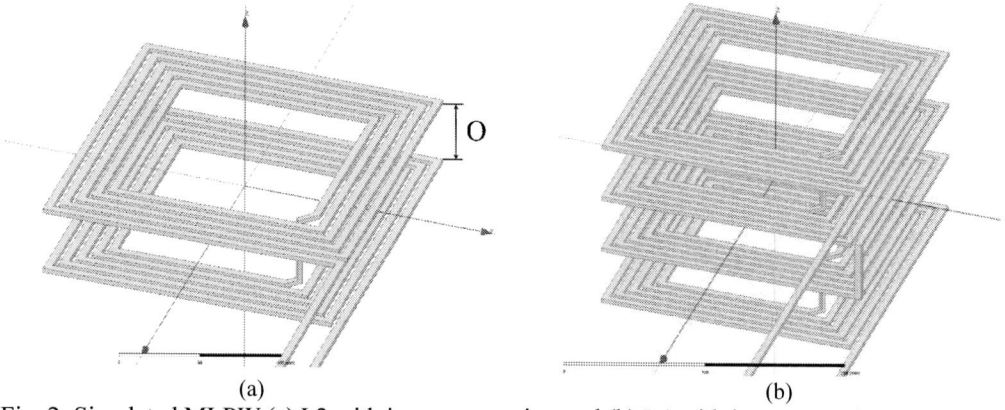

(a)   (b)

Fig. 2: Simulated MLPW (a) L2 with inner connection and (b) L4 with inner-outer-inner connection, with the distance between layers exaggerated.

It should be noted that for all simulations, D was varied with a step of 10 mm, resulting in 12 values, w was assigned three values, namely {3, 4, 5} mm, six values were selected for s, namely {0.1, 0.2, 0.5, 0.8, 1.0 2.0} mm, and three values for O, i.e., {0.8, 1.6, 3.2} mm, resulting in 648 simulations for each group (L2, L3, L4) and a total number of 1,944, for all the different combinations of the design dimensions.

Two-Layer Design

Two L1 windings with the same dimensions, connected in series and properly folded, as presented in Fig. 2a, can potentially even quadruple the total inductance, as it is indicated by the N^2 term in the equations. The exact amount of increase in inductance is determined by the coupling factor k, which can be calculated as

$$L_{L2} = 2L_{L1} + 2kL_{L1} \Rightarrow k = \frac{1}{2}\frac{L_{L2}}{L_{L1}} - 1, \tag{5}$$

where L_{L1} is the inductance for the L1PW of N_T turns and L_{L2} is the total inductance for the L2PW of $2N_T$ turns. In this case, k is calculated using the simulation results for L_{L1} and L_{L2}. After simulating the L2PW, with the variance of each parameter described previously, and their corresponding L1PWs, the dependance of k with respect to D and w, for $s = 0.5$ mm, is presented in Fig. 3a. Accordingly, extracting only the results for a single value of w, i.e., $w = 4$ mm, the dependence of k and D and s is illustrated in Fig. 3b. Each surface corresponds to a different value of O, namely 0.8 mm, 1.6 mm, and 3.2 mm, from top to bottom.

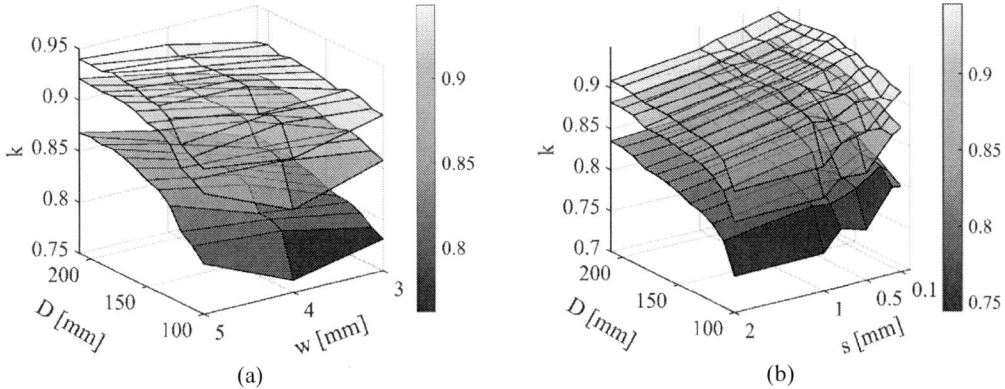

(a)   (b)

Fig. 3: For fixed number of turns $N = 10$ and $O = \{0.8, 1.6, 3.2\}$ mm (top to bottom), the coupling coefficient k, as a function of (a) D and w, for constant $s = 0.5$ mm, and (b) D and s, for constant $w = 4$ mm

As it is expected, k is inversely correlated to O, since less of the magnetic flux generated by each layer passes through the other, when O increases. Although O can be arbitrarily large (e.g., up to a few centimeters in wireless power transfer setups), the minimum allowable value is determined by the voltage difference of the vertically separated tracks and the dielectric properties of the insulating material (e.g., FR4). Another interesting observation is that k significantly increases with D. Regarding the other two parameters, k seems to increase with w and decrease with s. These effects are, however, minor compared to D. For this coreless case study, the coupling factor varies from 0.75 for $O = 3.2$ mm, $s = 2$ mm, $w = 3$ mm, and $D = 100$ mm, to 0.95 for $O = 0.8$ mm, $s = 0.1$ mm, $w = 5$ mm, and $D = 210$ mm.

The resulting simulated inductances, for a fixed number of $N_T N_L = 5x2 = 10$ turns, vary from 9 µH to 45 µH and depend strongly on D, with w and s having low impact on their values, as it can be observed in Fig. 4. Even though the coupling factor between two windings tends to increase with w, the total inductance of the winding decreases, since d becomes smaller (for the same D). Similarly, as s increases, both d and k decrease, and the total inductance follows the same trend too. The three surfaces in each subfigure of Fig. 4 correspond to O = $\{0.8, 1.6, 3.2\}$ mm (from top to bottom), and are very close to each other, with less than 5% difference for the same D, w, and s. This shows that a vertical separation of the layers in the range of up to 3.2 mm has very little impact on the total inductance of the winding. In order to observe large differences, it is mandatory to make O significantly larger, which in the context of MLPW, is not practical, but would still be relevant for WPT setups.

In Fig. 5 the Mean Absolute Errors (MAEs) between the estimations from (1) – (3) and the respective simulation results are presented, to show the accuracy of the equations, with respect to w and s. Each of the MAE% points represent the Mean Absolute Error % among a set of different D values (from 100 to 210 mm). It can be observed that for $O = 0.8$ mm, the equations provide accurate results, with MAE always less than 4%. In the worst case, for $O = 3.2$ mm, (1) and (2) provide an increased, albeit acceptable, MAE of less than 8% and 9%, respectively. Furthermore, the MAE created by these two approximations, increases with s, and decreases with w. This behavior is mainly attributed to the fact that the two equations relate to the current sheet approximation, and provide better results as the copper ring (sheet) becomes larger (i.e., larger D) and more dense (i.e., smaller s, larger w). The MAE of the Monomial approximation does not present a monotonic behavior, but it is also acceptable, as it is less than 6% in any case.

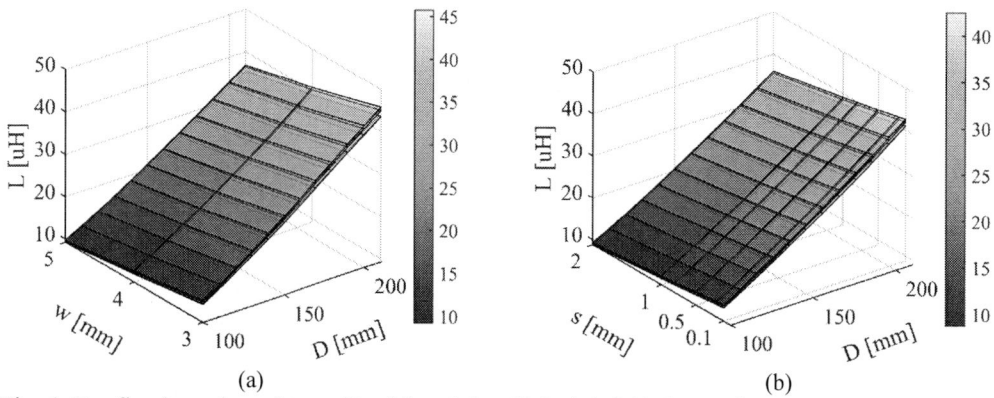

Fig. 4: For fixed number of turns $N = 10$ and $O = \{0.8, 1.6, 3.2\}$ (top to bottom) (a) the total inductance, as a function of D and w, for constant $s = 0.5$ mm, and (b) the total inductance, as a function of D and s, for a constant $w = 4$ mm.

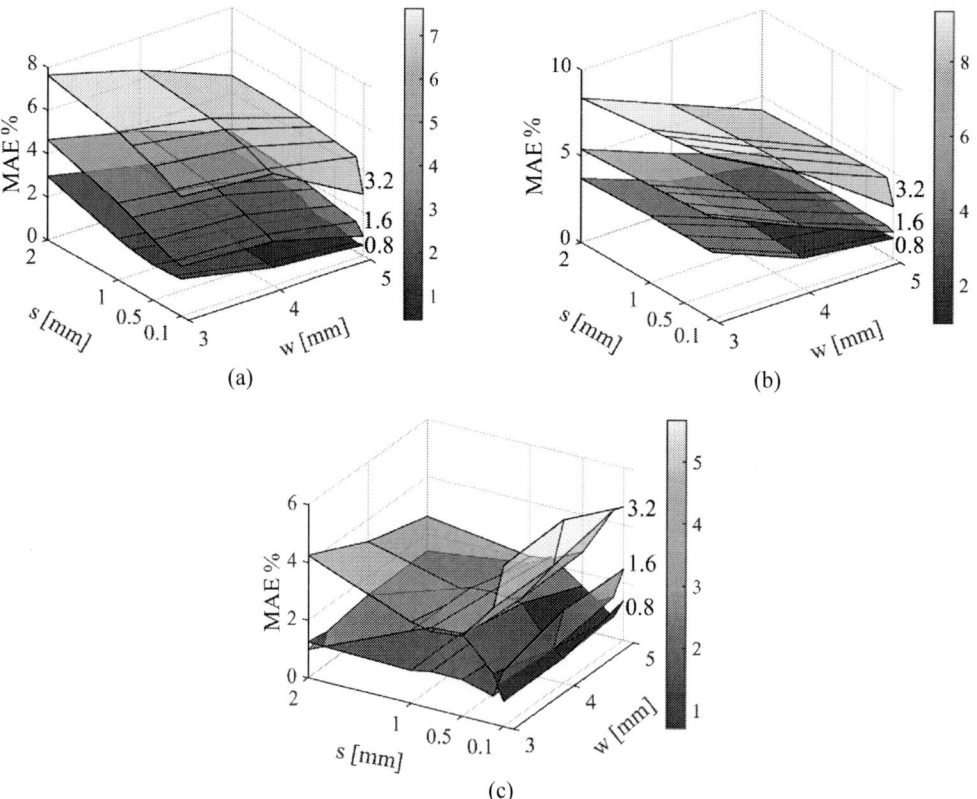

Fig. 5: MAE % for L2 and a fixed number of turns $N = N_T N_L = 10$ and $O = \{0.8, 1.6, 3.2\}$ (bottom to up), as a function of s and w, for (a) Wheeler's, (b) Rosa's and (c) the Monomial equations.

It should be noted that (1) and (2), in most L2 cases, provide an inductance estimation that is larger compared to (3), as can be seen in the Appendix for a selected number of windings. As these equations do not account for the effect of the coupling factor, it is reasonable to expect an overestimation of L, and therefore, increased MAE from (1) and (2) as O increases. In contrast, (3) presents a non-monotonic behavior. This behavioral difference can be mainly attributed to two reasons: In the first place, while (1) and (2) contain the N^2 term, the corresponding term in (3) is only $N^{1.78}$. This means that the inductance value resulting from (3) tends to increase less with each additional turn, compensating for the (less that one) coupling factor in multilayer architectures. This effect is pronounced especially in the cases of large vertical distances between the layers, e.g., for $O = 3.2$ mm. Additionally, (1) and (2) consider the average diameter $(D+d)/2$ raised to the power of one, and the filling factor $(D-d)/(D+d)$, raised to different power values. On the contrary, (3) does not consider the filling factor at all, and raises the average diameter to the power of 2.4. At the same time, the outer diameter D is raised to -1.21. This can also explain the non-monotonic behavior of the MAE when using (3). Due to these observations, it can be assumed that the Monomial equation, as expressed in (3), can provide a more accurate estimation of inductance in the case of MLPWs, especially as O increases.

Three- and Four-Layer Designs

Introducing more layers, with the proper orientation and inter-layer connections, results to windings with $N_T N_L$ turns, while the central aperture remains the same, with an area of d^2. The corresponding increase in the z-axis is negligible, since each new layer increases the profile by only a few mm. This leads to windings with exponentially higher inductance, as it has been discussed in the previous section.

The magnetic interaction between each layer defines three coupling factors, which can be calculated as in the L2 case, but their exact values do not contribute much to the analysis. It can be expected that the mutual flux linkage between any two layers follows the same trend as before and reduces as O increases. This means that, for example the middle layer in an L3 design presents higher inductance compared to the two outer layers, and hence experiences a greater voltage drop. This behavior shall be kept under consideration when the vertical distance of the layers is close to the breakdown voltage of dielectric material.

The resulting simulated inductances for different L3 design cases are presented in Fig. 6, with respect to D and w, for $s = 0.5$ mm and with respect to D and s, for $w = 4$ mm. As in the L2 cases, the inductance is strongly dependent and increases with D, while decreases with w and s but with a weaker dependence on these factors. In order to show that the relation of the N^2 term still holds, the resulting inductance of two designs with the same dimensions in L2 and L3 forms are compared. For $D = 100$ mm, $w = 5$ mm, $s = 1$ mm, and $O = 1.6$ mm, the inductance of the L2 case is 8.714 μH, while for the same dimensions the L3 case winding presents an inductance of 19.178 μH, which is approximately $(15/10)^2 = 2.25$ times more, confirming the N^2 term, at least when O is small enough.

The MAE between simulation and approximated results for the L3 designs for each of the three equations, with respect to w and s, with D varying from 100 mm to 210 mm, is presented in Fig. 7. As it is expected from the L2 case, the MAE increases with O for (1) and (2), resulting in inaccurate values for $O = 3.2$ mm (more than 10% off the simulated ones). The Monomial approximation, on the other hand, continues to provide accurate approximations (below 10% off the simulated values in all cases). It is interesting to observe that the correlation of the approximation accuracy to the vertical separation O is the opposite compared to the other two approximating equations, providing smaller errors when the vertical separation increases. This effect is the result of the inductance underestimation (attributed to $N^{1.78}$) as it was discussed previously.

Considering the N^2 relation to the inductance, it is easy to confirm that it holds by adding another layer to the design example discussed previously ($D = 100$ mm, $w = 5$ mm, $s = 1$ mm, $O = 1.6$ mm). The simulated inductance for the L4 design is 32.972 μH, which is 3.78 times larger from the L2 corresponding winding, hence, slightly smaller from the expected ratio of $(20/10)^2 = 4$. Studying the MAE for the results of each approximating formula in the L4 case, (1) and (2) provide relatively accurate results for

$O = 0.8$ mm, but exceed the 10% in MAE for higher O values, as can be seen in Fig. 8. In contrast, the $N^{1.78}$ term in (3) compensates the increase of the stray inductance in ML designs, providing very accurate results, with less than 8% MAE for the worst case. The inverse correlation to the increasing vertical spacing O can be stressed once again, according to the resulting MAE shown in Fig. 8c.

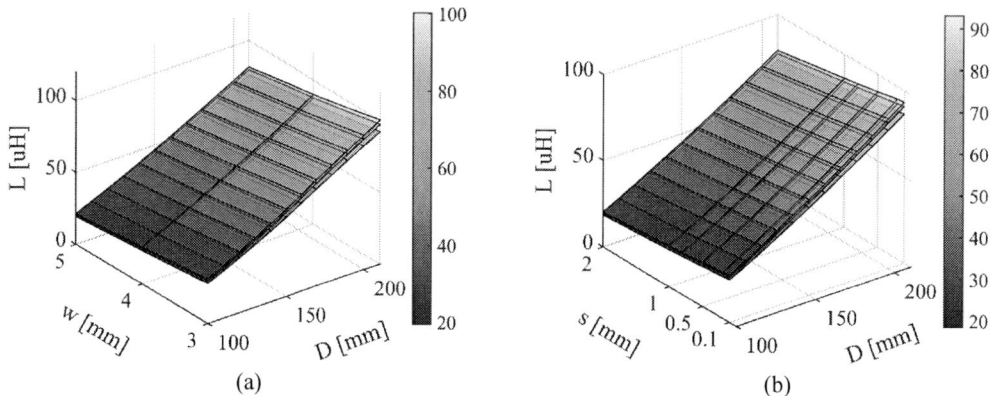

(a)   (b)

Fig. 6: For L3, with a fixed number of turns $N = 15$ and $O = \{0.8, 1.6, 3.2\}$ mm (top to bottom) (a) the total inductance, as a function of D and w, for constant $s = 0.5$ mm, and (b) the total inductance, as a function of D and s, for a constant $w = 4$ mm.

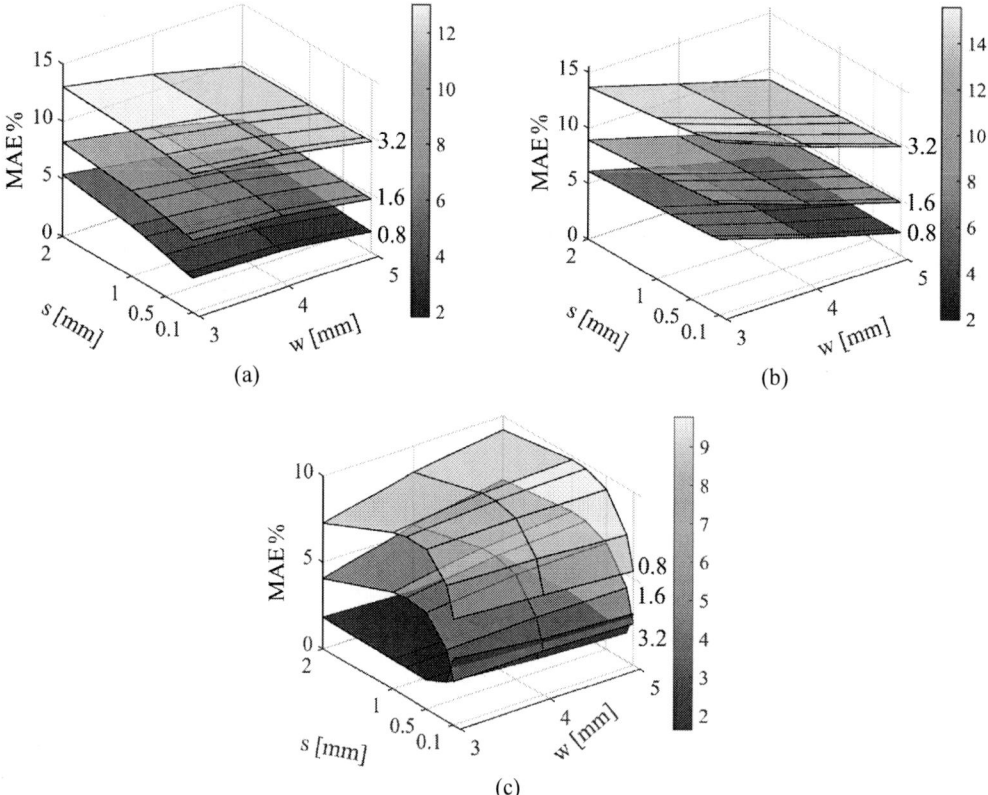

Fig. 7: MAE % for L3 and a fixed number of turns $N = N_T N_L = 15$ and $O = \{0.8, 1.6, 3.2\}$ mm for (a) Wheeler's, (b) Rosa's and (c) the Monomial equations.

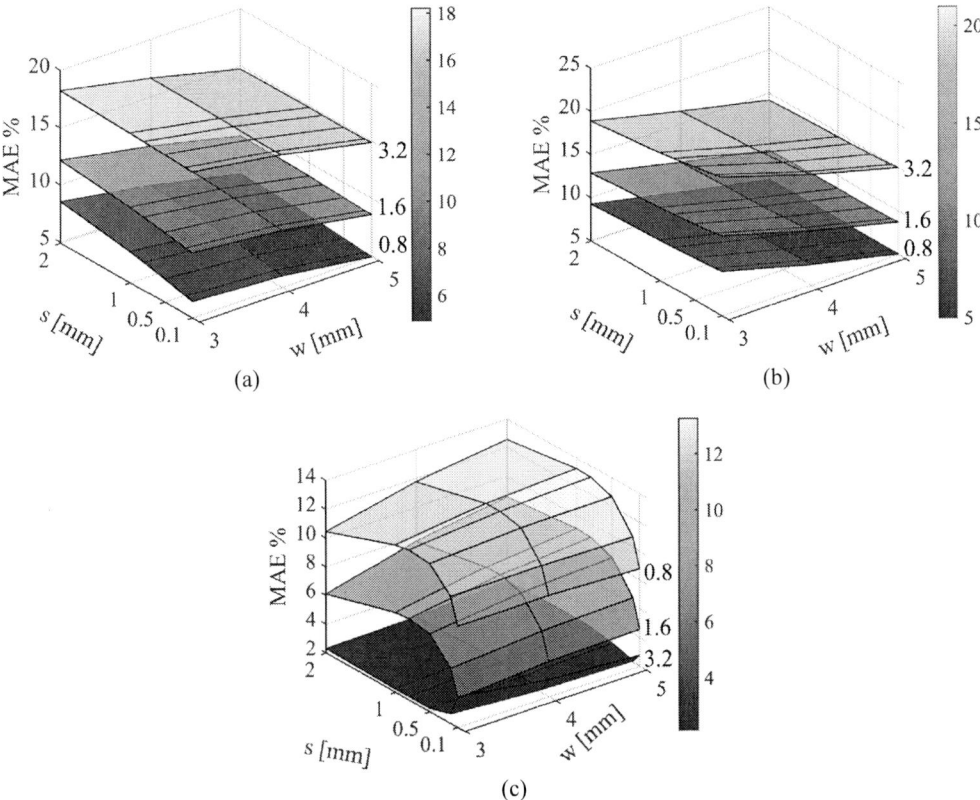

Fig. 8: MAE % for L4 and a fixed number of turns $N = N_T N_L = 20$ and $O = \{0.8, 1.6, 3.2\}$mm for (a) Wheeler's, (b) Rosa's and (c) the Monomial equations.

The estimation of inductance for windings with five layers or more cannot rely on (1) and (2), since the errors become unacceptable. Better estimations can be provided by (3), especially for cases with significant vertical separation between consecutive windings. For design cases with more than 5 layers the need for new estimation equations, which take into account O, can be considered.

Experimental Results

In order to verify the accuracy of both the equations and the simulation results, six MLPW have been printed and tested in the laboratory. The tests were conducted using a modified power amplifier for frequency, voltage, and current up to 50 kHz, 30 V and 2 A, respectively. In Fig. 9 the laboratory setup and one indicative case of an L2PW are illustrated. The results are presented in Table I, where the inductance is measured for 50 kHz, and is in good agreement, with less than 2.5 % error between the simulated and measured inductances. The error between each equation and measurement is presented in the three last columns, for Wheeler's, Rosa's, and the Monomial, respectively. It is less than 10% in any case, and less than 5% for the vast majority of the samples.

Considering the N^2 relation to the inductance, its validity can be highlighted once again for the MLPW described previously, but in this case from the measured values. The inductance of the L2PW is 8.912 μH, and the corresponding L3 and L4 cases are 19.564 and 33.512 μH, respectively, leading to: $(19.564/8.912) = 2.2 \approx 2.25$ and $(33.512/8.912) = 3.76 \approx 4$. Also, the reduction of the inductance as O increases (from 1.6 to 3.2 mm) is verified from rows 2 and 3, with a 4.3% reduction. Similarly, rows 4 and 5 indicate a 7.3% reduction. Finally, the strong dependence of inductance on d can be observed when comparing rows 1 and 2: for the same D, and with 2 mm difference (approximately 5%) in d, the inductance is increased by the same percentage, from 8.912 to 9.384 μH.

Table I: Experimental results for six indicative MLPW arrangements

N_L	D	w	s	N	d	O	WH	RS	MN	Sim	Meas	WH	RS	MN
		Dimensions [mm]						Equations [µH]					Error %	
2	100	4	2	10	44	1.6	9.377	9.284	9.302	9.202	9.384	-0.07	-1.08	-0.88
2	100	5	1	10	42	1.6	8.788	8.711	8.888	8.714	8.912	-1.41	-2.31	-0.27
2	100	5	1	10	42	3.2				8.485	8.532	2.91	2.05	4.01
3	100	5	1	15	42	1.6	19.774	19.601	18.291	19.178	19.564	1.06	0.19	-6.96
3	100	5	1	15	42	3.2				17.927	18.125	8.34	7.53	0.91
4	100	5	1	20	42	1.6	35.153	34.846	30.523	32.972	33.512	4.67	3.83	-9.79

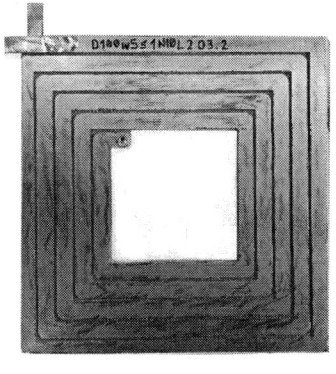

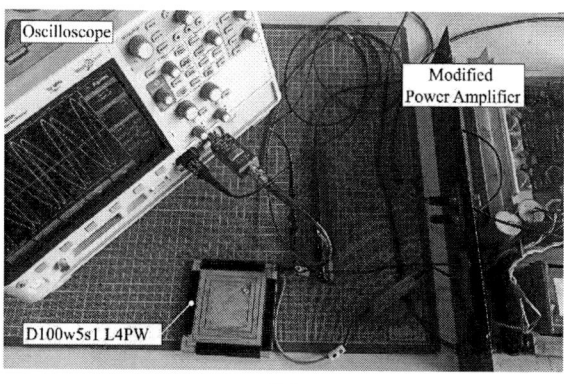

(a)   (b)

Fig. 9: (a) An indicative case of L2PW with D = 100 mm, w = 4 mm, s = 2 mm and O = 1.6 mm, and (b) the experimental setup for impedance and inductance measurement.

Conclusion

In this study, an effort to estimate the inductance for square-shaped MLPW has been made, using well-established approximation equations for L1 counterparts. A large number of MLPW have been simulated, with a variety of design dimensions (D, d, w, s, O, N_T, N_L), to quantify the accuracy of these approximations and its dependence on the number of layers and the vertical spacing between them. For the L2 case, all equations provide an acceptable level of accuracy, with less than 9% MAE, up to a reasonable O = 3.2 mm. The coupling factor in this case is relatively high, even without utilization of a ferrite core. For the L3 case, the Monomial provides accurate results, with less than 9% MAE for all values of O, while Wheeler's and Rosa's equations can be trusted only when layers are vertically close, with less than 1.6 mm spacing. For L4 designs, the accuracy of Wheeler's and Rosa's approximations is limited to small values of O, while the Monomial can still approximate relatively accurately, especially with increasing vertical separation.

References

[1] C. Buttay et al., "Application of the PCB-Embedding Technology in Power Electronics-State of the Art and Proposed Development," *3D-PEIM 2018 - 2nd Int. Symp. 3D Power Electron. Integr. Manuf.*, 2018, doi: 10.1109/3DPEIM.2018.8525236.

[2] Y. C. Liu et al., "Design and implementation of an integrated planar transformer for high-frequency LLC resonant converters," *Conf. Proc. - IEEE Appl. Power Electron. Conf. Expo. - APEC*, vol. 36, no. 5, pp. 2883–2890, 2021, doi: 10.1109/APEC42165.2021.9487046.

[3] Z. Zhang, C. Liu, M. Wang, Y. Si, Y. Liu, and Q. Lei, "High-Efficiency High-Power-Density CLLC Resonant Converter with Low-Stray-Capacitance and Well-Heat-Dissipated Planar Transformer for EV On-Board Charger," *IEEE Trans. Power Electron.*, vol. 35, no. 10, pp. 10831–10851, 2020, doi: 10.1109/TPEL.2020.2980313.

[4] K. W. Kim, Y. Jeong, J. S. Kim, and G. W. Moon, "Low Common-Mode Noise LLC Resonant

Converter with Static-Point-Connected Transformer," *IEEE Trans. Power Electron.*, vol. 36, no. 1, pp. 401–408, 2021, doi: 10.1109/TPEL.2020.3004168.

[5] M. K. Kazimierczuk, *High-Frequency Magnetic Components*. John Wiley & Sons, Ltd, 2014.

[6] A. M. Niknejad and R. G. Meyer, "Analysis, design, and optimization of spiral inductors and transformers for Si RF Ie's," *Phase-Locking High-Performance Syst. From Devices to Archit.*, vol. 33, no. 10, pp. 89–100, 2003, doi: 10.1109/9780470545492.ch8.

[7] H. A. Aebischer, "Inductance formula for rectangular planar spiral inductors with rectangular conductor cross section," *Adv. Electromagn.*, vol. 9, no. 1, pp. 1–18, 2020, doi: 10.7716/aem.v9i1.1346.

[8] C. Peters and Y. Manoli, "Inductance calculation of planar multi-layer and multi-wire coils: An analytical approach," *Sensors Actuators, A Phys.*, vol. 145–146, no. 1–2, pp. 394–404, 2008, doi: 10.1016/j.sna.2007.11.003.

[9] H. A. Wheeler, "Simple inductance formulas for radio coils," *Proc. Inst. Radio Eng.*, vol. 16, no. 10, pp. 1398–1400, 1928, doi: 10.1109/JRPROC.1928.221309.

[10] S. S. Mohan, M. D. M. Hershenson, S. P. Boyd, and T. H. Lee, "Simple accurate expressions for planar spiral inductances," *IEEE J. Solid-State Circuits*, vol. 34, no. 10, pp. 1419–1420, 1999, doi: 10.1109/4.792620.

[11] E. B. Rosa and F. W. Grover, "Formulas and tables for the calculation of mutual and self-inductance (Revised)," *Bull. Bur. Stand.*, vol. 8, no. 1, p. 1, 1912, doi: 10.6028/bulletin.185.

[12] T. Papadopoulos and A. Antonopoulos, "Formula Evaluation and Voltage Distribution of Planar Transformers Using Rectangular Windings," *2021 23rd Eur. Conf. Power Electron. Appl. EPE 2021 ECCE Eur.*, pp. 1–10, 2021.

APPENDIX

Table A: Inductance estimations of the three formulas, for an indicative number of L2 windings.

D	w	N	s	WH	RS	MN	s	WH	RS	MN
100	5.0	10	2.0	7.400	7.376	7.574	1.0	8.788	8.711	8.888
110	5.0	10	2.0	9.366	9.293	9.424	1.0	10.905	10.786	10.865
120	5.0	10	2.0	11.446	11.328	11.370	1.0	13.121	12.971	12.932
130	5.0	10	2.0	13.619	13.465	13.400	1.0	15.420	15.251	15.075
140	5.0	10	2.0	15.872	15.691	15.504	1.0	17.787	17.616	17.287
150	5.0	10	2.0	18.193	17.998	17.673	1.0	20.213	20.058	19.561
160	5.0	10	2.0	20.572	20.378	19.901	1.0	22.690	22.571	21.889
170	5.0	10	2.0	23.002	22.825	22.183	1.0	25.209	25.147	24.269
180	5.0	10	2.0	25.475	25.333	24.514	1.0	27.765	27.783	26.695
190	5.0	10	2.0	27.987	27.898	26.891	1.0	30.355	30.474	29.165
200	5.0	10	2.0	30.533	30.517	29.310	1.0	32.972	33.217	31.675
210	5.0	10	2.0	33.109	33.185	31.768	1.0	35.615	36.008	34.222
100	4.0	10	0.5	11.972	11.843	11.751	0.2	12.560	12.436	12.529
110	4.0	10	0.5	14.359	14.233	13.987	0.2	14.990	14.884	14.852
120	4.0	10	0.5	16.817	16.720	16.300	0.2	17.487	17.428	17.250
130	4.0	10	0.5	19.334	19.293	18.681	0.2	20.039	20.056	19.717
140	4.0	10	0.5	21.900	21.944	21.123	0.2	22.636	22.761	22.245
150	4.0	10	0.5	24.506	24.666	23.620	0.2	25.270	25.536	24.828
160	4.0	10	0.5	27.147	27.453	26.168	0.2	27.937	28.375	27.462
170	4.0	10	0.5	29.817	30.299	28.763	0.2	30.630	31.274	30.143
180	4.0	10	0.5	32.512	33.202	31.401	0.2	33.347	34.228	32.867
190	4.0	10	0.5	35.229	36.157	34.080	0.2	36.084	37.233	35.633
200	4.0	10	0.5	37.965	39.160	36.796	0.2	38.838	40.288	38.437
210	4.0	10	0.5	40.718	42.210	39.549	0.2	41.608	43.387	41.277

Cost and efficiency considerations in On-board Chargers

Marija Jankovic, Christian Felgemacher, Kevin Lenz, Aly Mashaly, Abdelmouneim
Charkaoui
ROHM SEMICONDUCTOR GMBH
Karl-Arnold-Straße 15, 47877
Willich, Germany
Tel.: +49 / (0) – 71172723722
E-Mail: marija.jankovic@de.rohmeurope.com
URL: http:// www.rohm.com

Keywords

Silicon Carbide (SiC), Efficiency, Automotive application, Battery charger, MOSFET, IGBT, Diode.

Abstract

Silicon Carbide (SiC) is an enabling technology for highly efficient power train applications such as
traction inverters and on-board chargers (OBC). SiC is foreseen as a dominating power device
technology in premium vehicles. However, in compact electric and hybrid vehicles a market share
with Silicon is also expected – setting high demands regarding efficiency and cost.

Introduction

Very tight regulations for CO_2 emissions are being introduced in many countries worldwide, driving
the development of electric and hybrid vehicles, that fully or partially use battery power and electric
motors for traction. Penetration of hybrid and electric vehicles (EVs) is tightly coupled with the
development of battery charging solutions. On the infrastructure side, the AC chargers developed first,
providing a single- or three-phase grid connection to the car and relying on AC/DC conversion taking
place inside the vehicle, in the on-board-charger (OBC). In this way, the charging power and the
efficiency of charging is limited by the OBC size and capability. Typical OBCs support 3.6 – 7.2 kW
single-phase charging and/or 11 – 22 kW three-phase charging. With the increase of the battery
capacity inside the vehicles, especially in full EVs, AC charging became too slow leading first to an
increase of OBC power rating, but later to the development of the DC chargers, capable of providing
DC power to the car. DC charging power and efficiency is defined with the infrastructure installed
power electronics and EV battery characteristics. Typical DC charging power ranges from 50 to 300
kW. DC charging therefore provides a reduction of charging time from a few hours to the portion of an
hour, resulting in a trend of reducing the OBC power rating. Reducing weight, volume, and cost of an
OBC and indirectly of an EV is an attractive trend for car manufacturers. Regardless of the power
rating there is a trend of increasing the OBC power density, from 2 kW/l typically achieved with Si
switches to 4-6 kW/l possible only with use of Wide-bandgap semiconductors [1, 2].

There are multiple reviews on OBC charging topologies and architectures [3,4]. The bidirectionality of
OBC also appears as a trend, especially with the development of smart homes and concepts such as
vehicle to grid (V2G), vehicle to vehicle (V2V) or vehicle to battery (V2B). The bidirectional concepts
are also analyzed in literature [3]. Since the OBC should be able to work with both single- and three-
phase grid, there are two main architecture concepts, modular and centralized, as presented in Fig. 1.

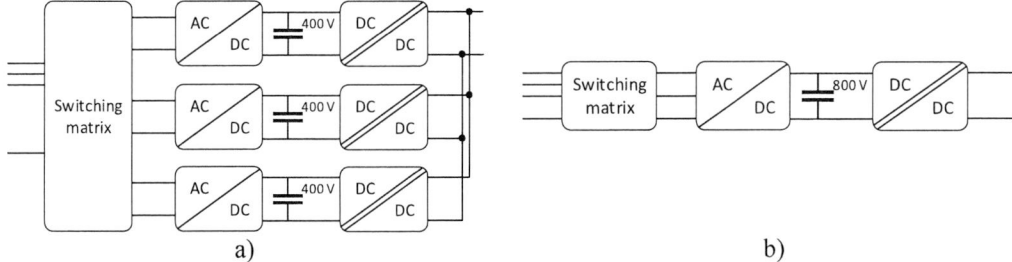

a)   b)

Fig. 1: Three-phase OBC Architecture concepts: a) Modular and b) Centralized.

Modular solutions use three identical blocks suitable for single-phase operation, typically rated at 3.6 kW or 7.2 kW each. These solutions allow the OBC manufacturers to have one-for-all solution blocks covering hybrid vehicles with only single-phase charging capability and EVs with both 3-phase and 1-phase charging capability.

On the other hand, centralized architectures use three phase AC/DC stages resulting in a higher DC link voltage and requiring 1200 V blocking voltage capability for the power semiconductors. This limits the choice on SiC-MOSFETS and Schottky barrier diodes and Si-IGBTs. Tight efficiency criteria and limited space usually limit the choice in this case to SiC. Centralized solutions require special strategies to enable single-phase operation. Sometimes this means that additional semiconductor components are added to the standard 3-phase PFCs.

This paper will first address the topologies and semiconductor technologies suitable for Modular and Centralized OBC architectures. Then it will discuss the simulation of 11 kW OBCs suitable for the European 3-phase 400 V Grid system based on the three most used topologies for two cases: predominantly SiC based solutions and predominantly Si based solutions. Finally, the AC/DC stage of a Modular solution suitable for a 3.6 kW 230 V phase to neutral grid voltage is investigated in more detail to assess the optimization of efficiency and cost. In this case the reduction of AC choke size can be achieved by using low switching losses devices and high switching frequencies. A compromise between the efficiency, semiconductor cost and AC choke could lie in the use of Hybrid IGBT which is addressed in the paper.

Topology and technology benchmark

Based on Fig. 1 various topologies can be considered a in the AC/DC and DC/DC stages. Modular solutions allow multiple choices of semiconductor technologies, since AC/DC stages and the primary side of a DC/DC stage need 650 V blocking voltage capability, which could be realized with SiC MOSFETs or Gallium Nitride (GaN) transistors as well as with Silicon technology, such as SJ-MOSFETs and Si-IGBTs. A first trend in OBC was to use a Boost PFC as a single-phase AC/DC stage and an LLC as a DC/DC stage. This topology normally requires a Si diode bridge and 650 V Si MOSFET (such as SJ-MOSFET) and FRD. Further trend towards higher efficiency and miniaturization, involved utilizing the SiC SBD as a Boost diode and interleaving. Although simple, this topology has limited efficiency and it is being replaced with other single-phase topologies such as the Totem pole topology. This topology is especially interesting when employing Wide-bandgap semiconductors. Suitable DC/DC stage topology, used in modular architecture is typically unidirectional LLC realized with a half- or full-bridge primary. Secondary side semiconductors are always rated according to the battery voltage.

Centralized topologies require blocking of 800 V DC link voltage, which results in deployment of 1200 V semiconductors for full voltage blocking or 650 V semiconductors if 3-level topologies manage the DC link splitting. 2-Level solutions use a 3-phase full bridge as an AC/DC stage and isolated DC/DC stage with 800 V input. In the case of both AC/DC stage and primary side of DC/DC stage 1200 V semiconductors are used. High efficiency requirements and miniaturization trends lead to the selection of SiC MOSFETs since IGBTs are too slow and lossy, especially for high resonant frequency LLC converters. If a 3-level AC/DC stage is used, the DC link is divided into two halves, ensuring lower blocking voltage of some semiconductors in AC/DC stage, and also allowing the use of two DC/DC stages, each with 400 V input. One popular AC/DC stage topology is Vienna rectifier, that requires 1200 V diodes and 650 V switches. Although perfectly suited for SiC semiconductors, it can deploy SJ-MOSFETs or GaN transistors as 650 V devices. Similarly, the primary side LLC switches can be realized with SiC, GaN or Si. One example of a GaN based Vienna rectifier with the split DC link fur multiple DC/DC converters is presented in [1]. This topology features high efficiency with relatively low cost. Unlike the 2-level full bridge topology, which is bidirectional, the Vienna rectifier is unidirectional and has only limited P, Q power controllability [5]. Given that some car manufacturers often specify P, Q controllability and bidirectionality Vienna rectifiers require additional active semiconductors to support this, resulting in increased cost.

Simulation – 11 kW OBC

A simulation study in PLECS has been performed to investigate the performance of different topologies and compare the solutions based on Si and SiC semiconductors for 11 kW unidirectional OBC feeding 800 V battery voltage. The semiconductors and switching frequency are selected in a way that their average junction temperature is less than 130 °C when case temperature is assumed to be 80 °C. Those thermal conditions correspond to typical coolant temperatures of 60 °C and good thermal contact case to coolant. Additionally, enough buffer to the datasheet defined maximum junction temperatures is ensured. The loss and thermal models are constructed based on the datasheets. The battery voltage varies between 600 V (empty battery) and 850 V (full battery) and the charging power is assumed to be 11 kW, since this is the case during 80 % of the charging time.

Three different topologies as shown in Fig. 2 are compared: 2-level based on full bridge and LLC, 3-level based on Vienna rectifier and LLC and single-phase solution, suitable for the modular architecture, based on Totem Pole and single-phase LLC. The target was to achieve an efficiency of above 96 % for all operating points.

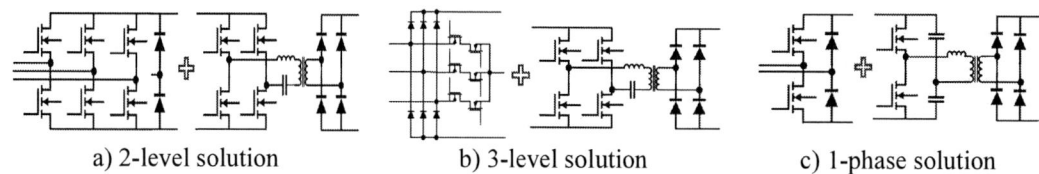

a) 2-level solution   b) 3-level solution   c) 1-phase solution

Fig. 2: Different OBC topologies: a) 2-level, b) 3-level and c) 1-phase solution suitable for modular architecture. DC link and AC side inductor are not presented here, but they are part of the design.

All three topologies are simulated based on SiC components, such as MOSFETs and Schottky barrier diodes, except for low frequency switched devices such as diodes in Totem pole (Fig. 2 a) and c)). Additionally, all three topologies are simulated based on Si devices, such as Si IGBTs and SJ-MOSFETs, where use of Si devices would not significantly compromise efficiency and SiC devices where Si device would significantly reduce efficiency. As an example, topology from Fig. 2. a) used Si IGBT at the PFC stage and SiC in LLC stage; topology from Fig. 2. b) uses only SiC diode in Vienna rectifier and all other components are based on Si, thanks to the use of split DC link and two 400 V input LLCs; topology from Fig. 2. c) is fully based on Si. The obtained efficiency profiles, in the 3-phase full power operating point are presented in Fig. 3 a). In those profiles only semiconductor losses obtained by simulation are considered.

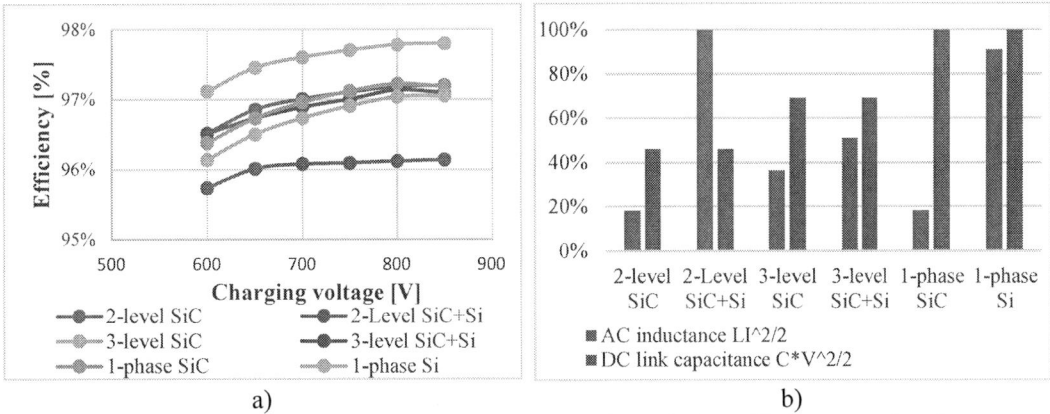

a)   b)

Fig. 3: Comparison of the 3 simulated solutions based on SiC and Si a) Efficiency vs. charging voltage; b) Normalized energy stored in AC inductance and DC link capacitance.

The 3-Level solution based on SiC has the highest efficiency due to the very efficient Vienna rectifier. The 2-level solution based on SiC, the single-phase solution based on SiC and the 3-level solutions based on either SiC or Si have similar efficiency profiles. Their efficiency peaks above 97 %. The Single-phase solution based on Si has only slightly lower efficiency then its SiC counterpart. The 2-Level solution based on SiC and Si that uses IGBTs in the PFC stage has more than 1 % lower efficiency than its SiC counterpart. This result indicates that in the case of 3-level solutions or single-phase solutions, Si devices could also lead to high efficiency operation.

Fig. 3. b) shows the normalized energy stored in the AC inductances and DC link capacitances. In this comparison a value of 100 % corresponds to the highest energy stored in AC inductance and DC link capacitance, respectively. For AC inductance we can see that the highest energy stored is in 2-level SiC+Si based solutions, whereas the highest energy stored in DC link capacitance is in 1-phase solutions suitable for modular architecture. The 2-Level SiC based solution has the lowest energy stored in AC inductances and DC link capacitances, and therefore lowest cost of those elements. 3-Level based solutions have moderate energy stored in AC inductances and DC link capacitances. Single-phase solutions store high energy in DC link capacitances. Although the efficiency of the Si based single-phase solution, considering semiconductor losses, is relatively high, this solution shows very high energy stored in AC inductance which might lead to the higher overall losses. The DC capacitances have significant volume in an OBC [6], and therefore where 3-phase OBC is implemented and OBC volume is critical, the use of centralized architecture and 2- or 3-level solutions might be preferred.

Modular OBC with Totem Pole topology

The Totem Pole topology, presented in Fig. 4. a), is very promising for modular OBC designs, since it features high efficiency, possibility for interleaved design and upgrade to a bidirectional solution. One example of 7.2 kW bidirectional OBC based on Totem Pole topology is presented in [2]. Switches Q1 and Q2 are fast switched, whereas the diodes D1 and D2 switch at 50 Hz. In general, use of Wide-bandgap devices guarantees high efficiency with high switching frequency, being in the range of hundreds of kHz for GaN and around 100 - 150 kHz for SiC. Use of Wide-bandgap devices means increase of semiconductor costs and decrease of the size and cost of passive elements. Given the single-phase nature of Totem Pole topology, the current and voltage on the DC link side is highly dominated by 100 Hz ripple, as presented in Fig. 4. b) requiring large DC link capacitances, typically realized with electrolytic capacitances. The capacitor bank is rated according to this ripple and therefore not significantly influenced by switching frequency. There are techniques of limiting the DC link capacitor bank such as active power decoupling, by adding additional circuit components [7]. However, [2] suggests that minimization of power density and volume is provided by proper arrangement of electrolytic capacitances and magnetic elements. Similar approach is done in the experimental converter presented in this paper. Switching frequency influences AC inductor sizing and therefore using the devices with lower switching losses, contributes to the OBC miniaturization.

Replacing the diodes D1 and D2 with Si MOSFETs used in synchronous rectification mode reduces the losses in 50 Hz operated leg and provides bidirectionality. In this paper we will focus on fast switching devices in a Totem pole topology (Q1 and Q2) and look how different semiconductor technologies influence efficiency and switching frequency.

A Hybrid IGBT, the name is used here for a discrete device comprising of a Si-IGBT and a co-packed SiC Schottky barrier diode, could be a compromise for AC inductor size vs. semiconductor cost in a Totem pole topology. The performance of hybrid IGBT, standard Si-IGBT with co-packed Si Diode and SiC MOSFET based solutions are compared in Totem pole topology.

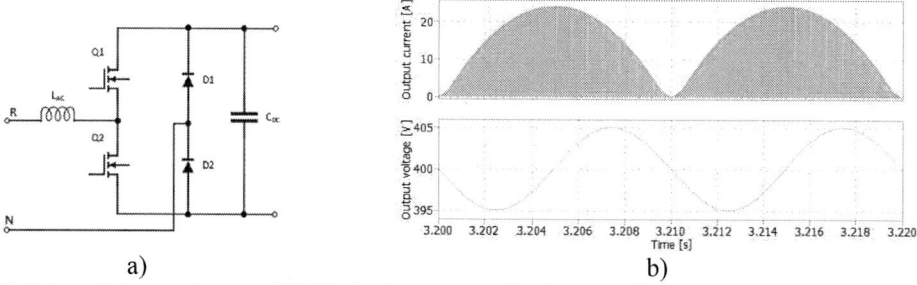

a)   b)

Fig. 4: a) Totem Pole topology and b) Typical current and voltage at the output.

Simulation based on Experimental evaluation – 3.6 kW Totem Pole

The simulation of a 3.6 kW Totem pole converter with 400 V output and 230 V single-phase grid has been performed. This is a typical power rating of a single phase OBC suitable of 3.6 kW single- and 11 kW three-phase charging. The same solution can be used in 7.2 kW single-phase charger by using two equivalent building blocks. Losses in the high frequency leg of a Totem pole are compared for different semiconductor components. Two different 30 A 650 V IGBT's are compared together with one SiC MOSFET. Both IGBT devices are based on the same IGBT die and different antiparallel diodes: 25 A Si diode (RGW60TS65DHR) and 25 A SiC diode (RGW60TS65CHR) [8, 9]. The IGBTs with antiparallel diodes are packed in TO-247-3L packages. SiC device is 45 mΩ 750 V MOSFET SCT4045DRHR packed in TO-247-4L [10]. TO-247-4L provides better switching characteristics compared to TO-247-3L package.

All three devices are firstly evaluated in the double-pulse test (DPT) experimental setup. This setup has a possibility to capture switching waveforms of the actively switched and commutating device in a half-bridge. DC link voltage is kept constant, and current is varied. DPT test is performed at 25 °C, 100 °C and 150 °C. Comparison of turn on and turn off behavior of all three devices is compared in Fig. 5 at 100 °C, 400 V DC link and 18 A. The turn on behavior (Fig. 5. a) shows a significant difference between the current overshoot and therefore turn on losses and diode reverse recovery losses by standard and hybrid IGBT (Fig. 6). The SiC MOSFET turns faster on than the IGBT devices. Fig. 5. b) shows significant difference in turn off speed of IGBT devices and SiC MOSFET. IGBT devices have significant tail current that amounts to high turn off losses.

Fig. 6 presents turn-on losses and reverse recovery losses (switching energies) calculated from transients captured by DPT at 100 °C. This highlights the main difference between standard IGBT co-packed with Si diode and Hybrid IGBT co-packed with SiC diode.

DPT test is performed at 25 °C, 100 °C and 150 °C to provide enough data for switching loss model in Plecs simulation environment. Inputs for the simulation model are switching energies calculated from measurements, as presented in Fig. 6. Conduction losses are modelled based on the datasheet information at 25 °C and 175 °C and thermal impedance is modelled with a 3rd order Cauer model. Case temperature in all the cases is set to 80 °C assuming the high performant water cooling in the automotive environment.

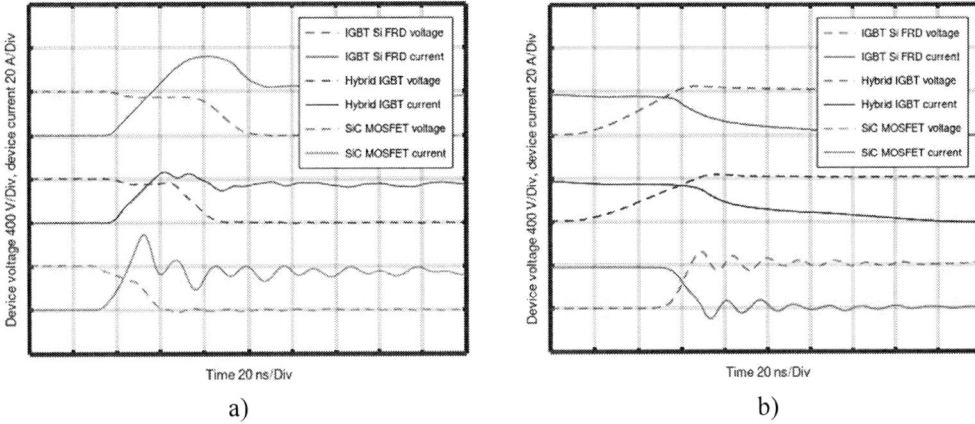

a)   b)

Fig. 5: Comparison of switching behavior by a) turn on and b) turn off. In the case of both IGBT devices driving condition is given with 15 V/0 V, $R_{G\_on} = R_{G\_off} = 10\ \Omega$. SiC MOSFET has driving condition of 18 V/0 V, $R_{G\_on} = 10\ \Omega$ and $R_{G\_off} = 3.3\ \Omega$.

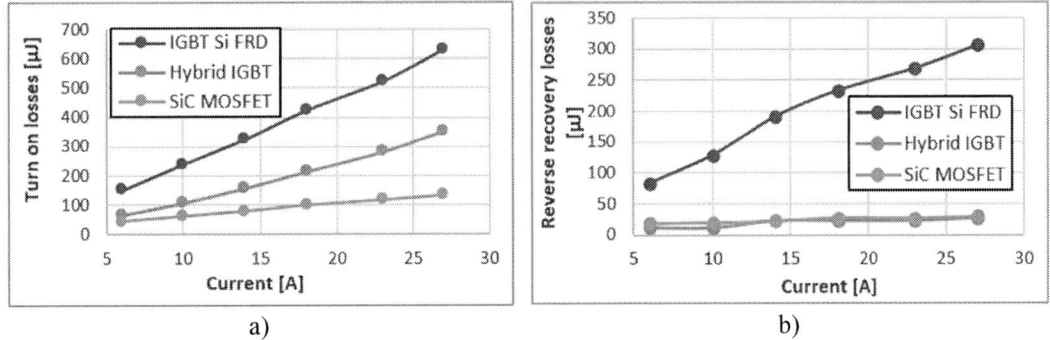

a)   b)

Fig. 6: Comparison of switching losses by a) turn on and b) reverse recovery. In the case of both IGBT devices driving condition is given with 15 V/0 V, $R_{G\_on} = R_{G\_off} = 10\ \Omega$. SiC MOSFET has driving condition of 18 V/0 V, $R_{G\_on} = 10\ \Omega$ and $R_{G\_off} = 3.3\ \Omega$.

Fig. 7. a) shows the semiconductor losses (both switching and conduction losses) in the switching leg normalized with the rated power at rated power operation. Unlike standard IGBT with Si diode, hybrid IGBT allows switching frequencies up to 90 kHz. Hybrid IGBT solution at 90 kHz and nominal power has 1 % less efficiency than the SiC MOSFET switched with the same switching frequency.

Fig. 7. b) shows normalized system parameters in three selected cases of switches: IGBT switched at 40 kHz, Hybrid IGBT switched at 60 kHz and SiC MOSFET switched at 100 kHz. Normalized system parameters are given as a percentage of the maximum of each parameter for all three cases (standard IGBT, Hybrid IGBT and SiC MOSFET). AC choke is selected to ensure 20 % peak-to-peak AC current ripple. As seen from the Fig. 7. b) SiC MOSFET based solution requires only 40 % and Hybrid IGBT 70 % of the AC choke inductance required in the case of standard IGBT based solution. Reduction of the inductance means reduction of stored energy and choke size, cost and weight and additionally reduction of losses in the AC choke. However, further increase of the switching frequency does not significantly reduce the required inductance value. For the selected switching frequencies and rated power operation, the losses in the switching leg of the Totem pole in a IGBT and Hybrid IGBT case are comparable. Losses in the switching leg bases on SiC MOSFET are 40 % lower, which results in 0.56 % higher system efficiency at nominal power. Finally, price of those switches is compared based on the internal price database. IGBT with Si diode costs approximately one third of SiC MOSFET price whereas expected Hybrid IGBT price development is between standard IGBT and SiC MOSFET price.

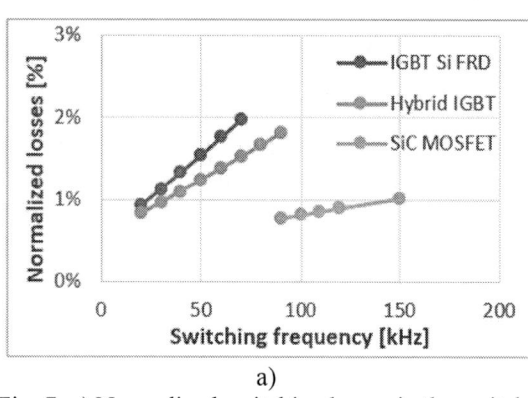
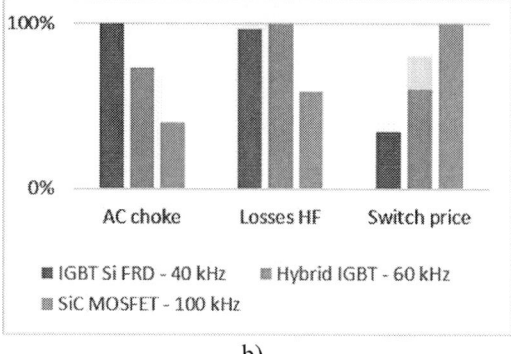

a)   b)

Fig. 7: a) Normalized switching losses in the switching leg at rated power operation and b) Normalized system parameters at the selected switching frequency: required AC choke for 20 % current ripple, losses in the switching leg and switch price.

The above analysis concludes the highest efficiency of SiC based Totem Pole even with the significantly higher switching frequency. Higher efficiency leads to reduction of cooling requirements which could lead to cost savings. Related to the cost targets, SiC MOSFET has the highest price but therefore the smallest AC choke. The size of the AC choke in the case of standard IGBT might not fit to OBC volume and weight requirements, and its price might compensate for the low semiconductor cost. Hybrid IGBT is an intermediate solution in both semiconductor price and choke size and cost. This analysis does not consider the EMI filtering size which would be briefly discussed in the next chapter.

Experimental results based on 45 mΩ 750 V SiC MOSFET Totem Pole

Experimental verification of a Totem Pole converter based on the 45 mΩ 750 V SiC MOSFET from the previous chapter has been performed to evaluate system efficiency. The experimental converter is bidirectional using Si SJ-MOSFET in synchronous rectification mode instead of the diodes in Fig. 4. a). Converter parameters are presented in Table I. The control algorithm consists of two control loops, slower proportional-integral controller of DC voltage and inner fast proportional-resonant AC current controller.

Table I: Converter parameters

Vin	Vout	Power	L_{AC}	f_{sw}	C_{DC}
230 V	400 V	3.4 kW	185 µH	100 kHz	2.24 mF

Fig 8. shows the photograph of the Totem Pole converter. In this case simple heatsink air cooling is used and therefore the converter is operated at somewhat higher case temperatures than in typical OBC application. As it can be seen, significant volume belongs to DC link capacitors that are independent on switching frequency, AC choke and EMI filters. AC choke is in this case minimized thanks to the 100 kHz switching frequency. AC side EMI filter has two stages as shown on Fig. 8. Use of Si switch technology would increase the size of the AC choke but might reduce the size of EMI filter. EMI filter design of the Totem Pole converter for all cases analyzed in the previous chapter is outside the scope of this paper.

1. AC choke
2. SiC MOSFETs
3. Heatsink
4. EMI filter AC side
5. DC link capacitors
6. EMI filter DC side

Fig. 8: Experimental Totem Pole converter.

Fig. 9 shows the efficiency behavior from input power source to the output including the control electronics, such as auxiliary power supply, and cooling fan. As shown the obtained efficiency of the overall system is higher than 98 % at the nominal power. The measured case temperature of the switching leg MOSFETs is about 110 °C which is 30 °C higher than in the case of simulation study. This would result in slightly higher losses compared to estimated 0.8 % at 100 kHz. However, considering that the overall system efficiency is higher than 98 % and there are losses associated with low frequency leg, traces, filters and control electronics, the obtained results match well high SiC based solution efficiency. Using a Hybrid IGBT would in this case mean reducing the switching frequency to 60 kHz, increasing AC choke (cost and volume), possibly reducing AC side EMI filtering, and reducing efficiency by roughly 0.6 % at the nominal power.

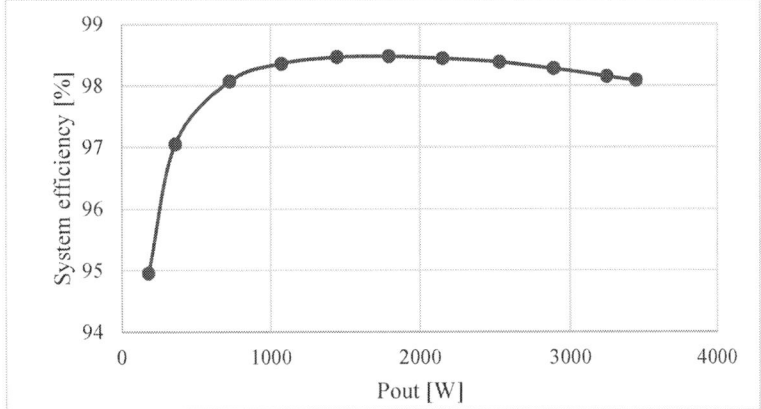

Fig. 9: Power analyzer obtained overall system efficiency of the experimental 3.4 kW Totem pole and 400 V output voltage.

Conclusion

This paper is a review of the current trends in OBC application, covering the topology and semiconductor technology trends. Centralized and modular architecture have been discussed together with the relevant topologies and selection of suitable semiconductors for those topologies. Three different topologies have been simulated to assess the efficiency and passive components sizing. Centralized 3-level Vienna based solution has the highest efficiency especially when based on SiC semiconductors. SiC based 2-level topology has the lowest requirement of passive elements providing the miniaturization of the system. Additionally, it features full P, Q controllability on the AC side, which makes it suitable for bidirectional solutions required for V2G/V2V/V2B trends. Therefore, SiC is the suitable choice for centralized OBC architecture, and it could be combined with Si in 3-level solutions.

Among Si based solutions, single-phase solution suitable for modular architecture has high efficiency based on Si but therefore the highest requirement for passive elements.

The second part of the paper discusses Totem pole, as an AC/DC stage of a single-phase solution suitable for modular architectures. This part focuses on the reduction of the AC choke by using a Hybrid IGBT with a co-packed SiC diode instead of a standard IGBT with Si diode. In Totem pole the highest efficiency, highest switching frequency and the smallest required AC choke is guaranteed with SiC MOSFET. Cost of a SiC MOSFET is higher than the cost of Si devices and the optimal solution depends on a trade of between cost, efficiency, and volume. Use of SiC MOSFET, on the other hand, reduces the cost of passive elements and cooling. However, Hybrid IGBT (IGBT with co-packaged SiC diode) could be suitable for more cost optimized solutions.

Finally, SiC based Totem pole converter has been built using 45 mΩ SiC MOSFETs in the fast switching leg. The experimental converter operated at 100 kHz confirmed high efficiency of above 98 % at the nominal power. The system design is very compact although it has been designed with forced air cooling, instead of more effective water-cooling typically present in OBC applications.

References

[1] M. Kasper, J. Azurza, G. Deboy, Y. Li, M. Heider, J. W. Kolar, "Next Generation GaN-based Architectures: From 240W USB-C Adapters to 11kW EV On-Board Chargers with Ultra-high-Power Density and Wide Output Voltage Range," 2022 IEEE Proceedings of the Conference on Power Electronics and Intelligent Motion (PCIM Europe 2022), Nuremberg, Germany, May 10-12, 2022.

[2] F. Vollmeier, A. Connaughton, I. Recepi, T. Langbauer, M. Pajnic, W. Konrad and C. Mentin, "Tiny Power Box - Exploiting Multiport Series Resonant Topologies for Very High-Power Density Onboard Chargers," 2022 IEEE Proceedings of the Conference on Power Electronics and Intelligent Motion (PCIM Europe 2022), Nuremberg, Germany, May 10-12, 2022.

[3] J. Yuan, L. Dorn-Gomba, A. D. Callegaro, J. Reimers and A. Emadi, "A Review of Bidirectional On-Board Chargers for Electric Vehicles," in IEEE Access, vol. 9, pp. 51501-51518, 2021, doi: 10.1109/ACCESS.2021.3069448

[4] I. Subotic and E. Levi, "A review of single-phase on-board integrated battery charging topologies for electric vehicles," 2015 IEEE Workshop on Electrical Machines Design, Control and Diagnosis (WEMDCD), 2015, pp. 136-145, doi: 10.1109/WEMDCD.2015.7194522.

[5] D. A. Molligoda, J. Pou, C. J. Gajanayake and A. K. Gupta, "Analysis of the Vienna Rectifier under Nonunity Power Factor Operation," 2018 Asian Conference on Energy, Power and Transportation Electrification (ACEPT), 2018, pp. 1-7, doi: 10.1109/ACEPT.2018.8610866.

[6] Baek, J.; Park, M.-H.; Kim, T.; Youn, H.-S. Modified Power Factor Correction (PFC) Control and Printed Circuit Board (PCB) Design for High-Efficiency and High-Power Density On-Board Charger. *Energies* 2021, *14*, 605. https://doi.org/10.3390/en14030605

[7] A. S. Morsy and P. N. Enjeti, "Comparison of Active Power Decoupling Methods for High-Power-Density Single-Phase Inverters Using Wide-Bandgap FETs for Google Little Box Challenge," in IEEE Journal of Emerging and Selected Topics in Power Electronics, vol. 4, no. 3, pp. 790-798, Sept. 2016, doi: 10.1109/JESTPE.2016.2573262.

[8] [online] https://fscdn.rohm.com/en/products/databook/datasheet/discrete/igbt/rgw60ts65dhr-e.pdf access on 13.6.2022.

[9] [online] https://fscdn.rohm.com/en/products/databook/datasheet/discrete/igbt/rgw60ts65chr-e.pdf access on 13.6.2022.

[10] [online] https://fscdn.rohm.com/en/products/databook/datasheet/discrete/sic/mosfet/sct4045drhr-e.pdf access on 13.6.2022.

A Novel Combined Control of Ground Current and DC-pole-to-Ground Voltage in Symmetrical Monopole Modular Multilevel Converters for HVDC Applications

Pablo Briff, Amit Kumar
GE Renewable Energy
Stafford, United Kingdom
Corresponding authors: {pablo.briff, amit.g.kumar}@ge.com
URL: http://www.ge.com

Abstract

This paper investigates the reduction in converter trips in the symmetrical Modular Multilevel Converter (MMC) by regulating the dc term of the valves' voltages, while allowing the dc pole-to-ground dc voltages to fluctuate within acceptable limits. Moreover, the control method proposed in this work serves two purposes: i) to regulate the pole-to-ground dc voltages of a symmetrical Modular MMC when the voltage of either of the poles at the dc side exceeds a predefined threshold; and ii) to control the ground current within acceptable limits. This results in an extended converter availability by avoiding unnecessary converter trips. The benefits of the proposed method is validated by simulation results for a symmetrical monopole MMC VSC-HVDC configuration.

Keywords: «modular multilevel converter», «ground current», «converter control».

Introduction

In HVDC power transmission networks, ac power is converted to dc power for transmission via overhead lines, under-sea cables and/or underground cables. This conversion removes the need to compensate for the ac capacitive load effects imposed by the transmission line or cable, and reduces the cost per kilometre of the lines and/or cables. Thus, dc transmission becomes cost-effective when power needs to be transmitted over a long distance [1], [2].

Fast and stable controllability of power is a major advantage of HVDC transmission over conventional ac transmission systems. Conventional thyristor-based line-commutated converter (LCC) HVDC systems have been a well-established technology for more than 60 years [3], especially when bulk power needs to be transferred over a long distance. However, when weak systems like wind or solar power plants need to be connected to the grid, Voltage Sourced Converter (VSC)-HVDC systems are the preferred solution [4]. During the past two decades, VSC-HVDC technology has undergone a series of evolutionary milestones until it finally reached the current state-of-art with the modular multilevel converter (MMC) technology. Most of these milestones were aimed at making the switching losses comparable to LCC HVDC systems, increase the rating of VSC converters and to reduce the harmonics emission of the converters [3].

A major topic in HVDC – being part of transmission infrastructure – is the availability of the converter stations. These are usually impaired by unplanned trips arising from protective actions. Examples of protection algorithms in VSC-HVDC are those which monitor the amount of ground current circulation and dc voltage to ground imbalance. While [5] addresses the issue of operation of MMC under dc pole imbalances, it does not investigate the problem of unexpected trips due to high ground current circulation and its impact on industrial applications.

More recent works such as [6–8] study the topic of dc voltage balancing in both pole and bipole configurations in great detail. However, these works do not address the issue of dc pole to ground voltage balancing combined with ground current control, and the prioritisation logic that must take place between these two controllers.

For a power transmission network, one of the major performance criteria are its reliability and availability. These form the basis of some of the key parameters an HVDC manufacturer considers while designing their VSC-HVDC scheme. At the same time, fitting a scheme with excessive design margins makes it uneconomical. By applying supplementary fast controls, the VSC-HVDC converter design can be optimised, while satisfying the reliability and availability targets.

This paper focuses on the symmetrical monopole MMC VSC-HVDC, which constitutes, at present, the most widely used dc circuit configuration in commercial HVDC projects. Compared to asymmetrical monopole or bipole, this configuration has the advantage that only two conductors (overhead transmission line or cable), each with only 50% of the total converter station dc voltage rating, are required.

This paper proposes a control method that serves two purposes: i) to regulate the pole-to-ground dc voltages of MMC when the voltage of either of the poles at the dc side exceeds a predefined threshold; and ii) to control the ground current within acceptable limits is also attained by regulating the dc term of the valves' voltages, while allowing the dc pole-to-ground dc voltages to fluctuate within acceptable limits. With the presented control method, the availability of the converter is extended by avoiding unnecessary converter trips. The potential of the method is validated by simulation results for a symmetrical monopole MMC for VSC-HVDC applications.

Ground Current and DC Pole to Ground Control

Model Statement

Fig. 1 shows a typical connection of an ac system to a VSC-HVDC, which is then connected to a dc line.

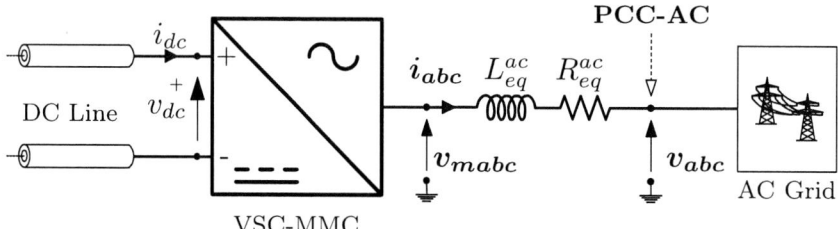

Figure 1: AC system connected to a VSC.

In particular, this paper focuses on the symmetrical monopole configuration of half-bridge (HB) MMC, where each valves' total voltage is in the range of $0 \rightarrow V_{dc}$ volts, where V_{dc} is the pole-to-pole voltage. A sample equivalent circuit of this topology configuration is shown in Fig. 2. While the concepts presented in this paper are applicable to both star-star and star-delta transformer connections, as an example, this paper investigates the application to the star-delta transformer configuration only. In a star-delta-connected transformer, the ground reference at the secondary side of the transformer is usually provided by inductors L_g with equivalent resistance R_g connected to ground. At the dc side, a virtual ground v_{gnd} with zero voltage potential is provided by means of the dc side equipment such as insulators, surge arresters, water cooling circuit, voltage divider and transmission line present at the positive and negative poles. Ultimately, an electrical connection between the solid ground provided by the grounding circuit and the virtual ground exists. Hence, a ground current i_g can flow at the star-point connection of the star-connected inductors.

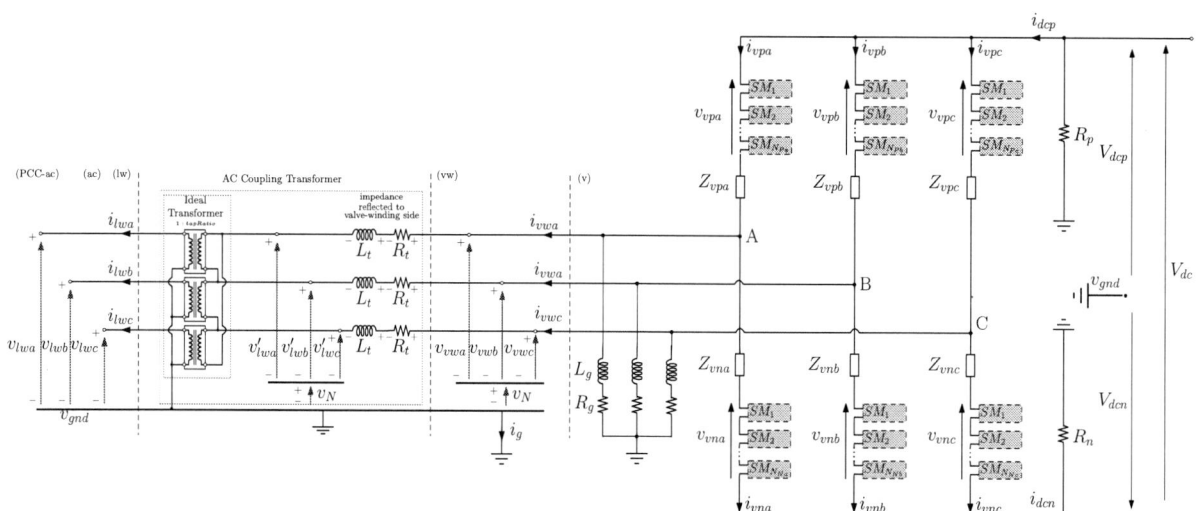

Figure 2: Single line diagram of MMC VSC-HVDC.

Tolerances in dc-side equipment such as surge arresters and insulators may alter the voltage reference at the V_{dcp}, V_{dcn}

terminals. Surge arresters and insulators in an overhead line usually present a dc resistance in the order of mega-ohms. If the value of such resistors varies with respect to each dc pole, an asymmetrical pole-to-ground dc voltage will be present since the pole-to-pole V_{dc} voltage will be unequally distributed by the resistive divider formed by the high resistance connected to each pole to ground.

Furthermore, if the pole-to-ground dc voltage exceeds a permitted maximum value, a protective action will be initiated, thus tripping the converter to avoid damage of equipment due to over-voltage.

An imbalance in the dc side conductance creates a current flow through the grounding circuit, and depending upon the grounding circuit resistance, this creates an imbalance in the pole to ground voltage, i.e. one of the pole-to-ground voltages increases with respect to the other pole. Other reasons for an imbalance of the pole-to-ground dc voltage include: manufacturing/ageing effect tolerance of voltage divider and surge arresters, difference in resistance of positive and negative pole conductor, dc current measurement error in the limbs, insulators degradation, to name a few. In addition to this, an imbalance in the pole-to-ground voltages will drive the ground current i_g. Such current can be measured accurately with a dc current transformer (DCCT) and it can be used as a measure of the amount of imbalance present at the dc poles. Moreover, if either the dc pole-to-ground voltage or the current through grounding circuit should exceed a permitted maximum value, then a protective action will be initiated.

Ultimately, the benefits of the proposed method relies on the extended operation of the converter by keeping both the pole-to-ground dc voltages and the ground current within acceptable limits. The main benefit is, therefore, an extension of the converter's availability, as well as the minimization of the liquidated damages due to excessive converter trips arising from excessive ground current circulation or over-voltage on the dc lines. What is more, these benefits are attained at no extra hardware costs.

Control Strategy

The pole-to-pole dc voltage is defined as $V_{dc} \triangleq V_{dcp} - V_{dcn}$. Under balanced operating conditions, the positive and negative valves' voltages, v_{vpx} and v_{vnx} respectively, for the MMC circuit shown in Fig. 2 are given by

$$
\begin{aligned}
v_{vpx} &= \frac{V_{dc}}{2} - v_x - v_{ctrl,x}^{\Sigma} \quad , \\
v_{vnx} &= \frac{V_{dc}}{2} + v_x - v_{ctrl,x}^{\Sigma} \quad ,
\end{aligned}
\tag{1}
$$

where $x = \{a, b, c\}$, $[v_a v_b v_c]^T$ is the three-phase equivalent converter's ac output voltage, and $v_{ctrl,x}^{\Sigma}$ is a common-model control voltage term that relates to the valves energy control. In general, the dc term of the valves voltage is $v_{zx,dc}$, where $z = \{p, n\}$.

The dc equivalent circuit of Fig. 2 is shown in Fig. 3. Consider the situation where the positive pole-to-ground dc voltage exceeds its maximum steady-state value by an amount ΔV_{dcp}, as shown in Fig. 4 . In order to reduce the pole-to-ground voltage on the positive dc rail, the dc voltages of the positive valves shall be reduced by ΔV volts, and typically $\Delta V \approx \Delta V_{dcp}$ volts.

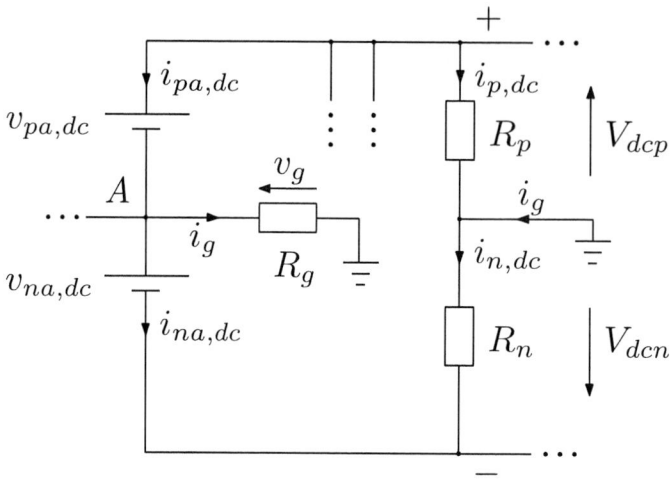

Figure 3: DC equivalent circuit of Fig. 2 for phase a.

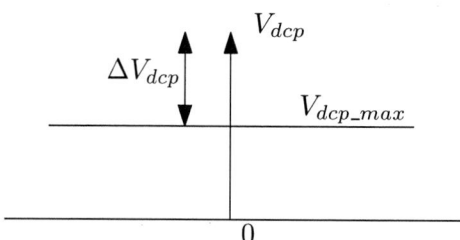

Figure 4: Pole-to-ground voltage V_{dcp} exceeding a threshold V_{dcp_max} by an amount ΔV_{dcp} at the positive dc pole.

An imbalance in the pole-to-ground dc voltages, defined as $\Delta V_{pn} \triangleq V_{dcp} - |V_{dcn}|$, is mainly due to an imbalance in the dc resistors R_p, R_n – these are usually in the order of mega ohms. If the ground current i_g is zero, then the voltage divider of R_p, R_n determines the values of V_{dcp}, V_{dcn} as follows

$$
\begin{aligned}
V_{dcp} &= V_{dc}\frac{R_p}{R_p + R_n}, \\
V_{dcn} &= -V_{dc}\frac{R_n}{R_p + R_n}.
\end{aligned}
\tag{2}
$$

Naturally, if $R_p \neq R_n$ then $V_{dcp} \neq |V_{dcn}|$, or equivalently $\Delta V_{pn} \neq 0$, unless some control action is performed. The imbalance in V_{dcp}, V_{dcn} can be influenced by controlling the currents $i_{p,dc}$ and $i_{n,dc}$. What is more, if $i_{p,dc} \neq i_{n,dc}$, then $i_g \neq 0$. This means that a dc pole-to-ground imbalance will generate a ground current, and vice versa. However, the amount of voltage imbalance at the dc rails may not be as significant – in terms of protective actions – as the current imbalance observed at the ground connection. This means that a ground current imbalance may be prioritised over an imbalance at the dc rails.

From Fig. 3, relation between the ground current i_g the dc pole-to-ground voltage imbalance $|V_{dcn}|$ is given by

$$
\begin{aligned}
V_{dcp} &= v_{pa,dc} + v_g, \\
V_{dcn} &= -v_{na,dc} + v_g, \\
v_g &= R_g i_g, \\
i_g &= i_{n,dc} - i_{p,dc}.
\end{aligned}
\tag{3}
$$

From (3), if $R_p > R_n$, then $V_{dcp} > V_{dcn}$. Meanwhile, $i_{n,dc}$ is increased by increasing i_g. The ground current i_g is increased by increasing v_g. And the voltage v_g is increased by increasing $v_{pa,dc}$. Consequently, this leads to decreasing $v_{na,dc}$ in order to keep V_{dc} constant.

Moreover, an important conclusion from (3) is that controlling both i_g and $|V_{dcn}|$ *simultaneously* is not possible. Only one of those variables can be controlled at any given time, and priority should be given to each of these controlled variables based on the protection levels defined by the converter design and its control strategy.

Defining the following

$$x = \begin{bmatrix} v_{pa,dc} \\ v_{na,dc} \\ V_{dcn} \\ v_g \\ i_{p,dc} \\ i_{n,dc} \\ i_g \end{bmatrix} \qquad b = \begin{bmatrix} V_{dc} \\ V_{dcp} \\ 0 \\ 0 \\ 0 \\ V_{dcp} \\ 0 \end{bmatrix} \qquad A = \begin{bmatrix} 1 & 1 & 0 & 0 & 0 & 0 & 0 \\ 1 & 0 & 0 & 1 & 0 & 0 & 0 \\ 0 & 1 & 1 & -1 & 0 & 0 & 0 \\ 0 & 0 & 0 & 0 & 1 & -1 & 1 \\ 0 & 0 & 0 & 1 & 0 & 0 & -R_g \\ 0 & 0 & 0 & 0 & R_p & 0 & 0 \\ 0 & 0 & -1 & 0 & 0 & -R_n & 0 \end{bmatrix}$$

$$(4)$$

then the system of equations for the circuit of Fig. 3 is

$$A \cdot x = b \tag{5}$$

The solution of the system (5) is found by simple inversion of the matrix, i.e. $x = A^{-1} \cdot b$.

Furthermore, in order to leave the ac converter voltage unaltered, all the positive valves' dc voltages shall be *decreased* an amount equal to ΔV volts. This is because, in a delta-connected transformer, the ac voltage output is a differential output of either the positive or negative valves' voltages; hence, shifting all the positive (or negative) valves voltages simultaneously and by the same amount will not affect the differential output.

What is more, in order to leave the dc voltage unaltered, all the negative valves voltages shall be *increased* an amount equal to ΔV volts. This is summarised in (6):

$$
\begin{aligned}
v_{vpx} &= \frac{V_{dc}}{2} - v_x + (-\Delta V), \\
v_{vnx} &= \frac{V_{dc}}{2} + v_x + (+\Delta V).
\end{aligned}
\tag{6}
$$

It is worth mentioning that the energy term in (1) has been neglected in (6), without loss of generality.

From (6), notice that $v_{vpx} + v_{vnx} = V_{dc}$, and $v_{vnx} - v_{vpx} = 2v_x + 2\Delta V$. Hence, the dc voltage remains unaffected, and thus the dc power is unaltered too. Meanwhile, the dc term of the differential voltage $(v_{vnx} - v_{vpx})/2$, i.e. ΔV, appears as common-mode voltage on the transformer's secondary windings with respect to ground. However, as the transformer is star-delta connected, there is no phase-to-phase dc voltage induced on the transformer windings. Hence, the difference $v_{vpx} - v_{vpy}$ contains no dc voltage, and the same holds true for $v_{vnx} - v_{vny}$, with $x, y = \{a, b, c\}$. Offsetting the valves voltages by ΔV volts has a small impact on equipment insulation levels, with the benefit of being able to control both the ground current and dc pole to ground voltages.

It must be noted that by altering (increasing or decreasing) the dc voltage term in each valve, the synthesized ac voltage peak, denoted as $\hat{v}_x$, is affected, as shown in Fig. 5. This is because, in a HB MMC topology, the maximum symmetrical swing of the synthesized ac voltage is limited by the dc term of each valve (excluding the application of triplen harmonic injection). For this reason, the proposed control strategy shall only be applied when either the relevant pole-to-ground dc voltage V_{dcp}, V_{dcn} or the ground current i_g has crossed a preset threshold associated with a preventive protective level. Applying the proposed method i normal operating conditions may have a negative impact on converter losses as well as it would limit its power delivery capability.

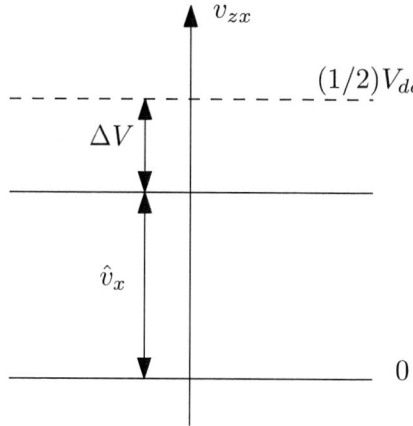

Figure 5: Reduced ac voltage swing due to a reduction in the valve's dc voltage for valve zx.

The control strategy proposed in this paper is shown in Fig. 6. This combines the control of pole-to-ground dc voltage and ground current with a prioritisation logic. The parameters of the deadband are settable by the control designer, either by offline or online methods.

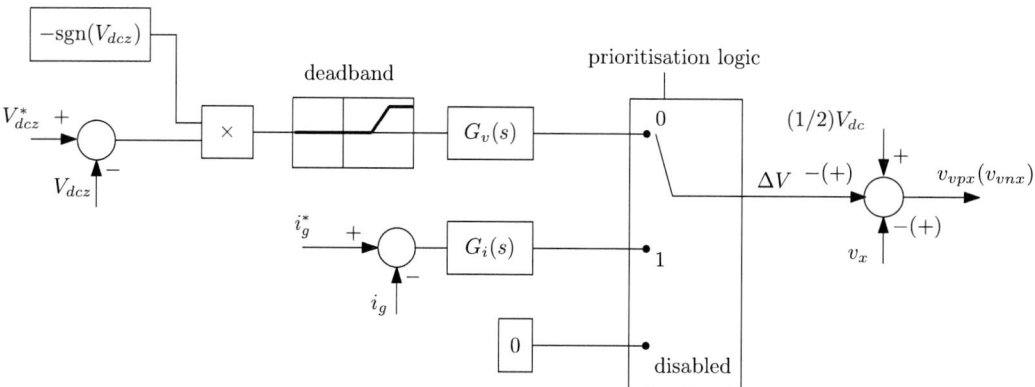

Figure 6: Strategy of pole-to-ground dc voltage feedback control using voltage and current measurements.

Fig. 6 shows how the positive (negative) valves voltages are modified when the pole-to-ground voltage V_{dcz} exceeds a predefined upper limit. The ground current measurement i_g allows the designer to control such current within acceptable limits by varying the dc term of the valves' voltages according to (6). The ground current reference i_g^* can be derived, e.g., by estimating the values of the resistors R_p, R_n and solving for (3). The controller gain $G_i(s)$ for the closed-loop control on i_g has the form of a negative (positive) resistance for over-voltages of the positive (negative) dc pole. More specifically, $G_v(s)$ and $G_i(s)$ are implemented using proportional-plus-integral (PI) controllers with harmonic filtering capability, in order to regulate the dc term of the error and hence bring its nominal value to zero. The prioritisation logic block implements enables the regulation of the ground current i_g over the relevant dc pole-to-ground voltage V_{dcz}, or vice versa; this block operates based on the measured quantities i_g and V_{dcz}, in combination with the converter's protection setting levels and trigger levels, V_{dcz_trig}, i_{g_trig}.

Simulation Results

In this section, the proposed control strategy is validated under different imbalance scenarios. The converter topology is a symmetrical monopole MMC with V_{dc} = 640 kV. The proposed control algorithm is implemented in MATLAB/Simulink.

The protection trip levels are shown in Table 1. The numerical results for various parameter combinations are shown in Table 2.

All figures and values shown below are approximate figures and values and are not definite in nature.

As it can be seen in Table 2, when R_g = 7.5 kΩ and there is a 20% imbalance between R_p, R_n, operating at the rated

Quantity	Trip level
V_{dcp}	1.05×320 kV
$|V_{dcn}|$	1.05×320 kV
i_g	250 mA

Table 1: Converter trip levels.

V_{dc}	R_p	R_n	R_g	$v_{pa,dc}$	$v_{na,dc}$	V_{dcp}	V_{dcn}	i_g
[kV]	[kΩ]	[kΩ]	[kΩ]	[kV]	[kV]	[kV]	[kV]	[mA]
640	300	$0.8R_p$	7.5	320	320	1.006×320	-318	252
640	300	$0.8R_p$	7.5	328	312	1.03×320	-310	195
640	300	$0.8R_p$	175	320	320	1.06×320	-300	115
640	300	$0.8R_p$	175	310	330	1.05×320	-304	146

Table 2: Application example for different parameter values of Fig. 3.

dc valves' voltage of $(1/2)V_{dc} = 320$ kV does not produce a significant voltage difference at V_{dcp}. However, in this case the ground current i_g goes beyond the trip limit of 250 mA (marked in red). Hence, the valves voltages shall be modulated as shown in the second row of the table to avoid the converter to trip.

The third row of the table shows an increase in the ground resistor $R_g = 175$ kΩ (shown in blue) for the same values of the remaining parameters. It can be noticed that now the voltage V_{dcp} has gone beyond the 5% allowable limit (shown in red), which can trip the scheme, thus reducing the system availability and reliability. To stabilise the system, the valves voltages shall be regulated as shown in the fourth row, where all parameters are controlled within acceptable limits.

To confirm the results shown above, a point to point VSC HVDC scheme has been simulated with an imbalance of $R_n = 0.8R_p$, and $R_p = 300$ kΩ. It is expected that this configuration of parameters trigger a ground current protection, and not a dc pole-to-ground protective action. Hence, ground current control takes priority over dc pole-to-ground voltage control in Fig. 6.

When the control of the ground current i_g is disabled, the ground current goes out of control, thus crossing the protection boundary of 250 mA, as shown in Fig. 7. When the i_g current control is activated, the ground current is controlled to a peak value below 7 mA, as shown in Fig. 8.

Based on the results obtained in this example, the adjustments required to control the ground current and/or pole-to-ground dc voltages are attained at minimal modification of the dc voltages at the valves ($< 3\%$ of the rated valves' dc voltage). Therefore, by using this technique no significant increase on the number of connected submodules per valves is expected. In summary, regulating the dc valves' voltages can serve the purpose of both regulating the pole-to-ground dc voltages as well as the ground current to avoid an unnecessary trip in the scheme.

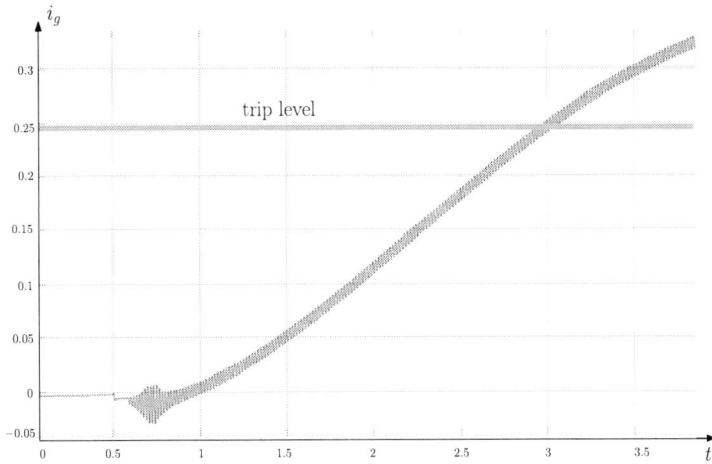

Figure 7: Uncontrolled ground current i_g as a function of time. The trip level is marked in green.

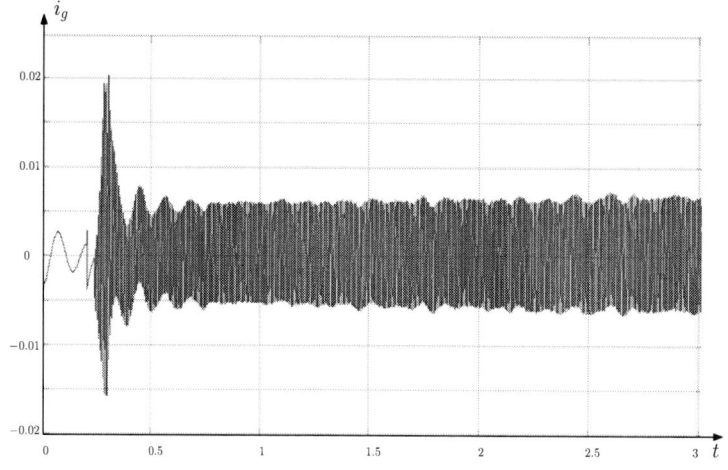

Figure 8: Controlled ground current i_g as a function of time.

Conclusions

This paper described a control method to regulate both the ground current and the pole-to-ground dc voltages of a symmetrical monopole modular multilevel converter, so that these do not exceed a pre-defined threshold. The paper has provided the mathematical modelling of the VSC HVDC converter voltage control, and a top level description of the proposed control algorithm. It has been discussed how, by controlling such voltages, the reliability and availability of the converter can be extended by avoiding unnecessary converter trips. To illustrate the effectiveness of the proposed control method, the control algorithm is applied to a typical VSC-HVDC system. The simulation results show the significance of the proposed method to improve the reliability and availability of the VSC HVDC system.

References

[1] O. Gomis-Bellmunt, J. Liang, J. Ekanayake, R. King, and N. Jenkins, "Topologies of multiterminal hvdc-vsc transmission for large offshore wind farms," *Electric Power Systems Research*, vol. 81, no. 2, pp. 271–281, 2011.

[2] I. Legorburu, K. R. Johnson, and S. A. Kerr, "Multi-use maritime platforms-north sea oil and offshore wind: Opportunity and risk," *Ocean & Coastal Management*, vol. 160, pp. 75–85, 2018.

[3] O. E. Oni, I. E. Davidson, and K. N. Mbangula, "A review of lcc-hvdc and vsc-hvdc technologies and applications," in *2016 IEEE 16th International Conference on Environment and Electrical Engineering (EEEIC)*. IEEE, 2016, pp. 1–7.

[4] A. Fernández-Guillamón, K. Das, N. A. Cutululis, and Á. Molina-García, "Offshore wind power integration into future power systems: Overview and trends," *Journal of Marine Science and Engineering*, vol. 7, no. 11, p. 399, 2019.

[5] A. Junyent-Ferre, P. Clemow, M. M. Merlin, and T. C. Green, "Operation of hvdc modular multilevel converters under dc pole imbalances," in *Power Electronics and Applications, 16th European Conference on*. EPE 14-ECCE Europe, 2014, pp. 1–10.

[6] Y. Zhao, W. Huang, and F. Bu, "Dc voltage balancing strategy of a bipolar-output active rectifier for more electric aircraft based on zero vector redistribution," *IEEE Access*, vol. 9, pp. 139 657–139 667, 2021.

[7] M. Wang, W. Leterme, G. Chaffey, J. Beerten, and D. Van Hertem, "Pole rebalancing methods for pole-to-ground faults in symmetrical monopolar hvdc grids," *IEEE Transactions on Power Delivery*, vol. 34, no. 1, pp. 188–197, 2018.

[8] M. Wang, J. Beerten, and D. Van Hertem, "Pole voltage balancing in hvdc systems: Analysis and technology options," in *2019 IEEE Milan PowerTech*. IEEE, 2019, pp. 1–6.

A PFC boost converter with reduced switching losses operating at a fixed switching frequency

Burkhard Ulrich
Reutlingen University
Alteburgstraße 150
72762 Reutlingen, Germany
Tel.: +49(7121) 271 – 7146
burkhard.ulrich@reutlingen-university.de
https://www.reutlingen-university.de

Keywords

«Power factor correction», «Switching losses», «Zero-voltage switching», «AC-DC converter», «Boost»

Abstract

A single-phase fixed-frequency operated power factor correction circuit with reduced switching losses is proposed. The circuit uses the combination of a boost converter with an added clamp-switch, a pulse wave shaping circuit, and a standard control IC to discharge the transistor's output capacitance prior to its turn-on. In this way, a very low-complexity control circuit implementation to reduce switching losses or even achieve complete zero-voltage switching without additional sensors is possible. Moreover, this operation method is achieved at a constant switching frequency, possibly simplifying the design of the EMI filter and the converter's inductor. Experimental test results for a 100 W prototype converter are presented to validate the feasibility of the proposed operating method and corresponding circuit structure.

Introduction

Soft-switching converters are commonly employed to reduce switching losses, mitigate EMI emissions, increase the operating frequency, and reduce the size of power conversion systems. E.g., for single-phase ac/dc converters in the lower to medium power range (i.e., below power levels of several hundred watts), critical conduction mode (CrM) or transition mode (TM) power factor correction (PFC) circuits, based on boost dc/dc converters, are commonly employed [1][2][3][4]. In these converters switching losses are reduced by turning the output diode off at zero current and afterward turning the main transistor on, in a quasi-resonant manner, at a low voltage. A drawback of this approach is that the converter operates at a varying switching frequency, depending on the output load and the momentary input voltage value. This variation in the switching frequency can complicate the inductor and input EMI filter design. Additionally, the minimum voltage at turn-on of the main switch depends on the momentary input to output voltage ratio. Therefore, zero-voltage switching (ZVS) is lost for high input voltages, as ZVS can be achieved in this type of converter only if $V_{in} < V_{out}/2$. For PFC circuits operating with input voltages at mains levels of 230 V (RMS) and a typical V_{out} of 380 V to 400 V, the ZVS operation is therefore limited to parts of the mains cycle. Also, this type of CrM converter often requires a detection circuit to achieve (partial) soft-switching.

Other approaches which allow a full ZVS operation independent of the input voltage level were presented in [5] and [6]. In [5], a four-switch buck-boost converter is proposed to achieve ZVS, and in [6], a triangular current mode (TCM) operation scheme for a bridgeless boost converter is presented. Both approaches rely on a varying switching frequency and more complicated control and sensor circuits which cannot be directly implemented using standard, off-the-shelf, PFC control ICs.

For the mitigation of some of the restrictions mentioned before, this article proposes a method to reduce switching losses or even achieve ZVS for single-phase PFC circuits, which allows the use of standard PFC control ICs and maintains a fixed operating frequency. This approach is based on a clamp-switch ZVS boost converter as presented in [7][8] and extends this operation method to single-phase active PFC systems.

Converter Circuit and Operation

Proposed Converter Circuit

A simplified schematic of the proposed single-phase PFC circuit, corresponding to the prototype presented later, is shown in Fig. 1. It is centered around the ZVS boost type dc/dc converter proposed in [7], which is used instead of a standard hard-switching boost topology to control the inductor current waveform and, therefore, the line current. Similar to the boost converter presented in [7], a clamp-switch network comprised of transistor T_{cl}, diode D_{cl}, and an additional clamp-capacitor C_{cl} in parallel to the converter's inductor L is added to a boost converter to reduce the switching losses and/or achieve soft-switching. Capacitor C_{cl} is added to the clamp-switch to adjust the ZVS range, as described in [7], allowing a ZVS operation even if $V_{out} < 2v_{in}$ and without any additional sensor circuitry.

Capacitors C_{in} and C_{out} are used in the same way as in a standard active PFC circuit. I.e., C_{in} is small in value, to only filter the high-frequency components in the inductor current i_L and capacitor C_{out} performs the low-frequency energy storage function at twice the mains frequency. Essentially the operation of this power stage can be traced back to the operating principle introduced in [7], with significant differences being here the continually varying rectified mains voltage as input and the added mains EMI filter. The latter also provides the required high-frequency differential mode filtering due to the large inductor current variation at the switching frequency in the proposed operating scheme.

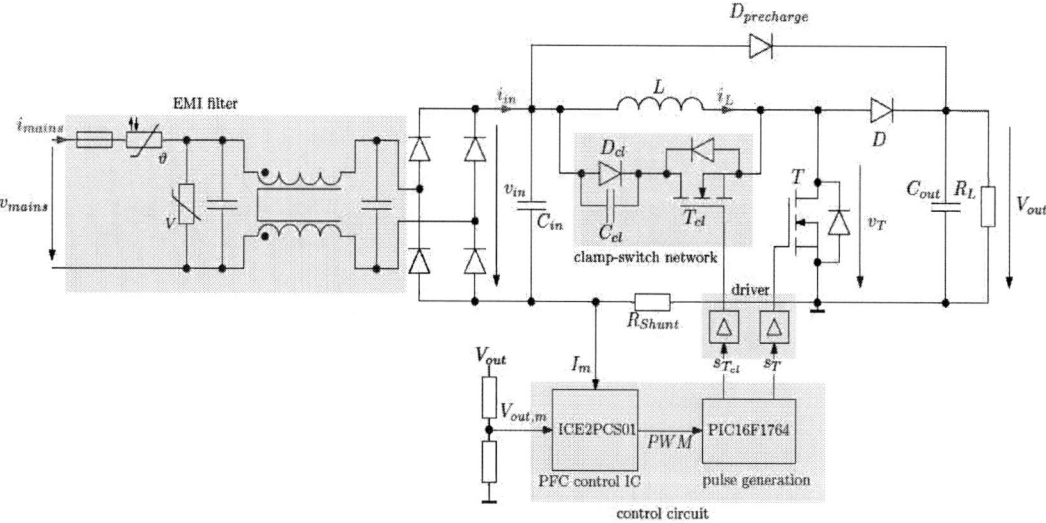

Fig. 1: Proposed clamp-switch single-phase boost type PFC rectifier

Basic Operation

Based on the more detailed description given in [7], a summary of the converter operation is presented in this paragraph.

The converter is designed to operate with discontinuous conduction mode (DCM) over the complete mains cycle if the clamp-switch is deactivated, by selecting a small inductance value L. The two transistors, T (main switch) and T_{cl} (clamp-switch), are driven with complementary pulse width modulated (PWM) signals at a fixed frequency, as shown in Fig. 2. Assuming a constant value of V_{in} during the switching period, eight different intervals can be identified. Here a possible ninth interval, where the body diode of the main transistor T conducts the current prior to T's turn-on, is omitted.

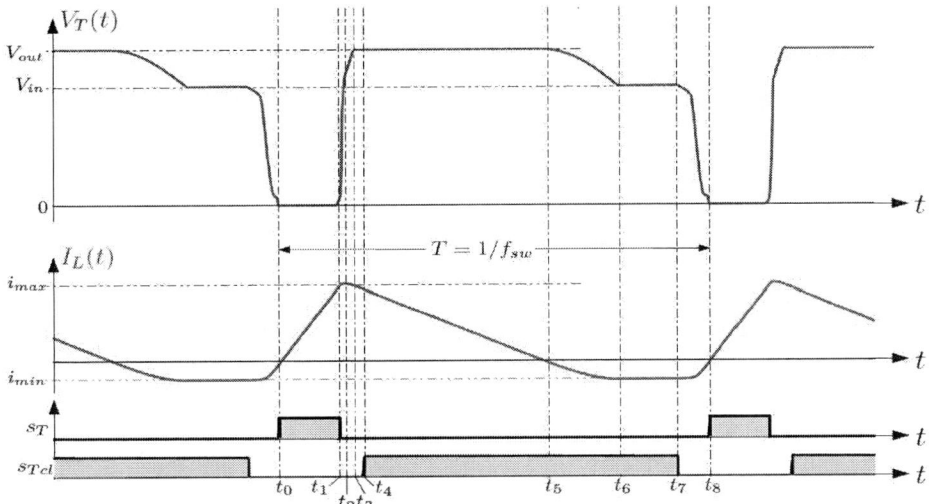

Fig. 2: Major simulated boost power stage waveforms during one high-frequency switching cycle

Interval 1 $[t_0 < t < t_1]$:
The switching cycle starts by turning on the main transistor at t_0, ideally a zero current. The turn-on time $t_1 - t_0$ is varied to control the primary power conversion operation, similar to a standard boost converter.

Interval 2 $[t_1 < t < t_2]$:
After T is turned-off at t_1, the inductor current will start to charge the total effective capacitance C_p at T's drain connection with the approximately constant maximum inductor current i_{max}. Voltage v_T will increase accordingly.

Interval 3 $[t_2 < t < t_3]$:
At t_2, the transistor voltage v_T reaches the value of V_{in}, causing the body diode of T_{cl} to turn on. This change in the switching state leads to an increase in the effective capacitance at the switch node. I.e., in the clamp-switch path, the resultant capacitance is no longer determined by the series connection of the capacitances of T_{cl} and C_{cl} (in parallel with D_{cl}'s capacitance C_{Dcl}) but then determined by resultant capacitance across the clamp diode D_{cl}. Therefore, the voltage slew rate of v_T will decrease.

Interval 4 $[t_3 < t < t_4]$:
The output diode D starts conducting at t_3 as the voltage v_T reaches V_{out}. The output capacitor will be charged, and the current i_L will decrease.

Interval 5 $[t_4 < t < t_5]$:
The clamp-switch T_{cl} is turned on at instant t_4 between t_3 and t_5. This process will not change the converter operation compared to the previous state but is necessary to allow a proper freewheeling during the following states. I.e., clamp-switch T_{cl} has to be turned on prior to t_5.

Interval 6 $[t_5 < t < t_6]$:
Diode D turns off at t_5 at the zero-crossing of i_L, comparable to the normal diode conduction state in a boost converter. Then, a resonant transition of v_T occurs, and inductor L resonates with the effective capacitance at the switch node during this interval. If the value C_{cl} is much larger than the other capacitances, C_{cl} will dominate this process and is used to set the minimum (negative) inductor i_{min} [7].

Interval 7 $[t_6 < t < t_7]$:
The resonant transition ends if the voltage v_T reaches the value V_{in}, as then diode D_{cl} gets forward biased. Therefore, the inductor current will freewheel, at a negative current value i_{min}, through the series connection of clamp-switch diode D_{cl} and transistor T_{cl}. In this way, energy is stored in the inductor to discharge later the transistor's effective capacitance at t_7.

Interval 8 $[t_7 < t < t_8]$:
At instant t_7, the clamp-switch T_{cl} turns off, and another resonant (discharge) transition of voltage v_T will be initiated. This will continue the discharge of the (parasitic) capacitances connected to the switch node prior to the turn-on of T. If this turn-on occurs after one-quarter of the resonant period, main switch T can be turned on at t_8 at the minimum voltage value $v_T(t)$. Thus, switching losses are reduced, and if v_T reaches approximately zero, even a ZVS turn-on is possible.

ZVS Condition at Varying Input Voltage

ZVS can be achieved in the converter if the minimum inductor current i_{min} has a sufficiently negative value to discharge the equivalent lumped parasitic capacitance C_p at the switch node during time interval six by an amount of V_{in}. I.e., the energy stored in the inductor at the end of the freewheeling phase at t_6 must be larger than the change in stored energy in the capacitor. Therefore, the ZVS condition can be written based on an energy balance as

$$L \cdot i_{min}^2 \geq C_p \cdot v_{in}^2 \tag{1}$$

In a PFC converter, the input voltage varies, during the line cycle, as

$$v_{in}(t) = \hat{v}_{in} \cdot |\sin(2\pi \cdot f \cdot t)| \tag{2}$$

This variation in $v_{in}(t)$ also changes the minimum inductor current value (c.f. i_{min} in Fig. 2) during the mains cycle. Using eq. (9) from [7] and substituting (2), $i_{min}(t)$ can be written as

$$i_{min}(t) = (\hat{v}_{in} \cdot |\sin(2\pi \cdot f \cdot t)| - V_{out}) \cdot \sqrt{\frac{C_{cl} + C_p}{L}} \tag{3}$$

As both $i_{min}(t)$ and $v_{in}(t)$, according to (1), determine if ZVS is possible, the condition to achieve a soft-switching will therefore vary across the mains cycle. The worst-case operating point regarding ZVS will be when $v_{in}(t)$ reaches its maximum value $\hat{v}_{in}$ at $t = 1/(2f)$. At this instant, the required energy to achieve ZVS, according to (1), has at its maximum value. In contrast, the current $|i_{min}(t)|$ will be at the lowest value, and therefore, the stored energy in L is also at its minimal value. The minimum current, in this case, will be

$$i_{min}\left(t = \frac{1}{2f}\right) = (\hat{v}_{in} - V_{out}) \cdot \sqrt{\frac{C_{cl} + C_p}{L}} \tag{4}$$

Substituting the (4) in (1) and simplifying yields the following ZVS condition for the converter

$$\left(1 - \frac{V_{out}}{\hat{v}_{in}}\right)^2 \left(\frac{C_{cl}}{C_p} + 1\right) - 1 > 0 \tag{5}$$

If (5) is solved for C_{cl}, the following design equation to select C_{cl} to achieve ZVS is derived:

$$C_{cl} > \left(\frac{1}{\left(1 - \frac{V_{out}}{\hat{v}_{in}}\right)^2} - 1\right) \cdot C_p \tag{6}$$

This last equation is valid for $V_{out} < 2\hat{v}_{in}$. If $V_{out} > 2\hat{v}_{in}$, then the evaluation of (6) leads to a negative value, indicating that a ZVS is in this case, at least in theory, possible even if C_{cl} is omitted. Therefore, by a suitable choice of C_{cl}, a ZVS operation over the whole mains cycle can (theoretically) be achieved. For a practical PFC converter at a nominal (European) mains voltage level of $\hat{v}_{in} = \sqrt{2} \cdot 230\ V$ and V_{out} in the range of 380 V to 400 V, the ratio $V_{out} / \hat{v}_{in}$ is 1.15...1.25. In this case, the required capacitance value C_{cl}, predicted by (6), would be about 16 to 44 times the value of C_p. Selecting such a large value increases the negative inductor current's absolute value and leads to increasing losses (e.g., inductor and semiconductor conduction losses), which could offset the loss reduction gained by achieving ZVS. Therefore, it is suggested to choose a smaller value for C_{cl} than predicted by (6), even if ZVS is only achieved over parts of the mains cycle.

The discussion of the ZVS range above assumes a lossless converter and linear capacitances values. In particular, the decrease of the stored energy in the inductor during the freewheeling phase due to voltage drop of the clamp-switch diode and transistor is neglected.

PFC Control Method and Line Current Distortion

A low-complexity control implementation for the converter circuit is proposed by using a standard PFC control IC and complementing it with a pulse waveform generator circuit. The latter converts the single-ended PWM drive signal of a standard PFC IC into two complementary drive signals with inserted dead times. (a suitable gate driver circuit could also perform this task). A possible scheme is presented in Fig. 3. Here, an IC employing a multiplier free control law (c.f. to [9-11]), which is commonly employed in continuous conduction mode (CCM) PFC circuits, is used.

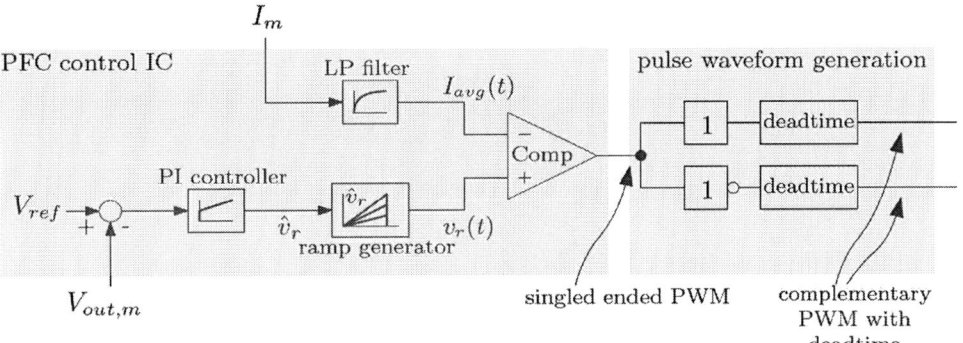

Fig. 3: Possible implementation of a PFC control scheme according to the proposed operation mode

In this control approach, a measured inductor current value $I_M = k_m \cdot I_L$ is first low-pass filtered to extract a value $I_{avg}(t)$ proportional to the moving average $\bar{\imath}_L$ of the inductor current. This value is compared to a sawtooth carrier signal with variable amplitude $\hat{v}_R$ to generate the main transistor's relative off time $d_{off} = t_{off} / T$. Therefore, $d_{off} = k \cdot I_{avg}$ is proportional to the average inductor current, and as it has been shown by Ben-Yaakov et al. in [9] (c.f. eq. (2) in [9]), the following applies for a PFC based on a CCM boost converter

$$\frac{v_{in}}{i_{in}} = \frac{d_{off}}{I_{avg}} \cdot V_{out} = k \cdot V_{out} \tag{7}$$

The factor k in (7) is the ratio of current measurement gain k_m to the amplitude $\hat{v}_R$ of the sawtooth,

$$k = \frac{k_m}{\hat{v}_R} \tag{8}$$

The converter will therefore show a resistive input characteristic in this case if V_{out} is approximately constant. The outer voltage loop controller determines the amplitude $\hat{v}_R$. Therefore, the average inductor current for a boost converter operating in CCM follows the rectified input voltage v_{in}, and a simple PFC functionality can be achieved [9-11].

The described operation method is intended to control a CCM boost converter; therefore, as the power stage is more comparable to a DCM converter, additional line current distortions will be introduced. The resulting distortions can be roughly estimated when evaluating the control law (7) for a DCM operating boost converter. Therefore, by neglecting the negative part of $i_L(t)$ and assuming an idealized DCM inductor current waveform during the high-frequency switching cycles, the moving average current $\bar{\imath}_L(t)$ is expressed as

$$\bar{\imath}_L(t) = \frac{1}{k} \cdot [\zeta(t, \hat{v}_{in}, k) + 1 - \sqrt{\zeta(t, \hat{v}_{in}, k) \cdot (2 + \zeta(t, \hat{v}_{in}, k))}] \tag{9}$$

with

$$\zeta(t, \hat{v}_{in}, k) = \frac{L \cdot f_{sw}}{k} \cdot \left(\frac{1}{\hat{v}_{in} \cdot |\sin(2\pi \cdot f \cdot t)|} - \frac{1}{V_{out}} \right) \tag{10}$$

Fig. 4 depicts the distorted average line current predicted by (9) for different values of k, representing different output load conditions. The distortion will be less severe in a practical converter, as the average value is reduced due to the negative part of $i_L(t)$. Besides this distortion, the converter will still show a high power factor, as shown later in the experimental results section.

It should be mentioned that the introduced distortion is not inherent to the clamp-switch operation method proposed here but would also occur in a DCM working boost converter as PFC if operated with a control scheme according to [9-11]. The selection of a different control scheme, e.g., average current mode control with mains derived reference signal and multiplier, could be used to avoid this distortion.

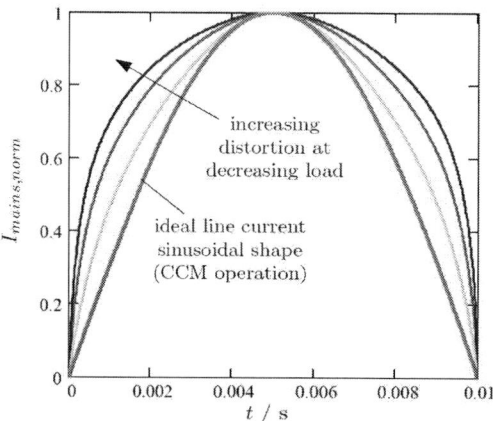

Fig. 4: Normalized estimated line current for different values k for multiplier free PFC control law

Prototype Converter and Experimental Results

For validation a *100* W prototype converter with $V_{out} = 375$ V and $V_{mains} = 230$ V ($f = 50$ Hz) operating at $f_{SW} \approx 240$ kHz has been built. The converter employs SiC cascode transistors due to the low stored energy E_{oss} in the output capacitance to reduce switching losses if a ZVS turn-on is not achieved. Some major prototype components are listed in Table 1. The PCB of the power stage is portrayed in Fig. 5; the additional control board, containing the PFC control IC and the wave shaping circuit, is not shown.

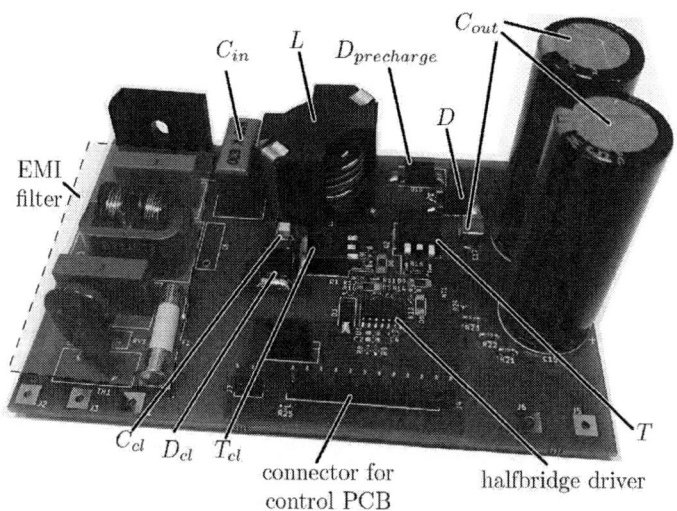

Fig. 5: Prototype PCB without external control board

Table I: Prototype Components

Component	Value and Manufacturer
Transistors T, T_{cl}	SiC FET UF3C065080B3 United SiC/Qorvo
Diodes D,D_{cl}	SiC Schottky Diode C6D04065E Wolfspeed/Cree
Inductor L	custom wound 75 µH core: RM10 material: N87 TDK-EPCOS
Halfbridge driver	2ED21814S06J Infineon
PFC control IC	ICE2PCS01 Infineon
Microcontroller (pulse shaping)	PIC16F1764 Microchip

The measured efficiency, power factor, and THD in the power range from 10 W to 100 W are shown in Fig. 6. The measured efficiency includes the losses in EMI filter and protection devices (i.e., fuse, common mode choke, varistor, inrush-current limiter). It, therefore, presents a realistic estimation of the total power stage losses, but the supply power for the control circuits is not considered.

Fig. 7 and 8 depict measured converter waveforms. On the left side in Fig. 7a) the measured mains current and voltage for an entire line cycle are shown, indicating that although the described line current distortion is visible, a PFC functionality with a high power factor of 0.982 is achieved. Figure 7b) shows the measured voltage across the main transistor T and the inductor current over a line cycle. The current waveform shows that the negative inductor current i_{min} varies over the mains cycle as predicted by Eq. (4). Magnified views of the transistor voltage and the inductor current for three different line voltage values are presented in Fig. 8 to illustrate the operation of the converter during a high-frequency switching cycle depending on the current time instant in the mains cycle. These show that a ZVS at low-line voltages, but for $V_{in} > V_{out}/2$, is achieved (c.f. Fig. 8a) and b)) and that at high line voltages (c.f. Fig. 8c), switching losses are reduced by turning on at the minimum value of v_T. In all cases, the converter is operated at a constant switching frequency, and parasitic ringing in the transistor voltage is reduced due to the clamp-switch operation.

As already mentioned, the ZVS range could be extended by using a larger C_{cl} capacitor (in the actual prototype $C_{cl} = 2.2$ nF), but this would also increase the losses in the clamp-switch network and the inductor, which offsets the effect of the reduced switching losses. Therefore, a ZVS operation for the entire line cycle of the converter could be possible, but a compromise is taken here to optimize the overall system efficiency.

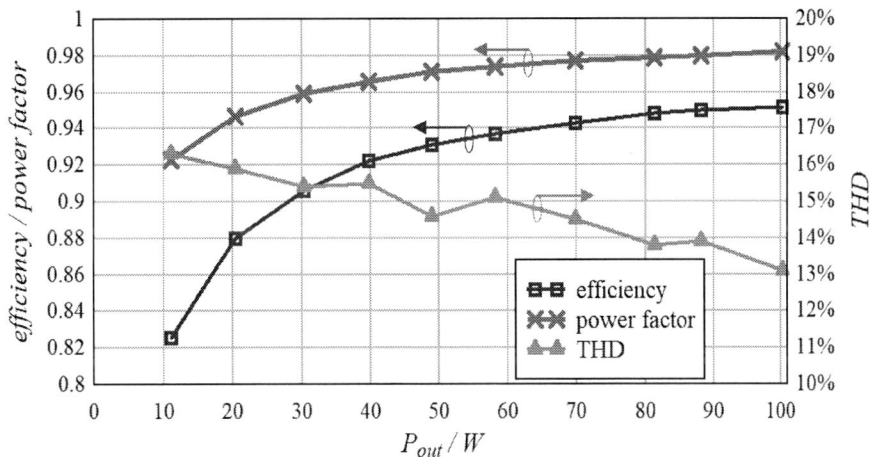

Fig. 6: measured efficiency, power factor and THD vs. P_{out}

A PFC boost converter with reduced switching losses operating at a fixed switching frequency   ULRICH Burkhard

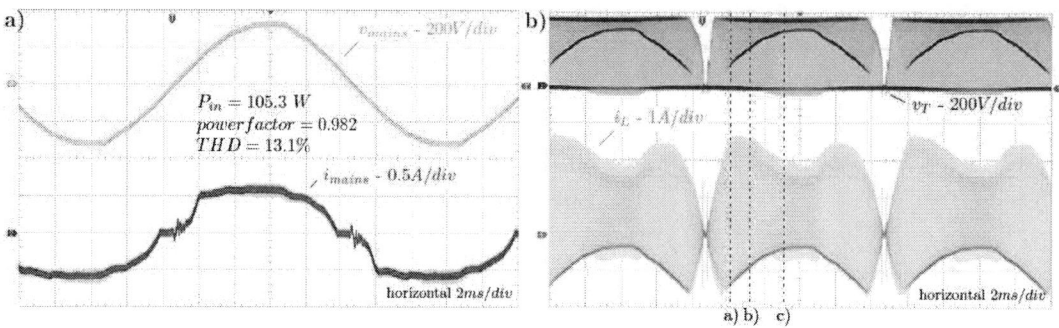

Fig. 7: a) mains voltage v_{mains} and current i_{mains} and b) switch voltage v_T and inductor current i_L over mains cycle

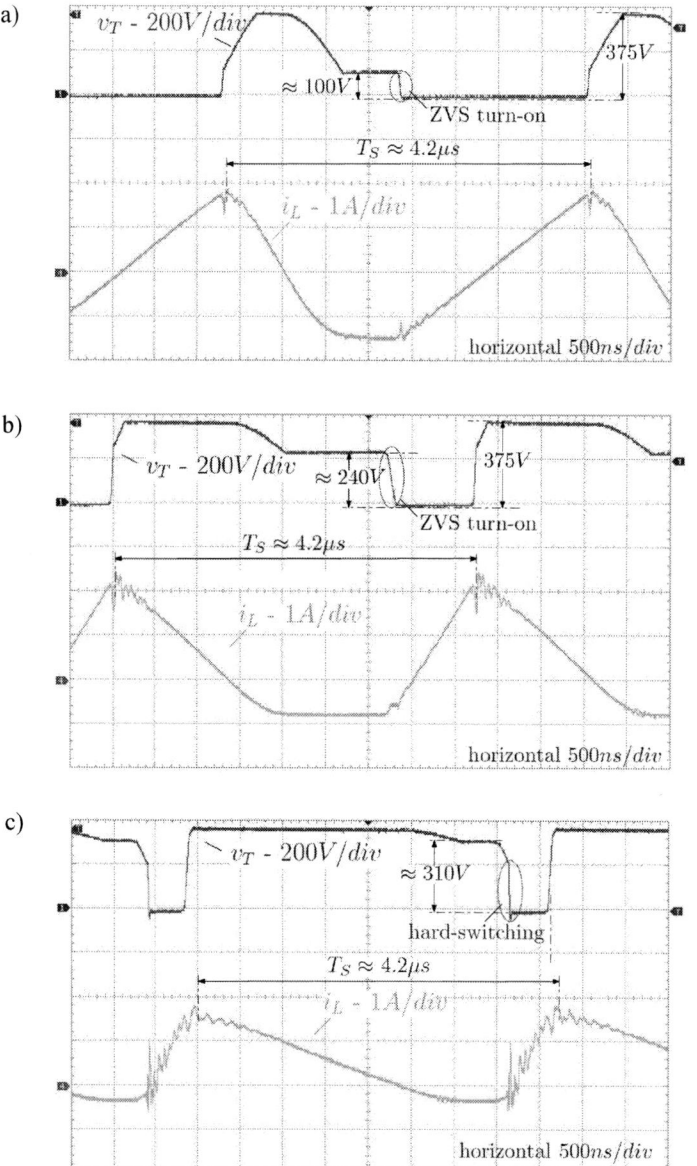

Fig. 8: Detail view of current i_L and voltage v_T for different V_{in} values a) 100 V, b) 240 V and c) 310 V corresponding to the instants marked with a), b) and c) in Fig. 7.

Conclusion

A method is presented to operate a single-phase boost converter PFC at a constant switching frequency to reduce switching losses and achieve ZVS at lower line voltages. The novelty of the approach presented here lies in the combination of the clamp-switch boost operating principle, which does not require any sensor circuitry to achieve reduced switching losses or ZVS with a control circuit based on a standard PFC control IC. This combined approach has not been reported in the literature. Also, it is shown that the multiplier-free approach for average current mode control [9] exhibits line distortions but is still usable in this application. Therefore, a low-cost and straightforward control circuit implementation is possible. The benefit of the proposed method is the utilization of a constant switching frequency and an operation similar to a DCM converter, which leads to a small required inductance value and can also help to ease the design of the inductor and EMI filter components.

Experimental results indicate the approach's feasibility and show that high efficiency and a high power factor at an output power level of 100 W are possible.

Although the implementation used here will introduce some additional line current distortions, this is due to the specific PFC control principle used and not inherent to the presented converter operation. A possible mitigation is using another standard IC for the control law implementation, e.g., an average current control mode PFC IC with input voltage sensing and a multiplier-based control algorithm.

References

[1] B. McDonald and B. Lough.: Power Factor Correction (PFC) Circuit Basics, Texas Instruments Power Supply Design Seminar 2020, accessed on 16.05.2022. [Online]. Available: https://www.ti.com/seclit/ml/slup390/slup390.pdf

[2] Z. Ye and B. Sun.: PFC efficency improvement and THD reduction at light loads with ZVS and valley switching, *2012 Twenty-Seventh Annual IEEE Applied Power Electronics Conference and Exposition (APEC)*, 2012, pp. 802-806

[3] J. Sun, X. Huang, N. N. Strain, D. J. Costinett and L. M. Tolbert.: Inductor Design and ZVS Control for a GaN-Based High Efficiency CRM Totem-Pole PFC Converter, *2019 IEEE Applied Power Electronics Conference and Exposition (APEC)*, 2019, pp. 727-733

[4] ST Microelectronics.: How to design a transition-mode PFC pre-regulator with the L6563S and L6563H – Application Note AN3027, 2011, accessed on 14.11.2021. [Online] Available: https://www.st.com/resource/en/application_note/cd00245012-how-to-design-a-transitionmode-pfc-preregulator-with-the-l6563s-and-l6563h-stmicroelectronics.pdf

[5] A. J. Hanson and D. J. Perreault.: A high frequency power factor correction converter with soft switching, 2018 IEEE Applied Power Electronics Conference and Exposition (APEC), 2018, pp. 2027-2034

[6] C. Marxgut, F. Krismer, D. Bortis and J. W. Kolar, "Ultraflat Interleaved Triangular Current Mode (TCM) Single-Phase PFC Rectifier," in IEEE Transactions on Power Electronics, vol. 29, no. 2, pp. 873-882, Feb. 2014

[7] B. Ulrich.: Improved Clamp-Switch Boost Converter with Extended ZVS range, *2021 IEEE Applied Power Electronics Conference and Exposition (APEC)*, 2021, pp. 1747-1754

[8] J. Prager and P. Vinciarelli.: Loss and noise reduction in power converters, US Patent US 6,522,108, 18-Feb-2003.

[9] S. Ben-Yaakov and I. Zeltser.: PWM converters with resistive input, in *IEEE Transactions on Industrial Electronics*, vol. 45, no. 3, pp. 519-520, June 1998

[10] Texas Instruments.: UCC28180 – datasheet, 2016, accessed on 16.05.2022. [Online]. Available: https://www.ti.com/lit/ds/symlink/ucc28180.pdf

[11] Infineon Technologies.: ICE2PCS01 – datasheet, 2011, accessed on 16.05.2022. [Online]. Available: https://www.infineon.com/dgdl/Infineon-ICE2PCS01-DataSheet-v02_03-EN.pdf?fileId=db3a304412b407950112b427caa43ccf

Predictive Control of Power Electronics Autotransformer for Mitigating Three-Phase Grid Current Unbalance in Railway Supply Systems

Tabish Nazir Mir*, Faysal Hardan, Masood Hajian, Tamer Kamel and Pietro Tricoli
Birmingham Centre for Railway Research and Education
University of Birmingham
United Kingdom
*Email: t.mir@bham.ac.uk

Acknowledgments

The authors express their gratitude to EPSRC Impact Acceleration Account (IAA) for supporting the work carried out in this manuscript through the follow-on fund.

Keywords

≪Railway power supply≫, ≪Power quality≫, ≪Current balancing≫, ≪Converter control ≫, ≪Grid-connected converter≫.

Abstract

In order to mitigate the unbalance of three-phase grid currents that is commonly prevalent in single-phase railway supply lines, a predictively controlled power electronics autotransformer is proposed in this paper. It comprises of a pair of back to back single-phase voltage source converters that are strategically connected between different phases of the three-phase grid. Using finite control set model predictive control, the converters are intelligently controlled with a regulated DC link, to draw balanced sinusoidal three-phase grid currents at nearly unity input power factor, even as the rail system connects as a single-phase load on the three-phase network. This in turn helps in maintaining the balance and quality of three-phase voltage supply at the point of common coupling.

Introduction

High speed railway traction systems typically use a 25 kV single-phase (1-ϕ) power supply that is generated through a step down transformer connected across two phases of a 132 kV three-phase (3-ϕ) transmission grid. This is popularly achieved through a 2×25 kV split-phase ($\pm$ 25 kV AC) arrangement, which is well proven to provide better efficiency [1]. Such a system is illustrated in Fig.1, where the overhead train line is at +25 kV (AC) with respect to the running rails, while there is a parellel negative feeder at -25 kV (AC). Regularly spaced auto-transformers connected between the split-phase network, continuously divert the returning rail currents along the negative feeder. Although the 2×25 kV railway supply system is widely prevalent world-over, it has significant drawbacks such as reactive power consumption and unequal loading of the three-phase grid [2]. An undesirable consequence of unbalanced grid currents is the deterioration in 3-ϕ voltage balance at the point of common coupling, which in turn has adverse effects on other end-user equipments connected to the same network [3].

A number of solutions have been proposed to maintain balanced 3-ϕ grid currents in 1-ϕ railway supply systems. These include passive solutions [4] as well as active solutions like rail power flow controllers [5, 6, 7]. This paper proposes an alternative 3×25 kV supply system coupled with a power electronics auto-transformer (PEAT), to mitigate grid current (and hence grid voltage) unbalance in railway power networks. The proposed system replaces the single-phase split-arrangement supply with a three-phase

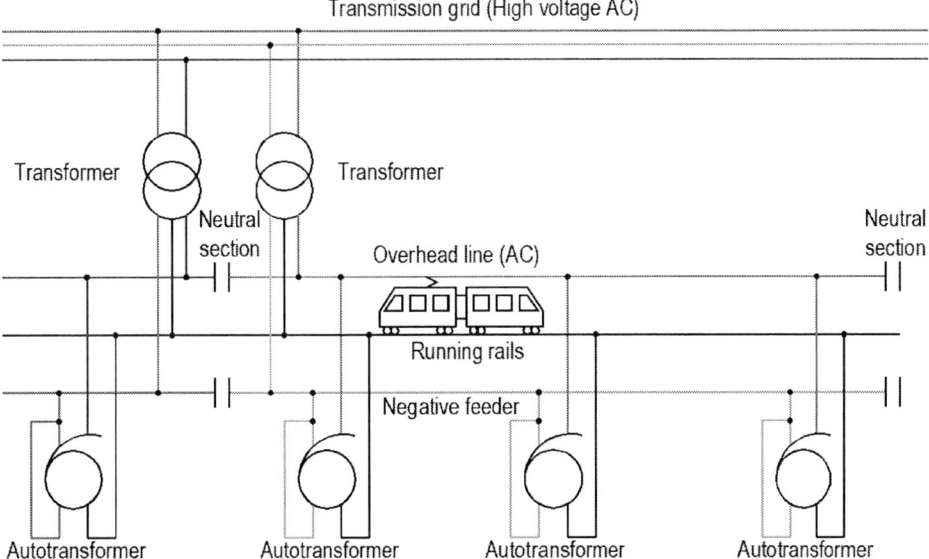

Fig. 1: Conventional state of the art 2×25kV railway power supply system with regularly spaced auto-transformers

transformer, which is cost-effective, compact and more efficient. While the rail load is supplied between two phases of the 3-φ transformer output, the compensating power electronics autotransformer (PEAT), comprising of back to back 1-φ voltage source converters (VSCs), is connected between different phase-pairs on its either end. The PEAT circuit is strategically operated to compensate for the single-phase train load on the three-phase network, ensuring balanced 3-φ power withdrawal from the grid, while circulating only a fraction of the total power. The circuit diagram of the proposed railway power supply and the PEAT connections are demonstrated in Fig.2.

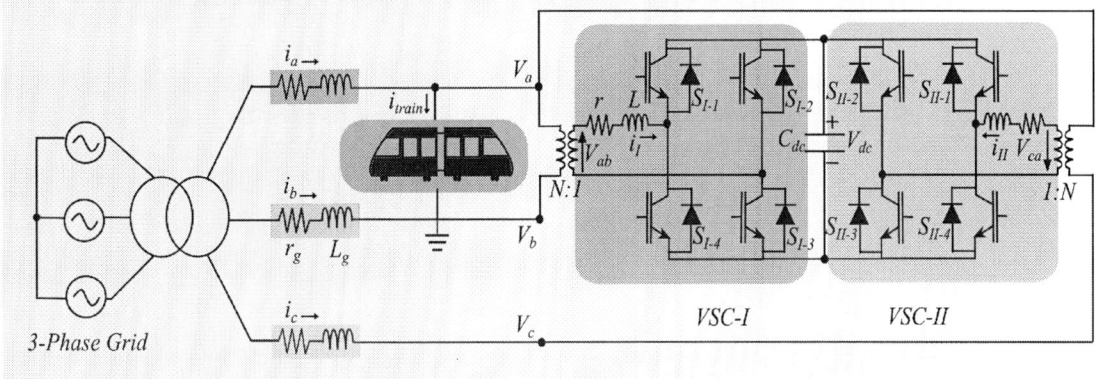

Fig. 2: Proposed 3×25kV railway power supply system with power electronics autotransformer

Control of Power Electronics Autotransformer

The back to back VSC pair forming the PEAT circuit is controlled to regulate the DC link voltage, compensate for the reactive component of source currents and to maintain balanced three-phase grid currents. In order to ensure fast and precise control of the power electronics auto-transformer, predictive current control is independently exercised for regulating the input currents of each VSC. Besides facilitating fast control dynamics with easy inclusion of system non-linearities, the use of predictive control also ensures that all objectives are accomplished with minimal sensing requirements, as the controlled currents are not measured, but mathematically predicted [8]. Due to these attractive features, predictive control is gaining rapid popularity in the control of power converters for various applications, including rail converters [9]. The reference currents for VSC-I and VSC-II, i.e i_I^* and i_{II}^* are generated as per various control objectives, followed by implementation of predictive current control for each VSC.

Reference Current Generation for PEAT

The first voltage source converter VSC-I, connected between the same phase pairs as the rail load, is controlled to ensure reactive power compensation, and to maintain the DC link voltage. On the contrary, VSC-II is responsible for drawing power balanced current in the otherwise unconnected phase of the 3-ϕ grid. Together, both converters work in tandem to maintain balanced 3-ϕ source currents.

For the control of VSC-I, the reference d-component of converter current (i_{Id}^*) is generated to regulate the DC link voltage, and the reference q-component of converter current (i_{Iq}^*) is derived to minimize the reactive component of the source currents. This is mathematically given as,

$$
\begin{aligned}
i_{Id}^* &= \left(k_p^d + \frac{k_i^d}{s} \right) \left(V_{dc}^* - V_{dc}^f \right) \\
i_{Iq}^* &= \left(k_p^q + \frac{k_i^q}{s} \right) \left(i_q^* - (N \times i_q) \right)
\end{aligned}
\tag{1}
$$

where, V_{dc}^* is the reference DC link voltage, V_{dc}^f is the measured DC link voltage after digital filtering, i_q^* is the reference q-component of source currents which is ideally set to zero, and i_q is the actual q-component of the source currents (i_{abc}). It must be noted that i_q is derived by using the phase angle from three-phase grid voltage to ensure reactive power minimization. Further noteworthy to mention is that the source currents (i_{abc}) are not directly measured, but estimated through measurement of train current and prediction of converter currents i_I and i_{II}. The use of predictive control significantly minimizes sensing requirements for undertaking current control.

From the dq components i_{Id}^* and i_{Iq}^*, the reference current i_I^* for VSC-I is generated as,

$$
i_I^* = i_{Id}^* \, sin(\theta_{ab}) + i_{Iq}^* \, cos(\theta_{ab})
\tag{2}
$$

where, θ_{ab} is the continuos phase of the converter side voltage, V_{ab}, obtained through a single-phase second order generalized integrator (SOGI) based phase locked loop.

For the control of VSC-II, the reference converter current (i_{II}^*) is derived from power balancing, followed by alignment with voltage of the unconnected phase (C) and subsequent reflection to the low voltage secondary (converter) side of the interfacing transformer. This is given as,

$$
i_{II}^* = \left[\left(\frac{V_a \cdot i_a + V_b \cdot i_b + V_c \cdot i_c}{V_a^2 + V_b^2 + V_c^2} \right) \times V_c \right] N
\tag{3}
$$

Once the reference currents i_I^* and i_{II}^* for VSC-I and VSC-II are produced, predictive current control of the converters is undertaken such that the VSCs are switched to draw currents as per their reference values.

Predictive Current Control

Finite set model predictive control (FS-MPC) is undertaken for VSC-I and VSC-II, to ensure that each converter traces its reference input current independently. The converter input currents, i_I and i_{II} are predicted one sample time ahead for all switching states of VSC-I and VSC-II, respectively. Switching models of the converters and discretized mathematical models of the interfacing inductors are used in generating the current predictions. The governing equations for VSC-I and VSC-II are respectively given as,

$$
\begin{aligned}
V_{ab} &= i_I r + L\frac{di_I}{dt} + [S_{I-1} + S_{I-3} - S_{I-2} - S_{I-4}]\frac{V_{dc}}{2} \\
V_{ca} &= i_{II} r + L\frac{di_{II}}{dt} + [S_{II-1} + S_{II-3} - S_{II-2} - S_{II-4}]\frac{V_{dc}}{2}
\end{aligned}
\tag{4}
$$

Upon discretization, the converter currents for the jth switching state in the $(k+1)^{th}$ sample time, are predicted for VSC-I and VSC-II respectively. These predictions are given as,

$$
\begin{aligned}
i_{I(k+1)}^{j} &= \frac{V_{ab} - i_{I(k)}r - [S_{I-1}^{j} + S_{I-3}^{j} - S_{I-2}^{j} - S_{I-4}^{j}]\frac{V_{dc}}{2}}{L}T_s + i_{I(k)} \\
i_{II(k+1)}^{j} &= \frac{V_{ca} - i_{II(k)}r - [S_{II-1}^{j} + S_{II-3}^{j} - S_{II-2}^{j} - S_{II-4}^{j}]\frac{V_{dc}}{2}}{L}T_s + i_{II(k)}
\end{aligned}
\tag{5}
$$

These current predictions are made for all four switching states of each converter and the corresponding errors in converter currents with respect to their reference signals are computed for each state. The errors for j^{th} state, for VSC-I and VSC-II are respectively given as,

$$
\begin{aligned}
e_I^j &= [i_I^* - i_{I(k+1)}^j]^2 \\
e_{II}^j &= [i_{II}^* - i_{II(k+1)}^j]^2
\end{aligned}
\tag{6}
$$

Independently for each converter, the switching state that yields the least error is finally switched in real time. It must be noted, that the control algorithm eliminates converter current measurement by propagating ahead the current prediction of the winning state, each time a switching decision is made.

Simulation Results

In order to ascertain the performance of power electronics autotransformer in mitigating 3-ϕ grid current unbalance in 1-ϕ railway supply systems, the proposed control algorithm is validated through simulation and analysis in Matlab/Simulink environment. The simulations are conducted for a 25 kV three-phase supply, with the 1-ϕ railway line connected between two of the three phases and a PEAT circuit connected across different phase pairs. Various system parameters used in the simulation are enlisted in Table-I.

Table I: Simulation Parameters

Parameter	Value
3-ϕ Secondary Side Supply Voltage	$25kV\,(L-L), 50Hz$
Grid Impedance $(L_g), (r_g)$	$10mH, 2\Omega$
Filter Parameters $(L), (r)$	$8mH, 0.5\Omega$
Transformer turns ratio $(N:1)$	$27.78:1$
DC Link Capacitance (C_{dc})	$8000\mu F$
DC Link Voltage (V_{dc})	$3000V$
Rail Load Emulating Resistance	$400\Omega/200\Omega$
Maximum Traction Power	$3.125MW$
Prediction Horizon/Sample Time (T_s)	$10\,\mu s$

The compensating performance of PEAT is well illustrated in Fig.3. The 3-ϕ source currents (i_{abc}), and their positive ($i_{abc}^{(+)}$), negative ($i_{abc}^{(-)}$), and zero sequence ($i_{abc}^{(o)}$) components, respectively are shown in Fig.3(a). At t=2.5 s, the PEAT circuit is disconnected to demonstrate unbalanced grid currents that are otherwise drawn by 1-ϕ railway lines. An unwelcome consequence of unbalanced grid currents, is the deterioration of voltage balance at the point of common coupling (PCC). To depict the promising impact of PEAT circuit, the PCC voltage (V_{abc}) and its sequence components, with and without the PEAT circuit are shown in Fig.3(b).

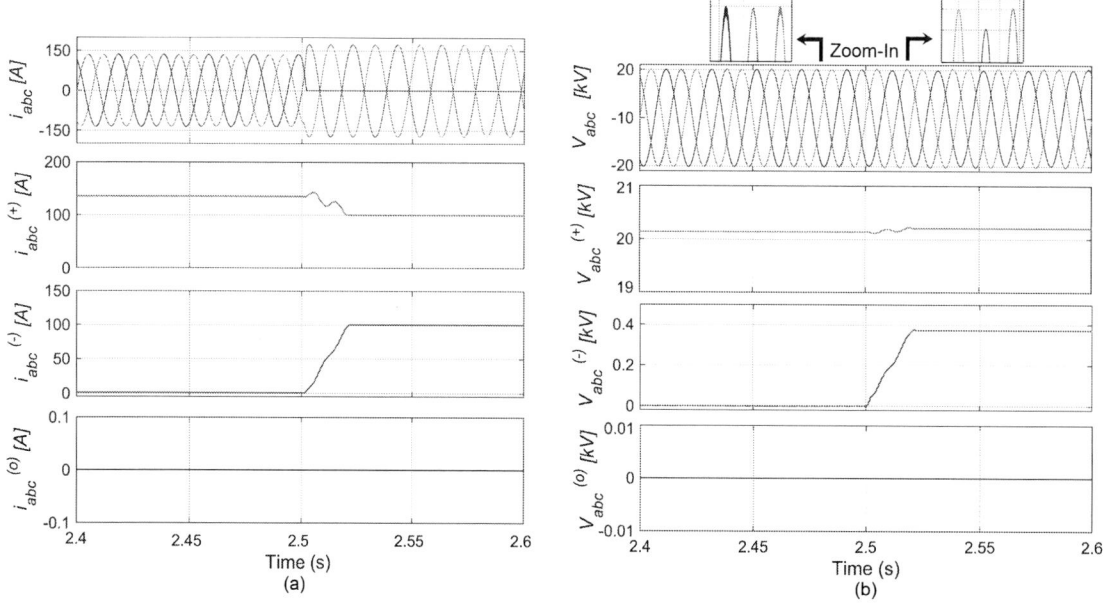

Fig. 3: System operation with and without power electronics autotransformer; (a) Three-phase source currents and their sequence components (b) Three-phase voltages at the point of common coupling and their sequence components

The dynamic response of the PEAT circuit with respect to an abrupt increase in the train load at t=3 s, is demonstrated in Fig.4. Fig.4(a) shows the 3-φ source currents (i_{abc}) and the sequence components of PCC voltage (V_{abc}), while Fig.4(b) shows the tracking performance of DC link voltage and converter currents i_I and i_{II} as the train load is abruptly increased. The DC link voltage reference is uniformly maintained at 3000 V.

The power quality performance of the proposed system is demonstrated in Fig.5. Fig.5(a) shows the source current (i_a) at nearly unity input power factor with respect to PCC voltage, V_a. Fig.5(b) shows the harmonic spectrum of grid current with a total harmonic distortion of only 0.58 %, which is well within the norms of IEEE-Std 519. It is well known that finite state model predictive control is a variable switching frequency algorithm, hence resulting in sporadic harmonic spectra.

Conclusion

From the aforementioned discussion, results and analysis, it is concluded that a 3×25 kV railway supply system when augmented with a power electronics autotransformer, can serve as a demonstrable alternative to conventional 2×25 kV railway supplies with passive autotransformers. While handling fractional power, the PEAT circuit can be effectively controlled to ensure that the 1-φ railway line appears as a balanced 3-φ load on the grid, with minimal harmonics and reactive power control.

References

[1] H. J. Kaleybar, H. M. Kojabadi, M. Brenna, F. Foiadelli and S. S. Fazel, "An active railway power quality compensator for 2×25kV high-speed railway lines," 2017 IEEE International Conference on Environment and Electrical Engineering and 2017 IEEE Industrial and Commercial Power Systems Europe (EEEIC / I&CPS Europe), 2017, pp. 1-6.

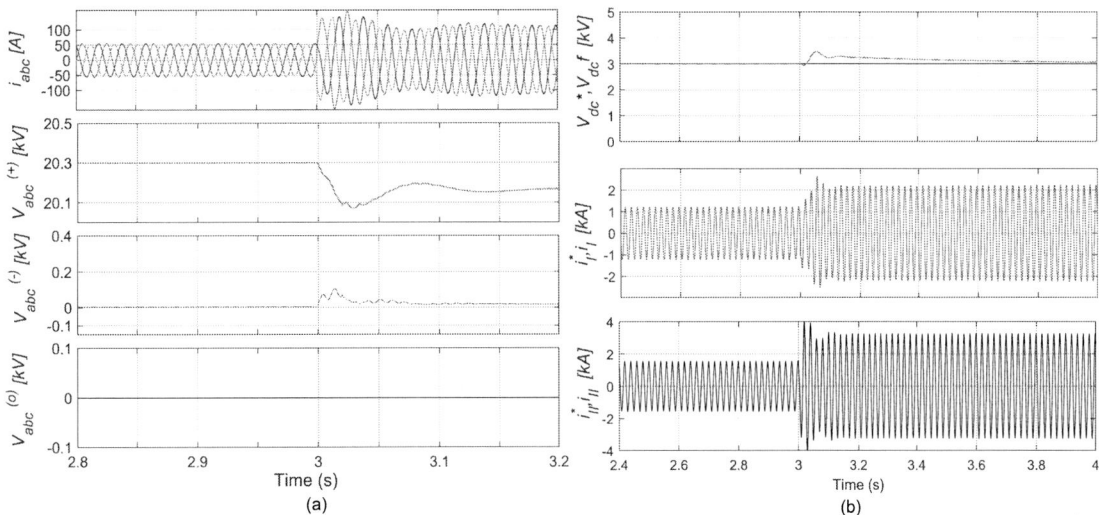

Fig. 4: Dynamic response of power electronics autotransformer during abrupt change in rail load; (a) Three-phase source currents (i_{abc}) and sequence components of voltage at the point of common coupling (V_{abc}). (b) Tracking response of DC link voltage (V_{dc}), and converter currents i_I and i_{II}.

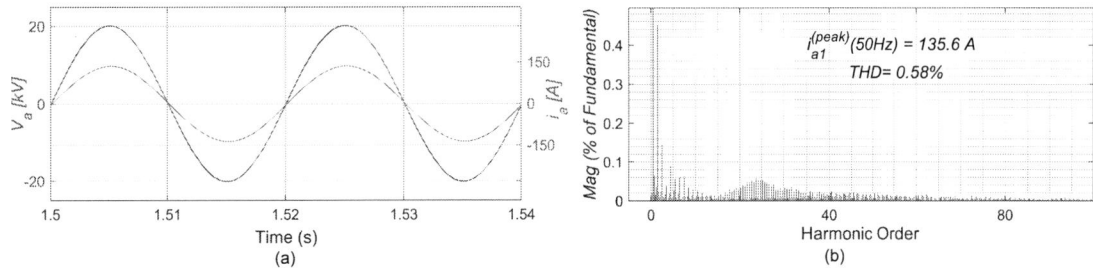

Fig. 5: Power quality performance of the power electronics autotransformer; (a) Source current (i_a) at nearly unity power factor with respect to voltage, V_a. (b) Harmonic spectrum of source current (i_a).

[2] S. M. Mousavi Gazafrudi, A. Tabakhpour Langerudy, E. F. Fuchs and K. Al-Haddad, "Power Quality Issues in Railway Electrification: A Comprehensive Perspective,"in IEEE Transactions on Industrial Electronics, vol. 62, no. 5, pp. 3081-3090, May 2015, doi: 10.1109/TIE.2014.2386794.

[3] V. Matta and G. Kumar, "Unbalance and voltage fluctuation study on AC traction system," 2014 Electric Power Quality and Supply Reliability Conference (PQ), 2014, pp. 303-308.

[4] G. Firat, Guangya Yang and H. A. H. Al-Ali, "A comparative study of different transformer connections for railway power supply-mitigation of voltage unbalance," 10th International Conference on Advances in Power System Control, Operation & Management (APSCOM 2015), 2015, pp. 1-6.

[5] S. T. Senini and P. J. Wolfs, "Novel topology for correction of unbalanced load in single phase electric traction systems," 2002 IEEE 33rd Annual IEEE Power Electronics Specialists Conference. Proceedings (Cat. No.02CH37289), 2002, pp. 1208-1212 vol.3.

[6] S. Hu et al., "A New Integrated Hybrid Power Quality Control System for Electrical Railway," in IEEE Transactions on Industrial Electronics, vol. 62, no. 10, pp. 6222-6232, Oct. 2015.

[7] S. Hu, S. Li, Y. Li, O. Krause and F. Zare, "A Comprehensive Study for the Power Flow Controller Used in Railway Power Systems," in IEEE Transactions on Industrial Electronics, vol. 65, no. 8, pp. 6032-6043, Aug. 2018.

[8] S. Borreggine, V. G. Monopoli, G. Rizzello, D. Naso, F. Cupertino and R. Consoletti, "A Review on Model Predictive Control and its Applications in Power Electronics," 2019 AEIT International Conference of Electrical and Electronic Technologies for Automotive (AEIT AUTOMOTIVE), 2019, pp. 1-6.

[9] N. Zhao, J. Liu, Y. Ai, J. Yang, J. Zhang and X. You, "Power-Linked Predictive Control Strategy for Power Electronic Traction Transformer," in IEEE Transactions on Power Electronics, vol. 35, no. 6, pp. 6559-6571, June 2020.

Parameter sensitivity of a MRAS-based sensorless control for AFPMSM considering speed accuracy and dynamic response at multiple parameter variations

Michael Brüns, Christian Rudolph, Tankred Müller
Hamburg University of Applied Sciences
Faculty of Engineering and Computer Science
Berliner Tor 21
Hamburg, Germany
Michael.Bruens@haw-hamburg.de; Christian.Rudolph@haw-hamburg.de;
Tankred.Mueller@haw-hamburg.de

Keywords

Sensorless control, Axial flux machines, Permanent magnet motor, Optimal control, Optimization method, Speed control, System identification, Real-time processing.

Abstract

In this paper, a novel approach is presented to tune the sensorless control based on a model reference adaptive system method (MRAS-based) in order to balance the variation of several system parameters. The method was used for speed and position estimation of a field-oriented controlled (FOC) wheel hub drive with an axial flux permanent magnet synchronous motor (AFPMSM). Parameter deviations of the control system are assumed to occur as disturbances. Their influences are reduced using an enhanced fundamental wave model of the AFPMSM. A model-based system engineering (MBSE) approach was chosen to compare simulation with experimental results. Following the simulation study, the MRAS-based method was implemented on a target system directly from the simulation models using code generation software. A performance evaluation of the tuning algorithm is focused on the accuracy of the calculated rotor position. The results demonstrate a resource-efficient MRAS implementation on a microcontroller (μC) suitable for operation with multiple parameter variations.

1 Introduction

The permanent magnet synchronous motor (PMSM) is a popular machine for modern drive solutions. It meets the requirements for high power density, high efficiency and sufficient robustness in the fields of mobile machinery, electromobility or some industrial applications at attractive system costs. But some drives require a particularly elevated energy density in addition of a short axial length. Therefore the choice of an AFPMSM could provide further advantages [1]-[2]. This type of permanent magnet synchronous machine is characterized by a design in which the air gap field is oriented axially, i.e. in the direction of the shaft, s. Fig. 1. Moreover, the machine can be extended using a modular design, for example as single stator double rotor or double stator, single rotor versions. The AFPMSM configuration enables a flat, cylindrical construction shape, which is ideal for integration as a wheel hub drive. However, the AFPMSM can be described by the same fundamental wave model as the PMSM.

Furthermore, sensorless control strategies are advantageous due to system simplifications increasing the reliability, reducing construction space and service costs of the drive system. Above all the redundant evaluation of a safety-critical system path becomes possible. In practice there are two main categories of sensorless control schemes subdivided in model-based sensorless control techniques for the medium and high-speed range and the saliency-based methods addressing the low speed range [3]. The MRAS-based speed estimators belong to the model-based group and present a closed loop method. Due to their advantages, such as straight forward design and comparative easy implementation on a μC they are recommended for low-cost applications, but can also be used in complex applications utilizing additional extensions [4]. Disadvantageous is the sensitivity on parameter variations of the real system, which can be reduced by an implementation of an optional parameter identification strategy [5]. Some publications release studies on the influence of parameters variations for the MRAS-based method by simulation changing particular parameters[6], [7]. However, in practice there are different parameter uncertainties at once. Therefore, this research focuses on the comparison of simulation and experimental results to identify an option to optimize the method under multiple parameter variations in order to increase the performance in practice.

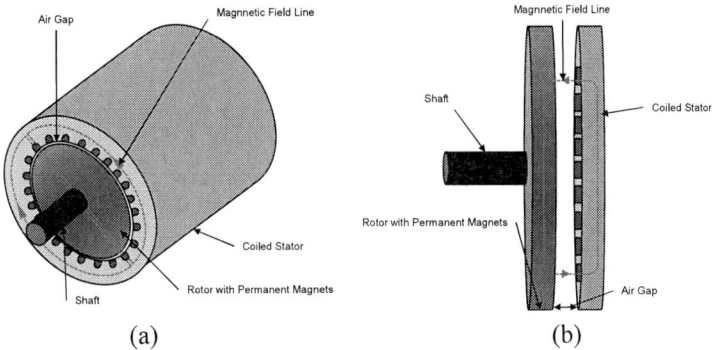

(a) (b)

Fig. 1: *Comparison of an interior-rotor PMSM and the AFPMSM:*
   a) structure of a radial flux permanent magnet synchronous motor with two poles
   b) structure of an axial flux permanent magnet synchronous motor with single stator and
       single rotor design, two-poles

2 MRAS-based sensorless control method

The MRAS method for speed and position estimation is based on the comparison of two independent machine models. Its adaptation mechanism is fed by the difference between state variables of the reference and the adjustable model. In the investigated MRAS-based method, the controlled machine itself serves as reference system and the measured stator current is used to adjust the model [4]. This creates a structure similar to an adaptive observer, but without the observer gain matrix as illustrated in Fig. 2.

The chosen method to design the adaptation mechanism is crucial for the stability of the sensorless control. There are three basic approaches [6]. One method is based on a local parameter optimization theory, missing sufficient stability conditions for operating in a wide field. A second method relies on the Lyapunov function and the third method is the one studied in this paper, based on the hyperstability theory design by V. M. Popov. There are indications that in practical systems the approach by Popov is superior in terms of response speed and convergence speed, since it yields a PI structure in contrast to a pure integral structure using the Lyapunov function [8]. However, both theories give a concept for sector stability of nonlinear feedback systems.

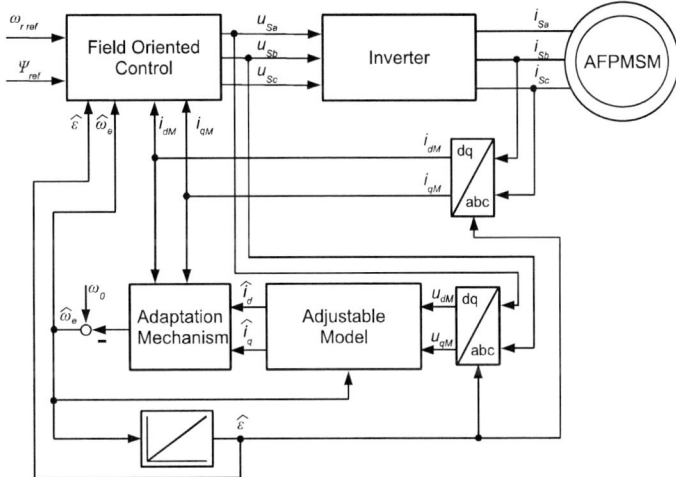

Fig. 2: *MRAS-based sensorless control structure for an AFPMSM*

2.1 Reference AFPMSM-model

The reference model can be described using the stator voltage equation of the AFPMSM. This is the well-known fundamental wave model [9] with the EMF induced by the permanent magnets, assuming a symmetrical winding distribution, neglecting saturation effects. Focused on machine parameter variation, it has to be considered which parts of the reference model can be imprecise. The machine parameters are dependent on the operating points, but in this approach, they are a part of the reference system, so they can be set as fixed. This includes the real values of the stator inductances L_d and L_q, the stator resistance R_s, the rotor magnetic flux Ψ_{PM}, the electrical rotation speed ω_e, the mechanical rotor speed ω, the flux-oriented angle ε, the number of pole pairs p and the stator currents and voltages i_d, i_q and u_d, u_q. However, the information we obtain about currents, voltages and the field or rotor angle by measurement and calculation are subject of errors in general, which are considered using error elements Δx_x. For example, the measured current consists of its actual real value and an unknown absolute error as follows:

$$\begin{bmatrix} i_{dM} \\ i_{qM} \end{bmatrix} = \begin{bmatrix} i_d \\ i_q \end{bmatrix} + \begin{bmatrix} \Delta i_d \\ \Delta i_q \end{bmatrix} \tag{1}$$

Subsequently, in the mathematical description of the reference AFPMSM-model these uncertainties are taken into account:

$$\begin{bmatrix} \dfrac{di_{dM}}{dt} \\ \dfrac{di_{qM}}{dt} \end{bmatrix} = \begin{bmatrix} -\dfrac{R_s}{L_d} & \omega_e * \dfrac{L_q}{L_d} \\ -\omega_e * \dfrac{L_d}{L_q} & -\dfrac{R_s}{L_q} \end{bmatrix} \begin{bmatrix} i_{dM} \\ i_{qM} \end{bmatrix} + \begin{bmatrix} \dfrac{1}{L_d} & 0 \\ 0 & \dfrac{1}{L_q} \end{bmatrix} \begin{bmatrix} u_{dM} \\ u_{qM} \end{bmatrix} + \begin{bmatrix} \dfrac{\Psi_{PM}}{L_q} \cdot sin(\Delta \epsilon) \\ -\dfrac{\Psi_{PM}}{L_q} \cdot cos(\Delta \epsilon) \end{bmatrix} \cdot \omega_e \tag{2}$$

2.2 Novel adjustable AFPMSM model approach considering parameter variation

The adjustable model is established in a similar way as the reference model, i.e. a fundamental wave model. Calculated or adjusted parameters of the adjustable model are indicated by the circumflex. Parameter deviations are taken into account using the theory of error. E.g. the stator resistance R_{sM} is described by its real value and an absolute error:

$$R_{sM} = R_s + \Delta R_s \tag{3}$$

Furthermore, an enhancement of the fundamental wave model using tuning parameters labeled $\hat{z}_d$ and $\hat{z}_q$ will be discussed. The mathematical description of the adjusted AFPMSM-model therefore includes influences by model parameter uncertainties and two tuning parameters $\hat{z}_d$ and $\hat{z}_q$:

$$\begin{bmatrix} \dfrac{d\hat{i}_d}{dt} \\ \dfrac{d\hat{i}_q}{dt} \end{bmatrix} = \begin{bmatrix} -\dfrac{R_{sM}}{L_{dM}} & \hat{\omega}_e \cdot \dfrac{L_{qM}}{L_{dM}} \\ -\hat{\omega}_e \cdot \dfrac{L_{dM}}{L_{qM}} & -\dfrac{R_{sM}}{L_{qM}} \end{bmatrix} \begin{bmatrix} \hat{i}_d \\ \hat{i}_q \end{bmatrix} + \begin{bmatrix} \dfrac{1}{L_{dM}} & 0 \\ 0 & \dfrac{1}{L_{qM}} \end{bmatrix} \begin{bmatrix} \hat{u}_d \\ \hat{u}_q \end{bmatrix} + \begin{bmatrix} 0 \\ -\dfrac{\Psi_{PMM}}{L_{qM}} \cdot \hat{\omega}_e \end{bmatrix} + \begin{bmatrix} \hat{z}_d \\ \hat{z}_q \end{bmatrix} \tag{4a}$$

$$\frac{d\underline{\hat{i}}_{dq}}{dt} = \underline{A}(\hat{\omega}_e) \cdot \underline{\hat{i}}_{dq} + \underline{B} \cdot \underline{\hat{u}}_{dq} + \underline{N}(\hat{\omega}_e) + \underline{\hat{Z}} \tag{4b}$$

2.3 Error model considering multiple parameter variations

The basis for the application of the hyperstability theory is the transformation of the structure depicted in Fig. 2 into a nonlinear feedback system of the following shape [4].

$$\underline{\dot{\delta}} = \underline{A}_E(\hat{\omega}_e) \cdot \underline{\delta} + \underline{B}_E \cdot \Delta \omega_e = \underline{A}_E(\hat{\omega}_e) \cdot \underline{\delta} - \underline{W}_E = \underline{A}_E(\hat{\omega}_e) \cdot \underline{\delta} + \underline{U}_E$$
$$\underline{y} = \underline{\delta} \tag{5}$$

In the first step, the mathematical transformation into an error model is done subtracting the Eq. (4) describing the adjustable model from the Eq. (2) describing the reference model, shown in Eq. (6).

$$\begin{bmatrix} \delta_d \\ \delta_q \end{bmatrix} = \begin{bmatrix} i_d - \hat{i}_d \\ i_q - \hat{i}_q \end{bmatrix} \tag{6}$$

Decoupling influences of parameter variations from the error model, four further assumptions have to be considered. The connected model parameters in Eq. (4a) should be expressed as shown for example below. A speed error is introduced.

$$\frac{R_{sM}}{L_{dM}} = \frac{R_s}{L_d} + \frac{L_d \cdot \Delta R_s - R_s \cdot \Delta L_d}{L_d \cdot L_{dM}} = \frac{R_s}{L_d} + \Delta \epsilon_{R_d} \tag{7}$$

$$\Delta \omega_e = \omega_e - \widehat{\omega}_e \tag{8}$$

$$-\frac{\pi}{2} \leq \Delta \epsilon \leq \frac{\pi}{2}, cos(\Delta \epsilon) \approx 1 - \frac{(\Delta \epsilon)^2}{2!} + \frac{(\Delta \epsilon)^4}{4!} \tag{9}$$

$$-\frac{\pi}{2} \leq \Delta \epsilon \leq \frac{\pi}{2}, sin(\Delta \epsilon) \approx \Delta \epsilon - \frac{(\Delta \epsilon)^3}{3!} + \frac{(\Delta \epsilon)^5}{5!} \tag{10}$$

The influence of parameter variations can now be estimated as disturbance and is considered using variable $Z = [z_d \ z_q]^T$. The mathematical description of the error model including influences by uncertainties and tuning parameters is carried out as follows:

$$\begin{bmatrix} \frac{d\delta_d}{dt} \\ \frac{d\delta_q}{dt} \end{bmatrix} = \begin{bmatrix} -\frac{R_s}{L_d} & \widehat{\omega}_e \cdot \frac{L_q}{L_d} \\ -\widehat{\omega}_e \cdot \frac{L_d}{L_q} & -\frac{R_s}{L_q} \end{bmatrix} \begin{bmatrix} \delta_d \\ \delta_q \end{bmatrix} + \begin{bmatrix} \frac{L_q}{L_d} i_q \\ -\frac{L_d}{L_q} i_d - \frac{\Psi_{PM}}{L_q} \end{bmatrix} (\Delta \omega) + \begin{bmatrix} z_d \\ z_q \end{bmatrix} - \begin{bmatrix} \hat{z}_d \\ \hat{z}_q \end{bmatrix} \tag{11}$$

The disturbance variable $\underline{Z}$ can be further subdivided into different parts, using the following equations.

$$\begin{bmatrix} z_d \\ z_q \end{bmatrix} = \underline{Z} = \underline{Z}_1 + \underline{Z}_2 + \underline{Z}_3 \tag{12}$$

$$\underline{Z}_1 = \begin{bmatrix} -\frac{R_s}{L_d} & \omega_e \cdot \frac{L_q}{L_d} \\ -\omega_e \cdot \frac{L_d}{L_q} & -\frac{R_s}{L_q} \end{bmatrix} \begin{bmatrix} \Delta i_d \\ \Delta i_q \end{bmatrix} - \begin{bmatrix} \frac{d\Delta i_d}{dt} \\ \frac{d\Delta i_q}{dt} \end{bmatrix} + \begin{bmatrix} \frac{1}{L_d} & 0 \\ 0 & \frac{1}{L_q} \end{bmatrix} \begin{bmatrix} u_{dM} - \hat{u}_d \\ u_{qM} - \hat{u}_q \end{bmatrix} \tag{13}$$

$$\underline{Z}_2 = \begin{bmatrix} \Delta \epsilon_{R_d} & -\widehat{\omega}_e \cdot \Delta \epsilon_{L_q} \\ \widehat{\omega}_e \cdot \Delta \epsilon_{L_d} & \Delta \epsilon_{R_q} \end{bmatrix} \begin{bmatrix} \hat{i}_d \\ \hat{i}_q \end{bmatrix} - \begin{bmatrix} \Delta \epsilon_d & 0 \\ 0 & \Delta \epsilon_q \end{bmatrix} \begin{bmatrix} \hat{u}_d \\ \hat{u}_q \end{bmatrix} - \begin{bmatrix} 0 \\ \Delta \epsilon_{\Psi_{PM}} \end{bmatrix} \cdot \widehat{\omega}_e \tag{14}$$

$$\underline{Z}_3 = \begin{bmatrix} \frac{\Psi_{PM}}{L_d} \left(\Delta \epsilon - \frac{(\Delta \epsilon)^3}{3!} + \frac{(\Delta \epsilon)^5}{5!} \right) \\ \frac{\Psi_{PM}}{L_q} \left(\frac{(\Delta \epsilon)^2}{2!} - \frac{(\Delta \epsilon)^4}{4!} \right) \end{bmatrix} \cdot \omega_e - \begin{bmatrix} 2 \frac{\Psi_{PM}}{L_d} \left((\Delta \epsilon) - \frac{(\Delta \epsilon)^3}{3!} + \frac{(\Delta \epsilon)^5}{5!} \right) \\ 0 \end{bmatrix} \cdot \widehat{\omega}_e \tag{15}$$

Eq. (13) comprises the influence of input and state vector values that are recorded incorrectly, Eq. (14) the effect of parameter variations and Eq. (15) a mix of both.

2.4 Adaptation Mechanism according to Popov's hyperstability

Assuming $\underline{Z} = \hat{\underline{Z}}$, the error model of Eq. (11) could expressed by Eq. (5) to obtain a nonlinear feedback closed-loop system, where the Popov theory can be applied to speed and position determination [4]. At first, it is necessary to prove the eigenvalues of matrix $\underline{A}_E$. The poles have to be located in the left-half complex plane.

$$\underline{H}_E = \left(s\underline{I} - \underline{A}_E(\widehat{\omega}_e) \right)^{-1} = \begin{bmatrix} s + \frac{R_s}{L_q} & \widehat{\omega}_e \cdot \frac{L_q}{L_d} \\ -\widehat{\omega}_e \cdot \frac{L_d}{L_q} & s + \frac{R_s}{L_d} \end{bmatrix} \cdot \frac{1}{\left(s + \frac{R_s}{L_d} \right)\left(s + \frac{R_s}{L_q} \right) + \widehat{\omega}_e^2} \tag{16}$$

$$Re\left\{ -\frac{R_s \cdot L_q + R_s \cdot L_d}{2 \cdot L_d \cdot L_q} \pm \sqrt{\left(\frac{R_s \cdot L_q + R_s \cdot L_d}{2 \cdot L_d \cdot L_q} \right)^2 - \frac{R_s^2}{L_d \cdot L_q} - \widehat{\omega}_e^2} \right\} \leq 0, \tag{17}$$

Eq. (17) is a condition showing that the poles are located in the left-half-plane for the investigated method. Further on, it must be verified that the Popov integral inequality shown below is true for the nonlinear feedback system [6].

$$\forall t_0 > 0, \gamma_0^2 \geq 0, \eta(0,t_0) = \int_0^{t_0} \underline{W_E}^T(\tau)\underline{y}(\tau)\,d\tau \geq -\gamma_0^2 \qquad (18a)$$

$$\eta(0,t_0) = \int_0^{t_0} \left[-\frac{L_q}{L_d}i_q(\omega_e - \widehat{\omega}_e) \quad \left(\frac{L_d}{L_q}i_d + \frac{\Psi_{PM}}{L_q}\right)(\omega_e - \widehat{\omega}_e) \right]\begin{bmatrix}\delta_d \\ \delta_q\end{bmatrix} d\tau \geq -\gamma_0^2 \qquad (18b)$$

The adaptive law for the stator frequency could be given in the form of a PI element by Eq. (19) [4], [6].

$$\widehat{\omega}_e = K_P\left[\frac{L_q}{L_d}i_q\delta_d - \left(\frac{L_d}{L_q}i_d + \frac{\Psi_{PM}}{L_q}\right)\delta_q\right] + \int_0^{t_0} K_i\left[\frac{L_q}{L_d}i_q\delta_d - \left(\frac{L_d}{L_q}i_d + \frac{\Psi_{PM}}{L_q}\right)\delta_q\right]d\tau \qquad (19)$$

2.5 Dynamic response of MRAS-based speed determination

Finding the forward transfer function of the MRAS-based speed determination, the dynamic response of the MRAS-System can be estimated:

$$G_{\omega s}(s) = \underline{c}^T \cdot \underline{H_E} \cdot \underline{b} \qquad (20)$$

$$G_{\omega s}(s) = \frac{\left[\frac{L_q}{L_d}i_q \quad \left(-\frac{L_d}{L_q}i_d - \frac{\Psi_{PM}}{L_q}\right)\right]}{\left(s+\frac{R_s}{L_d}\right)\left(s+\frac{R_s}{L_q}\right) + \widehat{\omega}_e^2} \cdot \begin{bmatrix} s+\frac{R_s}{L_q} & \widehat{\omega}_e\cdot\frac{L_q}{L_d} \\ -\widehat{\omega}_e\cdot\frac{L_d}{L_q} & s+\frac{R_s}{L_d} \end{bmatrix} \cdot \begin{bmatrix} \frac{L_q}{L_d}i_q \\ \left(-\frac{L_d}{L_q}i_d - \frac{\Psi_{PM}}{L_q}\right) \end{bmatrix} \qquad (21)$$

A closed loop block diagram represents the dynamic response of the calculated MRAS-based speed determination and can be described with Eq. (21), shown in Fig. 3. Moreover, its dynamic response is adjustable providing K_p and K_i as defined in Eq. (19).

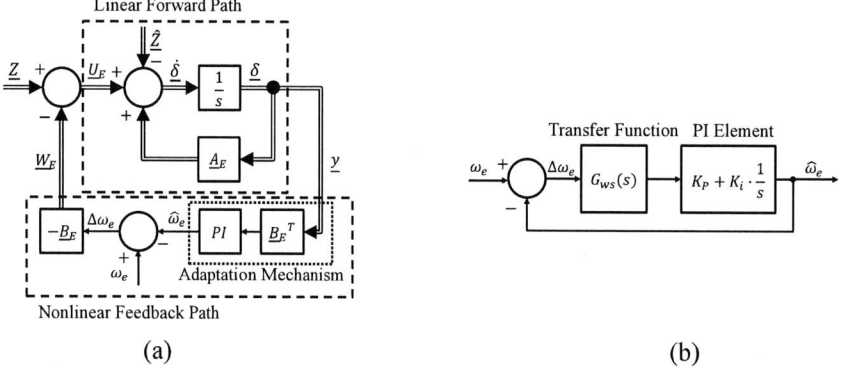

(a) (b)

Fig. 3: *MRAS-based closed loop block diagram for speed determination*
a) nonlinear and time varying feedback system considering parameter variation
b) block diagram of stator frequency determination

3 Model-based systems engineering approach

In order to compare simulation and experimental results, a model-based systems engineering (MBSE) approach was used, generating software for the µC directly from simulation models. The working procedure applied for this research can be subdivided into four main stages. In the first stage, we used Matlab Simulink 2022a to receive simulation results. Subsequently, in the second stage C-code for the control algorithm was generated by Embedded Coder 7.8 and Embedded Coder Support Package for Texas Instruments C2000 Processors 22.1.1.

On the next stage, the method could be verified using a Rapid Control Prototyping setup with serial communications interface (SCI)-based communication to acquire the data with the sampling time of PWM frequency of 15 kHz. After a functionality check of hardware components, the experimental verification began. Therefore, a wheel hub drive powered by a five-pole AFPMSM with nominal values P_n=350 W, I_n=10,1 A, M_n=1,11 Nm, n_n=3000 rpm and a gear ratio to the mounted wheel of i=8, was chosen.

3.1 Discrete model implementation

Even if the parameters are exactly defined, model variable deviations occur caused by the implementation. One reason is Euler's method for integration, realised by C code generation e.g. used for sampling operation on microcontrollers [10]. Euler's rule is a typical application to discretize observers for motor drives and also used in this research work. Therefore, the Eq. (4b) and Eq. (19) have to be rewritten into Eq. (22) and Eq. (26). The software code and the sampling operation is executed synchronously to the switching period T_{sw}.

$$\hat{\underline{i}}_{dq}(k+1) = \underline{A}_d(\hat{\omega}_e) \cdot \hat{\underline{i}}_{dq}(k) + \underline{B}_d \cdot \hat{\underline{u}}_{dq}(k) + \underline{N}_d(\hat{\omega}_e) + \hat{\underline{Z}} \tag{22}$$

$$\underline{A}_d(\hat{\omega}_e) = \begin{bmatrix} 1 - T_{sw} \cdot \dfrac{R_{sM}}{L_{dM}} & \hat{\omega}_e(k) \cdot T_{sw} \cdot \dfrac{L_{qM}}{L_{dM}} \\[3mm] -\hat{\omega}_e(k) \cdot T_{sw} \cdot \dfrac{L_{dM}}{L_{qM}} & 1 - T_{sw} \cdot \dfrac{R_{sM}}{L_{qM}} \end{bmatrix} \tag{23}$$

$$\underline{B}_d = \begin{bmatrix} \dfrac{T_{sw}}{L_{dM}} & 0 \\[3mm] 0 & \dfrac{T_{sw}}{L_{qM}} \end{bmatrix}; \; \underline{N}_d(\hat{\omega}_e) = \begin{bmatrix} 0 \\[2mm] -T_{sw} \cdot \dfrac{\Psi_{PMM}}{L_q} \cdot \hat{\omega}_e(k) \end{bmatrix}; \; \hat{\underline{Z}} = \begin{bmatrix} \hat{z}_d \\ \hat{z}_q \end{bmatrix} \tag{24}$$

$$fcn(k) = \frac{L_{qM}}{L_{dM}} i_{qM}(k)\delta_{dM}(k) - \left(\frac{L_{dM}}{L_{qM}} i_{dM}(k) + \frac{\Psi_{PMM}}{L_{qM}}\right)\delta_{qM}(k) \tag{25}$$

$$\hat{\omega}_e(k) = \hat{\omega}_e(k-1) + K_P \cdot fcn(k) - K_P \cdot fcn(k-1) + K_I \cdot fcn(k-1) + \hat{\omega}_e(0) \tag{26}$$

The goal is to compare the simulated and measured results regarding parameter variations. The implemented MRAS-based sensorless control for AFPMSM is depicted in a block scheme in Fig. 4.

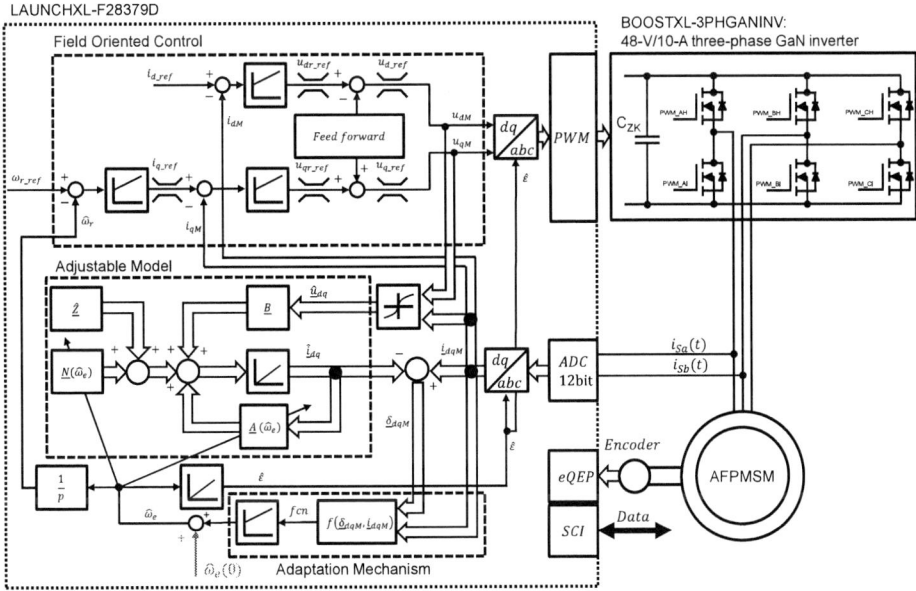

Fig. 4: Block scheme of MRAS-based sensorless control for AFPMSM

4 Simulation and measurement evaluation

Within this paper the sensitivity of parameter uncertainties is illustrated on the basis of the flux orientation angle error. In this way, the impact of parameter variations and measurement errors can be subdivided into static and dynamic disturbance effects, which can be represented by mean value and standard deviation. Furthermore, the co- and counter-coupling effects of parameter variation in the described error model approach should be assessed by tuning $\hat{\underline{Z}}$.

4.1 Simulation results

Simulation results show the influence of $\hat{\underline{Z}}_d$ and $\hat{\underline{Z}}_q$ interval steps at multi parameter variation. The difference between model and machine parameters was set to $L_{dM}=0,85 \cdot L_d$, $L_{qM}=1,10 \cdot L_q$, $R_{sM}=1,05 \cdot R_s$ and $\Psi_{PMM}=0,98 \cdot \Psi_{PM}$. Fig. 5 shows the estimated speed, the flux-orientation error for a set reference speed of 1000 rpm depends on interval steps of $\hat{\underline{Z}}$ and electric torque. The parameter set mentioned above led to problems in the implemented MRAS-based sensorless control method with $\hat{\underline{Z}} = 0$, presented in Fig. 5 at t=6,8 s. The sensorless startup is implemented without additional methods. This could produce higher torques or overshoots, depending on the initial conditions. Fig. 5 shows that the flux-orientation error can be tuned using $\hat{\underline{Z}}_d$ or $\hat{\underline{Z}}_q$ and is also be tuned at load. Setting $\hat{\underline{Z}}$ to a constant value has only a major effect to the standard deviation of flux-orientation error if the MRAS-based method runs in an unstable point. The mean value of flux-orientation error correlates to $\hat{\underline{Z}}$ presented in Fig. 5.

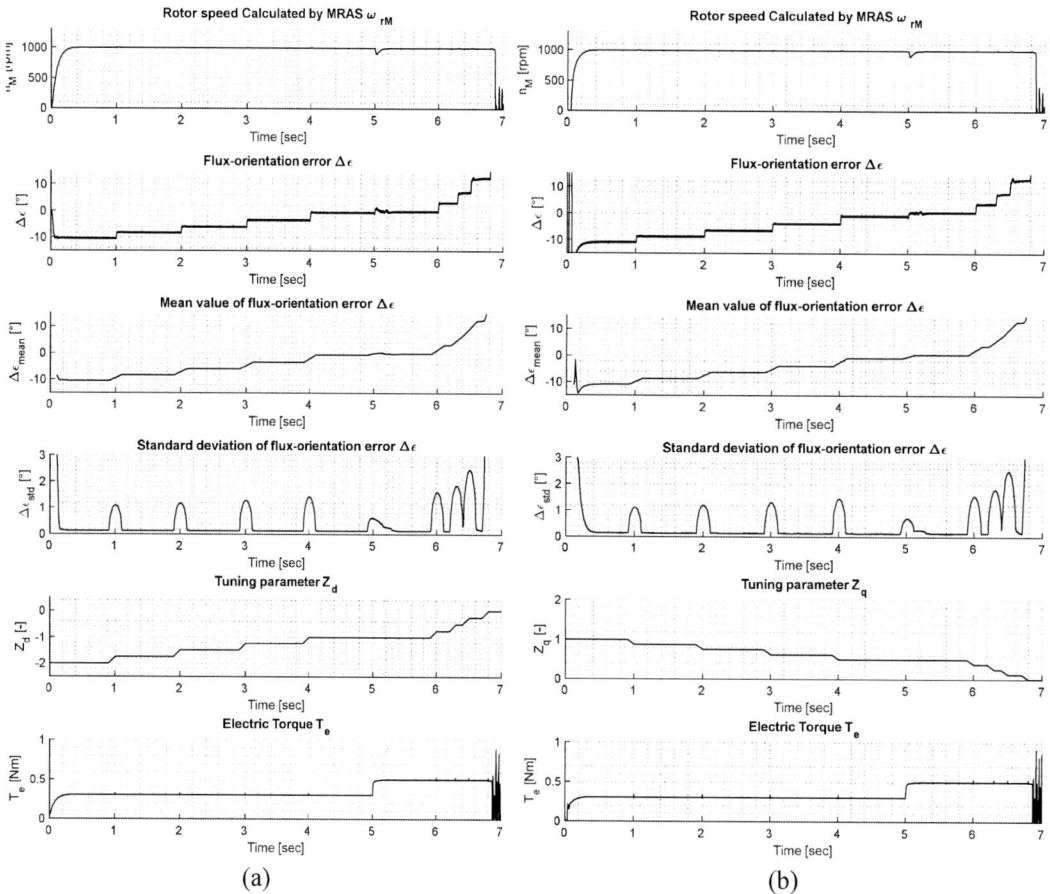

Fig. 5: *Flux orientation error at parameter variations. Stepwise modification of a tuning parameter and load*
a) stepwise modification of $\hat{\underline{Z}}_d$ at $\hat{\underline{Z}}_q=0$
b) stepwise modification of $\hat{\underline{Z}}_q$ at $\hat{\underline{Z}}_d=0$

Fig. 6a shows the influence of an implemented noise signal on the measurements of i_{sa} and i_{sb} with a boundary set to ±1 A. Fig. 6b illustrates the effect of an implemented dc-offset measurement error of 0,25 A_{DC} on i_{sa} and i_{sb}. Both have only a minor effect to the mean value of flux-orientation error compared to Fig. 5. However, the implemented noise signal and dc-offset have more effects on the standard deviation of flux-orientation error and could not be influenced by a constant $\hat{\underline{Z}}$ value.

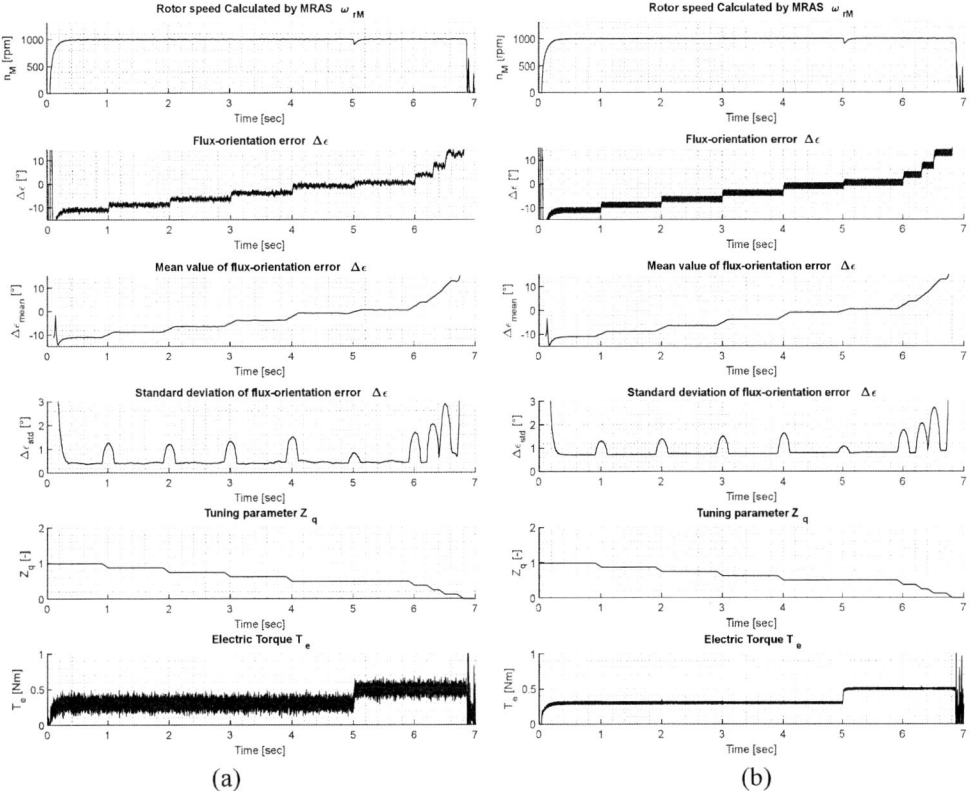

Fig. 6: Flux orientation error at parameter variations, stepwise modification of $\hat{\underline{Z}}_q$ at $\hat{\underline{Z}}_d$=0 and load
a) implemented current measurement noise
b) implemented current dc-offset

4.2   Measurement results

Measurement results show the influence of $\hat{\underline{Z}}_d$ interval steps at typical parameter variation. The machine parameters were identified by a parameter identification algorithm. The identified parameters are L_d=0,17 mH, L_q=0,17 mH, R_s=0,17 Ω and Ψ_{PMM}=0,0125 Wb. Fig. 5 shows estimated speed, flux-orientation error for a reference speed of 1000 rpm, depending on interval steps of $\hat{\underline{Z}}$ and electric torque. It can be assumed that there is an unknown parameter variation because of the flux orientation error measured at $\hat{\underline{Z}} = 0$, presented in Fig. 7 at t=0 s to 5 s. Furthermore, it can be assumed that there is also an unknown error due to the current measurements, because of the standard deviation of flux orientation error. Moreover, it could be observed a current noise boundary of ±0.6 A at nominal zero current. The sensorless startup is implemented without additional methods. This will produce higher torques or overshoots, depending on the initial conditions. Fig. 7 shows that the flux orientation error can be tuned using $\hat{\underline{Z}}$ and can also be tuned at load. Setting $\hat{\underline{Z}}$ to a constant value has only a major effect to the standard deviation of flux-orientation error if the MRAS-based method runs in an unstable point. The mean value of flux-orientation error correlates to $\hat{\underline{Z}}$ presented in Fig. 7.

Fig. 8 presents the measured flux-oriented angle error by a speed reversal process with and without a tuning parameter $\hat{\underline{Z}}_d$ at $\hat{\underline{Z}}_q$=0. The flux orientation angle can be optimized but has temporary an error peak evoked by reversion.

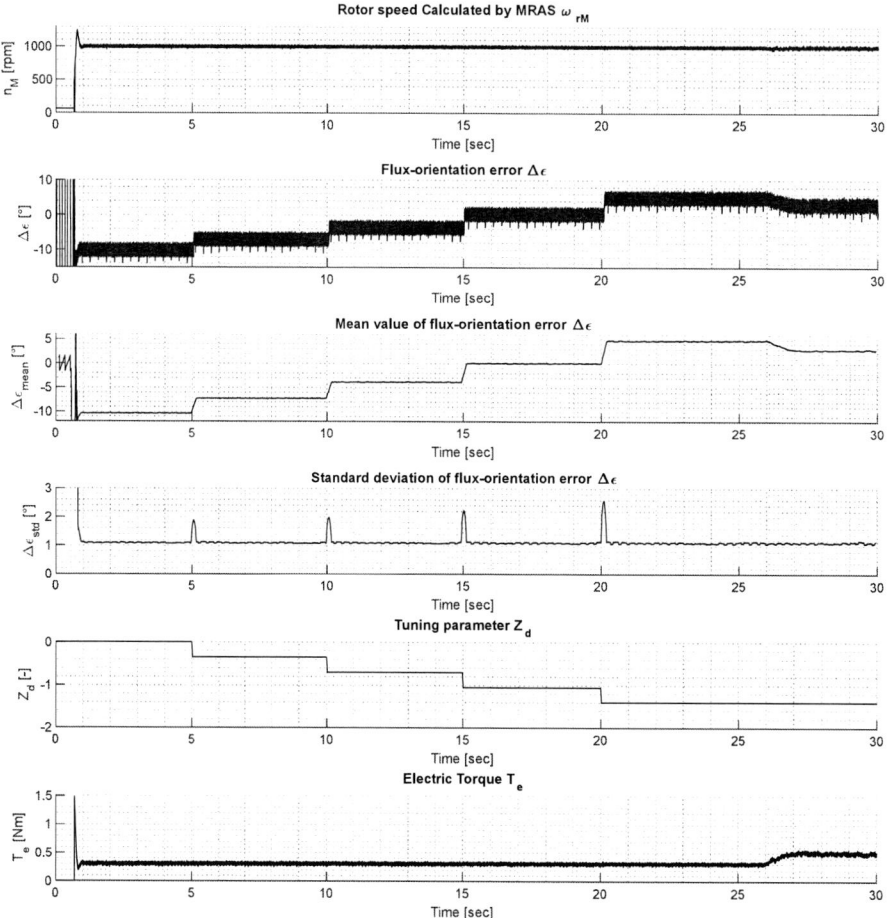

Fig. 7: *Measured flux orientation error angle at parameter variations, by stepwise modification of tuning parameter $\hat{\underline{Z}}_d$ at $\hat{\underline{Z}}_q=0$ and load*

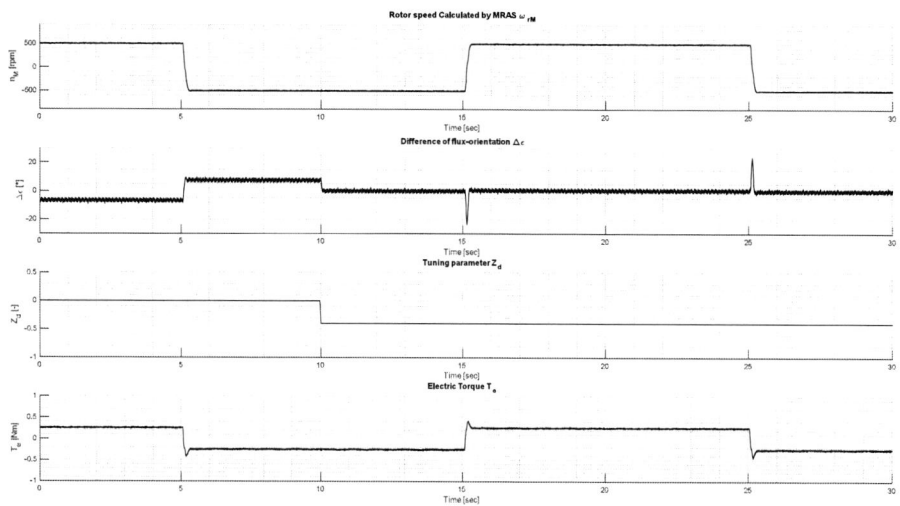

Fig. 8: *Measured flux orientation angle error at typical parameter deviations by reversion with and without adapted tuning parameter $\hat{\underline{Z}}_d$ at $\hat{\underline{Z}}_q=0$*

5 Conclusion

A MRAS-based sensorless control strategy designed for medium and high stator frequencies was investigated on a wheel hub drive with an AFPMSM in dependence of parameter sensitivity.

In a first step, the sensorless control method was introduced. Effects of multiple parameter variations and incorrectly recorded input and state vector values were described by a mathematical model. Dynamic response was also calculated. By enhancing the fundamental wave model with a tuning parameter, the effects of multiple parameter variations could be partially balanced. Furthermore, the sensorless method was simulated with an MBSE approach to be subsequently implemented on a target system.

By using the mean value and standard deviation of flux orientation error it could be shown that parameter variations have static and dynamic effects on the accuracy of the control. Setting the tuning parameter to a fixed value, the static effects could be reduced or increased. Thus, unstable operating points, that result from parameter variation, can be handled by well-chosen tuning parameters.

From simulation results it could be observed that a current measurement error has dynamic effects on the flux orientation error. This was shown using DC and noise signal offsets. The proposed sensorless control method works on a microcontroller, common in industrial applications.

References

[1] A. Cavagnino, M. Lazzari, F. Profumo, und A. Tenconi, „A comparison between the axial flux and the radial flux structures for PM synchronous motors", *IEEE Trans. Ind. Appl.*, Bd. 38, Nr. 6, S. 1517–1524, Nov. 2002, doi: 10.1109/TIA.2002.805572.

[2] K. Sitapati und R. Krishnan, „Performance comparisons of radial and axial field, permanent-magnet, brushless machines", *IEEE Trans. Ind. Appl.*, Bd. 37, Nr. 5, S. 1219–1226, Sep. 2001, doi: 10.1109/28.952495.

[3] G. Wang, M. Valla, und J. Solsona, „Position Sensorless Permanent Magnet Synchronous Machine Drives— A Review", *IEEE Trans. Ind. Electron.*, Bd. 67, Nr. 7, S. 5830–5842, Juli 2020, doi: 10.1109/TIE.2019.2955409.

[4] A. Khlaief, M. Boussak, und A. Châari, „A MRAS-based stator resistance and speed estimation for sensorless vector controlled IPMSM drive", *Electr. Power Syst. Res.*, Bd. 108, S. 1–15, März 2014, doi: 10.1016/j.epsr.2013.09.018.

[5] X. Liu, G. Zhang, L. Mei, und D. Wang, „Speed estimation and Parameters Identification simultaneously of PMSM based on MRAS", Bd. 12, S. 10, 2017.

[6] J. Zhao, X. Zhang, C. Lin, und G. Tian, „Simulation research of sensorless control of PMSM based on MRAS considering parameters variation and dead-time", in *2016 19th International Conference on Electrical Machines and Systems (ICEMS)*, Nov. 2016, S. 1–6.

[7] Y. Ni und D. Shao, „Research of Improved MRAS Based Sensorless Control of Permanent Magnet Synchronous Motor Considering Parameter Sensitivity", in *2021 IEEE 4th Advanced Information Management, Communicates, Electronic and Automation Control Conference (IMCEC)*, Juni 2021, Bd. 4, S. 633–638. doi: 10.1109/IMCEC51613.2021.9482131.

[8] K. Liu, Q. Zhang, Z.-Q. Zhu, J. Zhang, A.-W. Shen, und P. Stewart, „Comparison of two novel MRAS based strategies for identifying parameters in permanent magnet synchronous motors", *Int. J. Autom. Comput.*, Bd. 7, Nr. 4, S. 516–524, Nov. 2010, doi: 10.1007/s11633-010-0535-3.

[9] N. P. Quang und J.-A. Dittrich, *Vector Control of Three-Phase AC Machines*, Second Edition. Springer-Verlag Berlin Heidelberg, 2015. [Online]. Verfügbar unter: https://link.springer.com/book/10.1007/978-3-662-46915-6

[10] M. Comanescu, „Influence of the discretization method on the integration accuracy of observers with continuous feedback", in *2011 IEEE International Symposium on Industrial Electronics*, Juni 2011, S. 625–630. doi: 10.1109/ISIE.2011.5984230.

[11] B. Saunders, G. Heins, und F. De Boer, „Framework for sensitivity analysis of industry algorithms for sensorless PMSM drives", in *AUPEC 2011*, Sep. 2011, S. 1–6.

Synchronization Stability of a Grid Forming Converter Under the Effect of Current Limit in Voltage Dips with VI Based Current Limiting Method: Analysis and Solution

Siam Hasan Khan, Markel Zubiaga Lazkano, Pedro Izurza, Alain Sanchez-Ruiz, Javier Cañas
Aceña, Joseba Arza
Ingeteam R&D Europe S. L.
Parque Tecnológico de Bizkaia, Edificio 106, 48170
Zamudio, Spain
Tel.: +34-944039600
E-Mail: siamhasan.khan@ingeteam.com, markel.zubiaga@ingeteam.com,
pedro.izurza@ingeteam.com, alain.sanchez@ingeteam.com, javier.canas@ingeteam.com,
joseba.arza@ingeteam.com
URL: https://www.ingeteam.com/

Acknowledgements

The work presented in this paper is part of the "MISIONES: FLEXENER" project which is founded by the Spanish institution CDTI and supported by the Ministry of Science and Innovation under the grant agreement No MIG-20201002.

Keywords

« Grid forming », « Droop control », « Current limiter », « Virtual impedance », « Synchronization stability », « Fault ride-through »

Abstract

This article deals with the synchronization stability in voltage dips of a Grid Forming Converter (GFC) operating with Virtual Impedance based current limiter. In the event of a grid fault, if the converter can't evacuate the active power while limiting the current, stability issues arise. In this paper this effect is analyzed, and a solution is proposed.

Introduction

Currently most of the inverter-based resources (IBRs) available in transmission and distribution power system are based on Grid Following Converter (GFL) and it causes reduction to the total electrical grid inertia. With the high share of IBRs, the grid stability is an important phenomenon and the GFL doesn't have the ability to ensure it. The solution towards grid stability is the GFC which can operate in a stiff grid voltage while providing voltage and frequency support in normal and highly variable conditions [1]. The exchange of current between the converter and grid depends on the impedance that connects them (For two voltage sources) and for large voltage difference, the current can be higher than the nominal value. This high current can occur under the highly variable conditions such as short circuits, phase jumps and with connection of large loads. These highly variable conditions are very critical for the GFC, and it requires the use of current limiter. Over the years different current limiting methods have been proposed for GFC based on the current saturation control algorithm [2, 3]. But with these control algorithms, while limiting the maximum current, the system can become unstable because of the wind-up in the outer power loops [4]. For the synchronization during the grid faults, some authors have proposed switching the control to a PLL based current control [5]. With this, the main drawback is to implement a fault detection algorithm along with the setting of the triggering condition [4].

A new innovative approach has been developed known as Virtual Impedance (VI) where the effect of the physical impedance is emulated when the current output exceeds the nominal value [6, 7, 8, 9]. Compared to the current saturation control algorithm, the VI based current limiting method provides better performance in terms of limiting the current but the stability of GFC is still an issue [6] [9]. One of the main goals of this article is to highlight this issue for which a GFC with a VI is used as a base case.

Regarding the transient and synchronization stability, different analysis have been found such as Adaptive droop gain with respect to the current magnitude [6], Adaptive droop gain with respect to the AC voltage amplitude [6], Enhanced fault recovery using dynamic damping [7], synchronization with the virtual power angle generated from VI [8] etc. Among the studies regarding the synchronization stability, another control method that has been found is the use of virtual active power as feedback that is calculated by the voltage and unsaturated current measured at the point of common coupling (PCC) [10]. Based on this control method, The authors in ref [10] has solved the synchronization stability issue within the grid fault event, but in terms of Low Voltage Ride Through (LVRT) capability, the information with reactive power injection has not been clearly defined. This same idea of the Virtual active power has been introduced in this article as one of the techniques that incorporates the VI for the synchronization stability. With this control method, both the synchronization stability and LVRT capability has been studied.

Both LVRT capability with adequate reactive power injection and synchronization stability during and after the voltage dips are essential grid code requirements [11, 12, 13]. Considering these requirements, in this article, a new control structure has been proposed. Based on the VI current limiting method, a control subsystem is added which is the adaptive active power set point in the Active Power and Frequency droop (P-f droop). The main goal of this article is to solve the synchronization stability during and after a voltage dip event. To do so, three methods are compared, which are, the base case of the GFC with only VI, the virtual active power feedback (presented in [10]) and the proposed adaptive active power set point. Besides the stability, the performance during the LVRT in terms of reactive power is also analysed.

The stability analysis is carried out using Matlab-Simulink. For the GFC converter model, a three-level neutral-point-clamped (NPC) is considered arbitrarily as it is one of the most widely used converter topology, However, this article mainly focuses on the control strategy of the converter and the implications towards the converter characteristics are not within the scope of this article.

In this article firstly a widely used grid forming control with a VI based current limitation method is presented as a Base Case. Then the Virtual Active Power Control structure along with the proposed solution of Adaptive Active Power Set Point are presented. Afterwards, some voltage dips are simulated to highlight the stability issues with the Base Case. In order to prove the correct performance of the proposed solution, some comparative analyses have been performed between the Virtual Active Power Control and the proposed solution. Finally, a conclusion has been made based on the performed analysis.

Base Case: GFC Control Configuration With VI

The GFC control mechanism that has been used in this article as a Base Case is based on the droop control with a VI. This control structure has been widely used in different literature [6, 7, 8, 9] and even some prototypes have been developed [14, 15]. The control algorithm performed in this control structure are in dq reference frame and the equations of the voltage and frequency droop control used are provided below.

$$V_{d\_droop} = V_d^* - n_q \left(Q - Q^* \right) \tag{1}$$
$$f_{droop} = f^* - m_p \left(P - P^* \right) \tag{2}$$

The output frequency and voltage from this droop equations are f_{droop} & V_{d_droop} where the frequency output is used to generate the angle θ_{GFC} of the converter. The voltage output of the droop controller is fed into the VI. The outputs of the VI are the d and q reference voltages which are used to regulate the voltages V_d and V_q at the PCC. These regulated voltages are then converted to abc reference frame where the angle θ_{GFC} is used for the conversion. The overall structure of the GFC control is provided below.

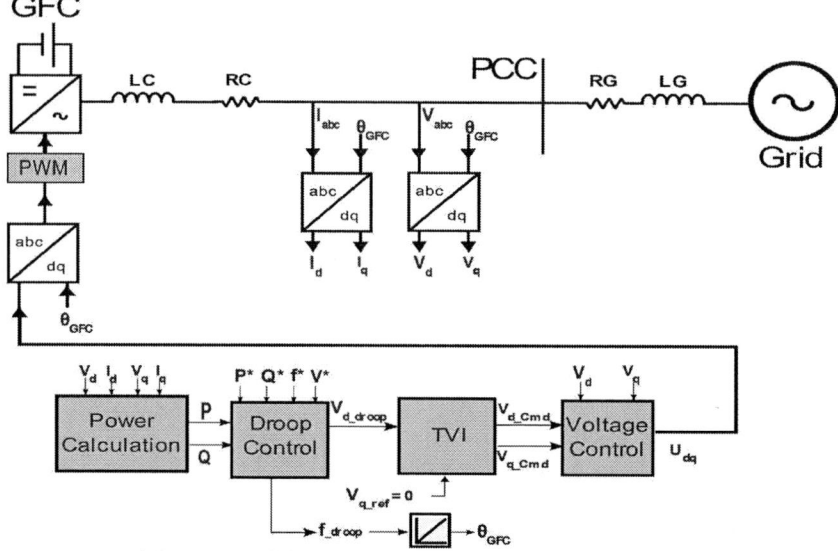

Fig. 1: Base case control Structure of the GFC

The voltage controllers used in this control method are PI based structures with anti-windup. The upper and lower saturation values for this anti-windup was provided as 1.2 PU and 0.85 PU respectively. The VI used in this control algorithm is to limit the fault current and it is only activated when the peak current of the converter current exceeds its threshold current value. The equation for calculating the peak current is provided below.

$$I_{Conv} = \sqrt{I_d^2 + I_q^2} \qquad (3)$$

The threshold current value is usually provided less than the maximum current value. The equation to determine the virtual impedance is provided below.

$$\begin{cases} X_{vi}\,(I_s) = K_{VI}\,|I_{Conv} - I_{thresh}|\ if\ I_{Conv} > I_{thresh} \\ \qquad X_{vi}\,(I_s) = 0, if\ I_{Conv} < I_{thresh} \\ \qquad R_{vi}\,(I_s) = X_{vi}/(\frac{\rho X}{R}) \end{cases} \qquad (4)$$

Here, K_{VI} is the reactance gain and $\frac{\rho X}{R}$ is the reactance to resistance ratio. With large value of $\frac{\rho X}{R}$, the VI has more inductive elements. In this control mechanism the values of K_{VI}, I_{thresh} and $\frac{\rho X}{R}$ are provided as 7.5, 1.1 PU and 4 respectively. After the inclusion of the VI, the d and q sequence voltage equation become like this,

$$V_{d\_cmd} = V_{d\_droop} - (I_d R_{vi}) + (I_q X_{vi}) \qquad (5)$$
$$V_{q\_cmd} = V_{q\_ref} - (I_q R_{vi}) - (I_d X_{vi}) \qquad (6)$$

Here, V_{d_cmd} and V_{q_cmd} are the voltages that are fed to the voltage controllers. V_{d_droop} and V_{q_ref} are the voltage references, where V_{d_droop} is coming from the droop and V_{q_ref}=0. $(I_d R_{vi}) - (I_q X_{vi})$ is the voltage drop due to the VI in the d sequence and $\left(I_q R_{vi}\right) + \left(I_q X_{vi}\right)$ is the voltage drop across the q sequence due to the VI.

The general parameters of the Converter and Grid used for the simulation purpose is provided in the table below.

Table I: General Parameters of the Converter and Grid

Parameter	Symbol	Unit	Value
Grid Voltage	V_{Grid}	Volt (Phase to Phase RMS)	600
Nominal Power	S_n	MW	2
Nominal Frequency	F_n	Hz	50
Grid Inductance	LG	mh	0.10313
Grid Resistance	RG	ohm	0.0022
Filter Inductance of the Converter	LC	mh	0.086
Filter resistance of the Converter	RC	ohm	0.01
Threshold Current	I_{thresh}	PU	1.1
Maximum Current Limit	I_Limit	PU	1.2
P-F Droop Constant	mp	-	0.02
Q-V Droop Constant	nq	-	0.05

The Base Case of the GFC with VI is not sufficient to provide post fault synchronization in the event of voltage dips [6] [9]. For this reason, the base case in this control structure is improved with two different control methods of virtual active power feedback and the proposed adaptive active power set point in the P-f droop. The detail description of these two control structures is provided below.

VI With Virtual Active Power Feedback (VI_P_Virt) [10]

The first method that has been used in this article, is the addition of virtual active power generated by the VI in the active power and frequency droop control which is explained in more detail in [10]. In the event of fault, the measured active power of the GFC includes the saturated current and with this the active power output becomes insensitive in the change of phase. For this reason, the problem of synchronization occurs. In order to counter this saturated current, the virtual active power can be used in the event of fault. The equation to calculate the Virtual Active Power is provided below.

$$P_{Virt} = \frac{V_{in\_Virtual} \times V_{out\_Virtual}}{X_{vi}} \sin(\delta) \qquad (7)$$

Here, $V_{in\_Virtual}$ is the modulas of V_{d_droop} and V_{q_ref}. $V_{out\_Virtual}$ is the modulas of V_{d_cmd} and V_{q_cmd} which are the outputs of the VI. X_{vi} is the inductance of the VI and δ is the angle difference between $V_{out\_Virtual}$ and $V_{in\_Virtual}$.

Proposed Control: VI With Adaptive Active Power Set Point in P-f Droop (VI_Adaptive_PSET)

From the GFC control perspective, the active power and frequency regulation of GFC can be made by changing the set point frequency or by changing the active power set point or by changing both. In order to provide fault ride-through capability by injecting more reactive power in the event of fault, a logic has been implemented in the active power set point of GFC. In this control method, the frequency set point of GFC is maintained as 1 PU and only the active power set point is changed. The equation for the active power set point is given as,

$$P_{Set_{Effective}} = P\_Set_{Gain} \times P_{Set} \tag{8}$$

Here, $P_{Set_{Effective}}$ is the effective active power set point that is given to the GFC. P_{Set} is the initial active power that is applied to the GFC. P_Set_{Gain} is the gain that varies based on the converter voltage and the condition for this is provided below.

$$\begin{cases} P_{Set_{Gain}} = 1, if\ V_{Conv_{Mod}} > 0.85\ PU \\ P_{Set_{Gain}} = 0.5 + \left(V_{Conv_{Mod}} - 0.85\right) \times K, if\ 0.5 < V_{Conv_{Mod}} \le 0.85 \\ P_{Set_{Gain}} = 0, if\ V_{Conv_{Mod}} \le 0.5\ PU \end{cases} \tag{9}$$

In equation (9) at 0.85 PU voltage, the value of $P_{Set_{Gain}}$ is considered as 0.5. Based on this the slope for the linear equation is found out to be, K= (0.5-0) / (0.85-0.5) = 1.4286.

Synchronization Assessment of the GFC During Voltage Dips

The IBRs should be connected to the grid in the event of voltage dips by providing voltage support by means of reactive power injection [16, 17, 18]. For that, the evacuation of active power during grid faults is an important factor for IBRs [11]. In general, this active power evacuation depends on many factors such as, the fault duration, fault depth, pre fault operating point of active power, Total system inertia etc [19]. This active power evacuation has a direct impact on the GFC in terms of both protection and stability as the GFC control uses the active power to synchronize with the grid.

It is well known fact that, in order to have stable operation, the phase angle difference between different busbars should not exceed 90° and the preferable phase difference is suggested to be ±30° [20]. During a grid fault event pole slipping may occur where the power generating unit loses synchronism to the transmission or distribution system it is connected to [21]. The main reason of the pole slipping in GFC is found out to be the P-F droop characteristic during fault [9]. If during a voltage dip event, the active power cannot be evacuated, the angle difference between the converter and the grid starts to increase significantly. For this, the whole system may lose synchronization and the angle difference will be found at the next stable regions where the equilibrium point lies at the ±2π periods [22].

In order to highlight these problems of active power evacuation, pole slipping and de-synchronization of the GFC, the base case with VI current limiting method is tested with 50% voltage dip for different time duration and with different active power generation point. The results are provided below.

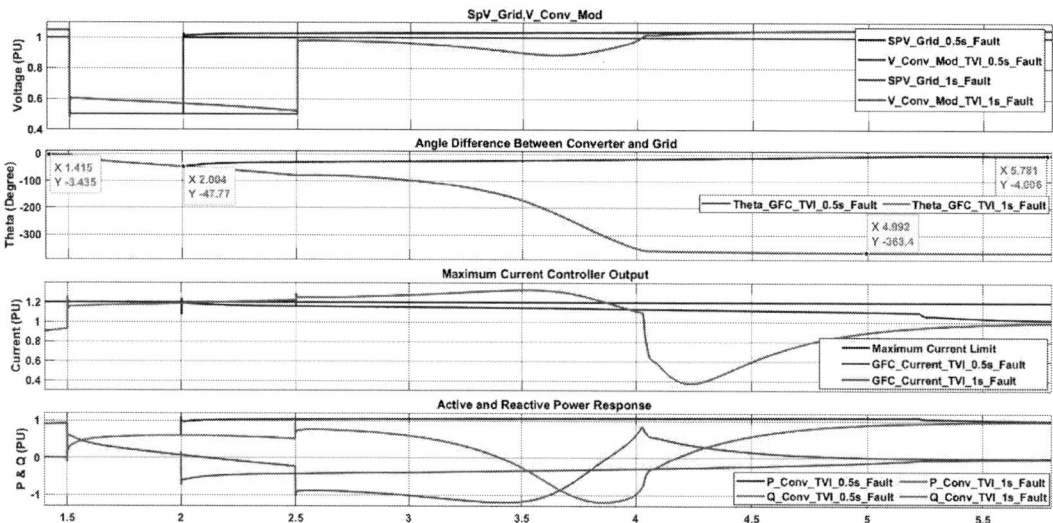

Fig. 2: Output of the Base Case GFC with VI for 500 ms and 1s Time Faults while GFC initially operating at P=1 & Q=0

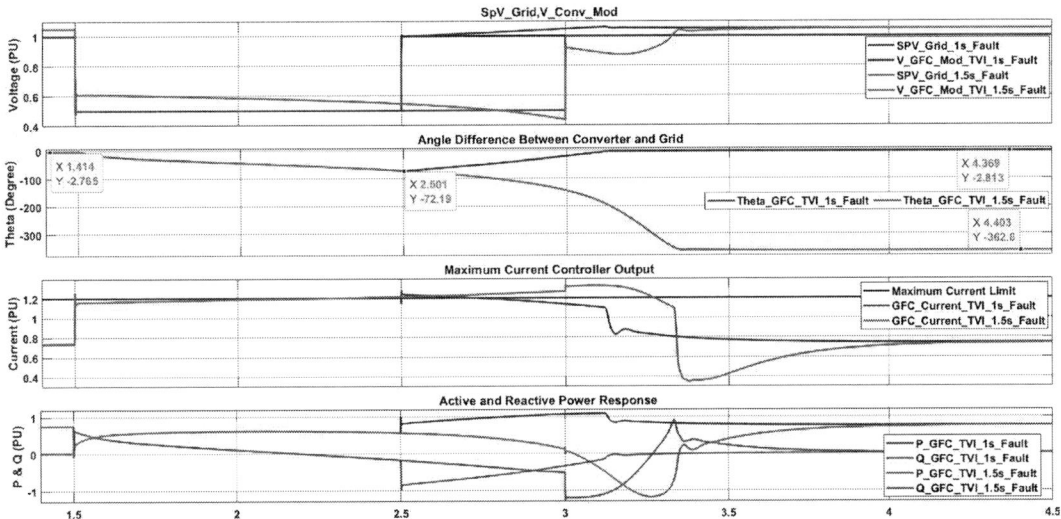

Fig. 3: Output of the Base Case GFC with VI for 1s and 1.5s Time Faults while GFC initially operating at P=0.75 & Q=0

From Fig. 2 it can be observed that, for 1s fault where GFC is operating at its rated active power (P=1 & Q=0), the VI operation for current limitation doesn't work properly as the active power cannot be evacuated and the impossibility to evacuate the required active power makes the de-synchronization process for the GFC. The angle difference with the grid surpasses 90° and finds another stable point. Similar to this, in Fig. 3 with 1.5 seconds of fault the de-synchronization process and the failure of the VI in terms of maximum current limitation are observed. From Fig. 2 with 500 ms fault and from Fig. 3 with 1 second of fault, it can be observed that the VI operation for maximum current limitation works (as the stability limit is not reached) but the post fault synchronization after the fault is very slow.

From the performed analysis it can be concluded that the use of a virtual impedance for limiting the current fails to provide LVRT capability as the angle stability is not guaranteed. Besides that, the reactive power injection is not adequate either. After an initial surge of reactive power at the beginning, decreases to zero. If the fault is long enough it also can reach negative values.

Along with the pre fault operating point of active power and the severity of the fault, the time duration for post fault resynchronization and desynchronization of GFC also varies depending on this emulated inertia [19]. Within the performed analysis, both the resynchronization and desynchronization time after the fault is more than 1 second which is not desirable in many grid codes [11]. So, it can be concluded that, the Base case only with VI cannot provide the LVRT capability and the synchronization stability.

Comparative Simulation Results of LVRT Capability and Synchronization Stability

Now a comparative analysis has been performed with the VI_P_VIRT control and proposed VI_Adaptive_PSET control in terms of LVRT capability and synchronization stability where the simulations are performed with a voltage dip of 50% for 1 second with GFC operating in its rated power (P=1 PU, Q=0 PU). The simulation results are provided below.

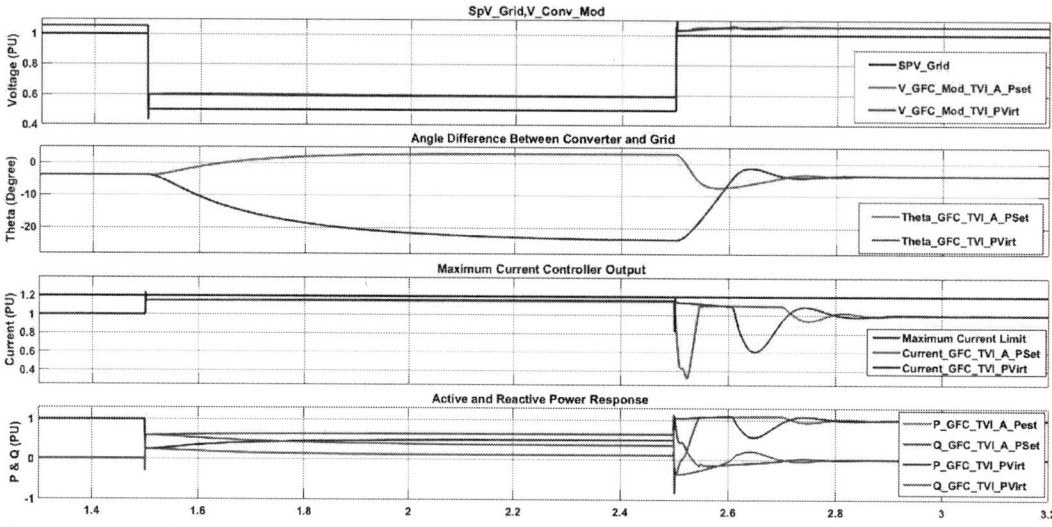

Fig. 4: Comparative results of the VI_P_Virt & VI_Adaptive_PSET for GFC (P=1, Q=0)

From the comparative results seen from Fig. 4 it can be concluded that the issue with the maximum current limitation and post fault synchronization seen with the base case (only the VI) is solved with both VI_P_VIRT and VI_Adaptive_PSET control. In terms of grid synchronization, the performance of the VI_Adaptive_PSET is better as it provides less angle difference between the GFC and the grid. Also, VI_Adaptive_PSET method provides more reactive power injection during faults. Therefore, with VI_Adaptive_PSET, both the LVRT capability and synchronization stability is achieved during the voltage dips.

Testing of the Proposed Solution with Different Level of Voltage Dips

It has been found that, the time GFC takes to return to its normal operation after the voltage dips is proportional to the severity of the event [23]. The dipper the voltage dip, the bigger will be the amount of curtailed active power. So, the depth of the fault is an important parameter that determines the stability. In order to validate this statement and to check the good performance of the proposed solution for all possible voltage depths, some simulations have been performed for a short period of time (140 ms) with different depth of voltage dips. The simulation results are provided below.

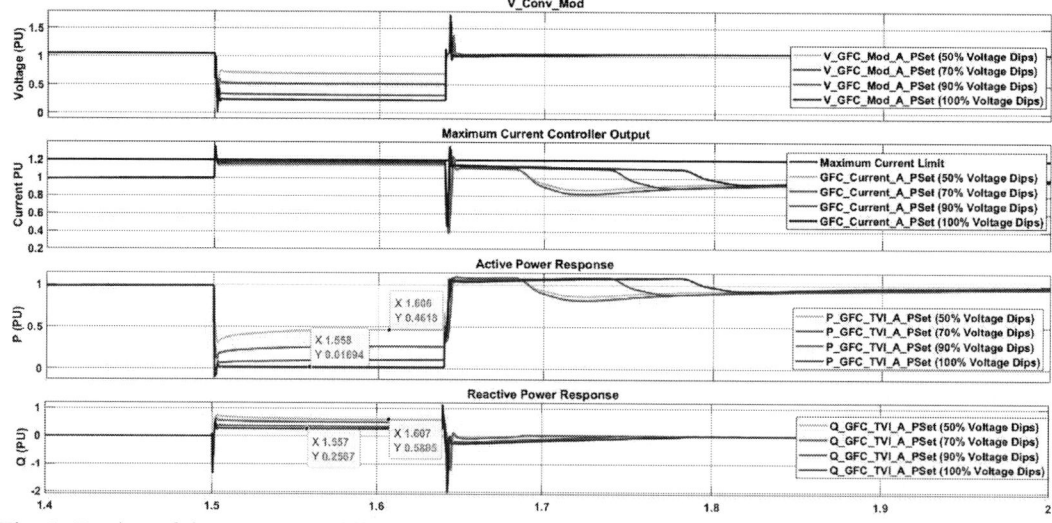

Fig. 5: Testing of the LVRT capability of GFM with VI_Adaptive_PSET control in different level of voltage dips for 140 ms (GFC initially operating at P=1, Q=0)

From Fig. 5, it can be observed that for all the voltage dips performed, the GFC quickly enters into the current limiting operation and in all cases the reactive power injection is significant, supporting the grid voltage. Although the transition from normal operation to the current limiting method is similar for all the voltage dip events, the transition from current limiting method to the normal operation at the end of voltage dip varies depending on the curtailment active power

Conclusion

The VI solves the current limiting problem, but the stability at some operation points was still an issue. For instance, when the active power cannot be evacuated, as analyzed in Fig. 2 & Fig. 3. To solve this issue many authors have provide new control methods. From all of them, one of the most promising from the authors point of view was the use of Virtual Active Power as feedback in the P-F Droop [10]. This control solved the stability issue, but as seen from the comparative results in Fig. 4, the voltage support during faults was still missing. From the performed comparative results in Fig. 4 & Fig. 5, the proposed Adaptive Active Power Set Point solution is proven to be effective for solving both issues, but not all the current limitation cases were considered. The effects with phase jump and rate of change of frequency (RoCoF) in high frequency change has not been examined for the current limitation in this article. This control method should be tested against those grid fault events also.

References

[1] Westman, Martin, and Ellen Nordén, "Modeling and comparative analysis of different grid-forming converter control concepts for very low inertia systems," 2020.

[2] Gkountaras Aris, Sibylle Dieckerhoff, and Tevfik Sezi, "Evaluation of current limiting methods for grid forming inverters in medium voltage microgrids," *In 2015 IEEE Energy Conversion Congress and Exposition (ECCE), pp. 1223-1230. IEEE, 2015*, 2015.

[3] Qoria, T., Cossart, Q., Li, C., Guillaud, X., Colas, F., Gruson, F. and Kestelyn, X, " WP3–Control and Operation of a Grid with 100% Converter-Based Devices. Deliverable 3.2: Local control and simulation tools for large transmission systems. MIGRATE Project," 2018.

[4] Paquette, Andrew D., and Deepak M. Divan, "Virtual impedance current limiting for inverters in microgrids with synchronous generators," in *IEEE Transactions on Industry Applications 51, no. 2 (2014): 1630-1638*, 2014.

[5] Cardozo, C.; Vernay, Y.; Denis, G.; Prevost, T.; Zubiaga, M.; Valera, J.J, "OSMOSE: Grid-Forming performance assessment within multiservice storage system connected to the transmission grid," *In Proceedings of the CIGRE Session 48, Paris, France, 25 August–3 September 2020.*

[6] Qoria, Taoufik, Francois Gruson, Frederic Colas, Guillaume Denis, Thibault Prevost, and Xavier Guillaud, "Critical clearing time determination and enhancement of grid-forming converters embedding virtual impedance as current limitation algorithm," in *IEEE Journal of Emerging and Selected Topics in Power Electronics 8, no. 2 (2019): 1050-1061*, 2019.

[7] Taul, Mads Graungaard, Xiongfei Wang, Pooya Davari, and Frede Blaabjerg, "Current limiting control with enhanced dynamics of grid-forming converters during fault conditions," in *IEEE Journal of Emerging and Selected Topics in Power Electronics 8, no. 2 (2019): 1062-1073.*

[8] Huang, L., Xin, H., Wang, Z., Zhang, L., Wu, K. and Hu, J., 2017, "Transient stability analysis and control design of droop-controlled voltage source converters considering current limitation," in *IEEE Transactions on Smart Grid, 10(1), pp.578-591.*

[9] Zubiaga, Markel, Carmen Cardozo, Thibault Prevost, Alain Sanchez-Ruiz, Eneko Olea, Pedro Izurza, Siam H. Khan, and Joseba Arza, "Enhanced TVI for Grid Forming VSC under Unbalanced Faults," *Energies,* vol. 14, no. 19, p. 6168, 2021.

[10] Paquette, A. D., "Power quality and inverter-generator interactions in microgrids," Doctoral dissertation, Georgia Institute of Technology, 2014.

[11] Kkuni, Kanakesh Vatta, and Guangya Yang, "Effects of current limit for grid forming converters on transient stability: analysis and solution," in *arXiv preprint arXiv:2106.13555 (2021).*

[12] Christiansen, Willi, and David T. Johnsen, "Analysis of requirements in selected Grid Codes," Prepared for Orsted-DTU Section of Electric Power Engineering, Technical University of Denmark (DTU), 2006.

[13] Bründlinger, Roland, "European codes & guidelines for the application of advanced grid support functions of inverters," In Sandia EPRI 2014 PV Systems Symposium-PV Distribution System Modeling Workshop, Santa Clara, CA. 2014.

[14] Fortmann, J., Pfeiffer, R., Haesen, E., van Hulle, F., Martin, F., Urdal, H. and Wachtel, S, "Fault-ride-through requirements for wind power plants in the ENTSO-E network code on requirements for generators," IET Renewable Power Generation 9, no. 1 (2015): 18-24.

[15] Denis, Guillaume, Thibault Prevost, Marie-Sophie Debry, Florent Xavier, Xavier Guillaud, and Andreas Menze, "The Migrate project: the challenges of operating a transmission grid with only inverter-based generation. A grid-forming control improvement with transient current-limiting control," in *IET Renewable Power Generation 12, no. 5 (2018): 523-529*, 2018.

[16] Pattabiraman, Dinesh, Robert H. Lasseter, and Thomas M. Jahns, "Transient stability modeling of droop-controlled grid-forming inverters with fault current limiting," *In 2020 IEEE Power & Energy Society General Meeting (PESGM), pp. 1-5. IEEE, 2020*.

[17] Paspatis, Alexandros G., and George C. Konstantopoulos, "Voltage support under grid faults with inherent current limitation for three-phase droop-controlled inverters," in *Energies 12, no. 6 (2019): 997*.

[18] Wang, Ren, Laijun Chen, Tianwen Zheng, and Shengwei Mei, "VSG-based adaptive droop control for frequency and active power regulation in the MTDC system," *CSEE Journal of Power and Energy Systems,* vol. 3, no. 3, pp. 260-268, 2017.

[19] QORIA, Taoufik, and Xavier Guillaud, "Grid-Forming Control Suitable for Large Power Transmission System Applications," *IEEE-TechRxiv,* vol. 14, p. 8, 2021.

[20] Boemer, Jens, "On stability of sustainable power systems: network fault response of transmission systems with very high penetration of distributed generation," Doctoral dissertation, Delft University of Technology, 2016.

[21] Raza, M., Peñalba, M. A., & Gomis-Bellmunt, O., "Short circuit analysis of an offshore AC network having multiple grid forming VSC-HVDC links," International Journal of Electrical Power & Energy Systems 102 (2018): 364-380.

[22] Ruberg, S, "MIGRATE–report on systemic issues," 2016.

[23] He X, Pan S, Geng H, "Transient Stability of Hybrid Power Systems Dominated by Different Types of Grid-Forming Devices," in *IEEE Transactions on Energy Conversion (2021)*.

[24] Energy Systems Integration Group's High Share of Inverter-Based Generation Task Force, "Grid-Forming Technology in Energy Systems Integration," ESIG-http://www.esig.energy/reports-briefs, Reston, VA, 2022.

Analytic calculation of touch and leakage currents of non-isolated EV chargers using a fast common mode calculation method and non-ideal passive component models

Christian Stutz, Sebastian Nielebock, Martin März
SIEMENS AG / FRIEDRICH-ALEXANDER-UNIVERSITY ERLANGEN-NUREMBERG
Erlangen, Germany
Tel.: +49 (1520) 3297094
E-Mail: christian.stutz@siemens.com; sebastian.nielebock@siemens.com;
martin.maerz@fau.de
URL: https://siemens.com; https://www.lee.tf.fau.de/

Keywords

«Battery charger», «Automotive component», «Charging infrastructure for EV's», «Failure modes», «Grid-connected inverter», «Non-isolated EV chargers», «Power converters for EV», «Simulation».

Abstract

Non-isolated charging systems suffer from tremendous touch currents due to the significant extension of the common mode coupling path by the electric vehicle. Since safety limits are very tough, the selection of converter topology, modulation scheme and filters with respect to touch current behavior is a very ambitious process. With the aim to ease integral system design, research is done for common mode analysis methods, eliminating the demand for long duration time domain simulations. A time saving method to analyze the touch current behavior in common mode domain is applied on a single stage charging system setup by a 3-level converter, filter system and a major coupling path. Furthermore, this paper improves that time-saving failure current prediction method by including the influence of non-ideal passive component models on the calculation result. All calculation results are compared to laboratory measurements. Border conditions from corresponding standards are used to define the required frequency range. Besides the touch current, which typically occurs during the failure condition of a broken protective earth conductor, the leakage current flowing in the protective earth conductor during normal operation is analyzed. Finally, this paper points out the influence of an internal common mode bypass via the filter system on the touch and leakage current.

1 Introduction

1.1 Common mode coupling path at non-isolated charging systems

Figure 1 shows a typical two-stage electric vehicle (EV) charging system comprising an AC-DC-Converter, its filter (1. Stage), a DC-DC-Converter (2. Stage) and the battery [1]. Capacitive coupling to the device case and chassis occurs due to parasitic effects ($C_{p,X}$) and filter capacitances (C_{CM}, $C_{Y,X}$).

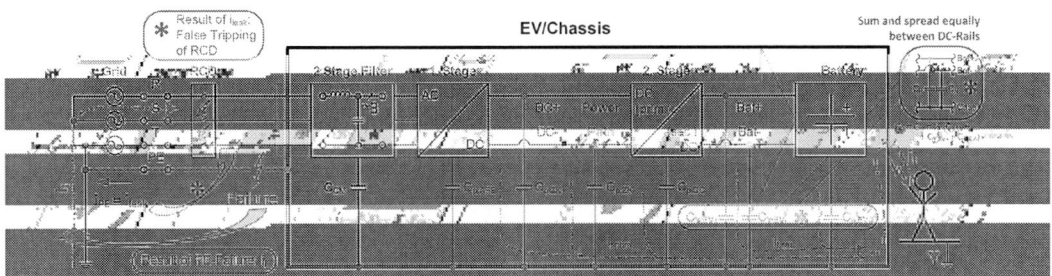

Figure 1: Example for a two-stage charging system

A major safety concern is the failure of the protective earth (PE) conductor. In this scenario, a person touching the chassis is coupled to all common mode voltage (CMV) sources within the power path referenced to ground, resulting in a certain touch current i_t flowing through the body. Isolated charging systems contain a transformer in the second stage, highly reducing the capacitive coupling due to the series connection of the transformer stray capacitance and the capacitive coupling of the charger secondary and EV (grey marking). Hence the major part of i_t equals $i_{t,prim}$ (orange marking). The sum of all secondary coupling capacitors, equally spread between both DC-rails and PE is represented by C_P and C_N (purple marking). Non-isolated chargers allow to omit the transformer with the advantage of volume, weight and cost reduction and an improvement in efficiency. On the other hand, C_P and C_N are added to the touch current path, hence i_t equals the sum of $i_{t,prim}$ and $i_{t,sec}$. Since the values of C_P and C_N may reach several μF, they will become the major coupling path and a significant increase of touch currents will occur [1]. Touch currents are limited to 3.5 mA at any operation scenario for Class 1 equipment by corresponding standards [10].

Beside the failure scenario of a broken PE conductor, the leakage current i_{leak} flowing through the PE conductor under normal operation is another important safety concern, since i_{leak} may lead to false tripping of the residual current device (RCD) at the supply side (grid side). As well as touch currents under fault conditions, leakage currents i_{leak} in the PE conductor at normal operation will increase in non-isolated charging systems since the coupling path is extended by C_P and C_N. The CMV driving these touch and leakage currents is influenced by the converter topology, DC link voltage, switching frequency, PWM, passive components and parasitic elements. This leads to a tremendous amount of combinations possible for system setup influencing i_t and i_{leak}. Hence, selecting and evaluating touch current behavior of certain combinations becomes a challenging process. With the aim to speed up that process and support integral system design, research is done to implement a workflow allowing flexibility in comparing different combinations and keeping time and simulation effort to a minimum.

1.2 Review on methods to analyze touch current behavior

Since touch and leakage currents are common mode (CM) currents, literature research is done regarding evaluation methods for CM behavior. In general, there are two main approaches within the results found: Either the system is transformed into CM equivalent circuits and results are analyzed within this reduced complexity (Group 1) or CM behavior is investigated by a full system simulation (Group 2). Figure 2 displays corresponding examples.

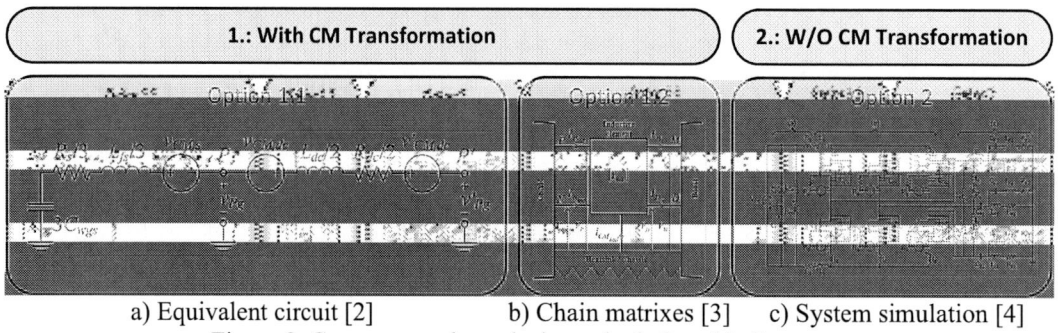

a) Equivalent circuit [2]   b) Chain matrixes [3]   c) System simulation [4]

Figure 2: Common mode analysis methods found in literature

When transforming the system into CM domain (Group 1), two options are common in literature. Option 1.1 transforms the system one-by-one into a CM equivalent circuit. This approach is commonly used for CM analysis [2], [3] and [4], as well as for filter design [5], [6] and [7]. This method allows full flexibility in terms of circuit design, but complexity of its transfer function corresponds with the nesting of the circuit. With Option 1.2, the system is described with the help of chain matrices [3]. Each matrix describes a certain subsystem, for example a CM choke or a capacitive DM filter, with the usage of the quadrupole theory. The approach has its strength in daisy chaining several chain matrices. Furthermore, this method allows to cover mode conversion effects. On the other hand, it requires a high initial mathematical effort. Option 2 (Group 2) uses a full system time domain simulation. It has its time demand in setting up the simulation model rather than in handling circuit equations. Option 2 allows full flexibility in terms of circuit design but has a high time demand in creating a model containing PWM, current regulation and grid synchronization. The precision of the result corresponds with the quality of the models used and the time step setting of the simulator. Hence, a single simulation run has a large time duration, especially when precise results are expected.

1.3 Selection of a touch current analysis method and creating the approach

Formulating the condition that time reduction and flexibility for CM analysis of certain system combinations are a priority, Option 1.1 becomes the method of choice. A workflow is implemented, calculating the CM behavior in time and frequency domain. The major steps of this workflow are summarized by Figure 3.

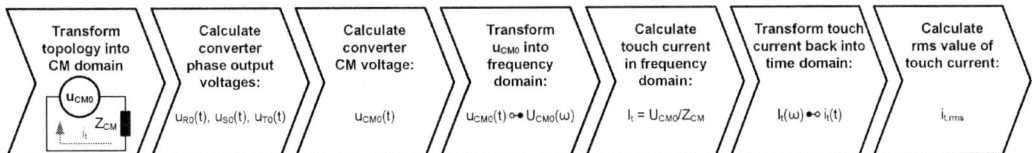

Figure 3: Workflow of the method for fast touch current prediction

Depending on the system specification, the second stage of a non-isolated charging system can be neglected. With respect to this fact, the implementation and validation of the workflow shown in Figure 3 is done for a single stage charging system working in normal operation and under the fault condition of a broken PE conductor. Since the literature cited focuses on filter design and CM analysis for electromagnetic compatibility (often in the context of drive applications), the aim of this work is to apply this approach on electrical safety topics of non-isolated charging systems, e.g. touch and leakage currents in the range of kHz. Furthermore, the research range is extended by investigating the influence of non-ideal passive component models on the calculation results.

2 Experimental setup

Figure 4 displays a 3-level neutral point clamped (NPC) converter used for the experimental validation. The converter operates with the switching frequency f_{sw} = 48 kHz, the PWM includes third harmonic injection. The rated DC-Link-Voltage is 700 V and the load resistor is set to a 11 kW operation point. The converter has a filter including two differential mode (DM) stages (C_1 & L_1 and C_2 & L_2,), as well as a CM choke L_{CM} and an additional CM path between C_2 and case/ground (G) via the CM capacitor C_{CM}. The switches are silicon carbide (SiC) half bridge modules. The Line Impedance Stabilization Network (LISN) is model "NNLK-8121" from Schwarzbeck. It is used to ensure a defined situation at the supply side.

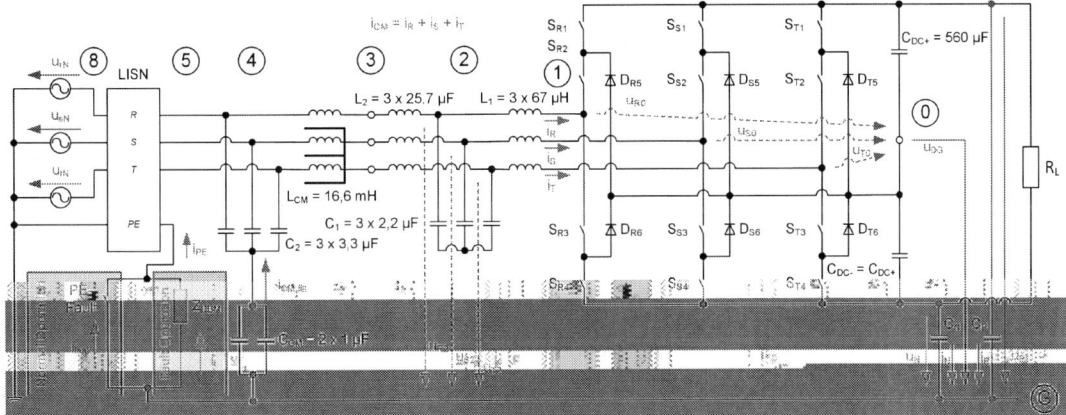

Figure 4: 3-Level NPC converter including its filter stages and a PE fault stimulation

3 Details on calculation method for common mode equivalent circuit and simulation

For the development of a single-phase CM equivalent circuit, definitions based on [2] are used:

$$u_{CM0} = \frac{u_{R0} + u_{S0} + u_{T0}}{3} \qquad (1)$$

$$i_{CM} = i_R + i_S + i_T. \qquad (2)$$

The following steps are executed to obtain the CM equivalent circuit of the experimental setup:
- Divide the entire topology into small sections.
- Define current and voltage expressions for each phase of the section.
- Transform equations for each section into CM domain.
- Daisy chain all sections to a complete CM equivalent circuit.

Figure 5 displays the entire CM equivalent circuit of Figure 4. The numbering of sections corresponds between both figures. A calculation example (green marking) is done for section 2-1-0-G-2 as follows.

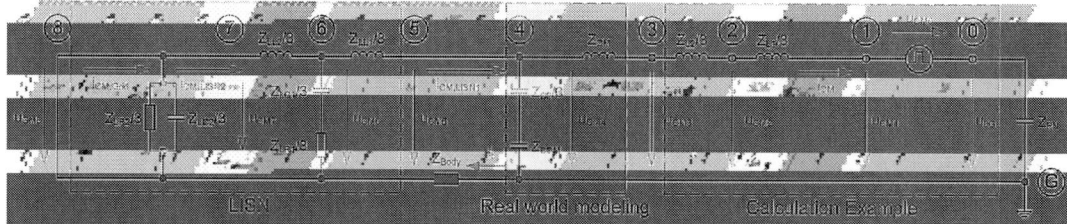

Figure 5: CM equivalent circuit of 3-level NPC converter

For each phase R, S and T of the setup shown in Figure 4, a voltage loop of this section is formulated:

$$Z_{L1} \cdot i_R + u_{R0} + u_{0G} - u_{R2} = 0 \qquad (3)$$
$$Z_{L1} \cdot i_S + u_{S0} + u_{0G} - u_{S2} = 0 \qquad (4)$$
$$Z_{L1} \cdot i_T + u_{T0} + u_{0G} - u_{T2} = 0. \qquad (5)$$

The summation of the three loops (3) to (5) equals to:

$$Z_{L1} \cdot (i_R + i_S + i_T) + (u_{R0} + u_{S0} + u_{T0}) + 3u_{0G} - (u_{R2} + u_{S2} + u_{T2}) = 0. \qquad (6)$$

Dividing (6) by 3 allows usage of the common mode definitions given in (1) and (2):

$$\frac{Z_{L1}}{3} \cdot i_{CM} + u_{CM0} + u_{0G} = u_{CM2}. \qquad (7)$$

Equation (7) shows, that the impedance Z_{L1} of the first DM choke L_1 has a contribution of one third of a single element value within the CM domain. The CMV driven by the converter is u_{CM0}. The connection point for the transformation of the next section on the ac-side is the CMV u_{CM2} at Point 2 referenced to ground. The transformation of the DC side can be daisy chained via the DC midpoint voltage u_{0G} referenced to ground. The entire CM equivalent circuit is achieved by transforming and daisy chaining the sections step-by-step. Their connection points are formulated as follows:

$$\frac{Z_{L2}}{3} \cdot i_{CM} + u_{CM2} = j\omega \frac{L_2}{3} \cdot i_{CM} + u_{CM2} = u_{CM3} \qquad (8)$$

$$Z_{CM} \cdot i_{CM} + u_{CM3} = j\omega L_{CM} \cdot i_{CM} + u_{CM3} = u_{CM4} \qquad (9)$$

$$\left(Z_{CCM} + \frac{Z_{C2}}{3}\right) \cdot \left(i_{CM,LISN1} - i_{CM}\right) = \left(\frac{1}{j\omega C_{CM}} + \frac{1}{j\omega 3 C_2}\right) \cdot \left(i_{CM,LISN1} - i_{CM}\right) = u_{CM4} \qquad (10)$$

$$Z_{Body} \cdot i_t + u_{CM4} = u_{CM5} \qquad (11)$$

$$\frac{Z_{LL1}}{3} \cdot i_{CM,LISN1} + u_{CM5} = j\omega \frac{L_{L1}}{3} \cdot i_{CM,LISN1} + u_{CM5} = u_{CM6} \qquad (12)$$

$$\left(\frac{Z_{LC1}}{3} + \frac{Z_{LR1}}{3}\right) \cdot \left(i_{CM,LISN2} - i_{CM,LISN1}\right) = \cdots$$
$$\cdots = \left(\frac{1}{j\omega 3 C_{L1}} + \frac{R_{L1}}{3}\right) \cdot \left(i_{CM,LISN2} - i_{CM,LISN1}\right) = u_{CM6} \qquad (13)$$

$$\frac{Z_{LL2}}{3} \cdot i_{CM,LISN2} + u_{CM6} = j\omega \frac{L_{L2}}{3} \cdot i_{CM,LISN2} + u_{CM6} = u_{CM7} \qquad (14)$$

$$\frac{u_{rN} + u_{sN} + u_{tN}}{3} = u_{CM8} = 0 \qquad (15)$$

$$i_{CM,Grid} = i_{CM,LISN2} \qquad (16)$$

$$Z_{PN} = \frac{1}{j\omega(C_P + C_N)}. \qquad (17)$$

According to (15), the grid is assumed to be ideal. Hence, the filter elements C_{L2} and R_{L2} of the LISN are shorted. Furthermore, L_{CM} is assumed to have ideal coupling. The resulting CM current i_{CM} at normal operation is achieved by applying Ohm's Law on the CMV u_{CM0} produced by the converter and the resulting CM impedance $Z_{CM,res}$. The touch current i_t is calculated using the current divider calculation rule, since i_t is the fraction of i_{CM} not flowing through the filter structure C_{CM} and C_2:

$$i_{cm} = -\frac{u_{CM0}}{Z_{CM,res}} \qquad (18)$$

$$i_t = \frac{\frac{Z_{C2}}{3} + Z_{CCM}}{\frac{Z_{C2}}{3} + Z_{CCM} + Z_{Body} + \left(\frac{Z_{LL2}}{3} \parallel \left(\frac{Z_{LC1}}{3} + \frac{Z_{LR1}}{3}\right)\right) + \frac{Z_{LL1}}{3}} \cdot i_{cm} = -Z_{CM,t} \cdot \frac{u_{CM0}}{Z_{CM,res}}. \qquad (19)$$

The calculation software "Maple" is used for factorizing the resulting impedance equations for $Z_{CM,res}$ and $Z_{CM,t}$. Matlab is used to compute all other calculation steps represented in Figure 3 and achieve results in frequency and time domain. The CMV pattern is created by sine-triangle comparison of three reference signals including third harmonic injection. The third harmonic amplitude is derived from the fundamental by multiplication with the factor 0.16667.

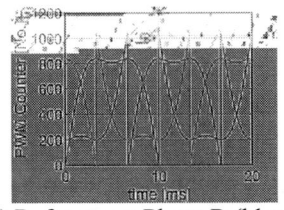

a) References: Phase R (blue),
Phase S (red), Phase T (yellow)

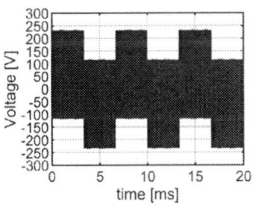

b) $u_{CM0} = f(t)$

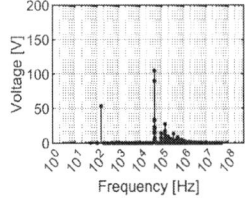

c) FFT of u_{CM0}

Figure 6: Creation of the converter common mode voltage u_{CM0}

The PWM FPGA code of the experimental setup is implemented one by one within the Matlab calculation workflow: The sampling frequency of the refence and triangle signals is 100 MHz. The base of the triangle signal is a 11 Bit counter. Counter and reference values are positive integers, hence negative values are flipped. Figure 6 shows the three-phase reference values (a), the resulting CMV (b) achieved by applying (1) on the phase output voltages (u_{R0}, u_{S0}, u_{T0}) of the converter and the FFT of the CMV (c) with its major components at the third harmonic, f_{SW} and its multiples.

4 Boundary condition: Human body model

For failure scenarios (break of PE conductor) resulting in a touch current i_t flowing through a person while touching the chassis, a human body model (HBM) is defined by corresponding standards [11], [12]. The equivalent circuit and its parameters can be found in Figure 7 a) and b). R_S and C_S are representing the skin, while R_B models the body itself. Since the human body is less sensitive for high frequency currents regarding startle response and the ability to let go, a first order filter utilized by R_1 and C_1 is added in parallel to R_B. For charging applications, evaluation of i_t is done on the basis of the peak value of the rated touch current $i_{t,rated}$, rather than on the raw touch current $i_{t,raw}$. The value of $i_{t,rated}$ is measured via the voltage u_2 of C_1 divided by 500 Ω [12].

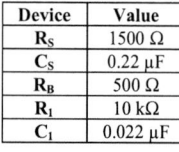

Device	Value
R_S	1500 Ω
C_S	0.22 μF
R_B	500 Ω
R_1	10 kΩ
C_1	0.022 μF

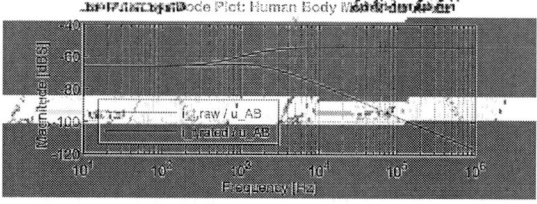

a) Equivalent circuit   b) Parameters   c) Impedance over frequency

Figure 7: Human body model [10] and [12]

In the bode plot of $i_{t,rated}$ shown in Figure 7 c), it can be clearly seen that the corner frequency is located between 2 kHz and 3 kHz. The bode plot finishes at 1 MHz, since the human body model is defined up to 1 MHz. This is the upper frequency limit for electrical safety defined in [12]. For further investigations of non-ideal modeling, two intermediate conclusions can be drawn:
- Frequency-dependent effects are relevant only in the frequency range from DC up to 1 MHz. Since [12] rates 1 MHz as corner frequency between electrical safety and electromagnetic compatibility, 1 MHz is used as upper frequency limit for investigations on fault operation (i_t) and normal operation (i_{leak}).
- Effects from 2 kHz onwards have a decreasing impact on $i_{t,rated}$ with a slope of -20dB/dec.

5 Passive components: Impedance curves, non-ideal models and expectations on their influence

The impedance curves over frequency of all passive components are measured with the impedance analyzer Bode 100. A summary of the measurement results is given in Figure 8.

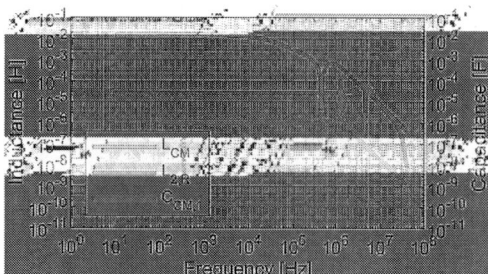

Figure 8: Passive component impedance curves

Table 1: Non-ideal models and their parameters

Element	L_{DM1}	L_{DM2}	C_{DM2}	C_{CM}
L / C	67 μH	26 μH	2.9 μF	0.9 μF
C_{par} / L_{ESL}	16 pF	45 pF	16 nH	17 nH
R_{ESR} [mΩ]	≤ 0.1 Ω	≤ 0.1 Ω	145 mΩ	159 mΩ
f_{res}	4.8 MHz	4.6 MHz	0.8 MHz	1.3 MHz
Equivalent Circuit				

Due to the limitation of space, only one impedance curve per element type (L_{CM}, C_{CM}, L_2) is presented. In general, the DM-inductors and all capacitors show very similar curves: Their impedance is very stable up to the resonance point and only small variations in resonance frequencies and nominal values occur. Corresponding values can be found in Table 1.

Equivalent circuit models covering the behavior of all DM inductances and all capacitors are taken from [8] and displayed at the last row of Table 1. The resonant points of both DM inductances are located at the area between 4 and 5 MHz, which is outside of the definition of the HBM, ending up at 1 MHz. Due to that reason, the influence of a non-ideal DM inductance model on the touch and leakage current is considered as neglectable.

The resonance points of the capacitors are located at the area of 1 MHz, which is the upper definition limit of the scope of this work. Hence the effect of non-ideal capacitor models on the calculation results is considered to be small. The DM capacitors C_1 are omitted since they have no contribution to the CM domain.

The impedance of the CM choke is unstable and changes over frequency, even in the range of kHz. Since the impedance of L_{CM} changes at the frequency range of f_{SW} and their multiples, a noticeable impact on the touch and leakage current prediction is expected and the focus in the results section is set on this device. For ease of implementation, the L_{CM} characteristic is implemented as a look-up table (LUT).

6 Results

The following sections compare prediction results using ideal and non-ideal models and laboratory measurement results. Since L_{CM} is expected to have the highest impact on the prediction result, the non-ideal prediction results are split into two groups using only a non-ideal model of L_{CM} and using non-ideal models of L_{CM}, C_{CM} and C_2. Due to the boundary conditions of Section 4, the frequency domain plots are cut off at 1 MHz. The converter operates with a DC-Link Voltage of $V_{DC} = 700$ V and $P_{charge} = 11$ kW. Due to the limitation of space, only one set point with $C_P = C_N = 1$ µF is presented. Results investigating the variation of C_P and C_N are given in Table 2.

6.1 Fault operation: Effects of non-ideal L_{CM}, C_{CM} and C_2 on $i_{PE} = i_t$ and i_{CM}

Figure 9 shows a comparison for the rated touch current $i_{t,rated}$ through the HBM. All calculation results only have one major component at 150 Hz due to the third harmonic injection of the PWM.

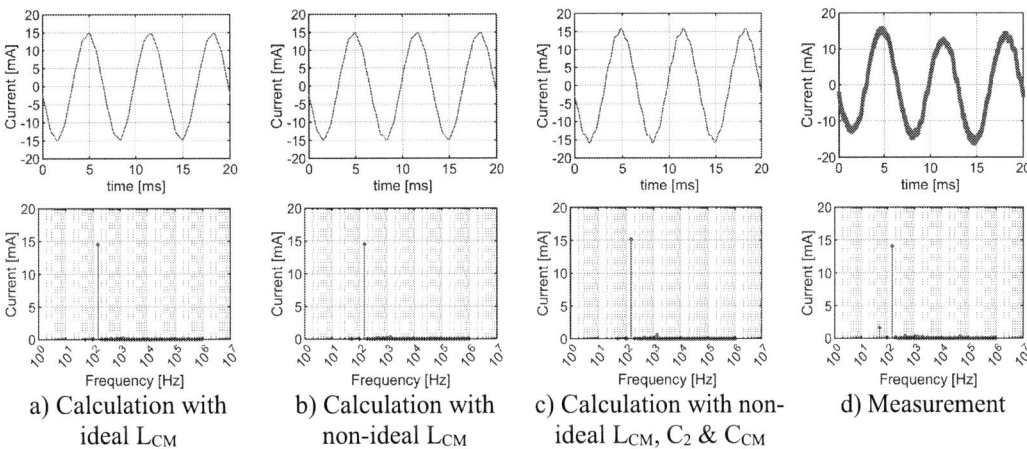

a) Calculation with ideal L_{CM} b) Calculation with non-ideal L_{CM} c) Calculation with non-ideal L_{CM}, C_2 & C_{CM} d) Measurement

Figure 9: Prediction and measurement of $i_{t,rated}$ through HBM, converter working under PE-Fault

The measurement matches almost perfectly with the calculations: There is one major component with an amplitude of approximately 15 mA at 150 Hz. But there is no sensitivity of $i_{t,rated}$ regarding the modeling of L_{CM} within the calculation. The usage of non-ideal capacitor models cause a small increase of $i_{t,rated}$. The non-ideal impedance of C_2 and C_{CM} slightly increases, since their actual values are smaller than their ideal nominal value, leading to a small shift of current from the path via C_2 and C_{CM} to the parallel path via the HBM. An additional unexpected component at 50 Hz occurs within the measurement. Since the unused neutral conductor port at the LISN shows an AC offset with several volts amplitude and 50 Hz frequency against PE, an asymmetry on the grid side is assumed to cause the additional 50Hz content in this and all other measurements.

The waveforms of the time domain plots of all other results only show small changes: The 150 Hz sine wave remains as a basis, only the amplitude and the width of the noise band varies in dependency of the operation point. Since space is limited only FFT plots are shown.

Figure 10 compares the CM current i_{CM} at the converter input between the prediction method with ideal and non-ideal models and the laboratory measurement result. Two basic trends apply for all results: All amplitudes increase significantly compared to Figure 9 and additional components at $f_{SW} = 48$ kHz and their multiples occur.

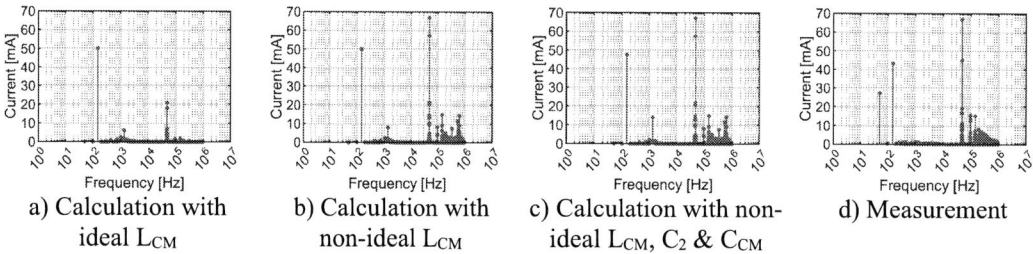

| a) Calculation with ideal L_{CM} | b) Calculation with non-ideal L_{CM} | c) Calculation with non-ideal L_{CM}, C_2 & C_{CM} | d) Measurement |

Figure 10: Prediction and measurement of i_{CM}, converter working under PE-Fault

When comparing the calculation results, it can be clearly seen that there is an increase of the f_{SW} component by a factor of 3 when using the non-ideal model of L_{CM}, whereas the 150 Hz component shows no significant change. This corresponds with the decreasing inductance of L_{CM}, starting with approximately 15 mH and decreasing to 5 mH at f_{SW}. The impedance curve of L_{CM} can be found in Figure 8. The 150 Hz component shows no change regarding to the L_{CM} model, since L_{CM} has a constant inductance at that frequency range. On the other hand, the 150 Hz component decreases slightly when using non-ideal capacitors models due to the slight increase of the resulting CM impedance. Adding the measurement results to the comparison it can be clearly seen that there is a good match between measurement and prediction when using non-ideal models regarding the major components of i_{CM} at 150 Hz and f_{SW}, whereas the prediction with the ideal modeling is inaccurate at f_{SW}. The major improvement within the calculation results is due the non-ideal L_{CM} model. On the other side, the non-ideal capacitors models have a small influence on the calculation result.

A comparison of i_{CM} to the previous results of $i_{t,rated}$ in Figure 9 shows that there is a reduction of the third harmonic of $i_{t,rated}$ by more than a factor of 3. Furthermore, $i_{t,rated}$ didn't show any components at f_{SW} and its multiples. In other words, during failure scenarios, the parallel internal CM path through C_{CM} and C_{DM2} is a very efficient current divider, keeping all high frequency content and a fraction of the third harmonic within the internal common mode loop rather than allowing that current to flow as i_t through the PE conductor and the HBM. Since the touch current is reduced to a problem of the third harmonic in this setup, further improvement could be achieved by minimizing or eliminating the third harmonic injection to pass the standard limit for the rated touch current of 3.5 mA.

6.2 Normal operation: Effects of L_{CM}, C_{CM} and C_2 on i_{CM} and $i_{PE} = i_{leak}$

The analysis is repeated during normal operation, hence the PE fault switch is closed (see Figure 4). Figure 11 displays the FFT plots of i_{CM}, achieved from calculation with ideal and non-ideal passive models and from laboratory measurement. When looking at the calculation results of Figure 11 a) to c), two major trends compared to fault operation in Figure 10 occur: First of all, the amplitude of the

150 Hz component increases significantly by a factor of 2, which can be explained by a decrease in the CM path impedance of the PE conductor due to the elimination of the HBM. Secondly, the influence of the non-ideal L_{CM} model appears the same as at failure mode: The f_{SW} component increases significantly within the calculation results, whereas the 150 Hz component remains the same.

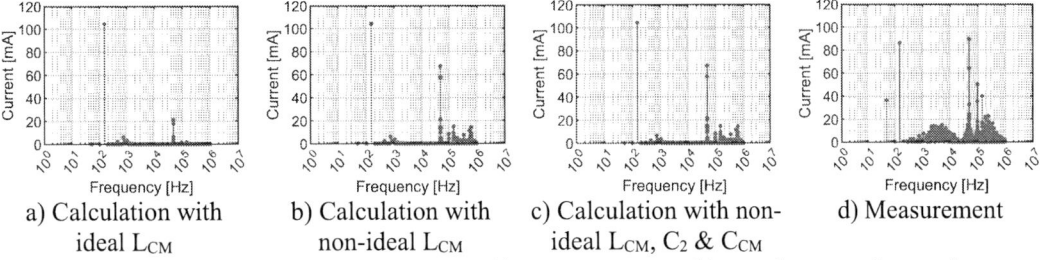

a) Calculation with ideal L_{CM} b) Calculation with non-ideal L_{CM} c) Calculation with non-ideal L_{CM}, C_2 & C_{CM} d) Measurement

Figure 11: Prediction and measurement of i_{CM}, converter working under normal operation

The non-ideal capacitor models have no visible effect on the calculation result. The calculation with the non-ideal models and the measurement show major components at 150 Hz and f_{SW} as well, but the amplitudes do not match perfectly.

The comparison of the PE conductor current i_{PE}, which is $i_{PE} = i_{leak}$ under normal operation condition, is displayed in Figure 12. The influence of the non-ideal L_{CM} model appears the same way as with i_{CM} in Figure 11: The f_{SW} component increases whereas the 150 Hz component remains at the same amplitude and the impact of the non-ideal capacitor models is minimal.

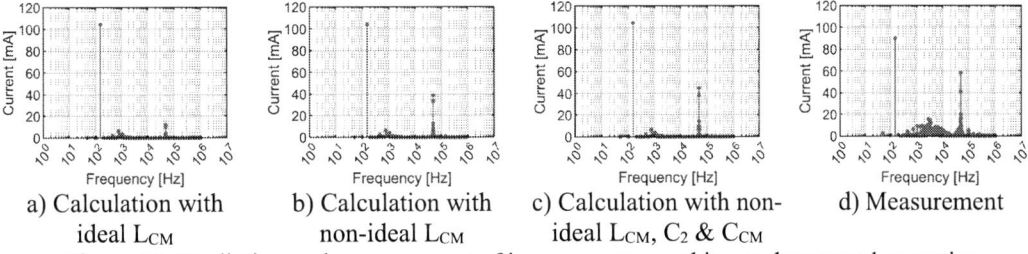

a) Calculation with ideal L_{CM} b) Calculation with non-ideal L_{CM} c) Calculation with non-ideal L_{CM}, C_2 & C_{CM} d) Measurement

Figure 12: Prediction and measurement of i_{PE}, converter working under normal operation

When comparing the frequency domains of i_{PE} in Figure 12 under normal operation and $i_{t,rated}$ under fault operation in Figure 9, several trends can be seen. First, the 150 Hz component increases by factor of 6 from 15 mA to approximately 105 mA. This can also be explained by the decrease of impedance of the ground loop due the elimination of the HBM. Secondly it can be clearly seen that the influence of the parallel CM path through C_{CM} and C_{DM2} decreases significantly under normal operation. Besides the 150 Hz component, i_{PE} has an additional major component at f_{SW}, whereas $i_{t,rated}$ is defined by the 150 Hz component itself. For normal operation, the CM path through the filter is less efficient and bypasses only parts of the high frequency CM currents from the ground loop.

For consumer households, it is assumed that the typical RCDs have a trip point of $i_{PE} \geq 30$ mA [9]. These RCDs provide protection against indirect touch and additional protection against direct touch [13]. Under this assumption, false tripping at normal operation is expected, since i_{PE} is above that limit by factor of 3 at 150 Hz. RCDs of type B and B+ have a decreasing sensitivity at higher frequencies. Their trip points at f_{SW} are at 8000 mA (Type B) and 330 mA (Typ B+) according to [13]. Therefore, false tripping due to the f_{SW} component is not expected.

Finally, Table 2 shows a comparison of RMS values between calculation and measurement results, when the coupling capacitor is increased. The calculations results are given for non-ideal models of L_{CM}, C_2 and C_{CM}. It can be seen that the dependency between the current in the PE conductor and the increasing capacitive coupling is significantly higher during normal operation (i_{leak}) than during a fault condition ($i_{t,rated}$). Furthermore, the difference between calculation and measurement is higher during normal operation. Both effects can be explained with the absence of the HBM and therefore the reduction of impedance of the ground loop during normal operation. Further research on the setup of the CM structure will be done with the aim of finding a CM structure allowing $i_{t,rated}$ being independent from C_P and C_N.

All calculation points have a difference between calculation and measurement in the range of single-digit values of mA, which underlines the precision of the chosen calculation approach.

Table 2: Comparison of RMS values between calculation using non-ideal models of L_{CM}, C_2, and C_{CM} and measurement

a) Comparison of $I_{t,rated,rms}$ during PE fault

C_P, C_N [nF]	I_{calc} [mA]	I_{meas} [mA]	ΔI [mA]
440	6.7	6.0	0.7
1000	10.7	9.3	1.4
3300	15.8	13.5	2.3

b) Comparison of $I_{PE,rms} = I_{leak,rms}$ during normal operation

C_P, C_N [nF]	I_{calc} [mA]	I_{meas} [mA]	ΔI [mA]
440	59.3	55.9	3.4
1000	87.7	97.5	-9.8
3300	267.8	259.0	8.8

7 Conclusion

This paper presents a fast and precise CM analysis method applied to electrical safety considerations of non-isolated EV chargers. The discrepancy of RMS values between measurement and calculation is in the range of single-digit mA. The runtime of the calculation for a certain setpoint is within a timeframe of 60-90 s. Both parameters underline the precision and speed of the chosen calculation approach.

For electrical safety considerations, a frequency limit of 1 MHz is defined in corresponding standards. This limit helps to reduce the number of passive elements considered for non-ideal component modeling. In this setup, the CM inductor was identified as an important element for non-ideal modeling due to its variable inductance over frequency in the range of kHz. The influence of non-ideal capacitor models was minimal.

For this given setup, the converter CM current i_{CM} turned out to be highly sensitive to the non-ideal model of L_{CM}, which is very important for filter design. On the other hand, the touch current $i_{t,rated}$ itself showed no sensitivity to detailed L_{CM} modeling. Further investigations will be done in an upcoming publication, considering whether the non-sensitivity of $i_{t,rated}$ for non-ideal passive component models can be generalized.

For failure scenarios, a parallel CM path (C_2, C_{CM}) within the filter system offers as a very powerful method to eliminate high frequency currents from the touch current path and to reduce i_t to a low frequency problem. The corresponding standard peak value limit for $i_{t,rated}$ is 3.5 mA. For this setup, a revision of the PWM (reducing or eliminating the third harmonic injection) is a practical approach to achieve the required standard.

Finally, this work clearly points out that the leakage current in the PE conductor under normal operation condition is another challenging situation of non-isolated charging systems. The 150 Hz component increases by factor six at normal operation compared to the touch current at fault condition. Furthermore, high frequency content is added to the spectrum of i_{leak}. False tripping of the RCD due to the tremendous 150 Hz component is expected.

References

[1] Y. Zhang et al.: Leakage Current Issue of Non-Isolated Integrated Chargers for Electric Vehicles, 2018 IEEE Energy Conversion Congress and Exposition (ECCE), Portland, OR, 2018, pp. 1221-1227, doi: 10.1109/ECCE.2018.8558133.

[2] A. D. Brovont, S. D. Pekarek: Equivalent Circuits for Common-Mode Analysis of Naval Power Systems, 2015 IEEE Electric Ship Technologies Symposium (ESTS), pp. 245-250, Jul. 2015, doi: 10.1109/ESTS.2015.7157897.

[3] C. Saber, D. Labrousse, B. Revol, A. Gascher: A Combined CM & DM Conducted EMI Modeling approach, Proc. of the 2017 International Symposium on Electromagnetic Compatibility - EMC EUROPE 2017, Sep. 2017, doi: 10.1109/EMCEurope.2017.8094616.

[4] D. Labrousse, B. Revol, C. Gautier, F. Costa: Fast Reconstitution Method (FRM) to Compute the Broadband Spectrum of Common Mode Conducted Disturbances, IEEE Transactions on Electromagnetic Compatibility, vol. 55, no. 2, pp. 248-256, Apr. 2013, doi: 10.1109/TEMC.2012.2219056.

[5] G. Mondal, J. Robinson and M. Finkenzeller: Modeling and Design of Common Mode and Differential Mode Filter for PWM Converters, 10th International Conference on Power Electronics and ECCE Asia (ICPE 2019-ECCE Asia), 2019, pp. 2191-2198, doi: 10.23919/ICPE2019-ECCEAsia42246.2019.8796878.

[6] D. O. Boillat, J. W. Kolar, J. Mühlethaler: Volume Minimization of the Main DM/CM EMI Filter Stage of a Bidirectional Three-Phase Three-Level PWM Rectifier System, 2013 IEEE Energy Conversion Congress and Exposition, Oct. 2013, pp. 2008-2019, doi: 10.1109/ECCE.2013.6646954.

[7] M. H. Hedayati, A. B. Acharya, V. John: Common-Mode Filter Design for PWM Rectifier-Based Motor Drives, IEEE Transactions on Power Electronics, vol. 28, no. 11, pp. 5364-5371, Nov. 2013, doi: 10.1109/TPEL.2013.2238254.

[8] G. Zschau: EMV-gerechte Schaltungsauslegung (LE), Vorlesungsskript, TU Dresden, 2009.

[9] Hofheinz et al.: Elektrische Sicherheit in der Elektromobilität, VDE-Schriftenreihe Normen verständlich 174, 2019, ISBN 978-3-8007-4882-2.

[10] IEC: IEC 61851-1, Electric vehicle conductive charging system – Part 1: General requirements, International Standard, Feb. 2017, ISBN 978-2-8322-3766-3.

[11] IEC: IEC 61851-23, Electric vehicle conductive charging system – Part 23: DC electric vehicle charging station, International Standard, Mar. 2014, ISBN 978-2-8322-1440-4.

[12] IEC: Methods of measurement of touch current and protective conductor current (IEC 60990:2016), German version EN 60990:2016, National Standard, Mar. 2017, ICS 13.260

[13] Siemens AG: Fehlerstromschutzeinrichtungen, Technik-Fibel, Feb. 2018, Purchase Order Number EMLP-T10158-00-00DE

Triple-Phase-Shift Controlled Dual Active Bridge Converter with Variable Input Voltage in Auxiliary Railway Supply

Martin Scohier, Olivier Deblecker, Carlos Valderrama
Electrical Engineering Division, Engineering Faculty, University of Mons
31, Bd. Dolez
Mons, Belgium
Phone: +3265374154
Email: martin.scohier@umons.ac.be
URL: https://www.epeu-umons.be/

Acknowledgments

The authors would like to thank C. Versèle, Alstom Belgium, for supporting this work.

Keywords

≪Railway power supply≫, ≪On-board auxiliary power supply system≫, ≪Dual Active Bridge (DAB)≫, ≪DC-DC≫, ≪Simulation≫

Abstract

Modern railway auxiliary supply uses a Dual-Active-Bridge whose input voltage, subject to variations, is regulated by means of a front-end boost stage. With the aim of gaining weight and simplicity, this paper proposes to implement Triple-Phase-Shift modulation in order to maintain high performance when this front-end stage is removed. Analytical models including a more accurate magnetic power losses calculation are developed and results in terms of efficiency are presented for comparison.

Introduction

Due to increased comfort and higher traveling speed demands, modern railways coaches require a continuous energy supply to auxiliary equipment such as air conditioning, lighting, pressure protection, etc. Due to the different voltage levels of the catenary across the world, the electric energy from the locomotive is transferred to the coaches via a supply line with the nominal voltage varying from 750 V to 3 kV in the case of a dc catenary [1]. The supply voltage of the consumers connected to such an electricity supply unit reaches from a few tenths of volts for battery charger to three-phase 400 V ac. The power level is typically within the range of few hundreds of kilowatts and a galvanic isolation is required for safety reasons.

Today, new auxiliary power supplies consist of a multiport isolated dc-dc converter to interface the dc input, the battery and the output dc-ac module (see Fig. 1). An input LC filter is used for harmonic rejection, supply line overvoltage mitigation, etc. Such a power supply is constrained by numerous design considerations. For instance, low volume and mass (i.e. a high power density), high efficiency and bidirectional power flow are required. To that end, in recent applications, a Multiple Active Bridge (MAB) converter has been adopted, with a front-end boost converter to adapt the supply line voltage, that is subject to wide variations (typically ±33% of the rated value for a 1.5 kV dc catenary [1]), to the desired (regulated) level. Note that, in this study, the battery port will not be considered in a first approach; hence, a Dual Active Bridge (DAB) converter [2],[3] is considered (instead of a MAB). The corresponding circuit diagram is shown in Fig. 2. The medium-frequency (MF) transformer provides the mandatory galvanic isolation between the high voltage side and the consumers.

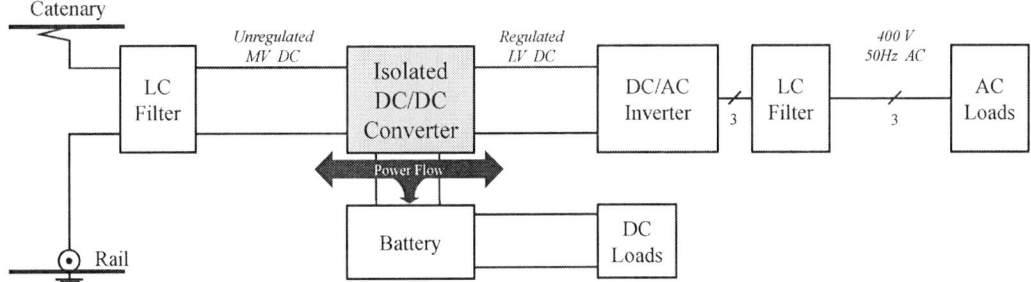

Fig. 1: Typical auxiliary power supply system with multiport isolated dc-dc converter

In this context, with a view of decreasing the mass and volume, the present paper aims to assess the feasibility of removing the 3-level boost converter shown in Fig. 2, hence simplifying the whole power conversion chain. The corresponding simplified topology of isolated dc-dc converter is shown in Fig. 3. This however leads to a nonregulated dc voltage at the input of the DAB, which makes the conventional Single Phase Shift (SPS) control strategy inappropriate for the considered application. Indeed, it is well established that despite the remarkable energy density of the DAB, most of its performance is highly dependent on the input/output voltage gain [2], [3],[4]. More concretely, poor efficiency and high current stress will occur as this gain departs from the MF transformer turns ratio. Therefore, in this contribution, it is proposed to adopt the Triple-Phase-Shift (TPS) modulation technique to overcome these limitations and extend the possibilities of the DAB converter. Several papers already addressed TPS controlled DAB converter operation focusing on various objectives (see, e.g., [2]-[6]). However, to our knowledge, addressing the application of a railway power supply with unregulated input voltage is a novelty.

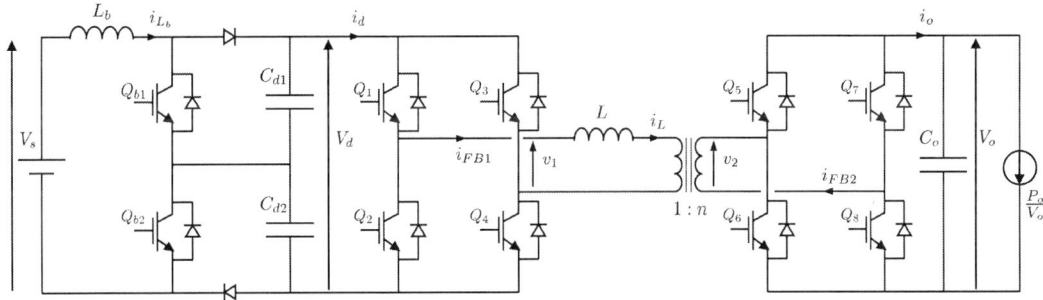

Fig. 2: Isolated dc-dc converter with front-end 3-level boost converter

Triple-Phase-Shift technique exploits the inner duty cycles of the primary- and secondary-side full bridge of the DAB converter (D_1 and D_2) in combination with the outer phase shift duty cycle D_3 in order to shape the inductor current i_L in the preferred way depending on the operating conditions. Fig. 4a shows some typical waveforms obtained with TPS by way of example. An overview of TPS modulation techniques can be found in [3], [4]. In this work, a backflow power minimization TPS strategy [7] is considered due to its effectiveness and because the modulation parameters (i.e. the duty ratios) can be quickly calculated using analytical formulae.

The remainder of this paper is organized as follows. The main assumptions and converter specifications (including the MF frequency transformers) are first presented. Then, analytical converter models are introduced followed by the main relationships used for power loss calculation. The results in terms of efficiency and power loss distribution obtained for the two isolated dc-dc converter solutions are presented in the next section. Finally, a discussion is being held to highlight the pros and cons of using TPS modulation strategy for the considered application.

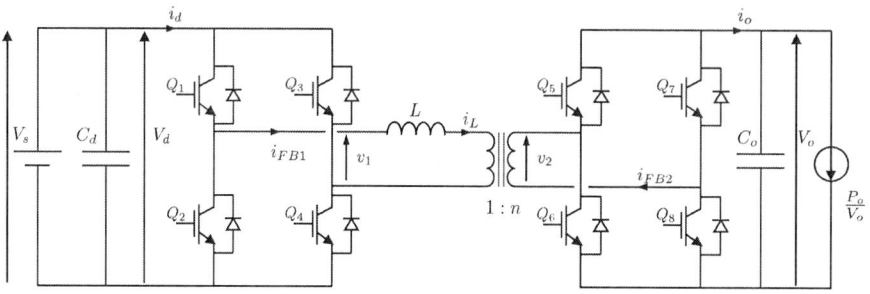

Fig. 3: Isolated dc-dc converter without front-end stage

Assumptions and specifications

Because output dc/ac module operation, filter design and battery management are not in the scope of this paper, assumptions are made to simplify the comparison between the two solutions of isolated dc-dc adaptation stage (namely with and without front-end boost converter) in steady-state. As already mentioned, the battery port is not considered. Moreover, it is initially assumed that the power flow is unidirectional from the dc voltage supply to a dc load (represented by a constant current source). The catenary voltage V_s is assumed constant within the permitted range. Input and output voltages of the DAB converter are considered as constant voltage sources assuming large capacitors.

Table I gives the main specifications and operating conditions used in this work. A 1.5 kV dc catenary voltage supply is chosen but it can vary from 1 kV to 2 kV [1]. In Fig. 2, the voltage V_d downstream of the 3-level boost converter is selected to be 2 kV, i.e. the upper limit of the supply line voltage. As IGBT modules are more mature at these voltage ratings and also due to data availability from manufacturer's datasheet, this technology is chosen here. Hence, the switching frequency f_s is set to 5 kHz.

Table I: Converter specifications

	Symbol	With boost converter	Without boost converter
Output Power Range	P_o	$10 \sim 150$ kW	$10 \sim 150$ kW
Supply dc voltage	V_s	$1 \sim 2$ kV	$1 \sim 2$ kV
DAB Input dc voltage	V_d	2 kV	$1 \sim 2$ kV
DAB Output dc voltage	V_o	750 V	750 V
Switching frequency	f_s	5 kHz	5 kHz

Transformer specifications

The absence of a front-end regulation circuit leads to a varying peak magnetic flux density in the core and increases rms currents in the windings. Since the MF transformer is one of the key elements of the converter, an accurate model is necessary for the comparison. First, the transformer turns ratio $n = V_o/V_d$ is set to match the DAB input and output voltage ratio under normal conditions (in this regard, V_d is equal to 2 kV or 1.5 kV according to whether the front-end boost stage is present or not). Applying a rule of thumb, the ac inductance L is determined with the SPS power transfer equation [2] ensuring that the maximum power is transferred with an outer phase shift D_3 of one eight (i.e. 45 degrees) of the switching period. In that equation, the DAB input voltage V_d is selected to be 2 kV or 1 kV (i.e. the lower bound of the supply dc voltage), depending on whether the the boost stage is present or not. The area product design method allows to determine both electrical and geometrical characteristics of the transformer for each configuration. A ferrite double-E shaped core is adopted here for the sake of simplicity. Indeed, the optimal dimensions of such a common shape are easily derived from the so-called area product, i.e. the product of the core cross section area A_{core} and the winding window area A_{wind} (see (1)). Moreover, in an effort to reduce the ac resistance value, aluminium Litz wire is considered in the windings. The area product can also be expressed as follows :

$$A_p = A_{core} \cdot A_{wind} = \frac{V_{1,rms}}{k_f f_s B_m} \frac{2I_{L,rms}}{J k_u} \tag{1}$$

where f_s, $V_{1,rms}$ and $I_{L,rms}$ depend on the converter specifications. The peak value of magnetic flux density B_m and the peak current density J are adjustable parameters. Because optimal design of the transformer is not in the scope of this project, such parameter values are derived from the works conducted in [8] for a similar power level and comparable converter specifications. Therefore, B_m is set to 0.15 T and J is taken equal to 3 A/mm^2. The form factor k_f is equal to 4 in the case of triangular shaped magnetic flux density. The utilization factor k_u of a Litz wiring is typically equal to 0.3. Furthermore, the transformer is assumed to be designed in a way to match its total leakage inductance with the desired ac link inductance L. As for the magnetizing inductance, based on previous railway auxiliary power supply projects, its value is fixed to limit the peak amplitude of the magnetizing current under 10% of the primary side maximum rms current ($L_m = V_{1,rms}/(k_f f_s \cdot 0.1 \cdot I_{L,rms})$). The mass of the transformer m_{TFO} is calculated from the core and winding volume along with their respective material mass densities. Hence, Table II shows the specifications adopted for the MF transformer in both isolated dc-dc converters.

Table II: Transformer specifications

	Symbol	With boost converter	Without boost converter
Transformer Turns Ratio	n	3/8	1/2
Primary side maximum rms Voltage	$V_{1,rms}$	2 kV	2 kV
Primary side maximum rms Current	$I_{L,rms}$	91 A	167 A
Core cross section	A_{core}	120 cm^2	163 cm^2
Core volume	V_{core}	10 dm^3	15 dm^3
Total dc resistance	R_{dc}	116 mΩ	55 mΩ
Total leakage inductance	$L = l_1 + l_2'$	500 μH	187.5 μH
Magnetizing inductance	L_m	10.9 mH	6 mH
Mass of the transformer	m_{TFO}	70 kg	111 kg

Modeling of the converter operation

In order to obtain the key waveforms of the two isolated dc-dc adaptation stages, analytical converter models valid under steady-state operation are implemented. The first model concerns the DAB converter when used with TPS modulation (noticing that SPS modulation is a specific case of TPS in which $D_1 = D_2 = 1$). It includes an accurate operation of the MF transformer. The second model relates to the 3-level boost converter possibly connected at the input of the DAB converter.

A lossless operation of the converter is assumed to obtain the current waveforms in each power switch and passive component. For each configuration of the dc-dc adaptation stage, the output power and input voltage are swept within their respective ranges of values. The duty cycles D_1, D_2 and D_3 are calculated according to the chosen modulation strategy. The voltages at the ports of the MF transformer can be defined as $v_1(t) = V_d \cdot s_1(t)$ and $v_2'(t) = V_o/n \cdot s_2(t)$, where s_1 and s_2 are the switching functions of the DAB primary and secondary side (see (2)). The inductor current is governed by (3) considering the simplified ac link circuit in Fig. 5a. The characteristic waveforms of the inductor current and the switch gate drive signals are shown in Fig. 4a. Here, the switching cycle starts when the power switch Q_1 turns on ($t_{2\to1} = 0$). Hence, considering TPS modulation, the switching instants are defined as follows over the first half of the switching period : $t_{4\to3} = D_1 T_s/2$, $t_{6\to5} = D_3 T_s/2$ and $t_{8\to7} = (D_3 + D_2)T_s/2$.

$$s_1(t) = \begin{cases} 1, & 0 < t < D_1\frac{T_s}{2} \\ 0, & D_1\frac{T_s}{2} < t < \frac{T_s}{2} \\ -1, & \frac{T_s}{2} < t < (1+D_1)\frac{T_s}{2} \\ 0, & (1+D_1)\frac{T_s}{2} < t < T_s \end{cases} \quad s_2(t) = \begin{cases} 1, & D_3\frac{T_s}{2} < t < (D_3+D_2)\frac{T_s}{2} \\ 0, & (D_3+D_2)\frac{T_s}{2} < t < (1+D_3)\frac{T_s}{2} \\ -1, & (1+D_3)\frac{T_s}{2} < t < (1+D_3+D_2)\frac{T_s}{2} \\ 0, & (1+D_3+D_2)\frac{T_s}{2} < t < D_3\frac{T_s}{2} + T_s \end{cases} \tag{2}$$

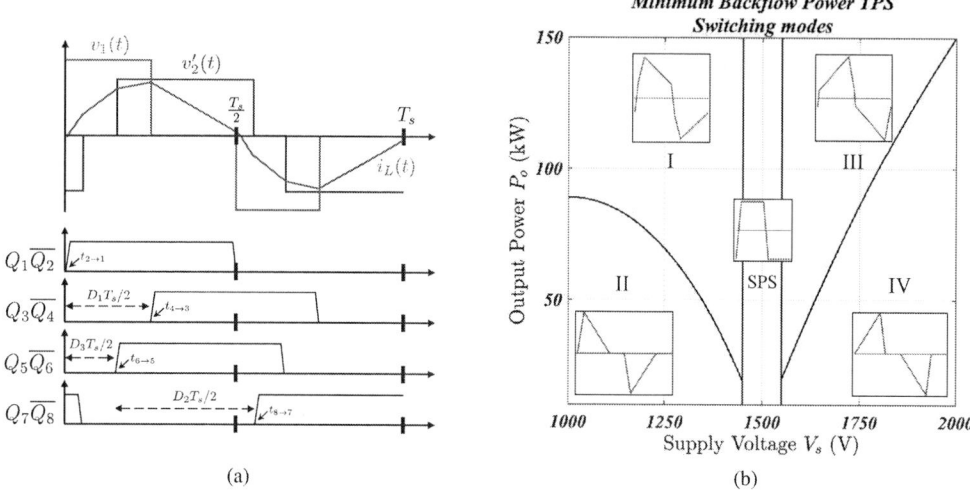

(a)   (b)

Fig. 4: (a) Characteristic waveforms using TPS modulation. (b) Minimum backflow power TPS strategy : existing modes and their inductor current typical waveform

$$L\frac{i_L(t)}{dt} = v_1(t) - v_2'(t) \tag{3}$$

The proposed TPS modulation strategy, aiming to minimize the reactive power in the ac link, switches between five modes depending on the output power and the ratio between input and output voltage of the DAB converter [7] (see Fig. 4b). First, SPS modulation is applied when the voltage is near its nominal value, i.e. when $V_s = 1500 \pm 50V$. Then, if the input voltage deviates from this range, the current is shaped into a discontinuous triangular waveform for light load conditions (modes II & IV). The other two modes (I & III) involve the calculation of optimal duty cycle values and are employed at higher power levels.

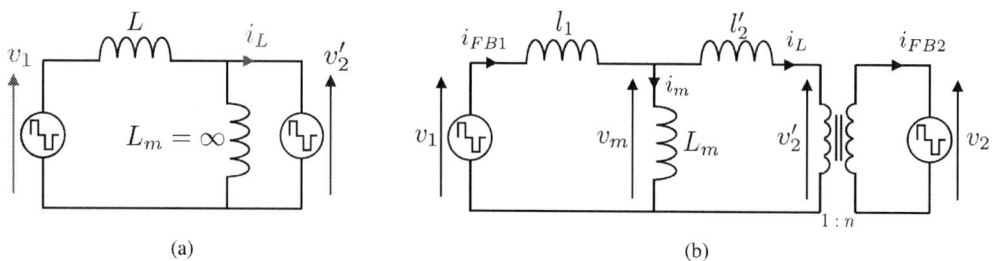

(a)   (b)

Fig. 5: (a) Lossless ac link circuit ($i_m = 0$). (b) Lossless ac link circuit ($i_m \neq 0$)

In this contribution, in an attempt to better describe the operation of the transformer, the magnetizing current is considered resulting in a more accurate model of the ac link (see Fig. 5b). The analytical expression of the voltage v_m across the magnetizing inductance L_m is given by (4), assuming an equal distribution of the leakage inductances and a large magnetizing inductance.

$$v_m(t) = \frac{v_1(t)l_2' + v_2'(t)l_1}{(l_1 + l_2' + \frac{l_1 l_2'}{L_m})} \approx \frac{v_1(t) + v_2'(t)}{2} \tag{4}$$

The magnetizing current is governed by :

$$L_m\frac{di_m(t)}{dt} = v_m(t) \tag{5}$$

Then, the ac currents of both full bridges, namely i_{FB1} and i_{FB2}, are determined assuming that the primary side full bridge (FB1) conducts the magnetizing current. Therefore, $i_{FB1}(t) = i_L(t) + i_m(t)$ and $i_{FB2}(t) = i_L(t)/n$. Depending on the direction of these currents and with the use of the switching functions in (2), the current waveforms can be obtained for each semiconductor device (transistor or anti-parallel diode).

In the front-end boost stage, whose operation is exposed in [9], the duty cycle depends on the input and output voltage ($D_b = 1 - V_s/V_d$). Similarly to the DAB, a switching function can be determined to reflect the state of the converter. Moreover, the current waveforms in each device are derived from the boost inductor current i_{L_b}. The average value is equal to P_o/V_s and the peak-to-peak current ripple is given by

$$\Delta I_{L_b} = (V_s - \frac{V_d}{2})\frac{D_b}{f_s L_b} \tag{6}$$

Power loss calculation

Due to the skin effect, the resistance of the Litz wire at the fundamental frequency of 5 kHz is assumed to be 1.5 times larger than the dc resistance [10]. Therefore, winding losses in the transformer can be calculated by means of $P_{TFO,wind} = 1.5R_{dc}I_{FB1,rms}^2$ where R_{dc} is the total dc resistance of the windings viewed from the primary side and $I_{FB1,rms}$ is the rms value of the ac current of FB1.

Magnetic power losses are usually evaluated with the Improved Generalized Steinmetz Equation (IGSE) because of its accuracy for non-sinusoidal induction [11]. Contrary to most existing works, this paper proposes to consider the exact time evolution of the magnetic flux density according to the governing equation (7), where N_1 is the number of turns of the transformer primary winding. The volume and the cross section area of the ferrite core are denoted V_{core} and A_{core}, respectively. Moreover, the steinmetz coefficients in (8) are selected from [11] and have the following values : $\alpha = 1.25$, $\beta = 2.35$ and $k = 16.9$.

$$\frac{dB(t)}{dt} = \frac{v_m(t)}{A_{core}N_1} \tag{7}$$

$$P_{TFO,core} = V_{core}\left(\frac{\Delta B}{2}\right)^{\beta-\alpha}\frac{k_i}{T_s}\int_0^{T_s}\left|\frac{dB}{dt}\right|^{\alpha}dt \quad \text{with} \quad k_i = \frac{k}{(2\pi)^{\alpha-1}\int_0^{2\pi}|\cos\theta|^{\alpha}d\theta} \tag{8}$$

The current and voltage waveforms calculated from the analytical converter models are used to select the IGBT modules. Their ratings are listed in Table III. The conduction losses of the devices (diodes and IGBTs) are calculated from their average and rms currents (denoted I_{avg} and I_{rms}, respectively) given the well-known formula

$$P_{cond} = V_{th}I_{avg} + rI_{rms}^2 \tag{9}$$

where V_{th} represents the threshold voltage and r the on-state resistance taken from the manufacturers datasheets.

Table III: IGBT module

	Voltage rating V_{CE}	Current rating I_{CE}
3-level Boost	1.7 kV	150 A
DAB Full Bridge 1	3.3 kV	200 A
DAB Full Bridge 2	1.7 kV	200 A

Due to the large number of switching devices, switching losses tend to decrease the power density of the DAB converter. Soft switching is thus desired within the entire range of operation. As the use of parallel snubber capacitors is commonly adopted to lower the turn off losses, these will not be considered. However, due to the presence of snubber capacitors, the voltage across each switch will decrease slowly, resulting in losses at turn on [12], [13]. To prevent this, the capacitor must fully discharge during the dead-time between the switch gate signals in each leg, allowing the anti-parallel diode to conduct and resulting in a quasi zero voltage turn on. This can be achieved if the dead-time is long enough and if the energy stored in the ac link inductance is sufficient. Moreover, it is generally adopted to simplify the ZVS turn-on condition to the sign of the inductor current at switching instants [4], [6]. Therefore, only hard switching loss at non-ZVS turn on is considered. For instance, equation (10) gives the average switching losses of the IGBT Q_1 over one switching period. The energy loss at turn on and turn off (E_{on} and E_{off}) are obtained from IGBT modules characteristics.

$$P_{swQ1} = \begin{cases} f_s E_{on}\left[i_{FB1}(t_{2\rightarrow1})\right], & \text{if } i_L(t_{2\rightarrow1}) < 0 \\ 0, & \text{if } i_L(t_{2\rightarrow1}) \geq 0 \end{cases} \tag{10}$$

The power switches in the 3-level boost converter operate in hard switching. Considering IGBT Q_{b1} as an example, the switching losses are calculated by using the following expression:

$$P_{swQb1} = f_s \left(E_{on}\left[i_{L_b}(t_{b2\rightarrow b1})\right] + E_{off}\left[i_{L_b}(t_{b1\rightarrow b2})\right] \right) \tag{11}$$

Finally, efficiency for each operating point of the converter within the admitted range is calculated using (12). $\sum P_{loss}$ stands for the total of losses including $P_{TFO,core}$, $P_{TFO,wind}$, $P_{DAB,cond}$, $P_{DAB,sw}$, $P_{Boost,cond}$ and $P_{Boost,sw}$.

$$\eta = \frac{P_o}{P_o + \sum P_{loss}} \tag{12}$$

Results

In this section, the simulations based on the above analytical models are conducted assuming three different cases. The first one (denoted as SPS case) concerns the DAB converter using SPS modulation directly connected to the varying voltage supply (as in Fig. 3) and exposes the limitations when the DAB converter faces varying operating conditions. The two other cases (simply denoted here as BOOST and TPS cases) address these limitations by using two different approaches which is the subject of a comparison. In the so-called BOOST case, a DAB converter with SPS modulation is used in combination with a front-end three-level boost stage (see the dc-dc converter configuration shown in Fig. 2). In the TPS case, the disadvantage of the non-regulated voltage at the input of the DAB converter is compensated by switching between modulation modes depending on the operating conditions. The isolated dc-dc converter operation over a switching period is simulated for a set of operating points, defined by values of the catenary voltage V_s and the output power P_o (within the ranges specified in Table I). Hence, the power losses can be calculated by exploiting the computed voltage and current waveforms at each operating point. In this way, Fig. 6 shows the ensuing efficiency maps for the three different cases considered here. Furthermore, Fig. 7 shows the efficiency curves under minimum, rated and maximum supply voltage conditions.

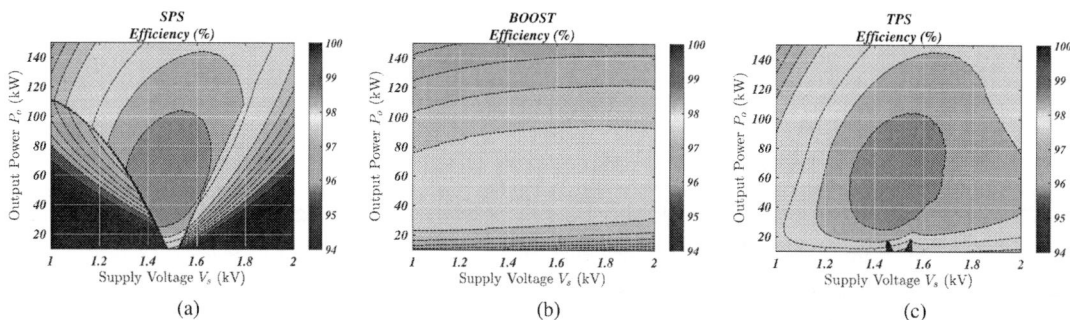

Fig. 6: Efficiency maps of the isolated dc-dc converter in the SPS, BOOST and TPS cases

It is first to be noted that the DAB converter with SPS modulation is subject to large power losses when the input voltage deviates from the nominal value (see Fig. 6a). This is even more pronounced in the case of light load conditions, given the high reactive power circulating in the circuit. Additionally, hard switching losses occur as the direction of the ac currents are no longer favorable for ZVS turn on. In Fig. 6b, the advantage of an input voltage regulation is exposed. The BOOST case is advantageous because the efficiency remains nearly unchanged (about 97%), whatever the operating point taken within the limits of variation of the output power and supply voltage. It can be observed in Fig. 6c that TPS case decreases the power losses in the critical areas of the SPS case while maintaining an overall higher efficiency than in the BOOST case. These high efficiency values represent an upper bound due to the approached calculation (based on analytical formulae) and because other losses exist in the converter (non-zero soft switching losses, losses in the filtering elements, etc.). At rated dc voltage supply, i.e. $V_s =$ 1.5 kV, the DAB converter with no front-end boost stage is clearly advantageous in terms of efficiency (see Fig. 7b). This is due to the remarkable performances of SPS modulation when the input and output dc voltage ratio matches the transformer turns ratio.

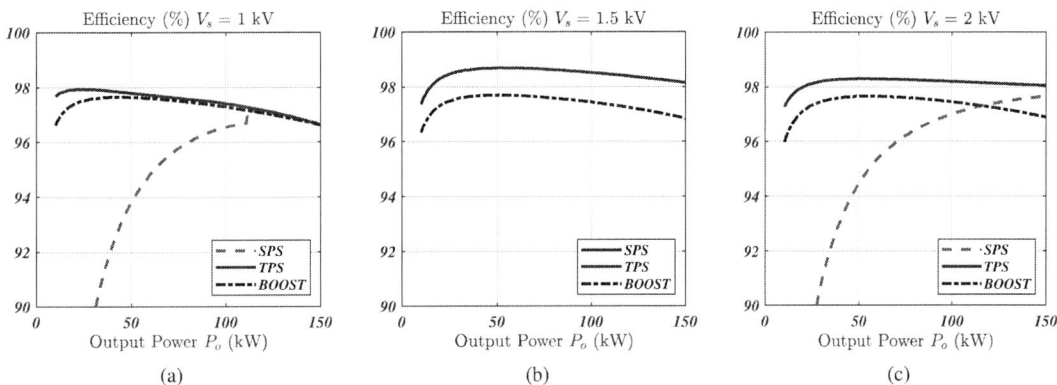

Fig. 7: Efficiency curves under different supply voltage conditions : (a) $V_s = 1$ kV, (b) $V_s = 1.5$ kV and (c) $V_s = 2$ kV

Fig. 8 shows the power loss distribution in the different components of the isolated dc-dc converter cases under different supply voltage ($V_s =$1 kV, 1.5 kV and 2 kV), at rated and light load conditions (i.e. $P_o =$ 150kW and 50kW, respectively). The DAB converter power losses mainly consist of conduction losses in the IGBT modules and in the transformer windings. Switching losses predominate in the three-level boost converter as only hard-switching occurs. On the other hand, soft switching is achieved in the DAB converter for the whole range of operation for the two compared dc-dc adaptation stage solution (see BOOST and TPS cases). As it can be observed in Fig. 8a and Fig. 8c, the TPS modulation in modes I and III (i.e. the modes that are used at high power level) shows slight improvement with regard to the SPS modulation.

Magnetic power losses in the ferrite core are relatively small due to the low switching frequency. Yet, they form a more important part of the transformer total losses when the load decreases. Due to their dependency on the DAB primary and secondary side voltage levels, such power losses tend to vary widely when the input voltage of the DAB is no longer regulated. Nevertheless, because of its ability to tune both inner duty cycles, applying TPS modulation strategy allows to reduce the peak magnetic flux density in the core resulting in fewer magnetic power losses than in the two other cases studied (SPS and BOOST).

The current waveforms in the filter capacitors can be computed as well from the analytical converter models. Considering the fundamental harmonic only ($=2f_s$), the removal of the front-end stage leads to an increase of the maximum rms value from 88 A to 167 A in the output filter capacitor C_o (for the BOOST and TPS cases, respectively). Therefore, a higher capacitance value (about twice as large) must be chosen in order to keep a comparable output voltage ripple. The same calculation for the input filter capacitor gives no variation regarding the maximum rms value of the current. However, the comparison

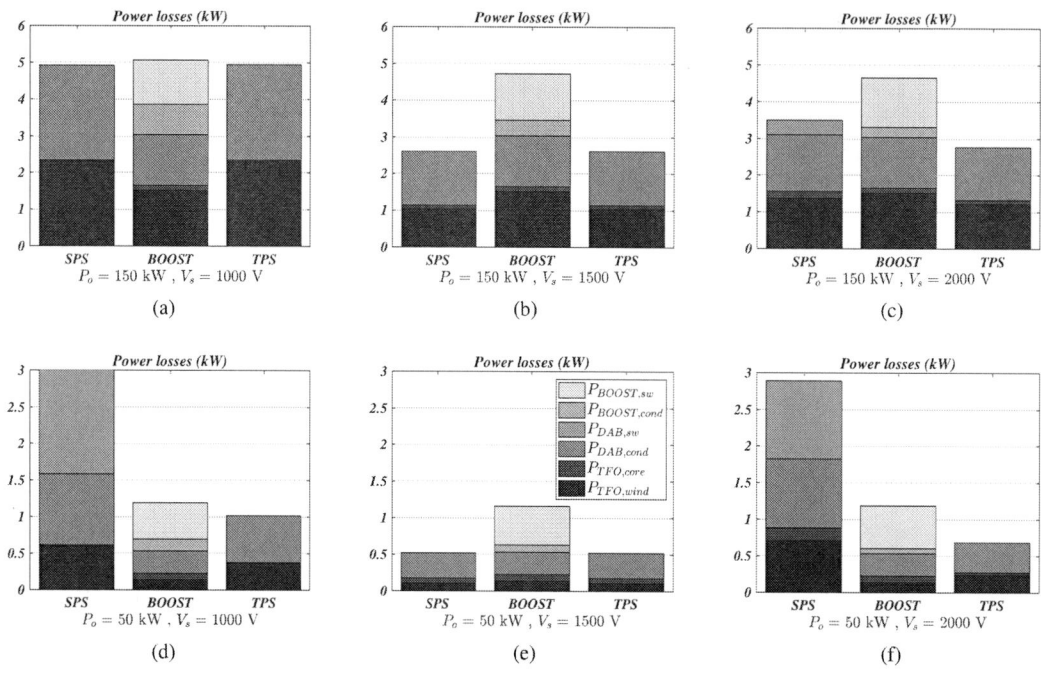

Fig. 8: Power loss distribution

analysis is not as straightforward as their role and topology in the two isolated dc-dc converter solutions are very different (see Fig. 2 and Fig. 3).

Discussion

As observed in previous section, for the considered specifications, the proposed isolated dc-dc converter with no front-end boost stage shows high performance at nominal voltage due to the optimal utilization of the SPS modulation and the elimination of the losses in the front-end stage. Moreover, the TPS modulation greatly improves the efficiency of the converter in case of severe voltage conditions. This is mainly due to the triangular shaped current which maintains minimal switching losses while decreasing the rms value of the ac link current at light load. Nevertheless, these results should be experimentally validated in future work.

From a component point of view, it is straightforward that removing the front-end stage will reduce the number of parts, in particular the front-end boost stage components. However, input voltage variation of the DAB converter will accordingly lead to higher thermal constraints and increased stresses on the remaining components, leaving the gain in terms of mass and volume uncertain. A higher capacitance value and rms currents in the output filter leads to a larger capacitor. Regarding the primary- and secondary-side full bridge of the DAB converter, while IGBT modules ratings are suitable for both configurations of isolated dc-dc converter, overall higher conduction losses as depicted in Fig. 8 will lead to the design of bulkier and heavier heatsinks. Furthermore, due to increased current at low supply voltage, the design method has resulted in a bigger and heavier transformer solution. The higher mass of the proposed MF transformer is compensated by lower losses and a wider cooling area, resulting in lower thermal management constraints. In a will to increase the power density, higher switching frequencies in combination with MOSFET SiC technologies will be increasingly used [3]. This will reduce the size of the passive elements as well as the overall power losses in the switches. In this context, the magnetic losses will become a more significant consideration, hence making the use of TPS modulation more relevant in such applications.

Finally, TPS modulation involves two extra degrees of freedom (D_1 and D_2) compared to SPS modulation strategy. It is therefore more difficult to implement in practice. Additionally, depending on the operating point, five different modes of operation exist which need to be switched accordingly from one to another (see Fig. 4b). Note that in this application, at high power level, the possibility of ignoring modes I and III to gain simplicity could be explored. On the other hand, the voltage controller (with inner current loop) necessary to regulate the output voltage of the 3-level boost stage is no longer necessary if the simplified configuration of isolated dc-dc converter is adopted, which is an asset.

Conclusion

This paper focused on a modern auxiliary power supply with bidirectional power flow capability. An isolated dc-dc converter is used to provide a (regulated) dc voltage at the input of the dc/ac module. Leaving apart the battery port, the topology of this converter is currently composed of a front-end 3-level boost stage and a DAB converter operated via SPS modulation strategy. The possibility of removing the front-end stage has been investigated by means of steady-state simulations conducted using analytical models, within the whole range of operation points in terms of output power and catenary supply voltage. In an effort to better compare the two dc-dc converter solutions, an analytical model of the MF transformer losses taking into account the magnetizing current has been employed. Based on simulation results, it was shown that using a minimum reactive power TPS strategy increases the overall efficiency of the isolated dc-dc converter. However, increased stresses is responsible for bulkier elements when removing the front-end circuit, with particular regard to the MF transformer. Moreover, it can be expected to simplify the control structure.

References

[1] BS EN 50163 Railway applications. Supply voltages of traction systems
[2] Krismer, Florian. "Modeling and optimization of bidirectional dual active bridge DC-DC converter topologies." (2010).
[3] B. Zhao, Q. Song, W. Liu and Y. Sun, "Overview of Dual-Active-Bridge Isolated Bidirectional DC–DC Converter for High-Frequency-Link Power-Conversion System," in IEEE Transactions on Power Electronics, vol. 29, no. 8, pp. 4091-4106, Aug. 2014, doi: 10.1109/TPEL.2013.2289913.
[4] N. Hou and Y. W. Li, "Overview and Comparison of Modulation and Control Strategies for a Nonresonant Single-Phase Dual-Active-Bridge DC–DC Converter," in IEEE Transactions on Power Electronics, vol. 35, no. 3, pp. 3148-3172, March 2020, doi: 10.1109/TPEL.2019.2927930.
[5] J. Huang, Y. Wang, Z. Li and W. Lei, "Unified Triple-Phase-Shift Control to Minimize Current Stress and Achieve Full Soft-Switching of Isolated Bidirectional DC–DC Converter," in IEEE Transactions on Industrial Electronics, vol. 63, no. 7, pp. 4169-4179, July 2016, doi: 10.1109/TIE.2016.2543182.
[6] Y. Tang et al., "RL-ANN Based Minimum-Current-Stress Scheme for the Dual Active Bridge Converter with Triple-Phase-Shift Control," in IEEE Journal of Emerging and Selected Topics in Power Electronics, doi: 10.1109/JESTPE.2021.3071724.
[7] S. Shao, M. Jiang, W. Ye, Y. Li, J. Zhang and K. Sheng, "Optimal Phase-Shift Control to Minimize Reactive Power for a Dual Active Bridge DC–DC Converter," in IEEE Transactions on Power Electronics, vol. 34, no. 10, pp. 10193-10205, Oct. 2019, doi: 10.1109/TPEL.2018.2890292.
[8] C. Versele, O. Deblecker and J. Lobry, "Multiobjective optimal design of high frequency transformers using genetic algorithm," 2009 13th European Conference on Power Electronics and Applications, 2009, pp. 1-10.
[9] M. T. Zhang, Yimin Jiang, F. C. Lee and M. M. Jovanovic, "Single-phase three-level boost power factor correction converter," Proceedings of 1995 IEEE Applied Power Electronics Conference and Exposition - APEC'95, 1995, pp. 434-439 vol.1, doi: 10.1109/APEC.1995.468984.
[10] Mohan, Ned, Tore M. Undeland, and William P. Robbins. Power electronics: converters, applications, and design. John wiley & sons, 2003.
[11] Hurley, William G., and Werner H. Wölfle. Transformers and inductors for power electronics: theory, design and applications. John Wiley & Sons, 2013.
[12] Naayagi, R., & Mastorakis, N. (2012). Performance verification of dual active bridge DC-DC converter. ACA'12 Proceedings of the 11th International Conference on Applications of Electrical and Computer Engineering, 1(October), 13–19.
[13] Sommer, F., Menger, N., Merz, T., & Hiller, M. (2021). Accurate Time Domain Zero Voltage Switching Analysis of a Dual Active Bridge with Triple Phase Shift. 2021 23rd European Conference on Power Electronics and Applications, EPE 2021 ECCE Europe, 1–9.

Loss characterization methodology for soft magnetic nano-crystalline tape materials in coupled inductors

David Bohne, Valentin Wagner, Patrick Deck, Christian P. Dick
TH KÖLN
Betzdorfer Straße 2
50679 Cologne, Germany
Phone: +49 221 8275-2262
Email: dbohne@th-koeln.de
URL: http://www.th-koeln.de/lea

Acknowledgments

The GaN-HighPower project on which this paper is based was funded by the German Federal Ministry for Economic Affairs and Climate Action under grant number 03EE1111F. The author is responsible for the content of this publication.

Keywords

≪loss characterization≫, ≪soft magnetic material≫, ≪four-wire measurement≫, ≪two- winding≫, ≪coupled inductors≫

Abstract

In this paper a four-wire magnetic core loss measurement test bench for waveforms typical for two-phase Coupled Inductors is presented. A low permeability nano-crystalline ribbon material is investigated. The advantage of the new test bench is its low measurement time of seconds for multiple operating points at good accuracy. Due to the improvements in preventing disturbances, the measurements are reliable and reproducible.

Introduction

General description

A big challenge in the improvement of power converters is the reduction of volume and weight. Cores used in filter inductors have a major impact on the power density. In [1] it is shown, that with inversely wound Coupled Inductors (CI) the core volume can be significantly reduced in N-phase interleaved converters. For an optimal design of CI, losses in deployed cores made of soft magnetic materials play an important role. These losses depend on temperature, AC flux frequency, DC bias, waveform and amplitude of the flux [2] [3]. Most accurate calorimetric methods for determining losses in magnetics are very time-consuming [4], this problem should be addressed with the developed test bench.

With power converters and their magnetic components improved to high power densities, core materials developed as well. In addition to low hysterisis and eddy current losses, the investigated nano-crystalline tape material features high saturation flux densities and low permeabilities [5]. These characteristics make it suitable for high frequency and high efficiency applications [6][7].

In this paper, a loss characterization methodology for the nano-crystalline tape material VITROPERM® 850FF [8] is introduced for special flux waveforms, which occur in the coupling and leakage path of coupled tape-wound cores. Fig. 1 depicts exemplary flux waveforms for such special setups as presented in [1].

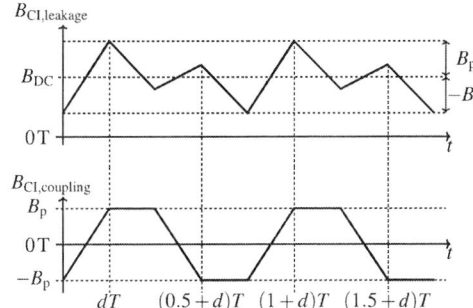

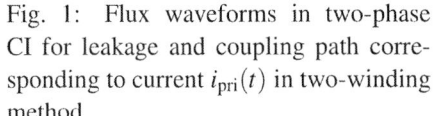

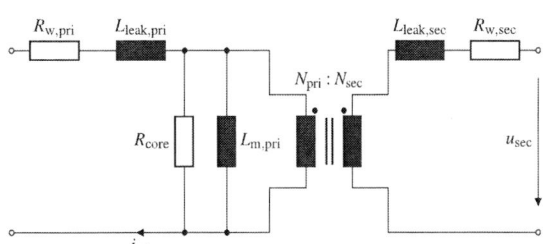

Fig. 1: Flux waveforms in two-phase CI for leakage and coupling path corresponding to current $i_{\text{pri}}(t)$ in two-winding method

Fig. 2: Equivalent circuit of core with primary and secondary side winding (leakage inductances $L_{\text{leak,pri}}$, $L_{\text{leak,sec}}$, winding resistances $R_{\text{w,pri}}$, $R_{\text{w,sec}}$, representative resistance for core losses R_{core})

Two-winding / four-wire method

For fast loss measurement, the electrical two-winding/four-wire method is widely used [9] [10]. Based on the equivalent circuit of the Device Under Test (DUT) (Fig. 2), core losses are represented by the resistor R_{Core} [11]. Loss measurement requires the primary side to be excited with waveforms for which losses are to be characterized. By measuring the current on the primary side and the voltage on the unloaded secondary side, core losses can be calculated with following equation:

$$P_{\text{Fe}} = \frac{N_{\text{pri}}}{N_{\text{sec}}T} \int_0^T u_{\text{sec}}(t) i_{\text{pri}}(t)\,\mathrm{d}t \tag{1}$$

This method allows hysteresis and eddy current losses to be measured separately from winding losses [3]. However, it suffers from parasitic effects in sensing the primary current as core losses are very sensitive to phase discrepancies between the measured signals [10] [12] [13]. With

$$P_{\text{Fe}} \sim \cos(\phi_{\text{actual}} + \phi_{\text{error}}) \tag{2}$$

this problem can be very severe, especially towards high frequencies.

Another common way to measure core losses is the thermal approach by means of a calorimeter [14], which doesn't face the challenges mentioned for high frequencies. The thermal approach is very time-consuming and with increasing frequency it becomes hard to distinguish core from winding losses [4] [2]. Compared to the two-winding method, the thermal approach is not preferred.

Developed test bench

The equivalent circuit of the developed test bench can be seen in Fig. 3. The waveforms shown in Fig. 1 are generated by two half bridges HB3 and HB4, while the supply voltages are provided by HB1 and HB2. A DC filter realized by a large capacitance ($C_{\text{f}} = 750\,\mu\text{F}$) is placed at the output to block any DC currents during measurement. This setup can be extended by placing a third winding on the DUT to superpose AC fluxes with DC bias. The switching patterns of HB3 and HB4 and the resulting voltage $u_{\text{DUT}}(t)$ and current $i_{\text{pri}}(t)$ are shown in Fig. 4 for leakage path and in Fig. 5 for coupling path waveforms in two-phase CI.

A coaxial shunt resistor R_{shunt} with very low ESL ($L_{\text{ESL}} \approx 35\,\text{pH}$) is chosen to measure the primary current $i_{\text{pri}}(t)$ and to avoid big errors caused by the phase discrepancy (2). The voltages $u_{\text{shunt}}(t)$ and $u_{\text{sec}}(t)$ are measured with differential probes.

The isolated measurement for $u_{\text{shunt}}(t)$ in Fig. 3 experiences interferences caused by common mode voltages, since the coupling capacitance to ground potential of the differential probe is much smaller compared to the coupling capacitance of DC sources ($C_{\text{CM,probe}}$ and $C_{\text{CM,source}}$ in Fig. 6).

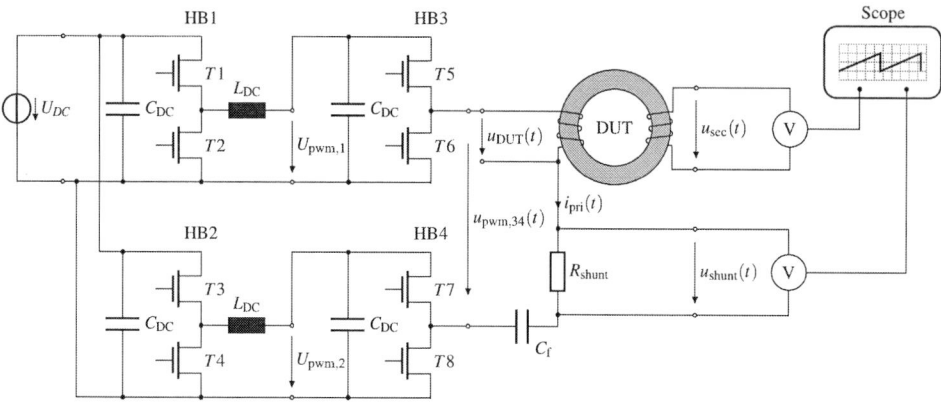

Fig. 3: Equivalent circuit of the developed test bench

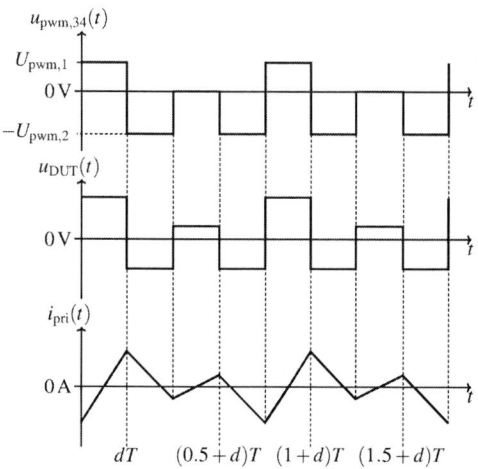

Fig. 4: Switchting pattern for the leakage path waveform in two-phase CI, $\overline{u_{\text{DUT}}} = 0\,\text{V}$

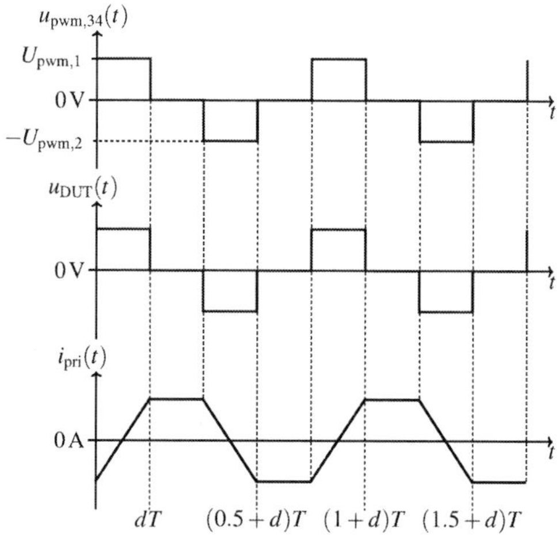

Fig. 5: Switchting pattern for the coupling path waveform in two-phase CI, $\overline{u_{\text{DUT}}} = 0\,\text{V}$

Grounding the shunt resistor leads to unsymmetrical common mode currents, resulting in an additional voltage over the shunt and thus, distorts the measurement. Without grounding the setup near the shunt resistor, unsymmetrical common mode voltages influence the measurement as well and prevent using higher measurement voltages since the probes used can not handle more than $U_{\text{CM,max}} = 60\,\text{V}$.

Various arrangements are able to suppress common mode currents despite grounding the setup on the primary and secondary side of the DUT. A common mode choke is placed in front of the shunt resistor and the output design is adapted by adding a second DC filter C_f and inductance L_f to achieve a symmetrical setup. Common mode currents are reduced even further by decoupling all power inputs from their supplies. This is done by opening relay contacts in each supply line during measurement. To store enough energy for one measurement, DC link capacitors are enlarged and a 12 V battery is placed at the logic power input as an auxiliary supply. The adapted test bench including the ground connections can be seen in Fig. 6.

To accelerate the measurement process, the scope and all half bridges are controlled by a PC performing all setups and calculations, including error analysis.

For each measurement, fifty periods of $u_{\text{shunt}}(t)$ and $u_{\text{sec}}(t)$ are recorded and losses are averaged to reduce the influence of noise. With the number of samples M for the amount of fifty periods recorded by the scope, the measured voltage $u_{\text{shunt}}(t)$ and the weight of the core m_{core}, the loss densitiy $P_{\text{Fe,dens}}$ is approximated by (3) based on (1) by the PC.

$$P_{\text{Fe,dens}} = \frac{N_{\text{pri}}}{N_{\text{sec}}R_{\text{shunt}}m_{\text{core}}M} \sum_{k=1}^{M} u_{\text{sec}}[k]u_{\text{shunt}}[k] \qquad (3)$$

Each measurement determines the loss density $P_{\text{Fe,dens}}$ specified for the maximum flux B_p induced in the core, which can be calculated by integrating the time-dependent flux from $t_1 = dT/2$ to $t_2 = dT$ (Fig. 1). With the equivalent core cross sectional area A_{Fe} and the sampling time $T_S = 100\,\text{ps}$ this results in:

$$B_p = \frac{T_S}{A_{\text{Fe}}N_{\text{sec}}} \sum_{k=1}^{M'} u_{\text{sec}}[k] = \frac{\rho\pi(r_{\text{in}} + r_{\text{out}})T_S}{m_{\text{core}}N_{\text{sec}}} \sum_{k=1}^{M'} u_{\text{sec}}[k] \qquad (4)$$

with samples according to

$$B_p = \frac{1}{A_{\text{Fe}}N_{\text{sec}}} \int_{dT/2}^{dT} u_{\text{sec}}(t)\,dt \qquad\qquad A_{\text{Fe}} = \frac{V_{\text{Fe}}}{l} \underset{l=2\pi\frac{r_{\text{in}}-r_{\text{out}}}{2}}{\overset{V_{\text{Fe}}=\frac{m_{\text{core}}}{\rho}}{=\!=}} \frac{m_{\text{core}}}{\rho\pi(r_{\text{in}} + r_{\text{out}})}$$

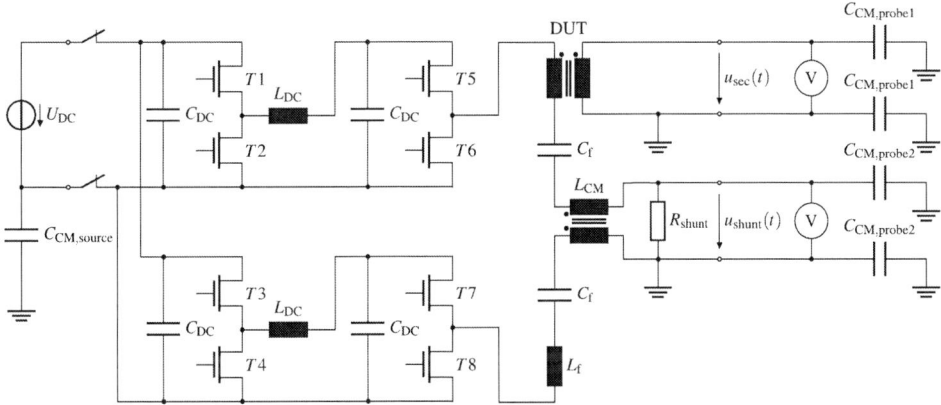

Fig. 6: Equivalent circuit of the extended test bench for common mode suppression

Error Analysis

Based on (3) and (4), various error sources need to be analyzed in this section. Due to non-sinusoidal waveforms, the error calculation is done by decomposing the two measured signals into their corresponding fourier series, neglecting the nonlinear dependence of core losses on both, flux and frequency. As can be seen in Fig. 12 later, more than 95 % of the total error can be explained with the proposed simplification based on the fourier decomposition. Thus, especially the share of the different loss contributors in Fig. 9 is valid in its main message. The remaining only 5 % difference of the losses in Fig. 12 mean that the calculated error bars have a very high confidence level. Therefore, this simplification is sufficiently accurate for error calculation and allows to derive the following calculations for sinusoidal waveforms.

Error sources

The total error is composed of following different sources:

1. weight m_{core}
2. value of shunt resistor R_{shunt}
3. amplitude and offset error in measured $u_{\text{shunt}}(t)/u_{\text{sec}}(t)$
4. phase ϕ_{actual}
5. radii $r_{\text{in}}/r_{\text{out}}$ of the DUT

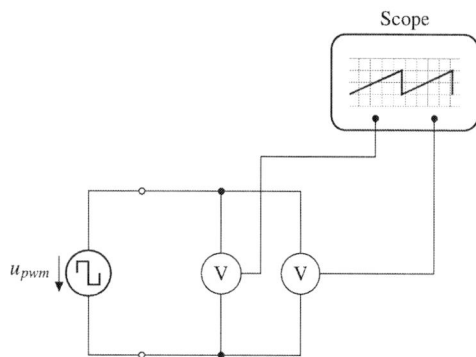

Fig. 7: Equivalent circuit of the mismatch determination between used probes

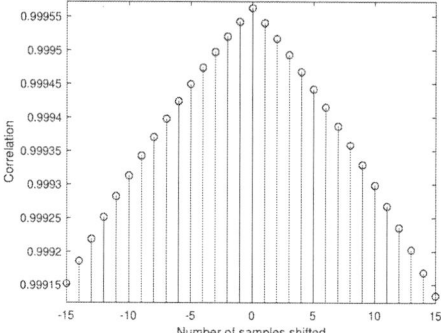

Fig. 8: Correlation of recorded signals depended on multiple shifts for mismatch determination

The limited accuracy of the scale used to weigh the cores leads to $\Delta m_{\mathrm{core}} = \pm 0.01\,\mathrm{g}$ and the resistance of the shunt resistor $R_{\mathrm{shunt}} = 0.2521\,\Omega$ has a tolerance of $\pm 0.2\,\%$.
The errors in the radii are considered as $\Delta r_{\mathrm{in}} = \Delta r_{\mathrm{out}} = \pm 50\,\mu\mathrm{m}$.

The signal amplitudes can be influenced by the probe's gain accuracy ($\pm 0.5\,\%$) and the scope. The scope's accuracy for AC signals is obtained by the frequency response of each channel with AC coupling and specified as $\pm 0.5\,\mathrm{dB} \approx \pm 6\,\%$ up to the cut-off frequency of $f_{\mathrm{cut-off}} = 1\,\mathrm{GHz}$. Both, the DC accuracy and offset error of the scope can be calculated as $\pm 0.5\,\%$ of the set full scale of each channel, which varies between measurements.

Phase discrepancies $\Delta\phi_{\mathrm{actual}}$ in this setup can be caused by the scope's sampling resolution, a mismatch between probes and the parasitic inductance of the shunt resistor. The sampling resolution results in a time delay of $\Delta t_{\mathrm{scope}} = 50\,\mathrm{ps}$ for the maximum sample rate of $10\,\mathrm{GS/s}$. The resulting phase error can be calculated with the period T of the corresponding frequency with following equation:

$$\Delta\phi_{\mathrm{scope}} = \frac{\Delta t_{\mathrm{scope}}}{T}2\pi \tag{5}$$

To specify the mismatch between the probes, a high frequency square wave signal is measured simultaneously with both probes (Fig. 7). One of the recorded signals is shifted in time by an integer number of samples ($1\,\mathrm{S} \hat{=} 100\,\mathrm{ps}$). For every shift the correlation of both signals is calculated. This procedure results in Fig. 8 showing a maximum correlation for no shift. Since shifting is done with the time discretization of the scope and the error caused by this time discretization has already been considered (5), the mismatch can be neglected.

The phase discrepancy caused by the inductance of the shunt $L_{\mathrm{ESL}} \approx 35\,\mathrm{pH}$ can be calculated with (6) by means of the angular frequency ω.

$$\Delta\phi_{\mathrm{shunt}} = \arctan\left(\frac{\omega L_{\mathrm{ESL}}}{R_{\mathrm{shunt}}}\right) \tag{6}$$

With (7) and $\Delta\phi_{\mathrm{actual}} = \pm(\Delta\phi_{\mathrm{scope}} + \Delta\phi_{\mathrm{shunt}})$, the overall phase error can be calculated.

$$\Delta P_{\mathrm{Fe,dens},\phi_{\mathrm{actual}}} = \frac{dP_{\mathrm{Fe,dens,weight}}}{d\phi_{\mathrm{actual}}}\Delta\phi_{\mathrm{actual}} = -\frac{N_{\mathrm{pri}}}{2N_{\mathrm{sec}}R_{\mathrm{shunt}}m_{\mathrm{core}}}\hat{u}_{\mathrm{shunt}}\hat{u}_{\mathrm{sec}}\sin(\phi_{\mathrm{actual}})\Delta\phi_{\mathrm{actual}} \tag{7}$$

It should be mentioned that the scope adapts the attenuation of each channel for resolutions above $100\,\mathrm{mV/div}$ to measure higher voltages and thereby, the time constant of the corresponding channel inevitably changes as well. However, for loss measurement it is crucial for the applied channels to be subject to the same time constant, otherwise this would result in severe errors.

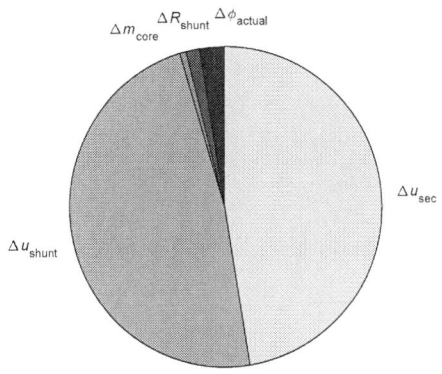

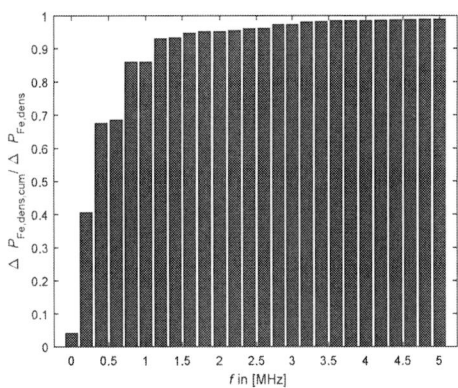

Fig. 9: Individual influences on total error of errors in phase discrepancy $\Delta\phi_{\text{actual}}$, weight Δm_{core}, value of shunt resistor ΔR_{shunt}, amplitude and offset of $u_{\text{shunt}}(t)/u_{\text{sec}}(t)$ for a typical measurement

Fig. 10: Cumulative error in loss density $\Delta P_{\text{Fe,dens,cum}}$ over frequency, standardized to total error $\Delta P_{\text{Fe,dens}}$, both calculated with FFT for a typical measurement

Calculation of total error

A Fast Fourier Transformation (FFT) is utilized for the total error calculation after each measurement. The resulting sinusoidal waveforms for each frequency simplify the further proceeding and allow to distinguish individual influences of all error sources and frequencies. Fig. 9 depicts the error distribution of an exemplary measurement with the cumulative loss density error over the first frequency components of the FFT in Fig. 10. It turns out that processing $u_{\text{shunt}}(t)$ and $u_{\text{sec}}(t)$ with the scope has the greatest impact on the total error. This can be explained by the scope's AC accuracy of $\pm 0.5\,\text{dB} \approx \pm 6\,\%$. The non-ideal DC filter deployed in the test bench leads to a small DC component of $u_{\text{shunt}}(t)$ causing the error component for $f = 0\,\text{Hz}$ in Fig. 10 with the scope's DC and offset error.

As can bee seen in Fig. 9, the error caused by phase discrepancies of $\Delta P_{\text{Fe,dens},\phi_{\text{actual}}}/\Delta P_{\text{Fe,dens}} < 3\,\%$ is small, although both signals measured entail very high frequencies caused by the approximately linear rising and falling current $i_{\text{pri}}(t)$. Fig. 11 shows $\Delta\phi_{\text{actual}}$ calculated by (5) - (6) and plotted up to the nyquist frequency of the scope. Despite very large phase errors for high frequencies, the influence on the total error remains relatively small due to the decreasing influence of high frequency components (Fig. 10).

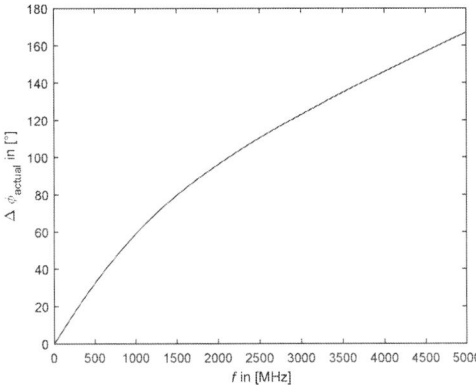

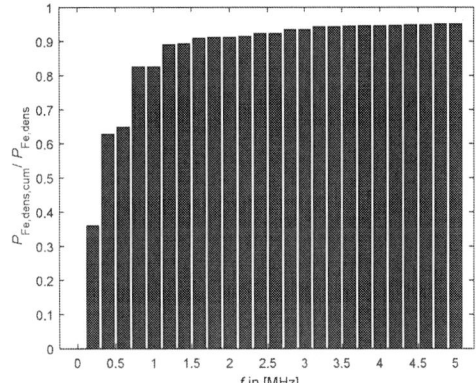

Fig. 11: Phase error $\Delta\phi_{\text{actual}}$ up to the nyquist frequency of the scope for a typical measurement

Fig. 12: Cumulative loss density $P_{\text{Fe,dens,cum}}$ over frequency calculated with FFT and standardized to total loss density $P_{\text{Fe,dens}}$ calculated by (3) for a typical measurement

The cumulative loss density of individual frequency components normalized to the loss density derived by (3) can be seen in Fig. 12 proving that the overall loss densitiy is mainly due to frequency components below $f < 5\,\mathrm{MHz}$ (here $> 95\,\%$ of losses can be explained). However, due to the nonlinear behavior of core losses mentioned above, the FFT approach isn't quantitatively accurate in calculating core losses [2] [15]. To maintain high accuracy in loss calculation, the FFT is just utilized for determining and analyzing the total error.

Depending on the waveform, the total error lies within a range of about $13\,\% < \Delta P_{\mathrm{Fe,dens}} < 16\,\%$ for the loss density and $6\,\% < \Delta B_{\mathrm{p}} < 9\,\%$ for the flux density. This is because of varying impacts of higher frequency components.

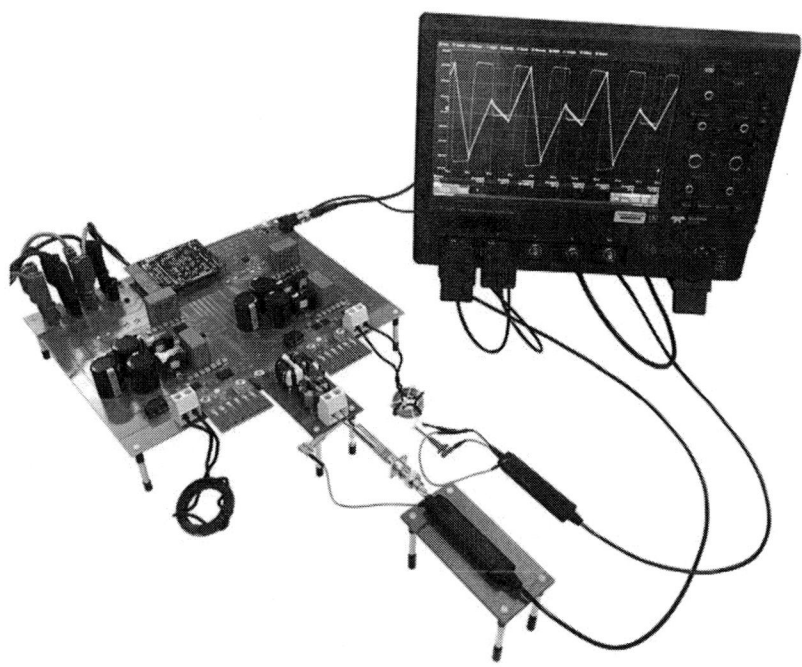

Fig. 13: Measurement setup with the scope (Lecroy HDO6104A), the differential probes (Lecroy DL10-HCM), the test core (Fig. 14) and the PCB

Experimental results

First verifications of the test bench proposed are performed by determining two hysteresis curves of a nano-crystalline tape wound core based on VITROPERM®850FF [8]. All information about the test core is compiled in Table I. Other cores such as powder cores can be inserted as well.

Table I: Data of test core

inner radius	r_{in}	$12.55\,\mathrm{mm}$
outer radius	r_{out}	$15.575\,\mathrm{mm}$
width	h_{core}	$6.15\,\mathrm{mm}$
mass	m_{core}	$10.53\,\mathrm{g}$
mass density	ρ	$7350\,\mathrm{kgm}^{-3}$
fill factor	η	$0.871\,56$
rel. permeability	μ_{r}	1500
layer thickness	b_{tape}	$18.183\,\mu\mathrm{m}$

Fig. 14: test core introduced in Table I with primary and secondary side winding

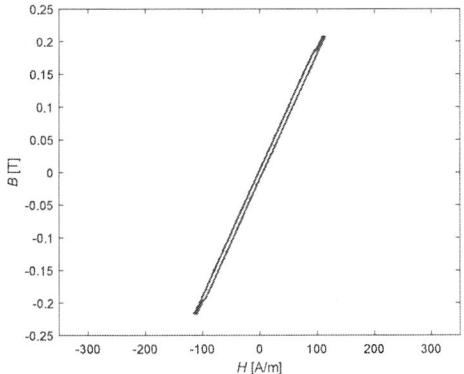
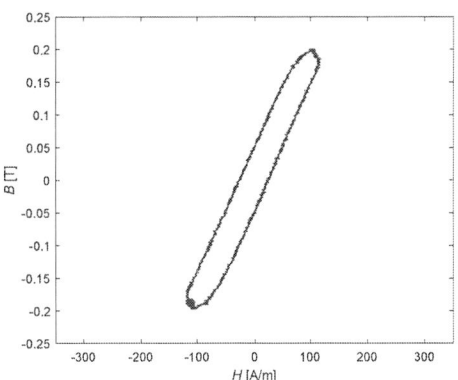

Fig. 15: Hysteresis curve of test core (Table I) for coupling path waveform with $B_p = 0.2\,\text{T}$, $d = 0.3$ and $f = 100\,\text{kHz}$

Fig. 16: Hysteresis curve of test core (Table I) for coupling path waveform with $B_p = 0.2\,\text{T}$, $d = 0.3$ and $f = 700\,\text{kHz}$

To avoid leakage flux in the air surrounding the core and to keep the coupling between the primary and secondary side high, windings are evenly distributed around the core (Fig. 14). This is important especially for cores with low permeability [3].

The material used has a nominal relative permeability of $\mu_r = 1500$, which corresponds to the following gradient of both hysteresis curves in Fig. 15 for $f = 100\,\text{kHz}$ and in Fig. 16 for $f = 700\,\text{kHz}$:

$$\mu_r = \frac{dB(t)}{dH(t)} \frac{1}{\mu_0} \approx 1500 \tag{8}$$

The small enclosed area of the B-H loop in Fig. 15 suggests very low hysteresis losses for the nano-crystalline tape material. For higher frequencies the enclosed area becomes significantly larger and with the proportionality given between the loop area and core losses, this can be directly attributed to a strong increase in losses.

There are different versions of the investigated nano-crystalline material varying in their relative permeability. As shown in Fig. 17, three cores are contrasted regarding their loss density and it can be seen, that with increasing permeability, losses slightly decrease, which was expected.

Fig. 18 shows measured loss densities $P_{\text{Fe,dens}}$ for three frequencies f over the flux density B_p, which

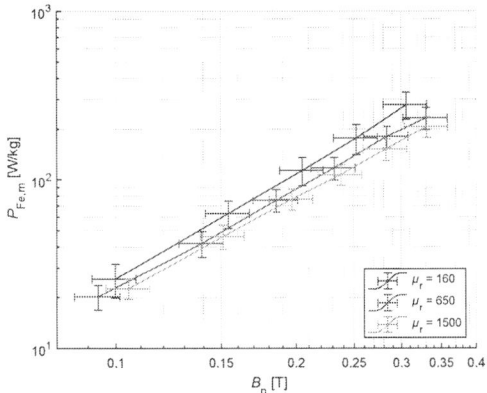
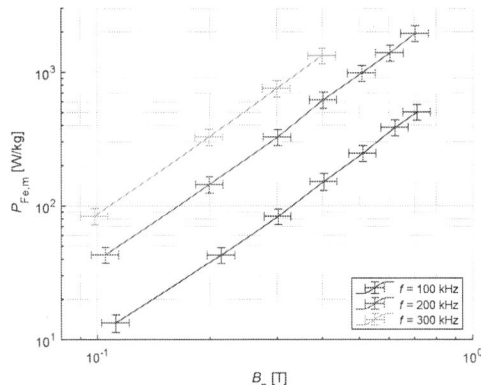

Fig. 17: Loss densities for three cores differing in their relative permeability μ_r over the flux density B_p specified for the leakage path waveform with $f = 150\,\text{kHz}$, $d = 0.4$ and $k = 0.5$

Fig. 18: Loss densities for the test core (Table I) for three different frequencies f over the flux density B_p specified for the coupling path waveform with $d = 0.3$

underlines the mentioned nonlinear dependence of core losses on both, the frequency and flux.

Core losses are proportional to $dB(t)/dt$. In order to achieve the same maximum flux B_p with smaller duty cycles, higher $dB(t)/dt$ are required. It follows that a duty cycle of $d = 0.5$ corresponds to minimal losses. As can be seen in Fig. 19, losses decrease with increasing duty cycle for $d < 0.5$ and would rise for $d > 0.5$ again.

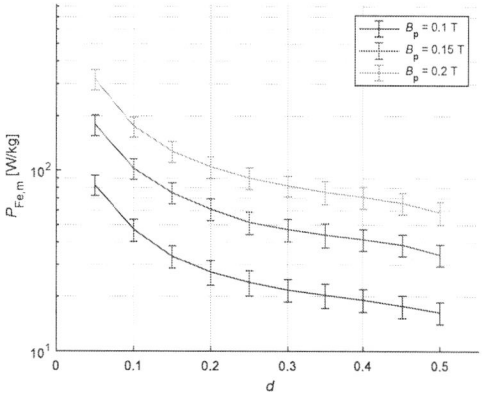

Fig. 19: Loss densities for the test core (Table I) for different duty cycles d specified for the leakage path waveform with $f = 150\,\text{kHz}$, $k = 0.4$ and different B_p

Fig. 20: Loss densities for the test core (Table I) for different coupling factors k specified for the leakage path waveform with $d = 0.3$, $f = 200\,\text{kHz}$ and different B_p

The total flux $B(t)$ in two-phase CI depends on the flux induced by the self inductance and on the $180°$-interleaved flux induced by the coupling inductance. A high coupling factor k means a pronounced $180°$-interleaved flux in the leakage path leading to higher losses (Fig. 20). The reason for that is the second harmonic of the fundamental frequency in $B(t)$ (compare Fig. 1) increasing with the coupling factor.

The leakage path waveform in Fig. 1 corresponds to triangular waveforms typically for single filter inductors in DC/DC converters for a coupling factor of $k = 0$ (compare Fig. 21). Fig. 22 depicts a comparison between loss densities measured with a coupling factor of $k = 0.4$ according to CI and $k = 0$ according to single inductors. CI show higher losses for duty cycles of $d < 0.5$ and $d > 0.5$ caused by the pronounced

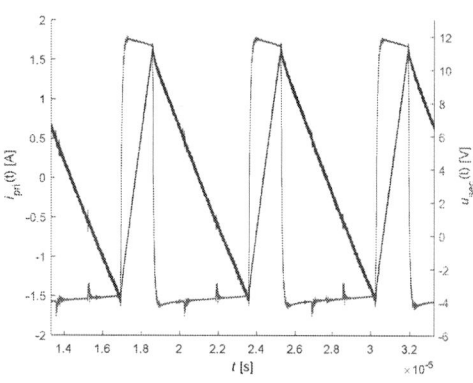

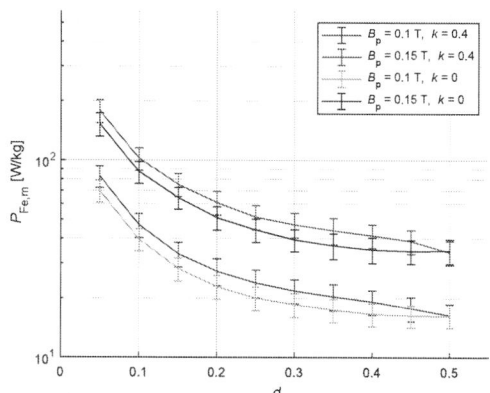

Fig. 21: Current $i_{\text{pri}}(t)$ and voltage $u_{\text{sec}}(t)$ corresponding to triangular waveforms typically for single inductors and measured for the leakage path waveform with $k = 0$, $f = 150\,\text{kHz}$, $B_p = 0.1\,\text{T}$ and $d = 0.25$

Fig. 22: Loss densities for the test core (Table I) for different duty cycles d specified for the leakage path waveform with $f = 150\,\text{kHz}$, different B_p and with $k = 0.4$ for CI and $k = 0$ according to single inductors

second harmonic of the fundamental frequency in $B(t)$ for $k > 0$.

With $d = 0.5$ the leakage path waveform in CI equals the triangular waveform in single inductors independent of k and so is the loss density.

Due to the automated measurement process controlled by a PC, it is possible to characterize core losses for an amount of twenty points of operation in less than $1.5\,\text{min}$, including error calculation for each measurement. The measurement duration varies depending on the frequency since lower frequencies require more samples per period to be processed.

Reproducing current shapes for loss measurements is crucial. For frequencies below $f < 100\,\text{kHz}$ large DC link capacitances with low ESL are necessary to maintain a constant DC link voltage of HB3 and HB4 in Fig. 6. The low inductance DC link capacitance deployed in this test bench is not sufficiently large to deliver such linear rising and falling current shapes (Fig. 4 and Fig. 5) at frequencies below $f < 100\,\text{kHz}$. Therefore, this test setup is currently limited to higher frequencies $f > 100\,\text{kHz}$ until the DC link capacity for HB3 and HB4 has been enlarged.

The GaN switches used, operate reliable up to frequencies $f \leq 900\,\text{kHz}$, which sets the upper frequency limit.

The maximum flux B_p which can be measured is dependent on the concrete core geometry but shouldn't exceed $B_\text{p} = 1\,\text{T}$ for the investigated material to avoid saturation.

Conclusion

The developed test bench shows good accuracy and a very high speed in measuring losses of soft magnetic materials. The automated test bench is easy to handle and fast in investigating different materials and waveforms, especially for two-phase CI.

The improvements for common mode suppression make the measurement results reliable and reproducible independent of the setup.

The low inductance DC Link capacitance needs to be enlarged for extending the frequency range for measurements to lower frequencies.

A calorimetric comparison is planned in near future.

References

[1] P. Deck and C. P. Dick: Material savings using coupled inductors in hard switched power-electronic building blocks: Modeling and experimental validation, 2016 18th European Conference on Power Electronics and Applications (EPE'16 ECCE Europe), 2016, pp. 1-10, doi: 10.1109/EPE.2016.7695346.

[2] M. Mu: High Frequency Magnetic Core Loss Study, Dissertation, February 2013.

[3] A. Stadler: Messtechnische Bestimmung und Simulation der Kernverluste in weichmagnetischen Materialen, Dissertation, Erlangen, 2009

[4] Z. Ma, J. Yao, Y. Li and S. Wang: Comparative Analysis of Magnetic Core Loss Measurement Methods with Arbitrary Excitations, 2019 IEEE Energy Conversion Congress and Exposition (ECCE), 2019, pp. 4125-4130, doi: 10.1109/ECCE.2019.8913150.

[5] G. Herzer, V. Budinsky, C. Polak: Magnetic properties of FeCuNbSiB nanocrystallized by flash annealing under high tensile stress, 2011, physica status solidi (b) 248, doi:10.1002/pssb.201147088.

[6] C. Jiang, X. Li, S. S. Ghosh, H. Zhao, Y. Shen and T. Long: Nanocrystalline Powder Cores for High-Power High-Frequency Power Electronics Applications, in IEEE Transactions on Power Electronics, vol. 35, no. 10, pp. 10821-10830, Oct. 2020, doi: 10.1109/TPEL.2020.2979069.

[7] J. Petzold: Advantages of softmagnetic nanocrystalline materials for modern electronic applications, Journal of Magnetism and Magnetic Materials, Volumes 242–245, Part 1, 2002, Pages 84-89, ISSN 0304-8853

[8] G. Herzer, V. Budinsky, C. Polak: Magnetic properties of nanocrystalline FeCuNbSiB with huge creep induced anisotropy, Journal of Physics: Conference Series, Volume 266 (2011) 012010, doi:10.1088/1742-6596/266/1/012010

[9] F. Dong Tan, J. L. Vollin and S. M. Cuk: A practical approach for magnetic core-loss characterization, in IEEE Transactions on Power Electronics, vol. 10, no. 2, pp. 124-130, March 1995, doi: 10.1109/63.372597.

[10] J. Muhlethaler, J. Biela, J. W. Kolar and A. Ecklebe: Core Losses Under the DC Bias Condition Based on Steinmetz Parameters, in IEEE Transactions on Power Electronics, vol. 27, no. 2, pp. 953-963, Feb. 2012, doi: 10.1109/TPEL.2011.2160971.

[11] D. Hou, M. Mu, F. C. Lee and Q. Li: New High-Frequency Core Loss Measurement Method With Partial Cancellation Concept, in IEEE Transactions on Power Electronics, vol. 32, no. 4, pp. 2987-2994, April 2017, doi: 10.1109/TPEL.2016.2573273.

[12] P. Y. Huang and T. Shimizu: High Power/Current Inductor Loss Measurement with Shunt Resistor Current-sensing Method, 2018 International Power Electronics Conference (IPEC-Niigata 2018 -ECCE Asia), 2018, pp. 2165-2169, doi: 10.23919/IPEC.2018.8507755.

[13] N. F. Javidi and M. Nymand: Error analysis of high frequency core loss measurement for low-permeability low-loss magnetic cores, 2016 IEEE 2nd Annual Southern Power Electronics Conference (SPEC), 2016, doi: 10.1109/SPEC.2016.7846098

[14] F. Schnabel and M. Jung: Calorimeter for exact determination of power losses, PCIM Europe 2019, International Exhibition and Conference for Power Electronics, Intelligent Motion, Renewable Energy and Energy Management

[15] M. Albach, T. Durbaum and A. Brockmeyer: Calculating core losses in transformers for arbitrary magnetizing currents a comparison of different approaches, PESC Record. 27th Annual IEEE Power Electronics Specialists Conference, 1996, pp. 1463-1468 vol.2, doi: 10.1109/PESC.1996.548774.

Substitution of Nanocrystalline Toroid by Laminated Ferrite Toroid in the Application of a Common-Mode Choke

Lukas Reißenweber, Fritz Wohlrath, Alexander Stadler
COBURG UNIVERSITY OF APPLIED SCIENCES AND ARTS
Friedrich-Streib-Str. 2
Coburg, Germany
E-Mail: lukas.reissenweber@hs-coburg.de
URL: http://www.coburg-university.de

Keywords

«Ferrite», «Impedance measurement», «EMC/EMI», «Filter optimization», «Filtering», «Nanocrystalline Core»

Abstract

In this work, it is investigated if a nanocrystalline toroid of a common-mode (CM) choke can be substituted by a laminated ferrite toroid. Small-signal measurement results of the complex impedance, of laminated toroidal cores of the material T38 with different diameters and layer thicknesses at 25 °C and 100 °C, are presented. Further, the calculated attenuation and the complex permeability are shown. Based on the obtained material data, a CM choke made of a laminated ferrite core is designed and analyzed. The focus is thereby on the comparison with the nanocrystalline CM choke.

Introduction

The high frequencies enabled by SiC- and GaN-semiconductors increase the efficiency of power electronic circuits and lead to a reduction of the component size of passive components, thus reducing the costs. However, the steep voltage rises also lead to undesirable amplitudes at higher frequencies. To meet electromagnetic compatibility (EMC) requirements, these must be damped if they cannot be avoided. The CM chokes made of ferrites have been more and more replaced since the development of the nanocrystalline cores. Ferrites are classified into MnZn and NiZn type, which is more suitable for high frequencies (higher Q-factor due to lower eddy-currents as a result of lower electrical conductivity). However, NiZn ferrites have a lower permeability ($\mu'_{r,25°C}$: 15 - 2300) and lower saturation flux density ($B_{sat.,25°C}$: 0.22 - 0.42 T) compared to MnZn ferrites ($\mu'_{r,25°C}$: 350 - 20000, $B_{sat.,25°C}$: 0.31 - 0.55 T) [1-3]. The nanocrystalline material is particularly suitable compared to ferrite due to the higher permeability ($\mu'_{r,25°C}$: 100 - 250000) and higher saturation flux density ($B_{sat.,25°C}$: 0.32 - 1.45 T) [4, 5]. Its real part of the permeability decreases above a certain frequency (around kHz) gradually as frequency increases. Compared to that, the ferrite drops abruptly around MHz (s. fig. 10). In addition, nanocrystalline cores have a low temperature dependence [4]. However, it is a more expensive material. A replacement by a less expensive material such as ferrite is therefore desirable. Investigations on laminated ferrite cores have shown that macroscopic eddy-current losses can be reduced and the impedance at higher frequencies can be increased [6-8].

Fundamentals of substitution of a nanocrystalline toroidal core

When designing a CM filter, it must be known what insertion loss or impedance is required. Typically, the spectrum of interference of the circuit is first measured without a filter. Then the permissible interference is subtracted according to the used standard [e.g. EN55011, EN55012, EN55025 or EN55032]. The result is the spectrum to be damped and so the required insertion loss or filter impedance is known [9, 10]. When developing a CM choke, dimensional effects (macroscopic eddy-currents and resonance effects) must be considered. The greater the difference between the inner and outer radius, the more inhomogeneous is the flux distribution over the cross-section. In addition, the occurring eddy-currents and their corresponding opposing fields are quadratically affected by the size of the cross-sectional area [6, 7, 11-14]. Fig. 1 a) shows the magnitude of the complex impedance normalized to the $f = 10$ kHz value and b) the argument of different sized toroidal cores (R16, R25 and R63) of the material T38 at $\vartheta = 25$ °C. It can be seen that due to the reasons mentioned above, the largest core already loses impedance at lower frequencies. Also, the argument of the larger cores first falls to zero and below.

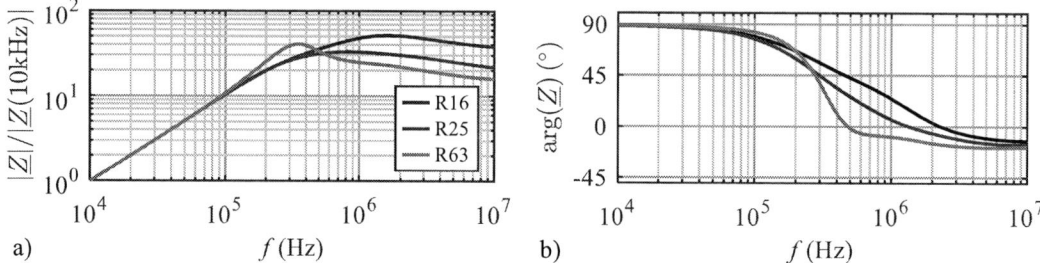

Fig. 1: a) Normalized magnitude, b) argument of different sized T38 toroidal cores at $\vartheta = 25$ °C.

The effect of the opposing fields caused by eddy-currents is illustrated in fig. 3, using a 2D axial symmetric magnetic FEM simulation. It shows the distribution of magnetic field strength H_φ in the core cross-section at different frequencies normalized to the highest occurring field strength $H_{\varphi,\max}$. Static values for the real part of the relative permeability $\mu_r' = 1 \cdot 10^4$, the electrical conductivity $\kappa = 6$ S/m and the real part of the relative permittivity $\varepsilon_r' = 1 \cdot 10^5$ are used. Such distributions are experimentally verified by measurements in [6, 7]. The upper row shows the field distribution of a R16 core as frequency increases. At $f = 1$ kHz, the field distribution is almost homogeneous, affected only by the different radii. At $f = 300$ kHz the field reducing effect occurs due to the opposing fields of the eddy currents. At $f = 1$ MHz, these opposing fields are so strong that negative field strength values occur in the center of the core. The effect is even more significant in the larger R63 core. This is clearly shown by the comparison of the field distribution at $f = 300$ kHz. The consequence is a reduction of the inductance (real part of complex permeability) with simultaneously increasing losses. The bottom row shows a laminated R63 core. Due to the smaller cross-sectional areas, the macroscopic eddy-currents are significantly reduced. This is shown by the nearly unchanged flux distribution with increasing frequency between $f = 1$ kHz and $f = 300$ kHz. Such a design of a larger ferrite core can therefore increase the frequency range at which it can be used [6-8]. This is examined here using toroidal cores of the sizes R16, R25 and R63 with different layer thicknesses made of the material T38. A detailed analytical analysis for calculation of core loss and structural effects is provided in [15, 16].

Investigations on laminated T38 toroids

Prototypes

Table I provides an overview of all investigated core samples and their number. $N_{\mathrm{lam.}} x h_{\mathrm{lam.}}$ is the name, where $N_{\mathrm{lam.}}$ is the number of layers and $h_{\mathrm{lam.}}$ the layers thickness. A selection of cores is shown in fig. 2.

Table I: Overview of the investigated core samples.

Core	$N_{\mathrm{lam.}} x h_{\mathrm{lam.}}$	Nbr. of samples	a (mm)	b (mm)	h (mm)	l_e (mm)	A_e (mm²)	$B_{\mathrm{sat.,25°C}}$ (T)	$B_{\mathrm{sat.,100°C}}$ (T)
R16	1x4mm	4	4.80	8.00	4.00	38.52	12.53		
	2x2mm	4							
	4x1mm	4							
R25	1x8mm	3	7.40	12.65	8.00	60.07	41.01	0.43	0.26
	2x4mm	3							
	4x2mm	3							
	8x1mm	2							
R63	1x20mm	3	19.00	31.50	20.00	152.09	244.74		
	2x8mm	1			16.00		195.79		
	6x2.5mm	2			15.00		183.56		
Nanocrystalline		2	10.00	15.00	15.00	76.43	59.19 (*ff* 80%)	1.10	

The layers of the sizes R16 and R25 were produced by milling and subsequent grinding. In order to exclude that the machining process has an influence on the material properties, test samples were additional manufactured using a diamond wire saw. Independent on the varying machining process the same results were obtained. The R63 layers were produced only by diamond wire saw. In order to obtain rectangular cross-sections, the chamfered upper and lower sides of the bulk cores were removed.

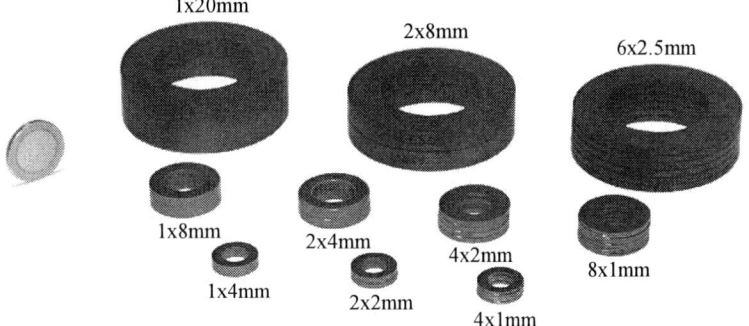

Fig. 2: Prototypes of bulk and laminated R16, R25 and R63 toroidal cores made of T38.

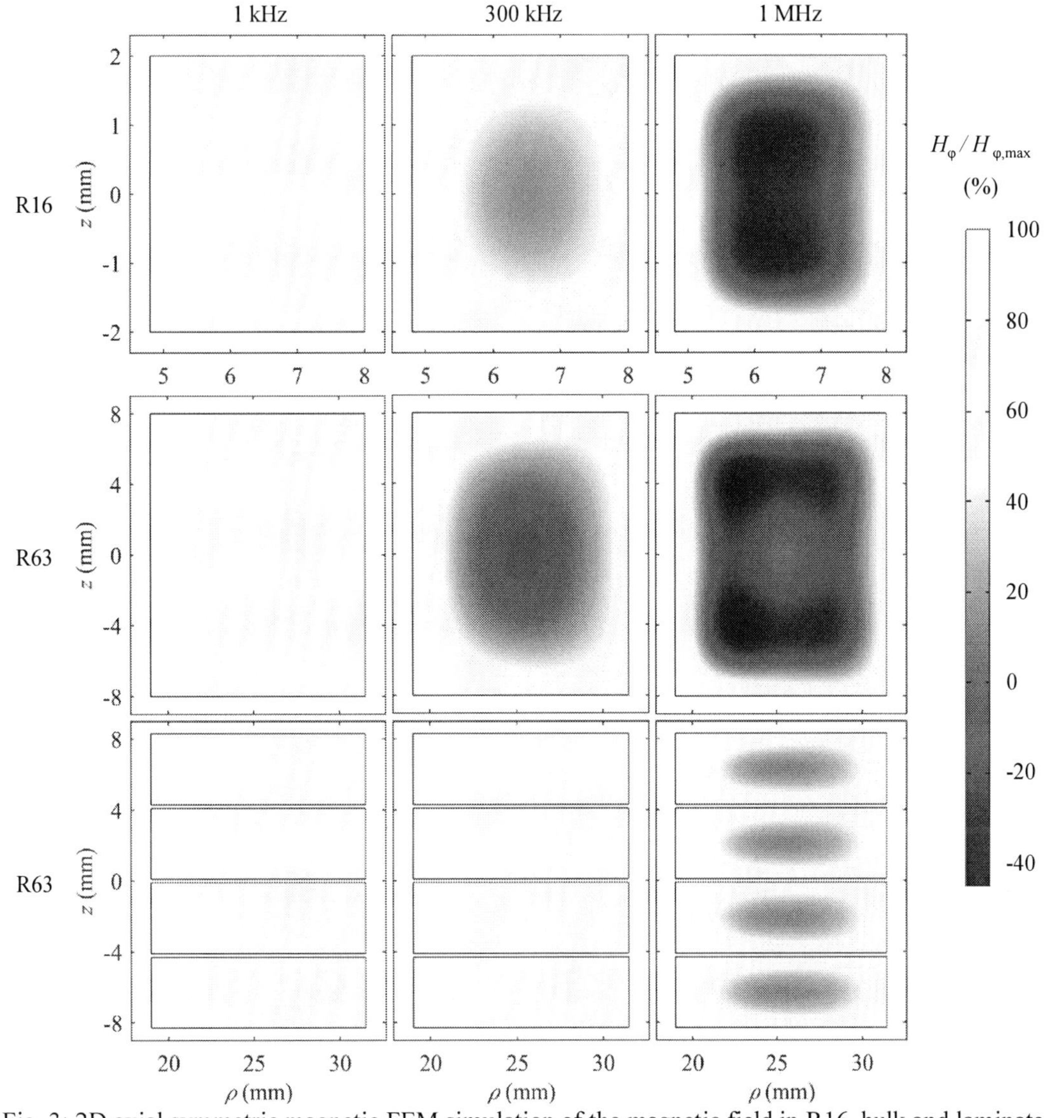

Fig. 3: 2D axial symmetric magnetic FEM simulation of the magnetic field in R16, bulk and laminated R63 toroidal cores at $f = 1$ kHz, $f = 300$ kHz and $f = 1$ MHz.

Adhesive strips serve as spacers between the layers. The fact that these have no influence on the measurement results was proven by measurements without them. All measurements were carried out in a coaxial cartridge manufactured for each core size according to the design in [17]. To eliminate the influence of premagnetization, the samples were demagnetized in advance. In the evaluation, the average values from all measurements of the respective size are shown.

Extension of the frequency range of complex permeability by lamination

The complex permeability of ferrites is influenced by a variety of parameters. These include the material composition, the sintering process, the geometry and the size of a core. The curves of complex permeability presented below are based on small-signal measurements. The equations from standard [IEC-60205] are used for the effective length l_e and the effective cross-section A_e:

$$l_e = 2\pi \frac{ab}{b-a} \cdot \ln\frac{b}{a} \tag{1}$$

$$A_e = h \frac{ab}{b-a} \cdot \left[\ln\frac{b}{a}\right]^2 \tag{2}$$

Therein, parameter a denotes the inner and parameter b the outer radius. The height h of the laminated cores is the sum of the number of layers and their height ($N_{lam.}xh_{lam.}$). The calculation of the complex permeability is based on the effective core parameters (1, 2), the magnetic field constant μ_0, the square of the number of turns N, the angular frequency ω and the real or imaginary part of the measured complex impedance $\underline{Z}$ [3, 4]:

$$\mu_r' = \frac{\text{Im}\{\underline{Z}\}}{\omega \cdot \mu_0 \cdot N^2 \cdot (A_e/l_e)} \tag{3}$$

$$\mu_r'' = \frac{\text{Re}\{\underline{Z}\}}{\omega \cdot \mu_0 \cdot N^2 \cdot (A_e/l_e)} \tag{4}$$

Fig. 4 a) shows the complex permeability of the core size R25 at $\vartheta = 25\,°C$. It can be seen that with decreasing layer thickness the real part of the permeability is more stable up to higher frequencies. Losses also decrease, but extend toward higher frequencies due to the longer lasting real part. Fig. 4 b) depicts the curves at $\vartheta = 100\,°C$. Fig. 5 a) shows the real part of the complex permeability of the bulk R16, R25 and R63 cores and their laminated versions with the thinnest layer thickness investigated. In fig. 5 b), the corresponding imaginary part of the cores are shown. Overall, the previously explained behavior is evident here. In the case of the R25 cores, the lamination achieves a behavior that is nearly similar compared to that of the R16 core. In addition, the mechanical machining of the cores always leads to a change in magnetic properties. We assume that this depends on a change in internal mechanical stresses. A decrease in the real part of the permeability of up to -27.7 % for the R16 cores and -27.6 % for the R25 cores was observed. For R63, an increase of 9.2 % was measured.

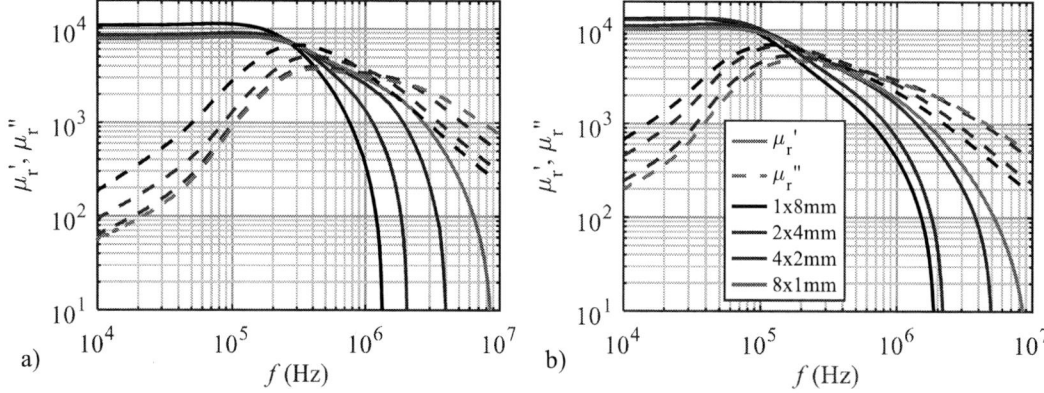

Fig. 4: Complex permeability of bulk and laminated R25 toroids made of T38 at a) $\vartheta = 25\,°C$ and
b) $\vartheta = 100\,°C$.

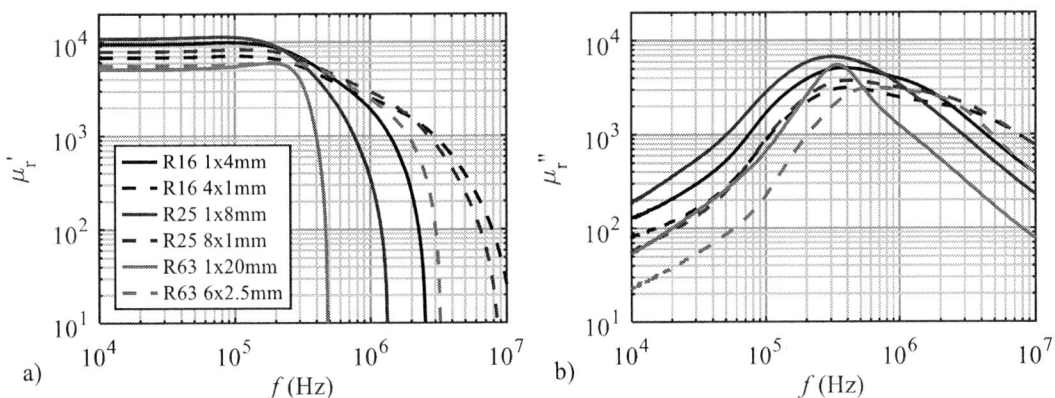

Fig. 5: Complex permeability of bulk and laminated R16, R25 and R63 cores made of T38 at $\vartheta = 25\ °\mathrm{C}$, a) real and b) imaginary part.

Distribution of permeability in axial-direction

When measuring the individual layers of the R63 cores with a layer thickness of $h_{\mathrm{lam.}} = 2.5$ mm, a distribution of the real part of the permeability was observed. Fig. 6 shows the averaged values of the two prototypes at $\vartheta = 25\ °\mathrm{C}$ and a frequency of $f = 10$ kHz. It can be seen that, with the exception of the chamfered top and bottom layer, the permeability decreases from S01 to S03 and then increases again from S04 to S06. In this case, the top and bottom layers deviate from the systematic. Such a distribution can already be introduced in the core by the manufacturing process (position of the parting plane of pressing tool). For the exact knowledge of the flux distribution, this has to be considered.

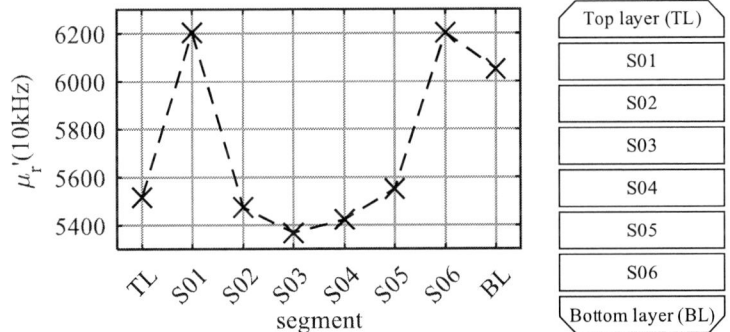

Fig. 6: Distribution of real part of complex permeability in the $h_{\mathrm{lam.}} = 2.5$ mm layers of R63 toroid after cutting and sketch for assignment.

Improved attenuation by increasing lamination

The comparison is based on equal effective core parameters (s. table I) for each core size. For the R63 cores, the curves were scaled (sc.) to $A_{\mathrm{eff}} = 195.79$ mm². Fig. 8 a) shows the magnitude of the complex impedance of the bulk and the laminated R25 cores at $\vartheta = 25\ °\mathrm{C}$. The laminated cores have a lower impedance up to about $f = 620$ kHz due to the reduced real part of the permeability. Above this, the impedance increases significantly. The thinner the layers, the more pronounced the improvement. The same can be seen in the measurements at $\vartheta = 100\ °\mathrm{C}$ in fig. 8 b), although the intersection point here is already about $f = 190$ kHz. Compared to the measurements at $\vartheta = 25\ °\mathrm{C}$ the impedances are higher at lower frequencies and lower at higher frequencies. The comparison of the attenuation of the laminated cores with respect to their bulk cores is made according to:

$$A(\mathrm{dB}) = 20\cdot \log\left(\frac{|Z_{\mathrm{lam.}}|}{|Z_{\mathrm{bulk}}|}\right) \tag{5}$$

In fig. 8 c) the attenuation of the laminated R25 cores compared to the bulk cores are depicted at $\vartheta = 25\ °\mathrm{C}$ and in fig. 8 d) for $\vartheta = 100\ °\mathrm{C}$. Due to the reduced real part of the permeability, the laminated cores initially show lower attenuation (up to $A = -3.0$ dB at $\vartheta = 25$ °C and $A = -2.4$ dB at $\vartheta = 100$ °C). At higher frequencies, the laminated cores show significantly higher attenuation (up to $A = 10.7$ dB at $\vartheta = 25$ °C and $A = 10.4$ dB at $\vartheta = 100$ °C). A comparable behavior is also seen in the measurements of the R16 cores in fig. 9 a) and c). For the R63 cores in fig. 9 b) and d), an improvement is obtained due to the increase of the real part of the permeability in the entire curves. The core with $h_{\text{lam.}} = 2.5$ mm layers shows a reduction of up to $A = -0.76$ dB in the range of $f = 150$ - 330 kHz. A maximum attenuation of up to $A = 14.7$ dB is achieved at $f = 3.6$ MHz.

CM choke made of laminated T38 toroid

The aim of the work is to substitute a nanocrystalline toroidal core by a laminated T38 ferrite core. This is done by an analytical approach with the measured and extrapolated material data. The geometry data of the nanocrystalline core are given in table I. For the effective core parameter A_e a filling factor (*ff*) of 80 % is considered. Also, the saturation flux densities of the nanocrystalline core (measured) and of the ferrite T38 (from datasheet [1]) are shown in table I. Fig. 7 a) illustrates two of the nanocrystalline cores in housings, b) the nanocrystalline core without the housing and c) a laminated R25 core with $h_{\text{lam.}} = 1$ mm layers of T38 scaled to the size of the nanocrystalline core.

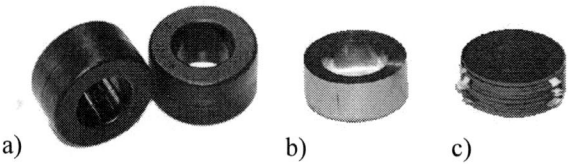

a)   b)   c)

Fig. 7: a) Nanocrystalline cores in housing, b) nanocrystalline core and c) R25 T38 core 8x1mm (sc.).

When designing a CM choke with a small number of turns, the equivalent series circuit of an inductor can be used as a first approximation. The complex impedance follows [3, 4, 14]:

$$\underline{Z} = R(\omega) + j\omega L \tag{6}$$

Here R represents the occurring losses and L the inductance of the arrangement. The aim is to achieve the highest possible amount of attenuation with the lowest possible material input. In addition, the saturation of the material must be considered [3, 4, 14]:

$$\hat{B} = \frac{L \cdot \hat{I}}{N \cdot A_e} \tag{7}$$

According to the occurring current $\hat{I}$, it must be ensured that the magnetic flux density $\hat{B}$ is below the saturation flux density $B_{\text{sat.}}$ (s. table I). The ferrite is designed to have the same radii as the nanocrystalline core. The height h and the number of turns N is adjusted so that the inductance at $f = 10$ kHz at $\vartheta = 25$ °C is comparable, considering the decrease of the real part of permeability. In addition to the complex impedance, the Q-factor, the quotient of the magnitude of the imaginary part of $\underline{Z}$ to its real part, is a decisive parameter - the smaller the losses, the higher the quality [3, 4, 14]:

$$Q = \frac{|\text{Im}\{\underline{Z}\}|}{\text{Re}\{\underline{Z}\}} = \frac{\omega L}{R} \tag{8}$$

For the complex permeability of the CM chokes made of ferrite, the data of the bulk and those of the R25 core with the thinnest layer thickness $h_{\text{lam.}} = 1$ mm are used. The material data of the R25 cores were chosen because they are closest to the desired geometry. Since the design is to be carried out up to a frequency of $f = 100$ MHz, the determined material data must be extrapolated. For this purpose, a modified debye approach described in [18] is used. Fig. 10 a) shows the result for the complex permeability of the nanocrystalline toroid at $\vartheta = 25$ °C. Fig. 10 b) depicts the extrapolated curves of the R25 core made of $h_{\text{lam.}} = 1$ mm layers. Based on the extrapolated material data, the CM chokes made of nanocrystalline material and T38 ferrite with and without lamination are designed at $\vartheta = 25$ °C and at $\vartheta = 100$ °C according to equations (6-8). For the nanocrystalline choke, one turn ($N = 1$) is used. To keep the height of the ferrite chokes as low as possible, $N = 2$ turns are used. Fig. 11 a) shows the magnitude of the complex impedance. The bulk ferrite choke exceeds the impedance of the nanocrystalline choke up to a frequency of $f = 8.1$ MHz at $\vartheta = 25$ °C and $f = 4.7$ MHz at $\vartheta = 100$ °C. After that, it is

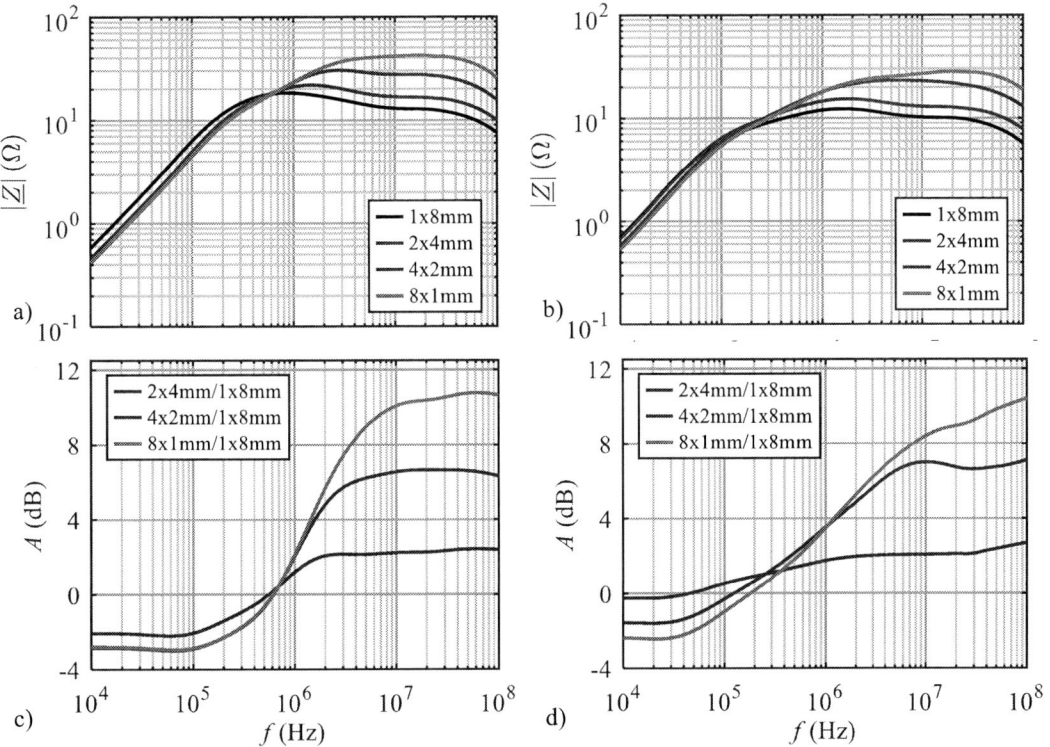

Fig. 8: Magnitude of complex impedance of bulk and laminated R25 T38 cores at a) $\vartheta = 25$ °C, b) $\vartheta = 100$ °C. Attenuation of laminated R25 cores related to bulk cores at c) $\vartheta = 25$ °C, d) $\vartheta = 100$ °C.

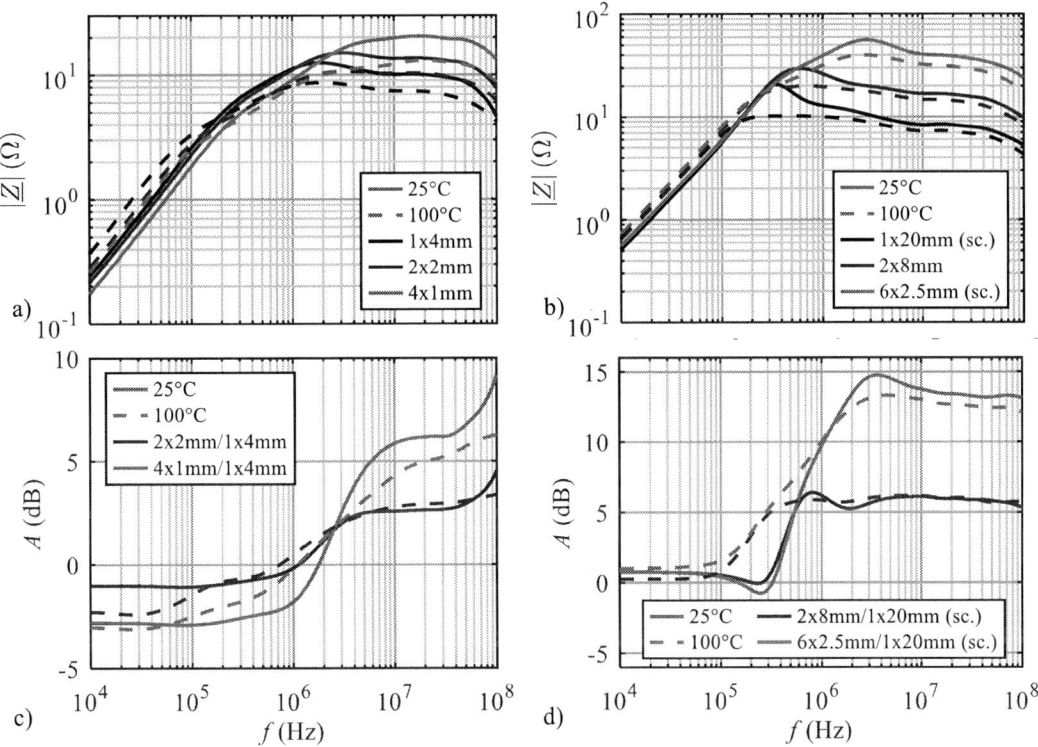

Fig. 9: Magnitude of complex impedance of bulk and laminated T38 cores at $\vartheta = 25$ °C/ 100 °C for a) R16, b) R63. Attenuation of laminated cores related to bulk cores c) R16, d) R63.

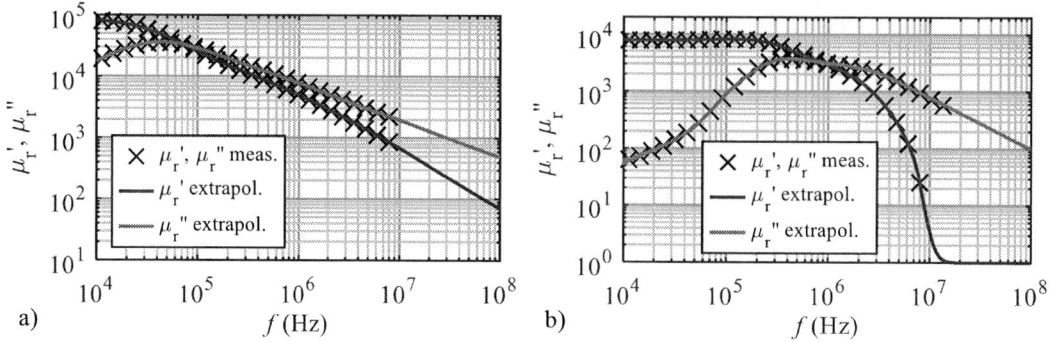

Fig. 10: Extrapolated complex permeability of a) nanocrystalline and b) T38 8x1mm core at $\vartheta = 25$ °C.

significantly lower. The laminated ferrite choke exceeds the impedance of the nanocrystalline choke in the entire frequency range. For the $\vartheta = 100$ °C curves, the ferrite chokes are above the curves of $\vartheta = 25$ °C up to $f = 100$ kHz and $f = 170$ kHz. At higher frequencies they are below. The laminated ferrite choke always shows a higher impedance compared to the nanocrystalline core. If we look at the Q-factor shown in fig. 11 b), we see that the ferrite chokes have a higher quality up to the lower MHz range. However, especially at the higher frequencies, the nanocrystalline choke has a higher quality. It should be noted that this is a simplified approach. Resonance effects due to wave propagation are not considered at this point. This can be seen when the curve of the magnitude of the complex impedance of the laminated core is compared to the curves in fig. 8 a) and b). In measurements, the impedance already drops slightly at frequencies above $f = 40$ MHz. Premagnetization also influences the material behavior. When looking at the calculated saturation currents in table II, the disadvantage of ferrite becomes apparent. If the saturation current at $\vartheta = 25$ °C is still above (10.4 % for the bulk and 50.5 % for the laminated ferrite) the nanocrystalline core, at $\vartheta = 100$ °C it is significantly below (-48.3 % bulk and -32.0 % for the laminated ferrite).

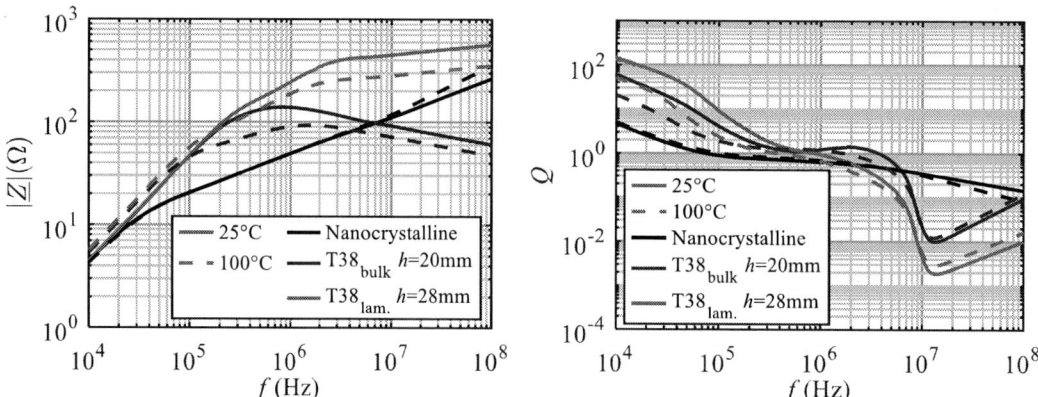

Fig. 11: a) Magnitude of complex impedance of CM chokes made of nanocrystalline material, bulk and laminated T38 ferrite at $\vartheta = 25$ °C/ 100 °C, b) quality factor of the CM chokes.

Conclusion

The substitution of a nanocrystalline core by a laminated ferrite core for a CM choke is possible. Compared to bulk ferrite cores, a significantly improved frequency behavior of the complex permeability and thus of the complex impedance could be achieved by a lamination. The developed laminated ferrite CM choke exhibits a higher impedance compared to the nanocrystalline core. However, the Q-factor is worse at higher frequencies. Also, lower saturation currents are obtained at higher temperatures. The mechanical machining of the cores influences its magnetic properties. A decrease of the real part of the permeability is detected for smaller cores (R16, R25). In contrast an increase for the R63 core were measured. It could be shown that for large cores there is a certain distribution of the real part of complex permeability in axial direction which has to be considered for the calculation of the exact distribution of the magnetic flux in the cross-section of the core.

Table II: Parameters of CM chokes.

Core type	N	h (mm)	$\hat{I}_{sat.,25°C}$ (A)	$\hat{I}_{sat.,100°C}$ (A)
Nanocrystalline	1	15	1.09	1.15
T38$_{bulk}$	2	20	1.20	0.60
T38$_{lam.}$	2	28	1.65	0.79

References

[1] TDK, "Ferrites and Accessories," EPCOS AG, Munich, 2017, www.epcos.com.

[2] Ferroxcube, "Soft Ferrites and Accessories," Ferroxcube International Holding B.V., Taiwan, 2013, www.ferroxcube.com.

[3] P. Zacharias, "Magnetische Bauelemente," Springer Fachmedien Wiesbaden GmbH, Wiesbaden, 2020, doi: 10.1007/ISBN978-3-658-24742-3.

[4] R. Hilzinger, W. Rodewald, "Magnetic Materials," Publicis Publishing, Erlangen, 2013, ISBN978-3-89578-352-4.

[5] M. Ferch, "Nanocrystalline core materials for modern power electronic designs," Magnetec GmbH, 2003, www.magnetec.de.

[6] M. Kącki, M. S. Ryłko, J. G. Hayes and C. R. Sullivan, "A Study of Flux Distribution and Impedance in Solid and Laminar Ferrite Cores," 2019 IEEE Applied Power Electronics Conference and Exposition (APEC), 2019, pp. 2681-2687, doi: 10.1109/APEC.2019.8722252.

[7] M. Kącki, M.S. Ryłko, J.G Hayes, C.R. Sullivan, E. Herbert, "Magnetic core dimensional effects - flux propagation in ferrites," PSMA Workshop during IEEE Applied Power Electronics Conference, 2018.

[8] J. Zhu, K. J. Tseng, P. Hing and C. F. Foo, "Effects of multi-segment structure in core loss reduction of MnZn ferrite," Proceedings IPEMC 2000. Third International Power Electronics and Motion Control Conference (IEEE Cat. No.00EX435), 2000, pp. 58-63 vol.1, doi: 10.1109/IPEMC.2000.885331.

[9] A. Nagel and R. W. De Doncker, "Systematic design of EMI-filters for power converters," Conference Record of the 2000 IEEE Industry Applications Conference. Thirty-Fifth IAS Annual Meeting and World Conference on Industrial Applications of Electrical Energy (Cat. No.00CH37129), 2000, pp. 2523-2525 vol.4, doi: 10.1109/IAS.2000.883177.

[10] S.-P. Weber, "Effizienter Entwurf von EMV-Filtern für leistungselektronische Geräte unter Anwendung der Methode der partiellen Elemente," Ph.D. dissertation, Faculty IV - Electrical Engineering and Computer Science of the Technical University of Berlin, 2007.

[11] G. R. Skutt, "High-Frequency Dimensional Effects in Ferrite-Core Magnetic Devices," Ph.D. dissertation, Virginia Polytechnic Institute and State University, Blacksburg, VA, 1996.

[12] G.R. Skutt, F.C. Lee, "Characterization of dimensional effects in ferrite-core magnetic devices," Power Electronics Specialist Conference, 1996.

[13] G. Hurley, T. Merkin, M. Duffy, "The performance factor for magnetic materials revisited: The effect of core losses on the selection of core size in transformers," IEEE Power Electronics Magazine, vol. 5, no. 3, page: 26-34, September 2018.

[14] E. C. Snelling: "Soft Ferrites, Properties and Application," London: Iliffe Books Ltd., 1969, ISBN-13: 978-0592027906.

[15] A. Stadler, M. Albach and A. Bucher, "Calculation of Core Losses in Toroids with Rectangular Cross Section," 2006 12th International Power Electronics and Motion Control Conference, 2006, pp. 828-833, doi: 10.1109/EPEPEMC.2006.4778502.

[16] W. Hauser and M. Albach, "Analytic model of structural effects in toroid cores with rectangular cross section," 2016 6th International Electric Drives Production Conference (EDPC), 2016, pp. 60-66, doi: 10.1109/EDPC.2016.7851315.

[17] A. Stadler, M. Albach and A. Lindner, "A Practical Method to Measure Electrical AC Conductivity of MnZn Ferrites Using Conventional Toroids," in IEEE Transactions on Magnetics, vol. 46, no. 2, pp. 678-681, Feb. 2010, doi: 10.1109/TMAG.2009.2030157.

[18] L. Reissenweber and A. Stadler, "Modeling of a Power Transformer including Higher Order Resonances," 2020 22nd European Conference on Power Electronics and Applications (EPE'20 ECCE Europe), 2020, pp. P.1-P.9, doi: 10.23919/EPE20ECCEEurope43536.2020.9215926.

Direct Active Stabilization of the DC-Link in Voltage-Source Converters

Matthieu Bertin*, Mohamad Koteich**
*Elsys Design, consultant for Schneider-Toshiba Inverter Europe
**Schneider-Toshiba Inverter Europe
Schneider Electric
27120 Pacy-sur-Eure, France
URL: https://www.se.com/drives

Keywords

≪Voltage-Source Converters≫, ≪DC-Bus≫, ≪Stability Analysis≫, ≪Motor Drives≫

Abstract

This paper presents an active stabilization method of the dc-link voltage in voltage-source converters. It is a direct compensation method, acting on the converter command voltages instead of the current loop reference, which allows for a wider bandwidth. The proposed method is compared to the state-of-the-art direct compensation method, i.e. filtering the dc-voltage used in the pulse-width-modulation process. The effectiveness of the proposed method over the complete operating range is demonstrated with theoretical analysis and simulation results.

Introduction

In voltage-source power converters, such as the Variable Speed Drive (VSD), undamped oscillations of the dc-link deteriorate the performances of the system and may damage the converter [1]. It has been shown that these oscillations can be caused by the control algorithm's design and tuning, which reduces the damping of the dc-link's LC circuit [2].

Most of the active stabilization methods proposed in the literature are based on the Constant-Power Load (CPL) assumption [1, 3, 4, 5]. Although this simplified model is well-suited for control design, more advanced design and tuning require a higher-order model to account for the load dynamics [2, 6, 7, 8]. For both the CPL model and the higher-order model, the stabilization mechanism consists of adding to the load power a component that is proportional to the oscillating dc-voltage signal δv_{dc}. This compensation can be applied indirectly at the input of the (current or power) controller [1, 8], in which case the stabilization controller's bandwidth is limited to the feedback system's bandwidth. Otherwise, it can be applied directly at the output of the controller (voltage command) [4, 9, 10], which allows for wider bandwidth.

In this paper, we only focus on the direct compensation method as it covers a wider range of applications, especially the slim-capacitor VSD [11], where the resonance frequency is in the range of several hundreds, up to several thousands, of hertz. One commonly used direct compensation method is to low-pass filter the dc-voltage used when normalizing the command voltages before the pulse-width-modulation (PWM) process [2, 10]. This method is simple and effective, yet it has not been thoroughly studied in the literature. In this paper, we present a detailed analysis of this method using the small-signal model. It is shown that it acts only on the amplitude of the output voltage, and that the compensation gain is proportional to both the modulation index and the oscillation δv_{dc}. When the output voltage is saturated, the performance deteriorates.

The main contribution of this paper is a new direct compensation method that exploits both the amplitude and the phase of the output voltage. It has two main advantages compared to the state-of-the-art methods:

1) it covers the whole operating range, including voltage saturation (and field-weakening), and 2) it is easier to tune using the virtual capacitance concept (one intuitive tuning parameter), independently of the modulation index.

After this introduction, the dc-link model and stabilization principle are presented in Section 1. The state-of-the-art dc-link voltage filtering method is presented and analyzed in Section 2. In Section 3, the proposed direct method is presented, analyzed, and compared with the state-of-the-art. The results of the theoretical analysis are confirmed by the simulation results.

1 DC-link model and stabilization principle

The equivalent model of the dc-link is nonlinear. It is described as:

$$L_{dc}\frac{di_{src}}{dt} = v_{src} - v_{dc} - R_{dc}i_{src} \tag{1}$$

$$C_{dc}\frac{dv_{dc}}{dt} = i_{src} - \frac{P}{v_{dc}} \tag{2}$$

In this paper, we take the slim-capacitance VSD as an application example. C_{dc} is the (slim) film capacitance, L_{dc} is the grid inductance referred to the dc-link, as there is no dc-choke in the VSD. R_{dc} is the sum of the grid resistance referred to the dc-link and the parasitic resistance due to unmodeled phenomena, such as commutation overlap in the input rectifier. P is the load power (motor and inverter). When the motor is controlled with tight bandwidth control loops, the inverter/motor system is often modeled as a CPL ($P = P_0$, and $\delta P = 0$). In small-signal, around the equilibrium, the model becomes:

$$L_{dc}\frac{d\delta i_{src}}{dt} = \delta v_{src} - R_{dc}\delta i_{src} - \delta v_{dc} \tag{3}$$

$$C_{dc}\frac{d\delta v_{dc}}{dt} = \delta i_{src} - \frac{\delta P}{v_{dc0}} + \frac{P_0}{v_{dc0}^2}\delta v_{dc} \tag{4}$$

Subscript 0 denotes the equilibrium point. The small-signal symbol δ can be omitted in the sequel for simplicity. The model described in equations (3), (4) is prone to instability. The stabilization principle relies on shaping the small-signal behavior of δP so it becomes proportional to the small-signal dc-link voltage $\delta P = K_a \delta v_{dc}$, to compensate for the negative incremental resistance $(-v_{dc0}^2/P_0)$ [2]. K_a is the active damping gain. In practice δv_{dc} is estimated by the high-pass filtered v_{dc}. As the high-pass filter (HPF) approximates the time-derivative in the large-signal domain, this strategy can be interpreted as adding a virtual capacitor C_v in parallel with C_{dc} [3].

Active Stabilization and the Concept of Virtual Capacitance

In a VSD, the power P is not controlled directly. Hence, to shape $P = P_0 + K_a\delta v_{dc}$, the compensation can be performed by acting on:

- the motor current i_s or torque T_e through the control loop (indirect compensation); or

- the output voltage's amplitude, frequency, or phase (direct compensation).

The performance of the active damping can be tuned using the gain K_a. To make this tuning intuitive, we can use the virtual capacitance (C_v) concept presented in [3]; the load power in small-signal is $\delta P = K_a\delta v_{dc}$, where δv_{dc} is equivalent to dv_{dc}/dt in large signal. Let $K_a = C_v v_{dc0}$, the power δP can be interpreted as the power of a virtual capacitance C_v mounted in parallel to C_{dc}. Therefore, the gain K_a can be interpreted as the amount of capacitance to be added to the dc-link.

General Form of Direct Compensation

Direct compensation does not rely on the controller structure or bandwidth. It covers a wider range of resonance frequency as it is limited by the bandwidth of the PWM process instead of the controller. Using complex-vector notation of the output voltage, we propose the following general direct compensation expression (δV_s, $\delta \omega_s$ and $\delta \phi_v$ are proportional to δv_{dc}):

$$\underline{v}_s = (V_s + \delta V_s)e^{j[(\omega_s + \delta \omega_s)t + (\phi_v + \delta \phi_v)]} \tag{5}$$

Where V_s, ω_s, and ϕ_v, are respectively the amplitude, angular frequency, and phase, of the command voltage. This general expression is thoroughly studied in section 3.

2 DC-link Voltage Filtering-based Stabilization

The voltage command at the output of the controller, $\underline{v}_s^* = V_s e^{j(\omega_s t + \phi)}$, is divided by v_{dc} to calculate the duty-cycles of the inverter switches. One common active stabilization method consists of low-pass filtering v_{dc} used in normalization. Taking $v_{dc}^* = LPF\{v_{dc}\} \approx v_{dc} - \delta v_{dc}$, the normalized voltage becomes:

$$\underline{m}_s = \frac{\underline{v}_s^*}{v_{dc}^*} = \frac{V_s}{v_{dc} - \delta v_{dc}}e^{j(\omega_s t + \phi)} = \frac{V_s/v_{dc}}{1 - \delta v_{dc}/v_{dc}}e^{j(\omega_s t + \phi)} \tag{6}$$

In Eq. (6), $\delta v_{dc} << v_{dc}$. Using the first-order approximation $1/(1-x) = 1+x$ for $x \to 0$, Eq. (6) becomes:

$$\underline{m}_s = \frac{V_s(1 + \delta v_{dc}/v_{dc})}{v_{dc}}e^{j(\omega_s t + \phi)}$$

Ideally, we want the applied output voltage, $\underline{v}_s = \underline{m}_s v_{dc}$, to be equal to the command voltage $\underline{v}_s^* = V_s e^{j(\omega_s t + \phi)}$. However, due to the dc-voltage filtering, it becomes:

$$\underline{v}_s = (V_s + m\delta v_{dc})e^{j(\omega_s t + \phi)} \tag{7}$$

where $m = V_s/v_{dc}$ is the modulation index. Note that Eq. (7) is equivalent to Eq. (5), if:

$$\delta V_s = m\delta v_{dc} \quad ; \quad \delta \omega = 0 \quad ; \quad \delta \phi = 0$$

Therefore, filtering the dc-voltage used in normalization is equivalent to adding a component to the output voltage amplitude that is proportional to both δv_{dc} and m. This explains why this is an effective stabilization method, but it also shows the rigidity of its tuning, as it depends on the operating point (m).

Simulation Results

To illustrate this analysis, simulations were performed in the Matlab/Simulink environment on a VSD model with a dc-link resonance at 1.2kHz ($C_{dc} = 12.5\mu F$, $L_{dc} = 1.4mH$). In a slim-capacitance VSD supplied with a 50Hz grid, the 1.2kHz harmonic is naturally present in the dc-voltage due to the three-phase rectification process. Hence, the resonance is continuously excited. Moreover, as a worst case scenario, the damping is neglected ($R_{dc} = 0$) in the simulation, which makes the resonance even more visible.

The following use case is considered: the VSD is driving a 4kW/50Hz induction motor, using Indirect Field-Oriented Control (FOC) with 50Hz bandwidth. A speed ramp is applied up to 150% rated speed, at 80% of rated torque. Fig. 1 shows the dc-link voltage behavior when no stabilization action is applied.

Fig. 2 shows the results of dc-link stabilization when using a first-order low pass filter (LPF) on the dc-link voltage signal at 600Hz. The resonance is effectively suppressed, however the damping cannot

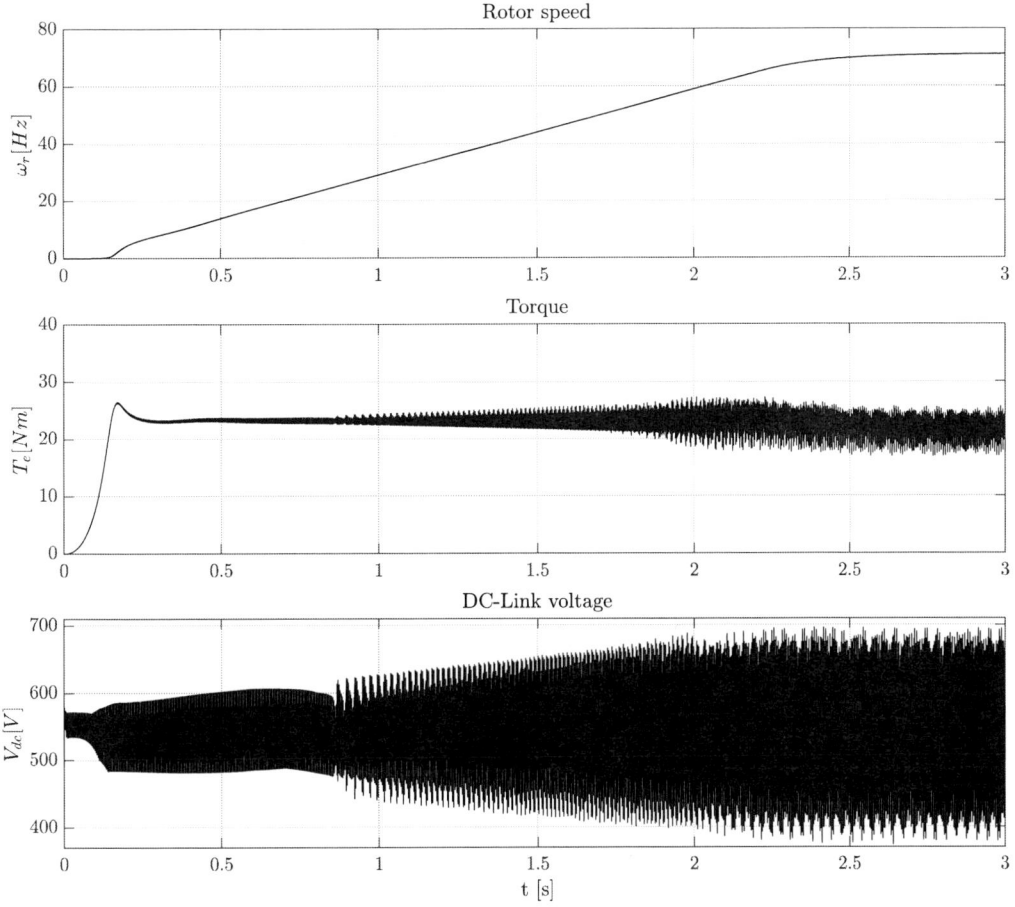

Fig. 1: DC-link resonance during acceleration

be precisely controlled at a specific operating point. The attenuation gain changes with the modulation index, which is clearly visible on the torque waveform, where the ripple amplitude increases with speed. It is possible however to roughly tune the performance by changing the filtering frequency. Choosing a higher filtering frequency, closer to the resonance, means that the naturally present harmonics in the system are not compensated in the modulation process. The consequence is lower torque ripple but higher dc-voltage ripple in all operating points.

3 Output Voltage-based Stabilization

As discussed in Section 1, direct compensation can be performed by adding the dc-link voltage ripple to the amplitude, the frequency, or the phase of the output voltage. Fig. 3 shows the implementation block diagram of the method within a VSD.

If we neglect the inverter losses, the load power can be expressed as:

$$P = \underline{i}_s^T \underline{v}_s = \frac{3}{2} V_s I_s \cos \varphi \tag{8}$$

where V_s is the peak phase voltage, I_s is the peak phase current, and $\cos(\varphi) = \cos(\phi_v - \phi_i)$ is the motor power factor. The current is the response to the voltage through the motor impedance. Therefore, when a δv_{dc} component is added to the voltage in small-signal, the current response is neglected (the higher the frequency, the higher the motor impedance). Note that if we seek a compensation mechanism that gives

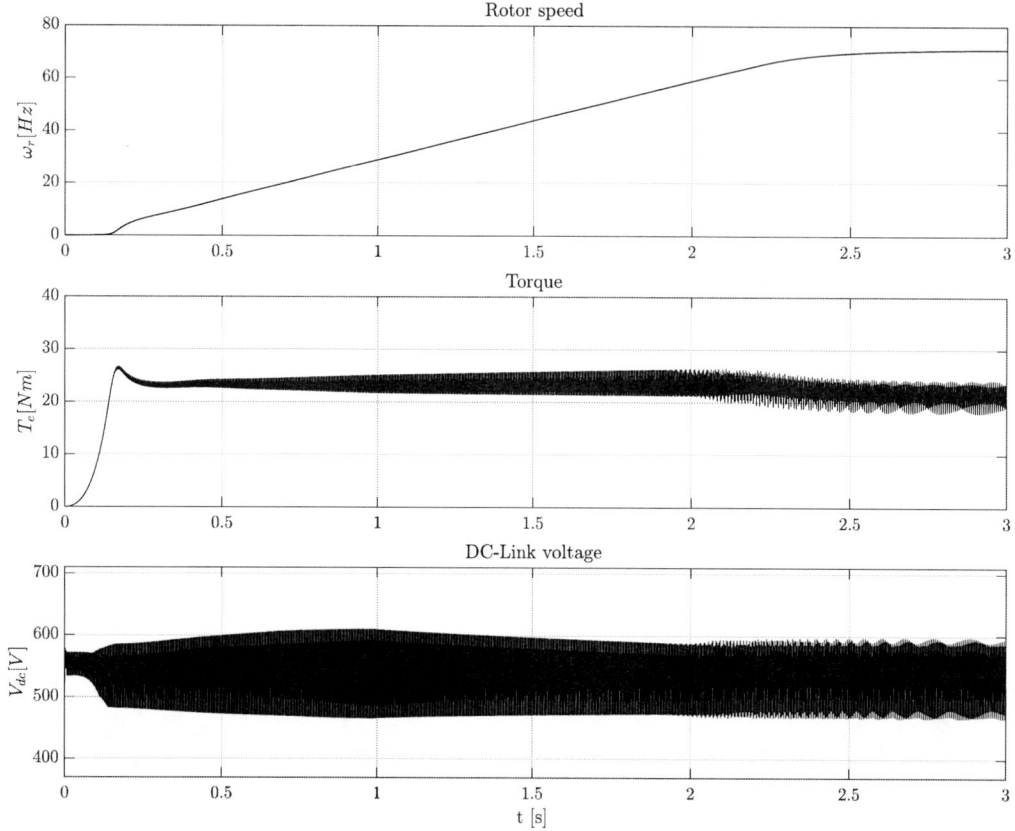

Fig. 2: DC-link stabilization by dc-link voltage filtering at 600Hz

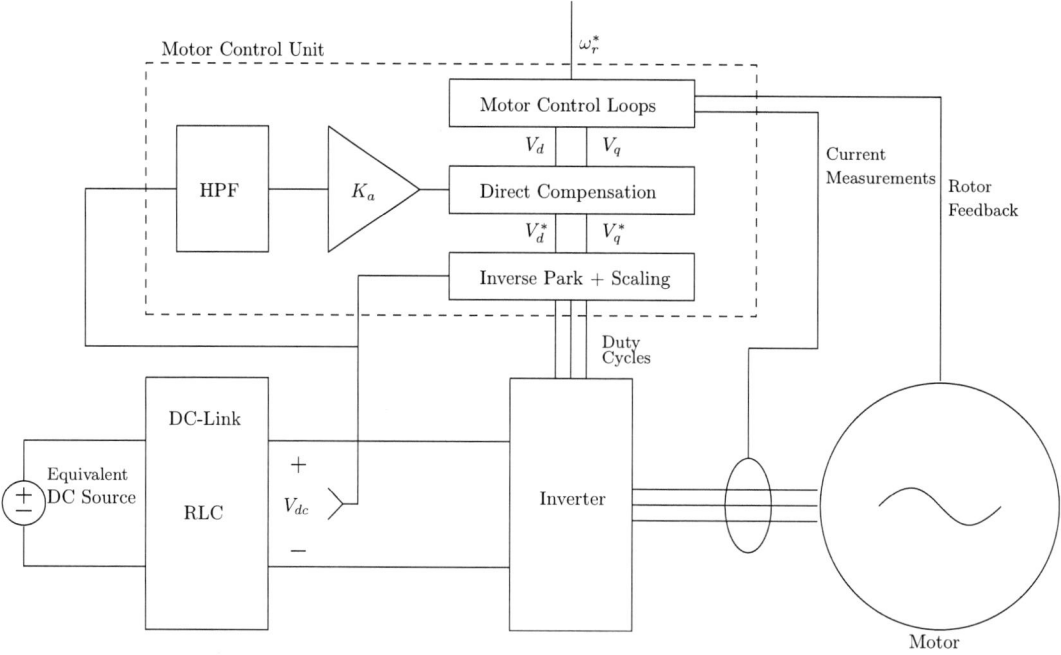

Fig. 3: Output Voltage Based Stabilization block diagram

$P = P_0 + K_a \delta v_{dc}$, only the voltage amplitude V_s and phase angle ϕ_v allow for it. Therefore, they will be thoroughly studied in the sequel. The frequency-based compensation is not in the scope of this study.

Amplitude-based Stabilization (ABS)

Using Eq. (5), amplitude-based stabilization means:

$$\delta V_s = K_v \delta v_{dc} \;\; ; \;\; \delta \omega_s = 0 \;\; ; \;\; \delta \phi_v = 0 \quad \Rightarrow \quad \underline{v_s} = (V_s + K_v \delta v_{dc}) e^{j(\omega_s t + \phi_v)}$$

K_v is the compensation gain applied to the voltage amplitude, which is different from K_a, the resulting compensation gain on the power. This method can be implemented in the synchronous reference frame by modifying v_d and v_q in a way that changes the voltage amplitude only. This can be done as follows:

$$v_d^* = v_d + K_v \delta v_{dc} \frac{v_d}{\sqrt{v_d^2 + v_q^2}} = v_d \left(1 + \frac{K_v \delta v_{dc}}{\sqrt{v_d^2 + v_q^2}} \right) \tag{9}$$

$$v_q^* = v_q + K_v \delta v_{dc} \frac{v_q}{\sqrt{v_d^2 + v_q^2}} = v_q \left(1 + \frac{K_v \delta v_{dc}}{\sqrt{v_d^2 + v_q^2}} \right) \tag{10}$$

The verification of the above calculation can be done by evaluating:

$$V_s = \sqrt{v_d^{*2} + v_q^{*2}} \;\; \text{and} \;\; \phi_v = \arctan(v_q^*/v_d^*)$$

Gain Tuning

The goal is to tune K_v to the value that results in the desired virtual capacitance in $K_a = C_v v_{dc0}$. From Eq. (8), the small-signal equation of the motor power can be derived as:

$$\delta P = \frac{3}{2}(V_{s0} \cos(\varphi_0) \delta v_s + V_{s0} \cos(\varphi_0) \delta i_s - V_{s0} I_{s0} \sin(\varphi_0) \delta \phi_v) \approx \frac{3}{2} I_{s0} \cos(\varphi_0) \delta v_s \tag{11}$$

With $\varphi_0 = \phi_{v0} - \phi_{i0}$. δi_s and $\delta \phi_i$ are neglected and $\delta \phi_v = 0$ by design, as only the amplitude is changed. To get $\delta P = K_a \delta v_{dc}$, with $K_a = C_v v_{dc0}$, and $\delta v_s = K_v \delta v_{dc}$, we can equate:

$$K_a \delta v_{dc} = \frac{3}{2} I_{s0} \cos(\varphi_0) \delta v_s \;\; \Leftrightarrow \;\; C_v v_{dc0} \delta v_{dc} \approx \frac{3}{2} I_{s0} \cos(\varphi_0) K_v \delta v_{dc} \;\; \Leftrightarrow \;\; K_v = \frac{2}{3} \frac{v_{dc0} C_v}{I_{s0} \cos(\varphi_0)}$$

Using this equation, and assuming $\cos \varphi_0 \approx 1$, the compensation gain tuning can be simplified:

$$K_v = \frac{2}{3} \frac{V_n C_v}{I_n} \tag{12}$$

where V_n, and I_n are respectively the rated phase voltage and the rated phase current. The gain can be easily changed by tuning the virtual capacitor's value.

Simulation results

Fig. 4 shows the ABS performance in simulation on the same setup as described in the previous section, using a 25µF virtual capacitance ($2 \times C_{dc}$), up to 150% of rated speed. The algorithm was implemented using equations (9) and (10) and a 600Hz HPF to approximate δv_{dc}. One can see that the ABS scheme has consistent performance over the whole speed range up to the voltage limitation. The damping can easily be changed by tuning K_v.

Fig. 5 shows the ABS performances with a varying gain $K_v = m$. As expected from theoretical results from section 2, this result is exactly similar to the one shown in Fig. 2.

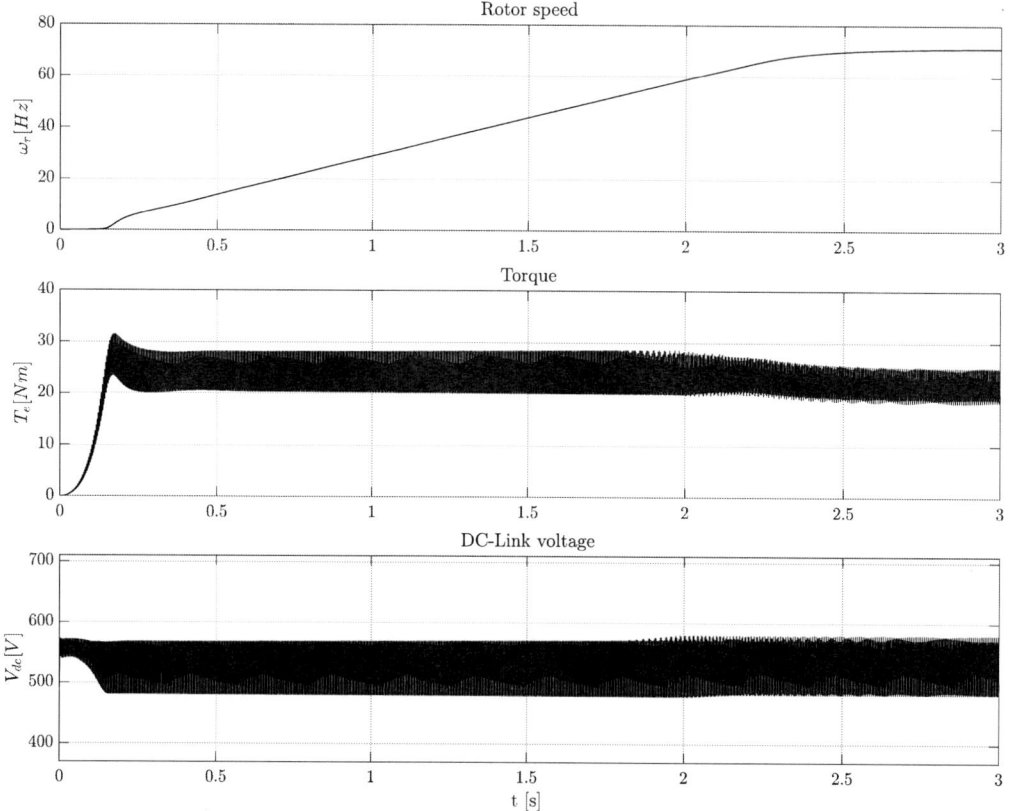

Fig. 4: ABS method performances, with $C_v = 25\mu F$ and 600Hz HPF

Phase-based Stabilization (PBS)

The ABS scheme can only be used outside of the voltage limitation region. Otherwise, the amplitude cannot be increased and the stabilization signal is not applied correctly. Nevertheless, it is possible to shape δP by acting on the voltage phase, by taking Eq. (5) and considering:

$$\delta V_s = 0 \; ; \; \delta\omega_s = 0 \; ; \; \delta\phi_v = K_\phi \delta v_{dc} \quad \Rightarrow \quad \underline{v_s} = V_s e^{j\left(\omega_s t + \phi_v + K_\phi \delta v_{dc}\right)}$$

This phase-based stabilization (PBS) method can also be implemented in the synchronous reference frame and expressed using the complex-vector notation, as follows:

$$\underline{v_s^*} = v_d^* + jv_q^* = \underline{v_s}e^{j\delta\phi} = \underline{v_s}e^{jK_\phi \delta v_{dc}} \tag{13}$$

In scalar form:

$$v_d^* = v_d \cos(K_\phi \delta v_{dc}) - v_q \sin(K_\phi \delta v_{dc}) \tag{14}$$
$$v_q^* = v_q \cos(K_\phi \delta v_{dc}) + v_d \sin(K_\phi \delta v_{dc}) \tag{15}$$

To simplify the implementation, the first order Taylor's approximation can be applied, as the angle oscil-

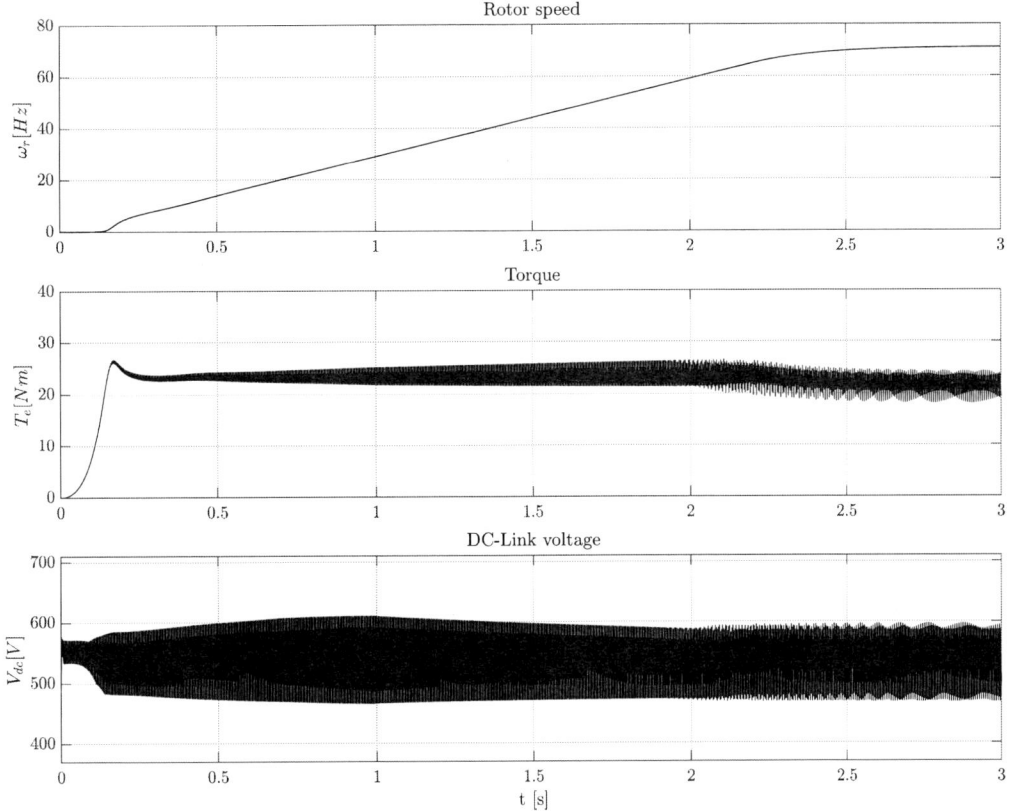

Fig. 5: ABS with $K_v = m$ and 600Hz HPF

lation should be small, $e^{jK_\phi \delta v_{dc}} \approx (1 + jK_\phi \delta v_{dc})$. This gives:

$$v_d^* = v_d - v_q K_\phi \delta v_{dc} \tag{16}$$
$$v_q^* = v_q + v_d K_\phi \delta v_{dc} \tag{17}$$

It is not advised to use the above approximation outside of the voltage limitation region. The voltage circle is smaller for lower voltage amplitudes, and the first order approximation loses precision. Performances with this method will be highest when the applied voltage is at its maximum value.

Gain Tuning

The gain K_ϕ links δP to $\delta \phi$. It can be calculated using the same approach of K_v and the virtual capacitance concept:

$$\delta P = \frac{3}{2}(V_{s0}\cos\varphi_0 \delta v_s + V_{s0}\cos\varphi_0 \delta i_s - V_{s0}I_{s0}\sin\varphi_0 \delta\phi_v) \approx -\frac{3}{2}V_{s0}I_{s0}\sin\varphi_0(\delta\phi_v) \tag{18}$$

with $\delta v_s = 0$ by design and δi_s neglected, which gives:

$$K_\phi = -\frac{2}{3}\frac{C_v v_{dc0}}{V_{s0}I_{s0}\sin(\varphi_0)} \approx -\frac{2}{\sqrt{3}}\frac{C_v}{I_n \varphi_0} \tag{19}$$

Note that K_ϕ is negative because of the inverse influence of the voltage phase on the electrical power in small-signal.

Full-range stabilization

Both amplitude-based stabilization (ABS) and phase-based stabilization (PBS) are combined to provide satisfactory performance over the complete operation range. To avoid a discontinuity when switching between methods, a progressive smooth transition has been implemented:

$$\underline{v}_s = (V_s + (1-\alpha)K_v \delta v_{dc})e^{j(\omega_s t + (\phi_v + \alpha K_\phi \delta v_{dc}))} \tag{20}$$

Where $\alpha \in [0,1]$ is a coefficient that is calculated based on the instantaneous value of V_s:

$$\begin{cases} \alpha = 0 , & V_s < V_1 \\ \alpha = \dfrac{V_s - V_1}{V_2 - V_1} , & V_1 < V_s < V_2 \\ \alpha = 1 , & V_s > V_2 \end{cases} \tag{21}$$

In a conventional VSD (with smooth dc-voltage), the voltage limitation region is reached close to the rated voltage, so V_1 could be set at 90% and V_2 at 95% for example. In slim-capacitor VSDs, the voltage oscillates continuously between 86% and 100% of the grid line-line voltage. In this case, V_1 is set to 80% and V_2 to 85% of the rated motor voltage.

Fig.6 shows the performances of the combined ABS-PBS method, up to 150% of rated speed. Results show that, compared to the ABS alone (see Fig.4), the combined method has superior performances in the voltage limitation region. The DC-link voltage resonance is suppressed and torque ripples are largely dampened in the field-weakening region.

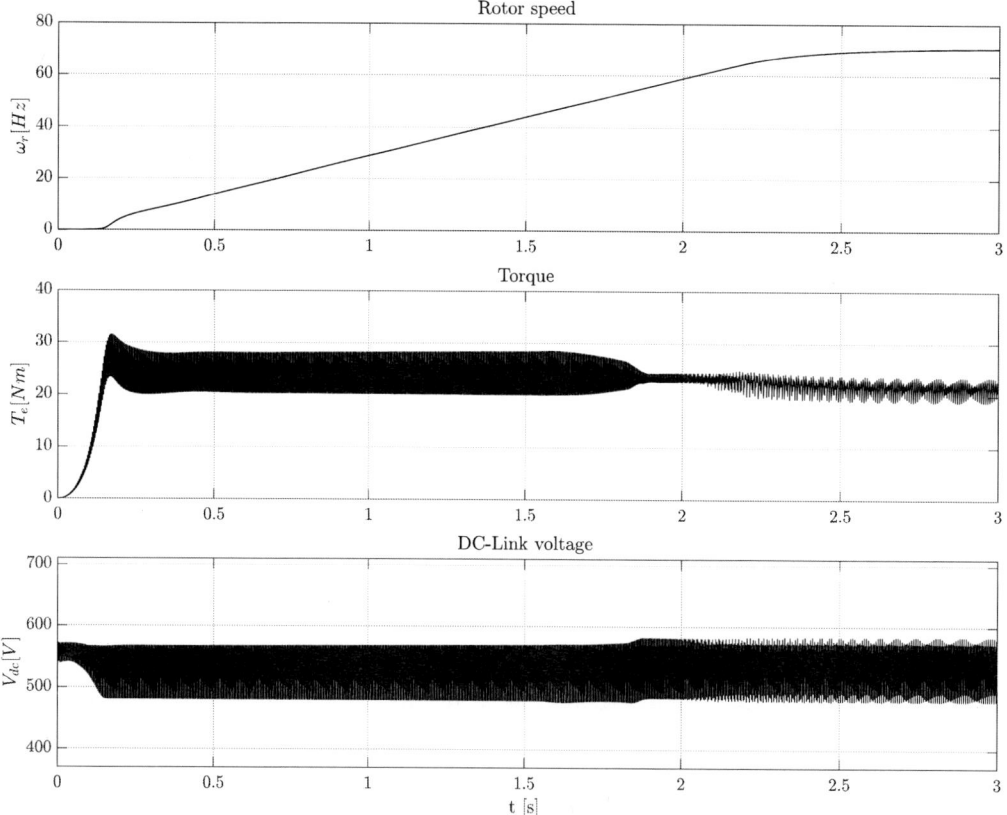

Fig. 6: Combined ABS-PBS method, with $C_v = 25\mu F$ and 600Hz HPF

4 Conclusion

In this paper, direct active stabilization of the dc-link oscillation has been studied. First, the dc-voltage filtering-based method was analyzed. It was shown that its small-signal behavior is equivalent to compensating the output voltage amplitude proportionally to the modulation index. This effective yet rigid method can be outperformed by the combination of two new direct methods: voltage amplitude-based stabilization and voltage phase-based stabilization. These methods are easy to implement in the synchronous reference frame, they require slightly more calculations than the voltage filtering method which is the cost of improved performance and superior tuning capability. This being said, the calculation burden of the proposed method remains low compared to other state-of-the-art methods.

The virtual capacitor concept is very useful to tune the stabilization gain. These methods were tested and compared in simulation. The frequency-based stabilization method was not analyzed in this paper, which can be the topic of future work.

References

[1] S. D. Sudhoff, K. A. Corzine, S. F. Glover, H. J. Hegner and H. N. Robey, "DC link stabilized field oriented control of electric propulsion systems," in IEEE Transactions on Energy Conversion, vol. 13, no. 1, pp. 27-33, March 1998

[2] M. Koteich, "Stability Analysis of a DC-link with Nonlinear Load – An Overview," 2021 23rd European Conference on Power Electronics and Applications (EPE'21 ECCE Europe), 2021, pp. P.1-P.11.

[3] D. Marx, P. Magne, B. Nahid-Mobarakeh, S. Pierfederici and B. Davat, "Large Signal Stability Analysis Tools in DC Power Systems With Constant Power Loads and Variable Power Loads—A Review," in IEEE Transactions on Power Electronics, vol. 27, no. 4, pp. 1773-1787, April 2012

[4] W. Lee and S. Sul, "DC-Link Voltage Stabilization for Reduced DC-Link Capacitor Inverter," in IEEE Transactions on Industry Applications, vol. 50, no. 1, pp. 404-414, Jan.-Feb. 2014

[5] K. Pietilainen, L. Harnefors, A. Petersson and H. -. Nee, "DC-Link Stabilization and Voltage Sag Ride-Through of Inverter Drives," in IEEE Transactions on Industrial Electronics, vol. 53, no. 4, pp. 1261-1268, June 2006

[6] O. Wallmark, S. Lundberg and M. Bongiorno, "Input Admittance Expressions for Field-Oriented Controlled Salient PMSM Drives," in IEEE Transactions on Power Electronics, vol. 27, no. 3, pp. 1514-1520, March 2012

[7] J. Koppinen and M. Hinkkanen, "Impact of the switching frequency on the DC-side admittance in three-phase converter systems," 2014 16th European Conference on Power Electronics and Applications, 2014, pp. 1-10

[8] H. Mosskull, "Stabilization of an induction motor drive with resonant input filter," 2005 European Conference on Power Electronics and Applications, 2005, pp. 10 pp.-P.10

[9] Y. A. I. Mohamed, A. A. A. Radwan and T. K. Lee, "Decoupled Reference-Voltage-Based Active DC-Link Stabilization for PMSM Drives With Tight-Speed Regulation," in IEEE Transactions on Industrial Electronics, vol. 59, no. 12, pp. 4523-4536, Dec. 2012

[10] T. Devos and P. Martin, "Avoiding DC link oscillations in drives," IECON 2020 The 46th Annual Conference of the IEEE Industrial Electronics Society, 2020, pp. 876-882

[11] M. Hinkkanen and J. Luomi, "Induction Motor Drives Equipped With Diode Rectifier and Small DC-Link Capacitance," in IEEE Transactions on Industrial Electronics, vol. 55, no. 1, pp. 312-320, Jan. 2008, doi: 10.1109/TIE.2007.903959.

Hardware-in-the-loop control of a modular induction motor drive in power electronics education

Jens Peter Kaerst

HAWK, University of Applied Sciences and Arts Hildesheim/Holzminden/Göttingen
Von-Ossietzky-Str. 99
D-37085 Göttingen, Germany
Phone: +49 (0) 551-3705-239
Email: kaerst@hawk.de
URL: http://www.hawk.de

Acknowledgments

Plexim GmbH has supported this work by granting the *PLECS academic software sponsorship*.

Keywords

≪Power Hardware-in-the-Loop≫, ≪Education tool≫, ≪Modular Converter≫, ≪Induction motor≫, ≪Vector control≫

Abstract

The development and the utilisation of a modular inverter for control of an induction or synchronous motor in a power electronics and electrical drives laboratory allows students to gain deep understanding. For easy deployment of own code and verification of analytical calculations a hardware-in-the-loop controller is used.

The laboratory hardware presented in [2] allows students to explore a variety of DC/DC-converter topologies. The most complex setup possible was an H-bridge designed for components with 200 V breakdown voltage at most. In order to extend the teaching concept presented in [2] to three-phase drives, a topology consisting of three inverter legs including an auxiliary voltage supply was developed extending the maximal component voltage to 600 V while retaining the modular structure.

In [2] the power semiconductors were controlled using an Arduino Nano, but its performance is by far insufficient for e.g. space vector modulation as commonly used in field-oriented control. Since the PLECS simulation software from Plexim GmbH is used in the power electronics and drive technology lectures, it was straightforward to use it in combination with hardware-in-the-loop (HIL) control. Plexim offers the RTBox CE as an inexpensive entry-level HIL-System. In the context of this publication, a RTBox CE is connected directly to the modular inverter driving a three-phase motor via an interface board. In order to apply a variable load the three-phase motor is connected in a motor test bench to a DC motor via a speed/torque measuring shaft. Actual values of rotor speed n and rotor angle γ are essential for field orientated control of the motor. Therefore the evaluation of the measuring shaft data is also performed by the RTBox.

The set-up, its possibilities and limitations are presented as examples within the scope of this work.

Introduction

University bachelor- and master-level power electronics and electrical drive education usually includes mandatory laboratory credits. Instead of using state of the art equipment from major companies it has been found that better outcome can be achieved using modular open frame components [1], [2], [4]. This allows for a deeper understanding of all involved hardware components including power electronic devices (MOSFET, IGBT), semiconductor material (Si, SiC, GaN), complementary gate drive, vector modulation and control, 3-phase AC machines (induction and synchronous motor), DC motor with external excitation, sensors (current, rotational speed, torque).

The most sophisticated part tends to be the shift from U/f-control to field orientated vector control of an induction motor. Therefor a hardware-in-the-loop (HIL) system using Plexim's PLECS as standalone simulator in conjunction with their entry-level HIL-system (RT Box CE) and PLECS-coder is used to control the power electronics and drives system in realtime [3]. Offline simulation of the entire system can be performed by students in advance to attending the laboratory session. This allows the students to implement and test control code prior to performing measurements which leads to a high level of understanding.

This paper is subdivided into three major parts. At first all hardware components used are described in detail. This includes both motors, the speed/torque measuring shaft, the modular inverter and the RTBox. The second part describes the implementation of the control with PLECS including data probing via the RTBox. Here, a few code examples are singled out for detailed discussion. Finally in the third part measurements both on system level and in detail are presented. Because of the open modular construction all voltages and currents can be easily accessed. In addition an exchange of e.g. Si $\leftrightarrow$ SiC freewheeling diode can be done on the fly, allowing to observe effects such as reverse recovery behaviour.

Hardware

Drive test stand

The drive test stand as depicted in Fig. 1 consists of a 3-phase AC-motor with slip rings in order to be operated either as induction motor or as synchronous motor, a rotational speed and torque sensor including an index impulse and a DC-Motor with separate excitation for recuperation.

Both motors can be operating using a single $0 \ldots 350\,\text{V}$ laboratory power supply. In this case by using recuperation only system losses have to be covered. The easiest way to vary the braking torque of the AC-motor is to short circuit the DC-motor as depicted in Fig. 1 and to vary the excitation current $0 \leq I_E \leq I_{E_N}$. This will be done throughout this paper.

The nominal values of the machinery are:

1. 3-phase AC-motor Δ: $U_1 = 230\,\text{V}$, $I_1 = 1.44\,\text{A}$, $P_N = 270\,\text{W}$, $n_N = 1360(1500)\,\frac{1}{\text{min}}$, $I_E = 4\,\text{A}$
2. DC-motor: $U_{A_N} = 220\,\text{V}$, $I_{A_N} = 1.8\,\text{A}$, $P_N = 270\,\text{W}$, $n_N = 2000\,\frac{1}{\text{min}}$, $I_E = 250\,\text{mA}$
3. Torque/speed measuring shaft: $T_{\text{min/max}} = \pm 5\,\text{Nm}$, $\sin/\cos$-output with $2 \times 360\,\frac{\text{pulses}}{\text{revolution}}$ with index impulse γ_0.

The torque/speed measuring shaft outputs the torque T as an analog voltage $U_T = \frac{1\,\text{V}}{\text{Nm}} \cdot T$, the rotation speed n as two logic level square waves $U_{\sin}$ and $U_{\cos}$ with $f_n = 6\,\frac{\text{min}}{\text{s}} \cdot n$ and the rotor angle reference γ_0 as open collector with $f_\gamma = \frac{n}{60\frac{\text{s}}{\text{min}}}$. The digital signals have 50% duty cycle and refer to $[n] = \frac{1}{\text{min}}$.

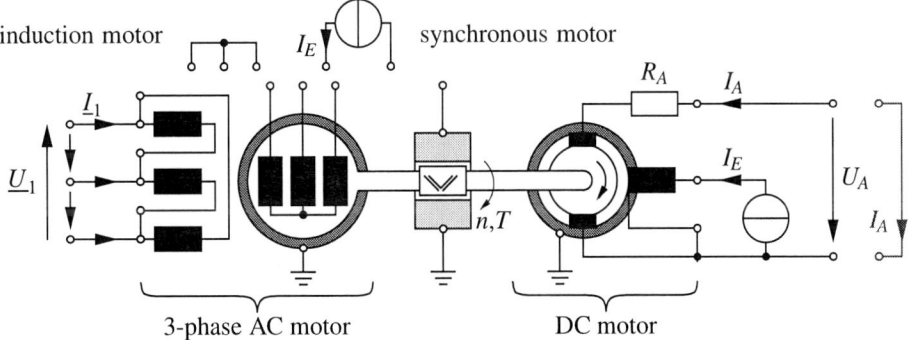

Fig. 1: Drive test stand, 3-phase AC motor, rotational speed and torque sensor, DC-Motor

Inverter

To allow students an easy recognition and understanding the modular power electronics concept presented in [2] is adopted. It consists of a half-bridge topology driven via a single driver (IR2131) incorporating a level-shift in conjunction with a bootstrap supply.

In order to maintain the modularity without compromising on issues as stray inductance an entire half-bridge including fault protection and phase current measurement is layouted on one PCB. This allows the use of 3 identical PCBs to ensemble the well known 3-phase voltage source inverter (VSI) topology. All clearances are chosen such to meet 600 V-requirements. All auxiliary voltages needed for drive and control of the inverter are generated via an additional PCB fed directly out of the DC-rail voltage.

Concluding the inverter hardware can be divided into the modular power electronics consisting of

- Backplane mainly populated with intermediate circuit capacitors and mounting all other PCBs
- 3× inverter leg including IGBTs and freewheeling diodes or MOSFETs and their gate drive, fault protection, fault status output and current measurement via a LEM current sensor
- Auxiliary power supply consisting of multiple DC/DC-converters generating +15 V and 5 V on DC-rail potential. In addition ±15 V isolated supply in order to decouple the HIL-System due to safety requirements. The operating voltage range is designed for $120\,\text{V} \leq U_{\text{DC}} \leq 350\,\text{V}$. It is measured via a resistive voltage divider.

The complete inverter is depicted in Fig. 2 in combination with the RTBox and the interface board allowing for direct HIL-control of the inverter. Fig. 3 a) and c) shows schematic excerpts of one of the inverter legs and the auxiliary power supply respectively.

As a side effect the DC/DC-converter includes several converter topologies as there are flyback with a planar transformer, buck and buck-boost. They can be integrated into measurement and simulation and are therefore part of the possible educational outcome.

Hardware-in-the-loop (HIL)-System for rapid control prototyping (RTBox CE)

Plexim GmbH has designed a real-time simulator specially for power electronics applications. It works in combination with PLECS (Piecewise Linear Electrical Circuit Simulation) and their PLECS-coder which generates real-time C code from a circuit or block diagram created with PLECS Blockset.

While the entry-level RTBox CE is used throughout this work, the described hardware is upward compatible. A brief summary of the RTBox CE key performance figures is included in following table:

Fig. 2: HIL-control of the 3-phase AC motor, operating as induction motor

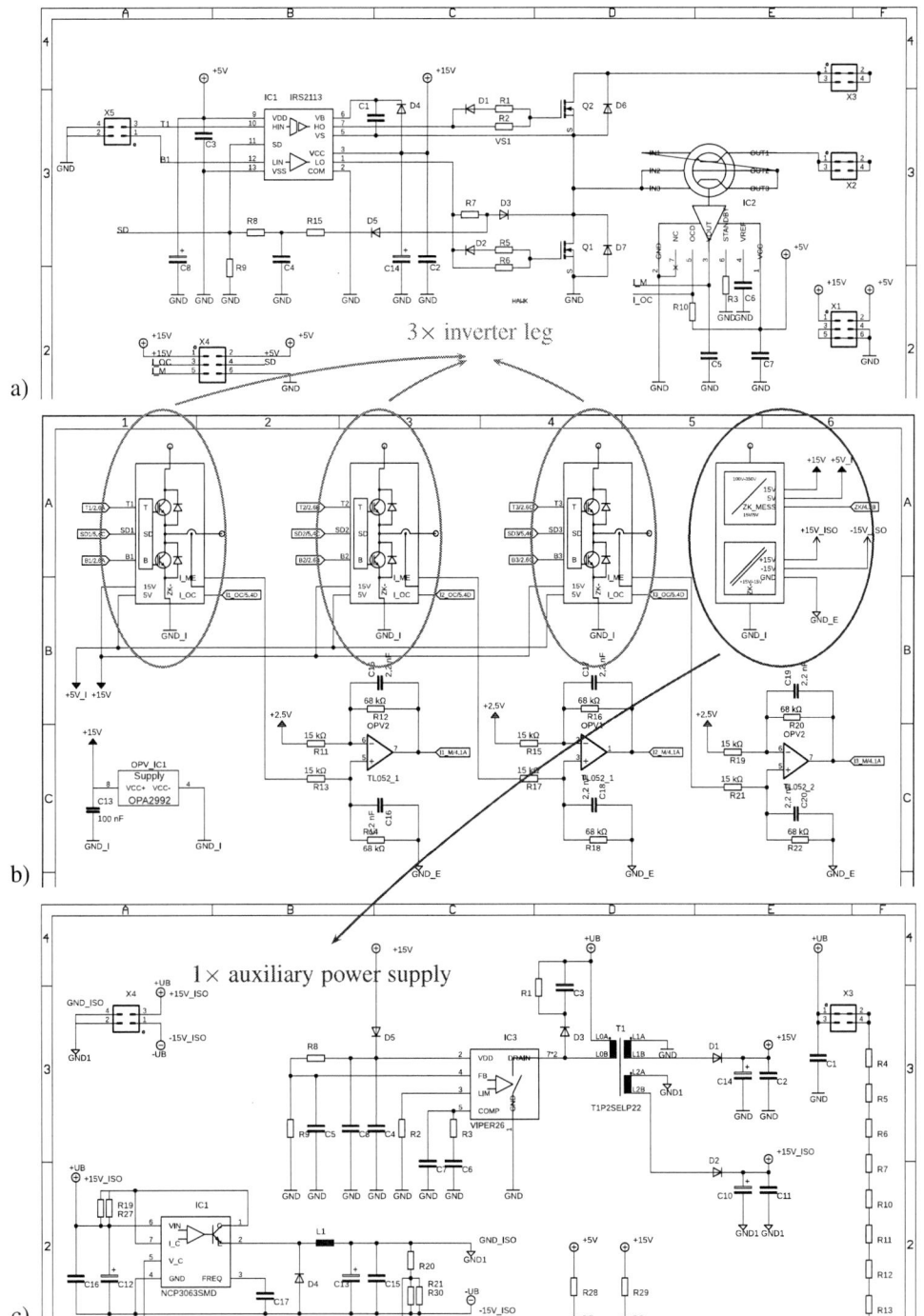

Fig. 3: Schematic excerpts, a) inverter leg, b) RTBox interface and c) auxiliary power supply

- Processor Xilinx Zynq Z-7030, CPU cores 2x ARM Cortex-A9, 1 GHz
- Analog inputs, 8 channels with 16 bit resolution, voltage ranges $\pm 5\,\text{V}$ or $\pm 10\,\text{V}$, differential input, simultaneous sampling with $2\,\text{Msps}$ at maximum
- Analog outputs, 16 channels with 16 bit resolution, voltage ranges single ended $5\,\text{V}$ or $10\,\text{V}$, differential $\pm 10\,\text{V}$ or $\pm 5\,\text{V}$, simultaneous update with $2\,\text{Msps}$ at maximum
- Digital inputs, 32 channels, logic level $3.3\,\text{V}$ ($5\,\text{V}$ tolerant)
- Digital outputs, 32 channels, logic level $3.3\,\text{V}$ or $5\,\text{V}$
- Connectivity, gigabit Ethernet

Inside the RTBox, signal ground (gnd) and protective earth (PE) are not connected directly, in order to avoid ground loops that might otherwise. The safety and grounding concept of the described setup is included in Fig. 4 giving an isolation and potential overview.

The performance of the RTBox CE is sufficient to run the code described in the software section of this paper with a discretisation step size as small as $T_S = 5\,\mu s$ which is sufficient. An even smaller step size may be chosen in case of less extensive code.

Interface board, RTBox to inverter

The interconnection between RTBox and inverter is attempted to be both reliable and structured in order to allow students to work safely and with reproducible outcome. Plexim offers similar interfaces for Texas Instruments LaunchPad evaluation boards. This concept has been adopted.

The main difficulty was to ensure safety by design. In case of any inverter malfunction it has to be ensured, that the maximum voltage levels between the RTBox inputs and outputs, all connected instruments such as oscilloscope and/or signal generator does not exceed $\pm 24\,V$ in order to prevent personal and equipment damage.

This goal was met by isolating the inverter ground (ignd) from the RTBox ground (gnd). Both grounds are connected at one point only via an inductor which avoids any DC voltage but decouples both grounds at higher frequency e.g. during transients. As stated above RTBox ground and protective earth (PE) are not connected inside the RTBox. This concept fails in case of connected measurement instruments with inputs or outputs referring to PE, which usually is the case. Therefor ground (gnd) and protective earth (PE) are connected on the interface board.

Digital signals are passed inbetween RTBox and inverter using magnetically isolated digital isolation (iCoupler ADuM225N, Analog Devices). This includes 3×2 gate drive signals and 2 overcurrent fault signals. All analog signals including $3\times$ phase current measurements and $1\times$ DC-rail voltage measurement use discrete differential amplifiers. An overview mainly showing the chosen isolation and potential scheme is included in Fig. 4. A foto of all PCBs is depicted in Fig. 5.

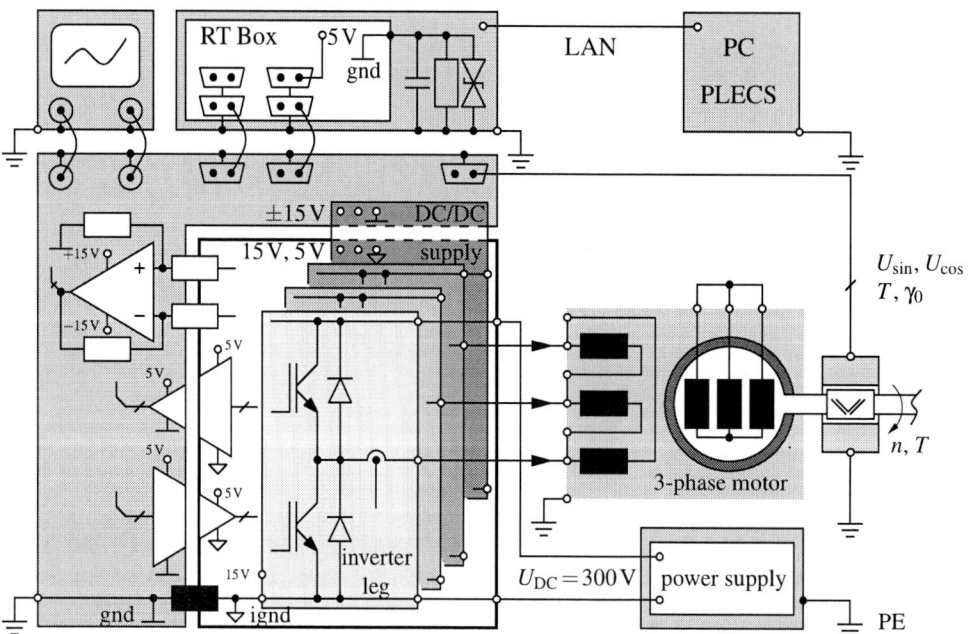

Fig. 4: Isolation and potential overview including RTBox with inverter, PC running PLECS, induction motor, torque/speed measuring shaft, laboratory equipment (power supply, oscilloscope, ...)

Fig. 5: Inverter, interface and backplane PCBs in top view

Two analog inputs and four analog outputs of the RTBox are accessible via PE referenced BNC connectors. Four digital inputs and outputs are connected to hardware switches and LEDs respectively. The RTBox uses four 37 pin D-Sub connectors for analog/digital in-/ and output.

In addition the digital and analog torque/speed measuring shaft outputs are inferfaces using a 15 pin D-Sub connector and processed via Schmitt trigger and the an active lowpass filter respectively.

Software

The control software is generated automatically using PLECS and PLECS-coder. Usually a board support package has to be developed in order to adapt to external microcontroller or DSP based control hardware. Obviously this is included for Plexim's own RTBoxes and for some DSPs. In the scope of this paper only the RTBox will be used.

The software development is done in 4 steps.
1. PLECS model of the drive test stand as depicted in Fig. 1. As this model can be reused, this has only to be done once.
2. PLECS model of the power electronics. This can also be reused by the students. At this point a clear interface naming and structure is essential, because in HIL simulation this will be the interface between RTBox and power electronics and electrical drives.
3. PLECS model for offline control of above components. This is the component developed by the students themselves. These 3 software components allow for offline simulation.
4. In order to deploy the offline control software for realtime control the PLECS-coder builds, downloads and starts the entire model for the chosen hardware.

The development of a a PLECS model including step 1 to 4 is shown in two examples.

Torque/speed measuring shaft

During offline simulation torque, speed and rotor angle can be probed using the AC motor, the DC motor or separate meters provided in PLECSs mechanical domain. The drive test stand for offline simulation including the inverter is depicted in Fig. 6 (top, blue rectangle).

In order to ensure compatibility to the real torque/speed measuring shaft the output signals as described in the hardware section have to be generated. PLECS allows for the implementation of C-code directly,

Hardware-in-the-loop control of a modular induction motor drive in power electronics education · KAERST Jens Peter

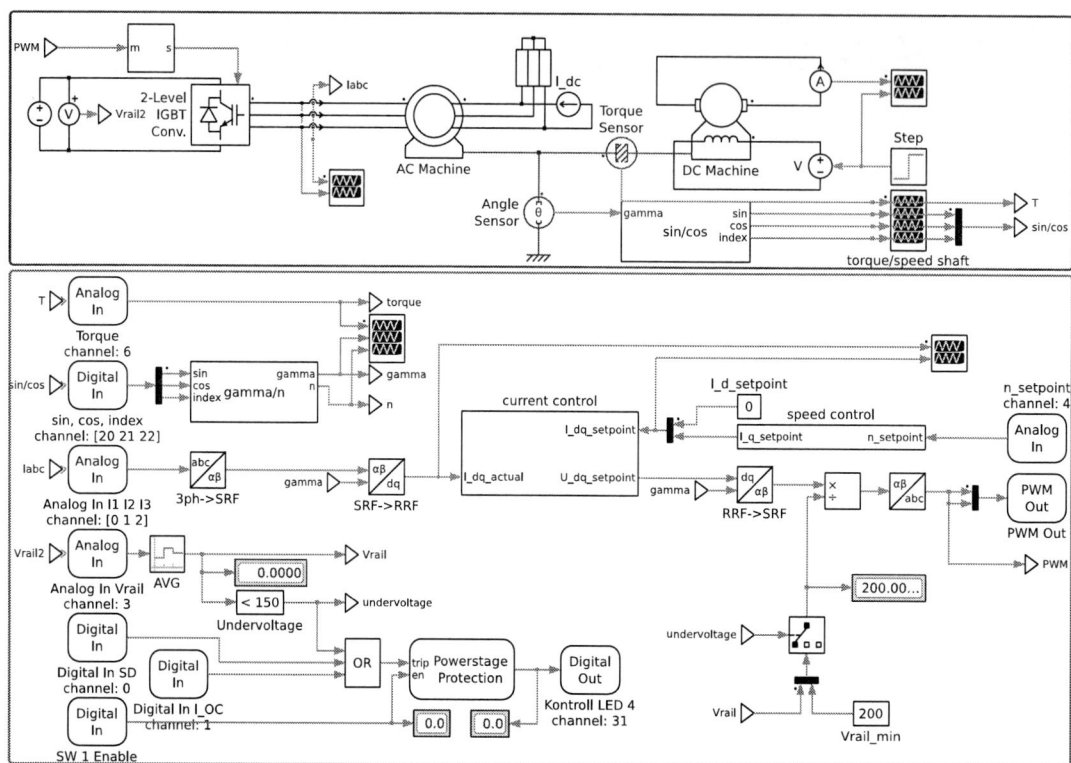

Fig. 6: PLECS, realtime/offline simulation including RTBox interfaces for drive output and measurement input including the torque/speed measuring shaft

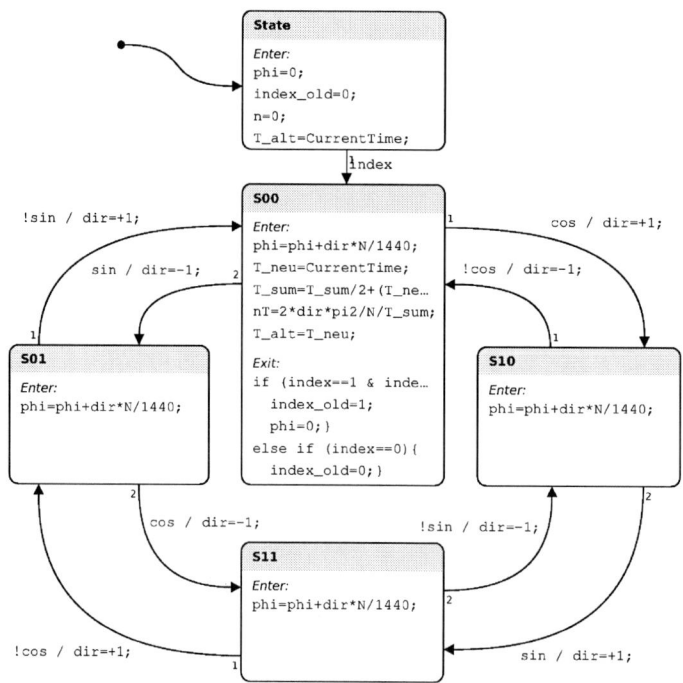

Fig. 7: PLECS, state-machine generating realtime outputs for rotor angle γ and speed n

defining input signals with arbitrary code of any complexity calculating the outputs signals. This has been used to generate the outputs U_{sin}, U_{cos} and γ_0 using $\gamma(t)$. The code is added into a user defined subsystem called *sin/cos*. This subsystem is also included into the user library.

During realtime simulation the torque/speed measuring shaft outputs are connected to the RTBox digital and analog inputs via the RTBox interface. Instead of using C-code a state-machine is used to calculate the rotor angle γ and the rotor speed n. The code of the single states of the state-machine also uses C-code. Fig. 7 shows the states including their enty- and exit-code. State-machines have been proven to be more transparent and easier to implement. The described state-machine is included into a user defined subsystem called *gamma/n*. It has also been added to the user library.

The PLECS model depicted in Fig. 6 allowes for developing and testing. For HIL simulation using the real drive test stand and the RTBox including interface and inverter, only the code depicted in Fig. 6 (bottom, red rectangle) is necessary. Fig. 8 shows a screenshot of PLECS running in external mode after building and deploying the model into the RTBox. The probe data is captured in the RTBox and transferred to the connected PC via LAN. On the right hand side T, γ and n are captured showing a load step $T = 0 \rightarrow 1.5\,\text{Nm}$. The trigger source and level can be set using the external mode of PLECS-coder.

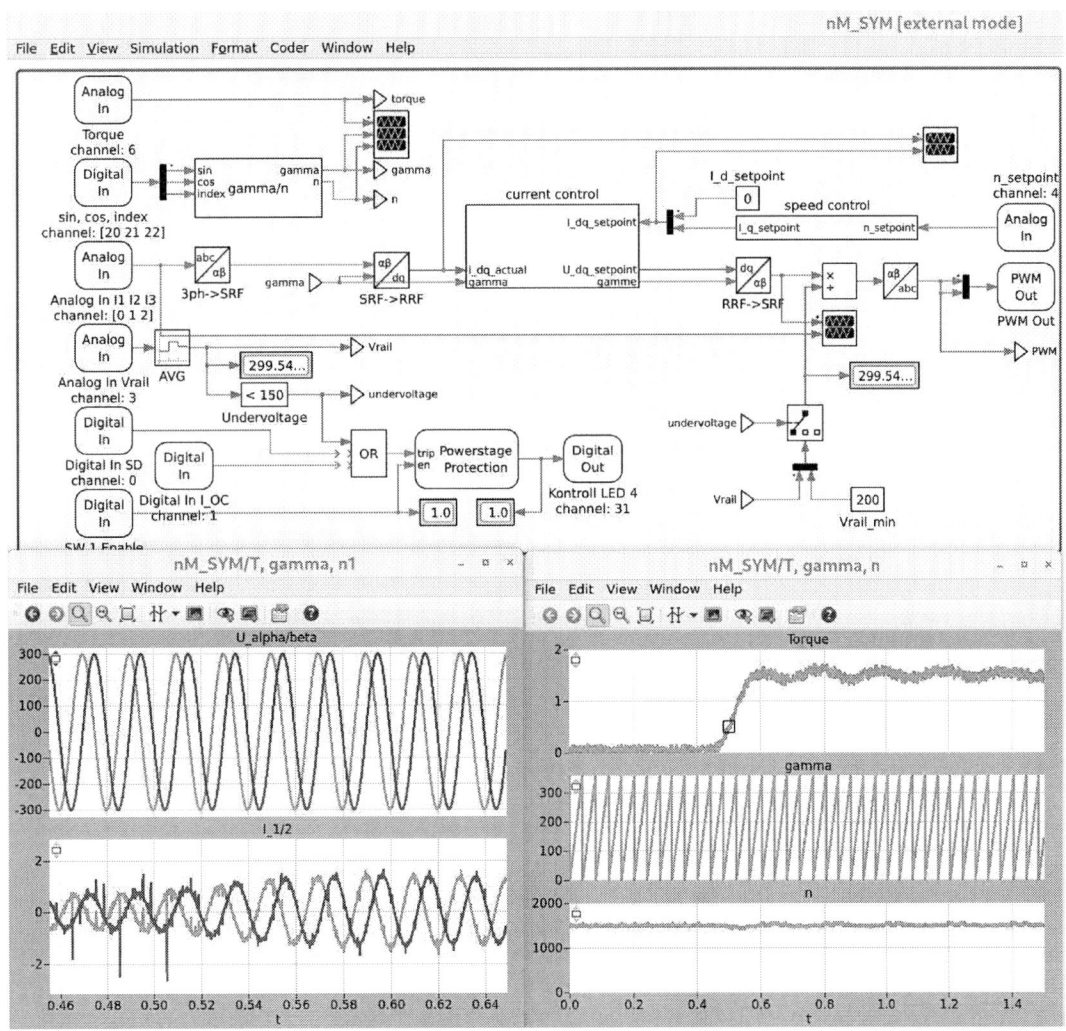

Fig. 8: PLECS running in external mode, PC connected to RTBox triggering and updating scope signals

Inverter drive

A second example is already included in Fig. 6. The AC motor, operated as synchronous motor, can be controlled via an analog speed setpoint which is fed into the RTBox interface e.g. using a signal generator. The speed control and the current control is modelled using continuous state block elements in equally named subsystems. In this case simple PI-controllers are used. The current control inputs the actual and the setpoint stator current as space-vector and outputs the desired voltage space-vector which has to be transformed back into stator-reference-frame. After Clarke transformation the inverter drive signals are generated using a pulse width modulation. Sadly the RTBox board-support-package does not provide a space-vector-modulation. Therefore only a symmetrical PWM has to be used via the *PWM Out*-block. An advantage of these blocks is their included fault protection using the *Powerstage Protection*-block ensuring safe shutdown in case of undervoltage or overcurrent.

The system response can be changed and observed easily. In this case the above mentioned load step $T = 0 \rightarrow 1.5\,\text{Nm}$ is imposed using the DC motor as eddy-current brake while maintaining the drive speed of $n = 1500\,\frac{1}{\text{min}}$. Again PLECS-coder in external mode is used to capture and to display realtime data. On the left hand side of Fig. 6 two line currents as measured in an inverter leg, transferred via a differential amplifier on the RTBox interface and sampled by the RTBox are displayed along with the stator voltage setpoint in stator-reference-frame $\vec{U}_1 = U_\alpha + \text{j} \cdot U_\beta$. Because all three line currents are sampled simultaneously (although two would be sufficient) the current can easily be transformed into a line current space-vecor $\vec{I}_1$ using either stator- or rotor-reference-frame. The later is needed for field orientated control.

This code was used to operate the AC-motor as synchronous motor performing the following measurement examples.

Measurements

Above PLECS model operating as HIL controller is capable of displaying measured data, variables etc. using the PLECS-scope via the PLECS-coder in external mode as depicted in Fig. 8. This option is limited to the sampling frequency which in turn is determined by the model complexity. In this case $f_S = 200\,\text{kHz}$ ($T_S = 5\,\mu\text{s}$) is chosen.

In order to observe fast switching events and to investigate e.g. the effect of SiC or GaN diodes the use of an oscilloscope is inevitable. Because of the open structure of the modular inverter it is possible to measure any signal of interest. This is shown in Fig. 9 comparing the effect of a line current commutating from the top diode into the bottom IGBT. The devices used are IBBTs IGP20N65H5, EmCon diodes IDP23E60 and the SiC Schottky diode IDH03G65C5 for comparison all from Infineon. The reduceced current peak due to the lack of reverse recovery charge Q_{rr} is eminent. This in turn reduces the turn on losses of the IGBT.

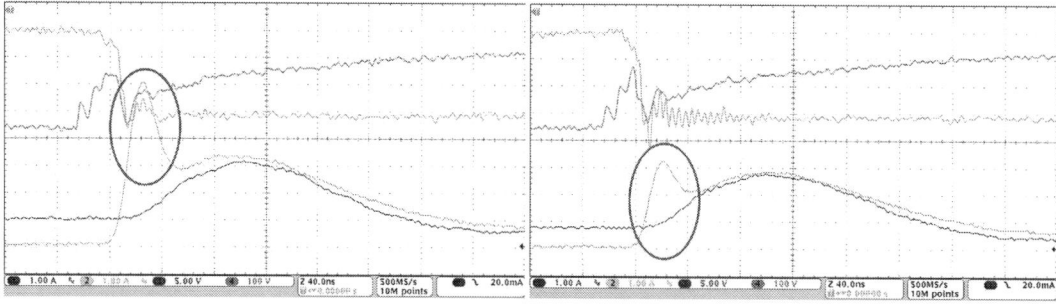

Fig. 9: Si versus SiC freewheeling diode (1) I_{line}, (2) I_{IGBT}, (3) U_{GE}, (4) U_{CE}

Of course line current and line voltage measurements as depicted in Fig. 10 are also possible. They allow students to verify offline simulation results and deepen their understanding. Because PLECS intends for system level simulations the effect of fast transients is observed best by performing measurements.

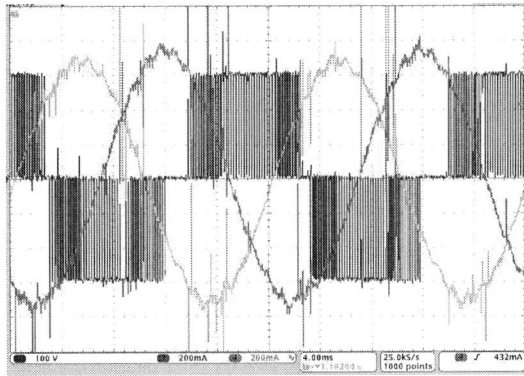

Fig. 10: Line voltage (1) U_{12} and line currents (3) I_{L1}, (4) I_{L2}

Conclusion

Modern simulation tools such as PLECS can automatically generate efficient code for either HIL-systems, microcontrollers or DSPs. Usually HIL-systems are used for real time simulation of e.g. the drive train in order to speed up control code development spire to finalisation of the drive train. For education purpose it is desirable to turn this around.

The graphical modeling of the control code and its deployment into a HIL-system allows students to concentrate on the underlying control principles such as calculations and transformations using space-vectors in different reference frames. The performance of the Plexims RTBox CE is sufficient for driving the described voltage source inverter including measuring the stator currents, the rail voltage and operating the torque/speed measuring shaft at a switching frequency of up to $f_S = 10\,\text{kHz}$.

Despite the lack of e.g. space-vector-modulation in the board support package the software has proven to be easy to use and the education outcome is superior to only performing measurements with commercial converters. Later do not allow for the deployment of own code and therefore the system has to be regarded as a black box which only accepts a set of parameters.

An additional advantage is the ability to perform both measurements on system and on component level. This allows for the investigation and the comparison of new components such as SiC or GaN devices.

The system described in this paper has already been used in a drive-systems master course generating a very promising outcome. This strategy will be expanded to more motor types and different control strategies in the near future.

References

[1] Michal Bonisławski and Marcin Hołub. Integrated test stand design for modern power electronics laboratory exercises. In *2017 19th European Conference on Power Electronics and Applications (EPE'17 ECCE Europe)*, pages P.1–P.6, 2017.

[2] Jens Peter Kaerst. Modular hardware for experimental investigation of dc/dc converter topologies. In *2016 18th European Conference on Power Electronics and Applications (EPE'16 ECCE Europe)*, pages 1–10, 2016.

[3] Sum Si Kai Kenny, R.T. Naayagi, Sze Sing Lee, and Cao Shuyu. Three phase vsi control system rapid prototyping with ti c2000 mcu and plecs coder. In *2021 5th International Conference on Green Energy and Applications (ICGEA)*, pages 42–46, 2021.

[4] Lucas Koleff, Gustavo Valentim, Victor Rael, Luciana Marques, Wilson Komatsu, Eduardo Pellini, and Lourenço Matakas. Development of a modular open source power electronics didactic platform. In *2019 IEEE 15th Brazilian Power Electronics Conference and 5th IEEE Southern Power Electronics Conference (COBEP/SPEC)*, pages 1–6, 2019.

Design and efficiency analysis of an LCL Capacitive Power Transfer system with Load-Independent ZPA

Francesco Musolino (*), Ahmed Abdullah (*), Mario Pavone (**),
Fabio Ferreyra (*), Paolo Crovetti (*)
(*) POLITECNICO DI TORINO (DET)
Corso Duca degli Abruzzi, 24, 10129
Torino, Italy
(**) STMicroelectronics Srl
Stradale Primosole, 50, 95121
Catania, Italy
Phone: +39 011 090 4166
Email: francesco.musolino@polito.it
URL: https://www.polito.it/

Keywords

≪Contactless Power Supply≫, ≪Current Source DC-DC≫, ≪DC-DC Converter≫, ≪Efficiency≫.

Abstract

This paper proposes a design procedure of an LCL compensation circuit for a capacitive power transfer (CPT) system. The design enables the achievement of load independent zero phase angle (ZPA) operation in order to increase the overall efficiency of the system by using a minimum number of compensation components. The proposed approach is supported and validated by circuital simulations and confirmed by the results of experimental tests carried out on a specifically designed prototype.

Introduction

Capacitive power transfer (CPT) is a method of delivering power without electric wires, through time-varying electric fields between two metallic plates forming a capacitance. The simple, light and cheap interface makes this technology an interesting alternative to the inductive power transfer (IPT) [1]. In addition, since electric fields can pass through metal materials without generating significant power losses, this technology is being actively studied for electric vehicle charging applications [2].

In charging applications, a constant current (CC) or constant voltage (CV) charging profile depending on the battery state of charge (SoC) is desirable to ensure its safety and durability [3]. A desirable feature in battery charger is the zero phase angle (ZPA) between the inverter output current and voltage, to reduce the volt-ampere rating and to enable soft-switching operation. This condition should be ensured for the entire charging profile, in other words, ZPA operation must be load independent [4].

According to [5], at least three passive components are required to obtain load independent ZPA operation with CC or CV output. In this article, an LCL compensation network is proposed, and a design procedure to obtain a CC output with load independent ZPA operation is given. The LCL network is able to improve the power delivered to the load and the power transfer efficiency [6]. In addition, a dc-dc efficiency analysis is performed [7], in which an analytical expression of the system efficiency as a function of the load is developed.

The proposed approach is validated both through circuital simulations and experimental tests. To this purpose, a prototype that include a half-bridge inverter and the LCL network designed to achieve ZPA as described in this paper is tested and measurement results are compared to those obtained by the simulations showing good agreement.

Load-independent ZPA operation analysis

Fig. 1 shows the proposed topology, which is composed of a half-bridge inverter (M_1, M_2), a diode bridge rectifier (D_1 through D_4) and a resonant network. This network consists of the LCL compensator (L_1, C_2 and L_2) and the interface capacitors (C_{int1} and C_{int2}).

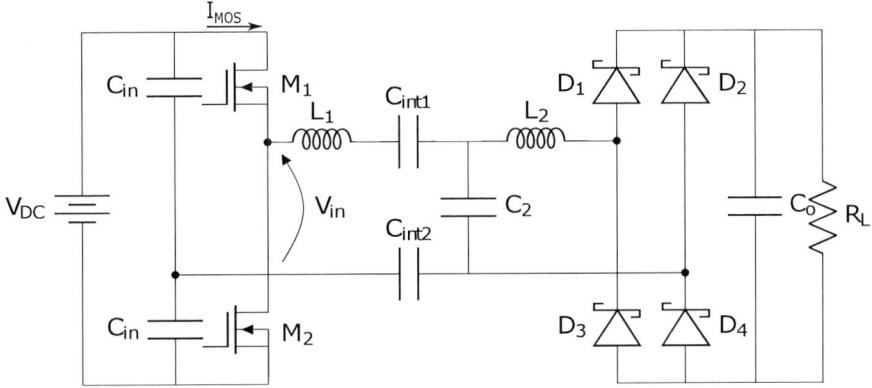

Fig. 1: CPT system with LCL compensator

This system can be simplified as shown in Fig. 2, using the fundamental harmonic approximation (FHA) [8] [9].

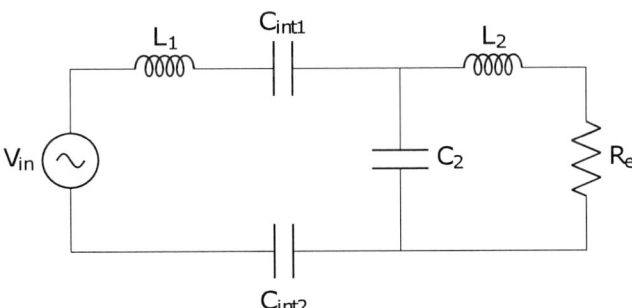

Fig. 2: Equivalent linear circuit

To this purpose, in what follows V_{in} is the fundamental component of the inverter output, which is assumed to be a pure sinusoidal with magnitude equal to:

$$V_{in} = \frac{2V_{DC}}{\pi} \tag{1}$$

and R_e is the equivalent resistance of the load with rectifier, calculated as:

$$R_e = \frac{8}{\pi^2}R_L \tag{2}$$

To obtain ZPA operation, the reactive part of the input impedance seen by V_{in}, $Z_{in} = R_{in} + jX_{in}$ must be zero. By circuit analysis, the reactance is found to be:

$$X_{in} = \frac{R_e^2 \dfrac{\omega_0 L_1^2}{L_1 + L_2}}{R_e^2 + \left(\dfrac{\omega_0 L_2^2}{L_1 + L_2}\right)^2} - \frac{1}{\omega_0 C_1} \tag{3}$$

C_1 is the series combination of the interface capacitances:

$$C_1 = \frac{C_{int1} \cdot C_{int2}}{C_{int1} + C_{int2}} \tag{4}$$

ω_0 is the resonance frequency:

$$\omega_0 = \frac{1}{\sqrt{C_2 L_p}} \tag{5}$$

where $L_p = L_1 \parallel L_2$.

According to (3), C_1 can be designed to achieve zero input reactance. Simply solving this equation for C_1 and selecting interface capacitors to obtain the required capacitance would not lead to a robust design, since the input reactance is load dependent, and consequently ZPA operation as well.

However, based on (3), load independent ZPA operation must be gauranteed when the following inequality is satisfied:

$$R_e^2 \gg \left(\frac{\omega_0 L_2^2}{L_1 + L_2}\right)^2 \tag{6}$$

If this is true, then the much smaller term is neglected with respect to R_e^2 in the denominator of (3), making the input reactance approximately load independent,

$$X_{in} \approx \frac{\omega_0 L_1^2}{L_1 + L_2} - \frac{1}{\omega_0 C_1} \tag{7}$$

$$C_1 = \frac{L_1 + L_2}{(\omega_0 L_1)^2} \tag{8}$$

Under such condition, for (8), X_{in} is effectively cancelled for any load and consequently load independent ZPA operation can be obtained for a particular frequency.

Design procedure

A design procedure for the CPT converter of Fig. 1 with load independent ZPA operation is proposed in this section. The specifications of the proposed system are given in Table I. The design of the system consists of calculating the components of the compensation network, to obtain the load independent ZPA operation previously explained and to fulfill the output requirements. By circuit analysis, the circuit is proven to work as a current source. Therefore, it makes sense to express the output requirements in terms of current rather than power, using $P_o = I_o^2 R_L$.

Simplified equations to calculate the component values can be obtained by considering the assumption $L_1 \gg L_2$. The design begins with the calculation of the inductance L_2, to obtain the required output current according to the following equation:

$$L_2 = \frac{4}{\pi^2} \frac{V_{DC}}{\omega_0 I_o} \tag{9}$$

The (6) is assumed to be:

$$R_e^2 = 10 \times \left(\frac{\omega_0 L_2^2}{L_1 + L_2} \right)^2 \tag{10}$$

Using (2), the (10) is solved for the inductance L_1 to guarantee ZPA condition for loads as low as $R_{\mathrm{L,min}}$:

$$L_1 = \frac{\pi^2 \sqrt{10}}{8} \frac{\omega_0 L_2^2}{R_{\mathrm{L,min}}} \tag{11}$$

Once both inductances are calculated, the equivalent capacitance of the capacitive interface is designed to compensate the input reactance, as shown in (8). Finally, C_2 is calculated to make the system resonant at the working frequency, expressed in (5):

$$C_2 = \frac{1}{\omega_0^2 L_\mathrm{p}} \tag{12}$$

Based on the specifications of Table I, the passive components computed using conventional design for $R_e = 12\,\Omega$ are shown in Table II. Applying the above procedure to get load independant ZPA operation to the same specifications of Table I, the optimal normalized passive components along with parasitics [10] are shown in Table III. The value of interface capacitors are not kept equal in order to have equivalent capacitance more closer to the computed value which is 55.04 pF.

Table I: System specifications

Parameter	Value
V_{DC}	60 V
f	1 MHz
P_o	4 W
R_L	15 Ω
$R_{\mathrm{L,min}}$	3 Ω

Table II: Conventional passive components

Component	Value
L_1	50 μH
L_2	7.74 μH
$C_{\mathrm{int1-2}}$	1436 pF
C_2	3.76 nF

Table III: Optimal normalized passive components

Component	Value
L_1	470 μH
DCR_{L_1}	3 Ω
L_2	10 μH
DCR_{L_2}	0.08 Ω
C_{int1}	100 pF
C_{int2}	120 pF
$ESR_{C_{\mathrm{int}}}$	0.05 Ω
C_2	2.7 nF
ESR_{C_2}	0.025 Ω
$R_{ds,on}$	0.03 Ω

Simulation results

The circuit shown in Fig. 1 has been simulated in MATLAB/Simulink for the components value specified in Table II and Table III and the corresponding results are reported in Fig. 3. This is observed in Fig. 3a where the phases of the input impedance Z_{in} for three different loads R_e, i.e., $10\,\Omega$, $12\,\Omega$ and $14\,\Omega$ are shown. The zero phase angle condition is achieved at different frequencies for each load. If the system works at a frequency of 1 MHz, only the curve for $R_e = 12\,\Omega$ achieves ZPA condition, meaning that only for this load soft switching is obtained, which is verified in Fig. 3c where the transistor currents are shown for the same three loads.

However, in Fig. 3b, there are zero phase input reactances at a frequency of 978 kHz for all three loads. The resonant frequency is obtained at 978 kHz instead of 1 MHz since the normalized component values are little different from the computed values. The load independent soft switching operation is achieved at this working frequency, as seen in Fig. 3d where the transistor currents for the same three loads are shown.

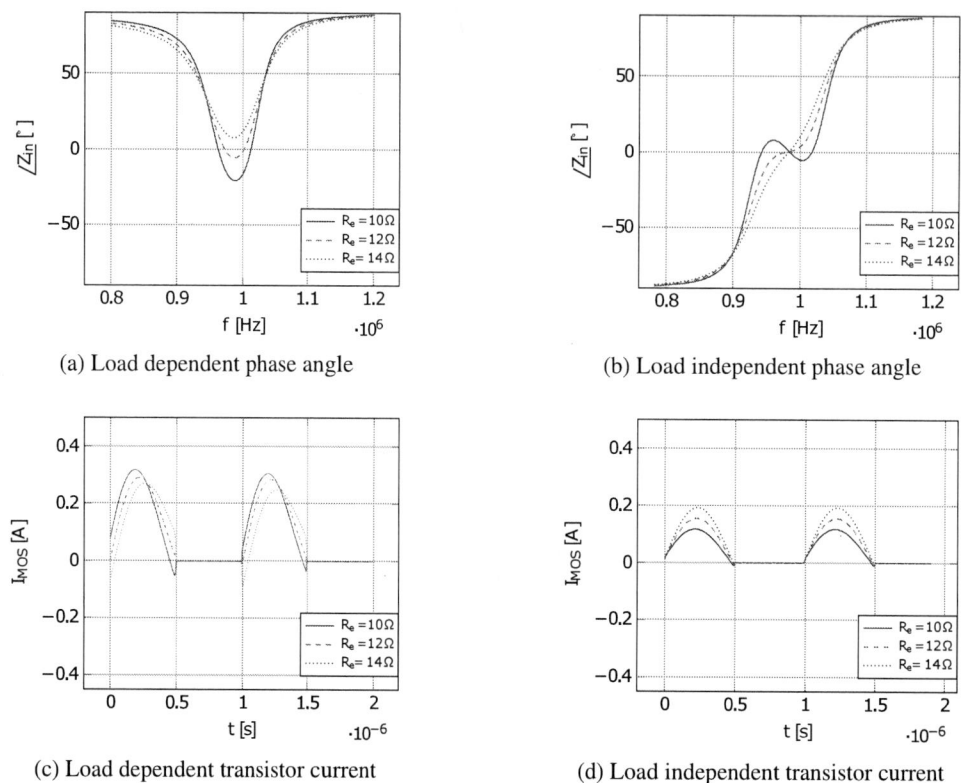

(a) Load dependent phase angle   (b) Load independent phase angle

(c) Load dependent transistor current   (d) Load independent transistor current

Fig. 3: Simulation results for load dependent and independent input reactance phase angle (a), (b), and load dependent and independent transistor current (c), (d)

Experimental setup and results

In order to validate the proposed design procedure, a prototype has been implemented and its performances have been experimentally characterized and compared to simulations. The schematic of the test setup is shown in Fig. 4a. It includes an EPC9006C development board [11] implementing the half-bridge with EPC2007C GaN enhancement mode field effect transistors (FET) (Q_1, Q_2). The input to the half-bridge is V_{DC} which can take a maximum value of 100 V. The board also contains the FET driver circuit, i.e., LM1553 gate-driver which is driven by V_{DD} = 7 V to 12 V and pulse-width modulation (PWM) circuit whose input is PWM_{IN} which can be maximum +6 V_{PP}, 6 MHz square-wave signal [11]. The LCL network components designed to obtain ZPA operation as explained in previous section, are mounted on a breadboard and connected to the EPC9006C development board as shown in Fig. 4a (V_{in} port). At the output, the equivalent resistance R_e described in the previous section (see eq.(2)) is considered instead of diode bridge rectifier.

The picture of the test setup is shown in Fig. 4b. The EPC9006C development board is mounted on the breadboard alongwith LCL network components specified in Table III. The potentiometer whose range is 0 to 50 Ω is placed and connected to the output through wires. The V_{DD} = 8 V to the driver circuit and V_{DC} = 60 V to the half-bridge inverter is provided by DC power supply [12]. The +5 V_{PP}, 1 MHz square-wave signal is given to PWM_{IN} by signal generator [13]. The input voltage and current (V_{in}, I_{in}) and respective output voltage and current (V_o, I_o) waveforms are taken on the oscilloscope [14] for various loads at different frequencies.

Since the real component values are slightly different from the nominal values, so the resonant frequency is obtained at 987 kHz instead of 978 kHz. At this frequency, the load is varied from 1 Ω to 30 Ω. It can be noted that the zero-phase angle operation is mantaned for all the loads above $R_{L_{min}}$. For three different loads, i.e., 10 Ω, 12 Ω and 14 Ω, the input phase is measured at all the frequencies ranging from

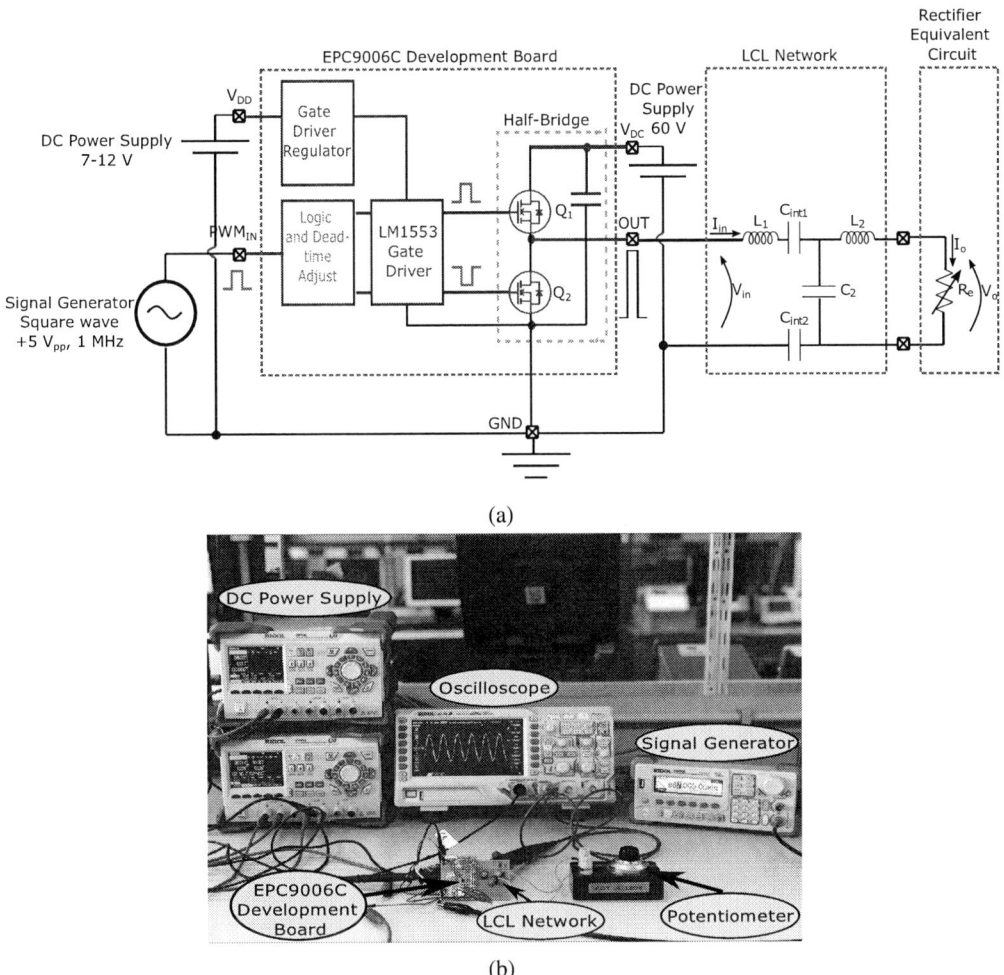

(a)

(b)

Fig. 4: Experimental test setup of the LCL CPT system with load-independent ZPA operation (a) Test setup schematic (b) test setup picture.

0.8 MHz to 1.2 MHz. The input phase versus frequency plot is generated and is reported in Fig. 5a. It is shown that, all three loads have almost zero-phase at working frequency. The phase difference increases for higher frequencies and symmetricaly decreases for lower frequencies.

Since, the transistor current (I_{MOS}) flows inside the EPC9006C board, so it can not be directly measured. However, we can measure the input current (I_{in}). The I_{MOS} is exactly the same as the I_{in} for the case when $V_{in} = 60$ V and zero when $V_{in} = 0$. Post processing of the current waveform I_{in} is perfomed in order to plot the I_{MOS} current and compare it to that shown in Fig. 3d. The resulting transistor currents are shown in Fig. 5b which verifies the ZPA operation for the same three loads ($10\,\Omega$, $12\,\Omega$ and $14\,\Omega$).

Efficiency analysis

In this section, the efficiency of the CPT system is studied. The efficiency depends on the losses of the devices, which are generated by parasitics specified in their datasheets, and are sometimes difficult to estimate. Therefore, to obtain a simple model, the following simplifications are considered:

- The losses generated by the passive components are represented by a series resistor (DCR for the inductors, ESR for the capacitors)
- The losses generated by the MOSFETs are modelled with a series resistor ($R_{ds,on}$) and by a constant contribution due to the gate capacitance

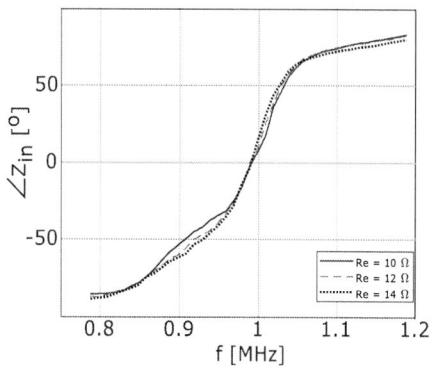

(a) Load independent phase angle

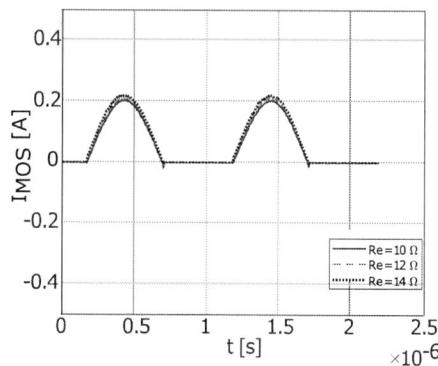

(b) Load independent transistor current

Fig. 5: Experimental results for load independent input reactance phase angle (a) and load independent transistor current (b)

- The losses generated by the Schottky diodes are caused by the forward voltage drop, V_F

In this way, the losses on all the devices are modelled using parameters specified in their datasheets. The goal is to obtain an analytical expression to represent the system efficiency as a function of a certain output quantity, such as the load. To this purpose, the efficiency is expressed as:

$$\eta = \frac{P_o}{P_o + \sum_i P_{\text{Loss,i}}} \tag{13}$$

If the output power is expressed as $P_o = I_o^2 R_L$, and the losses on the many devices are calculated with the simplifications considered, the efficiency can be expressed as follows:

$$\eta = \frac{R_L}{\alpha + R_L + \beta R_L^2} \tag{14}$$

where α and β are parameters that depend on component values and their parasitics, and are calculated by circuit analysis as shown below.

$$\alpha = \frac{\pi^2}{8} \left(\text{DCR}_{L_2} + \text{ESR}_{C_2} + 4V_F \frac{\omega_0 L_P}{V_{DC}} + V_{gs} Q_g f \left(\frac{\pi \omega_0 L_p}{V_{DC}} \right)^2 \right) \tag{15}$$

$$\beta = \frac{8}{(\pi \omega_0 L_p)^2} \left(\text{DCR}_{L_1} + 2\text{ESR}_{C_{\text{int}}} + R_{\text{ds,on}} \right) \tag{16}$$

With this model, the efficiency is represented as a function of the load, and efficiency versus load curves are developed. Selecting real components for Table III, the curves of Fig. 6 are generated both through computations and measurements. It is seen that the both curves are in close agreement to each other achieving a maximum of around 92% through computations (continuous line in Fig. 6) and 83% through measurements (dashed line). However, for the higher loads, there is a difference in the curves which is due to the fact that the proposed efficiency model doesn't take in to account the AC losses of the inductor which are high for higher loads, since the input current is high.

Conclusion

A design procedure for a CPT system using the LCL topology is explained in this article. This topology is able to provide a CC output with load independent ZPA operation, using just three passive components. In addition, the dc-dc efficiency is studied, and the system is expected to provide an efficiency of around 83% at nominal load. Effectiveness of the proposed procedure has been validated experimentally.

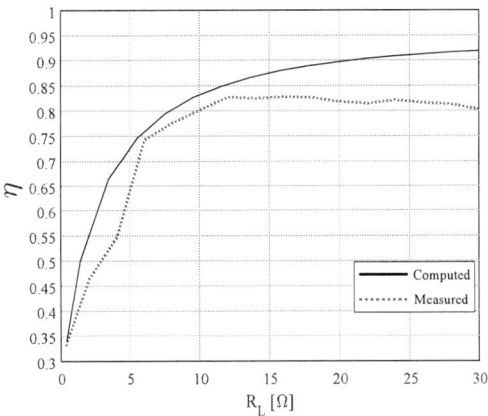

Fig. 6: Efficiency versus load

References

[1] M. M. El Rayes, G. Nagib, W. G. Ali Abdelaal, "A Review on Wireless Power Transfer," *International Journal of Engineering Trends and Technology (IJETT)*, vol. 40, no. 5, Oct. 2016.

[2] Chris Mi, "High Power Capacitive Power Transfer for Electric Vehicle Charging Applications" in *Proc. 6th International Conference on Power Electronics Systems and Applications (PESA)*, 2015, pp. 1-4.

[3] G. Buja, M. Bertoluzzo, and K. N. Mude, "Design and experimentation of WPT charger for electric city car," *IEEE Trans. Ind. Electron.*, vol. 62, no. 12, pp. 7436–7447, Dec. 2015.

[4] Su, YG;, Xie, SY.: "Capacitive Power Transfer System With a Mixed-Resonant Topology for Constant-Current Multiple-Pickup Applications," *IEEE Trans. Transp. Electrif.*, vol. 32, no. 11, pp. 8778-8786, Nov. 2017.

[5] Lu, J., Zhu, G., Lin, D., Zhang, Y., Jiang, J., Mi, C.C., "Unified Load-Independent ZPA Analysis and Design in CC and CV Modes of Higher Order Resonant Circuits for WPT Systems," *IEEE Trans. Transp. Electrif.*, vol. 5, no. 4, pp. 977-987, Dec. 2019,

[6] Theodoridis, M.P., "Effective Capacitive Power Transfer," *IEEE Trans. on Power Electronics*, vol. 27, no. 27, pp.4906-4913, Dec. 2012.

[7] Wu, Y, Chen, QH, Zhang, ZL, "Efficiency Optimization Based Parameter Design Method for the Capacitive Power Transfer System," *IEEE Trans. Transp. Electrif.*, vol. 36, no. 8, pp. 8774-8785, Aug. 2021.

[8] M. D. Bellar, T. S. Wu, A. Tchamdjou, J. Mahdavi, and M. Ehsani, "A review of soft-switched DC-AC converters," *IEEE Trans. Ind. Appl.*, vol. 34, no. 4, pp. 847–860, Jul./Aug. 1998.

[9] R. L. Steigerwald, "A comparison of half-bridge resonant converter topologies," *IEEE Trans. on Power Electronics*, vol. 3, no. 2, pp. 174-182, April 1988.

[10] Wire Wound Inductor 6065. Accessed: June 17, 2022. [online]. Available: https://eu.mouser.com/new/wurth-elektronik/we-ti

[11] EPC9006C Development Board. Accessed: June 17, 2022. [online]. Available: https://epc-co.com/epc/Products/DemoBoards/EPC9006C.aspx

[12] RIGOL DP832 DC Power Supply. Accessed: June 17, 2022. [online]. Available:https://www.rigolna.com/products/dc-power-loads/dp800

[13] RIGOL DG1022 Arbitrary Function Generator. Accessed: June 17, 2022. [online]. Available:https://www.rigolna.com/products/waveform-generators/dg1000z

[14] RIGOL DS1054 Digital storage Oscilloscope. Accessed: June 17, 2022. [online]. Available: https://www.rigolna.com/products/digital-oscilloscopes/1000z

A Pulse generator based on Transmission line Transformer for Insulation Aging Test

Xiao Yu, Khanh-Hung Nguyen, Peter Zacharias
UNIVERSITY OF KASSEL
Wilhelmshoeher Allee 71
D-34121 Kassel, Germany
Phone: +49 (561) 804-6477
Email: Xiao.yu@uni-kassel.de
URL: http://www.uni-kassel.de

July 11, 2022

Acknowledgments

This work has been funded in the frame of the ECPE Joint Research Programm ECPE Project 2020/PP03.

Keywords

≪Transformer≫, ≪Ferrite≫, ≪Pulsed power converter≫, ≪Aging≫, ≪GaN≫

Abstract

A pulse generator consisting of a transmission line transformer and GaN-based full bridge for generating high-repetitive impulse voltage stress with a high slew rate is proposed. GaN-half-bridges are used to generate fast switching edges, while the wide-band transmission line transformer is adopted to achieve the required voltage gain for an accelerated test without losing the interested high-frequency components contained in the fast edges. Another benefit of using the transmission line transformer is the spatial separation of pulse generation at room temperature and devices under test in a climate chamber, where the temperature for an accelerated test can be up to 150°C. The detailed operation principle, design considerations, simulation implementation, and experimental results for the proposed pulse generator are also presented. A peak-to-peak voltage of $400V \times 8$ and a maximal voltage slew rate of 250 V/ns have been achieved.

Introduction

New semiconductor technologies (e.g., 650 V GaN HEMTs) can generate high dv/dt during switching operations, resulting in significantly higher stress than that of conventional setups [1]. Together with high repetition rates, this stress can damage the insulation material and lead to partial discharges and breakdown. To generate the desired impulse voltage stress for the investigation of the long-term stability of relevant insulation materials under real geometries and environmental conditions, a pulse generator is needed, which should be able to provide a high charging/discharging current to the devices under test (DUTs) to achieve high dv/dt. On the other hand, for an accelerated lifetime test, the DUTs should be placed and stressed in a climate chamber, where the test temperature can be up to 150°C. No semiconductors can operate long hours under this ambient temperature. Therefore, the pulse generator should also enable a spatial separation between semiconductors generating the pulses and the DUTs. Furthermore, to fulfill a set of different test conditions, the slew rate, repetition rate, pulse width and amplitude of the generated voltage pulses should be easy and flexible to adjust.

In general, there are two basic methods for constructing pulse transformers, namely, ordinary transformers according to induction law and the transmission line type of transformer. The permeability drop and influences through stray capacitance and leakage inductance of windings limit the usage of ordinary transformers in the high-frequency range [2] [3]. The transmission line transformer (TLT) was first introduced by Guanella in 1944, and C.L.Ruthroff made another significant work in 1959 [4]. Due to their wide bandwidth and simple structure, TLTs are widely used for impedance matching for antennas and wideband voltage amplifiers in the high frequency and very high-frequency range [5] [6] [7].

For these reasons, a transmission line transformer-based pulse generator has been proposed and developed in this work. GaN-half-bridges are used to generate fast switching edges close to the real application scenarios. They are also responsible for the adjustable voltage amplitude, pulse width and repetition rate. At the same time, the transmission line transformer is adopted to achieve the required voltage gain for an accelerated test without losing the important high-frequency components contained in the fast edges. In the meantime, coaxial cables with a non-standard and lower characteristic impedance of 25Ω are chosen for a higher capacitive charging/discharging current so a fast rising/falling edge. The PTFE dielectrics of these coaxes allow the line ends to be placed and work in the climate chamber under high temperatures. In this paper, the relevant concepts and operation principle of the transmission line transformer has been presented. To extend the upper and lower bandwidth limits, a combination of ferrite materials is used.

The operating principle of transmission line transformer

The steep rising and falling edges of pulses contain a spectrum of harmonics with different intensities. The highest frequencies in the spectrum are determined by the slope of the pulse edges and are not related to the repetition frequency of the pulse train. A rule-of-thumb formula can be used to estimate the upper-frequency limit:

$$f_r = \frac{1}{\pi \cdot min(t_r, t_f)} \qquad (1)$$

Where t_r and t_f the rising time and falling time of a pulse, respectively.

The high-frequency effects become effective once the length of the transmission line approaches the order of magnitude of the wavelength of the highest frequency contained in the impulse spectrum [16]. Since the highest frequencies are still significantly higher than the upper-frequency limit f_r, the critical length l_{crit}, from which the line cannot be treated as electrically short, and the high-frequency effects should no longer be neglected, must therefore be significantly shorter than the wavelength λ corresponding to f_r:

$$\lambda = \frac{v}{f_r} = \frac{c \cdot \pi \cdot min(t_r, t_f)}{\sqrt{\varepsilon_r \mu_r}}, \ l_{crit} \approx \frac{\lambda}{10} \qquad (2)$$

Where ε_r and μ_r are, respectively, the relative permittivity and permeability of a medium in which the wave propagates. In fact, disturbances like overshoots occur already below this critical length but would not cause dramatic consequences. By using the high-frequency effects, a special design of transformers can be realized for high frequencies and pulse applications. These transformers use the propagation characteristics of electromagnetic waves in lines and are therefore also called delay line transformers [9]. Different forms of transmission lines such as a double line, a coaxial cable, a twisted pair, or a pair of wires wound on a ferrite core. In this work, only coaxial cables are considered.

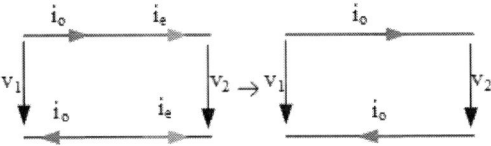
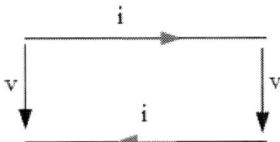

Fig. 1: Current Modes [3]   Fig. 2: Short line assumption

The following properties apply to an ideal TLT [3]:

- For a properly designed transmission line, only the current of the odd mode can flow through the line (see Fig. 1).
- All lines have the same length. The channels between two conductors of each line can be considered short compared to the intentionally delayed channels between conductors and the true ground.
- For a short line, the input voltage of a line is equal to its output voltage (see Fig. 2).
- Two different transmission lines are neither electrically nor magnetically coupled.
- The lines are connected in parallel on the input side and series on the output side (see Fig. 3).
- The order of a TLT is defined as the number of used lines.

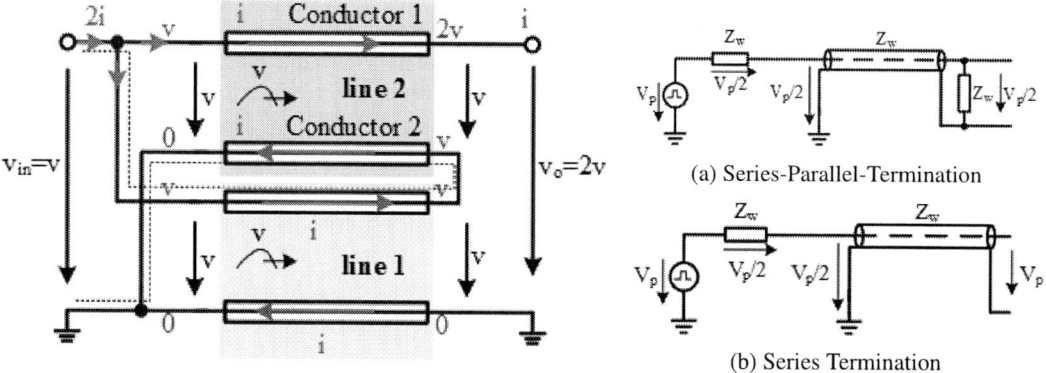

Fig. 3: Ideal TLT of 2nd order [9]

(a) Series-Parallel-Termination

(b) Series Termination

Fig. 4: Impedance matching

As shown in Fig.3, the ideal voltage gain and impedance ratio can be derived as:

$$n = \frac{v_o}{v_{in}} = \frac{i_{in}}{i_o} = 2, \ R_{in} = \frac{v}{2i} = \frac{1}{4} \cdot \frac{2v}{i} = \frac{1}{4} R_o = \frac{1}{n^2} R_o \tag{3}$$

The properly terminated transmission lines guarantee an undisturbed operation of TLT. Many methods are possible for a reflection-free termination (see Fig.4). The proposed pulse generator utilizes only the series termination, so in the case of a TLT with high order, no resistor needs to be dimensioned for the high output voltage. Furthermore, the incident voltage waves are totally reflected at the high impedance line ends due to a voltage reflection factor near +1 (capacitive load) and thus a voltage doubling occurs there and the voltage amplitude bounces back to V_p (see Fig. 4(b)). The reflected waves are eventually absorbed at the beginning of the line by the series resistor without causing further reflections.

Besides the propagation channels between the inner and outer conductor of the coaxial cables, the outer conductors and the ground form additional propagation channels, which prevent the normal operation of the transformer due to unwanted secondary reflections and short-circuit paths (see Fig.3 dot line) [10]. The transformer operates only when the secondary mode pulses are negligibly small, or be delayed a very long time, or both, without affecting the desired primary mode [2]. For these reasons, the usage of ferrite cores and ferrite sleeves is necessary. Due to the high permeability of the ferrites, the secondary mode pulses experience high impedance and larger delay in the channels between outer conductors and the ground plane. Another fact is that the highest frequency component in pulses can easily exceed 100 MHz, while the permeability of MnZn-ferrite (manganese zinc) drops rapidly from several megahertz. In order to provide enough impedance even in the high-frequency range, the cables are firstly wounded on the ferrite cores to obtain higher air inductance. Secondly, ferrite sleeves from NiZn (nickel-zinc) are installed on the coaxial cables, whose permeability is still far above zero in this frequency range (see Fig.5).

Compared to MnZn ferrites, the NiZn ferrites though have smaller permeability in the low-frequency range, but the feasible frequency range is much wider than that of MnZn ferrites and can be up to GHz-range. A combination of the two materials seems to be a good solution. It should be noted that, as shown in Fig.6, the higher the order of a cable, the higher the voltage stress on the installed ferrites so that the flux linkage. Therefore, the cables belonging to higher orders need more ferrites. In contrast, no ferrite

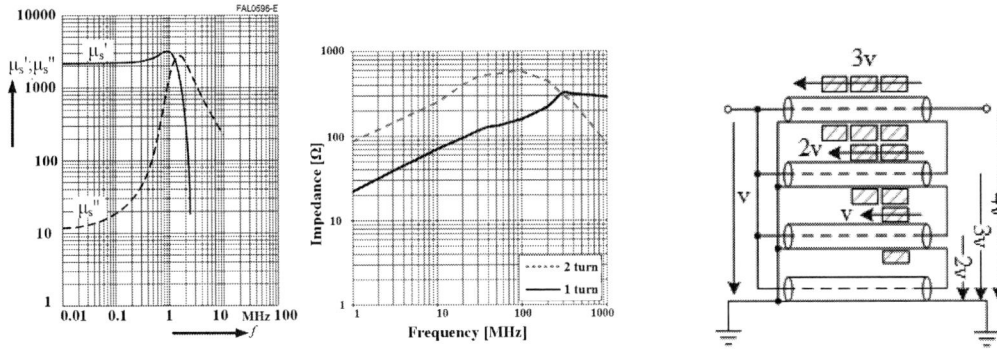

Fig. 5: Permeablitiy: MnZn (left) [11]; Impedance:NiZn (right) [12]   Fig. 6: Voltage on ferrites

is required for the cable at the bottom.

Basic circuit diagram and operation of proposed pulse generator

The pulse generator mainly consists of two GaN half-bridges, split DC link capacitors for blocking DC paths, resistors for impedance matching, and two TLTs of 4th order. The outer conductors of the bottom coaxial cable in each TLT are connected, while the DUT (device under test) is connected between two inner connectors of the top coaxial cable in each TLT. Fig. 7 shows the basic topology of the proposed pulse generator.

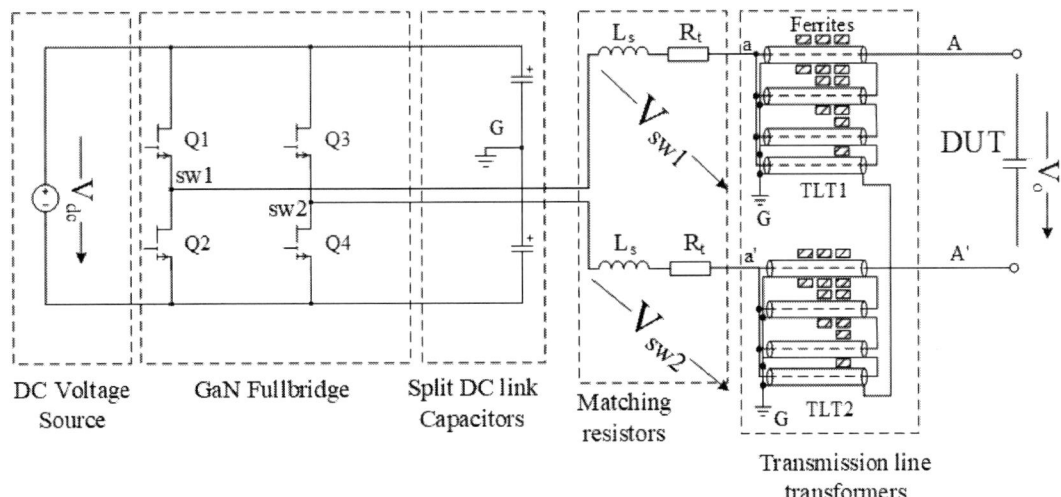

Fig. 7: Schematic of the proposed pulse generator

The relevant waveforms for the proposed pulse generator are shown in Fig.8. The GaN half-bridges generate complementary voltage pulses, V_{sw1} and V_{sw2}, with amplitude of $V_p = V_{dc}/2$. Taking the operation of TLT1 in Fig.7 as an example, since the matching resistors and the input impedance of TLT1 form a voltage divider, the amplitude of the incident voltage wave into the TLT1 is half of the pulse amplitude (see Fig.4). As the cables are parallel connected on the input side, a voltage wave with an amplitude of $V_p/2$ starts to travel between each cable's inner and outer conductor and towards the line end. Due to the same cable length, these voltage waves reach the serial connected line ends simultaneously. An Amplitude of $nV_p/2$ is induced at the output of TLT1. Taking the total reflection at the high impedance line ends into account, the amplitude of TLT1 output voltage V_{AG} equals nV_p. This also applies analogously to the wave propagation in TLT2, which results in an output voltage $V_{A'G}$ with an amplitude of $-nV_p$. Consequently, the amplitude of the voltage across the DUT equals the potential difference between two top inner conductors, which corresponds to $2nV_p$. The maximal change in voltage equals $4nV_p$, which occurs at the falling edge from $2nV_p$ to $-2nV_p$. It is noteworthy that to avoid shoot-through in half-bridges, a dead time t_d is inserted. During the dead time, because both switches in one half-bridge are open, exists no discharging path for charged cables, so the bridge voltages v_{sw1} and v_{sw2} are clamped. These voltages are allowed to change only when one switch in the corresponding half-bridge is on. Therefore, the dead time will not affect the output waveform.

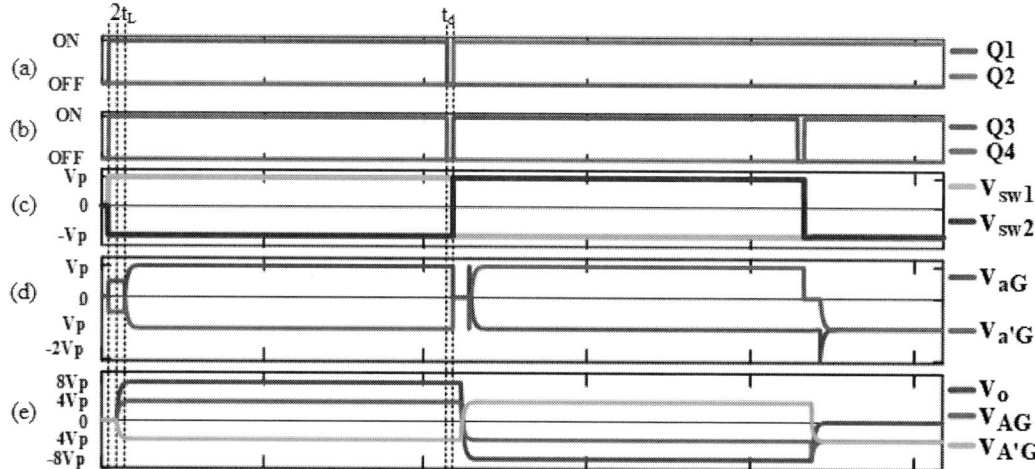

Fig. 8: Typical waveforms from simulation: (a), (b) Conductivity of switches ; (c) complementary voltage pulses ; (d) input voltages of TLTs ; (e) output voltage of each TLT and the voltage across the DUT

Simulation

The simulation model of the transmission line has been developed in [13]. For a lossless line, its two-port equations are derived as follows:

$$
i_a(t) = \frac{1}{Z_w} v_a(t) - i_a(t-t_L), \ i_a(t-t_L) := \frac{1}{Z_w} v_e(t-t_L) - i_e(t-t_L)
$$
$$
i_e(t) = \frac{1}{Z_w} v_e(t) - i_e(t-t_L), \ i_e(t-t_L) := \frac{1}{Z_w} v_a(t-t_L) - i_a(t-t_L)
$$

(4)

Where i_a and i_e the current flows into the line beginning and the line end, respectively, v_a and v_e the voltage across the input and output port of the transmission line, respectively. For a lossless line, applies:

$$
Z_w = \sqrt{\frac{L'}{C'}} \ (real), \ v = \frac{1}{\sqrt{L'C'}}
$$

(5)

It is assumed that the length of the line is equal to l. Then, the time for a voltage wave propagating from line beginning to the line end is:

$$
t_L = \frac{l}{v} = \frac{l}{\sqrt{L'C'}}
$$

(6)

Eq. 4 can be represented by the following equivalent circuit:

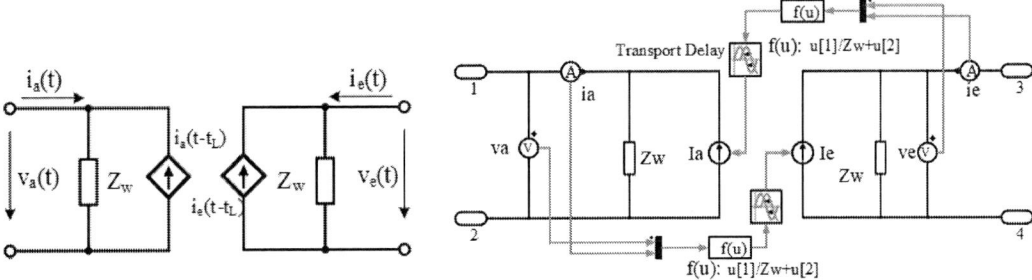

Fig. 9: Equivalent circuit for a lossless transmission line [13]

Fig. 10: Simulation realization of the model in PLECS

Specifications and design considerations

The required voltage slew rate range is from 100 to 250 V/ns, and a maximal peak-to-peak output voltage of 400 V×8 is prescribed for an accelerated aging test.

GaN full-bridge

GaN-HEMT transistors have smaller $R_{ds(on)}$ and need much fewer gate charges Q_G, which leads to a faster switching transient [1] and is beneficial for the generation of fast pulse edges. Commercial evaluation boards of GaN half-bridge offer high reliability, which also reduces the time consumption for designing gate driver and PCB layout of the full-bridge. Considering the limited project duration, the half-bridge evaluation board GS66508T-EVBDB2 from GaN-System has been chosen. The disadvantages of using evaluation boards are the fixed gate resistance and PCB layout of the bridges. It is worth mentioning that the method adopted in this work for adjusting the voltage slew rate is by introducing different loop inductance L_s (see Fig.7) instead of a changeable gate resistance through a potentiometer. However, the minimal achievable loop inductance $L_{s,min}$ so that the highest slew rate cannot be further reduced due to the fixed PCB layout of the evaluation board. For a well-designed and application-oriented GaN-full-bridge with changeable gate resistance, steeper edges, and more flexible adjustment of the slew rate are possible.

The operating voltage of the chosen GS66508T has been limited to 400 V for reliability reasons though a voltage rating of 650 V. To achieve the desired maximal voltage slew rate, the relation in Eq. 7 must be fulfilled:

$$\frac{dv_o}{dt}\Big|_{max} \approx \frac{\Delta V_{o,max}}{\tau_o} = \frac{4n\frac{400V}{2}}{2n \times 10^{-3}\frac{Z_w}{[\Omega]}\frac{C_{DUT}}{[pF]} \cdot ns} \geq 250\frac{V}{ns} \tag{7}$$

Where τ_o is the time constant of the output equivalent circuit, Z_w the surge impedance of the chosen coaxial cable, and C_{DUT} the capacitance of paralleled twisted pairs to be tested. The maximal allowable capacitance of DUTs can be derived as $C_{DUT,max} = 64pF$ for 25-Ω-coax. Taking the rising and falling time of V_{sw1} and V_{sw2}, which are decided by the switching speed of GaN switches, the matching resistance R_t and the loop inductance L_s, and the pulse droop due to finite secondary impedance into consideration, the total capacitance of paralleled DUTs should be lower than 64 pF. It is worth mentioning that the slew rate is theoretically independent of the order n, and TLTs are used to achieve the required peak-to-peak voltage.

Coaxial Cable

The desired voltage slew rate is mainly decided by loop inductance, the characteristic impedance of the cable, and the capacitance of DUT. The coaxial cables with lower characteristic impedance, e.g., 25 Ω, are preferred. Since the cable ends should be put in the climate chamber together with the DUT, where the temperature can be up to 150°C. The insulation materials with higher melting points must be chosen. Since no coaxial connector with 25 Ω is available in the market, the cables are directly mounted to the PCB. Considering the small size of these connection points, the impedance discontinuities caused by these direct connections can be ignored.

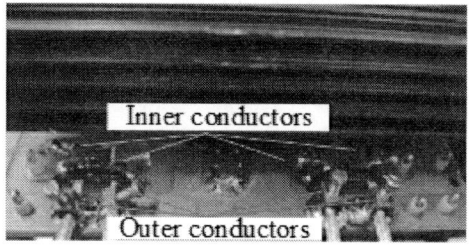

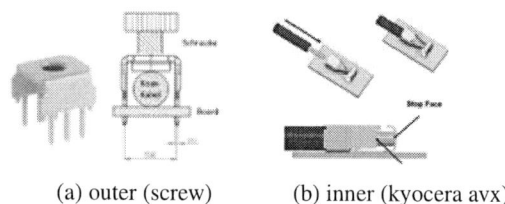

(a) outer (screw) (b) inner (kyocera avx)

Fig. 11: Screw mounted cables on PCB

Fig. 12: Alternative connectors

Matching networks

For each 4th order TLT, four resistors of 25 Ω are utilized. They are parallel connected to achieve the required termination resistance and reduce the introduced parasitic inductance. The effective current through each resistor can be derived as:

$$I_R(n.f_s, C_{DUT}) = \frac{1}{n}\sqrt{\frac{3n^3 C_{DUT} V_p^2}{2Z_w T_s} + \frac{2 \cdot t_L}{T_s}\left(\frac{nV_p}{2Z_w}\right)^2} = \frac{V_p}{Z_w}\sqrt{f_s\left(\frac{3n}{2}C_{DUT}Z_w + \frac{t_L}{2}\right)} \tag{8}$$

With a repetition rate of 140 kHz, propagating time of 25 ns, and a load capacitance $C_{DUT} = 64pF$, the calculation yields an RMS current of 0.445 A and power dissipation of 4.95 W in each resistor. It should be noted that current flows through those resistors only before the one-time reflection waves travel back to the line beginning, which lasts $2 \cdot t_L + t_r$ (see Fig.8 (d)).

Ferrite and ferrite sleeves

As already mentioned above, the task of ferrites is to provide sufficient secondary impedance between the outer conductor and the ground. Because different core shapes (ETD, PM, U, and toroid) are used at the same time, the derivation of the mathematical description of this impedance is already beyond the scope of the present work. The optimal number of installed ferrites cannot be estimated analytically, but the criterion is that as long as an improvement of the output waveform could be observed by increasing the number of ferrites, one more ferrite will be added.

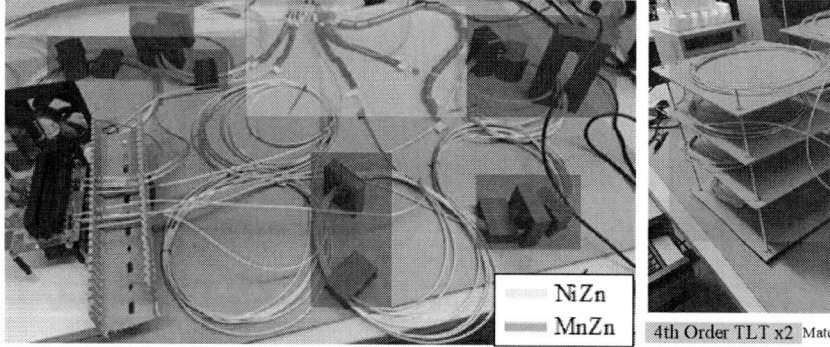

Fig. 13: Ferrites

Fig. 14: Prototype of pulse generator

PCB-Layouts for the full-bridge and impedance matching network

In order to enable an adjustable voltage slew rate, the length of interconnected lines between full-bridge and matching networks can be changed to introduce different values of loop inductance (see Fig.15(c), (d)). The design allows two different ways of connection. With the first variant (c), one can adjust the inductance by using different connection cables, while the smallest loop inductance is achieved via direct connection shown in 15(d).

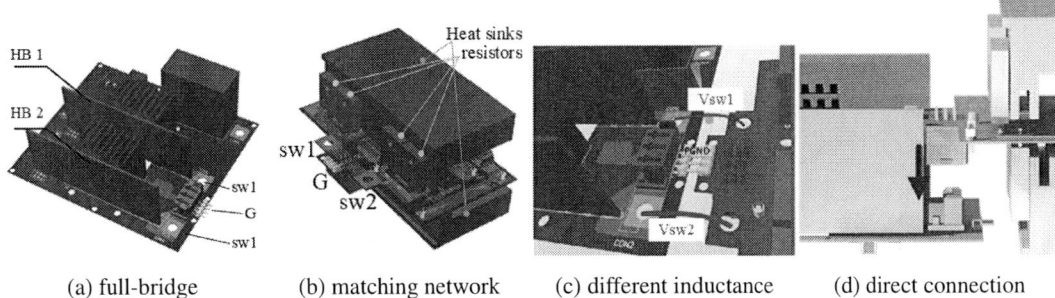

(a) full-bridge   (b) matching network   (c) different inductance   (d) direct connection

Fig. 15: Connection between half-bridges and matching networks

Measuring requirements and experimental results

For the measurement of fast transients or short switching times, an increased bandwidth of the measurement system is required. The lowest bandwidth required to measure a pulse with a rise time, t_r, or a falling time, t_f, can be estimated using the following rule of thumb [14] [16]:

$$BW\ [MHz] = \frac{0.35}{min(t_r, t_f)[ns]} \tag{9}$$

However, the specification of the bandwidth of a measuring device is usually given in such a way that the measuring signal is already slightly attenuated. Therefore, it is recommended to work with 3-5 times the calculated value. If it is not only about the display of a curve but also about the more precise measurement of the switching times or delay, even 10 times the calculated bandwidth is recommended [14]. For switching times in the 10 ns range, a bandwidth of at least 350 MHz is required.

It is also important to note that the effective bandwidth of the measurement system is determined not only by the oscilloscope but also by the probe used. The approximation of this effective bandwidth can then be made with the following formula:

$$BW_{sys} = \frac{1}{\sqrt{\frac{1}{BW_{osc}^2} + \frac{1}{BW_{probe}^2}}} \tag{10}$$

Where BW_{sys}, BW_{osc}, and BW_{probe} are the effective bandwidth of the measurement system, the bandwidth of the oscilloscope, and the voltage probe, respectively.

In the presented work, a prototype of the proposed pulse generator with two TLTs of the 4th order is constructed (see Fig.14). In Table I, the selected components for the pulse generator are listed. Forced air is used for the cooling of matching resistors and ferrites. No overheating has been detected during the tests on all components, which ensures a long time operation. Corona discharges were witnessed when the input voltage approached certain voltage values, which depend on the different isolation materials of different DUTs.

Table I: Selected components for the pulse generator

GaN half-bridge	GS66508T-EVBDB2	Ferrites	N87, N97
Coaxial cable	RG316/25 (PTFE), 5 m×8	Ferrite shelves	4W1500
Matching resistor	MP9100-25-1%, thick film		

A maximum voltage slew rate of 249.677 V/ns has been recorded in the measurement with $V_{dc} = 350V$ at no-load condition. It's worth noting that the slew rate during the falling time from 90% to 10% cannot always remain at this maximal value. This slew rate should theoretically be higher at the full voltage of

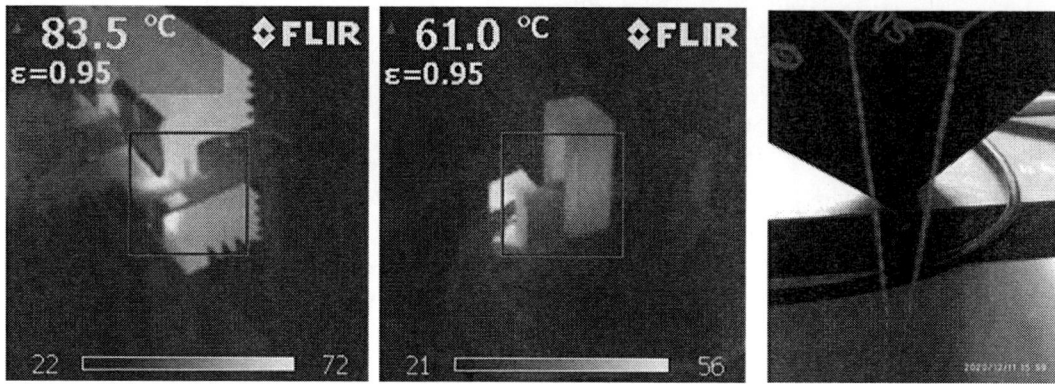

Fig. 16: thermography: resistors (left) ferrites (right)

Fig. 17: Corona discharge

400 V. The dv/dt can be adjusted by changing the DC voltage or the parasitic inductance. As shown in Fig.18, a pulse droop can be observed, and the measured voltage gain at $V_{dc} = 350\,V$ is around 6.8 and lower than the theoretical voltage gain of 8. This can be caused by the loss of TLTs, and the finite large secondary impedance [15], which reduce the lower bandwidth limit of TLTs.

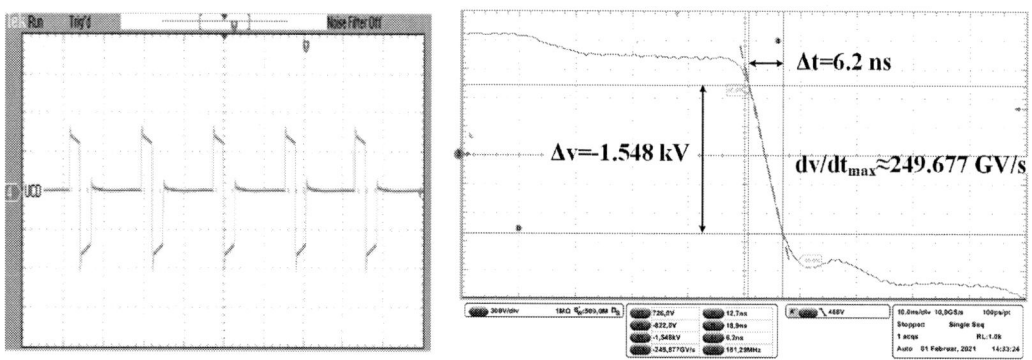

Fig. 18: Output voltage waveform

Fig. 19: Measured maximal voltage slew rate @ $V_{dc} = 350V$

Conclusion

A pulse generator based on TLT and GaN-based full bridge for generating high-repetitive impulse voltage stress with a high slew rate is built and tested. A maximal slew rate of 250 V/ns has been achieved. The generator can be operated at a repetition rate of 140 kHz without problems.

References

[1] GaN-System-Inc.: An Introduction to GaN Enhancement-mode HEMTs, 20 July 2021. [Online] Available: https://gansystems.com/design-center/application-notes/

[2] R. E. Matick:Transmission line pulse transformers—Theory and applications, Proceedings of the IEEE Vol 56 no 1, pp. 47-62, Jan. 1968

[3] W. A. Davis and K. Agarwal: Radio Frequency Circuit Design, John Wiley & Sons, pp. 105-121, 2001

[4] R. A. Mack and J. Sevick: Sevick's Transmission Line Transformers Theory and practice, 5th ed., SciTech Publishing, pp. 3, 2001.

[5] C. Trask: Transmission line transformers: Theory design and applications-part 2, High Frequency Electronics, pp. 46-52, Dec 2005.

[6] K. Yan: Corona plasma generation, Technische Universiteit Eindhoven, 2003.

[7] G. J. J. Winands: Efficient streamer plasma generation, Technische Universiteit Eindhoven, 2007.

[8] J. Horn and G. Boeck: Ultra broadband ferrite transmission line transformer, IEEE MTT-S International Microwave Symposium Digest, pp. 433-436, 15 July 2003.

[9] P. Zacharias: Magnetische Bauelemente, Springer Vieweg, pp. 602-611, 2020.

[10] P. N. Graneau, J. O. Rossi, and P. W. Smith: The operation and modeling of transmission line transformers using a referral method, Review of scientific instruments, pp. 3180-3185, 1999.

[11] TDK: Ferrites and accessories Siferrit material N87, September 2017. [Online] Available: https://www.tdk-electronics.tdk.com/en/529404/products/product-catalog/ferrites-and-accessories/ferrite-materials.

[12] Würth Elektronik: WE-AFB EMI Suppression Axial Ferrite 74270095, [Online] Available: https://www.we-online.com

[13] H. W. Dommel: Digital Computer Solution of Electromagnetic Transients in Single-and Multiphase Networks, IEEE Transactions on Power Apparatus and Systems, pp. 388-399, April 1969.

[14] Tektronix: ABC of the Probes, [Online] Available: https://www.tek.com/en/documents/whitepaper/abcs-probes-primer.

[15] C. Jiang et al.: A compact repetitive nanosecond pulsed power generator based on transmission line transformer, IEEE Transactions on Dielectrics and Electrical Insulation, pp. 1194-1198, 12 August 2011.

[16] Canavero, Flavio G. and Clayton R. Paul: Bandwidth of Digital Waveforms, 2010.

Design of a Single-Phase Common Mode and Differential Mode Inductor for Interleaved Converters

Jonathan Robinson, Gopal Mondal, Stefan Hänsel, Matthias Neumeister
SIEMENS AG
Erlangen, Germany
E-Mail: robinson.jonathan@siemens.com; gopal.mondal@siemens.com;
stefan.haensel@siemens.com, matthias.neumeister@siemens.com

Abstract

This paper analyses the design of single-phase interleaved inductors to provide inductance for differential mode (DM) circulating currents and common mode (CM) grid currents. The main design equations and a design optimization method are given. Comparison with other filter methods for equivalent performance shows reduced size with equivalent losses. Additional benefits using mixed core materials or due to phase shift in flux density are analyzed. Initial test results of a 120 kW converter demonstrate operation of the concept with a custom assembly that integrates the power electronics and coupled inductors in a combined heatsink.

1 Introduction

Interleaved converters for grid operation offer interesting potential to reduce the filter size while maintaining the same converter switching frequency and overall device losses. While in theory the concept can be used with any grid converter the real benefits from it are in applications requiring light weight converters in confined areas, such as mobility and offshore environments, especially at higher powers where parallel power modules would be normally required.

The main concept is shown in Fig. 1, where three interleaved converter modules are interconnected through the coupled DM inductor (L_{dmi}) and then connected to the grid through a 3-phase inductor (L_{cmi}).

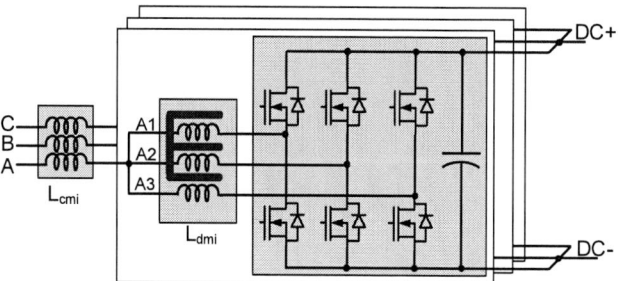

Fig. 1. Interleaved converter with coupled inductors to limit circulating current.

The dual-inductors have different purposes:

(a) L_{dmi}—this inductor limits circulating current between the parallel phases . By coupling the windings then only DM current (switching frequency circulating currents) will generate flux in the core. It is important that this inductor is connected across the interleaved phases (i.e. A1, A2, A3) and not between the grid phases.

(b) L_{cmi}—this inductor sees no switching frequency components (they are controlled to completely cancel at the common point between the interleaved sections) but is required for 50 Hz voltage regulation and filtering of higher order components (nf_{sw} where n is a multiple of the number of interleaved sections.

Note that CM and DM have been defined in the context of the coupled inductors and are different than the usual use of the terms. CM currents correspond to the main grid current (50 Hz) as well as harmonics of the order $n_{int}f_{sw}$. DM currents are generated from the out-of-phase switching, with the main harmonic at the switching frequency. These harmonics cancel at the grid side of the coupled inductor where the terminals are connected.

It is also possible to include additional capacitors and inductors to create higher order filters. The design of L_{cmi} and any additional filter components have an important advantage, that the harmonics they see are at a much higher frequency than the switching frequency of the converter. When the interleaved inductors see low frequency fundamental flux and high frequency circulating currents, then the core must be optimized for both, which is a difficult engineering task. Using coupled inductors, as shown in Fig. 1, splits the design into two parts—a high-frequency inductor design (L_{dmi}) and a low frequency inductor design (L_{cmi}).

However, it has been noted that peak flux densities of the interleaved section may be out of phase with the grid currents [1], and therefore it is possible to allow some additional fundamental flux to flow in the coupled inductors to maximize the utilization of the inductor. Example extensions to this concept have been proposed in [2], [3]. In [2] both a single-phase and a 3-phase inductor design were proposed and advantages of both were derived, however no detailed information was given on the design of the inductor nor a precise analyses on the benefits over a split inductor design. The design in [3] is similar to the single-phase design, although with only an additional core for common-mode flux and no additional winding (note that in [3], only a two-winding inductor is described, but the concept can be extended for additional interleaved sections). A basic electrical and magnetic circuit diagram of both concepts for an application with three interleaved sections is shown in Fig. 2.

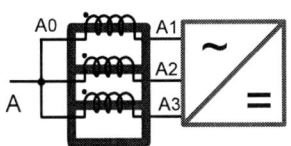
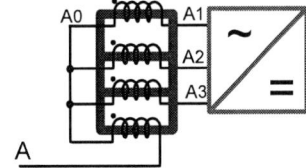

(a) Coupled inductor with common mode limb

(b) Coupled inductor with wound common mode limb

Fig. 2. Basic diagram of different CM/DM interleaved coupled inductors.

The additional winding gives an additional degree of engineering freedom (i.e. both air gap as well as the number of turns) for designing the inductor, but it is not clear what exactly the additional benefits of having this extra winding are in practice, nor is it clear if there are benefits over a split inductor design, which has the advantage that both inductors can be designed for either high-frequency or low-frequency flux. This paper studies the design of the single phase integrated CM and DM inductor to fully understand the design trade-offs and benefits of the concept. To simplify the analysis, only three interleaved sections will be considered, but all equations are easily adaptable for a different number.

2 Inductor Design Equations

The single-phase differential and common mode inductors are shown in Fig. 3. The DM flux (Fig. 3a) flows only between the interleaved windings since it is directly out of phase in other interleaved windings and get cancelled in the top and bottom limbs of the core and cannot then flow in the CM winding branch (W_{cm}). CM flux (Fig. 3b) will flow equally in the same direction of windings A1, A2, and A3, will then add up to 3x in the upper and lower limbs, and flow down the CM branch where winding W_{cm} is wound.

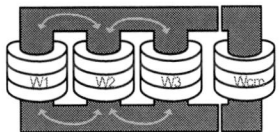
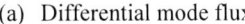
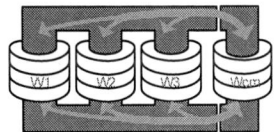

(a) Differential mode flux

(b) Common Mode flux

Fig. 3. Single phase differential and common mode inductor showing the different types of flux in the cores.

The inductance (L_{dmi}) and maximum flux density ($B_{dmi,max}$) for the differential section (windings W_1-W_3) are given by the basic inductance formulas as:

$$B_{dmi,max} = \frac{\mu_0 I_{dmi,pk} n_{dmi}}{l_{g,dmi}} \tag{1}$$

$$L_{dmi} = \frac{n_{dmi}^2 \mu_0 A_c}{l_{g,dmi}} \tag{2}$$

where n_{dmi} is the number of turns of the differential windings, A_c is the area of the core, and $l_{g,dmi}$ is the air gap. For the CM sections, similar equations for the maximum CM flux density ($B_{cmi,max}$) and CM inductance (L_{cmi}) can be derived as:

$$B_{cmi,max} = \frac{\varphi_{cmi}}{A_c} = \frac{\left(n_{dmi}\frac{l_{cmi}}{3} + n_{cmi}l_{cmi}\right)}{\frac{l_{g,dmi}}{3\mu_0} + \frac{l_{g,cmi}}{\mu_0}} \tag{3}$$

$$V_{cmi} = n_{dmi} \cdot \frac{1}{3}\frac{d\varphi_{cmi}}{dt} + n_{cmi}\frac{d\varphi_{cmi}}{dt} = \frac{\left(\frac{n_{dmi}}{3} + n_{cmi}\right)^2 \mu_0 A_c}{\left(\frac{l_{g,dmi}}{3} + l_{g,cmi}\right)} \cdot \frac{dI_{cmi}}{dt} = L_{cmi} \cdot \frac{dI_{cmi}}{dt} \tag{4}$$

where φ_{cmi} is the flux due to CM voltage across the winding, I_{cmi} is the CM current, and n_{cmi} is the number of turns for the CM winding, and $l_{g,cmi}$ is the total air gap for the CM core section. One aspect of (3) and (4) is that they use a common core area. When the core areas for the differential and CM sections are different then the equations can be modified as:

$$B_{cmi,max} = \frac{\mu_0 I_{cmi,pk}\left(\frac{n_{dmi}}{3} + n_{cmi}\right)}{\frac{l_{g,dmi}}{3}\frac{A_{c,cmi}}{A_{c,dmi}} + l_{g,cmi}} \tag{5}$$

$$L_{cmi} = \frac{\left(\frac{n_{dmi}}{3} + n_{cmi}\right)^2 \mu_0}{\left(\frac{l_{g,dmi}}{3A_{c,dmi}} + \frac{l_{g,cmi}}{A_{c,cmi}}\right)} \tag{6}$$

Note that the maximum CM current is given by the peak current at maximum power, while the maximum DM current can be obtained as described in [1].

Since the maximum DM current occurs around the zero crossing of the AC voltage (when the duty cycle is between 1/3 and 2/3—as shown in [1]) then the core utilization can be optimized for applications that require only active power at full power so that:

$$B_{dmcore,max} = B_{dmi,max} + k_r \frac{B_{cmi,max}}{3}\frac{A_{c,cmi}}{A_{c,dmi}} \tag{7}$$

and k_r is a factor varying from 2/3 to 1.

3 Optimized Design Methodology

Similar to other magnetic components, the optimum design is a trade-off between material cost, losses, and meeting the physical constraints of the materials, such as maximum temperature. In order to compare the designs in a basic way, then an optimization problem will be set up based on the dimension information shown in Fig. 4.

The following assumptions are made:

(a) The DM core will be considered square with side length d_c.
(b) The CM core will be considered rectangular with side length $d_c \cdot d_{c2}$.

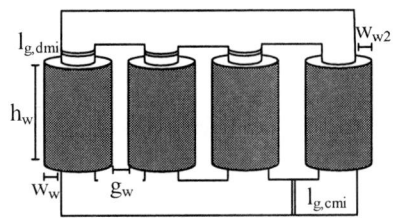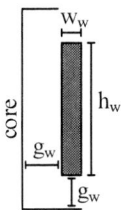

Fig. 4. Inductor dimensions for setting up optimization problem.

(c) The gaps between windings and between the windings and core (g_w) will be considered the same and equal to 2 mm.

(d) Dimensions of the winding (defined by w_w and h_w) will be increased by a factor $k=1/0.75$ to account for winding conductor insulation (this factor depends on the converter voltage).

(e) While the CM inductor is shown to the right of the interleaved inductors in Fig. 4, the ideal placement is in the middle, where the CM flux will then equally split in the different directions and reducing the required size of the upper and lower limbs. The design equations used here consider this ideal case.

The cost minimization function can then be defined as:

$$\underset{\underline{x}}{Min} = \{C_{cond} + C_{core} + C_{losses}\} \tag{8}$$

where the conductor cost (C_{cond}) depends on the cost of copper (C_{cu}) and conductor weight (M_{cu}):

$$C_{cond} = C_{cu}M_{cu} \tag{9}$$

the core cost (C_{core}) depends on the core material cost (C_{fe}) and core weight (M_{fe}):

$$C_{core} = C_{fe}M_{fe} \tag{10}$$

and cost of losses (C_{loss}) is calculated considering the present value of losses over N years, an interest rate i, and a cost per kWh of C_{kWh}:

$$C_{loss} = \frac{24 \cdot C_{kWh}E_{loss(kwh)}}{i}\left(1 - \frac{1}{(1+i)^N}\right) \tag{11}$$

It can be noted that C_u, C_{fe}, and C_{kWh} are effectively weighting functions that can be based on actual cost or modified to fulfill an objective (e.g. weight or loss minimization). The minimization variable $\underline{x}$, is defined as:

$$\underline{x} = \begin{bmatrix} w_w & l_w & l_{g,dmi} & d_c & n_{dmi} & J_{cond} & B_{max,dmi} & w_{w2} & l_{g,dmi} & n_{cmi} & B_{max,cmi} \end{bmatrix} \tag{12}$$

where J_{cond} is the current density of the conductor and other variables are defined previously and in Fig. 4. The mass of the conductors is the total of the DM and CM conductor mass:

$$M_{cu,dmi} = 3d_{cu}MLT_{dmi}n_{dmi}\frac{I_{rms,dmi}}{J_{cond}} \tag{13}$$

$$M_{cu,cmi} = d_{cu}MLT_{cmi}n_{cmi}\frac{I_{rms,cmi}}{J_{cond}} \tag{14}$$

where d_{cu} is the density of the conductor and MLT_{dmi} and MLT_{cmi} are the mean length per turn of the CM and DM windings, given by:

$$MLT_{dmi} = 4\left(d_c + 2g_w + \frac{w_w}{k}\right) \tag{15}$$

$$MLT_{cmi} = 4\left(\frac{d_c + d_{c2}}{2} + 2g_w + \frac{w_{w2}}{k}\right) \tag{16}$$

Similarly, the mass of the core can be calculated as the total of:

$$M_{fe,dmi} = d_{fe}d_c^2 \left(8\frac{w_w}{k} + 3\frac{h_w}{k} + 6d_c + 18g_w - 3l_{g,dmi} \right) \tag{17}$$

$$M_{fe,cmi} = d_{fe}(d_c \cdot d_{c2}) \left(2\frac{w_w}{k} + 2\frac{w_{w2}}{k} + \frac{h_w}{k} + 2d_c + 8g_w - l_{g,cmi} \right) \tag{18}$$

The losses are based on the total of the core and copper losses given by:

$$P_{cu} = I_{rms,dmi}^2 \left(\frac{r_{cu}J_{cond}}{I_{rms,dmi}} \right) MLT_{dmi}n_{dmi} + I_{rms,cmi}^2 \left(\frac{r_{cu}J_{cond}}{I_{rms,cmi}} \right) MLT_{cmi}n_{cmi} \tag{19}$$

where r_{cu} is the conductor resistivity and should be modified depending on the relevant frequency and type of wire [4]. The DM inductors contains the fundamental and all switching harmonics, while the CM contains the fundamental and the triplen harmonics (for three interleaved windings). The harmonics will depend on the PWM modulation and can be derived for different strategies as in [5].

Core losses can be estimated as:

$$P_{fe} = 0.5k_1 B_{max,dmi}^\beta f_{sw}^\alpha M_{fe,dmi} + 0.5k_1 \left(\frac{B_{max,cmi}}{3} \right)^\beta f_g^\alpha M_{fe,dmi} + 0.5k_1 B_{max,cmi}^\beta f_g^\alpha M_{fe,cmi} \tag{20}$$

where k_1, α, and β are Steinmetz constants that depend on the material, f_{sw} is the switching frequency of the interleaved converter, and f_g is the frequency of the grid. The flux in the differential part contains the non-triplen harmonics plus additional fundamental and triplen harmonics coming from the CM coupling, but split equally in the different branches. The CM core section should contain mainly the fundamental and triplen harmonics.

It should be noted that in (20) the losses are calculated using the Steinmetz equation considering separately the losses from the low frequency and high frequency harmonics. The result will not be completely accurate when there is a high percentage of low frequency flux but it is sufficient for a preliminary estimate.

The final step of the optimization is to ensure the maximum temperature of the inductor is within the limits of the materials, such as that described in [6]. In the example for this paper then the weighting function for losses was adjusted to keep the temperature rise in the core or conductor equal to 40°C.

4 Design Comparison

Using the equations given in Sec. 3, designs were made for a grid converter application consisting of three interleaved sections, and with a total rated power of 120 kW and connected to a 400V, 50 Hz grid and considering sinusoidal PWM with a switching frequency of 48 kHz. The actual filter requirements will depend on the grid and it is possible to use a complete inductive design or a higher order filter consisting of multiple components. An example filter is given in Sec. 4.4 although the complete design of the filter is outside the scope of this work. Instead this section focuses on studying the overall effects of varying the CM and DM inductor sizes on the inductor size. The tools and procedures developed here could then be applied as part of a complete optimization of the converter filter for any specific grid requirements.

The designs are based around a reference design considering a required 50 µH CM inductance and 200 µH DM inductance. Secs. 4.1 and 4.2 consider inductors with nanocrystalline cores (kOr 120), while Sec. 4.3 considers other materials and methods to optimize core utilization.

4.1 Differential Mode Inductance Variation

Variation of the DM inductance will directly impact the peak circulating currents. These currents flow through the devices, and therefore any large increase in circulating current will increase both conduction and switching losses. Fig. 5 shows changes in inductor size for a DM inductance varying from 50 to 500 µH, as well as the peak circulating current. The CM inductance was fixed at 50 µH.

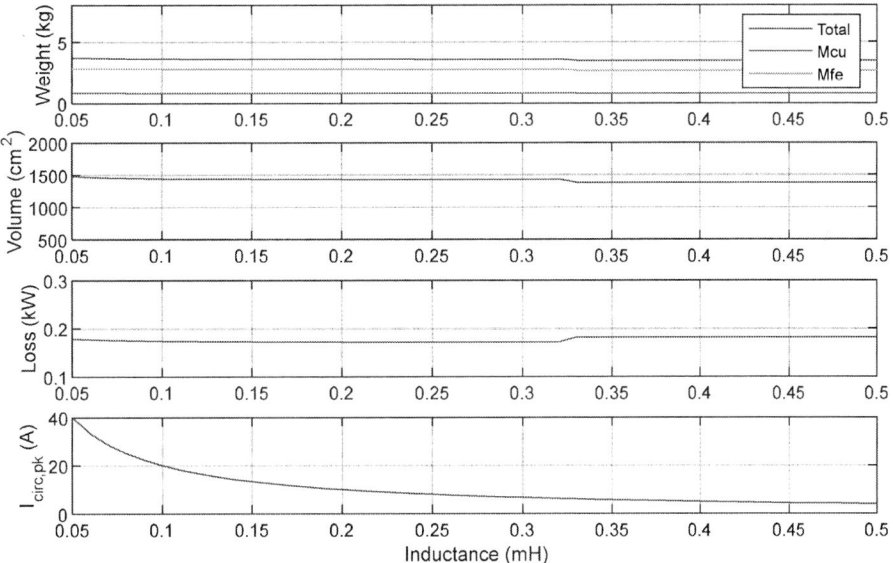

Fig. 5. Inductor size, weight and losses for varying differential mode inductance.

It can also be seen that the weight, volume and losses don't change much for different DM inductance. The inductance and peak current are inversely proportional so that from (1) and (2), increasing the inductance by reducing the air gap will result in a similar peak flux density. Eventually at high inductances then the air gap will reach a minimum value, the core reluctance will become more important, and then more turns will be required to further increase the inductance, resulting in increasing weight, volume, and/or losses.

Since converter losses generally increase with reduced inductance (due to higher ripple current) in the devices then there is an overall benefit to maximizing the DM inductance within construction limits.

4.2 Common Mode Inductance Variation

Variation of the CM inductance produces a more important change in the inductor size since there is no corresponding reduction of peak flux generating current such as in the differential design. Resulting weight, volume, and losses of the inductor are shown in Fig. 6 for a fixed DM inductance of 200 µH and CM inductance varying from 10 to 100 µH (equivalent to 0.2 – 1.5 % impedance range).

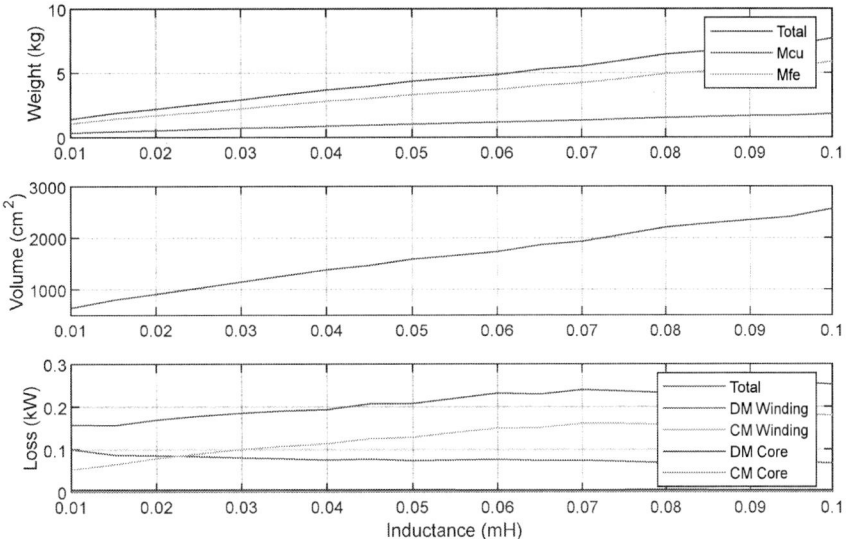

Fig. 6. Inductor size, weight and losses for varying common mode inductance.

Losses are dominated by winding losses, which must support the full current and all high frequency harmonics and the DM core, which has only high frequency flux. Losses slightly increase at low inductance since this low inductance leads to lower attenuation of triplen harmonics, which are only affected by the CM inductance. Core losses are very insignificant due to the fact that the nanocrystalline has excellent performance at high frequency and the actual percentage of high-frequency flux is fairly low, even in the DM part (the CM part has only a small amount of triplen harmonics).

This indicates a potentially important benefit from using a silicon steel core, which would allow higher flux densities and would have the effect of reducing overall inductor costs (estimated to be as high as 8x larger in [7]).

4.3 Core Optimization

A second key optimization possibility can be where it can be seen that the CM core will normally have only CM flux, which is mostly due to 50 Hz AC current. This low frequency flux would allow optimizing the inductor design by using a plated steel core in the CM branch, while using a higher frequency material only in the DM section.

Design results in Fig. 7 using a 0.2 mm plated Grain Oriented Electrical Steel show an improvement in weight, volume, and losses, except for low inductances where the higher switching harmonics results in increase core losses. Core losses in the CM core are still low, with only a small increase at low inductances due to the higher switching ripple frequencies.

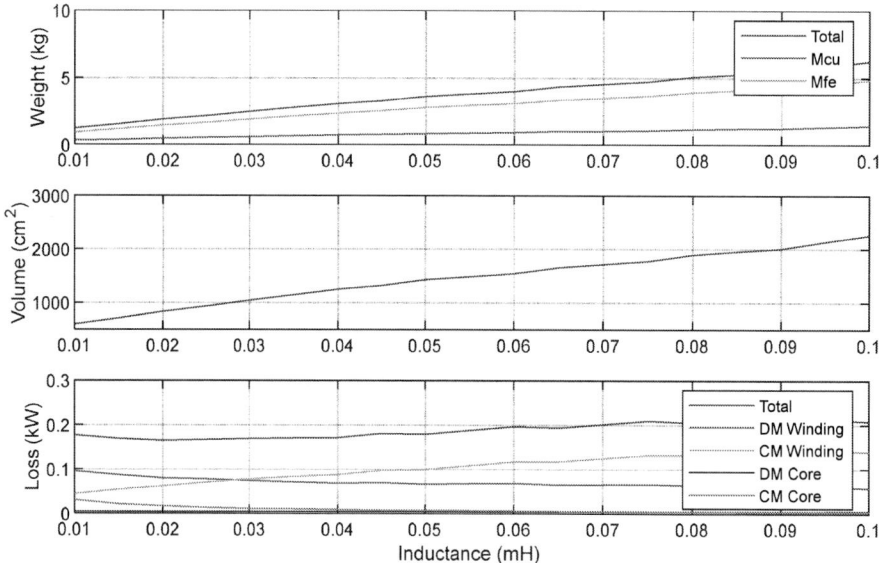

Fig. 7. Inductor size, weight and losses with a Si-Steel CM Core Section.

In fact further increases are possible if the entire core is made from Si-Steel (both DM and CM sections). The DM section of the core contains a large CM flux, so that the high frequency DM flux represents only a small percentage of the overall flux. It will result in higher losses but could offer a reduction in volume of >50%.

One final aspect of the interleaved converter is that the highest circulating current in the coupled inductors occurs at the zero crossing of the voltage. When the converter has mostly active power, then this implies that peak flux of the CM and DM parts occurs at different times, offering potential for overall core size optimization.

In the optimal case where k_r in (7) is equal to 2/3 then the resulting design shows a reduction similar to using the silicon steel CM core. The two techniques can both be applied resulting in an even smaller design, as shown in Fig. 8 compared with Fig. 6.

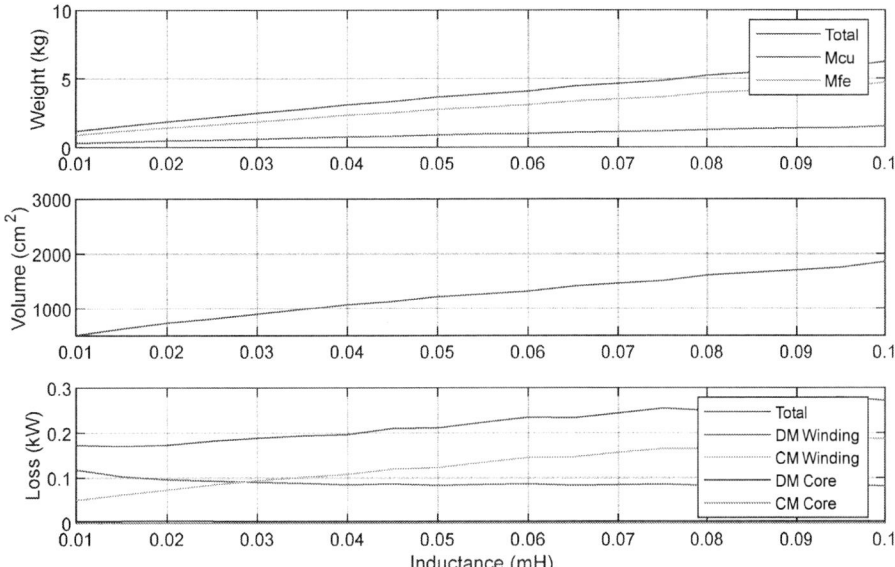

Fig. 8. Inductor size, weight and losses with a steel core and minimal reactive power.

4.4 Comparison with other Inductor Designs in a Grid Filter Application

In order to compare with other designs, then similar cost minimization functions were created to achieve 200 µH DM inductance and for varying CM inductance. Three cases are compared:

1. Case 1: 3-phase Inductor + 3x coupled inductor + CM Choke
2. Case 2: 3x coupled inductor with CM core (Fig. 2a) + CM Choke
3. Case 3: 3x coupled inductor with wound CM core (Fig. 2b) + CM Choke

A general illustration of the combined filter concepts is shown in Fig. 7.

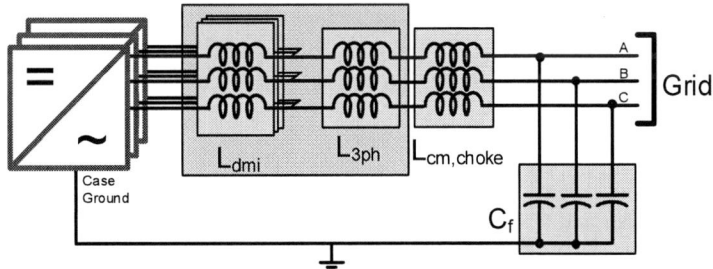

Fig. 9. Example filter concept circuit diagram.

The CM choke is for filtering 3rd harmonic (and other triplen harmonics) which are not attenuated by the DM or 3-phase inductors. For Case 1 it is set to 250 µH, for Cases 2 and 3, the total of the CM choke and CM inductance of the combined inductor kept constant at 250 µH. The CM filter capacitor (C_f) is not sized since it will be the same in all cases. As noted previously, the actual size requirements of the different components depend on the filter requirements to meet the grid code. Since all three cases are designed to the same requirement then it represents a valid comparison of the different filters, although the difference in benefits will vary for different applications.

Comparison of the volume is more difficult due to integration of separate components in Case 1 and therefore only the total weight of the core and conductor and full-load losses are shown. The optimizer was weighted toward minimum weight and in all cases the maximum temperature rise in the core or winding was kept to 40°C. Results are shown in Fig. 10, note that total weight and losses is of all core and windings for all inductor sections.

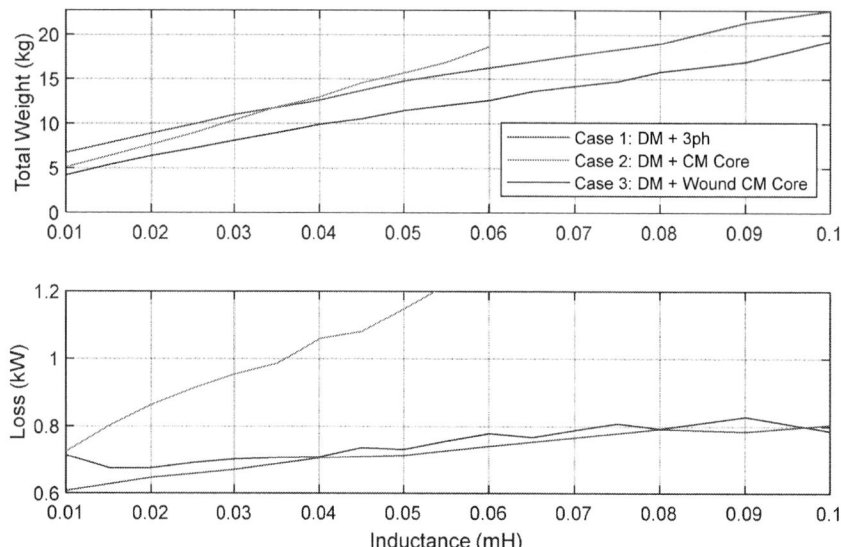

Fig. 10. Comparison of total inductor weight and losses.

The results show that the combined CM and DM (Case 3) design can potentially offer lower losses and weight compared to a conventional 3-phase design with separate coupled inductors. The unwound core design (Case 2: CMDM) achieves similar results to Case 3 at low inductance, but achieving higher CM inductance with only the additional core section results in higher losses in the DM winding (due to increased turns). It can also be seen that using separate inductors is still competitive, although overall weight is higher. An advantage of the separate inductors will be lower overall cost due to use of standard core sections.

It should also be noted that the CM inductance provides the main inductance at the grid frequency (50 Hz in our case) while providing very little benefit as a filter (non-triplen harmonics are cancelled after the DM inductor and triplen harmonics are not affected by the 3-phase inductor, only the CM choke). The resulting size is then mainly determined by the speed of the control or peak current during faults (which is also impacted by the control speed). A smaller inductance will offer a faster response, but also requires higher performance from the converter controller and sensors.

5 Experimental Results

A 3-phase interleaved converter was tested in the lab to verify the operation. The design uses a 200 uH coupled inductor with an additional 65 µH external inductor (type N CNW 854 / 250, not shown), corresponding to Case 1.

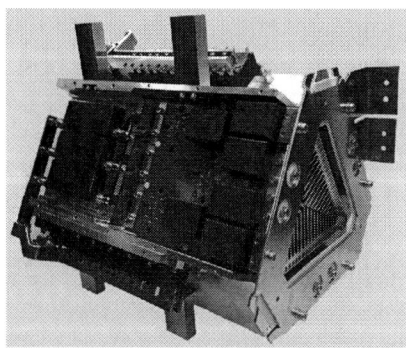

(a) Converter hardware

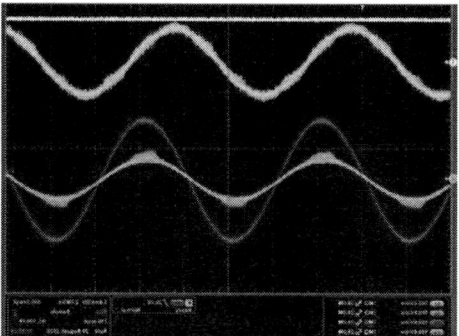

(b) Test results (Yellow: DC Link Voltage, Blue: AC voltage, Purple: AC grid current, Green: Coupled Inductor current

Fig. 11. Converter and test results.

It can be seen in Fig. 11a that the converter, DC link capacitors, and interleaved inductor were built into a common hardware design, resulting in a compact converter with a simple cooling system for both power devices and coupled inductor.

The resulting operation is shown in Fig. 11b where it can be seen that the highest ripple current in the coupled inductor occurs at the peak AC current. This is due to reactive power operation (the highest DM current is at the zero crossing of the voltage) and represents the case where no additional optimization of the core can be performed (such as STATCOM applications).

6   Conclusions

The analysis shows overall benefit for an inductor design combining the DM inductance with a wound CM limb. Splitting the inductors is still competitive and may have an overall lower cost, which may be the key criteria for some applications.

Comparing the combined inductors, losses in the unwound core design (Case 2) greatly increase for high CM inductances. When the core is wound (Case 3) then the extra degree of freedom allows the overall best design. Further improvements can be obtained by (a) using a high flux density core for the CM limb, (b) using a high flux density core for the whole inductor (trading off size with losses), (c) taking advantage of the phase shift in peak flux density between the DM and CM sections in applications with mostly real power.

While the results are interesting, test results are shown just with the split inductor case and work is ongoing on innovative methods to construct the combined inductor with the converter.

7   References

[1] G. Mondal, J. Robinson, S. Haensel, M. Neumeister, and S. Nielebock, 'On Optimal Construction of Interleaved Converter for minimum Circulating Current', in *2018 20th European Conference on Power Electronics and Applications (EPE'18 ECCE Europe)*, Sep. 2018, p. P.1-P.10.

[2] G. Mondal, J. Robinson, and S. Haensel, 'Design and Verification of Compact Inductor for Interleaved AC/DC Converter', in *2019 21st European Conference on Power Electronics and Applications (EPE '19 ECCE Europe)*, Sep. 2019, p. P.1-P.9. doi: 10.23919/EPE.2019.8915512.

[3] K.-B. Park, F. D. Kieferndorf, U. Drofenik, S. Pettersson, and F. Canales, 'Optimization of LCL Filter With Integrated Intercell Transformer for Two-Interleaved High-Power Grid-Tied Converters', *IEEE Trans. Power Electron.*, vol. 35, no. 3, pp. 2317–2333, Mar. 2020, doi: 10.1109/TPEL.2019.2926312.

[4] M. Bartoli, N. Noferi, A. Reatti, and M. K. Kazimierczuk, 'Modeling Litz-wire winding losses in high-frequency power inductors', in *PESC Record. 27th Annual IEEE Power Electronics Specialists Conference*, Jun. 1996, vol. 2, pp. 1690–1696 vol.2. doi: 10.1109/PESC.1996.548808.

[5] G. Holmes and T. Lipo, *Pulse Width Modulation for Power Converters: Principles and Practice.* 2003.

[6] M. Sippola and R. E. Sepponen, 'Accurate prediction of high-frequency power-transformer losses and temperature rise', *IEEE Trans. Power Electron.*, vol. 17, no. 5, pp. 835–847, Sep. 2002, doi: 10.1109/TPEL.2002.802193.

[7] D. Ruiz-Robles, J. Ortíz-Marín, V. Venegas-Rebollar, E. L. Moreno-Goytia, D. Granados-Lieberman, and J. R. Rodríguez-Rodriguez, 'Nanocrystalline and Silicon Steel Medium-Frequency Transformers Applied to DC-DC Converters: Analysis and Experimental Comparison', *Energies*, vol. 12, no. 11, 2019, doi: 10.3390/en12112062.

Steady-State Analysis and Comparison of SSFB, SDFB and DSFB MMC-based STATCOM

Mohamed Moez BELHAOUANE[1], Pierre VERMEERCH[1], François GRUSON[1], Pierre RAULT[2], Sébastien DENNETIERE[2], Xavier GUILLAUD[1]

[1]Univ. Lille, Arts et Métiers Institute of Technology, Centrale Lille, Junia, ULR 2697 - L2EP, F-59000 Lille, France.

E-Mail : mohamed-moez.belhaouane, pierre.vermeerch, xavier.guillaud @centralelille.fr, francois.gruson@ensam.eu

[2] RTE (Réseau de Transport d'Electricite), Research and Development Department of RTE, 92073, Paris La Défense, France.

E-Mail : pierre.rault, sebastien.dennetiere@rte-france.com

Acknowledgements

This research work has been supported by RTE, the French TSO (Transmission System Operator).

Keywords

« Static Synchronous Compensator (STATCOM) » « Modular Multilevel Converters (MMC)» « Double-Star Full Bridge » « Single-Star Full Bridge » « Single-Delta Full Bridge » « Energy requirement » « Losses estimation ».

Abstract

This work focuses on the steady-state analysis of three types of MMC based STATCOM. For a given STATCOM rating, Double-Star Full Bridge, Single-Star Full Bridge and Single-Delta Full Bridge have been compared in terms of design and losses. In this approach, the number of submodules is chosen according to the voltage and current ratings of semiconductor devices while the submodule capacitor value is obtained by following an energy storage criterion to maintain the submodule voltages within an acceptable voltage range.

Introduction

STATCOM are commonly considered for voltage regulation and control, reactive power compensation, improvement of steady state power transfer capacity, improvement of power quality such as flicker and harmonics, improvement of transient stability margin, damping power system oscillations and sub-synchronous oscillations, balancing loads of individual phases and eventually application including storage [1]. More recently, the massive penetration of renewable energy sources has increased the interest in HV STATCOM applications to strengthen the grid. The STATCOM-MMC family is usually classified into four topologies [2]:Single-Star Full Bridge (SSFB), Single-Delta Full Bridge (SDFB), Double-Star Full-Bridge (DSFB) and Double-Star Half-Bridge (DSHB). The single-star full-bridge and the single-delta full-bridge (FB) are based on three strings of multiple FB cells with star and delta connection, respectively. The double Star Full-Bridge (FB) and Half-Bridge (HB) topologies are based on two sets of star-configured converters in which the AC side of multiple FB and HB cells are cascaded to constitute each arm [3]. This topology is the same as an AC/DC converter station. Usually, this topology is seen in literature as unsuitable topology due its higher number of power devices. However, for STATCOM application, it seems to be interesting to investigate this structure compared to DSHB [4]. In terms of commercial application, some topologies have been implemented. For instance, the

SDFB topology has been adopted as STATCOM station and energy storage interface for an already operational project as illustrated in [5].

Nevertheless, few works in literature present an in-depth steady-state comparison study between the different topologies, presented above, for STATCOM application. For instance, in [2], only the difference in terms of the number of SMs has been evaluated for different topologies. In [6] a comparison of the two double star topologies, such as DSFB and DSHB, has been proposed. These topologies are compared in terms of steady-state operation, power losses and costs. The reference [7] compares the losses and topology cost of SDFB and DSHB topologies for STATCOM application based on a 100 MVA case study. In addition, regarding the converter losses, only the contribution of power semiconductor devices has been considered. In fact, few comparison studies related with the energy storage requirement ensuring a reliable operation of the topology are proposed in literature. In this sense, the authors in [8] propose some comparisons, which are supported by analytical results regarding minimum effective DC-bus voltage, number of SMs, current rating, energy storage requirements and operation during unbalanced conditions. However, only DSHB and SDFB topologies have been investigated for a 15 MVA MMC-STATCOM. Therefore, there is a lack of a more global comparative study between the main MMC topologies, about the main design criteria for a STATCOM application.

This paper aimed to provide a comprehensive approach on the design of MMC-STATCOM for high voltage application for different topologies:

   (i)   Proposition of steady-state analysis methodology allowing the design of SMs number per arm , the energy storage requirements and total losses including power electronic and passive high voltage equipment losses.
   (ii) Comparison of the three topologies in terms of arm design, storage energy requirments and semiconductor losses based on the elaborated criteria in **(i).** The comparison of the necessary components gives an insight of the total cost associated with each topology.
   (iii) Identification of the most efficient and reliable MMC-STATCOM topology that meets the defined criteria in the proposed study.

It can be demonstrated that the DSHB would induce an important energy storage requirement compared with the three other topologies [8]. So only SSFB, SDFB and DSFB will be studied in this paper.

The paper is outlined as follows. The configuration of the three main MMC-STATCOM topologies are detailed in section 2. The section 3 presents the steady-state analysis based on single star full-bridge configuration, considered as reference topology. The proposed methodology involves submodules number calculation and stored energy requirement per arm as well as total losses estimation. The expansion of the proposed design methodology on SDFB and DSFB Topologies is proposed in section 4. A comparative study between the studied configurations (i.e., SSFB, SDFB and DSFB) is performed to assess the steady-state performances of MMC-STATCOM topologies. The last section summarizes the main conclusions drawn from the analysis, showing the best topology of MMC-STATCOM based on the proposed criteria through this study.

STATCOM Topologies Description

As mentioned above, only the topologies with submodules (SM) in full-bridge configuration are considered. The circuit diagrams of the three main topologies are highlighted in Fig. 1. As shown, the DSFB topology presents 6 arms (Fig. 1c), while the SSFB and SDFB topologies presents 3 arms (Fig. 1a, Fig. 1b respectively). Each arm consisting of an inductor L_{arm}, an internal resistance R_{arm} and N-full-bridge cell capacitance C_{SM}. The converter is connected to the AC grid through the three-phase transformer connection impedance which is characterized by its transformation ratio and its leakage inductance (L_f, R_f). The grid side is modelled as a voltage source $v_{gj,j=a,b,c}$, with $i_{gj,j=a,b,c}$ are the three-phase grid current. The modulated voltages are designed by $v_{mj,j=a,b,c}$ (resp. $u_{mj,j=ab,bc,ca}$). As mentioned above, the SMs in SSFB, SDFB and DSFB topologies are of full bridge type, since both positive and negative voltages are required for the operation of the converter.

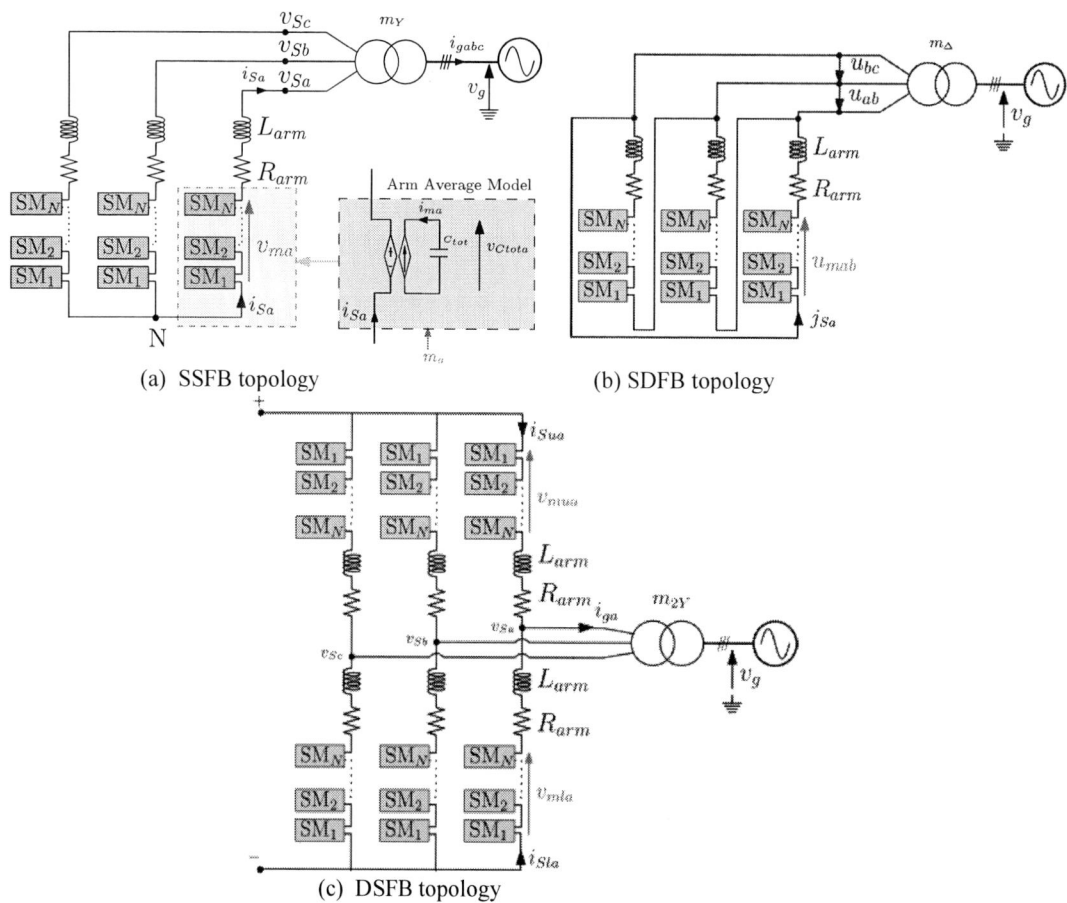

(a) SSFB topology (b) SDFB topology

(c) DSFB topology

Fig. 1: Classification of studied topologies

In addition, thanks to degrees of freedom, the Zero-sequence is considered to improve the design, a voltage v_{NO} is considered for star topology while a current i_0 for delta topology.

STATCOM Design: Example of Single-Star Full Bridge (SSFB) Topology

The same methodology is applicable for the three different topologies; however, it is only detailed on the SSFB configuration. First, the nominal arm current is obtained from the rating of power electronic devices which are connected in series. For instance, considering a 3300V-1500A, single switch IGBT - *Infineon FZ 1500R33HL3* [9], the arm current is defined as follow:

$$I_{SY} = I_{SY_{nom}} = \frac{1500}{\sqrt{2}} = 1061 \text{ Arms} \tag{1}$$

Considering a 300 MVA STATCOM connected to a 400kV grid, the nominal grid current is defined as follow:

$$I_{g_{nom}} = \frac{S_{nom}}{\sqrt{3} \times U_{nom}} = \frac{300MVA}{\sqrt{3} \times 400kV} = 433 \text{ Arms} \tag{2}$$

The interface transformer ratio is defined according to the ratio between the grid current and the valve current:

$$m_Y = \frac{I_{g_{nom}}}{I_{SY}} = \frac{433}{1061} = 0.408 \tag{3}$$

The valve nominal voltage line to line is directly derived from the transformer ratio:

$$U_S = m_Y U_g = 0.408 \times 400 = 163 \; kV \tag{4}$$

A. Number of submodules (N_{SM})

To determine the number of SMs per arm, a simplified phasor diagram (c.f., Fig. 2) can be used in per-unit system to determine the peak value of the modulated voltage that should be generated by the arm of SMs.

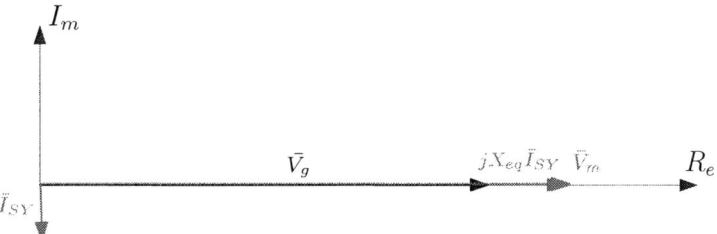

Fig. 2: Simplified phasor diagram by neglecting the resistive components

Therefore, depending on the topology, the operating point and the equivalent connection impedance (transformer leakage impedance and L_{arm}), the minimal number of SM (N_{SM}) can be expressed through the following equation when the STATCOM is injected the nominal power to the grid:

$$N_{SM} = \left\lceil \frac{\sqrt{2}}{V_{SM}} \left(\frac{m_Y U_g}{\sqrt{3}} + X_{eq} I_{SY_{nom}} \right) \right\rceil = \left\lceil \frac{\sqrt{2}}{1600V} \left(\frac{163kV}{\sqrt{3}} + 0.3 \times \frac{163^2}{300} \times 1061 \right) \right\rceil \approx 110 \tag{5}$$

where $X_{eq} = X_{arm} + X_T$ (i.e., X_{arm}=0.15 p.u and X_T =0.15 p.u) is the equivalent AC side impedance and V_{SM} is the submodule voltage (e.g., $V_{SM} = 1600V$). Thus, according to (5), the number of submodules is equal to $N_{SM} \sim 110$. In order to reduce the number of SM, a zero-sequence voltage injection (ZSVI) based on the following injection scheme [10] can be performed as

$$v_{NO} = - \frac{\max[v_{mj}] + \min[v_{mj}]}{2} \tag{6}$$

From (6), it results in a triangular waveform, naturally synchronous with STATCOM voltage v_S and its magnitude equals to $\frac{1}{4}$ of $\hat{V}_S$ [10]. Therefore, ZSVI allows the reduction of the required N_{SM} per arm by almost 15% for SSFB. Therefore, in our example the number of SMs can be reduced to 94.

B. Estimation of the stored energy requirement

The second step is to design the SM capacitors. The need of the internal energy storage depends on the acceptable SM capacitor voltage variation, caused by current oscillation over one network period. By assuming a perfect SM voltage balancing, it has been depicted in previous works [11], that an arm of SMs can be modelled using an average model that consists in an equivalent DC/DC converter. A more usual metrics to characterize the storage of energy is the constant H_c which corresponds the total energy divided by the nominal power [12].

$$H_c = \frac{\frac{1}{2} N_{arm} C_{tot} v_{Ctot}^2}{S_n} \tag{7}$$

with N_{arm} the number of arms, C_{tot} the total SMs capacitance (c.f. Arm Average Model in Fig. 1) and S_n the rated power. To estimate H_c, two methods have been proposed in literature [13]. The method developed in [14] has been used in this work. It is based on the limitation of the peak-to-peak voltage ripple defined as $\Delta v_{Ctot} = \hat{v}_{Ctot} - \check{v}_{Ctot}$, where $\hat{v}_{Ctot}$ and $\check{v}_{Ctot}$ are corresponding to the maximum and minimum voltages, respectively. The oscillations voltage ripple in the capacitors come from the waveform of the instantaneous power ($p_{C_{tot}}$). This power is calculated for the nominal value of the reactive power from the instantaneous steady state value of the voltage v_m and the arm current i_{arm}.

$$v_m = \left(\frac{m_Y U_g}{\sqrt{3}} + X_{eq} I_{SY}\right) \sqrt{2} \sin(\omega t) \tag{8}$$

$$i_{arm} = I_{SY} \sqrt{2} \sin\left(\omega t \pm \frac{\pi}{2}\right) = \frac{I_g}{m_Y} \sqrt{2} \sin\left(\omega t \pm \frac{\pi}{2}\right) = \frac{Q_n}{\sqrt{3} U_g m_Y} \sqrt{2} \sin\left(\omega t \pm \frac{\pi}{2}\right) \tag{9}$$

Based on (8) and (9), the instantaneous power expression in per-unit with $Q_n = S_n$ can be expressed as:

$$p_{Ctot\,pu} = \left(\frac{1}{3} + \frac{X_{eq}^{pu}}{3}\right) \cos\left(2\omega t \pm \frac{\pi}{2}\right) \tag{10}$$

By integrating (10), the peak-to-peak energy fluctuation Δw_{Ctot} is derived as:

$$\Delta w_{Ctot} = \frac{1}{\omega}\left(\frac{1}{3} + \frac{X_{eq}^{pu}}{3}\right) \tag{11}$$

If the average value of v_{Ctot} is assumed to be maintained around its nominal value V_{Ctot0}, then the total capacitor voltage can be expressed as:

$$v_{Ctot} = V_{Ctot0}\left(1 + \delta_{v_{Ctot}}\right) \tag{12}$$

where $\delta_{v_{Ctot}}$ is the voltage oscillation introduced by the AC component in the arm quantities. This oscillation is assumed to be evenly distributed around V_{Ctot0}. So, the peak-to-peak energy fluctuation can be express depending to the maximum and minimum energy variation as $\Delta w_{Ctot} = \hat{\delta}w_{Ctot} - \check{\delta}w_{Ctot}$, with:

$$\begin{cases} \hat{\delta}w_{Ctot} = \frac{1}{2} C_{tot} V_{Ctot0}^2 \left(1 + \delta_{v_{Ctot}}\right)^2 \\ \check{\delta}w_{Ctot} = \frac{1}{2} C_{tot} V_{Ctot0}^2 \left(1 - \delta_{v_{Ctot}}\right)^2 \end{cases} \tag{13}$$

Then, we obtain:

$$\Delta w_{Ctot} = \frac{1}{2} C_{tot} V_{Ctot0}^2 \left(1 + 2\delta_{v_{Ctot}} + \delta_{v_{Ctot}}^2 - 1 + 2\delta_{v_{Ctot}} - \delta_{v_{Ctot}}^2\right) = \frac{1}{2} C_{tot} V_{Ctot0}^2 4\delta_{v_{Ctot}} \tag{13}$$

According to (7) for STATCOM application (i.e., $S_n = Q_n$), it is possible to substitute the expression of C_{tot} in (14), to obtain:

$$\Delta w_{Ctot} = \frac{4 H_c Q_n \delta_{v_{Ctot}}}{N_{arm}} = \frac{2 H_c Q_n \Delta v_{Ctot}}{N_{arm}} \tag{15}$$

Thus, the value of H_c which guarantee the peak-to-peak voltage ripple Δv_{Ctot} is defined as:

$$H_c = \frac{N_{arm}\,\Delta w_{Ctot}}{2Q_n\,\Delta v_{Ctot}} \tag{14}$$

with $N_{arm} = 3$ for SSFB. From (16) and by considering X_{eq}^{pu} equals to 0.3 pu and $\Delta v_{Ctot} = 0.2\ pu$, we obtain:

$$H_c = 10.5\ \frac{kJ}{MVA} \tag{15}$$

In order to validate the stored energy requirement for SSFB configuration, obtained in (17), time-domain simulation based on Arm Average Model (see Fig. 1a) has been performed for maximal reactive power operating point. It can be checked that the peak-to-peak voltage ripple is equal to $\Delta v_{Ctot} = \hat{v}_{Ctot} - \check{v}_{Ctot} = 0.2\ pu$ as shown in Fig. 3.

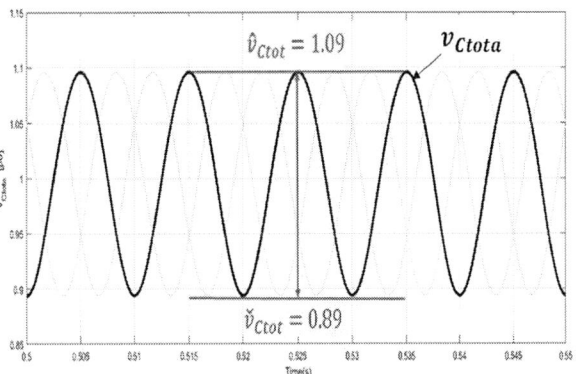

Fig. 3: H_C validation based on steady state time-domain response of v_{Ctota}

For simplicity of the presentation, the zero-sequence voltage has not considered in this steady-state modeling, but the methodology can be extended with this zero-sequence injection.

C. Semi-conductor loss estimation

The evaluation of losses in the MMC has been widely covered in the literature according to different methods, e.g., in [15]. In this study, power losses estimation relies on hybrid method combining the steady state analytical model and time domain simulation as depicted in Fig. 4, to get a fast estimation. The proposed semi-conductor loss estimation method is based on the previous work detailed in [11], extended to full-bridge SM configuration. From the sinusoidal waveform of the modulated voltage (v_m) and the arm current (i_{arm}), it is possible to simulate for an arm, the Capacitor-Balancing Algorithm (CBA) associated with the low level control, as shown in Fig. 4.

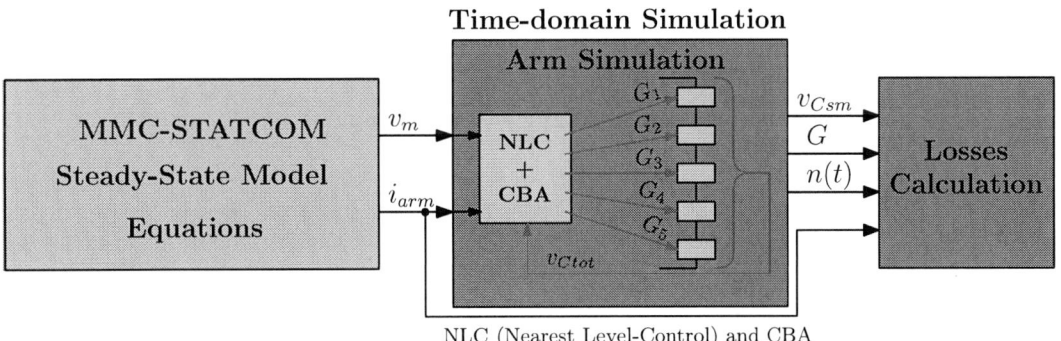

Fig. 4: Generic flow-chart of losses estimation method

For this study, the used CBA is based on the algorithm detailed in [11]. According to an execution time of $10\,\mu s$, the maximal voltage variation recorded is 320V between the most and least charged SM capacitors (i.e., $0.2\,pu$). This CBA execution time allows maintaining the switching frequencies around $\overline{f_{sw}} = 180\,Hz$.

The proposed method avoids performing an EMT simulation of the complete system, therefore reduces the computation effort and save time. Fig. 5a, shows the waveforms of a single-phase modulated voltage v_m and arm current i_{arm} in case of an injection of third harmonic. The SM capacitor voltages shown in Fig. 5b are obtained by the block which includes module selection and valve model. This block also provides the relevant inputs to the valve loss estimation which corresponds to SMs capacitors voltages, gate signals, and the arm currents.

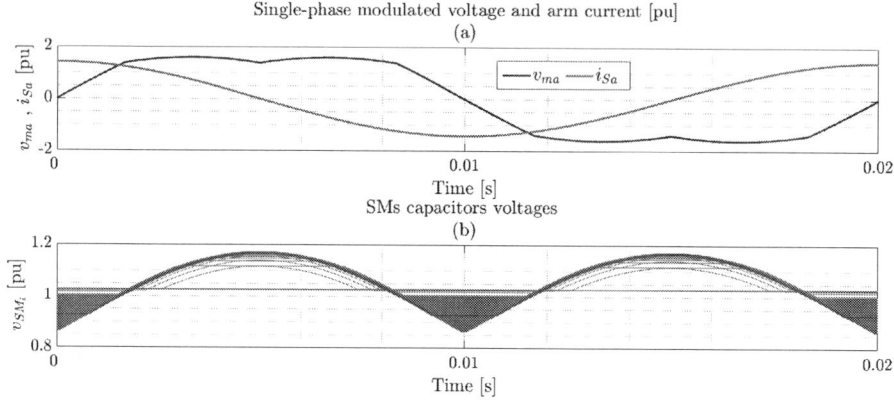

Fig. 5: Arm Voltage and current waveforms, and submodule voltages for phase (a)

As mentioned above, this algorithm discussed in [16], has been performed for HB-SM and adapted in this work to FB-SM, where the key elements are recalled in the following lines. As explained in [11], the semiconductor losses in the MMC can be derived into conduction and switching losses. From the datasheets of power electronic components (c.f., Infineon FZ 1500R33HL3), the VI static characteristics of IGBT and diode are approximated by polynomial functions of third order. In fact, the polynomial function corresponding to the on-state voltage characteristics versus the current for both IGBT and diode, denoted as $V_{CE(sat)}(i_{arm}, T_j)$ and $V_D(i_{arm}, T_j)$, respectively. In this work, for simplicity, the junction temperature T_j is kept constant equal to 125°C. In real life project, the temperature can be adapted to the operating point to get more accurate results. Thus, the conduction losses in the semiconductors of a given submodule can be calculated as the product of the arm current and its voltage drop as defined in [16]. Therefore, diode and IGBTs conduction losses of each FB-SMs are evaluated according to (18) and (19), respectively.

$$p_{c,Diodes\ D1 \to D4}(t) = f_{D1 \to D4} \cdot i_{arm} \cdot V_D(i_{arm}, T_j) \tag{16}$$

$$p_{c,IGBT\ T1 \to T4}(t) = f_{T1 \to T4} \cdot i_{arm} \cdot V_{CE(sat)}(i_{arm}, T_j) \tag{19}$$

where $f_{D1 \to 4}$ and $f_{T1 \to 4}$ are the Boolean functions, when the assigned valued is equal to one means that the current is following through the respective element and zero otherwise. The average conduction losses in the arm is calculated by integrating (18)-(19) over a fundamental period T using discrete-time integration.

Then, the switching losses in FB-SMs supposed to calculate energy which is lost at each commutation for IGBT (E_{ON} and E_{OFF}) and diode (E_{rec}). These energies are approximated by a polynomial description thanks to the switching energy characteristics of the module given from datasheets as curves in function of arm current (for a given T_j). So, based on the polynomial description of E_{ON}, E_{OFF} and E_{rec}, the energy lost at each commutation are expressed by the following equations.

$$IGBT_{Turn-ON} \to E_{ON}(i_{arm}, T_j)\frac{v_{Csmi}}{V_{CC}} \tag{20}$$

$$IGBT_{Turn-OFF} \to E_{OFF}(i_{arm}, T_j)\frac{v_{Csmi}}{V_{CC}} \tag{21}$$

$$Diode_{Turn-OFF} \to E_{rec}(i_{arm}, T_j)\frac{v_{Csmi}}{V_{CC}} \tag{22}$$

where v_{Csmi} is the submodule capacitor voltage value at a switching instant. V_{CC} is the voltage used by manufacturers to determine the energy lost during the switching (e.g., $V_{CC} = 1800\ V$). Then, the total switching losses at each commutation instant can be obtained, and the results are summed up and averaged over a fundamental period T.

The losses of passive element such as arm reactor and transformer are not presented in this paper but are added to total loss estimation.

The obtained power losses result according to the reactive power variation are given in Fig. 6, where the impact of ZSVI has been investigated. From Fig. 6a, in both capacitive and inductive modes, the SSFB without ZSVI exhibits almost similar levels of losses. AC high voltage Transformer with 99.75% of efficiency and 0.05% of winding resistances lead to 0.75 MW as losses in the transformer, below conduction losses at rated reactive power. The losses within the arm for a ratio $\frac{X}{R} \approx 200$ gives $P_{Rarm} \sim 0.23\ MW$, which are slightly below the switching losses. In fact, the switching losses in IGBTs are relatively low (i.e., $P_{SW} \sim 0.28\ MW$) compared to the total losses inside the converter. A second test on power losses estimation has been performed when using ZSVI as displayed in Fig. 6b. It is recalled that the zero-sequence voltage injection is based on (6) according to $\frac{1}{4}$ of maximal STATCOM voltage $\hat{V}_S$ as magnitude of v_{NO}. The details with respect to power losses based on ZSVI are highlighted through Table I. As it can be observed, the ZSVI allows reducing slightly the losses (mainly the conduction losses). Based on ZSVI, the total losses minimization is around $200\ kW$.

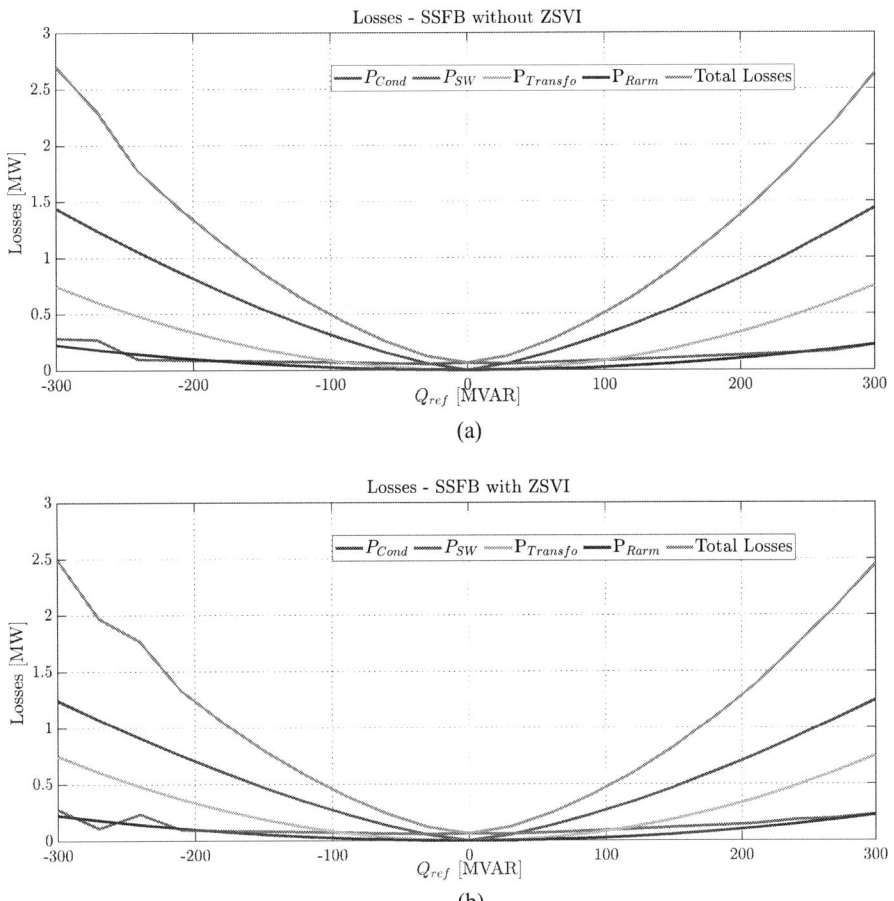

Fig. 6: Power losses in SSFB respect to reactive power reference Q_{ref} – (a): without ZSVI, (b): with ZSVI

Table I: Estimation of losses based on MMC-SSFB with ZSVI

	Absorption	Injection
Conduction Losses [MW]	**1.24**	**1.24**
Switching Losses [MW]	**0.28**	**0.24**
Transformer Losses [MW]	**0.75**	**0.75**
Arm Losses [MW]	**0.23**	**0.24**
Total Losses [MW]	**2.50**	**2.45**

Application of the proposed methodology to SDFB and DSFB Topologies and Comparative Results

The proposed methodology for the design of STATCOM topologies has also been applied to MMC-SDFB and MMC-DSFB configurations. To achieve a fair comparison, the same transistor is used for each topology. To obtain the same RMS current in the transistor, the transformer ratio m must be adjusted as shown in Table II.

Table II: Comparative results for a single arm in terms of m, N_{SM} and H_c

	MMC-SSFB	MMC-SDFB	MMC-DSFB
m	0.408	0.235	0.204
N_{SM}	110	110	55
$H_c(\frac{kJ}{MVA})$	10.5	10.5	10.5

From Table II, it can be displayed that the number of submodules per arm (N_{SM}) is the same for SSFB and SDFB. However, N_{SM} is twice less for the DSFB but since there are twice more arm, the total number of submodules is the same. When using the ZSVI, it is possible to decrease N_{SM} by 15% for SSFB and DSFB. On the SDFB topology, the zero-sequence current injection (ZSCI) has been assessed with a magnitude of $\frac{1}{6}$ of $\hat{I}_S$. It has been observed that ZSCI does not have a significant impact on N_{SM} for the delta configuration. In addition, the proposed design method of H_c has been performed also on the two other topologies. As displayed in Table II, the same value for the H_c index is obtained for each topology.

In terms of total losses, Fig. 7a shows quite similar results between the three topologies where no zero-sequence injection has been considered. Fig. 7b depicts the total losses by considering the ZSVI and ZSCI. It has been established that DSFB is more efficient compared to SSFB and SDFB. Indeed, a noticeable total loss reduction is observed for double-star topology respecting the total losses within the structure. Such difference involves around 16% of reduction (i.e., $\Delta P_{tot} \sim 410\ kW$) compared to SSFB and 19% (i.e., $\Delta P_{tot} \sim 503\ kW$) compared to SDFB.

Finally, from the proposed study, it is possible to conclude that DSFB is the most appropriate topology since it is more efficient while allowing losses minimization with ZSVI compared to SSFB and SDFB. In addition, using ZSVI, it is possible to decrease the total number of submodules by 15% and therefore minimize the cost of double-star topology.

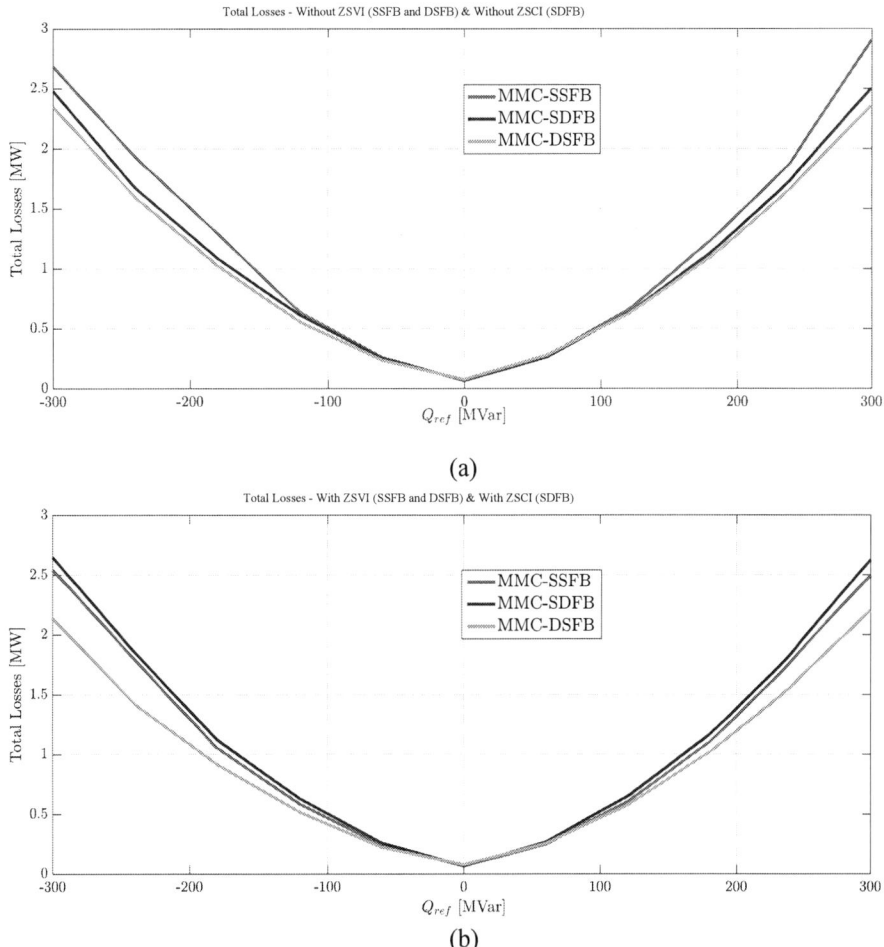

Fig. 7: Total losses estimation comparison respect to a variation of Q_{ref} – (a): without ZSI, (b): with ZSI

Conclusion

This paper gives a comprehensive steady-state analysis of the MMC-SSFB, MMC-SDFB and MMC-DSFB configurations in their application as STATCOM in transmission system. The proposed study focuses mainly on the stored energy requirements, submodules number design and total losses estimation to assess the design cost of each topology. A precise methodology based on a steady state analytical model of MMC-STATCOM has been performed. An in-depth comparison between the studied topologies is carried out to provide quantitative criteria to choose the MMC topology for STATCOM application. In addition, the impact of zero sequence voltage and current injection on the topology design has been discussed. The key points can be given: In terms of power losses, the DSFB is the best topology compared to SSFB and SDFB, especially when using ZSVI. A total loss reduction around 16% and 19% compared to SSFB and SDFB, respectively. The use of ZSVI induces also a 15% decrease of N_{SM} (i.e., the number of switches) in both single- and double-star topologies whereas the ZSCI has nearly no impact in case of SDFB. Then, it can be concluded that the double-star full bridge is the most appropriate topology for the MMC-STATCOM that meets the defined criteria in the proposed study.

As future works, the comparison between the investigated topologies may spread in dynamics to assess the dynamic performances under normal and distorted AC grid conditions. Furthermore, it seems interesting to extend the proposed design methodology for unbalanced and fault grid conditions using a steady state model of MMC-STATCOM based on the negative sequence components.

References

[1] "STATIC SYNCHRONOUS COMPENSATOR (STATCOM)", CIGRE Technical Brochure 144, Working Group 14.19, 2000.

[2] H. Akagi, "Classification, terminology, and application of the modular multilevel cascade converter (MMCC)", *IEEE Trans. Power Electronics*, vol. 26, n°.11, p. 3119–3130, 2011.

[3] S.G. Mian, P.D. Judge, A. Junyent-Ferr and T. Green, "A Delta-Connected Modular Multilevel STATCOM with Partially-Rated Energy Storage for Provision of Ancillary Services", *IEEE Transactions on Power Delivery*, vol. 36, n°15, pp. 2893-2903, 2021.

[4] E. Behrouzian, M. Bongiorno and H. Z. De La Parra, "Investigation of negative sequence injection capability in H-bridge multilevel STATCOM", *16th Conference on Power Electronics and Applications*, pp. 1-10, 2014.

[5] E. Spahic, C. P. Susai Sakkanna Reddy, M. Pieschel, and R. Alvarez, "Multilevel STATCOM with power intensive energy storage for dynamic grid stability - frequency and voltage support", *IEEE Electrical Power and Energy Conference (EPEC)*, pp. 73-80, 2015, London, UK.

[6] R. O. d. Sousa, D. d. C. Mendonça, W. C. S. Amorim, A. F. Cupertino, H. A. Pereira and R. Teodorescu, "Comparison of Double Star Topologies of Modular Multilevel Converters in STATCOM Application", *13th IEEE International Conference on Industry Applications*, pp. 622-629, 2018.

[7] H. A G. Tsolaridis, H. A. Pereira, A. F. Cupertino, R. Teodorescu, and M. Bongiorno, "Losses and cost comparison of DS-HB and DS-FB M based large utility grade STATCOM", *in 16th EEEI Conference*, pp. 1-6, June 2016.

[8] A. F. Cupertino, J. V. M. Farias, H. A. Pereira, S. I. Seleme and R. Teodorescu, "Comparison of DSCC and SDBC Modular Multilevel Converters for STATCOM Application During Negative Sequence Compensation", *in IEEE Transactions on Industrial Electronics*, vol. 66, n°13, pp. 2302-2312, March 2019.

[9] www.infineon.com/cms/en/product/power/igbt/igbt-modules/fz1500r33he3/

[10] O. Ojo, "The generalized discontinuous PWM scheme for three-phase voltage source inverters", *IEEE Transactions on Industrial Electronics*, vol. 51, n°16, pp. 1280-1289, 2004.

[11] P. Vermeersch, "Contribution to the Design and Control of the Extended Overlap-Alternate Arm Converter", *PhD. Thesis*, Ecole Centrale de Lille, L2EP, September 2021.

[12] H. Saad, "Modélisation et simulation d'une liaison HVDC de type VSC-MMC ", *PhD Thesis*, Université de Montréal, 2015.

[13] S. Heinig, K. Ilves, S. Norrga, and H. P. Nee, "On energy storage requirements in alternate arm converters and modular multilevel converters", *in Proceedings of the 18th European Conference on Power Electronics and Applications, EPE 2016 ECCE Europe*, 2016.

[14] M. M. Merlin and T. C. Green, "Cell capacitor sizing in multilevel converters: Cases of the modular multilevel converter and alternate arm converter", *IET Power Electronics,*, vol. 8, n°13, pp. 350-360, 2015.

[15] C. Oates and C. Davison, "A comparison of two methods of estimating losses in the modular multi-level converter", *14th European Conference on Power Electronics and Applications*, EPE'11, pp. 1–10, August 2011.

[16] J. Freytes, F. Gruson, P. Delarue, F. Colas, X. Guillaud, "Losses Estimation Method by Simulation for the Modular Multilevel Converter", *EPEC conference*, London, Ontario, October 2015.

[17] A. Hassanpoor, L. Angquist, S. Norrga, K. Ilves and H. Nee, "Tolerance Band Modulation Methods for Modular Multilevel Converters," *in IEEE Transactions on Power Electronics*, vol. 30, no. 1, pp. 311-326, Jan. 2015.

[18] "Power losses in voltage sourced converter (VSC) valves for high-voltage direct current (HVDC) systems – Part 2: Modular multilevel converters", *IEC 62751-2*, August 2019.

Current Distribution Control in Parallel Connected Power Converters with Continuous Output Voltage

Sabrina Ulmer, Andreas Brunner, Philipp Czerwenka, Gernot Schullerus, Ertugrul Sönmez

Reutlingen University
Electronics and Drives
Oferdingerstr. 50
72768 Reutlingen, Germany
+49 (7121) 271-7080

{sabrina.ulmer, andreas.brunner, philipp.czerwenka, gernot.schullerus, ertugrul.soenmez}
@reutlingen-university.de
www.electronics-and-drives.de

Acknowledgments

This work is supported by the German Federal Ministry of Education and Research.

Keywords

≪Paralleling≫, ≪Current sharing≫, ≪Converter control≫, ≪Inverter design≫, ≪Gallium Nitride (GaN)≫.

Abstract

This contribution presents a hardware-based current distribution control concept of parallel connected power converters with quasi-continuous output voltage as part of a scalable and modular power electronic system for motor control using wide bandgap semiconductors.

Introduction

The increased switching frequency of wide bandgap semiconductor switches leads to new hardware concepts for motor control. It enables the integration of filters into the power electronics to provide quasi-continuous voltages to a load thus avoiding the drawbacks of pulsed output voltages applied to motor terminals. Typically, LC filters are used to produce the quasi-continuous voltage at the terminals. However, such filter structure topologies exhibit a poor damping. This issue can be addressed by a dissipative damping as proposed in [1]. A more efficient and flexible alternative is to shape the filter transfer characteristic by a state feedback of the measured capacitor current [2] or the filter inductor current [3].

The sustainability of much higher switching frequencies also simplifies modular concepts. Power modules with continuous output quantities can be easily connected in parallel and/or in series to extend the current and/or voltage range to deal with the high power requirement and thus reducing the stress on the switches. For the connection of several power modules in parallel, several design objectives, such as equal current sharing, have to be considered. To ensure symmetric current sharing between the parallel power modules, different structures have been proposed. In [4] a comprehensive review based on a comparative analysis of current sharing control strategies of parallel-connected inverters modules is presented.

The active current distribution control can be divided into instantaneous average current sharing (IACS) control [5], master-slave control (MSC) [6] and circular chain control (3C) method [7]. In the MSC technique one power module is chosen as master acquiring voltage control mode while the remaining modules are selected as slaves embracing current control mode.

The current paper extends the idea for the parallel connection of power electronic modules from [8] and proposes a hardware-based current sharing control strategy for power modules connected in parallel such that a sufficient damping is ensured [3] and the motor current is equally distributed among the parallelized modules. In this contribution MSC is implemented. In contrast to conventional methods each power module controls its quasi-continuous output voltage.

The paper is organized as follows: First, the system model is introduced and the design of the control system is discussed in detail. Next, the simulation setup and results are illustrated. Then, preliminary measurement results for the implemented current distribution control are presented. Finally, conclusions are given.

System model

The power electronics module concept for controlling one motor phase considered here is based on so called *unit cells*. Each unit cell provides a quasi-continuous output voltage with a signal frequency f_S. By appropriately connecting unit cells, three-phase configurations can be obtained [3]. In addition, parallel [8] or series connection [9, 10] is feasible to increase the voltage and/or current range as illustrated in Fig. 1.

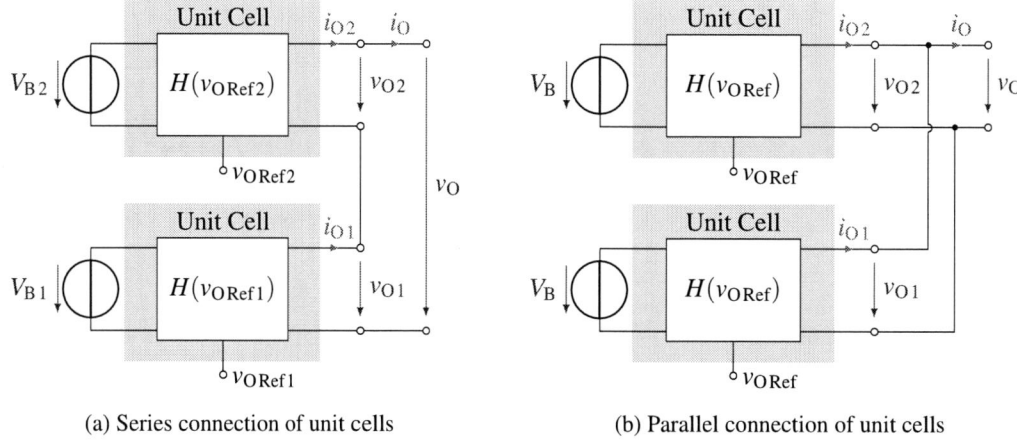

(a) Series connection of unit cells   (b) Parallel connection of unit cells

Fig. 1: Connection concept of unit cells

The internal structure of the unit cell used in this contribution is shown in Fig. 2. The output voltage v_O results from filtering the pulsed voltage v_{HB} with an LC filter. The maximum switching frequency $f_{SW\,max}$ is specified with 500 kHz at a duty cycle of 50 %. In this concept asynchronous delta-sigma ($\Delta\Sigma$) modulation is used. For active damping according to [3] an inductor current feedback with the transfer characteristic G_{FB} is applied. A motor winding is connected to the filter output terminals where the back-emf is not illustrated for simplicity.

The parallel connection of two unit cells is illustrated in Fig. 3. Both unit cells share the same DC voltage source V_B. For simplicity only the output filter structure is represented. The unit cells implemented in hardware will not exhibit exactly the same characteristic behavior with respect to each other due to parasitic and component influences resulting in an unsymmetrical current distribution among the unit cells. To avoid this problem, an external current controller based on a master-slave-principle was implemented as in [8]. In the new concept presented in this paper, a hardware-based solution is proposed. It is designed based on an extension of the state space model for each unit cell. Thus, a hardware-based current control is achieved and the motor current control scheme from a potential field oriented control is not affected.

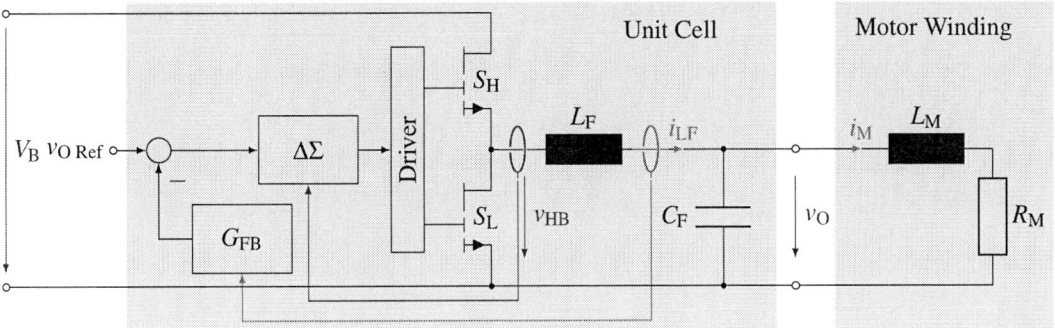

Fig. 2: Unit cell with motor winding

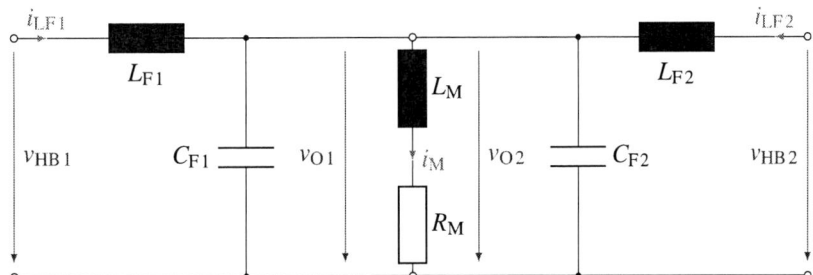

Fig. 3: Two parallel unit cells feeding one motor winding

By applying the mesh and nodal equations, as well as assuming $v_O = v_{O1} = v_{O2}$, the following system of equations can be created:

$$\dot{i}_{LF1} = -\frac{1}{L_{F1}}v_O + \frac{1}{L_{F1}}v_{HB1} \qquad (1)$$

$$\dot{v}_O = \frac{1}{C_{F1}+C_{F2}}(i_{LF1}+i_{LF2}-i_M) \qquad (3)$$

$$\dot{i}_{LF2} = -\frac{1}{L_{F2}}v_O + \frac{1}{L_{F2}}v_{HB2} \qquad (2)$$

$$\dot{i}_M = -\frac{R_M}{L_M}i_M + \frac{1}{L_M}v_O \qquad (4)$$

Based on the structure in Fig. 3, the state space representation of the system without inductor current feedback of i_{LFn} for $n = 1, 2$ is given by obtaining $x = [i_{LF1}\ i_{LF2}\ v_O\ i_M]^T$

$$\dot{x} = \begin{bmatrix} 0 & 0 & -\dfrac{1}{L_{F1}} & 0 \\[2ex] 0 & 0 & -\dfrac{1}{L_{F2}} & 0 \\[2ex] \dfrac{1}{C_{F1}+C_{F2}} & \dfrac{1}{C_{F1}+C_{F2}} & 0 & -\dfrac{1}{C_{F1}+C_{F2}} \\[2ex] 0 & 0 & \dfrac{1}{L_M} & -\dfrac{R_M}{L_M} \end{bmatrix} x + \begin{bmatrix} \dfrac{1}{L_{F1}} & 0 \\[2ex] 0 & \dfrac{1}{L_{F2}} \\[2ex] 0 & 0 \\[2ex] 0 & 0 \end{bmatrix} \begin{bmatrix} v_{HB1} \\[2ex] v_{HB2} \end{bmatrix} \qquad (5)$$

The following two subsections discuss the extension of (5) with respect to the control system objectives.

Integration of the inductor current feedback and current symmetrization

The feedback system illustrated in this section serves two objectives:
1. To introduce a sufficient damping for the LC filter.
2. To enable the symmetrization of the output currents of the two modules.

Although in principle, both functions could be implemented by one single feedback structure, they are considered separately here. The symmetrization function is not needed for single unit cells. The master unit cell does not need to fulfill the second objective either.

Damping is achieved by the transfer function G_{FB} illustrated in Fig. 2 where the time constant T_F and the gain k_I are determined following the procedure discussed in [11].

$$G_{FB}(s) = \frac{s T_F k_I}{1 + T_F s} \tag{6}$$

To integrate this feedback into (5), an additional state variable x_{Fn} with $n = 1, 2$ is introduced. An additional control variable v_{sn} with $n = 1, 2$ for current symmetrization for each unit cell is used. Then, we have for the control signals $v_{HB\,n}$ in (5) as illustrated in Fig. 4

$$v_{HB\,n}(t) = v_{ORef}(t) - x_{Fn}(t) + v_{sn}(t) \qquad \text{with} \qquad X_{Fn}(s) = G_{FB}(s) I_{LFn}(s) \tag{7}$$

where $X_{Fn}(s)$ and $I_{LFn}(s)$ are the Laplace domain representations of $x_{Fn}(t)$ and $i_{LFn}(t)$, respectively, and v_{ORef} is the reference input signal for both modules. Thus, we obtain using (5)

$$\dot{x}_{Fn} = -\frac{1}{T_F} x_{Fn} + k_I i_{LFn} \qquad \text{with} \qquad \dot{i}_{LFn} = -\frac{1}{L_{Fn}} v_O + \frac{1}{L_{Fn}}\left(v_{ORef} - x_{Fn} + v_{sn}\right) \tag{8}$$

$$\dot{x}_{Fn} = -\left(\frac{1}{T_F} + \frac{k_I}{L_{Fn}}\right) x_{Fn} - \frac{k_I}{L_{Fn}} v_O + \frac{k_I}{L_{Fn}} v_{ORef} + \frac{k_I}{L_{Fn}} v_{sn} \tag{9}$$

Integrating (9) into (5) for $n = 1, 2$ we obtain using $C_F = C_{F1} + C_{F2}$ and $x = \begin{bmatrix} i_{LF1} & i_{LF2} & v_O & i_M & x_{F1} & x_{F2} \end{bmatrix}^T$

$$\dot{x} = \begin{bmatrix} 0 & 0 & -\dfrac{1}{L_{F1}} & 0 & -\dfrac{1}{L_{F1}} & 0 \\[2mm] 0 & 0 & -\dfrac{1}{L_{F2}} & 0 & 0 & -\dfrac{1}{L_{F2}} \\[2mm] \dfrac{1}{C_F} & \dfrac{1}{C_F} & 0 & -\dfrac{1}{C_F} & 0 & 0 \\[2mm] 0 & 0 & \dfrac{1}{L_M} & -\dfrac{R_M}{L_M} & 0 & 0 \\[2mm] 0 & 0 & -\dfrac{k_I}{L_{F1}} & 0 & -\left(\dfrac{1}{T_F} + \dfrac{k_I}{L_{F1}}\right) & 0 \\[2mm] 0 & 0 & -\dfrac{k_I}{L_{F2}} & 0 & 0 & -\left(\dfrac{1}{T_F} + \dfrac{k_I}{L_{F2}}\right) \end{bmatrix} x + \begin{bmatrix} \dfrac{1}{L_{F1}} & 0 \\[2mm] 0 & \dfrac{1}{L_{F2}} \\[2mm] 0 & 0 \\[2mm] 0 & 0 \\[2mm] \dfrac{k_I}{L_{F1}} & 0 \\[2mm] 0 & \dfrac{k_I}{L_{F2}} \end{bmatrix} \begin{bmatrix} v_{s1} \\ v_{s2} \end{bmatrix} + \begin{bmatrix} \dfrac{1}{L_{F1}} \\[2mm] \dfrac{1}{L_{F2}} \\[2mm] 0 \\[2mm] 0 \\[2mm] \dfrac{k_I}{L_{F1}} \\[2mm] \dfrac{k_I}{L_{F2}} \end{bmatrix} v_{ORef}. \tag{10}$$

The model presented in Fig. 4 will be considered in the next section for current symmetrization.

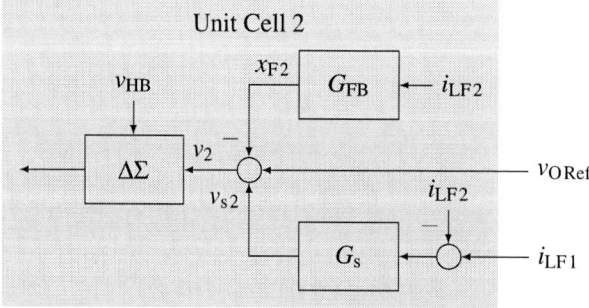

Fig. 4: New control signal generation

Current symmetrization

The controller structure for current symmetrization we will set $v_{ORef} = 0$ in (10). This is reasonable, as the symmetrization should work independently from the reference voltage. As already indicated by Fig. 4, the symmetrizing controller G_s considers the difference between the master current i_{LF1} from unit cell 1 and the slave current i_{LF2} in unit cell 2 which is expected to follow the master current.

A control design could be achieved to control the difference $\delta = i_{LF1} - i_{LF2}$ to zero based on the complete system (10). However, the complexity of this system does not allow the direct application of the standard design methods for designing G_s. Therefore, a simplified model will be used, where only the unit cell 2 is considered as process model and i_{LF1} is considered as a given reference value. The simplified model is given by obtaining $x_2 = [i_{LF2}\ v_O\ i_M\ x_{F2}]^T$

$$
\dot{x}_2 = \underbrace{\begin{bmatrix} 0 & -\dfrac{1}{L_{F2}} & 0 & -\dfrac{1}{L_{F2}} \\[2mm] \dfrac{1}{C_F} & 0 & -\dfrac{1}{C_F} & 0 \\[2mm] 0 & \dfrac{1}{L_M} & -\dfrac{R_M}{L_M} & 0 \\[2mm] 0 & -\dfrac{k_I}{L_{F2}} & 0 & -\left(\dfrac{1}{T_F} + \dfrac{k_I}{L_{F2}}\right) \end{bmatrix}}_{A_2} x_2 + \underbrace{\begin{bmatrix} \dfrac{1}{L_{F2}} \\[2mm] 0 \\[2mm] 0 \\[2mm] \dfrac{k_I}{L_{F2}} \end{bmatrix}}_{b_2} v_{s2} \tag{11}
$$

For objective 2, in the control signal generation in Fig. 2 an additional system input is included in the unit cell 2 as shown in Fig. 4. The controller G_s is designed such that i_{LF2} is sufficiently close to i_{LF1}. From the system matrices in (11) using the matrix $c^T = [1\ 0\ 0\ 0]$ and the identity matrix I a single-input-single-output (SISO) system transfer function can be computed

$$
G_2(s) = c_2^T (sI - A_2)^{-1} b_2 \qquad \text{such that} \qquad I_{LF2}(s) = G_2(s) V_{s2}(s) . \tag{12}
$$

The controller $G_s(s)$ with $V_{s2}(s) = G_s(s)(I_{LF1}(s) - I_{LF2}(s))$ (see Fig. 4) can then be designed using standard control theory methods based on the process model $G_2(s)$.

Controller design

The controller G_s will be designed using the root locus method. Other design methods may be applied, as well. This design uses the system parameters as given in Table I. Note, that the difference between R_{F2} and R_{F1} is used for illustrating the influence of parameter variations between the modules on the current sharing in the simulations. The parameters of the module will be used in correspondence to the datasheet values of the passive components.

Table I: Design parameter

Description M_1	Parameter	Setup	Description M_2	Parameter	Setup
Filter inductance	L_{F1}	$15\,\mu H$	Filter inductance	L_{F2}	L_{F1}
Resistance of L_{F1}	R_{F1}	$12\,m\Omega$	Resistance of L_{F2}	R_{F2}	$0.5\,R_{F1}$
Filter capacitance	C_{F1}	$1.36\,\mu F$	Filter capacitance	C_{F2}	C_{F1}
Motor inductance	L_M	$0.5\,mH$	Motor inductance	L_M	$0.5\,mH$
Resistance of L_M	R_M	$1\,\Omega$	Resistance of L_M	R_M	$1\,\Omega$

For the control system an *I*-Controller

$$G_s(s) = \frac{k_s}{s} \,.$$

was selected, resulting in the root locus plot of Fig. 5a. This root locus plot is used to determine k_s such that the poles have the desired position chosen by the designer. The displayed poles (red squares) of the closed loop transfer function $G_{i\,s}$ result from the choice $k_s = 1 \times 10^5$ in the root locus design process. Note, that the root locus plot just displays the dominant poles. Additional poles are located outside of the shown plot area and are not relevant for the system dynamics. Fig. 5b illustrates the frequency response of the closed loop transfer function $G_{i\,s}$. It is observed that the controller produces a reasonable behavior with respect to magnitude and phase of the closed loop in a frequency range up to 1 kHz.

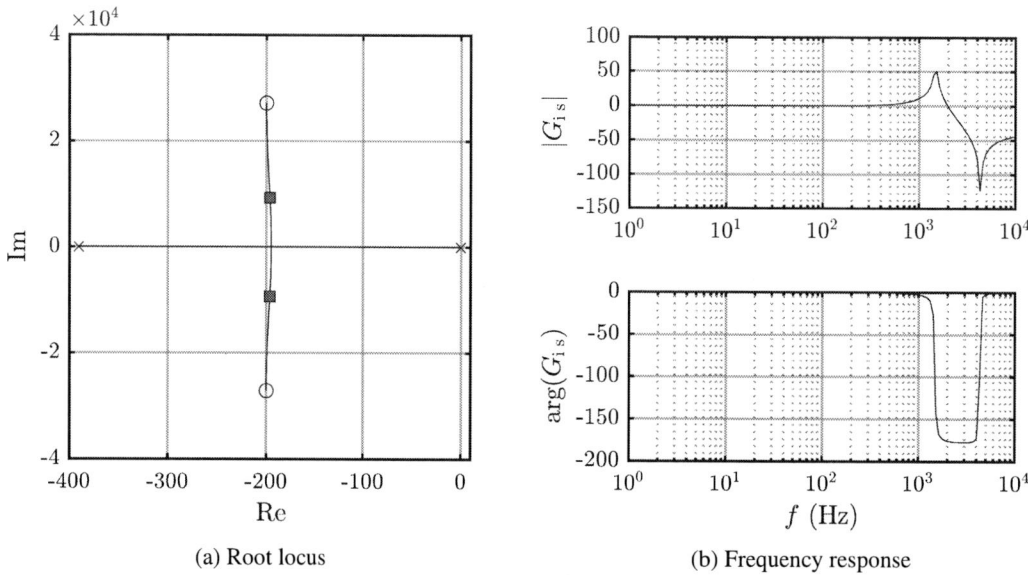

(a) Root locus (b) Frequency response

Fig. 5: Root locus and frequency response for closed current control loop

Simulation

A Simulink model for a parallel connection of two unit cells is shown in Fig. 6. Each of the cells M_1 (Master) and M_2 (Slave) exhibit the structure from Fig. 2 implemented in Simulink as shown in Fig. 7.

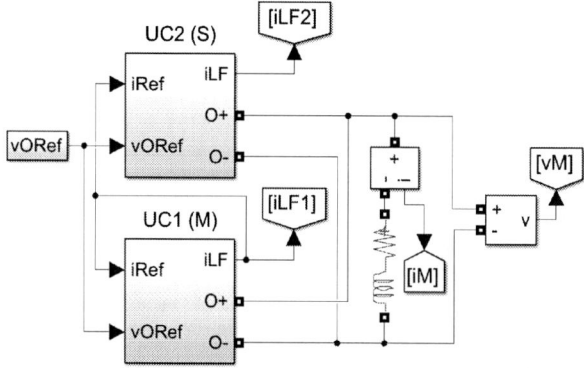

Fig. 6: Simulation model of parallel connected unit cells

The inductor current $i_{LF\,1}$ is the reference current $i_{Ref\,2}$ for unit cell M_2 and is fed back to M_1 just for consistency. Note, that the feedback path inside M_1 is deactivated. The load is represented by a motor winding. As this paper focuses on the properties of the power electronics rather than the motor behavior, the back-emf is not considered here and set to zero.

The unit cell simulation model shown in Fig. 7 can be divided into five main parts: $\Delta\Sigma$ modulation, power stage in half-bridge (HB) configuration, passive LC low-pass filter, inductor current feedback and current distribution control.

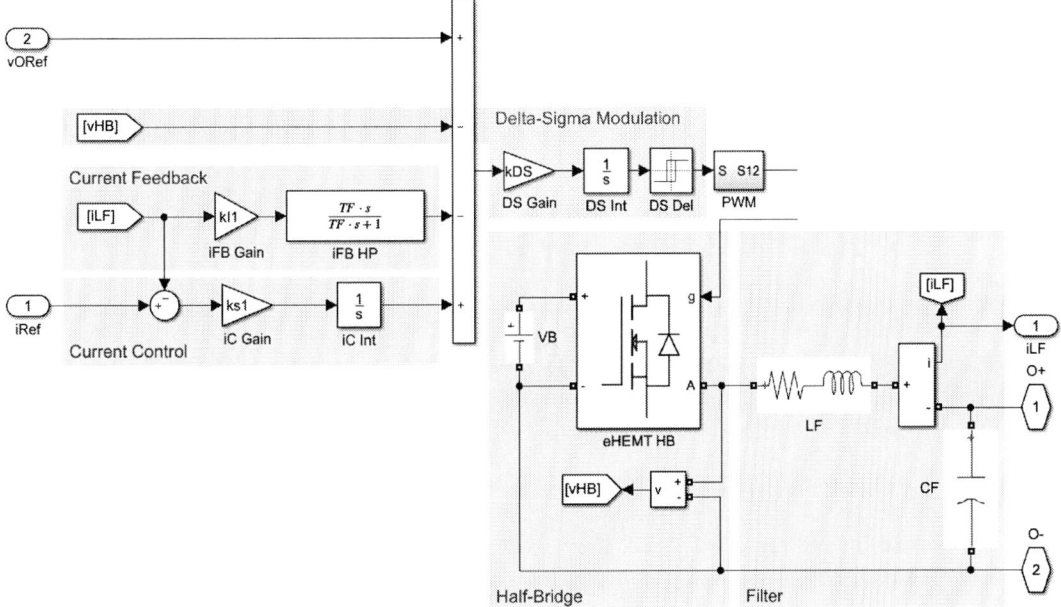

Fig. 7: Simulation model of one unit cell

The current feedback path based on the measured filter inductor current $i_{\mathrm{LF}\,n}$ ensures the filter damping as discussed for the design of a three-phase power stage in [3]. The additional high-pass filter in the current feedback path is used to additionally shape the transfer characteristic. Its design and implementation is discussed in [11].

The implementation of the current sharing control between the modules based on the feedback gain $k_{s\,i}$ is illustrated in Fig. 7 as well. The filter inductor current $i_{\mathrm{LF}\,2}$ is compared to the reference current $i_{\mathrm{LF}\,1}$ to calculate the input for the symmetrizing controller.

Assuming a sinusoidal reference input signal v_{ORef}, Fig. 8 illustrates the simulation results for the motor terminal voltage $v_{\mathrm{M}} = v_{\mathrm{O}}$ and the motor current i_{M} that are the same for the controlled and uncontrolled system.

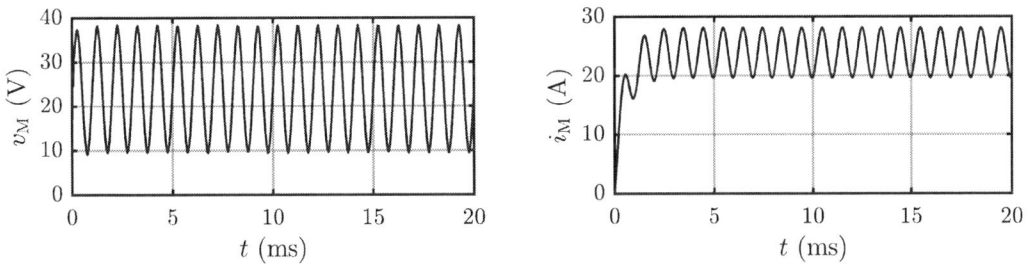

Fig. 8: Simulation results for motor signals v_{M}, i_{M}

Fig. 9 shows the inductor current signals $i_{\mathrm{LF}\,1}$, $i_{\mathrm{LF}\,2}$ of the two unit cells . In Fig. 9a the current control path is not activated, that is, $k_{s\,1} = k_{s\,2} = 0$. The current distribution among the unit cells is not equal due to the deviation from $R_{F\,2}$ as shown in Table I. By adjusting $k_{s\,2}$ to the designed value 1×10^5 an equal current distribution is achieved as shown in Fig. 9b.

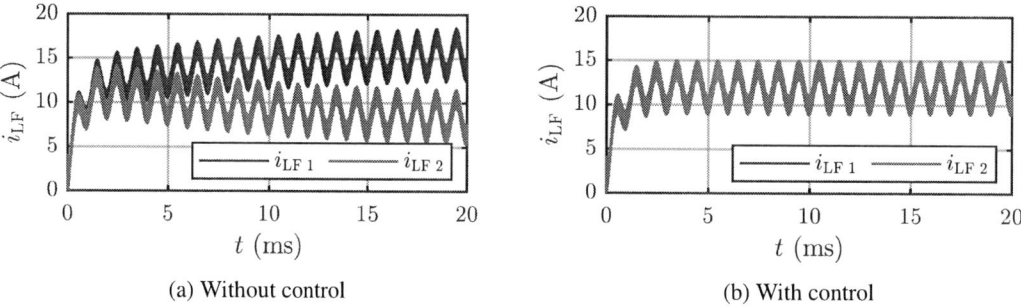

(a) Without control

(b) With control

Fig. 9: Simulation results illustrating current distribution control

Note, that the extension to several parallel unit cells is straightforward. Each unit cell receives the current output i_{LF} of its subsequent neighbor unit cell as current input i_{Ref}.

Measurement

In Fig. 10 the hardware setup for the parallelization of unit cells is presented. The representation of a single unit cell based on Fig. 2 and 4 is given in Fig. 10a. Note, that custom in-house developed plug-in HB modules based on *EPC2152* from *EPC Corp.*, that is, a GaN HB with monolithic integrated driver, are used. A second plug-in location is considered for the parallelization of two HB modules within a unit cell. The laboratory measurement setup is illustrated in Fig. 10b. Each unit cell is supplied with the same DC voltage source $V_B = 48\,\text{V}$ and the same reference input signal $v_{O\,Ref}$ provided by a waveform generator.

In these measurements only one system of parallel unit cells is connected to the terminals of a realistic motor winding. As the simulations illustrate, this leads to a high load current for the target applications. To avoid this situation an additional ohmic load is connected in series to the motor winding. The values of the motor winding ($L_M = 3.3\,\text{mH}$, $R_M = 3.2\,\Omega$) of a stepper motor *L32RFD-00N-NN-00* are determined by measurements. By energizing only one winding, the stepper motor is at stand still, which means no back-emf has to be considered.

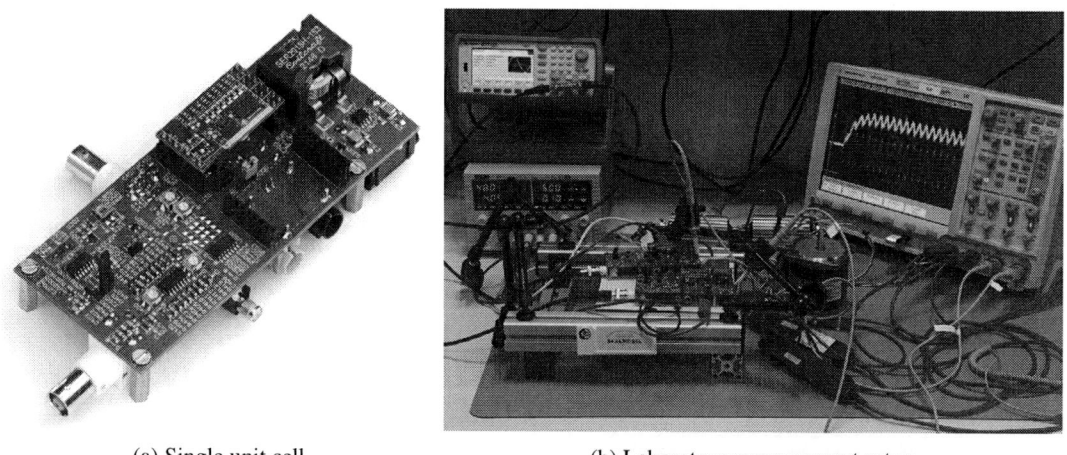

(a) Single unit cell

(b) Laboratory measurement setup

Fig. 10: Hardware setup for parallelization of unit cells

Fig. 11-12 show measurement results with two parallel connected unit cells and the current distribution control. The additional offset observed in the output current signals i_{OM}, i_{OS} at $t < 0$ is an effect of the utilized current probe and can be neglected. The enable signal indicates the simultaneous turn-on event for both unit cells in all measurements.

Fig. 11 illustrates the measurements obtained with a sinusoidal reference input signal $v_{O\,Ref}$. The measurements given in Fig. 11a on the left-hand side illustrate the symmetrical current distribution among the unit cells. The output voltage v_O is given on the right-hand side. The DC offset of 24 V of the output voltage occurs from working with a half-bridge.

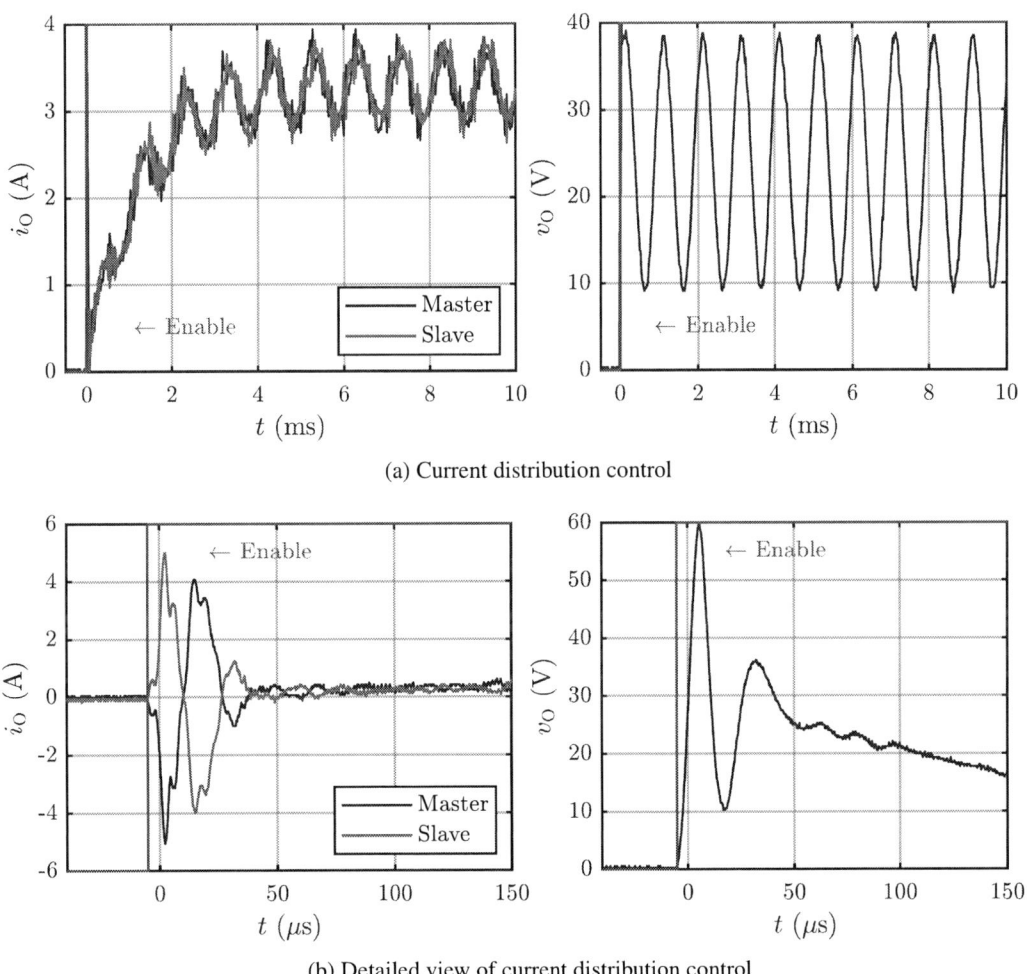

(a) Current distribution control

(b) Detailed view of current distribution control

Fig. 11: Measurement results for the output signals i_{OM}, i_{OS}, v_O of parallelized unit cells

A detailed view of the output signals v_O, i_{OM}, i_{OS} for two parallelized unit cells is illustrated in Fig. 11b. It is illustrated that the currents initially diverge. The output currents i_{OM}, i_{OS} increase rapidly in opposite directions as long as the current control is not in its operating point. Operating the parallelized unit cells without current control would lead to permanent damage due to the contrary eloping currents.

Finally, Fig. 12 shows a DC step $v_{O\,\Delta DC}$ of 10 V to illustrate the system dynamics. As expected, a symmetrical current distribution for the parallelized unit cells is achieved.

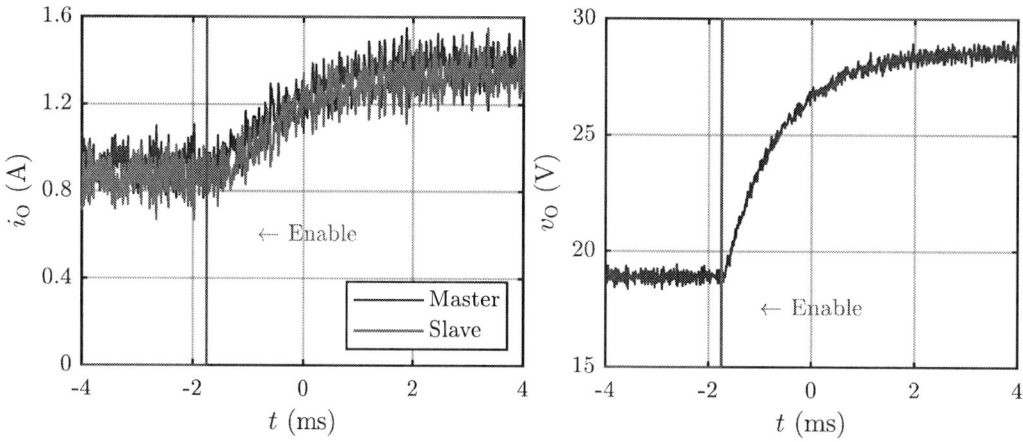

Fig. 12: Measurement results for the output signals $i_{O\,M}$, $i_{O\,S}$, $v_{O\,M}$ for a DC step $v_{O\,\Delta DC}$ of 10 V

Conclusion

This contribution presents a new hardware-based concept for current distribution control between two power electronic modules with quasi-continuous output voltage connected in parallel. The system model and the design of the control system are discussed in detail. Simulation and measurement results are given.

In future work, the extension to several unit cells will be demonstrated. In addition, the extension of the concept to a the combination of parallel and series connection of unit cells will be addressed. Furthermore, the impact of the controller on the modulation as well as the robustness of the controller will be a matter of research, finally resulting in a robust and flexible modular and scalable system.

References

[1] Stubenrauch F., Seliger N. and Schmitt-Landsiedel D.: Design and Performance of a 200kHz GaN Motor Inverter with Sine Wave Filter, 2017 International Exhibition and Conference for Power Electronics, Intelligent Motion, Renewable Energy and Energy Management (PCIM Europe)

[2] Maislinger F., Ertl H., Stojcic G., Lagler C. and Holzner F.: Design of a 100kHz Wide Bandgap Inverter for Motor Applications with Active Damped Sine Wave Filter, Journal of Engineering, Vol 2019, No 17

[3] Ulmer S., Walz-Lange A., Maatz A., Schullerus G., Soenmez E., Hennig E.: Active Filter Damping for a GaN-Based Three Power Stage with Continuous Output Voltage, 2021 23rd European Conference on Power Electronics and Applications (EPE ECCE Europe)

[4] Sinha A. and Jana K. C.: Comprehensive review on control strategies of parallel-interfaced voltage source inverters for distributed power generation system, IET Renewable Power Genereration, 2020, Vol. 14 Iss. 13, pp. 297-2314

[5] Sun. X, Lee Y.-S. and X. D: Modeling, Analysis, and Implementation of Parallel Multi-Inverter Systems With Instantaneous Average-Current-Sharing Scheme, IEEE Transactions on Power Electronics, 2003, Vol. 18 No. 3, 2003

[6] Prodanovic M., Green T. C. and Mansir H.: A Survey of Control Methods for Three-Phase Inverters in Parallel Connection, 8th International Conference on Power Electronics and Variable Speed Drives, 2000

[7] Wu T.-F., Chen Y.-K. and Huang Y.-H.: 3C Strategy for Inverters in Parallel Operation Achieving an Equal Current Distribution, IEEE Transactions on Industrial Electronics, 2020, Vol. 47 No. 2, 2000

[8] Ulmer S., Schullerus G. and Soenmez E.: A Modular and Scalable Power Electronics Device for the Control of Electric Drives, 2019 20th International Symposium on Power Electronics (Ee)

[9] Ulmer S., Schullerus G. and Soenmez E.: Active Damping in Series Connected Power Converters with Continuous Output Voltage, 2021 IEEE 19th International Power Electronics and Motion Control Conference (PEMC)

[10] Ulmer S., Schullerus G. and Soenmez E.: Active Damping in Series Connected Power Modules with Continuous Output Voltage, Power Electonics and Drives (PED), 2021 Vol 6 (41)

[11] Ulmer S., Schullerus G. and Soenmez E.: High Pass Design in Active Filter Damping, 2022 International Exhibition and Conference for Power Electronics, Intelligent Motion, Renewable Energy and Energy Management (PCIM Europe)

Optimized Pulse Pattern with Half-wave Symmetry for 5-Level Converter

Jonas Weires, Pedro Leal dos Santos, Steven Liu
Department of Electrical and Computer Engineering - LRS
TU Kaiserslautern
Gottlieb-Daimler-Straße 47
67663 Kaiserslautern, Germany
Email: sliu@eit.uni-kl.de

Keywords

≪Efficiency≫, ≪Harmonics≫ ,≪Optimization algorithm≫,≪Pulse Width Modulation (PWM)≫, ≪Grid-connected converter≫, ≪Multi-level converters≫

Abstract

In this paper, an Optimized Pulse Pattern is developed for a generic 5-Level Voltage Source Converter. The optimization process minimizes the current total harmonic distortion, while switching at the fundamental frequency. By relaxing the constraints to a half-period symmetry, a superior performance is obtained in comparison with existing solutions.

Introduction

Multilevel converters have become the enabling power conversion technology for enabling medium and high voltage in high power applications for the actual power grid and large motor drives. As a major research topic, several modulation techniques have been proposed to achieve the required output alternating voltage [1][2]. **Pulse Width Modulation (PWM)**-based techniques are preferred for multilevel converters with smaller number of voltages levels. At the expense of higher switching losses, low harmonic distortion is obtained. In opposite, other modulation techniques, as the **Nearest Level Control (NLC)**, switching at the fundamental frequency, fail to cope with required output voltage quality in most applications. In some applications, an additional Active Power Filter is required to mitigate harmonic content to ensure compliance with grid codes [3].

To counteract the problem of increasing harmonic distortion at low switching frequencies for converters with a low number of possible voltage levels the **Selective Harmonic Elimination (SHE)** has been developed. This technique focus on the elimination of a specific limited number of lower-order harmonics. Several formulations and algorithms have been developed in the last decades [4]. **Optimized Pulse Pattern (OPP)** is an alternative technique, eliminating not only specific harmonics, but minimizing the overall harmonic distortion [5]. OPP is based on an offline-calculation of the switching pattern with the help of optimization algorithms. Through this optimization process, the total harmonic distortion of the output current at low switching frequencies, down to the fundamental component, is greatly reduced. Consequently, both switching and conduction losses, due to lower harmonic distortion, are decreased and the overall system efficiency is increased. In high power applications even small improvements in transmission efficiency offer enormous potential for energy- and cost-savings.

The most common approach, i.e. **quarter-wave symmetry (QWS)**, imposes a strong symmetry constrain for every quarter of one period on the generated pattern. This restricts the range of solution and therefore the degree of freedom of the optimization. Investigations with reduced symmetry requirements have been done for **3-Level (3L)** patterns, in particular [6], while a derivation and an analysis for **5-Level (5L)** patterns with reduced symmetry have not been carried out yet. Whereby it already shows, that for

3L patterns, a reduction of the symmetry constrain leads to a significant improved harmonic distortion within certain modulation ranges [7]. In this paper the **half-wave symmetry (HWS)** is considered. The following work transfers the HWS concept to 5L patterns and a comparison with QWS 5L is provided. Furthermore an interesting variant of the HWS is derived which offers the performance of the QWS, but does not have any discontinuities within the switching angles and is thus ideally suited for integration into a linear controller concept.

Problem Formulation

An ideal multilevel three-phase converter is connected to a symmetrical three-phase grid, represented as in Fig.1. The grid is considered to be an ideal and symmetrical alternating voltage source of amplitude $\hat{U}_g$ and fixed angular frequency ω_1. Both filter and grid inductance with corresponding resistances are represented by L_{ac} and R_{ac}. For synchronous machines a further generalized OPP approach can be found in [8].

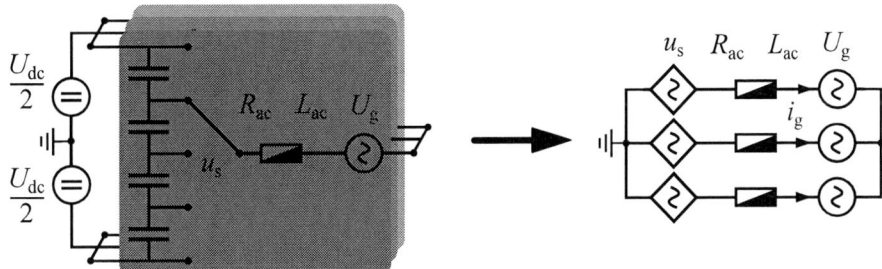

Fig. 1: Equivalent circuit of grid-connected converter

The optimization process intends to minimize the harmonic distortion of the output current i_g. The harmonic content of the output current is quantified by the **Total Demand Distortion (TDD)** and is used to derive the cost function:

$$I_{TDD} = \frac{1}{\sqrt{2}I_r} \cdot \sqrt{\sum_{h=2}^{h_{limit}} (\hat{i}_{g_h})^2} \tag{1}$$

Here I_r denotes the rated current of the converter (RMS) and h the harmonic order. The amplitudes of the current harmonic $\hat{i}_{g_h}$ can be expressed using the according output voltage harmonic amplitudes $\hat{u}_{s_h}$ of the converter:

$$\hat{i}_h = \frac{\hat{u}_{s_h}}{|Z_{ac_h}|} = \frac{\hat{u}_{s_h}}{\sqrt{R_{ac}^2 + (h\omega_1 L_{ac})^2}} \tag{2}$$

Neglecting the resistance ($R_{ac} \ll h\omega_1 L_{ac}$ for $h \gg 1$) and expressing the voltage harmonic amplitudes by normalized voltage amplitudes with $\hat{u}_{s_h} = \frac{U_{DC}}{2} \cdot \underline{\hat{u}}_{s_h}$ the equation can be written as:

$$I_{TDD} = \underbrace{\frac{U_{DC}}{\sqrt{2}I_r 2\omega_1 L_{ac}}}_{:=C} \cdot \sqrt{\sum_{h=2}^{h_{limit}} \left(\frac{\underline{\hat{u}}_{s_h}}{h}\right)^2} \tag{3}$$

The previously defined factor C contains only constant terms and therefore only scales the function. Furthermore, the square root is a strictly monotonically increasing function. Taking this into account and decomposing the normalized output voltage pattern with fourier series the cost function J can be derived:

$$J = \sum_{h=2}^{h_{limit}} \left(\frac{a_h^2 + b_h^2}{h^2}\right) \tag{4}$$

This general cost function is valid for all waveforms and symmetry conditions. Based on the respective waveform the fourier coefficients a_h and b_h are determined analytically and inserted into the objective function.

5L HWS OPP

The common QWS approach imposes not only QWS but also implicates HWS to the possible voltage pattern. Abolishing the strong QWS-constrain increases the degree of freedom for the optimization process. Thereby the HWS considers a half-wave period instead of only a quarter-wave period for optimization:

$$u_s(\pi + \omega t) = -u_s(\omega t) \tag{5}$$

The switching frequency of the switching units f_{SM} can be calculated by the number of switching angles per period $d_{2\pi}$ and the grid frequency $f_1 = \frac{\omega_1}{2\pi}$:

$$f_{SM} = \frac{d_{2\pi}}{2 \cdot (L-1)} \cdot f_1 \overset{5L}{=} \frac{d_{2\pi}}{8} \cdot f_1 \tag{6}$$

Following the symmetry reduction, the number of switching angles, that are available for optimization, are doubled for HWS in comparison to QWS:

$$d_{QWS} = \frac{d_{2\pi}}{4} \qquad d_{HWS} = \frac{d_{2\pi}}{2} \tag{7}$$

The comparison between 5L QWS and 5L HWS can be seen in Fig.2 for fundamental switching with eight switching angles per period $d_{2\pi} = 8$.

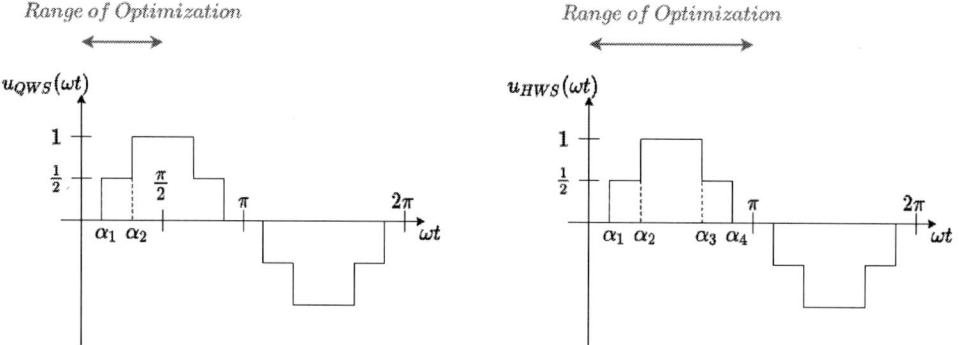

Fig. 2: Range of optimization QWS / HWS for fundamental switching $d_{2\pi} = 8$

Besides the number of switching transitions per period, the characteristic patterns are determined by the switching angles α_i with the corresponding switching transition Δu_{s_i} at the switching instance i. At each switching instance only one voltage step is allowed:

$$\Delta u_{s_i} \in \left\{ -\frac{1}{2}, \frac{1}{2} \right\} \tag{8}$$

Regarding the normalized pattern the output voltage is then allowed to take the values:

$$u_{s_i} \in \left\{ -1, -\frac{1}{2}, 0, \frac{1}{2}, 1 \right\} \tag{9}$$

In addition to the general reduction of the symmetry, the requirements for the possible voltage values are also greatly relaxed. Where the traditional QWS approach only allows positive voltage values in the first half-wave, now arbitrary voltage values are allowed e.g. negative values in the first positive half-wave. Additional patterns accomplished by HWS are shown exemplary for fundamental switching in Fig.3.

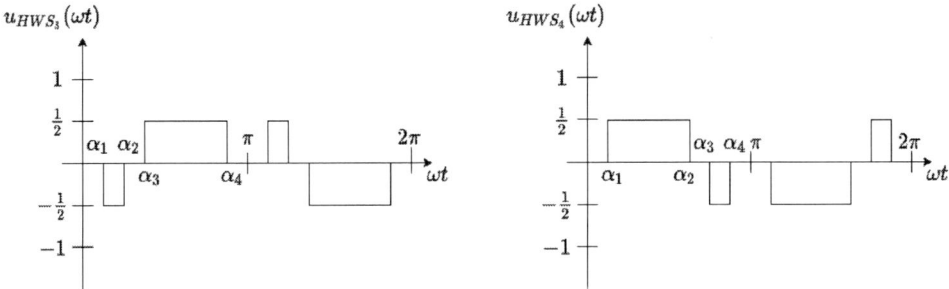

Fig. 3: Additional patterns 5L HWS $d_{2\pi} = 8$

To ensure that no further switching transitions at $\omega t = 0$ and $\omega t = \pi$ occur, the initial and end-value of the voltage pattern are restricted:

$$u_{s_0} = u_{s_{d_{HWS}}} = 0 \tag{10}$$

Due to the floating neutral point, common-mode voltages will not affect the current. Furthermore, as a result of the HWS, the amplitudes of even harmonics will be equal to zero. This leads to the final optimization problem, which considers two dimensions with switching angles α and switching transition Δu_s:

$$min\ J(\alpha, \Delta u_s) = \sum_{h=5,7,11,\ldots}^{h_{limit}} \left(\frac{a_h^2 + b_h^2}{h^2} \right) \tag{11}$$

$$subject\ to \quad a_1 = 0 \quad b_1 = m$$

$$0 \leq \alpha_1 \leq \alpha_2 \leq \ldots \leq \alpha_{d_{HWS}} \leq \pi$$

$$with \qquad \alpha = [\alpha_1 \ldots \alpha_{d_{HWS}}]^T$$

$$\Delta u_s = [\Delta u_{s_1} \ldots \Delta u_{s_{d_{HWS}}}]^T$$

The optimization problem is solved for fundamental switching i.e $d_{2\pi} = 8$ and the subsequent switching frequency with $d_{2\pi} = 12$. Assuming a grid frequency of $f_1 = 50Hz$ this leads to $f_{SM} = 50Hz$, $f_{SM} = 75Hz$

respectively. For low switching frequencies, and thus a low number of switching angles per period, it is suitable to reduce the two-dimensional optimization problem to a sum of one-dimensional optimization problems [9]. Therefor, all possible patterns are characterized by their combinations of transitions and determined in advance. Every combination of number of levels and number of switching angles per period leads to a characteristic set of possible patterns and the characterization has to be carried out again. In Table I an overview of the number of possible patterns for the considered combinations is provided.

Table I: Number of possible patterns

	3L QWS	5L QWS	5L HWS
$d_{2\pi} = 8$	1	2	4
$d_{2\pi} = 12$	1	2	14

Then for each pattern out of the set of possible patterns the global minimum of the non-linear non-convex optimization problem in the variable α is found. In a final comparison the best solution among all patterns is selected and thus the optimal waveform with the corresponding switching angles for each modulation index m is determined. The optimization problem in the variable α is solved numerically with the optimization tool fmincon of the Matlab software platform, which represents an implementation of **sequential quadratic programming (SQP)**. Due to the non-convex structure of the optimization problem locally converging methods like SQP are strongly dependent on the chosen initial value [8]. In order to be able to determine the global minimum, several calls of fmincon with different, randomly generated initial values are done. Finally the minima of all calls are compared and the best solution is selected. It should be noted here, that the minimum found is not necessarily a global minimum. However, if the number of calls is increased appropriately, the probability of finding the global minimum increases. In the optimization all harmonics up to the $100th$ order i.e. $n_{limit} = 100$ are considered. Furthermore, the modulation index m is represented with a resolution of $\Delta m = 0.01$ within the interval $m \in [0.01, 1.2]$. In Fig. 4 the concept of the complete algorithm is given and a comparison of the computation time for one iteration is provided. The computation time is normalized to the 3L QWS approach.

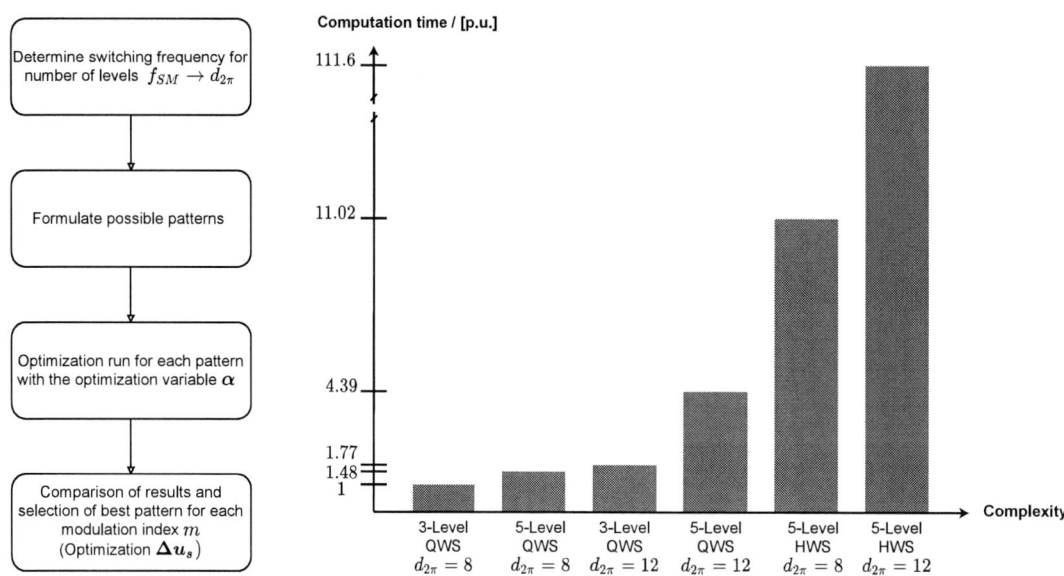

Fig. 4: Algorithm concept and computation time for one iteration

It can be seen that, with increasing degree of freedom, the complexity, and thus the calculation time, increases. Besides an increased calculation time per iteration, it becomes additionally more difficult to find the global minimum. Therefore the number of iterations, compared to 3L HWS [6], has to be increased significantly to ensure that the global minimum is found. An Intel® Core™ i7-4790 Processor with 16GB RAM was used to run the algorithm. The 5L HWS with $d_{2\pi} = 12$ needed an overall computation time of 222 hours.

The harmonic distortion is directly calculated with the use of the fourier coefficients and displayed as **Total Harmonic Distortion (THD)**. In both cases the HWS provides improved harmonic distortion in numerous modulation ranges in direct comparison to the QWS. The results are shown in Fig.5. For $d_{2\pi} = 8$ the THD is reduced by up to 12.95% at $m = 0.38$. Furthermore, significant percentage improvements of the THD are obtained in ranges that are particularly relevant for grid-connected power converters. For example, the THD is reduced at the modulation-index of $m = 0.62$ with 10.77%. The reduction in harmonic distortion is particularly evident in the range $m \in [0.82, 0.9]$. Here the already very low THD values can be further reduced by up to 31.63% at $m = 0.87$.

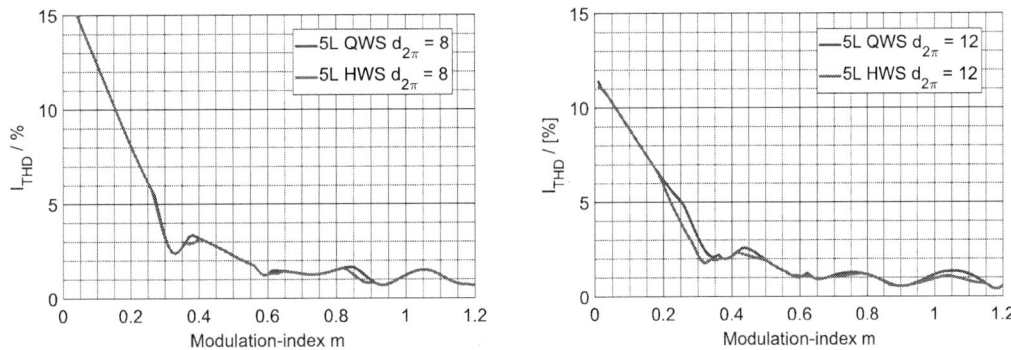

Fig. 5: Output current THD 5L QWS / 5L HWS

For $d_{2\pi} = 12$ it can be seen, that a reduction of the symmetry condition leads likewise to an improvement of the harmonic distortion over wide modulation ranges. Especially in the low modulation range there is a significant improvement by up to 25.68% at $m = 0.26$. Furthermore, within the range of $m \in [0.4, 0.5]$ there is a reduction of up to 14.39% at $m = 0.44$. At $m = 0.62$ an improvement of 14.92% is achieved. In the range of $m \in [0.68, 0.79]$ there is a reduction up to 12.11% at $m = 0.73$.

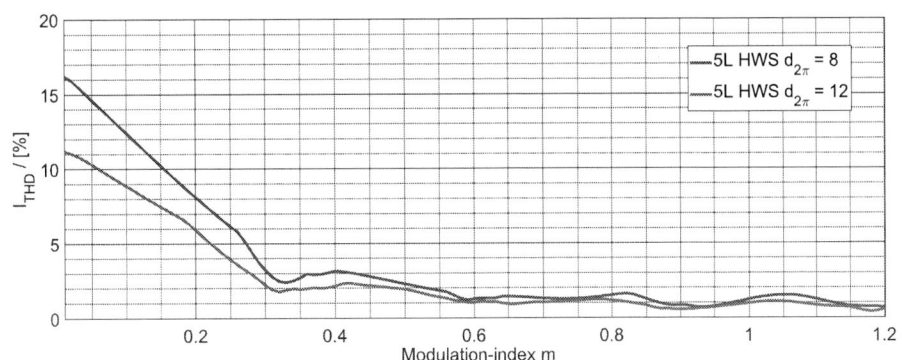

Fig. 6: Output current THD 5L HWS $d_{2\pi} = 8$ / 5L HWS $d_{2\pi} = 12$

In Fig. 6 the comparison for different number of switching angles per period for 5L HWS is given. It can be seen that an increased switching frequency leads to a reduced harmonic distortion. Overall, the 5L HWS leads to a significant reduction of harmonic distortion over numerous modulation ranges. Of particular importance is the lowest possible switching frequency of $f_{SM} = 50Hz$ with $d_{2\pi} = 8$. Here peak reductions of up to 31.63%, in particularly important operating ranges for power converters, can be achieved. As a result, the lowest possible switching frequency of $f_{SM} = 50Hz$ can be used without causing an excessive increase in THD. This leads to reduced switching losses and therefore improved overall system efficiency. For simple verification, the designed 5L HWS was simulated in open-loop conditions against an inductive load. The results for two different operation points are shown in Fig. 7.

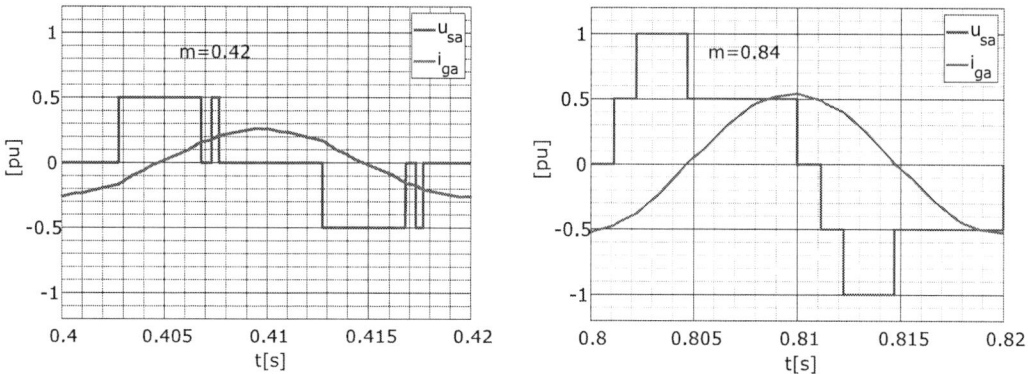

Fig. 7: Inverter voltage and output current 5L HWS $d_{2\pi} = 8$

5L HWS OPP with phase-shift

To further increase the degree of freedom and with it the solution range of the optimization the approach with allowed phase-shift is considered. It is basically based on the preceding approach. Hence, the computation regulation remains the same. The difference is given by adapted constraints. The fundamental component is now generally allowed to have a phase shift, i.e. both a_1 and b_1 can occur unequal zero. This leads to the constrain, that the amplitude of the fundamental component, resulting from the superposition of fundamental cosine and fundamental sine oscillation, must correspond to the modulation index:

$$\sqrt{a_1^2 + b_1^2} = m \tag{12}$$

The resulting phase-shift is determined after optimization and does not limit the optimization:

$$\tan(\phi_1) = \frac{a_1}{b_1} \tag{13}$$

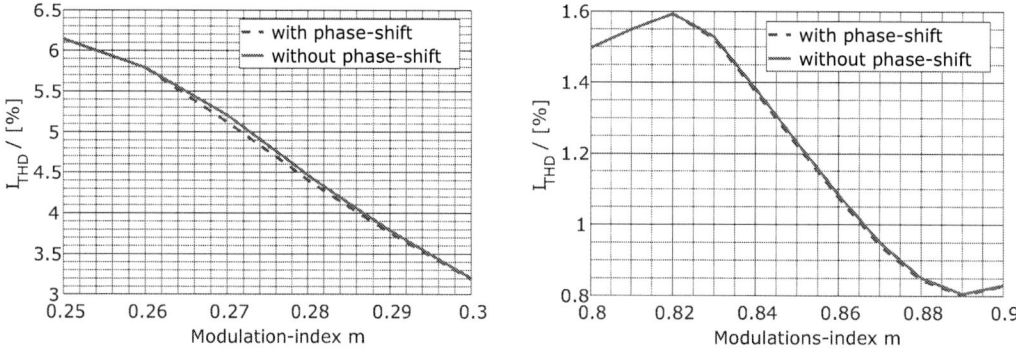

Fig. 8: Output current THD 5L HWS with / without phase-shift $d_{2\pi} = 8$

The optimization problem with allowed phase shift is now given by:

$$min\ J(\boldsymbol{\alpha}, \boldsymbol{\Delta u_s}) = \sum_{h=5,7,11,\ldots}^{h_{limit}} \left(\frac{a_h^2 + b_h^2}{h^2} \right) \tag{14}$$

$$subject\ to \quad \sqrt{a_1^2 + b_1^2} = m$$

$$0 \le \alpha_1 \le \alpha_2 \le \ldots \le \alpha_{d_{HWS}} \le \pi$$

$$with \quad \boldsymbol{\alpha} = [\alpha_1 \ldots \alpha_{d_{HWS}}]^T$$

$$\boldsymbol{\Delta u_s} = [\Delta u_{s_1} \ldots \Delta u_{s_{d_{HWS}}}]^T$$

This optimization problem is now solved for the case $d_{2\pi} = 8$. The relevant patterns remain the same and only the constraint has to be adapted accordingly. The results and the comparison with the approach without phase-shift are shown in Fig.8 for the relevant modulation ranges. A closer look reveals a further improvement of the THD in the represented modulation ranges. However, the resulting phase shift has to be considered and increases the requirement for the superimposed control system and therefore overall system complexity.

5L HWS OPP with reduced Discontinuities

Based on the general HWS approach (11) an alternative method has been developed which addresses the OPP inherent problem of discontinuities within the switching angles. These discontinuities in the switching angles, when varying the modulation index, prevent the use of linear controllers due to stability issues [10] and are therefore ideally avoided in advance. For 5L HWS with $d_{2\pi} = 8$ exemplary discontinuities can be seen in the left side of Fig.9.

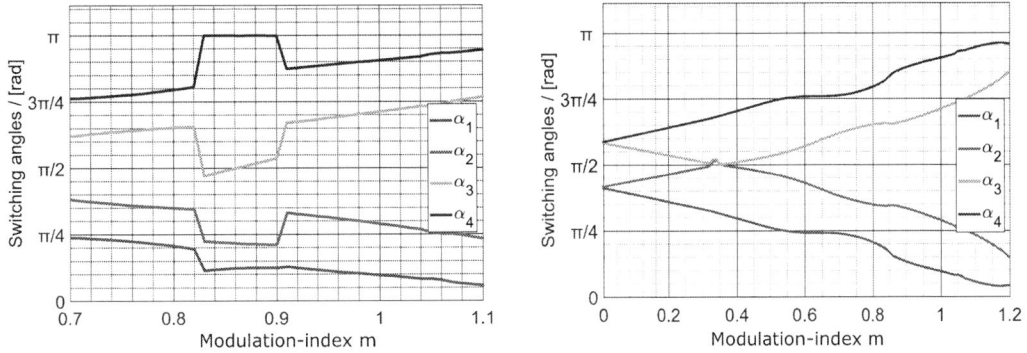

Fig. 9: 5L HWS with exemplary and reduced discontinuities $d_{2\pi} = 8$

Starting from the switching angles resulting for 5L HWS the ranges of discontinuities are regarded and specified. The possible range of values for each switching angle within the affected ranges are restricted, i.e. additional constrains are added to the optimization problem. These constrains have to be specified as minimal as possible to ensure the degree of freedom of the optimization remains as large as possible. For example for the discontinuity for α_4 in the range of $m \in [0.82, 0.91]$ the constrain is given that $\alpha_4 < \pi$. Following this constrain, the solution of the optimization problem where α_4 forces a discontinuity can not be selected anymore, however a large searching area for a alternative solution is maintained. This process is carried out for every discontinuity of each switching angle. After the restriction process the optimization problem with additional constrains is solved and the switching angles are checked until the desired continuous course is achieved. This approach offers a greater flexibility compared to a directly in the solving algorithm implemented post-optimization loop as proposed in [11]. The course of the switching angles after the process can be seen in the right side of Fig.9. The method results in a continuous and smooth progression of the switching angles without any discontinuities.

The harmonic distortion in comparison to 5L QWS is shown in Fig.10. It can be seen that the harmonic distortion of the 5L HWS with reduced discontinuities corresponds over a wide working range to the harmonic distortion of the 5L QWS. Thus the increased degree of freedom from the HWS can be used to compensate for emerging constraints due to reduced discontinuities. Furthermore the smooth progression of the switching angles significantly reduces the increased harmonic distortion that occur in transient processes as a result of discontinuities. Accordingly, the requirements for the superimposed control system are decreased and the use of linear controllers is allowed.

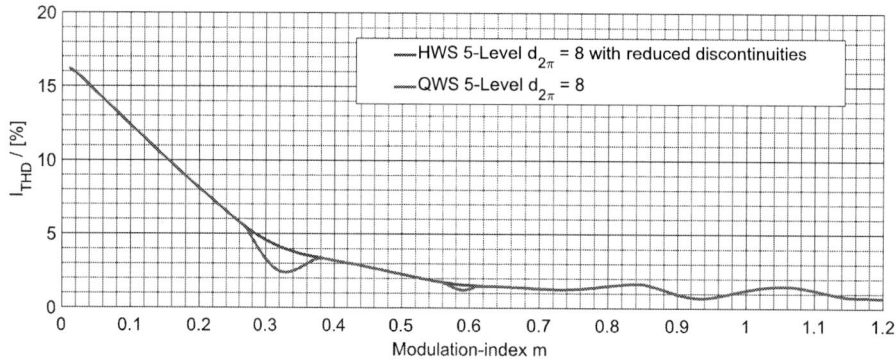

Fig. 10: Output current THD 5L QWS / 5L HWS with reduced discontinuities $d_{2\pi} = 8$

Conclusion

Reducing the symmetry constrain on the 5L OPP leads to significant improvements in harmonic distortion in several modulation ranges. Therefore 5L HWS OPP have superior performance compared to the common 5L QWS OPP. The increased degree of freedom can be used to avoid discontinuities under the expense of higher harmonic distortion. However, the harmonic distortion equals the performance of 5L QWS over a wide working range and therefore 5L HWS OPP can maintain the known harmonic distortion of 5L QWS with the advantage of smooth progression within the switching angles.

References

[1] A. A. Ferreira, C. C. Rodriguez and O. G. Bellmut: Modulation techniques applied to medium voltage modular multilevel converters for renewable energy integration: A review, Electric Power Systems Research, Volume 15, pp. 21-39, 2018

[2] P. dos Santos, M. Vazquez and S. Liu: Flexible and General Strategy of Space Vector Modulation for Multilevel Converters, PCIM Asia 2020 International Exhibition and Conference for Power Electronics, Intelligent Motion, Renewable Energy and Energy Management, pp. 1-7, 2020

[3] H. Wang and S. Liu: An optimal strategy with convex-concave constraints for power factor correction and harmonic compensation under different voltages, IEEE International Conference on Industrial Technology (ICIT), pp. 275-280, mar 2016

[4] M. S. A. Dahidah, G. Konstantinou and V. G. Agelidis: A Review of Multilevel Selective Harmonic Elimination PWM: Formulations, Solving Algorithms, Implementation and Applications, IEEE Transactions on Power Electronics, vol. 30, no. 8, pp. 4091-4106, aug 2015

[5] G. S.Buja and G. B. Indri: Optimal pulsewidth modulation for feeding AC motors, IEEE Transaction on Industrial Application, vol. IA-13, no. 1, pp. 38–44, jan 1977

[6] A. Birth, T. Geyer, H. du Toit Mouton, and M. Doring: Generalized three-level optimal pulse patterns with lower harmonic distortion, IEEE Transactions on Power Electronics, vol. 35, pp.5741 - 5752, jun 2020

[7] A. Birth, T. Geyer, and H. du Toit Mouton: Symmetry relaxation of three-level optimal pulse patterns for lower harmonic distortion, 2019 21st European Conference on Power Electronics and Applications (EPE 19 ECCE Europe), sep 2019

[8] N. Hartgenbusch, R. W. D. Doncker, and A. Thunen: Optimized pulse patterns for salient synchronous machines, 2020 23rd International Conference on Electrical Machines and Systems (ICEMS), nov 2020

[9] A. K. Rathore, J. Holtz, and T. Boller: Synchronous optimal pulsewidth modulation for low-switching-frequency control of medium-voltage multilevel inverters, IEEE Transactions on Industrial Electronics, vol. 57, pp. 2374 - 2381, jul 2010

[10] M. Vasiladiotis, A. Christe, and T. Geyer: Model predictive pulse pattern control for modular multilevel converters, IEEE Transactions on Industrial Electronics, vol. 66, pp. 2423 - 2431, mar 2019

[11] A. K. Rathore, J. Holtz, and T. Boller: Generalized optimal pulsewidth modulation of multilevel inverters for low-switching-frequency control of medium-voltage high-power industrial ac drives, IEEE Transaction on Industrial Electronics, vol. 60, no. 10, pp. 4215–4224, oct 2013

Characterization of Si-IGBT Crosstalk with a Concentration on Power Circuit Parasitic Elements and the Device Operation Point

Amir Azam Rajabian[1], Sadegh Mohsenzade[1], Javad Naghibi[2], Kamyar Mehran[2]

[1]Electrical Engineering Department, K. N. Toosi University of Technology, Tehran, Iran
[2]School of Electronics Engineering and Computer Science, Queen Mary University of London, London, UK
E-mails: amir.rajabian@email.kntu.ac.ir, s.mohsenzade@kntu.ac.ir,
s.naghibinasab@qmul.ac.uk, k.mehran@qmul.ac.uk

Keywords

«IGBT», «Parasitic elements», «Thermal stress», «Short circuit».

Abstract

Crosstalk is a serious issue in power electronics converters having a phase-leg in their structure. The crosstalk destructive effect can lead to failure of the device. Hence, several models are presented in the literature to analyze the crosstalk and prevent the undesired failures. However, these models need to be enhanced so as to be characterized in different aspects. In this paper, a new comprehensive model based on the last models in state-of-art has been presented. This model includes the parasitic elements of the power circuit and parasitic capacitances of Si-IGBT. Thus, the effect of variable elements can be investigated. In this paper, the effect of the high-side switch specifications such as off-sate voltage, conducting current, and turning-on time on the crosstalk has been figured out. It is found that the most critical condition in terms of the crosstalk is when the converter operates on the high voltage levels of DC-bus and light-load conditions. Moreover, the different values of parasitic inductance are considered in the model, and their effect on the crosstalk is evaluated. Furthermore, the experimental setup has been introduced in order to check the model accuracy. In order to study the effect of Si-IGBT parasitic capacitances, the ratio of C_{GC}/C_{GE} has been changed, and results have been presented. The case study device for the investigation of the crosstalk and experimental tests is IXGH60N60.

Introduction

Applications of high voltage/power insulated gate bipolar transistors (IGBTs) are very widespread in power converters. They are widely employed in traction systems [1], high-voltage DC/DC converters [2], pulsed power supplies [3], HVDC circuit breakers [4]-[5]-[6], and solid-state transformers [7]-[8]. The reliability of the power converters is very important since they play an important role in the industry. Prior studies indicate that 34 % of failures in power converters are related to IGBTs [9]. However, the power IGBTs are more rugged than their other counterparts in severe conditions such as short circuit faults [10], but there are many reliability issues threatening IGBTs during the operation. Hence, as an important topic, enhancing IGBTs' reliability and comprehension of the root of the failure in normal and harsh conditions have been the target of recent research.

When the IGBTs are used in a phase-leg structure, the shoot-through currents due to short-circuiting the DC-link can be fatal for the IGBTs. The phase-leg exists in many power converter topologies such as single/multi-phase inverters and DC/DC resonant converters. It should be carefully considered that even using intentional dead-time between the operation of IGBTs in the phase-leg, the risk of shoot-through currents is high. The origin of these shoot-through currents is crosstalk or parasitic turn-on. In crosstalk, the turning on of the upper device parasitically turns on the lower side device as depicted in Fig. 1. Consequently, for a while, both devices are conducting and the shoot-through current flows among the phase-leg. Crosstalk occurs due to the voltage change of the Miller capacitance of the low-side device in transition of the high-side device from the off-state to the on-state and vice-versa. Accordingly, when the high-side device changes state from the off to on, a positive voltage jump occurs for the low-side device gate-emitter voltage as shown in Fig. 1. When the peak value of the gate-emitter voltage (V_{GE1-P} in Fig. 1) exceeds the device gate threshold voltage (V_{TH}), it can operate in the active region, and saturated current can flow through the IGBTs in the phase-leg structure. The extra power loss due to crosstalk can increase the device's junction temperature. In some conditions, the value of the short circuit current can be several times the device's nominal current. This condition can potentially lead to device failure.

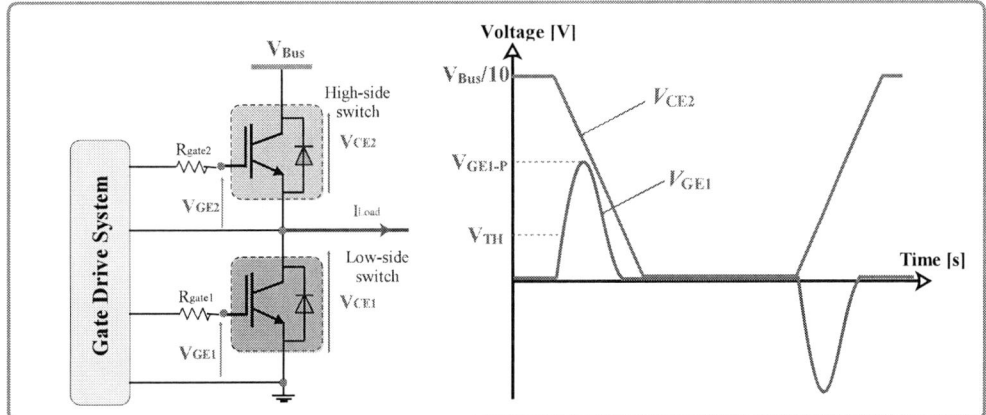

Fig. 1. The schematic of a phase leg and occurrence of a parasitic turn-on for the low-side switch.

To cease crosstalk and parasitic turn-on, several solutions have been reported in the prior research. Applying a negative bias in the off-state at gate-emitter terminals of the devices is proposed in [11] to alleviate the gate-emitter peak voltage in crosstalk. This method is promising but the cost added to the system for an isolated DC power supply is not acceptable in many applications. Moreover, this method cannot be served in boot-strap gate drive systems. In addition, negative bias solely cannot be effective in avoiding parasitic turn-on in many cases. Gate impedance regulation (GIR) is investigated in [12], [13], and [14]. These papers aim is to optimize the gate impedance to minimize the effect of Miller capacitance on the gate side in the crosstalk. These solutions are practical but the improvement is limited to the cases with specific conditions. Several novel gate drive systems also are proposed in [15], [16], and [17]. The focus of the mentioned solutions is on the gate side. The effect of the power path on the crosstalk has not been deeply investigated yet. More importantly, the behavior of the crosstalk strictly depends on the converter power rating and power path elements. Thus, before devising a crosstalk suppression scheme, a predictive approach for the determination of V_{GE1-P} in crosstalk seems to be mandatory.

This paper aims to characterize the crosstalk by considering power path parasitic elements and the converter operating point. It will be revealed that the power path elements have a dominant effect on the crosstalk. The results show a considerable dependency of the crosstalk of power path parasitic inductance. An optimum region for the power path parasitic inductance is obtained to minimize the effect of the crosstalk. This parasitic inductance depends on the PCB trace length and ESL of the DC-link capacitors. Hence, in the design phase, the practitioner has a practical guideline to design the power circuit for minimum crosstalk. In addition, the effects of the converter operating point as well as the high-side switch turning on speed on parasitic turn on are investigated. The theoretical discussions are validated using experiments. The case study device is IXGH60N60C a 600 V and 60 A discrete IGBT.

Characterization of the IGBT crosstalk

A. Existing approaches

There are several studies devoted efforts to characterizing the crosstalk using equivalent circuits. [18] provides (1) as a simple crosstalk model

$$V_{GE1} = R_G C_{GC} \frac{dV_{CE}}{dt} \left(1 - e^{\frac{-t}{R_G(C_{GC}+C_{GE})}} \right)$$

(1)

where R_G is the gate equivalent resistance, C_{GC} and C_{GE} are the junction capacitances of the device between gate-collector and gate-emitter terminals. Expression (1) also demonstrates the effect of the switch transient speed (dV/dt) and gate resistance (R_G) in order to investigate the crosstalk and the gate emitter voltage of the low-side switch (V_{GE1}). In addition, the parasitic capacitances of IGBT are considered. However, this model simplifies the power circuit and cannot study its effect.

The second model which can be considered as the most comprehensive one in the literature is discussed in [19] and is presented in Fig. 2. This model includes the power parasitic components. The shortcomings of this model are: 1- It does not consider the transient speed of the high-side device in the turning-on process. 2-It does not consider the mutual inductances between the gate circuit and power circuit.

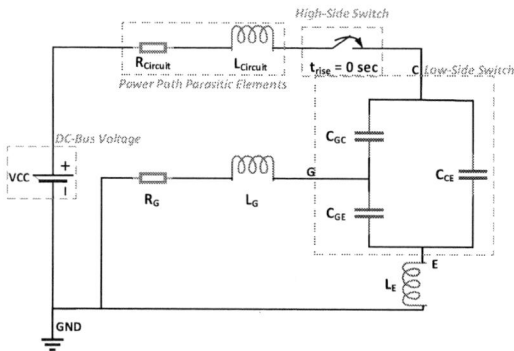

Fig. 2. The crosstalk model proposed in [19]

B. Proposed Model

Keeping in mind the shortcomings of the present crosstalk models, this paper tries to modify the model of [19]. These modifications are:

a) Considering the high-side switch turning-on transient time. Since this time depends on the device current and voltage, the analysis of the device operating point and its effect on the crosstalk can be carried out using the proposed model. In addition, the effect of the high-side switch gate drive system on the crosstalk can be well-studied.

b) The proposed model considers the mutual inductances of the power circuit and gate side circuit. Hence, the parasitic elements effect on the crosstalk can be studied more accurately.

The process of consideration of high-side device turning-on time is presented in Fig. 3-(a) and the proposed model schematic is presented in Fig. 3-(b). According to Fig. 3-(a), the time-domain equation of $V_{pulse}(t)$ can be written as

$$V_{pulse}(t) = \frac{V_{bus}}{t_{rise}} t \big(u(t) - u(t - t_{rise}) \big) + V_{bus} u(t - t_{rise})$$

(2)

Where V_{bus} is the DC-link voltage of the phase-leg, t_{rise} is the turning-on time of the high-side switch, and $u(t)$ is the Heaviside step function.

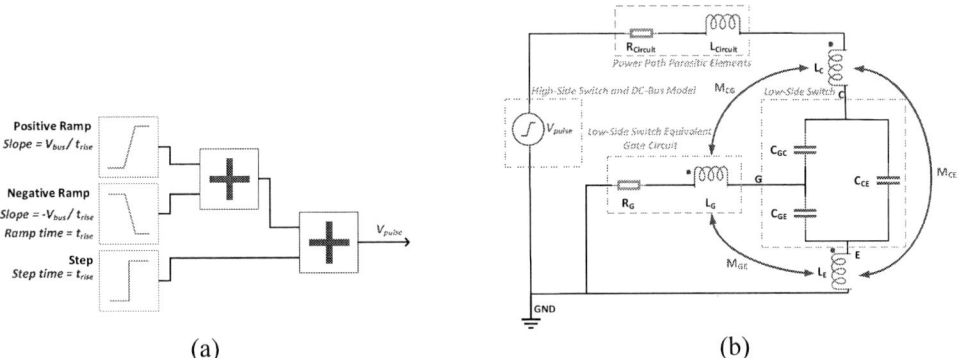

(a)   (b)

Fig. 3. The process of applying high-side device speed (a), and the full schematic of the proposed model (b).

In addition, considering Fig. 3-(b), the Laplace forms of the model equations can be written as

$$\frac{V_G}{R_G + sL_{GG}} + (V_G - V_E)sC_{GE} + (V_G - V_C)sC_{GC} = 0 \tag{3}$$

$$(V_E - V_G)sC_{GE} + \frac{V_E}{sL_{EE}} + (V_E - V_C)sC_{CE} = 0 \tag{4}$$

$$(V_C - V_G)sC_{GC} + (V_C - V_E)sC_{CE} + \frac{V_C - V_{pulse}}{R_{Circuit} + sL_{Circuit} + L_{CC}} = 0 \tag{5}$$

where V_C, V_G, and V_E are the collector, gate and emitter voltages of the device. In addition, regarding the investigations of [20], the mutual inductances of the model can be considered as

$$\begin{aligned} L_{GG} &= L_G + M_{CG} + M_{GE} - M_{CE} \\ L_{CC} &= L_C + M_{GC} + M_{CE} - M_{GE} \\ L_{EE} &= L_E + M_{CE} + M_{GE} - M_{GC} \end{aligned} \tag{6}$$

Regarding expressions (3) to (6), the equation of low-side switch in Laplace form can be expressed as

$$V_{GE1} = \frac{C_{GC}V_{Bus}e^{-s\times t_{rise}} \times (e^{s\times t_{rise}} - 1) \times (R_G + L_{GG}s)}{st_{rise} \times (D_4 s^4 + D_3 s^3 + D_2 s^2 + D_1 s + 1)} \tag{7}$$

The denominators of equation (7) are

$$\begin{aligned} D_4 &= (L_{Circuit} + L_{CC})\big((L_{EE} + L_{GG})(C_{CE}C_{GE} + C_{GC}C_{GE}) + C_{CE}C_{GC} \times L_{GG}\big) \\ &\quad + C_{GC}C_{GE} \times L_{EE}L_{GG} \\ D_3 &= C_{CE}C_{GE}(L_{EE}R_{Circuit} + L_{CC}R_G + L_{Circuit}R_G + L_{GG}R_{Circuit}) \\ &\quad + C_{CE}C_{GC}(L_{CC}R_G + L_{Circuit}R_G + L_{GG}R_{Circuit} + L_{EE}R_{Circuit}) \\ &\quad + C_{GC}C_{GE}(L_{CC}R_G + L_{Circuit}R_G + L_{EE}R_G + L_{GG}R_{Circuit}) \\ D_2 &= (L_{CC} + L_{Circuit})(C_{CE} + C_{GC}) + C_{GE}(L_{EE} + L_{GG}) + C_{GC}L_{GG} \\ D_1 &= R_{Circuit}(C_{CE} + C_{GC}) + R_G(C_{GC} + C_{GE}) \end{aligned} \tag{8}$$

According to (8), the time-domain of shoot-through current can be measured analytically, hence the effect of various elements can be investigated in the model.

C. Determination of the high-side switch turning-on time (t_{rise})

The turning-on time of the device depends on the device blocking voltage (V_{bus}) and conducting current (I_{Load}). According to investigations of [21] and [22], the turning-on time of the high-side switch can be written as

$$t_{rise} = C_{GC}R_G \frac{V_{bus}}{(V_{CC} - V_{GP})}; \tag{9}$$

, where V_{CC} is the applied on-state voltage of the gate driver, and V_{GP} is the Miller plateau voltage which can be expressed as

$$V_{GP} = V_{TH} + I_{Load}/gfs \tag{10}$$

, where V_{TH} is the gate threshold voltage. The parameter g_{fs} is the device transconductance and can be written as [23]

$$g_{fs} = \frac{\partial I_{CH}}{\partial V_{GE}} = \frac{1}{1 - \alpha_{PNP}} \frac{\mu_{ns} C_{ox} Z}{L_{CH}} (V_{GE} - V_{TH}) \tag{11}$$

, where μ_{ns} is the average electron mobility in the channel, C_{ox} is oxide capacitance per unit area, Z is the channel width, L_{CH} is the channel length.

C_{GC} is a nonlinear capacitance and depends on the device voltage in the transient time interval. [24] and [14] provide algorithms for consideration of this nonlinear capacitor. In this paper, we used the same method and consider the average value for the nonlinear gate-collector capacitance for the calculation of t_{rise}. According to (9) -(11), the turning-on time of the high-side device depends on the device operating point. On the other hand, t_{rise} is an important parameter in the gate-emitter voltage on the low-side device in crosstalk. Thus, the device operation point is very important to be considered in analyzing the crosstalk in the design phase and is feasible by the proposed approach.

D. Accuracy check

To check the accuracy of the proposed model, the output of the model is compared with a PSPICE simulation for a case study with the parameters of TABLE I.

TABLE I. MODEL PARAMETERS AND VALUES

C_{GC}	53.75 pF
C_{GE}	3614 pF
C_{CE}	204 pF
L_C	0.9 nH
L_G	3 nH
L_E	10 nH
M_{GC}	-5.71 nH
M_{GE}	-5.47 nH
M_{CE}	-10.3 nH
I_{Load}	20 A
$R_{Circuit}$	1 Ω
$L_{Circuit}$	50 nH
V_{bus}	400 V
Calculated t_{rise}	90 ns

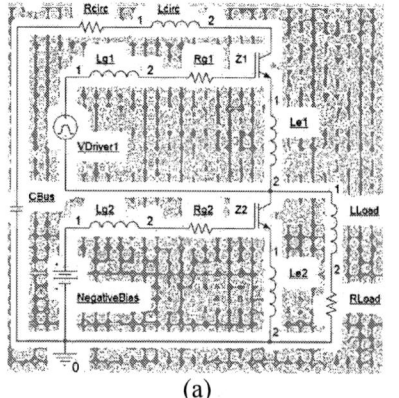
(a)

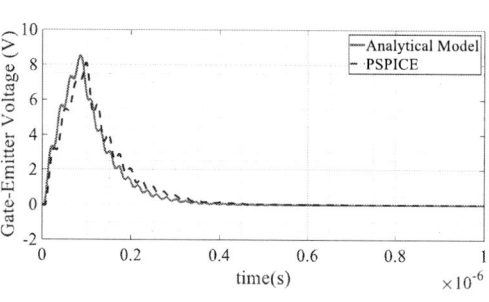
(b)

Fig. 4. The simulation circuit (a), and output waveforms (b).

The circuit used for the simulation of the crosstalk is presented in Fig. 4-(a) and the resultant waveforms are presented in Fig. 4-(b). According to Fig. 4-(b), the accuracy of the proposed model is acceptable and it can be employed to characterize the crosstalk.

Effect of the high-side switch operating point on Crosstalk

A. The high-side switch current effect on Crosstalk

To figure out the effect of high-side device current on the peak voltage of the gate-emitter, a variety of loads should be applied to the output. Moreover, in practice, there is always an inductance in the output. Thus, in the simulation of the case study, the load should be an RL type. The value of load inductance is, however, so less than its resistance 100 nH. In this study, the load resistance is changed to vary the load current. The dependency of the crosstalk voltage from the device current is presented in Fig. 5-(a). Fig. 5-(a) indicates that by increasing the load current, the peak of gate-emitter voltage decreases. This can be explained by the fact that the increment in the device current slows down the high-side device according to (9) to (11). Hence, the crosstalk effects are more dominant when the converter works under low power mode.

B. The high-side switch off-state voltage effect on Crosstalk

So as to investigate the other aspects of crosstalk, the effect of high-side off-state voltage should be investigated. The off-state voltage of the case study is 400 V. Thus, by decreasing the off-state voltage its effect can be studied. Fig. 5-(b) indicates the consequence of the off-state voltage of the high-side device on crosstalk. By increasing the off-state voltage, the crosstalk effect is dominant. This behavior is in accordance with (2) while the V_{bus} directly determines the input of the model. Thus the negative bias is necessary in order to prevent crosstalk in such conditions.

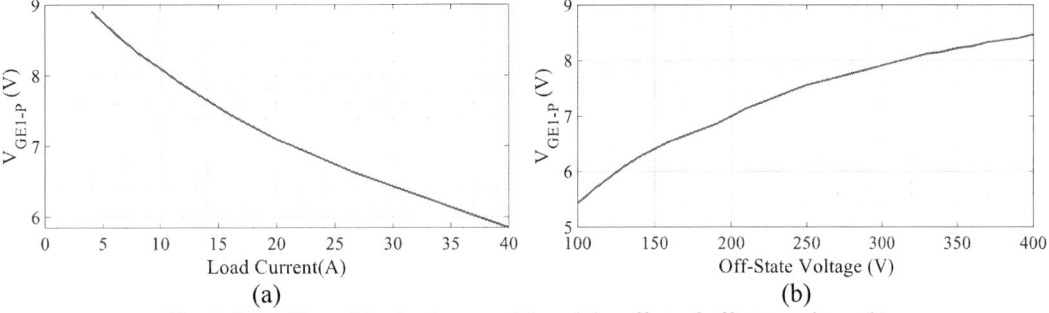

(a)   (b)

Fig. 5. The effect of the load current(a), and the effect of off-state voltage (b).

Effect of the power circuit parasitic inductance on crosstalk

In this paper, the concentration is on the parasitic elements of the power circuit. The PCB length and width are dominant factors for the circuit inductance. Fig. 6 presents the effect of the circuit inductance on the obtained crosstalk. There are some optimum regions where the obtained peak value of the V_{GE1} is less. On this basis, the PCB designer must take care of the length and width of the PCB track to minimize the effect of crosstalk. Moreover, in lower values of inductance, the maximum value of the curve is lower, and the distance between the DC-link and Si-IGBTs should be limited.

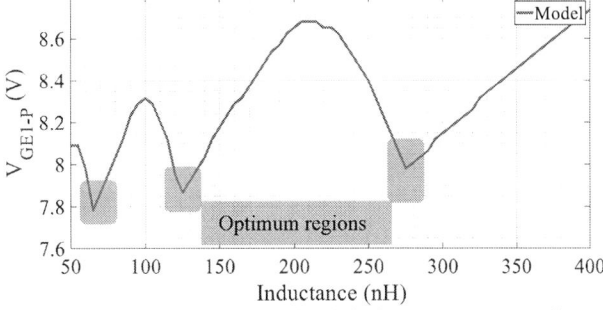

Fig. 6. The effect of the power circuit inductance on crosstalk.

Parasitic Capacitance Ratio C_{GC}/C_{GE} Effect

The values of the device parasitic capacitances are determined by the manufacturer and could be found in the device datasheet. Their values can affect the crosstalk, and hence, the change of C_{GC}/C_{GE} is studied. In the case study of this paper, the ratio is 0.014872. The value of this ratio is increased and decreased by 30%. Fig. 7 shows the effect of parasitic capacitances on crosstalk. By increasing this ratio, the gate-emitter voltage increases and may destroy the IGBT of the phase-leg. Moreover, the parasitic capacitance ratio does not have a tangible effect on the crosstalk time interval.

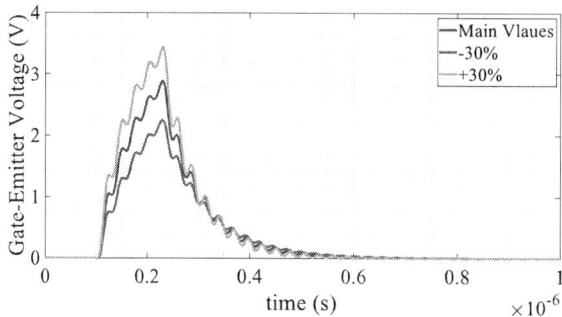

Fig. 7. The effect of C_{GC}/C_{GE} ratio on the crosstalk.

Experimental results

In order to verify the proposed crosstalk model, a phase-leg structure is implemented and experiments are conducted. The overview and block diagrams of the test setup are provided in Fig. 8-(a) and (b) respectively. The specifications of the experimental case are the same as the case used for simulation. The procedure of the test is as follows. The trigger unit provides a low-level signal for the input of the low-side device driver. Thus, the low-side device is in the off-state in the experiments. The trigger signal generates a 10 µs pulse width high-level signal for the high-side device. Before the arrival of this high-level signal, both switches are in the off-state. Since the resistance of R_{Load} is much less than the off-state impedance of the high-side switch, the voltage across the low-side switch is almost zero when both devices are in the off-state. After receiving the turning-on command, the high-side device voltage decays from V_{bus} to zero. At the same time, the collector-emitter voltage of the low-side device (V_{CE1}) raises from zero to V_{bus}. As deeply discussed in the paper, this voltage change causes a positive voltage jump at gate-emitter terminals of the low-side device (V_{GE1}). Fig. 9-(a), represents the experimental waveforms of this event in the high-side switch turning-on process. In the turning-off process of the high-side switch, V_{CE1} decays from V_{bus} to zero. Thus, V_{GE1} experiences a negative voltage change as occurred in Fig. 9-(b).

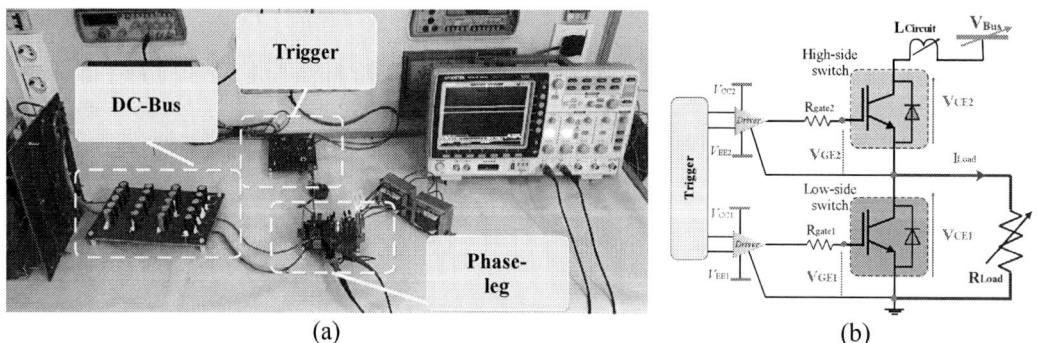

(a)	(b)

Fig. 8. The photo of the test bench (a), and the experimental setup block diagram (b).

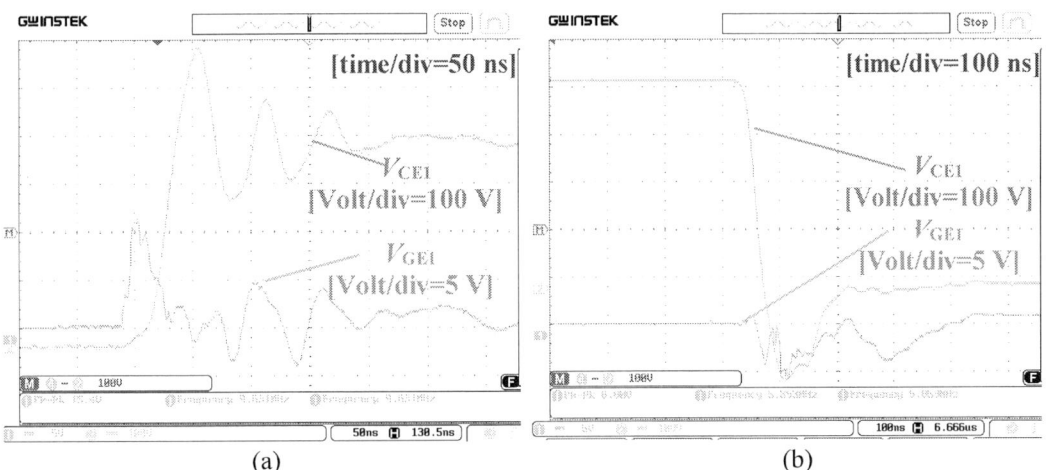

(a)   (b)

Fig. 9. The low-side switch waveforms in the high-side switch turning-on process (a), and in the high-side switch turning-off process (b).

In the experiments, the value of the power circuit inductance ($L_{circuit}$) is changed by adding external inductance. The value of the original circuit inductance is the sum of PCB traces inductance and the DC-link ESL. The inductance of PCB tracks is calculated based on the guidelines of [25] and [26] as 50 nH. In addition, the ESL of the DC-Link is 35 nH according to the component datasheet. Fig. 10-(a) presents the waveforms of V_{GE1} when the value of $L_{circuit}$ is changed. Fig. 10-(b) shows the normalized values of the low-side switch gate-emitter peak value (V_{GE1-P}) based on the existing maximum value. Considering Fig. 10, the optimum regions for $L_{circuit}$ values are also detectable in practice. This outcome is in accordance with the results achieved by the analytic model.

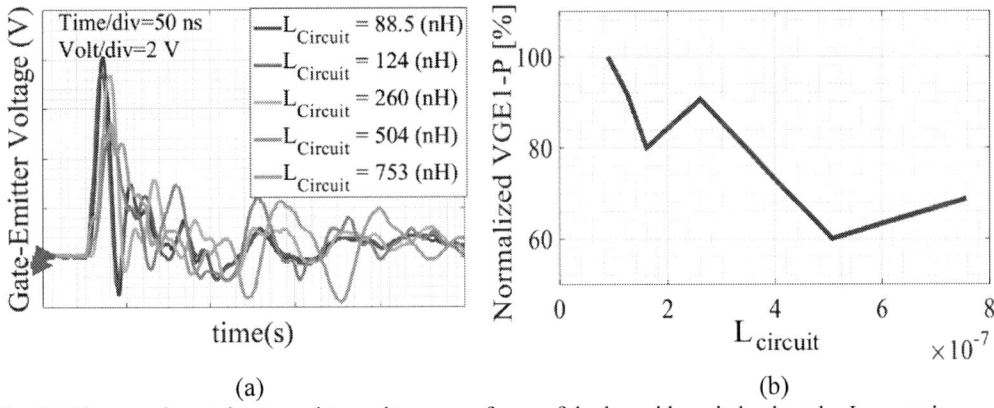

(a)   (b)

Fig. 10. The experimental gate-emitter voltage waveforms of the low-side switch when the $L_{circuit}$ varies.

In order to investigate the converter/device operating point on the crosstalk experimentally, several experiments are carried out. Firstly, the load current is changed and the V_{GE1-P} is measured. Fig. 11-(a) presents the obtained experimental results when the load current is varied. These results are in accordance with the results achieved from the crosstalk model. As can be seen, the V_{GE1-P} has higher values when the load current decreases. Thus, the effect of the crosstalk is more dominant in the light load conditions of the converter. As another important parameter, the off-state voltage of the device is changed in the experiments. The results of these experiments are provided in Fig. 11-(b). Regarding Fig. 11-(b), by increasing the off-state voltage of the device, the V_{GE1-P} value is increased. A saturated region is detectable in Fig. 11-(b) which can be explained by the non-linear behavior of C_{GC} which exhibits very low values when its voltage grows. Considering Fig. 11, the most dangerous condition for the converter in terms of crosstalk is when the converter is operating under high input voltage levels and light-load.

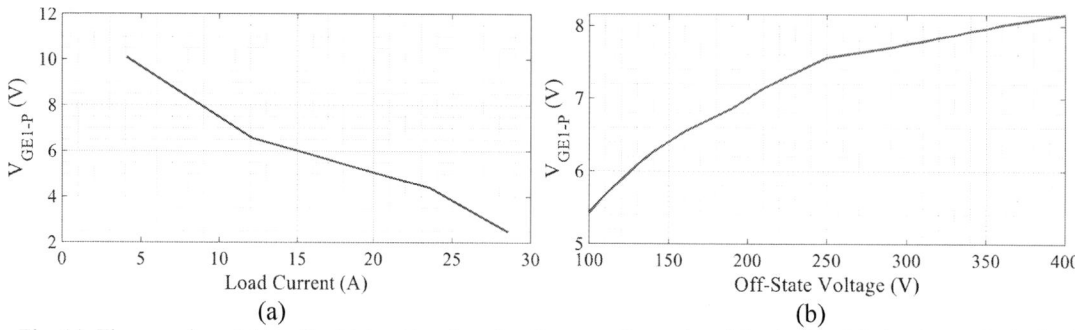

Fig. 11. The experimental results obtained by changing the operating point of the device. (a)-the change in the load current, and (b)- the change in the device off-state voltage.

Conclusion

Crosstalk remains a big challenge in power electronics converters including a phase-leg in their structure. Traditional models face problems in representing and analyzing the power path parasitic elements accurately. Thus, a new general model was introduced in this paper. The concentration is on the power path parasitic inductance and high-side switch current and off-state voltage. The crosstalk occurring in a converter arm is sensitive to the values of parasitic parameters. Hence, by changing the parasitic elements, their effect of them on crosstalk was analyzed. The considered switch nominal values are 600 V and 75 A. Therefore, the DC-link voltage in the case study is 400 V and the device current is 20 A. By increasing the output current, the peak value of the gate-emitter voltage decreases, which means crosstalk mostly has its highest values when there is a low power load in the output. The power path parasitic inductance is a significant element in crosstalk. Furthermore, there are several optimum regions in the curve due to the device model. Hence, the PCB designer must take ultra care of the power path inductance in order to mitigate the crosstalk effect on the arm. In addition, the parasitic capacitance ratio of IXGH60N60 is investigated. As a result, based on the application, the designer should opt for a device with a low parasitic capacitance ratio. Finally, the experimental results present so as to verify the proposed model and simulation results.

References

[1] S. Dieckerhoff, S. Bernet, and D. Krug, "Power loss-oriented evaluation of high voltage IGBTs and multilevel converters in transformerless traction applications," *IEEE Transactions on Power Electronics*, vol. 20, no. 6, pp. 1328–1336, Nov. 2005, doi: 10.1109/TPEL.2005.857534.

[2] M. Zarghani, S. Mohsenzade, A. Hadizade, and S. Kaboli, "An Extremely Low Ripple High Voltage Power Supply for Pulsed Current Applications," *IEEE Transactions on Power Electronics*, pp. 1–1, 2020, doi: 10.1109/TPEL.2020.2966682.

[3] S. Mohsenzade, M. Zarghani, and S. Kaboli, "A Voltage Balancing Scheme for Series IGBTs to Increase Their Expected Lifetime in Pulsed Load Applications," *IEEE Journal of Emerging and Selected Topics in Power Electronics*, vol. 9, no. 1, pp. 461–471, Feb. 2021, doi: 10.1109/JESTPE.2019.2958357.

[4] S. Mohsenzade, M. Zarghani, and S. Kaboli, "A Series Stacked IGBT Switch to Be Used as a Fault Current Limiter in HV High Power Supplies," *IEEE Journal of Emerging and Selected Topics in Power Electronics*, pp. 1–1, 2021, doi: 10.1109/JESTPE.2021.3061093.

[5] S. Mohsenzade, M. Zarghani, and S. Kaboli, "A Voltage Balancing Method for Series-Connected IGBTs Operating as a Fault Current Limiter in High-Voltage DC Power Supplies," *IEEE Transactions on Industrial Electronics*, vol. 68, no. 9, pp. 7895–7907, Sep. 2021, doi: 10.1109/TIE.2020.3009606.

[6] E. Deng *et al.*, "Research on the Multi-Physics Field-Circuit Coupling Model of Press Pack IGBT Considering the Application of Hybrid HVDC Breakers," *IEEE Journal of Emerging and Selected Topics in Power Electronics*, pp. 1–1, 2020, doi: 10.1109/JESTPE.2020.3019433.

[7] J. Zhang, J. Liu, S. Zhong, J. Yang, N. Zhao, and T. Q. Zheng, "A Power Electronic Traction Transformer Configuration With Low-Voltage IGBTs for Onboard Traction Application," *IEEE Transactions on Power Electronics*, vol. 34, no. 9, pp. 8453–8467, Sep. 2019, doi: 10.1109/TPEL.2018.2889107.

[8] X. Zhao *et al.*, "DC Solid State Transformer Based on Three-Level Power Module for Interconnecting MV and LV DC Distribution Systems," *IEEE Transactions on Power Electronics*, vol. 36, no. 2, pp. 1563–1577, Feb. 2021, doi: 10.1109/TPEL.2020.3007674.

[9] U.-M. Choi, F. Blaabjerg, and K.-B. Lee, "Study and Handling Methods of Power IGBT Module Failures in Power Electronic Converter Systems," *IEEE Transactions on Power Electronics*, vol. 30, no. 5, pp. 2517–2533, May 2015, doi: 10.1109/TPEL.2014.2373390.

[10] S. Mohsenzade, J. Naghibi, and K. Mehran, "Reliability Enhancement of Power IGBTs under Short-Circuit Fault Condition Using Short-Circuit Current Limiting-Based Technique," *Energies*, vol. 14, no. 21, Art. no. 21, Jan. 2021, doi: 10.3390/en14217397.

[11] W. W. T. Chan, K. O. Sin, P. K. T. Mok, and S. S. Wong, "A power IC technology with excellent cross-talk isolation," *IEEE Electron Device Letters*, vol. 17, no. 10, pp. 467–469, Oct. 1996, doi: 10.1109/55.537077.

[12] Z. Zhang, F. Wang, L. M. Tolbert, and B. J. Blalock, "Active Gate Driver for Crosstalk Suppression of SiC Devices in a Phase-Leg Configuration," *IEEE Transactions on Power Electronics*, vol. 29, no. 4, pp. 1986–1997, Apr. 2014, doi: 10.1109/TPEL.2013.2268058.

[13] Y. Li, M. Liang, J. Chen, T. Q. Zheng, and H. Guo, "A Low Gate Turn-OFF Impedance Driver for Suppressing Crosstalk of SiC MOSFET Based on Different Discrete Packages," *IEEE Journal of Emerging and Selected Topics in Power Electronics*, vol. 7, no. 1, pp. 353–365, Mar. 2019, doi: 10.1109/JESTPE.2018.2877968.

[14] H. Li, Y. Jiang, Z. Qiu, Y. Wang, and Y. Ding, "A Predictive Algorithm for Crosstalk Peaks of SiC MOSFET by Considering the Nonlinearity of Gate-Drain Capacitance," *IEEE Transactions on Power Electronics*, vol. 36, no. 3, pp. 2823–2834, Mar. 2021, doi: 10.1109/TPEL.2020.3016155.

[15] B. Zhang, S. Xie, J. Xu, Q. Qian, Z. Zhang, and K. Xu, "A Magnetic Coupling Based Gate Driver for Crosstalk Suppression of SiC MOSFETs," *IEEE Transactions on Industrial Electronics*, vol. 64, no. 11, pp. 9052–9063, Nov. 2017, doi: 10.1109/TIE.2017.2736500.

[16] M. Shen, "A simple gate assist circuit for SiC devices," in *2014 IEEE Conference and Expo Transportation Electrification Asia-Pacific (ITEC Asia-Pacific)*, Aug. 2014, pp. 1–6. doi: 10.1109/ITEC-AP.2014.6941017.

[17] Z. Zhang, J. Dix, F. F. Wang, B. J. Blalock, D. Costinett, and L. M. Tolbert, "Intelligent Gate Drive for Fast Switching and Crosstalk Suppression of SiC Devices," *IEEE Transactions on Power Electronics*, vol. 32, no. 12, pp. 9319–9332, Dec. 2017, doi: 10.1109/TPEL.2017.2655496.

[18] Z. Zhang, F. Wang, L. M. Tolbert, and B. J. Blalock, "Active Gate Driver for Crosstalk Suppression of SiC Devices in a Phase-Leg Configuration," *IEEE Transactions on Power Electronics*, vol. 29, no. 4, pp. 1986–1997, Apr. 2014, doi: 10.1109/TPEL.2013.2268058.

[19] S. Jahdi, O. Alatise, J. A. Ortiz Gonzalez, R. Bonyadi, L. Ran, and P. Mawby, "Temperature and Switching Rate Dependence of Crosstalk in Si-IGBT and SiC Power Modules," *IEEE Transactions on Industrial Electronics*, vol. 63, no. 2, pp. 849–863, Feb. 2016, doi: 10.1109/TIE.2015.2491880.

[20] K. Hasegawa, K. Wada, and I. Omura, "Mutual inductance measurement for power device package using time domain reflectometry," in *2016 IEEE Energy Conversion Congress and Exposition (ECCE)*, Milwaukee, WI, USA, Sep. 2016, pp. 1–6. doi: 10.1109/ECCE.2016.7855285.

[21] B. Ji *et al.*, "In Situ Diagnostics and Prognostics of Solder Fatigue in IGBT Modules for Electric Vehicle Drives," *IEEE Transactions on Power Electronics*, vol. 30, no. 3, pp. 1535–1543, Mar. 2015, doi: 10.1109/TPEL.2014.2318991.

[22] U. Karki, N. S. González-Santini, and F. Z. Peng, "Effect of Gate-Oxide Degradation on Electrical Parameters of Silicon Carbide MOSFETs," *IEEE Transactions on Electron Devices*, vol. 67, no. 6, pp. 2544–2552, Jun. 2020, doi: 10.1109/TED.2020.2990128.

[23] S. Mohsenzade, M. Zarghany, and S. Kaboli, "A Series Stacked IGBT Switch With Robustness Against Short-Circuit Fault for Pulsed Power Applications," *IEEE Transactions on Power Electronics*, vol. 33, no. 5, pp. 3779–3790, May 2018, doi: 10.1109/TPEL.2017.2712705.

[24] "Physical analysis and modeling of the nonlinear miller capacitance for SiC MOSFET | IEEE Conference Publication | IEEE Xplore." https://ieeexplore.ieee.org/document/8216240 (accessed Jun. 14, 2022).

[25] T. H. Hubing, T. P. Van Doren, and J. L. Drewniak, "Identifying and quantifying printed circuit board inductance," in *Proceedings of IEEE Symposium on Electromagnetic Compatibility*, Aug. 1994, pp. 205–208. doi: 10.1109/ISEMC.1994.385661.

[26] A. Azam Rajabian and S. Mohsenzade, "Investigating the Effect of the Power Path Parasitic Inductance on Si-IGBT Crosstalk Using a Comprehensive Model," in *2022 13th Power Electronics, Drive Systems, and Technologies Conference (PEDSTC)*, Feb. 2022, pp. 414–419. doi: 10.1109/PEDSTC53976.2022.9767324.

Impact of Higher Current Harmonics on Component Current Stress and Conduction Losses of Half-Bridge-Series-Resonant-Converters in Discontinuous Conduction Mode for High-Power Applications

Daniel Haake[1], Anton Grodnichev[1], Fabian Schnabel[1], Marco Jung[1,2]

[1] Fraunhofer Institute for Energy Economics
and Energy System Technology IEE
Joseph-Beuys-Straße 8
Kassel, Germany
Tel.: +49 561 7294 1589
daniel.haake@iee.fraunhofer.de
https://www.iee.fraunhofer.de/

[2] Hochschule Bonn-Rhein-Sieg
University of Applied Sciences
Grantham-Allee 20
Sankt Augustin, Germany
Tel. +49 2241 865 316
Marco.Jung@h-brs.de
https://www.h-brs.de

Acknowledgements

The authors acknowledge the support of the presented work by the German Federal Ministry for Economic Affairs and Climate Action within the project "MUSiCel: Mobile Umrichter und Energieübertragungslösungen auf SiC Basis für elektrische, leistungsstarke Land- und Baumaschinen" (FKZ 03EN2014E). Only the authors are responsible for the content of this publication.

Keywords

«DC power supply», «Isolated converter», «Resonant converter», «Silicon Carbide (SiC)», «ZCZVS converters»

Abstract

The half-bridge series resonant converter (HB-SRC) offers advantages with regard to power density compared to hard-switching topologies. To achieve soft-switching, the HB-SRC has to operate in discontinuous conduction mode (DCM). However, this operation influences component current waveforms and thus the component current stresses and losses in the converter, which are examined analytically and simulatively in this paper.

Introduction

Galvanically isolated DC-DC converters with high output power are used today in a wide variety of areas, from smart transformers in energy supply for AC and DC grids [1]-[3] to railway applications [4] and electrified agricultural machinery [5]. Series resonant converters (SRC) offer an advantage in achieving high power densities due to their low switching losses [5]. When silicon carbide power semiconductors are used, high switching frequencies can be achieved, which lead to an increase in power density. As shown in [1], [3]-[5], the half-bridge series resonant converter (HB-SRC) in discontinuous conduction mode (DCM), often referred to as half-cycle discontinuous-conduction-mode series resonant converter (HC-DCM-SRC [1], [3]), is particularly suitable due to the lower amount of required hardware in order to be able to implement the very compact systems with high performance. However, when the HB-SRC is operated to achieve soft switching, non-sinusoidal currents occur ([1], [3], [4]) which could lead to a massive increase in component current stress, particularly in high-power applications, and must therefore be taken into account throughout the design of the converter.

Soft switching Operation of the HB-SRC in DCM

The HB-SRC topology shown in Fig. 1 is examined. The associated waveforms of the currents and voltages in steady-state operation are shown in Fig. 2. For the following considerations, it is assumed that the input and output currents of the converter are approximately constant in steady-state operation. This applies, for example, in case that source and load of the HB-SRC are connected via inductances, for example through other converter stages [1] or sufficiently long cable connection [5].Furthermore, an energy flow from the HV-Side to the LV-Side is considered. It is also assumed that the transformer´s magnetization current i_{mag} is completely provided by the converter's HV-Side.

The HB-SRC topology (Fig. 1) consists of a primary-side half bridge with a split DC link at the converters input, a transformer, a resonance capacitance and a secondary-side full bridge, which is connected to the output capacitance of the converter. The equivalent circuit describing the transformer consists of an ideal transformer with the turn ratio N, a magnetizing inductance L_h, and a leakage inductance L_σ as described in [6]. The resonance tank of the HB-SRC is formed by the transformer's leakage inductance L_σ and the resonance capacitance C_{res}.

In the secondary-side full bridge, MOSFETs are used instead of diodes to allow a bidirectional power transfer through the converter. To reduce conduction losses, the full bridge works in synchronous rectification mode. The primary-side half bridge is controlled as described as follows. To ensure zero current switching (ZCS), the resonance frequency f_{res} is selected to be higher than the switching frequency f_{SW}, so that the resonant pulse of the current is completed before half the switching period is reached (see Fig. 2c), so-called "sub-resonant operation mode" [3]. After the resonant pulse is completed, the relatively small transformer magnetizing current i_{mag} continues to flow through the previously conductive semiconductors. If the magnetizing current is sufficiently large, it can be used to charge the output capacitances of the semiconductors and thus achieve zero voltage switching (ZVS) for turn on of the semiconductors (see [1], [7]).

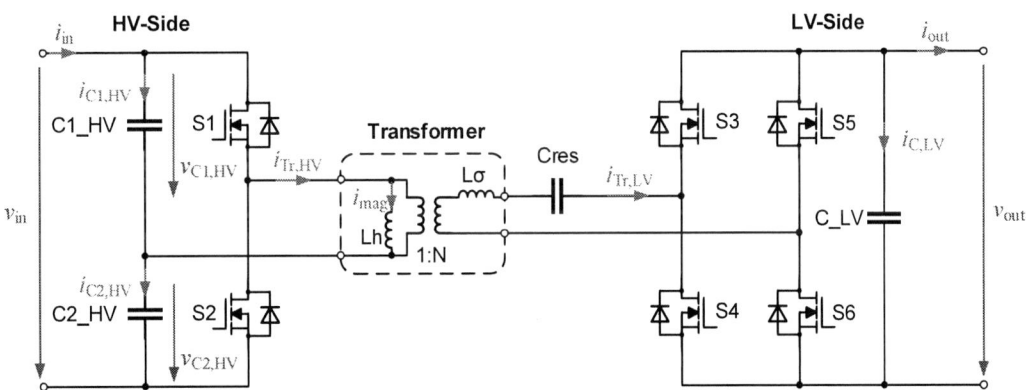

Fig. 1: HB-SRC topology

Due to the very small magnetization inductance L_h of the transformer that is required for this operation, the magnetizing current is limited by the smallest practically feasible inductance value. In order to still be able to provide soft switching when using power modules with large output capacitance values e.g. for high-power applications, a significant dead time between the gate signals of the upper and the lower switches is required. Large parasitic capacitances in the power switches can require a dead time of 10 % or higher in relation to the switching period T_{SW} to achieve proper soft switching operation. This is especially the case for converters that combine high switching frequencies with high output power. Therefore, a significant increase of the resonance frequency in relation to the switching frequency can sometimes be necessary (see Fig. 2c). The resonance tank of the converter, which consists of the resonance capacitance C_{res} and leakage inductance L_σ of the transformer, must be designed accordingly in order to set the desired resonance frequency, as described in [1].

However, increasing the resonance frequency while the switching frequency remains the same affects the shape of the currents through the components of the converter and thus their RMS values and frequency spectra. This leads to an increased current stress on the capacitors, the transformer and the power semiconductors. Although the power transmitted by the converter remains almost constant, the losses occurring in its components increase. These relationships are examined analytically and through simulations in the following sections.

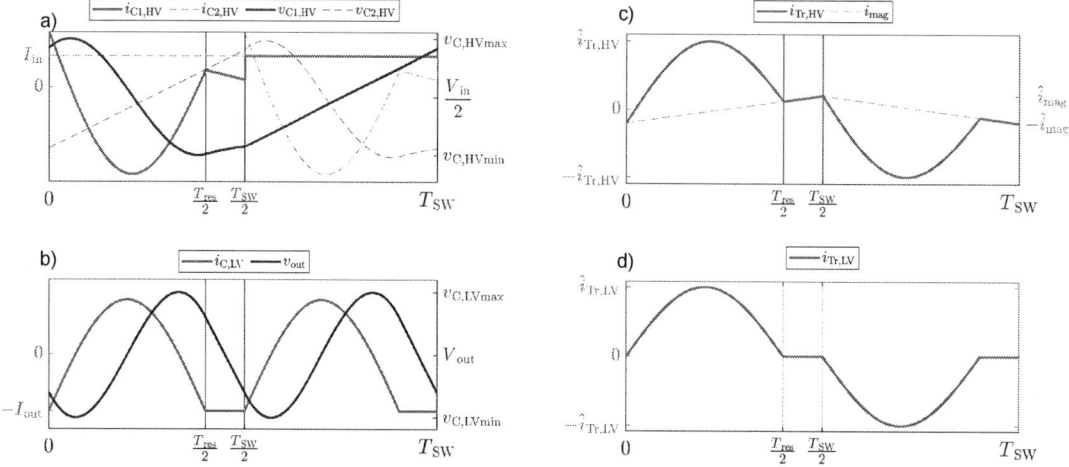

Fig. 2: Typical waveforms (qualitative) of HB-SRC: a) input capacitors (voltages and current), b) output capacitor (voltages and current), c) transformer primary side (current), d) transformer secondary side (current)

Semiconductor loss analysis

To analyze semiconductor losses, an analytical calculation method is derived and validated by simulations using PLECS. Assuming that the primary-side half-bridge always works in soft-switching operation and the secondary-side full bridge works as an ideal synchronous rectifier, consideration of the conduction losses is sufficient.

Table I: Simulation Parameters

Input voltage V_{in}	1900 V
Output voltage V_{out}	665 V
Output power P_{out}	250 kW
Switching frequency f_{SW}	50 kHz
Junction temperature T_j	125 °C

Using the waveform of the currents flowing throw the MOSFETs of the HV-Side (Fig. 3 left) and the LV-Side (Fig. 3 right), RMS values of these currents can be calculated analytically according to (1) and (3). With these RMS values, the conduction losses of the HV-Side and LV-Side MOSFETs could be calculated using (2) and (4), respectively. Note that the magnetization current i_{mag} is neglected for these calculations.

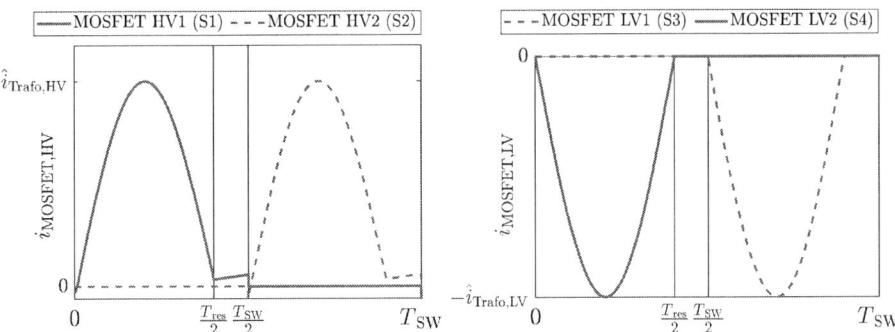

Fig. 3: MOSFET currents HV-Side (left) and LV-Side (right)

In Fig. 4 a) and c), calculated and simulated conduction losses for an application example of an HB-SRC for an electrified wired agricultural machinery presented in [5] are shown. The system parameters used are summarized in Table I and correspond to the application in [5]. In this application, SiC MOSFETs are used both on the primary side and on the secondary side. Energy flows from the HV-Side to the LV-Side of the converter.

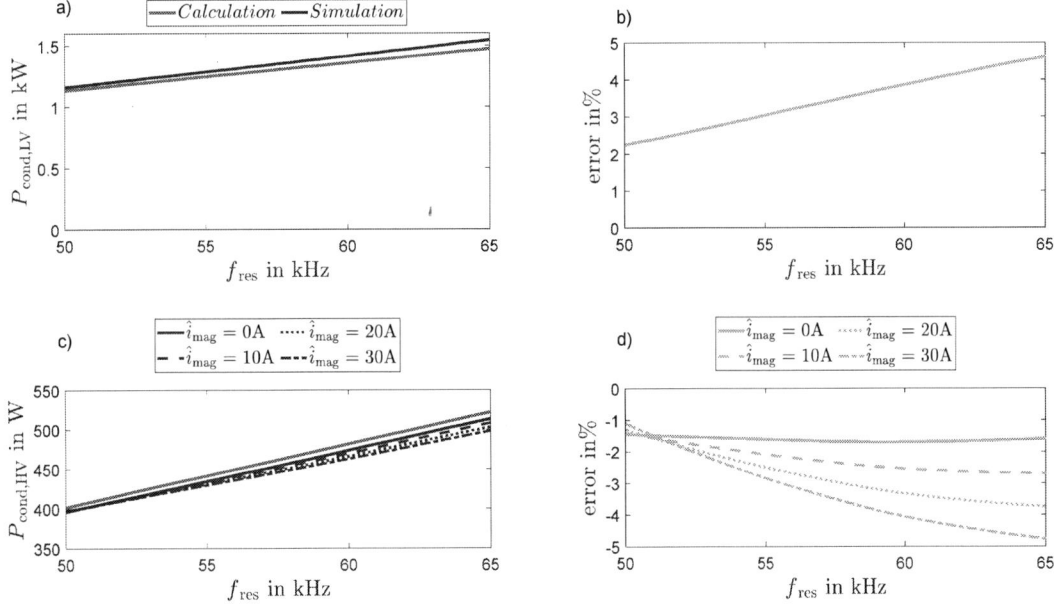

Fig. 4: Calculated and simulated conduction losses vs. resonance frequency f_{res}: a) HV-Side, b) HV-Side calculation error c) LV-Side and d) LV-Side calculation error

Fig. 4 shows the simulated power loss for the HV-Side (c) and the LV-Side (a). It can be seen that the conduction losses increase almost linearly with the resonance frequency for the HV-Side as well as for the LV-Side of the converter. This is true for both the calculated (red) and simulated (blue) conduction losses. For a general overview, the quantitative behavior of the losses is first described using the calculated values, before the deviations between the calculation and the simulation are discussed.

For example, with an increase of the resonance frequency f_{res} from 54 kHz to 62.5 kHz (15.7 %), the calculated conduction losses increase from 434 W to 502 W for the HV-Side module and from 1.224 kW to 1.417 kW for both LV-Side modules. This results in an increase in semiconductor conduction losses of 15.6 % and 15.8 %, respectively, for the same output power of the converter.

As shown in detail in Fig. 4 b) and d), there is a deviation between the calculation and the simulation. On the LV-Side, this is mainly due to the fact that the $R_{DS,on}$ used in the simulation is not constant but depends on the current through the respective MOSFET, which in this case leads to a higher calculation error as the resonance frequency increases. This is due to the fact that the amplitude of the current through the MOSFET increases with increasing resonance frequency and the deviation between the constant $R_{DS,on}$ used in the calculation and the $R_{DS,on}$ curve used in the simulation increases. This means that, for example, the error increases from 2.88 % at a resonance frequency of 54 kHz to 4.26 % at 62.5 kHz.

For the HV-Side however, neglecting the magnetization current i_{mag} leads to an additional error. The simulated losses for different magnetizing currents are compared with the calculated ones in Fig. 4 c). The corresponding calculation error can be seen in Fig. 4 d). With no magnetizing current ($\hat{\imath}_{mag} = 0$ A), the cause of the deviation is the same as for the LV-Side. However, an increasing magnetizing current leads to higher calculation errors. For example, with a resonance frequency of 62.5 kHz and a magnetizing current amplitude $\hat{\imath}_{mag}$ of 30 A, this leads to a deviation of -4.46 %, while without magnetizing current the calculation error is -1.66 %.

$$I_{\text{MOSFET,HV RMS}} = \sqrt{\frac{1}{T_{\text{SW}}} \cdot \int_0^{T_{\text{SW}}} i^2_{\text{MOSFET,HV}} \, dt} = I_{in} \cdot \frac{\pi}{2} \cdot \sqrt{\frac{f_{\text{res}}}{f_{\text{SW}}}}$$

$$\text{with } i_{\text{MOSFET,HV}} = \begin{cases} \hat{\imath}_{\text{Tr,HV}} \cdot \sin(\omega_{\text{res}} \cdot t) \, , & 0 < t \le \frac{T_{\text{res}}}{2} \\ 0 & , \frac{T_{\text{res}}}{2} < t \le \frac{T_{\text{SW}}}{2} \end{cases}$$

(1)

$$P_{\text{cond,HV}} = 2 \cdot R_{\text{DSon,HV}} \cdot I^2_{\text{MOSFET,HV RMS}}$$

(2)

$$I_{\text{MOSFET,LV RMS}} = \sqrt{\frac{1}{T_{\text{SW}}} \cdot \int_0^{T_{\text{SW}}} i^2_{\text{MOSFET,LV}} \, dt} = I_{out} \cdot \frac{\pi}{4} \cdot \sqrt{\frac{f_{\text{res}}}{f_{\text{SW}}}}$$

$$\text{with } i_{\text{MOSFET,LV}} = \begin{cases} -\hat{\imath}_{\text{Tr,LV}} \cdot \sin(\omega_{\text{res}} \cdot t) \, , & 0 < t \le \frac{T_{\text{res}}}{2} \\ 0 & , \frac{T_{\text{res}}}{2} < t \le \frac{T_{\text{SW}}}{2} \end{cases}$$

(3)

$$P_{\text{cond,LV}} = 4 \cdot R_{\text{DSon,LV}} \cdot I^2_{\text{MOSFET,LV RMS}}$$

(4)

Analytical Calculation of Passive Components Current Stress

The waveforms of the currents through the passive components (Fig. 2) are used for the analytical determination of the RMS current load and the frequency spectra of the component currents. The magnetization current i_{mag} of the transformer can usually be neglected in high-power applications due to its small magnitude compared to the current through the transformer.

$$I_{\text{C1,HV RMS}} = I_{\text{C2,HV RMS}} = I_{\text{C,HV RMS}} = \sqrt{\frac{1}{T_{\text{SW}}} \cdot \int_0^{T_{\text{SW}}} i_{\text{C1,HV}}^2 \, dt} = I_{in} \cdot \sqrt{\frac{\pi^2}{4} \cdot \frac{f_{\text{res}}}{f_{\text{SW}}} - 1}$$

$$\text{with } i_{\text{C1,HV}} = \begin{cases} I_{in} - \hat{\imath}_{\text{Tr,HV}} \cdot \sin(\omega_{\text{res}} \cdot t), & 0 < t \leq \frac{T_{\text{res}}}{2} \\ I_{in}, & \frac{T_{\text{res}}}{2} < t \leq T_{\text{SW}} \end{cases}$$

(5)

$$\hat{\imath}_{\text{C1,HV}}(n) = \hat{\imath}_{\text{C2,HV}}(n) = \hat{\imath}_{\text{C,HV}}(n) = 2 \cdot I_{in} \cdot \frac{\cos\left(\frac{\pi}{2} \cdot n \cdot \frac{f_{\text{SW}}}{f_{\text{res}}}\right)}{\left(n \cdot \frac{f_{\text{SW}}}{f_{\text{res}}}\right)^2 - 1}$$

(6)

The analytical expressions used to calculate the RMS values and the frequency spectra of the currents through the components the general equations for calculating a currents RMS value and the Fourier analysis of periodic signals according to [8] are applied to the current waveforms in Fig. 2. This enables the calculation of the RMS values and the frequency spectra of the current for the input capacitors according to (5) and (6), for the transformer (secondary side) according to (7) and (8) and for the output capacitor according to (9) and (10). It should be noted that the RMS values of the primary and secondary transformer currents differ only by the ratio of the input and output currents of the converter, while the frequency spectra only differ through the transformers turn ratio (when the magnetization current i_{mag} is neglected, as mentioned before). Therefore, it is sufficient to discuss the transformers secondary side current here.

$$I_{\text{Tr,LV RMS}} = \sqrt{\frac{1}{T_{\text{SW}}} \cdot \int_0^{T_{\text{SW}}} i_{\text{Tr,LV}}^2 \, dt} = \sqrt{\frac{2}{T_{\text{SW}}} \cdot \int_0^{\frac{T_{\text{SW}}}{2}} i_{\text{Tr,LV}}^2 \, dt} = I_{out} \cdot \sqrt{\frac{\pi^2}{8} \cdot \frac{f_{\text{res}}}{f_{\text{SW}}}}$$

$$\text{with } i_{\text{Tr,LV}} = \begin{cases} \hat{\imath}_{\text{Tr,LV}} \cdot \sin(\omega_{\text{res}} \cdot t), & 0 < t \leq \frac{T_{\text{res}}}{2} \\ 0, & \frac{T_{\text{res}}}{2} < t \leq \frac{T_{\text{SW}}}{2} \end{cases}$$

(7)

$$\hat{\imath}_{\text{Tr,LV}}(n) = I_{out} \cdot \frac{\sin\left(\frac{\pi}{2} \cdot n \cdot \left(\frac{f_{\text{SW}}}{f_{\text{res}}} + 1\right)\right) - \sin\left(\frac{\pi}{2} \cdot n \cdot \left(\frac{f_{\text{SW}}}{f_{\text{res}}} - 1\right)\right)}{\left(n \cdot \frac{f_{\text{SW}}}{f_{\text{res}}}\right)^2 - 1}$$

(8)

In Fig. 5 a) and b), the RMS values of the currents through the passive components are normalized to the input current i_{in} and the output current i_{out} of the converter, respectively, and plotted over the ratio of the resonance frequency f_{res} and the switching frequency f_{SW}. It can be seen that all current RMS values increase with an increasing ratio of resonance frequency to switching frequency. For example, if this ratio is increased from 1.05 to 1.3 (23.8 % increase), the RMS value of the current through the capacitors on the HV-Side normalized to the input current i_{in} increases by 17.8 % and for the transformer current normalized in the same way by 11.3 %. For the transformer current of the secondary side, normalized to the output current i_{out} of the converter, this leads to an increase of 11.3 %, while the LV-Side capacitor's normalized current increases by 43 %. It should be noted that the ohmic losses occurring in the components depend on the square of the RMS values of the currents (frequency dependency neglected) and thus even relatively small increases lead to an even higher increase in losses.

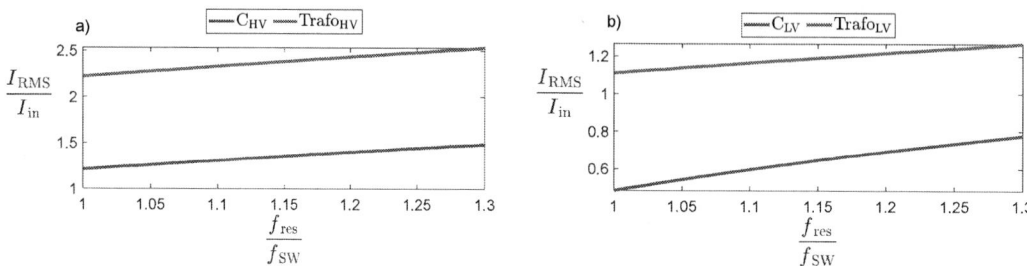

Fig. 5: Component RMS current vs. the ratio of f_{res} and f_{SW}: a) capacitor and transformer HV-Side, b) transformer and capacitor LV-Side

However, since the losses occurring in the passive components depend on the spectra of the currents (see [9], [11]), it is also necessary to consider the change of the spectra when the resonance frequency is increased. In particular, this effect applies to the transformer ([10], [11]). In Fig. 6 a) to c), the three most dominant harmonics, in the respective component currents are plotted over the ratio of resonance frequency f_{res} to switching frequency f_{SW}. In the current through the capacitors on the HV-Side, both even and odd harmonics occur, which all increase with increasing resonance frequency (Fig. 6 a)). Only odd harmonics occur in the transformer currents (Fig.6 b)). In particular, the third and fifth harmonics rise massively with increasing resonance frequency over the shown range. In the current through the capacitor on the LV-Side, only even harmonics appear (Fig. 6 c)).
In the area shown, only the first harmonic increases while the fourth and sixth harmonic even partially decrease.

$$I_{C,LV\ RMS} = \sqrt{\frac{1}{T_{SW}} \cdot \int_0^{T_{SW}} i_{C,LV}^2 \, dt} = \sqrt{\frac{2}{T_{SW}} \cdot \int_0^{\frac{T_{SW}}{2}} i_{C,LV}^2 \, dt} = I_{out} \cdot \sqrt{\frac{\pi^2}{8} \cdot \frac{f_{res}}{f_{SW}} - 1}$$

$$\text{with } i_{C,LV} = \begin{cases} \hat{\imath}_{Tr,LV} \cdot \sin(\omega_{res} \cdot t) - I_{out}, & 0 < t \le \frac{T_{res}}{2} \\ -I_{out} & , \frac{T_{res}}{2} < t \le \frac{T_{SW}}{2} \end{cases}$$

(9)

$$\hat{\imath}_{C,LV}(n) = 2 \cdot I_{out} \cdot \frac{\cos\left(\pi \cdot n \cdot \frac{f_{SW}}{f_{res}}\right)}{\left(2 \cdot n \cdot \frac{f_{SW}}{f_{res}}\right)^2 - 1}$$

(10)

As shown for the semiconductor losses, there is also a deviation between simulation and calculation for the passive component of the HV-Side due to the neglected magnetizing current. This affects the calculations of the RMS values and spectra of the HV-Side capacitors as well as the transformer's primary side currents.
To classify this calculation error quantitatively, numerical simulations in PLECS were carried out, again using the application example of an HB-SRC for an electrified wired agricultural machinery, in which $\hat{\imath}_{mag}$ was varied between 0 A and 30 A in 10 A steps. The resulting absolute (left) and relative (right) errors between simulations and calculations are shown in Fig. 7 for the RMS values (a) and the most dominant harmonics (b - c) of HV-Side transformer and capacitor currents.
As shown in Fig. 7, in case of no magnetization current ($\hat{\imath}_{mag} = 0$ A) the calculation error is zero for the whole investigated range of f_{res}, which is true for the RMS values as well as for the harmonics shown. Thus, it can be assumed that the calculation error is only caused by neglecting the magnetization current.

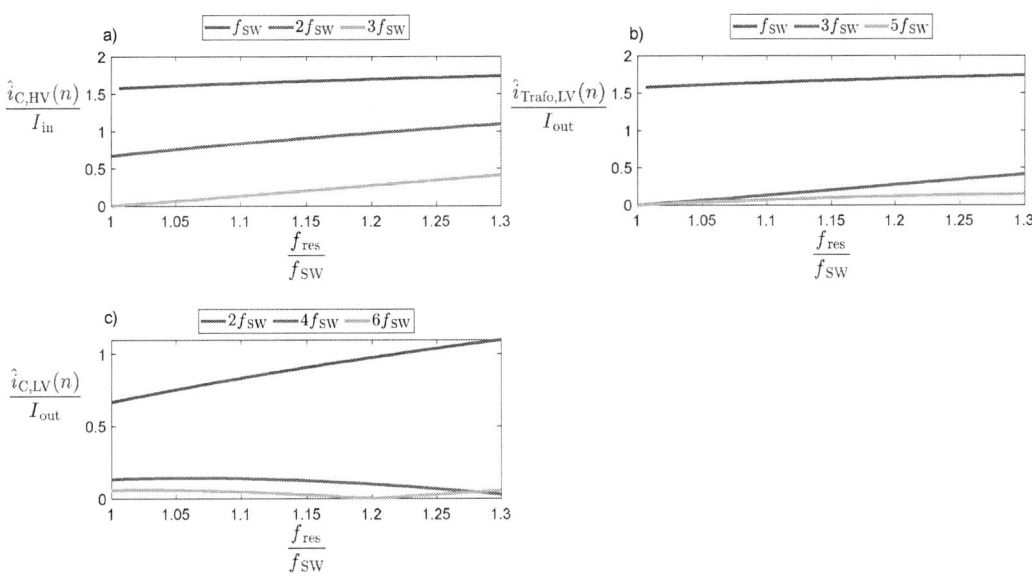

Fig. 6: Component current harmonics vs. the ratio of f_{res} and f_{SW}: a) capacitor HV-Side, b) transformer LV-Side, c) capacitor LV-Side

In Fig. 7 a) and b), the absolute and relative calculation errors of the RMS values and the first harmonics are plotted. For ratios of resonance frequency to switching frequency slightly above unity (approx. 1.05), the calculation error is positive and decreases to negative values with increasing resonance frequency f_{res} for both the transformer and capacitor currents (RMS values and first harmonic). With higher values of the magnetizing current amplitude $\hat{i}_{mag}$, the calculation error also increases, and this time almost linearly. Both observations also apply to the second harmonic of the capacitor current shown in Fig. 7 c).

For the third and the fifth harmonic of the transformer current as well as the third harmonic of the capacitor current, the calculation error is positive over the whole investigated range, as shown in Fig. 7 c) and d). First, the absolute error decreases to a minimum near the switching frequency ($f_{SW}/f_{res} = 1$) and then increases again with increasing resonance frequency, but with a declining slope. This leads to an approximately hyperbolic relative error curve.

Minimum required Capacitance for input and output Capacitor

The minimum capacitance required in order to not exceed a certain voltage ripple for both the input and the output side is a function of the switching and resonance frequency as well as the maximum power transmitted by the converter. The shifted charge can be calculated by integrating the current through the respective capacitor (see Fig. 2 a) and b)). For a specified voltage ripple, the minimum capacitance required for the capacitors on the HV- and LV-Side side can be calculated according to (11) and (12), respectively.

$$C_{HV} = \frac{\Delta Q_{C,HV}}{\Delta V_{C,HV}} = \frac{I_{in,max}}{\Delta V_{C,HV}} \cdot \left[\frac{1}{f_{SW}} \cdot \sqrt{1 - \left(\frac{1}{\pi} \cdot \frac{f_{SW}}{f_{res}} \right)^2} - \frac{1}{2 \cdot f_{res}} + \frac{1}{\pi \cdot f_{res}} \arcsin\left(\frac{1}{\pi} \cdot \frac{f_{SW}}{f_{res}} \right) \right] \quad (11)$$

$$C_{LV} = \frac{\Delta Q_{C,LV}}{\Delta V_{out}} = \frac{I_{out,max}}{2 \cdot \Delta V_{out}} \cdot \left[\frac{1}{f_{SW}} \cdot \sqrt{1 - \left(\frac{2}{\pi} \cdot \frac{f_{SW}}{f_{res}} \right)^2} - \frac{1}{f_{res}} + \frac{2}{\pi \cdot f_{res}} \arcsin\left(\frac{2}{\pi} \cdot \frac{f_{SW}}{f_{res}} \right) \right] \quad (12)$$

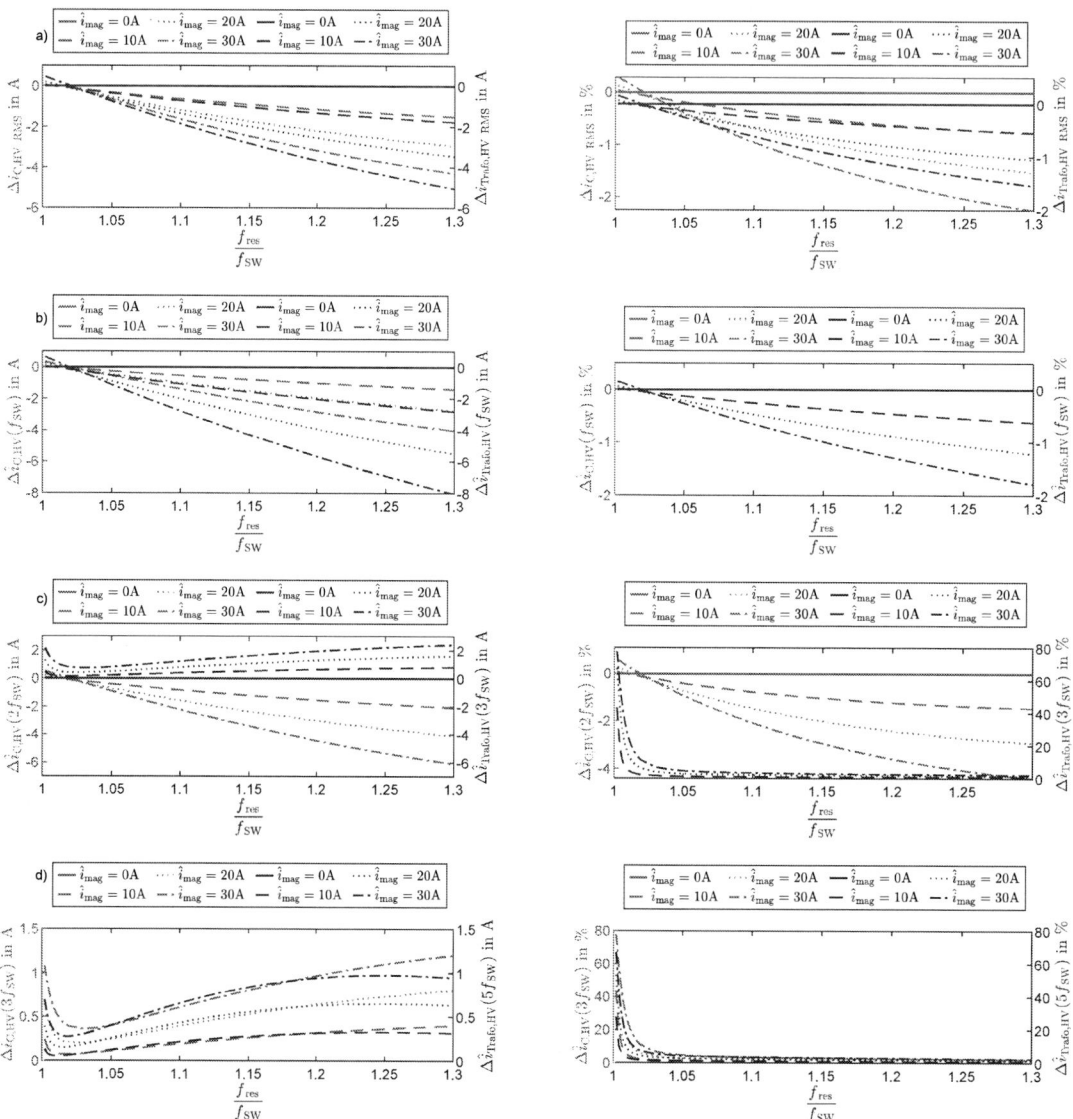

Fig. 7: Calculation error HV-Side capacitor and transformer vs. the ratio of f_{res} and f_{SW} (I_{out} = 376 A): a) RMS values, b) first harmonic (f_{SW}) c) capacitor second harmonic ($2f_{SW}$) and transformer third harmonic ($3f_{SW}$), d) capacitor third harmonic ($3f_{SW}$) and transformer fifth harmonic ($5f_{SW}$)

Conclusion

The influence of the ratio of the resonance frequency and the switching frequency on semiconductor conduction losses and component current stress in the discontinuous conduction mode (sub-resonant operation mode) of the HB-SRC has been investigated. For this purpose, the analytical equations to determine the semiconductor conduction losses as well as the RMS current loads and the frequency spectra of the respective currents in the passive components have been derived and analyzed. The calculation errors have been investigated for an application example using numerical simulations. Furthermore, analytical equations to calculate the minimum necessary capacitance values for the capacitors on the HV- and LV-Side were derived.

References

[1] D. Rothmund, J. Huber, J. Kolar: "Operating Behavior and Design of the Half-Cycle Discontinuous-Conduction-Mode Series-Resonant Converter with Small DC Link Capacitors", 2013 IEEE 14th Workshop on Control and Modeling for Power Electronics (COMPEL)

[2] R. Unruh, F. Schafmeister, N. Fröhleke, J. Böckler: "MMC-Topology for High-Current and Low-Voltage Applications with Minimal Number of Submodules, Reduced Switching and Capacitor Losses", 2019 PCIM Europe

[3] G. Ortiz, H. Uemura, D. Bortis, J. Kolar, O. Apeldoorn: "Modeling of Soft-Switching Losses of IGBTs in High-Power High-Efficency Dual-Active-Bridge DC/DC Converters", in IEEE Transactionson Electron Devices, VOL. 60, NO. 2, February 2013

[4] C. Stackler, A. Fouineau, P. Ladoux, F. Morel, F. Wallart, P Dworakowski, N. Evans: "NPC assessment in insulated DC/DC converter topologies using SiC MOSFETs for Power Electronic Traction Transformer", 2019 20th International Symposium on Power Electronics (Ee)

[5] D. Tatusch, A. Gorodnichev, D. Haake, F. Schnabel, J. Friebe, M. Jung: "Hardware and control design considerations for a mobile 1 MW Input-Series Output-Parallel (ISOP) DC-DC converter in Medium Voltage range", IEEE Energy Conversion Congress and Expo 2021 (ECCE 2021)

[6] T. Guillod, D. Rothmund, J. W. Kolar: "Active Magnetizing Current Splitting ZVS Modulation of a 7 kV/400 V DC Transformer", IEEE Transactions on Power Electronics, VOL 35, No. 2, February 2020

[7] J. Huber, G. Ortiz, F. Krismer, N. Widmer, J. W. Kolar: "η-ρ Pareto Optimization of Bidirectional Half-Cycle Discontinuous-Conduction-Mode Series-Resonant DC/DC Converter with Fixed Voltage Transfer Ratio", 2013 Twenty-Eighth Annual IEEE Applied Power Electronics Conference and Exposition (APEC)

[8] F. Zach: "Leistungselektronik: Ein Handbuch Band 1", 4. Auflage, 2010 Springer

[9] Y. Yang, K. Ma, H. Wang, F. Blaabjerg: "Instantaneous thermal modeling of the DC-link capacitor in PhotoVoltaic systems", 2015 IEEE Applied Power Electronics Conference and Exposition (APEC)

[10] N. Kimura, K. Nakao, T. Morizane: "Loss Analysis and Temperature Measurement of Middle Frequency Transformer Applied for Solid State Transformer", 2019 8th International Conference on Renewable Energy Research and Applications (ICRERA)

[11] S. Wang, D. Dorell: "Copper Loss Analysis of EV Charging Coupler", IEEE Transactions on Magnetics VOL 51, Issue 11, November 2015

Control of a Zero-Voltage Switching Isolated Series-Resonant Power Circuit for Direct 3-phase AC to DC Conversion

Yusuf Kosesoy, Remco Bonten, Henk Huisman, Jan Schellekens
Eindhoven University of Technology
Eindhoven, The Netherlands
y.k.kosesoy@tue.nl

Acknowledgement

This project has received funding from the ECSEL Joint Undertaking (JU) under grant agreement No 101007281. The JU receives support from the European Union's Horizon 2020 research and innovation programme and Austria, Germany, Slovenia, Netherlands, Belgium, Slovakia, France, Italy, Turkey.

Keywords

≪Soft Switching≫, ≪ZVS Converter≫, ≪Resonant Converter Control≫, ≪Fast Transient Response≫.

Abstract

A novel control method is presented for a fully zero-voltage switching series-resonant isolated 3-phase AC to DC converter. The control is derived from first principles such as energy and charge conservation. The approach results in operation with a high power factor at the AC grid side and is beneficial in terms of EMI due to soft switching and low dv/dt across the switches even when extremely fast WBG devices such as SiC or GaN are used.

Introduction

The application of the newest generation of Wide-Bandgap (WBG) devices in classical, hard-switching power circuits leads to extremely steep voltages at the switching nodes and as a consequence to large amounts of electromagnetic interference (EMI). With such circuits, switching times in the order of a few nanoseconds are observed [1], and voltage slopes up to several tens and even above one hundred volts per ns [2]. These very steep slopes give rise to a large amount of both differential and common-mode EMI. Typically, EMI in power converters is attenuated by means of filters, which represent between 20-40% of the volume and weight of the total assembly [3]. With the steep slopes caused by hard-switching WBG devices, the design of such filters becomes increasingly difficult, as parasitic elements in the filter components become more relevant. Due to the presence of inter-winding capacitance in inductors and parasitic series inductance in capacitors, a filter which was designed to show low-pass behaviour actually becomes high-pass at the frequencies related to the very fast switching, thereby losing (part of) its function.

It can be expected that due to the developments described above, the relative volume, weight and cost of EMI filters will have to increase, which is not a very attractive outlook: the gain in efficiency and size of the core power processor due to using WBG devices is at least partially lost due to increased filter size and losses. Therefore, in this paper, we will present an approach which attempts to avoid hard-switching and thereby eliminate the root cause for the high-frequency EMI. Similar approaches (ZVS) have already been shown for some classes of DC-DC converters such as full/half bridge driven resonant circuits [4]-[6], DAB circuits [8, 7] and SR circuits [9]. The approach will be illustrated by means of a direct three-phase AC to DC converter. The circuit of this converter was proposed in [10], but a suitable control law to facilitate soft switching and experimental results are still missing.

Circuit operation

The proposed circuit is shown in Figure 1. It consists of a 4-phase voltage selector using 8 MOSFET devices (Q1..Q8) which are pair-wise connected in anti-series, a resonant tank (C_{res}, L_{res}), an isolation transformer also for scaling and a diode rectifier bridge supplying the load. The AC supply is represented as three ideal voltage sources ($V_R \cdots V_T$) combined with the neutral ($V_Z = 0$). In the sequel we will assume the AC supply to be balanced, in particular that the sum of the AC voltages is zero.

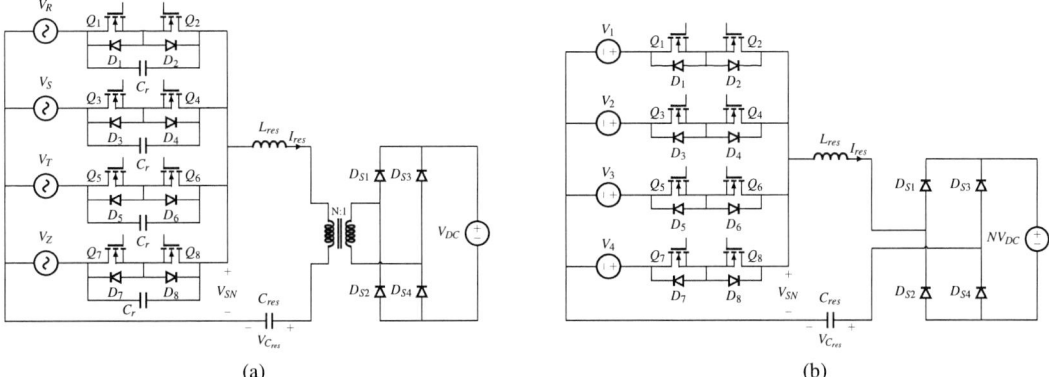

(a)   (b)

Fig. 1: Direct three-phase to DC converter (a) proposed circuit; (b) simplified circuit for control method.

We will address the operation of the circuit as shown in Figure 1a during one cycle of the resonant tank, i.e. from one rising zero crossing of the resonant current I_{res} to the next. To simplify the analysis, we will assume that the voltages both at the supply and load sides are constant for the duration of the resonant cycle, and can therefore be represented by DC sources. Also, the transformer, which is assumed to be ideal and has turns ratio N, is removed with a corresponding change in naming for the DC source representing the load ($V_{DC} \to NV_{DC}$). The effect of the device capacitances is represented by capacitances C_r. Note that it is possible to place extra capacitance here to further reduce the value of dV/dt. However, the influence of these capacitances and losses in the circuit are ignored for the analysis in this research.

The voltages at the supply side (including the neutral voltage) will be sorted from most positive to most negative, and receive a numerical subscript for the analysis to follow. Once these simplifications/modifications have been implemented, the circuit can be depicted as shown in Figure 1b. Due to the sorting of the voltages, V_1 and V_2 will always be non-negative, and V_3 and V_4 non-positive.

To obtain zero voltage switching (ZVS) in the half-cycle where $I_{res} > 0$, the generalized supply voltages need to be applied starting with the most positive (i.e. V_1) and ending with the most negative (V_4). For the other half-cycle, where $I_{res} < 0$, the inverse applies: this half-cycle needs to start at V_4 and end at V_1. The behaviour over a complete resonant cycle can be depicted as shown in Figure 2.

As shown in Figure 2a, the complete resonant cycle is composed of 8 sub-intervals. During each of these, one of the AC side voltages $V_1 \cdots V_4$ is applied to the left of the resonant tank which represents V_{SN}. The interval boundaries are defined by the switching on/off of the MOSFETs at the AC side and the diodes at the load side.

In the switching modulation, at the rising zero crossing of resonant current, the MOSFETs Q1, Q3, Q5 and Q7 are turned on under zero current switching (ZCS) conditions, but only the phase with the highest voltage, phase V_1, provides energy to the resonant circuit through the diode D_2 while the diodes D_4, D_6 and D_8 are blocking and prevent any short circuit between the phases. Afterwards, the MOSFET in parallel the conducting diode, Q_2, can be turned on under ZVS conditions. The amount of charge to be delivered from each phase for one full resonant cycle is calculated using the energy balance approach at the beginning of the cycle. When the calculated amount of charge for the phase V_1 is provided, the MOS-FET Q_1 is turned off. The second highest phase, phase V_2, can provide energy over the MOSFET Q_3 only after the V_{SN} voltage drops to the voltage level of V_2, through the D_4 diode and the ZVS conditions

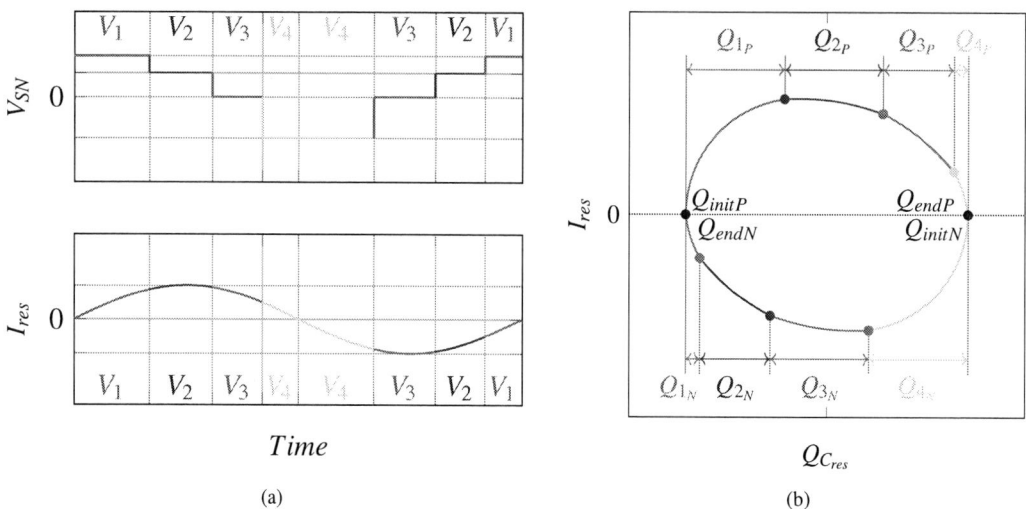

Fig. 2: Example of waveforms for two positive and one negative phase voltages (a) applied voltage to the resonant tank and resonant current, (b) state-plane diagram.

are satisfied for the Q_4 MOSFET to be turned on. In this simplified model, the change on V_{SN} occurs instantaneously as the node capacitance, C_r, is ignored. This procedure continues for all the phase transitions in the positive half cycle. A similar procedure applies for the negative half-cycle. Consequently, soft switching is achieved throughout the circuit operation.

The operation of the circuit is also depicted in Figure 2b in the state plane spanned by the resonant charge $Q_{C_{res}}$ (i.e. the charge stored in C_{res}) and the resonant current I_{res} by using the approach presented in [11]. Note that when resonant capacitor C_{res} is linear, a representation using the voltage $V_{C_{res}}$ across it would look exactly the same, apart from the obvious scaling with value C_{res}. However, a description using charges leads to somewhat simpler expressions, and is also applicable for the case when C_{res} shows non-linear behaviour, as would be the case for some varieties of ceramic capacitors.

The frequency of the inductor current, to be pointed out, is slightly higher than the resonant frequency, but it is not a control parameter, but a consequence of the control technique. The excess frequency of the inductor current can be observed on the state plane from the inflection at each phase transition. Also, the frequency of the inductor current increases as the load decreases which could be possible to observe in the state plane since there is going to be more inflection for light loads.

Some important variables, to be used in the derivation which follows, are shown there: the initial charge on C_{res} for the positive half-cycle is depicted as Q_{initP}, similar for the end of the half-cycle and for the negative half-cycle. As we consider a single cycle starting at a positive zero crossing for I_{res}, we find $Q_{initN} = Q_{endP}$. However, it is not necessarily the case that $Q_{initP} = Q_{endN}$, this holds only when steady-state operation is assumed.

The charges provided or absorbed by the AC supply per sub-interval have been depicted as $Q_{1P} \cdots Q_{4N}$. For the positive and negative half-cycles respectively we find the following relations:

$$\begin{aligned}
Q_{DC_P} &= Q_{endP} - Q_{initP} = Q_{1P} + Q_{2P} + Q_{3P} + Q_{4P}, \\
Q_{DC_N} &= Q_{endN} - Q_{initN} = Q_{4N} + Q_{3N} + Q_{2N} + Q_{1N}.
\end{aligned} \tag{1}$$

Note that all charges with subscript $..N$ are negative. Due to the diode rectifier at the load side, these are also the net charges per interval which flow (after rectification) to the load.

The energy balances for the positive and negative half-cycles can respectively be formulated as:

$$Q_{1P}V_1 + Q_{2P}V_2 + Q_{3P}V_3 + Q_{4P}V_4 - Q_{DC_P}NV_{DC} = \frac{Q_{endP}^2 - Q_{initP}^2}{2C_{res}},$$

$$Q_{1N}V_1 + Q_{2N}V_2 + Q_{3N}V_3 + Q_{4N}V_4 + Q_{DC_N}NV_{DC} = \frac{Q_{endN}^2 - Q_{initN}^2}{2C_{res}}. \tag{2}$$

The net charge per full resonance cycle supplied by phase V_1 equals $Q_1 = Q_{1P} + Q_{1N}$, and similar for the other phases. By adding the two parts of (2), the energy balance for the full cycle can be expressed as:

$$Q_1V_1 + Q_2V_2 + Q_3V_3 + Q_4V_4 + (Q_{DC_N} - Q_{DC_P})NV_{DC} = \frac{Q_{endP}^2 - Q_{initP}^2 + Q_{endN}^2 - Q_{initN}^2}{2C_{res}}. \tag{3}$$

Steady state

Under steady state conditions ($Q_{initP} = Q_{endN}$ and $Q_{initN} = Q_{endP}$), the right-hand side of (3) vanishes, and using (1), (3) can be further simplified to:

$$Q_1V_1 + Q_2V_2 + Q_3V_3 + Q_4V_4 = 2Q_{DC}NV_{DC} \tag{4}$$

where due to the steady state we can use $Q_{DC} = Q_{DC_P} = -Q_{DC_N}$. Note that either V_2 or V_3 is always zero and the corresponding (neutral) phase is only conducting current until the net charge reaches its final value after the active phases. To obtain a maximum power factor, the net charge per phase is chosen linearly proportional to the supply voltage as:

$$Q_1 = KV_1, Q_2 = KV_2, Q_3 = KV_3, Q_4 = KV_4 \tag{5}$$

in which K is a constant with the dimension of capacitance. Applying this to the energy balance (4), we obtain:

$$KV_1^2 + KV_2^2 + KV_3^2 + KV_4^2 = 2Q_{DC}NV_{DC} \rightarrow K = \frac{2Q_{DC}NV_{DC}}{V_1^2 + V_2^2 + V_3^2 + V_4^2} \tag{6}$$

To avoid the flow of reactive current, combinations of charge and voltage which lead to negative power flow will be discarded. For this purpose it is now suitable to distinguish two cases:

Case 1 (12Z4 sequence)
The case relates to the situation where two of the AC supply voltages are positive and the remaining one negative. In this case it follows that $V_1 > 0, V_2 > 0, V_3 = 0, V_4 < 0$. Hence, as $Q_{4P} > 0$, the interval related to V_4 can be discarded/skipped for the positive resonant half-cycle, and similarly V_1 and V_2 can be skipped for the negative half-cycle. In other words this implies that $Q_{1N} = 0$, $Q_{2N} = 0$, $Q_{4P} = 0$.

Case 2 (1Z34 sequence)
The other case applies to the opposite situation: one supply voltage is positive and the two others negative. In this case we find $V_1 > 0, V_2 = 0, V_3 < 0, V_4 < 0$. Here, both V_3 and V_4 can be skipped for the positive half-cycle, and V_1 for the negative half-cycle. This implies that $Q_{1N} = 0$, $Q_{3P} = 0$, $Q_{4P} = 0$. As a result, the general state plane diagram as depicted in Figure 2b changes to the two varieties depicted in Figure 3.

In order to simplify the expressions, the average charge, Q_{AV}, is defined as in (7).

$$Q_{AV} = (Q_{endP} + Q_{initP})/2 \tag{7}$$

Using (6) and (7), (2) can be rewritten for case 1 as in (8):

$$(V_1^2 + V_2^2)\frac{2Q_{DC}NV_{DC}}{V_1^2 + V_2^2 + V_3^2 + V_4^2} - Q_{DC}NV_{DC} = \frac{Q_{DC}Q_{AV}}{C_{res}},$$

$$(V_3^2 + V_4^2)\frac{2Q_{DC}NV_{DC}}{V_1^2 + V_2^2 + V_3^2 + V_4^2} - Q_{DC}NV_{DC} = -\frac{Q_{DC}Q_{AV}}{C_{res}}. \tag{8}$$

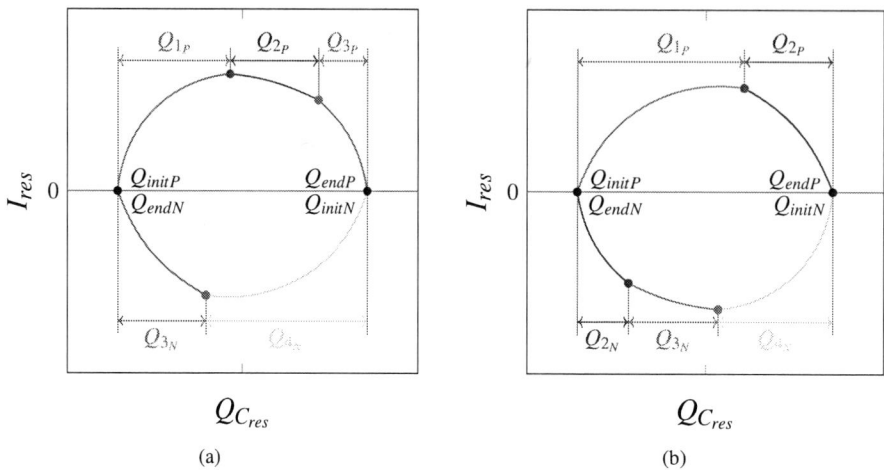

Fig. 3: Example of state-plane diagrams, (a) case 1 and (b) case 2.

For case 2 exactly the same result is found. After some manipulation of (8) we find for the average charge Q_{AV}:

$$Q_{AV} = \frac{V_1^2 + V_2^2 - V_3^2 - V_4^2}{V_1^2 + V_2^2 + V_3^2 + V_4^2} N V_{DC} C_{res}. \tag{9}$$

Note that (9) holds for both case 1 and 2, and that Q_{AV} will be continuous at the transitions between these cases. Furthermore, (9) implies that the state plane trajectory will in general not be centered around (0,0), as would be the usual case for most resonant converters.

Transient operation

In this section we will derive equations for the transient operation, in which Q_{endN} is not necessarily equal to Q_{initP}. As a first step, the energy balances per half cycle in (2) are re-used. Just as for the steady-state derivation, we will assume that the charge displacements are made linearly proportional to the applied AC-side voltage. Also here, intervals which lead to negative power flow at the AC side will be discarded. Hence we find the charge displacements for the case 1 and case 2, respectively:

$$Q_{1P} = K_P V_1, Q_{2P} = K_P V_2, Q_{4N} = K_N V_4,$$
$$Q_{1P} = K_P V_1, Q_{3N} = K_N V_3, Q_{4N} = K_N V_4. \tag{10}$$

with K_P, K_N again proportionality constants with the dimension of capacitance. Using these constants, (2), leaving out the undesired intervals, and simplifying the notation by using (7) leads to:

$$K_P(V_1^2 + V_2^2) - Q_{DC_P} N V_{DC} = \frac{Q_{DC_P} Q_{AV}}{C_{res}} \rightarrow K_P = \frac{Q_{DC_P}(N V_{DC} + \frac{Q_{AV}}{C_{res}})}{V_1^2 + V_2^2},$$

$$K_N(V_3^2 + V_4^2) + Q_{DC_N} N V_{DC} = \frac{Q_{DC_N} Q_{AV}}{C_{res}} \rightarrow K_N = \frac{Q_{DC_N}(-N V_{DC} + \frac{Q_{AV}}{C_{res}})}{V_3^2 + V_4^2}. \tag{11}$$

As the transient operation should ideally converge to the steady-state operation, the steady-state value for Q_{AV} as shown in (9) will be used to calculate K_P and K_N in (11). For the two cases addressed before, the charge levels at which switching actions and commutations need to start and end can be derived, as shown in Table I. Similar to steady-state operation, after the active phases, the neutral line is conducting until the charge reaches its final value. Note that in both cases the intervals corresponding to V_4 and V_1 are always skipped for $I_{res} > 0$ and $I_{res} < 0$, respectively. Initial and final charge levels are expected to be the same in steady-state. In an implementation, the value of Q_{initP} can be obtained by sampling the value of the voltage across C_{res} at the zero crossing of I_{res} while the values for $Q_{endP} = Q_{initN}$ and Q_{endN} are calculated using ($Q_{initN} = Q_{AV} + 0.5 Q_{DC}$) and ($Q_{endN} = Q_{AV} - 0.5 Q_{DC}$).

Charge level #	Case 1 (12Z4)	Case 2 (1Z34)
$Q_{comm}(1)$	Q_{initP}	Q_{initP}
$Q_{comm}(2)$	$Q_{comm}(1) + K_P V_1$	$Q_{comm}(1) + K_P V_1$
$Q_{comm}(3)$	$Q_{comm}(2) + K_P V_2$	Q_{endP} (skip V_3)
$Q_{comm}(4)$	Q_{endP} (skip V_4)	Q_{endP} (skip V_4)
$Q_{comm}(5)$	Q_{initN}	Q_{initN}
$Q_{comm}(6)$	$Q_{comm}(5) + K_N V_4$	$Q_{comm}(5) + K_N V_4$
$Q_{comm}(7)$	Q_{endN} (skip V_2)	$Q_{comm}(6) + K_N V_3$
$Q_{comm}(8)$	Q_{endN} (skip V_1)	Q_{endN} (skip V_1)

Table I: Charge levels for switching actions per case.

Simulation tests

To verify the intended operation of the circuit and its control, a simulation model was built using the PLECs blockset [12] in the Simulink environment [13]. In Figure 4, results for steady state operation when supplying a 48V load with 1 kW are shown where the input line voltage is 400 V_{AC}, 50 Hz. The transformer ratio is selected as 4 to provide an easy operating range for the 400 V_{AC} input voltage of the converter.

The natural resonant frequency is set to 5000 Hz. This very low resonant frequency value is used to increase visibility in the figures, in practice a much higher frequency can be used. The red waveforms are representing the averaged value of the current (blue) waveforms and as shown in Figure 4b the averaged phase current looks sinusoidal. The averaged phase current in Figure 4b and phase voltage in Figure 4a are in the same phase which indicates that, after removal of the HV part by a low-pass filter, unity power factor is successfully achieved. The average of neutral line current, I_Z, is zero in the Figure 4c as expected.

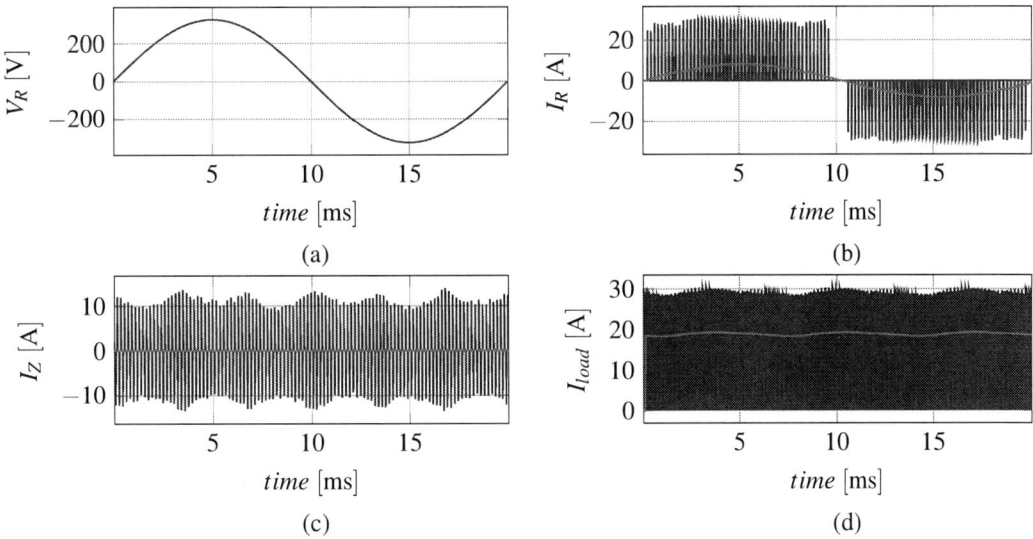

Fig. 4: Simulation results: (a) phase R voltage (b) phase R current and its average (c) zero line current and its average (d) load current and its average.

A more detailed spectral analysis is presented in Figure 5 for the normalized phase R current. The fundamental 50 Hz component is shown as first order in the Figure 5. The calculated total harmonic distortion using simulation results is 2.21%. Compared to the fundamental component, other components are quite insignificant. The results are similar for the other phases.

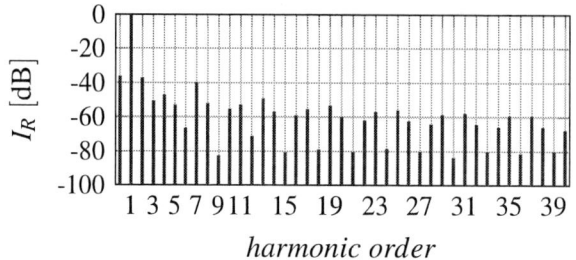

Fig. 5: Spectral analysis of the phase R current for the first 40 harmonic components.

Conclusion

Analytic derivations and simulation results are presented for the proposed novel control method which is applied to a fully zero-voltage switching series-resonant isolated 3-phase AC to DC converter. The simulation results show that the proposed control method is a promising approach to operate a fully zero-voltage switching series-resonant isolated 3-phase AC to DC converter, achieving unity power factor and low total harmonic distortion.

References

[1] J. Choi, D. Tsukiyama and J. Rivas, "Evaluation of a 900 V SiC MOSFET in a 13.56 MHz 2 kW resonant inverter for wireless power transfer," 2016 IEEE 17th Workshop on Control and Modeling for Power Electronics (COMPEL), 2016, pp. 1-6.

[2] H. Kim, A. Anurag, S. Acharya and S. Bhattacharya, "Analytical Study of SiC MOSFET Based Inverter Output dv/dt Mitigation and Loss Comparison With a Passive dv/dt Filter for High Frequency Motor Drive Applications," in IEEE Access, vol. 9, pp. 15228-15238, 2021.

[3] J. L. Schanen, A. Baraston, M. Delhommais, P. Zanchetta, and D. Boroyevitch, "Sizing of power electronics emc filters using design by optimization methodology," 2016 7th Power Electronics and Drive Systems Technologies Conference (PEDSTC), pp. 279–284, 2016.

[4] H. Huisman, I. de Visser and J. Duarte, "Optimal trajectory control of a CLCC resonant power converter," 2015 17th European Conference on Power Electronics and Applications (EPE'15 ECCE-Europe), 2015, pp. 1-10, doi: 10.1109/EPE.2015.7309101.

[5] R. Bonten, J. M. Schellekens, B. Vermulst, F. Clermonts and H. Huisman, "Improved Dynamic Behaviour for the Series-Resonant Converter using Bidirectional Charge Control," in IEEE Transactions on Power Electronics, doi: 10.1109/TPEL.2022.3169710.

[6] R. W. T. Bonten, J. M. Schellekens, H. Huisman and C. G. E. Wijnands, "A Comparative Evaluation of Series-Resonant, Bidirectional Optimal Trajectory Controlled Isolated DC-DC Converters," 2019 21st European Conference on Power Electronics and Applications (EPE '19 ECCE Europe), 2019, pp. P.1-P.10, doi: 10.23919/EPE.2019.8915383.

[7] G. E. Sfakianakis, J. Everts, H. Huisman, T. Borrias, C. G. E. Wijnands and E. A. Lomonova, "Charge-based ZVS modulation of a 3–5 level bidirectional dual active bridge DC-DC converter," 2016 IEEE Energy Conversion Congress and Exposition (ECCE), 2016, pp. 1-10, doi: 10.1109/ECCE.2016.7854914.

[8] R. W. T. Bonten, J. M. Schellekens and H. Huisman, "Optimal Utilization of the Dual-Active Bridge Converter with Bidirectional Charge Control," 2021 22nd IEEE International Conference on Industrial Technology (ICIT), 2021, pp. 452-457, doi: 10.1109/ICIT46573.2021.9453656.

[9] H. Huisman, "A three-phase to three-phase series-resonant power converter with optimal input current waveforms. I. Control strategy," in IEEE Transactions on Industrial Electronics, vol. 35, no. 2, pp. 263-268, May 1988, doi: 10.1109/41.192658.

[10] F. P. Kusumah and J. Kyyra, "Successive injections modulation of a direct three-phase to single-phase ac/ac converter for a contactless electric vehicle charger," The Journal of Engineering, vol. 2019, no. 17, pp. 4106–4110, 2019.

[11] R. Oruganti, Fred C. Lee, "Resonant Power Processors, Part I State Plane Analysis", IEEE Trans. On Industry Applications, vol. IA-21, No. 6, Nov./Dec. 1985, pp. 1453-1460.

[12] Plexim engineering software documentation, Plecs blockset packages, 2002. [Online]. Available: https://www.plexim.com/documentation.

[13] Simulink documentation, Simulation and model-based design, 2020. [Online]. Available: https://www.mathworks.com/products/simulink.html.

Design of a Robust Voltage Control for Inverters with LC Filter based on the Internal Model Control

Frederik Stallmann and Axel Mertens
Institute for Drive Systems and Power Electronics
Leibniz University Hannover
Welfengarten 1
Hannover, Germany
Phone: +49 511 762 2217
Email: frederik.stallmann@ial.uni-hannover.de
URL: https://www.ial.uni-hannover.de/en/

Lukas Fräger
BLOCK
Transformatoren-Elektronik GmbH
Max-Planck-Straße 36
Verden (Aller), Germany
Phone: +49 4231 678 434
Email: lukas.fraeger@block.eu
URL: https://www.block.eu/en_US/

Acknowledgments

This paper is funded by the German Federal Ministry of Economic Affairs and Climate Action (BMWK) pursuant to a decision of the German Parliament in the project STIM (Smart Transformers as Power Supply for the Future Mechanical Engineering Industry). Funding number: 03EN2010E.

Keywords

≪Voltage Source Inverter (VSI)≫, ≪Robust Control≫, ≪Microgrid≫, ≪Converter Control≫, ≪Uninterruptible Power Supply (UPS)≫.

Abstract

Voltage-controlled inverters with an LC filter are widely used in uninterruptable power supplies and droop-controlled inverters. However, as inverters usually have an operational lifetime of several years, important plant parameters such as the inductance and capacitance of the LC filter may diminish due to aging. In this paper, deviations in the LC filter parameters and their effect on the stability and robustness of the control are analyzed. An internal model control is designed for the inverter and a robust stability analysis is carried out. Furthermore, different controller designs for increased robustness are discussed and validated by simulations. Finally, the robust internal model control developed, is implemented in an FPGA-based system and used in an experimental setup to investigate the controller performance.

Introduction

Many industrial and microgrid applications rely on power electronic converters to supply local loads. The control of these converters determines their technical capabilities. While grid-feeding, current-controlled inverters are usually used in distributed generation units (DG) to inject a defined current into the grid, a different control concept is needed for industrial and islanded microgrids. In contrast, voltage-controlled inverters can provide a well-defined voltage to the load grid, are capable of black start and are usually referred to as grid-forming inverters [1]. When it comes to parallel operation of grid-forming inverters, the reference value for the voltage control is typically given by an outer power-related control, like a droop control or virtual synchronous generator control [2][3][4]. For these types of inverters, passive LC filters are a common choice for damping harmonics, caused by the switching devices. As detailed information about the connected loads is not usually available, the LC filter is considered as a plant model for the control design. Deviations in the filter capacitance and inductance values due to component tolerances and aging pose a threat to the system stability. Damping in the form of a small resistive component in the LC filter is a reasonable assumption, as inductors and capacitors, as well as connectors, possess parasitic resistances. For the LC filter, this means that the gain will not be infinite at the resonant frequency, which enables a stable control design without additional damping. The assumption of a parasitic resistance, and the knowledge of its value, thus have a significant impact on the control design. The parallel operation of inverters does not only require a stable voltage control, but also the consideration of impedance and passivity in certain frequency ranges [5]. Stability issues are usually observed around the nominal frequency due to the phase-locked loop or the power-related control, or around the critical frequency due to the time delay and LC resonance [2][6][7][8]. One design option for robust control is the H_∞ method, which yields good results, but involves the numerical solution of the Riccati equation and thus the use of sophisticated MATLAB algorithms [9][10]. A more comprehensible way of designing a robust control is the internal model control, which can be easily designed as it only takes into account the LC filter parameters and the sampling time constant [11]. The aim of this paper is to provide a straightforward voltage control design that can be easily adapted for different systems and is robust against possible parameter changes in the LC filter. Three control variations will be shown to increase the robustness and improve the stability of the system [12]. Moreover, the control will be tested, using an experimental setup in passive load and grid-connected scenarios.

Voltage Control of Inverters with LC Filter

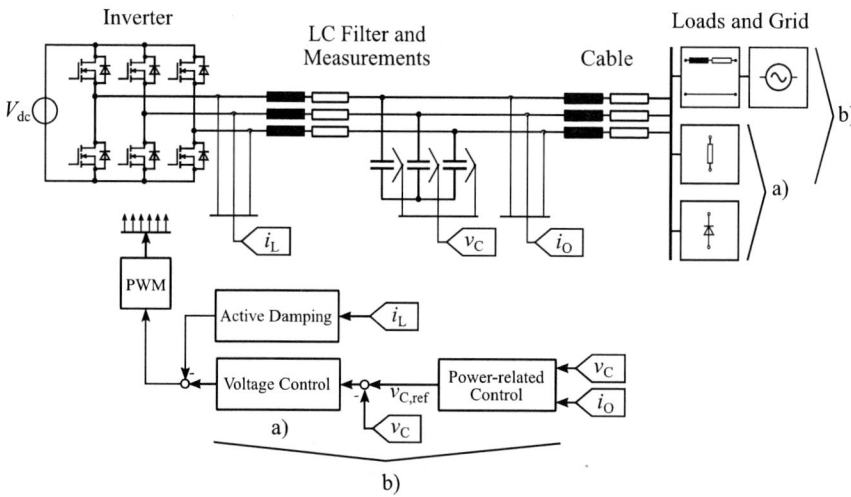

Fig. 1: Schematic of the inverter, LC filter, control system, loads and grid for the experimental setups a) voltage-controlled and b) droop-controlled operation

The basic outline of the system including the inverter, the filter, the digital control, a grid and loads is shown in Fig. 1. Apart from the capacitor voltage, measurements of the inductor and/or output current are typically available. The control may also include a superimposed power-related control to provide the reference voltage and enable power sharing in parallel operation or grid-connected operation. Active damping, depending on the filter parameters, might also be necessary, but is considered optional in this paper. The block diagram for the single-loop voltage control is shown in Fig. 2. The system in Fig. 1 can be divided into the voltage controller $K(s = j\omega)$ and the plant, which includes the sample-and-hold delay $G_D(s)$, the inductance L_f and capacitance C_f of the LC filter, as well as a parasitic resistance R_f in series with the inductor.

Internal Model Control

The basic concept of the internal model control (IMC) for the LC filter plant is shown in Fig. 2. As the name indicates, the control concept is based on a plant model $G_{P,M}$ of the LC filter and the delay. The plant model is used to estimate the capacitor voltage and feed it back to the control error signal. The controller K_{IMC} is used to control the remaining voltage error and minimize the influence of the disturbance signal. The transfer functions for the plant and plant model are given in Eq. (1)-(3) and

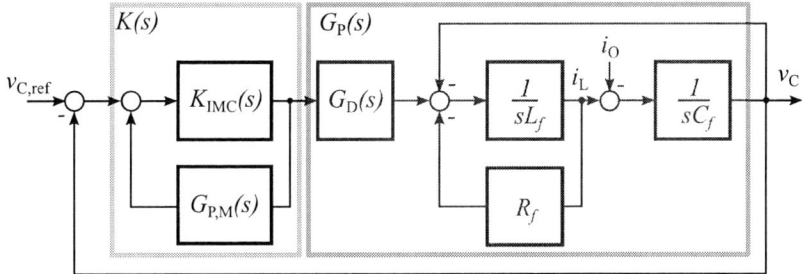

Fig. 2: Internal model control for inverter with LC filter

Eq. (4)-(6). The indices M and N indicate model and nominal quantities, respectively.

$$G_P(s) = G_D(s)G_{LC}(s) \qquad (1)$$

$$G_{LC}(s) = \frac{1}{L_f C_f s^2 + C_f R_f s + 1} \qquad (2)$$

$$G_D(s) = e^{-sT_D} \qquad (3)$$

$$G_{P,M}(s) = G_{D,M}(s)G_{LC,M}(s) \qquad (4)$$

$$G_{LC,M}(s) = \frac{1}{L_{f,N}C_{f,N}s^2 + C_{f,N}R_{f,N}s + 1} \qquad (5)$$

$$G_{D,M}(s) = \frac{T_D^2 s^2 - 6T_D s + 12}{T_D^2 s^2 + 6T_D s + 12} \qquad (6)$$

The effectiveness of this controller can be demonstrated by determining the closed-loop transfer function of the control system and assuming a perfect plant model. In Eq. (7) it can be seen that the denominator of the closed-loop transfer function $G_{cl}(s)$ of the system in Fig. 2 becomes equal to 1 if $G_P(s)$ and $G_{P,M}(s)$ are equal.

$$G_{cl}(s) = \frac{G_P(s)K_{IMC}(s)}{(G_P(s) - G_{P,M}(s))K_{IMC}(s) + 1} \stackrel{\text{for } G_P = G_{P,M}}{=} G_P(s)K_{IMC}(s) \qquad (7)$$

The resulting closed-loop transfer function simplifies the design of a controller $K_{IMC}(s)$. In [11], the reciprocal of the minimal phase part of the plant model is taken as controller transfer function and adjusted by an additional low-pass filter $G_F(s)$ with order n to increase robustness and make the controller a proper transfer function. In this case, the minimal phase part of the plant is the LC filter transfer function, modeled as $G_{LC,M}(s)$.

$$K_{IMC}(s) = \frac{1}{G_{LC,M}(s)} \cdot G_F(s) \qquad (8) \qquad G_F(s) = \left(\frac{\omega_{LP}}{s + \omega_{LP}}\right)^n \qquad (9)$$

As the LC filter is a second-order system, a second-order low-pass filter is needed to achieve a proper controller transfer function and suppress high frequency noise. The cut-off frequency of the low-pass filter is dependent on the required robustness of the system on the one hand, and on the desired control bandwidth on the other hand. The controller bandwidth depends on the cut-off frequency and order of the low-pass filter. As lower-order harmonics should be controllable, a control bandwidth of 750 Hz is suggested.

Robust Stability Analysis

As stated above, the deviations in the filter inductance L_f and capacitance C_f and their influence on the control system stability will be analyzed in this section. The derived IMC control will be analyzed in terms of robustness using the complementary sensitivity function $T_M(s)$ and the multiplicative model uncertainty function $\delta_{P,M}(s)$. $T_M(s)$ resembles the closed-loop transfer, but is based on the nominal plant model and not the real plant. The IMC controller K_{IMC} and the model feedback can be condensed into $K(s)$.

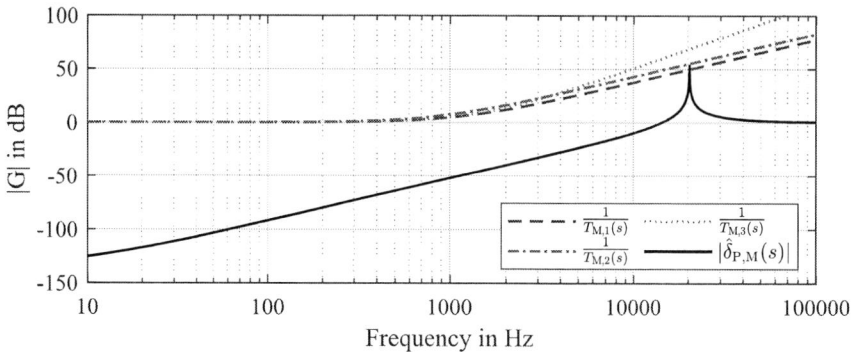

Fig. 3: Robustness analysis of $T_{M,1}(s)$ with $G_{F,1}(s)$, $T_{M,2}(s)$ with $G_{F,2}(s)$, $T_{M,3}(s)$ with $G_{F,3}(s)$

$$T_M(s) = \frac{G_{P,M}(s)K(s)}{G_{P,M}(s)K(s)+1} \qquad (10) \qquad K(s) = \frac{K_{IMC}(s)}{1 - K_{IMC}(s)G_{P,M}(s)} \qquad (11)$$

The multiplicative model uncertainty $\delta_{P,M}(s)$ is a measure of the difference between the nominal plant and the real plant, including parameter deviations.

$$G_P(s) = G_{P,M}(s)(1 + \delta_{P,M}(s)) \qquad (12) \qquad \delta_{P,M}(s) = \frac{G_P(s)}{G_{P,M}(s)} - 1 \qquad (13)$$

For a robust and stable system, the condition $\left|\frac{1}{T_M(s)}\right| > |\hat{\delta}_{P,M}(s)|$ must hold, where $\hat{\delta}_{P,M}(s)$ is the most significant deviation between the nominal plant and real plant [11].

The robust stability analysis is carried out for an unloaded system, as a resistive load could provide additional damping. The parameters are given in Table I and Table II. For $|\hat{\delta}_{P,M}(s)|$, a deviation in the filter inductance and capacitance (Table II) of 30% each is assumed. The performance of the control is validated using simulations in MATLAB/Simulink with the toolbox PLECS. First, the control with $G_F(s) = G_{F,1}(s)$ (Eq. (8)) is considered. In Fig. 4, the simulation results are shown. Before 0.02 s, the LC filter parameters are equal to their nominal values $L_{f,N}$ and $C_{f,N}$ and the system is stable. After 0.02 s, the filter inductance and capacitance are reduced by 30 % each, leading to an instability. This agrees with the gain plot in Fig. 3. The solid black line representing $|\hat{\delta}_{P,M}(s)|$ lies above the reciprocal complementary sensitivity function $\frac{1}{T_{M,1}(s)}$ at around 20.3 kHz and thus indicates a potentially unstable system. The same analysis and validation are carried out for $T_{M,2}(s)$ and $T_{M,3}(s)$. Using the respective filter transfer functions $G_{F,2}(s)$ and $G_{F,3}(s)$, stability can be guaranteed for the described deviations in the LC parameters. However, in case of $G_{F,2}(s)$, the bandwidth of the system was significantly reduced. In the case of $T_{M,3}(s)$, the bandwidth was kept at the same frequency while achieving robust performance. However, the higher order of the filter and more complex design indicate a greater effort to implement the control on a real system. The simulation results in Fig. 4 validate the robust performance of the system, as both filters $G_{F,2}(s)$ and $G_{F,3}(s)$ achieve a stable system.

Table I: Parameters of the filter transfer functions $G_F(s)$; $f = \frac{\omega}{2\pi}$

	$G_{F,1}(s)$	$G_{F,2}(s)$	$G_{F,3}(s)$
Order	2	2	3
f_{LP}	1164 Hz	854 Hz	1451 Hz
f_{BW}	750 Hz	550 Hz	750 Hz

Table II: Parameters of the example control system

P_N	V_N	f_s	$L_{f,N}$	$C_{f,N}$	$R_{f,N}$
10 kW	400 V	100 kHz	250 μH	0.5 μF	50 mΩ

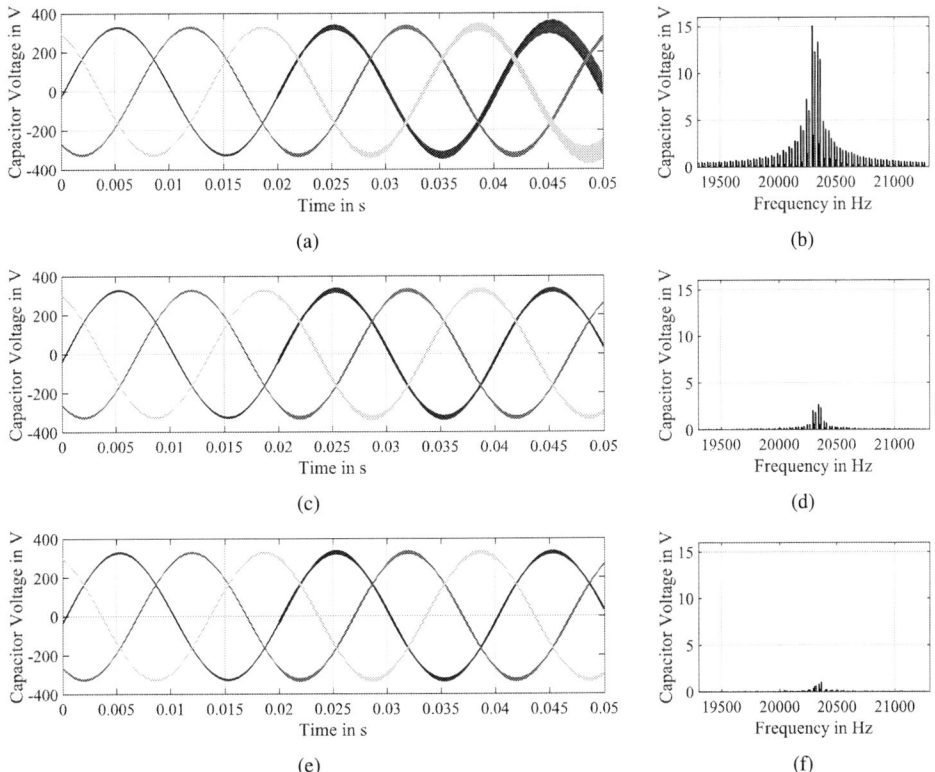

Fig. 4: Simulation results in the time domain and spectrum for Δt from 0.03 s to 0.05 s using (a)-(b) $G_{F,1}(s)$, (c)-(d) $G_{F,2}(s)$, (e)-(f) $G_{F,3}(s)$

Impedance Modeling

To evaluate the controller performance in parallel operation with other inverters or connected to the grid, the impedance-based approach is a suitable methodology [5]. The first step when performing the stability analysis is to derive the impedance characteristic of the whole inverter, including the control, LC filter and digital delay. The resulting impedance is expressed as a continuous transfer function $Z(s)$ and shown for the three different controller parametrizations in the Bode plot in Fig. 5.

$$Z(s) = \frac{L_{f,M} \cdot s + R_{f,M}}{C_{f,M} \cdot L_{f,M} \cdot s^2 + C_{f,M} \cdot R_{f,M} \cdot s + G_D(s) \cdot K(s) + 1} \tag{14}$$

Together with the closed-loop transfer function, this results in the full small signal description of the capacitor voltage.

$$V_C(s) = G_{cl}(s) \cdot V_{C,ref}(s) - Z(s) \cdot I_O(s) \tag{15}$$

The impedance of all three systems has an inductive characteristic until the resonant frequency is reached. For the lower frequencies from 10 Hz to 2 kHz, the impedance even has a non-passive inductive characteristic. This is not usually considered an advantageous feature, as non-passive regions can lead to instabilities when connected to a grid or parallel inverters. However, as the impedance of the IMC has a non-passive inductive behavior, only a capacitive grid impedance may lead to an instability. As grid impedances are usually ohmic-inductive, instabilities are not expected. Nevertheless, an improvement

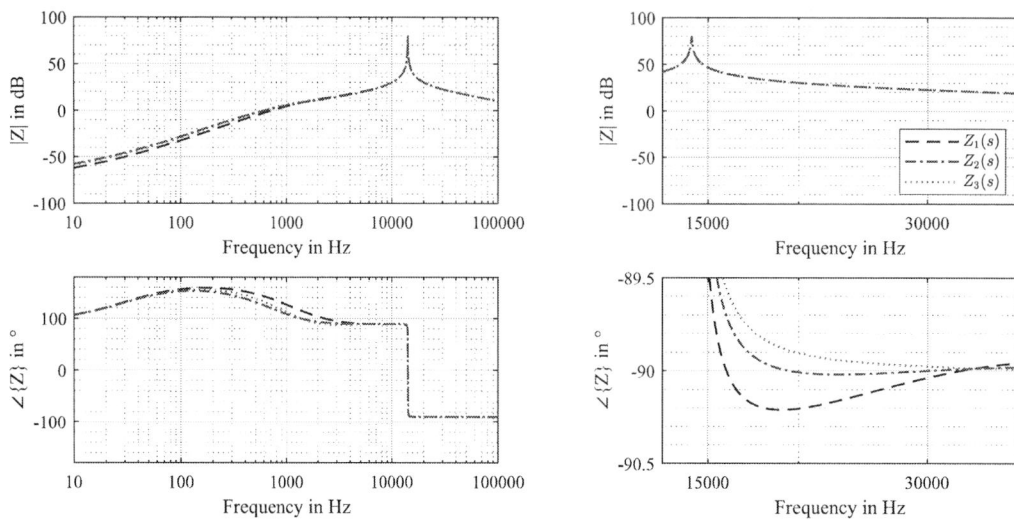

Fig. 5: Impedance amplitude and phase of the inverter system for controllers K_1, K_2, K_3; left: from 1 Hz to 100 kHz, right: zoomed in from 13 kHz to 30 kHz

towards increased passivity by using active damping would be beneficial.

Another impedance characteristic is the very low amplitude at frequencies below 1 kHz. Especially when a large share of non-linear loads, like three-phase rectifiers, are connected to the inverter, the low impedance amplitude leads to a low voltage distortion.

On the right side of Fig. 5, a close-up of frequencies around the resonance is shown. For the first two controller parametrization, a small non-passive capacitive characteristic can be observed. However, the third controller parametrization does not show this characteristic and instead stays passive, because the forward gain of the voltage controller becomes small enough that it does not excite the resonant frequency. At this point, the passive damping of the resistance $R_{f,N}$ is sufficient to damp the inverter system.

Implementation and Experimental Results

To implement the control presented in the previous chapter, the continuous transfer functions were discretized using the bilinear method. The discrete transfer function was then simplified and rearranged into the direct form II (DF-II).

$$G(s) = \frac{B_0 + B_1 s + ... B_m s^m}{A_0 + A_1 s + ... A_n s^n} \qquad (16)$$

bilinear method: $s = \dfrac{2}{T} \cdot \dfrac{z-1}{z+1}$ $\qquad (17)$

$$G(z) = \frac{b_0 + b_1 z^{-1} + ... b_m z^{-m}}{a_0 + a_1 z^{-1} + ... a_n z^{-n}} \qquad (18)$$

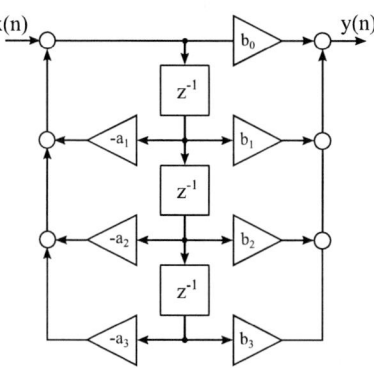

Fig. 6: IIR-filter in DF-II used for the discrete implementation

Table III: Parameters of the experimental setup

P_{max}	V_{LL}	V_{dc}	f_s	$L_{f,N}$	$C_{f,N}$	$R_{f,N}$
500 W	53 V	200 V	100 kHz	250 µH	1 µF	50 mΩ

$R_{load,DC}$	$R_{load,AC}$	L_g	k_P	k_Q	f_{Droop}	f_{LP}
30 Ω	16 Ω	3.5 mH	0.02	0.0005	50 Hz	850 Hz

The controller was then implemented by using the block diagram of a third-order generalized infinite impulse response (IIR) filter as shown in Fig. 6 and filling in the resulting coefficients $\vec{a} = [a_0...a_n]$ and $\vec{b} = [b_0...b_m]$. As mentioned before, the controller operates in the natural reference frame. However, because of the controller's high DC gain and a small steady DC component of the measurement signals, an additional high-pass filter was implemented to filter out any unwanted DC components. The high-pass filter is considered to be negligible at frequencies higher than 1 Hz. Note that operation in the dq frame is also possible with this controller.

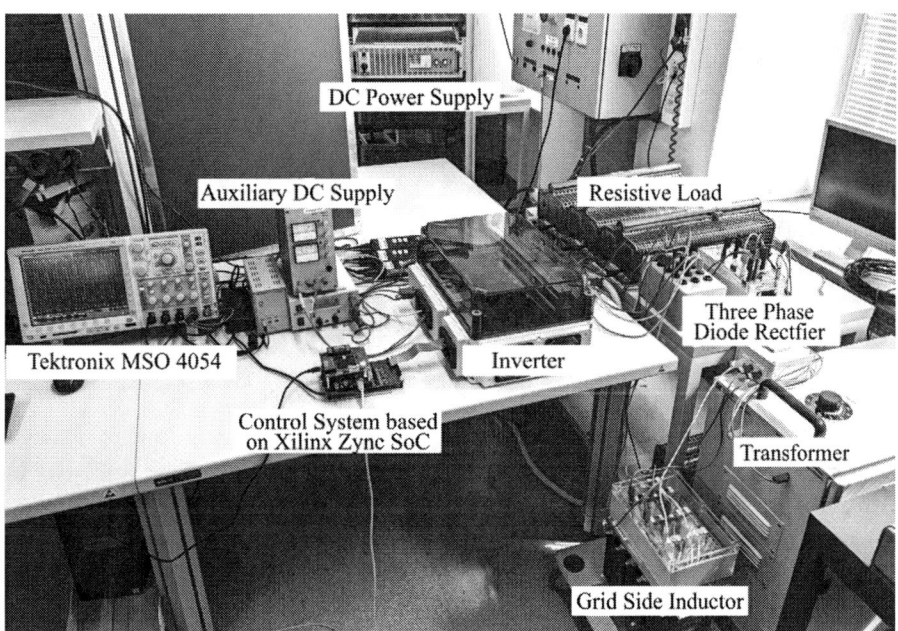

Fig. 7: Experimental setup

To validate the proposed control, the IMC with a third order low-pass filter was implemented on an FPGA using a Xilinx Zynq 7020. The experimental setup presented in Fig. 7 consists of the inverter, the control hardware (Xilinx Zync 7020), a DC source (EA-PSI 91599-30), an oscilloscope (Tektronix MSO 4054 and LeCroy HDO8108), differential probes (TT SI 8052), current probes (Keysight N2783B), load resistors, a three-phase bridge rectifier, a grid-side inductor, a transformer and additional auxiliary DC sources.

Measurements were carried out using two different experimental setups. The first one is shown in Fig. 1a) and consists of the three-phase inverter with the LC filter, a three-phase resistive load and a three-phase rectifier. This setup is used to test the IMC as a pure voltage control with only loads. The second setup includes a transformer with an inductor representing the grid. Furthermore, the three-phase resistive load is connected in this setup. The parameters of the experimental setup can be seen in Table III.

The results of the measurement with a resistive and non-linear load are shown in Fig. 8. It can be seen that for most frequencies the voltage distortion due to the rectifier harmonics is below 1% of the fundamental, which is due to the low inverter impedance of the inverter system. In applications with a high number of non-linear loads, these low distortions are satisfactory. In Fig. 9, a load jump is performed. First, only

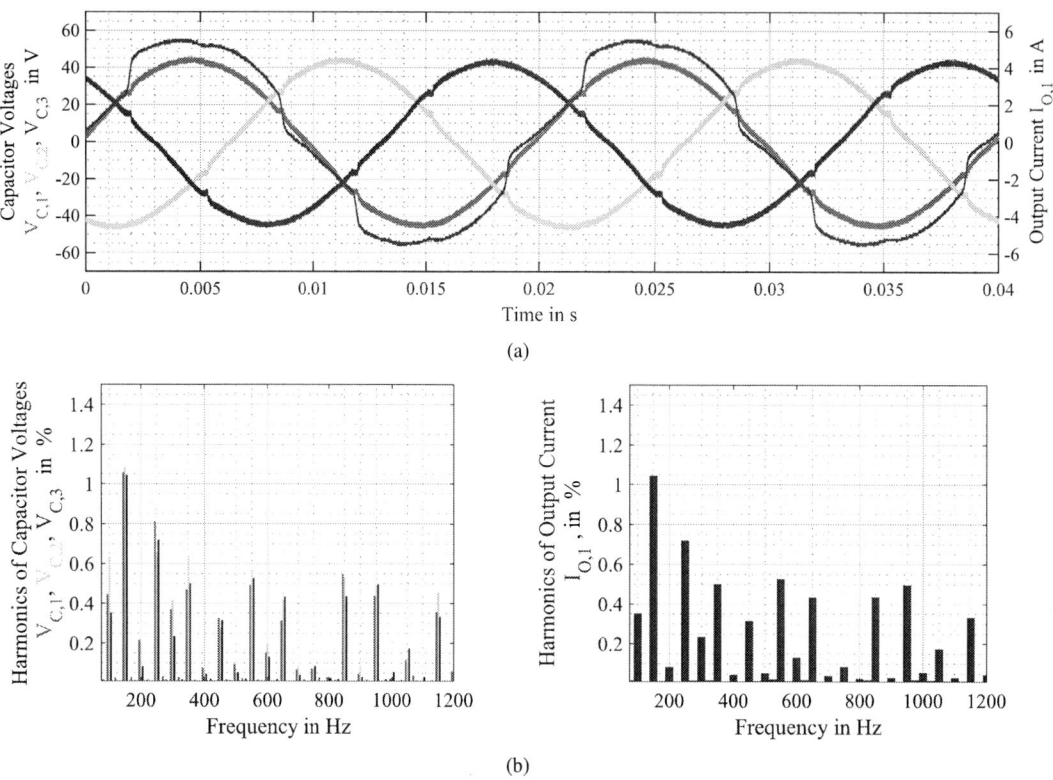

(a)

(b)

Fig. 8: Measurement of the capacitor voltages and output current $I_{O,1}$ with a resistive and non-linear load (see Fig. 1a)); (a) Time domain plot, (b) FFT of voltages (left) and current (right)

the three-phase rectifier load is connected. At 0.014 s, the three-phase resistive load is connected. This sudden connection leads to a visible dip and distortion in the capacitor voltages. The dip has a magnitude of approximately 10 V. However, the control quickly recovers and reaches its reference. Especially in industrial applications, load jumps can occur regularly. Therefore, fast recovery after a distortion is an advantageous feature.

The second experimental setup used can be seen in Fig. 1b). The setup includes the transformer, grid-side inductor and three-phase resistive load. This setup is used to validate the performance of the IMC in combination with a droop control and connected to a grid. The droop control can be expressed as

$$\theta_{ref} = ((P_{ref} - P) \cdot k_P \cdot \frac{\omega_{Droop}}{s + \omega_{Droop}} + \omega_N) \cdot \frac{1}{s} \tag{19}$$

$$\hat{V}_{ref} = (Q_{ref} - Q) \cdot k_Q \cdot \frac{\omega_{Droop}}{s + \omega_{Droop}} + \hat{V}_N. \tag{20}$$

Before the droop-controlled operation can be started, the inverter and control first need to be synchronized with the grid. For this purpose, the PWM is initially disabled, while the grid frequency, angle and voltage amplitude are obtained by measuring the voltage across the capacitors and feeding it to a phase-locked loop (PLL). When the estimated frequency and voltage amplitude of the PLL reach a steady value and are synchronized with the frequency and amplitude of the droop control, the PWM is enabled and the operation can be started.

The measurement results are shown for the capacitor voltages, output currents, active powers and reactive

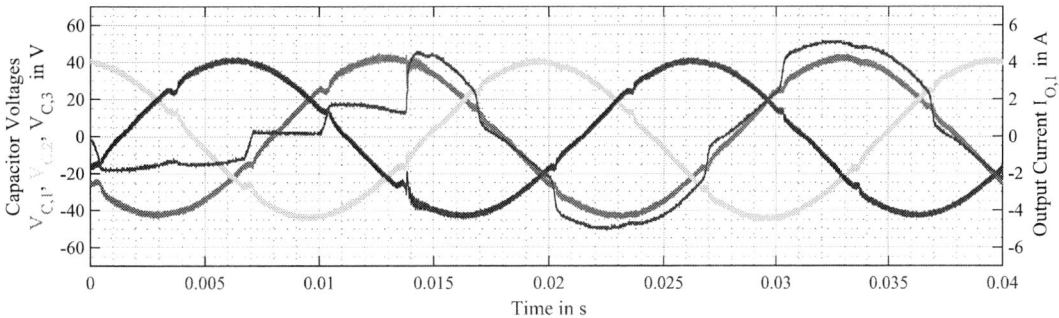

Fig. 9: Measurement of the capacitor voltages and output current $I_{O,1}$; three-phase rectifier load, three-phase resistive load jump at $0.014\,\mathrm{s}$

powers. Until $0.2\,\mathrm{s}$, the system operates in a steady state with an active power reference of $300\,\mathrm{W}$. At $0.2\,\mathrm{s}$, a reference jump to $500\,\mathrm{W}$ is performed. It can be seen that the new reference is reached at around $0.3\,\mathrm{s}$. The distortions which are visible at both powers are likely due to the slightly unbalanced voltages caused by an imperfectly symmetrical grid inductor and transformer. Moreover, the inverter is performing below its rating, which might lead to an unsuitable measuring range for the voltage and current sensors. Nonetheless, the measurements show that the IMC is suitable for the operation in a grid-connected scenario in combination with a droop control.

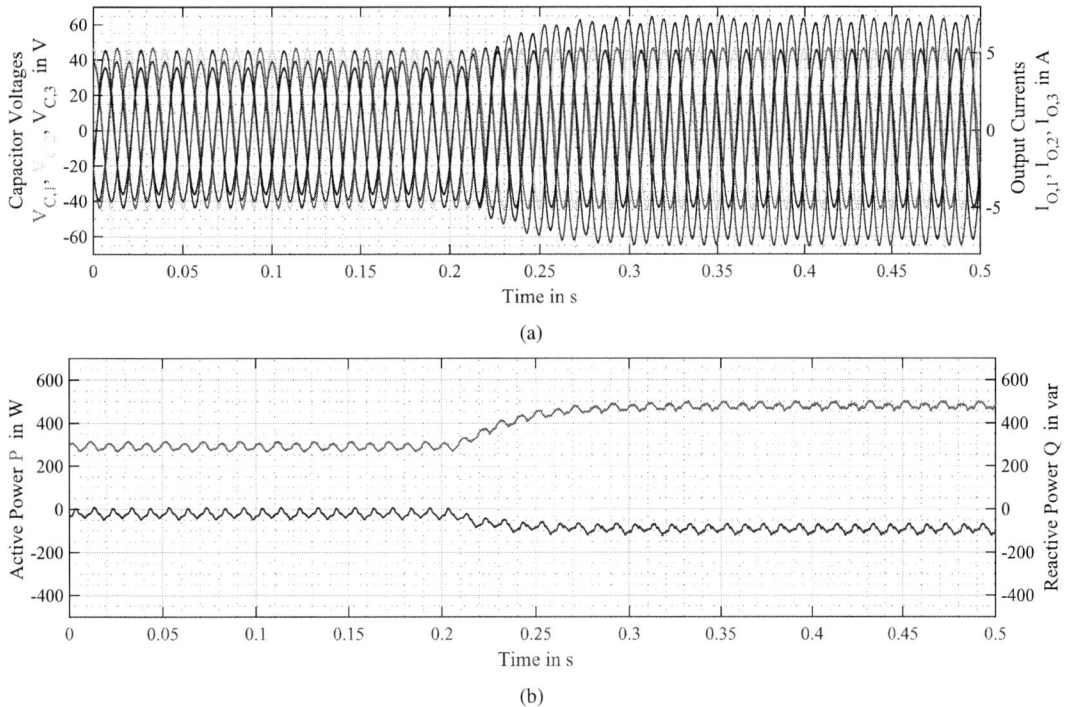

Fig. 10: Measurement of (a) the capacitor voltages & output currents and (b) active & reactive powers; using droop-controlled, grid-connected operation with an active power reference jump at $0.2\,\mathrm{s}$ (see Fig. 1 b))

Conclusion

In this paper, a robust voltage control based on the IMC was designed for a three-phase inverter with an LC filter. Three different controllers, modifying the low-pass filter, were used to increase the robustness of the IMC. The controls developed were analyzed using the model uncertainty method for a specified LC filter parameter deviation. It was found that two of the three controller variations are robust and that a trade-off between controller complexity and bandwidth should be considered. Simulations in Simulink/PLECS validated the analytical approach. Furthermore, the impedance characteristics of the IMC-controlled systems were derived to evaluate the stability in grid-connected or parallel operation. Moreover, the IMC was implemented on an FPGA-based hardware and tested in passive load and grid-connected scenarios. The measurement results show small voltage distortions when supplying a non-linear load, while also proving stable in combination with a droop control. In conclusion, the robust IMC seems to be a suitable and easily parametrized solution for voltage-controlled inverters in standalone, parallel or grid-connected operation.

References

[1] J. Rocabert, A. Luna, F. Blaabjerg, and P. Rodríguez. Control of Power Converters in AC Micro-grids. *IEEE Transactions on Power Electronics*, 27(11):4734–4749, November 2012.

[2] M. Dokus and A. Mertens. Sequence Impedance Characteristics of Grid-Forming Converter Controls. In *2020 IEEE 11th International Symposium on Power Electronics for Distributed Generation Systems (PEDG)*, pages 413–420, September 2020. ISSN: 2329-5767.

[3] J.M. Guerrero, L. Hang, and J. Uceda. Control of Distributed Uninterruptible Power Supply Systems. *IEEE Transactions on Industrial Electronics*, 55(8):2845–2859, August 2008. Conference Name: IEEE Transactions on Industrial Electronics.

[4] P. Unruh, M. Nuschke, P. Strauß, and F. Welck. Overview on Grid-Forming Inverter Control Methods. *Energies*, 13(10):2589, May 2020.

[5] J. Sun. Impedance-Based Stability Criterion for Grid-Connected Inverters. *IEEE Transactions on Power Electronics*, 26(11):3075–3078, November 2011.

[6] L. Harnefors, X. Wang, A.G. Yepes, and F. Blaabjerg. Passivity-Based Stability Assessment of Grid-Connected VSCs—An Overview. *IEEE J. Emerg. Sel. Topics Power Electron.*, 4(1):116–125, March 2016.

[7] X. Wang, P. C. Loh, and F. Blaabjerg. Stability Analysis and Controller Synthesis for Single-Loop Voltage-Controlled VSIs. *IEEE Transactions on Power Electronics*, 32(9):7394–7404, September 2017. Conference Name: IEEE Transactions on Power Electronics.

[8] F. Stallmann and A. Mertens. Sequence Impedance Modeling of the Matching Control and Comparison with Virtual Synchronous Generator. In *2020 IEEE 11th International Symposium on Power Electronics for Distributed Generation Systems (PEDG)*, pages 421–428, September 2020. ISSN: 2329-5767.

[9] T.-S. Lee, S.-J. Chiang, and J.-M. Chang. H_∞ loop-shaping controller designs for the single-phase UPS inverters. *IEEE Transactions on Power Electronics*, 16(4):473–481, July 2001. Conference Name: IEEE Transactions on Power Electronics.

[10] S. Yang, Q. Lei, F. Z. Peng, and Z. Qian. A Robust Control Scheme for Grid-Connected Voltage-Source Inverters. *IEEE Transactions on Industrial Electronics*, 58(1):202–212, January 2011. Conference Name: IEEE Transactions on Industrial Electronics.

[11] J. Lunze. *Regelungstechnik 1: Systemtheoretische Grundlagen, Analyse und Entwurf einschleifiger Regelungen*. Springer Berlin Heidelberg, Berlin, Heidelberg, 2020.

[12] X. Wang, F. Blaabjerg, and P. C. Loh. Virtual RC Damping of LCL-Filtered Voltage Source Converters With Extended Selective Harmonic Compensation. *IEEE Transactions on Power Electronics*, 30(9):4726–4737, September 2015.

Influence of Power Semiconductor Device Variations on Pulse Shape of Nanosecond Pulses in a Solid-State Linear Transformer Driver

Raffael Risch, Anliang Hu and Jürgen Biela
Laboratory for High Power Electronic Systems, ETH Zurich
Email: risch@hpe.ee.ethz.ch
URL: http://www.hpe.ee.ethz.ch

Keywords

≪Wide bandgap devices≫, ≪Silicon Carbide (SiC)≫, ≪Pulsed power≫, ≪Modelling≫, ≪Statistics≫

Abstract

Power semiconductors show a significant variation in their electrical characteristics attributed to fluctuations during their fabrication process. This can lead to critical voltage and current imbalances in ultra-fast switching multi-cell topologies/pulse generators. This paper explores this problem based on a statistical model of a SiC MOSFET and Monte-Carlo simulations.

1 Introduction

Silicon carbide (SiC) MOSFETs with high breakdown voltages ($> 1.2\,\text{kV}$) are increasingly used in new pulse generator designs [1], [2]. One reason for this is the fast switching speed of SiC MOSFETs, which enables the generation of very short high-voltage pulses with pulse widths of only a few nanoseconds. Such short pulses could not be achieved with silicon-based IGBT devices in the past. Pulse generators with nanosecond high-voltage pulses are used in various fields, such as the generation of transient plasmas or in injection/extraction systems of particle accelerators.

To achieve high output voltages of several kilovolts, multi-cell topologies such as the solid-state Marx generator [3] or the linear transformer driver (LTD) [4], as depicted in Fig. 1(a), are frequently used. In these topologies, many devices are typically connected in series and/or in parallel. Differences in the characteristics (as e.g. current characteristics, device capacitances or internal gate resistance) of the devices — mainly caused by tolerances during the fabrication process of the semiconductor chips [5] — lead to an unsynchronized switching of the devices. This results in imbalanced device voltages and currents that might eventually cause device failures due to overcurrents in some devices.

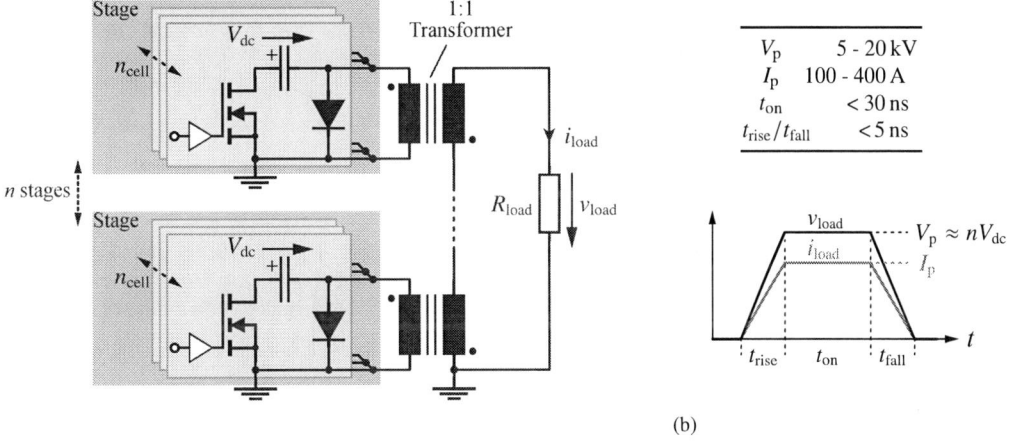

(a)   (b)

Fig. 1: (a) Solid-state LTD consisting of n series-connected stages. Each stage has n_{cell} SiC MOSFETs connected in parallel. (b) Pulse specifications of the target design.

The influence of device mismatches is even more pronounced when the individual SiC MOSFETs are operated at very high switching speeds. This is often necessary in very fast pulse applications in order to be able to meet the demanding pulse specifications. Therefore, besides harmful voltage and current imbalances, potential device mismatches can also have a significant impact on the output voltage pulse shape. For very fast pulse generators, the achievable rise and fall times of the output pulse are typically of primary interest.

In order to compensate for steady-state and transient imbalances, passive or active synchronization methods can be employed [6]–[9]. However, these methods are limited in their bandwidth and their operation is prone to malfunctioning at very fast switching speeds due to the large influence of parasitics. In addition, they usually slow down the switching speed of the SiC MOSFETs. Hence, these methods are generally not applicable for pulse generators, that aim for ultrafast nanosecond switching times.

Having only very limited possibilities to actively influence the balancing of device voltages and currents, it is crucial to be able to estimate the influence of device variations on the circuit performance. Missed performance goals or potential risks for device failures can be addressed by pre-screening semiconductor chips [10] and defining boundaries for the maximum allowed spread in the device characteristics.

Characteristic measurements of a large number of SiC MOSFETs showing the approximate range of device variations as well as methods for systematically analysing the variations are discussed in [10]–[12]. Methods to statistically analyse device variations are developed in [13], [14], primarily focusing on the current characteristics of the MOSFET. Furthermore, the effects of device variations on the current distribution of parallel-connected SiC MOSFETs are analysed. In [15], a basic analysis of the transient current waveforms is carried out. [16] investigates the unequal switching losses of parallel-connected SiC MOSFETs and the resulting thermal imbalances using statistical methods.

However, an analysis of the current and voltage distribution in multi-cell topologies with several mismatched SiC MOSFETs connected in series and/or parallel is missing. Especially when the switches are operated at very high switching speeds. Therefore, this paper investigates the behaviour of a multi-cell topology, using the example of the solid-state LTD illustrated in Fig. 1(a), under the influence of device variations of SiC MOSFETs. The parametrisation and target specifications of the load pulse are shown in Fig. 1(b). The focus is on the achievable output voltage switching times t_rise and t_fall and their dependence on the device variations.

Estimating the circuit performance under the influence of device variations requires statistical modelling of the device properties. The statistical modelling approach presented in this paper consists of three steps. First, a suitable device model for the SiC MOSFET is selected. A behavioural model is chosen since only the terminal behaviour of the MOSFET is of interest. In a second step, the variation of the device properties is statistically modelled. This is done based on measurements of different SiC MOSFET samples. In the last step, the statistical models of the SiC MOSFET are implemented in a circuit simulation and Monte Carlo (MC) simulations are performed to investigate the circuit performance of the LTD under the influence of varying device characteristics.

Section 2 of this paper explains the selected behavioural device model for the SiC MOSFET. The statistical modelling of the SiC MOSFETs is described in 3. Finally, section 4 presents the results of the Monte-Carlo simulations.

2 Behavioural Device Modelling of SiC MOSFET

Since only the terminal behaviour of the MOSFET is of interest, a behavioural model is chosen. In behavioural models, the underlying equations only model the electrical behaviour of the SiC MOSFET at its terminals. The equations do not directly describe the physical mechanisms by which the MOSFET operates. The basis for the modelling is usually a series of measurements of the current-voltage behaviour and the impedance behaviour at the terminals of the MOSFET. Based on these measurements, the parameters of the model equations can be determined in such a way that the resulting model behaviour closely matches the measurement curves.

Fig. 2(a) shows the behavioural SiC MOSFET model used in this paper. It contains the following submodels: The voltage-controlled current source i_ch, the voltage-dependent capacitances C_gs, C_dg and C_ds, the internal gate resistance $R_\text{g,int}$ and the body diode D_b.

The electrical behaviour of the submodels is described based on mathematical functions. In this context, four categories of mathematical functions can be distinguished [17]: Continuous functions, segmented functions based on case distinctions, look-up tables or combinations of the three. Due to the large number of simulations required for the statistical investigation in this work, good/fast convergence and short simulation times of the model are

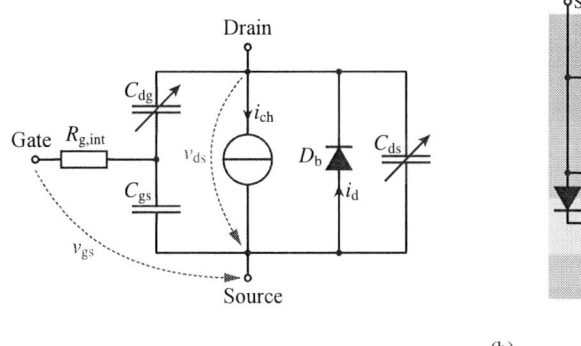 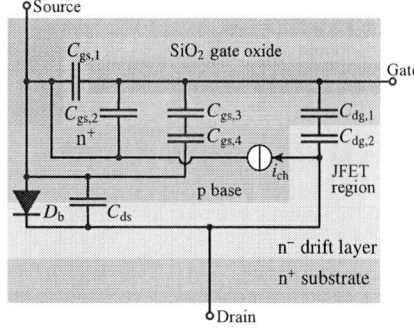

(a) (b)

Fig. 2: (a) Behavioural SiC MOSFET model consisting of: Voltage-dependent current source i_{ch}, voltage-dependent device capacitances C_{gs}, C_{ds} and C_{dg}, internal gate resistance $R_{\mathrm{g,int}}$ and body diode D_{b}. (b) Cell structure of a planar SiC MOSFET indicating the submodels of the behavioural model.

important. Segmented functions and look-up tables tend to have worse convergence properties than continuous functions [17]. Therefore, continuous functions are used in the following.

In addition, the main focus of this paper is on the simulation of very short voltage pulses. Therefore, the self-heating of the switch is neglected. As a consequence, the temperature dependence of the switch characteristics is not modelled and only the switch characteristics at room temperature $T \approx 25\,^{\circ}\mathrm{C}$ are considered.

2.1 Voltage-Controlled Current Source

The channel current i_{ch} of the MOSFET is mainly determined by the applied gate-source voltage v_{gs} and the drain-source voltage v_{ds}. Accordingly, the current function is formulated as a function of these two voltages. The continuous equation (1) is used, which is presented in [18]. It approximates the static current characteristics of the device using the fitting parameters $k_1 - k_{10}$. Essentially, the model equation consists of a multiplication of the terms $i_{\mathrm{ds,tran}}$ and $i_{\mathrm{ds,out}}$. The term $i_{\mathrm{ds,tran}}$ describes the transfer characteristics and the term $i_{\mathrm{ds,out}}$ models the output characteristics of the SiC MOSFET.

$$i_{\mathrm{ch}}(v_{\mathrm{ds}}, v_{\mathrm{gs}}) = i_{\mathrm{ds,tran}} \cdot i_{\mathrm{ds,out}} = k_1 \underbrace{\left[1 + \tanh\left(k_2(v_{\mathrm{gs}}+k_3) + k_4(v_{\mathrm{gs}}+k_5)^2 \right) \right]}_{i_{\mathrm{ds,tran}}} \cdot \underbrace{\frac{p(v_{\mathrm{gs}})v_{\mathrm{ds}}}{1 + q(v_{\mathrm{gs}})v_{\mathrm{ds}}}}_{i_{\mathrm{ds,out}}} \qquad (1)$$

The parameters $p(v_{\mathrm{gs}})$ and $q(v_{\mathrm{gs}})$ are exponential functions defined by (2) and (3). They are added to the current equation in order to take into account the dependency of the output characteristics on the gate-source voltage v_{gs}. As a result, a better fit of the output characteristics can be achieved. Compared to the original equation in [18], the fitting parameter multiplied with the exponential function in (2) is omitted because this parameter is not independent of the other fitting parameters and therefore, does not influence the resulting fitting curve.

$$p(v_{\mathrm{gs}}) = \exp(k_6 v_{\mathrm{gs}}) + k_7 \qquad (2)$$
$$q(v_{\mathrm{gs}}) = k_8 \exp(k_9 v_{\mathrm{gs}}) + k_{10} \qquad (3)$$

2.2 Device Capacitances

The gate oxide and the p-n junctions within a MOSFET cell result in distributed capacitances as shown in Fig. 2(b). These capacitances are charged and discharged during each switching operation. Thus, they also have a major influence on the switching behaviour and have to be taken into account for the transient modelling of the SiC MOSFET in addition to the current characteristics.

The distributed capacitances are typically modelled by the three lumped capacitances C_{gs}, C_{dg} and C_{ds}. The distributed capacitances can be related to the lumped capacitances as follows: $C_{\mathrm{gs}} = C_{\mathrm{gs,1}} + C_{\mathrm{gs,2}} + C_{\mathrm{gs,3}} + C_{\mathrm{gs,4}}$ and $C_{\mathrm{dg}} = C_{\mathrm{dg,1}} + C_{\mathrm{dg,2}}$ [19].

The capacitances C_{dg} and C_{ds} are largely dependent on the p-n junctions within the MOSFET cell. As a result, the capacitance values of C_{dg} and C_{ds} strongly decrease with larger reverse voltage because the widths of the depletion regions increase. In addition, both capacitance curves, but especially the capacitance curve of C_{dg}, show two different decay rates at low reverse voltages. This is caused by different doping concentrations in the JFET and the drift region of the MOSFET, resulting in two different expansion rates of the depletion region [20].

The decreasing characteristics in the capacitance curves of C_{dg} and C_{ds} are modelled by a sum of exponential functions with negative exponents. The sudden change of decay rates at low voltages is modelled by an additional sum of inverse exponential functions. In summary, (4) is used to model the capacitance curves of C_{dg} and C_{ds}.

$$C(v_{ds}) = \sum_{i=1} a_i e^{b_i v_{ds}} + \sum_{j=1} \frac{c_j}{1 + e^{d_j(v_{ds}-e_j)}} + f \tag{4}$$

In contrast to C_{dg} and C_{ds}, the gate-source capacitance C_{gs} is mainly determined by the oxide capacitance. This value is largely independent of the drain-source voltage and is therefore assumed to be constant.

2.3 Gate Resistance & Body Diode

A lumped gate resistor $R_{g,int}$ is used to model the distributed resistance of the gate structure of the SiC MOSFET. Furthermore, the Shockley equation in (5) is used to model the current behaviour of the body diode.

$$i_d(v_{sd}) = I_s \left(e^{(v_{sd} - R_s i_d)/NV_T} - 1 \right) \tag{5}$$

The fitting parameters are: The saturation current I_s, the series-resistance R_s and the emission coefficient N. V_T is the thermal voltage at $T = 25\,°C$.

2.4 Fitting of Model Parameters

A series of static current measurements and impedance measurements at different values of v_{gs} and v_{ds} are used to fit the model parameters. In order to achieve high model accuracy over the entire operating range, the measurement points used for fitting the model parameters should also cover the entire operating range. However, this raises an issue regarding the measurement of the static current characteristics. At higher drain-source voltages, the self-heating of the MOSFET distorts the measurement results. For this reason, the current characteristics is only measured up to a maximum drain-source voltage of $10\,V$. To increase the modelling accuracy over the whole operating range, the measurement range of the curve tracer would need to be extended, e.g. by the method explained in [21].

The fitting of the model parameters to the measurement curves represents a non-linear optimisation problem that is solved using the least squares method. Specifically, the Nelder-Mead simplex algorithm is applied to find a solution to the minimisation problem in (6).

$$\min_{\boldsymbol{p}} \|y_m(\boldsymbol{p}, \boldsymbol{v}) - y_{meas}\|_2^2 \tag{6}$$

Here, y_m represents the model functions (1)–(5), each consisting of $\boldsymbol{p}$ modelling parameters. y_{meas} are the measurements of the modelled quantities at the measurement voltages $\boldsymbol{v}$.

2.5 Model Implementation in Circuit Simulation

SPICE is used to implement the circuit model. The current equation (1) is implemented based on a voltage-dependent current source. Regarding the implementation of voltage-dependent capacitances in SPICE, there have been various methods presented in literature [22]. With respect to a capacitance-based implementation, there are two general ways: Either using the intrinsic *ddt* function in SPICE, which calculates the discrete time derivative of the capacitor voltage, or using the intrinsic voltage-current relation of a capacitance. Using the intrinsic voltage-current relation of a capacitance has several advantages, which are summarized in [22]. Therefore, this method is chosen. It is shortly described in the following.

The circuit for modelling a voltage-dependent capacitance is depicted in Fig. 3. It consists of a controlled current source G_C, a voltage-controlled voltage source E_C and a reference capacitance C_{ref}. C_{ref} is used to generate the time derivative of the capacitor voltage. The capacitance current i_C is directly determined by G_C, which is equal to the reference current i_{ref} multiplied by the expression $C(v_C)$ for the capacitance. Furthermore, choosing E_C to be equal to v_C/C_{ref} results in equation (7).

$$i_C = G_C = C(v_C)i_{ref} = C(v_C)C_{ref}\frac{dE_C}{dt} = C(v_C)\frac{dv_C}{dt} \tag{7}$$

EPE'22 ECCE Europe

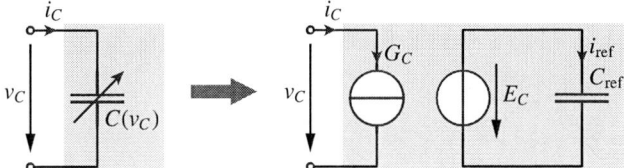

Fig. 3: Circuit used in SPICE to implement voltage-dependent capacitances, specifically C_{dg} and C_{ds}.

For the implementation of the body diode, the intrinsic diode model of SPICE is used. This model directly implements the Shockley equation (5).

3 Statistical Parameter Modelling

Tolerances during the manufacturing cause each switch to have slightly different properties. The switch properties are described by the fitted parameters of the behavioural model derived in section 2. For each switch, this results in N_{param} model parameters. Consequently, the fitting of a sample set with N_{sample} switches results in an N_{param}-dimensional point set consisting of N_{sample} points, where each point corresponds to a modelled switch. The point set determined in this way reflects the variation of the switch properties of the N_{sample} samples.

This set of points can be modelled using statistical equations. This allows to generate an arbitrarily large theoretical sample set, which has the same statistical properties as the measured sample set. Thereby, the correlation between submodels is neglected. Furthermore, not all model parameters have the same influence on the variation of the switch properties. Usually, the switch variation can be modelled with sufficient accuracy by taking into account only a few dominant parameters, as will be shown in the next subsections.

In the following, the statistical modelling of the submodels for the current characteristics, the device capacitances and the internal gate resistance is explained. The variation of the parameters of the diode equation is neglected. The basis for the statistical modelling is the measurement of 10 SiC MOSFETs of the type C3M0075120D. The characteristic measurements were carried out with a B1506A power device analyzer from Keysight.

3.1 Statistical Modelling of Current

The current characteristic of a MOSFET is essentially determined by the structure of the MOSFET cell. The most influential parameters are the epitaxial layer, the channel length, the interface traps and the inversion layer [5]. Variations of these parameters mainly affect the threshold voltage and the transconductance of the MOSFET.

The transconductance essentially describes the current amplification. In the current equation (1), this corresponds to the parameter k_1. The threshold voltage on the other side describes the gate-source voltage at which the MOSFET channel becomes electrically conductive. With reference to the transfer characteristic of a MOSFET, a change in the threshold voltage primarily causes a shift in the characteristic curve along the v_{gs}-axis. The threshold voltage does not explicitly appear as a parameter in the current equation. However, the same effect can be modelled with the parameter k_3.

To determine the statistical distribution of k_1 and k_3 as well as the value of the other parameters of the current equation, the following two steps are performed:

1) First, the mean values of the model parameters $k_1 - k_{10}$ are determined. For this purpose, all measurement curves of the N_{sample} samples are taken into account at the same time for fitting $k_1 - k_{10}$.
2) Subsequently, the statistical distributions of the dominant parameters k_1 and k_3 are determined. For this purpose, the previously determined mean values are first assigned to the other parameters. Then k_1 and k_3 are individually fitted to the N_{sample} characteristic current measurements. The resulting distributions of k_1 and k_3 consist of N_{sample} values, each of which can be fitted by Gaussian distributions [14]. This results in (μ, σ) pairs for both k_1 and k_3, where μ is the mean and σ is the standard deviation of the respective statistical distribution.

The parameters determined in this way are summarised in Table I. To check the accuracy of this procedure, the measured variation of the current characteristic is re-simulated based on 100 Monte Carlo simulations. The results are depicted in Fig. 4(a) and show good agreement with the measured current variation.

Table I: Fitting parameters of SiC MOSFET C3M0075120D. The distributions of statistical parameters are highlighted in bold. They are indicated with (μ, σ) pairs, where μ is the mean value and σ is the standard deviation.

Current model

k_1	$(-5.95, 49.6e-3)$	k_2	$264e-3$	k_3	$(-8.0, 137e-3)$	k_4 $21.4e-3$	k_5	-7.78
k_6	$-32.2e-3$	k_7	-1.71	k_8	32.3	k_9 $-631e-3$	k_{10}	$75.9e-3$

Gate resistance & Body diode model

$R_{\mathrm{g,int}}$	$(10.02, 129.4e-3)$	I_{s}	$102.6e-3$	R_{s}	$14.7e-3$	N	41.04

Capacitance model C_{gs}

C_{gs}	$(1.184e-9, 8.6384e-12)$

Capacitance model C_{ds}

a_1	$(618.1e-12, 3.6e-12)$	b_1	$(-94.7e-3, 1.9e-3)$	c_1	$11.7e-12$	d_1	-1.553	e_1	5.45
a_2	$120.1e-12$	b_2	-1.13	c_2	$23.2e-12$	d_2	$-164.3e-3$	e_2	31.22
a_3	$94.71e-12$	b_3	$-5.47e-3$	c_3	$115.1e-12$	d_3	$-239.9e-3$	e_3	12.12
a_4	$121.8e-12$	b_4	$-27.3e-3$	c_4	$-57.9e-12$	d_4	-3.27	e_4	9.45
f	$-48.8e-12$								

Capacitance model C_{dg}

a_1	$(167.6e-12, 7.4e-12)$	b_1	$(-257.6e-3, 7.6e-3)$	c_1	$29.9e-12$	d_1	-2.75	e_1	-75.99
a_2	$181.3e-12$	b_2	-1.86	c_2	$-30.9e-12$	d_2	-15.78	e_2	9.03
f	$3.86e-12$								

3.2 Statistical Modelling of Device Capacitances

Process fluctuations during chip manufacturing also cause a variation of the capacitance curves of C_{gs}, C_{dg} and C_{ds}. The variation of the capacitance curves can essentially be described by a scattering of the DC value $C(v=0)$ and in the case of C_{dg} and C_{ds} by a scattering of the capacitance change dC/dv.

With respect to the capacitance equation (4) used for C_{ds} and C_{dg}, the DC value and the capacitance change are mainly determined by the sum of the exponential terms with the model parameters a_i and b_i. Therefore, the variation of the capacitance curves is modelled based on these parameters. Usually, several exponential terms are necessary to achieve an accurate fit of the capacitance curves. In the following, only the parameters of one exponential term are used to describe the variation of the capacitance curves, as this already allows for a sufficiently accurate representation of the variation. The remaining parameters are considered constant. Furthermore, C_{gs} is modelled by a voltage-independent capacitance as already described in section 2. Consequently, the statistical parameter in this case is the capacitance value C_{gs} itself.

The procedure for determining the statistical distribution of the dominant capacitance parameters is the same as for the statistical modelling of the current. The resulting parameters are summarised in Table I. The variation of the capacitance characteristics has been again re-simulated based on 100 MC simulations. The results are depicted in Fig. 4(c)-(e) and show good agreement with the measured variation.

3.3 Statistical Modelling of Internal Gate Resistance

The internal gate resistance $R_{\mathrm{g,int}}$ primarily depends on the film thickness and the resistivity of the gate electrode material [20]. It is modelled by a single constant parameter $R_{\mathrm{g,int}}$. Therefore, $R_{\mathrm{g,int}}$ itself is modelled as a statistical variable. The resulting statistical distribution is also listed in Table I. Fig. 4(b) shows the re-simulation, which also shows good agreement with the measured variations.

4 Monte-Carlo Simulation of the LTD

The parallel and/or series connection of switches with different characteristics leads to imbalanced voltage and current distributions within the circuit during switching operations. In a pulse generator, asymmetrical voltage and current distributions have an influence on the pulse shape of the output voltage. In very fast pulse generators, the influence on the rise and fall time of the output voltage is of particular interest. To determine an expected

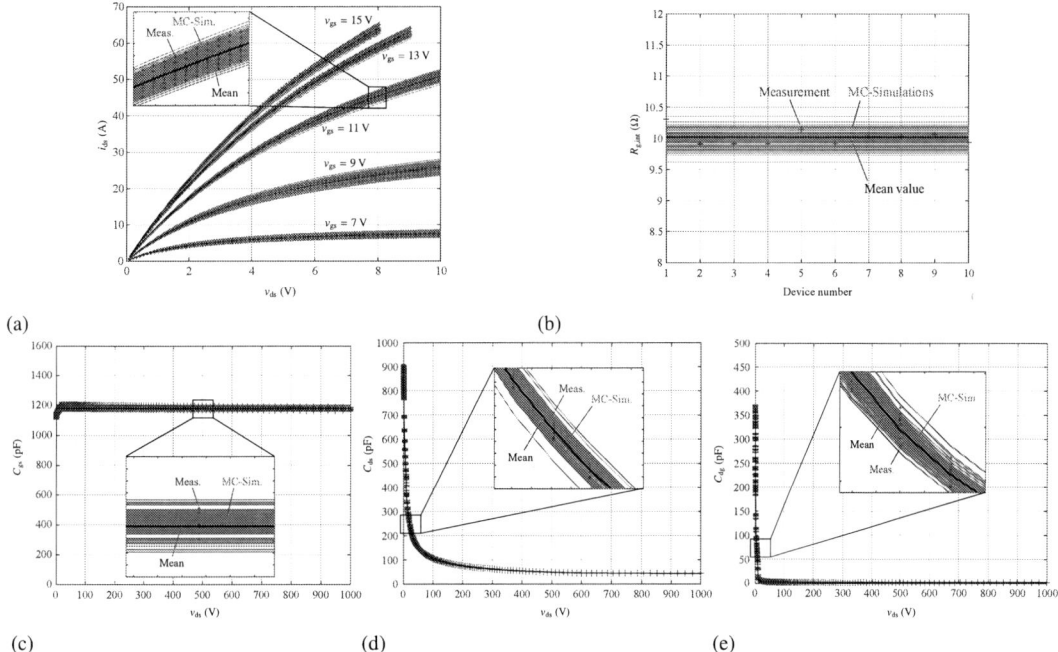

(a)

(b)

(c)

(d)

(e)

Fig. 4: (a) Output characteristics of current, (b) $R_{g,int}$, (c) C_{gs}, (d) C_{ds} and (e) C_{dg}. The measurement points are marked with blue crosses. The fitted mean value is indicated with a black line. The simulation of the measured spread is shown in red.

value for the rise and fall time under the influence of device variations, a suitable statistical analysis of the circuit is required.

A simple worst-case analysis does not provide any information about the expected value and statistical distribution of the circuit variables. The conclusions drawn from such an analysis are therefore often too restrictive. In contrast, a statistical analysis based on MC simulations is more meaningful because it also takes probabilities into account. The basis for a statistical analysis of a circuit topology is the statistical modelling of the switch, which is discussed in section 3. The behaviour of the topology can then be modelled by independently sampling the statistical equations of each switch.

4.1 Circuit Model & Simulation Parameters

To assess the influence of variations of the switch parameters on the switching behaviour of a pulse generator, the LTD shown in Fig. 1(a) is investigated. A more detailed model of a single stage of the LTD consisting of n_{cell} parallel-connected switching cells is shown in Fig. 5. Component parasitics are highlighted in green. Parasitics from the layout of the switching cell and the mechanical arrangement are highlighted in blue. The transformer core is modelled with a core resistance R_c and a magnetising inductance L_m. The capacitance of the DC link is chosen to be large enough such that the voltage drop due to the discharge of the capacitors during the flat pulse can be neglected. Therefore, the DC link capacitors are replaced by a constant voltage source. A detailed explanation of the modelled parasitics including their values can be found in [23]. The gate driver circuit is modelled by the voltage source v_{drv} and a series resistor $R_{on(drv)}$. The simulation parameters used for the MC simulations are listed in Table II. With these parameters an output voltage amplitude of 5 kV and a load current amplitude of 100 A for the LTD results. Further, it is assumed that synchronised gate voltages are applied to the switches, i.e. that all gate voltages are equal. Considerations of variations in the gate driver circuits or jitter effects in the trigger signals is out of the scope of this paper and are not taken into account.

4.2 Simulation Results of Sample Measurements

In the following, the influence of variations of the switch parameters on the synchronization of the switches and on the rise and fall time of the output pulse are investigated. The variations in the device characteristics are based on the measured samples of section 3.

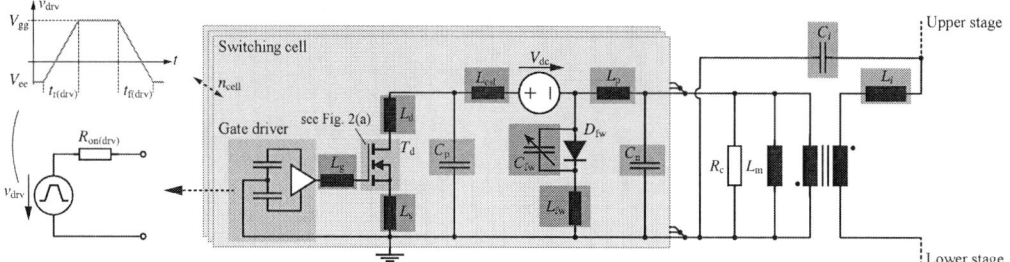

Fig. 5: Circuit model of the i-th LTD stage including stray inductances and capacitances.

Table II: Parameter values used for the MC simulations. Detailed parameter values of the parasitics are given in [23].

Topology							
V_{dc}	720 V	n	7	n_{cell}	4	R_{load}	50 Ω

Gate driver							
V_{gg}	18 V	V_{ee}	−5 V	$t_{r(drv)}$	10 ns	$t_{f(drv)}$	10 ns	$R_{on(drv)}$	0.6 Ω

The synchronicity of the SiC MOSFET switching is determined by looking at the turn-on and turn-off times. In this context, the turn-on, respectively turn-off time is defined as the time instant when the drain current of the respective switch reaches 10 % of the nominal switch current. Subsequently, the maximum deviation of the turn-on times Δt_{on} and the turn-off times Δt_{off} of all switches are calculated for each MC simulation. For a better understanding, Δt_{on} and Δt_{off} are illustrated in Fig. 6(a) where, as an example, the switch currents of all 28 SiC MOSFET are plotted for varying current characteristics. The rise and fall time of the output pulse is measured between 10 % and 90 % of the voltage amplitude. The times t_{rise} and t_{fall} are illustrated in Fig. 6(b), where the output voltage of the example simulation is shown.

In the following, the influence of variations in the current characteristics, the device capacitances and the internal gate resistance on Δt_{on}, Δt_{off}, t_{rise} and t_{fall} is investigated. The results are shown in Fig. 7. The expected values are added to each histogram. During the MC simulations of one submodel, the parameters of the other submodels are kept constant. Each histogram is based on 1000 MC simulations.

The parameter variation of the MOSFETs leads to an unsynchronised switching in the range of a few hundred picoseconds. As a result, the current distribution between parallel-connected SiC MOSFETs is no longer balanced and a single MOSFET has to switch a current that is larger than the nominal current. The switching times typically scale with the switched current. As a result, the expected values of the switching times are higher

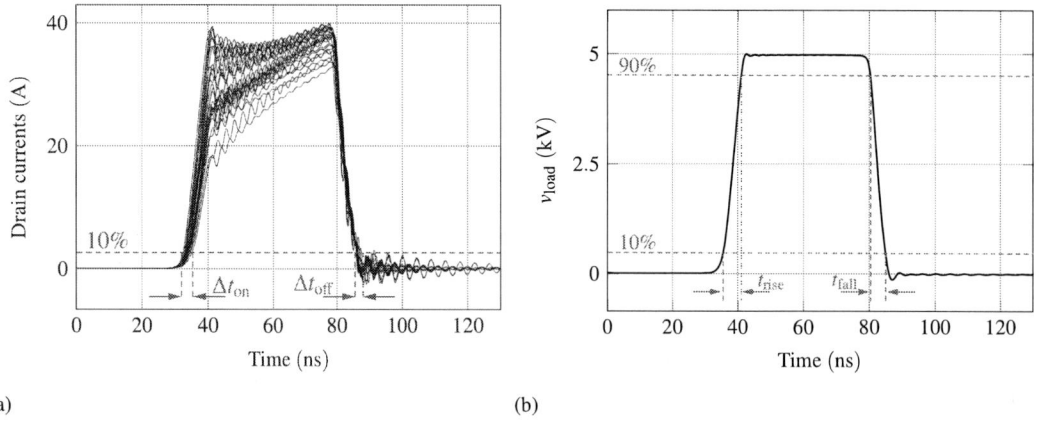

(a)   (b)

Fig. 6: Simulation example for a varying current characteristics: (a) Drain currents of the 28 SiC MOSFETs and (b) the output voltage. Also indicated are the investigated key figures.

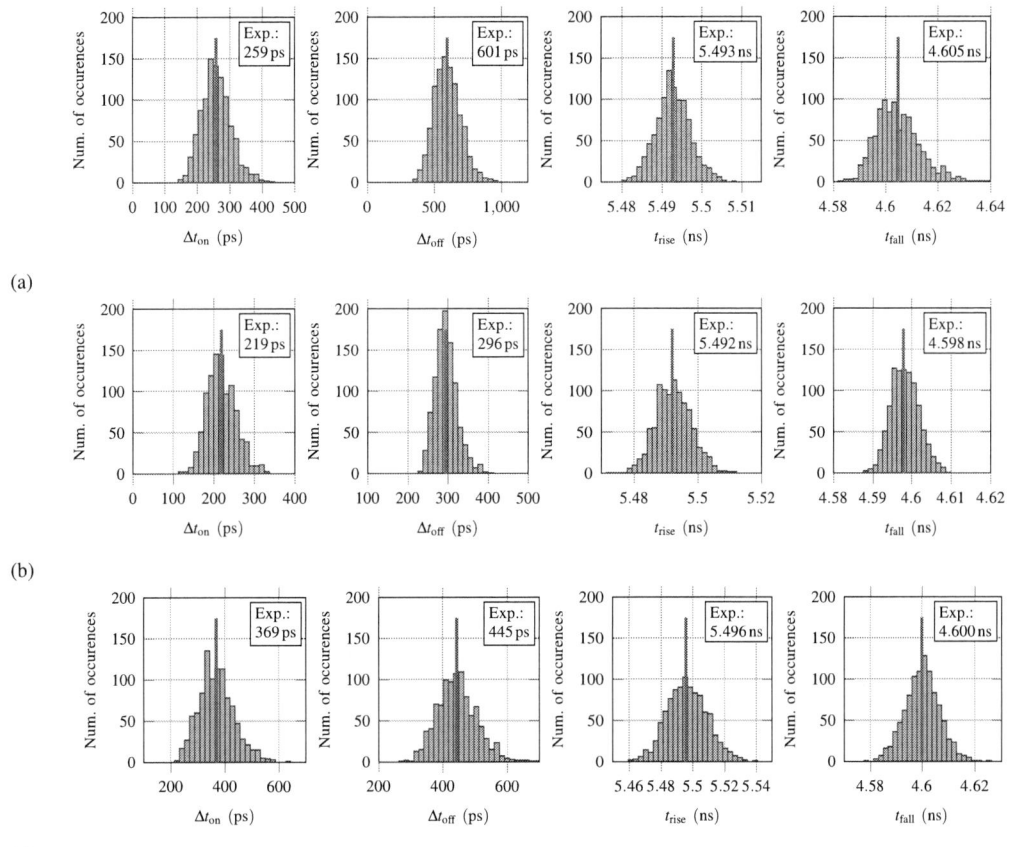

(a)

(b)

(c)

Fig. 7: Histograms of Δt_{on}, Δt_{off}, t_{rise} and t_{fall} for parameter variations in (a) the current model, (b) the device capacitances and (c) the internal gate resistance. Each histogram is based on 1000 MC simulations. The expected values are also indicated and marked with a red line.

than the values for perfect synchronisation, which are $t_{\text{rise}} = 5.491\,\text{ns}$ and $t_{\text{fall}} = 4.598\,\text{ns}$. However, one has to mention that the switching times only increase by a small amount, partly because the parameter variations of the measured MOSFETs are rather small.

4.3 Dependence on Degree of Variation

The simulation results in the previous subsection have shown that under the influence of device parameter variations, the switches no longer switch synchronously and thus the expected rise and fall time of the output voltage of the LTD increase. This section examines the dependency of the synchronicity and the rise and fall time on the degree of variation, expressed by the respective standard deviations of the submodels.

As a basis σ_{base} for each submodel, the standard deviations extracted in section 3 for the 10 MOSFET samples are used. To increase the degree of variation, the standard deviations are successively increased by a multiple of the respective σ_{base} of each submodel. For each value of the standard deviation, 1000 MC simulations are carried out and the respective expected values are calculated.

The results are summarised in Fig. 8. The Δt_{on} and Δt_{off} increase with a larger degree of variation. Similarly, t_{rise} and t_{fall} slightly increase with a larger range of parameter variation because the transient current distribution is more unevenly distributed. For instance, to limit the increase of the rise/fall time to 5%, the parameter variation should be confined to approximately $6\sigma_{\text{base}}$.

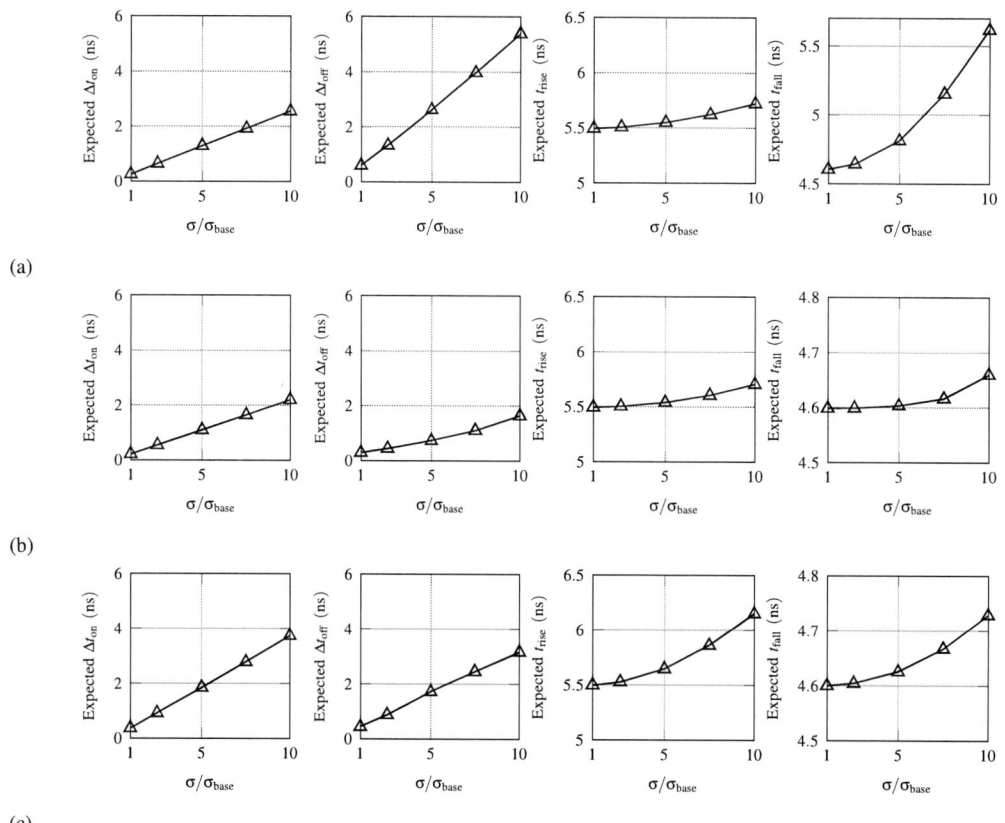

(a)

(b)

(c)

Fig. 8: Expected values of Δt_{on}, Δt_{off}, t_{rise} and t_{fall} for different standard deviations of the parameter variation of (a) the current model, (b) the device capacitances, and (c) the internal gate resistance. Each simulation point is based on 1000 MC simulations.

5 Conclusion

For a comprehensive assessment of the pulse performance of a nanosecond solid-state LTD, parameter variations of the employed power semiconductors have to be taken into account. For this purpose, statistical modelling of the SiC MOSFETs is applied.

In a first step, the SiC MOSFET is modelled using a behavioural model with separate submodels for the current characteristics, the device capacitances and the internal gate resistance. Subsequently, the dominant model parameters, which have the largest influence on the device variations, are identified for each submodel. The statistical distribution of these dominant parameters are then calculated based on characteristic measurements of 10 SiC MOSFETs. In a last step, MC simulations of the LTD are performed in order to investigate the influence of device variations on the synchronicity of the SiC MOSFETs as well as the rise and fall times of the output voltage.

The simulation results show that device variations lead to a unsynchronised switching of the SiC MOSFETs, which can easily reach a few hundred picoseconds and increases approximately linearly with increasing range of variation. This results in an imbalanced current distribution among parallel-connected switches, which can eventually lead to temperature differences among the switches. As a consequence, some switches might experience an overstress, potentially reducing the lifetime and the reliability. As a further consequence of the current imbalance, the expected values of the output voltage rise/fall times of the sample MOSFETs are slightly higher than for perfect synchronisation. With increasing degree of device variation, the expected values of the rise/fall times further increases by several hundred picoseconds.

References

[1] L. M. Redondo, A. Kandratsyeu, and M. J. Barnes, "Marx Generator Prototype for Kicker Magnets Based on SiC MOSFETs", *IEEE Trans. Plasma Sci.*, vol. 46, no. 10, pp. 3334–3339, Oct. 2018.

[2] L. Pang, T. Long, K. He, Y. Huang, and Q. Zhang, "A Compact Series-Connected SiC MOSFETs Module and Its Application in High Voltage Nanosecond Pulse Generator", *IEEE Trans. Ind. Electron.*, vol. 66, no. 12, pp. 9238–9247, Dec. 2019.

[3] T. Huiskamp and J. J. Van Oorschot, "Fast Pulsed Power Generation With a Solid-State Impedance-Matched Marx Generator: Concept, Design, and First Implementation", *IEEE Trans. Plasma Sci.*, vol. 47, no. 9, pp. 4350–4360, Sep. 2019.

[4] F. Yu, T. Sugai, A. Tokuchi, and W. Jiang, "Development of Solid-State LTD Module Using Silicon Carbide MOSFETs", *IEEE Trans. Plasma Sci.*, vol. 47, no. 11, pp. 5037–5041, Nov. 2019.

[5] J. Mueting, P. Natzke, A. Tsibizov, and U. Grossner, "Influence of Process Variations on the Electrical Performance of SiC Power MOSFETs", *IEEE Trans. Electron Devices*, vol. 68, no. 1, pp. 230–235, Jan. 2021.

[6] T. Wang, H. Lin, and S. Liu, "An Active Voltage Balancing Control Based on Adjusting Driving Signal Time Delay for Series-Connected SiC MOSFETs", *IEEE J. Emerg. Sel. Top. Power Electron.*, vol. 8, no. 1, pp. 454–464, Mar. 2020.

[7] C. Yang, Y. Pei, Y. Xu, *et al.*, "A Gate Drive Circuit and Dynamic Voltage Balancing Control Method Suitable for Series-Connected SiC MOSFETs", *IEEE Trans. Power Electron.*, vol. 35, no. 6, pp. 6625–6635, Jun. 2020.

[8] Y. Wen, Y. Yang, and Y. Gao, "Active Gate Driver for Improving Current Sharing Performance of Paralleled High-Power SiC MOSFET Modules", *IEEE Trans. Power Electron.*, vol. 36, no. 2, pp. 1491–1505, Jun. 2020.

[9] C. Zhao, L. Wang, F. Zhang, and F. Yang, "A Method to Balance Dynamic Current of Paralleled SiC MOSFETs With Kelvin Connection Based on Response Surface Model and Nonlinear Optimization", *IEEE Trans. Power Electron.*, vol. 36, no. 2, pp. 2068–2079, Feb. 2021.

[10] J. Ke, Z. Zhao, P. Sun, H. Huang, J. Abuogo, and X. Cui, "Chips Classification for Suppressing Transient Current Imbalance of Parallel-Connected Silicon Carbide MOSFETs", *IEEE Trans. Power Electron.*, vol. 35, no. 4, pp. 3963–3972, Apr. 2020.

[11] J. Ke, Z. Zhao, Q. Zou, J. Peng, Z. Chen, and X. Cui, "Device Screening Strategy for Balancing Short-Circuit Behavior of Paralleling Silicon Carbide MOSFETs", *IEEE Trans. Device Mater. Rel.*, vol. 19, no. 4, pp. 757–765, Dec. 2019.

[12] B. Zhao, Q. Yu, P. Sun, Y. Cai, and Z. Zhao, "Device Screening Strategy for Suppressing Current Imbalance in Parallel-Connected SiC MOSFETs", *IEEE Trans. Device Mat. Rel.*, vol. 21, no. 4, pp. 556–568, Dec. 2021.

[13] H. Tsukamoto, M. Shintani, and T. Sato, "A study on statistical parameter modeling of power MOSFET model by principal component analysis", in *Proc. IEEE Int. Conf. Microelect. Test Struct.*, Mar. 2019.

[14] ——, "Statistical Extraction of Normally and Lognormally Distributed Model Parameters for Power MOSFETs", *IEEE Trans. Semicond. Manufact.*, vol. 33, no. 2, pp. 150–158, May 2020.

[15] H. Li, S. Munk-Nielsen, X. Wang, *et al.*, "Influences of Device and Circuit Mismatches on Paralleling Silicon Carbide MOSFETs", *IEEE Trans. Power Electron.*, vol. 31, no. 1, pp. 621–634, Jan. 2016.

[16] A. Borghese, M. Riccio, A. Fayyaz, *et al.*, "Statistical analysis of the electrothermal imbalances of mismatched parallel SiC power MOSFETs", *IEEE J. Emerg. Sel. Top. Power Electron.*, vol. 7, no. 3, pp. 1527–1538, Sep. 2019.

[17] B. W. Nelson, A. N. Lemmon, B. T. DeBoi, *et al.*, "Computational Efficiency Analysis of SiC MOSFET Models in SPICE: Static Behavior", *IEEE Open Journal Power Electron.*, vol. 1, pp. 499–512, Nov. 2020.

[18] H. Li, X. Zhao, K. Sun, Z. Zhao, G. Cao, and T. Q. Zheng, "A Non-Segmented PSpice Model of SiC MOSFET With Temperature-Dependent Parameters", *IEEE Trans. Power Electron.*, vol. 34, no. 5, pp. 4603–4612, May 2019.

[19] R. Stark, A. Tsibizov, N. Nain, U. Grossner, and I. Kovacevic-Badstuebner, "Accuracy of Three Interterminal Capacitance Models for SiC Power MOSFETs Under Fast Switching", *IEEE Trans. Power Electron.*, vol. 36, no. 8, pp. 9398–9410, Aug. 2021.

[20] J. Ke, Z. Zhao, P. Sun, H. Huang, J. Abuogo, and X. Cui, "Influence of Device Parameters Spread on Current Distribution of Paralleled Silicon Carbide MOSFETs", *J. Power Electron.*, vol. 19, no. 4, pp. 1054–1067, Jul. 2019.

[21] A. Endruschat, T. Heckel, H. Gerstner, C. Joffe, B. Eckardt, and M. Maerz, "Application-related characterization and theoretical potential of wide-bandgap devices", in *Proc. IEEE Workshop Wide Bandgap Power Devices Appl.*, Oct. 2017.

[22] B. W. Nelson, A. N. Lemmon, S. J. Jimenez, *et al.*, "Computational Efficiency Analysis of SiC MOSFET Models in SPICE: Dynamic Behavior", *IEEE Open Journal Power Electron.*, vol. 2, pp. 106–123, Feb. 2021.

[23] R. Risch and J. Biela, "Solid-State Marx Generator vs. Linear Transformer Driver: Comparison of Parasitics and Pulse Waveforms for Nanosecond Pulsers", in *Proc. IEEE Pulsed Power Conf.*, Dec. 2021.

Optimal design of integrated motor drives - Comparison of topologies (2L/3L/modular), PWM variants, and switch technologies (Si/SiC/GaN)

Thilo Bringezu and Jürgen Biela
Laboratory for high power electronic systems (HPE), ETH Zürich
bringezu@hpe.ee.ethz.ch

Keywords

≪Optimal converter design≫ ≪Electrical drive≫ ≪System integration≫ ≪Integrated motor drive≫

Abstract

In this paper, a new design procedure for the optimal design of an integrated motor drive is presented, including an extended iron loss model. The design procedure is based on a multi-objective optimization of power density, efficiency, and cost. In the optimization, a large design space is covered, including the inverter topology, the PWM scheme, the chip technology (Si/SiC/GaN), the winding scheme, the chip area/cost, and the switching frequency. In addition to power density/efficiency/cost, the system reliability is investigated. Considering a 1.5 kW IMD as example, the optimal design in terms of efficiency and cost is achieved using a modular topology, GaN HEMTs, and a 9-phase motor winding. This design enables an efficiency increase of +2.26 % at 36 % higher cost compared to the cost-optimal design that is achieved with the standard 2L-topology, Si IGBTs, and a 3-phase motor winding.

1 Introduction

Integrated motor drives (IMDs) combine an electric motor and its driving inverter in one mechanical unit [1]. A basic question during the design of such IMDs is what is the optimal inverter topology and what are the optimal topology-related design parameters such as the PWM scheme and the chip technology (Si/SiC/GaN) in terms of power density/efficiency/cost.

Existing literature already covers several topology comparisons for inverters in general [2–4], and for IMDs in particular [1, 5, 6]. According to [1], modular topologies are advantageous due to their potential for higher fault-tolerance, smaller size, and lower cost. In [5], the focus is on different topologies for the rectifier whereas for the inverter always the same 2L-topology is considered and the motor

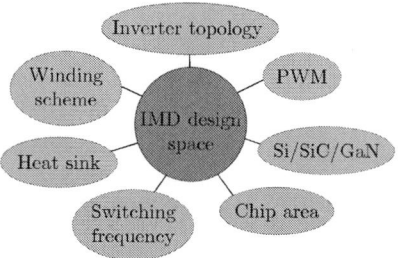

Figure 1: Design space considered for the optimal design of IMDs.

has a 3-phase winding. However, [5] considers neither different topologies for the inverter, nor multiphase motor windings. In addition, a comprehensive design procedure that enables a holistic comparison of modular inverter topologies and standard 2L- and 3L-topologies is missing. A limited approach of such a comparison is given by [6], wherein standard 2L- and 3L- inverter topologies are compared to different modular inverter topologies such as a *2L-2S-VSI* (series connection of two 2L-modules) or a *3L-2P-VSI* (parallel connection of two 3L-modules). However, the approach in [6] is limited because the chip area/size is not optimized and the analysis only focuses on efficiency, i.e. power density and cost are not included in the comparison.

To fill this gap, this paper presents a design procedure for a holistic comparison of modular converter topologies and standard 2L- and 3L-topologies. In this context, "holistic" means that multiple design goals (power density/efficiency/cost/reliability) and a relatively large set of topology-related design parameters is considered. An overview of the considered design parameters (i.e. the design space), is shown in Fig. 1. Therein, the variable *winding scheme* indicates that also multiphase motors are considered in this paper.

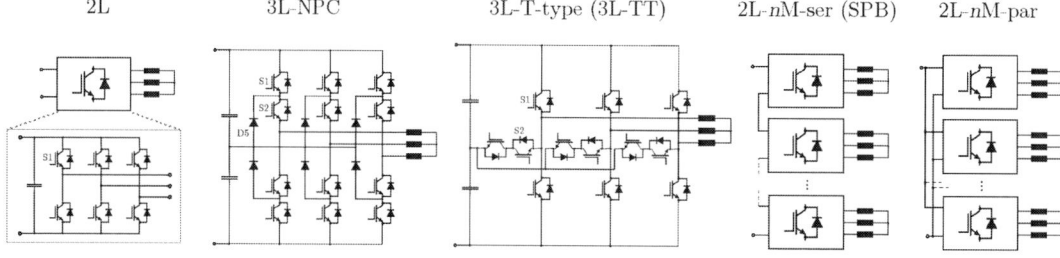

Figure 2: Inverter topologies considered for the IMDs. The topologies 2L, 3L-NPC, and 3L-TT are standard non-modular topologies that are connected to the 3-phase motor. The topologies 2L-nM-ser & 2L-nM-par are modular topologies that consist of $n \times$ 2L-modules connected in series (2L-nM-ser) or parallel (2L-nM-par) and that are connected to multiphase motors.

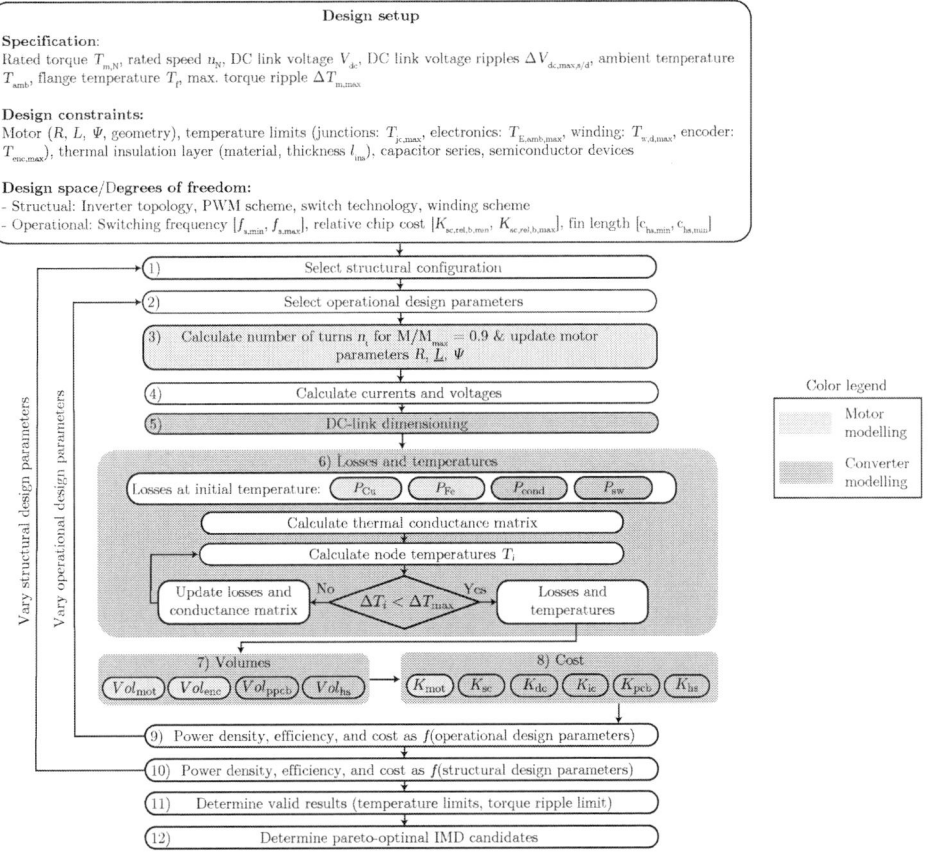

Figure 3: Procedure for the optimal design of an IMD in terms of power density, efficiency, and cost.

In the following, first the proposed design procedure is presented in section 2. The models used for the design procedure and for evaluating the system reliability are given in section 3. In section 4, the results of applying the design procedure to an exemplary system specification of a 1.5 kW high-torque, low-speed motor are presented and discussed. In section 5, the reliability of the considered converter topologies is analysed and compared.

2 Design procedure

Fig. 3 shows the proposed procedure for the IMD design for a fix DC link voltage and a given PMSM, where only the number of turns of the stator winding is varied. The considered integration concept is an axially integrated IMD, which is shown in Fig. 4 and explained in detail in [7]. In the following, the design setup and the different steps of the design procedure are explained.

2.1 Design setup - Specification, constraints, and design space

The start of the design procedure in Fig. 3 is to initialise the system specification, the design constraints, and the design space. The system specification comprises the DC link specification, the motor specification, and the ambient temperatures (air/flange). The design constraints for the different parameters are determined by the selected motor, the semiconductor devices, the capacitor series, the electronics temperature rating, and the thermal insulation layer.

The overall design space is split into structural degrees of freedom and operational degrees of freedom

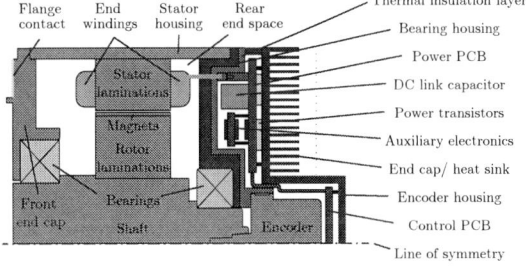

Figure 4: Inverter integration concept [7].

(DOFs). The structural DOFs in Fig. 3 define the basic system configuration, i.e. they describe a combination of an inverter topology, a PWM scheme, a semiconductor technology, and a winding scheme. The considered system configurations are summarized in Tab. I. The procedure could be extended to other configurations as well.

Table I: Considered system configurations. In case a certain *PWM-/SC-/W*-variant is only combined with particular topologies, these topologies are indicated in parenthesis.

Topology *Top*	Modulation scheme *PWM*	Semiconductor technology *SC*	Winding scheme *W*
1: 2L	1: Sin-PWM	1: Si	1: 3-phase ($Top = 1/2/3$)
2: 3L-NPC	2: OC-PWM ($Top = 1/4/5$)	2: SiC	2: 9-phase ($Top = 4/5$)
3: 3L-TT	3: NPB-PWM ($Top = 2/3$)	3: GaN	
4: 2L-3M-ser			
5: 2L-3M-par			

The selected topologies are shown in Fig. 2. This selection of topologies is based on the motivation to compare standard non-modular topologies (2L, 3L-NPC, 3L-TT) to modular topologies (2L-3M-ser, 2L-3M-par), that are particularly developed/presented for IMDs in literature [6, 8]. As PWM schemes, standard Sinus-PWM (Sin) and advanced schemes such as Optimal Clamped-PWM (OC) and Neutral Point Balanced-PWM (NPB) are considered. Regarding semiconductor technologies, Si IGBTs, SiC MOSFETs, and GaN HEMTs are considered with an overall voltage blocking capability of 1200 V for DC link voltages up to 800 V. Concerning the winding scheme, non-modular topologies are combined with a standard 3-phase winding and modular topologies are combined with a non-phase-aligned 3x3-phase winding, called 9-phase in the following. This winding scheme is chosen, because for the same topology, it is already known to perform better than phase-aligned 3x3-phase windings due to a higher winding factor [9].

The operational DOFs in Fig. 3 are design parameters that are linked to the relevant trade-offs in the design goals (system power density ρ_{sys}, system efficiency η_{sys}, system cost K_{sys}). These trade-offs are:

1. With increasing switching frequency the switching losses increase, but harmonic motor losses and the torque ripple are decreased in the standard case of 3-phase motors. With multiphase-motors like 9-phase motors, the switching frequency also affects the harmonic motor losses and the torque ripple but the dependency is more complex due to the fact that multiphase motors have multiple inductances for different harmonic sequences.

2. Increasing the heat sink size decreases the power density, but increases efficiency due to the temperature dependence of the conduction losses ($R_{ds,on}$) and the motor losses.

3. Increasing the chip area/chip cost decreases conduction losses, but increases the system cost. Depending on the $E_{on}(I)/E_{off}(I)$-curves, increasing the chip area/chip cost can furthermore increase switching losses due to the larger parasitic capacitances that have to be charged/discharged during switching.

The relative chip cost $K_{sc,rel,b} := K_{sc}/K_{sc,b}$ is used as an indicator for the chip area, where $K_{sc,b}$ is a chip cost budget that is constant for all system configurations to ensure a fair comparison.

2.2 Calculation steps of the design procedure

The procedure in Fig. 3 starts in step 1) with an outer loop (brute force) over the structural DOFs and continues, in step 2), with an inner loop over the operational DOFs. In each iteration, which starts in step 3), the number of turns n_t is adapted to a relative modulation index M/M_{max} of 90 % to achieve a good DC link utilization with 10 % margin for the control. In step 3) also the motor resistance R, the inductance matrix $\underline{L}$, and the PM flux linkage Ψ are calculated. In step 4)-5), these motor parameters and the modulation index are used to calculate the switched currents and voltages, and the required DC link capacitance. In step 6), the system losses and temperatures are calculated in an iterative routine. Therein, all considered loss components (copper, iron, conduction, switching) are calculated, which are the input values of a lumped parameter thermal network (LPTN) model. Then the LPTN is solved resulting in the node temperatures of the LPTN. For this resulting temperature distribution, the motor copper losses and the inverter conduction losses are recalculated and the LPTN is solved again. This process is repeated until the losses and the temperatures converge. Thereafter, the system volume and costs are calculated in step 7)-8). The results are used in step 9)-11), where the system power density, the system efficiency, and the system costs are calculated and the designs are identified, which meet the temperature limits and the torque ripple limit defined in the design setup. Finally, the pareto-optimal designs are determined from the set of valid designs in step 12). In the following, the models, that are used in the design procedure, are described.

3 Modelling of IMDs

In this section, the models used in the design procedure in Fig. 3 are presented, separately for the calculation of electrical quantities (steps 3-5 in Fig. 3), losses and temperatures (step 6), volumes (step 7), and costs (step 8). In addition, a model used for calculating the system reliability is given. Models taken from literature are mainly referenced, whereas new models (iron loss model & redundancy model) are explained in detail.

3.1 Electrical model

Step 3 of the design procedure in Fig. 3 uses basic motor design equations to adapt the number of turns of a given reference motor to the required DC link voltage and to calculate the motor parameters (R, L, Ψ) accordingly [10]. In step 4, the motor/converter voltages and currents are calculated based on known models for non-salient 3-phase PMSMs [11] (6.4) and for non-salient 9-phase PMSMs [12] (5.9). As part of step 4 in Fig. 3, the torque ripple $\Delta T_\mathrm{m} = \max(T_\mathrm{m}) - \min(T_\mathrm{m})$ is calculated based on the transient torque $T_\mathrm{m}(t)$ that is calculated with (1). In (1), N_mod is the number of three-phase modules, $T_{\mathrm{m},i}(t)$ is the transient torque, $v_{\mathrm{bemf,r/s/t},i}(t)$ are the motor back-emfs, and $i_{\mathrm{r/s/t},i}(t)$ are the currents flowing in the i-th 3-phase winding module.

$$T_\mathrm{m} = \sum_{i=1}^{N_\mathrm{mod}} T_{\mathrm{m},i(t)} = \frac{1}{\omega_\mathrm{mech}} \sum_{i=1}^{N_\mathrm{mod}} \left(v_{\mathrm{bemf,r},i} i_{\mathrm{r},i} + v_{\mathrm{bemf,s},i} i_{\mathrm{s},i} + v_{\mathrm{bemf,t},i} i_{\mathrm{t},i}\right) \tag{1}$$

The model for the DC link dimensioning in step 5 of the design procedure, as well as the analytical duty cycle formulas for different modulation schemes are taken from [13].

3.2 Loss model

The inverter losses comprise conduction and switching losses. Both loss types are calculated using the equations given in [14]. For the conduction losses, this includes a scaling of the $R_\mathrm{ds,on}$ of selected reference devices which could be based on the chip area. Instead of such chip area based scaling, $R_\mathrm{ds,on} \propto 1/I_\mathrm{r}$ is assumed, where I_r is the rated current of the semiconductor device, and I_r is scaled with the relative semiconductor costs $K_\mathrm{sc,rel,b}$ in the considered model. For the switching losses, the loss energies $E_\mathrm{T,on}$, $E_\mathrm{T,off}$, and $E_\mathrm{D,off}$ of the reference devices at a reference operating point are taken from the datasheet/application note. For the 3LTT-topology, the switching energies are scaled with the same factors that are also used in [13] (Tab. 2.6) to account for the difference between the switching energies occurring in a 3LTT-converter and the loss energies given in the datasheets that are based on a 2L-topology.

In the motor losses, copper losses and iron losses are included. For the copper losses, a frequency and temperature dependent stator resistance $R_\mathrm{ac}(f,T)$ is used, taking the skin effect and the proximity effect into account based on [15] (6.43). For the iron losses, a new model for calculating the iron losses in laminated steel in the frequency domain is applied. The model is given by the following formulas for the iron loss density p_m.

$$p_\mathrm{m} = \bar{K}_\mathrm{h,dc} K_\mathrm{h,0}(f) \hat{B}^2 f + \bar{K}_\mathrm{c,dc} K_\mathrm{c,0}(f) (\hat{B}f)^2 + \bar{K}_\mathrm{e,dc} K_\mathrm{e,0}(f) (\hat{B}f)^{1.5} \tag{2}$$

$$\bar{K}_{h/c/e,\mathrm{dc}} = \{\frac{1}{N} \sum_{k=1}^{N} K_{h/c/e,\mathrm{dc}}(H_\mathrm{dc}(t_k)), f > 10f_1; 1, f < 10f_1\} \tag{3}$$

Therein, $\hat{B}$ is the peak flux density, f is the frequency, f_1 is the fundamental frequency, $K_{h/c/e,0}(f)$ are frequency dependent weighting factors, and $\bar{K}_{h/c/e,\mathrm{dc}}$ are DC-bias dependent weighting factors. This approach is a variation of the classic Bertotti equation and combines the idea of fitting the Bertotti loss terms (hysteresis, eddy current, excess) to frequencies in the range of PWM harmonics ($\approx 10..100\,\mathrm{kHz}$) [16] (3.22) with the idea of fitting the loss terms to DC-biased excitation [17] (3.7.1). In electric motors the "DC-bias" is given by the fundamental wave excitation and hence (slowly) varies over time. This is taken into account with (3) by calculating the effective DC-bias weighting factors $\bar{K}_{h/c/e,\mathrm{dc}}$ from an averaging of the DC-bias weighting factors at the actual dc excitation $K_{h/c/e,\mathrm{dc}}(H_\mathrm{dc})$, where H_dc is the DC magnetic field strength. The parameter functions $K_{h/c/e,0}(f)$ and $K_{h/c/e,\mathrm{dc}}(H_\mathrm{dc})$ are found from a curve fitting of measured iron loss curves where sinusoidal excitation is used.

3.3 Volume model

The considered system volume contributions are the motor volume Vol_motor, the volume of the PCB populated with the DC link capacitors Vol_ppcb, the inverter heat sink volume Vol_hs, and a constant volume of the encoder housing Vol_enc. For the DC link dimensioning, the model from [9] is used and the heat sink volume is obtained from an offline-optimization of a finned heat sink based on [7].

3.4 Cost model

For the inverter costs, the cost model from [18] is used that takes into account the chip cost, consisting of the power semiconductor cost K_sc and the cost of auxiliary ICs K_ic, the capacitor cost K_dc, the PCB cost K_pcb, and

Table II: Formulas for calculating the converter reliability for the converter topologies shown in Fig. 2. The used reliability functions of the power semiconductor devices S1/S2/D5 from Fig. 2 are given by $R_\mathrm{S1} = e^{-\lambda_\mathrm{S1} \cdot t}$, $R_\mathrm{S2} = e^{-\lambda_\mathrm{S2} \cdot t}$, and $R_\mathrm{D5} = e^{-\lambda_\mathrm{D5} \cdot t}$. The value pair $\{k,n\}$ defines the k-out-of-n redundancy for the modular topologies with $k \in [0, n-1]$.

Topology	2L	3L-NPC	3L-TT	2L-nM-ser-k / 2L-n-par-k
Reliability	$R_\mathrm{S1}^6 =: R_\mathrm{2L}$	$R_\mathrm{S1}^{12} \cdot R_\mathrm{D5}^6$	$R_\mathrm{S1}^6 \cdot R_\mathrm{S2}^6$	$\sum_{i=0}^{k} \binom{n}{i} (R_\mathrm{2L})^{(n-i)} (1 - R_\mathrm{2L})^i$

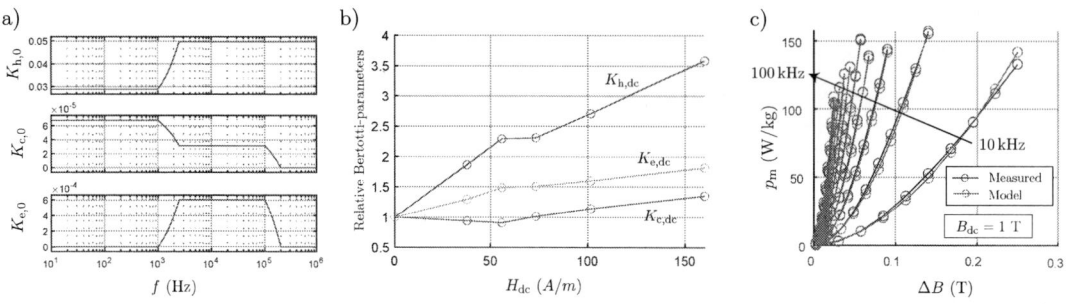

Figure 5: a) Frequency dependent loss parameters and b) DC-bias dependent loss parameters for the parametrization of the iron loss model. c) Measurement vs. model with DC-bias.

the heat sink cost K_{hs}. The motor costs K_{mot} are approximated based on the costs of the copper windings, the iron laminations, and the magnets. The motor costs are assumed to be constant (constant copper fill factor). Concerning the parametrization of the cost model, the specific prices from [18] are used for K_{dc}, K_{pcb}, and K_{hs}, whereas K_{sc} & K_{ic} are parametrized based on the cost study presented in section 4.1.

3.5 Thermal model

For the thermal model, a lumped parameter thermal network (LPTN) is used, where the motor part is based on [19] and the converter part is based on [7]. Solving the LPTN is an iterative process, as shown in Fig. 3, because the copper losses, the conduction losses, and the thermal resistance of the finned heat sink are temperature dependent.

3.6 Reliability model

2L-nM-ser- & 2L-nM-par-converters can be designed with a k-out-of-n-redundancy, which means that up to k 2L-modules can fail while the system can still operate (in part-load). On the one hand, such redundancy improves system reliability. On the other hand, the higher component count of the modular topologies compared to a 2L-converter degrades system reliability due to the higher risk of component failure. Given these contrary effects, the question is, if the modular topologies 2L-nM-ser & 2L-nM-par overall have an advantage over the 2L- and 3L-topologies in terms of reliability. To answer this question, a reliability model is built that is based on the following three definitions according to [20].

1. The *reliability* $R(t)$ of an item is defined as the probability that an item can perform a required function under given conditions for a given time interval $[0,t]$.

2. The *failure rate* $\lambda(t)$ of an item is defined as the average number of failing items per time unit normalized to the total number of still good/working items at the time instant t.

3. The *Mean-Time-To-Failure MTTF* is defined as $\int_0^\infty R(t)dt$.

To calculate the reliability of a system consisting of multiple items, the reliabilities of these items are combined in one of the following two ways depending on the type of redundancy: 1) If the items are non-redundant, their reliability functions (i.e. probabilities of working) are multiplied. 2) If the items are redundant, the overall reliability corresponds to the sum of the probabilities of all working combinations. This concept is applied to all considered

Table III: Design space limits for the operational DOFs in Fig. 3.

Switching frequency f_s	$1\,kHz \ldots 100\,kHz$
Relative semiconductor cost $K_{sc,rel,b}$	$0 \ldots 1$
Heat sink fin length c_{hs}	$0\,mm \ldots 30\,mm$

Table IV: Benchmark system parameters.

System configuration {*Top-PWM-SC-W*}	2L-Sin-Si-3ph
Switching frequency f_s	$26.28\,kHz$
Relative semiconductor cost $K_{sc,rel,b}$	0.0764
Heat sink fin length c_{hs}	$1\,mm$
System power density $\rho_{sys,ref}$	$0.1291\,kWL^{-1}$
System efficiency $\eta_{sys,ref}$	$86.05\,\%$
System cost $K_{sys,ref}$	$190.45\,CHF$
Motor cost K_{mot}	$138.18\,CHF$

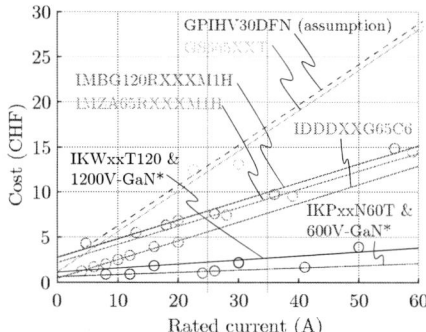

Figure 6: Chip cost as a function of rated current/chip size based on distributor prices for an order volume of 1000 pieces.

Table V: System specifications and design constraints of a general-purpose, high-torque, low-speed IMD.

Rated power P_N	1.5 kW	Max. junction temperature $T_{j,max}$	125 °C/150 °C
Rated torque $T_{m,N}$	48 Nm	Max. electronics ambient temperature $T_{e,amb,max}$	80 °C
Rated speed $n_{n,N}$	300 rpm	Max. winding temperature $T_{w,max}$	130 °C
DC link voltage V_{dc}	800 V	Max. encoder temperature $T_{enc,max}$	80 °C
Max. static DC link voltage ripple $\Delta V_{dc,max,s}$	$0.01 V_{dc}$	Max. torque ripple $\Delta T_{m,max}$	0.5 Nm
Max. dynamic DC link voltage ripple $\Delta V_{dc,max,d}$	$0.125 V_{dc}$	Min. PCB height $T_{pcb,min}$	15 mm
Max. load step $\Delta i_{dc,out}/I_{mod}$	1	Insulation layer thickness l_{ins}	0 mm
Ambient temperature T_{amb}	40 °C	Semiconductor cost budget $K_{sc,b}$	190.2 CHF
Flange temperature T_f	65 °C		

converter topologies (Fig. 2), where the items are given by the power semiconductor devices. Considering a constant failure rate over time for each device, the reliability of a single device, e.g. switch S1 of the 2L-topology, is given by $R_{S1} = e^{-\lambda_{S1} \cdot t}$ [20]. The resulting formulas for calculating the converter reliability are summarized in Tab. II. The reliability results for a specific number of modules and for a specific failure rate are given in section 5.

4 Application of the design procedure

The design procedure presented in Fig. 3 is applied to the optimization of a 1.5 kW IMD. In the following, first the design setup is given. Then, the optimization results are presented and discussed.

4.1 Design setup - Specifications, constraints, and design space

The considered system specification, design constraints, and the design space limits of the operational DOFs are summarized in Tab. V & III. The parameters of the benchmark system used for the definition of the relative system cost $K_{sys,rel} := K_{sys}/K_{sys,ref}$ are given in Tab. IV. Tab. VI shows the selected reference power semiconductor devices and the cost parameters obtained from a cost study of the corresponding device series. The results of this study are shown in Fig. 6, indicating that the considered WGB devices (SiC/GaN) are 5 to 10 times more expensive than conventional Si IGBTs of the same current rating. However, as the production of GaN devices is based on silicon substrate, the cost of GaN is expected to reach the level of Si devices in the long run. Therefore, GaN HEMTs are considered in two "cost versions": 1) Based on present costs, denoted as GaN, and 2) with future costs which are assumed to be equal to the costs of Si IGBTs as indicated in Fig. 6. These devices are denoted as GaN* in the following.

The parameters used in the iron loss model are given in Fig. 5a & b. These parameters are obtained from a curve fitting of the iron loss density curves shown in Fig. 5c, which originate from iron loss measurements performed with a BH-analyzer (IWATSU SY-8219) and a DC-bias tester (IWATSU SY-960).

The parameters used for the semiconductor loss model are shown in Fig. 7a-d. Fig. 7a & b show the switching energies of the reference devices which decrease from Si IGBTs to SiC MOSFETs to GaN/GaN* HEMTs, as expected. Compared to MOSFETs and HEMTs, IGBTs have an on-state resistance with a strong dependency on the current. Therefore, in order to compare the conduction loss behavior of MOSFETs, HEMTs, and IGBTs on the device level, an effective on-state resistance $R_{ds,on,eff}$ is defined as the on-state resistance at a typical operating current which is assumed to be equal to $1/3$ of the rated current. Fig. 7c & d show that, for the same cost per transistor, the static losses increase from GaN* HEMTs to Si IGBTs to SiC MOSFETs to GaN HEMTs. The fact that the effective on-state resistance of GaN* HEMTs (i.e. GaN HEMTs with future expected costs) is lower than the one of the considered Si IGBTs is due to the following two reasons: 1) As previously mentioned, it is assumed that, for the same cost per device, GaN* HEMTs have the same rated current as the considered Si IGBTs (Fig. 6). 2) Due to that assumption, the effective on-state resistance of the reference devices is scaled to the same rated current (using $R_{ds,on} \propto 1/I_r$), for which the considered GaN HEMTs have a lower $R_{ds,on,eff}$ than the Si IGBTs.

4.2 Optimization results

In this section, the pareto-optimal IMD designs resulting from the design procedure in Fig. 3 are analysed. Therefore, in a first part, the pareto-front is shown and an overall optimal system configuration is suggested. In a second

Table VI: Reference power semiconductor devices, selected as the devices with the largest current rating within each device series with the same package dimensions. Given parameters: Rated current $I_{r,25°C}$, cost parameters a_{sc} & b_{sc}, and maximum junction temperature $T_{j,max}$. As the switching energies $E_{on/off}$ of the 1200V GaN HEMT are not provided in literature, it is assumed that these scale with $V_{ds,max}$ like the switching energies of SiC MOSFETs, i.e. $E_{on/off,T6} = E_{on/off,T3} \cdot (E_{on/off,T5}/E_{on/off,T2})$.

ID	Device type	$I_{r,25°C}$ (A)	Name	a_{sc} (CHF/A)	b_{sc} (CHF)	$T_{j,max}$ (°C)	Source of $E_{on/off}$
T1	600 V Si IGBT	41	IKP20N60T	0.0443	1.097	175	Datasheet
T2	600 V SiC MOSFET	59	IMZA65R027M1H	0.2016	2.22	150	Application note [21]
T3	600 V GaN HEMT	30	GS66508T	0.7976	2.083	150	Datasheet
T4	1200 V Si IGBT	75	IKW40T120	0.0802	2.026	150	Datasheet
T5	1200 V SiC MOSFET	56	IMBG120R030M1H	0.2063	2.778	175	Datasheet
T6	1200 V GaN HEMT	30	GPIHV30DFN	0.7976	3.083	150	Assumption
D1	600 V SiC Diode	51	IDDD20G65C6	0.2073	0.4457	175	-

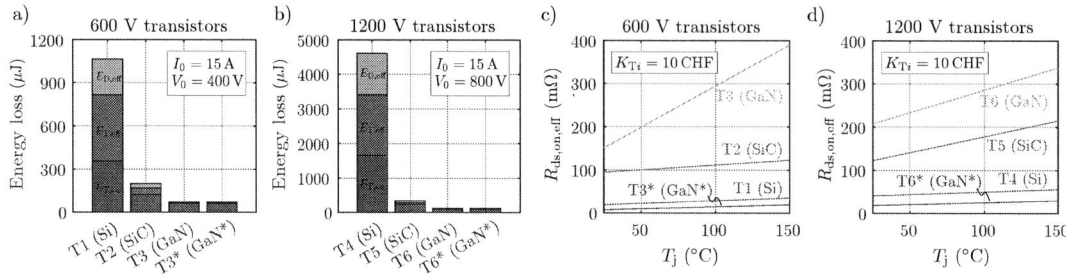

Figure 7: a) & b) Switching loss energies of the reference devices from Tab. VI, based on the scaling of one $E_x(I)$-value pair from the datasheet/application note which is scaled to the switched current I_0 and voltage V_0. c) & d) Effective on-state resistance of the reference devices over the junction temperature, scaled to the same cost per device $K_{T,i}$. The values in a)-d) are used to parametrize the semiconductor loss model described in section 3.2.

part, the considered semiconductor technologies, PWM schemes, and topologies are compared with regard to the maximum system efficiency at approximately equal system cost and volume.

4.2.1 Investigation of the pareto-optimal designs

Fig. 8a shows the pareto-front of the pareto-optimal IMD designs in the $(\rho_{sys}, \eta_{sys}, K_{sys,rel})$-space resulting from the design procedure in Fig. 3, where only the present costs of semiconductor technologies Si/SiC/GaN are considered, i.e. GaN* is not considered. Fig. 8a, shows six pareto-optimal {*Top-PWM-SC-W*}-configurations (i.e. system configurations in terms of the topology, the PWM scheme, the semiconductor technology, and the winding scheme). These configurations are color-coded by means of the colored configuration names in Fig. 8a & b. The overall optimal/best configuration depends on how the design objectives $(\rho_{sys}, \eta_{sys}, K_{sys,rel})$ are weighted. This is shown in Fig. 8b which displays the color of the configuration that minimizes the scalar cost function $f_s = w_\rho(-\rho_{sys}/\rho_{sys,ref}) + w_\eta(-\eta_{sys}/\eta_{sys,ref}) + w_K(K_{sys}/K_{sys,ref})$, as a function of the weighting factors w_ρ & w_η with the constraint $w_\rho + w_\eta + w_K = 1$. To illustrate how to read Fig. 8b, an exemplary point at $w_\rho = 0.1$, $w_\eta = 0.2$, and $w_K = 1 - (w_\rho + w_\eta) = 0.7$ is indicated in Fig. 8b. The color at that point belongs to the configuration *2L-OC-Si-3ph*, indicating that this is the overall optimal configuration for the weighting factors at the considered exemplary point.

Fig. 8a & b show that the cost-optimal system configuration is *2L-OC-Si-3ph*, the most efficient configuration is *2L3Mpar-OC-GaN-9ph*, and the highest power density is reached by all configurations (in the case of no cooling fins). For a compromise between costs and efficiency, the least expensive IMD of the most efficient system configuration, i.e. of the configuration *2L3Mpar-OC-GaN-9ph*, is suggested as the overall optimal IMD (*IMD$_{sg}$* in Fig. 8a). Fig. 8a shows that the suggested IMD enables an efficiency increase of +2.26 % (i.e. percentage points) at the expense of 36 % higher costs compared to the cost-optimal IMD (*IMD$_{co}$* in Fig. 8a).

The same analysis is performed now including the future expectation of GaN HEMT costs (GaN*). The result is shown in Fig. 8c & d, indicating that only two GaN*-based system configurations are on the pareto-front. Hence, for the considered specification, GaN HEMTs are expected to outperform Si IGBTs and SiC MOSFETs in the future. As shown in Fig. 8c & d, the cost-optimal configuration is now *2L-OC-GaN*-3ph* and the most efficient configuration is *2L3Mpar-OC-GaN*-9ph*. The suggested IMD (*IMD$_{sg}^*$* in Fig. 8c) now enables 0.65 % more efficiency for 23 % more costs, compared to the future expected cost-optimal IMD (*IMD$_{co}^*$* in Fig. 8c).

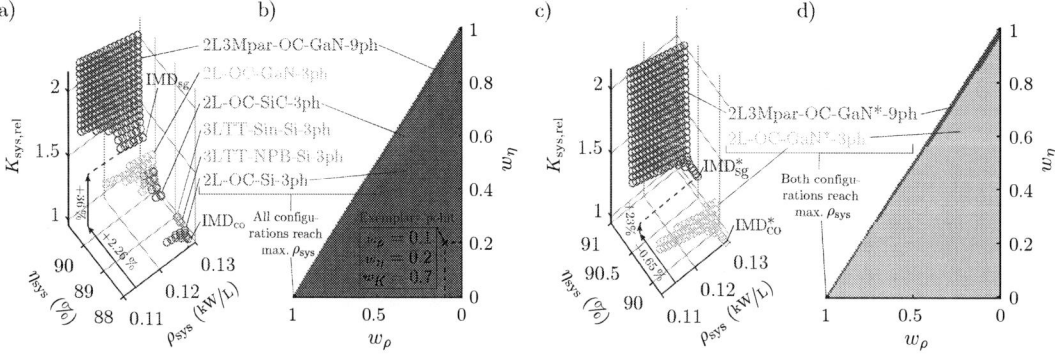

Figure 8: a) Pareto-front of the pareto-optimal IMD designs resulting from the design procedure in Fig. 3 and from the system specification in Tab. V, considering present semiconductor prices (i.e. Si/SiC/GaN). b) Overall optimal system configuration as a function of the weighting factors w_ρ & w_η of the cost-function $f_s = w_\rho(-\rho_{sys}/\rho_{sys,ref}) + w_\eta(-\eta_{sys}/\eta_{sys,ref}) + w_K(K_{sys}/K_{sys,ref})$ with $w_K = 1 - (w_\rho + w_\eta)$. c) & d) Analog results to a) & b), considering expected future GaN prices (i.e. Si/SiC/GaN*).

4.2.2 Comparison at equal cost and volume

While the previous analysis revealed the best system configurations for the given system specification, in a next step, the impact of the different semiconductor technologies, PWM schemes, and topologies on the system performance is investigated. For that purpose, all system configurations defined in Tab. I are compared with regard to maximum system efficiency/minimum system losses $P_{\text{sys,loss,opt}}$ at approximately equal system cost $K_{\text{sys,rel}} = 1.8 \pm 0.03$ and approximately equal system volume $Vol_{\text{sys}} = 12.89\,\text{L} \pm 0.17\,\text{L}$. These intervals for $K_{\text{sys,rel}}$ & Vol_{sys} are chosen so that all considered system configurations can be compared. In other intervals some configurations might not have a solution because the ranges of $K_{\text{sys,rel}}$ & Vol_{sys} are different for each configuration.

Fig. 9 shows the resulting cost distribution, volume distribution, and optimal power loss distribution as bar plots, as well as a table listing the corresponding optimal design parameters ($f_{\text{sw,opt}}$, $K_{\text{sc,rel,b,opt}}$, $c_{\text{hs,opt}}$), the efficiency gain compared to the reference system (Tab. IV), and the optimum type (*opttype*). The optimum type indicates

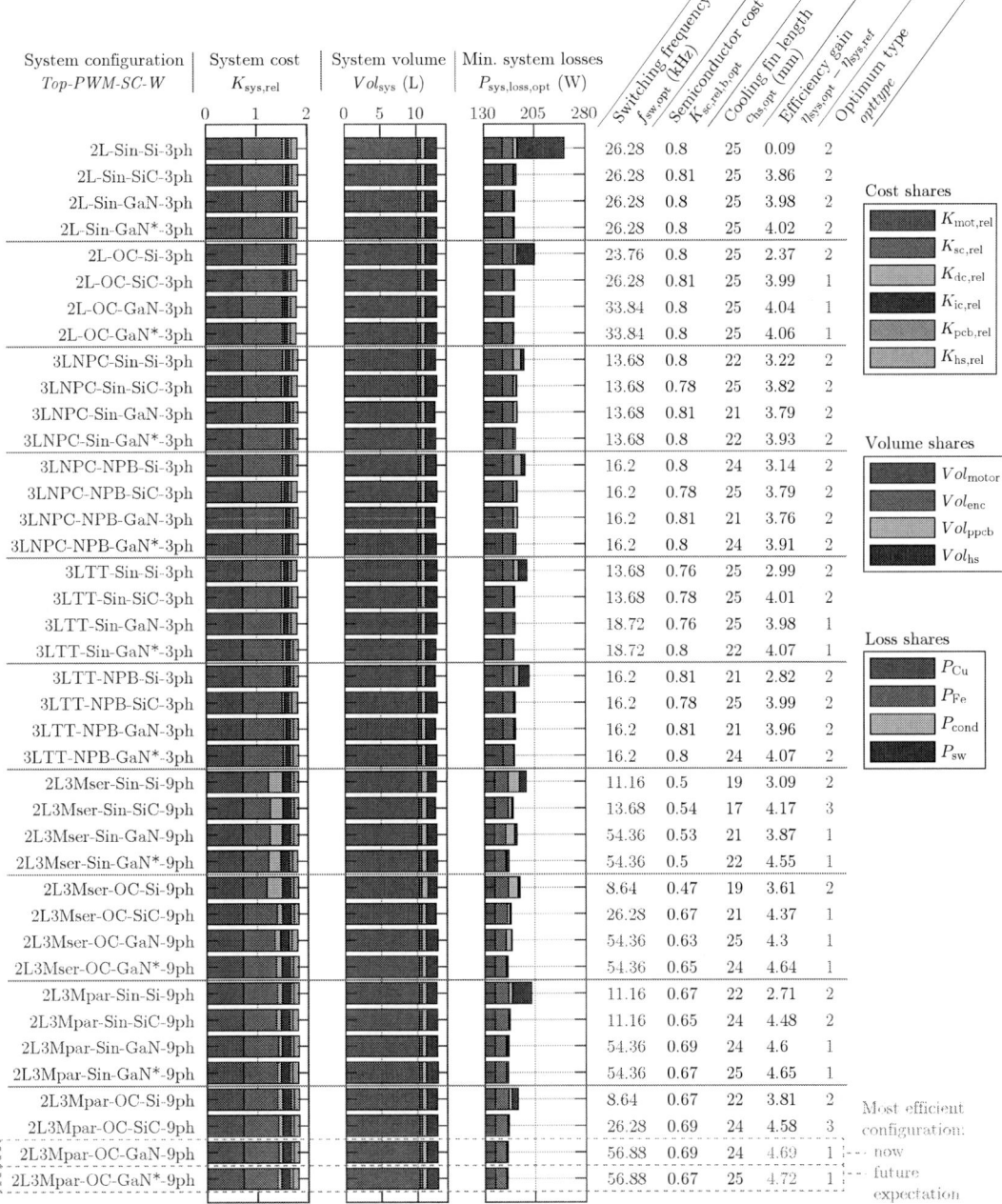

Figure 9: System performance resulting from the design procedure in Fig. 3 at approximately equal system cost $K_{\text{sys,rel}}$, volume Vol_{sys}, and minimized losses $P_{\text{sys,loss,opt}}$ for all system configurations defined in Tab. I. Bar plots: Distribution of system cost, volume, and losses. Table on the right: Optimal design parameters ($f_{\text{sw,opt}}$, $K_{\text{sc,rel,b,opt}}$, $c_{\text{hs,opt}}$), efficiency gain with regard to the reference system (Tab. IV), and the optimum type that indicates what determines the optimum design parameters.

what determines the optimal design parameters as explained in the following. The optimal semiconductor costs $K_{sc,rel,b,opt}$ and the optimal cooling fin length $c_{hs,opt}$ are determined by the upper limits of the already mentioned cost and volume intervals for all considered system configurations. However, different cases are identified for the optimal switching frequency $f_{sw,opt}$: For $opttype = 1$, $f_{sw,opt}$ results from an optimal balance between the switching losses and the harmonic motor losses. That is, for switching frequencies lower than $f_{sw,opt}$, the harmonic motor losses increase faster than the switching losses decrease, so that the system efficiency decreases. For $opttype = 2$, $f_{sw,opt}$ lies on the boundary defined by the torque ripple constraint $\Delta T_m < \Delta T_{m,max}$. That is, for switching frequencies lower than $f_{sw,opt}$, the torque ripple exceeds $T_{m,max}$. For $opttype = 3$, $f_{sw,opt}$ lies on the boundary defined by the upper cost limit of this analysis $K_{sys,rel} < 1.83$. That is, for switching frequencies lower than $f_{sw,opt}$, the cost exceeds its limit (as larger and hence more expensive capacitors are required to keep the DC link voltage ripple constant). In the following, the semiconductor technologies, modulation schemes, and topologies are compared, based on the results shown in Fig. 9.

Table VII: Comparison of semiconductor technologies. The most efficient variant is printed in bold.

Topology-PWM	$\eta_{sys,opt,ref}$ Si	$\eta_{sys,opt} - \eta_{sys,opt,ref}$		
		SiC	GaN	GaN*
2L-Sin	86.14	3.77	3.89	**3.93**
2L-OC	88.41	1.63	1.68	**1.70**
3LNPC-Sin	89.26	0.61	0.57	**0.72**
3LNPC-NPB	89.19	0.65	0.62	**0.77**
3LTT-Sin	89.04	1.02	0.99	**1.08**
3LTT-NPB	88.86	1.18	1.15	**1.26**
2L3Mser-Sin	89.13	1.09	0.79	**1.46**
2L3Mser-OC	89.66	0.76	0.68	**1.03**
2L3Mpar-Sin	88.76	1.77	1.89	**1.94**
2L3Mpar-OC	89.86	0.77	0.88	**0.91**

Table VIII: Comparison of PWM schemes. The most efficient variant is printed in bold.

Topology (2L)	$\eta_{sys,opt,ref}$ Sin-PWM	$\eta_{sys,opt} - \eta_{sys,opt,ref}$ OC-PWM
2L	90.07	**0.04**
2L3Mser	90.59	**0.10**
2L3Mpar	90.70	**0.07**

Topology (3L)	Sin-PWM	NPB-PWM
3LNPC	**89.98**	-0.02
3LTT	**90.12**	0.00

1) Comparison of semiconductor technologies (Si/SiC/GaN/GaN*)

Based on the efficiency gain given in Fig. 9, Tab. VII compares the maximum efficiency achieved with different semiconductor technologies (Si/SiC/GaN/GaN*) for each combination of topology and PWM scheme, where Si IGBTs are used as a reference. The following conclusions are drawn from Tab. VII: 1) For the considered specification, Si IGBTs are the least efficient, which is expected due to the high switching loss energies (Fig. 7a & b). 2) In general, GaN HEMTs are expected to be the most efficient technology in the future (at the same cost & volume). 3) In general, and in particular for SiC vs. GaN for the considered system specification, which semiconductor technology is more efficient cannot be determined just based on the device level characteristics. Instead, whether one technology is more efficient than the other depends on the device level characteristics (Fig. 7), the used combination of topology and PWM scheme (which influences the P_{sw}/P_{cond}-ratio), and the optimal switching frequency that depends on the design constraints such as the torque ripple constraint.

2) Comparison of modulation schemes

Based on the efficiency gain given in Fig. 9, Tab. VIII compares the maximum efficiency achieved with different PWM schemes for each topology (taking the maximum over Si/SiC/GaN/GaN*), where Sin-PWM is used as a reference. The following conclusions are drawn from Tab. VIII: 1) For the 2L-topologies (2L, 2L3Mser, 2L3Mpar) with the considered system specification, OC-PWM is more efficient than Sin-PWM, as expected, because OC-PWM avoids switching of the phase-leg with the largest current, thereby reducing switching losses. 2) For the 3LNPC-topology and the considered specification, NPB-PWM leads to slightly higher switching losses due to an additional space vector used in every switching period. 3) For the 3LTT-topology with the considered system specification, Sin-PWM & NPB-PWM reach approximately the same efficiency because the loss-related disadvantage of the additional space vector of NPB-PWM is compensated by a lower switching frequency of 3LTT-NPB-GaN*-3ph compared to 3LTT-Sin-GaN*-3ph in Fig. 9. 4) In general, the most efficient PWM scheme cannot be determined just based on a simple comparison at equal switching frequency but the most-efficient PWM scheme depends also on the optimum switching frequency and hence also on the design constraints.

3) Comparison of topologies

Based on the efficiency gain given in Fig. 9, Tab. IX compares the maximum efficiency achieved with different topologies (taking the maximum over all combinations of Si/SiC/GaN/GaN* and PWM schemes). The following conclusions are drawn from Tab. IX: 1) For the considered specification, the topology 2L3Mpar has the best efficiency with +0.66 % higher efficiency than the 2L-topology. 2) For the considered specification, the modular topologies (2L3Mpar/2L3Mser) enable an efficiency increase of +0.65 % compared

Table IX: Comparison of topologies. The most efficient variant is printed in bold.

$\eta_{\text{sys,opt,ref}}$ 2L	3LNPC	$\eta_{\text{sys,opt}} - \eta_{\text{sys,opt,ref}}$ 3LTT	2L3Mser	2L3Mpar	$\eta_{\text{sys,opt,ref}}$ 2L/3LNPC/3LTT	$\eta_{\text{sys,opt}} - \eta_{\text{sys,opt,ref}}$ 2L3Mser/2L3Mper
90.11	-0.13	0.01	0.58	**0.66**	90.12	**0.65**

to the non-modular topologies (2L/3LNPC/3LTT), which is due to the higher winding factor of the 9-phase winding that reduces the motor losses. Indeed, Fig. 9 shows lower copper losses P_{Cu} for all configurations with modular topologies compared to the configurations with standard/non-modular topologies. The fact that the modular topologies can be combined with 9-phase windings with a higher winding factor than 3-phase windings is a general advantage of these topologies.

5 Reliability analysis

The considered converter topologies are also compared with regard to the reliability based on the model given in section 3.6 for the reliability related system specification summarized in Tab. X. For simplicity, the same failure rate of $\lambda = 500$ fit is assumed (which is chosen based on Fig. 6 in [22]) for all power semiconductor devices, where 1 fit corresponds to 1 *failure in time* with a reference time interval of 10^9 hours. The 2L-nM-ser-topology is analyzed with 3 modules ($n = 3$) and for three different redundancy configurations (no redundancy ($k = 0$), 1-out-of-3-redundancy ($k = 1$), and 2-out-of-3-redundancy ($k = 2$)) to investigate the effect of redundancy on the reliability. The 2L-nM-par-topology is not separately listed, as the model equations in Tab. II already indicate that both modular topologies (2L-nM-ser & 2L-nM-par) have the same reliability.

Tab. X shows the results for the absolute mean time to failure $MTTF$ and for the relative mean time to failure with regard to the 2L-topology, $MTTF_{\text{rel,2L}}$. Note that the $MTTF$ depends on λ, whereas the $MTTF_{\text{rel,2L}}$ is independent of λ. The resulting values for the $MTTF$/$MTTF_{\text{rel,2L}}$ show that the modular topologies without any redundancy ($k = 0$) are as reliable as the 3L-NPC but less reliable than 2L and 3L-TT. With a 1-out-of-3-redundancy, the reliability of the modular topologies increases, as expected, but is still lower than the one of the 2L-topology. This shows that for a 1-out-of-3-redundancy, the higher risk of component failure due to more components of the modular topologies outweighs the redundancy effect on the reliability. Only for a 2-out-of-3-redundancy, the modular topologies are more reliable in terms of the $MTTF$ than the 2L-topology. It should be noted that for 2L-nM-ser and for a given DC link voltage, a higher redundancy level implies a higher required voltage rating of the power semiconductor devices, whereas for 2L-nM-par there is no such additional design requirement.

Table X: Reliability related system specification and results of the reliability analysis.

Topology	Device failure rates (fit)	Redundancy	$MTTF$ (years)	$MTTF_{\text{rel,2L}}$
2L	$\lambda_{\text{S1}} = 500$	-	38.1	1
3L-NPC	$\lambda_{\text{S1}} = \lambda_{\text{D5}} = 500$	-	12.7	$1/3$
3L-TT	$\lambda_{\text{S1}} = \lambda_{\text{S2}} = 500$	-	19.0	$1/2$
2L-nM-ser	$\lambda_{\text{S1}} = 500$	$\{k,n\} = \{0,3\}$	12.7	$1/3$
2L-nM-ser	$\lambda_{\text{S1}} = 500$	$\{k,n\} = \{1,3\}$	31.7	$5/6$
2L-nM-ser	$\lambda_{\text{S1}} = 500$	$\{k,n\} = \{2,3\}$	69.8	$1\,1/6$

6 Conclusion

The following conclusions are drawn:

- A procedure for the optimal design of IMDs in terms of ρ_{sys}, η_{sys}, and K_{sys} is presented. The design procedure incorporates a new iron loss model and enables the comparison of modular converter topologies (2L3Mser/2L3Mpar) to standard/non-modular topologies (2L/3LNPC/3LTT) covering a large design space including different PWM variants, semiconductor technologies (Si/SiC/GaN), and winding schemes.

- For a 1.5 kW IMD, the cost-optimal system configuration is 2L-OC-Si-3ph. Compared to this, the suggested design has the configuration 2L3Mpar-OC-GaN-9ph and enables an efficiency increase of $+2.26\%$ at 36% higher cost. Assuming a utilization of 90% and an energy price of 0.1 CHF/kWh, the higher costs are compensated by the lower energy consumption after approximately 2 years.

- For a 1.5 kW IMD, different topologies, PWM schemes, and semiconductor technologies are compared with regard to maximum efficiency at equal system cost and volume. Modular topologies reach higher efficiencies than the standard topologies due to the general advantage that modular topologies can be combined with multiphase windings. The optimal PWM scheme and the optimal semiconductor technology cannot be predicted in general, as they depend on the system specification, such as the torque ripple constraint.

- A model for comparing the system reliability of standard topologies and of modular topologies is presented. The modular topologies 2L3Mser and 2L3Mpar only have an advantage over the standard 2L- and 3L-topologies, if a 2-out-of-3-redundancy is achieved.

Acknowledgement

This research is financially supported by B&R Industrie-Automation AG.

References

[1] R. Abebe *et al.*, "Integrated motor drives: state of the art and future trends," *IET Electric Power Applications*, vol. 10, no. 8, pp. 757–771, Sep 2016.

[2] J. Wyss and J. Biela, "Optimal design of bidirectional PFC rectifiers and inverters considering 2l and 3l topologies with Si, SiC, and GaN switches," *IEEJ Journal of Industry Applications*, vol. 8, no. 6, pp. 975–983, Nov 2019.

[3] B. Welchko, T. Lipo, T. Jahns, and S. Schulz, "Fault tolerant three-phase AC motor drive topologies: A comparison of features, cost, and limitations," *IEEE Trans. on Power Electronics*, vol. 19, no. 4, pp. 1108–1116, Jul 2004.

[4] S. Norrga, L. Jin, O. Wallmark, A. Mayer, and K. Ilves, "A novel inverter topology for compact EV and HEV drive systems," in *39th Annual Conference of the IEEE Industrial Electronics Society (IECON)*, Nov 2013.

[5] C. Klumpner, F. Blaabjerg, and P. Thogersen, "Evaluation of the converter topologies suited for integrated motor drives," in *38th Annual Meeting on Conf. Record of the Industry Applications Conference (IAS)*. IEEE, Oct 2003.

[6] M. Ugur, H. Sarac, and O. Keysan, "Comparison of inverter topologies suited for integrated modular motor drive applications," in *IEEE 18th International Power Electronics and Motion Control Conference (PEMC)*, Aug 2018.

[7] T. Bringezu and J. Biela, "Cooling limits of passively cooled integrated motor drives," in *23rd European Conference on Power Electronics and Applications (EPE ECCE Europe)*, Sep 2021.

[8] J. Wang, Y. Li, and Y. Han, "Integrated modular motor drive design with gan power fets," *IEEE Trans. on Industry Applications*, vol. 51, no. 4, pp. 3198–3207, Jul 2015.

[9] T. Bringezu and J. Biela, "Comparison of optimized motor-inverter systems using a stacked polyphase bridge converter combined with a 3-, 6-, 9-, or 12-phase PMSM," in *22nd European Conference on Power Electronics and Applications (EPE ECCE Europe)*, Sep 2020.

[10] B. P. Germar Müller, Karl Vogt, *Berechnung Elektrischer Maschinen*. WILEY-VCH, 2008.

[11] R. D. Doncker, D. W. Pulle, and A. Veltman, *Advanced Electrical Drives*. Springer Netherlands, 2011.

[12] A. A. Rockhill, "On the modelling and control of high phase order synchronous machines," Ph.D. dissertation, University of Wisconsin-Madison, 2012.

[13] J. Wyss, "Multi-domain optimization for highly efficient, active rectifiers in drive systems," Ph.D. dissertation, ETH Zürich, 2018.

[14] J. Biela, M. Schweizer, S. Waffler, and J. W. Kolar, "SiC versus Si— Evaluation of potentials for performance improvement of inverter and DC–DC converter systems by SiC power semiconductors," *IEEE Trans. on Industrial Electronics*, vol. 58, no. 7, pp. 2872–2882, Jul 2011.

[15] J. Biela, "Optimierung des elektromagnetisch integrierten serien-parallel-resonanzkonverters mit eingeprägtem ausgangsstrom," Ph.D. dissertation, ETH Zürich, 2005.

[16] M. T. Kakhki, "Modeling of losses in a permanent magnet machine fed by a pwm supply," Ph.D. dissertation, University of Laval, 2016.

[17] J. Muehlethaler, "Modeling and multi-objective optimization of inductive power components," Ph.D. dissertation, ETH Zurich, 2011.

[18] R. Burkart and J. W. Kolar, "Component cost models for multi-objective optimizations of switched-mode power converters," in *IEEE Energy Conversion Congress and Exposition*, Sep 2013.

[19] D. Staton, A. Boglietti, and A. Cavagnino, "Solving the more difficult aspects of electric motor thermal analysis in small and medium size industrial induction motors," *IEEE Trans. on Energy Conversion*, vol. 20, no. 3, pp. 620–628, Sep 2005.

[20] A. Birolini, *Reliability Engineering: Theory and Practice*. Springer, 2017.

[21] Infineon, "An_1907_pl52_1911_144109, coolsic™ mosfet 650 v m1 trench power device," Dec. 2021. [Online]. Available: https://www.infineon.com/dgdl/Infineon-MOSFET_CoolSiC_650V_SiC_trench_power-ApplicationNotes-v04_00-EN.pdf?fileId=5546d4626fc1ce0b016ff6c9a86f6e08

[22] A. Akturk, R. Wilkins, J. McGarrity, and B. Gersey, "Single event effects in Si and SiC power MOSFETs due to terrestrial neutrons," *IEEE Trans. on Nuclear Science*, vol. 64, no. 1, pp. 529–535, Jan 2017.

AUTHOR INDEX

Abdalrahman, Adil 2241, 3282, 3757
Abdullah, Ahmed.. 554
Abedini, Hossein... 865
Aceña, Javier Cañas.. 484
Adabi, Jafar... 2537
Addin, Ali Sharaf.. 1824
Afonso, Luciana C. 4018
Aganza-Torres, Alejandro... 1328
Agarwal, Ritika.. 3615
Agirrezabala, Eneko 3327
Aguglia, D. 1955
Ahmed, Emad M.. 1015
Aiello, Giuseppe 2628
Aillerie, Michel.. 315
Aizpuru, I. 2903
Aizpuru, Iosu 325, 3327, 3574, 3750
Akuru, Udochukwu B. 2958
Al-Haddad, Kamal 1025
Alaluss, Mohamed 1424
Alatise, Olayiwola 1497, 2477
Albrecht, Fabian 2726
Aldarmon, Mohamed 2574
Ali, Mohammad 2392, 3022
Ali, Ramy 390
Ali, Rana Asad... 698
Allard, Bruno 169, 3862
Allioua, Abdelmoumin 2835
Alvarez, Asier... 279
Alvarez-Herault, Marie-Cecile 1147
Alves, Wendell Da Cunha... 1046
Alvi, Muhammad H 1692
Aly, Mokhtar 1015
Andersen, Michael A. E. 1561
Ando, Y. 1785
Andresen, Jan 1684
Ansari, Sajad A. 3440
Antonopoulos, Antonios 297, 432
Anzola, J. 2903, 2967
Anzola, Jon 3574
Apostolidou, Nena 1796
Appel, Tobias... 1121
Apte, Pramod 2773
Arabsalmanabadi, Bita... 1025
Arias, Manuel 152
Arrizabalaga, Antxon... 325
Arrozy, Juris 681
Arruti, Asier... 3574, 3750
Artal-Sevil, J. S.. 2903, 2967

Arza, Joseba 484, 2011
Asllani, Besar 2515
Asoodar, Mohsen 2843
Atzler, Frank 3391
Aunsborg, Thore Stig 825
Aviñó, Oriol 2715
Ayarzaguena, Ibán.. 1765, 3336
Aztiria, Jon 325
Baars, Nico 2788
Babin, Anthony 3696
Baburske, Roman 1424
Bacha, S. 2422, 3179
Bacha, Seddik... 3140, 3928
Bacheti, Gabriel Gaburro 421
Bachmann, Matthias... 3501
Badenhop, Niklas.. 145, 1939
Baek, Seung-Hyuk 2877
Bagaber, Bakr... 3037, 3711
Baimel, D. 3254
Baimel, N. 3254
Bak, Claus Leth.. 2504
Bakhos, Gianni 3928
Bakran, Mark-M.................................... 805, 1036, 2744
Bakri, Reda 1046
Balachandran, Arvind 1456
Balasubramanian, Sridhar 2030
Ballestín-Bernad, V.. 2903, 2967
Banana, Shady.. 1064
Banavath, Satish Naik 730
Banda, Joseph.. 187, 289
Barba, V. 1975
Barbi, Eli 3254
Barg, Sobhi 361
Barman, Subhranil.. 2462
Barón, Kevin Muñoz 2698
Bashar, Erfan 2477
Basic, Duro 125
Basler, Michael 242
Basler, Thomas.................................... 1424, 1713, 1733, 3373
Bauer, Luca 971
Bauer, Pavol...................................... 1319, 3607, 3729
Baumann, Michael 1167
Baumann, Timm Felix 2355
Bäumler, Christian 1733
Bayer, Markus.. 115
Bayhan, Sertac 3518
Bayram, Islam Safak 3518
Beck, Simon... 1434, 2038

Beckemeier, Christian..2327
Beczkowski, Szymon Michal..2661
Beineke, Stephan 3501
Beiranvand, Hamzeh..................... 833, 3092, 3846, 3966
Belhaouane, Mohamed Moez 582
Benchaib, Abdelkrim 3928
Bendfeld, Christian 1620
Benech, Philippe..169
Bensetti, Mohamed 3883
Bergmann, Lukas...1036
Bergveld, Henk Jan..3796
Bermejo, Jose Manuel..1765, 3336
Bernal, Carlos..3327
Bernal-Agustín, J. L..2967
Bernal-Ruiz, Carlos 3750
Bernichon, Thomas..3920
Bertilsson, Kent 361
Bertin, Matthieu 534
Beukes, Johan 3112
Beye, Mamadou Lamine...2736
Beza, Mebtu 1187
Bezerra, Vinicius Freire...2689
Bhatnagar, Pallavee 3804
Bhattacharya, Arghyadip 178
Bhoi, Sachin Kumar...3031
Biadene, Davide...865
Biela, Juergen 1402
Biela, Jürgen 651, 662, 933, 1391, 1434, 2038, 2544
Bieler, Arne 1121
Bier, Anthony 922, 2736
Billa, Laxma R..2301
Bimmel, Luc 2736
Binder, Andreas 2316
Bitsi, Konstantina 3246
Blaabjerg, Frede................... 2110, 2182, 2496, 2504, 2939
Blanes, J. M. 3382, 3401
Blank, Thomas..232
Blanquez, Francisco R. 2189, 2451
Blasco-Gimenez, Ramon 2189, 2451
Blasuttigh, Nicola 3846
Blatsi, Zoe..2824, 3813
Blömeke, Alexander 4025
Böcker, Joachim 2276, 2432, 2754, 3625, 3686
Bockholt, Jan 1286
Boettcher, Norman...1128
Bohllaender, Marco 4016
Bohne, David 514
Boige, Francois 944
Boisson, Guillaume Piquet...960
Bolzoni, A..1371
Bongiorno, Massimo..1187
Bonten, Remco 634

Böorngen, Hannes 1754
Borcherding, Holger...2852
Börngen, Hannes 3362
Boroyevich, Dushan 2806
Bosch, Swen...2219
Bosga, Sjoerd G. 3246
Bouscayrol, Alain...2175
Boutleux, Emmanuel 251
Boutry, Arthur 2515
Brabetz, Ludwig...2383
Branco, Cesar Augusto Santana Castelo 2948
Braun, Gerrit 2205
Braz, Cesar 1445
Briff, Pablo..451
Bringezu, Thilo 662
Brinker, Tobias...2977
Brogioli, Doriano Constantino 833
Brommer, Volker 2726
Bronstein, S. 3254
Brooks, Michael 279
Brückner, Thomas 1824
Brulin, Pierre-Yves 3831
Brunner, Andreas 593
Brunner, Frank 3775
Brüns, Michael 474
Bruyere, Antoine 1046
Bruyere, Paul..960
Bucarey, Victor 1074
Budo, Kohei 213, 351
Bueno, Emilio José...421
Bueno-Mariani, Guilherme 3272
Bugarski, Stevan 2334
Bünte, Andreas...380
Burgos, Rolando...1692, 2806
Burgos-Mellado, Claudio 1074, 3429
Burkart, Ralph M. 203
Burke, Richard 3696
Bushra, Rehnuma...2392, 3022
Busquets-Monge, Sergio 2715
Buticchi, Giampaolo 3014
Buttay, Cyril...2049, 2515
Byen, Byengjoo 1207
Caarls, Esin Ilhan 681
Cabrera, Michel..169
Cacciato, Mario 2628
Caillierez, Antoine 3883
Cajander, D. 1955
Cakal, Gokhan 3947
Caldognetto, Tommaso 865
Camargo, Renner Sartório..421
Camurca, Luis 3101
Can, Görkem 3092

Cano, Tania C.	335
Cao, Jingming	3215, 3225
Cao, Yongtao	2003
Cappelle, Jan	1300
Cárcamo, Alberto	1083
Carcouet, S.	843
Carpita, Mauro	1543
Carrasco, Miguel	370
Casado, P.	3382, 3401
Castellazzi, Alberto	689, 2156, 2285, 2402, 2893, 3084
Castelli-Dezza, Francesco	1476
Castro, Ignacio	335
Catalán, Pedro	2011
Catellani, Stéphane	922, 990
Ceccarelli, Lorenzo	681
Chakraborty, Sajib	2101, 3031
Chang, Che-Wei	1692
Charkaoui, Abdelmouneim	442
Chatterjee, Kishore	178, 2462
Chen, Zhe	2011
Chen, Zhu	3235
Chevalier, Florian	3582
Chida, Makoto	1580
Chinthavali, Madhu Sudhan	344
Chiumeo, Riccardo	3206
Choksi, Kushan	344
Choudhury, Soham	1966
Chub, Andrii	730
Cimetiere, Xavier	1046
Clerc, Guy	251
Clerici, Alessio	3206
Cobaleda, Diego Bernal	2581
Cogitore, Bruno	1216
Colmenero, Manuel	2189, 2451
Cosso, Simone	2919
Coumont, Martin	1966
Crovetti, Paolo	554
Cui, Yi	3986
Czerwenka, Philipp	593
Dahmen, Christopher	1824, 1855
Damian, Ioan Catalin	2266
Damm, Gilney	3590
Danielsson, Christer	2843
Dargahi, Vahid	2073
Davari, Pooya	2496
Davidson, Jonathan N.	3440
De Bernardinis, Alexandre	315
De Carne, Giovanni	3014
De Cesaris, Ivan	223
De Donato, Giulio	1569
De Doncker, Rik W.	709, 1266, 2119, 3599, 3676, 3740, 3766, 3893
De Lillo, Liliana	3450
De Matos, Jose Gomes	2948
De Oliveira, Eduardo Facanha	2441
De, Dipankar	689
Deb, Arkadeep	1497
Deblecker, Olivier	504
Deboy, Gerald	3984
Deck, Patrick	514
Deckers, Martijn	2795
Degaa, Laid	3696
Delette, Gérard	922
Deng, Kai	3235
Dennetiere, Sébastien	582
Derammelaere, Stijn	3344
Despouys, Olivier	2486
Dick, Christian P.	514
Dickmann, Stefan	758
Dieckerhoff, Sibylle	1466, 2596, 2607, 2644, 3775
Dieng, A.	2092, 2930
Dierks, Rebecca	1533
Dietrich, Tim-Hendrik	1094
Disselkamp, Simon	2912
Domae, Shinichi	3084
Domes, Daniel	2744
Domes, Konrad	1137
Dong, Chaoyu	3215, 3225
Dong, Dong	1692, 2515
Dong, Jianning	1319
Dong, Tenghui	3084
Dorner, Oscar	1177
Dos Santos, Pedro Leal	604
Dragicevic, Tomislav	2496, 2939, 3429
Drexler, Christoph	411, 1167
Driesen, J.	3655
Driesen, Johan	2795
Drimizi, Youssef	2869
Drissi, Khalil El Khamlichi	3786
Duarte, Jorge L.	681, 798
Duarte, Jorge	2788
Duchamp, Jean-Marc	169
Dujic, Drazen	2049
Dumtzlaff, Jacob	1865
Duquesne, Thierry	3582
Dürbaum, Thomas	88, 307
Duun, Sune Bro	825
Dworakowski, P.	2422
Dworakowski, Piotr	2049
Ebel, Thomas	3130
Ebner, Kathrin	4015
Eckart, Martin	3646
Eckel, Hans-Guenter	3460

Eckel, Hans-Günter...... 11, 59, 70, 980, 1294, 1703, ... 1744, 1885, 1895, 2308, 4003
Eckstein, Mattea 1277
Effenberger, Thomas 1754
Eggers, Malte 1466
Ehlich, Martin 2852
El Baghdadi, Mohamed 2101, 2293, 3031
El Sherif, Alaa 3796
El-Refaie, Ayman 719, 1692
Ellinger, Thomas 2885
Emmers, G. 3655
Emmers, Glenn 2795
Empringham, Lee 3450
Encarnação, Lucas Frizera 421
Endo, Yusuke 2285
Epping, Daniel 749
Erckrath, Tobias 1350, 1620
Eremia, Mircea 2266
Eriksson, Lars 1456
Erlbacher, Tobias 1128
Ernst, Alexander 3149, 3159
Es-Seghier, Hajar 922
Escoffier, René 990
Etoz, Burhan 1497
Faber, Samuel 307
Falchi, Daniele 2486
Faramehr, Soroush 3822
Farhangi, Shahrokh 787
Fauth, Leon 2003, 2638, 3838
Fayolle-Lecocq, Murielle..................................... 990
Fazli, Nastaran 11
Fehr, Hendrik 49, 3391
Felgemacher, Christian 442, 4004
Fernández, Arturo 152
Ferreyra, Fabio 554
Festerling, Tobias 1237
Finney, Stephen 80, 3470, 3813
Fischer, Katharina 1674, 1804
Fischer, Manuel 749
Fischer-Baeumer, Rico 1137
Fölkel, Lorandt 279
Formentini, Andrea 2919, 3975
Forouzesh, Mojtaba 1590, 1601
Forsstrom, Ville 3301
Förster, Nikolas 2432
Foster, Martin P. 3353, 3440
Foteinopoulos, Georgios 1985
Fräger, Lukas 145, 641, 1939, 2588, 2773
Frank, S. R. 3411
Franzki, Jonas 261
Frey, David 1147
Fricke, Tobias 1247

Fricke, Torben 1381
Friebe, Jens .... 1914, 2003, 2327, 2392, 2588, 2638, ... 2655, 2689, 2773, 2977, 3022, 3059, 3545, 3838
Fritze, Eric.. 758
Fröhling, Sören.. 1674
Fuchs, Simon.. 1434, 2038
Fuhrmann, Jan 980
Fukunaga, Shuhei 108
Ganeshpure, Dhanashree Ashok....................... 3729
Gao, Xiang 3014
Gaona, Daniel.. 2441
Garces, Santiago Ramos..................................... 3344
Garcia, Raul Murillo 2355
Garrigós, A.. 3382, 3401
Gaubert, Jean-Paul 1525
Gauthier, Jean-Yves 3862
Gavelle, Mathieu 2618
Gehl, Adrian 2912
Geiss, Michael.. 2554
Gemma, Filippo 3975
Geng, Weiwei... 3722
Geng, Xiaomeng 2596, 2644, 3775
Gennaro, Francesco 2628
Gensior, Albrecht......................... 49, 370, 3391
Gerges, Tony 169
German, Ronan 2175
Germishuizen, J. J... 3318
Geury, Thomas.. 2101
Gholami, M. 3179
Gholami, Mehrdad 3140
Ghumman, Sukhjit S 2763
Gieraths, Antje 767
Gierschner, Magdalena...................................... 1294
Gierschner, Sidney 11
Gillon, Frédéric 1046
Girona-Badia, Jaume 3704
Glaser, Martin 4020
Gleissner, Michael............................... 805, 1036
Gnärig, Lasse 370
Goetz, Stefan.............. 1025, 1064, 1197, 3636, 3665
Gohler, Katherina 1804
Gohrmann, Kai 1137
Golev, Victor.. 1286
Goller, Maximilian 1733
Gomes, Lucas Vinícius De Araújo....................... 3059
Gomes, Zariff Meira.. 3590
Gómez, Alexis A............................... 1765, 3336
Gomis-Bellmunt, Oriol 2486, 3704
Gonzalez, Jose Ortiz.. 1497
Gonzalez-Hernando, Fernando............................ 3938
Gonzalez-Torres, Juan-Carlos 3928
Götz, Georg Tobias 709

Gräber, Hendrik 2977
Grabs, Volker 97
Gradinger, Thomas B. 203
Grant, Thomas 2301
Grass, Norbert 2366
Grau, Vivien 854
Gremme, Florian 4021
Griepentrog, Gerd 160, 2780, 2835
Grodnichev, Anton 624
Groke, Holger 3169
Groon, Fabian 3092
Groten, Jonas 279
Gruson, François 582
Guerrero, Bruno 944
Gui, Qiuye 49
Guillaud, Xavier 582
Günes, Ece Olcay 1361
Gupta, Kirti 2110
Gupta, Krishna Kumar 3615, 3804
Gutierrez, Alonso 2618
Haag, Felix 2726
Haake, Daniel 624
Haarer, Jörg 971, 1237, 1277
Habersetzer, Antoine 4015
Hably, A. 3179
Hably, Ahmad 3140
Hackl, Philipp 39
Haederli, Christoph 3282
Häfner, Ying-Jiang 2241, 3282, 3757
Hagedorn, Maximilian 1875
Hajar, K. 3179
Hajar, Khaled 3140
Hajian, Masood 468
Hakkila, Akseli 297
Hald, Alex 380
Hameyer, Kay 3005, 3235
Hammes, David 11
Handt, Karsten 2607
Hanf, Michael 3169
Hanisch, Lucas Vincent 261
Hanisch, Lucas 1094
Hänsel, Stefan 572
Hansen, Sandra 3966
Hanson, Alex J. 1722
Hanson, Jutta 1966
Hao, Chuantong 80, 3470
Hardan, Faysal 468
Harmand, Souad 2996
Hasan, Md. Mahamudul 3031
Hasler, J. P. 1371
Hassan, Tayssir 1466
Hatori, K. 1785

Hatori, Kenji 777
Hattori, Takato 739
Hauenschild, Philipp 1506
Haug, Martin 279, 698
Hayes, John G. 2470
Hegazy, Omar 2101, 2293, 3031
Heide, Daniel 3711
Heien, Christian 1294
Heimler, Patrick 1713
Hein, Yves 1294
Helmholdt-Zhu, Ting 97, 854
Hembel, Ahmed 3947
Henke, Markus 261, 1094, 2030
Henkenjohann, Jonas 1684
Henn, Jochen 3599
Henneberg, Dustin 2885, 3491
Herbold, Johannes 749
Hernando, Marta M. 1083, 1765, 3336
Herzog, Hans-Georg 952
Heydari, Rasool 2682
Hikihara, Takashi 108
Hiller, M. 3411
Hiller, Marc 115, 999
Hillmer, Hartmut 2383
Hilt, Oliver 2596, 2644, 3775
Himker, Niklas 1631
Himmelmann, Patrick 999
Hiraki, Eiji 2164
Hirning, David 971, 1237, 1277, 3536
Hissel, Daniel 315
Hjerrild, Jesper 2504
Hoerner, Michael 1754
Hofer, Heimo 1445
Hofer, Matthias 2251
Hoff, Bjarte 3198
Hoffmann, Klaus F. 758, 2726, 3188
Hoffmann, Madlen 3262
Hoffstadt, Thorben 1157
Hofmann, Viktor 195, 400
Hofmann, Wilfried 3957
Hofstetter, Patrick 195, 400
Hölscher, Jonas 2432
Holtje, Pauline 1665
Holzke, Wilfried 3149, 3159, 3169
Horn, Markus 2383
Hortans, Magnus 3309
Hoshi, Nobukazu 1776, 1844
Hosseinabadi, Farzad 3031
Hosseini, Elham 1025
Hou, Jingning 3722
Houwen, Simon 3344
Hridya, I 187

Hu, Anliang 651
Hu, Bin 2182
Hu, Xiaowei 3722
Huang, Jiasheng 1561
Huerta, Gabriel Ramos 1226
Huesgen, Till 2230
Huisman, Henk 634, 673, 681
Hutzler, Michael 1445
Idir, Nadir 2996, 3582, 3822
Igic, Petar 3822
Iida, Masaki 2164
Iman-Eini, Hossein 787
Imgart, Paul 1187
Incurvati, Maurizio 223, 268
Inoue, Michiko 3420
Iraola, Unai 3327
Ishihara, Mastaka 2164
Itoh, Jun-Ichi 902, 1104, 2127
Ittamveettil, Hridya 289
Izurza, Pedro 484
Jaber, Hamzeh J. 2156, 2285, 3084
Jacques, Dries 3344
Jagannath, Sriram 3362
Jahdi, Saeed 1497, 2477
Jain, Anekant 3615
Jain, Sanjay K. 3615, 3804
Jamal, Adeel 2780
Jaman, Shahid 3031
Jankovic, Marija 442
Jayathurathnage, Prasad 1947
Jena, Kasinath 3804
Jenhani, Firas 1343
Jeong, Byunghwang 1207
Jeschke, Sebina 3235
Jha, Kapil 187, 289
Jia, Hongjie 3215, 3225
Jia, Ming 1266
Joebges, Philipp 1266
Johansson, N. 1371
Johnson, C. Mark 3450
Jonsson, Tomas 1456
Jordà, Xavier 2715
Jørgensen, Asger Bjørn 825, 1641, 2661
Jöst, Dominik 4025
Jovanovic, Raka 3518
Juchem, Ralf 4023
Judge, Paul 80
Junemann, Lennart 1665
Jung, Marco 624, 1515, 1611, 1620
Junghans, Christoph 3460
Junyent-Ferre, Adria 2574
Kabbara, Wassim 3883

Kacetl, Jan 1197, 3636, 3665
Kacetl, Tomáš 1197, 3636, 3665
Kacki, Marcin 2470
Kadem, Karim 3590
Kaerst, Jens Peter 544
Kaiser, Jeremias 307
Kallfass, Ingmar 2698, 3565
Kamel, Tamer 468
Kaminski, Nando 2230, 3149, 3169
Kamm, Simon 2698
Kampen, Dennis 145, 1939, 2588
Kamper, Maarten J. 2958
Karakasli, Vefa 2835
Karamanakos, Petros 297, 1476, 1754
Karau, Fabian 3292
Karnehm, Dominic 767
Karwatzki, Dennis 195
Kasten, Henning 3501
Kayser, Felix 59, 4003
Keilmann, Robert 891
Kempchen, Malte 2912
Kemper, Philipp 749
Kennel, Ralph 1754, 2366, 3362
Kerekes, Tamas 1933
Keshavarzi, Davood 1064
Khader, Meriem 2655
Khan, Basit Ali 2537
Khan, Mohammed Ali 135
Khan, Nameer 3796
Khan, Siam Hasan 484
Khanzadeh, Babak 2344
Khenfri, Fouad 3831
Kiehnle, Philip 999
Kiffe, Axel 1157
Kikuchi, Naoto 1104
Kim, Dong-Uk 1207
Kim, Sungmin 1207, 2877
Kinzer, Dan 3987
Kirsch, Andreas 380
Kitagawa, Wataru 739
Kjærsgaard, Benjamin Futtrup 825
Klee, Matthias 1515
Klever, Severin 3676
Klötzer, Sebastian 4011
Knebusch, Benjamin 1665, 3048
Ko, Youngjong 3014
Kobayashi, Hiroyasu 1580
Kocewiak, Lukasz 2504
Koch, Jan-Niklas 2852
Koczy, Dawid 3149
Kohlhepp, Benedikt 88, 307
Kojima, Tetsuya 3740

Kondo, Keiichiro 1580
Kondratenko, Dmytro 1906
Kopp, Tobias 912
Kormska, Tomáš 1114
Körner, Patrick 2021
Korthauer, Bastian 3625
Kosesoy, Yusuf 634
Kostka, Benedikt 1649
Kostynski, Daniel 3855
Koteich, Mohamad 534
Kouro, Samir 1015
Koutroulis, Eftychios 1985
Kowal, Julia 4014
Kragl, Robert 2554
Krick, Alexander 3989
Krigar, Tim 2375
Krishnamoorthy, Harish Sarma 730
Krüger, Helge 3966
Krümpelmann, Marcel 1631
Kubulus, Pawel Piotr 2661
Kuder, Manuel 767
Kumar, Amit 451
Kumar, Kaushik Naresh 1486
Kumar, Manish 3511
Kuperman, A. 3254
Kuprat, Johannes 3067
Kuring, Carsten 2596, 2644, 3775
Kurrat, Michael 912
Kurukuru, V S Bharath 135
Kusaka, Keisuke 1104, 2127
Kusche, Stephan 3704
Kusebauch, Manuel 3491
Küster, Pierre 411
Kwak, Jaedon 2893
Kyyrä, Jorma 1947
La Mantia, Fabio 833
Labonne, A. 3179
Labonne, Antoine 3140
Labrousse, D. 843
Lacerda, Vinícius Albernaz 3704
Laclaverie, Julien 944
Laforet, David 1445
Lamar, Diego G. 335, 1083, 1765, 3336
Lange, Jarren 2276
Lange, Yannic 2644
Langfermann, Sascha 1939, 2588
Lanzarotto, D. 2564
Larrañaga, Uxue 3938
Larrazabal, Igor 1765, 3336
Larsson, Anders 1456
Lataire, Philippe 2293
Laumen, Michael 3766

Lauri, Andrea 865
Laza, Saioa Burutxaga 370
Lazkano, Markel Zubiaga 484
Le Leslé, Johan 2526
Le Métayer, Pierre 2049
Lee, Jaehong 2877
Lee, Seung-Hwan 2877
Lee, Yonghwa 2402
Lefebvre, Bruno 2515
Lefevre, Guillaume 2526
Legay, Florian 3529
Lehn, Peter W. 1995, 2084, 2145, 2763
Leifert, Torsten 4013
Lemaire-Semail, Betty 2175, 2996
Lembeye, Yves 1216
Lenz, Kevin 442
Lenzen, Patrick 2413
Leuer, Michael 3292
Leuzzi, Riccardo 3975
Lévy, PE 843
Lewicki, Arkadiusz 1906
Lexow, Daniel 1744
Li, Feifei 3235
Li, Ke 3822
Li, Marui 3215, 3225
Li, Qiang 3722
Li, Weihan 4025
Li, Xiang 2301
Li, Xupeng 3373
Li, Zheming 2744
Liang, Mincui 3786
Lichtenstein, Timo 1674
Liebfried, Oliver 2726
Liegmann, Eyke 1754, 3362
Lievre, Aurelien 2175
Lin, Siqi 1914, 2638
Lin-Shi, Xuefang 3862
Lindemann, Georg 3555
Linder, Stefan 3992
Lippold, Florian 1506
Liserre, Marco 421, 833, 3014, 3067, 3092, 3101, 3846, 3966
Liu, Chao 1561
Liu, Steven 604
Liu, Xing 1733, 3373
Liu, Yan-Fei 1590, 1601
Liu, Yining 1947
Llanos, Jacqueline 3429
Löfgren, Jonas 3920
Lombard, Philippe 169
López, Abraham 152
Lorenz, Andreas 814

Lorenz, Erwin 1167
Lorenz, Malte 1875
Lorenz, Oscar 873
Loudot, Serge 3883
Lu, Xuyang 3822
Lu, Yizhou 883
Luan, Shaokang 3309
Luckert, Franz 2706
Luecke, Stefan 3075
Luh, Matthias 232
Luo, Fang 344, 2860
Lusardi, Federico 3975
Lutsch, Michael 88
Lutz, Josef 1713
Lutzen, Hauke 2230
Ma, Wenhao 80
Maamri, Nezha 1525
Maibach, Philippe 3282
Maier, Robert W. 2744
Maitra, Abhishek 1424
Mallwitz, Regine 891, 912, 1094, 1247, 1506
Mambetow, Arthur 145
Manthey, Tobias 2655, 2689, 3059
Marca, Ygor Pereira 798
Marcaide, Inko 3920
Marcault, Emmanuel 2618
Marchesoni, Mario 2919
Margreiter, Thomas 223
Margueron, Xavier 1046
Marks, Hendrik 2030
Marquardt, Rainer 1855
Marroquí, D. 3382, 3401
Martin, Jérémy 990, 2736
Martinez, Wilmar 1914, 2197, 2581
Martinez-Garcia, Herminio 1056
Martinez-Padron, Daniel S. 1256
Martnez, Wilmar 2638
Marx, Philipp 1237, 1277, 3536
März, Martin 493, 3262
Mashaly, Aly 442
Mashayekh, Ali 767
Mathúna, Cian Ó. 4006
Mattavelli, Paolo 865
Matthies, David 3159
Maussion, Pascal 2869
Maynard, X. 843
Mazuela, Mikel 325, 3327, 3574
Meddour, Aissam Riad 3696
Mehran, Kamyar 614, 3353
Mehrasa, M. 3179
Mehrasa, Majid 3140
Meier, Hans 2021

Meinert, Janus Dybdahl 825
Meissner, Michael 758, 3188
Mellor, Phil 2477
Mendoza-Araya, Patricio 1177, 1226
Meng, Qingchao 933
Menzel, Steffen 3169
Merlin, Michael M. C. 2824, 3813
Merlin, Michael .... :......80, 3470
Mersche, Stefan 115
Mertens, Axel 641, 1350, 1533, 1631, 1649, 1665,
..1684, 1865, 1875, 2003, 2066, 2392, 2706, 3022, 3037, 3048,
3075, 3555, 3711
Miaja, Pablo F. 152
Mijatovic, Nenad 2496, 2939
Miller, T. J. E. 3318
Minami, Masataka 2285
Mir, Tabish Nazir 468
Mirza, Abdul Basit 344
Mirzadeh, Mina 1350
Mirzaeva, Galina 3903
Miskiewicz, Rafal 1486
Mistretta, C. 1975
Mita, Salvatore 2628
Mo, Wai Keung 3130
Möckel, Andreas 3391
Moench, Stefan 242
Mogorovic, Marko 203
Mohanta, MK Kharabela 689
Möhlenkamp, Georg 3993
Mohsenzade, Sadegh 614, 3353
Moldenhauer, Deniz-Heinz 2205
Mondal, Gopal 572
Mondzik, Andrzej 3804
Monmasson, Eric 1256
Mönninghoff, Sebastian 3005
Montero, E. Rodriguez 1834
Morales-Paredes, Helmo K. 1074
Morand, Julien 2526
Morel, F. 2422, 2564
Morey, Philippe 1543
Morshed, Muhammad 2301
Motte-Michellon, Denis 1216
Mouselinos, Theodoros P. 1551
Moussa, Hassan 3590
Movagharnejad, Hedieh 3048
Mu, Yunfei 3215
Müller, Jonas 2230
Müller, Tankred 474
Munk-Nielsen, Stig 825, 1641, 2661, 3309
Muñoz-Carpintero, Diego 1074, 3429
Muruaga, Endika Bilbao 3529
Musolino, Francesco 554

Mustafeez-Ul-Hassan 2860
Musumeci, S. 1975
Muyllaert, Koenraad 2383
Mysore, Madhu Lakshman 1424
Naeve, Tomasz 1445
Nagayasu, Kiwa 2164
Naghibi, Javad 614, 3353
Nahalparvari, Mehrdad 2843
Najjar, Mohammad 2682
Nakamura, Keiichi 777
Nakamura, Taketsune 3084
Nami, Ashkan 2241, 3757
Nannen, Hauke 160
Nassurdine, B. Mohamed......................................843
Nayak, Khirod Kumar....................................2241, 3757
Nayampalli, Vishwas Acharya................................1703
Nazeri, Ahmad Ali 1309, 1336, 1343, 2670, 3871
Neal, Harley 2301
Nee, Hans-Peter 2843
Nehmer, Dominik 1036
Neira, Sebastian 2824, 3813
Neuland, Tanja 3991
Neumann, Christian 1895
Neumann, Ingmar 1445
Neumeister, Matthias 572
Nguyen, Allen 1722
Nguyen, Khanh-Hung 562, 1309
Nguyen, Van-Sang......................................922, 990
Nguyen, Xuan Viet Linh 169
Nian, Heng 2182
Niasar, Mohamad Ghaffarian................................3729
Nie, Shuang 2145
Niedernostheide, Franz-J. 2744
Niedernostheide, Franz-Josef................................1424
Nielebock, Sebastian.....................................493, 2607
Niemetz, Michael 2021
Niggemann, Oliver 3545
Nikowitz, Mario 2251
Nishio, Atsushi 351
Nishitani, Yota 3420
Nishizawa, Shin-Ichi 1128
Noboru, Wakana 777
Noisette, Philippe 3910
Nooshabadi, Morteza Tadbiri 787
Nordström, Lars 883, 1006
Nymand, Morten 2682
O'Donnell, Terence 390
O'Driscoll, Seamus 4006
Obernolte, Urs 854
Odeh, Charles 1906
Okada, Ryohei 1776, 1844
Olbrich, Markus 2912

Oliveira, Hercules Araujo 2948
Orbay, Raik 3920
Orchard, Marcos...3429
Orfanoudakis, Georgios I. 1985
Örgüt, Osman 1361
Orlik, Bernd 3149, 3159, 3169
Ortega, David 1765, 3336
Ortiz-Gonzalez, Jose 2477
Orts, C. 3382, 3401
Oshnoei, Arman 2939
Ota, Ryosuke 1776, 1844
Ouyang, Ziwei..1413, 1561
Owzareck, Michael 1939, 2588
Oyarbide, Estanis 3327
Paasch, Kasper M...3130
Pace, Loris 3582
Páez, J. D..2422
Pagnani, Daniela 2504
Panigrahi, Bijaya Ketan................................2110, 3511
Papadopoulos, Georgios.......................................1391
Papadopoulos, Theofilos 432
Papafotiou, George..2788
Papanikolaou, Nick 1796, 2257
Papastergiou, Konstantinos 2355
Pascal, Yoann 3067
Pasquier, Christophe 3786
Passalacqua, Massimiliano....................................2919
Passmore, Brandon 4005
Pathmanathan, Mehanathan 1995, 2084, 2145, 2763
Patin, Nicolas 1256
Patti, Dario 2628
Patzelt, Nikolaus 1923
Paul, Arup Ratan 178
Pauls, Denis...2441
Pavone, Mario 554
Pedroso, Douglas 335
Peftitsis, Dimosthenis...................................1486, 2355
Pelletier, Sebastien 223
Penczek, Adam 3804
Peng, Hujun 3235
Péra, Marie-Cécile 315
Pereda, Javier 2824
Pereira, Thiago...................3014, 3092, 3101, 3846
Perez, Gaëtan 960
Perez-Cebolla, Francisco Jose.........................3574, 3750
Peroutka, Zdenek 1114
Perpiñá, Xavier 2715
Perrin, Rémi 2526
Perrin, Remi 3272
Petritz, Andreas 279
Petzoldt, Jürgen 2885, 3491
Peyghami, Saeed 2939

Pfeiffer, Jonas 411, 1167
Pfost, Martin 2375, 2413
Phanse, Ajinkya 1722
Phulpin, Tanguy 3883
Pichon, Pierre-Yves 2526
Pickert, Phil Leon 1381
Piepenbrock, Till 2432
Pietrzak-David, Maria 2869
Pigott, John 3796
Pinheiro, José Renes 3590
Piqué, Gerard Villar 3796
Piróg, Stanislaw 3804
Placzek, Julius M. 833
Plat, Arnaud 3862
Plötz, Till-Mathis 980
Pogulaguntla, Aditya 730
Pohlmann, Sebastian 767
Polezhaev, Vladimir 2230
Ponick, Bernd 1381, 1665, 3048, 3711
Poormohammadi, Fereshteh 2795
Pöschke, Florian 3704
Pouresmaeil, Edris 2537
Pouresmaeil, Kaveh 2788
Pouresmaeil, Mobina 2537
Pramanick, Sumit 1658, 3511
Pree, Elias 1445
Prenleloup, Pierre 3529
Prieto-Araujo, Eduardo 2486, 3704
Puls, Simon 2852
Puschmann, Frank 749
Qin, Zian 3607
Quabeck, Stefan 3893
Quade, Katharina Lilith 4025
Quay, Rüdiger 242
Rabkowski, Jacek 1486, 3938
Rädel, Uwe 2885, 3491
Radha, Krishna Moorthy 344
Rafiq, Aamir 1658
Raggini, Diego 3206
Raghavendra, I Venkata 730
Rahmani, Mehdi 2496
Raison, Bertrand 1147
Rajabian, Amir Azam 614
Ramdane, Brahim 1216
Ramirez, Fernando 289
Rasekh, Navid 3120
Rasool, Haaris 2101, 2293
Raßmann, Rando 1286
Rathjen, Kai-Uwe 758
Rault, Pierre 582
Ravyts, Simon 1300
Raya, Mariana 2715

Razi, R. 3179
Razi, Reza 3140
Regnat, Guillaume 2526
Rehlaender, Philipp 2432, 2754, 3625
Reimann, René 3159
Reincke-Collon, Carsten 370, 3391
Reindl, Andrea 2021
Reiner, Richard 242
Reißenweber, Lukas 525
Reitmeier, Dominik 2211
Remón, Daniel 1083
Rettner, Cornelius 4019
Reyes-Chamorro, Lorenzo 3429
Reynaud, Jean-François 3529
Ribeiro, Luiz Antonio De Souza 2948
Richard, Lucas 1147
Rickert, Kai 115
Rigbers, Klaus 4023
Rigogiannis, Nick 2257
Ringbeck, Florian 4025
Risch, Raffael 651
Rizoug, Nassim 3696, 3831
Robinson, Jonathan 572
Rocha, Gabriel Silva 2948
Roche, Jan-Philipp 3545
Rodríguez, Alberto 335, 1083, 1765, 3336
Rodriguez, Daniel C. 3893
Rodriguez, Joan Marc 2574
Rodriguez, José 1015
Roes, Maurice G. L. 798
Roes, Maurice 2788
Roß, Tilo 3391
Rossi, Mattia 1476
Rothenburger, Max 2383
Roth-Stielow, Jörg 971, 1237, 1277, 3536
Rouphael, Rosalie 1525
Rudolph, Christian 474
Rueß, Manuel 3565
Rufer, Alfred 30
Ruppert, Lukas A. 3766
Ruthardt, Johannes 971
Rylko, Marek S. 2470
Sadarnac, Daniel 3883
Saeidi, Mahmoud 1336, 1343, 3871
Safdarzadeh, Omid 2316
Sah, Gyanendra Kumar 1885
Sahan, Benjamin 1137
Sahin, Ilker 1361
Sahoo, Subham 2110, 2182
Sahu, Malaya Kumar 2241, 3757
Sahu, Silpashree 689
Said, Nasri 2618

Saito, Wataru	1128
Sakai, J.	1785
Salehi, Navid	1056
Samples, Ben	4005
Sanchez, Juan	873
Sanchez-Ruiz, Alain	484
Santos, Francisco	3101
Sanusi, Bima Nugraha	1413
Sanz-Alcaine, José Miguel	3750
Sarlioglu, Bulent	3947
Sato, Kota	1580
Sato, Takashi	3420
Sauer, Dirk Uwe	4012, 4025
Sauerland, Henning	3159
Sawicki, Jean–paul	315
Scarcella, Giuseppe	1569
Scelba, Giacomo	1569, 2628
Schäffner, Philipp	279
Schafmeister, Frank	2432, 2754, 3625, 3686
Schanen, Jean-Luc	787
Schanen, JL	843
Schefer, Hendrik	891, 912, 1094
Schellekens, Jan	634
Schierle, Guido	3188
Schiestl, Martin	223, 268
Schillinger, Tobias	3646
Schillingmann, Henning	2030
Schlegel, Christian	1923
Schlegel, Ludwig	3957
Schmid, Markus	268
Schmidhuber, Michael	411, 1167
Schmies, Dominik	2276
Schmitz, Laurids	3599
Schnabel, Fabian	624, 1515
Scholjegerdes, Moritz	3005
Schön, André	814
Schrödl, Manfred	2251
Schueltzke, Jens	1167
Schuerhuber, Robert	39
Schuhmann, Thomas	3646
Schullerus, Gernot	593, 2334
Schulte, Horst	3704
Schulz, D.	3411
Schulze, Gerold	2383
Schulze, Hans-Joachim	1424
Schumann, Christian	2058
Schumann, Sven	4022
Schümann, Ulf	1286
Schupp, Jan	3309
Schütt, Michael	1885, 2308
Schwarz, Babette	1381
Schwendemann, R.	3411

Scohier, Martin	504
Scrimizzi, F.	1975
Sebastián, Javier	1765, 3336
Seibel, Axel	1515, 1620
Seitz, Arne	4015
Seliger, Norbert	22
Semail, Eric	2996
Sen, Paresh C.	1590, 1601
Sepehr, Amir	2537
Serdyuk, Yuriy	2344
Sergentanis, Grigorios	3450
Serra, Amiron Wolff Dos Santos	2948
Seybold, Felix	3536
Shahparasti, Mahdi	2682
Sharma, Kanuj	2698
Shawky, Ahmed	1015
Shen, Chengjun	2477
Shen, Xiaobing	1914, 2197
Shinoda, Kosei	3928
Shintani, Michihiro	3420
Shousha, Mahmoud	279, 698
Shuqin, Wang	1815
Siala, Sami	125
Siemaszko, Daniel	3910
Siemieniec, Ralf	1445
Sievers, Markus	3855
Singh, Rupam	135
Singh, Shashank Shekhawat	279
Singh, Sukhjit	2084
Skala, Aleksander	3804
Skibin, Stanislav	3301
Soeiro, Thiago Batista	1319, 3729
Solomentsev, Michael	1722
Solovyov, Vyacheslav	2860
Soltau, N.	1785
Soltau, Nils	777
Sönmez, Ertugrul	593, 2334
Soundararajan, Ajeeth Phrassanna	3729
Soupremanien, Ulrich	922
Spieler, Matthias	1692
Sprunck, Sebastian	1611
Sreekanth, T	730
Stadler, Alexander	525
Stadlober, Barbara	279
Staiger, Jochen	2219
Stala, Robert	3804
Stalleicken, Frederik	2607
Stallmann, Frederik	641
Stärz, Ronald	223, 268
Stathis, Spyridon	1402
Staubach, Christian	1137
Steckler, P. B.	2564

Stefanski, L.	3411
Steffen, Jonas	1515
Steinhart, Heinrich	2219
Štengl, Josef	1114
Stevic, Marija	2985
Stewart, Joshua	2806
Steyn, Kyle	3112
Stille, Karl Stephan	2276
Stock, Alexander	1
Stöckl, Thomas	952
Stone, David A.	3440
Strunk, Robin	1350
Stul, Koen	1300
Stutz, Christian	493
Suberski, Martin	2885, 3491
Sujeeth, Arjun	2628
Sullivan, Charles R.	2470
Svensson, Jan R.	1187
Tabrizi, Gholamreza	1611
Takamori, Taro	1128
Takayama, Hajime	108
Takeshita, Takaharu	213, 351, 739
Talla, Jakub	1114
Tang, Chengjun	2813
Tang, Zhongting	1933
Tashakor, Nima	1025, 1064, 1197, 3636, 3665
Tatakis, Emmanuel C.	1551
Tegtmeier, Bernd	1674
Teske, Peter	1466
Thiringer, Torbjörn	2344, 2813, 3920
Thoma, Jürgen	2554
Thönelt, Nick	1713
Thönnessen, André	3676
Tian, Fanghao	2581
Tillmann, Philipp	3740
Tiwari, Arvind Kumar	289
Tiwari, Arvind	187
To, Pham Ha Trieu	59, 70, 4003
Tornello, Luigi Danilo	1569
Torres, C.	3382, 3401
Torrico, Grover	361, 1815
Tournez, Florian	2175
Tran, Dai Duong	2293
Tran, Manh Tuan	2101, 2293
Tresca, Giulia	3975
Trescases, Olivier	3796
Tricoli, Pietro	468
Trochimiuk, Przemyslaw	1486
Tschepp, Andreas	279
Turrisi, Gaetano	1569
Tzanakis, Athanasios	3920
Uicich, Simon	3862
Ulbing, Alexander	3855
Ulmer, Sabrina	593, 2334
Ulrich, Burkhard	459
Umetani, Kazuhiro	2164
Unruh, Peter	1620
Unruh, Roland	3686
Urkizu, June	325
Vaccaro, Luis	2919
Vaessen, Peter	3729
Vagg, Christopher	3696
Vagnon, Eric	2515
Vahid, Sina	719
Vala, Sama Salehi	344
Valderrama, Carlos	504
Valenzuela, Rodrigo Alonso Alvarez	814
Van Cappellen, Leander	2795
Van Mierlo, Joeri	2101
Van Oosterwyck, Nick	3344
Van Tuan, Mai	351
Vandenbussche, Thomas	1300
Vanfretti, Luigi	3928
Vanwalleghem, Bart	3344
Vasiladiotis, Michail	1923
Vatamanu, Lucian	1046
Vázquez, Aitor	1083
Vázquez, Francisco	1765, 3336
Velasco-Quesada, Guillermo	1056
Velazco, Diego	251
Vellvehi, Miquel	2715
Venkataramanan, Giri	3480
Venugopal, Ravinder	2985
Verdier, Jacques	169
Vermeerch, Pierre	582
Veroni, Alessandro	3206
Vershinin, K.	2564
Viana, Caniggia	1995, 2084
Viarouge, I.	1955
Viarouge, P.	1955
Vidal-Albalate, Ricardo	2189
Videau, Nicolas	944
Videt, Arnaud	3822
Villar, Irma	3529, 3938
Vitorino, Montiê Alves	2689, 3059
Vogelsberger, M.	1834
Volzer, Benjamin	2554
Von Hoegen, Anne	3740
Wada, Keiji	1128
Wagner, Valentin	514
Wakelin, Bruce	3309
Wallart, Francois	251
Wallscheid, Oliver	2276, 2432
Waltereit, Patrick	242

Wang, Chu 3722
Wang, Jun 2136, 3120
Wang, Kangan 3014
Wang, Rui 673, 1641
Wang, Xiaoya 3722
Wang, Xin 315
Wang, Yanbo 2011
Wang, Yangang 2301
Waradzyn, Zbigniew 3804
Watanabe, Hiroki 1104
Wattenberg, Martin 873
Weicker, Martin 2316
Weires, Jonas 604
Weiser, Mathias C. J. 3565
Weiss, Xavier 1006
Wenzel, Johannes C. 2066
Werlig, Christian 3966
Weyh, Thomas 767
Wicht, Bernhard 2912
Wieczorek, Nick 3775
Wiemer, Adrian 2544
Wiesemann, Julius 1865
Wiesner, E. 1785
Wiesner, Eugen 777
Wijnands, Korneel 673, 798, 2788
Wilkowski, Matt 4008
Willer, Felix 3838
Willich, Viktor 3555
Wohlrath, Fritz 525
Wolbank, T. 1834
Wolf, Mihaela 2596, 2644, 3775
Wolfstädter, Simon 4017
Wölk, Alexander 279
Wouters, Hans 2197
Woywode, Oliver 758
Wu, Weimin 1985
Wu, Xiangqiang 1933
Wu, Yuxuan 2860
Wunsch, Bernhard 3301
Würfl, Joachim 2596, 2644, 3775
Würsig, Andreas 3966
Xia, Peizhou 3470
Xiao, Qian 3215, 3225
Xiao, Xiong 1966
Xie, Jun 2885, 3491
Xie, Lihong 2136
Xu, Huihui 709
Xu, James 3796
Xu, Qianwen 883, 1006
Xu, Wei 2136
Xu, Zhongqing 912
Xu, Zixiao 2182

Yadav, Sachin 3607
Yamaguchi, Masamichi 2127
Yamashita, Shota 213
Yamauchi, Kohei 2119
Yang, Huoming 1466
Yang, Jiajun 3014
Yang, Juefei 2477
Yang, Yinghui 3993
Yang, Yongheng 2257
Yaqoob, M. 1815
Yasuda, Takumi 902
Yeganeh, Mohammad Sadegh Orfi 2496, 2939
Yu, Guangyao 1319
Yu, Xiao 562, 1309, 2383
Yu, Xiaodan 3225
Yuan, Xibo 2136, 3120
Zacharias, Peter 411, 562, 1309, 1328, 1336, 1343,
.... 2383, 2670, 3871
Zacher, Benjamin H. 2058
Zampardi, Giorgia 833
Zanchetta, Pericle 3975
Zatocil, Heiko 160
Zdanowski, Mariusz 3938
Zhang, Bo 1733
Zhang, Shimin 709
Zhang, Yaqian 2182
Zhang, Zhe 1561
Zhang, Zhuoqi 1776
Zhang, Ziqian 39
Zhao, Hongbo 1641, 3309
Zheng, Zhixue 315
Zhetessov, Aidar 3480
Zhu, Zi-Qiang 2958
Ziani, Adel 944
Ziegler, Philipp 971, 1237, 1277, 3536
Zilic, Rufad 1336
Zocher, Markus 2366
Zolfi, Pouya 719
Zou, Zhixiang 3014
Zsurzsan, Tiberiu Gabriel 1561

IEEE
445 Hoes Lane
Piscataway, NJ 08854-4141

ISBN 978-1-6654-8700-9